ADVANCES IN FRUIT BREEDING

Advances in Fruit Breeding

edited by Jules Janick and James N. Moore

1975
Purdue University Press
West Lafayette, Indiana

Library of Congress Catalog Number 73-76916
International Standard Book Number 0-911198-36-9
Printed in the United States of America
Twenty-five dollars

Dedication

This book is dedicated to Dr. George M. Darrow, premier breeder of strawberries and blueberries, whose research accomplishments, spanning a career of 46 years with the United States Department of Agriculture, are unparalleled in the history of fruit cultivar improvement.

Born at Springfield, Vermont, February 2, 1889, George Darrow graduated from Middlebury College in Vermont in 1910. Going on to Cornell University, he received the A. B. degree in horticulture the following year, and that same year was employed by the USDA.

For two years he was assigned to studying problems in the handling and transporting of fruits, and worked in various parts of this country. He was then reassigned to study production practices and climatic influences on fruit production, particularly including berry crops. During that assignment, he prepared farmers' bulletins on berry growing in different geographic areas of the United States.

Recognizing the importance of obtaining better cultivars, he initiated breeding work with strawberries in 1919. Hybridizations were made in a greenhouse in Washington, D.C., and seedlings grown in field plots at nearby Glenn Dale, Maryland. Cooperative testing was also arranged, particularly with North Carolina State University. From that early work came several outstanding selections, the most valuable being 'Blakemore,' which for more than two decades was the most widely grown strawberry cultivar in the United States.

During this time, Darrow was also taking graduate work in plant physiology and genetics at Johns Hopkins University in Baltimore. He received the Ph.D. degree there in 1927.

The fruit research of the USDA, including the strawberry breeding, was moved to Beltsville, Maryland, in 1932. The first strawberry plots were on heavy soil and many plants died from what was later found to be red-stele root rot. That disease, little recognized previously, was found to be rather widespread throughout the northern half of the United States, and development of cultivars resistant to it became important. Darrow and his associates devised screening techniques for the disease, which involved growing the seedlings in a cool greenhouse in beds of wet, infested soil. This proved very successful as a means of eliminating susceptible seedlings and became a routine part of the breeding program.

In the 1940s, the importance of virus diseases in strawberries also became recognized. Mainly transmitted by aphids, the disease infected many new cultivars even before they were distributed to growers; they soon became weakened and "ran out." Pathologists instituted a search for "disease-free" plants of important cultivars and in many cases found them. Dr. Darrow was a leader in convincing strawberry nurserymen to propagate cultivars with these clean plants, grow them isolated from other strawberries, and control aphids by systematic spraying. With the cooperation of state inspectors and nurserymen, this program has resulted in production of relatively disease-free planting stock of more vigor and productiveness.

The virus problem also required revision of breeding and testing methods to avoid infection of breeding material. Growing such material isolated from other strawberries, with aphid control, became standard practice.

Frederick Coville had initiated breeding work to improve the native blueberry in 1908. Following Coville's retirement from the USDA in 1937, Dr. Darrow was assigned leadership of that program. He expanded the work and developed cooperation with several states and individual growers. Darrow made crosses in the greenhouse at Beltsville, distributed seedlings to cooperators, and later assisted in their evaluation. From this work of Coville, Darrow, and recently, D. H. Scott, has come nearly all cultivars of cultivated blueberries, now a $35 million industry with over 30,000 acres (12,000 ha) planted.

In strawberries, it is estimated that 30% of the area grown in the United States is of cultivars introduced from the USDA breeding work which Darrow initiated and conducted for nearly 40 years. Another 40% is of cultivars derived in part from breeding material originated in that program, but developed by others. Certainly, the accomplishments of Dr. Darrow in the field of cultivar improvement of strawberries and blueberries is without precedent.

Dr. Darrow's interest in berries, however, has not been limited to breeding and genetics. In addition, he extensively studied climatic factors and adaptation. An early study by Darrow and G. F. Waldo showed that day length and temperature were controlling factors in the initiation of strawberry flower buds as well as runner formation. He traveled to Central America, Ecuador, and Chile to study the behavior of strawberries under the varying day lengths and temperatures prevailing in those areas. He was particularly interested in the species *Fragaria chiloensis*, one of the parent species of present-day cultivars.

Why has Dr. Darrow's work been so successful? With a firm foundation in genetics, he is an all-around plantsman, a keen observer interested in every facet of the plant and its environment. Added to this has been the magnitude of the work which involved hundreds of thousands of progeny plants, both of strawberry and blueberry. Further, he has been a prolific writer, authoring many research papers and popular articles climaxed by his excellent and authoritative book, *The Strawberry: History, Breeding, and Physiology*. And at 84, he is still active, contributing a chapter to this present collection!

Dr. Darrow's generosity and unselfishness have led to the willingness on the part of cooperators in all parts of the United States to work with him. He has shared

GEORGE McMILLAN DARROW

material freely, and has been as unbiased in evaluating the work and products of others as in evaluating his own material. He has always given freely of his time to discuss problems with his colleagues and has been especially willing to counsel young fruit breeders just entering their professional careers. Through the years his optimism and enthusiasm has been an inspiration to many young breeders, as well as to his peers. These admirable qualities, coupled with innate ability and long hours of arduous work, have led to the worldwide recognition that is now his.

John R. Magness

Contents

v Dedication

John R. Magness, United States Department of Agriculture (retired), Takoma Park, Maryland

xiii Preface

James N. Moore, Professor, Department of Horticulture and Forestry, University of Arkansas, Fayetteville, Arkansas
Jules Janick, Professor of Horticulture, Department of Horticulture, Purdue University, West Lafayette, Indiana

1 TEMPERATE FRUITS

3 Apples

A. G. Brown, Principal Scientific Officer, Department of Applied Genetics, John Innes Institute, Norwich, England

38 Pears

Richard E. C. Layne and Harvey A. Quamme, Research Scientists (Pomology), Agriculture Canada, Research Branch, Research Station, Harrow, Ontario, Canada

71 Strawberries

Donald H. Scott, Research Horticulturist, Fruit Laboratory, United States Department of Agriculture, Beltsville Agricultural Research Center, Beltsville, Maryland
F. J. Lawrence, Research Horticulturist, United States Department of Agriculture, Western Region, Corvallis, Oregon

98 Brambles

D. K. Ourecky, Associate Professor, Department of Pomology and Viticulture, New York State Agricultural Experiment Station, Geneva, New York

130 Grapes

John Einset, Professor of Pomology and Viticulture, and *Charlotte Pratt*, Research Associate, Department of Pomology and Viticulture, New York State Agricultural Experiment Station, Geneva, New York

154 Blueberries and Cranberries

Gene J. Galletta, Professor of Horticultural Science and Genetics, Department of Horticultural Science, North Carolina State University, Raleigh, North Carolina

197 Currants and Gooseberries

Elizabeth Keep, Fruit Breeder, Fruit Breeding Section, East Malling Research Station, East Malling, Maidstone, Kent, England

269 Minor Temperate Fruits

George M. Darrow, United States Department of Agriculture (retired), Glenn Dale, Maryland

285 Peaches

Claron O. Hesse, Professor of Pomology, Department of Pomology, University of California, Davis, California

336 Plums

John H. Weinberger, Research Leader, United States Department of Agriculture, Western Region, Fresno, California

348 Cherries

Harold W. Fogle, Research Horticulturist, Fruit Laboratory, United States Department of Agriculture, Beltsville Agricultural Research Center, Beltsville, Maryland

367 Apricots

Catherine H. Bailey and L. Fredric Hough, Research Professors of Pomology, Department of Horticulture and Forestry, Rutgers University, New Brunswick, New Jersey

385 TEMPERATE NUTS

387 Almonds

Dale E. Kester, Professor of Pomology, and *Richard A. Asay,* Staff Research Associate, Department of Pomology, University of California, Davis, California

420 Pecans and Hickories

George D. Madden, Research Horticulturist, United States Department of Agriculture, United States Pecan Station, Brownwood, Texas
Howard L. Malstrom, Plant Physiologist, United States Department of Agriculture, Southeast Fruit and Nut Crops Research Station, Byron, Georgia

439 Walnuts

Harold I. Forde, Staff Research Associate, Department of Pomology, University of California, Davis, California

456 Filberts

H. B. Lagerstedt, Research Horticulturist, United States Department of Agriculture, Western Region, Corvallis, Oregon

490 Chestnuts

Richard A. Jaynes, Associate Geneticist, Connecticut Agricultural Experiment Station, New Haven, Connecticut

505 SUBTROPICAL FRUITS

507 Citrus

R. K. Soost, Chairman, and *J. W. Cameron*, Professor of Horticulture, Department of Plant Sciences, University of California, Riverside, California

541 Avocados

B. O. Bergh, Specialist and Lecturer, Department of Horticultural Science, University of California, Riverside, California

568 Figs

W. B. Storey, Professor of Horticulture, Department of Plant Sciences, University of California, Riverside, California

591 Author Index

603 Subject Index

Preface

The art of plant breeding has been practiced continually since primitive man gave up nomadic food gathering to settle in permanent settlements around dependable sources of food. Our early ancestors selected practically all the useful species we now use and propagated those individuals and clones which expressed the most desirable phenotypes. This has been a truly remarkable achievement—our crop plants may be man's most important heritage of the past. Plant breeding continued to be largely based on empirical knowledge until the "discovery" in the beginning of this century of Gregor Mendel's great 1865 paper on inheritance which gave rise to the science we know today as genetics. While the breeder's art is still critically important, knowledge of the "laws" of genetics has transformed plant improvement into a truly scientific discipline in its own right. Progress in crop improvement has brought tremendous changes in this century. The impact of this new discipline on the fruit industry can be seen in the superior cultivars being grown in orchards, vineyards, and plantations. And it has become increasingly apparent that the success of the fruit industry in any area is usually directly related to the success of fruit breeding programs to develop superior phenotypes in that area. This understanding has led to renewed interest in fruit breeding worldwide.

Breeders of fruit species are faced with certain unique problems not common in other agricultural crops. The long generation cycle, requiring from two to ten years depending on species, greatly curtails the exploitation of genetic recombination. The long juvenile period results in costly demands on field space and magnifies misjudgments in parental selection and breeding approach. In addition, most fruit species have been maintained in a highly heterozygous condition and large seedling populations are required. The polyploid nature of many fruit species is a complicating factor in genetic analyses; segregation ratios for even monogenic characters become complex in tetraploid, hexaploid, or octoploid plants. Finally, the vegetative propagation of fruit crops has made it possible to perpetuate unique combinations which has tended to over-emphasize the occurrence of chance rather than purpose in breeding.

The current scientific approach to fruit breeding, augmented by ever-increasing basic knowledge in genetics, statistics, physiology, and other disciplines, has led to many new concepts and techniques within the last four decades. Much of this information is scattered throughout various horticultural publications; there has been no fully comprehensive compilation of this subject in a single volume since the United States Department of Agriculture's 1937 *Yearbook of Agriculture* (*Better Plants and Animals* 2). The 1937 *Yearbook* has been a valuable contribution to fruit breeders, but it is now seriously outdated. It is our purpose in compiling *Advances in Fruit Breeding* to bring up-to-date information on the breeding of temperate and subtropical fruit and nut crops. Our original intention to include fruit crops of the tropics proved to be unnecessary in light of the 1969 collection, *Outlines of Perennial Crop Breeding in the Tropics* (Miscellaneous Papers 4, Landbouwhogeschool, Wageningen, the Netherlands).

To accomplish our aims we have solicited outstanding breeders of specific fruit crops as authors. These scientists have not only reviewed the available scientific literature but have included many observations and data of their own not previously recorded. They have tried to use a world approach wherever possible combined with their own unique experiences as practical fruit breeders. Although there are common features in a discussion of fruit crops, many of which are related, we have thought it most useful to present this information on a species-by-species basis. While this has resulted in some duplication, we feel that this approach will prove to be most useful.

We are grateful to the contributors for taking the time to explain and record for others their knowledge and experience. We also are appreciative of the assistance of the following people who reviewed portions of the manuscript: R. L. Andersen, C. H. Bailey, B. H. Barritt, O. A. Bradt, R. S. Bringhurst, C. W. Campbell, M. N. Dana, D. F. Dayton, C. C. Doughty, A. D. Draper, H. W. Fogle, D. T. Funk, J. R. Furr, R. A. Hamilton, C. J. Hearn, C. O. Hesse, L. F. Hough, R. W. Jones, R. J. Knight, Jr., H. G. Kronenberg, H. B. Lagerstedt, R. C. Lamb, K. O. Lapins, R. E. C. Layne, J. W. Lesley, G. D. Oberle, H. P. Olmo, E. N. O'Rourke, D. K. Ourecky, John Popenoe, Charlotte Pratt, Piero Romisondo, D. H. Scott, W. B. Sherman, G. L. Slate, G. F. Solignat, L. P. S. Spangelo, Darrell Sparks, J. B. Storey, W. B. Storey, M. M. Thompson, T. K. Toyama, Ernesto Vieitez, J. P. Watson, R. D. Way, J. H. Weinberger, John Welke, and D. Wilson.

It is our hope that this book will be a useful reference source for professional fruit breeders as well as amateurs. We especially intend this for students of fruit breeding, and for young plant breeders entering the important and exciting profession of developing new fruit cultivars for the enjoyment and benefit of mankind.

James N. Moore
Jules Janick

Temperate Fruits

Apples

by A. G. Brown

The apple is undoubtedly the most ubiquitous of all fruits. It has been cultivated in Europe and Asia from the earliest times, being known to the Greeks and Romans and mentioned by Theophrastus in the third century B.C. Since then it has been spread by man into almost all parts of the world. This has been possible because of the great amount of genetic variability found in the apple which has allowed adapted types to be selected for different environments. Selection continues for new types to extend apple culture into colder regions. Orchards are now found in Siberia and northern China where winter temperatures fall to −40 C. Apples now account for approximately 50% of the output of deciduous tree fruit, and the annual world production, although it varies from year to year according to conditions in the main centers of production, is now over 23 million tons (Anon., 1970).

The greater part of world production is of dessert and cooking apples although the dual-purpose apple is gaining in importance. The high fruit quality now being demanded makes it necessary to have an outlet for the lower-quality fruit, and this means they must be suitable for processing in various ways or for the production of apple juice. In Europe a large tonnage of apples is grown specifically for the manufacture of cider.

At one time each country and area had its own particular local cultivars; this is still so, but only to a small extent for, with expanding production, the search for high yield and good quality has led to a few cultivars being grown throughout the world where climatic conditions are similar. The result is that the most widely grown cultivars are 'Golden Delicious' and the red sports of 'Delicious', both of American origin.

Another example is the increasing popularity in Britain of 'Mutsu' which originated in Japan and 'Granny Smith' from Australia. Apple breeding which was primarily to produce cultivars to meet national needs is now aimed at world-wide distribution, particularly since the introduction of breeders' rights.

Since the apple is a long-lived tree and propagated vegetatively, cultivars known hundreds of years ago are still extant. In the large world collection we have a great museum of trees showing the development and improvement of the apple and also containing a vast reservoir of desirable genes available when needed. These old cultivars, however, are not used extensively by breeders since their seedlings generally produce fruit very inferior in quality to the cultivars being grown today. In general most breeding is between the best recent cultivars. To introduce some character from the old cultivars requires a lengthy program which few are prepared to undertake. It is of utmost importance that these collections are maintained, for they represent a vast gene pool which may well have to be drawn upon at some time in the future.

Origin and Early Development

The apple, in company with most of the important fruits—pears, plums, peaches, apricots, cherries, strawberries, raspberries, and blackberries—cultivated in the temperate regions of the world, belongs to the Rosaceae or rose family. The apples, pears, quinces, medlars, and some less well-known genera have been classified into the subfamily Pomoideae or the pome fruits. These are characterized by having fruits consisting of two to five carpels enclosed in a fleshy covering. The genus *Malus* has, according to most authorities, 25 to 30 species and several sub-species of so-called crab apples many of which are cultivated as ornamental trees for their profuse blossom and attractive fruits. Many of the species intercross freely and, since self-incompatibility is common, trees grown from seeds obtained from a botanic garden or arboretum where collections of *Malus* are grown are frequently interspecific or intervarietal hybrids. It is therefore very difficult to be certain of the authenticity of specific names.

The cultivated apple is usually referred to as *Malus pumila* Mill. but undoubtedly *M. sylvestris* Mill. and other species have been involved in its evolution. *Malus baccata* Borkh. has been used in attempts to induce winterhardiness and, more recently, disease resistance from *M. floribunda* Sieb., *M. micromalus* Mak., *M. prunifolia* Borkh., *M. atrosanguinea* Schneid., and other species has been introduced into the cultivated apple by a backcross and selection program.

Malus pumila is generally considered to be the parent of most of our cultivated apples and is endemic in the area from the Balkans and southeastern Russia eastwards through the Transcaucasus, Iran, Turkestan, and north to the Altai Mountains in central Russia. Almost certainly *M. sylvestris*, which is distributed over much of Europe and into western Turkestan, was involved in the early development of the apple as was *M. baccata* from eastern Asia including eastern Siberia, Manchuria, eastern Mongolia, China, Japan, and the Himalayas. The center of origin or of greatest diversity of the apple is in southwestern Asia, and Vavilov (1931) in his explorations found many wild apples in woods in the Caucasus and in Turkestan bearing fruits in a wide range of sizes and some of quite good quality.

Cultivation of the apple seems to have been practised from a few centuries B.C. by the Greeks and Romans and, as a result of their travels and invasions, to have been spread by them throughout Europe and Asia. Later cultivation was concentrated around the medieval religious houses. Cultivars were selected and propagated in very early times, for budding and grafting were known at least 2,000 years ago. By the end of the thirteenth century, many named cultivars were known, and from this time we get the names of 'Pearmain' and 'Costard'. At the time when the first settlers were getting established in the New World, many hundreds of apples were grown in Europe and soon many of these were dispatched to America together with seeds from which new apples were raised.

Apples in those days arose either as selected wildings from the hedgerow or were grown from seeds of selected apples. It was not until Thomas Andrew Knight (1759–1835) began creating new apples by deliberate hybridization that fruit breeding on a scientific basis began. Knight was the first to apply the discovery of sex in plants to practical ends. "New varieties of every species of fruit will generally be better obtained by introducing the farina (pollen) of one variety of fruit into the blossom of another, than by propagating from single kind," he wrote (1806). He emasculated the seed parent and made deliberate crosses. Knight had shown the way and, although many more chance seedlings were to find a place among the best cultivars, consistent improvement has been made in many directions by scientific breeding.

Modern Breeding Objectives

The main breeding objective is aimed at producing good quality fruit as cheaply as possible. In order to achieve this the first requirement is high yield—in the region of 70,000 to 90,000 kg/ha (1,500 to 2,000 bushels/acre). To produce such crops, breeders must give attention to the efficiency of the foliage, structure of the tree to carry such weight, and to annual production. The quality must be such that it pleases the maximum number of consumers; that is, fruit must be firm and juicy, for while the soft apple may have minority appeal it is not desirable for large scale commercial production. Flavor is quite a problem since "good" flavor is a peculiarly personal thing. Since the fruit must have a more or less universal appeal, fruits with distinct, particularly aromatic flavors are only for the connoisseur. A pleasant but undistinguished flavor is regrettably the answer—following the maxim that most will like that which has nothing to dislike. The present trend is for self colors, good red, yellow, or green, while the rather nondescript striped fruits, with the exception of 'Cox's Orange Pippin', are not popular. All-over russet has a certain appeal but apart from this a peculiar prejudice exists against even the smallest trace of russet, especially on some cultivars, leading to a worldwide quest for a russet-free 'Golden Delicious', for example. Size is of considerable importance and the aim is for apples up to 75 mm in diameter, and although some prefer smaller, large fruit commands a premium.

Good storage qualities are essential for large scale production and, with correct harvesting and optimum storage conditions, the fruit should keep almost the year round. Breeding early ripening cultivars which have a season of a month or two must have a future only for small scale production.

Disease and pest resistance are of considerable importance in that maximum yield can only be achieved on disease- and pest-free trees. Many of the pests and diseases can be satisfactorily controlled by frequent applications of costly sprays. Breeders must concentrate on the production of new cultivars resistant to all the important diseases and to pests, for although growers may be quite content to control the diseases with sprays, the time has come when some insecticides and fungicides are considered unsafe. Legislation can quickly prohibit pesticide use and breeders should be prepared for such eventualities.

In some countries certain cultivars are grown especially for cooking and processing but the tendency seems to be for dessert apples to be also used for processing. Very little breeding is being done to produce new cooking apples although there is some demand from processors. There is probably still a place for breeding good specialty apples for those who want something other than the mass produced cultivars.

The apple as grown today is a composite tree usually consisting of a rootstock and a scion but occasionally a rootstock, interstock, and scion. The rootstock needs to be improved as well as the cultivar, and so the aim is to produce rootstocks with the potential for maintaining good growth and reproductive ability in a virus-free condition.

Breeding Techniques

Floral Biology

The flowers of apple cultivars and seedlings vary considerably in size, petal shape, and in color from white to deep pink. They are produced at the same time as the leaves, and are borne in cymose clusters on fairly short pedicels, usually on spur type growth but in some instances from the terminal or lateral buds of the season's growth. The flower consists of five petals, a calyx of five sepals, about 20 stamens and the pistil which divides into five styles. The ovary has five carpels each containing two ovules so that in most cases the maximum seed content is ten. A few cultivars have more than ten and the maximum in these cases is probably 20.

Pollination

Pollination is the mechanical transfer of pollen from the anthers to the stigmas and is usually a prerequisite to the fertilization of the ovules and the development of seeds and fruit. Most apples require cross-pollination and this necessitates the transfer of pollen from the flowers of one tree to the stigmas of another tree of a different cultivar. In the orchard, pollination is carried out by insects, particularly bees, in their search for pollen and nectar as they move from flower to flower and from tree to tree. Most flowers are admirably designed to facilitate this process but in a few genotypes the stamens are long and the styles short which allows the bees to visit the flowers without making contact with the stigmas. This character may lead to a considerable loss of the potential crop.

In making controlled crosses the pollen has to be transferred by the breeder. The pollination may be made on small trees planted in pots or small tubs and transferred to a cool greenhouse to flower and fruit. In this way the trees are protected from the weather and, provided the house is screened against insects, conditions are ideal for controlled cross pollination. The number of pollinations that can be made in this way is limited by the size of the trees and by the size of the collection of potential parents that can be kept in this way. There is much more freedom of choice of cultivars and size of tree in the orchard and large-scale hybridizations can only be done on trees grown in the open.

Pollen collection: The collection of pollen is the first part of the process of hybridization. Flowers are collected preferably at the balloon stage just before the petals expand, and before the anthers dehisce, although circumstances often demand that they are collected a little earlier than this. The flowers are taken into the laboratory or greenhouse and the anthers removed with a small comb or rubbed on a screen into petri dishes or similar containers. These are put into a warm or sunny position and the anthers will soon dehisce and liberate the pollen. Anthers collected too soon tend to shrivel and fail to dehisce. The pollen can be used immediately. Since all cultivars do not flower at the same time and six weeks may elapse between the flowering of the earliest and the latest, it is often difficult to synchronize the flowering of both parents. When parents do not flower at the same time, pollen may be collected from the earlier parent, and stored until required. The pollen is put into small vials very loosely stoppered with cotton-wool, put into a desiccator with calcium chloride, and kept cool in a domestic-type refrigerator. When it is desired to keep pollen for a year or more it should be prepared in the same way but put in a deep-freeze unit at −15 C. If proper conditions are maintained and the pollen dry it will remain viable for some years.

In planning a breeding program one may wish to use certain cultivars not available locally. With the cooperation of other breeders or fruit growers it is a simple matter to have pollen sent from any distance by ordinary air mail letter. The pollen is collected in the usual way and when dry it is put into a cellophane packet, sealed and dispatched. This is much more satisfactory than using rigid containers. This saves the tedious wait for scion wood to be sent, grafted, and for the young trees to flower.

Quality: Quality of the pollen may be tested after staining with acetocarmine-glycerol jelly (Marks, 1954) and determining the proportion of good appearing grains microscopically. In order to determine the viability of the pollen a sample can be germinated on a drop of sugar solution (2.5 to 20%). Most recommend 10% sucrose in aqueous solution or 5% agar and a temperature of 20 to 25C for good pollen tube formation in order to calculate the percentage of germination. The addition of 10 ppm boron to the medium increases germination (Thompson and Batjer, 1950).

Emasculation: Before crossing, the flowers of the "female" or seed parent must be emasculated. This consists of the removal of the stamens before the anthers dehisce. This serves two purposes: firstly, it prevents self-pollination and contamination, and secondly, it exposes the stigmas and facilitates cross-pollination. Even when trees are grown under glass or screened from insects and the parent tree is self-incompatible it is helpful to emasculate. With trees growing in the open, emasculation is essential because the stamens and petals attract insects which bring unwanted pollen and contaminate the crosses. When hybridizations are being made for genetical research as well as for breeding, then very strict precautions need to be taken to prevent contamination. If, however, the crosses are part of a breeding program with the sole purpose of producing new improved cultivars, then 100% control is not essential. One way of preventing the bees from being attracted to the flowers is to remove the petals, sepals, and stamens with the nails of thumb and index finger just before the buds have reached the balloon stage. One can become quite expert at this with a little practice. Only the styles are left sticking up from the receptacle and ovary. There are other methods including the use of button-hole scissors with a notch cut in the blade which has proved to be very efficient (Barrett and Arisumi, 1952).

Apple trees produce flowers in great abundance and only a small proportion of them can set and produce fruit. It is therefore quite unnecessary to emasculate all the flowers and as a rule two flowers per cluster are ample and the others should be removed. Any unemasculated flowers producing fruit can be identified, as the fruit from emasculated flowers have no calyx.

Experiments have shown that insects do not visit flowers when the stamens and petals have been removed and there is then no need to protect them from insects (Visser, 1951). Fruit set is improved, however, if the emasculated flowers are protected with bags since the desiccation of wounded floral parts is lessened. In windy and exposed situations, such protection is desirable.

Cross-pollination: Pollination is achieved by dipping a small, soft brush into a vial containing the pollen and lightly brushing the pollen onto the stigmas. The brushes are cleaned by washing in absolute alcohol and drying before using for a cross involving pollen from another parent. Pollination can be done equally well by dipping the finger tip into the pollen and touching the stigmas to transfer the pollen. Either way is quick and effective. Some pollinate flowers immediately after emasculation; others prefer to emasculate and return in a few days and make the pollinations when the stigmas are more receptive. Both methods are effective but the latter seems to give slightly better results.

The whole tree may be pollinated with one parent or a number of crosses may be made on different parts of the tree, providing limbs are tagged carefully to prevent mix-ups.

The fruit should be harvested when ripe, and seeds removed from the mature fruit. Fruits which have been attacked by birds or otherwise damaged quickly rot, so the seeds should be removed immediately after the fruit is harvested to avoid trouble during stratification. Apple seeds will not germinate unless stratified and this involves keeping the seeds in moist conditions and subjecting them to a period of cold. This allows the after-ripening process to proceed, during which changes take place in the embryo. When this process has advanced sufficiently, the seeds can be placed in warm conditions and germination proceeds. After-ripening will proceed at temperatures between 0 and 10 C, but the optimum is from 3 to 5 C; the time required may be anything from six to 14 weeks and depends to some extent on the temperature. Abbott (1955) has shown that the temperatures for after-ripening are critical with a threshold temperature of about 17 C at which the seeds are held and their capacity to germinate is constant. Above 17 C the process goes into reverse and the longer they are held above this temperature the longer will be the period of cold required to bring them to a condition to germinate.

In practice, seeds should not be stored dry but stratified on removal from the fruit. They can be put in damp sand or peat and kept in cold conditions out of doors or in cold frames. Modern materials, however, provide easier and more certain methods. Seeds may be placed thickly on a moist, 6mm (¼ inch) thick, plastic foam sheet in a tray and covered with another sheet, another layer of seeds, a sheet of foam, and so on until the tray is full. The tray is put into a polythene bag to prevent drying out and put into a refrigerator or cold room and kept at 3 to 5 C. After six weeks the seed should be examined from time to time, for even at that temperature some seeds will start to germinate. When this is noticed the tray should be removed and placed in warmer conditions and as the seeds germinate they should be planted in boxes or similar containers and given optimum conditions for growth.

Rots may be troublesome during stratification particularly if the seeds are not clean when put in the trays. *Rhizoctonia solani* Kühn, the principal culprit, grows on the outside of the testa and in due course penetrates and causes the embryo to rot, quickly spreading from seed to seed. It is always advisable to surface sterilize the seeds in calcium hypochlorite solution (10 g in 140 ml water and filtered) for five minutes, and then wash them before putting them in the trays.

Juvenility

Apple trees, like many other woodland trees, undergo different phases of development between the germination of the seed and the adult fruiting tree. The first is the juvenile phase, during which plants may differ considerably from mature adult trees. The leaves are smaller, usually more finely serrate, and the shoots are thinner and are often produced at right or obtuse angles to the main stem. During this phase no flowers are produced. The onset of the adult phase is marked by the development of flower buds. Between these two phases there is undoubtedly a transition phase when the lower part of the plant is still juvenile and the upper part adult. Cuttings from most adult apple trees are extremely difficult, if not impossible to root. Cuttings from young seedlings root readily but as the seedlings age rooting is more difficult to achieve, suggesting that the transition is gradual. The duration of the juvenile phase is very variable and may be anything from three to ten or more years depending on the genetics of the seedlings and the cultural practices employed.

Over the years many investigations have been carried out into ways of shortening the duration of the juvenile phase (Kemmer, 1953; Murawski, 1955; Visser, 1964). Methods used to induce adult trees to flower such as pruning, root-pruning, bark-ringing, etc., have been tried. These methods, and indeed any method which restricts growth in the very young seedling, do not shorten the duration of the juvenile phase but in fact lengthen it and thus delay the onset of fruiting.

Cultural methods which check the growth of seedlings are only effective in hastening flowering when a certain stage of development has been reached. Thus Way (1971) was able to induce seedlings to flower by bark-ringing when they were four years old, yet three-year-old seedlings failed to respond. Four bark-ringing techniques were used, all equally effective.

There are other cultural methods which can effectively shorten the juvenile phase. One is to grow the seedling as fast and vigorously as possible. This can best be achieved by avoiding any check to growth in the early stages of development and where this is not possible to keep disturbance to a minimum. Seedlings grown under glass, closely planted, watered, and fed, may reach 3m in height in the first season's growth compared with 1m for seedlings planted out in the field. These may be planted in their permanent positions when they go dormant. They should continue to grow freely and, since the seedlings first come out of the juvenile condition at the top of the seedling, they should be left unpruned until they flower and fruit (Brown, 1964; Zimmerman, 1971). In the other method Tydeman and Alston (1965) have shown that the juvenile phase can be considerably shortened by budding seedlings onto dwarfing rootstocks. Buds were taken from the upper part of the main shoot in late summer of the second year's growth and worked on the dwarfing rootstocks 'M.9' and 'M.27', grown as closely planted cordons with all lateral growth pruned back annually in late summer. Nine years after germination 88% of 902 budded seedlings had fruited compared with 49% of those on their own roots. A possible disadvantage of this method is the risk of contamination of the seedling from a virus-infected rootstock. Where rootstocks are virus free, as in the apomictic seedlings of *M. sikkimensis* Koehne, this does not arise and a similar shortening of juvenility is achieved (Campbell, 1961).

Early Selection

As soon as the seeds have germinated and the first true leaves are showing, the long process of seedling evaluation may begin. It is possible to select for resistance to some of the important diseases at this stage. In fact this is the best time to select for resistance to diseases where the initial infection is through young leaves, such as scab (*Venturia inaequalis* [Cooke] Wint) and cedar-apple rust (*Gymnosporangium juniperi-virginianae* Schw.). Inoculations with spore suspensions at this early stage show infections on the young leaves in a few weeks and the susceptible seedlings can be discarded. Selection for scab resistance alone can reduce the size of the progeny by 50%. Some other diseases such as mildew (*Podosphaera leucotricha* [Ell. & Everh.] Salm) are better left until the seedlings are in their second year of growth before being assessed for resistance. It is important to be able to discard inferior seedlings at the earliest stage of growth.

Many growth characters can be evaluated in the early years of development of the seedlings such as tree weakness, the production of long spindly shoots, and other defects in general growth which, no matter how good the fruit might be, would prevent the seedling from ever making a satisfactory tree. Any such defects should be enough to eliminate the seedling. There are also some correlated characters which can be detected, such as late leafing, which is associated with late flowering, and early selection can be made when this is one of the breeding aims.

Preselection

A search is always being made for characters which can be assessed in the juvenile stage which are correlated with fruit characters in the mature tree. Any positive correlation of this kind would be of considerable value in allowing seedlings with undesirable fruit characters to be discarded at an early stage. Such correlations must fulfill two particular requirements: 1) the degree of correlation must be of a fairly high order and 2) the character in the juvenile plant must be reasonably easy to recognize, for any time-consuming series of measurements or complicated chemical analyses greatly reduces the value of the correlation as a practical method of preselection. In order to establish correlations between leaf and fruit characters, the tests are often carried out on mature trees so that the leaves and fruit are available during the same season. Where correlations appear to be established in this material, there is no proof that the same correlation will hold when juvenile leaves are used.

A number of such correlations have been claimed but few are of any real value to the breeder. Nybom (1959) found a correlation between the pH of the leaf sap and the pH of the fruit juice. The leaves of the sweet type usually have pH 5.7 and the acid types 5.5–5.6. This has not proved to be a reliable method as the differences are so small that the error between samples from the same seedling are often greater than between seedlings. Another correlation that is often claimed is between red fruit color and anthocyanin pigmentation of other parts of the tree such as one-year-old shoots, leaf petioles, etc. The distribution of anthocyanin in the tree depends on so many factors and there are so many exceptions to these correlations as to make this of no value as a method for predetermining fruit color. Only in progenies from trees which are pigmented throughout the tissue with purplish red anthocyanin, as in *Malus pumila niedzwetzkyana* Schneid., can red-leaved seedlings be selected which will produce trees with red leaves and shoots and fruit with purplish-red skin and flesh. Some useful correlations have been established between late leafing and late flowering, between length of juvenile period and fruiting age in the vegetative propagants, and others which will be discussed under the appropriate character headings.

Evaluation of Fruit

When the seedlings have emerged from the juvenile phase and produce flowers and fruit, the most important part of seedling evaluation begins. It is assumed that the seedlings have already been screened for characters such as weak growth, poor habit, small leaves, and other undesirable features and that these have been eliminated or at least discounted. The next stage is to assess the fruit of the seedlings for quality. There are many criteria by which the fruit can be judged and to fall below the standard in any of these will automatically eliminate the seedling. Size is one which can be easily assessed. If the fruits are consistently below 60mm in diameter they are too small for the present-day standard for a commercial dessert apple. Color, too, is important and the preference is for clear yellow, bright red, or bright green apples. These should be free from russet or nearly so, although there may be a place for some completely russet apples. Dull-looking apples even with many good attributes are not worthy of further consideration for the public buys by sight. Shape is not very important and, provided that the fruits are not peculiarly shaped or very irregular in outline, they will be accepted by the consumer and grower alike. Flavor is of prime importance. Many seedlings have fruit with very little acid and in consequence they are very sweet and insipid. This type is quite unacceptable. Fruit may have a strongly aromatic or distinct aniseed-like flavor and, while these may prove to be good home garden apples, they are not acceptable as commercial apples for large scale production because such flavors are not universally liked. The ideal is sub-acid with a pleasant but not too distinct flavor. The connoisseur who likes the subtle flavor of some apples will no doubt grow his own. The texture of the flesh should be firm and crisp with plenty of juice. Those who prefer a soft apple are very much in the minority.

When the fruits of the seedlings have been examined and those that meet all the requirements have been selected, the next stage of selection is for yield. For this a number of trees need to be propagated on fairly dwarfing rootstocks and grown in properly conducted trials with the best standard cultivars as controls and crop weights taken and grading assessed. At the same time, the selection can be framework-grafted onto an existing tree so that fairly large samples of fruit can be seen and can be subjected to storage trials and examined for susceptibility to storage diseases. Where possible, the selected seedlings which show real promise should also be grown on trial in a number of localities to see if their behavior is consistent under varying conditions. In these trials, only those selections giving very heavy annual yields need be considered further.

During the whole process of selection it is important to be absolutely ruthless and to eliminate from further consideration any seedling which does not reach the required standard in every respect.

Much time and trouble is taken unnecessarily in carrying out trials on selected seedlings which never make the grade and which, if properly evaluated, would have been eliminated earlier. One good rule is that if there is any doubt that a seedling is good enough then it should be discarded. There must be no doubt!

Records

A most important part of any breeding program is the keeping of records. What records are kept, and how detailed they need to be, depend entirely on how much information one wants to extract from them for future use. Records should obviously be kept of the parentage of crosses and the seed yield from the fruit harvested. This can be a guide to good seed parents since some parents are much better in this respect than others. Detailed records can be kept of tree, flower, and fruit characters of all the seedlings but very careful consideration should be given to the characters to be recorded. Field recording is very time consuming and therefore only characters about which further information is required should be recorded. Seedling records can be used in studies of the inheritance of characters and in calculating the contribution to specific characters that a parent is making to its progenies. This information can be used to estimate the effect a given parent will have in combination with other parents. Unless the records are used in this way it is largely a waste of time to collect them.

There are a number of ways of making field records but whatever method is used it is essential that all the seedlings are recorded in the same way for the same characters, otherwise the results cannot be analyzed. The "chatty" description is quite useless. One common method is to have printed or duplicated sheets on which the characters are listed with the possible categories for each, such as "Fruit shape: oblong, conical, roundish, oblate." The appropriate one is underlined or ticked. This usually requires one sheet per seedling. Another method is to use a numerical classification where for example, for size: "1 = very small; 2 = small; 3 = medium; 4 = large; 5 = very large." Such a method saves much time and space providing the characters are printed across the top of the sheet above the appropriate columns. The details of a seedling are contained in one line across the sheet.

More and more use is being made of computers and there is no doubt that they can play a large part in breeding research and planning. Field data can be analyzed, meaningful genetic studies can be made, the inheritance of various characters can be calculated, correlations may be made, and much useful information can be extracted. Data can be recorded in the field with the aid of a tape recorder or on punch cards using one of several different systems (Fogle and Barnard, 1961). Before embarking on a recording system of this kind it is advisable to consult the computer operator as analysis may be easier if all the characters are recorded in one digit classes (0–9).

Records should also be kept of data and information relating to cultivars and selected seedlings used, or likely to be used, as parents. These are most conveniently kept on punch cards and information on a vast number of characters can be recorded on one card. Data on fruit size, shape, season, resistance or susceptibility to pests and diseases, and so forth can be collected both from field records and from published information. By this method all the parents having any particular character can quickly be withdrawn from the pack.

Mutations and Chimeras

A bud sport or mutation is a variant of an inherited character arising in a cell from which a bud eventually develops; and this in due course produces a shoot and later a branch which differs, usually in only one character, from the plant on which it was produced. These mutations can affect any part of the plant. In the apple, quite naturally, mutations affecting the fruit are those most commonly observed; many others no doubt arise which pass unnoticed.

There are two important types of mutations which occur in the apple: 1) those which produce single gene differences in some character of the tree or fruit and 2) those which alter the ploidy. Both types occur naturally but the mutation rate is very low. Mutations affecting the appearance of the fruit are easily recognized and are in consequence the type most often found. Increases in the amount of anthocyanin in the outer cell layers of the fruit skin are the most common and red sports of many of the popular cultivars have arisen. Some cultivars such as 'Delicious', 'Rome Beauty', 'Winesap', and 'Cox's Orange Pippin' are very prone to produce mutants of this type while other widely grown cultivars never seem to mutate. All the red sports of one cultivar are not necessarily the same, for apart from differences in the area of the fruit covered, the intensity of the pigmentation in the three outer cell layers can differ (Dayton, 1959). The mutation is likely to be limited to one cell layer only and therefore the plant is a chimera; since the gametes are formed from the subepidermal layer (Crane and Lawrence, 1952), only mutants in that layer will differ from the original in their breeding behavior.

A bright red fruit is considered desirable and when these sports are discovered they should be propagated and tried to see if they are superior to the original clone. Russet sports also occur from time to time and some of these are also desirable providing the russet is not rough and the coverage is complete. Mutations of this character can be produced in either direction and non-russet sports can arise from completely russet apples.

Important mutations affecting a tree character which have been observed and selected in recent years are the so-called spur or compact types which produce, as the names imply, dwarf, freely spurring trees. These are being sought in all the important commercial apples. Presumably any character can be affected and a watch should be kept for such mutations for such important things as resistance to diseases and pests. There is also no doubt that mutations can occur which may not be very obvious and which are inferior to the original clone, and careful observation should be made on stock trees so that no inferior mutants are unwittingly propagated.

The rate at which single gene mutations occur can be increased by irradiation by x-rays, gamma rays, or thermal neutrons. Bishop (1959) produced two dark red sports of 'Cortland', two sports of 'Sandow' with less color and more russet, and a 'Golden Russet' with considerably less russet than normal. Lapins (1965) and others have produced compact mutations by means of irradiation. The effective dosage is 3–5 kr for dormant scions and 2–4 kr for summer buds when x-rays are used and $3.9 - 15.6 \times 10^{12}$ thermal neutrons per cm^2. Mutants from irradiated material do not readily show themselves and the normal growth if allowed to develop will suppress the mutant. When dormant scions have been treated and grafted, the basal ten buds or so of the shoot emerging from each treated bud should be removed and budded onto a dwarfing rootstock. The mutants are likely to show in the second vegetative generation.

The other mutations important in apple breeding are the "large" or "giant" sports (Einset and Imhofe, 1947, 1949, 1951). These are not due to single gene differences but to a change in ploidy. These giant sports are tetraploids with double the number of chromosomes of the parent tree. These sports are usually recognized first by their large fruit, which are sometimes twice as large as their diploid counterpart. The fruits, apart from the increase in size, are usually flatter and more irregular in shape. The growth is vigorous, the shoots are thick, and tend to be short-jointed, and the leaves are thicker, broader, and rounder, while the tree habit tends to be flatter with more widely angled branches.

Giant sports of many of our most widely grown cultivars are now known and the majority of these have arisen spontaneously and have been discovered in orchards by observant growers. One peculiarity, although an expected one, is that when these mutations occur all three outer cell layers are not usually affected so that practically all of these tetraploids are periclinal chimeras and can be grouped according to the arrangement of the 2x and 4x tissues. Group 1, in which the epidermis is diploid and L-2 and L-3 and the rest of the plant is tetraploid, is designated 2-4-4-4; Group 2 as 2-2-4-4, and Group 3 as 2-2-2-4. Other combinations of cell layers are possible (Derman, 1951, Pratt, Ourecky and Einset, 1967), but in spontaneous mutations, 4-4-4-4 are very rarely found (Batra, Pratt and Einset 1963).

The breeding behavior of these sports differs according to the position of the 4x tissue. Only when L-2 is tetraploid will the tree breed as a tetraploid. Thus many giant sports, apart from having tetraploid growth characters, genetically behave as diploids. It is, however, possible to produce homogenous tetraploid plants from 2-2-4-4 sports by inducing shoots to grow from endogenous tissue. The method is to grow one-year-old trees of the giant form in large pots, cut them down to 30 cm, and remove all the buds and any growth that appears in the region of the removed buds or from the rootstocks. This encourages the formation of sphaeroblasts in the internodal regions. These will in due course crack and produce adventitious buds which develop into shoots. Having been developed from the region of the phloem, these shoots should be 4-4-4-4 (Derman, 1948).

Although practically all the 4x sports have arisen spontaneously, it is possible to encourage the formation of tetraploids by the use of colchicine. There have been no spectacular successes with this treatment as there have been in many other plants. The method is to use one-year-old trees which are cut back to encourage a vigorous shoot. When this is growing rapidly the colchicine should be applied to the growing point either in solution or in lanolin or glycerine. The leaves at the growing point are stripped back and the colchicine applied in one application if in lanolin, or daily for three or four days if in solution. The strength used is 1% colchicine. The point of application needs to be marked and later, when the shoot has grown, further buds are removed above the treatment point and budded onto dwarf rootstocks much in the same way as for radiation mutants (Hunter, 1954).

This colchicine method has also produced a number of other mutants showing differences in color, russet, and time of fruit ripening. The explanation for this is not obvious but may have been brought about by a segregation of chimerical tissue (Brown, 1966).

Breeding Systems

The majority of cultivated apples are diploids (2n = 34). There have been suggestions (Darlington and Moffett, 1930) that they are complex polyploids, being partly tetraploid and partly hexaploid with the basic number of x = 7, which is common in Rosaceae. The hypothesis, based on the associations and behavior of the chromosomes, is that the 34 chromosomes are made up of four sets of seven chromosomes and six sets of three chromosomes.

The basic chromosome numbers of 8 and 9 also occur in Rosaceae, and Sax (1931) put forward the theory that all the Pomoideae were allopolyploids derived from a doubled hybrid between two remote ancestral types, one with the basic number of 8 and the other 9. Derman (1949) while agreeing in general with Sax, considered that, because of the great diversity between genera in the Pomoideae, each genus must have arisen in this way independently. Be that as it may, they are functionally diploids and, although most of the characters which have been studied are polygenically controlled, there are a few under single gene control which have diploid segregation.

Among the cultivars there are also triploids (2n = 51). These have arisen naturally from the fertilization of unreduced gametes. Among the cultivated apples triploids appear to be more common than one would expect, accounting for about 10% of the commonly grown cultivars, yet only appearing in populations from diploid parents at the rate of approximately 0.3%. Some of these apples are of considerable importance such as 'Baldwin', 'Gravenstein', 'Rhode Island Greening', 'Blenheim Orange', and more recently 'Mutsu', to name but a few. It would appear that there is more chance of a triploid seedling being outstanding than a diploid. They are more vigorous and tend to have larger fruits which might account for some of their selective advantage. Triploids produce very poor pollen and this leads to pollination problems in the orchard since it requires a diploid to pollinate the triploid, and another diploid to ensure fruit on the diploids. The triploids produce very few good seeds; nevertheless they are sufficiently fertile to produce a good crop of fruit if properly pollinated. They are more or less useless as parents for further breeding since if they are selfed or crossed they produce few seeds, practically all of which produce aneuploid seedlings which are weak and seldom develop into trees.

In more recent years tetraploids (2n = 68) have arisen or at least been discovered as large fruited sports on trees of diploid cultivars. These usually occur as chimeras with one or more layers of diploid tissue over the tetraploid. Where only one diploid layer occurs, the tree behaves as a tetraploid in the formation of gametes; with more than one diploid layer, it behaves as a diploid. In growth and fruit the tetraploid seems to have no advantage over the diploid counterpart having thick, rather short-jointed growth, and large, somewhat misshapen fruit. By crossing those which breed as tetraploids with diploids, large progenies of triploids can be produced. These are not quite the same as triploids derived from an unreduced diploid egg cell fertilized by haploid pollen, and it is believed that the latter are superior to those derived from tetraploids x diploids.

The explanation for the superiority of these

natural triploids given by Knight and Alston (1969) is that in triploids produced from 4x × 2x parents, the gametes from both parents have undergone meiosis and, in consequence, have a random re-assortment of genes. Natural triploids, however, usually arise from the fertilization of an unreduced egg cell by a haploid pollen grain. The pollen has undergone full segregation but the maternal gamete, on the other hand, will have undergone the minimum of reassortment and is, therefore, contributing to the resulting triploid two-thirds of the genomic constitution almost unchanged from a parent that was probably a very successful cultivar.

The usual problem in this type of breeding is that so many seedlings have to be grown to get a reasonable number of triploid seedlings from which to select. The method used by Knight and Alston is to save seeds from discarded fruit from a pack-house. The female parent is known and if the pollinator in the orchard is known, so much the better. By selecting the largest seeds by eye from the mass of seeds saved, growing seedlings, and selecting for leaf size and vigor, a number of triploids can be obtained without growing the thousands of seedlings that would otherwise have been necessary.

Tetraploids have been produced from triploids pollinated by diploids. The number of good seeds produced from such crosses is very small and most of these produce aneuploid seedlings. Laubscher and Hurter (1960) made root-tip counts of 884 seedlings from triploid cultivars growing in a mixed orchard of diploids. In all, eight tetraploids were obtained or one in 110 seedlings. All triploids do not respond in the same way and some produce many more tetraploids than others (Einset, 1952). The tetraploids are presumed to arise from unreduced 51-chromosome egg cells fertilised by 17-chromosome pollen, and it is suggested that tetraploids produced in this way may be better parents for breeding than those from colchicine-treated diploids. The first tetraploid cultivar to have been bred in this way is the Swedish 'Alpha 68'.

One disadvantage of many of these polyploid seedlings is that they have very long juvenile periods (some of the natural triploids are said to be better in this respect) and there is some evidence that this is correlated with slowness to come into bearing after propagation of adult material.

Hexaploids have also been obtained by colchicine treating triploids and 6-3-3 and 3-6-6 chimeras have been produced of the cultivars 'Paragon' and 'Stayman' (Derman 1965).

Sterility and Incompatibility

There are two main causes of unfruitfulness in the apple—sterility and incompatibility. The former, in the form of generational sterility, can be brought about by the failure of any of the processes concerned with the development of pollen, embryo-sac, embryo, and endosperm. This is very clearly seen in the formation of the gametes in the triploids due to chromosome imbalance which leads to very poor seed set and to a very small percentage of good pollen being produced. Even in diploids, however, there is often quite a lot of aborted pollen and some cultivars are much worse than others in this respect. This may be due to lethal genes and several suggestions have been put forward to explain this occurrence. Gagnieu (1951) concludes that the segregation suggests a simple disomic inheritance of four different and possibly allelomorphic genes p^1, p^2, p^3, p^4.

Sexual incompatibility which is due to the failure of the pollen, although functional, to grow down the style and bring about fertilization, is widespread in the apple. Self-incompatibility is particularly common, although cases of cross-incompatibility are also known. Apples have an incompatibility system whereby the pollen tube growth is arrested in the style. Pollen germination is not noticeably lower in incompatible pollinations than in compatible ones and is not related to subsequent pollen tube growth.

Practically all apple cultivars are self-incompatible to some extent: some are completely so and even those which appear to be self-compatible set more fruit with higher seed content when pollinated with a cross-compatible cultivar, showing that even in such cases some degree of self-incompatibility is present. The figures from Crane and Lawrence (1952) of percentage of fruit set from selfing about 50 cultivars ranged from 0 to 9.6%. In some the number of flowers pollinated was rather small and so some results may be somewhat exaggerated. Brittain (1933) selfed 'Cox's Orange Pippin', 'Golden Russet', and 'Northern Spy' on a large scale—several thousand flowers of each—and the mean fruit set was about 1.5% although when crossed they set about 7%. Cross-incompatibility is not very common but a number of cases have been reported, usually between closely related kinds. Sports of a cultivar are cross-incompatible with the original form and with the other sports but this is more strictly self-incompatibility.

Triploid cultivars behave in much the same way as diploids, varying in the degree of self-incompatibility and also producing a considerable increase in fruit set when pollinated with diploids, but varying very considerably when crossed with other triploids. Natural tetraploids are also very variable, some being possibly self-fertile and others only partly so.

In a series of crosses (Brown, unpublished), the following results were obtained: 2x × 4x gave 9% fruit set; 4x × 2x gave 3% fruit set; 3x × 4x gave 7% fruit set. Other information suggests that 4x × 4x sets fruit freely.

Apomixis

Some plants, while appearing to produce seeds in the normal fashion, actually reproduce vegetatively from unfertilized eggs. This form of reproduction, known as apomixis, is found in many plants of hybrid origin, some of which never produce true seeds as a result of fertilization of the ovules; others are usually apomictic but can on occasions produce sexual hybrids.

Facultative apomixis is characteristic of a number of *Malus* species which are probably of hybrid origin but do not appear to occur among the cultivated apples. The apomictic species which have been investigated are polyploids. *Malus sikkimensis* (Hook) Koehne is a triploid; *M. coronaria* (L.) Mill., *M. hupehensis* (Pamp) Rehd., *M. lancefolia* Rehd., *M. platycarpa* Rehd., and *M. toringoides* (Rehd.) Hughes are known in both triploid and tetraploid forms; *M. sargenti* Rehd. is known in diploid, triploid, and tetraploid forms, and *M. sieboldii* (Reg.) Rehd. is known in diploid, triploid, tetraploid, and pentaploid forms. It is probable that the diploid forms of *M. sargenti* and *M. sieboldii* are sexual.

Under normal circumstances these species reproduce themselves freely by apomictic "seeds" but most of them can produce sexual hybrids if crossed with sexual diploids. Sax (1959) has shown that in controlled crosses the tetraploid form of *M. sargenti* when pollinated with sexual diploids produced maternal tetraploids and hybrid triploids and pentaploids; the triploid var. *rosea* produces maternal triploids and a few tetraploids. Seedlings from these apomictic species are not necessarily identical and a certain amount of variation can be found. The importance of this character in *Malus* species is that the seedlings of some are sufficiently uniform to be used as rootstocks which are virus free. There are certain problems of stock-scion incompatibility and sensitivity to virus which will be dealt with in the section on apple rootstock breeding.

The inheritance of apomixis is not very straightforward. Sax (1959) considered apomixis to be a dominant trait in the F_1 hybrids from controlled crosses, whereas the evidence from more recent work by Schmidt (1964) shows it to be recessive but varying according to the ploidy of the species. Triploid hybrids of apomictic x amphimictic forms of *M. sieboldii* were in the ratio of 6 amphimicts: 1 apomict, but in crosses between high chromosome number apomicts and low chromosome number amphimicts, the proportion of apomicts was higher and increased proportionally to the difference in chromosome number.

Parthenocarpy

The apple usually has ten ovules but it is not necessary for them all to be fertilized and develop into seeds for a fruit to be produced. Often a single seed is sufficient for the development of the fruit; thus fruitfulness may still be maintained even when a high degree of generational sterility is present. It is even possible in some cultivars and in certain conditions for fruit to develop parthenocarpically without fertilization and without seeds. Fruits which arise in this way vary according to the cultivar; in some they are small and often misshapen, and in others they are quite normal in size and appearance. There is evidence, however, that these fruits tend to ripen earlier and do not keep so well in storage as seeded fruits.

Recommendations have been made for cultivars which produce parthenocarpic fruits consistently to be planted because of their ability to produce a good crop in years when the flowers are damaged by late spring frosts or when conditions are unfavorable for pollination (Thiele, 1956). Ewert, as long ago as 1909, was advocating the planting of seedless apples as a safeguard against failure of fertilization. Others have suggested breeding for this character but it is doubtful if any has been undertaken. Some quite heavy-cropping cultivars have been described which produce parthenocarpic fruit but none seems to have been grown to any extent.

Parthenocarpy can be induced in some cultivars by spraying with various hormones.

Inbreeding

Any program of inbreeding by selfing is very difficult to pursue with apples because of the high degree of self-incompatibility. A great many flowers have to be selfed to produce a few fruits from which only a few seeds are obtained. The average fruit set from selfing a great many cultivars is about 2% and the number of seeds per fruit averages one to two. Only about 30% of the seeds germinate and a great many of these are weak and die. The end result is that only a very small proportion of the seeds produce viable seedlings capable of flowering and fruiting. In these selfed progenies it is possible to get a few good seedlings which, if not completely self-incompatible, can be selfed again. By selecting the most vigorous and most self-fertile it has been possible to carry this process through three selfed generations at the John Innes Institute, Norwich, England, and it may be possible to carry on long enough to achieve some degree of homozygosity without completely losing vigor.

It is important when making crosses between apparently widely different cultivars to consider their parentage and pedigree, if known. Otherwise, disappointing results may occur through unintentional close inbreeding. This is illustrated in Fig. 1. If these crosses had been planned taking only the fruit characters and vigor of the parents into consideration without studying their parentage, the poor results would have been entirely unexpected.

Looking at the parentage of new selections and cultivars, it is significant that a very large proportion of them are derived from a backcross to one of the parents, usually 'Cox's Orange Pippin', 'Delicious', 'McIntosh', or one of the well-known cultivars, or from crossing two half-sibs. It appears that a limited amount of inbreeding is helpful in intensifying certain desirable characters without having unduly harmful effects on the vigor of all the seedlings. In such crosses, however, inbreeding depression does occur and large progenies should be grown to overcome the deficiency of vigorous seedlings.

Outbreeding and Backcrossing

Most breeding programs are based on outcrossing, usually between cultivars of known merit, with the object of combining the good qualities of both parents in some of the progeny. Much of the present breeding is aimed at specific objectives, and

FIG. 1. Mortality and average height of apple seedlings after six years with various degrees of inbreeding

Mortality	Average height			
3.2%	2.62 m	Ellison's Orange	Cox's Orange Pippin	
			Calville Blanche	
		No. 413	Antonowka	
			Worcester Pearmain	
29.5%	2.00 m	Merton Beauty	Ellison's Orange	Cox's Orange Pippin
				Calville Blanche
			Cox's Orange Pippin	
		Merton Charm	McIntosh	
			Cox's Orange Pippin	
55.0%	1.56 m	Merton Beauty	Ellison's Orange	Cox's Orange Pippin
				Calville Blanche
			Cox's Orange Pippin	
		Cox's Orange Pippin		

parents are selected which are expected to impart some specific character to the offspring. In such cases a certain amount of backcrossing is used to introduce one specific character into another background as in single gene resistance to disease. Dominant single gene resistance in a *Malus* species, for example, can be transferred to the cultivated apple by a modified backcross procedure to avoid inbreeding (Hough et al., 1953). The method involves crossing the wild species with a large-fruited cultivar. The resistant F_1s are heterozygous and the best are selected and backcrossed to a good cultivar and their progeny yields 50% resistant seedlings. The best of these are again backcrossed to a good cultivar and so on until all the good qualities of the cultivated apple are recovered and the resistance from the species retained. This form of backcrossing avoids inbreeding by alternating different cultivars for the "recurrent" quality parent and eliminates loss of vigor and incompatibility problems.

Characters Controlled by Single Genes

In the early days of genetics, when the simple segregations from single gene differences were being rediscovered, it was thought that plant breeding would be revolutionized, and so it was for some plants. But when fruit trees were crossed the results did not fit any of the simple segregations and no single gene characters were known for a long time. All the important characters such as fruit shape, size, and color were shown to be inherited quantitatively and in those days this was not easy to understand and genetics was thought to have little to offer the fruit breeder.

One of the first characters to be found in the apple which was controlled by a single dominant gene was the complete anthocyanin pigmentation originating in *Malus pumila niedzwetzkyana* Schneid. (Lewis and Crane, 1938). Within the red-leaved seedlings derived from crossing *M.p.-niedzwetzkyana* with normal green apples, there is considerable variation in the intensity of the pigmentation, suggesting that other factors are involved. Nevertheless, in this case one can make a positive classification into those which are pigmented and those which are not. Since then a number of single gene characters have been found, particularly in relation to disease resistance and such minor characters as albinism and lethals.

Where a character is controlled by a dominant gene, it is very easy to transfer it into other apples. The number of generations required depends entirely on the quality of the apple in which the desired character is found. If it is already within the cultivated apple, then possibly only one generation is required; if on the other hand it is in a species of *Malus*, then several generations, possibly five or six, of crossing and backcrossing may be required to eliminate the undesirable characters of the species and retain the good qualities of the large-fruited cultivars and the desired character of the species.

When the character is recessive, and most mutations are, then the breeding becomes much more complicated and protracted. When a cultivar carrying a recessive character is crossed with a normal, the F_1 is normal and the recessive character reappears in the F_2 resulting from sib crossing. Thus only one-fourth of the seedlings are available for selection every two generations and even with a short juvenile period one generation will take about five years. Where one or more outcross generations may be required to achieve success, such breeding programs become quite uneconomical.

Characters Controlled by Polygenes

Most of the qualities that go to make a good apple—size, shape, cropping, etc.—are under polygenic control, which means that when two cultivars are intercrossed there will be a wide and continuous range of expression of all these characters in the seedlings. They will not segregate into discrete categories. Most polygenic characters behave independently and the range of variation is related to the expression of the characters in the parents and the progeny mean is always related to the parental mean. In some, given a strictly additive system where there is no dominance, the progeny mean and the parental mean will be the same; in others it may be above and in others below the parental mean. The extent of the deviation and its direction will depend on dominance and on the difference between the dominant and recessive phenotype.

When the number of seedlings is plotted against the character values it forms a simple distribution curve about the mean. In cases where no dominance is present, and the parental values are known, it is easy to predict the progeny mean and the proportion of the progeny that will be above and below the parental values. Characters which behave in this way appear to be those which, in the evolution of the apple, have not been subjected to great selective pressure. Fruit shape is such a character where there has been no selection at either end of the scale for particularly flat apples or tall ones. There are cases where there is a limit in the values of expression. Thus if July is considered the earliest that an apple can ripen because this is the minimum time from flowering in which a fruit can develop, then July is a fixed low value. The progeny from two early-ripening cultivars will show a one-sided distribution because of the barrier at July and will only produce July and later ripening seedlings but none earlier. In such cases the progeny mean will be higher than the parental mean.

It frequently happens that the parental mean is considerably greater than the progeny mean although the distribution forms the same simple curve. This occurs with characters which have been subjected to extreme selective pressure over many generations in evolution. Where selection is always at the extreme end of the curve, the progeny mean will tend towards the middle of the range of variation for the character. An example of this is fruit size where strong selection has been away from the small-fruited crabs and in this case the progeny mean may be as much as 40% less than the parental mean (Brown, 1960).

In addition to having characters controlled by either a single major gene or by polygenes it is possible to have one system superimposed on the other. This can present a rather confusing picture until the situation is realized. Such a situation exists in the inheritance of malic acid concentration in the fruit where medium to high acid is dominant over a very low acid type, but within both types there is a range of variation typical of polygenic control (Brown and Harvey, 1971).

Parental Contributions

Where detailed fruit and tree records of seedlings of many progenies have accumulated over many years, it is possible to use these data to calculate the contributions made by individual parents to their progenies. This assumes that the data for the different characters have been recorded in such a way that it is possible to classify the seedlings into categories suitable for analysis. Gilbert (1967) describes a method of analyzing a series of biparental crosses which have some parents in common by which the main effects of each parent on each of the characters can be calculated by treating the data as an incomplete diallel cross. Since the combining abilities are additive, the calculated main effects for each of the parents allows predictions to be made with considerable accuracy as to the progeny mean of untested combinations of parents.

Selection of Parents

Since most of the breeding of apples is based on selection from progenies produced by crossing extremely heterozygous parents, a great deal of consideration must be given to parent selection; otherwise, breeding becomes haphazard and the outcome quite unpredictable. If the major aim is to introduce a particular character which is controlled by a major gene or genes then the choice of at least one parent is limited to cultivars carrying these genes. However, most of the important characters are polygenically controlled, and for each, there will be a considerable range of variation and the breeder seeks the optimum expression of each. Fortunately the breeding behavior of highly heterozygous parental lines, with respect to measurable characters having a low level of genetic dominance, is fairly predictable. As has been shown, the mean value for the progenies falls with reasonable consistency around the mean of the two parents. It should be the aim to try to combine parents which between them have all the chosen characters present at as near the optimum expression as possible. Thus if polygenic mildew resistance is desired, then one and preferably both parents should have a high degree of resistance, in order that as many resistant seedlings as possible are available from which to select for other characters.

Having chosen the parents, the breeder must determine the size of the progeny required to be reasonably certain of obtaining a new apple having all the attributes visualized when planning the cross. Using actual records, Williams (1959) calculated that the percentage of desirable seedlings that can be expected as the main product of an apple breeding program for polygenically controlled characters is seldom more than 40% and for every additional character the figure rapidly decreases. Thus for a program in which the main objective is polygenically controlled mildew resistance, size of fruit, season of maturity, flavor, and color of skin, a reasonable estimate would be 40, 20, 20, 10, and 10% respectively.

Assuming that these characters behave independently, a progeny of 6,250 would, on the average, yield one seedling possessing a combination of the five traits at an acceptable level of expression, and at least one desirable one would appear by chance in two out of three such progenies. If the frequency of success is to be increased from two in every three progenies to say 19 in every 20, then the progeny size has to be increased to 18,750, using the formula $((N-1)/N)^N = 1/20$ where $(N-1)/N$ is the rejected fraction of the progeny ($N =$ family size).

In addition to these definable characters there are others for which allowance must be made, as well as some which cannot be readily included in the calculation, such as minor characters of flavor or texture which so often decide the ultimate success or failure of a selection. Therefore, if success is to be assured at a fairly high level of probability and, assuming that the probability of failure is held at one progeny in three, an adequate progeny would consist of 31,250 individuals.

To observe 30,000 seedlings would be practically impossible if the seedlings are to be evaluated at the fruiting stage. Nevertheless, in any serious breeding program, provided that great care is taken in the selection of the parents, progenies as large as practicable should be aimed at. Instead of growing all the seedlings to fruiting, as many as possible should be eliminated in the early stages of growth. For example, a high proportion might be eliminated because of an unacceptable level of mildew resistance by the second season of growth.

Breeding of Specific Plant Characters

Vigor

The study of the inheritance of vigor in progenies from crosses between apple cultivars is not easy and Spinks (1936) concluded that there is no correlation between the size and vigor of the seedlings and that of the parents. Watkins and Spangelo (1970) in a diallel analysis of seedling vigor (expressed as plant height) found one set of parents which showed 100% additive variability. The progeny mean would equal the mid-parent value. From another set of parents the additive variance was 50–70% which would result from considerable divergence of the progeny mean from the mid-parent value. This is not surprising because there are several factors which can greatly influence the vigor of seedlings. The chief difficulty arises in trying to assess the vigor of the parent cultivars. All cultivars are budded or grafted onto rootstocks, and one of the functions of the modern rootstock is to control the vigor of the scion. Thus by appropriate selection of rootstock, the scion cultivar can be dwarf at one extreme and large and vigorous at the other. Within reason, therefore, provided a seedling is not a weakling, the rootstock can be looked to for any desired variation in vigor.

Certain aspects of the parents and breeding procedure can, however, have a considerable effect on the vigor of the progeny. One factor already mentioned is inbreeding which leads to a very considerable loss of vigor, resulting in many seedlings being too weak to survive. The other cause of extreme weakness in seedlings is the use of parents which produce progenies practically all of which are aneuploids. This occurs in crosses involving triploid parents either intercrossed or in combination with diploids or tetraploids. Such seedlings are very weak and any which survive to flower are likely to be completely sterile.

Cold Hardiness

In many parts of the world probably the most important character of the tree is winterhardiness, for in severe winters the trees of many cultivars are seriously injured and, in extreme cases, even killed. There is, therefore, a continual need for apples of quality which are more winter-hardy so that they may escape damage in conditions of severe cold in existing apple growing areas, and also that the cultivation of the apple can be extended into colder areas. There are considerable differences in cold hardiness between different cultivars and it is quite possible to breed for increased hardiness. The breeding and selection for cold hardiness in deciduous fruit crops is reviewed by Stushnoff (1972).

It is important to use laboratory methods of testing cold hardiness because field observation under natural conditions is not always reliable. There are so many factors which produce variable microclimates in the orchard or nursery that it becomes impossible to know which parts of the orchard, or tree, have been exposed to what temperatures. This can be overcome by using a portable low temperature chamber in which trees or blocks of trees can be subjected to artificial cold stresses similar to those occurring in severe winters (Scott and Spangelo, 1964). A number of laboratory methods have been devised which are fairly satisfactory. There may be some problems in testing the parental cultivars as there seem to be conflicting reports on the effect the rootstock has on the hardiness of the tree grafted upon it. Most agree that the hardiness of the scion is independent of the rootstock, although the hardiness of the rootstock may be affected by the scion—in some cases increased and in others considerably decreased (Westwood, 1970).

The principle of one widely used method is that shoots or other parts of the plant, after having been frozen and soaked for a certain time in pure water, will exude electrolytes into the water by exosmosis from the cells which have been killed by the freezing. Living cells do this to a lesser extent. The amount of electrolyte exuded may be estimated by measuring the electrolytic conductivity of the water. The electrolytic readings can be used in various ways to estimate the amount of frost injury and as a measure of cold hardiness (Stuart, 1941; Wilner, 1960). By comparing the frozen section and the control section of the same shoot, Nybom et al. (1962), who described the whole method in detail, were able to calculate the relative conductivity of parents and seedlings and thus have a basis for comparison.

Lapins (1962) describes a method whereby one-year-old shoots of cultivars and seedlings were exposed to very intense cold. They were put in a freezer at 0 C. and lowered at 2° per hour to −43 C. where they were left for eight hours. The method of assessing the cold hardiness was to estimate the recovery of the shoots after three weeks in a growth chamber at 27 C (day) and 15 C (night). The estimates ranged from 1 = no recovery to 10 = no injury. This method is claimed to be more sensitive than the readings of electrolytical conductivity of the exosmosis liquid of shoot sections. A practical sample size when selecting young apple seedlings is five to seven shoots per tree.

Detailed information on the inheritance of hardiness is somewhat sparse, but from progenies estimated by each of the above methods it is evident that hardiness is under polygenic control and the frequency distribution of the trees in the progenies form a normal curve about the mean. Where it was possible to compare the progenies with the parent values the progeny mean was near but a little above the mid-parent value. This indicates that the progeny as a whole was a little less hardy than would be expected if there were no dominance—about 10% in the very limited data available. Nevertheless, in all the progenies there were some seedlings more hardy than either parent. These results are similar to those obtained by Watkins and Spangelo (1970) who did a diallel analysis of a number of expressions of winterhardiness and found approximately 100% additive variance.

Winter Chilling Requirements

Most apple trees require a certain amount of chilling during the dormant season for their proper development. In sub-tropical climates this requirement is not always met and prolonged dormancy or delayed foliation results. Cultivars with low chilling requirements occur locally but most are of poor quality. Oppenheimer (1968) in Israel has crossed some of these with good quality cultivars and from the progenies selected seedlings and introduced some cultivars with low chilling requirements and much improved quality. Early leafing is the criterion for selection and seedlings that broke bud within three weeks of the earliest were grown. This character appears to be polygenically controlled and about 30% of the F_1 progeny and 40% of the back-crosses to 'Delicious' and 'Jonathan' were sufficiently early to be retained for further test. It may be possible to eliminate the chilling requirement since some of the earliest to leaf were earlier after warm winters than after cooler ones.

Season of Flowering

Since cross-pollination is, to a great extent, essential for maximum fruit production in the apple, the time of flowering is important because cultivars whose flowering times coincide must be planted together. Cultivars which flower either very early or very late present a problem of providing efficient pollinators when planning an orchard. It is therefore convenient if new cultivars flower midseason or late midseason to coincide with most of the widely grown cultivars. Late flowering is considered important to avoid disastrous spring frost damage.

The object of many breeding programs is to produce cultivars which flower so late that practically all danger to the blossoms from late frost is past. Such breeding has been going on for some years but the fruit quality of the very late flowering cultivars which are available as parents is such that probably more than one generation will be required to get really good dessert quality into very late flowering sorts. This is not necessarily the complete answer as there is a problem of combining late flowering with early ripening which is difficult because of the minimum time required from flowering to the development of full fruit size and maturity.

The inheritance of time of flowering is polygenically controlled and a fair estimate of the mean flowering date of the progeny can be obtained from the mid-flowering time of the parents. The seedling distribution about the mean appears to be normal and the spread is such that seedlings flowering earlier and later than either parent will normally occur unless, of course, both parents are at the very extreme limit of the range, either early or late, and then distribution will be mainly away from the extreme. From the small amount of data available, it appears that the progeny mean tends to be slightly earlier than the parental mean (Tydeman, 1944).

There is a considerable amount of evidence from the results of Tydeman (1964) and Murawski (1967) that there is a very close correlation between the time of leaf emergence and season of flowering. In the analysis of Murawski the coefficient of regression was $b = 0.42$. In a breeding program where late flowering is one of the principal aims, this useful correlation allows seedlings which would be expected to be early and midseason flowering to be discarded, probably in their second year of growth, and to retain only those seedlings which are late in coming into leaf.

Another contributary factor to crop losses due to late frost is the susceptibility of the flowers to injury. The degree and amount of injury is affected by many environmental conditions, such as air drainage in an orchard, whether the orchard is under arable cultivation or grass, the position of the blossoms on the tree, and particularly the stage of development of the flower bud. In general it is in the later stages of development that flowers are most tender—the styles being most susceptible to damage. Fortunately, not all the flower buds on a tree are at the same stage of development at one time and frequently some escape damage while others are frosted.

There are undoubtedly some cultivars which are less susceptible to damage than others. Assuming that the evidence from pears (Brown, 1954) also applies to apples, tetraploid mutants are likely to be much more susceptible to frost damage than their respective diploid form. It appears that triploids and tetraploids, with larger cells, are more prone to damage than diploids. The hardiness of the tree to winter injury and the hardiness of the flowers to spring frost damage are inherited independently.

Duration of the Juvenile Period

In any breeding program the length of the juvenile period of the seedlings is of prime importance: 1) it determines the number of years the trees have to be grown and occupy nursery space before the results can be seen and assessed, and 2) the length of the juvenile period is correlated with fruiting precocity after propagation.

Apart from the effects that cultural practices have on the length of the juvenile period, the within-progeny differences and between-progeny differences are due to heredity. Some parents contribute to very short juvenile periods with many of their seedlings flowering in three or four years from seeding. Others contribute to very long juvenile periods; 'Northern Spy' progenies, for example, may take ten or more years to fruit.

The inheritance of this character is quantitative and follows the usual pattern, except that it is not easy to measure the parental values since the actual length of the juvenile period of the parents is not usually known. If sufficient data from seedlings are available, the parental contribution to the seedlings can be calculated. Visser (1970) has shown that there is a good correlation between the vegetative phase of a cultivar (i.e., the time from propagation to fruiting) and the contribution it makes to the length of the juvenile period of its progeny.

From this information about the parents it is possible to predict whether the mean juvenile period of a progeny will be short, medium, or long. There is now a considerable amount of evidence to show that seedlings must attain a certain size before they reach the stage at which they can flower and fruit. Therefore, within a progeny, those seedlings which are most vigorous are likely to be those which will attain the optimum size in the shortest time. Visser suggests that by selecting the more vigorous seedlings and discarding the weaker ones the breeder is also selecting for short juvenile period and consequently for precocious fruiting.

All this information applies to diploids. When

considering triploids and tetraploids, while the same principles apply, different standards must be applied. In general, the polyploids have much longer juvenile periods and the seedlings, although they may attain much greater size than diploids, may take several more years to fruit.

Spur Types

The spur type of tree is a fairly recent development which has been selected from mutant forms among a number of widely-grown cultivars. The spur types are characterized by compact habit, reduced internode length, limited side branching on shoots, and prolific development of fruit spurs, producing trees which are precocious in bearing and crop more heavily and regularly in the early years. The spur types are sometimes referred to as dwarfs or compact mutants but it is possible to have compact trees which do not develop spurs freely.

New spur types can be produced either by inducing mutations in existing cultivars or by breeding. Mutations can be induced by irradiating dormant scions. The optimum X-ray dose for apples appears to be 3 krad. The treated scions are then grafted onto rootstocks, and the selection for compact types can begin during the first season's growth. The selection criterion is that the length/diameter ratio of the shoots be less than normal, and the shoots have shorter internodes. Selection for short internodes alone is not reliable. Selections are budded onto rootstocks. The trees are cut back into the original graft and the lateral shoots cut back to induce latent buds to develop and the selection method repeated. This process may be repeated in the third year. These compact selections can be further selected for free spurring and precocious fruiting. These methods are described in detail by Visser et al. (1971).

Spur types which have been used in breeding do not all behave in the same way. Lapins (1969) reports that a spontaneous compact or spur mutant of 'McIntosh' when crossed with 'Golden Delicious' produced a progeny 43.9% of which had compact growth while a radiation-induced spur mutant of 'McIntosh' also crossed with 'Golden Delicious' did not produce any compact seedlings. This suggests that some of the mutants are chimeras and it depends in which cell layer the mutation occurs whether the tree breeds as a spur type or as a normal.

Decourtye (1967) investigated the inheritance of dwarfness which appears superficially to be the same as the compact type: all dwarfs need not necessarily be spur types. The results showed conclusively that the dwarf character is controlled by a single recessive gene, n, and that a number of cultivars, including 'Golden Delicious' were heterozygous Nn.

Fruit Size

Fruit size is undoubtedly one of the most critical characters in the selection of apple seedlings; if fruits do not attain the required standard size, the seedling is automatically eliminated from further consideration. Fruit size is a somewhat variable character which can be influenced by environmental conditions, but if the seedlings and parents are grown under comparable conditions there should be little difficulty in establishing the standard.

There are a number of ways of expressing size and the most common way in the field is based on fruit diameter. The generally accepted size is about 65mm diameter with a tendency towards larger fruit, up to 75mm. In analyzing the inheritance of size, the unit used by Brown (1960) was "estimated volume":

$$\text{Estimated volume (ml)} = \left(\frac{\text{Height (mm)} \times \text{radius (mm)}^2}{1{,}000}\right) \times 2.7.$$

While not as accurate as measurement by displacement, this method does permit the use of records of fruit outlines. By classifying the seedlings into size groups it is possible to show that this character is polygenically controlled and by plotting the percentage of the progeny in each of the size groups a normal distribution curve about the mean results.

The contribution made by parents to size is not quite as straightforward as some other characters. In some progenies a small percentage of the seedlings have fruits larger than the larger parent and quite frequently 50% or more of the seedlings have fruit smaller than the smaller parent.

The parental contribution to size can be calculated if data are available from a large number of progenies, by treating them as an incomplete diallel using the method described by Gilbert (1967). This can be much more accurate than trying to estimate the progeny mean from the parental mean.

Fruit size in the apple has been greatly increased by rigorous selection over the years from the small wild types to the size of the present-day commercial apple. Where extreme selection pressure of this kind occurs there is a tendency for the progeny mean to be smaller than the parental mean. This is borne out by the analyses of many crosses where the mean fruit size of the progenies is smaller than the average size of the parents by about 34%. This means that many cultivars which are acceptable themselves produce progenies in which the fruit size of the majority of the seedlings is below commercial size. Some combinations of cultivars are unlikely to yield any or very few seedlings with fruit of acceptable size. In general, the larger the fruits of the parents the greater will be the proportion of the seedlings producing fruit of acceptable size. It may well be sound policy to select very large-fruited seedlings, not necessarily acceptable as cultivars, for use as parents to ensure an increase in the proportion of seedlings with good fruit size from which selection for other desirable characters can be made.

Fruit Shape

The fruits of the apple vary in shape from very flat oblate to oblong. If the shape is translated into height/diameter and expressed as a percentage then the very flat apples are about 65 and the very tall ones about 100. It is very rare indeed for the height to exceed the diameter. This procedure does not take into account those conic fruits which taper to the eye nor to other irregularities of contour such as ribbing. Wilcox and Angelo (1937) have shown that the tapering conical character is rarely observed in oblate apples (1%) and is most common in the round oblong (32%). By dividing the range of ratio percentages from 65 to 100 into 5% intervals eight shape classes are formed and, by classifying the fruits of seedlings into these categories, it is possible to plot the frequency distribution among the eight classes for the different progenies. It has been shown (Brown, 1960) that the majority of the seedlings fall between the parental values although the distribution frequency extends beyond the parents and forms a normal distribution curve about the mean. Although progenies vary the average distribution seems to be about 5.4 shape classes.

By studying a fairly large number of progenies it is possible to calculate the parental contribution to shape of the different cultivars used. However, this does not seem to be necessary for in 35 progenies studied there was a remarkable similarity between the actual progeny mean, the estimated progeny mean (calculated from the sum of the calculated main effects of the parents), and the mid-parent value. Thus in the selection of the cultivated apple, apart from gross irregularities of contour, any shape from flat oblate to tall conic has been acceptable and fruit shape has not been subjected to great selective pressure. It is possible by knowing the height/diameter ratio of the fruit of the parents to predict with considerable accuracy the progeny mean and the range of shape classes expected.

Ribbing

The records (Spinks, 1936) of a study of the amount of ribbing of seedling fruits in a number of small progenies suggest that the progenies from angular-fruited parents will have a preponderance of angular-fruited seedlings. Practically all apple progenies will have seedlings with angular, slightly angular, and not angular fruit, but the proportion in each category will differ according to the character of the parents.

Percentage of progeny distribution
(calculated from the results of Spink, 1936)

Cross	angular	intermediate	smooth
Smooth x smooth	2	38	60
Intermediate x smooth	7	57	36
Angular x smooth	19	58	23

Season of Ripening

An important character in many breeding programs is the season at which the fruit ripens. This is the time when the fruit is in prime condition for use. This may be at harvest time for early ripening cultivars or after a period of cool storage for those ripening later. Improved early-ripening and late-ripening cultivars are sought in most countries. The results obtained by Brown (1960) show that when the time of ripening of seedlings and cultivars is recorded by months on a scale with August = 0 and March = 7, it is possible to plot the percentage frequency distribution of ripening time within the progenies. The results show a polygenic pattern of inheritance and a normal distribution curve about the mean. The parental mean does not equal the progeny mean but shows fairly consistent divergence towards earliness. The progenies from crosses between early-ripening parents have a small range of distribution which is no doubt due to the limit of earliness being fixed by the minimum time from flowering to fully developed fruit. Thus when very early ripening parents are crossed, the progeny mean is quite close to the parental mean. On the whole, however, the progeny mean is earlier than the parental mean and the difference between them increases with later-ripening parents. In crosses between very late ripening parents the progeny mean may be more than two months earlier. In some 35 progenies the mean was just over two weeks earlier than the parental mean and the average distribution within the progenies was 4.8 months.

It is thus comparatively easy to produce progenies of early-ripening and midseason seedlings although in the early-ripening seedlings quality is more difficult to attain. To increase the lateness of ripening on the other hand is much more difficult for even by crossing very late ripening parents only a small proportion of the progeny will be late-ripening, and therefore very large progenies are required to give a fair chance of selecting one with the other good qualities required.

Fruit Color

Fruit color is determined primarily by the ground color of the skin and secondly by anthocyanin pigmentation which is superimposed. The ground color of the immature fruit is dark green and as the fruit matures one of three things can happen: 1) the green may fade until it has completely disappeared and the ground color of the fruit will then be in the range from very pale cream to deep yellow, 2) the green may fade but not completely, producing ground colors in the greenish-yellow to yellowish-green range; and 3) the green may not fade at all, leaving a green ground color. Results suggest that the ground color is polygenically controlled and that the yellow range (which is related, at least to some extent, to the flesh color) and the green range may well be controlled independently.

Anthocyanin production in the fruit skin may be present or absent. If absent, then the fruit is either yellow or green and there is considerable interest in self-colored apples of this kind. If pigmentation is present, it can take several forms from small red flecks to bold stripes and from a faint blush to solid red. The intensity of the color can vary from very pale to very deep red and the area can be from practically nil to complete coverage. A study of all aspects of fruit color is complex and often confusing because the expression of all these characters can be affected by the state of maturity of the fruit, by the general environment, and by the micro-environment within the area of the tree.

The production of anthocyanin is normally dominant over the lack of it and it seems that the majority of colored apples are heterozygous since most when intercrossed produce a few seedlings without coloring (Crane, 1953). The coloring most often takes the form of striping, but blushes may also occur. These two characters are distinct and one can get stripes, stripes on a blush, and blush alone. Wilcox and Angelo (1936) found that by classifying seedlings into striped and non-striped (which included the blushed and unpigmented fruit), striping behaved more or less as a single major gene producing in fair-sized progenies 100% striped, 1:1, and 3:1 segregations of striped and non-striped. The blush may be quite pronounced but in some cases is quite ephemeral, being extremely light sensitive. Quite a number of normally yellow apples will, if exposed to full sun, develop a slight blush and it is possible for a few presumed yellow cultivars to produce a few seedlings with colored fruit. Normally when blushed apples are intercrossed, the offspring will be either blushed or non-pigmented but not striped (Spinks, 1936).

The shade of red which develops depends to a very great extent on the ground color. The most brilliant red is produced when the ground color is almost white and the dullest brown when the ground color is green. The area of the fruit covered and the intensity of the color are again inherited quantitatively and in most families there is a continuous gradation in color and the average amount in the progeny is proportional to the color of the two parents.

The anthocyanin pigment is in solution in cells of the epidermis and two sub-epidermal layers, but not all cells are pigmented. Some apples have no pigment in the epidermis and the intensity of color depends on the proportion of cells in each layer that contain pigment (Dayton, 1959). Since overall red is a very desirable character, red sports from many cultivars have been selected which have either greater coverage or the red is more intense. Most of these sports have arisen naturally but some have been induced by irradiation. It is natural when breeding red-colored apples to select the red sports as parents in preference to the original cultivar, believing the red sports to be better parents. Bergendal (1970) has crossed 'Golden Delicious' with six cultivars and their red sports and grown a progeny from each. In all classes the proportion of seedlings in each of the ground color classes was the same from original and sport. In the amount of red, the results from four of the pairs showed no difference whereas in the other two the results were quite different.

Crossed with Golden Delicious	% Fruits more than 50% red
Safstaholm	28
Red Safstaholm	64
Delicious	32
Richared	84

Mišić and Tešović (1970) examined the pigmentation of the three other outer-cell layers of the fruit and showed that the number of pigmented cells in the epidermis of 'Delicious' and 'Richared' are the same but that in the sub-epidermis (layer 2) there are twice as many pigmented cells in 'Richared' than in 'Delicious', confirming the results of Dayton (1959). The mutation must have taken place in layer 2 and in consequence affected the gametes. This indicates that a histological examination of the three outer-cell layers of the fruit skin of cultivars and their red sports would indicate whether a red sport will produce more red-fruited seedlings or will breed the same as the original or, if there is a choice of red sports, which to choose.

Fruit Flesh Color

The flesh of the apple varies in color from white through cream to pale yellow; it may be greenish white or it may even be tinged with red. These categories are not discrete and one merges with another. There is no strong preference for one flesh color over another but the preferred clear yellow and red skins tend to be more attractive on fruit with white or creamy white flesh.

In order to analyze progenies and show the mode of inheritance, flesh color has been put into four categories: white, cream, yellow, and green. Table 1 has been compiled from unpublished results together with those of Spinks (1936), and shows the percentage of seedlings in each of the four color classes obtained from crosses. Many progenies were involved and these figures show the average behavior of different parental types. All the progenies from one type of cross (white x white, for example) do not produce exactly the same proportion in each category. This is to be expected because in the classification there is no clear division between one color and another. Nevertheless the pattern is fairly clear and from these figures a rough guide can be obtained (Table 2) from which to predict the expected behavior of parents with respect to color. Adding together any two parental contributions will give the percentage of seedlings expected in each color class and a good fit to the results in Table 1 is obtained. The value for green flesh is based on too few seedlings to be accurate.

TABLE 1. Distribution of seedlings in the flesh color classes among progenies from various parental types

Cross	No. of seedlings	Frequency distribution (%)			
		yellow	cream	white	green
White x white	85	1	47	44	8
White x cream	85	6	54	34	6
White x yellow	198	29	38	25	8
Cream x cream	134	17	57	16	10
Cream x yellow	782	38	43	5	14
Yellow x yellow	219	70	24	1	5
Yellow x green	136	20	48	5	27

TABLE 2. Approximate percentage of contribution of parental types to flesh color in seedlings

Parental phenotype	Seedling phenotype			
	yellow	cream	white	green
White	0	24	22	4
Cream	8	29	8	5
Yellow	35	12	0	3
Green	0	31	0	19

The purplish-red flesh which can be incorporated into the cultivated apple from *Malus niedzwetzkyana* is controlled quite independently from the flesh colors already mentioned. The red is superimposed on the normal flesh colors, almost completely obscuring them, and probably accounts for the differences in redness found between seedlings. This character, which has value for ornamentals but not for commercial apples, is controlled by a single dominant gene (Lewis and Crane, 1938).

Russet

Russet on the fruit is a very variable character. Some cultivars have fruit which are completely covered with russet, others have fruit entirely free, and between these two extremes are intermediate types. Many have russet confined to the stalk cavity, others have it around the calyx, and others have it in patches over the fruit. These are inherited characteristics of the cultivar but external factors can also affect or produce russet patches on the fruit; the most frequent causes are low temperatures, humidity, and spray damage. These last defects are of course inherited weaknesses and are distinct from the normal russet character. 'Golden Delicious' is susceptible to such effects of external conditions. Cultivars which are completely covered with russet are very popular in some countries; a little russet in the stalk cavity is also acceptable but patchy russet is not very popular. It is strongly disliked on apples like 'Golden Delicious', but tolerated on 'Cox's Orange Pippin'. Some associate a certain amount of russet with good aromatic flavor.

Information on the inheritance of russet on the fruit is somewhat scant but some results are available from small progenies where russeting was classified as none, slight (where the russet was in the stalk cavity or calyx only), moderate (where it was in patches over the fruit), and complete (Brown, unpub.). In two progenies where two completely russeted cultivars were crossed, approximately 50% of the seedlings were completely russet and the others either slight to moderate. Similarly, when 'Cox's Orange Pippin', a moderate russet, was crossed with the completely russet 'Egremont Russet', 50% of the seedlings had completely russet fruit. Yet 'McIntosh', which would be classified as slightly russet, when crossed with 'Egremont Russet' produced no completely russet seedlings in a rather small progeny. 'Cox's Orange Pippin' crossed with the completely russet 'Golden Russet' produced only 25% completely russet seedlings. In a number of progenies where moderate russets were crossed with slightly russet parents, a few completely russet seedlings were produced. It is obvious from these results that more than one factor is involved since all full russets do not behave in the same way. Where the aim is to produce russet cultivars, russet x russet should produce progenies with at least 50% of the seedlings completely russet.

Fruit Flavor

Fruit flavor, one of the most important criteria in the selection of apple seedlings, is difficult to analyze since the constituents of flavor are a complex combination of acids, sugars, tannins, and aromatic substances. The appreciation of flavor is something that is personal and what one may consider a very good flavor another may not.

The basis of flavor is acidity and sweetness and it is the balance between these, irrespective of aroma, that primarily determines the acceptability of the fruit. Apples that are high in acid and low in sugar are quite unpalatable, being too acid; likewise, apples high in sugar and low in acid are equally unpalatable, being too sweet and insipid. The acid in the mature fruit is almost entirely malic acid and is measured either as percentage of malic acid in the fruit juice or as the pH of the juice. The main sugars are fructose, sucrose, and glucose and are conveniently measured as percentage of total sugars in the fruit juice (Brown and Harvey, 1971).

Acidity and sweetness are inherited independently and from an analysis of over 100 cultivars the majority of successful dessert apples are in the groups of medium acid/medium sugar, medium acid/high sugar, and low acid/medium sugar.

TABLE 3. Distribution of dessert and culinary apple cultivars in sugar acid groups

Malic acid (%)	Type of cultivar	Sugars (%) Low 9.0-11.4	Medium 11.5-13.4	High 13.5-16.0
High (1.0-1.6)	dessert	0	0	0
	cooking	2	14	4
Med. (0.5-0.9)	dessert	6	25	16
	cooking	4	0	0
Low (0.0-0.4)	dessert	6	12	1
	cooking	0	0	0

68 dessert cultivars, 24 cooking cultivars

The inheritance of sweetness, measured as the sugar concentration in the fruit juice, shows a quantitative pattern with a normal distribution of progenies about the mean which is very close to the mean value of the two parents. The inheritance of acidity is, however, more complicated in that two patterns are involved. A single gene control with medium to high acidity being dominant over very low acidity is superimposed on a quantitative pattern. The very low acid seedlings, often referred to as "sweets," have from 0.1 to 0.3% malic acid in the fruit. In progenies where one or both parents are homozygous acid (*MaMa*) all the seedlings will have normal acidity. In other progenies, a quarter or a half of the seedlings, depending whether their origin is *Mama* x *Mama* or *Mama* x *mama*, will be of the sweet type. These very low acid types are very sweet and insipid and are on the whole undesirable and have to be discarded. They are only tolerated in a few very early ripening cultivars.

In progenies where one or both parents are homozygous dominant, the distribution of acidity is normal with the progeny mean approximating the parental mean. In progenies where sweets are segregating and the sweets are excluded, the mean of the remaining acid portion is somewhat higher than the parental mean. When this is known, however, allowance can be made for this and a fair estimate of the progeny mean for malic acid concentration can be made from the parental mean.

An unusual feature of this "sweet" character is that because of its unpleasant flavor it would have a negative selection value, yet few cultivars are known which are homozygous for the dominant acid type and the majority of the cultivars studied are heterozygous. This suggests that this is an example where selection for quantitative economic characters has favored selection of heterozygotes for a neutral character (Williams and Brown, 1956).

By knowing the sugar concentration and the malic acid concentration in the fruits of cultivars, parents can be selected which will produce progenies in which the majority of the seedlings will combine desirable sugar and acid contents and if the *Ma* genotype is known the number of "sweets" which will have to be discarded can be predicted. If there is a selective advantage in the heterozygotes, then *MaMa* x *mama* progenies would all be heterozygous and such crosses would be worth considering. A comparison between a number of cultivars and their color sports and their tetraploid sports showed no major differences in the sugar or acid concentration in their fruits.

In a study of aromatic substances it has been shown by gas chromatography that between 20 and 30 components are present but the precise role played by individual components or by various combinations in flavor is not yet understood (Stackenbrock, 1962).

Disease Resistance

Apple scab: Caused by the fungus *Venturia inaequalis* (Cke) Wint., it is probably the most serious disease of the apple wherever it is grown. As well as attacking the leaves and shoots, it attacks and disfigures the fruit and makes it virtually unsalable. It can be controlled chemically but at considerable expense. The best way to combat the disease is to breed resistant cultivars and this has been attempted on a large scale for many years in many countries and the final goal of producing new cultivars virtually field immune to this disease is now being reached.

Two forms of resistance are available, one polygenic and the other monogenic. The most desir-

TABLE 4. Host reaction to *Venturia inaequalis* biotypes

	Isolates of *Venturia inaequalis*										
Cultivars	A19	A70	A30	A57	A72	A20	W13	A4	A7	A1	A14
Mrs. Lakeman's Sdg.	−	−	+	−	−	−	−	−		−	−
White Transparent	+	+	−	−	+	−			−	−	−
Wealthy	−	+	+	+	−		−	−			
Miller's Seedling	+	+	+	−	−	−	−	+	−	−	−
Epicure	−	+	+	−	+	−	+	−	−		
Ontario	+	+	+	+			−		+		+
Early Harvest	−	+	+	+	+	−	+	−	+	+	−
Lord Lambourne	+	+	+	+	+	+	+	+	+	+	−

+ = sporulating freely, − = no apparent infection

able type of resistance is likely to be obtained from a combination of both.

As with most diseases some cultivars are much more resistant than others and this allows selection for high, but not complete, resistance. A great many biotypes of the pathogen exist and when cultures are grown from single spores and tested on young leaves of cultivars some give positive and some negative results as shown above in Table 4. A "very susceptible" cultivar is susceptible to a great many biotypes while a "resistant" cultivar is resistant to most biotypes but is nevertheless susceptible to some. Under certain conditions or in certain localities the so-called resistant cultivar could become badly infected by encountering biotypes to which it is susceptible. This form of resistance is therefore not reliable.

A much more reliable type of polygenic resistance is found in the cultivar 'Antonowka', which is claimed to be field resistant to all known races of the fungus. This type of polygenic resistance is also found in selections of some species, notably *M. baccata* Borkh., *M. sargenti* Rehd., *M. sieboldii* (Reg) Rehd., and *M. zumi calocarpa* Rehd. (Shay et al., 1962). Forms of 'Antonowka' and selections from its progenies have been used extensively in breeding programs, particularly in Europe.

The breeding of scab-resistant apples based on the single dominant gene resistance found in some of the *Malus* species is almost a classical piece of research and breeding based on collaboration and cooperation, and is a pattern for other large-scale breeding. The program started early in this century in Illinois when a cross was made between a form of *Malus floribunda* carrying the gene V_f and 'Rome Beauty', and in 1926 two sibs were intercrossed to produce an F_2 population of 38 trees. These segregated into highly susceptible and highly resistant in the ratio of 1:1. Two of these seedlings have formed the basis on which an enormous breeding program for scab resistance has been carried on, mainly in the United States and Canada but also in Europe. *M. micromalus*, *M. atrosanguinea*, *M. prunifolia*, and other species have been introduced into the program but the most advanced material carries the *floribunda* resistance. The resistance being dominant, a simple backcross program of resistant selections to susceptible cultivars produces seedlings 50% of which are resistant in each generation (Hough et al., 1953; Shay et al., 1953). The stage has now been reached when many selections—potentially good commercial cultivars—are on trial and 'Prima' and 'Priscilla' have already been named (Dayton et al., 1970, and Williams et al., 1972).

FIG. 2. Scab infection on apple leaves. Top to bottom: type 2 irregular chlorotic lesions and no sporulation; restricted sporulating lesions; extensive and abundantly sporulating lesions.

There is a large pool of scab resistance genes in *Malus* and so far six loci have been identified and five physiologic races of the pathogen have been recognized which can infect certain of the original sources of resistance (Hough et al., 1970).

The resistance is scored by the symptoms on young leaves grown under warm glasshouse conditions. The scores are: 0 = no macroscopic evidence of infection; 1 = pin-point pits, no sporulation; 2 = irregular chlorotic or necrotic lesions and no sporulation; 3 = few restricted sporulating lesions; and 4 = extensive, abundantly sporulating lesions. Since the original classification another class, 'M', has been added, which indicates a mixture of necrotic, nonsporulating, and sparsely sporulating lesions (Shay and Hough 1952*b*). This has been the accepted method of scoring and only class 4 is considered as field susceptible—all the others are classified as field resistant and do not show symptoms of infection when grown outdoors. Using this definition of resistance segregation in a B_1 progeny will be 1 resistant: 1 susceptible.

The method of testing is to put germinating seeds into flats or very small pots and grow in a glasshouse at a temperature of 18 to 20 C. When the first true leaves are showing between the cotyledons the seedlings are sprayed with a spore suspension of mixed inoculum of the fungus and covered with polythene sheeting for 48 hours to ensure that the leaf surface does not dry. They are inoculated again after ten days as a safeguard and when the sheets come off for the second time, or soon after, sporulation should be evident on the susceptible seedlings. The susceptible seedlings can be discarded at this stage and only the resistant ones grown. Satisfactory infection can only be produced on very young leaves. The inoculum is prepared by washing the conidia from laboratory cultures of *Venturia* collected from many sources to give a spore suspension of a representative mixture of the pathogen. It has been shown that resistance to leaf infection is correlated with resistance to fruit infection.

Breeding for scab resistance and the extensive research on the pathogen and the resistance mechanism is fully reviewed by Williams and Kuc (1969).

Mildew: Caused by the fungus *Podosphaera leucotricha* (Ell. & Everh.) Salm., mildew is one of the most important diseases of apple. Although it does not damage the fruit to any extent, it debilitates the tree, and can kill seedlings. This disease, which attacks the foliage and young shoots, has a very marked effect on the crop and the quality of the fruit produced. Mildew varies in its intensity according to the climate but is a disease which is present wherever apples are grown. Some cultivars are very susceptible; others are highly resistant and between these two extremes there is a complete range. It is very doubtful if complete immunity can be found within the cultivated apple. It is possible, however, to select some which are very highly resistant and only rarely produce any infected shoots.

It has been shown (Brown, 1959) that resistance is inherited quantitatively with no evidence of dominance and the influence of the parents on the progeny is very marked. It is important to be able to estimate the value of the different cultivars when used as parents. From a survey of many progenies it was found that the parental contribution to mildew resistance of the progenies could be calculated accurately if a considerable amount of data were available (Brown, 1959). It was also found that the results from selfed families gave a good estimate of the resistance contribution of cultivars. This information was obtained by scoring the degree of susceptibility of the seedlings in the selfed progeny and calculating the progeny mean. By getting the mid-value of the selfed progeny means of two cultivars to be used as parents, a very close fit to their hybrid progeny mean is obtained. Because of self-incompatibility it is not always possible to produce selfed progenies. There is, however, evidence that direct assessment of mildew resistance among cultivars provides a sufficiently reliable guide to their breeding behavior.

Polygenic resistance to mildew is not easy to assess in young seedlings. The differences in resistance are not clear until the seedlings are two years old.

McIntosh and Lapins (1966) have found large differences in susceptibility to mildew among 21 clones of 'McIntosh' grown from dormant scions x-rayed at doses from 3.75 to 5 krad. Most showed normal susceptibility but a few were very susceptible and some quite resistant. The changes are attributed to mutations.

In addition to polygenic resistance, it is possible to find single gene immunity within some of the species of *Malus*. Knight and Alston (1969) have found single gene (Pl_1) dominant immunity in *Malus robusta* Rehd. and *M. zumi* Rehd. (Pl_2).

FIG. 3. Degrees of susceptibility of apple mildew. Left to right: all shoots severely infected; a few shoots severely infected; shoots mildew free.

These species also confer a short juvenile period. Several generations will be required to eliminate the undesirable characters of the species but if seedlings are selected for short juvenile period also, the time required will be kept to a minimum.

Apple canker: The cause of branch or stem encircling cankers, *Nectria galligena* Bres seems to be becoming more widespread and creates considerable concern among fruit growers. On heavy soil in particular, trees seem prone to attack. To date, few investigations have been carried out on this disease, but at least it is known that there are considerable differences between cultivars in their susceptibility. Some cultivars, particularly among the cider apples, are claimed to be very highly resistant (Wormald, 1955) as are 'M.1' and 'M.12' among the rootstocks (Moore, 1960). Where degrees of resistance exist there is always the possibility that breeding and selection will yield a high degree of resistance combined with good commercial qualities.

A method for testing the susceptibility of seedlings and cultivars has been described by Alston (1970*b*). Leaves are removed from the current year's shoots in the fall and a drop of spore suspension (180,000 spores/ml) is applied to each fresh leaf scar. Applications of the spore suspension are repeated twice just prior to leaf-fall. The lesions formed following infection can be recorded the following June. It may be important to collect the fungus for test purposes from quite a wide area, as there is evidence that isolates show considerable differences in virulence.

Fireblight: A very serious bacterial disease caused by *Erwinia amylovora* (Burr.) Winslow et al., fireblight can infect most of the Pomoideae. It can be a devastating disease of pears and, although it is less severe on apples, no cultivar is immune and outbreaks can be very serious. The bacterium attacks flowers, fruit, shoots, and branches, and the susceptibility of any tree depends on the condition of the growth, time of flowering, and environmental conditions at the time when infective material is present. This can lead to conflicting reports on the susceptibility or resistance of a cultivar although this may also be due to the existence of different biological races (Nonnecke, 1948).

No source of immunity is known in *Malus* and all cultivars of the apple are susceptible if inoculum from a virulent culture of the bacterium is injected into succulent tissue. There can, however, be considerable resistance to the rapid advance of the disease into the tissue.

There are a number of ways of assessing the susceptibility of seedlings but most involve inoculation at the tip of actively growing terminal shoots and, after a given time, measuring the distance the disease has invaded the shoots. To compare resistance, one method is to measure the length of the current shoot at the time of inoculation and express infection as the percentage of the shoot which becomes infected. Where infection goes into two-year-old wood, the figure will be in excess of 100. From the results of Moore (1946) it appears that resistance is polygenically controlled. Calculating susceptibility by the above method, the results ranged from 0% to over 400% and by taking 0–20% as being resistant the following figures were calculated from Moore's results.

TABLE 5. Fireblight resistance in apple progenies

Status of parents	Resistant seedlings in progenies (%)
Susceptible x susceptible	3.0
Susceptible x moderate	13.6
Resistant x susceptible	30.8
Resistant x moderate	39.2
Resistant x resistant	43.5

It would seem therefore that by the careful choice of parents new cultivars could be bred with a high degree of resistance although not with immunity.

Collar rot: Sometimes called crown rot, it is caused by the fungus *Phytophthora cactorum* (Leb. and Cohn) Schroet. and is a disease which attacks the bark of apple trees at or near ground level. It is capable of killing quite large trees by girdling the main stem. It is possible for either the rootstock or the scion cultivar or both to be attacked. Since infection comes from zoospores in the soil and is only likely to reach up the stem to the height of rain or irrigation splashes, it is possible to avoid infection of the scion by grafting about 60cm above soil level. It is therefore very important to have resistant rootstocks, and desirable to have resistant cultivars.

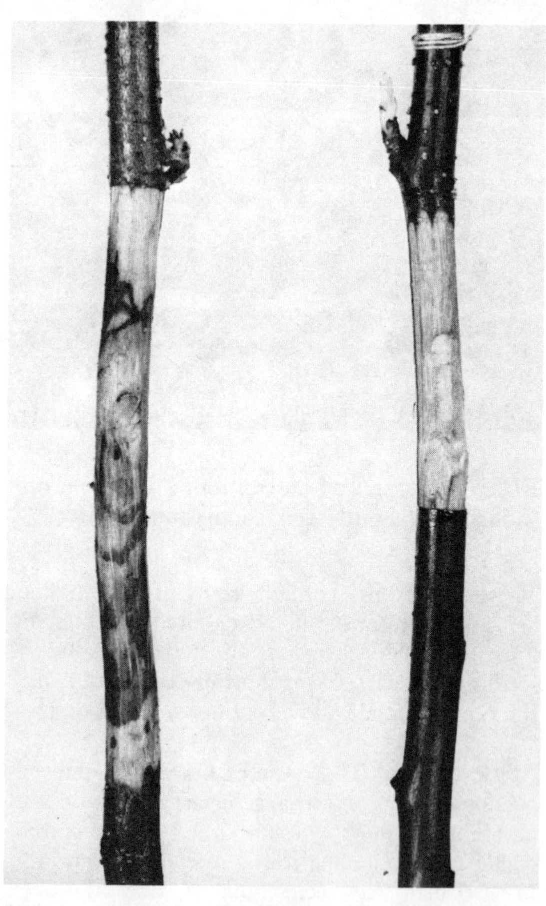

FIG. 4. Collar rot of apples shown on shoots after inoculation with isolate P.39. Bark has been removed to show area of infection on 'Cox's Orange Pippin' (left), which is very susceptible, and 'MM.103' (right), which is resistant.

There would appear to be two classes of resistance, polygenic and monogenic. Polygenic resistance is evinced by the wide variation found among the cultivars from very susceptible to very resistant, and within progenies as shown by the variation in seedling response to inoculation which shows a very marked influence of the parents. The results of McIntosh and Mellor (1954) using a mixed inoculum to infect trees suggests that resistance is partially or completely dominant. Their mean disease rating for resistant x resistant progenies was 7.2%; resistant x susceptible 23.5%; and susceptible x susceptible 93.3% of seedlings infected with collar rot.

The other type of resistance is monogenic and Alston (1970a) has found this in the cultivar 'Northern Spy' where resistance is primarily controlled by a single dominant major gene (*Pc*). The very high degree of resistance to collar rot in several of the Malling-Merton rootstocks is presumably derived from 'Northern Spy'.

Apart from differences in susceptibility in the host plant there are also great differences in pathogenicity between isolates of the fungus, and some cultivars which are normally highly resistant can be severely attacked by certain isolates.

The methods of testing cultivars and seedlings usually depend on culturing the fungus in the laboratory and then inoculating with mycelium-infected agar discs into the stem of growing trees or into cut-off shoots in controlled conditions in the glasshouse or laboratory and after a given time measuring the infection (Sewell and Wilson, 1959). A very rapid method for the pre-selection for collar rot resistance is described by Watkins and Werts (1971) where young seedlings two to three weeks old are grown in a peat-sand mixture which is inoculated by flooding with a zoospore suspension. The effects of attack are seen after one week and the complete results after two weeks. By this method it is only the resistant ones that are left to grow on for further testing.

Apple rust: The most important rust disease which attacks the apple is the cedar-apple rust caused by *Gymnosporangium juniperi-virginianae* Schw. which has as its alternate host the red cedar (*Juniperus virginiana* L.). There are considerable differences among cultivars in their resistance to this disease and resistance is present in a number of good-quality cultivars. Resistance was considered to be dominant and simply inherited, with such cultivars as 'Arkansas Black' and 'McIntosh' being homozygous for resistance; 'Delicious', 'Winesap', and 'Wolf River' heterozygous; and 'Jonathan' and 'Rome Beauty' fully susceptible (Moore, 1940; Shay and Hough, 1952a).

More recently Mowry (1964) has found that his results could not be explained by a single gene and suggested that inheritance is more complicated and is controlled by two genes, designated G and *J*, and that susceptibility is controlled by the two recessive genes, with duplicate epistatic reaction between pairs. It is thus not possible to be certain of the genotype of a cultivar by its phenotypic behaviour.

Apple blotch: Mowry and Dayton (1964) have analyzed the progenies from controlled crosses for susceptibility to this disease caused by *Phyllosticta solitaria* E. & E. They have found that susceptibility is controlled by two completely dominant genes with duplicate recessive epistatic interaction between the gene pairs. Because of this it is not possible to predict with any accuracy the genotype of a cultivar from its expressed apple blotch phenotype.

Storage rots: Apart from the diseases which attack the fruit on the tree, very considerable losses can occur from fruit rotting diseases while the fruit is in storage, and species of *Gloeosporium*, particularly *G. perennans* Zellar & Childs, are responsible for much of it. In a survey of over 200 cultivars, Alston (1967) found that some are very susceptible and a few (about 5%) were highly resistant, with the cultivar 'Cravert Rouge' most resistant of all. A small progeny from 'Cox's Orange Pippin', which is very susceptible, crossed with 'Cravert Rouge' produced seedlings with a fair degree of resistance. When compared with a progeny of 'Cox' selfed which had a mean of 4.7 lesions per fruit, the progeny from the cross with 'Cravert Rouge' had a mean of 0.8 lesions. Also among the most resistant cultivars in this survey were 'Jonathan' and 'Jonared', and it is known that a number of derivatives of 'Jonathan' also show good resistance to this pathogen.

The method of estimating the susceptibility is to dip the fruits in a spore suspension (50,000 spores/ml) and store in paper bags at 5 C for up to five months. Five or more fruits of each cultivar need to be sampled.

Virus diseases: These diseases present a number of problems to the fruit-grower and the propagator since many (like rubbery wood and flat limb) are quite serious and may disable the tree, while others affect the efficiency of the tree, and most cause a reduction in crop. Viruses pass freely across graft unions so that it is important to have virus-free rootstocks as well as virus-free commercial cultivars. The response of apple cultivars to viruses differs widely. Some viruses are present in many cultivars without showing any symptoms. With others, such as vein banding mosaic, the severity of the symptoms varies from one cultivar to another, some being very severe, some moderate, some slight, and some showing no symptoms at all (Luckwill, 1953). Other viruses produce very severe symptoms or none at all. Many of the viruses have been present but undetected in cultivars until fairly recent times when indicator plants have been found which produce visible symptoms. How many more may be present awaiting the discovery of other indicators is anyone's guess.

Since so many cultivars carry viruses without showing symptoms it would be an almost impossible task to try to breed for resistance. There is some evidence that sensitivity may be passed to offspring. In the scab breeding program 'Russian R.12740-7A', which is an indicator for chlorotic leaf spot, passed on the sensitive reaction to some of its seedlings (Mink and Shay, 1962).

The present state of our knowledge suggests that it is much simpler to inactivate the viruses in important material by heat therapy than to attempt to breed for resistance.

Pest Resistance

Apart from the thorough investigations into woolly aphis resistance and its inheritance, information on breeding for resistance to insect attack is somewhat sparse. There is, however, considerable evidence that sources of resistance to many important pests exist.

Woolly aphis (*Eriosoma lanigerum* Hausm.) is a very widespread pest of apples which is capable of attacking not only the above-ground stems of the tree causing cankers but also the roots where, under certain conditions, large colonies can become established. The above-ground infestations can be controlled by spraying but those on the roots cannot and can be quite a serious problem on nursery stocks. There is a need for resistant rootstocks as well as resistant scions.

Crane et al. (1936) found that there are degrees of resistance and that inheritance from some sources is quantitatively controlled. Among the cultivars reputed to be immune or highly resistant, 'Northern Spy' was found to be the most resistant. Using this cultivar in crosses with Malling rootstocks, many new rootstocks were bred which were highly resistant to woolly aphis—first the Merton series and later the Malling-Merton series, which provide a range of rootstocks imparting various degrees of vigor.

Knight et al. (1962) have shown that there are well authenticated but conflicting reports on the resistance and susceptibility of various cultivars which confirm the existence of biological races of woolly aphis. So far none of the races have been capable of successfully colonizing 'Northern Spy' whose resistance they have shown to be under the control of a single dominant gene (*Er*). In breeding cultivars resistant to woolly aphis, 'Northern Spy' is a parent which confers resistance not only to woolly aphis but also to collar rot.

One inoculation technique is to repeatedly transfer insects by a camel's hair brush to the shoots of seedlings to see if colonies develop. The most effective method, however, is to tie pieces of heavily infested shoots carrying about 200 aphids to each seedling. The susceptibility is determined by the amount of colonization of the seedling shoot which may be intense, slight, or nonexistent.

Resistance to other aphids is also to be found, and resistance to the rosy leaf curling aphis (Dysaphis devecta Wlk.) is controlled by single dominant genes. Alston (1970c) has found three biotypes of the insect and at least three resistant genes in the host. 'Cox's Orange Pippin' carries Sd_1 and is resistant to biotypes 1 and 2; 'Northern Spy' has Sd_2 and is resistant to biotype 1 only; and a selection from *Malus robusta* with Sd_3 is resistant to all three biotypes.

M. robusta also carries the gene Sm_h which confers hypersensitivity to the rosy apple aphis (*Dysaphis plantaginea* Pass.) (Alston and Briggs, 1970).

Minor Characters

There are a few characters found among seedlings which are apparently of no value to the breeder nor to the plant. Among those commonly found is albinism; albino seedlings are completely devoid of chlorophyll and very soon die. This is controlled by the recessive gene c_1. A similar character is the pale green lethal (Brown, 1958, and Klein et al., 1961) which is also detected in young seedlings. Shortly after germination the lethals are indistinguishable from normal seedlings but as the cotyledons expand the lethals become pale green and at the second leaf stage seedlings wither and die. This character is also controlled by a single recessive gene c_2 and is estimated to be present in 50% of cultivars.

Albinos and pale green lethals cannot exist as mature trees and therefore cannot be involved in any hybridizations, and yet these characters have become so numerous among cultivars in the heterozygous state that it appears that selection for quantitative economic characters has favored the heterozygotes. It would therefore seem desirable in planning crosses to try to ensure that progeny will have as high a ratio of heterozygotes as possible.

Rootstocks

The rootstock is a very important part of the apple tree and its genetics for its particular function should be considered with as much care as that of the scion cultivars. In some respects breeding rootstocks is easier in that they do not have to comply with the very high standards of fruit quality demanded for commercial apples. Nevertheless they do have to fulfill many special requirements and the testing period is long and difficult, for it is only when the tree worked on the rootstock has reached full cropping that the full potential of the rootstock can be assessed.

There are a number of special requirements for any rootstock. It must be easy to propagate either vegetatively or by seeds; it must produce a good clean upright stem, easy to bud or graft; it must have a root system which will provide adequate anchorage to support the tree grafted on it without staking; it should not exhibit any stock-scion incompatibility; it should not produce any suckers in the orchard; it should induce early and heavy cropping; and of course it should, where possible, be resistant to the important pests and diseases. Some may have to be selected for additional specific requirements such as a high degree of winter-hardiness for some areas.

At present there is a wide range of vegetatively-propagated apple rootstocks which impart different degrees of vigor, ranging from those producing dwarf, compact trees to those producing very large, vigorous trees. There is a need for a greater range in winter-hardy stocks and for resistance to many pests and diseases.

One of the problems of vegetatively-propagated rootstocks is virus diseases which may be present in the rootstocks and then passed to the scion. One possible way of overcoming this is to raise rootstocks from seeds. This practice has been in use for a great many years and one of the reasons for the change to clonal stocks was the great lack of uniformity among the seedlings since those used as rootstocks displayed the enormous variability found in most apple seedling populations. In recent years apomictic seedlings which are much more uniform have been bred from some *Malus* species. These too present problems in showing quite an amount of stock-scion incompatibility, and in some, extreme sensitivity to virus infection from the scion. They do have the advantages that propagation is simple and the seedlings are uniform and free from virus. A number of species have been tried, particularly *M. Sieboldii*, *M. sargenti*, *M. hupehensis*, *M. sikkimensis*, *M. toringoides*, and hybrids between some of them.

M. hupehensis and *M. sikkimensis* are autonomous apomicts and can produce apomictic seeds without pollination which produce seedlings of the maternal type. Other species are facultative apomicts and a small percentage of the seedlings may be hybrids. Since some of these require pollination to produce seeds, they can be pollinated with a homozygous red-leaved plant and the hybrids immediately identified by the pigmentation of the leaves. Many show a high degree of incompatibility with standard cultivars but much of this is due to extreme sensitivity to latent viruses in the scion. When virus-free clones are used as scions much of the incompatibility disappears. Some *M. Sieboldii* hybrids have shown a good affinity for virus-infected scions and appear tolerant of latent virus infection (Schmidt, 1970).

Testing vegetatively-propagated rootstocks is not easy and takes a long time. In the initial stages each has to be tested to determine its ability to root and to produce a good stem, its rate of multiplication, and whether it has all the normal qualities that make a good healthy plant. If disease or pest resistance is being incorporated, this can be determined at this stage. Beakbane and Thompson (1947) have shown that there is a strong correlation between the relative area of bark to wood in transverse section of lateral roots and the size of the tree at maturity—those with most bark being most dwarfing and most likely to induce early cropping. The same criteria apply to a great extent to apomictic seedling rootstocks with the necessity for additional tests for uniformity or the easy identification of hybrids, and tests for incompatibility. The ultimate tests of the selected stocks are the proportion of trees produced when grafted or budded, their anchorage, cropping, and general behavior. Many years may elapse before the rootstock's behavior can be fully assessed.

Achievements and Prospects

The length of time which elapses between the sowing of the seeds, the testing and selection of the seedlings, and the periods of trial to the launching of a new cultivar can easily amount to 20 or more years. So while there is no doubt that great advances have been made in apple breeding, the full benefits are still to be seen. New scab- and mildew-resistant cultivars are about to be introduced and many others having special attributes are on trial and a number of new rootstocks which show considerable improvements on the older ones are now in use.

The use of various forms of irradiation to induce mutations in existing cultivars has already produced some interesting sports. These techniques are likely to be used to a greater extent.

What of the future? Since new cultivars take so long to produce it is obvious that the breeder should not only be concerned with present-day problems but should be looking ahead and trying to envisage what will be required from some of the seedlings he is now producing. It is doubtful if fruit standards will change a lot. Where the greatest changes are likely to occur will be in cultural techniques particularly as a result of mechanization and the breeder may well be called upon to produce cultivars suited to different management regimes.

Hand labor is very expensive and machines for harvesting apples have already been tried but not very successfully with the present type of orchard tree. The possibilities are greater when the trees are grown as a hedge and with apples bred specifically for processing. It seems likely that in due course the breeder and the engineer will get together and devise a machine and a tree which will be compatible.

A great deal of attention will have to be given to breeding for resistance to pests and diseases, for it is almost certain that much more attention will be given to the possible pollution of the soil by frequent applications of poisonous sprays in the orchard and there could be a move to reduce the amount of spraying permitted.

Physiological disorders cause severe losses of fruit and there are indications that it will be possible to breed apples which are not prone to these troubles. The work of Faust et al. (1971) shows that apple cultivars differ in their ability to accumulate calcium in the fruit. The effect of high calcium content is to decrease the incidence of a number of physiological disorders. The analysis of many seedlings indicates genetic control. This is likely to be an important character in future breeding programs. Similar investigations into variations in mineral, hormone, and allied substances in plant tissue are likely to be used as aids to apple breeding.

How will breeders meet these problems? At present most are engaged by government departments or universities and are working on their own individual problems. Apple breeding being such a slow process, there is a great need for cooperation and collaboration on a world-wide scale, particularly on the exchange of breeding material, possibly as pollen, and an exchange of information. There is a need for a common policy in methods of recording information about seedlings so that the results can be classified, fitted to a computer program, and be analyzed to extract the maximum amount of information. This would also allow results from different workers to be compared. The Apple Breeders Cooperative in the United States and Canada which arose out of collaboration in the scab breeding program has shown that such cooperation can work to everyone's advantage and others could do well to follow their example.

Literature Cited

A great many more references could be cited. These will be found in the excellent abstract bibliography of Robert Knight (1963).

Abbott, D. L. 1955. Temperature and the dormancy of apple seeds. *Rpt. Int. Congr. Scheveningen.* p.746–753.

Alston, F. H. 1967. Varietal response to *Gloeosporium perennans* in the apple. *Rpt. E. Malling Res. Sta. 1966.* p.132–134.

———. 1970a. Resistance to collar rot, *Phytophthora cactorum* (Leb and Cohn) Schroet. in apple. *Rpt. E. Malling Res. Sta. 1969.* p.143–145.

———. 1970b. Response of apple cultivars to canker, *Nectria galligena. Rpt. E. Malling Res. Sta. 1969.* p.147–148.

———. 1970c. Integration of major characters in breeding commercial apples. *Proc. Angers Fruit Brdg. Sym.* p.231–248.

Alston, F. H., and Briggs, J. B. 1970. Inheritance of hypersensitivity to rosy apple aphis *Dysaphis plantaginea. Canad. J. Gen. Cyt.* 12:257–258.

Anon. 1970. *Fruit—a review of production and trade relating to fresh, canned, frozen and dried fruit.* London: Commonwealth Secretariat.

Batra, S., Pratt, C., and Einset, J. 1963. Chromosome numbers of apple varieties and sports 4. *Proc. Amer. Soc. Hort. Sci.* 82:56–63.

Barrett, H. C., and Arisumi, T. 1952. Methods of pollen collection, emasculation and pollination in fruit breeding. *Proc. Amer. Soc. Hort. Sci.* 59:259–262.

Beakbane, A. B., and Thompson, E. C. 1947. Anatomical studies of stems and roots of hardy fruit trees. 4. The root structure of some new clonal apple rootstocks budded with Cox's Orange Pippin. *J. Pomol.* 23:206–211.

Bergendal, P. O. 1970. On the inheritance of red color in crosses between 'Golden Delicious' and various red apple mutants. *Proc. Angers Fruit Brdg. Sym.* p.181–184.

Bishop, C. J. 1959. Radiation-induced fruit colour mutations in apples. *Can. J. Gen. Cyt.* 1:118–123.

Brittain, W. H. 1933. Apple pollination studies in the Annapolis Valley, N.S., Canada. 1928–32. *Can. Dept. Agr. Bul.* 162.

Brown, A. G. 1954. Frost susceptibility of 2x and 4x fertility pear. *45th Ann. Rpt. John Innes Inst.* p.7.

———. 1958. Pale green lethal in apple varieties. *48th Ann. Rpt. John Innes Inst.* p.6–7.

———. 1959. The inheritance of mildew resistance in progenies of the cultivated apple. *Euphytica* 8:81–88.

———. 1960. The inheritance of shape, size and season of ripening in progenies of the cultivated apple. *Euphytica* 9:327–337.

———. 1964. Fruit breeding. *55th Ann. Rpt. John Innes Inst.* p.37–38.

———. 1966. New fruits from old. p.10–24 In *Fruit present and future.* London: Royal Hort. Soc.

Brown, A. G., and Harvey, D. M. 1971. The nature and inheritance of sweetness and acidity in the cultivated apple. *Euphytica* 20:68–80.

Campbell, I. 1961. Shortening the juvenile phase of apple seedlings. *Nature* 191:517.

Crane, M. B., Greenslade, R. M., Massee, A. M., and Tydeman, H. M. 1936. Studies on the resistance and immunity of apples to woolly aphis, *Eriosoma lanigerum* (Hausm.). *J. Pomol.* 14:137–163.

Crane, M. B., and Lawrence, W. J. C. 1952. The genetics of garden plants. 4th ed. London: Macmillan.

Crane, M. B. 1953. The genetics and breeding of tree fruits. *Rpt. 13th Int. Hort. Cong.* London, p. 687–695.

Darlington, C. D., and Moffett, A. A. 1930. Primary and secondary chromosome balance in *Pyrus. J. Gen.* 22:129–151.

Dayton, D. F. 1959. Red color distribution in apple skin. *Proc. Amer. Soc. Hort. Sci.* 74:72–81.

Dayton, D. F., Mowry, J. B., Hough, L. F., Bailey, C. H., Williams, E. B., Janick, J., and Emerson, F. H. 1970. Prima—an early fall red apple with resistance to apple scab. *Fruit Var. Hort. Dig.* 24:20–22.

Decourtye, L. 1967. A study of some characters controlled by simple genetic means in apple trees. (*Malus sp.*) and pear trees (*Pyrus communis*) (in French). *Ann. Amelior. Plantes* 17:234–266.

Derman, H. 1948. Chimeral apple sports and their propagation through artificial buds. *J. Hered.* 39: 235–242.

———. 1949. Are the pomes amphidiploid? A note on the origin of the Pomoideae. *J. Hered.* 40:221–222.

———. 1951. Ontogeny of tissues in stem and leaf of cytochimeral apples. *Amer. J. Bot.* 38:753–760.

———. 1965. Colchiploidy and histological imbalance in triploid apple and pear. *Amer. J. Bot.* 52:353–359.

Einset, J. 1952. Spontaneous polyploidy in cultivated apples. *Proc. Amer. Soc. Hort. Sci.* 59:291–302.

Einset, J. and Imhofe, B. 1947. Chromosome numbers of apple varieties and sports. *Proc. Amer. Soc. Hort. Sci.* 50:45–50.

———. 1949. Chromosome numbers of apple varieties and sports 2. *Proc. Amer. Soc. Hort. Sci.* 53:197–201.

———. 1951. Chromosome numbers of apple varieties and sports 3. *Proc. Amer. Soc. Hort. Sci.* 58:103–108.

Ewert, R. 1909. New investigations about the parthenocarpy of fruit trees and some other fruit-bearing plants (in German). *Ladw. Jahrb.* 38: 767–839.

Faust, M., Shear, C. B., Oberle, G. B., and Carpenter, G. T. 1971. Calcium accumulation in fruit of certain apple crosses. *HortScience* 6:542–543.

Fogle, H. W. and Barnard, E. 1961. Punched cards as aids in evaluating seedlings in the field. *Fruit Var. Hort. Dig.* 15:47–50.

Gagnieu, A. 1951. Production of pollen in the apple tree; the possibility of lethality in the single factor gene (in French). *Nat. Rech. Agron.* (Paris) (Ser. B.) 1:455–496.

Gilbert, N. 1967. Additive combining abilities fitted to plant breeding data. *Biometrics* 23:45–49.

Hough, L. F., Shay, J. R., and Dayton, D. F. 1953. Apple scab resistance from *Malus floribunda* Sieb. *Proc. Amer. Soc. Hort. Sci.* 62:341–347.

Hough, L. F., Williams, E. B., Dayton, D. F., Shay, J. R., Bailey, C. H., Mowry, J. B., Janick, J., and Emerson, F. H. 1970. Progress and Problems in breeding apples for scab resistance. *Proc. Angers Fruit Brdg. Sym.* p. 217–230.

Hunter, A. W. S. 1954. Tetraploidy in vegetative shoots of the apple induced by the use of colchicine. *J. Hered.* 45:15–16.

Kemmer, E. 1953. Over the primary and the fertile stages of apple orchards (in German). *Züchter*, 23: 122–127.

Klein, L. G., Way, R. D., and Lamb, R. D. 1961. The inheritance of a lethal factor in apples. *Proc. Amer. Soc. Hort. Sci.* 77:50–53.

Knight, R. L. 1963. *Abstract bibliography of fruit breeding and genetics to 1960. Malus and Pyrus.* London: Commonwealth Agric. Bur.

Knight, R. L., and Alston, F. H. 1969. Developments in apple breeding. *Rpt. E. Malling Res. Sta. 1968.* p.125–132.

Knight, R. L., Briggs, J. B., Massee, A. M., and Tydeman, H. M. 1962. The inheritance of resistance to woolly aphid, *Erisoma lanigerum* Hsmnn. in the apple. *J. Hort. Sci.* 37:207–218.

Knight, T. A. 1806. Observations on the means of producing new and early fruits. *Trans. Hort. Soc. London.* 1:30–37.

Lapins, K. 1962. Artificial freezing as a routine test of hardiness of young apple seedlings. *Proc. Amer. Soc. Hort. Sci.* 81:26–34.

———. 1965. Compact mutants of apple induced by ionizing radiation. *Can. J. Plant Sci.* 45:117–124.

———. 1969. Segregation of compact growth types in certain apple seedling progenies. *Can. J. Plant Sci.* 49:765–768.

Laubscher, F. X., and Hurter, N. 1960. The chromosome numbers of seedling progeny from triploid apples. *S. Afr. J. Sci.* 3:31–39.

Lewis, D., and Crane, M. B. 1938. Genetical studies in apples 2. *J. Genet.* 37:119–128.

Luckwill, L. C. 1953. Virus diseases of fruit treees 4. Further observations on rubbery wood, chat fruit and mosaic in apples. *Rpt. Long Ashton Res. Sta.* p.40–46.

McIntosh, D. L., and Lapins, K. 1966. Differences in susceptibility to apple powdery mildew observed in McIntosh clones after exposure to ionizing radiation. *Can. J. Plant Sci.* 46:619–623.

McIntosh, D. L., and Mellor, F. C. 1954. Crown rot of fruit trees in British Columbia. 3. Resistance trials on apple seedlings obtained from controlled crosses. *Can. J. Agr. Sci.* 34:539–541.

Marks, G. E. 1954. An acetocarmine-glycerol jelly for use in pollen fertility counts. *Stain Tech.* 29:277.

Mink, G. I., and Shay, J. R. 1962. Latent viruses in apple. *Purdue Univ. Agr. Expt. Sta. Res Bul.* 756.

Mišič, P., and Tešovič, Z. 1970. Distribution of anthocyanin in the skin of apple varieties Jonathan, Red Delicious and their dark red mutants. *J. Sci. Agr. Res. Boegrad.* 23:121–126.

Moore, M. H. 1960. Apple rootstocks susceptible to scab, mildew and canker for use in glasshouse and field experiments. *Plant Path.* 9:84–87.

Moore, R. C. 1940. A study of inheritance of susceptibility and resistance to apple cedar rust. *Proc. Amer. Soc. Hort. Sci.* 37:242–244.

———. 1946. Inheritance of fireblight resistance in progenies of crosses between several apple varieties. *Proc. Amer. Soc. Hort. Sci.* 47:49–57.

Mowry, J. B. 1964. Inheritance of susceptibility to *Gymnosporangium juniperi-virginianae*. *Phytopathology* 54:1363–1366.

Mowry, J. B., and Dayton, D. F. 1964. Inheritance of susceptibility to apple blotch. *J. Hered.* 55:129–132.

Murawski, H. 1955. Investigations of stages of apple seedings as a basis for fruit research (in German). *Arch. Gertenb.* 3:255–273.

———. 1967. A contribution to apple cultivation research. 10. Results of cultivation of apple species with late leafing-out blooming (in German). *Der Züchter* 3:134–139.

Nonnecke, I. 1948. Fireblight resistance across the prairies in *Malus*. *Proc. W. Can. Soc. Hort. Sci.* p.31–32.

Nybom, N. 1959. On the inheritance of acidity in cultivated apples. *Hereditas* 45:332–350.

Nybom, N., Bergendal, P. O., Olden, E. J., and Tamas, T. 1962. On the cold resistance of apples. *Proc. Eucarpia Fruit Conf.* Arnhem. 1961. p.66–73.

Oppenheimer, C., and Slor, E. 1968. Breeding of apples for a sub-tropical climate. 2. *Theor. and Appl. Gen.* 38:97–102.

Pratt, C., Ourecky, D. K., and Einset, J. 1967. Variation in apple cytochimeras. *Amer. J. Bot.* 54:1295–1301.

Sax, K. 1931. The origin and relationship of the Pomoideae. *J. Arnold Arboretum.* 12:3–22.

———. 1959. The cytogenetics of facultative apomixis in *Malus* species. *J. Arnold Arboretum* 40:289–297.

Schmidt, H. 1964. A contribution to the cultivation of apomictic apple rootstocks. 1. Zygogenetic and embryologic research (in German). *Z. Pflanzenz.* 52: 27–102.

———. 1970. A contribution to the cultivation of apomictic apple rootstocks. 3. Their affinity with cultured species in relation to virus content (in German). *Z. Pflanzenz.* 63:214–226.

Scott, K. R., and Spangelo, L. P. S. 1964. Portable low temperature chamber for winterhardiness testing of fruit trees. *Proc. Amer. Soc. Hort. Sci.* 84:131–136.

Sewell, G. W. F., and Wilson, J. F. 1959. Resistance trials of some apple rootstock varieties to *Phytophthora cactorum* (L. and C.) Schroet. *J. Hort. Sci.* 34:51–58.

Shay, J. R., Dayton, D. F., and Hough, L. F. 1953. Apple scab resistance from a number of *Malus* species. *Proc. Amer. Soc. Hort. Sci.* 62:348–356.

Shay, J. R., and Hough, L. F. 1952a. Inheritance of cedar rust resistance in apples. *Phytopathology.* 42: 19.

———. 1952*b*. Evaluation of apple scab resistance in selections in *Malus*. *Amer. J. Bot.* 39:288–297.

Shay, J. R., Williams, E. B., and Janick, J. 1962. Disease resistance in apple and pear. *Proc. Amer. Soc. Hort. Sci.* 80:97–104.

Spinks, G. T. 1936. Apple breeding investigations. 1. Results obtained from certain families of seedlings. *Rpt. Long Ashton Res. Sta. 1936*. p.19–49.

Stackenbrock, K. H. 1962. The effect of environment on the aroma of the apple. *Rpt. Eucarpia Conf.* (1961) p.50–56.

Stuart, N. W. 1941. Cold hardiness of seedlings from certain apple varieties as determined by freezing tests. *Proc. Amer. Soc. Hort. Sci.* 38:315.

Stushnoff, C. 1972. Breeding and selection methods for cold hardiness in deciduous fruit crops. *HortScience* 7:10–13.

Thiele, I. 1950. Fruitfulness of maiden pome trees as a grower's project (in German). *Der Züchter.* 26: 241–243.

Thompson, A. H., and Bakjer, L. P. 1950. The effect of boron in the germinating medium on pollen germination and pollen tube growth for several deciduous tree fruits. *Proc. Amer. Soc. Hort. Sci.* 56:227–230.

Tydeman, H. M. 1944. A preliminary account of experiments in breeding early and mid-season dessert apples. *Rpt. E. Malling Res. Sta. 1943*. p.34–42.

———. 1964. The relation between time of leaf break and of flowering in apple seedlings. *Rpt. E. Malling Res. Sta. 1963*. p.70–72.

Tydeman, H. M., and Alston, F. H. 1965. The influence of dwarfing stocks in shortening the juvenile phase of apple seedlings. *Rpt. E. Malling Res. Sta. 1964*. p.97–98.

Vavilov, N. I. Wild progenitors of the fruit trees of Turkistan and the Caucasus and the problems of the origin of fruit trees. *Proc. 9th Int. Hort. Cong.* p.271–286.

Visser, T. 1951. Floral biology and crossing technique in apples and pears. *Meded. Dir. Tuinb.* 14:707–726.

———. 1964. Juvenile phase and growth of apple and pear seedlings. *Euphytica* 13:119–129.

———. 1970. Environmental and genetic factors influencing the juvenile period in apple. *Proc. Angers Fruit Brdg. Sym.* p. 101–115.

Visser, T., Verhaegh, J. J., and de Vries, D. P. 1971. Pre-selection of compact mutants induced by X-ray treatment in apple and pear. *Euphytica* 20:153–207.

Watkins, R., and Spangelo, L. P. S. 1970. Components of genetic variance for plant survival and vigor of apple trees. *Theor. Appl. Gen.* 40:195–203.

Watkins, R., and Werts, J. M. 1971. Pre-selection for *Phytophthora cactorum* (Leb. and Conn.) Schroet. resistance in apple seedlings. *Ann. Appl. Biol.* 67: 153–156.

Way, R. D. 1971. Hastening the fruiting of apple seedlings. *J. Amer. Soc. Hort. Sci.* 96:384–389.

Westwood, M. N. 1970. Rootstock-scion relationships in hardiness of deciduous fruit trees. *HortScience* 5: 418–421.

Wilcox, A. N., and Angelo, E. 1936. Apple breeding studies 1. Fruit color. *Proc. Amer. Soc. Hort. Sci.* 33:108–113.

———. 1937. Apple breeding studies 2. Fruit shape. *Proc. Amer. Soc. Hort. Sci.* 34:9–12.

Williams, E. B., Janick, J., Emerson, F. H., Dayton, D. F., Mowry, J. B., Hough, L. F., and Bailey, C. 1972. 'Priscilla,' a fall red apple with resistance to apple scab. *Fruit Var. Hort. Dig.* 26:34–35.

Williams, E. B., and Kuc, J. 1969. Resistance in *Malus* to *Venturia inaequalis*. *Ann. Rev. Phytopath.* 7: 223–246.

Williams, W. 1959. Selection of parents and family size in the breeding of top fruits. *Rpt. 2nd Congr. of Eucarpia*. p.211–213.

Williams, W., and Brown, A. G. 1956. Genetic response to selection in cultivated plants; Gene frequencies in *Prunus avium*. *Heredity* 10:237–245.

Wilner, J. 1960. Relative and absolute electrolytic conductance tests for frost hardiness of apple varieties. *Can. J. Plant Sci.* 40:630–637.

Wormald, H. 1955. *Diseases of fruit and hops*. London: Crosby Lockwood.

Zimmerman, R. H. 1971. Flowering in crab apple seedlings: Methods of shortening the juvenile phase. *J. Amer. Soc. Hort. Sci.* 96:404–411.

Pears

by Richard E. C. Layne and Harvey A. Quamme

The pear originated in prehistoric times as a fruit crop. Cultivar development has been continuous since early days and a high level of improvement has been achieved. Pears are now grown in all temperate regions of the world.

The European pear, the most widely grown type, is considered by many to be the most delectable of all tree fruits. It combines a buttery juicy texture with unsurpassed delicacy of flavor and aroma. The Oriental pears, which arose independently in China and Japan, are preferred there for their uniquely different texture and flavor. But these pears are not grown extensively in other parts of the world.

Second only to apples in world production of deciduous tree fruits, pears are consumed fresh, cooked, dried, or as preserves. In Europe some are used for making perry or pear wine. In 1968 the world production of pears was 7.3 million metric tons (mt) (United Nations, 1969). Europe produced the most (4.5 million mt), followed by Asia (1.7 million mt), North America (597,000 mt), South America (223,000 mt), Oceania (172,000 mt), and Africa (88,000 mt). The major producers in Europe were Italy (1.36 million mt), West Germany (599,000 mt), and France (493,000 mt). In North America, the major producers were the United States (559,000 mt) and Canada (38,000 mt). In South America they were Argentina (112,-000 mt) and Brazil (55,000 mt). In Africa the major producer was the Republic of South Africa (68,000 mt). In Asia, China was the largest producer (890,000 mt), followed by Japan (476,000 mt). In Oceania, Australia was the largest producer (150,000 mt), followed by New Zealand (22,000 mt).

It is noteworthy that production in Italy alone exceeded that of North and South America, Africa, and Australia combined. Italy is also the biggest exporter of pears, followed in descending order by Argentina, France, Australia, South Africa, the Netherlands, and the United States (USDA, 1970).

European pear breeders have been particularly successful in developing cultivars of high quality, large size, and attractive appearance that thrive best in France, Italy, and Belgium. North American breeders have not been as accomplished in developing high-quality cultivars, perhaps because they have had to place greater emphasis on breeding for disease resistance and cold hardiness, which has usually been at the expense of optimum quality.

Pyrus is genetically diverse and the gene pool available to breeders is well documented (Knight, 1963). Greater cooperation among pear breeders in all parts of the world should lead to more rapid development of cultivars with a broader range of adaptation than those presently available, while still meeting the needs of specific countries and areas of production. Hopefully, greater efforts will follow in pear breeding and genetics with more worldwide cooperation among breeders.

Origin and Early Development

Taxonomy of *Pyrus*

The genus *Pyrus* is a member of the Rosaceae family and has been further classified into the subfamily Pomoideae. *Pyrus* contains more than 20 species, all indigenous to Europe and Asia (Rehder, 1967). Many of these species are similar and often classification is difficult.

In Europe, North America, South America, Africa, and Australia *P. communis* L. is the main species of commerce. In Europe *P. nivalis* Jacq., the snow pear, is also grown to a limited extent for making perry. In some parts of North America hybrids of *P. communis* and *P. pyrifolia* (Burm.) Nak. are grown for processing. In southern and central China and in Japan, *P. pyrifolia*, the sand pear, is the main cultivated species, but in northern China and Japan *P. ussuriensis* Max., hybrids of *P. pyrifolia* and *P. ussuriensis*, and *P. Bretschneideri* Rehd., which is reputed to be a hybrid of the cultivated types and *P. betulaefolia* Bge., are grown (Kikuchi, 1946). In southern China (Pieniazek, 1966) and northern India (Mukherjee et al., 1969), selections of *P. Pashia* P. Don. are cultivated. Other species are used for rootstock and ornamental purposes, and few are grown for their fruit.

The Evolution of *Pyrus*

Pomoideae is distinguished by a basic chromosome number of 17 compared with 7 to 9 for the other sub-families of Rosaceae, and it has been suggested that the Pomoideae group arose as an amphidiploid of two primitive forms of Rosaceae, one having a basic chromosome number of 8 and the other 9 (Sax, 1931, and Zielinski and Thompson, 1967). All species of *Pyrus* that have been examined are diploid ($2n = 34$, $x = 17$). A few polyploid cultivars of *P. communis* exist. Speciation in *Pyrus* has proceeded without a change in chromosome number (Zielinski and Thompson, 1967).

The genus *Pyrus* is believed to have arisen during the Tertiary period in the mountainous regions of western China which are now exceedingly rich in forms of Pomoideae. Dispersal is believed to have followed the mountain chains both east and west. Speciation probably involved geographic isolation of *Pyrus* populations by mountain ranges and adaptation to colder and drier environments (Rubzov, 1944).

These differentiated populations provided early man with much variability from which to select desirable genotypes to grow in his gardens. Vavilov (1951) proposed that the domestication of the pear involved a process similar to that which could be still seen in the Caucasus Mountain region where wild pears grow abundantly. He observed all phases of pear culture, from wild groves through a transition in which the better-fruited trees were left after the forest was cleared, to a phase in which the better-fruited forms were grafted onto poorer trees.

It is evident that the evolution of the cultivated forms also involved interspecific hybridization. Rubzov (1944) reported that modern cultivars of *P. communis* exhibit characteristics of at least three species, *P. elaeagrifolia* Pall., *P. salicifolia* Pall., and *P. syriaca* Boiss. The most improved group of cultivars cultured in northern China also appears to belong to a hybrid complex involving *P. ussuriensis* and *P. pyrifolia*. The species *P. Bretschneideri*, in fact, appears to be a cultivated form involving hybridization of *P. betulaefolia* and the cultivated types of *P. pyrifolia* (Kikuchi, 1946).

Vavilov (1951) listed three centers of diversity for cultivated pear which he regarded as centers of origin: 1) the Chinese center, where forms of *P. pyrifolia* and *P. ussuriensis* are grown; 2) the central Asiatic center, comprising northwest India, Afghanistan, and the Soviet republics of Tadjikistan and Uzbekistan, and western Tian-Shan, where *P. communis* is grown, and 3) the Near Eastern center in the Caucasus Mountains and Asia Minor where *P. communis* also is grown. The Near Eastern center is of special importance because it is believed that the domesticated forms of *P. communis* from which the modern cultivars are derived originated there. The central Asiatic center is thought to be a secondary center of diversity where *P. communis* hybridized with *P. heterophylla*, *P. Boisseriana*, and *P. Korshinskyi* (Zukovskij, 1962).

History of Improvement

Europe

According to Hedrick et al. (1921), the earliest written record of pear culture in Europe is that of Homer of ancient Greece who wrote around 1000 B.C. that pears were one of the "gifts of the gods." By the time of Theophrastus (ca. 300 B.C.), pear culture was well established in Greece and cultivars with distinct names were propagated by grafting and cuttings.

The ancient Romans also contributed greatly to the knowledge of pear growing. Cato (235–150 B.C.) described methods of pear culture similar to those practiced today and described six cultivars grown in his time. Somewhat later, Pliny (23–79 A.D.) described 35 cultivars of pear grown in Rome. The range in fruit characters of the cultivars described by these authors is similar to the range of fruit characters found in today's commercially-grown cultivars.

In central and eastern Europe, pears were also grown but few cultivars were developed that attained the importance of the Belgian and French cultivars. Pears grown in Germany and described by Cordus (1515–1544) already had all the fruit characters possessed by modern cultivars, with the exception of good texture.

By the time of Charlemagne, pear culture in France was well established, and by the sixteenth and seventeenth centuries, France was a major pear-producing country. In 1628 an amateur fruit collector, Le Lectier, had more than 254 pear cultivars in his garden. By the early 1800s there were 900 cultivars of pears growing in France, but most were rather crisp-fleshed.

During the eighteenth century pear culture and improvement in Belgium made major strides through the work of Hardenpont, Van Mons, and others. In that century probably more progress was made in pear improvement in Europe than in all preceding centuries. Hardenpont developed at least 12 cultivars by selecting from large populations of open-pollinated seeds of cultivars that were being grown at that time. Van Mons alone originated or distributed more than 400 cultivars of which at least 40 have had lasting merit. Belgian breeders selected the cultivars with soft, melting, buttery flesh including some that are still important today, such as 'Beurre Bosc', 'Beurre d'Anjou', 'Flemish Beauty', and 'Winter Nelis'.

The pear is not native to England but is believed to have been introduced before the Roman conquest. Commercial pear culture in England probably began by 1200 A.D. In 1826 a catalog of the Royal Society of London listed 622 cultivars. According to Hooker (1818), the cultivar 'Williams Bon Chretien' ('Bartlett') grew from a seedling around 1796 and was widely dispersed in England by 1816. This is now the most important cultivar grown in the world. 'Conference' is another cultivar selected in the nineteenth century that is still commercially important.

The history of pear improvement in Europe deals with improvement in only two species: *P. communis* (the common European pear), and *P. nivalis* (the perry pear), which is used for making wines. The history of improvement and present status of perry pears in England is well reviewed by Luckwill and Pollard (1963). Perry-making has been a traditional rural industry in the West Midlands for more than 400 years. Perry pears are also important in France. Most improvements in perry pears in France and England were made from selection among open-pollinated seedlings as opposed to selection from controlled hybridization.

Most of the cultivars developed in Europe during the golden age of pear improvement (1750–1850) were derived from selection among open-pollinated seedlings of common cultivars. According to Vávra and Orel (1971), the possibility of improving fruit trees by controlled hybridization was discussed as early as 1840 at an agriculture and forestry congress in Brno, Czechoslovakia, but they found no record of this procedure being used by pear breeders at that time. However, Knight (1806) in England reported the value of controlled hybridization in obtaining new combinations of characters in pears as well as other fruits as early as 1806. Mendel, the father of genetics, is credited with hybridizing pears at Brno in the latter half of the nineteenth century. It appears from his hybridization plans that he was trying to develop late-ripening cultivars with superior flesh quality. Apparently he had some degree of success because he received an award in 1883 by the Heitzing Society of Gardeners for new cultivars of pear that attracted attention for their beauty, shape, and color (Vávra and Orel, 1971).

Asia

Several pear species are cultivated in Asia for their fruit. The most important of these are: *P. pyrifolia, P. ussuriensis, P. Bretschneideri,* and *P. Pashia*. According to Pieniazek (1966), pears have been cultivated in China for more than 2,000 years. Kikuchi (1946) states that separation of edible from wild sorts in China began about 1100 B.C. European pears (*P. communis*) are not appreciated in China and 95% of the pears produced are derived from the above-mentioned Asiatic species and their hybrids. Over the past centuries of culture, many cultivars have been developed which vary considerably in size, shape, taste, texture of flesh, color of skin, and many other fruit and tree characters. The fruits lack the smooth, buttery texture of European pears, tend to have crisp-textured flesh and some have abundant stone cells in the flesh. The fruits tend to have little aroma but the flesh is often sweet and juicy.

Cultivars of the Ussurian pear (*P. ussuriensis*) are the most cold hardy of all pear cultivars. They are grown in northeastern China and produce small, globose, medium-sized fruit with persistent calyxes. The flesh is not too gritty but the taste is bland. Two of the best cultivars are 'Sian-sui-li' and 'An-li' (Pieniazek, 1966). The Chinese white pears (*P. Bretschneideri*) grow further south than the Ussurian pears. Although they are not as hardy as the cultivars of *P. ussuriensis*, they are hardier than other Asiatic species. The fruits are of medium size, vary from pyriform to obovate in shape, and have the best texture and flavor of the Oriental pears. 'Yar-li' and 'Pingo-li' are two important cultivars in this group (Pieniazek, 1966).

The Chinese or Japanese sand pear (*P. pyrifolia*) is widely distributed in China. Several hundred cultivars of this species are grown. The size of their fruits varies from very small to very large. The skin is usually russeted, the flesh has many stone cells, and the flavor is bland. They are sweet and juicy and well liked by the Orientals. The cultivar 'Twentieth Century', of Japanese origin, is considered to be the best of this group. The sand pears are not cold hardy and are adapted to warm climates.

In southwestern China and northern India some cultivars derived from *P. Pashia* are grown commercially. They are adapted to a hot, humid climate. The fruits are medium-sized and of mediocre quality by European standards. Some cultivars such as 'U-li' have dark-brown, almost black flesh (Pieniazek, 1966).

In Japan, pears were cultivated as early as the eighth century (Kajiura, 1966), but commercial orchards as we now know them were not planted until after 1868. The sand pear (*P. pyrifolia*) is the most widely-grown type in Japan and Korea. The most important cultivar in Japan is 'Twentieth Century,' as in China. Cultivars of *P. ussuriensis* are also grown in northern Japan where *P. pyrifolia* is susceptible to winter injury. *P. communis* cultivars, mainly 'Bartlett,' are also grown on a small scale in Japan.

North America

Pears are not native to North America. They were introduced by the early French and English settlers to eastern Canada and the eastern United States. According to Hedrick et al. (1921) the earliest reference to pear culture in New England was in 1629. The Franciscan monks introduced pears to the West Coast, and extended cultivation from British Columbia to lower California. Pears were found under cultivation in California by Vancouver in 1792.

During the seventeenth and eighteenth centuries, only cultivars derived from *P. communis* were grown in North America. A significant event in the history of pear improvement was the introduction in the early 1800s from Europe of the Chinese sand pears (*P. pyrifolia*). The first cultivar derived from *P. communis* x *P. pyrifolia* named 'Le Conte' appeared in 1846. 'Kieffer' was introduced in 1873 and 'Garber' in 1880 (Hedrick et al., 1921). These were the first pear cultivars to be developed fom interspecific hybridization between the European and Oriental types. The fruits were of lower quality than the European cultivars being grown, but the trees were more resistant to fire blight caused by *Erwinia amylovora* (Burr.) Winslow et al. It is not known whether these cultivars were derived from controlled crosses or from chance outcrosses, although the latter is more probable. It was not long before controlled crosses were being made between *P. communis* and *P. pyrifolia*, with the hope of obtaining cultivars with fruit of the European type and the blight resistance of *P. pyrifolia*.

Another significant event in the history of pear improvement in North America was the introduction of hardy cultivars of *P. communis* from Russia with the first shipment arriving in 1879. The fruit quality was poor, however, and the trees were susceptible to fire blight. Their main value was as a source of cold hardiness in further breeding (Magness, 1937). The introduction of *P. ussuriensis* by Patten around 1867 in Iowa was another historic event. According to Magness (1937) chance hybrids that occurred between *P. ussuriensis* x *P. communis* proved to be very cold hardy.

Further introductions of pear species were made by Reimer (1925) from Europe and the Orient. The superior blight resistance of some of the Oriental introductions set the stage for a major effort in breeding pears for fire blight resistance by the U. S. Department of Agriculture and at several experiment stations in the United States and Canada, beginning in the early 1900s.

Although pears are also grown in South America, Australia, New Zealand, and Africa, comparatively little effort has been given to pear breeding in these regions and very few cultivars have been developed that have attained commercial importance comparable to the European cultivars. The most noteworthy is 'Packhams Triumph', which was developed in Australia and introduced commercially around 1900. It is second in importance only to 'Bartlett' in Australia and ranks second in new plantings in New Zealand (Cole, 1966).

The efforts of current pear improvement programs in various parts of the world will be discussed in following sections.

Modern Breeding Objectives

Pear breeders have many objectives in common, because new cultivars must possess certain general pomological characteristics in order to be commercially acceptable. New cultivars must have adequate climatic adaptation to thrive in the regions where they are to be grown. In some regions the emphasis may be on cold hardiness, whereas in others it may be drought or heat tolerance.

In addition to climatic adaptation, disease resistance is very important in some regions. Thus, in the eastern and southern parts of North America, the disease fire blight (*E. amylovora*) is widespread in occurrence and devastating in effect, and the prime objective is improved resistance to fire blight.

In all pear breeding programs improvement of fruit quality is an important objective, although the attributes that constitute good quality in some areas may be different in others. This is especially the case with European pears (*P. communis*) in comparison with Oriental pears (*P. ussuriensis, P. pyrifolia, P. Bretschneideri,* and *P. Pashia*). Attributes that constitute good quality among European pears, such as soft, buttery flesh, are not desirable in Oriental pears, in which crisp, breaking flesh is preferred instead. Cultivars that are well adapted with adequate disease and pest resistance have little potential unless they also possess the appearance and quality attributes desired by the consumer.

Breeding programs differ most in the relative emphasis given to specific characters because the need for improvement in these characters varies from one region to another.

There are a number of important attributes required for new cultivars to be successful if grown in North America, east of the Rocky Mountains. A level of fire blight resistance equivalent to 'Kieffer' or better has the most priority. Early bearing and consistent cropping are important. An overlapping bloom period and compatibility with major pollenizer cultivars are important. Medium vigor and adaptability to a high density system of culture are desirable. Resistance to cold is important and resistance to insects such as pear psylla and to other diseases such as powdery mildew, leaf spot, and scab are desirable. Fruit size is important and should exceed 6 cm in length and 5 cm in diameter. A pyriform shape is preferred although other shapes may be acceptable. When ripe, the skin should be golden yellow and bright with or without a red blush, rather than green or greenish yellow. There may also be a place for some cultivars with a bright red skin for the fresh market. The skin should be relatively free of russet, but some completely russeted types would be acceptable for the fresh market provided the russet was uniform and attractive. The skin should also resist bruising from handling during harvesting, grading, storing, and ripening. White flesh is preferred with absence of stone cells under the skin or in the flesh. If some stone cells are present, they are more acceptable if located in the core. The core should be small in relation to the fruit. Sweet, juicy, aromatic flavor and smooth, buttery texture with low fiber is highly desirable. Uniformity of maturity, good storage quality, and uniform ripening are important. Good processing quality as canned halves, as diced pieces for fruit cocktail, or as finely ground purée for baby food is important, especially for cultivars intended to replace 'Bartlett'. Season of maturity is important to ensure continuity of supply for fresh and processing outlets. New cultivars are needed that ripen from August to October and that have all or most of these characters combined with fire blight resistance.

Active breeding programs in Canada and the United States share many of these objectives in cultivar improvement because the requirements of the industry in both countries are the same or similar. Progress in breeding for these and other specific characters will be discussed more fully in a later section.

Modern approaches to breeding pear rootstocks also include several major objectives, the most important of which is to develop rootstocks that induce similar size control and precocity in the scion cultivar as 'Quince A', but which are more compatible, winterhardy, and disease resistant. Indeed, it may be better to develop dwarfing *Pyrus* rootstocks than to attempt further improvement of the quince rootstocks because more incompatibility is encountered between intergeneric than intrageneric grafts. Furthermore, some of the *Pyrus* species are also more winterhardy, disease resistant, drought tolerant, and have better anchorage than the quince rootstocks. The main drawback in developing dwarfing *Pyrus* rootstocks is the low frequency with which dwarfing potential is found in *Pyrus* populations. Progress in breeding pear rootstocks will be discussed in a later section.

Breeding Techniques

Hybridization

Floral biology: Pear blossoms are white, rarely pinkish, and borne in umbel-like racemes. The flower consists of five petals and sepals and 20 to 30 stamens with anthers that are usually red. Styles vary from two to five and are free but closely constricted at the base. The ovary has five locules with two ovules per locule with a maximum seed set of ten. Gayner (1941) found that the flowers in the lower positions in the cultivar 'Conference' were larger and heavier than those in the upper positions except for the king blossom. A greater percentage of the larger basal flowers set fruit compared with the smaller more terminal ones. The terminal flower, despite its large size, had a low percentage set.

This feature has been observed with other cultivars and is used as a basis for selecting blossom clusters for emasculation. The king blossom is usually allowed to open and perhaps even the second blossom; then one or more of the remaining blossoms are emasculated when they are at the balloon stage of development. Magness (1937) notes that only one or two blossoms per cluster should be emasculated and pollinated; otherwise, fruit set may be reduced. The remaining flowers should be removed at the time of emasculation.

Draganov et al. (1964) found that the lowest mean daily temperature for the start of flowering in pears was 9 C. The length of the flowering period depended upon the mean daily temperature and atmospheric humidity. The higher the mean temperature and the lower the humidity the shorter the flowering period. Brown (1943) found that pear cultivars usually flowered in the same order each year although the dates varied widely from year to year. Flower development proceeded only at temperatures above 6 C. Differences among cultivars studied by Brown and Draganov et al. may account for the differences in base temperatures reported.

Pollen collection and storage: Pear blossoms open very quickly if the weather is warm and dry. It is preferable, therefore, to collect and store pollen several weeks in advance of the bloom period to ensure that the pollen required is available when needed. It will retain its viability for two to three weeks at room temperature. Fortunately, pear pollen is quite easy to collect and store. It can also be preserved for more than a year without significant loss of viability using simple procedures and commonly available laboratory equipment.

The method we use in our laboratory is in general use with some variations by other pear and apple breeders in North America. Basically the method is as follows: branches 1 to 1.5 m in length are collected at any stage from the delayed dormant to the tight cluster stage of blossom development and forced in the laboratory or greenhouse with the cut ends in water at ca. 23 C. When collected at the dormant stage in late spring, about two weeks of forcing are required before the blossoms are ready for harvest. At the tight cluster stage two to three days of forcing are usually sufficient. The blossoms are harvested just before the anthers begin to dehisce. The inverted blossoms are rubbed over a wire mesh grid (1.5 mm^2) placed over a sheet of clean paper. The anthers fall through the grid onto the paper. The edges of the paper are then folded inward to avoid spillage of pollen. The paper trays are allowed to dry at room temperature ca. 23 C or are placed in an incubator at ca. 25 C to hasten dehiscence of the anthers and release of the pollen. This usually occurs within 24 hours. The pollen is then poured into glass vials which are labeled, stoppered, then placed over anhydrous $CaSO_4$ in a dessicator and stored in a refrigerator at ca. 5 C. The indicating type of $CaSO_4$ is preferred since it changes color from blue to pink when hydrated. When pink, the dessicant should be replaced, or dried in an oven for 24 hours to restore its drying ability.

When the pollen is required in the orchard the vials containing pollen are placed in larger vials containing anhydrous $CaSO_4$ which are stoppered to prevent absorption of moisture from the air. They are then placed over ice in a styrofoam insulated cooler before being taken to the orchard. The vials of pollen are removed from the cooler when required for pollination and replaced immediately after use. On return to the laboratory they are again placed in the dessicator at ca. 5 C and stored. Pollen collected, stored, and handled in this way will maintain its viability for more than a year.

Pollen viability can be quickly checked by determining its stainability in acetocarmine using standard procedures, or by germinating it in a liquid medium containing 10 ppm boron and 15–20% sucrose at room temperature.

Other methods of pollen collection, preservation, and storage have been tried (Johri and Vasil, 1961). Visser (1955) found that pear and apple pollen remained fairly viable for two years stored at 2–4 C and 10% relative humidity. Pollen stored at −20 C and −190 C retained its viability for two to three years. After pre-drying, pollen longevity was generally increased at low humidity and temperatures below 0 C. King and Hesse (1938) found that pear pollen could be stored for 550 days at 0 C and 25% relative humidity and still retain high viability as indicated by 80% pollen germinability. Pear pollen has also been successfully preserved by freeze-drying and vacuum-drying methods (Visser, 1955).

Emasculation and pollination: Techniques of emasculation and pollination vary from one institution to another. To make sure there is no chance of foreign pollination, the flower clusters may be bagged or the whole tree enclosed with a muslin tent. The cover is put on before the flower clusters open and is taken off after the stigmas have turned completely brown. On the other hand, many breeders do not consider that these elaborate precautions are necessary to prevent foreign pollination in pear if the calyx, corolla, and stamens are removed before the flowers open. The removal of the corolla is all that is needed to prevent bee visitation and foreign pollination. Studies in which the uncovered flowers were emasculated and left unpollinated verify this conclusion (Visser, 1951). There is the possibility that wind pollination may occur on some varieties if the pistils are exposed, and this should be considered if genetic studies are planned.

Emasculation is accomplished by using finger nails, scalpel, tweezers, or small scissors specially modified with a notch in the blades and an adjustable screw to control the amount of closure (Fig. 1). These scissors are used to cut below the sepals; then with a slight pull the flower except for the pistil is removed. This method is faster and less injurious than other methods. Usually, three blossoms per cluster at the balloon stage of development are emasculated. Open and immature flowers are cut off with the emasculating scissors or finger tips.

Pollen is applied to the stigmas by glass rods, finger, camel hair brush, or velvet-covered cork stopper. Glass rods make an efficient use of pollen when it is in short supply. They are also easily and quickly cleaned by immersing in 70% ethyl alcohol before a different pollen is used. Although the seed set varies widely from one cross to another, our experience has been that approximately two flowers must be emasculated and pollinated to produce one seed.

Seed stratification and germination: Pear seeds require stratification, i.e., a period of low temperature exposure in a moistened state to overcome their internal dormancy or rest. This requirement may be met by sowing the seeds in a seed bed outdoors in the fall or by holding them at low temperature in cold storage for 60 to 90 days at 0 to 7 C in a moist, well-aerated medium (Hartman and Kester, 1969). The optimum temperature and time is dependent on the species stratified. Species from warm winter climates require a shorter stratification period than those from cold winter climates, as the former species have a higher optimum temperature (7 to 10 C) than the latter (3 to 5 C) (Westwood and Bjornstad, 1968).

At Harrow, we stratify pear seeds in moist, finely-ground peat moss placed in polyethylene bags. In order to prevent damping off and molds, ferbam (ferric dimethyl-dithio-carbamate) is added to the water used to moisten the peat moss. Other fungicides and seed disinfectants may also be used with satisfactory results.

The seeds are planted in individual peat pots (6 cm in diam.) containing a mixture of equal parts sand and peat moss (U.C. Mix-C, Matkins and Chandler, 1957) which has been steam sterilized. At 20 C germination occurs within seven to ten days. When the seedlings attain a height of 15 cm, they are transferred to 10 cm fiber pots. A conventional potting mixture of soil, sand, and peat moss (2:1:1) is used which is also steam sterilized before use.

Inducing early fruiting: Seedlings of many plants, including pear, are juvenile following germination and cannot be induced to flower. This growth phase is termed the juvenile phase and is distinct morphologically and physiologically from the adult phase of trees propagated by scions or cuttings from the fruiting or adult portions of another tree (Doorenbos, 1965). Pear seedlings remain unfruitful six years or longer, and thus any means of shortening the juvenile period is vitally important to

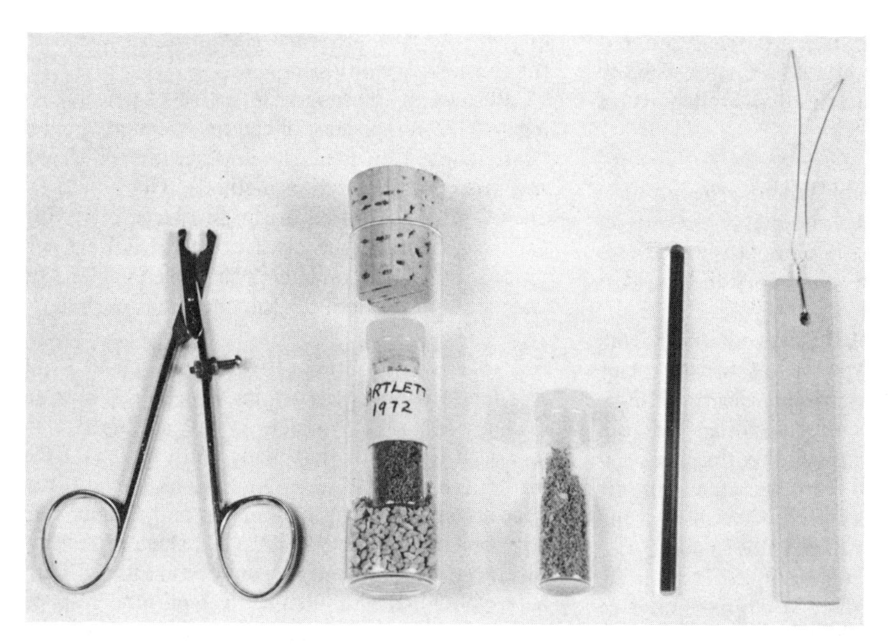

FIG. 1. Equipment used in emasculating and pollinating pears. Left to right: emasculating scissors with notched blades and adjustable screw; pollen in small stoppered glass vial enclosed in larger cork-stoppered vial containing anhydrous $CaSO_4$; plastic-stoppered glass vial with pollen; glass rod for applying pollen to stigma; small wooden label with wires for identifying crosses and fastening to pollinated branches.

obtaining early fruit evaluation and shortening the time to produce a generation.

The length of the unfruitful phase appears to depend on the amount of growth the seedling makes. Therefore, the most effective way of shortening the juvenile period is to grow the seedlings as fast as possible (Zimmerman, 1972). Weather, soil, and environmental conditions which promote growth tend to decrease the time to fruiting (Visser, 1964). For early fruiting it seems advisable to grow pear seedlings at high levels of fertility and to avoid excessive pruning and overcrowding. Perhaps, while the seedlings are growing in the greenhouse, supplemental lighting and enrichment of the atmosphere with CO_2 may offer a further means of speeding up growth and shortening time to fruiting.

Grafting onto bearing trees, ringing, scoring, bark inversions, root pruning, and sprays with growth retardants may stimulate earlier flowering of older seedlings which are already in transition from the juvenile to the adult phase, but are not effective on more juvenile seedlings (Zimmerman, 1972). Grafting pear seedlings at the time of transition on quince rootstocks may also shorten the time to fruiting. This technique has only limited utility, however, because of variable compatibility of pear on quince and the limited availability of virus-free quince rootstocks.

Seedling evaluation: Establishment of valid selection criteria based on real needs, objectivity in selection, and ruthlessness in eliminating seedlings of questionable value are mandatory if success is to be achieved.

Some seedling characters, such as disease and insect resistance, when expressed at an early stage of growth permit early seedling selection to be practised. In general, the more seedlings that can be eliminated at this stage the better, since it greatly reduces the number to be grown in the nursery or field. Screening for multiple disease and pest resistance may be possible at this stage.

There are other advantages of early selection. The seedlings can be grown in a greenhouse or controlled environment chamber because at this stage space requirements are minimal. It is thus possible to more effectively control environmental factors, such as temperature, humidity, and light, which are important for reducing the environmental effect on genetic discrimination. Thus, efficiency in selection is greatly increased, especially for characters like disease and pest resistance where effective screening requires control of certain environmental factors. Furthermore, if greenhouse and controlled environment chambers are used, the tests can be scheduled at the convenience of the breeder.

Some characters which are expressed in the juvenile stage are correlated with economically important characters expressed in the adult stage. Early selection may be practiced for such correlated characters. Some correlated characters in *P. communis* include: seedling vigor with precocity (Zimmerman, 1972), early flowering before leafbreak with high yield (Moruju and Slusanschi, 1959), seed size with germinability and seedling vigor (Schander, 1955) and fruit color with foliage color. Other useful correlations might be found by further study of seedling and fruiting characters.

Selection at the time of fruiting involves consideration of such fruit characters as precocity and productivity, fruit size and quality, optimum harvest maturity, ripening uniformity, and storage and processing qualities. Storage and processing tests should be conducted only on the most promising seedlings. Furthermore, only a few of the best seedlings in the best progenies should be advanced to the second stage of testing.

Because of the long-term nature of pear breeding and the importance of careful selection, greater efforts should be made to improve the efficiency and precision of selection methods. This would reduce the time required in the first stage of testing and also improve the chances that seedlings advanced into the second stage have been adequately screened and selected for the characters desired.

Advanced testing and cultivar introduction: Fruit breeders can be criticized for generally doing an inadequate and incomplete job of advanced testing, with the result that many cultivars have been introduced that did not warrant commercial introduction in that they had major faults that were not detected during the advanced stage of testing. Advanced testing of pears could be improved with better designed and adequately replicated tests to establish the range of adaptation of new selections, to compare performance with standard cultivars, and to remove personal bias by using objective tests and measurements where possible. More information is required with pears, however, to determine the optimum number of experimental units per plot and the optimum number of replications, locations, and years required to adequately test advanced selections.

The procedure in advanced testing followed by many breeders involves three stages of testing: 1) as closely planted seedlings, 2) as grafted trees at normal orchard spacings in one location, 3) as grafted trees at normal orchard spacings at a number of locations.

The use of three separate stages of testing is optional. However, it is possible to adequately determine cultivar potential of selections by the end of the second stage of testing, if the second and third test are combined and conducted at several locations simultaneously.

The procedure that we are using should make it possible to release a new cultivar 18 to 20 years after the cross is made instead of 25 to 30 years or longer which is usually the case. In addition, there will be an ample supply of virus indexed, true-to-name budwood available to satisfy the requirements of commercial nurseries at the time of introduction.

The scheme in use at Harrow is outlined as follows. The same year that a selection is made, budwood of that selection is entered into a virus indexing scheme that will lead to certified, virus-tested, true-to-name trees in ten years. Simultaneously, open-pollinated seedlings of 'Bartlett' and 'Old Home' and virus-free clones of 'Quince A' are budded in sufficient quantity to produce trees for regional tests by cooperating experiment stations and selected growers using both replicated and non-replicated plantings. By pooling the resources of cooperating experiment stations, precise estimates can be obtained for those characters which show a large environmental response. Experienced growers can also provide valuable assistance in the assessment of advanced selections for commercial potential and their participation in such testing programs should be encouraged. They can provide useful information on such characters as productivity, harvesting, handling and storage ability, and consumer acceptance.

Mutation Induction

Although bud mutations arise naturally, mutagenic agents can be used to increase their frequency of occurrence (Granhall et al., 1949, and Bishop, 1966). Ionizing radiation, either 3 to 8 krads gamma rays (Decourtye and Lantin, 1971) or 4.5 to 7 krads X-rays (Visser et al., 1971) applied to dormant scions, have been used to induce mutations in pear.

After irradiation, special techniques are further required to identify and isolate the mutated buds. The irradiated dormant scions are first grafted onto rootstocks. The primary effects of the irradiation treatment such as abnormal leaves and bifurcations seen on the shoots growing on the scions are not an indication of the presence of mutations. Decourtye and Lantin (1971) found it necessary to propagate a second vegetative generation (V_2) in order to select for bud variations such as bloom date, ripening time, fruit color, and russeting. Buds situated high on the shoot were taken to ensure that they were formed after the mutagenic treatment. Final testing was done in the V_3 generation. Visser et al. (1971) found that they could select for mutants with the compact growth habit among shoots produced on the irradiated scions in the V_1 generation. The shoots on the scions were cut back after the first and second growing season and new shoots that were thicker than normal for their length were selected to produce a V_2 generation. These shoots are morphologically the same as the V_2 plants produced by Decourtye and Lantin. The compact habit was produced by 0.5% of the surviving V_2 trees.

Polyploid Induction

Polyploids have occurred spontaneously, such as tetraploid forms of 'Bartlett' and 'Anjou' and some natural triploid seedlings like 'Beurre Diel', 'Beurre d'Amanlis', 'Cantillac', and 'Pitmaston Duchess' (Moffett, 1934). These have been called "giant sports" and can be recognized by their larger than normal shoots and fruits and increased leaf thickness. Other triploids, such as 'Merton Pride' have been derived by crossing a diploid cultivar with a tetraploid ('Merton Pride' = 'Glou Morceau' [2x] x 'Bartlett' [4x]) (Anon., 1959).

Dermen (1947, 1965) was able to produce triploids by crossing a colchicine-produced tetraploid with several diploid cultivars. Colchicine treatment of a triploid pear selection produced a complete hexaploid (6x) and also produced chimeras with a 6-3-3 and a 3-6-6 constitution in the first three histogenic layers of the apical meristem. The totally hexaploid (6x) and chimeral hexaploid 3-6-6 would breed true as a hexaploid and could be useful in breeding.

Polyploids can also be produced by crossing diploids with hexaploids. But Dowrick (1958) discovered a feature of the diploid cultivar 'Beurre Bedford' that makes it particularly useful in generating large progenies of triploid and tetraploid offspring if the proper parental combinations are made. He found that as a result of abnormal gametogenesis, nuclei of pollen varied in polyploidy from 1x to 4x. Young pollen grains had four nuclei due to failure of cell wall formation following meiosis. At the first pollen grain mitosis the four spindles fused giving rise to mature pollen with 4x generative nuclei in 81.4% of the cases. In the remainder, one or more nuclei remained free and a variable number of 1x, 2x and 3x nuclei resulted. Where no fusion occurred the mature pollen grain contained four generative and four vegetative nuclei. When 'Beurre Bedford' was used as a pollen parent and crossed with a diploid, the majority of offspring were diploid, but when crossed with triploid or tetraploid females, the offspring were mainly triploid and tetraploid, respectively. Embryos whose chromosome numbers differed from the maternal parent usually aborted. He postulated that embryo breakdown was caused by a lethal mechanism controlled by the balance of chromosome sets of the male and female sex nuclei. This unusual feature of 'Beurre Bedford' offers the opportunity of using it as a pollen parent in crosses with selected triploid and tetraploid cultivars. Other diploid pear cultivars may also possess this mechanism.

Breeding Systems

The plant breeding strategy used in improving specific characters in pears is influenced by a number of considerations which dictate the approach that may be most efficient and fruitful for the characters in question. The same breeding strategy for pear may not be useful in all situations, but the best approach may vary depending on the type of character and its inheritance. The presence of incompatibility and sterility barriers to crossing, inbreeding effects, and the occurrence and frequency of mutants also influence the selection of an appropriate breeding strategy.

Incompatibility and Sterility

As previously mentioned, all pear species studied are diploid and intercrossable. There are no major sterility barriers to interspecific hybridization in *Pyrus*. In fact, several cultivated forms appear to be derived from interspecific hybridization.

Outcrossing in pear is ensured by gametophytic incompatibility (Crane and Lawrence, 1952). Cross incompatibility also occurs in pears (Crane and Lewis, 1942) but this is rare. Naturally, most bud sports derived from the same cultivar are cross incompatible.

Diploid pears are mostly self incompatible. This includes cultivars of *P. communis* as well as other pear species (Crane and Lewis, 1942 and Westwood and Bjornstad, 1971). A few cultivars will form a limited number of seeds when selfed. Apparently the self-incompatibility system breaks down when the chromosome numbers are doubled (Crane and Lewis, 1942); thus tetraploid cultivars are partly or fully self compatible. Inbreeding to increase homozygosity in diploid pear cultivars by self pollination is thus severely restricted. Other forms of inbreeding at the diploid level such as backcrossing and sib-mating are more successful for increasing homozygosity.

Several types of sterility also impose restrictions on the crosses that can be made: 1) generational sterility due to the failure to obtain normal development of the pollen, embryo-sac, embryo, and endosperm; and 2) morphological sterility due to suppression or abortion of the sex organs (Crane and Lawrence, 1952). These forms of sterility are important in only a few cases. 'Magness', for example, can be used only as a seed parent because it is completely male sterile. Generational sterility is greater in triploid than in diploid pears. It is generally not an important consideration in breeding at the diploid level, however.

Polyploidy

It was previously mentioned that the genus *Pyrus* was thought to have arisen from a doubled hybrid between two ancestral species with basic chromosome numbers of $x = 8$ and $x = 9$. All pear species examined are diploid ($2n = 34$, $x = 17$). The pear likely arose as an allopolyploid but it behaves as a diploid and the monogenically inherited characters show typical diploid segregation (Crane and Lewis, 1940).

Most *Pyrus* cultivars are diploid but polyploid

cultivars do exist, especially in *P. communis.* A number of cultivars are triploid (2n = 51, x = 17), a few are tetraploid (2n = 68, x = 17), and hexaploid forms have also been produced (2n = 102, x = 17).

Although polyploidy may offer some advantages in pear breeding, such as greater vigor, larger fruit size, and higher self compatibility in the case of the tetraploids, it also has the disadvantage of usually greater generational sterility than diploids with resultant reductions in fruit set. In the case of triploid cultivars, two diploid pollenizer cultivars are required—one to pollinate the triploid, the other to allow reciprocal pollination between diploids because the triploid produces a high percentage of sterile pollen. Only one pollenizer is required if both cultivars are diploid (Crane and Lawrence, 1952).

Polyploidy may have some utility in disease and pest resistance breeding, for example, where the best source of resistance may be present in unadapted derivatives of a wild species that have small, lower quality fruit. In this case it usually requires several generations of backcrossing to parents with large fruit and high quality to improve size and quality in the offspring. If the diploid wild species were crossed with a better quality tetraploid cultivar such as 'Bartlett' (4x) or 'Anjou' (4x), it might be possible to obtain triploids in the F_1 generation that have more size and perhaps better quality than one might obtain if the diploid forms of 'Bartlett' or 'Anjou' were used as parents. If cultivars are not achieved by this approach, however, the triploid seedlings are difficult to use as parents in succeeding generations.

Mutations

Mutations occur spontaneously or they can be induced by ionizing radiations and chemical mutagenic agents. Except in the case of pollen, genetic changes usually involve a sector of tissue or a whole bud or shoot which can affect any character. These mutated forms can then be maintained by asexual propagation from the shoots on which they occur and if beneficial receive varietal status. Some of the mutants breed true and others do not depending on the location of the mutated tissue and the complexity of the chimeras produced.

Mutations affecting fruit color and russet are easily recognized. Spontaneous mutations ('sports') have resulted in completely red fruits ('Starkrimson' from 'Clapp's Favourite' and 'Red Anjou' from 'Anjou') and completely russeted fruits ('Russet Bartlett' from 'Bartlett') being produced. Other mutants also occur that may affect less visible characters and consequently may not be recognized. Mutagenic agents have been used with tree fruits to increase the frequency of mutations affecting such characters as skin color and russet, disease resistance, compact (spur)-type growth, self-fertility, season and muturity, and bloom (Bishop, 1966, Nybom and Koch, 1965).

It should be noted that the time required to produce and adequately test superior clones from a mutation breeding program is not much different from that required from a conventional hybridization program. Careful study of the mutated clones is important because of the danger of accompanying minor inherent defects which may be also induced by radiation treatment. It is also important that the mutant be stable on repropagation.

Mutation induction may be a promising fruit breeding method for characters such as compact growth, skin color, and russet changes, but does not seem to be effective for improving characters such as fruit size and quality. It may be useful in breaking up closely linked genes and in inducing greater variability where it may be limiting. It may also be useful in breaking down self incompatibility to induce self fertility, or in breeding for disease resistance. Its place in pear breeding is not yet established but it appears to be no substitute for conventional hybridization.

Hybridization

Cultivar improvement in pear by seedling selection and clonal propagation has been continuous for over 2,000 years. The Belgian breeders (1750–1850) successfully practiced a form of mass selection from which many outstanding cultivars were developed. It was not until the late eighteenth century that Knight (1809) recognized the value of controlled hybridization as a method of cultivar improvement. Since then, controlled hybridization has been the main method used by pear breeders to develop new cultivars.

Controlled hybridizations are usually made between parents that have the best phenotypes for the characters desired. One of the dangers of mating "the best with the best" is that only a few outstanding cultivars are used as parents, such as 'Bartlett' and 'Comice', and this leads to a narrow genetic base. Furthermore, inherent in the strategy of crossing only "the best with the best" is the tendency to repeat the same crosses. There is a limit to the genetic progress that can be expected from continually sampling within the same progenies (Andersen, R. L. 1970).

A better approach might be to create a population with an ever-increasing concentration of desirable genes by employing recurring cycles of selection and intermating of the best selections. Anderson, working from theoretical considerations, proposed that the most appropriate system for improvement of tree fruits such as pears is recurrent mass selection. This method involves selecting a source population which contains all the traits desired for cultivar improvement. The population is intermated and the best seedlings are selected to form the second source population. These seedlings are intercrossed and the best selections from their offspring form the next source population. The cycle can be repeated many times. Each cycle of recurrent mass selection should result in progressive population improvement. The testing of selections for cultivar status is a separate phase of the breeding program that is conducted after each cycle.

Watkins (1970), however, cautions against relying solely on procedures based on quantitative genetics because these procedures may produce poorly defined genetic stocks. In his opinion, such an approach is only warranted if a careful search fails to show the presence of major genes for the characters in which improvement is desired.

Sometimes better sources of cold hardiness, disease resistance, or other characters are present in wild seedlings of *P. communis* or even in other *Pyrus* species. The breeder may wish to introduce these characters into his material to effect cultivar improvement. Often a number of undesirable characters such as small fruit size and poor quality may accompany the transfer of desired characters. To overcome this, backcrossing is often employed. Sometimes a strict form of backcrossing is used in which crosses are made to a single recurrent parent. But often a modified form of backcrossing is employed in which more than one recurrent parent of similar phenotype is used. This is done to avoid excessive inbreeding and resultant inbreeding depression. Backcrossing has been frequently used in programs to improve fire blight resistance.

Parental Selection, Progeny Testing, and Progeny Size

Parental selection is most often based on the phenotype of the parents. If the character to be improved is highly heritable and the genetic variance is additive then genetic advance can be achieved. On the other hand, if dominance and epistatic variance is high then less genetic advance can be made by selecting parents on the basis of their phenotypes and selection based on progeny performance may be more useful.

The progeny test is especially useful if the characters can be assessed at an early stage. It is less useful if it is necessary to await fruiting to assess progeny performance. In the case of fire blight resistance, for example, Layne et al. (1968) found that selections which had potential as parental sources of resistance could be crossed to a common susceptible tester parent and the progeny screened for fire blight resistance the following year. In this case, relatively small progenies of 50 to 100 seedlings were sufficient to test each potential parent. Those that were found to have the greatest prepotency for transmitting resistance could then be crossed to appropriate cultivars the following year, but larger progenies could then be secured to increase the chances of obtaining offspring in which resistance was combined with high quality fruit of suitable size. The diallel cross can also be used effectively in a progeny test to determine the general and specific combining ability of the parents involved. This can improve precision in parental selection.

Progeny size depends on the breeding system that is being used. It also depends on the characters being selected, their mode of inheritance, and whether the parents having these characters are homozygous or not.

Breeding for Specific Characters

Tree Characters

Growth habit: The modern trend in tree fruit production is toward dwarf trees which bear early, and which are easier to prune, spray, and harvest. Yield per unit area of dwarf trees grown at close spacings is also greater than the standard type of tree grown at wide spacings because of the more efficient interception and use of light (Tukey, 1964).

Dwarfing can be induced by using a dwarfing rootstock, but some cultivars, such as 'US 309' (personal correspondence, T. van der Zwet, USDA, Beltsville, Md.) have a natural dwarf or compact growth habit. These dwarf and compact cultivars can be used in further breeding. Another promising approach is that of using ionizing radiation to produce compact and dwarf mutants which has already been outlined in a previous section.

Precocity: Besides the importance of a short juvenile period to ensure early selection of the fruit and shortening of the generation time, it is important to the development of cultivars that are more precocious and productive when propagated on rootstocks (Visser and DeVries, 1970). Because the length of the juvenile period is heritable (Zielinski, 1963; unpublished data, R. H. Zimmerman), selection for precocity is feasible. The selection of parents to use in breeding for precocity may be based on the length of the vegetative period (Visser, 1967) or on progeny performance. Seedlings exhibiting juvenile growth may be identified by the presence of thorns, irregular leaf margins, and wide angled branches (Fritzsche, 1948) and it is possible that absence of these features might be used for early selection of seedlings with a short juvenile phase.

Self compatibility: While incompatibility may be a hindrance to the plant breeder wishing to carry out selfing, it is an important pomological character because it requires the orchardist to plant cultivars together which are mutually compatible and which flower at the same time in order to obtain good cross pollination and fruit set. Fortunately few cultivars are incompatible when crossed with each other (Crane and Lewis, 1949), but the availability of efficient pollinizers is an important consideration in the release of any new pear cultivar.

The development of self compatible cultivars may be possible through radiation of the pollen to induce mutations for a self fertility allele (Lewis and Crowe, 1954). Self compatible cultivars would eliminate the need for pollinizer cultivars in orchard plantings.

Parthenocarpy: Many pear cultivars are capable of producing fruit parthenocarpically, that is, without fertilization (Gorter and Visser, 1958). Parthenocarpy has obvious advantages: 1) fruit is produced even though pollination conditions are unfavorable and 2) it may be possible to dispense with the pollinizer cultivar altogether. In fact, in certain valleys and coastal regions of California, 'Bartlett' is grown with few or no pollinizer cultivars and productivity is maintained by parthenocarpy. As much as 85 to 99% of the fruit in these orchards are seedless. The main problem encountered is that the seedless fruits lack flavor, soften earlier and have lower soluble solids accumulation compared to the seeded fruits (Griggs et al., 1960). In most other regions, however, 'Bartlett' requires cross pollination to maintain good production.

The development of cultivars that produce parthenocarpic fruit seems possible. But because of the variable expression of this trait in different environments and the lower quality of seedless fruit, it has low priority in most breeding programs.

Productivity: In their effort to improve other characters such as quality and disease resistance, pear breeders may have a tendency to overlook productivity. Productivity has become increasingly more important as production costs continue to rise and financial returns to the grower decline. It is therefore important that new cultivars produce large crops of marketable fruit throughout the life of the orchard. Not only should the accumulated yield be high, but it is also important that the cultivar bear consistently from year to year. Biennial bearing is a highly undesirable trait.

There are large differences in productivity of seedlings. They must be noted and recorded. Quite often there may be several seedlings with similar over-all characters. In this case the breeder should select the one with the greatest yield for further breeding or for second test.

Fruit Characters

Flavor: Flavor in pears is often difficult to judge. It is highly variable, depending on the time of picking and conditions of ripening. Time of picking is especially important with early maturing seedlings in which the best flavor is ephemeral. On the other hand many of the winter pears require a period of low temperature near 0 C to develop their best flavor. For this reason it may be necessary to ripen one sample of pears immediately after picking and another after a period of cold storage.

Flavor is dependent on a delicate balance of sugars, acids, tannins, and aromatic compounds. The main acids contributing to low pH are malic and citric acids. Sweetness depends on the amounts of fructose, glucose, and sucrose. In general the best dessert pears tend to have high sugar content. Acidity, on the other hand, tends to vary as much among good dessert pears as it does among those of lower quality (Visser et al., 1968). The contribution of the tannins and aromatics to flavor is not as well understood as sugars and acids but these compounds are nonetheless components of flavor. Astringency often located in the skin may be caused by this group of compounds.

Juiciness also affects flavor (Visser et al., 1968). Sugar tends to taste sweeter to the taster if the fruit is juicy than if it is dry.

Acidity and sweetness seem to be inherited independently, both in a quantitative way (Visser et al., 1968 and Zielinski et al., 1965). Juiciness, however, was reported by Zielinski et al. (1965) to be inherited as a single gene with juiciness dominant over dryness.

Texture: Fruit texture is a collective term for those attributes of the fruit that determine how it feels in the mouth. Flesh firmness is highly dependent on ripeness, but even at optimum ripeness it varies from soft and buttery to crisp and breaking.

In most pear growing regions, the soft, buttery texture typical of *P. communis* cultivars is preferred. In China and Japan, where the main cultivars are from *P. pyrifolia* and *P. ussuriensis*, the crisp, breaking texture is preferred. Thus, the breeding and selection criteria for flesh texture are quite different in different regions, depending on the species that are cultivated and the preferences that exist.

Textural properties that are desired in European-type pears include: a tender breaking skin, a soft buttery flesh, absence of fibers in the flesh, and absence of grit (stone cells). Grit is more tolerable if located in the core than under the skin or in the flesh. It is particularly undesirable if pears are made into purée for baby food. Grittiness is a characteristic of *P. pyrifolia* and its hybrids and accounts for the name 'sand pears'.

Very little is known about the inheritance of flesh texture in pears. Westwood and Bjornstad (1971) reported that in interspecific hybrids of wild forms of *Pyrus*, stoneless flesh was dominant to stony. Golisz et al. (1971) on the other hand, found that in crosses of cultivars of *P. communis* with *P. ussuriensis*, the majority of offspring had numerous stone cells typical of the *P. ussuriensis* cultivars used as parents. Zielinski et al. (1965) presented data indicating that presence of stone cells is dominant to their absence among certain crosses of *P. communis* cultivars.

Color: Skin color of pear is dependent upon the number of pigments present and their amounts. The basic ground color is due to the presence of chlorophyll (green) and carotenoids (yellow) pigments. The change in ground color from green to yellow that accompanies ripening results from the increase in carotenoids and the breakdown of chlorophyll (Spencer, 1965). When the fruit is fully ripe the ground color can vary from dark green to deep yellow depending on relative amounts of chlorophyll and carotenoids.

The red pigmentation which may be present on all or part of the fruit in different intensities is due to the presence of anthocyanins, namely cyanidin-3-galactoside and cyanidin-3-arabinoside (Francis, 1970) in one or more of the skin layers. Bright red-colored pear fruits have considerable eye appeal and it seems possible that in the future more effort will be made to selecting red-fruited bud sports and seedlings.

Many of the components of skin color appear to be under the control of single major genes. Zielinski et al. (1965) suggested that the basic ground color in several progenies which he studied was under the influence of a major gene with yellow dominant to green. The inheritance of blush vs. non-blush in these same progenies indicated that the blushed condition was recessive. A genetic mutation of 'Bartlett' designated 'Cardinal Red' (*C*) by Zielinski (1963), which produced a deep red pigmentation of the fruit, was inherited as a single dominant gene. Brown (1966) came to a similar conclusion using 'Max Red' which is another red sport of 'Bartlett'. Not all red mutants, however, are capable of transmitting red fruit coloration to their progenies. For example, the cultivar 'Starkrimson' is a red mutant of 'Clapp's Favorite', in which the mutation only affects the epidermis. The germ layer is normal and the mutant fails to transmit the red character to its progenies (Dayton, 1966).

Corking of the skin produces a brown russeted appearance which is seen on many cultivars. Many people do not find this appearance unattractive and for the fresh market trade russeted fruit might be acceptable, but when the fruit is processed whole for purée, russeting results in a brown colored product which is unacceptable. The presence of large and dark colored lenticels on the skin is also objectionable for the same reason.

Wellington (1913) first proposed that russeting was controlled by a single gene but Kikuchi (1930) concluded that two genes were involved, *R* and *I*. *R* gives russet brown and *r* smooth green. *I* partially inhibits cork formation so that it does not extend over the whole fruit. Both of these workers based their conclusion on the segregation within progenies that involved *P. pyrifolia* or hybrids of *P. pyrifolia* and *P. communis*. In the *P. communis* progenies studied by Zielinski et al., (1965) and Crane and Lewis (1949) the inheritance of russeting appeared to be more complex.

Flesh color of pears is generally creamy-white to white, but green, yellow, pink, and almost black flesh occur. Less is known about the inheritance of flesh color than of skin color, but a white flesh vs. green segregates as if a single gene were controlling color with white dominant and green or cream behaving as alternate alleles (Zielinski et al., 1965). Evidence from a progeny of 'Sanquinole' (red flesh) crossed with 'Conference' (white flesh) suggests that red flesh is dominant to white (Brown, 1966).

Size: Fruit size is a highly variable character because it is influenced by many environmental factors as well as by fruit set. Consequently, more than one season may be required to adequately assess this character. Considerable genotypic variation exists in *Pyrus* for fruit size. Normal size in some clones of *P. betulaefolia* and *P. Calleryana* may be as small as 1 cm in diameter while in some clones of *P. communis* and *P. pyrifolia* it may exceed 12 cm in diameter. The optimum size of a cultivar depends somewhat with the use to which it is put. Dessert cultivars should generally be 5 to 10 cm in diameter. Pears intended for processing also have certain size requirements. The size of the can influences the maximum size of fruit that can be used. Thus, in North America where the processing industry is geared to 'Bartlett', a fruit diameter greater than 5 cm is required for packing as halves or quarters. If, on the other hand, the fruit is used for purée, a smaller size is acceptable. Generally speaking, a cultivar producing fruits smaller than 5 cm in diameter would be undesirable as a dual-purpose cultivar in North America. Fruit size, at least among *P. communis* cultivars, appears to be under polygenic control with the mean fruit size of the progeny being smaller than either parent (Crane and Lewis, 1949 and Zielinski et al., 1965).

Shape: Although many of the common cultivars of pear are pyriform, fruit shape can vary from oblate to round to pyriform to conical. The inheritance of fruit shape also appears to be under polygenic control (Crane and Lewis, 1949, and Zielinski et al., 1965). It appears that round and obovate shapes may be dominant to pyriform and turbinate shapes because they are recovered with much higher frequency in the progeny (Zielinski et al., 1965). While shape is not as critical as size, texture, and flavor, the pyriform shape is generally preferred in North America and should be considered in seedling selection.

Ripening season: Cultivars which ripen at different times are required to extend the picking season and market period. Early maturing cultivars are especially needed for the fresh market trade. But early pears tend to overripen quickly, are subject to internal breakdown, and quality is ephemeral. The breeding of an early pear that does not quickly overripen has been attempted, but without success (Morettini, 1967). Crane and Lewis (1949) found that the mean season of ripening of all F_1 families they studied was much earlier than that of either parent. Very few seedlings were found that were later than the later parent. In breeding for earliness or lateness, therefore, both parents should be either early or late to ensure a greater proportion of the offspring with a ripening period that is desired. Ripening season appears to be polygenically inherited.

Keeping quality: The ability to keep in cold storage without developing internal breakdown allows marketing over a prolonged period of time. Cultivars with good keeping qualities, such as 'Anjou' and 'Bosc', are commonly held in cold storage and marketed during the winter when prices tend to be high.

Keeping quality is difficult to evaluate with precision. The usual method is to hold fruit samples in storage (−1 to +3 C), and periodically ripen them in a ripening room, examining them for internal breakdown, texture, and flavor. However, considerable differences are known to exist among cultivars for resistance to internal breakdown and long storage ability (Howlett, 1957). A noteworthy cultivar is 'Pingo-li' (*P. Bretschneideri*) which can be harvested at the end of September and kept a year in common storage without spoilage (Pieniazek, 1966).

Evidence obtained from segregating progenies suggests that keeping quality is under polygenic control (Zielinski et al., 1965).

Processing quality: Many breeding programs have as one of their major aims the development of cultivars for the processing industry—canning as halves, quarters, or diced with other fruit, purée, perry-making, or drying of fruit slices. Fruit characteristics important in processing quality have already been discussed under flavor, texture, color, and size.

Laboratory tests which can be carried out on small samples of fruit are important for the evaluation of quality when seedlings first begin to fruit. Most laboratory procedures try to simulate mass production procedures. Thus, seedlings and advanced breeding lines can be screened for processing ability as an aid to ultimate selection for cultivar introduction. Special taste panel techniques are usually employed which allow precise organolyptic evaluation (Larmond, 1967). The use of special instruments such as the hand refractometer for measuring soluble solids, the Hunter colorimeter for color measurement of halves and purée and the Ottawa Texture Measuring System (Voisey, 1971) for textural qualities allow more objective evaluation of processing qualities.

Because so many components are involved in processing quality, its inheritance is difficult to analyze. It may be difficult to breed specifically for processing quality, but it is possible to effectively select good processing types in seedling progenies of pear.

Disease Resistance

Fire blight: This is the most serious disease of pear in North America, and has influenced pear production more than any other single factor. The disease has confined the major production areas to the protected river valleys of the Atlantic seaboard in Nova Scotia and New York; to a narrow region bordering the Great Lakes in New York, Ohio, Michigan, and Ontario; and to mild, dry valleys of the Pacific region extending from British Columbia to Washington, Oregon, and California (Anderson, 1956). Commercial pear growing in the warm humid regions of the southeastern, southern, and central United States has been virtually abandoned due to the unusual severity of fire blight. The disease is a constant threat, even in the climatically favored production areas. It is difficult, expensive, and laborious to control even when all recommended practices are followed. Blossoms, actively growing shoots, and immature fruits are most readily infected, but the trunks and roots may become infected as well.

Fire blight was noticed as early as 1780 in New York (Denning, 1794). It spread westward with the settlers until it is now present in all pear-growing regions of North America. The disease is also present in Japan and New Zealand, parts of South America, and more recently in Northern Europe (van der Zwet, 1968 and 1970).

Breeding for fire blight resistance in pear began in the nineteenth century after introduction of the Chinese sand pears (*P. pyrifolia*) to the eastern United States (Hedrick et al., 1921). The cultivars 'Le Conte', 'Kieffer', and 'Garber' were derived from interspecific hybridization and were grown because they were substantially more resistant to fire blight than the European cultivars of *P. communis* but they were inferior in terms of fruit quality. It was not long after their introduction that work was initiated in the United States to breed high quality pears resistant to fire blight (Magness, 1937).

The stage was set for an intensive effort at breeding for fire blight resistance following the pioneer work of Reimer (1925), who introduced a number of pear species and species hybrids from the Orient and tested them for resistance in comparison with hundreds of European cultivars of *P. communis*. Sources of resistance identified by

Reimer (1925) and by others (Hough, 1944; Carpenter and Shay, 1953; Lamb, 1960; and Thompson et al., 1962) led to a more rational approach to blight resistance breeding with a better choice of parental material and more effective screening methods.

Major blight resistance breeding programs in the United States now include those at the U.S. Department of Agriculture (Brooks et al., 1967); New Brunswick, New Jersey (Hough and Bailey, 1968); Geneva, New York (Lamb, 1960); and West Lafayette, Indiana (Janick et al., 1962). In Canada, the major program is at Harrow, Ontario (Layne, 1968 and Layne et al., 1968). To our knowledge the only breeding program outside of North America where crosses are being made and seedlings are being screened for fire blight resistance is at the East Malling Research Station in England (Alston, 1971). The major concern in England is breeding for fire blight avoidance by crossing parents that have little or no tendency to produce secondary blossoms, a major point of entry for blight infection in England. The second aspect involves using blight resistant backcross derivatives of *P. ussuriensis* x *P. communis* from North American breeding programs in crosses with high quality *P. communis* cultivars.

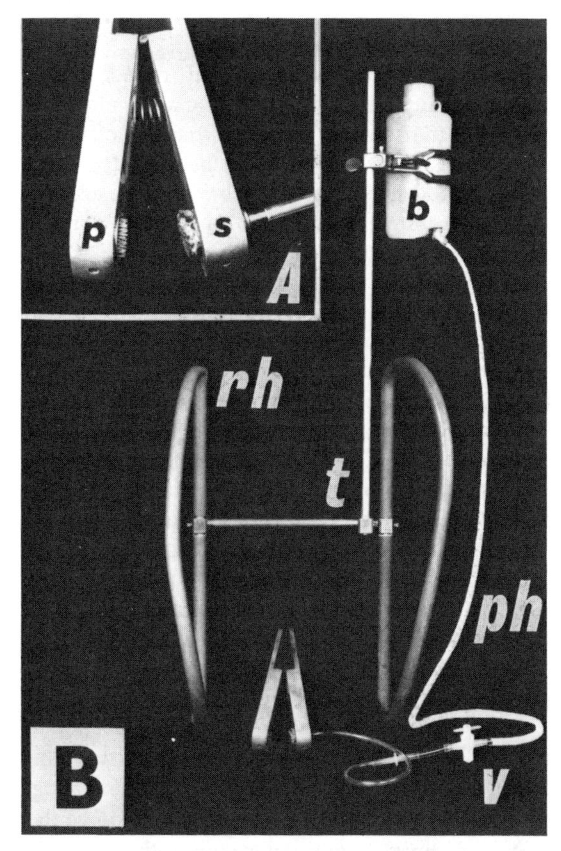

FIG. 2. Simple apparatus for efficient inoculation of pear seedlings with *Erwinia amylovora*. In picture (A), aluminum forceps with hinge and spring attachments have a florist's pin holder (p) and a sponge (s) in opposite wells. In (B), back pack equipment made of aluminum tubing (t) and a rubber hose (rh) carries a plastic inoculum bottle (b) which is attached to the inoculator by a plastic hose (ph). The flow of inoculum is regulated by a valve (v) (van der Zwet and Oitto, 1973).

A limiting factor in breeding for fire blight resistance in pear was the lack of an efficient, reliable seedling screening technique that would give a high level of infection but still distinguish levels of resistance in the offspring with reproducible results. The most common method basically involves inoculating actively growing seedlings about five to six months old with a virulent standardized suspension of the bacterium by needle injection into the stem tissue near the shoot apex (Carpenter and Shay, 1953; Thompson et al., 1962; Layne et al., 1968; and van der Zwet, 1970). Inoculated seedlings are then usually placed in a chamber at 95–100% relative humidity for three days, are returned to the greenhouse, and after approximately two months, the length of blight is measured. Degree of resistance can then be expressed as a percentage of total shoot length. Following inoculation, seedlings may be grouped into 10 percentile classes based on percent of shoot blighted, where class 1 = trace to 10% blight and class 10 = 90.1 to 100% blight (Layne et al., 1968). The bacterial isolates selected should be carefully tested for virulence and a mixed inoculum containing several isolates may be used.

Recently, van der Zwet and Oitto (1973) have developed a new method to efficiently inoculate large numbers of pear seedlings. Tips of succulent seedling terminals are slightly injured with a set of metal pins in a clamp while simultaneously the bacterial cell suspension is released through a sponge connected to a plastic bottle carried on a back pack made of aluminum tubing and rubber hose (Fig. 2). Inoculated seedlings are not moved but remain on the greenhouse bench and are in-

TABLE 1. Distribution of potency for transmitting fire blight resistance in 3 species of *Pyrus* determined from progeny tests at Harrow.
(Layne, Hough, and Bailey, unpublished)

Species	No. of resistant selections tested	Distribution (no.) by potency rating				
		Poor	Poor to Fair	Fair	Fair to Good	Good
P. communis	24	6	6	4	3	5[W]
P. pyrifolia	25	6	4	8	4	3[X]
P. ussuriensis	9		2	3	3	1[Y]

[W] Magness, Maxine, NJ490340020 (P. ussuriensis #76 x Bosc), Old Home, Purdue 80-51 (Early Sweet x Old Home).
[X] NJ501948213 (Beierschmidt x Meigetsu), Purdue 110-9 (Beierschmidt x Tenn. 34S603).
[Y] Purdue 77-73 ([Bartlett x P. ussuriensis 76] x Comice).

cubated under a tent-like structure of plastic sheeting over a wood and wire frame, and the enclosed area is maintained at a temperature of 21–27 C and a relative humidity of 85–100%. Six to eight weeks after inoculation, seedlings are rated for their degree of blight resistance.

At Harrow, progeny testing using our screening method has proven invaluable in identifying the best sources of resistance to use in further breeding. The progeny test proved to be a reliable way of sorting out the parents that were most prepotent in transmitting resistance to their offspring. Many resistance sources have been tested since 1963, and many different hybridization schemes have been tested in breeding for fire blight resistance.

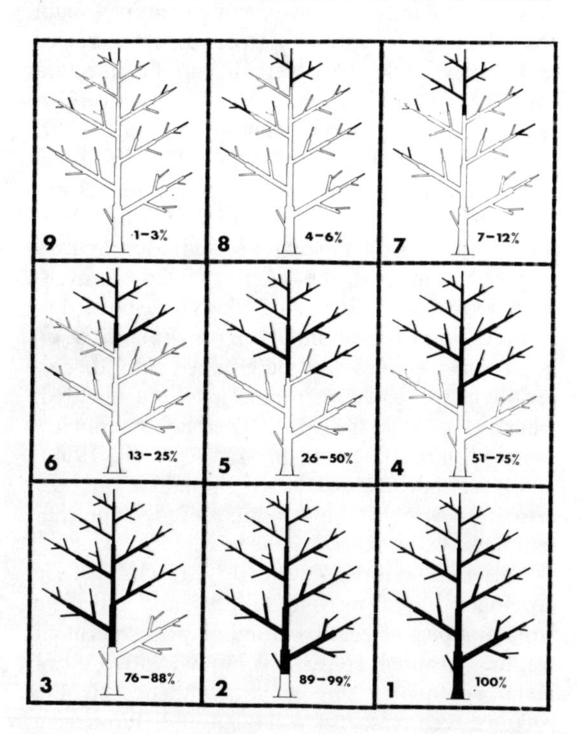

FIG. 3. The USDA scoring system for rating the severity of fire blight in the field, based on age of tissue and percentage of tree infected (van der Zwet et al., 1970).

Layne et al. (1968) and subsequent data by the same authors (unpublished) have shown that various types of simple and complex hybridization schemes can be used successfully in breeding for fire blight resistance in pear. The crossing schemes were devised to utilize resistance present in *P. communis* (C), *P. pyrifolia* (P), and *P. ussuriensis* (U) and to transfer this resistance to high quality *P. communis* cultivars. They used simple intraspecific and interspecific crosses: CxC, CxP, CxU; complex interspecific crosses: (CxP) x (CxP) and (UxC) x (CxP); interspecific backcrosses to different *P. communis* cultivars: (UxC) x C and (PxC) x C; and more modified back crosses: C x (C x [CxP]) and (PxC) x (C x [CxP]). No single breeding system appeared to be clearly superior to another in terms of the proportion of resistant offspring recovered, however. It appeared that progeny performance was influenced more by the specific parental combinations used than by the species source of resistance or the breeding scheme tested. In terms of recovery of resistant offspring with high quality fruit, however, the simple CxC breeding system and the modified backcross system just mentioned produced the greatest proportion of resistant offspring with good quality fruit.

A number of resistance sources derived from *P. ussuriensis*, *P. pyrifolia*, and *P. communis* have been progeny tested at Harrow (Table 1) to determine the best sources of resistance to use as parents in further breeding. The efficacy of the parents as transmitters of resistance was rated subjectively depending on the proportion of resistant seedlings recovered in the progeny, and their level of resistance (Layne, Hough and Bailey, unpublished data).

As a group, the *P. ussuriensis* sources were slightly more efficient than the *P. pyrifolia* sources which in turn were slightly better than the *P. communis* sources, but important exceptions were present in each case. The breeding value of a parent is determined not only by its ability to transmit resistance, however, but also by the other tree and fruit characteristics it possesses. Thus, the *P. communis* sources that have been identified as fair to good transmitters of resistance would have greater breeding value than comparable parents from the other two species sources, because they usually had superior fruit size, appearance, and quality.

Because of the inferior fruit quality associated with *P. ussuriensis* and *P. pyrifolia*, and because their hybrids with *P. communis* usually have inferior fruit quality compared to most *P. communis* cultivars, only the most advanced backcross selections with the best quality fruit should be used in further breeding. Some of these advanced backcross selections possess greater cold hardiness and *Fabrea* leaf spot resistance than most *P. communis* cultivars in addition to fire blight resistance.

In addition to the species we have tested, there are others which have resistance to fire blight. Hartman (1957) lists additional cultivars and species including *P. Calleryana*, *P. betulaefolia*, *P. phaeocarpa*, *P. Fauriei*, and *P. variolosa*. He mentions five Chinese cultivars of *P. ussuriensis* that are immune under Oregon conditions—'Ba Li Hsiang', 'Chien Pa Li', 'Huang Hsiang Sui Li', 'Hung Guar Li', and 'Ta Tau Huang'.

In our progeny tests we have not encountered any true immunity in the seedling we have screened. The previously mentioned cultivars have not been progeny tested to determine whether this immune-type response will be transmitted to the offspring and in what frequency.

Recent ratings of large numbers of pear cultivars and pear species have been published by Oitto et al. (1970) and Mowry (1964) following unusually severe epiphytotics in two regions—Beltsville, Maryland, and Carbondale, Illinois, respectively—where environmental conditions are usually optimum for the disease. The resistant individuals listed by them and those listed by Hartman (1957) should be progeny tested to determine their efficacy as parents for future breeding.

The rating system devised by van der Zwet et al. (1970) for estimating blight resistance in the field following natural infection appears to be a useful and comparatively easy system (Fig. 3). It should be adopted by all pear breeders for rating natural infection in the field in cultivar and species collections and in seedling populations. The system would also aid in comparing the relative resistance of pear cultivars and species in different locations.

The inheritance of resistance to fire blight in *Pyrus* is not yet established. The difficulty in defining phenotypic resistance is due to a number of factors which affect its expression including: 1) age, vigor, succulence, and kind of tissue infected; 2) temperature and humidity relationships during the pre- and post-infection period; 3) inoculum purity and concentration; 4) virulence of isolates; and 5) method of inoculation (Carpenter and Shay, 1953; Lamb, 1960; Thompson et al., 1962; Layne et al., 1968; and van der Zwet and Oitto, 1973).

Layne et al. (1968) found that when most of the factors affecting phenotypic expression of fire blight resistance were controlled, seedling segregation for resistance was continuous with each of the ten resistance classes represented in the progeny. Continuous segregations were obtained regardless of the resistance phenotypes of the parents or the species sources of resistance being tested. The proportion of offspring in each class was significantly influenced by parental phenotypes, however, and to a lesser extent by the species source of resistance. Because most segregation distributions were continuous and a number of them showed an approximately normal distribution, fire blight resistance was concluded to be polygenically inherited. Because a specific type of inheritance pattern that was characteristic of any of the three species they studied was not detected, it was concluded that the same or similar genes for resistance may be present in each species. In a few cases, with each species source of resistance, they found some seedling progenies with a segregation pattern that suggested major gene inheritance. Thompson et al. (1962) also found evidence for major gene and polygenic inheritance in studies with the same *Pyrus* species, although they thought that monogenic resistance was present only in *P. ussuriensis*.

A further complication in breeding for fire blight resistance is the recent finding by van der Zwet and Oitto (1972) that some cultivars are shoot resistant and trunk susceptible, others are shoot susceptible but trunk resistant, and others still are shoot and trunk susceptible or shoot and trunk resistant. It may be necessary, therefore, to screen seedling for shoot resistance in the greenhouse, plant the resistant seedlings in the orchard, then screen again for shoot and trunk resistance as they begin to fruit. This will avoid the danger of releasing a cultivar like 'Magness' which is shoot resistant but trunk susceptible. It would also serve to correlate the resistance of seedlings in the juvenile and adult stages of development.

Leaf blight and fruit spot: This disease, caused by the fungus *Fabraea maculata* Atk., occurs in most areas of the world where pears are grown. Susceptible cultivars are often defoliated by midsummer. Infected fruits are worthless since they become disfigured, cracked, and misshapen. In the nursery the disease can be serious, causing defoliation and stunted growth of nursery trees (Anderson, 1956).

In the Soviet Union, resistance to *Fabraea* has been recognized in some cultivars of *P. communis* including 'Beurre Diel', 'Phelps', 'Louise d'Avranches', 'Clapp's Favourite', and 'Bartlett' (Kovalev, 1963). Resistance has been observed in *P. ussuriensis* seedlings in Illinois (Hough, 1944), and in crosses of *P. ussuriensis* x *P. communis* (Thompson et al., 1962). Zaleski et al. (1959) report from Poland that *P. caucasica* seedlings when inoculated were the most resistant; intermediate resistance was shown by *P. salicifolia* x *P. communis* and *P. elaeagrifolia* x *P. communis;* and *P. communis* seedlings were the most susceptible. Since sources of resistance to this disease have been identified, breeding for resistance should be possible although we are not aware of any major programs where this work is underway. It is selected against, however, especially in regions where natural epiphytotics occur among seedlings and advanced test selections.

Pear scab: This disease occurs in most regions where pears are grown, but it is more serious in Europe than in North America. Scab is caused by the fungus *Venturia pirina* Aderh. which attacks leaves, fruit and twigs. It renders fruit unmarketable because of blemished appearance and deformed shape. The disease is caused by a fungus that is similar to apple scab (Anderson, 1956). Host resistance is somewhat confusing because the fungus has a number of biotypes which appear to have a fairly narrow range of distribution. Thus, cultivars that are resistant in one region may be susceptible in another where they are exposed to another biotype.

Siebs (1955) found that the resistant cultivars he studied had leaves with smaller epidermal cells and a higher hydroquinone content than susceptible cultivars. Hydroquinone was highly toxic to the scab fungus. Further tests are needed with a wide range of biotypes to determine whether hydroquinone content may provide a possible laboratory screening test for resistance.

Stanton (1953) performed controlled inoculation tests on two pear progenies using single biotypes of the pathogen. Resistance in the host was dominant to susceptibility in every case. He obtained some simple 3:1 and 1:1 ratios of resistant: susceptible which suggest that a single major gene controls resistance to pear scab in some cases. Other complex ratios were also obtained which suggest more complex inheritance. Brown (1960) concluded that scab resistance in pear was polygenically inherited. His progenies were not inoculated under controlled conditions with single biotypes of the pathogen, however, and further tests of this type are needed before we can be certain that quantitative inheritance is involved. If separate genes in the host condition resistance to individual biotypes of the pathogen, for example, progenies exposed to a mixture of biotypes in the field may appear to be segregating in a quantitative fashion when in fact relatively few major genes may be involved.

A more careful search should be undertaken with pears to determine whether major genes can be found that control resistance to pear scab. The major gene approach in breeding for scab resistance in apples has been very successful, and it should be tried with pears if suitable genes can be found. Breeding for scab resistance in pears is likely to be more successful if the seedlings are screened in the greenhouse under controlled conditions. The use of the greenhouse permits biotypes that are prevalent in different locations to be used. Resistant seedlings screened in this way would be expected to have a broader genetic base for resistance than those screened only for resistance to single biotype, or those exposed to only a narrow range of biotypes in the field.

Other diseases: Sources of resistance have been identified for other diseases of pear. Resistance for these diseases therefore could be incorporated into breeding programs in areas where they are a problem. Some of the diseases include: *Monilia* fruit rots (Mittmann-Maier, 1940); blackspot, caused by *Alternaria kikuchiana* Tanaka (Hiroe et al., 1958); leaf spot, caused by *Mycosphaerella sentina* (Fckl.) Schroet (Wenzl, 1935, and Lefter, 1959); powdery mildew, caused by *Podosphaera leucotricha* (Ell. & Ev.) Salm. (Fisher, 1922); some pear virus diseases, such as pear bud drop (Trifonov, 1970); stony pit (Morvan, 1961); ring spot mosaic (Kegler, 1960); and mycoplasma diseases, such as pear decline (Hibino and Schneider, 1970).

Host resistance in the long run is the least expensive and most ecologically safe method of disease control. Pear breeders should be on the constant alert to identify sources of disease resistance present in species and cultivar collections and in seedlings and should report the results to assist other breeders elsewhere. In addition, every effort should be made to breed and select for multiple disease resistance, where practicable.

Pest Resistance

Pear psylla (*Psylla pyricola* Foerster): This is the major insect pest of pears in the northeastern United States and in Ontario, Canada. Infested leaves turn brown and often drop and the fruits drop prematurely or are small and of poor quality. Foliage feeding also suppresses root growth and reduces tree vigor. In addition to the damage pear psylla causes by itself, it also transmits a mycoplasma which is the causal agent of pear decline (Jensen et al., 1964). Fortunately, pear species vary considerably in their resistance to pear psylla and breeding for resistance is possible. Westigard et al. (1970) found that Asiatic species of *Pyrus* were generally less attractive to oviposition than those from Asia Minor, North Africa, or Europe. Attractiveness of *P. pyrifolia* and *P. communis* cultivars was greater than that of *P. ussuriensis*. Susceptibility of *Pyrus* was based on differences between total egg deposition and resulting nymphal populations. Using the nymph/egg ratio, it was found that the Asiatic species were more resistant than those from Asia Minor or Europe. The most resistant species were *P. Fauriei* and *P. Calleryana*. Because these species have small, hard, inedible fruit, they may not be as useful in breeding as resistant selections of larger fruited *P. communis* and *P. ussuriensis* cultivars.

Harris and Lamb (1972) report the use of *P. ussuriensis* as a source of psylla resistance. They found that 60% of one progeny of *P. communis* x *P. ussuriensis* was resistant to psylla. Furthermore, resistance in the offspring was positively correlated with large fruit size, and not correlated with poor quality or stone cells. The fact that psylla resistance appears to be inherited independently will simplify any breeding program utilizing this source of resistance.

Woolly pear aphid (*Eriosoma pyricola* Baker and Davidson): This pest infests pear root and devitalizes nursery trees and young trees in the orchard. Nursery trees are seldom killed but may be stunted and unmarketable. In Oregon, orchard trees growing on susceptible rootstocks may be killed if they become dry in late summer and if aphid infestations are high. Westwood and Westigard (1969) have found that at least eight pear species were resistant and six were either susceptible or variable in resistance. They developed a technique of infesting root pieces that could be used in a rootstock breeding and selection program to develop pear rootstocks resistant to woolly aphid. Immunity to wooly apple aphid (*Eriosoma lanigerum* Hausm) has also been reported by Kahn (1955) for two pear rootstocks used in India—'Shiara' (*P. communis* var. *shiara*) and 'Shegal' (*P. Pashia*).

Pear leaf blister mite (*Eriophyes pyri* Pagenstecher): Host resistance to this mite has been reported by Briton-Jones (1923) and Susa (1967). Susa noted that resistant cultivars had dull, oval leaves with thinner cuticles and epidermis than leaves from susceptible cultivars. In addition, tissue respiration was higher, sugar content was lower, and osmotic pressure of cell sap was higher in resistant versus susceptible cultivars. These differences may provide the basis for a selection procedure that could be used for mite resistance. A wider range of resistant and susceptible varieties should be tested, however, to establish which of these tests may be most useful in selecting for pear blister mite resistance.

Root-knot nematodes (*Meloidogyne* spp.): Resistance to root-knot nematodes was reported by Tufts and Day (1934) among seedlings of some cultivars of *P. communis* including 'Easter Beurre' and 'Beurre Hardy'. Seedlings of *P. variolosa* were free of infestation and were apparently immune. Screening for nematode resistance should be incorporated into pear rootstock breeding and selection programs, especially in areas where nematode infestations are high and adversely affect tree vigor and longevity. Screening in the young seedling stage in the greenhouse should be possible.

Adaptability

Cold hardiness: Pears are grown in many parts of the world where the winter temperatures are sufficiently severe to cause cold injury to shoots, fruit spurs, trunks, and roots, and may even cause death of whole trees. In northern parts of Europe, Asia, Canada, and the United States, the need for cold hardy cultivars is especially important. Spring frost during bloom is also a constant threat in some pear growing regions, particularly in southern Europe and California. Thus, resistance of blossoms to frost injury is also an important consideration in the adaptability of pear cultivars to particular regions. In recent times, the greatest efforts in breeding and selecting for cold hardiness in pear have been made in the Soviet Union and North America, especially in the steppe and prairie regions where the winters are severe. Magness (1937) reviewed the early work in the United States to improve cold hardiness of pears, especially the work in Minnesota and Iowa where *P. ussuriensis* seedlings imported from northern Russia, and hybrids of *P. ussuriensis* x *P. communis* were tested. From this work several pear cultivars have been developed that are sufficiently cold hardy to survive the harsh prairie winters, but none of them are grown elsewhere to any extent. They should be important sources of cold hardiness in further breeding in other regions, however. Some of these cultivars include 'Patten', 'Harbin', and 'Bantam'.

In Canada, breeding and testing for cold hardiness in pear has been done by public and private plant breeders in Ontario, Manitoba, Saskatchewan, and Alberta. Few *P. communis* cultivars have withstood the harsh prairie winters. The most promising cultivars that combine cold hardiness with fair size and quality are hybrids of *P. communis* x *P. ussuriensis*. These selections have better size and quality than the *P. ussuriensis* parents and greater hardiness than the *P. communis* parents. The most promising cultivars for the prairies include 'Golden Spice', 'Olia', 'David', 'John', 'Peter', 'Philip', 'Pioneer 3', 'Tait Dropmore', and 'Tioma' (Morrison et al., 1965). None of these cultivars have as high quality as the cultivars grown in the main pear growing regions. Nevertheless, they are valuable sources of germ plasm for improving cold hardiness in pear, since they are able to withstand winter temperatures which are often as low as −30 to −40 C and they have superior fruit quality to their *P. ussuriensis* parents.

Cold hardiness of pear cultivars, species, and interspecific hybrids have been assessed in various European countries, especially after unusually severe winters (Ludin, 1942; Anjou, 1954; Enikeev, 1959; Zavoronkov, 1960; Matjunin, 1960; and Sansavini, 1967). In general, as was found in North America, the cultivars of *P. communis* are less hardy than those of *P. ussuriensis*. Greatest progress in breeding for cold hardiness has been made by crossing hardy selections of *P. ussuriensis* with good quality cultivars of *P. communis* and selecting for hardiness combined with good fruit characters in the offspring.

The difficulties in breeding for cold resistance is the same as that encountered in breeding for fire blight resistance. Most of the sources of cold resistance lack suitable fruit size and quality. It is difficult to improve cold hardiness and at the same time retain good fruit quality in one generation. The progenies of high quality cold susceptible parents crossed with low quality cold resistant parents tend to have small fruit and poor quality, although they often have adequate cold hardiness. Some cultivars, however, appear to be better parents than others, e.g., 'Bessumianka' for transmitting cold resistance, fruit size, and quality; and cultivated forms of *P. ussuriensis*, 'Hua-gei,' and 'Angko-li' for cold hardiness (Golisz et al., 1971).

Methods of breeding and selecting for cold hardiness in deciduous fruit crops was recently reviewed by Stushnoff (1972). There are many test procedures available for testing and evaluating cold hardiness of pear scions and rootstocks that could easily be integrated into pear breeding programs where greater cold hardiness is an objective in breeding.

Inheritance of cold hardiness has not been studied in pear, but analogies with apple show that inheritance of cold hardiness is polygenic and additive with little evidence for epistatis or dominance being important components of the genetic variance for winterhardiness. If we can assume that the same applies to pears, then the best breeding system would involve crossing the hardiest selections from each successive generation, using parents that have also been selected phenotypically for other important pomological characters, but avoiding inbreeding (Watkins and Spangelo, 1970).

Breeding for blossom frost resistance in pear has been approached in two ways. The first is to breed for late flowering to reduce the chance of frost occurring during the bloom period. This approach is valid since late flowering is not correlated with late ripening, and it would be possible to breed cultivars that were early, midseason, or late ripening, regardless of the bloom period of the parents (Baldini, 1949).

The other approach is to hybridize cultivars which have apparent resistance of blossoms to frost injury. This approach has pitfalls, however, because of the large error associated with proper parental classification of resistance to blossom frost due to microclimate differences. Another pitfall is that injury from frost may not always result in reduced fruit set. Perraudin (1955) showed that the yield of 'Louise Bonne' on which 70% of the blossoms were killed was six times greater than that of 'Beurre Giffard' on which only 39% of the blossoms were killed. Parthenocarpic fruit set following frost damage and varietal differences in parthenocarpy also reduce accuracy in parental classification. For example, he found that 'Conference' with 96% of the flowers killed produced a high yield, but 98% of the fruit were parthenocarpic.

Floral organs also vary in their sensitivity to frost (Simovski et al., 1968). Thus, breeding for frost resistance in pear may not be too feasible when one considers the major sources of error that are involved in parental classification and seedling selection.

Drought resistance: Pears are frequently grown in arid to semi-arid areas where drought resistance of scions and rootstocks are important features of adaptability. Cultivars vary widely in their drought tolerance. Pruss and Eremeev (1969) assessed the drought tolerance of 99 pear cultivars of Russian, European, Australian, and American origin and classified them into three classes of tolerance—high, medium, and low. Kuznetzov (1941) found that pears grafted on *P. salicifolia* thrived better in arid soils than those on *P. communis* rootstocks. In addition to drought tolerance, *P. salicifolia* was found to be tolerant to extreme temperature changes and saline soils, and resistant to pear scab, woolly aphid, and many other diseases and pests. It appears, therefore, that one can breed for greater drought resistance in both scions and rootstocks.

Rootstocks

General Considerations

Almost all pear cultivars are reproduced asexually by budding or grafting on a rootstock of a different genotype. The development of improved rootstocks is an important phase of pear breeding. Some cultivars like 'Bartlett' can be propagated on their own roots by softwood or hardwood cuttings (Westwood and Lombard, 1966), but rooting is generally difficult to achieve and the trees produced are often unthrifty.

Seedlings are commonly used as rootstocks because of the ready availability of seeds and ease of propagation, but clonal propagation is sometimes used to produce certain rootstocks which have highly desirable characters such as the capacity to dwarf the scion cultivar and also induce precocity. Although seedling rootstocks have been predominantly used for propagating pear, the trend towards dwarf trees and high density plantings has resulted in an increased use of clonally propagated, dwarfing rootstocks in many pear growing areas especially in western Europe (Tukey, 1964).

Pyrus Rootstocks

Wherever pears are grown, the common practice has been to use the seed extracted as a byproduct of the fruit to produce seedling rootstocks. As a result most of the species cultivated for their fruit have also been used for rootstock purposes. The important characteristics of pear rootstocks and the localities where they are grown are summarized in Table 2.

In Europe, the perry mills have provided much of the seed for *P. communis* production. Before World War I this seed was also imported into North America, but during and after the war imported seed was not available and the American nurseries turned to the local canneries for a source

TABLE 2. Important characteristics of the commonly used pear rootstocks including resistance to diseases, pests, physiological disorders and environmental stress.

Rootstock species	Locality of use	Scion vigor	Black end	Pear decline	Fire blight	Woolly pear aphid	Cold	Soil adaptability
P. communis 'Bartlett' sdlg. 'Winter Nelis' sdlg. 'Bosc' sdlg.	North America New Zealand Australia	vigorous	R	MR	S	S(V)	R	tolerates a wide range of soil textures and soil moisture
P. nivalis perry pear sdlg.	Europe and formerly North America	vigorous	R	MS	S	—	R	tolerates a wide range of soil textures and soil moisture
P. Calleryana D-6 strain	southern United States & China Australia	vigorous	R	MR	R	R	S	tolerates a wide range of soil textures and moisture D-6 strain is drought tolerant
P. betulaefolia	central China northern Italy	vigorous	R	R	MR	R	MS	tolerates a wide range of soil texture and soil moisture
P. ussuriensis	northern Asia and Great Plains region of North America	vigorous	MS	S	R	R	HR	sensitive to excessive moisture
P. pyrifolia	Japan	vigorous	S	S	MR	S(V)	MR	tolerates a wide range of soil texture and soil moisture sensitive to excessive moisture
P. communis x *P. pyrifolia* 'Kieffer' sdlg.	Australia	vigorous	MS	MR	—	R	—	—
P. Pashia	northern India	vigorous	R	MR	MS	S(V)	S	—
Cydonia oblonga 'Quince A' 'Provence'	Europe and North America	dwarf	R	R	S	R	MS	trees do not tolerate poorly drained or alkaline soils
Authority		Batjer et al. (1967)	Batjer et. al (1967) Westwood (pers. comm.)	Westwood et al. (1971)	Reimer (1925) Westwood and Lombard (1966)	Westwood and Westigard (1969)	Reimer (1925) Westwood and Lombard (1966)	Batjer et al. (1967)

HR = highly resistant; MR = moderately resistant; MS = moderately susceptible; S = susceptible; and S(V) = generally susceptible but variable.

of seed. At the present time most of the pear trees in North America are grown on seedlings of 'Bartlett', 'Winter Nelis', and 'Bosc' (Batjer et al., 1967). Seedlings mainly derived from 'Bartlett' are also commonly used in New Zealand and Australia (Cole, 1966).

P. communis seedlings in general produce vigorous trees that are adapted to a wide range of climates and soil types. They are mostly resistant to pear decline and oak root rot caused by *Armallaria mellea* Vahl. ex Fr., but are susceptible to fire blight in the nursery and in the orchard where the disease enters the crown through root suckers (Batjer et al., 1967). To overcome the blight problem, seedlings and clones of the resistant progeny 'Old Home' x 'Farmingdale' are sometimes used. The cultivar, 'Old Home', is fire blight resistant and easy to root, and is also used to a limited extent as a clonal rootstock (Westwood and Lombard, 1966).

Seedlings of some of the Oriental species have been tried as pear rootstocks, especially in North America because of their fire blight resistance. *P. Calleryana* has been the most successful of the Oriental rootstocks tested to date and is grown to a limited extent in southern regions of the United States. In northern regions it is susceptible to winter injury (Batjer et al., 1967). In Australia, a special rootstock derived from *P. Calleranya*, 'D-6', is used which is resistant to drought (Cole, 1966).

P. pyrifolia and *P. ussuriensis* are less useful as rootstocks mainly because they are prone to the physiological disorder "black end" or "hard end" which spoils the fruit because of hardening of the calyx end. Seedlings of 'Kieffer' (*P. communis* x *P. pyrifolia*) which were formerly used in Australia because of their resistance to pear root aphid also have a tendency to produce black end (Selimi, 1966).

Susceptibility to pear decline has been another weakness of *P. pyrifolia* and *P. ussuriensis* rootstock (Batjer et al., 1967). *P. ussuriensis*, however, remains an important rootstock for northern Asia and the great plains region of North America where extreme winterhardiness is required.

The trend in modern tree fruit production is to grow dwarf trees. To date no satisfactory dwarfing rootstock has been produced within *Pyrus*. However, L. A. Brooks, Daybreak Nurseries, Oregon (personal correspondence) has reported some success in obtaining dwarfing selections within the progeny 'Old Home' x 'Farmingdale'. At Oregon State University the dwarf species *P. Fauriei* is under observation to assess its rootstock potential. The selection of dwarfing rootstocks within *Pyrus* is an important objective in pear rootstock development in several other programs, including the Station de Recherches d'Arboriculture Fruitière in France, the East Malling Research Station in England, and the Research Station at Harrow in Canada.

Quince Rootstocks

Quince, *Cydonia oblonga* L., a genus closely related to *Pyrus*, has been used for centuries in Europe as a dwarfing rootstock for pear. When pear is grafted on quince the tree may be reduced 15 to 60% of the size it would attain when grafted on a standard seedling rootstock. Time to fruiting is also shortened. Quince is readily propagated by cuttings or mound layering making it possible to select and then maintain a range of dwarfing clones.

Many seed strains and clones of quince have been developed in Europe, but on the basis of morphological and phenological evidence, they fall into one of two main groups: the Angers quinces developed in northeastern France, and the Provence quinces in southern France (Brossier, 1965a). The clones and seedling rootstock designated 'Pillnitz', 'Orleans', and 'Fontenay' quinces fall into either one of these two groups (Tukey, 1964).

Within strains and clones commonly used in Europe at the beginning of the twentieth century, Hatton (1935) selected a series of clones designated 'Quince A', 'B', and 'C'. 'Quince A' has become the leading pear rootstock planted in Europe and is widely used in other regions where quince rootstocks are planted. 'Quince B' and 'Quince C' have not performed as well as 'Quince A' (Parry, 1966). They are known to be infected with the sooty ring and stunt viruses, respectively, and this may have contributed to their poor performance (Posnette et al., 1962). Clones of these quince rootstocks freed of virus by heat treatment are now available.

Several problems besides virus disease are encountered with the use of quince rootstocks. Quince is graft incompatible with many common pear cultivars. Barrett (1959) lists the cultivars that are compatible and incompatible with quince. To overcome the incompatibility problem it is often necessary to first graft a compatible pear cultivar, such as 'Beurre Hardy' or 'Old Home' on the quince rootstock and then graft the incompatible cultivar onto the compatible one. Bud sports of some of the incompatible cultivars such as 'Bartlett' have been discovered that are more compatible with quince than the normal type (Posnette and Cropley, 1962). Selection within the 'Provence' quince at the Station de Recherches d' Arboriculture Fruitière also has produced clones that are more graft compatible than 'Quince A' (Brossier, 1965b).

Another problem encountered with quince rootstocks in the colder regions of North America and Europe is that they are susceptible to winter injury. Several clones reported to be more winter hardy than 'Quince A' have been selected in Poland and Czechoslovakia (Millikan and Pieniazek, 1967), but none is in general use.

Quince is also susceptible to fire blight and oak root rot, lacks tolerance to wet soils, and often presents problems of poor anchorage in the soil. But in spite of all these problems quince remains the most useful dwarfing rootstock for pear yet developed.

Other Genera

Several genera other than quince in the Pomoideae are graft compatible with *Pyrus* and have been used with some success for dwarfing pear. *Crataegus Oxyacantha* L., English hawthorne, has been reported to have dwarfing effects and to be especially tolerant of cold, wet soils (Tukey, 1964). Olden (1960) reported that *Amelanchier canadensis* (L.) Med., Juneberry, promoted earlier fruiting than quince, although anchorage was inferior. *Sorbus aucuparia* L., the European mountain ash, is also compatible enough to be used as a rootstock and produces dwarf trees. But to date these species have found little use in pear production and are considered essentially as novelties.

Modern Approaches to Breeding Pear Rootstocks

Dwarfing germ plasm may be found but it requires a careful search of the available sources and the use of effective test procedures. At the Station de Recherches d'Arboriculture Fruitière, d'Angers, France, a direct approach to selection is being tested. A vigorous cultivar, 'Beurre Hardy', is budded on test seedlings or grafted on test clones. Growth data are taken on the scions for one or two years. Reduction of growth in the scion is then considered as a measure of dwarfing ability of the rootstock. Dwarfing rootstocks identified in this way are then rooted by soft wood cuttings and further tested for compatibility and dwarfing ability with other test varieties. Other indicators of dwarfing such as the relative bark thickness used as a measure of dwarfing in apple may be useful for pear rootstock selection, but they remain untested (Beakbane and Thompson, 1939).

Ease of rooting is an important character because it permits clonal propagation of rootstocks and the preservation of characters such as dwarfing. Although many selections of quince root easily, ease of rooting is less common within *Pyrus*. Several cultivars, namely 'Old Home' (Hartman et al., 1963), 'Magness', 'Moonglow' (McFaddin et al., 1968), and un-named selections from Turkey (Sykes, 1972) are known to be easy to root. Ease of rooting of both *Pyrus* and quince may be tested as a matter of routine by placing softwood cuttings under mist or by rooting hardwood cuttings (Hartman and Kester, 1969).

Incompatibility appears to be a problem involved in the use of intergeneric grafts. The conventional method used to test for incompatibility is to graft a susceptible scion cultivar on the rootstock to be tested and wait for the incompatibility reaction to appear. This may involve a considerable period of time because the scion may fail years after the graft was made. In the case of quince, R. M. Samish, the Hebrew University of Jerusalem, Rehovot, Israel (personal communication), suggested that a biochemical test be used to speed up selection for types of quince compatible with pear. This test is based on the studies of Gur et al. (1968) who found that incompatibility of pear with quince results from toxic substances produced by the quince rootstock and released in the scion by the hydrolysis of a cynogenic glycoside, presumed to be prunasin. The incompatible cultivars are more capable of cleaving the cynanogenic glycocide than compatible ones due to the presence of larger amounts of the compounds, arbutin and methyl arbutin, which promote the cleavage reaction. The incompatibility of a clone may be determined by the amount of arbutin or methylarbutin present or by the rate of cleavage.

The improvement of winterhardiness of both *Pyrus* and quince rootstocks is an important objective in several rootstock breeding programs. The greatest opportunity for selection of cold hardy genotypes probably exists with certain species of

Pyrus, e.g., *P. ussuriensis*, but selection within *Cydonia* might also lead to further improvements in winterhardiness. The use of test chambers to induce freezing stress in excised roots or on the plant *in situ* is more advantageous in selecting winter-hardy genotypes than depending on the normal winter climate (Budagovskij, 1966).

Drought is a problem in several areas of pear production. Drought tolerant rootstock selections have been made from *P. Calleryana* (Oldham and Way, 1965), and *P. salicifolia* (Kuznetzov, 1941). Further selection and testing for drought tolerance is required.

Although disease and insect resistance might be lower on the list of priorities in many rootstock breeding programs, testing and selection for disease and insect resistance should be considered. Techniques used in breeding and testing for disease and pest resistance which apply to rootstock improvement have been discussed earlier.

Achievements and Prospects

Scion Cultivars

Centuries of selection have resulted in the development of several broadly adapted cultivars of high quality including 'Bartlett', 'Anjou', 'Comice', 'Bosc', and 'Passe Crassane' that are successfully grown in many parts of the world. In North America where fire blight has been a constant menace, only limited progress has been made in breeding blight resistant cultivars. None of the cultivars presently released seem destined to replace or even compete favorably with the susceptible cultivars that are the mainstay of the industry. However, some new selections from several breeding programs appear to have sufficient resistance combined with good quality to compete with susceptible varieties at least in eastern North America. It is possible that it may become necessary to breed for fire blight resistance in Europe and Asia if the disease continues to spread in these regions. Certainly, breeding for fire blight resistance must continue to be the major objective at least in North America.

Improved cold hardiness has extended the range of pear growing into northern regions of North America, Europe, and Asia where pear growing was not previously possible. There is, however, a continuing need for better cultivars in these harsh climates to provide fruit which otherwise is limited and expensive.

It is difficult for breeders working with long-lived perennials such as pears to quickly adjust their breeding programs to meet new needs in the industry which are sometimes created by other economic or technological developments. Because certain new trends in the pear industry are now becoming evident, the emphasis in breeding and selection will have to be changed to ensure that the cultivars developed are adapted to the new cultural and management systems under which they will probably be grown.

Trends in modern orchard mangagement point to a much more intensive culture of tree fruits than ever before. Trees are being grown at increasingly higher densities per unit area, usually in hedgerows or on wires. To reduce management costs and to shorten time to bearing, pears are being grown on dwarfing rootstocks where the climate will permit.

High density culture is creating many new requirements on cultivars that were not important when they were developed. Adaptability to high density, intensive culture will require cultivars that come into production early and show good annual productivity. They should compete effectively for sunlight and nutrients to grow well in a restricted space. Greater disease and pest resistance will be required since the intensive system of culture is generally more favorable for rapid increase and spread of diseases and pests. Cold hardiness, drought, and frost tolerance will also be required to reduce the losses caused by the increased environmental stress of high density culture.

Another trend that may become important before long is that pears will be harvested mechanically both for fresh use and for processing. The exact design of the machines are not known yet, but the principles on which they will probably operate will involve mechanical separation of the fruit from the tree by shaking, and a device for catching the falling fruit and loading it into bulk storage bins. Pear cultivars to be adapted to such a harvesting system should produce fruit of uniform size and maturity that would separate fairly easily on shaking but have sufficient toughness of skin to resist rupture and bruising during harvesting and bulk handling. Greater emphasis will probably be placed on tree shape and mechanical strength of trunks and limbs as well.

Some of the pesticides that have been used have harmful effects on man and his environment that were not previously recognized. It is possible that greater restrictions will be placed on their use and manner of application. If safe methods of pest control are not quickly found, there will be a greater need to develop disease and pest resistant cultivars.

Some of the problems foreseen will probably be solved by other technological developments, but pear breeders cannot afford to ignore these real or potential problems, nor fail to direct their energies toward their solution. The crosses being made today will not become cultivars for 20 or more years. Now is the time to consider breeding and selection for cultivars that would improve their adaptability to a high density system of culture, mechanical harvest, and restricted use of pesticides.

Rootstocks

Little intensive breeding work has been carried out to improve pear rootstocks. Superior seedlings and clones of some *Pyrus* species and clones of *Cydonia* have been selected, however, which represent definite improvements in such characters as adaptability, disease and pest resistance, and dwarfing ability.

Present trends in the pear industry towards intensive culture with high density plantings and possible mechanical harvesting indicate that greater demands will be made for improved pear rootstocks than ever before. To be adapted to the intensive system of culture, pear rootstocks should have certain properties, including broad adaptability to varying climates and soils, resistance to major diseases and pests, good anchorage and compatibility with scion cultivars, dwarfing ability, and capacity to induce precocious fruiting.

No rootstock cultivars are available that possess all of the characters just mentioned, although the genetic resources exist in *Pyrus*, and to a lesser extent in *Cydonia*, to make the required improvements possible.

Genetic progress should be greater with *Pyrus* since better sources of some of the characters have been identified in *Pyrus* than in *Cydonia*. In addition, fewer instances of graft incompatibility occur with *Pyrus* than *Cydonia* rootstocks. The need for a hardy, size-controlling rootstock for pear that induces precocity in the scion cultivar is particularly urgent.

Literature Cited

Alston, F. H. 1971. Fire blight *Erwinia amylovora* resistance. *In* R. L. Knight. Fruit breeding. *Rpt. E. Malling Res. Sta.* 1971. p. 103–104.

Anderson, H. W. 1956. *Diseases of fruit crops.* New York: McGraw Hill.

Anderson, R. L. 1970. Some quantitative genetic considerations pertinent to breeding asexually. Ph.D. dissertation, University of Minnesota.

Anjou, K. 1954. Winter injury of applies and pears at Balsgard, 1953 (in Swedish). *Sverig. Pomol. Foren. Arsskr.* 54:139–147. (*In* R. L. Knight. 1963. Abstract bibliography of fruit breeding and genetics to 1960 *Malus* and *Pyrus*. E. Malling: *Commonwealth Agr. Bur. Tech. Comm.* 29. p. 7. Abstr. 35.)

Anonymous, 1959. Description of Merton fruits. *Rpt. John Innis Inst.* 1958. p. 41.

Baldini, E. 1949. Spring frost damage to fruit trees in the spring of 1949 (In Italian). *Riv. Ortoflorofrut-tic. Ital.* 33:78–86. (*Plant Brdg. Abstr.* 19:865. 1949.)

Barrett, H. C. 1959. Compatibility of certain pear-quince combinations. *Fruit Var. Hort. Dig.* 14:3–6.

Batjer, L. P., H. A. Schomer, E. J. Newcomer, and D. L. Coyier. 1967. *Commercial pear growing.* USDA Handbook 330.

Beakbane, A. B., and E. C. Thompson. 1939. Anatomical studies of stems and roots of hardy fruit trees. 2. The internal structure of roots of some vigorous and some dwarfing apple rootstocks, the correlation of structure with vigor. *J. Pomol.* 17:141–149.

Bishop, C. J. 1966. Radiation induced mutations in vegetatively propagated tree fruits. *Proc. 17 Int. Hort. Congr.* 3:15–25.

Briton-Jones, H. R. 1923. Pear leaf blister. (*Taphrina bullata*, Tul.) *Rpt. Long Ashton Res. Sta.* p. 89–90.

Brooks, H. J., T. van der Zwet, and W. A. Oitto. 1967. The pear breeding program of the United States Department of Agriculture. *Chron. Hort.* 7:34–35.

Brown, A. G. 1943. The order and period of blossoming in pear varieties. *J. Pomol.* 20:107–110.

———. 1960. Scab resistance in progenies of varieties of the cultivated pear. *Euphytica* 9:247–253.

———. 1966. Genetical studies in pears 5. Red mutants. *Euphytica* 15:425–429.

Brossier, J. 1965a. The selection of rootstocks for pears from natural populations of quinces. 1. A study of natural populations of quinces (In French). *Ann Amél. Plantes.* 15:263–326.

———. 1965*b*. The selection of rootstocks for pears from natural populations of quinces. 2. The selection of quince rootstocks for pears (In French). *Ann. Amél. Plantes.* 15:373–404.

Budagovskij, V. I. 1966. The use of artificial methods of freezing roots for testing frost-resistance of rootstocks (In Russian). p. 37–43. *In* V. G. Trushechkin, G. I. Tarakanov, and N. P. Nicolaenko, eds. *Reports of the Soviet Scientists to the 17 International Congress on Horticulture*. Moscow: Ministry of Agriculture of the U.S.S.R.

Carpenter, T. R., and J. R. Shay. 1953. The differentiation of fire blight resistant seedlings within progenies of interspecific crosses of pear. *Phytopathology* 43:156–162.

Cole, C. E. 1966. The fruit industry of Australia and New Zealand. *Proc. 17 Int. Hort. Congr.* 3:321–368.

Crane, M. B., and W. J. C. Lawrence. 1952. *The genetics of garden plants*. 4th. ed. New York: Macmillan.

Crane, M. B., and D. Lewis. 1940. Genetical studies in pears. 2. A classification of cultivated varieties. *J. Pomol.* 18:52–60.

———. 1942. Genetical studies in pears. 3. Incompatibility and sterility. *J. Genet.* 43:31–43.

———. 1949. Genetical studies in pears. 5. Vegetative and fruit characters. *Heredity* 3:85–97.

Dayton, D. F. 1966. The pattern and inheritance of anthocyanin distribution in red pears. *Proc. Amer. Soc. Hort. Sci.* 89:110–116.

Decourtye, L., and B. Lantin. 1971. Methodology in induced mutagenesis for apple and pear trees (In French). *Ann. Amél. Plantes.* 21:29–44.

Denning, W. 1794. On the decay of apple trees. *N. Y. Soc. Prom. Agric. Arts & Manuf. Trans.* 2:219–222.

Dermen, H. 1947. Polyploid pears. *J. Hered.* 38:189–192.

———. 1965. Colchiploidy and histological imbalance in triploid apple and pear. *Amer. J. Bot.* 52:353–359.

Doorenbos, J. 1965. Juvenile and adult phases in woody plants. p. 1222–1235. *In* W. Ruhland, ed. *Encyclopedia of plant physiology*. vol. 15, part 1, Differentiation and development, Berlin:Springer-Verlag.

Dowrick, G. J. 1958. Abnormal gametogenesis and embryo abortion in the pear variety Beurre Bedford (*Pyrus communis*). *Z. Indukt. Abstamm. Vererblehre* 89:80–93.

Draganov, D., and others. 1964. The relationship between the flowering time of some fruit species and the air temperature (In Bulgarian). *Grad. Lozar. Nauka* 1(2):3–12. (*Hort. Abstr.* 35:2823, 1965.)

Enikeev, H. K. 1959. The results of interspecific hybridization of fruit trees and soft fruits (In Russian). *Agrobiology*, Moscow, 6:924–928. (*Plant Brdg. Abstr.* 31:3751, 1961.)

Fisher, D. F. 1922. An outbreak of powdery mildew (*Podosphaera leucotricha*) on pears. *Phytopathology* 12:103.

Francis, F. J. 1970. Anthocyanin in pears. *HortScience* 5:42.

Fritzsche, R. 1948. The juvenile forms of apple and pear trees and their bearing on rootstock and variety breeding (In German). *Ber. Schweiz Bot. Ges.* 58: 207–267. (*Hort. Abstr.* 28(3):1626, 1948.)

Gayner, F. C. H. 1941. Studies in the non-setting of pears. 5. The size of flower in relation to its position in the truss. *Rpt. E. Malling Res. Sta.* 1940. p. 41–43.

Golisz, A., A. Basak, and S. W. Zagaja. 1971. Pear cultivar breeding. *In* S. A. Pieniazek. Studies on some local Polish fruit species, varieties and clones and on those recently introduced to Poland with respect to their breeding value and other characters. Nov. 1, 1966, to Oct. 31, 1971. *Res. Inst. Pomology*, Skierniewice, Poland.

Gorter, C. J., and T. Visser. 1958. Parthenocarpy of pears and apples *J. Hort. Sci.* 33:217–227.

Granhall, I., A. Gustafsson, F. Nilsson, and E. J. Olden. 1949. X-ray effects in fruit trees. *Hereditas* 35:269–279.

Griggs, W. H., L. L. Claypool, and B. T. Iwakiri. 1960. Further comparison of growth, maturity and quality of seedless and seeded Bartlett pears. *Proc. Amer. Soc. Hort. Sci.* 76:74–84.

Gur, A., R. M. Samish, and E. Lifchitz. 1968. The role of the cyanogenic glycoside of the quince in the incompatibility between pear cultivars and quince rootstocks. *Hort. Res.* 8:113–134.

Harris, M. K., and R. C. Lamb. 1972. Pear breeding for resistance to the pear psylla. *HortScience* 7:327. (Abstr.)

Hartman, H. 1957. Catalog and evaluation of the pear collection at the Oregon Agricultural Experiment Station. *Oregon Agr. Expt. Sta. Tech. Bul. 41.*

Hartmann, H. T., W. H. Griggs, and C. J. Hansen. 1963. Propagation of own rooted Old Home and Bartlett pears to produce trees resistant to pear decline. *Proc. Amer. Soc. Hort. Sci.* 82:92–102.

Hartmann, H. T., and D. E. Kester. 1969. *Plant propagation, principles and practices*, 2nd. ed. Englewood Cliffs, N.J.: Prentice-Hall.

Hatton, R. G. 1935. Rootstocks for pears. *Rpt. E. Malling Res. Sta.* 1934. p. 75–86.

Hedrick, U. P., G. H. Howe, O. M. Taylor, E. H. Francis, and H. B. Tukey. 1921. The pears of New York. *N. Y. Dept. Agr. 29th Ann. Rpt.* vol. 2, part 2.

Hibino, H., and H. Schneider. 1970. Mycoplasmalike bodies in seive tubes of pear trees affected with pear decline. *Phytopathology* 60:499–501.

Hiroe, I., S. Nishimura, and M. Sato. 1958. Pathochemical studies on *Alternaria kikuchiana*. On toxins secreted by the fungus (In Japanese). *Trans. Tottori Soc. Agric. Sci.* 11:291–299. (*Plant Brdg. Abstr.* 29:3036, 1959.)

Hooker, W. 1818. Account of a new pear, called Williams' Bon Chretien. *Trans. Hort. Soc.* London 2: 250–251.

Hough, L. F. 1944. The new pear breeding project. *Trans. Ill. Hort. Soc.* 78:106–110.

Hough, L. F., and Catherine H. Bailey. 1968. Star, Lee and Mac—three blight resistant fresh market pears from New Jersey. *Fruit Var. Hort. Dig.* 22:43–45.

Howlett, F. S. 1957. Preliminary evaluation of new and uncommon pear varieties including comparison with standard sorts. *Ohio Agr. Expt. Sta. Res. Bul.* no. 790.

Janick, J., S. S. Thompson, and E. B. Williams. 1962. Fireblight resistance in pear. *Proc. 16 Int. Hort. Congr.* 3:244–247.

Jensen, D. D., W. H. Griggs, C. Q. Gonzales, and H. Schneider. 1964. Pear decline virus transmission by pear psylla. *Phytopathology* 54:1346–1351.

Johri, B. M. and I. K. Vasil. 1961. Physiology of pollen. *Bot. Rev.* 27:326–368.

Kajiura, M. 1966. The fruit industry of Japan, South Korea and Taiwan. *Proc. 17 Int. Hort. Congr.* 4:403–425.

Kegler, H. 1960. Studies on pome fruit viruses. 2. Ring spot mosaic virus of pears (In German). *Phytopath. Z.* 37:379–400.

Khan, K. A. W. 1955. Studies on stock immune to woolly aphis of apple (*Eriosoma lanigerum* Hausm.) *Punjab Fruit J.* 19(70/1):28–35. (*Hort. Abstr.* 26: 162, 1956.)

Kikuchi, A. 1930. On skin colour of the Japanese pear, and its inheritance (In Japanese). *Contr. Inst. Plant Ind.* 8:1–50.

———. 1946. Speciation and taxonomy of Chinese pears. *Collected Records of Hort. Res.* 3, p. 1–8. Kyoto Univ. (Trans. by K. Park; ed. by M. N. Westwood.)

King, J. R., and C. O. Hesse. 1938. Pollen longevity studies with deciduous fruits. *Proc. Amer. Soc. Hort. Sci.* 36:310–313.

Knight, R. L. 1963. Abstract bibliography of fruit breeding and genetics to 1960 *Malus* and *Pyrus*. E. Malling: *Commonwealth Agr. Bur. Tech. Comm.* 29.

Knight, T. A. 1806. Observations on the means of producing new and early fruits. *Trans. Hort. Soc. London.*

Kovalev, N. V. 1963. Leaf blight of pears (In Russian). *Zasc. Rast.* Vred. *Bolez.* 8(11):58. (*Hort. Abstr.* 34:2545, 1964.)

Kuznetzov, P. V. 1941. The role of *Pyrus salicifolia* Pall. in the development of fruit growing in arid regions (In Russian). *Sovetsk. Bot.* No. 1–2:103–107. (*Plant Brdg. Abstr.* 13:p. 359, 1943.)

Lamb, R. C. 1960. Resistance to fire blight of pear varieties. *Proc. Amer. Soc. Hort. Sci.* 75:85–88.

Larmond, Elizabeth. 1967. Methods for sensory evaluation of food. *Can. Dept. Agr. Publ.* 1284.

Layne, R. E. C. 1968. Breeding blight-resistant pears for southwestern Ontario. *Can. Agr.* 13(3):28–29.

Layne, R. E. C., Catherine H. Bailey, and L. F. Hough. 1968. Efficacy of transmission of fire blight resistance in *Pyrus*. *Can. J. Plant Sci.* 48:231–243.

Lefter, G. 1959. The behaviour of some pear varieties in response to infection by the fungi *Mycosphaerella sentina* (Fuck.) Schroet. and *Endostigme pirina* (Aderh.) Sydow (In Italian). *Gradina Via Livada* 8:38–41. (*Plant Brdg. Abstr.* 30:809–810, 1960.)

Lewis, D., and L. K. Crowe. 1954. The induction of self-fertility in tree fruits. *J. Hort. Sci.* 29:220–225.

Luckwill, L. C., and A. Pollard, eds. 1963. *Perry pears.* Bristol: Univ. Bristol Press.

Ludin, Y. 1942. Hardiness of fruit trees in the winter of 1941–42 (In Swedish). *Fruktodlaren* 6:168–171. (*In* R. L. Knight, 1963. Abstract bibliography of fruit breeding and genetics to 1960 *Malus* and *Pyrus*. E. Malling: *Commonwealth Agr. Bur. Tech. Comm.* 29. p. 262. Abstr. 1289.)

Magness, J. R. 1937. Progress in pear improvement. In *USDA Yearbook of Agriculture.* 1937:615–630.

Matjunin, N. F. 1960. The question of frost resistance in fruit growing (In Russian). Sadovodstvo, Moscow. 11:31–33.

Matkin, O. A., and P. A. Chandler. 1957. Preparation and uses of soil mixes and fertilizers. p. 73. *In* E. F. Baker, ed. *The U.C. system of producing healthy container-grown plants.* Univ. Calif. Manual No. 23.

McFaddin, N. J. Jr., H. J. Sefick, and G. E. Stembridge. 1968. The rooting of three varieties of pears by cuttage. *S.C. Agr. Expt. Sta. Bul. Res. Ser.* 110.

Millikan, D. F. and S. A. Pieniazek. 1967. Superior quince rootstocks for pear from east Europe. *Fruit Var. Hort. Dig.* 21:2.

Mittmann-Maier, G. 1940. The susceptibility of apple and pear varieties to monilia fruit rots (In German). *Gartenbauwiss.*, 15:334–361. (*Hort. Abstr.* 12:89, 1942.)

Moffett, A. A. 1934. Cytological studies in cultivated pears. *Genetica* 15:511–518.

Morettini, A. 1967. The quest for very early pear varieties that do not over-ripen (In Italian). *Frutticoltura* 29:575–579. (*Hort. Abstr.* 38:4875, 1968.)

Morrison, J. W., W. A. Cumming, and H. J. Temmerman. 1965. Tree fruits for the prairies. *Can. Dept. Agr. Publ.* 1222.

Moruju, G., and H. Slusanschi. 1959. The study of the correlation between the processes of growth and fruiting at the commencement of shooting in some apple and pear varieties (In Romanian). *Lucar. Sti. Inst. Cerut. Hort.-Vit.* Baneasa-Bucuresti, 1957. p. 317–330. (*Plant Brdg. Abstr.* 30:3069, 1960.)

Morvan, G. 1961. The virus diseases of pear (In French). *Pomol. Franc.* 3:375–381.

Mowry, J. B. 1964. Maximum orchard susceptibility of pear and apple varieties to fire blight. *Plant Dis. Rptr.* 48:272–276.

Mukherjee, S. K., S. S. Randhawa, A. M. Gosami, and P. H. Gupta. 1969. Description of wild species of pome and stone fruits. 1. *Pyrus*. *Indian J. Hort.* 26: 130–133.

Nybom, N. and A. Koch. 1965. Induced mutations and breeding methods in vegetatively propagated plants. *In* The use of induced mutations in plant breeding. *Rad. Bot.* 5 (Suppl.):661–678.

Oitto, W. A., T. van der Zwet, and H. J. Brooks. 1970. Rating of pear cultivars for resistance to fire blight. *HortScience* 5:474–476.

Oldén, E. J. 1960. Preliminary nursery and rootstock studies with some dwarfing pear stocks (In Swedish). *Sver. Pomol. Foren. Arsskr.* 61:103–114. (*Hort. Abstr.* 31:3877, 1961.)

Oldham, Margaret, and K. K. Way. 1965. Drought resistance of D6 pear rootstocks. *Agr. Gazette.* Nov. 1965, p. 689–690.

Parry, M. S. 1966. Dwarfing quince rootstocks for pears. *Rpt. E. Mall. Res. Sta.* 1965. p. 83–87.

Perraudin, G. 1955. The susceptibility of fruit trees to

late frosts (In Italian). *Rev. Romande Agr. Vitic.* 11:87–88. (*Hort. Abstr.* 26:344, 1956).

Pieniazek, S. A. 1966. Fruit production in China. *Proc. 17 Int. Hort. Congr.* 4:427–456.

Posnette, A. F., and R. Cropley. 1962. Further studies on a selection of Williams' Bon Chrétien pear compatible with Quince A rootstocks. *J. Hort. Sci.* 37: 291–294.

Posnette, A. F., R. Cropley, and L. D. Wolfswinkel. 1962. Heat inactivation of some apple and pear viruses. *Rpt. E. Malling Res. Sta.* 1961, p. 94–96.

Pruss, A. G. and G. N. Eremeev. 1969. Drought resistance in pear varieties of diverse geographical origin (In Russian). *Trudy Prikl. Bot. Genet. Selek.* 40: 56–67. (*Hort. Abstr.* 41:358, 1971.)

Rehder, A. 1967. *Manual of cultivated trees and shrubs.* 2nd. ed. New York: Macmillan.

Reimer, F. C. 1925. Blight resistance in pears and characteristics of pear species and stocks. *Oregon Agr. Expt. Sta. Bul.* 214.

Rubzov, G. A. 1944. Geographical distribution of the genus *Pyrus* and trends and factors in its evolution. *Amer. Nat.* 78:358–366.

Sansavini, S. 1967. Studies on cold resistance in pear varieties (In Italian). *Riv. Ortoflorofruttic. Ital.* 51: 407–416. (*Hort. Abstr.* 38:2614, 1968.)

Sax, K. 1931. The origin and relationships of the Pomoideae. *J. Arnold Arbor.* 12:3–22.

Schander, H. 1955. On the causes of differences in weight in the seeds of pome fruits (apple and pear). 1. The relationship between seed and fruit (In German). *Z. Pflanzenz.* 34:255–306. (*Plant Brdg. Abstr.* 26:643, 1956.)

Selimi, A. 1966. Rootstocks for pears in the Goulburn valley. *Victoria Hort. Dig.* 10:21–23.

Siebs, E. 1955. Investigations on scab resistance in pears. 3. An indication of the chemical basis for resistance to leaf scab (In German). *Phytopath. Z.* 23: 37–48.

Simovski, K., B. Ristevski, and R. Spirovska. 1968. Effects of negative temperatures and temperature fluctuations on pears (In Macedonian). *Annu. Fac. Agric. Sylvic* Skopje, Agric. 21:125–129. (*Hort. Abstr.* 40:5700, 1970.)

Spencer, Mary. 1965. Fruit ripening. In *Plant biochemistry.* J. Bonner and J. E. Varner, eds. p. 793–823. New York: Academic Press.

Stanton, W. R. 1953. Breeding pears for resistance to the pear scab fungus, *Venturia pirina* Aderh. 2. The study of field resistance on selected pear seedlings and the inheritance of resistance in seedling pear families under controlled conditions. *Ann. Appl. Biol.* 40:192–196.

Stushnoff, C. 1972. Breeding and selection methods for cold hardiness in deciduous fruit crops. *HortScience* 7:10–13.

Susa, V. I. 1967. The resistance of pears to pear leaf blister mite (In Ukrainian). *Zborn. Nauk. Prac'. Umans'k. Sil's'-Kogasp. Inst.* 15:268–272. (*Hort. Abstr.* 40:507, 1970.)

Sykes, J. T. 1972. Tree fruit resources in Turkey. In *Plant Genetic Resources Newsletter.* 27:Feb. 1972, p. 17.

Thompson, S. S., J. Janick, and E. B. Williams. 1962. Evaluation of resistance to fire blight of pear. *Proc. Amer. Soc. Hort. Sci.* 80:105–113.

Trifonov, D. 1970. The susceptibility of some pear varieties to the virus disease pear bud drop (In Bulgarian). *Grad. Lozar. Nauka.* 7(5):37–43. (*Hort. Abstr.* 41:3370, 1971.)

Tufts, W. P., and L. H. Day. 1934. Nematode resistance of certain deciduous fruit tree seedlings. *Proc. Amer. Soc. Hort. Sci.* 31 (Suppl.):75–82.

Tukey, H. B. 1964. *Dwarfed fruit trees.* New York: Macmillan and London: Collier-Macmillan Ltd.

United Nations. 1969. *Production yearbook.* New York: Food & Agr. Org. of U.N. 23:186–187.

U.S. Department of Agriculture. 1970. *Agricultural statistics.* Washington: U.S. Govt. Printing Office. p. 241–244.

van der Zwet, T. 1968. Recent spread and present distribution of fire blight in the world. *Plant Dis. Rptr.* 52:698–702.

———. 1970*a*. Evaluation of inoculation techniques for determination of fire blight resistance in pear seedlings. *Plant Dis. Rptr.* 54:96–100.

———. 1970*b*. New outbreaks and current distribution of fire blight of pear and apple in northern Europe. *FAO Plant Prot. Bul.* 18:83–88.

van der Zwet, T. and W. A. Oitto. 1972. Further evaluation of the reaction of "resistant" pear cultivars to fire blight. *HortScience* 7:395–397.

———. 1973. Efficient method of screening pear seedlings in the greenhouse for resistance to fire blight. *Plant Dis. Rptr.* 57:20–25.

van der Zwet, T., W. A. Oitto, and H. J. Brooks. 1970. Scoring system for rating the severity of fire blight in pear. *Plant Dis. Rptr.* 54:835–839.

Vavilov, N. I. 1951. *The origin, variation, immunity and breeding of cultivated plants.* New York: Ronald Press.

Vávra, M., and V. Orel. 1971. Hybridization of pear varieties by Gregor Mendel. *Euphytica* 20:60–67.

Visser, T. 1951. Floral biology and crossing technique in apples and pears (In Dutch). *Meded. Dir. Tuinb.* 14:707–726.

———. 1955. Germination and storage of pollen. *Meded. Landb. Hogesch.* Wageninen 55:1–68.

———. 1964. Juvenile phase and growth of apple and pear seedlings. *Euphytica* 13:119–129.

———. 1967. Juvenile period and precocity of apple and pear seedlings. *Euphytica* 16:319–332.

Visser, T., and D. P. De Vries. 1970. Precocity and productivity of propagated apple and pear seedlings as dependent on the juvenile period. *Euphytica* 19: 141–144.

Visser, T., A. A. Schaap, and D. P. De Vries. 1968. Acidity and sweetness in apple and pear. *Euphytica* 17:153–167.

Visser, T., J. J. Verhaegh, and D. P. De Vries. 1971. Pre-selection of compact mutants induced by x-ray treatment in apple and pear. *Euphytica* 20:195–207.

Voisey, P. W. 1971. The Ottawa texture measuring system. *Can. Inst. Food Tech. J.* 4:91–103.

Watkins, R. 1970. Fruit breeding methodology—major

gene and quantitative genetics. *Proc. Angers Fruit Brdg. Symp.*, 1970. p. 251–263.

Watkins, R. and L. P. S. Spangelo. 1970. Components of genetic variance of plant survival and vigor of apple trees. *Theor. Appl. Genet.* 40:195–203.

Wellington, R. 1913. Inheritance of the russet skin in the pear. *Science* 37:156.

Wenzl, H. 1935. Observations on the susceptibility of pear varieties to *Mycosphaerella sentina* (In German). *Z. Pflkrankh.* 45:305–316. (*In* R. L. Knight, 1963. Abstract bibliography of fruit breeding and genetics to 1960 *Malus* and *Pyrus*. E. Malling: *Commonwealth Agr. Bur. Tech. Comm.* 29. Abstr. 2225.)

Westigard, P. H., M. N. Westwood, and P. B. Lombard. 1970. Host preference and resistance of *Pyrus* species to the pear psylla, *Psylla pyricola* Foerster. *J. Amer. Soc. Hort. Sci.* 95:34–36.

Westwood, M. N., and H. O. Bjornstad. 1968. Chilling requirements of dormant seeds of 14 pear species as related to their climatic adaptation. *Proc. Amer. Soc. Hort. Sci.* 92:141–149.

———. 1971. Some fruit characteristics of interspecific hybrids and extent of self-sterility in *Pyrus*. *Bul. Torrey Bot. Cl.* 98:22–24.

Westwood, M. N., and P. B. Lombard. 1966. Pear rootstocks. *Ann. Rpt. Oregon Hort. Soc.* 58:61–68.

Westwood, M. N., and P. H. Westigard. 1969. Degree of resistance among pear species to the woolly pear aphid, *Eriosoma pyricola*. *J. Amer. Soc. Hort. Sci.* 94:91–93.

Westwood, M. N., H. R. Cameron, P. B. Lombard, and C. B. Cordy. 1971. Effects of trunk and rootstock on decline, growth, and performance of pear. *J. Amer. Soc. Hort. Sci.* 96:147–150.

Zaleski, K., J. Wierszyllowski, and Z. Rebandel. 1959. Observations and experiments on leaf blight of pear (*Fabraea maculata* Atk., *Entomosporium maculatum* Lev.) and its biology and control in seedlings during nursery production (from 1948 to 1954) (In Polish). *Prace Kom. Nauk. Roln.* Lésn., Poznán 5(1):46. (*Plant Brdg. Abstr.* 30:723, 1960.)

Zavoronkov, P. A. 1960. Breeding winter-hardy pear varieties (In Russian). *Sadovodstvo*, Moscow, No. 11: 28–31. (*Plant Brdg. Abstr.* 31:5311, 1961)

Zielinski, Q. B. 1963. Precocious flowering of pear seedlings carrying the Cardinal Red color gene. *J. Hered.* 54:75–78.

Zielinski, Q. B., F. C. Reimer, and V. L. Quackenbush. 1965. Breeding behaviour of fruit characteristics in pears, *Pyrus communis* L. *Proc. Amer. Soc. Hort. Sci.* 86:81–87.

Zielinski, Q. B., and M. M. Thompson. 1967. Speciation in *Pyrus*: chromosome number and meiotic behaviour. *Bot. Gaz.* 128:109–112.

Zimmerman, R. H. 1972. Juvenility and flowering in woody plants: a review. *HortScience* 7:447–455.

Zukovskij, P. M. 1962. Cultivated plants and their wild relatives. *Commonwealth Agr. Bur.* (Abr. trans. by P. S. Hudson.)

Strawberries

by Donald H. Scott and F. J. Lawrence

The cultivated strawberry is grown throughout the world and rates high on the list of preferred foods. Delicate of flavor and rich in vitamins and minerals, it is a valuable food in the diets of millions of people around the globe, but only within the past 25 to 50 years has it attained this extensive popularity. Fifty years ago only people in a few countries were acquainted with strawberries, and then only as a rare delicacy. The change has resulted from breeding work that has provided cultivars that produce abundant fruit under widely different environmental conditions. Strawberry yields and fruit quality are greatly influenced by the interaction of photoperiod and temperature. In addition, length of rest period, disease resistance, tolerance to different soil conditions, winterhardiness, high-temperature tolerance, and inherent vigor of growth, all influence the behavior of plants. As a result, plants of a particular cultivar may grow well and be very satisfactory in one area, but may be unsatisfactory in another area where environmental conditions are different. This is usually referred to as regional or environmental adaptation of a cultivar. The cultivated strawberry displays a wide variation in adaptation to environmental conditions and, therefore, is an excellent subject for genetic engineering. Today we see the results of such engineering in the range from certain cultivars that survive the bitter cold winters at Fairbanks, Alaska, to others that perform well in the long, hot season at Pretoria, South Africa. The numerous breeding programs in the United States and throughout the world are based to a large extent on the need for improved cultivars to fit prescribed environmental conditions. Only by the breeding and selection of material under specific environmental conditions can genetic engineering attain the full expression of the characteristics sought by the breeder. Thus the need for breeding programs at many places is evident.

Origin and Early Development

The cultivated strawberry, *Fragaria* x *ananassa* Duch., is a result of hybridization of two native American species, *F. chiloensis* (L) Duch. and *F. virginiana* Duch. Duchesne in his classical book *L'Histoire Naturelle des Fraisiers* (1768) documented the beginnings of the cultivated strawberry. In recent years, Darrow (1966) and Lee (1964) have provided further historical information. According to their accounts, we are indebted to Amedee Francois Frezier, a French army officer, for the first development of the cultivated strawberry. In 1714 he returned to France from a foreign mission where he had seen the large-fruited *F. chiloensis* at Concepcion, Chile. He tenderly nursed five plants on the long six-month trip from Concepcion to France. Upon arrival in France, he gave two plants to the cargomaster of the ship, one plant to the King's Garden, Paris, one to his superior at Brest, and kept one himself. Within a few years an industry developed at Brest, where Frezier's pistillate plants of *F. chiloensis* were interplanted by chance with staminate plants of *F. virginiana*, and improvement of cultivated strawberries began. Improvement during the next 200 years was largely from efforts of amateur private breeders, and progress was slow. However, during the past 50 years, strawberry breeding has been conducted by several state experiment stations and the United States Department of Agriculture, and progress has been rapid. The cultivated strawberry as we know it today is a greatly improved fruit compared with that of 50 to 100 years ago. Much of the improvement is a result of the last 25 years of work.

Darrow (1966) lists eleven wild species of strawberries, in four chromosome groups with seven chromosomes as the base number: five diploids, two tetraploids, one hexaploid, and three octoploids. The cultivated strawberry *F.* x *ananassa* is also an octoploid. Brief descriptions of the species follow.

Of the diploids (2n=14), (1) *F. vesca* L., the wood strawberry, is circumpolar and the most extensively distributed of all species in the genus. It occurs in North America, northern Asia, northern Africa, and Europe. Plants are erect and runnering. They have thin light-green leaves borne on slender petioles, and produce inflorescences about the same length as the leaf petioles. Flowers are bisexual, and fruits are long ovate, bright red in color, and usually highly aromatic. Many forms exist, but the most prominent is *F. semperflorens* Duch., an everbearing type, runnered or runnerless (the alpine form). (2) *F. viridis* Duch., native in Europe and in eastern and central Asia, is found in meadows and along edges of forests. Plants are slender, upright; with few runners without nodes; with deep-green leaves; small inflorescence, bisexual flowers larger than *F. vesca*. Fruit is small, firm, pink to red in color, and aromatic. (3) *F. nilgerrensis* Schlecht. is native in southeast Asia. Plants are vigorous, spreading, with strong runners; pubescent, dark-green heavily veined leaves; small inflorescence, large bisexual flowers; small, nearly round, pale-pink, tasteless fruit with many seeds. (4) *F. daltoniana* J. Gay occurs in a small area of the Sikkim Himalayas at elevations of 10,000 to 15,000 feet. Plants are vigorous with slender runners; leaves are petiolulate, with few indentations on the margins; flowers are solitary; fruit may be elongate or fusiform, 2 to 2½ cm long, bright red, nearly tasteless. (5) *F. nubicola* Lindl. ex Lacaita is native in the Himalayas at 5,000 to 13,000 feet elevation. The plants closely resemble *F. vesca* with slender runners, but they are dioecious rather than hermaphroditic. *F. iinumae* Makino may merit species rank although it resembles *F. daltoniana*. Its plants are vigorous, erect, with slender filiform runners; leaflets are shortly petiolulate and coarsely dentato-serrate. Scape is one or few, erect, one or two-flowered. Fruit is ovoid, 1½ cm long, achenes imbedded in pits on fruit. This plant is found in central and northern Japan in alpine mountains.

Tetraploids (2n=28) include (6) *F. moupinensis* (Franch). Card. It occurs in eastern Tibet, Yunnan, and western China. Plants closely resemble *F. nilgerrensis*, with short runners; leaves are trifoliate, with smaller leaflets on the petiole below; flower stems longer than leaf petioles, with two to four flowers per inflorescence; fruit is small, similar to *F. nilgerrensis*. This species is not well studied. (7) *F. orientalis* Losinsk is native in western Siberia, Mongolia, Manchuria, and Korea. Plants are small, upright with long, slender runners; leaves are ovate, nearly sessile, light green, deeply serrate margins; inflorescences with few large flowers; fruit is soft, conical to round, and slightly aromatic.

The hexaploid, 2n=42, is (8) *F. moschata* Duch., occurring from northern and central Europe eastward into Russia and Siberia. Plants are dioecious, vigorous, tall, freely runnering; leaves are large, heavily veined, rugose, pubescent, dull green; inflorescence longer than leaf petioles, large flowers; fruit is dark red, soft, irregular-globose to ovoid, aromatic or musky flavor, larger than *F. vesca*. Cultivated forms usually are perfect-flowered.

Octoploids (2n=56) include (9) *F. virginiana* Duch., the meadow strawberry of eastern North America, found from Louisiana and Georgia northward to the plains states and Hudson Bay. Plants are slender, tall, profusely runnering, usually prolific, dioecious; leaves are medium thick, medium to dark green, with deeply dentate margins; flowers imperfect, large, the staminate larger than the pistillate, usually borne on inflorescenses shorter than the leaf petioles; fruit is soft, with deeply imbedded seeds, nearly round, 1 to 1½ cm in diameter, light to deep red surface, white flesh, tart, aromatic. Plant and fruit characters are variable, with many desirable characters to use in breeding. This species produces fertile hybrids in crosses with the cultivated species, *F.* x *ananassa*, and other octoploids. (10) *F. chiloensis* (L.) Duch. is found along the North American beaches from Alaska to near Los Angeles, California, along the beaches of Chile and inland and in the Andes Mountains from Concepcion southward to below Coyhaique, on eastern slopes in Argentina, and on the mountain tops in Hawaii. Species characteristics are highly variable. Plants are low, spreading, very vigorous, usually dioecious with rare hermaphrodites, runner prolifically; runners are long usually, heavy-textured, very pubescent; leaves are thick, leathery, usually dark-green and very glossy except some South American types that are gray-green and not glossy; inflorescence is variable with few to many flowers, staminate flowers are very large, pistillate very small, and hermaphrodites large; fruit is dull reddish-brown, white flesh, firm, mild flavored, round to oblate, 1½ to 2 cm in diameter, although some South American clones have large fruit. The species hybridizes easily with other octoploids, but most of hybrid seedlings have low fertility when grown in the eastern United States (Scott et al., 1962). (11) *F. ovalis* (Lehn.) Rydb occurs from the mountains of northern New Mexico northward to Alaska and westward to the West Coast states. This is another species with highly variable characteristics that are useful in breeding. Plants are slender, upright, with leaves that resemble *F. virginiana* but with a bluish-green sheen; dioecious, runner profusely; inflorescence usually short, with several flowers. The fruit is nearly round, 1 cm in diameter, pink, with deeply imbedded seeds, and flavorful. It hybridizes freely with other octoploid species. Staudt (1962) does not give species rank to this material, but the distinct fruit and plant characteristics and the isolated areas where it occurs argue for species status. (12) *F.* x *ananassa* Duch. The common cultivated strawberries with large fruits belong in this species (Darrow, 1966). Staudt (1959) lists *F. mandschurica* as a new species from Manchuria.

Bringhurst et al. (1964), using paper-chromatography, made a study of phenolic compounds of diploid, tetraploid, hexaploid, and octoploid *Fragaria* species. All had distinctive spot patterns, and they confirmed phylogenetic relationships as determined by cytogenetic studies. Breeding studies with the distinctive compounds of Alpine *F. vesca* and *F. orientalis* at the tetraploid level indicate discrete conventional segregation for the genes conditioning the compounds.

Up to the present time *F. chiloensis*, *F. virginiana*, and *F. ovalis* are probably the only wild species involved in the improvement of the cultivated strawberry.

History of Improvement

Private Breeders

Although the cultivated strawberry originated in the vicinity of Brest, France, about 1750, little was done to obtain improved cultivars for the next 60 years. Then in 1817, Thomas A. Knight, an English amateur horticulturist with an avid interest in cultivar improvement, raised his first strawberry seedlings. These were from controlled crosses that are presumed to have *F. virginiana* and *F. chiloensis* parentage and resulted in the 'Downton' and 'Elton' cultivars. 'Downton' was later used extensively in strawberry breeding.

Michael Keen, a market gardener near London, was also interested in strawberry improvement, and he, too, grew seedlings, of which his 'Keens Imperial' was the first to be named. It was from seed of the 'White Chili'. Seedlings of 'Keens Imperial' yielded a selection named 'Keens Seedling' in 1819. 'Keens Seedling' became the progenitor of many of our modern cultivars. The success of 'Keens Seedling' encouraged many amateurs throughout Europe and the United States to breed strawberries. Notable were the works of John Wilmot, John Williams, J. Barnes, Samuel Bradley, and Thomas Laxton, all of England; Gabriel Pelvilain, J. L. Jamin, F. Gloede, and others of France; C. M. Hovey, Marshall P. Wilder, Arthur B. Howard, James Wilson, Albert Etter, and others in the United States.

Famous cultivars from the early breeding work of amateurs were 'Downton' (1820) and 'Elton' (1828) from Thomas Knights' work, 'Keens Seedling' (1821) from Michael Keen, 'Hovey' (1834) from the work of C. M. Hovey, and 'Wilson'

(1850) originated by James Wilson, which was used to establish large-scale production of commercial strawberries.

Public Agencies

The New York Experiment Station at Geneva was the first public institution to do strawberry breeding in the United States. Work began in 1889, but a continuing program did not begin until 1924. In the meantime, N. E. Hansen, South Dakota Agricultural Experiment Station, began work in 1907, and C. C. Georgeson, in Alaska, began in 1905. Both were interested in breeding for winterhardiness. In 1920 George M. Darrow of the USDA and Carl Schuster of the Oregon Agricultural Experiment Station began breeding for general improvement of horticultural characters (Darrow et al., 1933; Darrow et al., 1934). Within a few years several state experiment stations initiated breeding programs (Darrow, 1937). At the present time 19 state experiment stations and the USDA are engaged in breeding work. Publicly supported breeding programs are also active in Canada, South America, Europe, Asia, Africa, and the USSR. Progress in origination of improved cultivars has been fairly rapid in recent years, but much remains to be done (Scott, 1962).

Modern Breeding Objectives

There are two general classes of objectives. One deals with genetics and inheritance studies of characters, and the other with applied breeding for improvement of cultivars. The characters that are of interest in most breeding programs are similar, but the relative importance placed on them differs among breeders, depending on the conditions within a region. Thus, resistance of cultivars to red stele root rot is an important character in the northeastern, central, and northwestern United States, but not in the southeastern United States nor in southern California where the disease rarely occurs.

Plant characters that are included in the objectives are: yield, vigor, fruiting habit (everbearing, day-neutral, or single-crop), time of ripening, winterhardiness, blossom frost-hardiness, high-temperature tolerance, length of rest period, concentrated ripening, disease resistance (root diseases, foliage diseases, virus tolerance), and mite resistance. Yield, vigor, and fruiting habit are of primary universal interest in breeding programs, but the other plant characters may not be considered, depending upon the problems.

Fruit characters included in objectives are size, symmetry, shape, flesh firmness, skin firmness, color of surface, color of flesh, gloss, flavor, ease of capping (removal of calyx), vitamin content, soluble-solids content, acidity, and resistance to rot. Of these characters, size and flavor are important in all programs, whereas some of the other characters are of minor concern as objectives.

In all applied programs, an overlying objective is origination of cultivars that are adapted to regional environmental conditions, as noted earlier.

Breeding Techniques

Three flower types exist among octoploid species —pistillate, which is devoid of anthers; staminate, with nonfunctional pistils; and hermaphrodite (perfect) or complete, as first noted by Duschesne (1768). Plants may be dioecious. Modern cultivars have hermaphrodite flowers only and therefore they must be emasculated for controlled cross-pollinations.

The blossom of the strawberry is composed of many pistils, each with its own style and stigma, attached to a receptacle which, on fertilization of the pistils, develops into a fleshy "fruit." The true fruits of the strawberry are the nutlets (achenes) containing one seed each that are found on the surface of the receptacle and that have resulted from the fertilization and development of the pistils. A single blossom may have from fewer than 50 to more than 400 pistils that develop seeds, depending on the size of the blossom and its position on the cluster. Primary flowers have the largest number of pistils and secondary, tertiary, or later flowers have progressively fewer (Darrow, 1966).

Emasculation usually removes anthers, petals,

and sepals in one operation to leave the pistils fully exposed for pollination. A scalpel or tweezers or thumb nails can be used to do the emasculating, but careful emasculation is important to avoid breaking an anther and causing self-pollination. Emasculation should be done one to three days before anthesis to prevent self-pollination. If humidity is low, anthers will dehisce at least a day before anthesis.

Emasculated flowers must be protected from foreign pollen by bagging or by covering with a layer of cotton, or the plants with emasculated flowers kept isolated from all plants that have open blossoms (Fowler and Janick, 1972). Much of the hybridization work is now done in greenhouses using potted plants, but crosses can be made in the field if precautions are taken to avoid contamination.

Pollen is collected either by removing individual anthers from the blossoms one to two days before anthesis and placing them in vials to dehisce, or by detaching flowers from the clusters, removing petals and sepals, and placing the blossoms in smooth paper-lined shallow vessels such as petri dishes for anthers to dry and dehisce. Pollen will remain viable for several days when stored at 5 to 10 C. Pollen can be transferred to stigmas with small camel's hair brushes or on a finger tip or a rubber-tipped stirring rod. Care must be taken to protect nearby emasculated flowers if a different pollen is used with a camel's hair brush.

Fruits are ripe in 25 to 30 days after pollination under normal temperature conditions, and seeds can be separated from the pulp at once. The fruit can be threshed in water with a 10- to 15-second spin in a food blender (Morrow et al., 1954) and the pulp floated off in water to leave the seeds, or the fruit can be mashed on absorbent paper and the seeds scraped off after the residue has dried.

Seeds retain high viability for 23 or more years when stored at 1 to 4 C (Scott and Draper, 1970). Parents differ in their seed germination capacity (Henry, 1935). Germination and emergence of seedlings is irregular, with some seedlings emerging in ten days and others in the same progenies not emerging until 60 to 90 days after planting (Scott and Ink, 1955; Gupta, 1962). Time of emergence can be hastened by after-ripening the seed for two and one half to three months (Bringhurst and Voth, 1957), and by exposing the seed to light (Scott and Draper, 1967). Young seedlings are susceptible to damping-off diseases, and therefore either sand, sterilized soil, or shredded sphagnum is used as a germination medium.

Young seedlings must be handled carefully during their early growth. They are usually grown in the seed trays for six weeks and then transplanted into soil in flats or pots, where they are grown another six to eight weeks before being set in the field. Field plantings of first-test seedlings may be set in the hill system or in small matted blocks. Both systems have advantages. The hill system requires much less space than a matted block, and the fruit is relatively easy to examine. The matted block provides several plants for observation and later transfer to new plots and an extensive leaf and root area for disease readings. The hill system is used particularly in breeding programs where the commercial crop is grown in hill systems, such as California, Louisiana, and Florida.

If young seedlings are being screened for disease resistance in a greenhouse or cold frame, they can be transplanted directly from the seed tray into benches or flats containing sand or soil or into small pots.

Artificial-inoculation techniques have been developed to screen for resistance to verticillium wilt (Wilhelm, 1955; Bringhurst et al., 1967; Bowen et al., 1966), leaf spot (Nemec and Blake, 1971), anthracnose (Horn et al., 1972), and red stele. All methods require that the disease organisms be produced in aseptic cultures for use in inoculation. Races of the fungi exist in all of the diseases mentioned. Young seedlings are not satisfactory for screening for resistance to verticillium wilt (Wilhelm, 1955).

Seedlings are commonly screened in the field by natural inoculation for resistance to powdery mildew, leaf spot, leaf scorch, anthracnose, and red stele. Screening for verticillium wilt resistance is usually done in the field by artificially inoculating plants as they are planted in the spring. Tests for virus tolerance usually require that plants be maintained in the field for one or more years in areas where viruses are present and the vectors are abundant.

Symptoms of foliage diseases are most evident late in fall and again about the time fruit begins ripening in the spring. Red stele symptoms may appear early in spring soon after new growth begins, but are most evident about the time fruit begins to ripen. Wilting from verticillium may occur within two to three weeks after plants are set in the field, but usually the symptoms are most noticeable about 12 to 15 weeks after planting. Anthracnose infection is most severe late in summer.

Many of the characters of seedlings and selections are recorded by subjective ratings on a scale of 1 to 10 (some breeders use 0 to 9), with 10 being the most desirable quality and 1 the least desirable. Ratings of 6 and higher indicate degrees of acceptability, and 5 or lower unacceptability (Morrow et al., 1949). Cultivars may be used as standards for comparison, as in Table 1.

Selections may be evaluated either subjectively or objectively, in single plots or in replicated plots. The most frequently used objective measurements are yield and fruit size by weight, and fruit firmness by Instron readings or by a pressure-plunger tester. Leaf-spotting diseases have been studied by counting the number of spots on leaves collected at random or by scoring for degrees of infection (Spangelo and Bolton, 1953; Nemec and Blake, 1971). However, most characters are given subjective ratings. Although subjective ratings are influenced by the judgment of the breeder, the agreement of ratings among breeders is usually very consistent. For example, all breeders will rate the fruit of 'Dixieland' 10 for firmness and will rate 'Catskill' 3 to 4 for firmness.

Data sheets or data cards are used for recording large quantities of data, and these can be handled directly by computers for analysis. Plot replication is essential, particularly if the differences among means are narrow. Randomized blocks are used most frequently, but latin squares and lattices are also used (Allard, 1960). Duncan's multiple-range test is frequently used to determine significant differences among means (Duncan, 1955). Koch (1953) and Aalders et al. (1968) have described methods for determination of appropriate plot sizes for strawberries.

TABLE 1. Subjective method for rating firmness and leaf scorch characters of strawberries at Beltsville, Maryland

Rating[z]	Firmness	Leaf scorch
10	Dixieland	
9	Blakemore	Albritton
8	Tennessee Beauty	Blakemore
7	Albritton	Surecrop
6	Pocahontas	Dixieland
5	Robinson	Jerseybelle
4	Suwannee	Vesper
3	Catskill	*F. ovalis*
2	*F. ovalis*	*F. chiloensis*
1	*F. virginiana*	

[z] 10 indicates most desirable and 1 the least desirable expression of a character.

Breeding Systems

Strawberry selections and cultivars are propagated vegetatively by runner plants rather than by seed. Most cultivars and selections are highly heterozygous, so that their seedlings show wide variability. Such heterogeneity no doubt has influenced the plant-breeding system that is used for cultivar improvement, since it is not necessary to have pure lines for propagation of a cultivar.

Outbreeding and Inbreeding

The outcross method of breeding has been used most often for the improvement of cultivars (Darrow, 1966). In such a system, all characters are segregating in each generation of seedlings, and selection pressure must be exerted simultaneously on each character. The goal is to select the seedlings that have a combination of very desirable characters, with the expectation that the plants with the new combination of characters will be superior to those of one or both of the parents. A wide parentage base is necessary to avoid inbreeding or line breeding, which often results in loss of vigor and reduction in yield.

Inbreeding has been used as a method for origination of new cultivars, but it has not been used as extensively as outcrossing. Jones and Singleton (1940) and Anstey and Wilcox (1950) were among the early breeders to study inbreeding. They developed second- and third-generation inbred lines, which were then combined to produce seedlings for selection as cultivars. Jones introduced three cultivars, but none were retained. Morrow and Darrow (1952) found that one generation of selfing produced selections that were weaker, less productive, and smaller fruited than their parents. However, crosses of S_1 selections yielded desirable seedlings. The 'Albritton' cultivar was originated from a cross of two S_1 selections ('NC 1065' from 'Southland' selfed x 'NC 1053' from 'Massey' selfed). More recently Craig et al. (1963) have used recurrent reciprocal selection as a means of reducing variability among seedlings in a progeny and raising the general level of the characteristics. Spangelo et al. (1971*b*) grew progenies from S_5 inbreds of 'Howard 17' ('Premier') with four tester cultivars and obtained some progenies that were superior in yield to 'Howard 17' crossed with the same testers. They outlined a method of inbreeding for improvement of characters of cultivars.

One of the difficulties in working with inbred material is a lack of vigor. Inbred seedlings are much weaker than seedlings obtained from outcrosses (Jones and Singleton, 1940; Morrow and Darrow, 1952), and inbred selections frequently are difficult to keep alive. They also may succumb quickly to virus infection because of their inherent weakness. However, inbreeding is a valuable tool when used in genetic studies or when there is need to concentrate genes for a few particular characters (Spangelo et al., 1971a; Morrow and Darrow, 1952).

Interspecific Hybridization

Interspecific hybridization has as an ultimate goal the improvement of the cultivated strawberry either by incorporation of single characters from native species or by full genome complements. Characters sought from species with lower than the octoploid chromosome number perhaps can be incorporated into cultivars through polyploidy and the addition of full sets of genomes. Results from such work are achieved slowly and require a sustained program for many years.

All of the octoploid *Fragaria* species are interfertile, and all have contributed desirable characters to the improvement of the cultivated strawberry. In recent times some of the notable contributions of characters of octoploid species that have been studied genetically are: winterhardiness of *F. ovalis* (Powers, 1945), in which hardiness was found to be partially dominant; blossom low-temperature hardiness of *F. virginiana* (Darrow and Scott, 1947), where hardiness was dominant; resistance to red stele root rot of *F. chiloensis* (Waldo, 1953; Stembridge and Scott, 1959; Pepin and Daubeny, 1966), in which resistance is partially dominant; resistance of *F. chiloensis* to verticillium wilt (Bringhurst et al., 1966); virus tolerance of *F. chiloensis* (Darrow, 1966); and the day-neutral fruiting character from *F. ovalis* (Powers, 1954; Bringhurst and Voth, unpublished). However, the incorporation of these characters into cultivated strawberries is usually complicated by the presence of deleterious characters such as small fruit size, partial fertility, soft flesh, black seeds, susceptibility to leaf diseases, and dull fruit color (Scott et al., 1962). Consequently the use of the native octoploid species in a breeding program requires at least three to four generations of seedlings from outcrosses or backcrosses to the cultivated species to obtain commercial material. Also, large populations are needed to provide for segregation among the many characters involved in a practical breeding program. However, Powers (1944) found, in crosses of *F. ovalis* and *F.* x *ananassa*, that meiotic behavior was the same as in a diploid. Evans and Jones (1967) and Mok and Evans (1971), in meiotic studies of interspecific hybrids among diploids and from analysis of nine cultivars, found evidence of tetrasomic inheritance and they emphasize that this needs to be considered in interpretation of genetic data.

Hybridization is successful between some diploid species, but not others, as indicated in Tables 2 and 3 (Dowrick and Williams, 1959; Fadeeva, 1966). Some hybrids are fertile, others sterile. Fadeeva (1966), in a study of *Fragaria* genomes, found a strong genetic similarity among the ge-

TABLE 2. Interspecific crosses among diploid species of *Fragaria* (after Dowrick and H. Williams, 1959)

Seed parent	Pollen parent				
	vesca	*viridis*	*nubicola*	*nipponica*	*nilgerrensis*
vesca	1	1	1	3	1, 2, 3[z]
viridis	1, 4[z]	1	2	1	4
nubicola	4	1	4	1	4
nipponica	4	1	1	4	4
nilgerrensis	4	4	4	3	1

1 = full seed set, viable F_1 plants
2 = full seed set, embryo lethal
3 = full seed set, lethal after germination
4 = no seed set
z = depends on form of *vesca* used as parent

TABLE 3. Cross fertility of species of the genus *Fragaria*
(after Fadeeva, 1966)

Species	*F. vesca* 2x	*F. vesca* 4x	*F. viridis* 2x	*F. nipponica* 2x	*F. nilgerrensis* 2x	*F. orientalis* 4x	*F. moschata* 6x	*F. virginiana* 8x	*F. chiloensis* 8x	*F.* x *ananassa* 8x
F. vesca 2x	f	f,s	f,s		m,s	s	s,m	s	s,f	s,f
F. vesca 4x	f,s	f	f			s,f	s,f			s,f
F. viridis 2x	f,s		f		m,s	s	s,f			s
F. nipponica 2x				f			s			
F. nilgerrensis 2x	m,s	m	m,s		f					
F. orientalis 4x	s,f		s,f			f	s,f			
F. moschata 6x	n,m	n, m,s	s,f	f		s,f	f	n	n	n
F. virginiana 8x			s				s	f	f	f
F. chiloensis 8x			s				s	f	f	f
F. x *ananassa* 8x	s,f	s,f	n,s				s,f	f	f	f

f = fertile; s = sterile hybrids; m = sprouts die; n = seeds do not germinate.

nomes of the species and also found homology, as demonstrated by fertility of hybrids from some species crosses.

Attempts to incorporate characters directly from diploid, tetraploid, and hexaploid species into the octoploid cultivated species have failed. Usually hybrid seedlings can be obtained from such crosses, but they are sterile or too low in fertility to be used in further breeding (Jones, 1966; Bringhurst and Khan, 1963; Yarnell, 1931*a*, 1931*b*). However, by doubling the chromosome numbers of such sterile hybrids, new fertile forms may be created that will be valuable in breeding improved cultivars. Some results along these lines are indicated in the section on polyploidy.

Polyploidy

Four different chromosome groups occur naturally in *Fragaria* species, as described previously, to form a polyploid series: diploid, tetraploid, hexaploid, and octoploid.

Islam (1961) detected a haploid of *F. vesca*, a diploid species, and his meiotic studies indicated that the seven chromosomes behaved largely as univalents.

From a study of sex types and their behavior in triploid, tetraploid, pentaploid, and hexaploid plants derived from *F. vesca* (2x), *F. viridis* (2x), and *F. orientalis* (4x), Staudt (1967, 1968) concluded that *F. vesca* genomes are the basic units in tetraploid and hexaploid species. Induced doubling of chromosomes of a triploid from *F. vesca* x *F. orientalis* yielded a hexaploid very similar to *F. moschata* in its sex expression.

Fertile decaploid hybrids from *F.* x *ananassa* and *F. vesca* have been reported from several studies of polyploidy in *Fragaria* (Scott, 1951; Bauer and Bauer, 1967; Ellis, 1962; Bringhurst and Gill, 1970). The decaploids were obtained in different ways, but all resulted from either unreduced- or equivalent-type gametes that occurred as a last step in the development of the hybrid. In one instance, a 4x *F. vesca*, obtained by colchicine doubling of diploid *F. vesca* (Dermen and Darrow, 1939), was crossed with 8x *F.* x *ananassa* to produce a 6x hybrid. The 6x hybrid was crossed with 8x *F. ananassa*, and a small percent of the seedlings were fertile 10x hybrids. Examination of the 6x hybrid revealed the production of unreduced gametes (Scott, 1951). A similar method was used by Bauer (1960) to obtain decaploids. A second method involved hybridization of 4x *F. vesca* ($2n = 28$) with an induced 16x *F.* x *ananassa* ($2n = 112$) to obtain a highly fertile 10x hybrid ($2n = 70$) (Ellis, 1962).

Of particular significance in the origin of *Fragaria* polyploids is the finding of natural polyploids, which were used in further studies (Bringhurst and Senanayake, 1966; Bringhurst and Gill, 1970). In a coastal area of California where 8x *F. chiloensis* and 2x *F. vesca* were growing contiguously, three different natural hybrids were detected: a partially fertile pentaploid from normally reduced gametes, a partially fertile hexaploid assumed to be from an unreduced gamete of *F. vesca* combined with a normally reduced gamete of *F. chiloensis*, and a highly fertile enneaploid (9x) from an unreduced gamete of *F. chiloensis* with a normally reduced gamete of *F. vesca*. When these three different hybrids were pollinated naturally or by hand-pollination with *F. chiloensis*, many of the resulting seedlings had chromosome numbers that could come only from unreduced and in some cases from double unreduced gametes. The pentaploid yielded mostly 9-ploid seedlings (from unreduced 5x plus the 4x from *F. chiloensis*), the hexaploid yielded more than 50% 10x seedlings (from unreduced 6x plus the 4x from *F. chiloensis*), and the 9x yielded mostly aneuploids with chromosome numbers of 57 to 60 and a few with $2n = 56$ and $2n = 63$ euploids. Because of the relatively high fertility of the enneaploid seedlings, the authors postulated that enneaploids may be a means for introgression of *F. vesca* genomes into *F. chiloensis*. Presumably the same introgression might occur with cultivated strawberry.

Senanayake and Bringhurst (1967) determined from a cytological analysis of pentaploids derived from native species that two pairs of genomes in *F. chiloensis* and *F. virginiana* are homologous. Octoploids share a common genome with *F. vesca*. They proposed that the genome formula for octoploids should be AAA'A' BBBB.

Islam (1960) theorizes that the octoploid species, all of which are of American origin, probably arose from unreduced gametes of tetraploid species not yet discovered in America.

Further studies with polyploidy have shown that most 9x hybrids obtained from crosses of 8x cultivated x 10x selections are sterile, whereas 10x selections x 10x selections yield mostly fertile seedlings (Scott, 1951). When two of the 10x selections were crossed with 6x *F. moschata*, all of the 570 hybrid seedlings were sterile and offered no possibility for further breeding (Draper and Scott, unpublished).

Ellis (1963) treated pentaploids from 4x *F. vesca* x 6x *F. moschata* with colchicine and obtained decaploids. The decaploids were crossed with *F. moschata*, and allo-octoploid (8x) plants were obtained that resembled the hexaploid parent and were dioecious. He speculated that 'Belle Bordelaise' could have been a similar allo-octoploid and could have entered into the ancestry of some of the nineteenth-century cultivars of *F.* x *ananassa*. By treating seed of 8x parents with colchicine just as germination began, Hull (1960) obtained 16x seedlings, some of which were fertile and as large as 8x seedlings.

Improvement at the decaploid level may be a possibility. Some of the decaploid seedlings are aromatic, with a musky flavor similar to that of *F. vesca*, but most of the economic characters of the decaploids such as yield, size of fruit, firmness, flesh quality, and general appearance of the fruit are much inferior to those of modern octoploid cultivars. Future species hybridization should give particular attention to unreduced gametes as bridges to obtain fertile hybrids.

Intergeneric Hybridization

Fragaria and *Potentilla* have been hybridized, but most of the hybrids die at an early age. Mangelsdorf and East (1927) obtained two hybrid seedlings from a cross of *F. vesca* x *Potentilla nepalensis* Hook, but they lived for only a few weeks. Jones (1955) was able to hybridize *F. vesca* with *P. anserina*, *P. anglica*, *P. erecta*, *P. reptans*, and *P. rupestris*. Crosses with *P. anglica*, *P. erecta*, and *P. reptans* produced seedlings, but all died in a few weeks.

More recently Ellis (1962) used four forms of *Fragaria* and six species of *Potentilla* in a hybridization study. Results of the hybridization are given in Table 4. Seedlings were obtained from seven of the 16 intergeneric crosses. Some seedlings in four progenies survived to maturity: *F.* x *ananassa* x *P. fruticosa* hybrids were pentaploid and sterile; *F.* x *ananassa* x *P. palustris* were heptaploid, slightly female-fertile but male-sterile, and very vigorous; 10x hybrid *Fragaria* x *P. fruticosa* were hexaploid and sterile; 4x *F. vesca* x *P. fruticosa* were triploid and sterile. When chromosomes of a heptaploid seedling from *F.* x *ananassa* x *P. palustris* were doubled with colchicine, fertility was improved as compared with the heptaploid. These results indicate a close relation between *Fragaria* and *Potentilla* and offer possibilities of new sources of characters for improvement in *Fragaria*.

In a cross of *F. moschata* (6x) x *Potentilla fruticosa* (2x), Asker (1970) obtained 7 hybrids from 124 seeds. The hybrid plants were 4x and resembled *F. moschata*, but were weaker. Three plants flowered, but all were sterile. Bivalents and trivalents occurred at meiosis, which resulted in formation of abnormal tetrads and poor pollen grains. Crosses of *F. moschata* x *P. erecta* (4x) and *P. palustris* (6x) did not germinate.

Senanayake and Bringhurst (1967) concluded from cytological examination of a hybrid of *F. chiloenis* x *P. glandulosa* L. ($2n = 14$) that *P. glandulosa* is not closely related to *F. chiloensis* as indicated by the low level of bivalents observed.

Barrientos and Bringhurst (1973) reported a tetraploid seedling (haplo-octoploid) from a cross of 'Tioga' x *Potentilla anserina* L pollen. Two plants survived out of 34 seeds that germinated and one of the two plants had 28 chromosomes, the tetraploid number for *Fragaria*. The plant produced runners and flowers, but did not set fruit under greenhouse conditions. The use of *P. anserina*, or other genera, to stimulate seed development in octoploid strawberries may be a means of obtaining haplo-octoploid plants ($2n = 28$) in quantity

TABLE 4. *Fragaria-Potentilla* intergeneric hybridization (from Ellis, 1962)

Seed parent / Pollen parent		% pollinations giving fruit	Avg. no. of seeds/fruit	% germination	% seedlings surviving to maturity	Chromosome nos. of surviving hybrids
F. x ananassa 2n=56						
P. fruticosa	2n=14	90	67	35	4	35
P. erecta	2n=28	75	42	60	0	—
P. reptans	2n=28	0	—	—	—	—
P. sterilis	2n=28	0	—	—	—	—
P. palustris	2n=42	100	126	77	50	49
F. vesca-F. x ananassa (allopolyploid hybrid 2n=70)						
P. fruticosa	2n=14	90	30	28	20	42
P. erecta	2n=28	0	—	—	—	—
P. reptans	2n=28	0	—	—	—	—
P. sterilis	2n=28	0	—	—	—	—
P. palustris	2n=42	0	—	—	—	—
P. anglica	2n=56	64	17	56	0	—
F. vesca 2x 2n=14						
P. fruticosa	2n=14	72	41	0	—	—
P. palustris	2n=42	0	—	—	—	—
F. vesca 4x						
P. fruticosa	2n=14	100	36	8	5	21
P. palustris	2n=42	0	—	—	—	—

Mutation

Rybakov (1966) obtained changes in yield, fruit quality, morphological characteristics, winterhardiness, and time of ripening in clones of two cultivars that had been treated with 0.001 to 0.1% solutions of ethylenimine or with gamma irradiation at dosages of 80 to 6,000 r. Although the results were incomplete, they indicated that strawberry tissues are relatively sensitive to such treatments.

Breeding of Specific Plant Characters

Most of the plant and fruit characters of the cultivated strawberry are inherited quantitatively, with several genes for each character. In some studies of genetic variance, it was found that additive, dominant, and epistatic components may be acting in the expression of particular characters. Such types of inheritance frequently create complex breeding problems when the maximum expression of a combination of characters is desired.

Yield

Yield, or the quantity of fruit produced by a plant, is the most important single character in all breeding programs. Yield may be the result of the expression of a combination of characters such as number and size of fruit, vigor of the plant, hardiness, and disease resistance of the plant.

Most of the studies that have been conducted on inheritance of yield have included other characters also. Morrow et al. (1958), in a study of eight characters involving 40 parents, determined that sufficient genetic variation existed for most of the characters, including yield, to allow considerable improvement. From the same source of data, Comstock et al. (1958) found that epistatic variance accounted for a sizeable part of the total genetic variance for five characters, but that additive and dominance variances were present also. The five characters investigated were size of plant area, number of berries, total weight of berries, average weight per berry, and weight of berries per unit plot size. In this study, and in others where matted beds are used, measurements for yield may be confounded with plant competition which could reduce additive variance.

Watkins et al. (1970); Spangelo et al. (1971a); and Bedard et al. (1971) studied the genetic variance components of several characters in 64 progenies from crosses involving 32 parents. They found that nonadditive variance comprised about 50% of the total genetic variance, and that much of this was epistatic variance. Table 5 gives results of the study dealing with economic characters. Since epistatic variance is so important, Watkins et al. (1970) postulate that progress may best be obtained by a two-step breeding procedure, which would involve small-scale testing of progenies, followed by growing large progenies of the best ones. Spangelo et al. (1971*b*) found that heritability estimates were high for certain yield components such as average berry weight, berries per flower stem, yield per flower stem, and flower stem number. Bedard et al. (1971) studied the correlations among some fruit and plant characters. Total berry number and total marketable yields were positively correlated with average berry weight, berries per flower stem, yield per flower stem, leaf area, and petiole diameter, but negatively correlated with stolon number and flower stem number. They concluded that no genetic barrier was detected to prevent combining high yield with good quality of berry.

Watkins and Spangelo (1971) did not obtain increased yields when only additive variance was present. They point out that, when several characters are involved in a breeding program, crosses should be designed to exploit all the genetic variance, whether it be additive, dominant, or epistatic.

Aalders and Craig (1968) used seven inbred lines in diallel crosses to study inheritance of yield, appearance, quality, and mildew resistance. Significant general combining ability occurred for all four characters, and specific combining ability existed for yield, appearance, and mildew incidence. They concluded that seed-propagated strawberries would produce acceptable yields, but that berry characteristics might be unacceptable.

Hansche et al. (1968) determined the genetic

TABLE 5. Significance of the general (gca) and specific (sca) combining ability for 20 commercially important strawberry characters (from Watkins et al., 1970)

Character	gca	sca	gca/sca[z]
Total yield	***	**	2.4
Marketable yield	***	***	2.0
Early yield	***	***	2.3
Early marketable yield	***	***	2.3
Late yield	***	***	2.1
Late marketable yield	***	***	2.1
Berry number	***	***	3.5
Average berry weight	***	***	2.2
Berries/flower stalk	***	***	1.4
Yield/flower stalk	***	***	1.2
Flower stalk number	***	***	1.2
SD of picks	***	*	2.7
Three highest picks	***	***	2.0
Mildew resistance	***	***	1.7
Leaf scorch resistance	***	***	1.6
Appearance	***	*	1.1
Processing (external)	***	*	1.4
Processing (internal)	**	**	0.9
Vigor	***	N.S.	—
Composite	***	**	2.0

*, **, *** = Significant at the 5%, 1%, and 0.5% probability levels, respectively.
N.S. = Not significant.
[z] Calculated only when both gca and sca were significant.

and environmental variances, heritabilities, and genetic and phenotypic correlations associated with fruit size, firmness, yield, and appearance. Extensive genetic variability was found associated with all the characters. Yield and fruit firmness were highly heritable, fruit size moderately heritable, but fruit appearance nonheritable. Genetic correlation existed between fruit size and yield, indicating that plants with large berries have a genetic potential for large yield. Little nonadditive variance but much additive variance was associated with yield in the material under study. Since the plants were grown in the hill system without competition from runner plants, the variances were well defined. The yield performance of the parents served as indicators for improving the mean yields of the seedlings indicating strong additive variance.

Hondelmann (1965) concluded from a study of breeding for yield that the genetic potential for increased yield is still large, and that flower number and fruit size are the two most important components of yield.

Disease Resistance

Inheritance of resistance to six fungus diseases has been studied. These are leaf spot, caused by *Mycosphaerella fragariae* (Tul.) Lindau; leaf scorch, caused by *Diplocarpon earliana* (Ell. & Everh.) Wolf; powdery mildew, caused by *Sphaerotheca macularis* (Fr.) Magn.; verticillium wilt, caused by *Verticillium albo-atrum* Reinke & Berth. and *V. dahlia;* red stele root rot, caused by *Phytophthora fragariae* Hickman; and anthracnose, caused by *Colletotrichum fragariae* Brooks. The genetics of botrytis fruit rot resistance have not been determined, although the disease is very destructive and causes extensive losses in some seasons.

Leaf spot: Leaf spot is a problem throughout most areas of the United States and many other countries. Several races of the fungus exist, with different ones occurring in different regions. Breeding for resistance, therefore, has been on a regional basis, and cultivars that are resistant in one region may be susceptible in a different region. No parent has been found that is resistant to all races (Nemec, 1971). Nemec (1969), in Illinois, developed a screening technique for young seedlings. 'Dabreak', from Louisiana, is the most resistant cultivar. Other cultivars resistant to leaf spot are 'Albritton', 'Blakemore', 'Earlibelle', 'Fairfax', 'Fairmore', 'Headliner', 'Howard 17', 'Klonmore', 'Massey', 'Midland', 'Missionary', and 'Surecrop'. 'Blakemore', 'Fairmore', 'Klonmore', 'Massey', and 'Missionary' were observed to transmit high resistance to their seedlings in North Carolina (Morrow and Darrow, 1941). Sato et al. (1965) found that 'Tohoku No. 1' transmitted a fairly high degree of resistance in crosses with other cultivars and in its S_1 in Japan.

Leaf scorch: Leaf scorch is most destructive in the southeastern United States, but it can be damaging in other places also. Breeding for resistance has been an objective in work in Illinois, Maryland, and North Carolina. Resistant cultivars have been originated in connection with general programs on breeding for improved characters. Material has been screened by natural infection, and ratings have been on a scale of either 1 to 6 (Spangelo and Bolton, 1953) or 1 to 10 (Morrow et al., 1949). Nemec and Blake (1971), in a study of 16 progenies, noted that 'Sunrise' x self yielded 97% of the seedlings in the most resistant class, as compared with 'Albritton' x self as the next highest in resistance with 72%, and only 13% for 'Redglow' x 'Headliner', the progeny with the least resistance. Resistance of the progenies could not always be predicted from the classification of the parent cultivars. The same behavior was observed previously by Drain and Fister (1938). Cultivars with high resistance are 'Albritton', 'Blakemore', 'Catskill', 'Earlibelle', 'Fairfax', 'Fletcher', 'Howard 17', 'Redcoat', and 'Sunrise'.

Powdery mildew: Powdery mildew is widely prevalent on strawberry leaves and frequently infects the fruit of cultivars that are very susceptible. Plant vigor is reduced when mildew infection is severe. Cultivars range in reaction from those that are immune, through those with different degrees of resistance, to some that are very susceptible (Darrow et al., 1954). In an extensive study involving 22 progenies, Daubeny (1961) found that, of the nine parents used, 'Puget Beauty' gave the highest degree of resistance in the progenies. His data indicated that resistance was dominant in 'Puget Beauty', but that 'Siletz', which is very resistant itself, did not possess dominant genes for resistance. Harland and King (1957) noted that resistance in diploid *F. vesca* selections is controlled by two genes. In seedling progenies that involved *F. chiloensis* and *F. virginiana* parentage, Hondelmann (1967) observed that resistant seedlings had dark-green glossy foliage similar to that of *F. chiloensis,* and that susceptible ones had light-green, tender foliage. Hsu et al. (1969), in quantitative studies on inheritance of resistance to powdery mildew among 32 parents of North American origin, found nonadditive variance more important than additive, and considerable epistatic variance was present. Segregation appeared to depend on two additive genes for resistance and one pair of epistatic genes for susceptibility.

Verticillium wilt: Verticillium wilt has been recognized as a serious disease of strawberries for at least 40 years (Thomas, 1932). It is widespread and is especially infectious on strawberry plants grown in soil previously occupied by solanaceous crops. Although fumigation treatments with chloropicrin or a combination of methyl bromide-chloropicrin control the disease (Wilhelm et al., 1961), they are expensive, not always fully effective, and may be a hazard near inhabited areas.

Resistant cultivars are the most practical solution to the problem.

Wilhelm (1955) found, by using a special inoculation technique to screen seedlings, that verticillium wilt resistance is dominant and is inherited quantitatively. He found three classes of seedlings: resistant, with no visible symptoms but plants invaded by the fungus; tolerant, with infection symptoms but plants vigorous; and susceptible, in which plants were severely stunted or dead.

Bringhurst et al. (1961) found that more than one race of the fungus existed, and that strawberry cultivars differed in reactions to the races. None of the cultivars tested were immune, but some were highly resistant. In a further study of inheritance of resistance, Bringhurst et al. (1967) determined that resistance is mainly additive, with a heritability estimate of 0.73 ± 0.03.

Newton and van Adrichem (1958) grew seedlings of seven different *Fragaria* species with a screening test and found that all seedlings of *F. vesca*, *F. bracteata*, and *F. virginiana* were susceptible, but 6% or more of *F. ovalis* and *F. chiloensis* were resistant. Van Adrichem and Orchard (1958) found resistant seedlings in some *F. chiloensis* progenies from Chile. Van Adrichem and Bosher (1962) reported 'Sierra' the only cultivar out of 26 that was highly resistant to wilt. 'Climax', 'Marshall', 'Perle de Prague', 'Redcrop', 'Sierra', and 'Temple' produced high percentages of resistant seedlings when selfed. Varney et al. (1959) found a clone of *F. virginiana* that was highly resistant to verticillium wilt. Wilhelm (1955) found a clone of *F. chiloensis* in California to be highly resistant, and Bringhurst et al. (1966) found 6% of the *F. chiloensis* plants to be resistant in 11 of 14 sites along the California coast where collections were made.

Bowen et al. (1966), in an extensive study of the breeding behavior of parents in relation to verticillium wilt, obtained a large percent of highly resistant seedlings only in progenies where both parents were resistant. Resistant x susceptible crosses produced a low percentage of highly resistant seedlings, and the resistant parents differed in the progeny-mean disease rating. More resistant seedlings occurred in the cross of resistant x resistant than in S_1 progenies of either of the resistant parents. Susceptible x susceptible and S_1 progenies of susceptible parents yielded a low percentage of resistant seedlings, indicating that several genes govern susceptibility.

Barnes et al. (1966) noted that 'Howard 17' transmitted more resistance to seedlings than 'Robinson', even though both are resistant themselves.

Jordan (1971) developed a simple screening method for determining resistance of strawberry plants, by immersing rooted plants for an hour in a conidial suspension of the fungus and then planting in a light sandy soil. Soil temperatures of 20 and 25 C were best.

Cultivars reported as very resistant to verticillium wilt are 'Aberdeen', 'Blakemore', 'Catskill', 'Cavalier', 'Fletcher', 'Gala', 'Gem', 'Guardian', 'Hood', 'Howard 17', 'Juspa', 'Redchief', 'Redgauntlet', 'Robinson', 'Salinas', 'Siletz', 'Sunrise', 'Surecrop', 'Talisman', 'Tennessee Beauty', 'Vermilion', and 'Wiltguard'.

Red stele root rot: Red stele root rot is a destructive disease of strawberry plants under cool, moist soil conditions in many countries. It was first known as Lancashire Disease, for the Lancashire district in Scotland where it was first observed and studied in 1920. The original source of infection of red stele in strawberry is unknown, but Moore et al. (1964) found that five of nineteen species of *Potentilla* were susceptible to red stele when inoculated experimentally. They speculate that *Potentilla* might serve as a natural host of the disease.

Once land becomes contaminated with the red stele fungus, the only practical solution is use of resistant cultivars.

Existence of races of the fungus has complicated red stele resistant breeding programs, which must be aimed at the origination of cultivars resistant to several races. Ten races have been detected in the USA, eleven in the UK, six in Canada, and six in Japan (Morita, 1968). Recent work has indicated the desirability of breeding for field resistance rather than immunity to specific races (Reid, 1966; Gooding, 1972). Gooding (1972) found that field resistance and immunity to specific races were not necessarily correlated, but that field resistance appeared to be positively correlated with a low percentage of roots infected with red stele. His results suggest that 'Aberdeen'-type parentage contributed race immunity, but not field resistance, whereas 'Frith' parentage contributed both kinds of resistance.

Screening techniques have been developed to

eliminate susceptible seedlings at an early age (Waldo et al., 1947; Draper et al., 1970; Muscle and Fay, 1971). The simplest method is that of Draper et al., (1970) in which four- to six-week old seedlings are dipped into a slurry of mycelium-agar-water and planted immediately into sand. Cool temperatures of 15 to 18 C are best for good inoculation and development of the disease and should be accompanied by frequent watering or sprinkling of the plant beds to sponsor secondary infection of roots. Seedlings are grown in the beds for five months and then dug, the roots are washed, and susceptible seedlings are discarded. The resistant seedlings are planted in the field, where they fruit the following spring.

Inheritance studies have revealed that in all instances resistance is partially dominant, and quantitatively inherited, and that resistant parents are heterozygous for the character (Scott et al., 1950, 1962; Stembridge and Scott, 1959). Since races of the fungus exist, parents differ in ability to transmit resistance to seedlings (Reid, 1953, 1959; Scott et al., 1962; Daubeny, 1963; Daubeny and Pepin, 1965), depending on the genetic constitution of the resistant parents and the race or races of the fungus that are present. Thus, Stembridge (1961), in an inheritance study involving races A-1, A-2, A-3, A-4, and A-5 of the fungus and several resistant parents, found that 'Aberdeen' (resistant) x 'Fairfax' (susceptible) yielded 48% resistant seedlings when screened with A-1. 'Midland' (susceptible) x 'Md 683' ('Frith' resistant) yielded only 18.4%, but the cross of 'Aberdeen' x 'Md 683' yielded 59.5% resistant seedlings, indicating that 'Aberdeen' and 'Frith' types of resistance operated independently and were additive. This interaction was further substantiated in a cross of 'Surecrop' x 'Stelemaster', where both parents possess both types of resistance, that yielded 69.5% resistant seedlings. None of the cultivars transmitted resistance to Race A-5, but the 'Yaquina' clone of *F. chiloensis* in crosses with susceptible parents gave 28% of seedlings resistant to Race A-5.

Kronenberg et al. (1971), in a field test for resistance, compared European and American resistant cultivars and found most of them equally vigorous. Resistant European cultivars were 'Crusader', 'Gorella', 'Juspa', 'Lavo', 'Rabunda', 'Redgauntlet', 'Precosana', 'Talisman', 'Tamella', and 'Templar'. Resistant American cultivars were 'Guardian', 'Hood', 'Md US 2700', 'Md US 2929', 'Midway', 'Puget Beauty', 'Redchief', 'Siletz', 'Sparkle', 'Sunrise', and 'Surecrop'. 'Guardsman', a Canadian cultivar, was also resistant. 'Gorella' x 'Siletz'; and 'Gorella' x 'IVT 6032' (both parents resistant) gave progenies with 85% of the seedlings resistant, whereas 'Senga Sengana' x 'IVT 6032' (susceptible x resistant) gave 51% resistant seedlings, and 'Senga Sengana' S_1 (susceptible) had 10% resistant.

Suzuki et al. (1962) screened seedlings for red stele resistance from crosses of 'Temple' with 13 Japanese parents. From 19 to 65% of the seedlings were resistant in the different progenies, indicating that some of the Japanese parents transmitted resistance as well as 'Temple'. Diseased roots were used as inoculum for the screening tests.

Four basic sources of resistance to red stele have been classified for most of the known American races of the fungus (Table 6), but their breeding behavior against all races has not been determined. The sources are 'Aberdeen', an American domestic cultivar of unknown parentage; 'Md 683', a cultivated type containing Scottish parentage ('Scotland BK 46') and probably derived from 'Frith'; selected clones of *F. virginiana,* including 'Little Scarlet'; and selected clones of North American *F. chiloensis.*

TABLE 6. Sources of resistance in strawberries to ten American races of *P. fragariae*

Source	A-1	A-2	A-3	A-4	A-5	A-6	A-7	A-8	A-9	A-10
F. chiloensis										
Del Norte	R	S	R	S	R	S	R	R	R	S
Yaquina	R	R	R	R	R	R	R	R	S	S
F. virginiana										
Sheldon	R	R	R	R	S	–	–	–	–	–
N 3953	R	R	R	–	R	–	–	–	–	–
Aberdeen	R	R	S	R	S	S	R	S	R	R
Frith										
Md 683	R	S	R	R	S	R	S	R	R	R

R = resistant; S = susceptible; – = unknown

Montgomerie (1967) catalogued the resistance of sixteen European and American cultivars for immunity or susceptibility to six American and six Canadian races of the red stele fungus and related these to British races (Table 7). This serves as a guide in selection of certain cultivars to use as parents in breeding for resistance to red stele.

To combine resistance to the various races of

the fungus and simultaneously to maintain a high level of desirable plant and fruit characters is a complex breeding problem. Despite the complexity, several resistant cultivars have been originated that are being grown extensively. These include 'Cambridge Favourite', 'Early Midway', 'Gorella', 'Guardian', 'Hood', 'Midway', 'Redchief', 'Redgauntlet', 'Redglow', 'Shuksan', 'Sparkle', 'Sunrise', 'Surecrop', and 'Talisman'.

Other cultivars that have resistance to red stele are 'Cambridge Vigour', 'Cheam', 'Columbia', 'Fairland', 'Guardsman', 'Juspa', 'Marmion', 'Puget Beauty', 'Redcrop', 'Siletz', 'Stelemaster', 'Templar', 'Temple', 'Totem', and 'Vermilion'.

Anthracnose: Anthracnose (*Colletotrichum fragariae* Brooks) attacks the crowns and runners of strawberry plants under the warm, humid conditions that prevail during the summer in southeastern United States. Horn et al. (1972) identified three races of the fungus in eight isolates. They developed a screening technique for seedlings by growing the organism on an oatmeal-agar medium and spraying an aqueous conidial suspension on the crowns of plants when temperatures ranged from 25 to 35 C. In breeding for resistance, progenies from crosses and selfs of cultivars and selections were classified according to the percentage of plants that survived inoculation. Although only a few seedlings were tested in each cross, it was sufficient to indicate that some Louisiana selections transmitted considerably more resistance than others.

Cactorum crown rot: In some areas of Europe, *Phytophthora cactorum* (Leb. & Cohn) Schroet. causes severe damage in strawberry plants as a crown rot (Molot and Nourrisseau, 1970). Development of the disease is enhanced by wet, warm weather in the fall and spring. Breeding for resistance to the disease has been started in France, as there appear to be differences among cultivars in resistance, although the differences are not large (Risser, 1972).

Virus tolerance: Breeding for virus tolerance has been an important objective in California, Oregon, and Washington, where large aphid populations disseminate viruses rapidly, and susceptible plants deteriorate quickly.

In breeding for virus tolerance, plants have been screened in field plots (Waldo, 1960; Daubeny et al., 1972) or by transmission tests in a greenhouse (Miller, 1959). Miller and Waldo

TABLE 7. Pathogenicity of American and Canadian races of *Phytophthora fragariae* (from Montgomerie, 1967)

Race	Huxley	*F. chiloensis* Del Norte	Auchincruive No. 6	52AC18	Perle de Prague	Climax	Redgauntlet	Juspa	Sparkle	Aberdeen	Talisman	Auchincruive No. 11	53Q13	Cambridge Vigour	Templar	Siletz	Probable British race
A-1	+	•	+	−	−	+	+	+	+	−	−	−	−	−	−	−	?
A-2	+	+	+	+	+	−	−	−	−	−	−	−	−	−	−	−	B66-3
A-3	+	+	•	+	+	+	+	+	•	+	+	+	+	+	+	−	B66-10
A-4	+	+	+	+	+	−	−	−	−	−	−	−	−	−	−	−	B66-3
A-5	+	•	+	+	+	+	+	•	+	+	+	+	+	+	+	+	B66-11
A-6	+	+	+	+	+	+	+	•	+	+	+	+	+	+	+	+	B66-11
(M-1)	+	•	+	+	+	−	−	−	•	−	−	−	−	−	•	−	B66-3
C-1	+	+	•	+	+	−	−	−	•	−	−	−	−	−	−	−	B66-3
C-2	+	+	+	+	+	−	−	−	•	−	−	−	−	−	−	−	B66-3
C-3	+	•	+	+	+	−	−	−	•	−	−	−	−	−	−	−	B66-3
C-4	+	•	+	−	−	+	+	+	•	−	−	−	−	−	−	−	? (=A-1)
C-5	+	•	+	+	+	−	−	•	•	−	−	−	−	−	•	•	?
C-6	+	•	•	−	−	+	+	•	•	−	−	−	•	•	•	•	?

+ = Pathogenic; − = nonpathogenic; • = not tested.

(1959) reported a high degree of tolerance in several clones of *F. chiloensis* and indicated their value for a breeding program. Daubeny et al. (1972) noted large differences in virus tolerance among cultivars exposed to natural infection. Among named cultivars, 'Cheam', 'Northwest', 'Shuksan', and 'Totem' were most tolerant. 'Northwest' appeared to transmit tolerance to its selections. However, no inheritance studies of virus tolerance have been reported. With both methods it is essential to maintain virus-free breeding stock in isolation.

Aphids are not so prevalent on strawberry plants in the eastern United States as on the Pacific Coast, and breeding programs in the eastern United States do not include specific screening for virus tolerance. Reliance is placed on nursery propagation of virus-free stocks.

Soil-retained viruses are problems in Europe, where cultivars are known to differ in their resistances to the different ones. However, cultivars have not been compared directly for virus tolerance, and no work has been reported on breeding for tolerance to the soil-retained viruses. Soil-retained viruses have not been detected in the United States.

Virus-tolerant American cultivars are 'Blakemore', 'Columbia', 'Dunlap', 'Earlibelle', 'Fresno', 'Howard 17', 'Northwest', 'Robinson', 'Sequoia', 'Shasta', 'Shuksan', 'Siletz', 'Surecrop', 'Tennessee Beauty', 'Tioga', 'Totem', and 'Trumpeter'.

Fruiting Habit

Three types of fruiting habit are now recognized in strawberry plants: single-crop or "June" fruiting habit, which is characteristic of most of the cultivars; everbearing, in which plants fruit two or more times per season, and the summer flowering is governed by long photoperiods; and day-neutral, which does not become dormant as the day length shortens and flowering and runner production are continuous from early spring until growth is stopped by cold temperatures late in fall. The last two types are similar, but distinct (Bringhurst and Voth, unpublished).

The inheritance and breeding behavior of the everbearing characteristic has been the subject of several studies (Clark, 1938; Powers, 1954; Moulton and Johnston, 1955; Brown and Wareing, 1965; and Ourecky and Slate, 1967). Ouercky and Slate (1967), on the basis of 46 progenies, determined that complementary dominant genes governed inheritance, and that they segregated in an octoploid manner. Thus the percentage of everbearing seedlings in a progeny depends on the genetic constitution of the everbearing parent (Table 8). In general, 'Geneva' transmitted the everbearing character to a higher percent of seedlings than 'Streamliner', 'Ogallala', or 'Arapahoe'. When progenies were retained for a second year, a higher percentage of the seedlings displayed the everbearing character. The same results have been obtained in our work (Scott and Draper, unpublished).

Powers (1954) intercrossed and selfed three everbearing and seven non-everbearing parents in all combinations and recorded the seedlings that were everbearers. Progenies with everbearing parentage differed greatly in the percentage of ever-

TABLE 8. Percentage of everbearing seedlings obtained by crossing non-everbearing x everbearing strawberries, 1962 (from Ourecky and Slate, 1967)

Parentage Non-everbearing	x everbearing	Total no. seedlings	Percentage everbearing
Fletcher	x Geneva	57	70.2
Fortune	x Geneva	73	17.8
Midland	x Geneva	160	50.0
Surecrop	x Geneva	138	46.4
New York 254	x Geneva	181	49.2
New York 318	x Geneva	401	43.6
Redglow	x Geneva	134	54.5
Frontenac	x Geneva	114	46.5
Suwannee	x Geneva	124	41.1
Average			46.6
Fletcher	x Streamliner	96	28.1
Fortune	x Streamliner	212	39.7
Midland	x Streamliner	283	42.4
Surecrop	x Streamliner	153	36.6
New York 254	x Streamliner	143	30.8
New York 318	x Streamliner	257	14.4
Average			32.0
Fletcher	x Ogallala	135	39.3
Fortune	x Ogallala	231	36.0
Midland	x Ogallala	338	37.0
Surecrop	x Ogallala	211	31.3
New York 254	x Ogallala	115	29.6
New York 318	x Ogallala	308	24.0
Frontenac	x Ogallala	71	21.1
Average			31.2
Midland	x Arapahoe	304	31.3
Frontenac	x Arapahoe	112	22.3
New York 318	x Arapahoe	208	18.8
Average			24.1

New York 254 = (Tennessee Shipper x Fairfax)
New York 318 = (Streamliner x Fairfax)

bearing seedlings, depending on the parentage. Some progenies segregated 3 everbearing : 1 non-everbearing; some were 1 : 1; some were 37.5 : 62.5; and some fit ratios of 56.25 : 43.75. He postulated dominant and recessive genes for everbearing, with two or more complementary dominant genes of unequal potency and at least four recessive genes governing expression of the character.

Bedard (1968) grew progenies of three June-bearers and two everbearers in a 5x5 diallel. Specific combining ability occurred for plant vigor, height, spread, fruit weight, and fruit number. General combining ability was nonsignificant for all five characters. Nonadditive variance exceeded additive variance, and epistatis was the main component of the genetic variance.

Brown and Wareing (1965) studied everbearing in diploids of wild-type *F. vesca*, 'Baron Solemacher', and 'Bush White', the last two being everbearers. Everbearing was governed by a single recessive gene and acted independently from a single recessive gene for nonrunnering.

Winterhardiness

Winter injury is a serious problem in many strawberry areas, and cultivars differ widely in their hardiness. As an example, Daubeny et al. (1970) found the 'Northwest' cultivar to be the least winter-hardy of 29 selections and cultivars on test in Washington, and 18 kinds were hardier than 'Northwest' in British Columbia. An unusual degree of winterhardiness exists in clones of *F. ovalis* (Hildreth and Powers, 1941). Powers (1945) determined the inheritance of several characters involving cultivars crossed with winter-hardy *F. ovalis*. He found that extreme winterhardiness was partially dominant over nonhardiness, and that most of the important economic characters were independently inherited, so that undesirable linkages were not a problem. Winterhardiness, large fruits, very vigorous plants, high runner production, and good fruit quality were recombined in about 1% of the double-cross seedlings. Eight cultivars were introduced from the work, but only 'Arapahoe' and 'Ogallala' everbearers have persisted. However, the outstanding winterhardiness of the material furnishes an excellent source for further breeding.

Generally, cultivars that originate in the north-central and northwestern United States and northern Europe are more winter-hardy than those originated in southern areas.

Blossom Hardiness at Low Temperatures

Late spring frosts are a hazard in strawberry-growing areas, especially where plant growth begins in early spring and blossom-hardiness of early flowering cultivars is especially important. Early flowering is essential for early ripening (Wilson and Giamalva, 1954; Zych, 1966; Ferguson, 1971). Darrow and Scott (1947) reported that a high percentage of seedlings obtained from crosses of two very hardy, early flowering clones of *F. virginiana* x 'Midland' (an early, blossom-tender cultivar) possessed hardiness. Some of the seedlings were as hardy as the hardy parent, and their fruit ripened early.

Wide differences exist among cultivars in low-temperature blossom-hardiness. 'Earlidawn' and 'Howard 17' are two early cultivars that possess blossom-hardiness, whereas 'Midland' and 'Stelemaster' (both early ripening) and 'Redstar' and 'Robinson' (both late ripening) are very susceptible to low temperatures at flowering.

Time of Flowering and Time of Ripening

As noted previously (Wilson and Giamalva, 1954; Zych, 1966), time of bloom and time of ripening are rather closely correlated. Cultivars that bloom early generally ripen early, and those that bloom late ripen late.

Peterson (1953), in a study of time of bloom, time of ripening, and rate of fruit development of strawberry seedlings, found that all characters were inherited quantitatively without heterosis. Specific tests of combining ability were not necessary to determine breeding behavior for these characters, but the most efficient progeny test to establish the breeding value of parents was use of selfed progenies.

In Powers' (1945) work with *F. ovalis*, the F_1 hybrids were nearly as early flowering as the early parent, and earliness was partially dominant. Scott et al. (1972) noted that earliness of *F. virginiana* behaved as a partial dominant in crosses with cultivars.

Short Rest Period

Cultivars differ substantially in the duration of their rest periods. Usually cultivars that originate in and are adapted to warm southern areas such as the southern United States, the Mediterranean region, and Africa, have the shortest rest periods, and plants are capable of growing and ripening fruit during the short days of winter. Examples of such cultivars are 'Missionary', 'Florida Ninety', 'Dabreak', 'Fresno', 'Sequoia', 'Tioga', 'Festival', 'Parfait', and apparently 'Benizuri', 'Fukuba', and 'Yachivo' of Japan. The 'Missionary' arose as a chance seedling near Norfolk, Virginia; 'Florida Ninety' was from breeding work in Florida; 'Dabreak' was from breeding work in Louisiana; 'Fresno', 'Sequoia', and 'Tioga' were selected from breeding work in southern California; and 'Festival' and 'Parfait' originated from breeding work in South Africa (Steyn, 1960). The short-rest-period character appears to be heritable, but it has not been studied genetically.

Concentrated Ripening

A recent objective in breeding work is concentrated ripening of fruit for possible once-over machine harvesting. Some selections and cultivars have been noted that have about half the crop ripe and in usable condition at one time (Denisen et al., 1969; Lawrence, 1970). Lawrence (1970) found that crosses yielding the highest percentage of concentrated-ripening seedlings had at least one early-season parent.

Mite Resistance

Red spider mites, *Panonychus ulmi* Koch, and two-spotted mites, *Tetranychus urticae* Koch, can cause serious damage to strawberry plants under certain conditions favorable for mite infestation. Infestations usually occur late in fall and during the fruiting period.

In seedling fields severely infested with mites, clones have been observed that are affected less by mites than others. Chaplin et al. (1968) selected two seedling clones that were resistant to mites. These were used as parents to produce F_1 and S_1 progenies as a first step in a breeding program. Later work (Chaplin et al., 1970) included backcrosses, outcrosses, and combinations of S_1 lines. In F_1 crosses between mite-resistant and mite-susceptible parents and in backcrosses to the resistant parents, the progeny means were intermediate between those of the parents. The F_1 progeny mean was nearer the mean of the resistant parent than that of the susceptible one, indicating partial dominance of resistance. In progenies of selfs, backcrosses, and crosses of some S_1 selections, the progeny means were nearer the mean of the susceptible parent than that of the resistant, but in such crosses there was no way to separate mite susceptibility from inherent weakening of the seedlings caused by inbreeding effects. Apparently many genes govern resistance. Further search for resistance among cultivars and clones of species appears desirable.

The long fruiting season in California provides favorable conditions for mite infestation on strawberry plants which lead to severe damage. Kishaba et al. (1972) screened 98 strawberry cultivars and selections, most of which were derivatives of either a resistant or a susceptible parent, for mite resistance over a two- to three-year period. Wide differences in mite resistance were detected. 'Sierra', 'Shasta', and several 'Shasta' derivatives were susceptible, whereas 'Lassen' and some of its derivatives were very resistant. The most susceptible selection was an F_1 of 'Shasta' x South American *F. chiloensis*, and the most resistant an F_1 of 'Lassen' x *F. ovalis*. Mite resistance appeared to be associated with near absence of pubescence on the leaves and petioles.

Blossom Sterility

Blossom sterility, as discussed here, means the failure of blossoms with normal-appearing pistils to set fruit. Three broad classes of blossoms exist in cultivated strawberries: pistillate, which are completely devoid of anthers; hermaphrodite; and staminate, which appear hermaphroditic but do not have functioning pistils. When pistillates are crossed with hermaphrodites, usually about half of the seedlings are pistillate and half hermaphrodites (Cummings and Jenkins, 1923; Cristoferi et al., 1967) (Table 9). However, when hermaphrodites are selfed, a low percentage of pistillates is obtained. This indicates that the pixtillate character is heterozygous, and that modifying genes are present in the hermaphrodite class.

Scott et al. (1962) noted two different types of blossom sterility, depending on the parentage involved in the progenies. In crosses of perfect-flowered parents, some staminate seedlings were obtained. The percentage of staminate seedlings depended on the parental combinations. Other seedlings in the same progenies appeared fully fertile and productive. Evidently one or more recessive genes control the staminate character. When cultivars were crossed with staminate clones of *F. chiloensis*, the resulting seedlings generally had low fertility, with a few that were sterile. A wide range in fertility occurred among seedlings within any one progeny.

TABLE 9. Breeding behavior for flower type in crosses between 'Warfield' and perfect-flowered cultivars (from Cummings and Jenkins, 1923)

Parents		Number of progeny				Percentage hermaphrodite
Pistillate	Hermaphrodite	Generation	Total	Hermaphrodite	Pistillate	
Warfield x	Barrymore	first	196	94	102	48.0
		second	181	164	17	90.6
		third	1	1	0	100
Warfield x	Dicky	first	217	96	121	44.2
		second	196	177	19	90.3
Warfield x	Gandy	first	349	190	159	54.4
		second	1,906	1,891	15	99.2
		third	298	287	11	96.3
Warfield x	Golden Gate	first	57	29	28	50.9
		second	440	401	39	91.1

The mode of inheritance of the different flower classes has not been determined, but it appears to be complex.

Variegation

Variegation (June yellows or transient yellows) is an abnormal yellow-green mottled appearance of the leaves of strawberry plants. Symptoms vary among seedlings from light green-yellow mottling to intense green and yellow mottling. Variegation can occur early in the growth of a seedling, or may not occur until the plant is several years old. Notable examples of the delayed appearance of the condition are 'Auchincruive Climax' and 'Blakemore' cultivars, which had gained extensive prominence before variegation appeared and weakened the plants so that they were no longer satisfactory. Variegation has become an increasingly important problem because of the use of parents carrying the disorder (Darrow, 1955). These have included 'Auchincruive Climax', 'Blakemore', 'Dixieland', 'Howard 17', 'Klonmore', 'Madame Moutot', 'Midland', 'Vermilion', and others.

Investigations by Williams (1955), Wills (1962), and Misic (1965) largely agree that transmission does not fit a Mendelian pattern but that plasmagene control could account for behavior of the disorder. Williams obtained large differences in variegation among 'Blakemore' and 'Howard 17' progenies, depending on the condition of the parent plants, age of seedlings, and whether grown in a glasshouse or cold frame (Table 10). Therefore, he postulates that a cytoplasmic effect is operating as a plasmagene, which interferes with chlorophyll development. All attempts to transmit a causal virus have been unsuccessful.

Nonrunnering

In *F. vesca* the runnering versus nonrunnering character is simply inherited, with runnering dominant and a 3:1 segregation in the F_2 (Richardson, 1914; Brown and Wareing, 1965). Brown and Wareing found the nonrunnering and "bushy" habit associated or controlled by the same gene.

In cultivated strawberries, the breeding behavior of the runnering character is complex. Corbett (1963) and Corbett and Meader (1965), in an extensive study of the character, found that nonrunnering appeared to be heritable. Nonrunnering was closely correlated with the everbearing character. With few exceptions, the nonrunnering plants were everbearers. When nonrunnering plants were selfed, their seedlings produced runnering and nonrunnering plants. Some parents produced nearly twice as many nonrunnering seedlings as others. Inbreeding to the S_3 generation gave a significant increase in the percentage of nonrunnering seedlings among progenies, but no homozygous nonrunnering genotype was obtained.

TABLE 10. The percentages of yellows-affected seedlings in 1952 crosses and number of seedlings scored (from Williams, 1955)

Cross	1953				1954			
	Glasshouse		Cold frame		Glasshouse		Cold frame	
	%	No.	%	No.	%	No.	%	No.
Blakemore								
y x y	90.1	212	70.3	91	98.6	190	90.5	89
g x g	5.0	285	4.2	120	16.2	267	10.3	116
y x g	4.7	129	2.3	130	32.0	125	19.5	118
g x g	15.3	124	8.7	127	64.2	123	53.3	122
Blakemore and NS								
y x NS	0.0	204	0.0	123	30.2	199	18.9	122
g x NS	0.0	166	0.0	90	0.0	160	0.0	90
NS x y	0.0	235	0.0	131	50.4	225	39.9	131
NS x g	4.1	121	2.4	125	27.7	119	11.2	125
Howard 17								
g x g	1.6	125	0.8	118	2.5	120	0.8	117
y x y	58.3	103	64.1	92	77.9	86	72.6	73

y = yellows affected, g = normal green, NS = nonsusceptible variety.
The female parent is placed first in the crosses.

Breeding Specific Fruit Characters

Size

Large fruit size is sought in all practical breeding programs. Size differs within a fruit cluster, the primary berry being the largest, and berries at the secondary position and more inferior positions on the inflorescence being progressively smaller. Sherman and Janick (1966) and Janick and Eggert (1968) reported that the decline in size of fruits from the primary to inferior positions was similar in all the cultivars under study and reported a high correlation between fruit size and number of achenes. This infers that fruit size can be selected on the basis of position in the cluster. However, Moore et al. (1970) found that the relative decline in size of fruit from the primary to secondary to tertiary positions was much greater for large-fruited cultivars than for small-fruited ones.

Size of fruit is inherited quantitatively (Powers, 1945; Baker, 1952; Comstock et al., 1958; Scott, 1959; Sherman et al., 1966; Spangelo et al., 1971a; Hansche et al., 1968). Baker (1952) observed heterosis for fruit size in some progenies of crosses of cultivars. Scott (1959) and Scott et al. (1972) found small fruit size of *F. virginiana* and of a North American clone of *F. chiloensis* were partially dominant in crosses with standard cultivars. Sherman et al. (1966), in a study of diallel combinations involving seven parents, one of which was the same small-fruited parent used by Scott (1959), reported that the highest estimate of general combining ability was obtained when the small-fruited species tester was included in the analysis. When the small-fruited tester was deleted from the analysis, specific combining ability was more important than general combining ability. The small-fruited tester contributed very few or no genes for large size and, therefore, small fruit size was partially dominant. Among progenies from crosses of large-fruited parents, much of the variance appeared to be epistatic.

Oydvin (1965) noted that 'Abundance', 'Gorella', 'Senga Precosa', 'Senga Sengana', and 'Ydun', when used as parents, gave significant differences in combining ability for yield, fruit size, and number of fruits per plant. 'Ydun' had greatest combining ability for yield, 'Gorella' for fruit size, and 'Abundance' for number of fruit. Fruits were especially large on seedlings of 'Senga Sengana' x 'Gorella' and 'Ydun' x 'Gorella'. Diallel crosses indicated that high correlations occurred between parents and means of progenies for size of fruit and yields.

Hansche et al. (1968) reported that an important part of the genetic variance for fruit size was epistatic, and there was a fairly high value for heritability. In their material, a correlation coefficient of 0.65 ± 0.07 was found between fruit size and yield, indicating that selection for yield may result in an increase in fruit size.

Considerable improvement in fruit size has been achieved by breeders in recent years. Cultivars that are notable for large fruit size are 'Belrubi', 'Catskill', 'Gorella', 'Guardian', 'Robinson', 'Sequoia', 'Tioga', 'Titan', and 'Vesper'.

Flesh Firmness and Skin Toughness

Satisfactory handling quality depends on firm flesh and tough skin of the fruit. Both firmness of flesh and toughness of skin are greatly influenced by temperature and humidity during fruit development and ripening. Relatively warm temperatures and high humidity result in softer fruit than cool temperatures and low humidity. Cultural conditions will also influence flesh firmness and skin toughness. Both subjective and objective ratings have been used in evaluation of the characters (Scott, 1959; Hansche et al., 1968; Ourecky and Bourne, 1968). Hansche et al. (1968) found fruit firmness, as measured by a pressure-puncture tester, to be highly heritable, but negatively correlated with verticillium wilt resistance. Estimates of genotypic correlation between firmness and yield and between firmness and size were near zero, indicating no genetic relationship between firmness and fruit size or yield. Morrow and Darrow (1941) noted that a soft-fruited parent transmitted the soft-fruit character to a high percentage of its seedlings, which indicated many genes for soft fruit. Sato et al. (1965) reported that a progeny of 'Chiyoda' x 'Florida Ninety' had the firmest progeny-mean value of 3.58 ± 0.096 on a scale of 1 to 5, and that 'Tohoku No. 1' x 'Howard 17' had the lowest value of 1.90 ± 0.136, in a comparison of 20 progenies.

Ourecky and Bourne (1968) found that an Instron machine, which measures compressive and tensile characters, can be adapted to measure skin toughness and flesh firmness. They obtained very good reproducible results in a study of 64 cultivars and selections. Several New York selections had the toughest skins, and among cultivars 'Tennessee Shipper' and 'Tioga' were the toughest. 'NY 844' led the list for firm flesh. Among cultivars, 'Tennessee Shipper' was the firmest, with 'Pocahontas' and 'Gorella' second and third.

'Blakemore' was the first very firm-fruited cultivar to be originated, and it established a new standard for firmness. Several cultivars now are notable for firm flesh: 'Albritton', 'Apollo', 'Blakemore', 'Dixieland', 'Earlibelle', 'Holiday', 'Tennessee Beauty', 'Tennessee Shipper', and 'Tioga'.

Frozen-Pack Processing Characters

An increasing quantity of fruit is processed by freezing in the United States and other countries. Characteristics of high-quality fruit for freezing include a uniform red flesh color, firmness, high flavor with tartness, and ease of capping or plugging.

Flavor rating is a variable characteristic that is influenced by environmental and cultural conditions and by personal preference. Overcash et al. (1943), Slate (1943), and Darrow (1966) reported the importance of 'Fairfax' as a parent for flavor. Overcash et al. (1943), in a breeding study of 14 progenies involving 2,660 seedlings, rated only 7% of the seedlings as having high flavor, and 50% of these highly flavored seedlings had 'Fairfax' as a parent.

Sistrunk and Moore (1971) reported that 'Earlibelle', 'Md US 2713', and 'Ark 5018' had outstanding flavor in frozen pack. They also noted that 'Earlibelle', 'Md US 2713', and 'Ark 5018' had bright intense-red flesh color as fresh fruit and rated high after freezing in slice wholeness, color, flavor, and general appearance. They were also high in soluble-solids content and total acidity, which characters appear to be closely correlated with frozen-pack quality. High acidity sponsors retention of flesh color. 'Earlibelle' was a good parent in transmission of quality characters.

Murawski (1968), in a breeding test of eight parents involving 23 progenies, observed that progenies tended to express characters much like those of the parents for external and internal color of the fruit. Both characters were quantitatively inherited.

Some cultivars cap or plug more easily than others. Usually, but not always, easy capping is correlated with soft flesh. 'Tioga' and 'Redchief' are examples of firm-fleshed cultivars that cap easily. Inheritance of the character has not been determined.

Vitamin C Content

The fruits of some cultivars are rich sources of vitamin C, with some having more than 100 mg/100 g of fresh tissue, but cultivars differ widely in this respect. Breeding studies have established that vitamin-C content is genetically controlled. Hansen and Waldo (1944) found that 'Marshall', 'Clarke', 'Progressive', and *F. chiloensis* transmitted relatively high vitamin-C content to their seedlings whereas other parents did not. Selections with highest vitamin content had double the amount of those with the lowest. Darrow et al. (1947) compared the vitamin-C content of seedlings in six progenies from parents having high, medium, and low vitamin-C content. Some seedlings in progenies from two high-vitamin parents had higher vitamin C than either parent. Anstey and Wilcox (1950) compared seedlings from inbred lines and from outcrosses in 16 progenies for vitamin-C content of fruit. Three of the inbred parents had high vitamin content, and two had medium. A high correlation existed between the mean vitamin-C content of fruits of a progeny and the mean of the parents. High vitamin C of the fruit was partially dominant.

Soluble Solids and Acidity

Soluble-solids content of the fruit is a quality factor that appears to be fairly consistent among cultivars. Variations between years may occur, but rankings of cultivars within years are usually similar.

As mentioned previously, high total acidity favors retention of bright-red flesh color in frozen pack.

Both soluble solids and acidity are controlled genetically. From a comparison of the soluble-solids content of six progenies derived from parents with high, medium, and low soluble solids, Duewer and Zych (1967) determined that high content of soluble solids was dominant in all except one of the progeny means. A wide range in soluble-solids content occurred among seedlings within a progeny, so that selection for high soluble-solids content should be fairly simple. Their data for titratable acidity indicated that low titratable acidity was partially dominant. Again, there was a wide range among seedlings within progenies for titratable acidity.

Botrytis Fruit Rot Resistance

Fruit rot caused by *Botrytis cinerea* Pers. ex Fr. is the most destructive disease that attacks strawberries, when environmental and cultural conditions are favorable for infection. Significant differences in the quantity of rotted fruits among cultivars have been reported in many studies, one of the most recent being that of Barritt et al. (1971). They reported 12% rot in 'Shuksan', which had the least, to 52% in 'WSU 1217', which had the most of 14 cultivars in the test. Selections with 'Columbia' and 'Molalla' parentage were less susceptible than those with other parentage. 'Valentine' was reported in Sweden to transmit fruit-rot resistance to seedlings (Koch, 1963; Nybom, 1968).

Koch (1963) in a two-year comparison of 18 selections derived from a cross of 'Senga Sengana' x 'Valentine' noted that some selections had a low incidence of botrytis fruit rot while others were relatively high. The lowest incidence of fruit rot was 1.7% and the highest 17.8%. Considerable variation occurred between years. However, in all such differences in the measured fruit rot, there is a question of whether the fruit is genetically resistant to rot or whether other factors are involved, such as density of leaf canopy, placement of fruit on the ground, firmness of flesh, or waxiness of the skin. More work is needed along lines similar to that of Irvine and Fulton (1959) who reported that mycelial growth of *Botrytis* was inhibited when sterilized juice of ripe fruit of 'Empire' and 'Sparkle' was placed in the culture. Juice from 'Albritton,' 'Howard 17,' 'Earlidawn,' and 'Robinson' did not inhibit growth of the fungus.

Actual inheritance or breeding behavior of selected parents has not been determined.

Achievements and Prospects

The rapidity with which newer cultivars have replaced older ones is a testimonial to the success of present-day strawberry breeders. Several of the cultivars being grown commercially have been introduced in recent years. Some of the most recent introductions will no doubt replace older cultivars even more rapidly than in the past, because of their desirable combinations of characteristics. However, the needs are evident for more extensive work on inheritance of characters, breeding behavior of parent material, incorporation of new characteristics, and the practical origination of improved cultivars, which is the final objective in most breeding programs. Although newer cultivars are far superior to their predecessors, no region in the United States, and probably nowhere else, has cultivars with completely satisfactory combinations of plant and fruit characteristics.

For the future, the greatest challenge is in breeding for resistance to such diseases as red stele, verticillium wilt, leaf spot, leaf scorch, mildew, anthracnose, virus infections, botrytis fruit rot, mites, and nematodes. Sources of resistance to the last three troubles are largely unknown, and it may be necessary to obtain them via wide interspecific hybridization or intergeneric hybridization. Such breeding work requires specific information on the pathogenicity of the organisms and suitable screening techniques for evaluation of breeding material. Programs of this type are costly to operate and require special facilities and highly trained personnel, but they produce pest-resistant cultivars that reduce risks and increase production per unit of space.

Breeders must constantly search species and closely related genera for new desirable genes to incorporate into the cultivated strawberry. American breeders are especially fortunate because of the wealth of native octoploid species near at hand for their use. However, such material is being used very little at the present time. An extensive collection of clones of native octoploid species is needed, which would be evaluated at several places for several years in a search for new characters. No doubt there are genes for large fruit size in some clones of *F. chiloensis* from South America that are not present in our cultivars. Parents are needed to originate cultivars that retain high flavor in fruit under unfavorable environmental conditions. These are only two examples of many that could be listed that await the strawberry-genetics engineer to incorporate into cultivars of the future.

Viruses are a constant problem in most strawberry-breeding programs, as they cause degeneration of stocks. Frequently virus symptoms in cultivars are not evident and can be detected only by indexing to sensitive indicator plants. For descriptions and illustrations of viruses of strawberries, see Frazier (1970).

Descriptions of cultivars, with excellent illustrations in color, are given in Baldini and Branzanti (1964), Darrow (1966), and Bazzocchi et al. (1972). All three volumes are excellent references for strawberry breeders.

Literature Cited

Aalders, L. E., and D. L. Craig. 1968. General and specific combining ability in seven inbred strawberry lines. *Can. J. Gen. Cyt.* 10:1–6.

Aalders, L. E., D. L. Craig, and H. B. Cannon. 1968. Field plot technique for the evaluation of strawberry cultivars. *HortScience* 3:234–235.

Allard, R. W. 1960. *Principles of plant breeding.* New York: Wiley.

Anstey, T. H., and A. N. Wilcox. 1950. The breeding value of selected inbred clones of strawberries with respect to their vitamin C content. *Scient. Agr.* 30:367–374.

Asker, Sven. 1970. An intergeneric *Fragaria* x *Potentilla* hybrid. *Hereditas* 64:135–139.

Baker, R. E. 1952. Inheritance of fruit characters in the strawberry: a study of several F_1 hybrid and inbred populations. *J. Hered.* 43:9–14.

Baldini, E., and E. C. Branzanti. 1964. *Monograph of the principal non-everbearing cultivars of strawberries* (in Italian). Bologna: Istituto di Coltivazioni Arboree Dell'Universita di Bologna.

Barnes, E. H., J. E. Moulton, and J. A. Ignatoski. 1966. *Verticillium*-resistant selections of strawberries from crosses of named varieties. *Mich. Agr. Expt. Sta. Quart. Bul.* 48:514–520.

Barrientos, F., and R. S. Bringhurst. 1973. A haploid of an octoploid strawberry cultivar. *HortScience* 8:44.

Barritt, B. H., L. Torre, and C. D. Schwartze. 1971. Fruit rot resistance in strawberry cultivars adapted to the Pacific Northwest. *HortScience* 6(3):242–244.

Bauer, R. 1960. Aspects of polyploid breeding in the genus *Fragaria* (in German). *Institut Agronomique et des Stations de Recherches de Gemblaux. Hors. serie* 2:994–1006.

Bauer, R., and A. Bauer. 1967. New ways in strawberry breeding (in German). *Der Erwerbsobstbau* 9:83–85.

Bazzocchi, R., E. C. Branzanti, G. Cristoferi, and P. Rosati. 1972. *Monograph of the principal non everbearing cultivars of strawberries* vol. 2. (in Italian). Rome: Consiglio Nazionale Delle Ricerche.

Bedard, P. R. 1968. Strawberry breeding studies involving diallel crosses between everbearing and June-bearing varieties. Ph.D. dissertation, Univ. of Minnesota. *Diss. Abstr.* 28:4830B.

Bedard, P. R., C. S. Hsu, L. P. S. Spangelo, S. O. Fejer, and G. L. Rousselle. 1971. Genetic, phenotypic, and environmental correlations among 28 fruit and plant characters in the cultivated strawberry. *Can. J. Gen. Cyt.* 13:470–479.

Bowen, H. H., L. F. Hough, and E. H. Varney. 1966. Breeding studies of verticillium wilt resistance in *Fragaria x ananassa* Duch. *Proc. Amer. Soc. Hort. Sci.* 93:340–351.

Bringhurst, R. S., and T. Gill. 1970. Origin of *Fragaria* polyploids. 2. Unreduced and double-reduced gametes. *Amer. J. Bot.* 57(8):969–976.

Bringhurst, R. S., and D. A. Khan. 1963. Natural polyploids *Fragaria chiloensis—F. vesca* hybrids in coastal California and their significance in polyploid *Fragaria* evolution. *Amer. J. Bot.* 50:658–661.

Bringhurst, R. S., and Y. D. A. Senanayake. 1966. The evolutionary significance of natural *Fragaria chiloensis* x *F. vesca* hybrids resulting from unreduced gametes. *Amer. J. Bot.* 53:1000–1006.

Bringhurst, R. S., and V. Voth. 1957. Effect of stratification on strawberry seed germination. *Proc. Amer. Soc. Hort. Sci.* 70:144–149.

Bringhurst, R. S., P. E. Hansche, and V. Voth. 1967. Inheritance of verticillium wilt resistance and the correlation of resistance with performance traits of the strawberry. *Proc. Amer. Soc. Hort. Sci.* 92:369–375.

Bringhurst, R. S., D. Khan, and V. Voth. 1964. Biochemical relationships among *Fragaria* species. *Amer. J. Bot.* 51:688.

Bringhurst, R. S., S. Wilhelm, and V. Voth. 1961. Pathogen variability and breeding verticillium wilt resistant strawberries. *Phytopathology* 51:786–794.

———. 1966. Verticillium wilt resistance in natural populations of *Fragaria chiloensis* in California. *Phytopathology* 56:219–222.

Brown, T., and P. F. Wareing. 1965. The genetical control of the everbearing habit and three other characters in varieties of *Fragaria vesca*. *Euphytica* 14: 97–112.

Chaplin, C. E., L. P. Stoltz, and J. G. Rodriguez. 1968. The inheritance of resistance to the two-spotted mite *Tetranychus urticae* Koch in strawberries. *Proc. Amer. Soc. Hort. Sci.* 92:376–380.

———. 1970. Breeding behavior of mite-resistant strawberries. *J. Amer. Soc. Hort. Sci.* 95:330–333.

Clark, J. H. 1938. Inheritance of the so-called everbearing tendency in the strawberry. *Proc. Amer. Soc. Hort. Sci.* 35:67–70.

Comstock, R. E., T. Kelleher, and E. B. Morrow. 1958. Genetic variation in an asexual species, the garden strawberry. *Genetics* 43:634–646.

Corbett, E. G. 1963. The breeding behavior of the nonstoloniferous strawberry, *Fragaria* x *ananassa* Duch. Ph.D. dissertation, Univ. of New Hampshire. Ann Arbor, Mich.: Univ. Microfilms.

Corbett, E. G., and E. M. Meader. 1965. Non-stoloniferous strawberry. *J. Hered.* 46:237–241.

Craig, D. L., L. F. Aalders, and C. J. Bishop. 1963. The genetic improvement of strawberry progenies through recurrent reciprocal selection. *Can. J. Gen. Cyt.* 5:33–37.

Cristoferi, G., D. S. Francassini, and A. Zocca. 1967. Research on strawberries 12. Studies on inheritability of characters (in Italian, English summary). *Rivista della Ortoflorofruttic.* 51:3–26.

Cummings, M. B., and E. W. Jenkins. 1923. Sterility of strawberries: strawberry breeding. *Bul. Vt. Agr. Expt. Sta.* No. 232. 61 pp.

Darrow, G. M. 1937. Strawberry improvement. In *USDA Yearbook of Agriculture.* 1937:445–495.

———. 1955. Leaf variegation in strawberry—a review. *Plant Dis. Reptr.* 39:363–370.

———. 1966. *The strawberry. History, breeding and physiology.* New York: Holt, Rinehart and Winston.

Darrow, G. M., and D. H. Scott. 1947. Breeding for cold hardiness of strawberry flowers. *Proc. Amer. Soc. Hort. Sci.* 50:239–242.

Darrow, G. M., D. H. Scott, and A. C. Goheen. 1954. Relative resistance of strawberry varieties to powdery mildew at Beltsville, Maryland. *Plant Dis. Rptr.* 38: 864–866.

Darrow, G. M., G. F. Waldo, and C. E. Schuster. 1933. Twelve years of strawberry breeding: a summary of the strawberry breeding work of the United States Department of Agriculture. *J. Hered.* 24:391–402.

Darrow, G. M., G. F. Waldo, C. E. Schuster, and B. S. Pickett. 1934. Twelve years of strawberry breeding. 2. From 170,000 seedlings, seven named varieties: a summary of the crosses made and an evaluation of their effectiveness as breeding material. *J. Hered.* 25: 451–462.

Darrow, G. M., M. S. Wilcox, D. H. Scott, and M. C. Hutchins. 1947. Breeding strawberries for vitamin C. *J. Hered.* 38:363–365.

Daubeny, H. A. 1961. Powdery mildew resistance in strawberry progenies. *Can. J. Plant Sci.* 41:239–243.

———. 1963. Effect of parentage in breeding for red stele resistance of strawberry in British Columbia. *Proc. Amer. Soc. Hort. Sci.* 84:289–294.

Daubeny, H. A., R. A. Norton, C. D. Schwartze, and B. H. Barritt. 1970. Winterhardiness in strawberries for the Pacific Northwest. *HortScience* 5:152–154.

Daubeny, H. A., R. A. Norton, and B. H. Barritt. 1972. Relative differences in virus tolerance among strawberry cultivars and selections in the Pacific Northwest. *Plant Dis. Rptr.* 56:792–795.

Daubeny, H. A., and H. S. Pepin. 1965. The relative resistance of various *Fragaria chiloensis* clones to *Phytophthora fragariae*. *Can. J. Plant Sci.* 45:365–368.

Denisen, E. L., R. Garren, J. N. Moore, and E. J. Stang. 1969. Cultural practices and plant breeding influences for strawberry harvest mechanization. *Rural Manpower Center Rpt.* 16:469–495.

Dermen, H., and G. M. Darrow. 1939. Colchicine-induced tetraploid and 16-ploid strawberries. *Proc. Amer. Soc. Hort. Sci.* 36:300–301.

Dowrick, G. J., and H. Williams. 1959. Strawberry breeding. John Innes Hort. Inst. *Ann. Rpt.* 50:9.

Drain, B. D., and L. A. Fister. 1938. Some strawberry breeding progeny data. *Proc. Amer. Soc. Hort. Sci.* 35:60–66.

Draper, A. D., D. H. Scott, and J. L. Maas. 1970. Inoculation of strawberry with *Phytophthora fragariae*. *Plant Dis. Rptr.* 54:739–740.

Duchesne, A. N. 1768. Paris: *Natural history of strawberry plants* (in French).

Duewer, R. G., and C. C. Zych. 1967. Heritability of soluble solids and acids in progenies of the cultivated strawberry (*Fragaria* x *ananassa* Duch.). *Proc. Amer. Soc. Hort. Sci.* 90:153–157.

Duncan, D. B. 1955. Multiple range and multiple F tests. *Biometrics* 11:1–42.

Ellis, J. R. 1962. *Fragaria-Potentilla* intergeneric hybridization and evolution in *Fragaria*. *Proc. Linnean Society of London* 173:99–106.

———. 1963. The allopolyploid hybrid between *Fragaria vesca* and *F. moschata*. Genetics Today. The Hague: *Proc. 11 Int. Cong. of Gen.* 1:209–210.

Evans, W. D., and J. K. Jones. 1967. Incompatibility in *Fragaria*. *Can. J. Gen. Cyt.* 9:831–836.

Fadeeva, T. S. 1966. Communication 1. Principles of genome analysis (with reference to the genus *Fragaria*). *Genetika* 2(1):6–16.

Ferguson, J. H. A. 1971. Earliness of flowering and fruiting of forty strawberry varieties: a statistical study. *Euphytica* 20:362–370.

Fowler, C. W., and J. Janick. 1972. Wide crosses and pollen contamination in strawberry. *HortScience* 7: 566–567.

Frazier, N. W. 1970. *Virus diseases of small fruits and grapes handbook*. Berkeley: Univ. of Calif. Press.

Gooding, J. J. 1972. Studies on field resistance of strawberry varieties to *Phytophthora fragariae*. *Euphytica* 21:63–70.

Gupta, P. K. 1962. The occurrence of indigenous auxins and inhibitors and their effect on the germination of strawberry seed. Ph.D. dissertation, Kansas State Univ.

Hansche, P. E., R. S. Bringhurst, and V. Voth. 1968. Estimates of genetic and environmental parameters in the strawberry. *Proc. Amer. Soc. Hort. Sci.* 92: 338–345.

Hansen, E., and G. F. Waldo. 1944. Ascorbic acid content of small fruits in relation to genetic and environmental factors. *Food Res.* 9:453–461.

Harland, S. C., and E. King. 1957. Inheritance of mildew resistance in *Fragaria* with special reference to cytoplasmic effects. *Heredity* 11:257.

Henry, E. M. 1935. The germination of strawberry seeds and the technique of handling the seedlings. *Proc. Amer. Soc. Hort. Sci.* 32:431–433.

Hildreth, A. C., and L. Powers. 1941. The Rocky Mountain strawberry as a source of hardiness. *Proc. Amer. Soc. Hort. Sci.*, 38:410–412.

Hondelmann, W. 1965. Investigations on breeding for yield in the garden strawberry, *Fragaria ananassa* Duch. (in German). *Pflanzensuchtung* 54:46–60.

Hondelmann, W. 1967. Field observations on mildew attacks on strawberry seedlings (in German). *Nachrichtenbl. Deutch. Pflanzensuchtung*. 19:137–139.

Horn, N. L., K. R. Bernside, and R. B. Carver. 1972. Control of the crown rot phase of strawberry anthracnose through sanitation, breeding for resistance, and benomyl. *Plant Dis. Rptr.* 56:515–519.

Hsu, C. S., R. Watkins, A. T. Bolton, and L. P. S. Spangelo. 1969. Inheritance of resistance to powdery mildew in the cultivated strawberry. *Can. J. Gen. Cyt.* 11:426–438.

Hull, J. W. 1960. Development of colchicine induced 16-ploid breeding lines in *Fragaria*. *Proc. Amer. Soc. Hort. Sci.* 75:354–359.

Irvine, T. B., and R. H. Fulton. 1959. A study of laboratory methods to determine susceptibility of strawberry varieties to grey mold fruit rot. *Phytopathology* 49:542. (Abstr.)

Islam, A. S. 1960. Possible role of unreduced gametes in the origin of polyploid *Fragaria*. *Biologia (Lahore)* 6:189–192.

———. 1961. The haploid strawberry, *Fragaria vesca* L., and the significance of haploidy in phylogeny and plant breeding. *Scientist* 4:1–21.

Janick, J., and D. A. Eggert. 1968. Factors affecting fruit size in the strawberry. *Proc. Amer. Soc. Hort. Sci.* 93:311–316.

Jones, D. F., and W. R. Singleton. 1940. The improvement of naturally cross pollinated plants by selection in self fertilized lines. 3. Investigations with vegetatively propagated fruits. Strawberry and raspberry hybrids. *Bul. Conn. Agr. Expt. Sta.* 435:325–347.

Jones, J. K. 1955. Cytogenetic studies in the genera *Fragaria* and *Potentilla*. Ph.D. dissertation, University of Reading.

———. 1966. Evolution and breeding potential in strawberries. *Sci. Hort.* 18:121–130.

Jordan, V. W. L. 1971. A method of screening strawberries for resistance to verticillium wilt. *Plant Path.* 20:167–170.

Kishaba, A. N., V. Voth, A. F. Howland, R. S. Bringhurst, and H. H. Toba. 1972. Two-spotted spider mite resistance in California strawberries. *J. Econ. Entom.* 65:117–119.

Koch, Annelise. 1963. Valentine, a noteworthy parent in strawberry breeding (in German). *Der Zuchter* 33:352–354.

Koch, E. J. 1953. Plot technique in small fruits. *Proc. Amer. Soc. Hort. Sci.* 62:14–20.

Kronenberg, H. G., C. P. J. Lindeloof, and L. M. Wassenar. 1971. Strawberry breeding for red core resistance in the Netherlands. *Euphytica* 20:228–234.

Lawrence, F. J. 1970. Breeding strawberries in Northwestern USA for mechanical harvesting of fruit. *Proc. Ore. Hort. Soc.* 61:107–109.

Lee, Vivian. 1964. Antoine Nicholas Duchesne—first strawberry hybridist. *Amer. Hort. Mag.* 43:80–88.

Mangelsdorf, A. J., and E. M. East. 1927. Studies on the genetics of *Fragaria*. *Genetics* 12:307–339.

Miller, P. W. 1959. An improved method of testing the tolerance of strawberry varieties and new selected seedlings to virus infection. *Plant Dis. Rptr.* 43: 1247–1249.

Miller, P. W., and G. F. Waldo. 1959. The virus tolerance of *Fragaria chiloensis* compared with the Marshall variety. *Plant Dis. Rptr.* 43:1120–1131.

Misic, P. D. 1965. Occurrence of June yellows and symptoms of white streaks among strawberry hybrid seedlings. *J. Sci. Agr. Res. Yugoslavia* 18:152–159.

Mok, D. W. S., and W. D. Evans. 1971. Chromosome associations at diakinesis in the cultivated strawberrey. *Can. J. Gen. Cyt.* 13:231–236.

Molot, P. M., and J. G. Nourrisseau. 1970. Research on the death of strawberry caused by *Phytophthora cactorum* (L. and C.) Schroet (in French). *Ann. Phytopathol.* 2:117–137.

Montgomerie, I. 1967. Pathogenicity of British isolates of *Phytophthora fragariae* and their relationship with American and Canadian races. *British Mycol. Soc. Trans.* 50:57–67.

Moore, J. N., G. R. Brown, and E. D. Brown. 1970. Comparison of factors influencing fruit size in large-fruited and small-fruited clones of strawberry. *J. Amer. Soc. Hort. Sci.* 95(6):827–831.

Moore, J. N., D. H. Scott, and R. H. Converse. 1964. Pathogenicity of *Phytophthora fragariae* to certain *Potentilla* species. *Phytopathology* 54:173–176.

Morita, H. 1968. Physiologic races of *Phytophthora fragariae. Shizuoka Agr. Expt. Sta. Bul.* 13.

Morrow, E. B., R. E. Comstock, and T. Kelleher. 1958. Genetic variances in strawberry varieties. *Proc. Amer. Soc. Hort. Sci.* 72:170–185.

Morrow, E. B., and G. M. Darrow. 1941. Inheritance of some characteristics in strawberry varieties. *Proc. Amer. Soc. Hort. Sci.* 39:262–268.

———. 1952. Effects of limited inbreeding in strawberries. *Proc. Amer. Soc. Hort. Sci.* 59:269–276.

Morrow, E. B., G. M. Darrow, and J. A. Rigney. 1949. A rating system for the evaluation of horticultural material. *Proc. Amer. Soc. Hort. Sci.* 53:276–280.

Morrow, E. B., G. M. Darrow, and D. H. Scott. 1954. A quick method of cleaning berry seed for breeders. *Proc. Amer. Soc. Hort. Sci.* 63:265.

Moulton, J. E., and S. Johnston. 1955. The breeding value of 35 strawberry varieties tested in Michigan. *Proc. Amer. Soc. Hort. Sci.* 66:246–253.

Murawski, H. 1968. Research on the heritability among strawberry varieties for height of flower stems, powdery mildew resistance, fruit color, flesh color, and berry shape (in German). *Arch. Gartenb.* 16:293–318.

Muscle, H., and Florence E. Fay. 1971. A method for screening strawberry seedlings for resistance to *Phytophthora fragariae. Plant Dis. Rptr.* 55:471–472.

Nemec, S. 1969. Determination of leaf spot races in southern Illinois strawberry plantings. *Plant Dis. Rptr.* 53:94–97.

———. 1971. Studies on resistance of strawberry varieties and selections to *Mycosphaerella fragariae* in southern Illinois. *Plant Dis. Rptr.* 55:573–576.

Nemec, S., and R. C. Blake. 1971. Reaction of strawberry cultivars and their progenies to leaf scorch in southern Illinois. *HortScience* 6(5):497–498.

Newton, W., and M. C. J. van Adrichem. 1958. Resistance to *Verticillium* wilt in F_1 generations of *Fragaria. Can. J. Bot.* 36:297–299.

Nybom, N. 1968. A report from the Balsgard Breeding Institute. *Fruit Var. Hort. Digest* 22:520.

Ourecky. D. K., and M. C. Bourne. 1968. Measurement of strawberry texture with an Instron machine. *Proc. Amer. Soc. Hort. Sci.* 93:317–325.

Ourecky, D. K., and G. L. Slate. 1967. Behavior of the everbearing characteristics in strawberries. *Proc. Amer. Soc. Hort. Sci.* 91:236–241.

Overcash, J. P., L. A. Fister, and B. D. Drain. 1943. Strawberry breeding and the inheritance of certain characteristics. *Proc. Amer. Soc. Hort. Sci.* 42:435–440.

Oydvin, J. 1965. The inheritance of yield and berry size in five strawberry varieties (English summary). *Meld. Norg. Landbrtogsh.* 44:1–11.

Pepin, H. S., and H. A. Daubeny. 1966. Reaction of strawberry cultivars and clones of *Fragaria chiloensis* to six races of *Phytophthora fragariae. Phytopathology* 56:361–362.

Peterson, R. M. 1953. Breeding behavior of the strawberry with respect to time of blooming, time of ripening, and rate of fruit development. Ph.D. dissertation, Univ. of Minnesota.

Powers, L. 1944. Meiotic studies of crosses between *Fragaria ovalis* and *F. ananassa. J. Agr. Res.* 69:435–448.

———. 1945. Strawberry breeding studies involving crosses between the cultivated varieties (*Fragaria* x *ananassa*) and the native Rocky Mountain strawberry (*F. ovalis*). *J. Agr. Res.* 70:95–122.

———. 1954. Inheritance of period of blooming in progenies of strawberries. *Proc. Amer. Soc. Hort. Sci.* 64:293–298.

Reid, R. D. 1953. Breeding strawberries resistant to red core root rot. London: *Int. Hort. Congr. 1952.* 13(2):739–750.

———. 1959. Strawberry breeding. *Ann. Rpt. Scottish Hort. Res. Inst.* 6:46–48.

———. 1966. Strawberry breeding. *Ann. Rpt. Scottish Hort. Res. Inst.* 12:67–70.

Richardson, C. W. 1914. A preliminary note on the genetics of *Fragaria. J. Genetics* 3:171–177.

Risser, G. 1972. Breeding strawberries at the National Institute for Agronomic Research, Montfavet, France. *Jardins de France, N.S.* 6:16–17.

Rybakov, M. N. 1966. Variability of garden strawberry clones induced by gamma rays and ethylenimine (in Russian with English summary). *IXV. Timirjazev Sel'-hoz Akad* 2:36–44.

Sato, T., T. Hanaoka, T. Takai, and S. Henmi. 1965. Studies on the breeding of strawberries adapted to the northern part of Japan. 2. Expression of some characters in S_1 and F_1 populations. *Bul. Hort. Res. Sta.* series C, no. 3, Morioka.

Scott, D. H. 1951. Cytological studies on polyploids derived from tetraploid *Fragaria vesca* and cultivated strawberries. *Genetics* 36:311–330.

———. 1959. Size, firmness, and time of ripening of fruit of seedlings of *Fragaria virginiana* Duch. crossed with cultivated strawberry varieties. *Proc. Amer. Soc. Hort. Sci.* 74:388–393.

———. 1962. Breeding and improvement of the strawberry in the United States of America—a review. *Hort. Res.* 2:35–55.

Scott, D. H., G. M. Darrow, W. F. Jeffers, and D. P. Ink. 1950. Further studies on breeding strawberries for resistance to red stele disease. *Trans. Peninsula Hort. Soc.* 40:1–9.

Scott, D. H., A. D. Draper, and L. W. Greeley. 1972. Interspecific hybridization in octoploid strawberries. *HortScience* 7:382–384.

Scott, D. H., and A. D. Draper. 1967. Light in relation to seed germination of blueberries, strawberries, and *Rubus. HortScience* 2(3):107–108.

———. 1970. A further note on longevity of strawberry seed in cold storage. *HortScience* 5(5):439.

Scott, D. H., and D. P. Ink. 1955. Treatments to hasten the emergence of seedlings of blueberry and strawberry. *Proc. Amer. Soc. Hort. Sci.* 66:237–242.

Scott, D. H., R. J. Knight, and G. F. Waldo. 1962.

Blossom sterility of strawberry seedlings, in relation to other characteristics. *J. Hered.* 53:187–191.

Scott, D. H., G. E. Stembridge, and R. H. Converse. 1962. Breeding studies with *Fragaria* for resistance to red stele root rot (*Phytophthora fragariae* Hickman) in the United States of America. *Int. Hort. Congr.* 16(3):92–98.

Senanayake, Y. D. A., and R. S. Bringhurst. 1967. Origin of *Fragaria* polyploids 1. Cytological Analyses. *Amer. J. Bot.* 54:221–228.

Sherman, W. B., and J. Janick. 1966. Greenhouse evaluation of fruit size and maturity in strawberry. *Proc. Amer. Soc. Hort. Sci.* 89:303–308.

Sherman, W. B., J. Janick, and H. T. Erickson. 1966. Inheritance of fruit size in strawberry. *Proc. Amer. Soc. Hort. Sci.* 89:309–317.

Sistrunk, W. A., and J. N. Moore. 1971. Strawberry quality studies in relation to new variety development. *Ark. Agr. Expt. Sta. Bul.* 761.

Slate, G. L. 1943. A second report on the best parents in strawberry breeding. *Proc. Amer. Soc. Hort. Sci.* 43:175–179.

Spangelo, L. P. S., and A. T. Bolton. 1953. Suggested infection scales for roguing strawberry seedlings susceptible to *Mycosphaerella fragariae* and *Diplocarpon earliana*. *Phytopathology* 43:345–347.

Spangelo, L. P. S., C. S. Hsu, S. O. Fejer, P. R. Bedard, and G. L. Rousselle. 1971*a*. Heritability and genetic variance components for 20 fruit and plant characters in the cultivated strawberry. *Can. J. Gen. Cyt.* 13:443–456.

Spangelo, L. P. S., C. S. Hsu, S. O. Fejer, and R. Watkins. 1971*b*. Inbred line x tester analysis and the potential of inbreeding in strawberry breeding. *Can. J. Gen. Cyt.* 13:460–469.

Staudt, G. 1959. Cytotaxonomy and phylogenetic relationships in the genus *Fragaria*. 9 *Int. Bot. Congr. Proc.* 2:377.

———. 1962. Taxonomic studies in the genus *Fragaria*. Typification of *Fragaria* species known at the time of Linnaeus. *Can. J. Bot.* 40:869–886.

———. 1967. The genetics and evolution of heterosis in the genus *Fragaria* 2. Species hybridization of *F. vesca* x *F. orientalis* and *F. viridis* x *F. orientalis* (in German). *Z. Pflanzenzuchtg.* 58:309–322.

———. 1968. The genetics and evolution of heterosis in the genus *Fragaria* 3. Research with hexaploid and octoploid kinds (in German). *Z. Pflanzenzuchtg.* 59:83–102.

Stembridge, G. E. 1961. A study of the genetic resistance of strawberry plants to several physiologic races of the red stele fungus, *Phytophthora fragariae*, Hickman. Ph.D. dissertation, Univ. of Maryland.

Stembridge, G. E., and D. H. Scott. 1959. Inheritance of resistance of strawberry to the common race of the red stele root rot fungus. *Plant Dis. Rptr.* 43:1091–1094.

Steyn, P. A. L. 1960. Strawberries in South Africa. *Bul. Dept. Agr. Tech. Serv. Union S. Afr.* 372:26.

Suzuki, J., A. Kotani, and H. Shimoda. 1962. Studies on breeding strawberries for resistance to root rot diseases. Okitsau Hort. *Res. Sta. Bul.* 1:74–87.

Thomas, H. E. 1932. Verticillium wilt of strawberries. *Calif. Agr. Expt. Sta. Bul.* 530.

Van Adrichem, M. C. J., and J. E. Bosher. 1962. A search for *Verticillium* resistance in strawberry varieties. *Can. J. Plant Sci.* 42:365–367.

Van Adrichem, M. C. J., and W. R. Orchard. 1958. Verticillium wilt resistance in the progenies of *Fragaria chiloensis* from Chile. *Plant Dis. Rptr.* 42:1391–1393.

Varney, E. H., J. N. Moore, and D. H. Scott. 1959. Field resistance of various strawberry varieties and selections to *Verticillium*. *Plant Dis. Rptr.* 43:567–569.

Waldo, G. F. 1953. Sources of red stele root disease resistance in breeding strawberries in Oregon. *Plant Dis. Rptr.* 37:236–242.

———. 1960. Breeding new strawberry varieties: a progress report. *Proc. Ore. State Hort. Soc.* 75:133–135.

Waldo, G. F., G. M. Darrow, W. F. Jeffers, J. B. Demaree, and E. M. Meader. 1947. Breeding strawberries for resistance to red stele root disease. *Proc. Amer. Soc. Hort. Sci.* 49:219–220.

Watkins, R., and L. P. S. Spangelo. 1971. Genetic components from full, half and quarter diallels for the cultivated strawberry. *Can. J. Gen. Cyt.* 13:515–521.

Watkins, R., L. P. S. Spangelo, and A. T. Bolton. 1970. Genetic variance components in cultivated strawberry. *Can. J. Gen. Cyt.* 12:52–59.

Wilhelm, S. 1955. Verticillium wilt of the strawberry with special reference to resistance. *Phytopathology* 45:387–391.

Wilhelm, S., R. C. Storkan, and J. E. Sagen. 1961. Verticillium wilt of strawberry controlled by fumigation of soil with chloropicrin and chloropicrin-methyl bromide mixtures. *Phytopathology* 51:744–748.

Williams, H. 1955. June yellows: A genetic disease of the strawberry. *J. Gen.* 53:232–243.

Wills, A. B. 1962. Genetical aspects of strawberry June yellows. *Heredity* 17:361–372.

Wilson, W. F., Jr., and M. J. Giamalva. 1954. Days from bloom to harvest of Louisiana strawberries. *Proc. Amer. Soc. Hort. Sci.* 63:201–204.

Yarnell, S. H. 1931a. Genetic and cytological studies on *Fragaria*. *Genetics* 16:422–454.

———. 1931*b*. A study of certain polyploid and aneuploid forms in *Fragaria*. *Genetics* 16:455–489.

Zych, C. C. 1966. Fruit maturation times of strawberry varieties. *Fruit Var Hort. Dig.* 20:51–53.

Brambles

by D. K. Ourecky

To the pomologist, members of the genus *Rubus* (Tourn.) L., which includes raspberries, blackberries, and dewberries, are called brambles. The genus is composed of a highly heterozygous series of species and species-hybrids. Few plant genera are in such confusion as to nomenclature and identity. Between 400 and 500 species have been reported, mainly from cold and temperate regions of the Northern Hemisphere, some being found in the tropical mountainous regions in the Southern Hemisphere and beyond the Arctic Circle and on many oceanic islands (Focke, 1910–1914). Taxonomy has been greatly hindered by inadequate and scanty herbarium specimens from many regions of the world; however, L. H. Bailey preserved an enormous quantity of specimens. Many natural and induced forms have been called hybrids on the basis of guesses as to species' parentage (Darrow, 1937).

Brambles grow in shrubby, rigidly erect, arching, or trailing forms. Most produce biennial canes, but a few species consist of perennial canes. Most are deciduous but a few are evergreen. Reproduction varies from apomictic to sexually fertile. Raspberries (subgenus *Idaeobatus*) are distinguished from blackberries (subgenus *Eubatus*) in that the mature fruit separates from the torus. Ploidy in *Rubus* $x = 7$ ranges from a diploid number of $2n = 14$ to dodecaploid, $2n = 84$. Most raspberries are diploid ($2n = 14$) except for a few triploids (Anonymous, 1940; Einset, 1947; Pratt et al., 1958) and natural (Darrow, 1937) and induced (Hull, personal communication) tetraploids, while blackberries cover a continuous range in ploidy and include numerous species hybrids.

The bramble industry is scattered in many countries around the world. The largest commercial plantings are found in Poland, Scotland, Yugoslavia, in the states of Washington and Oregon in the United States, and in British Columbia, Canada. Estimates given at the *Rubus* symposium in Scotland in 1967 indicated that the world cultivated area devoted to *Rubus* increased from 16,800 ha in 1955 to 21,450 ha in 1966. In the United States the area was over 24,300 ha in 1900 with only 7,700 ha reported in 1965. The greatest reduction has occurred in New York where 5,000 ha were reported in 1900 and today the area has dropped to less than 810 ha. Much of this decline is related to the poor quality of nursery stock as a result of virus problems and high costs associated with hand harvest. British Columbia is considered to be the top raspberry production area in the world. The commercial planting is between 810 ha and 1220 ha, but production averages over 9000 kg per ha, compared to 4500 to 5600 kg per ha in Scotland and 6700 to 7800 kg in Oregon.

The development of indexed virus-free stocks and the development of mechanical harvesters have created a renewed interest in *Rubus* culture.

Raspberries: *Ideobatus*

Origin and Speciation

Eastern Asia is considered to be the center of origin for the *Idaeobati* with 195 species recognized by Focke (1910–1914) and Haskell (1954).

The major species which produce edible fruits include: *R. idaeus* L. var. *vulgatus*, (European red raspberry); the American species *R. occidentalis* L. and *R. leucodermis* Torr. & Gr. (black raspberries), *R. odoratus* L. (thimbleberry), *R. spectabilis* Pursh. (salmonberry), and *R. idaeus* var. *strigosus* (red raspberry); the Asiatic species, *R. phoenicolasius* Maxim. (wineberry), *R. ellipticus* Sm. (Golden Evergreen raspberry), *R. illecebrosus* Focke, *R. kuntzeanus* Hemsl. (Chinese raspberry), *R. nivens* Thumb., *R. coreanus* Miq., and *R. parvifolius* Nutt.; the Hawaiian species *R. macraei* A. Gray and *R. hawaiiensis* A. Gray; and the South American species *R. glaucus* Benth.

The first raspberries were introduced into cultivation in Europe over 450 years ago. The original European raspberry belongs to the genus *R. idaeus* L. var. *vulgatus* Arrhen. Rozanova (1941) recognized five subspecies of *R. idaeus*. No subspecies have been identified in *R. idaeus* L. var. *strigosus* (Michx.) Maxim., the native American red raspberry, because of the wide variability. Present cultivars have been derived from hybridization between these two botanical varieties. Darrow (1937) presented a list of American cultivars of which he considered ten to be from pure *strigosus*. Sixteen were from *vulgatus*, nine were hybrids between the two varieties, and seven were from *R. strigosus* x *R. occidentalis* L. parentage.

The two botanical varieties differ in that *vulgatus* fruits are dark and long-conic with few or no glandular hairs, while *strigosus* fruits are round and light red with numerous glandular hairs. *Strigosus* canes are more hardy, slender, and erect with a reduced number of prickles.

Many investigators consider *R. idaeus* var. *vulgatus* and *R. idaeus* var. *strigosus* to be separate, distinct species; however, hybrids show little or no sterility in the F_1. The F_1 from crosses with *R. occidentalis* are highly variable in fertility, with extreme variability in the F_2 in fertility and plant growth habit. No complete study has been made on the interrelationships of *Idaeobati* species. Rietsema (1955) noted that there was some fertility in a cross between *R. idaeus* and *R. spectabilis*, but backcrossing to *R. idaeus* failed.

Williams et al. (1949) and Williams (1950) found that the F_1 hybrids between *R. occidentalis* and the four species *R. parvifolius*, *biflorus*, Buch.-Ham., *coreanus*, and *kuntzeanus*, were completely sterile, while crosses between *R. idaeus* and the four species were slightly fertile. Rietsema (1955) also found that crosses between *R. idaeus* and *R. illecebrosus*, *lasiostylus* Focke, and *phoenicolasius* were easy to make, but not between *illecebrosus*, *lasiostylus*, and *phoenicolasius*. These results show that *R. idaeus* may be an intermediate species. Gruber et al. (1962) have summarized the relationships between the *Idaeobati* and other subgenera.

The black raspberry (*R. occidentalis* L.) is indigenous only to North America and is only slightly variable taxonomically. Specific status has been assigned to its hybrid derivatives with *R. idaeus-strigosus*. These crosses and their derivatives produce the purple raspberry, so-called *R. neglectus* Peck. However, repeated backcrossing and selection produce hybrids indistinguishable taxonomically from one or the other original parent.

History of Improvement

At the beginning of the Christian era, Pliny wrote of wild raspberries as having come from Mount Ida, a statement which no doubt led Linnaeus to give the plant its botanical name.

The first record of cultivated raspberries dates from around 1548 and is attributed to Turner, an English herbalist, who said "they growe in certayne gardines in Englande." The first extensive writing was a chapter devoted to the raspberry by Parkinson in 1629. Other early accounts of their history are given by Hedrick (1925).

William Prince of Flushing Landing, N.Y., in 1771 was the first to sell raspberry plants commercially in the United States. By 1790 four cultivars were sold, two derived from European species and two from American species. Prince also published many detailed early accounts on raspberries.

Red raspberry cultivars were imported into the United States from Europe in the late eighteenth century because they produced larger fruits with higher quality than the American forms. During much of this same period, many private and commercial growers in the United States were selecting clones from the wild, importing seeds and cultivars, and making crosses among the more promising forms to serve their own particular objectives (Darrow, 1967). Much of this activity was prior to the development of modern concepts of genetics. Unfortunately, the few remaining records of original sources of plants or specific crosses made are mostly inadequate.

A few well-documented records of early breeding work are extant. The efforts of Downing, Brinckle, and Fuller contributed greatly to the early history of raspberry culture in the United States (Hedrick, 1925) and the extensive writings by and about Luther Burbank, the papers of E. Brainerd, A. K. Pieterson, W. O. Focke, and L. H. Bailey, and letters by J. H. Logan and B. M. Young are some of the more prominent. These and many of their less literary-minded contemporaries were excellent plantsmen and produced many cultivars of blackberries and raspberries as well as information on breeding which are standards of excellence even today. Somewhat overlapping these developments in time were the beginnings of state and federally sponsored breeding programs in *Rubus:* Canada began in the 1880s; Minnesota, New York, and South Dakota followed shortly; and Texas, Washington, and the John Innes Institute (England) all developed active programs by the end of the first decade in the 1900s. A precise date has not been established for the beginning of formal programs in the USSR, but one at the Michurin Institute may predate that of Canada.

The comprehensive monograph, *The Small Fruits of New York* (Hedrick, 1925), is a standard reference covering description, performance, and potential value of many species, including over 415 named red raspberries and their hybrids, 94 black raspberries, and 193 blackberries and dewberries. Most of these named cultivars originated as chance seedlings and had limited distribution.

By 1937 many new superior cultivars had been released from breeding programs, exhibiting increased hardiness, fruit size, quality, processability, disease and insect resistance. Many species also were observed to be valuable sources for specific breeding characteristics (Darrow 1937).

Black raspberry breeding has produced a number of superior cultivars which are hardier, larger-fruited, and less susceptible to powdery mildew (*Sphaerotheca humuli* [DC.]) Burr. and leaf spot (*Sphaerulina rubi* Demaree & M. S. Wilcox). The major breeding programs have been in New York, North Carolina, and Maryland. The breeding programs at Maryland and North Carolina have been aimed at producing better cultivars adapted to the South while northern programs have concentrated on hardiness and fruit size. The 'Mysore' black raspberry was a seedling selected at the south Florida station from *R. albescens* Roxb., a subtropical species from Asia. Drain (1956) used both *R. albescens* and *R. crataegifolius* Bunge. in extensive crosses with the black raspberry. Waldo (1934) and Williams (1945) used such Asiatic species as *R. coreanus*, *R. biflorus*, and *R. kuntzeanus* while Slate (Darrow, 1937) used *R. phoenicolasius;* however no commercial cultivars have yet been released from these interspecific hybrids.

Purple raspberries are first generation hybrids between black and red raspberries. Purple raspberries are grown commercially in central New York where they have obtained recognition. The New York State Agricultural Experiment Station has developed the 'Sodus', 'Marion', and 'Clyde' cultivars. Nowhere else in the United States have large plantings of purple raspberries been made. Other cultivars which have been released include 'Purple Autumn' from Illinois, 'Success' from New Hampshire, and 'Amethyst' from Iowa. Many superior purple clones have been selected in the breeding programs aimed at transferring fruit and plant characteristics from reds into black and vice versa.

Darrow (1937, 1967) has summarized developments in *Rubus* breeding, taxonomy, and cytology. Knight and Keep (1958) and Knight et al. (1972) have published extensive world-wide abstract bibliographies on *Rubus*. Information on name, origin, parentage, and most valuable characteristics of cultivars from 1920–70 can be found in *The Register of New Fruits and Nut Varieties* (Brooks and Olmo, 1972), while more recent breeding lists are featured annually in *HortScience*, published by the American Society for Horticultural Science.

Modern Breeding Objectives

Good vigor, high productivity, increased hardiness, large fruit size, high quality, and disease and insect resistance are important objectives in *Rubus* breeding. Most of these characteristics are quantitatively inherited, and may have numerous components, each of which may be polygenic. Productivity, for example, involves flower fertility or ability to set fruit, which in turn is dependent upon hardiness, and photoperiodic and chilling requirements. Drupelet size, number of drupelets per fruit, number of fruits per cluster, number of clusters per plant, ability to develop and mature the crop, and other factors affect productivity.

Climatic adaptation is a continuing problem, especially in peripheral areas of production. Survival at low temperatures is sought in the northern United States (Schwartze, 1935), Canada, and northern Europe (Jennings et al., 1967a), while resistance to leaf diseases and to fluctuating winter temperatures is required in the more southern regions (Hull, 1961; Overcash, 1972; Williams, 1945).

The most important plant characteristics are good vigor, strong primocane (first year cane) formation, and erect sturdy canes which require no trellising.

The major fruit characteristics sought include large fruit size, mainly for fresh market; firmness, for handling and mechanical harvesting; concentrated ripening, for mechanical harvesting or long season of ripening for "pick-your-own" operations; and good quality fruits with small bright colored seeds which are attractive in jams.

Raspberries normally fruit on biennial canes which grew the previous year and in which the terminal and axillary buds became dormant. Fruit buds are initiated in the fall from early September until January in New York, and during the following spring fruiting laterals develop, fruit is produced, and the cane dies. In autumn-fruiting or fall-bearing cultivars, fruit is produced on laterals formed basipetally on the current season's growth, ranging from a few clusters at the tip of the cane to numerous laterals along the full length of the primocane. A spring crop may be obtained from the same cane from laterals which did not grow and fruit the previous fall. The full potential value of this characteristic has not been exploited. The production of a crop on the current season's growth may begin to ripen in Geneva, New York, on August 1 and extend until frost, depending upon the clone. Special clones have been selected which produce a complete canopy of fruiting laterals. This type of growth habit does not require a special trellising system, providing the canes are erect and not arching and can easily be mechanically harvested by an over-the-row harvester. The primocanes are more easily manipulated by the harvester than one-year-old woody canes. Selected clones may ripen fruit over a short period of time or fruit and flowers may be produced from early August until frost. 'Heritage' (Ourecky, 1969), for example, begins to ripen fruit around August 20 in Geneva, New York, and around August 1 in Beltsville, Maryland. During most years all the fruit may be harvested before frost. Late-fruiting selections have been tested in Maryland to extend the season from 'Heritage' to frost. After frost, all cane growth may be removed, thus sacrificing the next year's summer crop. Complete sanitation can be practiced and the need for hardier canes is not required. The rows can be kept at a desired width by cultivation, making it possible to completely mechanize raspberry growing by eliminating removal of old canes and spring topping.

Breeding Techniques

Floral biology: The *Idaeobati* are characterized by having biennial canes which usually require a dormant period before fruiting. Fall-bearing raspberries are exceptions in that fruit is produced basipetally on the current season's primocanes. Jennings (1964*b*) found that fully grown canes of 'Malling Jewel' initiated flower buds without the onset of dormancy when grown for six weeks at 7 C with a nine-hour daylength. In the north temperate zone, flower buds are differentiated in autumn; in early cultivars beginning in mid-September (Haltvick and Struckmeyer, 1965); and in very late cultivars in November. Flower buds differentiate basipetally (Waldo, 1934) within a primocane and flowers basipetally within an inflorescence. Initiation of the torus, sepals, and stamens occurs soon after flower bud formation but further development may be delayed until spring. Meiosis in anthers takes place in April and about four weeks later in the gynoecium.

Many species of *Rubus* produce secondary or tertiary buds, in which differentiation occurs very late. These buds are arranged in descending order below the primary and vary in degree of differentiation depending upon location within the cane. These buds enable a species to recover from frost damage or offer the possibility of selection for high secondary crop potential (Keep, 1969*b*).

Jennings (1964*a*) suggested the presence of population differences with respect to time of budburst and plant height. There are also ecological differences within the same population or clone, with plants from more exposed locations being later to start growth in the spring and producing shorter canes with shorter internodes. In the raspberry, *R. idaeus* L., Crane and Lawrence (1931) observed four flower types: hermaphrodite, female, male, and neuter. Crosses between types indicated that two factors were involved. They further concluded that genetic differentiation of sex in *R. idaeus* was similar to that observed in *R. chamaemorus* L. and suggested that this condition could be one from which alternative chromosome types might originate as the result, not the cause, of sexual differentiation.

Pollination: Anthers may be recovered from emasculated flowers when large amounts of pollen are required or the material is scarce. Pollen may be stored in a desiccator with calcium chloride. With fairly large quantities of anthers the addition of a wide-mouthed shallow dish of fresh concentrated sulfuric acid speeds drying and release. Storage of a sealed desiccator in a refrigerator at 5 C prolongs the pollen viability up to four or more weeks. An incandescent light placed 60 cm above an open petri dish of anthers will produce enough heat to speed drying and release. Shallow seamless metal salve tins are excellent containers for individual collections of pollen, and pollen may be added as the season progresses.

For making crosses out of season, dormant plants are dug up in the fall and placed in cold storage at − 2 C to 1 C. After a rest period of six to ten weeks, depending upon the species, the plants can be removed and potted in 15–30 cm clay pots. A potting medium with good drainage is advised as extreme moisture fluctuations affect *Rubus* growth. Plants are generally placed in a cool greenhouse at 18–21 C for flowering. Plants can be moved to other temperature ranges to delay or speed flower development.

With most cold-stored deciduous species, the time required for greenhouse flowering is 30 to 45 days. Some semi-evergreen species may require a considerably longer period. These may be brought in correspondingly earlier or they may be used as seed parents, utilizing stored pollen from the other earlier-flowering parent. Other means of securing uniform flowering in the greenhouse seem to be relatively ineffective or impractical with *Rubus*. The author prefers to make the majority of crosses in the field.

Jennings (1964*b*) found that fully grown canes of 'Malling Jewel' maintained for six weeks at 7.5 C and nine-hour daylength resulted in initiation of flower buds without the onset of dormancy. Loss of dormancy could also be obtained by treating dormant canes for 30 hours with ethylene chlorhydrin vapor. Both methods have been used to produce out-of-season fruits for experimental purposes.

A razor blade or scalpel cut, ringing the base of a full-sized bud, removes sepals, petals, and stamens together. The *Idaeobati* often shed pollen early and should be emasculated before the separation of the sepals, or else accidental self-pollination may occur. In any case, both razor blade or scalpel and fingertips should be rinsed in alcohol (70–95%) and dried before emasculating a different clone. The lateral branches with the emasculated flowers are best covered with a white paper bag. They are most receptive after two to three days. Two, or preferably three, pollinations at daily intervals are required for best seed production; however, most breeders pollinate only once if flowers are in good condition.

The author prefers to cover three or four flowering laterals of the pollen parent with a white paper bag, first removing all open flowers and developing fruits. White bags are preferred as they allow diffuse light to fall on the enclosed laterals, stimulating flower development. Brown bags exclude light, while plastic bags affect fruit set because of high temperature. When emasculated flowers are receptive, these bags are removed and fresh flowers containing large quantities of pollen are removed by forceps from the bagged male parent, placed in a petri dish, and subsequently used as "brushes" to apply pollen. The bags may be replaced and pollen recollected later.

Salve tins also make excellent pollen applicators. While closed and held upside down, a sharp tap on the edge of a hard surface causes pollen to adhere to the inside of the lid. The lid is touched to the mature stigmas and obvious streaks are apparent when the pollen is transferred. Pollen may also be applied by camel's hair brush, finger, or other means, but larger quantities may be required.

Seed set and germination are strongly influenced by the seed parent (Jennings, 1971*b*, 1971*c*). Thus, it is wise to make reciprocal crosses whenever possible to assure a desired genetic combination.

Seed germination: Mature fruits may be collected and refrigerated until a sufficient quantity has been obtained. A laboratory blender with a variable powerstat is useful for separating the seeds from the pulp. The pulp mixture is poured into a container to which a larger quantity of water has been added; the sound seed settles rapidly and the pulp can be decanted (Morrow et al., 1954).

The cleaned seed is rinsed thoroughly, air-dried, and stored in the refrigerator at 1–5 C until time for planting. If properly dried and stored, seeds of most *Rubus* species will remain viable for several years.

Pregermination treatment is necessary in most cases to assure prompt and uniform sprouting of seeds of most temperate and arctic species. Dormancy is due to three conditions: impermeability of the seed coat to air and water, mechanical resistance of the seed coat to swelling of the embryo, and the requirement of the embryo for certain physiological changes which occur only under proper conditions of air, moisture, and temperature.

The breaking down of the seed coat is most readily achieved by scarification with concentrated sulfuric acid (Scott and Ink, 1957; Jennings and Tulloch, 1965; Heit, 1966). The seed coat must be absolutely dry, or the heat generated by the treatment will kill the embryo. It also is necessary to work with small lots of seed and to stir frequently for the same reason. Plunging beakers into an ice bath during treatment helps to reduce heat generated by the acid. Depending on seed size, treatments should last 15 to 20 minutes (Jennings and Tulloch, 1965; Heit, 1966) for *Idaeobati*. After rapidly flooding with water (to avoid heating) the treated seed is washed ten minutes in running tap water and the remaining acid is neutralized in a saturated water solution of sodium bicarbonate. The seed is then rinsed again and spread out on absorbent paper to surface dry before planting. Drying should be only long enough to facilitate handling, but not long enough to desiccate the embryo within the charred seed coat.

Germination has been increased by pretreatment of seed with 500 ppm gibberellic acid after a chilling requirement and before sowing (Jennings and Tulloch, 1965).

The author has found that seeds from full-sized green fruits germinate as readily as those from ripe fruits after scarification and after-ripening.

Either moist vermiculite or moist, finely shredded sphagnum moss is a satisfactory planting medium. The seeds should not be covered with more than 2–8 mm of the medium. There is evidence that light favorably affects germination of at least some species (Jennings and Tulloch, 1965; Scott and Draper, 1967). The flats or pots of planted seed are held in cold storage for three to four months at 2–5 C for after-ripening. Care must be taken to maintain an adequately moist (but not wet) condition throughout the period or secondary dormancy may result, which is difficult to overcome. When the after-ripening process is complete, the containers are held in the greenhouse at about 21 C for germination and early growth.

Just as the second true leaf begins to develop, seedlings may be "pricked out" into individual pots or containers. The author uses peat pots. It is difficult to transplant very young seedlings as the main root is often injured, while a more fibrous root system on older seedlings makes them easier to establish. A fungicide drench after pricking out is beneficial, often eliminating the damping-off disease complex.

Seedling evaluation: As soon as the cotyledons have expanded, populations segregating for thornlessness may be screened. Thornlessness from certain European species is associated with a total lack of stalked glands on the cotyledons. A hand lens in bright light makes identification relatively simple and saves the labor and space requirements of carrying the total population to the stage of thorn development. First-test seedlings, in a two- to four-leaf stage, may also be inoculated with various disease organisms or infested with aphids in the greenhouse. Resistant or immune seedlings may then be transplanted to the field for selection and evaluation. This method would eliminate the necessity of growing large populations of susceptible seedlings.

Unless the genetics of a particular trait is known, it is difficult to predict the size of population needed to achieve a specific objective. The amounts of space and labor available may be overriding limitations. Recessive characteristics occur much less frequently in polyploids than in diploids and selection of these characteristics requires larger populations. Often large populations are required to combine or separate certain combinations of characters while for some cytogenetic investigations only very few plants may suffice. For most crosses where the breeding behavior of the parents is not fully known, populations of 100 to 200 individuals generally give an idea of the potential value of a given cross. Such populations are often referred to as "first-test" seedlings.

Except for unusual or specific reasons, no data are taken on seedling populations at this station until the year of selection. Hardiness, fruitfulness, and other clonal characteristics are not accurately exhibited by one- or two-year old seedlings because seedlings do not produce as vigorous a growth the first year in the eastern United States as they do in the West. Therefore, selections are made the third year instead of the second year, except for precocious seedlings, or seedlings for cytogenetic studies or unusual hybrids. In the year of selection, records are made on fruit and plant characteristics, and insect and disease ratings.

After preliminary selections have been recorded and labeled, the unwanted seedlings should be destroyed as soon as possible. This serves a general sanitation measure and reduces labor and the cost of spray materials. Selected plants are moved the following dormant season to a semi-permanent nursery for future observation and detailed recording of characteristics. These are called "second-test seedlings." At this stage some breeders propagate a few apparently outstanding selections to avoid accidental loss of a clone and/or to provide for short-row tests as soon as possible. After one or more years of further observation, more selections will be found to warrant propagation for short-row tests. These plantings should be considered separately from the original tests because of age difference.

At this station, suckering-type clones are grown in hedgerows. Each clone is planted to develop a hedge 5–6 m in length with 3 m between rows. For yield data, a 1.3 m sector within each hedge is staked out for picking. (This constitutes a 1/1,000-acre plot and the resulting data are readily transformed into yield per acre). Only three replications are generally used because of labor requirement, but four would be more satisfactory for statistical analysis.

If the clone "passes" these preliminary and secondary trials, it usually is propagated for test by other breeders and testers, most often federal or state agricultural experiment stations in areas where the clone would be expected to succeed. While time and facilities of the cooperator often do not allow for detailed record taking, general comments on adaptability are invaluable to the original breeder in his decision for possible release of the clone as a new cultivar.

Besides screening for horticultural characteristics on second test, selections are usually rated for susceptibility to insects and diseases.

Plant Breeding Systems

Genetic structure: The chromosome morphology of *Rubus* species and secondary pairing at meiosis, in polyploid forms, indicates that the chromosome complements are made up of seven basic chromosomes ($x = 7$). *Rubus* chromosomes range from ca. 1.5 to 2.0 μ in length. Analysis of pachytene chromosomes in *R. parvifolius* (Bammi, 1965) showed that they are highly differentiated; centromeres were easily located by chromatic regions on either side and each chromosome ended with well-defined telochromomeres. A comparison could not be based on absolute length, but was reliable if based on arm ratios and chromatic patterns. In diploid *R. idaeus* and *R. ulmifolius* Schott., two satellite chromosomes were observed (Heslop-Harrison, 1953).

Chromosome numbers of *Rubus* have been published from various research centers of the world (Komarov and Juzepczuk, 1939; Komarov et al., 1941; Einset, 1947; Heslop-Harrison, 1953; Darlington and Wylie, 1955; Britton and Hull, 1957a; Jinno, 1958; Bolkhovskikh et al., 1969).

Rozanova (1939) identified three races of *R. idaeus* as being autotetraploids based on good pollen production and quadrivalent formation at meiosis.

Fertility and sterility: Fertility in raspberry is affected by many factors, including genetic factors, chromosome number, and virus infection. The degree of fertility affects the number of drupelets formed, which in turn determines the size of the fruit and to some extent the coherence between drupelets. "Crumbly fruit" is a term applied to an aggregate fruit which lacks coherence between drupelets because it has a reduced drupelet set and it breaks apart when picked. "Crumbly fruit" has been shown to be a genetic defect due to a mutation. In 'Malling Jewel' and possibly 'Latham', Jennings (1967*b*) observed a mutation of a dominant allele in a heterozygote to the double recessive. This gene apparently is linked with a balanced lethal system. Daubeny et al. (1967) found that a clone of 'Sumner' with crumbly fruit exhibited reduced drupelet numbers due to retarded embryo sac development and the lack of fertile pollen.

In another survey by Daubeny (1969), percentage of drupelet set in self-pollinated clones varied from 11% to 75% compared to 84% to 94% when open-pollinated. In one of the selections the total number of apparently normal pollen grains was low, indicating partial male sterility as a possible cause of low self fertility. In various clones a reduction of self fertility usually followed self or open pollination; reductions in set following self pollination mostly seemed to involve self incompatibility, while reductions in open pollination involved reduced numbers of functional embryo sacs (Daubeny, 1971).

Sterility may be associated with virus infection in some cultivars. Daubeny et al. (1970) studied the effect of four viruses on drupelet set in four red raspberry cultivars. They found a significant reduction in drupelet set, expressed as crumbliness, only on those plants which showed relatively severe symptoms of decline from tomato ring spot virus infection. 'Newburgh' and possibly 'Latham' appeared to be less tolerant than 'Willamette' but more tolerant than 'Fairview' and 'Sumner'.

A triploid raspberry clone (Pratt et al., 1958) had reduced and unreduced embryo sacs. Upon crossing this clone with a 2x cultivar, variable percentages of triploid seedlings were obtained. This was explained in that the triploids were derived parthenogenetically from 2x eggs or by fertilization of diploid eggs by haploid pollen. In another population of 3x x 4x, the majority of the seedlings were tetraploid. This is an unusual case of apomixis in *Idaeobatus*, since most of the species are diploid.

Fertility in crosses involving diploid and autotetraploid raspberries were studied in relation to embryo, fruit, and seed development by Topham (1967*a*, 1967*b*, 1970*a*, 1970*b*). Chromosome doubling was found to reduce fertility of both male and female gametes, especially in the pollen. Seed size and embryo development in crosses between plants of different ploidy depend on a balance between early growth following the stimulus of fertilization and later slower growth due to a weakness in the endosperm. Both diploid and tetraploid ovules appeared to be more stimulated by pollen from tetraploid clones than from diploid clones, while embryo differentiation differed according to the direction of the cross. Radicles tended to be abnormally large when diploid pollen was used and smaller than usual with haploid pollen.

Jennings and Topham (1971) found that artificial pollen dilution reduced set and rate of fruit ripening. Observations on numbers of pollen grains germinating in the styles revealed maternal and paternal interactions. Genetic factors have also been observed to affect fruit development. In one of two diallel crosses, Jennings (1971*a*) observed that seed set was influenced by interactions between pollen and maternal effects, the interactions being reduced with auxin application. Self pollination, however, greatly improved seed set. In the second diallel, maternal effects had the most influence on seed set. The maternal parents differed in rate of fruit ripening but the pollen parent had some effect on fruit ripening. The results also indicated that the factors determining the production of small empty seeds differ from those determining the set of large ones. The crosses also showed that the size and shape of different parts of the seeds were influenced by maternal effects and interaction between maternal and genetic factors (Jennings, 1971*b*).

Self incompatibility in 11 of 23 *Rubus* species was due to stylar inhibition of self pollen tubes (Keep, 1968*b*). Crosses with a wild *R. idaeus* selection demonstrated the presence of S_1 and S_2 incompatibility alleles. However, nearly all raspberry cultivars are self fertile.

For many years, *Rubus* breeders have used, as a general rule, a dark-fruited clone as the seed parent and a lighter-fruited clone as the pollen parent in crosses involving red, black, and purple raspberries or blackberries in order to secure seed set. Zych (1965) substantiated this practice in controlled greenhouse crosses where he found that red x black and purple x black combinations were completely incompatible. Red x purple, purple x purple, and purple selfed produced fairly normal fruit but with enough irregularity to indicate some incompatibility. All reciprocal combinations were fully compatible. In the incompatible crosses, the pollen tubes did not grow beyond the upper portion of the style. No genetic explanation has been formulated. In contrast, Lawrence (1966) observed that the red raspberry could be used as a seed parent in some crosses with purples and some purples could be used as seed parents in crosses with the black raspberry.

Inbreeding, outbreeding, and backcrossing: Inbreeding, outcrossing, and backcrossing are the major breeding systems used by *Rubus* breeders. Test crosses for the determination of lines with general or specific combining ability have been made over the years. 'Lloyd George', for example, has shown good combining ability and therefore appears in the lineage of most presently grown cultivars. Oydvin (1970) summarized the important parents and lines of 129 North American and European cultivars. He observed that 32% had had 'Lloyd George' as one parent, 37% were less closely related, and 31% were unrelated. The 'Preussen' cultivar has not been used in breeding in North America but ranks with 'Lloyd George' as a parent among West European cultivars.

Inbreeding does not change gene frequency in a population, but brings about homozygosity so that recessives are expressed and selection pressure may be applied. Since many of the desirable quantitative characters such as large fruit size, high flavor, and good texture appear to be inherited as recessive gene complexes, a certain amount of inbreeding may be desirable. In contrast, many of the major genes for desirable plant and fruit characteristics, including insect and disease resistance, are dominant and do not have to be homozygous in each parent line.

Jennings (1963) classified raspberry cultivars into four groups based on the extent of outcrossing in their background or amount of inbreeding which they will tolerate. These four groups, in order of decreasing inbreeding tolerance, are 1) outstanding chance seedlings from the wild; 2) clones which arose from the crossing between a seedling or cultivar and the unrelated local population; 3) clones derived from controlled crosses between cultivars or seedlings of different origin within the variety of *R. idaeus;* 4) clones derived by controlled crosses between cultivars or other species of *Rubus*. He suggested that upon incorporating major genes into breeding lines, while retaining and stabilizing the good characteristics, the best success would be more easily obtained in clones which give a good phenotypic performance following inbreeding. It was concluded, however, that F_1 hybrids would be easier to produce and may have the advantage of possessing greater developmental stability. Jennings (1967a) further suggested that the presence of lethal genes may help raspberry populations to maintain heterozygosity for particular chromosome segments and to avoid inbreeding depression.

Rietsema (1946) found that his inbred red raspberry lines, after several generations, became very weak. Most breeders also find a considerable amount of variation upon inbreeding. Some lines are extremely weak and unproductive while others are little affected. Inbred lines of the black raspberry are generally uniform and vigorous showing little inbreeding depression.

Diallel crosses between nine cultivars of red raspberry (Ure, 1963) were studied to determine general and specific combining ability with respect to vigor and winterhardiness. Progenies with low mean vigor were found to have common parentages. Despite a high degree of heterozygosity, results indicated that vigor was quantitatively inherited, while certain parents transmitted a high degree of hardiness to progenies.

Backcrossing and modified systems have been used to transfer a few characteristics from one parent into a desirable genotype. Keep and Knight (1968) have used several species as sources for specific genes. These characteristics include resistance to the virus vector, *Amphorophora rubi* Hottes, cane spot, spinelessness, cane pubescence, autumn fruiting, dwarfing habit, flavor, and firmness of fruit. The resulting synthesized parents are then used as intermediates in crossing with advanced promising material for production of commercial cultivars.

Interspecific hybridization: Although *R. idaeus* var. *vulgatus* and *R. idaeus* var. *strigosus* have been regarded by many breeders as separate and distinct species, hybrids between them show little or no sterility in F_1 and subsequent generations; thus, any differences may be considered as sub-specific or varietal.

Crosses between *R. idaeus* and *R. occidentalis* show a wide range of sterility in certain combinations, indicating that both are distinct species. In the wild, *R. idaeus* var. *strigosus* and *R. occidentalis* have assimilated genes from each other by introgressive hybridization. Controlled crosses between the red and black raspberries have produced many purple cultivars such as 'Sodus', 'Marion', 'Clyde', and 'Amethyst'.

Hybrids between *R. idaeus* and *R. spectabilis* are easy to make and show some fertility, although backcrossing to the red raspberry fails (Rietsema, 1955). Crosses between *R. occidentalis* and four Asiatic species, *R. parvifolius, R. biflorus, R. coreanus,* and *R. kuntzeanus* all gave completely sterile F_1 plants; however, the same four species crossed with *R. idaeus* gave some fertility in the F_1 and could be successfully backcrossed (Williams, 1950). Rietsema (1955) found that, in general, crosses were easy to make between *R. idaeus* and

R. illecebrosus, *R. lasiostylus*, and *R. phoenicolasius*. Subsequent backcrosses to *R. idaeus* were successful with *R. phoenicolasius* but not with *R. illecebrosus*. These results suggest that *R. idaeus* is intermediate in status between the Asiatic group and *R. occidentalis*. Keep (1958) found that the F_1 plants from *R. palmatus* Thunb. (2x) x *R. idaeus* (2x) were almost completely sterile and that upon backcrossing partially fertile F_1 plants to *R. idaeus*, triploid plants were obtained as a result of fertilization of 2x ovules with haploid *R. idaeus* pollen.

Black raspberry cultivars are relatively homozygous resulting from crosses between similar clones. *R. albescens* and *R. crataegifolius* have been crossed with named black raspberry cultivars to develop clones adapted to southern areas (Drain, 1956). F_1 hybrids between *R. occidentalis* and *R. leucodermis* (Ourecky and Slate, 1966) are extremely vigorous and productive. Soft, dull fruit and susceptibility to winter injury are dominant characteristics from *R. leucodermis*. Repeated backcrossing has not produced superior clones. Keep and Knight (1968) found *R. occidentalis* to be an outstanding source of fruit firmness and quality, source of resistance to strains 1, 2, and 3 of *Amphorophora rubi*, and possessing a tip-layering method of propagation. *R. kuntzeanus*, *R. biflorus*, and *R. parvifolius* are being used as sources of resistance to leaf spot (*Mycosphaerella confusa*) and fluctuating winter temperatures which damage canes (Hull, 1961). *R. parvifolius* has proved to be the most promising parent in breeding red raspberries adapted to southern regions (Hull, 1961; Overcash, 1972).

R. cockburnianus Hemsl. has been used in breeding for increased yield and high flower number per lateral and *R. crataegifolius* for erect sturdy canes (Keep and Knight, 1968).

Records of hybridization between *Idaeobati* and subgenera other than *Eubati* are relatively rare. Darrow (1955*b*) has presented a lengthy discussion and review on crossing blackberries with raspberries, while Einset and Pratt (1954) made numerous crosses at different levels of ploidy.

Polyploidy and mutation breeding: Polyploidy has played only a minor role in *Idaeobati*, being confined, when it does occur, to autopolyploidy. In contrast, allopolyploidy has proved to be an important evolutionary mechanism in *Eubati*, where autopolyploidy is relatively rare. Hybrids between diploid *Eubati* and *R. idaeus* are nearly sterile but allotetraploids derived from such F_1 hybrids show complete autosyndesis and are fertile. Genome differentiation between *Idaeobati* and *Eubati* appear to have arisen by continuous gene mutation affecting processes such as chromosome pairing, rather than by major structural changes in the chromosomes (Vaarama, 1953).

'LaFrance' and 'Erskine Park' are two old fall-bearing cultivars which are triploid, but the fall-bearing characteristic is not a direct expression of polyploidy. Today all commercial fall-bearing cultivars are diploid.

Chromosome doubling in *R. idaeus* reduces pollen and ovule fertility with a more pronounced effect on the pollen in which the ability to induce set is impaired (Topham, 1967*b*).

Polyploidy has been induced in a number of red raspberry clones for the specific purpose of using the synthetic clone as an intermediate in breeding. Hull (1961) found that *R. parvifolius* proved promising in breeding, but 2x F_1 and F_2 progenies from crosses with diploid red raspberries did not possess adequate resistance to leaf-spot (caused by *Mycosphaerella confusa* Wolf), and were only partially fertile. Crossing colchicine-induced tetraploid *R. parvifolius* (2n=28) with 4x raspberry clones provided a means of transferring desirable traits.

Mutation breeding seems to be especially useful in changing single, simply inherited characteristics in highly developed genic systems. Many plant and fruit characteristics, including insect and disease resistance and linkage relationships, have yet to be determined in *Rubus;* therefore, much remains to be accomplished through breeding before the induction, identification, and isolation of mutants add to the improvement of present cultivars. Several populations of seed irradiated by the author yielded mutant types but no proof of origin could be made because of the heterozygosity within the progeny.

Few documented reports have been made on spontaneous bud sports in raspberries. 'Cuthbert', a red raspberry, originated as a chance seedling in 1865. 'Golden Queen', an amber-fruited cultivar, originated as a bud sport from 'Cuthbert'. Jennings (1966a) obtained a mutant of 'Malling Jewel' which had larger fruiting laterals, flowers, and fruits. The mutation was identified as a single dominant gene and has been useful in breeding.

Breeding for Plant Specific Characters

The identification of specific genes and mode of inheritance is essential to any efficient breeding program. Most of the studies on inheritance have been carried out by the East Malling Research Station and the Scottish Horticultural Research Institute. Table 1 summarizes the genes which have been identified for plant, leaf, flower, and fruit characteristics and for disease and pest resistance, along with the species source and research reference. Table 2 summarizes the linkage relationships known to date while Table 3 is a general list of current cultivars with specific characteristics. Table 3 refers to red raspberry except where noted.

TABLE 1. List of genes identified in *Rubus*, subgenus *Idaeobatus* (Adapted from Keep et al., 1970, and Knight et al., 1972)

Gene Symbol	Original Symbol if Different	Gene Effect	*Rubus* Species	Reference
		PLANT		
B		Waxy bloom on stem	*idaeus*	Keep, 1964, 1968*a*
Bd_1Bd_2		Accessory buds	*idaeus*	Keep, 1968*c*
C		Growth pigmented	*idaeus*	Crane and Lawrence, 1931
d_1d_2		Sturdy dwarf	*idaeus*	Jennings, 1967*a*; Keep, 1969*a*
d_3d_4		Crumpled dwarf	*idaeus*	Keep, 1969*a*
d_5	fr	Frilly dwarf	*idaeus*	Knight et al., 1959
H		Pubescent cane	*idaeus*	Crane and Lawrence, 1931; Knight, 1962
no		Dwarf; probably identical with d_1d_2	*idaeus*	Rietsema, 1939
s		Spineless cane, eglandular cotyledons	*idaeus*	Knight and Keep, 1964
sk_1		Suckering	*idaeus*	Knight and Keep, 1960
sk_2sk_a		When homozygous, epistatic to Sk_1sk_1	*idaeus*	Knight and Keep, 1960
Tr		Tip-rooting	*occidentalis*	Knight and Keep, 1960
w		Pollen tube inhibitor	*idaeus*	Oydvin, 1968
wh		Lethal affecting H:h segregation	*idaeus*	Jennings, 1967*a*
ws		Lethal affecting S:s segregation	*idaeus*	Jennings, 1967*a*
wt		Lethal affecting T:t segregation	*idaeus*	Jennings, 1967*a*
x		Red hypocotyl	*idaeus*	Lewis, 1939
ch_1	g	Pale green leaf	*idaeus*	Lewis, 1939
sl		Simple leaf	*idaeus*	Jennings, 1967*a*
S^1S^2	S_1S_2	Incompatibility	*idaeus*	Keep, 1968*b*
S^f	S_f	Self-fertility (no pollen reaction)	*idaeus*	Keep, 1968*b*
sx_1	f	Female-sterile, obtuse foliage	*idaeus*	Crane and Lawrence, 1931
sx_2	m	Male-sterile, normal leaf	*idaeus*	Crane and Lawrence, 1931
sx_3	d	Sepaloid	*idaeus*	Keep, 1964
		FRUIT		
Bl		Fruit color: BlBlT-black; BlBltt-apricot; BlblT-purple	*occidentalis*	Britton et al., 1959
L_1		Large fruit and calyx	*idaeus*	Jennings, 1966*a*
l_2		Miniature fruit	*idaeus*	Jennings, 1966*b*
P		pT fruit red, spines tinged; PT fruit red, spines dark to base	*idaeus*	Crane and Lawrence, 1931; Oydvin, 1968
t		Fruit color: pt fruit yellow, spines green; Pt fruit apricot, spines green	*idaeus*	Britton et al., 1959; Jennings, 1963, 1967*a*
		DISEASE AND PEST RESISTANCE		
AB		Resistance to *A. rubi*; probably identical with A_{K4a}	*idaeus*	Baumeister, 1961
A_1		Resistance to *A. rubi* strains 1 and 3	*idaeus*	Knight et al., 1960
A_2		Resistance to *A. rubi* strain 2	*idaeus*	Knight et al., 1960
A_1A_3		Resistance to *A. rubi* strain 1	*idaeus*	Knight et al., 1960

Gene Symbol	Original Symbol if Different	Gene Effect	*Rubus* Species	Reference
A_3A_4		Resistance to *A. rubi* strain 2	*idaeus*	Knight et al., 1960
A_5		Resistance to *A rubi* strain 1	*idaeus*	Knight et al., 1960
A_6		Resistance to *A rubi* strain 1	*idaeus*	Knight et al., 1960
A_7		Resistance to *A rubi* strain 1	*idaeus*	Knight et al., 1960
A_8		Resistance to *A. rubi* strains 1, 2, 3, 4	*idaeus*	Knight, 1962
A_9		Resistance to *A. rubi* strains 1, 2, 3, 4	*idaeus*	Knight, 1962
A_{10}		Resistance to *A. rubi* strains 1, 2, 3, 4	*occidentalis*	Knight et al., 1972
A_{L503}		Resistance to *A. rubi* strains 1, 2, 3, 4; possibly identical with A_{10}	*occidentalis*	Knight et al., 1972
A_{K4a}	?AB	Resistance to *A. rubi* strains1, 2, 3, 4	*idaeus*	Knight et al., 1972
$A_{cor.1}$		Resistance to *A. rubi* strains1, 2, 3, 4	*coreanus*	Knight et al., 1972
$A_{cor.2}$		Resistance to *A. rubi* strain 2	*coreanus*	Knight et al., 1972
Ag_1		Resistance to *A. agathonica*	*idaeus*	Daubeny, 1966
b		Lack of bloom on cane confers *Elsinoe* avoidance	*idaeus*	Jennings, 1962*b*
H		Cane pubescence confers *Botrytis* and *Didymella* avoidance	*idaeus*	Knight, 1962
Iam		Immunity to *arabis* mosaic virus	*idaeus*	Jennings, 1964c
Irr		Immunity to raspberry ring spot virus; common Scottish strain	*idaeus*	Jennings, 1964c
I_{tb}		Immunity to tomato black ring virus	*idaeus*	Jennings, 1964c
s		Spinelessness confers *Elsinoe* avoidance	*idaeus*	Jennings, 1962*b*
Sp_1Sp_2		Resistance to *Sphaerotheca macularis*	*idaeus*	Keep, 1968a
sp_3		Resistance to *S. macularis*	*idaeus*	Keep, 1968a

Adaptation: Most *Rubus* species are indigenous to the cooler temperate regions of the Northern Hemisphere, with few species found in the tropical and subtropical regions. 'Southland' (Hull, 1961) and 'Dormanred' (Overcash, 1972) are two new red raspberry cultivars which produce well farther south than most northern red raspberry cultivars. Both cultivars were derived in part from the Asiatic species, *R. parvifolius*. The possibility of transgressive segregation should not be overlooked in breeding, as neither *R. idaeus* var. *strigosus* nor *R. parvifolius* is fully winter-hardy in southern regions. Many progenies from crosses and backcrosses between *R. idaeus* and *R. biflorus, R. coreanus, R. kuntzeanus*, and *R. parvifolius* are vigorous and tolerant of high summer temperatures and drought (Hull, 1961).

Breeding for hardiness in raspberries should include, in addition to a high degree of cold resistance, such factors as 1) the ability to harden rapidly; 2) longer rest or deeper dormancy in order to avoid out-of-season bud activity, and 3) ability to reharden if initial cold resistance is lost (Brierley and Landon, 1946). Most seedling populations are screened by subjection to natural field temperatures. Specially designed cold chambers have not yet been fully tested for screening of seedling populations. Attempts have been made to use electrical impedance (Craig et al., 1970) to measure hardiness but no correlations were found. Van Adrichem (1970) observed that cane elongation and leaf drop were associated with winterhardiness. Jennings et al. (1967*b*) found that hardiness appeared to be inherited as a dominant character.

In second-test and trial plantings, breeders have noted selection and cultivar responses to soil type and moisture content.

Plant habit: Vigorous, erect, sturdy canes are essential in new cultivars. This type of growth habit does not require a trellis and can be easily harvested mechanically. Dwarfing (Keep, 1968a) is a potential source of erect sturdy cane growth habit, although large mechanical harvesters may require a taller growth habit.

Two components of yield are large fruit size and number of fruits per lateral. Keep and Knight (1968) have been using *R. cockburnianus* as a source of high flower number per lateral. Yield may also be increased by using such sources as 'Glen Clova' which have a high frequency of nodes bearing more than one fruiting lateral.

Keep (1969*b*) found accessory axillary buds in 21 out of 32 *Rubus* species, including wild *R.*

TABLE 2. Gene linkages in *Rubus*, subgenus *Idaeobatus*
(Adapted from Knight, Parker, and Keep, 1972)

Linked genes, cross-over values determined	Reference
B —11[z]— Sx_3 —17— T —15— Ch_1 ——— X ——— Sp_3 ——— W	
28	Keep, 1968a
23	Crane & Lawrence, 1931
20	Crane & Lawrence, 1931
32	Crane & Lawrence, 1931
36	Crane & Lawrence, 1931
25	Knight et al., 1959
A_1 ——— D_5	

Linked genes, no cross-over values determined	Reference
A_3 or 4_4-A_5	Knight et al., 1960
H-mildew resistance	Keep, 1968a
H-dwarf	Jennings, 1967a
H-wh	Jennings, 1967a
Irr-Iam-I_{tb}	Jennings, 1964c
S-T	Jennings, 1967a
S-ws	Jennings, 1967a
T-dwarf	Jennings, 1967a
T-wt	Jennings, 1967a

[z] Cross-over values

idaeus. In disbudding experiments, secondary laterals showed various degrees of flowering, suggesting the possibility of selecting for a high secondary crop potential. In the red raspberry cultivar, 'Lloyd George', accessory bud formation is controlled by two complementary genes, Bd_1 and Bd_2. The expression is influenced by homozygosity, minor genes, and environment (Keep, 1968c).

Fall bearing: The full potential of the fall-bearing characteristic has not been investigated. Complete mechanization is possible with this characteristic thus eliminating removal of old canes and topping. An extensive breeding program for fall bearing has been underway at the Geneva station since 1934 (Slate, 1940; Slate et al., 1966; Slate and Watson, 1964).

Haskell (1960*b*) observed 3:1 ratios in open-pollinated populations indicating that the fall-bearing habit was controlled by a single recessive gene; however, Keep (1961) assumed that autumn fruiting was under complex genetic control and influenced largely by temperature and day length. Slate (1940) also observed that the environment had a direct effect on the expression of the fall- or autumn-fruiting character. In a temperate climate the fall crop must mature early. Many of the old fall-bearing cultivars were triploid and tetraploid and were of little value because of the low proportion of viable pollen they produced. 'Lloyd George', a *R. idaeus* source, was used extensively in breeding for fall-bearing types because of its large fruit, but it was very late ripening. 'Ranere', a *R. idaeus* var. *strigosus* source, was used as a source of earliness despite the small fruit size.

Earliness was obtained from a wild selection of *R. idaeus* var. *strigosus* (Slate and Watson, 1964); however, small fruit size was characteristic. Selections were obtained which ripened on August 15, three weeks earlier than most clones (Slate et al.,

1966; Ourecky, 1969). The bearing surface or degree of branching was increased from a few fruits at the tip of the cane to numerous lateral branches produced basipetally over 60 cm or more of the cane.

Insect resistance: The most serious disease problem in raspberries is caused by viruses. Once a plant is infected it will not recover and consequently all future propagules will be diseased. Virus-free tip cuttings may be obtained by special heat treatment techniques. Today, most states have virus-free raspberry programs. Raspberry viruses are spread by certain aphids. *Amphorophora rubicola* Oestl. is of some importance in spreading mosaic in some areas. The raspberry cane aphid, *Amphorophora sensorita* Mason, occurs in large colonies on the lower portions of the stems of black and purple raspberries but it does not readily transmit mosaic or other known viruses (Bennett, 1932).

Various sources of resistance (Baumeister, 1961) have been found to *Amphorophora rubi* Kltb., the European raspberry virus vector, with the identification of numerous dominant genes (Knight et al., 1960; Keep et al., 1970). The cultivar 'Chief' was found to have three dominant genes, A_5, A_6, and A_7, each conferring strong resistance to strain 1 of *A. rubi*. It also carries A_2, A_3, and A_4 for resistance to strain 2. A_2 is dominant, conferring full resistance by itself, while A_3 and A_4 are dominant complementaries and A_5 is linked with A_3 or A_4 with a cross-over value of 10%. Gene A_1, from 'Baumforth A' (Knight et al., 1960), conferred resistance to strains 1 and 3 of *A. rubi* but when combined with A_3 gave resistance to strain 2. Therefore, the combinations of A_1A_2 or A_1A_3 may may be used to control three strains of *A. rubi* in Great Britain.

Other sources of resistance to various strains of *A. rubi* include 'Klon 4a' described by Baumeister (1961) which appears to carry a single dominant gene for resistance to four strains (Keep et al., 1970; genes A_8 and A_9 from *R. strigosus* (Knight, 1962) which confer resistance to strains 2, 3, and 4; *R. occidentalis* clone 'L 503', resistance in which is controlled by a single dominant gene probably identical with A_{10}; and *R. coreanus* clone 'L 646', which carries two genes, one giving resistance to strains 2, 3, and 4 and the other to strain 2 (Keep et al., 1970). The relationship between ten major genes for resistance and four aphid strains are summarized by Keep et al. (1970).

The North American strain of the raspberry aphid has been designated as *Amphorophora agathonica* Hottes. Daubeny and Stace-Smith (1963) and Daubeny (1972) have observed a number of selections and cultivars which are immune to *A. agathonica*. Daubeny (1972) also described immune selections in which immunity was derived from sources distinct from 'Lloyd George'; namely, 'Pyne's Royal' and 'Burnetholm', as well as the tetraploid 'La France'.

In controlled crosses, Daubeny (1966) found immunity was determined by a single dominant gene. Several populations suggested that two gene pairs were present in certain parents. He also noted some divergent ratios suggesting linkage with a semi-lethal gene.

Kennedy et al. (1973) screened 22 cultivars and three selections for resistance to *A. agathonica* and *A. rubicola*. 'Canby', 'Lloyd George', and 'New York 632' were immune to *A. agathonica*; 'Washington' and 'Malling Exploit' demonstrated a high level of resistance. No immunity to *A. rubicola* was observed but several cultivars did exhibit some tolerance.

Disease resistance: Jennings (1962*b*) found that the presence or absence of prickles, hairs, wax, and pigmentation on canes influenced the disease incidence of several fungus diseases. *Didymella applanata* (Niessl.) Sacc. (spur blight) was less frequent on spine-free, wax-free, and nonpigmented canes. *Botrytis cinerea* Pers. ex Fr. was similar but slightly greater on wax-free canes.

'Lloyd George' and 'Burnetholm' are examples of two cultivars which are heterozygous for three genes governing resistance to *Sphaerotheca humuli* (DC.) Burr. (powdery mildew). Sp_1 and Sp_2 are dominant complementaries and sp_3 a recessive. Resistance has been shown to be epistatic to susceptibility whether of Sp_1, Sp_2, or sp_3sp_3 origin (Keep, 1968a). Sp_3 is linked with the gene T for fruit color with a cross-over value of 25%. The gene order in this linkage group has been designated as B-Sx_3-T-Sp_3. Evidence was also presented indicating linkage with H (hairy canes).

Daubeny et al. (1968) discuss various methods which provide a more complete assessment of parental resistance. An analysis based on the model of discontinuous variation supported the hypothesis that segregation was controlled by two additive genes for resistance and one epistatic for susceptibility, while the model for continuous variation indicated that inheritance was additive with significant genetic interactions.

Inheritance of immunity to raspberry ringspot, arabis mosaic, and tomato black ring virus following graft inoculation of raspberry seedlings was studied by Jennings (1964*c*). Immunity was dominant to susceptibility to each virus but more than one gene was involved in each case. It was not possible to determine whether the second gene was dominant complementary or linked recessive affecting viability of immune segregates because of aberrant ratios and possible lethal genes.

Breeding for Specific Fruit Characteristics

Size: Large fruit size is important where hand picking operations are involved, but less important for mechanical harvesting where yield is foremost. Fruit size, as measured in grams, may vary from less than 2 g in *R. idaeus* var. *strigosus* to more than 8 g in hybrid selections. In many breeding programs, large fruit size was originally obtained from 'Lloyd George'. Toyama (1961) found that fruit weight, drupelet number, drupelet weight, seed weight, percentage of light seed, and percentage of seediness were quantitatively inherited. Correlation results indicated that small fruit, large seeds, and a higher percentage of light seeds were partially dominant. As fruit size increased, seed weight tended to increase and the percentage of seediness decreased. Intensive efforts to breed for small seeds and lower percentages of seediness were not considered worthwhile.

A mutant of 'Malling Jewel' has been obtained (Jennings, 1966*a*) with larger fruiting laterals, flowers, and fruits. Breeding results indicated that this characteristic is due to a single gene mutation. The dominant mutant L_1 slightly depresses the amount of reproductive second year's growth. 'Norfolk Giant' and 'Baumforth B' are heterozygous for the recessive gene l_2 which causes a reduction in vegetative growth and the development of miniature fruits or blind-fruiting laterals (Jennings, 1966*b*).

Heterosis may influence fruit size (Haskell, 1960*a*); however, large-fruited inbred lines have been obtained. Haskell (1960*a*) also noted that there was no correlation between fruit size or torus shape and ease of picking.

Firmness: Outstanding fruit firmness has been transferred from *R. occidentalis* to several red raspberry selections (Keep and Knight, 1968). At the Geneva station, 'Newburgh' has been a good source of firmness along with 'Heritage'. Most of the fruit firmness in 'Heritage' is attributed to the small, compact, coherent drupelets (Ourecky, 1969). Nybom (1962) substantiated the firmness of 'Newburgh' by means of transmitted vibrations. He listed 24 raspberry cultivars in descending order of firmness.

In crosses between *R. occidentalis* and *R. leucodermis* (Ourecky and Slate, 1966) soft fruit was found to be dominant to firm fruit; however, no data were presented on mode of inheritance.

Haskell (1960*a*) found that texture or firmness depended upon the ability of drupelets to adhere and not collapse under slight pressure.

Fruit rot: Daubeny and Pepin (1969) evaluated 27 red raspberry cultivars for relative susceptibility to fruit rot. The percentage of rotted fruit was observed 72 and 96 hours after harvest at temperatures between 21 and 26 C. The least susceptibile cultivars were 'Carnival' and 'Matsqui', while the most susceptible were 'Fairview', 'Sumner', 'Latham', and 'Malling Jewel'. 'Willamette' showed intermediate susceptibility. Barritt (1971) confirmed the low susceptibility of 'Matsqui' and also found 'Meeker' and 'Cuthbert' to show low susceptibility. Resistance to fruit rot may not exist.

Fruit color: Various interpretations have been made of the inheritance of fruit color. Crane and Lawrence (1931) observed that red- and tinged-spined raspberries had red fruit, while green-spined forms have yellow fruits. It was postulated that the genotype *PT* gave red spines and red fruits, *pT* tinged spines and red fruits, with *Ptt* and *ptt* giving green spines and yellow fruits. *P* was designated as intensifying the color. The spine color of *P* and *pt* is green but *Ptt* fruits had a very faint tinge of red when fully mature while *pptt* fruits were clear yellow.

In the results of crosses between apricot-fruited black raspberries and red raspberries, Britton et al. (1959) postulated that the lack of development of red pigment was controlled by the gene *t* in both red and blacks. This gene was epistatic to the gene or genes which controlled the formation of the black pigments in black raspberries. They set forth the following genotypes: red raspberry, *bl bl T–*; apricot-fruited red raspberry, *bl bl tt*; black raspberry, *Bl BL T–*; apricot-fruited black raspberry, *Bl Bl tt*; purple raspberry, *Bl bl T*–; and apricot-fruited purple raspberry, *Bl bl tt*.

A genetic and biochemical study on fruit color (Lawrence, 1966) indicated that the inheritance of fruit color was quantitative and under the control of a number of factors.

Processing quality: The chemical composition, freezing, and canning suitability are important characteristics, especially with the advent of mechanical harvesting. Most raspberry breeders submit samples of fresh and frozen fruit to taste panel analysis. Color, texture, and flavor are scored organoleptically on a 1 to 10 or 1 to 5 scale. In studying flavor, Haskell (1960*a*) used five classes. He found that upon backcrossing to a well-flavored clone, the frequency of well-flavored seedlings increased and that of acid types decreased. Segregation for all classes occurred in sib and self populations. Bright red-fleshed fruit is preferred because light-colored fruit turns slightly pink upon freezing while dark-colored fruit is unattractive.

Additional characteristics: The cultivar 'Cuthbert' has been considered as a standard of quality in the United States. Readily available cultivars with plant and fruit characteristics useful in breeding are listed in Table 3. Many breeders have a large collection of genetic material which transmit certain characteristics in breeding even though the exact mode of inheritance is not known. Information on old cultivars and characteristics may be found in numerous references (Darrow, 1937; Hedrick, 1925; Schwartze, 1934).

In black raspberry breeding, lack of genetic variability is a limiting factor. If breeding is done within the species *R. occidentalis*, no major advances can be made in fruit size, production, and disease resistance. It is nearly impossible to separate the present cultivars based on plant and fruit characteristics. Drain (1952) has presented detailed data on the inheritance of fruit and seed characteristics in crosses involving *R. occidentalis* cultivars. Cold injury was found to be less severe on seedlings of *R. albescens* and *R. crataegifolius* than on the black raspberry cultivars 'Bristol' and 'Cumberland' (Drain, 1956). In *R. occidentalis* x *R. albescens* or *R. occidentalis* x *R. crataegifolius* crosses, seedlings were found to have larger seeds, heavier bloom, less juice, greater acidity, and greater firmness than clones of *R. occidentalis*. Future progress in black raspberry breeding can only be made by utilizing other species.

Achievements and Prospects

The raspberry breeding programs throughout the United States have contributed greatly to the production of cultivars which are vigorous, productive, hardy, and adapted to specific soil and climatic regions (Darrow, 1937, 1967). These new cultivars have greatly increased production, extended the range of adaptation, and provided a wide variety of high quality fruit. Many characteristics are available in cultivars and species (Table 3), such as large fruit size, firmness, high flavor, small seeds, productivity, vigor, erect growth habit, different seasons of ripening, and resistance to various insects and diseases. The task still remains to incorporate many of these desirable characteristics into future cultivars. Much remains to be solved in understanding various genetic and physiological relationships.

One important goal is development of cultivars suited to mechanization. Mechanical harvesters have been developed for raspberries and many of those developed for other crops such as blueberries and grapes could be adapted for raspberries. These new cultivars should have a very erect cane growth habit, resistance to cane breakage, and exhibit concentrated ripening. The plants should be very productive with firm high-quality fruit. Most mechanically harvested fruit will require processing while a limited quantity may be suited for the fresh market. Because of high production costs, mechanization is essential for quick removal of ripe fruit. Along with mechanization will come the necessity of mechanical pruning. The potential utilization of the fall-bearing characteristic in raspberry has not yet been fully realized. A raspberry crop could be harvested on the current season's growth, in repeated or once-over operations, and after frost all canes could be removed. This would permit better sanitation, reduce insect and disease problems, and eliminate potential loss of crop on winter-killed canes, as well as eliminate the removal of old fruiting canes and topping of primocanes as dictated by conventional methods. Cultivars with sturdy erect primocanes would eliminate the need for trellises.

Through intensive breeding efforts, new fall-bearing raspberry clones have been developed which produce fruiting laterals down two-thirds of the current season's cane with production ap-

TABLE 3. Cultivars of *Idaeobatus* with specific characteristics

Characteristic	Cultivar	Species	Reference[z]
ADAPTATION			
Northern regions	Latham, Madawaska, Chief, Newburgh	*R. idaeus* var. *strigosus*	
Central regions	Reveille, Citadel, Sentry, Scepter, Pocahontas, Cherokee	*R. idaeus* var. *strigosus*	
Southern regions	Southland, Dormanred	*R. albescens* *R. parvifolius*	Hull, 1961; Overcash, 1972; Sherman and Sharp, 1971
Light soils	Cuthbert, Puyallup	*R. idaeus*	
Heavy soils	Sumner, Newburgh, Fairview, Malling Exploit, Malling Enterprise	*R. idaeus*	
Drought resistance	Latham, Marcy	*R. idaeus* var. *strigosus*	
GROWTH HABIT			
Erect cane	Heritage, Durham, Viking, Canby	*R. cockburnianus* *R. crataegifolius*	Keep and Knight, 1968
Many fruits per lateral	Glen Clova	*R. cockburnianus*	Keep and Knight, 1968
Reduced prickles to glabrous cane	Canby, June, Viking, Carnival	*R. idaeus*	
Accessory buds	Lloyd George	*R. idaeus*	Keep, 1969*b*
Fall-bearing	September, Fall Red, Heritage, Fall Gold, Indian Summer, Durham, Zeva Herbsternte, Southland, Tennessee Autumn Red	*R. idaeus* *R. arcticus* *R. strigosus* *R. stellatus*	Slate and Watson, 1964
INSECT RESISTANCE			
Aphid immunity *Amphorophora rubi*	Malling Landmark, Canby, Carnival, Creston, Lloyd George, Miranda, Mitra Rideau, Trent	*R. occidentalis; idaeus; strigosus; coreanus*	Baumeister, 1961;. Keep et al., 1970
Amphorophora agathonica	Lloyd George, Canby, Haida, Pyne's Royal, Burnetholm	*R. idaeus*	Converse and Bailey, 1961; Daubeny, 1966; Kennedy et al., 1973
Mite resistance	Preussen, Milton, Viking, Newburgh, Pyne's Royal	*R. idaeus*	
FRUIT			
Size			
Large	Marcy, Preussen, Hilton, Pyne's Royal, Malling Exploit	*R. idaeus*	Jennings, 1966*a*
Small	Latham, Chief	*R. strigosus*	
Firmness	Heritage, Pyne's Royal, Newburgh	*R. occidentalis*	Keep and Knight, 1968
Rot susceptibility			
Low	Matsqui, Carnival, Meeker, Cuthbert, Malling Jewel	*R. idaeus*	Barritt, 1971; Daubeny and Pepin, 1969
Intermediate	Willamette		
High	Fairview, Sumner, Latham, Malling Jewel		
Good fresh quality	Cuthbert	*R. idaeus; strigosus*	Schwartze, 1934

Characteristic	Cultivar	Species	Reference[z]
Good freezing quality	Fairview, Willamette, Meeker, Taylor, Malling Giant, Malling Promise	*R. idaeus*	
Early ripening	Early Red, June, Sunrise, Reveille, Chief, Trent, Malling Promise	*R. strigosus; idaeus*	
Late ripening	Latham, Taylor, Cuthbert	*R. idaeus*	
Small drupelets	Heritage, Malling Jewel	*R. idaeus*	
Small seed	Heritage	*R. idaeus*	
Yellow fruit	Amber, Fall Gold	*R. idaeus*	
DISEASES			
Spur blight resistance	Ottawa, Newburgh, Marcy, Viking, Pyne's Royal, Preussen	*R. idaeus; R. occidentalis*	Jennings, 1962a; 1962*b*
Mildew resistance	Malling Promise, Lloyd George	*R. idaeus*	Keep, 1968*a*
Mildew susceptibility	Latham, Carnival, Ottawa	*R. idaeus* var. *strigosus*	
Anthracnose resistance	Ottawa	*R. coreanus; biflorus; kuntzeanus; nivens; innominatus; idaeus*	Jennings, 1962a; 1962*b*
Verticillium resistance	Cuthbert	*R. idaeus* var. *strigosus; R. biflorus*	
Root rot resistance	Fairview, Sumner, Newburgh, Matsqui (Tolerant)	*R. idaeus*	
Yellow rust resistance	Lloyd George, Chief	*R. idaeus*	
Yellow rust immunity	Black raspberries	*R. occidentalis*	
Cane spot resistance		*R. coreanus; R. glaucus*	
Leaf spot resistance	Citadel, Pyne's Royal	*R. biflorus; kuntzeanus; innominatus; parvifolius; inoperatus; phoenicolasius*	
Soil-borne viruses resistance	Lloyd George, Malling Landmark	*R. idaeus*	Jennings, 1962*b*; 1964c
Green mottle resistance	Latham, Chief	*R. idaeus* var. *strigosus*	

[z] Literature cited, Darrow (1937), and records of New York State Agricultural Experiment Station, Geneva, N.Y.

proaching one-half that of summer-fruiting cultivars. The use of growth regulators to induce fruiting laterals and loosen and ripen fruit for mechanical harvesting are new areas of research.

There will always be a demand for hand-picked fruit either by "pick-your-own" operations, roadside fruit stands, or speciality and gourmet shops. For these outlets, large fruit size, high quality, good production, and extended season of ripening are important. Firmness is also important if fruit is to be sorted or fancy packed.

Another important goal of *Rubus* breeding will be in the area of insect and disease resistance. Sources of immunity and resistance to raspberry aphids have been identified and are being used in breeding to reduce spread of viruses by these vectors. The development and use of virus-free nursery stock has contributed greatly to the improvement of raspberry plantings. Until a large number of immune cultivars are available for planting, a strict inspection service and release of indexed virus-free material will be necessary.

New programs are needed to screen cultivars and species for resistance to such serious pests as *Lygus pratensis oblineatus* Say (tarnished plant bug), *Oberea bimaculata* Oliv. (cane borer), and *Tetranychus urticae* Koch (two-spotted spider mite). Differences in susceptibility to *Alternaria humicola* Oud. (Alternaria fruit rot), *Botrytis cinerea* (grey mold), *Rhizopus stolonifer* (Ehr. ex Fr.) Lind. (Rhizopus fruit rot), *Didymella applanta* (spur blight), and *Elsinoe veneta* (anthracnose) have been noted among cultivars and species but new sources of resistance have yet to be found.

The tremendous variation in the genus *Rubus* offers an unlimited opportunity for research and breeding. Greater cooperation between the plant breeders, entomologists, plant pathologists, and agricultural engineers will be required for development of future cultivars.

Blackberries: *Eubatus*

Origin and Speciation

In their original habitat, species of *Eubatus*, which include the cultivated blackberries, were usually found along the edges of forests, clearings, and streams. Retreating glaciers in Europe and North America opened large areas to colonization by the genus. More recently, man's clearing of forests in North America served a similar purpose (Darrow, 1937). The then relatively distinct species moved in, sexual and apomictic seeds were randomly distributed by birds and fauna, and a vast natural breeding program followed. Rapid vegetative propagation by suckering and tip layering aided rapid establishment and spread.

The extensive intercrossing and several levels of polyploidy, coupled with the highly variable phenotypic response to site and other factors led taxonomists to compile extensive descriptions of supposedly different species at different locations. In many instances the same type was assigned different specific names or the same name was applied to quite different forms. Even now there is not agreement on criteria to distinguish closely related "species."

European contributions on origin and speciation include the works of Sudre (1908), Focke (1910–1914), Gustafson (1942), and Watson (1958). Sudre (1908) confined his study to the European blackberries in which he tried to determine the original indigenous species by studying their distribution and relationship based on pollen production. He described some 110 species. Focke (1910–1914) divided the genus into subgenera, gave a description of some 132 species, and discussed the origin of the European blackberry.

Gustafsson (1942) presented evidence supporting the hypothesis that there were two major centers of origin for the blackberry, namely eastern North America and Europe. He divided the subgenus into two complex groups, *Moriferi veri* and *Corylifolii*. The *Moriferi veri* complex consists of five primary diploid species. These species have intercrossed, undergone chromosome doubling and apomixis, and given rise to a complex of hybrids which are widespread in Europe and Asia (Gustafsson, 1942).

Watson (1958) has also documented the basic work on species in Europe.

North American contributions to speciation include a report made by Brainerd and Peitersen (1920) from a detailed study of New England blackberries. Brainerd and Peitersen (1920) concluded that much of the observed variation was due to environmental effects. All of the species studied were nearly or completely self-sterile based on percentage of aborted pollen and inter-crossing. The two extreme types of blackberries, *R. allegheniensis* Porter and *R. baileyanus* Britton, showed less sterility and other hybrid characteristics than any of the intermediate species and forms. They concluded that due to intercrossing a large number of so-called species are either hybrids or crypto-hybrids brought about by crosses between two or more elementary species.

Bailey (1932) found classification difficult but assigned specific rank to innumerable hybrids; in contrast, Gustafsson (1942) defined a few primary species to which he related hybrid forms.

Darrow (1937) classified the wild blackberries of North America into five major groups: 1) the erect or nearly erect types like 2x 'Early Harvest' and 4x 'Eldorado' of the eastern United States from Florida to Canada and from the Atlantic coast to the Midwest; 2) the eastern trailing blackberries whose canes do not have red hairs, similar to 'Lucretia', with a distribution similar to the erect types; 3) the southeastern trailing types which have red hairs on the canes, found along the Atlantic and Gulf coasts from Delaware to Texas; 4) the trailing blackberries of the Pacific Coast from which 'Logan' was derived; and 5) the semi-trailing evergreen forms on the Pacific Coast.

In general, the *Eubatus* species in eastern Asia, Africa, and to a lesser extent in Latin America appear to be fairly distinct, although highly variable. More sophisticated and detailed investigations may well modify this opinion.

Vaarama (1949) and Thompson (1961) presented substantial evidence that differentiation between the chromosomes of various *Rubus* genomes has taken place mainly through gene mutation and minor structural changes. However, breeders and cytologists have demonstrated the unlikelihood of successful distant crosses within the genus, even at the diploid level where chromosome pairing pressure would be at its greatest. J. Hull (personal communication) has found evidence that at least one and possibly up to six of the seven chromosome pairs of the 2x hybrids between *R. rusticanus inermis* Willd. ('Burbank Thornless') and *R. argutus* Link. ('Early Harvest') are heterozygous for major translocations.

History of Improvement

Blackberry cultivation began in the early nineteenth century, mainly in North America, where most cultivars have been developed. Selections which were introduced from the wild include 'Lawton', 'Snyder', 'Eldorado', and the trailing form 'Lucretia'. Among the first efforts to improve cultivars resulted in the planting of 'Wilson Junior', a seedling of 'Wilson Early' and 'Minnewaski' (Darrow, 1937). In the western United States, many species and polyploid forms contributed to blackberry improvement. 'Aughinbaugh' was one of the first trailing blackberry selections to be grown commercially.

In 1881 J. H. Logan selected from seedlings of 'Aughinbaugh' the cultivars 'Logan' and 'Mammoth'. The exact origin of the 'Loganberry' is obscure; however, it exhibits regular chromosome pairing and it behaves like a true species. Many hybrids between it and wild blackberries are fertile while crosses with the raspberry are sterile. It has been used extensively in many blackberry breeding programs involving crosses at high levels of ploidy.

'Black Logan' and several other cultivars were selected from 'Logan'. Near the turn of the century, Luther Burbank introduced 'Phenomenal' which was widely grown despite its faults.

Some of the early state blackberry breeding programs include: Texas, beginning in 1909; New York, where the first crosses were made in 1912; North Carolina, where breeding began in 1926; and Rhode Island, beginning in 1929. The first blackberry breeding work of the United States Department of Agriculture began in 1919. Most of the early European programs were directed toward understanding speciation.

The Texas program in 1912 produced hybrids between a selection of *R. trivialis* Michx. and the 'Brilliant' red raspberry. The first and second generations were quite sterile, but selections made in the third generation gave rise to the 'Nessberry' which was then released in 1921. The 'Nessberry' was drought-resistant and high in flavor, but the fruit did not separate as readily from the plant as did either the raspberry or the blackberry.

The first blackberry crosses in New York were aimed at producing productive, hardy, erect cultivars. 'Eldorado' and 'Brewer' were used extensively as parents.

In 1926 crosses between 'Young' and 'Lucretia' were made in North Carolina in an attempt to obtain productive, disease resistant, highly flavored, thornless clones, 'Cameron' was released because of its vigor, high flavor, thornlessness, and resistance to *Elsinoe veneta* (anthracnose) and *Sphaerulina rubi* (leaf spot).

The Rhode Island program in 1929 sought to develop thornless clones by crossing *R. canadensis* L., 'Austin Thornless', and 'Cory' with 'Snyder', 'Eldorado', 'Alfred', and 'Lucretia'.

Other objectives of the early breeding programs in the eastern United States were to combine the fruit size and hardiness of the eastern cultivars with the productivity and vigor of 'Himalaya'.

Until the 1940s many of the cultivars in production were chance seedlings selected from the wild. Since 1950 the following experiment stations have released a number of promising cultivars: Florida, 'Flordagrand' and 'Oklawaha'; Georgia, 'Flint', 'Early June', 'Gem', and 'Georgia Thornless'; Oregon, 'Olallie', 'Marion', and 'Aurora';

North Carolina, 'Carolina' and 'Williams'; New Jersey, 'Jerseyblack'; New York, 'Bailey', 'Hedrick', and 'Darrow'; Maryland, 'Raven' and 'Ranger'; and Texas, 'Brazos'.

The USDA blackberry breeding program began in 1919 at Atlanta, Georgia. The seedlings which were produced were raised at Beltsville, Maryland, where only two grew to maturity, one of which was named 'Brainerd' (Darrow, 1937). 'Brainerd' resulted from a cross between 'Himalaya' x 'Georgia Mammoth'. It was grown extensively on the West Coast, replacing 'Himalaya'.

Most of the breeding for thornless blackberries has been carried out by the USDA at Beltsville, Maryland, and Carbondale, Illinois. In 1966 the thornless cultivars 'Smoothstem' and 'Thornfree' were released. These cultivars are extremely vigorous and productive but not hardy in northern regions. Various sources of thornlessness have been used in conjunction with manipulation of ploidy levels (Hull and Britton, 1968) for transfer of the character.

Southern breeding programs have concentrated on the use of indigenous species to extend *Rubus* growing farther south. The USDA program in Oregon is aimed at combining the high flavor of the native trailing blackberry with the size, firmness, productivity, and hardiness of commercial cultivars.

About 1926 B. M. Young of Morgan City, Louisiana, introduced the 'Youngberry' which was a hybrid between 'Phenomenal' and 'Austin Mayes'. A few years later the 'Boysenberry' was introduced; it is similar to the 'Youngberry' but its parentage is unknown.

The cultivars 'Evergreen' and 'Himalaya' were introduced from Europe after the turn of the century and today may be found widely naturalized on the West Coast.

At present, active blackberry breeding programs are in progress in New York, Oregon, Arkansas, Texas, Georgia, and Florida. Numerous crosses have been made between the *Eubati* and *Idaeobati* (Einset and Pratt, 1954) but they succeed best at the polyploid levels. A cross between a 2x blackberry and a 2x red raspberry is almost completely sterile, while at the tetraploid level the hybrids are moderately fertile and some combinations are fully fertile. This breeding approach was undertaken to develop cultivars similar to the 'Boysen', 'Logan', and 'Youngberries', but with greater hardiness. The blackberry-raspberry hybrids are very difficult to pick. The torus does not separate from the plant readily as in blackberries nor do the drupelets separate from the torus readily as in raspberries. Backcrossing to the raspberry has produced seedlings with poor fruit color, low quality, and reduced vigor. Backcrossing to the blackberry has had similar results. Because of discouraging results many breeders have not pursued this line of breeding. The cultivars 'Mahdi' and 'Kings Acre' ($2n = 21$), 'Veitchberry' ($2n = 28$), and 'Lawtonberry' ($2n = 49$) were developed in England. None of these hybrids, despite good quality, were grown commercially.

Modern Breeding Objectives

The major breeding objectives include selection for a vigorous erect growth habit, climatic adaptation, productivity, and large, attractive, high-quality fruit containing small seeds (Sistrunk and Moore, 1973).

Erect, rigid canes are desirable for mechanical harvesting and can be easily pruned by use of pneumatic shears or a tractor-mounted cutting bar, while trailing types require a considerable amount of hand work. In addition, fruiting laterals must be well above the ground. The fruits must be firm, easily removed, and of good quality. Concentrated ripening would reduce the number of harvests required, but several cultivars would be needed to extend the season.

In the northern United States, hardiness is of great importance. Several aspects of hardiness must be considered: 1) retention of cold resistance once attained; 2) deep or sufficiently long rest period so that fluctuating temperatures do not stimulate growth, making the plant susceptible to injury by low spring temperatures; and 3) ability to reharden if initial cold resistance is lost (Brierley and Landon, 1946). Natural selection for hardiness has not occurred in many northern regions because plants are protected by a sufficient snow cover often lasting until late spring. Selected clones of *R. canadensis* (3x) and *R. allegheniensis* (2x) have been used extensively as sources of hardiness. Since many *Rubus* species do not grow well in regions of the southern United States, programs have been initiated to combine the adaptability of native species with the commercial characteristics of temperate species. The search for low chilling period for uniform bud break after warm winters and a high heat requirement for growth after dormancy are among the greatest challenges to southern breeders

(Sherman and Sharpe, 1971). Most of the native southern species are small-fruited with low quality.

Breeding for thornlessness is of great importance in trailing types which require trellising. When erect types are mechanically harvested, thorns are of less importance except for possible injury to the fruit. However, thornlessness would be of value for use in local and "pick-your-own" fresh market operations. Most sources of thornlessness are not hardy in northern regions, thus requiring the transfer of this character to hardy clones.

Summer maturation of fruit is desirable in areas such as the northwestern United States, where school children pick fruit. 'Evergreen' and 'Himalaya' ripen too late for such labor.

Breeding Techniques

Most of the techniques described are used at the Geneva station; however, each breeder has developed specific techniques based on experience and facilities.

Floral biology: The time of flower bud differentiation within *Eubatus* is more variable than *Idaeobatus*. The cultivars 'Evergreen', 'Austin Thornless', 'Santiam', and the species *R. macropetalus* Hook. differentiate flower buds in early October in Oregon, whereas 'Himalaya' does not until January or February (Waldo, 1934). Greater variations have been reported in Scotland (Gruber et al., 1962). Eastern blackberry cultivars begin to differentiate flowers in late August but little change takes place between September and March. Blackberry-raspberry hybrids, 'Youngberry' and 'Loganberry', differentiate flowers in early autumn.

Emasculation and pollination: Blackberry flowers within a fruiting lateral are more uniform in stage of development than those of the red raspberry. From 20 to 60 flowers are produced in a few initiated leaf axils. Sequence of flowering is basipetal within the cluster with opening covering a two- to three-day period in contrast to a raspberry lateral which may flower over two to three weeks. The anthers are larger than those of the raspberry and easily removed from the flowers with forceps or scalpel and placed in a petri dish. Large quantities of pollen are obtained by merely shaking a petri dish containing thoroughly dried anthers. The petri dish lid can easily be used to apply pollen. Pollen can be stored for many years, −5 C and 10–20% relative humidity being optimum.

Blackberry flowers are emasculated when they reach full size but before the petals separate. A cut is made at the base of the bud between the torus and sepals with a razor blade or sharp scalpel. The emasculated flowering laterals of a cane are immediately covered with a white paper bag, avoiding bending which lends to breakage, and tied to the main stem to prevent breakage by strong winds. At Geneva, after fruit is set, the paper bag is generally removed and replaced with a muslin bag which allows greater light penetration, protects the fruit until harvest, and catches over-ripe fruit.

Seed germination: Blackberry seeds are handled in much the same way as that described for raspberries. Blackberry seed is difficult to germinate and requires 40 to 60 minutes of scarification with concentrated sulfuric acid (Heit, 1966; Heit and Slate, 1950) followed by an after-ripening period of three to five months. Scott and Ink (1957) found that germination improved by holding the seeds at 21–24 C for seven to eight weeks before after-ripening. Light may also beneficially affect germination (Scott and Draper, 1967).

At Geneva, seed may be sown on flats of ground sphagnum moss and over-wintered in cold frames, or placed in bags of moist moss in the refrigerator for after-ripening and later sown. Most seeds germinate in eight to 15 days in a greenhouse at 21 C. The use of ground sphagnum moss facilitates pricking out and reduces root damage. The seedlings may be transplanted to peat pellets or into clay or peat pots filled with soil.

Seedling evaluation: Blackberry seedlings are evaluated in the same manner as red raspberries. Seedlings may be screened at an early stage for thornlessness by examining cotyledons for stalked glands. In New York, seedlings are frequently grown in the field for three years prior to selection. Depending upon the cross or species involved, some seedlings may be selected in the second or fourth year.

After selection, the seedling is dug and root cuttings are made for establishment of second-test plots. The author generally places the root cuttings in a nursery for one year prior to test plot planting.

For nonsuckering types usually grown on trellises ('Boysenberry', 'Loganberry', 'Thornfree', etc.), three plants of a clone are set 3 m apart in the row. Yield records are taken from a sector from the base of the middle plant to a point 1.33 m in one direction including any overlapping canes from adjacent plants. This constitutes a 1/1000th-acre plot; the results are handled as previously described for raspberry.

Propagation: In addition to conventional methods of propagation (tip-layering, root suckers, and crown division) most *Eubati* and some *Idaeobati* can be propagated by leaf-bud or stem cuttings in an intermittent mist bed. If the material is in short supply, a single node cutting with attached leaf will suffice, but two- or three-node cuttings are preferred. Semi-mature wood of the current season's growth can be used in conjunction with a rooting agent, such as indole-butyric acid. Care should be taken so that the cuttings never wilt before being placed in the mist bed. Most species will be adequately rooted for potting in five to seven weeks. Dormant cuttings of some species will root if taken in early winter.

Cultivar, selection, and breeding line stocks are propagated primarily from root cuttings, grown one year in the nursery and then field planted in replicated trials.

Exotic species may be obtained from *Rubus* breeders throughout the world, but international restrictions severely limit the importation and exportation of materials. The United States Plant Quarantine Division usually requires methyl bromide fumigation of imported plants but this is generally lethal to *Rubus* plants. Great Britain requires for importation an official certification against certain diseases, often unrelated to *Rubus*, which is practically impossible to obtain.

Many unusual species often require special conditions for optimum growth and flower development, especially light, temperature, and humidity. The desirability of making crosses in the greenhouse in mid-winter further complicates the problem. Many temperate species can be grown for two to three years in large pots in a lath house or cooled greenhouse during the spring, summer, and fall and then transferred to cold storage (2–3 C) for holding and completion of their rest period. Two or three months at this temperature satisfies the cold requirement of most temperate species.

Arctic and tropical species do not respond to the same treatment. Both can be overwintered in a cold greenhouse at 4–7 C. The arctic species are very difficult to maintain through the hot summer months even with heavy shade and evaporative cooling systems. Certain facilities have been developed for maintaining and propagating parental stocks. A fan and pad evaporative cooled greenhouse or a lath house with 50% shade can serve as an effective holding area.

Pollen may be imported but deteriorates rapidly in transit. Seeds are more easily shipped but are of unknown genetic constitution because of frequent open pollination and high heterozygosity. Plant materials, collected by plant explorers, can be obtained from the USDA Plant Introduction Division.

Breeding Systems

Sexual reproduction: The large number of species, wide range in ploidy, and plant habit and adaptation offer the breeder an unlimited opportunity for improving blackberries. In all diploid *Rubus* species examined (Crane and Thomas, 1939; Pratt and Einset, 1955), reproduction was found to be sexual, but in polyploids reproduction was variable or apomictic.

Einset (1951) observed that many American and European polyploid species are pseudogamous, producing a high proportion of apomictic seedlings upon selfing, crossing with plants of different chromosome numbers, or upon open pollination. The apomictic seedlings may be produced aposporously from unreduced eggs. *R. procerus* P. J. Muell. ($2n = 28$) and *R. laciniatus* (West.) Willd. ($2n = 28$) are pseudogamous facultative apomicts, and F_1 hybrids of them are believed to be sexually formed (Bammi and Olmo, 1966). Pollination of 'Oregon Evergreen' and 'Himalaya' with pollen from American cultivars produced various proportions of hybrids and the remainder were maternal apomicts (Darrow and Waldo, 1933).

Many cultivars such as 'Hedrick', 'Bailey', 'Darrow', 'Raven', and 'Ranger' have been developed through crossing of superior clones. Few improvements remain to be made in this line of breeding without incorporating new characteristics from American and European species. New sources of hardiness, disease and insect resistance, small seed size, flavor, season of ripening, climatic adaptation, and thornlessness have been observed in wild species and are now being incorporated into breeding lines (Table 4). 'Smoothstem' and 'Thornfree' blackberries originated from crosses involving the cultivar 'Merton Thornless' while 'Oklawaha' and 'Regal Ness' were produced from direct species crossing. Recombination of characteristics between wild species for the development of new and

TABLE 4. Sources of desirable characteristics in *Eubatus*

Characteristic	Cultivar	Species	Reference[z]
PLANT			
Hardiness	Snyder, Eldorado, Darrow, Alfred	*R. allegheniensis* Porter *R. canadensis* L. *R. odoratus* L. *R. caesius* L.	Darrow, 1937; Yeager and Meader, 1958
Thornlessness	Smoothstem, Thornfree, Austin Thornless, Whitford Thornless, Merton Thornless	*R. canadensis* L.	Hull, 1968; Scott and Ink, 1966; Sherman and Sharpe, 1971; Yeager and Meader, 1958
Drought resistance	Nessberry		Darrow, 1937
Erect cane	Early Harvest, Eldorado, Darrow, Lawton	*R. argutus* Link. *R. allegheniensis* Porter	 Darrow, 1937
Adaptation to southern United States	Smoothstem, Thornfree, Brazos, Flordagrand, Oklawaha, Raven, Lawton, Humble	*R. biflorus* Buch.-Ham. *R. ellipticus* Sm. *R. kuntzeanus* Hemsl. *R. trivialis* Michx.	Sherman and Sharpe, 1971
Disease resistance			
Orange rust	Eldorado, Evergreen, Snyder		Darrow, 1937
Double blossom	Himalaya, Brainerd	*R. trivialis* Michx.	Sherman and Sharpe, 1971
Leaf spot	Himalaya, Evergreen, Early Harvest, Boysen, Young	*R. coreanus* Miq. *R. biflorus* Buch.-Ham. *R. parvifolius* Nutt. *R. morifolius* Hort. *R. wrightii* A. Gray *R. albescens* Roxb.	Darrow, 1937
Anthracnose	Young, Boysen	*R. coreanus* Miq. *R. biflorus* Buch.-Ham. *R. kuntzeanus* Hemsl. *R. parvifolius* Nutt. *R. innominatus* S. Moore	Darrow, 1937
Verticillium			
Resistant	Logan, Mammoth, Cascade, Marion, Olallie, Merton Thornless	*R. ursinus* Cham. & Schlecht	Wilhelm and Thomas, 1954
Immune	Himalaya, Evergreen		
Large flower	Himalaya, Evergreen, Chandler	*R. thysiger* Banning & Focke *R. nitidioides* W.	Darrow, 1937
FRUIT			
Early ripening	Advance, Lucretia, Young, Early Harvest	*R. borreri* Bell Salter *R. pungens* Chambers. *R. morifolius* Hort.	Darrow, 1937
Late ripening	Himalaya, Evergreen, Thornfree, Smoothstem	*R. cuneifolius* Pursh. *R. rusticanus* Merc.	Darrow, 1937 Scott and Ink, 1966
Firmness	Evergreen, Mersereau		
Large size	Boysen, Young, Logan, Brazos, Mammoth	*R. glaucus* Benth. *R. caucasicus* Focke *R. anatolicus* Focke	Darrow, 1937
Small seed	Early Harvest, Chehalem, Whitford Thornless	*R. glaucus* *R. cuneifolius* Pursh.	Darrow, 1937; Sherman and Sharpe, 1971
Flavor	Darrow, Young, Boysen, Logan, Zielinski	*R. macropetalus* Hook. *R. allegheniensis* Porter *R. arcticus* L. *R. chamaemorus* L.	Darrow, 1937

[z] Some observations are based on author's experiences and files of the New York State Agricultural Experiment Station, Geneva, New York.

unusual types of blackberries has not been fully investigated.

Only limited inbreeding has been done in blackberries. Species vary greatly in their response to inbreeding, some exhibiting little depression while others are greatly affected. The author has found that seed from selfed or inbred lines do not germinate as rapidly as those from outcrosses.

Backcrossing is more frequently used to incorporate desirable characteristics from wild species. Frequently the desired characteristic is found in different species at a different level of ploidy. For example, "dominant," in contrast to "recessive," thornlessness occurs in 'Austin Thornless'. This cultivar is 8x or 2n = 56, while most eastern erect blackberry cultivars are 4x or 2n = 28. In order to incorporate this unique source of thornlessness into an adapted cultivar a series of crosses is required; for example, 'Austin Thornless' (8x) X 'Eldorado' (4x) = F_1 (6x); selected F_1 (6x) X 'Early Harvest' (2x) = (4x) hybrids. The selected 4x hybrids can then be crossed with other tetraploid cultivars. A percentage of all progenies will be thornless from which selection can be made for vigor, fertility, hardiness, and other important characteristics. Analogous crosses substituting 2x and both natural and artificially induced 4x red raspberries for the blackberry forms indicated above show considerable promise for developing 'Logan' type 4x and 6x thornless blackberry-raspberry hybrids. Genes from exotic species are also incorporated in the same way (Hull, personal communication).

Sterility in the F_1 and later generations may adversely affect a practical breeding program. Such is the case with the derivatives of *R. rusticanus inermis* ('Burbank Thornless' 2x) x *R. argutus* ('Early Harvest' 2x), in which the best selections are too sterile for introduction after five generations of selection for fertility (Hull, personal communication). However, derivatives of tetraploids of similar origin ('Merton Thornless' 4x) x (colchicine-induced 4x 'Early Harvest') produce some fertile seedlings in the second generation and selection in subsequent generations has developed lines which are fully fertile. Thornless derivatives of 'Merton Thornless' x 'Brainerd' (4x) or 'Eldorado' (4x) have been introduced commercially (Scott and Ink, 1966).

As the chromosome number of hybrids approaches the 6x (2n = 42) level and higher, parental differences in chromosome number and genetic constitution become less critical for fertility of the hybrid. An extreme example of this was an aneuploid hybrid (2n = 39) of the cross 'Youngberry' (2n = 49) x 'Eldorado' (2n = 28). This seedling and all others of the same progeny that were 6x or above were fertile and nonapomictic (Hull, personal communication). Those between 2n = 39 and 2n = 42 were variable in fertility and those below 2n = 39 were sterile. Aneuploidy can be tolerated at these higher polyploid levels. For example, many seedlings of 'Boysenberry' (2n = 49), 'Youngberry' (2n = 49), and 'Cascade' (2n = 63) have been satisfactory parents when crossed with other parents of high chromosome numbers. Conversely, aneuploid hybrids at lower than the 5x level have been mostly sexually sterile, although some will form fruit apomictically.

The Pacific Coast blackberries, *R. ursini* Cham. and Schlecht., are dioecious as compared to the eastern and European species which are perfect flowered. Crosses between pistillate and perfect-flowered types gave 1:1 sex ratios (Vaarama, 1953).

In *R. arcticus* at least five incompatibility alleles were involved in 16 self-incompatible clones (Tammisola and Tyynanen, 1970).

Polyploidy: Natural polyploidy in the subgenus *Eubatus* ranges from 2x to 12x. Allopolyploidy has proved to be a major evolutionary mechanism in *Eubatus* while autopolyploidy is relatively rare. Gustafsson (1943) observed that 87% of the *Moriferi veri* complex were tetraploid with few 2x, 3x, and 5x types. Tetraploids also predominate in the *Corylifolii* complex, but other polyploid forms are found.

Einset (1947) found no significant differences in chromosome size among a large number of species but noted that eastern species generally were lower in chromosome number than Pacific Coast species. Chromosomes ranged in length from 1–4 μ. In polyploids (Heslop-Harrison, 1953) it has been shown that there is no correlation between number of satellites and degree of ploidy.

During meiosis in the *Moriferi veri* complex, Gustafsson (1943) observed a high proportion of bivalents, with univalent and multivalent formation being rare. Einset (1947) found various degrees of meiotic irregularity in North American species. The high percentage of multivalents observed in tetraploid species suggests that both auto- and allo-polyploidy are involved.

Britton and Hull (1957*b*) found a considerable amount of mitotic instability in both the *Eubati* and *Eubati* x *Idaeobati*. They also observed that root cuttings from single plants gave rise to mitotically stable plants with differing phenotypes, with generally low aneuploid chromosome numbers, and that leaf-bud cuttings produced plants similar to the phenotype of the mother plant, thus perpetuating the mitotic instability.

Most of the genome differentiation in *Rubus* appears to have occurred by continuous gene mutation (Vaarama, 1953; Thompson, 1961); however, other evidence indicates that gross structural changes may have occurred (Hull, personal communication).

The manipulation of polyploidy has been important in *Rubus* breeding. Shoemaker and Sturrock (1959) found that in order to combine the characteristics of diploid species *R. trivialis* and *R. cuneifolius* Pursh. with the flavor, firmness of fruit, and erect plant habit of many polyploids, 3x hybrid seedlings had to be treated with colchicine to produce fertile 6x plants.

Tetraploid seedlings of 'Nessberry' x 'Lawton' which are semi-erect, productive, with attractive high-quality fruits were treated with colchicine to obtain octoploids for crossing with 'Austin Thornless' (2n = 56) to obtain progenies segregating for thornlessness. Hexaploid and higher ploidy clones are often highly fertile and possess excellent fruit quality, but many of these plants have a high chilling requirement and are not adapted to southern areas. Therefore, by inducing polyploidy in lower-chromosome, indigenous species adapted to warm climates, hexaploid forms may be used in breeding (Hull, personal communication).

The following method has proved successful in inducing chromosome doubling of blackberries. Following an after-ripening period of four to six months at 2–3 C, seeds which have germinated enough to crack the endocarp are soaked for six to nine hours in an aqueous solution of 0.2% colchicine plus 5% glucose at 30 C. They are washed thoroughly and sown immediately (Hull and Britton, 1958; Shoemaker and Sturrock, 1959). Colchiploid seedlings are generally cytochimeras. Fruit size varies directly with the ploidy of the third histogenic layer in otherwise comparable colchiploid plants (Hull and Britton, 1958). Many induced tetraploid blackberries and raspberries have shown promise or merit as potential cultivars in addition to serving as parents in wide crosses.

Sanders and Hull (1970) found that the addition of 2% dimethyl sulphoxide to a 0.2% colchicine solution used on germinating seeds resulted in higher percentage of induced polyploids than colchicine alone but concentrations higher than 2% were phytotoxic.

Methods for detection of induced polyploidy have been developed. Transmitted blue light accentuates foliage characteristics (Hull and Britton, 1956). A quick cytological method has been developed for making chromosome smears from flower buds (Haskell and Paterson, 1964) and root-tips (Tun and Haskell, 1961).

Apomixis: Evolution in the *Eubati* has resulted from sexual reproduction, vegetative propagation, and various degrees and types of apomictic processes. Apomixis has not been observed in diploid species (Crane and Thomas, 1939; Pratt and Einset, 1955) but is associated with various levels of polyploidy. Polyploids may reproduce sexually, apomictically, or partly sexually and nonsexually (Crane and Thomas, 1939).

Jennings et al. (1967*a*) and Dowrick (1966) found that by using *R. laciniatus*, a pseudogamous apomict, as the seed parent, the proportion of non-hybrids differed with different pollen parents and was positively correlated with the ploidy level of the pollen parent. From these studies they postulated that the genotypes obtained from different crosses resulted from selection pressure on the embryo sac determined by its relative compatibility with the genetic constitution of the pollen parent.

In European species crosses, Berger (1953) found that apospory was more frequent in hybrids than in the parents. He also observed that in all the apomictic forms examined except *R. procerus* and the completely apomictic species *R. caesius* L., structures resembling embryo sacs were occasionally formed from binucleate tapetal cells of the anther and in some cases developed into 8-nucleate embryo sacs. Dowrick's (1966) results differ markedly for he found that in *R. caesius*, *R. calvatus* Lees., and *R. laciniatus* (all tetraploids), megaspore mother cells always underwent normal meiosis forming a *Polygonum* type of embryo sac and that there was no evidence for the formation of aposporic embryo sacs. He suggested that maternal seedlings arose from restitution of the first division of the egg cell or by automixis. Pratt and Einset (1955) studied the development of embryo sacs in a wide range of polyploid American blackberry species. Apospory was very rare in diploid *R. allegheniensis*, but common in the polyploids, especially odd-ploid species. The origin of embryo sacs could be determined at the one-nucleate stage when the remains of megaspores could be seen.

Pseudogamy refers to the situation in which pollination without fertilization of the egg is required for seed set. Darrow and Waldo (1933) reported that European blackberries were pseudogamous and American blackberries, except for *R. canadensis*, were not. However, Einset (1951) found a high proportion of apomictic seedlings from a wide range of polyploid American species which apparently were produced pseudogamously.

Breeding for Specific Characters

The selection of parents for hybridization requires a previous knowledge of the combination of characteristics desired and breeding behavior of the parents. The best individuals of a given type are often not the best parents. In *Rubus* most clones are highly heterozygous and even an extremely desirable combination of characters often is fragmented during meiosis or masked during later gene recombination in the hybrid. Only experience or advice from other breeders can help in selecting those clones which are effective in transmitting combinations of desirable genes for specific characteristics and do not transmit undesirable characteristics. Studies on the inheritance of specific characteristics have been limited by high levels of polyploidy and complex apomictic processes.

Several sources of thornlessness have been studied (Hull, 1968). Dominant epistatic thornlessness was observed in 'Austin Thornless' (8x); recessive thornlessness in 'Burbank Thornless' (2x), and 'Merton Thornless' (4x). 'Whitford Thornless' (2x) also transmits recessive thornlessness to its progeny, but apparently of a different nature from that of 'Burbank Thornless'. The thornlessness in 'Burbank Thornless' and 'Merton Thornless' is associated with eglandular cotyledons, making it easy to separate newly germinated seedlings in segregating populations. A difference of one or more translocations was found between the genomes of 'Burbank Thornless' and 'Early Harvest' thus limiting the use of 'Burbank Thornless' as a source of thornlessness (Hull, 1968). *R. canadensis* (3x) clones vary in hardiness and amount of thorns and may be useful for breeding by producing haploid and diploid gametes (Craig, 1960). Thornlessness in 'Merton Thornless' is inherited as a simple recessive giving a tetraploid ratio of 35 thorny to one thornless. This cultivar was used in the development of 'Smoothstem', a selfed seedling from a first backcross of 'Eldorado' to 'Merton Thornless', and 'Thornfree', a selection from ('Brainerd' x 'Merton Thornless') x ('Merton Thornless' x 'Eldorado').

A number of thornless blackberry sports have been described by Darrow (1955a). Many thornless sports are sterile or chimeral, giving thorny seedlings when used in breeding but thornless plants when tip-rooted. The thornless mutation appears to be in the first histogenic layer.

Yeager and Meader (1958) have used *R. canadensis* for thornlessness, *R. odoratus* for hardiness and disease resistance, *R. arcticus* L. and *R. chamaemorus* L. for flavor, and *R. pungens* Cambess. and *R. morifolius* Hort. for earliness. Others (Williams, 1945; Williams et al., 1949) have used *R. parvifolius* L. for large fruit size, yield, and disease and heat resistance; *R. glaucus* for large fruit size and quality; and *R. biflorus*, *R. coreanus*, *R. ellipticus*, *R. glaucus*, and *R. kuntzeanus* for disease resistance and adaptation to southern conditions.

'Flordagrand' (2x) and 'Oklawaha' (2x) are well adapted to southern regions of the United States. These are early, productive, disease-resistant and large-fruited cultivars selected from an open-pollinated population of the F_1 of 'Regal-Ness' x *R. trivialis*. Double-blossom disease resistance has also been obtained from *R. trivialis* and used in crossing with 'Brazos' (Sherman and Sharpe, 1971).

'Mysore', a black raspberry (*R. albescens*) which is not hardy and has low quality, has proved to be a source of high heat tolerance.

The cultivars 'Burbank Thornless', 'Chehalem', 'Himalaya', 'Merton Thornless', 'Oregon Evergreen', 'Logan', 'Black Logan', 'Cascade', 'Mammoth', 'Ollalie', and 'Phenomenal' have been reported to be resistant to *Verticillium albo-atrum* Reinke & Berth. (Verticillium wilt) (Wilhelm and Thomas, 1954).

Large seed size, as exhibited by 'Smoothstem' and 'Thornfree', is objectionable because it makes mastication difficult. Sherman and Sharpe (1971) found a wide genetic variation in seed size among blackberry crosses. 'Whitford Thornless' x 'Flordagrand' produced a large proportion of progeny with small seeds.

A number of promising cultivars have been produced from crosses between parents of differing high chromosome numbers. For example, 'Santiam' is a selection from a cross between an 8x wild trailing western blackberry x 'Himalaya' (4x); 'Cascade' was selected from 'Zielinski' (12x) x 'Logan' (6x) and 'Aurora' from ('Zielinski' [12x] x 'Logan' [6x]) x ('Logan' [6x] x 'Austin Thornless' [8x]).

Future commercial cultivars of the blackberry must be considered in relation to their adaptability to mechanical harvest (Sistrunk and Moore, 1973). Several characteristics have been identified which are important in selecting for adapted types. Plants must be rigidly erect, without support, with fruit borne well above the ground in order to be machine collectable. Fruit must abscise readily with agitation at the optimum stage of fruit maturity. Concentrated fruit maturity would be advantageous, reducing the number of times a planting would need harvesting. Fruit firmness is essential for machine handling. Very early and very late ripening types are needed to expand the picking season and allow more acreage to be harvested by each machine. Since much of the mechanically harvested fruit is being processed, good processing quality is paramount.

A list of species and cultivars with specific plant and fruit characteristics is presented in Table 4.

Achievements and Prospects

The large number of blackberry species, variable chromosome numbers, and complex breeding behavior have reduced to some extent rapid advances in blackberry cultivar improvement. Cultivars have been developed for southern regions of the United States; however, many of them need a smaller seed size and added resistance to various diseases. New thornless cultivars have been developed which are extremely productive, but again seed size is objectionable, they are not winter-hardy in northern regions, and they require extensive hand labor and trellising. Northern cultivars are erect and productive but additional hardiness is essential. They do not perform well in southern regions and are extremely thorny. In general, present cultivars are more vigorous, productive, and hardy than old cultivars and selections from the wild. Many desirable characteristics have been observed among species (Table 3), particularly such things as large fruit size, firmness, high quality, small seed, productivity, thornlessness, growth habit, and tolerance to insects and diseases. The characteristics available in these species offer great potential for the development of new cultivars.

Blackberries have been successfully harvested mechanically, producing a more uniform product than hand harvesting. The greatest improvement is needed in the area of plant growth habit which could facilitate both mechanical harvesting and pruning. The fall-bearing characteristic is also present in blackberries; however, only a few fruits are produced at the tip of the cane. The breeding of fall-bearing raspberries began at a similar level and today highly branched, productive clones have been developed which ripen fruit on the current season's cane.

Breeding blackberries resistant to various insects and diseases has not progressed very rapidly. Sources and modes of inheritance have yet to be identified as only differences in susceptibility have been noted. Future progress will require a team effort and the utilization of diverse species.

Literature Cited

The author wishes to thank Dr. J. W. Hull for his contribution to the text and Miss Charlotte Pratt for manuscript editing.

Anonymous. 1940. Pomology report. *Rpt. John Innes Hort Inst. 1939* p. 3–5.

Bailey, L. H. 1932. The blackberries of North America. *Gentes Herbarium* 2:269–423.

Bammi, R. K. 1965. Cytogenetics of *Rubus*. 4. Pachytene morphology of *Rubus parvifolius* L. chromosome complement. *Can. J. Gen. Cyt.* 7:254–258.

Bammi, R. K., and H. P. Olmo. 1966. Cytogenetics of *Rubus*. 5. Natural hybridization between *R. procerus* P. J. Muell. and *R. lacinatus* Willd. *Evolution* 20: 617–633.

Barritt, B. H. 1971. Fruit rot susceptibility of red raspberry cultivars. *Plant Dis. Rptr.* 55:135–139.

Baumeister, G. 1961. Investigations on the resistance of different raspberry varieties to the virus vectors *Amphorophora rubi* (Kalt.) and *Aphis idaei* (v.d. Goot) (in German). *Züchter* 31:351–357.

Bennett, C. W. 1932. Further observations and experiments with mosaic diseases of raspberries, blackberries, and dewberries. *Mich. Agr. Expt. Sta. Bul.* 125.

Berger, X. 1953. Investigations on the embryology of partially apomictic *Rubus* hybrids (in German). *Ber. Schweiz. Bot. Ges.* 63:224–266.

Bolkhovakikh, Z., V. Grif, T. Matvejeva, and O. Zakharyeva. 1969. *Chromosome numbers of flower plants* (in Russian). (Acad. Sci. U.S.S.R., V. L. Komarov Bot. Inst.) Leningrad: Science.

Brainerd, E., and A. K. Peitersen. 1920. Blackberries of New England—their classification. *Vt. Agr. Expt. Sta. Bul.* 217.

Brierley, W. G., and R. H. Landon. 1946. Some relationships between rest period, rate of hardening, loss of cold resistance, and winter injury in the 'Latham' raspberry. *Proc. Amer. Soc. Hort. Sci.* 47: 224–234.

Britton, D. M., and J. W. Hull. 1957a. Chromosome numbers of *Rubus*. *Fruit Var. Hort. Dig.* 11:58–60.

———. 1957b. Mitotic instability in *Rubus*. *J. Hered.* 48:11–20.

Britton, D. M., F. J. Lawrence, and I. C. Haut. 1959. The inheritance of apricot fruit-color in raspberries. *Can. J. Gen. Cyt.* 1:89–93.

Brooks, R. M., and H. P. Olmo. 1972. *Register of new fruit and nut varieties*. 2nd ed. Berkeley: Univ. of Calif. Press.

Converse, R. H., and J. S. Bailey. 1961. Resistance of some *Rubus* varieties to colonization by *Amphorophora rubi* in Massachusetts. *Proc. Amer. Soc. Hort. Sci.* 78:251–255.

Craig, D. L. 1960. Studies on the cytology and the breeding behavior of *Rubus canadensis* L. *Can. J. Gen. Cyt.* 2:96–102.

Craig, D. L., D. A. Gass, and D. S. Fensom. 1970. Red raspberry growth related to electrical impedance studies. *Can. J. Plant Sci.* 50:59–66.

Crane, M. B., and W. J. Lawrence. 1931. Inheritance of sex, color, and hairiness in the raspberry, *Rubus idaeus* L. *J. Gen.* 24:243–255.

Crane, M. B., and P. T. Thomas. 1939. Segregation in asexual (apomictic) offspring in *Rubus*. *Nature* 143:684.

Darlington, C. D., and A. P. Wylie. 1955. *Chromosome atlas of flowering plants*. London: Allen & Unwin.

Darrow, G. M. 1937. Blackberry and raspberry improvement. In *Better plants and animals 2. USDA Yearbook of Agriculture*. 1937:496–533.

———. 1955a. Nature of thornless blackberry sports. *Fruit Var. Hort. Dig.* 10:14–15.

———. 1955b. Blackberry-raspberry hybrids. *J. Hered.* 46:67–71.

———. 1967. The cultivated raspberry and blackberry in North America—breeding and improvement. *Amer. Hort. Mag.* 46:203–218.

Darrow, G. M., and G. F. Waldo. 1933. Pseudogamy in blackberry crosses. *J. Hered.* 24:313–315.

Daubeny, H. A. 1966. Inheritance of immunity in the red raspberry to the North American strain of the aphid, *Amphorophora rubi* Kltb. *Proc. Amer. Soc. Hort. Sci.* 88:346–351.

———. 1969. Some variations in self-fertility in the red raspberry. *Can. J. Plant Sci.* 49:511–512.

———. 1971. Self-fertility in red raspberry cultivars and selections. *J. Amer. Soc. Hort. Sci.* 96:588–591.

———. 1972. Screening red raspberry cultivars for immunity to *Amphorophora agathonica* Hottes. *HortScience* 7:265–266.

Daubeny, H. A., P. C. Crandall, and G. W. Eaton. 1967. Crumbliness in the red raspberry with special reference to the 'Sumner' variety. *Proc. Amer. Soc. Hort. Sci.* 91:224–230.

Daubeny, H. A., J. A. Freeman, and R. Stace-Smith. 1970. Effects of virus infection on drupelet set of four red raspberry cultivars. *J. Amer. Soc. Hort. Sci.* 95:730–731.

Daubeny, H. A., and H. S. Pepin. 1969. Variations in susceptibility to fruit rot among red raspberry cultivars. *Plant Dis. Rptr.* 53:975–977.

Daubeny, H. A., and R. Stace-Smith. 1963. Note on immunity to the North American strain of the red raspberry mosaic vector, the aphid *Amphorophora rubi* Kalb. *Can. J. Plant Sci.* 43:413–414.

Daubeny, H. A., P. B. Topham, and D. L. Jennings. 1968. A comparison of methods for analyzing inheritance data for resistance to red raspberry powdery mildew. *Can. J. Gen. Cyt.* 10:341–350.

Dowrick, G. F. 1966. Breeding systems in tetraploid *Rubus* species. *Gen. Res.* 7:245–253.

Drain, B. D. 1952. Some inheritance data with black raspberries. *Proc. Amer. Soc. Hort. Sci.* 60:231–234.

———. 1956. Inheritance in black raspberry species. *Proc. Amer. Soc. Hort. Sci.* 68:169–170.

Einset, J. 1947. Chromosome studies in *Rubus*. *Gentes Herb.* 7:181–192.

———. 1951. Apomixis in American polyploid blackberries. *Amer. J. Bot.* 38:768–772.

Einset, J., and C. Pratt. 1954. Hybrids between blackberries and red raspberries. *Proc. Amer. Soc. Hort. Sci.* 63:257–261.

Focke, W. O. 1910–1914. Species *Ruborum* (in German). *Monographiae Generis Rubi Prodromus*. Stuttgart.

Gruber, F., R. L. Knight, and E. Keep. 1962. Fruitbreeding: berries. *Rubus* L. Sub-genera *Idaeobatus* Focke and *Eubatus* Focke. 1. Systematics. 2. Floral biology and seed formation. *Handbuch der Pflanzenzüchtung* 6:477–487.

Gustafsson, A. 1942. The origin and properties of the European blackberry flora. *Hereditas* 28:249–277.

———. 1943. The genesis of the European blackberry flora. *Acta. Univ. Lund* 39:1–199.

Haltvick, E. T., and B. E. Struckmeyer. 1965. Blossom bud differentiation in red raspberry. *Proc. Amer. Soc. Hort. Sci.* 87:234–237.

Haskell, G. 1954. The history and genetics of the raspberry. *Discovery* 15:241–246.

———. 1960a. Biometrical characters and selection in cultivated raspberry. *Euphytica* 9:17–34.

———. 1960b. The raspberry wild in Britain. *Watsonia* 4:238–255.

Haskell, G., and E. B. Paterson. 1964. Quick preparation of corolla chromosomes from flower buds. *Nature* 203:673–674.

Hedrick, U. P. 1925. *The small fruits of New York*. Albany: J. B. Lyon.

Heit, C. E. 1966. Propagation from seed. 7. Germinating six hard-seeded groups. *Amer. Nurseryman* 125: 10–12, 37,39–41, 44–45.

Heit, C. E., and G. L. Slate. 1950. Treatment of blackberry seed to secure first year germination. *Proc. Amer. Soc. Hort. Sci.* 55:297–301.

Heslop-Harrison, Y. 1953. Cytological studies in the genus, *Rubus* L. 1. Chromosome numbers in the British *Rubus* flora. *New Phytol.* 52:22–39.

Hull, J. W. 1961. Progress in developing red raspberries for the south. *Fruit Var. Hort. Dig.* 16:13–14.

———. 1968. Sources of thornlessness for breeding in bramble fruits. *Proc. Amer. Soc. Hort. Sci.* 93: 280–288.

Hull, J. W., and D. M. Britton. 1956. Early detection of induced internal polyploidy in *Rubus*. *Proc. Amer. Soc. Hort. Sci.* 68:171–177.

———. 1958. Development of colchicine-induced and natural polyploid breeding lines of the genus *Rubus* (Tourn.) L. *Md. Agr. Expt. Sta. Bul.* A-91.

Jennings, D. L. 1962*a*. Some aspects of breeding for disease resistance in the raspberry. *Proc. 16th Int. Hort. Cong.* 3:87–91.

———. 1962*b*. Some evidence on the influence of the morphology of raspberry canes upon their liability to be attacked by certain fungi. *Hort. Res.* 1: 100–111.

———. 1963. Some evidence on the genetic structure of present-day raspberry varieties and some possible implications for further breeding. *Euphytica* 12: 229–243.

———. 1964*a*. Some evidence of population differentiation in *Rubus idaeus* L. *New Phytol.* 63:153–157.

———. 1964*b*. Two further experiments on flower-bud initiation and cane dormancy in the red raspberry (var. 'Malling Jewel'). *Hort. Res.* 4:14–21.

———. 1964*c*. Studies on the inheritance in the red raspberry of immunities from three nematode-borne viruses. *Genetica* 34:152–164.

———. 1966*a*. The manifold effects of genes affecting fruit size and vegetative growth in the raspberry. 1. Gene L_1. *New Phytol.* 65:176–187.

———. 1966*b*. The manifold effects of genes affecting fruit size and vegetative growth in the raspberry. 2. Gene l_2. *New Phytol.* 65:188–191.

———. 1967*a*. Balanced lethals and polymorphism in *Rubus idaeus*. *Heredity* 22:465–479.

———. 1967*b*. Observations on some instances of partial sterility in red raspberry cultivars. *Hort. Res.* 7:116–122.

———. 1971*a*. Some genetic factors affecting fruit development in raspberries. *New Phytol.* 70:361–370.

———. 1971*b*. Some genetic factors affecting the development of endocarp, endosperm, and embryo in raspberries. *New Phytol.* 70:885–895.

———. 1971*c*. Some genetic factors affecting seedling emergence in raspberries. *New Phytol.* 70:1103–1110.

Jennings, D. L., D. L. Craig, and P. B. Topham. 1967*a*. The role of the male parent in the reproduction of *Rubus*. *Heredity* 22:43–55.

Jennings, D. L., and P. B. Topham. 1971. Some consequences of raspberry pollen dilution for its germination and for fruit development. *New Phytol.* 70: 371–380.

Jennings, D. L., and B. M. M. Tulloch. 1965. Studies on factors which promote germination of raspberry seeds. *J. Expt. Bot.* 16:329–340.

Jennings, D. L., B. M. M. Tulloch, and P. B. Topham. (Reported by C. North). 1967*b*. Plant breeding: raspberry. *13th Ann. Rept. Scot. Hort. Res. Inst. 1966*. p. 45.

Jinno, T. 1958. Cytogenetic and cyto-ecological studies on some Japanese species of *Rubus*. 2. Cytogenetic studies on some F_1 hybrids. *Jap. J. Gen.* 33:201–209.

Keep, E. 1958. Cytological notes. *Rpt. E. Malling Res. Sta. 1957*. p. 75–78.

———. 1961. Autumn-fruiting in raspberries. *J. Hort. Sci.* 36:174–185.

———. 1964. Sepaloidy in the red raspberry, *Rubus idaeus* L. *Can. J. Gen. Cyt.* 6:52–60.

———. 1968*a*. Inheritance of resistance to powdery mildew, *Sphaerotheca macularis* (Fr.) Jacwewski in the red raspberry, *Rubus idaeus* L. *Euphytica* 17: 417–438.

Keep, E. 1968*b*. Incompatibility in *Rubus* with special reference to *R. idaeus* L. *Can. J. Gen. Cyt.* 10: 253–262.

———. 1968*c*. The inheritance of accessory buds in *Rubus idaeus* L. *Genetica* 39:209–219.

———. 1969*a*. Dwarfing in the raspberry, *Rubus idaeus* L. *Euphytica* 18:256–276.

———. 1969*b*. Accessory buds in the genus *Rubus* with particular reference to *R. idaeus*. *Ann. Bot.* 33:191–204.

Keep, E., and R. L. Knight. 1968. Use of the black raspberry (*Rubus occidentalis* L.) and other *Rubus* species in breeding red raspberries. *Rpt. E. Malling Res. Sta. 1967*. p. 105–107.

Keep, E., R. L. Knight, and J. H. Parker. 1970. Further data on resistance to the *Rubus* aphid, *Amphorophora rubi* Kltb. *Rpt. E. Malling Res. Sta. 1969*. p. 129–131.

Kennedy, G. G., G. A. Schaefers, and D. K. Ourecky. 1973. Resistance in red raspberry to the aphids *Amphorophora agathonica* Hottes and *Aphis rubicola* Oestlung. *HortScience* 8:311–313.

Knight, R. L. 1962. Heritable resistance to pests and diseases in fruit crops. *Proc. 16th Int. Hort. Cong.* 3:99–104.

Knight, R. L., J. B. Briggs, and E. Keep. 1960. Genetics of resistance to *Amphorophora rubi* (Kalt.) in the raspberry. 2. The genes A_2-A_7 from the American variety, 'Chief.' *Gen. Res.* 1:319–331.

Knight, R. L., and E. Keep. 1958. Abstract bibliography of fruit breeding and genetics to 1955—Rubus and Ribes—A survey. *Tech. Comm.* 25. Commonwealth Bur. Hort. Plantation Crops. East Malling, Maidstone, Kent, England.

———. 1960. The genetics of suckering and tip rooting in the raspberry. *Rpt. E. Malling Res. Sta. 1959*. p. 57–62.

Knight, R. L., and E. Keep. 1964. Soft fruit breeding. *Rpt. E. Malling Res. Sta. 1963*. p. 158–160.

Knight, R. L., E. Keep, and J. B. Briggs. 1959. Genetics of resistance to *Amphorophora rubi* (Kalt.) in the raspberry 1. The Gene A_1 from 'Baumforth A.' *J. Genet.* 56:261–280.

Knight, R. L., J. H. Parker, and E. Keep. 1972. Abstract bibliography of fruit breeding and genetics. 1956–1969. *Rubus* and *Ribes*. *Tech. Comm.* No. 32.

Commonwealth Bur. Hort. Plantation Crops. East Malling, Maidstone, Kent, England.
Komarov, V. L., and S. V. Juzepczuk (eds.). 1939. *Flora of the U.S.S.R.* (in Russian). Vol. 9. Moscow: Acad. Scientiarum.
Komarov, V. L., B. K. Schischkin, and S. V. Juzepczuk (eds.). 1941. *Flora of the U.S.S.R.* (in Russian). Vol. 10. Moscow: Acad. Scientiarum.
Lawrence, F. J. 1966. Genetic-biochemical studies of fruit color in *Rubus* species hybrids. *Diss. Abstr.* 26:6272.
Lewis, D. 1939. Genetical studies in cultivated raspberries. 1. Inheritance and linkage. *J. Gen.* 38:367–379.
Morrow, E. B., G. M. Darrow, and D. H. Scott. 1954. A quick method of cleaning berry seed for breeders. *Proc. Amer. Soc. Hort. Sci.* 63:265.
Nybom, N. 1962. A new principle for measuring firmness of fruits. *Hort. Res.* 2:1–8.
Ourecky, D. K. 1969. 'Heritage', a new fall-bearing red raspberry. *Fruit Var. Hort. Dig.* 23:78.
Ourecky, D. K., and G. L. Slate. 1966. Hybrid vigor in *Rubus occidentalis-R. leucodermis* seedlings. *Proc. 17th Int. Hort. Cong.* 1: Abstr. 277.
Overcash, J. P. 1972. 'Dormanred' raspberry: new variety from Mississippi. *Miss. St. Univ. Agr. For. Expt. Sta. Bul.* 793.
Oydvin, J. 1968. Inheritance of some morphological varietal characteristics and yield characters in raspberry *Rubus idaeus* L. (in Norwegian). *Melt. Norg. LandbrHogsk.* 47:1–20 (*Pl. Brdg. Abst.* 39:739).
———. 1970. Important breeding lines and cultivars in raspberry breeding (in Norwegian). *St. Forsokag. Njos.* 1970:1–42.
Pratt, C., and J. Einset. 1955. Development of the embryo sac in some American blackberries. *Amer. J. Bot.* 42:637–645.
Pratt, C., J. Einset, and R. T. Clausen. 1958. Embryology, breeding behavior and morphological characteristics of apomictic, triploid *Rubus idaeus* L. *Bul. Torrey Bot. Club* 85:242–254.
Rietsema, I. 1939. A solution of the mosaic problem in raspberries (in Dutch). *Landbouwk. Tijdschr.* 51: 14–25.
———. 1946. Radbond and Gertrude's raspberries (in Dutch). *Meded. Dir. Tuinb. 1946.* p. 387–390.
———. 1955. Interspecific hybrids in the genera, *Ribes* and *Rubus* (in Dutch). *Rpt. Int. Hort. Cong.* 14: 712–715. (*Plant Brdg. Abstr.* 27:4307).
Rozanova, M. A. 1939. Role of autopolyploidy in the origin of Siberian raspberry (in Russian). *C. R. (Doklady) Acad. Sci. U.S.S.R.* 24:58–60.
———. 1941. Distant hybridization within the genus *Rubus* (in Russian). *Vestnik Gibridizatssi* 2:21–50.
Sanders, J., and J. W. Hull. 1970. Dimethyl sulfoxide as an adjuvant of colchicine in the treatment of *Rubus* seeds and shoot apices. *HortScience* 5:111–112.
Schwartze, C. D. 1934. The genetic constitution of certain red raspberry varieties in relation to their breeding behavior. *Proc. Amer. Soc. Hort. Sci.* 30:113–116.
———. 1935. Rest period response and cold resistance in the red raspberry in relation to the breeding of hardy varieties. Ph.D. dissertation, State College of Washington.
Scott, D. H., and A. D. Draper. 1967. Light in relation to seed germination of blueberries, strawberries and *Rubus*. *HortScience* 2:107–108.
Scott, D. H., and D. P. Ink. 1957. Treatment of *Rubus* seeds prior to after-ripening to improve germination. *Proc. Amer. Soc. Hort. Sci.* 69:261–267.
———. 1966. Origination of 'Smoothstem' and 'Thornfree' blackberry varieties. *Fruit Var. Hort. Dig.* 20:31–33.
Sherman, W. B., and R. H. Sharpe. 1971. Breeding *Rubus* for warm climates. *HortScience* 6:147–149.
Shoemaker, J. S., and T. T. Sturrock. 1959. Chromosome relations in blackberries. *Proc. Fla. St. Hort. Soc.* 72:327–330.
Sistrunk, W. A., and J. N. Moore. 1973. Progress in breeding blackberries. *Ark. Farm Res.* 22(3):5.
Slate, G. L. 1940. Breeding autumn-fruiting raspberries. *Proc. Amer. Soc. Hort. Sci.* 37:574–578.
Slate, G. L., D. K. Ourecky, and J. Watson. 1966. Autumn-fruiting raspberry breeding. *Proc. 17th Int. Hort. Cong.* 1:278 (Abstr.).
Slate, G. L., and J. Watson. 1964. Progress in breeding autumn-fruiting raspberries. *Farm Res.* (N.Y.) 30(2):6–7.
Sudre, H. 1908. *Illustrated monograph of European* Rubus (in Italian). volumes 1–6. Paris.
Tammisola, J., and A. Tyynanen. 1970. Incompatibility in *Rubus arcticus* L. *Hereditas* 66:269–278.
Thompson, M. M. 1961. Cytogenetics of Rubus. 2. Cytological studies of the varieties 'Young', 'Boysen' and related forms. *Amer. J. Bot.* 48:667–673.
Topham, P. B. 1967*a*. Fertility in crosses involving diploid and autotetraploid raspberries. 1. The embryo. *Ann. Bot.* 31:673–686.
———. 1967*b*. Fertility in crosses involving diploid and autotetraploid raspberries. 2. Fruit and seed development. *Ann. Bot.* 31:687–697.
———. 1970*a*. The histology of seed development in diploid and tetraploid raspberries (*Rubus idaeus* L.). *Ann. Bot.* 34:123–135.
———. 1970*b*. The histology of seed development following crosses between diploid and autotetraploid raspberries (*Rubus idaeus* L.). *Ann. Bot.* 34:137–145.
Toyama, T. K. 1961. The breeding behavior of selected varieties of the red raspberry especially with respect to certain fruit characters which may be related to seediness. *Diss. Abstr.* 22:384.
Tun, N. N., and G. Haskell. 1961. A squash technique for small chromosomes of *Rubus* and *Ribes*. *Hort. Res.* 1:62–63.
Ure, C. R. 1963. A study of the parental value of nine red raspberry varieties with respect to combining ability and inheritance of vigor and hardiness. *Diss. Abst.* 23:2671.
Vaarama, A. 1949. Cytogenetic studies on two *Rubus arcticus* hybrids (in Swedish). *Maataloust. Aikakausk.* 20:67–79.
———. 1953. Cytology of hybrids between Pacific

Rubi ursini types and the eastern blackberry. *Amer. J. Bot.* 40:565–570.

Van Adrichem, M. C. J. 1970. Assessment of winterhardiness in red raspberries. *Can. J. Plant Sci.* 50:181–187.

Waldo, G. F. 1934. Fruit bud formation in brambles. *Proc. Amer. Soc. Hort. Sci.* 30:263–267.

Watson, W. C. R. 1958. *Handbook of the* Rubi *of Great Britain and Ireland.* Cambridge, England: Cambridge University Press.

Wilhelm, S., and H. E. Thomas. 1954. Blackberries resistant to wilt. *Calif. Agric.* 8(1):8, 12.

Williams, C. F. 1945. Breeding raspberries for North Carolina. *Prog. Rpt. Res. and Farming* (N.C.) 3 (2):9, 12.

———. 1950. Influence of parentage in species hybridization of raspberries. *Proc. Amer. Soc. Hort. Sci.* 56:149–156.

Williams, C. F., B. W. Smith, and G. M. Darrow. 1949. A Pan-American blackberry hybrid. *J. Hered.* 40:261–265.

Yeager, A. F., and E. M. Meader. 1958. Breeding better fruits and nuts. *N.H. Agr. Expt. Sta. Bul.* 448.

Zych, C. C. 1965. Incompatibility in crosses of red, black and purple raspberries. *Proc. Amer. Soc. Hort. Sci.* 86:307–312.

Grapes

by John Einset and Charlotte Pratt

World production of grapes exceeds that of any other fruit. While the major portion is consumed as wine and spirits, grapes are also used in quantity as fresh fruit and dried as raisins.

The economic importance of grape products has stimulated a large volume of work which is concerned with both the basic taxonomic and cytogenetical problems encountered by breeders and by the demand for cultivars adapted to varied climates and uses. This chapter emphasizes the accomplishments and controversies in these areas, particularly those reported since 1937 (Snyder, 1937). Secondary sources of information, especially in the historical and taxonomic aspects, have been used wherever available, in order to reduce the volume of references. Since such sources are scanty in the technical and cytogenetical areas, a large number of primary sources have been cited for these subjects. From these sources, the student may obtain detailed procedures and evaluate the biological soundness of the present information in order that further advances in grape breeding may be made.

It may appear that this review places undue emphasis on grapes that are not purely *Vitis vinifera* L. since these probably make up a very small portion of the world production. However, it is in this area that the most active breeding programs have been, and are being, pursued. The non-vinifera sources of germ plasm have provided opportunity to incorporate into vinifera resistance to insects, diseases, and environmental extremes that have on occasion threatened to destroy segments of the industry. It has also resulted in the extension of grape growing into areas where the vinifera does not succeed because of uncontrollable pests or temperature extremes.

Origin and Early Development

The cultivation of the grape is a very ancient art. Documents concerning viticulture and wine-making in Egypt take us back some 5000 to 6000 years. Certainly before the beginning of the Christian era, grapes and wine had considerable importance to the Middle Eastern and Mediterranean peoples. As pointed out by Amerine and Singleton (1965), not only was wine important but fresh grapes were, too. These had a high caloric value, being about 20 to 25% sugar. Dried grapes could have a sugar content as high as 80%, and they constituted one of the few sources of sweet materials which could be easily stored and transported.

In the area of the earliest human civilization,

in a region somewhere between the Black and Caspian seas, the grape still grows wild. This region is considered by plant taxonomists to be the original home of the Old World grape, *Vitis vinifera* L. (Snyder, 1937).

From its early beginnings in the Near East the vine spread first around the Mediterranean region. Grape culture next extended inland from the coastal areas. It reached up the Rhone River valley of France and as far north as the Rhine and Moselle valleys at the time of the Romans. In the fifteenth century viticulture became established in Madeira and the Canary Islands. Later it spread to South Africa, Australia, and South America. The first wine grapes were brought to California from Mexico late in the eighteenth century. During the first half of the next century grape growing and wine-making became established in California and rapid expansion took place from 1860 to 1900.

The botanical family, the Vitaceae, is made up of 11 genera and about 600 species, widely distributed in the tropics and sub-tropics with ranges extending into the temperate regions. The genus of greatest economic importance and the only one containing food plants is *Vitis*. This genus is divided into two subgenera, *Muscadinia* Planch., whose members have the somatic chromosome number of 40, and *Euvitis* Planch., the bunch grapes, all of whose species have 38 somatic chromosomes. In *Muscadinia*, *V. rotundifolia* Michx. is indigenous to Florida and the south coast of the United States and will be considered in a later section.

De Lattin (1939), in a study of the origin and world distribution of *Vitis*, has grouped the many species of *Euvitis* into nine sections based essentially on earlier classifications. He has plotted the limits of distribution of these groups on a world map (Fig. 1, p. 148). De Lattin includes 18 species of North American *Euvitis* in his groups. Bailey (1934) includes 28 American species, and his groupings and their designations differ from those of de Lattin. Galet (vol. 1, 1956) states that about 20 species of *Vitis* can be found in the United States and Mexico. Some of the confusion is due to the lack of agreement among systematic botanists and ampelographers as to what constitutes good species, extreme variants and hybrid forms (Barrett et al., 1969; Levadoux et al., 1962). The species of *Euvitis* are interfertile and separated only by geographic, phenologic, and ecologic barriers; Levadoux et al. (1962) would place them in a single coenospecies with 38 somatic chromosomes.

The Asiatic groups (Fig. 1) are made up of 10 to 15 species native to a vast area in eastern Asia, China, Japan, and south into Java. *V. amurensis* Rupr. is perhaps most commonly known. This species is not cultivated but in some areas of northeastern China the edible fruit is collected for use as fresh fruit, juice, and jelly (Pieniazek, 1967). In Russia the species has been crossed with *V. vinifera* L. as a source of resistance to frost and cold (Negrul', 1969; Snyder, 1937).

The European and middle Asian group consists of one species, *V. vinifera* L., which will be discussed in some detail in a later section.

The American species which are of interest to breeders are those that have entered into the origin of bunch grape cultivars. These include the so-called American-type slip-skin grapes like 'Concord,' 'Catawba,' 'Delaware,' and also the French hybrid grapes, mostly identified as numbered selections of Seibel, Seyve-Villard, Landot, Ravat, and several other hydridizers. Another role of the American species has been as rootstocks and as parents of rootstocks that show resistance to various soil pests and also tolerate unfavorable soil conditions, notably high lime content. The species considered by Galet (vol. 1, 1956) to merit description will be listed here:

V. rupestris Scheele, the sand grape of southern Missouri and Illinois, Arkansas, Kentucky, Tennessee, Oklahoma, and eastern and central Texas, is outstanding for its high vigor, phylloxera resistance, and the ease of rooting of its cuttings. A fault is its sensitivity to high lime soils, which induce chlorosis of the vine.

V. riparia Michx. or *V. vulpina* L., the riverbank grape, has a wide range from Canada to Texas and west to Great Salt Lake. It has high resistance to phylloxera, to most fungus diseases, and to winter cold. It roots well and is easy to propagate but is not tolerant of high lime soils.

V. monticola Buckl., the sweet mountain grape of Texas, attracted the attention of viticulturists because of its high resistance to lime in the soil. It is resistant to phylloxera and to certain diseases. However, it propagates with difficulty and has been little used in hybridization.

V. cordifolia Michx., the frost grape, ranges widely from the Great Lakes to Florida. It is vigorous and phylloxera resistant but does not tolerate high lime soils and thus is not suitable as a rootstock in many parts of France.

V. aestivalis Michx., the summer grape, is found from New England to Georgia and westward to the Mississippi River. Many American cultivars have been derived from it. It lacks phylloxera resistance but does carry resistance to fungus diseases and has desirable fruit characteristics. As a rootstock it has not played an important role.

V. lincecumii Buckl, or *V. linsecomii* Buckl. (Bailey, 1934), the post oak grape of the Southwest, is a large-fruited southwestern form of aestivalis, according to Galet (vol. 1, 1956). It is reported to have healthy foliage and moderate phylloxera resistance.

V. bicolor Leconte, also called *V. argentifolia* Bailey, is found in northeastern North America from Canada to northern Georgia and is similar to aestivalis in plant characteristics.

V. candicans Engelm., is the mustang grape of Texas, Arkansas, Oklahoma, Louisiana, and Mexico. It has good resistance to phylloxera and is well adapted to hot, dry conditions. The species is not used directly as a rootstock because of poor take on grafting, poor rooting, and sensitivity to high lime content in soils.

V. cinerea Engelm., the pigeon grape of central and southeastern United States, has good resistance to phylloxera and fungus diseases but is poorly adapted to high lime soils. Another fault is its poor rooting and high percentage of graft failure.

V. berlandieri Planch., the Spanish grape or winter grape of Texas and northern Mexico, has high phylloxera resistance and also a high degree of resistance to lime chlorosis, approaching that of *V. vinifera* and *V. monticola*. Its major faults are poor take obtained on grafting and poor rooting of cuttings.

V. labrusca L., the fox grape, is found from New England to northern Georgia, westward to Indiana and bordering the Ohio River. The species has provided cold resistance, large-berried fruits, and strong distinctive flavor, but it has little resistance to phylloxera.

The breeding characteristics of these and some other species are listed in tabular form by Snyder (1937).

History of Improvement

V. vinifera has given rise to the many thousands of cultivated clones that are grown today in most of the grape areas of the world. Botanists separate the wild vinifera into two subspecies. The first is ssp. *sylvestris* Gmel. of south and central Europe, northwestern Africa, western Turkey, and Palestine. The second, ssp. *caucasia* Vav., is found in Bessarabia, South Russia, Armenia, Caucasia, Anatolia, Iran, Turkestan, and Kashmir, according to de Lattin (1939). In these subspecies occur the vast range of fruit characteristics that are found in the purely vinifera wine, dessert and raisin grapes of the world. The cultivated grape is designated ssp. *sativa* D.C.

The Russian worker Negrul' has further subdivided the subspecies *sativa* into "proles," groups of cultivars indigenous to different geographical areas, viz., proles *pontica* Negr. in the basin of the Black Sea and proles *orientalis* Negr. in Asia. The latter comprises subproles *caspica* Negr. of central Asia and subproles *antasiastica* Negr., large-berried table grapes (Peliakh, 1963).

Levadoux (1956) summarizes the geographical distribution and the morphological and biological characteristics of these proles and lists some French cultivars belonging to them. He also gives a table of synonyms. As results of isolation of eastern and western populations during the glacial period, climatic adaptations of populations during the precultural (neolithic) period, and cultivation by man, populations originally derived from cultivated forms but growing wild today vary from those which have always been wild.

The origin of many of the most important vinifera cultivars is lost in the distant past. The exhaustive four-volume work of Galet (1956–1964) includes descriptions of 229 wine grapes, with more than 1500 synonyms. These include the major cultivars of the vineyards of France and other countries. Sixty of these cultivars each occupy more than 1000 ha (2500 acres) of French vineyards. The origin of the great majority of these cultivars is described as ancient, very ancient, or from time immemorial. The 'Muscat Frontignan' was known to the Greeks and the Romans; the 'Syrah' or 'Petite Sirah' was supposedly brought by the

Roman legions from Syracuse to the Rhone valley; 'Chenin Blanc' was known with certainty in the year 845 A.D. in Anjou. Very few of the vinifera wine types are of relatively recent origin. A rare exception appears to be the 'Müller-Thurgau', a cultivar grown mainly in Alsace and in Germany, which was obtained from a cross of 'Riesling' x 'Sylvaner' made in 1891 by the Swiss, H. Müller-Thurgau. The Alsacian hybridizer C. Oberlin also produced a number of 'Riesling'-type cultivars. 'Goldriesling', planted to some extent before 1919, is the best known of his introductions.

The story of the origin of table grape cultivars is somewhat different. Man has certainly consumed grapes as fresh fruit even before the first wine was made. Most early vineyards in Europe included a number of different cultivars. Sorts more desirable for fresh use were planted with the wine grapes. These provided fruit for the vineyardist and his family and perhaps some for local sale. Several grapes that are major wine cultivars in Europe today are also among the most important in the fresh market. Outstanding examples are 'Golden Chasselas', 'Cinsaut', and 'Muscat of Alexandria'. However, toward the end of the nineteenth century, the production of table grapes as such became a more special and commercialized enterprise. The French, with the English, were the early leaders in selecting desirable fresh fruit cultivars. In France, the names of Vibert, Moreau-Robert, Courtiller, P. Besson, and P. Giraud should be recognized. Hungarian workers, like J. Mathiasz, Stark, Krasznay, P. Kocsis, and the Italians, like A. Pirovano, Bogni, B. Bruni, G. Dalmasso, F. Paulsen, and V. Prosperi, originated many table cultivars that have entered commercial channels.

Galet's list of table grapes (vol. 4, 1964) includes about 175 cultivars. A total of 53 of these are selections of Pirovano, then director of the Institute of Fruit Culture at Rome. Most of these were produced during the first quarter of the twentieth century. Galet's list also includes three introductions from the United States Department of Agriculture program at Fresno, California: 'Cardinal,' introduced in 1939; 'Blackrose,' introduced in 1940; and 'Calmeria,' introduced in 1939 by Snyder. 'Perlette,' introduced by Olmo in 1936 from the program at the University of California at Davis, is also included. The most commercially important table grape introductions made during the twentieth century from planned breeding programs are 'Muscat-Italia' ('Pirovano 65'), 'Cardinal,' and 'Perlette'.

Early Improvement of Native Grapes In the United States

Much has been written about attempts by early colonists to establish vinifera grapes in eastern North America. Determined efforts were made starting early in the seventeenth century and continuing to the latter part of the next century. We know today that the resulting failures were due to lack of resistance to native diseases and soil pests and the low winter temperatures in the more northern areas. It was gradually recognized that the wild native species could contribute resistance to these conditions. The reader is referred to Hedrick (1908) and to Snyder (1937) for comprehensive accounts of the early improvement of the American native grapes which today are the foundation of the grape industry in the areas of the Great Lakes, the Finger Lakes, and the Northwest.

Between 1800 and 1850 the 'Catawba', 'Isabella', and 'Concord' cultivars came on the scene. These and many others introduced during this period were either produced by amateur breeders, or were chance seedlings or selections from the wild. Most of them are interspecific hybrids and have vinifera in their background. The most commonly incorporated native species was V. *labrusca*. One of the earliest hybridizers was E. S. Rogers (1826–1899) of Massachusetts who crossed a large-fruited red labrusca with two vinifera cultivars, 'Black Hamburg' and 'White Chasselas', to obtain the so-called Rogers' Hybrids. Several of these were named, and one of them, 'Agawam,' is still commercially grown today. T.V. Munson of Texas did notable work in botanical studies and hybridizing of grapes. He named and introduced many cultivars especially suited to southern conditions. Many of his introductions have V. *lincecumii* in their parentage as well as V. *labrusca*, V. *aestivalis*, V. *riparia*, V. *rupestris*, V. *champini* Planch., V. *cinerea*, and V. *vinifera*.

Improvement of American Grapes in Europe

Hybridization of native American species assumed importance in France when the devastations of phylloxera made necessary the grafting of vinifera on resistant roots in the last quarter of the nineteenth century (Snyder, 1937). The phylloxera is an insect (*Phylloxera vitifoliae* Fitch), indigenous to eastern and central United States, that feeds on roots of vines and also forms galls on the leaves of some species. This insect was carried to France before 1860, probably on rooted vines imported because they were resistant to powdery mildew (*Uncinula necator* Burr.). In order to combine the most desirable characteristics of different species, interspecific crosses were made between various American species and between vinifera and the American species. Some work was done by official institutions but much by private breeders, nurserymen like M. Contassot, G. Couderc, and A. Seibel. An extension of the early work resulted in the production through breeding of the so-called direct producers or French hybrids that would combine the resistance of the American species with the fruit qualities of V. *vinifera*. Many breeders contributed to this effort, including F. and M. Baco, Bertille Seyve, P. Castel, F. Gaillard, V. Ganzin, Humbert, A. Jurie, E. Kuhlmann, V. Malègue, A. Millardet, C. Oberlin, Peage, A. Roy-Chevrier, V. Rouget, and others. The work has continued into recent years and more names have been added, such as Burdin, A. Galibert, Joannès Seyve, P. Landot, Meynieu, A. and M. Perbos, J.-F. Ravat, Rudelin, and Seyve-Villard (Barrett, 1956). Some measure of the importance of the direct producers to the industry in France may be gained from the fact that in 1961 they occupied approximately 370,000 ha compared to 900,000 for the viniferas (Galet, vol. 3, 1962).

Modern Breeding Objectives

The common objectives of most grape breeding programs are to produce locally adapted, high yielding cultivars with quality that is desirable for the intended use. An important special objective has been the combination of adaptability to environmental extremes and high fruit quality. Grapes are generally grown in the Northern Hemisphere between 20° and 51° N latitude (Snyder, 1937). The most northern extent of cultivars of V. *vinifera* is in the Rhine valley in Germany. The southern range extends into India. In the Southern Hemisphere, grape growing is mainly carried on between 20° and 40° S latitude. The major factors that limit vinifera production to these boundaries toward the poles are the length of the growing season, which must be sufficient to mature the fruit and the canes, and the winter cold. With vinifera, temperatures lower than about −20 C for even a relatively short period result in injury. The limits in the equatorial direction may be determined by disease, the lack of sufficient chilling to induce dormancy, and the excessively high temperatures in the growing season. However, in some areas in India, grape cultivars of both vinifera and American hybrid origin are grown with no real dormant season.

Northern species, especially V. *riparia* and V. *labrusca*, are the major sources of short season, cold hardy sorts. Southern species, such as V. *lincecumii*, V. *aestivalis* var. *bourquiniana*, and especially V. *rotundifolia*, provide tolerance to hot conditions.

The most damaging grape pests are indigenous to the New World and, because the vinifera cultivars have little or no inherent resistance, they created havoc when introduced into the vineyards of Europe during the nineteenth century. The damage caused by *Phylloxera vitifoliae*, the sources of resistance, and how this resistance has been incorporated in rootstocks and in hybrid direct producers was discussed earlier in this chapter.

The first serious disease problem, and probably the most common vine disease in all grape-growing regions today, is the powdery mildew or oidium caused by the fungus *Uncinula necator* Burr. This disease was brought from North America to Europe between 1800 and 1850 and caused severe economic losses in the period 1850–54, until it was found that sulfur gave good control. Twenty to thirty years later downy mildew caused by another fungus, *Plasmopara viticula* Berl. and Toni., became a serious problem. Soon after, black rot, *Guignardia bidwellii* Ellis, appeared in the vineyards of Europe from eastern North America. Sources of resistance to all these diseases may be found in the many native American species. Another disease that can be serious under warm, wet conditions is anthracnose, *Elsinoë ampelina* Shear, one of the oldest European vine diseases and indigenous to Europe and North Africa. Cultivars vary greatly in their susceptibility to the disease, even within the vinifera. Pierce's disease, a destructive disease caused by a bacterium-like organism (Goheen et al., 1973) and transmitted by insect vectors, has been the principal cause of grapevine decline in Florida and other southern vineyard areas. Apparent resistance to the disease has been found in *Vitis smalliana* Bailey, *V. simpsoni* Mun. and *V. champini* (Stover, 1960).

Plant parasitic nematodes of several kinds may attack grape roots. In California, according to Raski et al. (1965), all cultivars and the most commonly used commercial rootstocks may be injured. The importance of the root-knot nematode has been recognized for years. Many other species can also cause serious injury. The problem becomes more acute as growers plant grapes on old agricultural lands formerly used for vineyards, tree crops, or nematode-susceptible vegetable or field crops. The most serious problem is encountered with young vines that may be stunted or killed by high nematode populations in the soil. Some species of nematodes may also carry and transmit such virus diseases as fanleaf of grape which may persist in the soil for many years. Inherent resistance to nematodes is found in certain species native to the south-central states of the United States. The rootstocks 'Dog Ridge' and 'Salt Creek' are nematode-resistant selections of *Vitis champini* and *V. doaniana*, respectively (Loomis and Lider, 1971). Hybrids of these and certain other species have also demonstrated their value as resistant stocks (Weinberger and Harmon, 1966).

Breeding Techniques

Once the objectives of a grape breeding program are determined, and the sources of the characteristics to be combined are available, controlled pollinations must be made and the resulting seeds harvested and grown. The traditional techniques of breeding depend upon basic knowledge of flower development and seed germination. A few new techniques seem to be potentially useful to grape breeders. The theories of Michurin and Lysenko on the inheritance of certain types of adaptation to the environment are being applied in Russia for the selection and pre-conditioning of parents and for influencing the phenotypic expression of hybrid progenies (Pogossian, 1963).

Flower Development

The development of the grape flower is condensed from the detailed review by Pratt (1971). Flower clusters are formed in buds developing during the summer preceding bloom, but flowers develop mostly in the spring. The parts of the flower differentiate in the following order: five sepals, a petal and a stamen jointly arising from each of five primordia, and lastly the pistil which appears as two papillae which later fuse along their margins. The petals are free at first, but later join along their edges to form the cap or calyptra which covers the flower bud. The ovary has typically two locules, each containing two ovules. The papillate glandular epidermis of the stigma continues down the canal of the style. This stigmatoid tissue secretes a sugar solution on the receptive stigma and in the stylar canal. The pollen tube grows intercellularly on the stigma, in the stigmatoid tissue, onto the surface of the attachment of the ovule, down the micropyle, and between the cells of the nucellus into the embryo sac. The development of the pollen begins earlier than that of the ovules.

There are three main types of flowers (Pratt, 1971). In hermaphroditic or perfect flowers the stamens are erect with the anthers producing functional pollen and the pistil is functional. In pistillate flowers, found on female vines of dioecious species or cultivars, the pistil is well developed and functional but the stamens are more or less reflexed and the pollen is generally sterile. Staminate flowers found on male vines have erect stamens with viable pollen and a more or less aborted pistil.

Anthesis occurs most frequently between 6 and 9 A.M. with a rising air temperature, and may also occur from 2 to 4 P.M. (Pratt, 1971). The method of pollination has been debated, but probably most hermaphroditic flowers are self-pollinated. Pollen tubes have been observed in styles within three hours after pollination at 16–27 C (Mayer, 1964). The consequences of fertilization appear two to three days after pollination.

Inheritance of Sex

There are several hypotheses for the inheritance of flower type in hybrids between *V. vinifera* (mostly perfect flowers) and American species (mostly dioecious). Breider and Scheu (1938), studying several species, propose that in *V. vinifera* sex is inherited in a simple Mendelian ratio, but in some other species modifying genes are present. Males are *XY* or *YY* and females *XX*. Hermaphrodites arise from a mutation in the Y-chromosome. Oberle (1938) proposes two dominant linked genes, *So* (suppression of ovules), and *Sp* (development of normal pollen). Loomis (1960) supports this hypothesis. Barrett (1966) has emphasized the quantitative action of these genes, producing in a male vine various degrees of development of the pistil. Negi and Olmo (1970, 1971a), in a detailed study of a wild vine of *V. vinifera* which is predominantly male but also spontaneously partially hermaphroditic, propose another scheme for inheritance of sex: males are $Su^f\ Su^f$ or $Su^f\ Su^m$ and females are $Su^m\ Su^m$, partial hermaphroditism being controlled by minor modifying genes in response to unknown environmental factors. Moore (1970) has pointed out that a permanent change from maleness to hermaphroditism, such as one in *V. rotundifolia* (Detjen, 1917), must be the result of mutation, while temporary increases in the proportion of hermaphroditic flowers in a cluster are the result of environmental factors. Actually, the difference between these two types of natural sex conversion may be quantitative rather than qualitative and based upon a similar mechanism. Doazan and Rives (1967) discuss Oberle's theory, but support Levadoux' (1946) hypothesis of three alleles with a decreasing order of dominance, *M* (male flowers), *H* (hermaphroditic flowers), and *F* (female flowers). Avramov et al. (1967) hypothesize a single pair of genes, S^h, a dominant gene for hermaphroditic flowers, and S^f, a recessive gene for pistillate flowers. Wagner (1967*b*) proposes a pair of genes for flower type, that for hermaphroditic flowers being dominant.

Chemicals Stimulating Development Of Pistil in Male Flowers

Several synthetic and natural cytokinins, 6-substituted adenine derivatives, when applied to a prebloom inflorescence of a male plant of a number of species and interspecific hybrids of grape stimulate the development of the pistil of many flowers so that ovules can be fertilized and develop into viable seeds (Doazan and Cuellar, 1970; Hashizume and Iizuka, 1971; Moore, 1970; Negi and Olmo, 1966, 1970, 1971*b*). The most responsive time for application is when the megaspore mother cell is being formed (Negi and Olmo, 1970, 1971*b*). Moore (1970) has postulated that the flower type produced in an inflorescence depends on the balance of inhibitors of male and female organs under the influence of modifying genes. The addition of exogenous cytokinins at an early stage in the development of the ovule can alter this balance so that the pistil and its ovules continue to develop. Negi and Olmo (1971*b*) have emphasized that the causal factors of both natural and induced sex conversion operate in individual flower buds, rather than on the inflorescence or on the shoot as a whole. Kender and Remaily (1970) found that clusters of a male vine treated with (2-chloroethyl)phosphonic acid (ethephon) produced some hard-seeded berries; ethephon was thought to promote hermaphroditism through an anti-gibberellin action. No material has yet been found to convert female flowers to hermaphroditic flowers by stimulating erect growth of stamens and production of viable pollen by anthers.

Progenies raised from selfing of induced hermaphroditic flowers gave a ratio of 3 males : 1 female (Doazan and Cuellar, 1970; Hashizume and Iizuka, 1971), similar to that obtained from a spontaneously converted vine (Negi and Olmo, 1971a). Male plants can be either homozygous or heterozygous for maleness; their male progenies from selfing gave a ratio of 1 homozygote : 2 heterozygotes (Doazan and Cuellar, 1970).

Induction of femaleness in male clones has several possible applications besides its use to test theories of inheritance of sex in grapes (Moore, 1970). Converted males can be used in selfing or as female parents in crossbreeding for rootstocks. In certain progenies of interspecific hybrids in which only male seedlings reach reproductive age these seedlings can be used for further hybridization (Negi and Olmo, 1971*b*).

Pollen Collection, Storage and Testing

Pollen production by hermaphroditic grape flowers is reported to be comparable to that of apple, 1242–3790 grains per anther in three cultivars (Oberle and Goertzen, 1952). The production of viable pollen by hermaphroditic cultivars varies with the clone (Olmo, 1942*a*) and in some cultivars with the degree of pruning (Shaulis and Oberle, 1948; Winkler, 1926, 1927).

Controlled pollination may be done by cutting

a previously bagged, freshly blooming cluster and tapping it lightly against an emasculated cluster, which is immediately bagged again (Snyder, 1931), or by applying dry pollen with a camel's hair brush (J. H. Weinberger, personal communication). An atomizer has been devised to blow pollen onto emasculated flowers (Barrett and Arisumi, 1952).

Pollen is collected from newly opened flowers. Winkler (1926) allowed anthers to dehisce in vials. Barrett and Arisumi (1952) stripped flowers from the cluster, dried them on a glass plate, then sifted out the pollen. Olmo (1942*b*) harvested clusters on which half of the flowers had opened and tapped the clusters against a clean glass plate. The dry pollen was scraped up with a razor blade and put into small vials or gelatin capsules. The equipment was cleaned with alcohol to kill unwanted pollen.

Grape pollen is stored to use on a female parent of widely different blooming date or for other special purposes. It has been kept for four years at low temperature (−12 C optimum) and low relative humidity (28% optimum) maintained by the appropriate mixture of sulfuric acid and water in a desiccator; pollen which showed a germination percentage of 6% or better gave as good a set in the field as fresh pollen (Olmo, 1942*b*). Other investigators have reported successful storage up to a year at −12 to 8 C and 0–56% relative humidity (Bamzai and Randhawa, 1967; Gollmick, 1942; Nagarajan et al., 1965; Nebel and Ruttle, 1936; Weaver and McCune, 1960).

In vitro testing of grape pollen has been done in connection with storage or pruning experiments rather than as a physiological study. While acetocarmine is useful to indicate the proportion of aborted grains, it does not appear to indicate reliably the viability of the pollen, but no data have been presented. Germination of pollen *in vitro* is said to be a reliable method of predicting the performance of the pollen in setting fruit, but here again no quantitative data are available (Nagarajan et al., 1965).

The usual germination method is to add grape pollen to a hanging drop culture of water with 20% sucrose at 25–30 C or at unspecified temperatures (Olmo, 1942*b*; Weaver and McCune, 1960; Winkler, 1926). Boric acid (5–20 ppm) has been added to increase germination (Bamzai and Randhawa, 1967). Incubation times are given as six to 12 hours (Bamzai and Randhawa, 1967; Olmo, 1942*b*). Agar (0.5–2.0%) with 5–10% sucrose and incubation for two to 24 hours at 26 C have been used (Gollmick, 1942; Nebel and Ruttle, 1936; Mayer, 1964). Some of these investigators stain pollen tubes with chrome-cresol-green or lacmoid; other stains can be used.

Emasculation

Since most hermaphroditic grapes are self-fertile, the buds must be emasculated for use in controlled crosses. The cap and the stamens are removed by fine-pointed forceps. A pair of eyebrow tweezers with broad ends can be notched with a file and the ends bent slightly inwards. The calyptra is grasped between the notches and removed with one motion (J. H. Weinberger, personal communication). Another emasculation tool was devised by Barrett and Arisumi (1952); small, sharp-pointed scissors were notched on the inside of the blades and the degree of closure regulated by a thumb screw.

Attempts to induce male sterility by chemical growth regulators have shown that such substances as gibberellin, 6-benzylaminopurine, maleic hydrazide, napthaleneacetic acid, and 2,3,5-triiodobenzoic acid have rendered pollen impotent, but have also retarded ovule development and reduced seed set (Iyer and Randhawa, 1966; Mohamed, 1968; Weaver and McCune, 1960). The latter also show that pollen germination was progressively reduced by increasing concentrations of gibberellin added to the medium. Two applications of maleic hydrazide at 500 ppm to the inflorescence before meiosis appear to be the most promising method for grape breeding (Iyer and Randhawa, 1965).

Seed Germination

Grape seeds often germinate poorly. Research on germination and seedling growth has been largely the by-product of breeding programs, rather than systematic physiological studies. Many of the experiments reported here are fragmentary.

Vitis seeds are usually extracted from the berries at or soon after fruit maturity by manual pressing or in a laboratory blender operated at low speed to avoid chipping the seeds. Seeds collected two weeks after veraison and held at 10 C for three weeks showed nearly as good germination as those harvested a month later (Rives, 1965).

The seeds have a hard seed coat of variable thickness (Pratt, 1971). Removal of part of the seed coat by sulfuric acid favored germination

(Rives, 1965), if the treatment was not continued too long (Scott and Ink, 1950). Ramirez' (1968) data show that seeds extracted by a blender with sharp-edged blades germinated better than those extracted by hand; the author thinks that the seeds were being scarified.

Without stratification, freshly harvested seeds germinated better than dry seeds (Harmon and Weinberger, 1959; Scott and Ink, 1950). Reports on the occurrence of dwarf, rosetted seedlings from unchilled seeds are at variance with each other (Flemion, 1937; Scott and Ink, 1950). Stratification outdoors in winter or at 0–10 C under moist conditions for about three months is essential for quick, uniform, high germination (Flemion, 1937; Harmon and Weinberger, 1959; Rives, 1965; Scott and Ink, 1950). Séchet (1962) shows that the percentage of germination of seeds of one *V. vinifera* cultivar decreased in both dry and moist storage during one year.

Soaking of unstratified seeds in solutions of growth regulators had only slight effect on seed germination (Chadha and Randhawa, 1967; Randhawa and Negi, 1964; Rives, 1965). Endogenous IAA-like substances have been shown to increase during stratification (Kachru et al., 1969). Reports as to the amount of growth by the embryo during stratification are conflicting (Pratt, 1971).

Grape seeds have been usually germinated in soil, sand, or mixtures containing sand in a greenhouse (Harmon and Weinberger, 1959; Ramirez, 1968; Scott and Ink, 1950; Snyder, 1931; Snyder and Harmon, 1937). Rives (1965) devised a polyethylene grid placed under constant mist in a greenhouse; this facilitated counting of germinated seeds. Other workers germinated seeds in incubators at alternating temperatures (Flemion, 1937) or at 25–30 C (Séchet, 1962; Wagner, 1967a).

The differences in germination between treatments are often less than those between cultivars used as the female parents (Olmo, 1942a; Scott and Ink, 1950). The environment of the female parent has also been thought to influence the degree of germination (Rives, 1965). Wagner (1962) reports that he could nearly triple the yield of seedlings per selfed inflorescence by enclosing a vine in a glasshouse from mid-March to pollination time. His data suggest that this treatment favored the development of the embryo and endosperm rather than of the pollen or pollen tube. Seeds resulting from self pollination germinate more poorly than those from cross pollination (Olmo, 1942a).

Shortening of the Juvenile Period

Another handicap for grape breeders is that under ordinary growing conditions a seedling vine fruits in three to six years (Roos, 1968; Snyder, 1931). This so-called juvenile period has been experimentally shortened to two years by forcing vegetative growth in nutrient sand cultures in a greenhouse and pruning to stimulate the fruitful buds into growth (Huglin and Julliard, 1964; Wagner, 1967a). Another method of hastening fruiting is to force vegetative growth of seedlings and in the first season to insert buds or grafts from them into bearing vines (Negi and Olmo, 1971a; Roos, 1968; Snyder and Harmon, 1937). Selection of seedlings for certain characteristics, such as sex and muscat flavor, can be made before setting the vines in the vineyard, thus reducing the space and labor required (Wagner, 1967a).

Breeding Systems

Changes in the phenotypes of grapes may occur by natural or controlled inter- and intraspecific hybridization with consequent segregation of characteristics. Cytogenetical barriers to sexual reproduction may exert their effect before fertilization, resulting in sterile pollen or ovules, or after fertilization, resulting in seed abortion.

Crosses between species within *Euvitis* (2n = 38) (Darlington and Wylie, 1956) may be readily made. These species show very little difference in karyotype (Raj and Seethaiah, 1969; Vatsala, 1961). Meiosis is generally regular, except for a few lagging chromosomes (Hilpert, 1958; Raj and Seethaiah, 1969; Shetty, 1959). The principal problems are those of dioeciousness in some species and sterility of pollen and ovules of some hybrids, which will be treated under appropriate headings. Crosses between *Euvitis* and *Musca-*

dinia, which differ in chromosome number, are made with difficulty and will be discussed under muscadine grapes.

Inheritance of Some Characteristics

The inheritance of specific characteristics other than flower type, which has already been discussed, has been worked out for a few characteristics which are readily analyzed. The grape is not a convenient plant for genetical analysis due to its long life cycle, large number of chromosomes, partial sterility of ovules, and low seed germination. De Lattin (1957) lists 53 genes, six of them belonging to a polyallelic series. These genes affect anthocyanin formation, chlorophyll formation, hairiness, shape and structure of leaves, number of cotyledons, tendrils, habit, and vigor. He assumes that linkage between some of these characteristics must occur, since there are more genes listed than the haploid number of chromosomes. These genes segregate in the F_2 according to Mendelian ratios. Some of these genes are indirectly of practical significance. For example, de Lattin (1950) shows a relationship between anthocyanin inheritance, lethality, and sugar content of berries.

The genetics of fruit color have been recently reviewed by Barritt and Einset (1969). They propose two pairs of genes with epistatic action: *B*, a dominant gene for black fruit, and *R*, a dominant gene for red fruit. White-fruited grapes are considered to be recessive for both genes.

Other genes have been proposed since de Lattin's list of 1957. Resistance to Pierce's disease requires three dominant genes, Pd_1, Pd_2, and Pd_3 (Mortensen, 1968). Stenospermocarpic fruit is governed by a pair of genes, *S*, with modifying factors (Stout, 1936, 1939). However, Weinberger and Harmon (1964) present evidence that seedlessness, here apparently meaning stenospermocarpy, behaves as a recessive character in vinifera crosses. Two gene pairs govern anthocyanin development: G controlling diglucosides or g controlling monoglucosides, and *O* being specific for triphenols or o being specific for diphenols (Durquety, 1957). According to Wagner (1967*b*), muscat flavor is controlled by five complementary dominant genes. No specific gene for resistance to three important pests has been identified. Resistance to root attack by phylloxera is controlled by several genes (Boubals, 1966). V. *vinifera* is considered by Boubals to be homozygous for sensitivity; other species crossed with vinifera show various degrees of partial dominance of resistance. Resistance to downy mildew in *Vitis* depends on two genic systems: a single gene for necrosis of the stomatal tissues at the time of infection, for which V. *vinifera* is homozygous recessive and resistant species homozygous dominant; several genes for inhibition of growth of mycelium in the host (Boubals, 1959; Coutinho, 1963). Resistance to powdery mildew depends on a separate polygenic system (Boubals, 1961).

Sterility

Most grapes with hermaphroditic flowers and erect stamens, which include most vinifera cultivars, are self-fertile (Beach, 1898). Some cultivars set few or no fruits because of the failure of either pollen or ovules at various stages (Pratt, 1971). Nonfunctional pollen may be the result of breakdown at two points in its development. Abortion of pollen results from irregularities in chromosome distribution at meiosis. It occurs in different degrees in both functional and nonfunctional anthers. In contrast, the sterile pollen found in pistillate flowers with reflexed stamens lacks furrows and germ pores and shows postmeiotic and mitotic disturbances. The absence of germination or fruit setting by pollen of female cultivars has apparently been established.

In flowers of some male vines, the development of the ovule may be arrested at various stages from initiation to mature embryo sac, in proportion to the size of the ovary. In other male cultivars, most ovules and sacs appear normal but are only half the usual size. Not all the ovules in an ovary of a normally seeded cultivar may contain a mature and functional embryo sac at anthesis. Some ovules may be arrested before meiosis. In other nonfunctional ovules even in the same cultivar, the embryo sac fails after meiosis, either with arrest of the sac in its early stages or with degeneration of the egg apparatus of the mature sac. A cultivar characterized by an excessive proportion of nonfunctional ovules is said to be *coulard.*

Seedlessness

Some cultivars bear so-called seedless berries and are popular as table grapes or as raisins. These berries are of two types, those in which the seeds abort while still small and soft (stenospermocarpic), and those in which the seeds do not develop at all (parthenocarpic) (Pratt, 1971). The pollen of such cultivars has generally been found to be viable and has been used in controlled crosses to produce seedless types (Stout, 1936, 1939; Weinberger and Harmon, 1964).

Somatic Mutation

Besides the segregation of characteristics due to sexual reproduction, natural and induced mutations are important sources of variation in grapes. Sporting or mutation is interpreted as the result of a change in a gene in an initial cell of the organ in which it appears. In another view, a clone, especially if it has been cultivated widely for a long time, may have accumulated so many mutations, expressed or latent, that it is somatically heterogeneous (Rives, 1961). The appearance of a new characteristic may be the expression of a preexisting gene change. The distribution of the new characteristic within the plant, if it is not uniform, has been used to study the morphological development of the organ in question (Breider, 1953).

The new characteristic may be of economic significance. Superior clones of some established cultivars have been selected, tested and released (Antcliff, 1967–1969; Olmo, 1964; Rives, 1961). Selection of high-yielding forms should prove advantageous, as would also the suppression of low-yielding clones such as those described by Fleming (1960), Pratt and Einset (1961), Steuk (1945), Winkler (1950), and Woodham and Alexander (1966). Seedless mutants of such seeded cultivars as the following may be of interest as table grapes: 'Catawba' (Steuk, 1945), 'Concord' (Nitsch et al., 1960), 'Emperor' (Olmo, 1940; Winkler, 1950), 'Habshi' (Muthuswamy and Abdul Khader, 1959), and 'Muscat of Alexandria' (Snyder and Harmon, 1940). A seeded mutant was found in the seedless 'Black Corinth' (Harmon and Snyder, 1936; Snyder and Harmon, 1935). Mutants with phylloxera resistance or cold hardiness (Scherz, 1940a) occur but not much use seems to have been made of them.

The most commonly observed mutation is a change in color of a few or all of the berries within a cluster, cane, or vine (e.g., Müller-Stoll, 1950; Breider, 1953). Rives (1961) proposes that such mutants may be periclinal chimeras with the mutation for color in apical meristem layer 1 and that reversions to the original type may occur frequently.

Other reported mutations are variegated leaves (e.g., Antcliff and Webster, 1962), change in hairiness of leaves (Scherz, 1940a), replacement of functional flowers by more or less leafy structures (e.g., Breider, 1962), and small berries (Snyder and Harmon, 1935). Breider (1964) has cataloged 17 spontaneous mutations and their comparative frequency in *Vitis vinifera*.

The induction of mutations by irradiation of dormant buds or cuttings with X or gamma rays or thermal neutrons has been attempted by many workers (Antcliff, 1963–1965, 1965–1967; Kelperis and Daris, 1963; Milosavljević and Mijajlović, 1965; Olmo, 1960; Pratt, 1959; Reichert, 1955; Shimotsuma, 1962). A dose which permits survival of more than 50% of the plants and which induces mutations is about 2000 rads. Since root formation is inhibited by irradiation (Pratt, 1959), rooted cuttings were treated (Pratt, 1959), or the lower node of the two-bud cutting received a lower dose because it was shielded (Antcliff, 1963–1965) or was further from the source of radiation than the upper node (Reichert, 1955).

Grapes have been treated with a chemical mutagen (ethyl methane sulfonate) and mutations have been obtained (Das and Mukherjee, 1968; Sarasola, 1966).

Vegetative growth from irradiated grape buds shows distinctive features, such as distortion of leaves, inhibition of shoot growth, or death of the terminal bud. They have been closely observed as indicators of the direct impact of radiation on the plant (Kelperis and Daris, 1963; Pratt, 1959; Reichert, 1955; Shimotsuma, 1962). Such primary effects are seen only in the first year after irradiation and are distinct from mutations.

The mutated phenotypes, which have been carried on by somatic cell multiplication, often appear first as sectors (Reichert, 1955). They include chlorophyll defects in leaves, change in shape, size or hairiness of the leaf, polyploidy of the epidermis at least, partial sterility due to asynapsis of chromosomes, change in size or seediness of berries, early maturity of fruit and wood, and change in aroma of fruit (Antcliff, 1965–1967; Breider, 1964; Das and Mukherjee, 1968; Olmo, 1960; Reichert, 1955). Breider (1964) reports that a radiation-induced form of 'Perle' is commercially cultivated in Germany for its cold hardiness. A radiation-induced, partially sterile form of the seedless 'Perlette' seems to have commercial promise because its loosely filled clusters do not require as much hand-thinning as does the original 'Perlette'

(Olmo, 1960). Mutation rates in general appear low, but Antcliff (1965–1967) reports than in irradiated 'Sultanina' 33% of the plants from shielded cuttings showed mutations, mostly in vegetative characteristics, but less than 5% of the totally irradiated cuttings were mutated.

Rather few mutations and no new chimeral rearrangements have been recovered from irradiated grapes. The selection of a mutable cultivar appears important, since there seem to be differences in cultivars in this capacity (Reichert, 1955). Even a single dormant bud of grape contains many shoot meristems in various stages of development (Pratt, 1959). They would be expected to vary widely in radiosensitivity and capacity to recover from damage. The irradiated terminal meristem seems to have limited capacity to resume growth (Pratt, 1959). These facts may make the expression and isolation of induced mutants especially difficult in grapes.

Polyploidy

Polyploidization of grapes has been enthusiastically explored by grape breeders as a method of obtaining self-fertile interspecific hybrids with muscadines or large-berried forms of desirable cultivars.

Polyploid forms of grape cultivars have appeared spontaneously in many places (Einset and Lamb, 1951; Einset and Pratt, 1954; Furusato et al., 1955; Gargiulo, 1957, 1969; Golodriga et al., 1970; Olmo, 1936, 1942*c*, 1952; Ourecky et al., 1967; Scherz, 1940*b*; Wagner, 1958). The polyploid branch may arise from a latent bud near a pruning wound (Olmo, 1952*b*). Polyploids are thought to arise from somatic chromosome doubling (endopolyploidy) in initial cell(s) of a shoot meristem. They may be wholly tetraploid or tetraploid with a diploid epidermis (2x–4x) (Thompson and Olmo, 1963).

Detection of polyploid branches is usually by their large berries. Some early reports of large-berried forms are given by Dorsey (1917). Many suspected polyploids have been confirmed by chromosome counts in pollen mother cells (Nebel, 1929), in root tips (Olmo, 1937), or in shoot tips (Einset and Lamb, 1951; Einset and Pratt, 1954; Ourecky et al., 1967; Rives and Pouget, 1959; Thompson and Olmo, 1963). The latter method gives a direct reading of a periclinal chimera. Recently, determination of DNA in the nuclei of the cells in the outer two layers of the apical meristem has been found to give a better separation of 2x and 4x nuclei than nuclear size (Sauer and Antcliff, 1969). Since the characteristics of polyploids have become known, a suspected polyploid can be classified by the size of its stomates (an indicator of the ploidy of layer 1 of the apical meristem) and the size of the pollen grains (an indicator of the ploidy of layer 2).

Tetraploids have been used in breeding to obtain 3x and 4x progeny. The 4x and 2x–4x breed like tetraploids (Gargiulo, 1967; Olmo, 1952*b*). Meiotic disturbances are reported by Alley (1957), Hilpert (1959), and Olmo (1952*b*). Pollen germination of tetraploid vines is lower than that of the original diploid (Alley, 1957; Rives and Pouget, 1959; Wagner, 1958). Polyploids are said to produce fewer functional ovules than diploids but no data were given (Olmo, 1952*b*). The high percentage of empty seeds ("floaters") in mature berries indicates late post-fertilization breakdown (Alley, 1957). In crosses with diploids, more viable seeds are produced by using the tetraploid as the pollen parent rather than as the seed parent (Alley, 1957; Olmo, 1952*b*). Two cultivars from a cross of 4x x 4x have been named in California (Olmo and Koyama, 1962*b*).

Obvious morphological differences between diploids and spontaneous tetraploids have been cataloged by many authors (Einset and Lamb, 1951; Einset and Pratt, 1954; Gargiulo, 1957; Olmo, 1936, 1942*c*, 1952*b*; Rives and Pouget, 1959; Scherz, 1940*b*; Wagner, 1958). Ourecky et al. (1967) conclude that the consistent differences (larger flowers, pollen, seeds, berries, and stomates in tetraploids rather than in diploids) are those dependent on the large size of tetraploid cells; other differences, such as pollen germination and fruit and seed set, may vary with cultivar or with environmental conditions. No consistent difference was found between 2x and 4x vines in physiological characteristics, such as vitamin content of the berry juice (Smith and Olmo, 1944), concentration of cell sap in leaf tissue (Alleweldt, 1961), or transpiration, CO_2 assimilation, and structure of leaves (Geisler, 1961).

To induce polyploidy in grapes most workers have used several applications of a colchicine solution on the shoot tip of a growing bud, but Almeida (1952) applied a heat shock to the inflorescence to produce 3x and 4x seedlings. Colchicine has been generally used as an aqueous solution of 0.25–0.50% of the drug with 5–10% glycerine (Das and Mukherjee, 1967; Gargiulo, 1960; Lelakis, 1957), sometimes with the addition of a wetting agent and preservative (Dermen, 1954). A lower concentration of colchicine has sometimes been used (de Lattin, 1940). The colchicine solution has been injected into the bud (Zuluaga and Gargiulo, 1954). Increase in number of polyploid shoots has been obtained by adding gibberellic acid (10 ppm) to the colchicine solution applied to etiolated shoots (Iyer and Randhawa, 1965). The growth of the induced tetraploid cells is thought to be stimulated by the application of kinetin (1 ppm) seven days after the colchicine treatment (Narasimham and Mukherjee, 1968, 1969*b*).

The early detection of polyploid shoots has been done by visual inspection. Growth of the treated shoot is retarded (Das and Mukherjee, 1967; de Lattin, 1940; Dermen, 1954). The leaves are larger and greener (Lelakis, 1957), or have a dark green mosaic or sectors of 4x cells (Dermen, 1954; Gargiulo, 1960), larger, more numerous stomates (Das and Mukherjee, 1967; Dermen, 1954; Lelakis, 1956, 1957), and a U-shaped sinus (Dermen, 1954).

Selected branches are then more carefully checked for internal polyploidy by chromosome counts in root tips (Das and Mukherjee, 1967; Lelakis, 1956). Narasimham and Mukherjee (1968) used squashes of stem tips for chromosome counts and nuclear volumes. Only Thompson and Olmo (1963) have sectioned shoot tips of colchiploid grapes to determine the presence of a chimera. They found 2x–4x, 4x, and 4x–2x types, the latter being peculiar to the colchiploids, perhaps because of the method of selection.

The morphological characteristics of colchicine-induced tetraploids are described as being similar in general to those of spontaneous polyploids (Bauer, 1968; Dermen and Scott, 1962; Narasimham and Mukherjee, 1970; Zuluaga and Gargiulo, 1954). Larger size of organs here also depends on increased cell size (Das, 1970). Narasimham and Mukherjee (1969a, 1970), working with a group of tetraploids characterized by empty seeds, found that in the development of the berry the lag period (2) before the period of rapid berry enlargement (3) is shorter, that the seeds abort by the beginning of period 3, and that the berries mature earlier than those of the corresponding diploids. These colchiploids behave like spontaneous tetraploids in crosses. Triploids which have been doubled with colchicine to produce 6x plants grow more poorly than the original forms (Bauer, 1968).

One colchiploid grape, derived from 'Loretto', was introduced by Dermen because of its large berry size (Anonymous, 1956). Neither spontaneous or induced polyploids have become economically important for several reasons, notably their irregular bearing (Olmo, 1936, 1937, 1942c).

Breeding for Specific Characters

The more general breeding objectives such as adaptability to environmental extremes and resistance to diseases and pests have been discussed in an earlier section. Although some of the most widely grown grapes are dual or multipurpose in their use, most cultivars serve specific needs. The extreme variation found in fruit characteristics makes the grape the most fascinating of all fruits for the breeder. Based on use, cultivars may be grouped as table or dessert grapes, raisin grapes, wine grapes, and juice grapes.

Dessert Grapes

Grapes that are used fresh as food or for table decoration are generally called table or dessert grapes. Attractive appearance, such as large size of berry, bright color, and unusual shape, is of primary value and often determines the price. When the fruit is stored for lengthy periods or shipped for considerable distances, firm flesh, tough, fairly thick skin, and good adherence of the berry to the pedicel are very important. Fairly loose bunches of even size with uniformly sized berries and long shelf life are important. According to Winkler (1949), less than a dozen cultivars are grown extensively in the world as table grapes and those possess to a marked degree the qualities listed above. Outstanding examples are 'Flame Tokay', 'Emperor', 'Malaga', 'Almeria', and 'Alphonse Lavallcc', all of which are grown in California, South Africa, Australia, Argentina, and elsewhere. All these are pleasant to the taste, but lacking in high or distinctive flavor.

General objectives of table grape breeding are better eating quality and extension of the season either with early maturing types or with late storage types (Snyder and Harmon, 1952).

Seedless grapes have gained in use as table grapes in recent decades. The 'Thompson Seedless' owes its popularity as a table grape mainly to its seedlessness. The objectives of breeding programs with seedless dessert grapes have been to extend the season and to improve the eating quality. 'Perlette' and 'Delight', introduced in 1948 by the University of California at Davis, are very early maturing, seedless cultivars (Olmo, 1948*b*). Other seedless introductions by Olmo are 'Beauty Seedless', 'Emerald Seedless', and 'Ruby Seedless' (J.H. Weinberger, personal communication).

Seedless grapes are also used largely in canned combinations of fruits for fruit salad and fruit cocktail. 'Perlette' and 'Delight' were both found to be superior to the standard 'Thompson Seedless' in canning tests (Olmo, 1948*b*).

Raisin Grapes

Raisin grapes are grapes that produce an acceptable dried product. Raisins are mainly used as sweets and in baking, and, according to Winkler (1949), should possess certain characteristics, namely, a soft texture, little tendency to become sticky, seedlessness, a marked pleasing flavor, and either large or very small size. The raisins of commerce are produced mainly from three cultivars which collectively meet most of these requirements: 'Sultanina', 'Black Cornith', and 'Muscat of Alexandria'.

Wine Grapes

A wine grape is one that is capable of producing an acceptable wine in some locality (Winkler, 1949). Mature fruit of all named grapes will ferment into a kind of wine but only a limited number of cultivars produce standard or higher quality wines. Table or dry wines of quality are produced from grapes of relatively high acidity and moderate sugar while dessert or sweet wines are traditionally made of grapes high in sugar and moderately low in acid. High quality wines are made from cultivars recognized as contributing distinctive flavor, bouquet, and general excellence to the product. Outstanding examples of such cultivars are 'White Riesling', 'Chardonnay', 'Cabernet Sauvignon', and 'Pinot noir', when these are grown under the climatic conditions favorable to the development of these special characteristics. Different cultivars and different conditions of climate are required for the quality sought in such dessert wines as muscatel, sherry, and port.

The main breeding objective is to combine high yield with high quality of the product for the conditions under which the cultivar is to be grown. A few examples of the results of such breeding are given here.

Under the high-temperature ripening conditions of the Mediterranean area, and similarly in most of California's vineyard regions, even the best wine grapes are often too deficient in acid to produce the best quality table wines. Quality cultivars that produce a juice with the desired sugar: acid ratio under these conditions have been a major objective of the breeding work in California (Olmo, 1952a) and in Portugal (Almeida, 1957–1958). High acidity is available from several American species as well as from vinifera.

Specific flavors, aromas, or bouquets of wines are derived from certain cultivars and are often highly prized in the market. Olmo (1948a) has combined the 'Cabernet Sauvignon' aroma with the high production and excellent vineyard performance of 'Carignane' to produce the 'Ruby Cabernet'.

A deficiency of color for red wines of various types has also been a problem in the warmer areas. The most common source of color has been 'Alicante Bouschet', an old cultivar with intense red color in the juice but producing a poor wine. 'Rubired' and 'Royalty' (Olmo and Koyama, 1962a) are results of the University of California, Davis, program that combine high color with improved wine quality and excellent vineyard performance.

Juice Grapes

Grapes used for sweet or unfermented juice retain a fresh grape flavor when the juice is preserved by pasteurization or rendered sterile by ultrafiltration or by a high concentration of SO_2. In America, where pasteurization has been employed to preserve grape juice, the strongly flavored American cultivars, particularly 'Concord', retain their characteristic flavor while most vinifera cultivars lose their fresh flavor and acquire a cooked taste (Winkler, 1949). This fact largely accounts for the general use of 'Concord' for juice in the United States. In Switzerland and other central European countries juices preserved by close filtration are obtained from the traditional wine cultivars, such as 'Golden Chasselas', 'White Riesling', and also from a number of the French hybrids.

Programs of breeding improved juice types have been confined largely to the United States and have the general objective of producing cultivars of 'Concord' juice type adapted to local conditions. The 'Concord' ripens very irregularly under conditions of high temperatures, especially high night temperatures in the period from beginning of berry color to maturity (Shaulis, 1966). For this reason it is not grown successfully in warm areas.

Breeding of Rootstocks

The ideal rootstock must combine a number of characteristics. The plant itself should be vigorous, hardy to cold, and resistant to diseases and pests. With a minimum of care it should produce large quantities of wood for grafting. The cuttings should root readily and profusely and make strong unions with the scion cultivar. The roots should be resistant to injury from phylloxera, nematodes, and other soil pests, and also to soil conditions of high lime, salinity, and drought.

Rootstocks are of major importance in countering the attack of the root louse, phylloxera, which is now commonly present in vineyard soils, but which is not an economic problem in all areas. The major sources of resistance have been derived from V. *riparia*, V. *rupestris*, and V. *berlandieri*, and, while some pure species have been used, in France many important phylloxera-resistant stocks are hybrids of riparia-rupestris, berlandieri-riparia and berlandieri-rupestris. Other stocks whose resistance is somewhat less but nevertheless sufficient, are hybrids of berlandieri-vinifera and riparia-rupestris-vinifera. Such combinations as vinifera-rupestris and vinifera-riparia have been discarded because of difficulties in propagation and hybrids of aestivalis, monticola, and labrusca have been found to be insufficiently resistant to the insect (Galet, 1968).

Lime-induced chlorosis of the vine became a problem in many vineyard areas of France and elsewhere in Europe when American species were used as rootstocks. While the vinifera is quite tolerant to high lime content many of these species are not. The vineyard soils of California rarely contain more than 5% of calcium—too low a proportion to cause an iron deficiency problem, according to Winkler (1962), while in France many of the vineyard soils are of much higher lime content than those of California. V. *berlandieri* has been the most important source of tolerance to high lime and has been crossed with other American species, as well as with vinifera. Outstanding examples of lime-tolerant stocks are '41B' of Millardet, a cross of the vinifera 'Chasselas' x V. *berlandieri*, and '99' ('99R') of Richter, a cross between selections of berlandieri and V. *rupestris*.

In droughty soils the hybrids of berlandieri and rupestris have given the best performance while riparia-rupestris and riparia-berlandieri combinations have shown little adaptation to dry conditions, according to Galet (1968).

The role of nematodes in suppressing growth and in virus transmission has been realized in recent years. A range of resistance has been found in the rootstocks. The chief sources of resistance have been V. *champini* and other species native to the south-central United States.

In the interior of California a nematode- and phylloxera-resistant rootstock ('Couderc 1613' ['Solonis' x 'Othello']) is principally used. In 1965 'Harmony', a selection of a cross between seedlings of 'C. 1613' and 'Dog Ridge' (V. *champini*) from a rootstock breeding program of the United States Department of Agriculture at Fresno, California, was introduced for improved resistance, vigor, and ease of propagation (Weinberger and Harmon, 1966).

The vigor of the vine resulting from a given rootstock-scion combination as it grows in a given soil may be a major factor in deciding what rootstock will be used. Some stocks impart more vigor than others. This may be related to the degree of resistance or tolerance to unfavorable soil conditions (Galet, 1968). The selection of stocks is dependent on the cultivar to be grown and the use to which the crop will be put. For wine grapes to be used for the more ordinary wines and where no special premium is paid for fruit from relatively light-cropping vines, maximum vigor and maximum crops are the goal. For early table grapes and for high quality wines, less vigorous stocks are used. These result in smaller vines, better leaf and fruit exposure, and higher quality as measured by fruit appearance and composition in table grapes, and by early high sugar, color, and other fruit components in wine grapes (Robinson et al., 1970).

Muscadines

Muscadine grapes are derived from *Vitis rotundifolia*, a species adapted to a humid, warm climate and resistant to many diseases and pests (Olmo, 1971). They differ from the vinifera in dioeciousness and in bark, leaf, and cluster characteristics (Patel and Olmo, 1955). They differ from all the bunch grapes in having 40, rather than 38, somatic chromosomes.

The early cultivars were selected directly from the wild (Reimer, 1909), such as 'Scuppernong' which occurs in several strains differing in season of ripening and fruit characteristics (Woodroof, 1934). The inheritance of the characteristics of *V. rotundifolia* was worked out early in the breeding programs (e.g., Dearing, 1917; Detjen, 1917; Loomis et al., 1954; Williams, 1923). Objectives of controlled hybridization have been greater vigor, larger clusters and berries, and especially hermaphroditic flowers (Loomis and Williams, 1957). Since most cultivated forms of *V. rotundifolia* are pistillate, hermaphroditic cultivars, rather than sterile male vines, are recommended as pollinators (e.g., Crops Research Division, 1961).

Polyploidy in muscadine grapes has been reported in experiments with gamma rays on several cultivars (Fry, 1963). Fry obtained 2x–4x chimeras on most of his plants by suitable treatments. They showed the usual characteristics of polyploids. There are no reports available yet on the use of these tetraploids in breeding.

Hybridization of *Euvitis* species with *V. rotundifolia* has been carried on for over a century. The objective is to combine the pest and warm climate resistance of *V. rotundifolia* with the large cluster and desirable berry qualities of *V. vinifera* (Olmo, 1971) and other bunch grapes.

Two serious obstacles stand in the way of achieving this goal. Pollen of *V. rotundifolia* will fertilize the egg of *V. vinifera*, but the reciprocal cross is less successful (Dearing, 1917; Jelenković and Olmo, 1968; Patel and Olmo, 1955). It does not appear to be a matter of incompatibility of pollen tube and style, but failure occurs just before fertilization (Patel and Olmo, 1955). Partially fertile F_1 hybrids ($2x=39$) can cross reciprocally between themselves or with *V. vinifera*, but with *V. rotundifolia* only when the latter is used as the male parent (Jelenković and Olmo, 1968).

The second obstacle is the sterility of most F_1 hybrids between *Euvitis* and *V. rotundifolia* (Dunstan, 1962, 1964, 1967; Jelenković and Olmo, 1968; Patel and Olmo, 1955). It is caused by abnormal pairing and irregular distribution of chromosomes in PMC at meiosis, as in many hybrids of distantly related parents (Jelenković and Olmo, 1968; Patel and Olmo, 1955), and it probably also arises from differences in genomes (Nesbitt, 1966). Ovule fertility is greater than pollen fertility (Jelenković and Olmo, 1968). Nesbitt (1966) notes the lack of synchronization of stages of meiosis in anthers; four megaspores are usually formed, but since he did not study later stages, he assumes that abortion of ovules occurs after meiosis. Some of the sterile hybrids have been proposed as disease-resistant rootstocks (Davidis and Olmo, 1964).

Various methods have been tried to overcome the sterility of *Euvitis*-rotundifolia hybrids. Some parents, especially certain selections of *V. vinifera*, have produced more fertile F_1 individuals than others (Dearing, 1917; Jelenković and Olmo, 1968). Dunstan (1962, 1967) found several examples in which a bunch grape which was itself an interspecific hybrid produced more fertile offspring than a purely *V. vinifera* female parent. A second method of recombining genomes of *Euvitis* and *V. rotundifolia* in a more fertile hybrid is to backcross a partially fertile F_1 with one of the parents for one or more generations (e.g., Dunstan, 1962, 1964, 1967; Fry, 1968; Jelenković and Olmo, 1969*a*; Olmo, 1971). A third method of increasing fertility is that of doubling the chromosome number of parent species and F_1 hybrids (e.g., Dermen, 1954). This does not change the crossability pattern of the parents or of the original F_1 hybrids (Jelenković and Olmo, 1969*b*), but does increase berry and bunch size (Dermen, 1964). Chromosome pairing is better in the tetraploids, but there appears to be no relationship between the degree of pairing and fertility (Jelenković and Olmo, 1969*b*). Spontaneous tetraploidy appeared in the progeny of a *V. vinifera* x *V. rotundifolia* cross and was accompanied by self-fertility and increased bunch size (Dermen et al., 1970). However, no tetraploid has as yet been successful in commercial cultivation (Olmo, 1971).

A triploid vine from a complex hybrid female x a male selection of *V. rotundifolia* produced a few seeds when used as a pollen parent on a selection of *V. vinifera*; its meiotic pairing of chromosomes was very irregular (Patel and Olmo, 1957).

Achievements and Prospects

Grape breeding programs are carried on in many countries of the world. These programs are varied in scope and size. Some are concerned with rootstocks, others with wine, dessert, and raisin types. Most of these programs are located at viticultural institutes and plant breeding stations listed by Mao (1961, 1965, 1967) and the International Society for Horticultural Science (1966), such as the following centers in Europe: In Czechoslovakia active breeding programs with vinifera are indicated for two institutions, one at Bratislava and the other at Znojmo. In France programs are reported at Pont de la Maye, Colmar, Marseillan Plage, and Montpellier. In Germany the institutes at Geisenheim, Landau, and Wurzburg are deeply involved in breeding programs concerned with cultivar and rootstock improvement. In Hungary programs are carried on by the Ministry of Agriculture and the College of Horticulture and Viticulture in Budapest. In addition two other programs are listed at the Establishment "Mathiasz Janos" at Kecskemet-Miklostelep and at the Agricultural College at Keszthely. Almeida at the Centro Nacional de Estudos Vitivinicolas at Lisbon, Portugal, has introduced a number of table grapes as well as wine types since 1960 (Almeida and Grácio, 1969). A grape breeding program is indicated at the Institut Agronomique, "N. Balcescu" in Bucharest, Romania, and another at the Viticultural Research Institute at Sremski Karlovici, Yugoslavia. The Bulgarian grape breeding is carried on at Pleven (Stoev, 1970). The U.S.S.R. has breeding programs at several stations (Negrul', 1969; Williams, 1971).

Such programs are not limited to Europe. In Africa, breeding programs are conducted at two institutions in Algeria. Work is in progress in the Union of South Africa at the Pretoria Horticultural Research Station and at the Fruit and Food Technology Research Station at Stellenbosch where both wine and table grape breeding is carried on. Japan has grape programs, which are reported to include breeding, at six different locations. Two such programs are reported for India and one for Pakistan. Australia in recent years has activated a major project designed to increase and improve the existing range of planting material through the creation of new wine and seedless cultivars through hybridization (Antcliff, 1969–1971). In South America Argentina has listed active breeding programs at the Agricultural Institute at Castelar and the Agricultural Experiment Station at General Roca. Brazil has a very large program at Campinas, S.P. (H.P. Olmo, personal communication).

According to a survey of fruit breeders in North America, made in 1970 by the Fruit Breeding Committee of the American Society for Horticultural Science, active breeding programs were carried on in 11 states: Arkansas, California, Florida, Georgia, Missouri, New York, North Carolina, Oklahoma, South Carolina, South Dakota, and Virginia. In addition, the USDA has active programs. A sizeable program is carried on in Canada, at Vineland, Ontario. Of the programs in the states, one has been terminated in the last decade, two others will likely be discontinued soon, while four programs have received rather strong additional support in this period.

Breeding with pure vinifera is carried on only in California. The program of the USDA at Fresno was started in 1923 with objectives of improving all types of grapes (Snyder and Harmon, 1952), but work on wine and juice grapes has been discontinued (J. H. Weinberger, personal communication). Results of this program have already been cited. The University of California program at Davis was started in 1931, originally to improve seedless cultivars and table grapes. As has been described, the program expanded to include wine types and juice types. Recent work has also included the muscadines (Olmo, 1971).

An expanded program has been initiated in Arkansas in the last decade. Major emphasis has been put on table grape breeding as offering the greatest possibility for improvement, since the juice and wine industries are quite well served by the present cultivars. Much vinifera has been involved in crosses; hardiness to cold is less of a problem than disease under Arkansas conditions.

A breeding program at Leesburg, Florida, was established in the mid-1940s to combine the resistance and longevity of native Florida species (*Vitis coriacea* Shuttleworth, *V. simpsoni* Mun., and *V. smalliana* Bailey) with the acceptable flavors and desirable fruit qualities of vinifera and its hybrids (Mortensen, 1971; Stover and Mortensen, 1963).

Some introductions resulting from this program have been named.

Earlier programs in Georgia were concerned entirely with improvement of vine and fruit qualities of *V. rotundifolia*. Now that fruitful hybrids have been obtained between the bunch grapes and the muscadines, a program has been initiated to combine the desirable characteristics of both groups. At present emphasis is also placed on obtaining vines adapted to mechanical harvest and machine pruning (Lane, 1969).

In Missouri grape breeding has been carried on since 1933 at Mountain Grove to improve the quality, vigor, and disease resistance of the American grapes.

The New York program at Geneva was started in 1888. The major objective through the years has been to combine vinifera fruit quality with cold hardiness, vigor, productiveness, and disease resistance in dessert, juice, and wine types. The emphasis in the last decade has been to approach vinifera-type quality in table wines. American, vinifera, and French hybrid cultivars have been used extensively as parents.

An expanded program of muscadine breeding was activated at North Carolina after 1960. This goes back many years to a joint program at Raleigh between the University of North Carolina and the USDA. The department's muscadine work in North Carolina was started at Willard about 1907.

The USDA project at the United States Horticultural Field Station at Meridian, Mississippi, from 1941 through 1965 obtained much of its basic breeding material from the early program at Willard.

An earlier program at Stillwater, Oklahoma, to breed juice and dessert grapes adapted to local conditions is being phased out.

A project of breeding bunch grapes for the Southeast is carried on at the South Carolina Agricultural Experiment Station at Clemson.

In South Dakota the program was started early in the century by N. E. Hansen to produce grapes that would stand the extreme winters. The work continues today through the crossing of high-quality cultivars with the native hardy sorts.

The Virginia program started in 1930 at Blacksburg to improve the standard American cultivars by incorporating resistance to black rot and downy and powdery mildew, and by improving texture and flavor, but the program may be discontinued in the near future.

The USDA, in addition to the vinifera work in California and the earlier muscadine programs in North Carolina and Mississippi already mentioned, has carried on a program of bunch grape breeding at Beltsville, Maryland, which has dealt largely with disease resistance, primarily to black rot.

The grape breeding program at the Horticultural Research Institute of Ontario at Vineland, Ontario, Canada, was initiated in 1913, at first with the objectives of originating improved fresh fruit cultivars for shipping and storing. Later, as more of the crop was used by wineries, more emphasis was placed on wine types. Dessert-wine and table-wine selections have been introduced, using as parents mainly American and French hybrids and but few vinifera cultivars.

These many programs will certainly contribute new sorts that will improve the cultivar picture as it exists today. The more thorough understanding of the cytology, genetics, and breeding behavior of *Vitis* that has been gained since the turn of the century presents us with useful guidelines for future work. The great diversity of plant material with which to work and the interfertility of most of the many and varied species offer great promise. Well-planned programs of adequate size and scope will surely yield desirable fruit quality in vines that will bear successfully beyond the present commercial geographical ranges and harvest seasons. Increasing attention will be given to the incorporation of inherent pest resistance with the aim of eventual elimination of chemicals for pest control.

The increasing occurrence of leaf symptoms related to oxidant pollution of the air and specifically ozone injury (Richards et al., 1958) which may sharply reduce grape yields (Thompson and Kats, 1970) emphasizes the need to select tolerant individuals in breeding populations. Richards et al. (1958) found marked differences in ozone injury among different cultivars. Cultivars have also been found to vary in their tolerance to injury from atmospheric contamination of 2,4-dichlorophenoxyacetic acid (2,4-D) (Abmeyer, 1969).

Mechanical harvesting by machines that shake the berries or cluster parts off the vine has progressed more rapidly in some areas than in others. In New York more than 85% of the crop was thus harvested by 1971. The berries of some cultivars can be shaken off and recovered by catching frames with little berry or juice loss. Other cultivars are harvested with much greater difficulty. This is related to the cluster structure and compactness, which are characteristics now even more important than formerly in the selection of seedlings.

Pruning of the vine is the last major cultural operation that has not been mechanized. The natural growth habit of the vines as well as the training system will prove to be critical in the solution of this problem. Here, as with so many characteristics, a wide range of growth habits in *Vitis* species is available to the breeder.

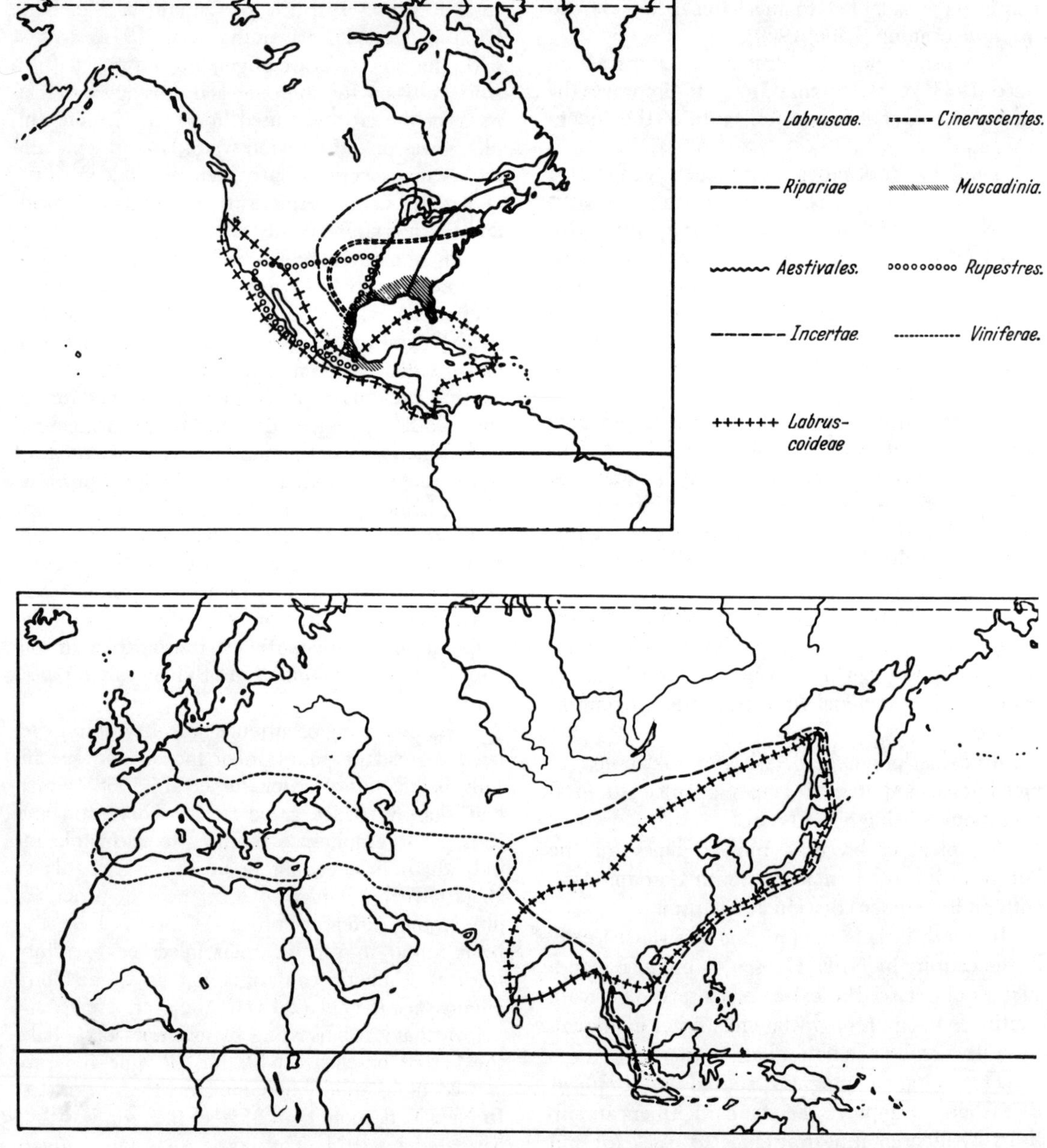

FIG. 1. The distribution of the sections of the genus Vitis (de Lattin, 1939).

Literature Cited

Abmeyer, E. 1969. Tolerance of several grape cultivars to injury from atmospheric contaminations of 2,4-D. *Fruit Var. Hort. Dig.* 23:53.

Alleweldt, G. 1961. Investigations on the cell sap concentration of grapes. 2. The osmotic value of diploid and tetraploid grapes as well as of a few interspecific hybrids (in German). *Vitis* 3:48–56.

Alley, C. J. 1957. Cytogenetics of Vitis. 2. Chromosome behavior and the fertility of some autotetraploid derivatives of *Vitis vinifera* L. *J. Hered.* 48: 194–202.

Almeida, J. L. F. de. 1952. Advance note on the production of polyploids in *Vitis vinifera* L. *Agron. Lusit.* 14:173–174. (*Plant Brdg. Abstr.* 23:2983).

——— 1957–1958. Improvement of grape vines. Preliminary note (in Portuguese). *A Junta Nacional Vinho* 9:5–16.

Almeida, J. L. F. de, and A. M. Grácio. 1969. Macrozones of table grapes in continental Portugal (in Portuguese). *Vin. Port. Doc.* Ser. 1, 4:2:1–168.

American Society for Horticultural Science Fruit Breeding Committee. 1970. Register of fruit breeders of North America. (Mimeo.)

Amerine, M. A., and V. L. Singleton. 1965. *Wine. An introduction for Americans.* Berkeley: Univ. of Calif. Press.

Anonymous. 1956. New tetraploid grape. *Fruit Var. Hort. Dig.* 11:3.

Antcliff, A. J. 1963–1965. Selection following irradiation. *Rpt. 1963–1965 C.S.I.R.O. Sect. Hort. Res.* p. 33–34.

———. 1965–1967. Selection following irradiation. *Rpt. 1965–1967 C.S.I.R.O. Div. Hort. Res.* p. 10–11.

———. 1967–1969. Clonal selection. *Rpt. 1967–1969 C.S.I.R.O. Div. Hort. Res.* p. 39–40.

———. 1969–1971. Hybridization of vines. *Rpt. 1969–1971 C.S.I.R.O. Div. Hort. Res.* p. 25.

Antcliff, A. J., and W. J. Webster. 1962. Bruce's sport —a mutant of the Sultana. *Australian J. Expt. Agr. Anim. Husb.* 2:97–100.

Avramov, L. et al. 1967. Inheritance of flower type in some grape varieties (*Vitis vinifera* L.) *Vitis* 6:129–135.

Bailey, L. H. 1934. The species of grapes peculiar to North America. *Gentes Herbarum* 3:150–244.

Bamzai, R. D., and G. S. Randhawa. 1967. Effects of certain growth substances and boric acid on germination, tube growth and storage of grape pollen (Vitis ssp.) *Vitis* 6:269–277.

Barrett, H. C. 1956. The French hybrid grapes. *Nat. Hort. Mag.* 35:132–144.

———. 1966. Sex determination in a progeny of a self pollinated staminate clone of Vitis. *Proc. Amer. Soc. Hort. Sci.* 88:338–340.

Barrett, H. C., and T. Arisumi. 1952. Methods of pollen collection, emasculation and pollination in fruit breeding. *Proc. Amer. Soc. Hort. Sci.* 59:259–262.

Barrett, H. C., S. G. Carmer, and A. M. Rhodes. 1969. A taximetric study of interspecific variation in Vitis. *Vitis* 8:177–187.

Barritt, B. H., and J. Einset. 1969. The inheritance of three major fruit colors in grapes. *J. Amer. Soc. Hort. Sci.* 94:87–89.

Bauer, O. 1968. Polyploid Vitaceae, experimental production and comparative studies of polyploid grapes and their initial forms (in German). *Mitteilungen* (Klosterneuberg, Austr.) (Ser. A.) 20:77–78. (Abstr.)

Beach, S. A. 1898. Self-fertility of the grape. N.Y. (*St.*) *Agr. Expt. Sta., Geneva, Bul.* 157:397–441.

Boubals, D. 1959. Contribution to the study of the causes of resistance of Vitaceae to downy mildew (*Plasmopara viticola* [B. and C.] Ber. and de T.) and their inheritance (in French, English summary). *Ann. Amélior, Plantes* 9:5–233.

———. 1961. Study of the causes of resistance of Vitaceae to vine powdery mildew—*Uncinula necator* (Schw.) Burr.—and their inheritance (in French, English summary). *Ann. Amélior. Plantes* 11:401–500.

———. 1966. Inheritance of resistance to radicicolous Phylloxera in the vine (in French, English summary). *Ann. Amélior. Plantes* 16:327–347.

Breider, H. 1953. The somatic mutations of grape and their relations with the structure of branches (in French). *Prog. Agr. Vitic.* 139:43–47; 70–73.

———. 1962. Spontaneous flower mutations in grape (in German). *Züchter* 32:100–102.

———. 1964. On the exploitation for breeding and the practical utilization of X ray-induced somatic mutations in long-lived and vegetatively propagated cultivated plants (represented by investigations on *Vitis vinifera*) (in German, English summary). *Mitteilungen* (Klosterneuberg, Austr.) (Ser. A) 14: 165–171.

Breider, H., and H. Scheu. 1938. The determination and inheritance of sex in the genus Vitis (in German). *Gartenbauwissenschaft* 11:627–674.

Chadha, K. L., and G. S. Randhawa. 1967. Studies on grape seed germination—a review. *Indian J. Hort.* 24:181–187.

Coutinho, M. P. 1963. Notes on the genetical aspect of the resistance of vines to *Plasmopara viticola* (in Portuguese, French summary). *Agron. Lusit.* 25: 355–365. (*Plant Brdg. Abstr.* 37:5033).

Crops Research Division, Agricultural Research Service. 1961. Muscadine grapes, a fruit for the South. *USDA Farmers' Bul.* 2157.

Darlington, C. D., and A. P. Wylie. 1965. *Chromosome atlas of flowering plants.* New York: Hafner.

Das, P. K. 1970. Studies on some morphological and anatomical characteristics of induced tetraploid grapes (*Vitis vinifera* L.). *Technology* 7:30–33.

Das P. K., and S. K. Mukherjee. 1967. Induction of autotetraploidy in grapes. *Indian J. Gen. Plant Brdg.* 27:107–116.

———. 1968. Effect of gamma radiation and ethyl

methane sulphonate on seeds, cuttings and pollen in grapes. *Indian J. Gen. Plant Brdg.* 28:347–351. (*Biol. Abstr.* 52:31585; *Plant Brdg. Abstr.* 40: 8503.)

Davidis, U. X., and H. P. Olmo. 1964. The *Vitis vinifera* x *V. rotundifolia* hybrids as phylloxera resistant rootstocks. *Vitis* 4:129–143.

Dearing, C. 1917. Muscadine grape breeding. *J. Hered.* 8:409–424.

De Lattin, G. 1939. On the origin and distribution of grapes (in German). *Züchter* 11:217–225.

———. 1940. Spontaneous and induced polyploidy in grapes (in German). *Züchter* 12:225–231.

———. 1950. On the lethality of the anthocyanin gene in grapes (in German). *Naturwissenschaften* 37: 428–429.

———. 1957. On the genetics of grapes. Present results of factor analysis in the genus Vitis (in German). *Vitis* 1:1–8.

Dermen, H. 1954. Colchiploidy in grapes. *J. Hered.* 45: 159–172.

———. 1964. Cytogenetics in hybridization of bunch and muscadine-type grapes. *Econ. Bot.* 18:137–148.

Dermen, H., F. N. Harmon, and J. H. Weinberger. 1970. Fertile hybrids from a cross of a variety of *Vitis vinifera* with *V. rotundifolia. J. Hered.* 61: 269–271.

Dermen, H., and D. H. Scott. 1962. Potentials in colchiploid grapes. *Econ. Bot.* 16:77–85.

Detjen, L. R. 1917. Inheritance of sex in *Vitis rotundifolia. N.C. Agr. Expt. Sta. Tech. Bul.* 12.

Doazan, J.-P., and V. Cuellar. 1970. Artificial modification of sex expression in the genus Vitis (in French, English summary). *Ann. Amélior. Plantes* 20:79–86.

Doazan, J.-P., and M. Rives. 1967. On the genetical control of sex in the genus Vitis (in French, English summary). *Ann. Amélior. Plantes* 17:105–111.

Dorsey, M. J. 1917. The inheritance and performance of clonal varieties. *Proc. Amer. Soc. Hort. Sci.* 13: 41-71.

Dunstan, R. T. 1962. Some fertile hybrids of bunch and muscadine grapes. *J. Hered.* 53:299–303 (Corrigendum 54:25 [1963]).

———. 1964. Hybridization of Euvitis x *Vitis rotundifolia*: backcrosses to muscadine. *Proc. Amer. Soc. Hort. Sci.* 84:238–242.

———. 1967. Inheritance of inflorescence in bunch grape-muscadine hybrids. *J. Hered.* 58:235–237.

Durquety, P. M. 1957. Note on the genetics of grape anthocyanins (in French). *Prog. Agr. Vitic.* 147: 309–315.

Einset, J., and B. Lamb. 1951. Chimeral sports of grapes. *J. Hered.* 42:158–162.

Einset, J., and C. Pratt. 1954. "Giant" sports of grapes. *Proc. Amer. Soc. Hort. Sci.* 63:251–256.

Fleming, H. K. 1960. Concord grape vines transmit productivity. *Penn. State Univ. Coll. Agr. Prog. Rpt.* 217.

Flemion, F. 1937. After-ripening at 5° C. favors germination of grape seeds. *Contrib. Boyce Thompson Inst.* 9:7–15.

Fry, B. O. 1963. Production of tetraploid muscadine (*V. rotundifolia*) grapes by gamma radiation. *Proc. Amer. Soc. Hort. Sci.* 83:388–394.

———. 1968. Hybridization of Euvitis x *Vitis rotundifolia. Proc. Assoc. Southern Agr. Workers* 65:179 (Abstr.) (*Plant Brdg. Abstr.* 39:1364).

Furusato, K., K. Ishibashi, and Y. Ohta. 1955. Polyploid grape varieties (in Japanese). *Ann. Rpt. Nat. Inst. Gen. Japan* 6:70–71. (*Plant Brdg. Abstr.* 27:3826).

Galet, P. 1956–1964. *Varieties and vineyards of France* (in French). Montpellier: Déhan.

———. 1968. *Treatise on practical ampelography* (in French). Montpellier: Déhan.

Gargiulo, A. 1957. Spontaneous tetraploid mutation in Barbera d'Asti (in Spanish). *Vitis* 1:156–158.

———. 1960. Artificial induction of polyploidy in *Vitis vinifera* with colchicine (in Spanish). *Vitis* 2:181–189.

Gargiulo, A. A. 1967. Barbera d'Asti, a tetraploid cultivar excellent for blending wine (in Spanish). *Vitis* 6:364–365.

Geisler, G. 1961. Investigations on transpiration, CO_2 assimilation, respiration and leaf structure of spontaneous tetraploid mutants of *Vitis vinifera* in comparison with the diploid sources (in German). *Züchter* 31:98–106.

Goheen, A. C., G. Nyland, and S. K. Lowe. 1973. Association of a rickettsialike organism with Pierce's disease of grapevines and alfalfa dwarf and heat therapy of the disease in grapevines. *Phytopathology* 63:341–345.

Gollmick, F. 1942. On the longevity of grape pollen (in German). *Angew. Bot.* 24:221–232. (*Biol. Abstr.* 24:10438).

Golodriga, P. Y., P. V. Korobets, and S. G. Topale. 1970. Spontaneous tetraploid mutants of grape (in Russian, English summary). *Tsitol. Genet.* 4:24–29. (*Biol. Abstr.* 51:93466).

Harmon, F. N., and E. Snyder. 1936. A seeded mutation of the Panariti grape. *J. Hered.* 27:77–78.

Harmon, F. N., and J. H. Weinberger. 1959. Effects of storage and stratification on germination of vinifera grape seeds. *Proc. Amer. Soc. Hort. Sci.* 73:147–150.

Hashizume, T., and M. Iizuka. 1971. Induction of female organs in male flowers of Vitis species by zeatin and dihydrozeatin. *Phytochemistry* 10:2653–2655.

Hedrick, U. P. 1908. The grapes of New York. *N.Y. (St.) Agr. Expt. Sta., Geneva, Rpt.* 1907.

Hilpert, G. 1958. Investigations on early meiotic stages of *Vitis vinifera* L. (in German). *Vitis* 1:218–223.

———. 1959. Investigations of pachytene in *Vitis vinifera* L. (in German). *Vitis* 2:79–83.

Huglin, P., and B. Juilliard. 1964. On obtaining very vigorous, early fruiting grape seedlings and their significance for genetic improvement of grapes (in French, English summary). *Ann. Amélior. Plantes* 14:229–244.

International Society for Horticultural Science. 1966. *Horticultural research international. Directory of horticultural research institutes and their activities in 23 countries.* Wageningen: Centre Agr. Publ. Doc.

Iyer, C. P. A., and G. S. Randhawa. 1965. Increasing colchicine effectiveness in woody plants with special reference to fruit crops. *Euphytica* 14:293–295.

———. 1966. Induction of pollen sterility in grapes (*Vitis vinifera*). *Vitis* 5:433–445.

Jelenković, G., and H. P. Olmo. 1968. Cytogenetics of Vitis. 3. Partially fertile F_1 diploid hybrids between V. *vinifera* L. x V. *rotundifolia* Michx. *Vitis* 7:281–293.

———. 1969*a*. Cytogenetics of Vitis. 4. Backcross derivatives of V. *vinifera* L. x V. *rotundifolia* Michx. *Vitis* 8:1–11.

———. 1969*b*. Cytogenetics of Vitis. 5. Allotetraploids of V. *vinifera* L. x V. *rotundifolia* Michx. *Vitis* 8:265–279.

Kachru, R. B., E. K. Chacko, and R. N. Singh. 1969. Physiological studies on dormancy in grape seeds (*Vitis vinifera*). 1. On the naturally occurring growth substances in grape seeds and their changes during low temperature after ripening. *Vitis* 8:12–18.

Kelperis, J. P., and B. T. Daris. 1963. A contribution to research on the effects of gamma rays on vine tissues (in Greek, English summary). *Delt. Inst. Ampel.* (Athens) 2:7–15. (*Hort. Abstr.* 34:6502).

Kender, W. J., and G. Remaily. 1970. Regulation of sex expression and seed development in grapes with 2-chloroethylphosphonic acid. *HortScience* 5:491–492.

Lane, R. P. 1969. Muscadine grape breeding program. *N. Amer. Grape Brdrs. Conf.* 1969. (Clarksville, Ark.) (Mimeo.) (Abstr.)

Lelakis, P. 1956. On a rapid method of control of the success of induction of polyploidy in woody plants (in French). *Prog. Agr. Vitic.* 145:27–28.

———. 1957. Induction of polyploidy in *Vitis vinifera* L. by application of colchicine (in French). *Ann. École Nat. Agr. Montpellier* 30:3–60. (*Plant Brdg. Abstr.* 28:3403).

Levadoux, L. 1946. Study of the flower and sexuality in grapes (in French). *Ann. École Nat. Agr. Montpellier* N.S. 27:1–89.

———. 1956. Wild and cultivated populations of *Vitis vinifera* L. (in French, English summary). *Ann. Amélior. Plantes* 6:59–118.

Levadoux, L., D. Boubals, and M. Rives. 1962. The genus Vitis and its species (in French, English summary). *Ann. Amélior. Plantes* 12:19–44.

Loomis, H. N. 1960. Notes on a staminate grape selection as a parent. *J. Hered.* 51:212–213.

Loomis, N. H., and L. A. Lider. 1971. Nomenclature of the 'Salt Creek' grape. *Fruit Var. Hort. Dig.* 25: 41–43.

Loomis, N. H., and C. F. Williams. 1957. A new genetic flower type of the muscadine grape. *J. Hered.* 48:294,304.

Loomis, N. H., C. F. Williams, and M. M. Murphy. 1954. Inheritance of flower types in muscadine grapes. *Proc. Amer. Soc. Hort. Sci.* 64:279–283.

Mao, Y. T. 1961. *World list of plant breeders*. Rome: F.A.O.

———. 1965. *World list of plant breeders*. Suppl. 1. Rome: F.A.O.

———. 1967. *World list of plant breeders*. Suppl. 2. Rome: F.A.O.

Mayer, G. 1964. Investigations into the causes of different germinabilities of various Vitis species (in German, English summary). *Mitteilungen* (Klosterneuberg. Austr.) (Ser. A.) 14:118–132.

Milosavljević, M., and R. Mijajilović. 1965. Investigations on the radiation sensitivity of grape buds (in German). *Vitis* 5:88–93.

Mohamed, A. H. 1968. The effect of prebloom treatments of some chemicals on the induction of pollen sterility and on fruit set in some grape varieties. Ph.D. dissertation, Rutgers. *Diss. Abstr.* 29:1229–B.

Moore, J. N. 1970. Cytokinin-induced sex conversion in male clones of Vitis species. *J. Amer. Soc. Hort. Sci.* 95:387–393.

Mortensen, J. A. 1968. The inheritance of resistance to Pierce's disease in Vitis. *Proc. Amer. Soc. Hort. Sci.* 92:331–337.

———. 1971. Breeding grapes for central Florida. *HortScience* 6:149–153.

Müller-Stoll, W. R. 1950. Mutant color changes in wine grapes (in German). *Züchter* 20:288–291.

Muthuswamy, S., and J. B. M. M. Abdul Khader. 1959. A note on the occurrence of a seedless bunch in Habshi, a seeded variety of grape (*Vitis vinifera* Linn.). *S. Indian Hort.* 7:22–23. (*Plant Brdg. Abstr.* 33:2151).

Nagarajan, C. R., S. Krishnamurthi, and V. N. M. Rao. 1965. Storage studies with grape pollen. *S. Indian Hort.* 13:1–14. (*Hort. Abstr.* 36:4374).

Narasimham, B., and S. K. Mukherjee. 1968. Induction, isolation and performance of autotetraploids in grapes. *Nucleus* (Calcutta) Suppl., p. 295–313. (*Plant Brdg. Abstr.* 41:6191).

———. 1969*a*. Early maturity and seed abortion in tetraploid grapes. *Vitis* 8:89–93.

———. 1969*b*. Use of kinetin in recovery of tetraploids from grape shoots treated with colchicine. *Sci. Cult.* 35:266–267. (*Am. J. Enol. Vitic.* 21:168 (Abstr.); *Plant Brdg. Abstr.* 40:6086).

———. 1970. Seed fertility in tetraploid grapes and their crosses with diploids. *Vitis* 9:177–183.

Nebel, B. 1929. On the cytology of Malus and Vitis (in German). *Gartenbauwissenschaft* 1:549–592.

Nebel, B. R., and M. L. Ruttle. 1936. Storage experiments with pollen of cultivated fruit trees. *J. Pom. Hort. Sci.* 14:347–359.

Negi, S. S., and H. P. Olmo. 1966. Sex conversion in a male *Vitis vinifera* L. by a kinen. *Science* 152:1624–1625.

———. 1970. Studies on sex conversion in male *Vitis vinifera* L. (*sylvestris*). *Vitis* 9:89–96.

———. 1971*a*. Conversion and determination of sex in *Vitis vinifera* L. (*sylvestris*). *Vitis* 9:165–279.

———. 1971*b*. Induction of sex conversion in male Vitis. *Vitis* 10:1–19.

Negrul', A. 1969. Genetics and improvement of grapes. Soviet Reports (in French). *Bul. Off. Int. Vin* 42: 479–488.

Nesbitt, W. B. 1966. Behavior of F_1 hybrids of Euvitis P. cultivars and *Vitis rotundifolia* Mich. and some backcross individuals during microsporogenesis and megasporogenesis. Ph.D. dissertation, Rutgers. (*Diss. Abstr.* 27:4205–B).

Nitsch, J. P., et al. 1960. Natural growth substances in Concord and Concord Seedless grapes in relation to berry development. *Amer. J. Bot.* 47:566–576.

Oberle, G. D. 1938. A genetic study of variations in floral morphology and function in cultivated forms of Vitis. *N.Y. (St.) Agr. Expt. Sta., Geneva, Tech. Bul.* 250.

Oberle, G. D., and K. L. Goertzen. 1952. A method for evaluating pollen production of fruit varieties. *Proc. Amer. Soc. Hort. Sci.* 59:263–265.

Olmo, H. P. 1936. Bud mutation in the vinifera grape. 2. Sultanina gigas. *Proc. Amer. Soc. Hort. Sci.* 33: 437–439.

———. 1937. Chromosome numbers in the European grape (*Vitis vinifera*). *Cytologia. Fujii Jubilee Vol.*, p. 606–613. (*Plant Brdg. Abstr.* 9:479).

———. 1940. Somatic mutation in the vinifera grape. 3. The Seedless Emperor. *J. Hered.* 31:211–213.

———. 1942*a*. Choice of parent as influencing seed germination in fruits. *Proc. Amer. Soc. Hort. Sci.* 41:171–175.

———. 1942*b*. Storage of grape pollen. *Proc. Amer. Soc. Hort. Sci.* 41:219–224.

———. 1942*c*. Breeding new tetraploid grape varieties. *Proc. Amer. Soc. Hort. Sci.* 41:225–227.

———. 1948*a*. Ruby Cabernet and Emerald Riesling. Two new table-wine grape varieties. *Calif. Agr. Expt. Sta. Bul.* 404.

———. 1948*b*. Perlette and Delight. Two new early maturing seedless table grape varieties. *Calif. Agr. Expt. Sta. Bul.* 705.

———. 1952*a*. Wine grape varieties of the future. *Amer. Soc. Enol. Proc. 3rd Ann. Open Mtg.* p.45–51.

———. 1952*b*. Breeding tetraploid grapes. *Proc. Amer. Soc. Hort. Sci.* 59:285–290.

———. 1960. Plant breeding program aided by radiation treatment. *Calif. Agr.* 14:7:4.

———. 1964. Improvement in grape varieties. *Wines Vines* 45:2:23,25.

———. 1971. Vinifera rotundifolia hybrids as wine grapes. *Amer. J. Enol. Vitic.* 22:87–91.

Olmo, H. P., and A. Koyama. 1962*a*. Rubired and Royalty. New grape varieties for color, concentrate, and port wine. *Univ. Calif. Div. Agr. Sci. Calif. Agr. Expt. Sta. Bul.* 789.

———. 1962*b*. Niabell and Early Niabell. *Univ. Calif. Div. Agr. Sci. Calif. Agr. Expt. Sta. Bul.* 790.

Ourecky, D. K., C. Pratt, and J. Einset. 1967. Fruiting behavior of large-berried and large-clustered sports of grapes. *Proc. Amer. Soc. Hort. Sci.* 91:217–223.

Patel, G. I., and H. P. Olmo. 1955. Cytogenetics of Vitis: 1. The hybrid *V. vinifera* x *V. rotundifolia*. *Amer. J. Bot.* 42:141–159.

———. 1957. Interspecific triploid hybrid in grape. *Caryologia* 9:340–352. (*Plant Brdg. Abstr.* 27: 4454).

Peliakh, M. A. 1963. History of vine growing in the U.S.S.R. (in French). *Bul. Off. Int. Vin* 36:1389–1405.

Pieniazek, S. A. 1967. Fruit production in China. *Proc. 17th Int. Hort. Congr.* 4:427–452.

Pogossian, S. A. 1963. Selection methods for grapes in the U.S.S.R. (in French). *Bul. Off. Int. Vin* 36: 1267–1279.

Pratt, C. 1959. Radiation damage in shoot apices of Concord grape. *Amer. J. Bot.* 46:103–109.

———. 1971. Reproductive anatomy in cultivated grapes—a review. *Amer. J. Enol. Vitic.* 22:92–109.

Pratt, C., and J. Einset. 1961. Sterility due to premeiotic ovule abortion in Small-clustered Concord and normal Concord grapes. *Proc. Amer. Soc. Hort. Sci.* 78:230–238.

Raj, A. S., and L. Seethaiah. 1969. Karyotype analysis and meiotic studies in three varieties of grape (*Vitis vinifera* L.). *Cytologia* 34:475–483. (*Biol. Abstr.* 51: 105135).

Ramirez, O. C. 1968. Comparative embryogenesis of Erie, Concord and Golden Muscat grape varieties as related to the germinability of seeds. Ph.D. dissertation, Rutgers. (*Diss. Abstr.* 29:1230-B).

Randhawa, G. S., and S. S. Negi. 1964. Preliminary studies on seed germination and subsequent seedling growth in grape. *Indian J. Hort.* 21:186–196. (*Hort. Abstr.* 35:7372).

Raski, D. J., W. H. Hart, and A. N. Kasimatis. 1965. Nematodes and their control in vineyards. *Calif. Agr. Expt. Sta. Ext. Service Cir.* 533.

Reichardt, A. 1955. Experimental investigations on the effect of X rays on vegetative reproduction of old grape varieties (in German). *Gartenbauwissenschaft* 20:355–413.

Reimer, F. C. 1909. Scuppernong and other muscadine grapes: origin and importance. *N.C. Agr. Expt. Sta. Bul.* 201.

Richards, B. L., J. T. Middleton, and W. B. Hewitt. 1958. Air pollution with relation to agronomic crops: 5. Oxidant stipple of grape. *Agron. J.* 50: 559–561.

Rives, M. 1961. Genetic bases of clonal selection in grape (in French). *Ann. Amélior. Plantes* 11:337–348.

———. 1965. The germination of grape seeds. 1. Preliminary experiments (in French, English summary). *Ann. Amélior. Plantes* 15:79–91.

Rives, M., and R. Pouget. 1959. Chasselas Gros Coulard —a tetraploid mutant (in French). *Vitis* 2:1–7.

Robinson, W. B., J. Einset, and N. J. Shaulis. 1970. The relation of variety and grape composition to wine quality. *Proc. N.Y. St. Hort. Soc.* 115:283–287.

Roos, T. J. 1968. A shortening of the generation interval in the grape and the importance thereof in the breeding program. *Decid. Fruit Grower* 18:11–13.

Sarasola, J. A. 1966. A method to induce mutations by treating detached buds of grape vines with ethylmethanesulfonate. *Bol. Genét. Inst. Fitotec.* (Castelar, Argentina) 2:41–45.(*Plant Brdg. Abstr.* 38: 1373).

Saurer, W., and A. J. Antcliff. 1969. Polyploid mutants in grapes. *HortScience* 4:226–227.

Scherz, W. 1940*a*. Grape mutations, their significance

and exploitation for breeding (in German). *Wein Rebe* 22:73–86. (*Biol. Abstr.* 16:15781).

———. 1940*b*. On somatic mutations of *Vitis vinifera* —variety "Moselriesling" (in German). *Züchter* 12:212–225.

Scott, D. H., and D. P. Ink. 1950. Grape seed germination experiments. *Proc. Amer. Soc. Hort. Sci.* 56: 134–139.

Séchet, J. 1962. Influence of low temperatures on gerination of grape seeds (in French). *C. R. Acad. Sci.* 255:2653–2655.

Shaulis, N. 1966. Light intensity and temperature requirements for Concord grape growth and fruit maturity. *Proc. 17th Int. Hort. Cong.* 1:589. (Abstr.)

Shaulis, N., and G. D. Oberle. 1948. Some effects of pruning severity and training on Fredonia and Concord grapes. *Proc. Amer. Soc. Hort. Sci.* 51:263–270.

Shetty, B. V. 1959. Cytotaxonomical studies in Vitaceae. *Bibl. Genet.* 18:167–272.

Shimotsuma, M. 1962. Irradiation experiments with grapes. *Seiken Zihô. Rpt. Kihara Inst. Biol. Res.* (Yokohama) 14:102–103.

Smith, M. B., and H. P. Olmo. 1944. The pantothenic acid and riboflavin in the fresh juice of diploid and tetraploid grapes. *Amer. J. Bot.* 31:240–241.

Snyder, E. 1931. A preliminary report on the breeding of vinifera grape varieties. *Proc. Amer. Soc. Hort. Sci.* 28:125–130.

———. 1937. Grape development and improvement. *USDA Yearbook of Agriculture* 1937:631–664.

Snyder, E., and F. N. Harmon. 1935. Three mutations of *Vitis vinifera*. *Proc. Amer. Soc. Hort. Sci.* 33: 435–436.

———. 1937. Hastening the production of fruit in grape hybridizing work. *Proc. Amer. Soc. Hort. Sci.* 34:426–427.

———. 1940. "Synthetic" zante currant grapes. *J. Hered.* 31:315–318.

———. 1952. Grape breeding summary 1923–1951. *Proc. Amer. Soc. Hort. Sci.* 60:243–246.

Steuk, W. K. 1945. Variations of the Catawba grape. *Ohio Agr. Expt. Sta. Bimonthly Bul.* 30:232:31–33.

Stoev, K. 1970. Criteria of endorsement of selected clones and new wine and table grape varieties with the object of maintaining common records (in French). *Bul. Off. Int. Vin* 43:370–376.

Stout, A. B. 1936. Breeding for hardy seedless grapes. *Proc. Amer. Soc. Hort. Sci.* 34:416–420.

———. 1939. Progress in breeding for seedless grapes. *Proc. Amer. Soc. Hort. Sci.* 37:627—629.

Stover, L. H. 1960. Progress in the development of grape varieties for Florida. *Proc. Fla. St. Hort. Soc.* 73:320–323.

Stover, L. H., and J. A. Mortensen. 1963. Four promising grape selections. *Proc. Fla. St. Hort. Soc.* 76: 341–345.

Thompson, C. R., and G. Kats. 1970. Antioxidants reduce grape yield reductions from photochemical smog. *Calif. Agr.* 24:9:12–13.

Thompson, M. M., and H. P. Olmo. 1963. Cytohistological studies of cytochimeric and tetraploid grapes. *Amer. J. Bot.* 50:901–906.

Vatsala, P. 1961. Chromosome studies in Ampelidaceae. *Cellule* 61:191–206. (*Plant Brdg. Abstr.* 31:5404).

Wagner, E. 1958. On spontaneous tetraploid mutants of *Vitis vinifera* L. (in German). *Vitis* 1:197–217.

———. 1962. Production of Vitis pollen for crossing purposes. 1. Effects of forcing grapevines (in German). *Vitis* 3:117–129.

Wagner, R. 1967*a*. Preliminary selection of grape seedlings under glass (in French, English summary). *Ann. Amélior. Plantes* 17:159–173.

———. 1967*b*. Study of some segregations in progenies of Chasselas, Muscat Ottonel and small-berried Muscat (in French). *Vitis* 6:353–363.

Weaver, R. J., and S. B. McCune. 1960. Further studies with gibberellin on *Vitis vinifera* grapes. *Bot. Gaz.* 121:155–162.

Weinberger, J. H., and F. N. Harmon. 1964. Seedlessness in vinifera grapes. *Proc. Amer. Soc. Hort. Sci.* 85:270–274.

———. 1966. Harmony, a new nematode and phylloxera resistant rootstock for vinifera grape. *Fruit Var. Hort. Dig.* 20:63–65.

Williams, C. F. 1923. Hybridization of *Vitis rotundifolia*. Inheritance of anatomical stem characters. *N.C. Agr. Expt. Sta. Tech. Bul.* 23.

Williams, H. A. 1971. Soviet wine making. *Wines Vines* 52:12:15–17.

Winkler, A. J. 1926. The influence of pruning on the germinability of pollen and the set of berries in *Vitis vinifera*. *Hilgardia* 2:107–124.

———. 1927. Improving the fruiting of the Muscat (of Alexandria) grape by less severe pruning. *Proc. Amer. Soc. Hort. Sci.* 24:157–163.

———. 1949. Grapes and wine. *Econ. Bot.* 3:46–70.

———. 1950. Pitfalls of vine propagation. Amer. Soc. *Enol. Proc. 1st Ann. Open Mtg.* p. 5–15.

———. 1962. *General viticulture*. Berkeley: Univ. Calif. Press.

Woodham, R. C., and D. M. Alexander. 1966. Reproducible differences in yield between sultana vines. *Vitis* 5:257–264.

Woodroof, J. G. 1934. Five strains of the Scuppernong variety of the muscadine grapes. *Proc. Amer. Soc. Hort. Sci.* 32:384–385.

Zuluaga, P. A., and A. Gargiulo. 1954. Polyploidy in *Vitis vinifera* L. (in Spanish). *Rev. Fac. Cienc. Agr.* (Mendoza) 4:1–13. (*Plant Brdg. Abstr.* 27:2241).

Blueberries and Cranberries

by Gene J. Galletta

The blueberry is the most recent of the major fruit crops to be brought under cultivation, having been domesticated entirely in the twentieth century. The cultivated blueberries offer a most dramatic example of the results of fruit crop breeding and selection. All blueberries were harvested from wild plants prior to the first shipment of fruit of Dr. F. V. Coville's hybrid seedlings from Miss Elizabeth White's farm at Whitesbog, New Jersey, in 1916 (Coville, 1921). The introduction of the 'Pioneer', 'Cabot', and 'Katherine' cultivars from Coville's breeding program in 1920 (Coville, 1937) served as the basis for an entirely new agricultural industry, which was adapted to the utilization of acid, imperfectly drained soils that had been previously classed as agriculturally worthless. This industry has continued to thrive and expand with the continuing development of newer and better cultivars; more than 45 highbush and 11 rabbiteye cultivars which originated from controlled pollinations have been named to date. There are over 8100 ha (20,000 acres) of cultivated blueberries in the United States at present, and there is considerable European interest in expanding blueberry culture, based initially on the use of adapted American cultivars. The future outlook is good for a greatly increased world production of blueberries, but the realization of this expansion will be dependent on further developments in blueberry breeding and genetics.

The large or American cranberry (*Vaccinium macrocarpon* Ait.) has long been prized for its acid red fruit, which is high in Vitamin C content. Henry Hall of Dennis, Massachusetts, on Cape Cod, started the culture of this native American crop (which is also adapted to acidic, organic lowland soils like its highbush blueberry relatives) in about 1816 (Peterson et al., 1968). Cranberries are now grown on approximately 8700 ha in the United States; they are important in Massachusetts, Wisconsin, New Jersey, Washington, and Oregon. Production in 1970 was over 90,700 mt, up from 58,955 mt in 1960 (Marucci, personal correspondence, 1972). The gross value of the crop was about $25 million in 1969 (Norton, 1969). Canadian production (in Nova Scotia, Quebec, and British Columbia) in 1967 totaled about 1,338 mt (Hall et al., 1969). Cranberry culture has shown promise in recent years in experimental trials in Poland, Austria, Germany, and Russia (Soczek and Scholz, 1969; Klein, 1971; Liebster, 1971; and Kolupaeva, 1971).

Consumption of cranberries was once limited to Thanksgiving and Christmas fare in the form of jellies and sauces. Starting in the early 1960s, new products, such as cranberry juice, grape-berry juice, cran-apple juice, and cranberry-orange relish began to be vigorously promoted. In 1968 the industry voted to accept a marketing order which permits withholding part of the crop each year to stabilize prices. Although crop area has remained essentially the same since the early 1960s, production has risen 50 to 75% due to improved cultural practices such as weed control, fertilizer management, and water harvesting in eastern areas, and total dollar value has tripled in the period from 1963 to 1971, due largely to the impact of new products, especially juice. Cranberries are now consumed the year around and they are being exported in quantity. The industry today appears sound and healthy.

Blueberries

Origin and Early Development

The *Vacciniaceae* has been treated variously taxonomically as a suborder of the order *Ericaceae* (Chapman, 1883) and as a family of the order *Ericales* (Small, 1933), but usually as a subfamily of the family *Ericaceae* (Robinson and Fernald, 1908; Sleumer, 1941). Camp (1942a) observed that the *Vacciniaceae*, as usually defined, contained two tribes, the apparently monogeneric *Gaylussacieae* (huckleberries) and the *Vaccinieae* (whortleberries, bilberries, blueberries, cranberries). Camp further indicated that the genus *Vaccinium* was very old, and that he felt that segments of it were already highly differentiated in the Cretaceous Period of the Mesozoic Era. It is of interest to note that the majority of species in the subfamily *Vacciniaceae* are tropical in distribution (Camp, 1942a), and it appears on morphological grounds that the deciduous temperate forms of *Vaccinium* may be part of, or derived from, certain evergreen tropical groups.

Camp (1942a), drawing largely on the experience of his cooperative investigations (with G. M. Darrow, E. B. Morrow, and F. B. Chandler) of species of the subgenus *Cyanococcus* of eastern North America, made the following observations in regard to speciation in *Vaccinium*: 1) a lack of fundamental sterility barriers between *Vaccinium* species of the same ploidy level; 2) a high incidence of polyploidy in the genus ($x = 12$), with many natural tetraploid ($2n = 48$) and hexaploid ($2n = 72$) species; 3) individuals of many species are functionally self-unfruitful, which promotes the incidence of interspecific hybrid swarms in the genus; 4) intolerance of dense shade and alkaline soil which restricts their habitats, and encourages speciation through ecological separation; and 5) result of migrations caused by geologic events or changed distribution patterns as a consequence of the antiquity of *Vaccinium*. These events permitted formerly disjunct species to come together, hybridize, and recede, and disturbed areas permitted colonization opportunities for hybrid segregants. In addition, *Vaccinium* is an excellent primary colonizer of disturbed areas, such as logged or burned-over forests and abandoned agricultural land, and *Vaccinium* seed is very widely distributed in nature by birds.

Camp's studies in the *Ericales* were terminated by wartime conditions, but he published a series of tentative taxonomic monographs on several groups within the subfamily *Vacciniaceae* (*Gaylussacieae*, 1941; *Euvaccinium*, 1942b; *Oxycoccus*, 1944; and polypetalous forms of *Vaccinium*, with C. L. Gilly, 1942). This series culminated in Camp's monumental monograph on the cluster-fruited or "true" blueberries (1945, subgenus *Cyanococcus*). Camp hoped that this incomplete work "will serve to begin the clarification of a floristically and horticulturally important group which has long been nomenclaturally confused." In this work, Camp grouped the North American *Vacciniaceae* (north of Mexico) as follows:

Gaylussacieae

1) *Gaylussacia*—the huckleberries—have a ten-celled ovary and fruit with ten one-seeded nutlets. There are six or more species (Robinson and Fernald, 1908) in the eastern United States and Canada.

Vaccinieae has a four- to five-celled ovary. Its fruit is usually many seeded, each carpel containing two to 15 small seeds.

2) *Vaccinium ovatum* Pursh—California or box-blueberry—is an evergreen species with tropical relatives found on the Pacific Coast.
3) *Vaccinium uliginosum* L. and *V. occidentale* A. Gray—the whortleberries—are arctic species descending through Canada to the extreme northern United States. They are also found in Europe and Asia.
4) Subgenus *Euvaccinium*—the bilberries—grow in the western United States and Canada, with some members in Europe and Asia. (See also J. H. Shultz [1944], who describes

and gives chromosome counts and Washington distributions for nine species of bilberries, including the European species *V. myrtillus* L., the two whortleberry species, *V. ovatum* Pursh, and *V. quadripetalum*, an *Oxycoccus* representative.)

5) *Polycodium*—the deerberries—range from Florida to Maine and westward to central Canada and Texas (Small, 1933, recognized ten species).
6) *Batodendron*—the sparkle- or farkleberry *V. arboreum* Marsh—appears as one species from North Carolina to Florida and west to Kentucky, and from Indiana through the Ozarks to Texas. It has Mexican relatives.
7) *Oxycoccus*—the cranberries—are circumboreal in distribution and have several species.
8) *Hugeria*—the southern mountain cranberry *V. erythrocarpum* Michx.—grows in the Southern Appalachians, Virginia, and West Virginia, and south to North Carolina and Tennessee. It has relatives in China.
9) *Vitis-Idaea*—the ling- or linberry, *V. vitis-idaea* L.—is found in the northern United States, Canada, Europe, and Asia.
10) *Herpothamnus*—the creeping blueberry, *V. crassifolium* Andr.—is endemic to sandy areas from Virginia to South Carolina, and has Mexican and South American relatives.
11) *Cyanococcus*—the true or cluster-fruited blueberries—is predominantly an eastern North American group with Asiatic relatives. Camp recognized and provided keys, species descriptions, natural distributions, taxonomic correlations, frequent species hybrids encountered, and a tentative phylogenetic pattern for nine diploid species, 12 tetraploid species, and three hexaploid species.

The improved cultivated blueberries have all been derived from species hybrids within the *Cyanococcus* subgenus. Eck (1966) has very nicely summarized the *Vaccinium Cyanococcus* species for ploidy level, taxonomic equivalents, and morphological characters. Table 1 lists the general habit, habitats and distributions, and the purported origins of the polyploid species of the 24 *Cyanococcus* species recognized by Camp (1945).

History of Improvement

Commercially grown blueberries occur in three major groups: 1) the wild lowbush types of rocky uplands in southern Canada, the northeastern and Great Lakes areas of the United States and as far south as Pennsylvania and West Virginia; 2) the cultivated highbush species hybrids adapted to organic lowland situations from the Carolinas north to southern Maine, and in similar areas in Ohio, Indiana, Illinois, Michigan, Washington, Oregon, Canada, and Europe; and 3) the hybrid clones developed within the rabbiteye species which perform well in coastal plain and piedmont soils from northern Florida through the Gulf Coast states as far north as central Arkansas in the west and the Carolinas and southern Virginia in the east.

Excellent general and regional summaries of blueberry breeding parents, objectives, methods, and improvement progress within and among the above cultivated groups are found in the following chronologically arranged sources: Coville, 1937; Darrow, 1960*a*; Sharpe and Darrow, 1960; Brightwell, 1960; Galletta, 1960; Whitton, 1960; and Knight, 1960; Moore, 1965 and 1966; Sharpe, 1966; Brightwell, 1966; Galletta, 1966; Stushnoff and Hough, 1966; Johnston, 1966; Bailey, 1966; Aalders, 1966; and Draper, et. al., 1966; Johnston and Moulton, 1968; Sharpe and Sherman, 1971; and Brightwell, 1971*a* and 1971*b*.

Lowbush blueberries: The commercial lowbush blueberry barrens are found to consist of basically four morphological types (Camp, 1945; Aalders and Hall, 1962*b*, 1963*b*; Aalders, 1966; Hall and Aalders, 1961a, 1968; and Whitton, 1964). These types are: 1) the acid, blue-fruited, heavily pubescent diploid growing usually 20–40 cm tall and possessing fruit 4–7 mm in diameter. This species (*V. myrtilloides* Michx.) has been variously referred to as the Canada, velvet-leaf, hairy-leaf, or sourtop blueberry. This species predominates in land newly cleared from woodlots (Barker et al., 1964); 2) the common lowbush blueberry or sugarberry (*V. angustifolium* Ait.), a sweet, bright blue-fruited, glabrous plant growing from 5–40 cm tall, depending on soil fertility and moisture conditions. This species was thought to be diploid by Camp, but the Canadian studies of Hall and Aalders and the Maine studies of Whitton, cited above, indicate that the species is tetraploid. This is the predominant lowbush blueberry form in land recently cleared from abandoned hayfields (Barker et al., 1964); 3) an uncommon low-growing species confined to localized areas in Newfoundland and

TABLE 1. Habitats and general distributions of the 24 living cluster-fruited blueberry species of eastern North America (subgenus *Cyanococcus* of genus *Vaccinium*) and the proposed origins of the polyploid species (modified from Camp, 1945).

Species	Habitat	Presumed origin
DIPLOIDS (2n = 24)		
V. *myrtilloides* Michaux	Lowbush,[a] muskey or upland barrens, Mont. and Brit. Columb. to Labrador, S to N.Y., Ind., and in mts. to W. Va.	
V. *angustifolium* Aiton[b] (V. *boreale* Hall and Aalders)	Lowbush, open rocky upland, dry sandy areas and swamp hummocks. Labrador and Newfoundland W to Minn., S to N.J., and in mts. to Va.-W. Va.	
V. *vacillans* Torrey	Lowbush, upland dry woods, rocky ledges and abandoned farm lands. Minn. and Ont. to Me., S to Mo., Tenn., and Ga.	
V. *pallidum* Aiton	Lowbush—halfhigh, dry upland woods and brushy mt. areas. N. Ga. and Ala. N to W. Va., Ky., Pa. and Ark. Ozarks to Mo. & Kans.	
V. *darrowi* Camp	Lowbush, sandy areas, La. to Fla.	
V. *tenellum* Aiton	Lowbush, upland open forests and meadows. SE Va. to Ga. and Ala., N. Fla., and Miss.	
V. *elliottii* Chapman	Highbush, river basins. SE Va. to Fla., W to La., Ark. & Tex.	
V. *caesariense* Mackenzie	Highbush, coastal plain swamps, inland boggy areas, occasional open wooded slopes. N. Fla. to S. Me., and central N.Y.	
V. *atrococcum* Heller	Highbush, swamps & rivers, occasional dry ground. W. Tenn. & Ark. S to the gulf, E to N. Fla., N to central N.Y. & Me. (possibly in Great Lakes & S. Canada).	
TETRAPLOIDS (2n=48)		
V. *brittonii* Porter ex Bicknell[c] (V. *angustifolium* var. *nigrum*)	Lowbush, rocky uplands, dry sandy areas, swamp borders. Newfoundland to Minn. S to N.J. & in mts. to W. Va.	Unknown
V. *lamarckii* Camp[b] (V. *angustifolium* Aiton)	Lowbush, ecological requirements of V. *boreale*. Newfoundland W to Minn., S in uplands to Va.-W. Va.	Autotetraploid form of V. *boreale*
V. *myrsinites* Lamarck	Lowbush to halfhigh, dry sandy areas, Gulf Coast and Fla., Ga., and S.C.	Allotetraploid forms from *darrowi-tenellum* hybridizations
V. *virgatum* Aiton	Lowbush—halfhigh, ecological requirements of *tenellum*. Ga.-Fla. to Tex.-Ark.	Autotetraploid of V. *tenellum*
V. *hirsutum* Buckley	Halfhigh, dry ridges and mtn. meadows. N.C. and Tenn.	Unknown
V. *fuscatum* Aiton	Highbush, sandy flatwoods and bottomlands, along streams and lakes. S. Ga. to central Fla.	Allotetraploid from *darrowi-atrococcum* hybridizations
V. *simulatum* Small	Highbush, open mtn. slopes and meadows. N. Ala. and Ga. to Ky. and Va.	Autotetraploid of V. *vacillans*
V. *altomontanum* Ashe	Halfhigh, open woods and rocky uplands at high elevations. N. Ga. and Ala. to Va., Tenn. and Ky.	Autotetraploid of V. *pallidum*
V. *australe* Small	Highbush, coastal plain marshes and swamps, inland bogs. SE Ala. & N. Fla. to N.J. (Similar phenotypes accompanying *corymbosum* N to Me., N.Y. and S. Canada)	Autotetraploid of V. *caesariense*
V. *marianum* Watson	Highbush, isolated lowland populations. N.C. to N.Y.	Allotetraploid from *caesariense-atrococcum* hybridizations
V. *arkansaum* Ashe	Highbush, sandy lake and stream margins, swamps, occasionally in open flatwoods or on low ridges. N. Fla. W to Tex. and Ark.	Autotetraploid of V. *atrococcum*

Species	Habitat	Presumed origin
TETRAPLOIDS (2n=48)		
V. *corymbosum* Linnaeus	Highbush, swamps and bogs, lake and stream margins, moist sandy areas, hillside seepages. Mich. to Nova Scotia S to glacial boundary.	Allotetraploid hybrid complex involving germplasm from V. *simulatum, brittonii, angustifolium, australe, marianum,* and *arkansanum*
HEXAPLOIDS (2n=72)		
V. *ashei* Reade	Highbush, streams, lake margins, open woods, abandoned fields. Ga., Ala., Fla., S.C. (?), Miss. (?)	Allohexaploid hybrid complex involving germplasm from V. *virgatum, arkansanum, australe, fuscatum* & *myrsinites*
V. *amoenum* Aiton	Highbush, uplands, open woods, stream margins. S.C. to Fla. W to Tex. and Ark.	Autohexaploid of V. *virgatum*
V. *constablaei* Gray	Highbush, mtn. tops and upper slopes. W. N.C. and E. Tenn.	Allohexaploid from *simulatum-altomontanum* hybridizations

[a]Lowbush = less than 50 cm, halfhigh = 50 cm — 1 m, highbush = over 1.5 m tall.
[b]V. *angustifolium* Aiton is the better taxonomic binomial for the tetraploid common lowbush blueberry and V. *boreale* for the diploid lowbush (Hall and Aalders 1961a).
[c]Hall and Aalders (1961*a*) consider V. *brittonii* the var. *nigrum* of the common tetraploid lowbush species V. *angustifolium*.

Maine. This species is of the *angustifolium* phenotype but differs from it in being diploid, having a different degree of branching, shorter flowers, and smaller pollen grain diameters. In a uniform environment, this "ground hurts" (named V. *boreale* by Hall and Aalders) is also shorter and bears smaller leaves and fruit; and 4) a dark-fruited tetraploid species (V. *brittonii* Porter ex Bickn. or V. *angustifolium* var. *nigrum* [Wood] Dole). Growing usually 15–35 cm tall, it is an infrequent variant. It also differs from the common lowbush type in the presence of a waxy bloom on the leaf surface, no hairs on the leaf midvein, and a zigzag shoot growth pattern. Aalders and Hall (1963*b*) considered this type as a taxonomic form of V. *angustifolium* because the leaf bloom character appeared to be inherited as a single semi-dominant gene. However, the leaf midvein hairs and zig-zag shoot pattern were not consistently correlated with the leaf bloom character.

Improvement of the lowbush blueberry has been initiated with the selection of horticulturally superior wild clones at various times by personnel with the United States and Canada Departments of Agriculture and with the Maine, Michigan, Wisconsin, Minnesota, and West Virginia Agricultural Experiment Stations. Recently, the author has grown and is selecting in North Carolina representative lowbush types from seed collected in Maine, Wisconsin, West Virginia, and Minnesota.

Kender (1966) listed the objectives of such lowbush selection as large fruit size, good blue color, fine flavor, heavy productivity, self-fruitfulness, late blooming, uniform ripening, disease resistance, vigorous rhizome growth, easy propagation, and upright, vigorous, tall stems.

Two problems exist in the further improvement and establishment of superior lowbush cultivars. One is the existence of a considerable amount of pollen and ovule sterility in different clones, as shown by Aalders and Hall and associates (1961, 1961*b*, 1963*a*, 1966). This sterility would lead to varying degrees of self- and cross-unfruitfulness. The second is the establishment of clonal propagations in the field. Difficulties in propagating selected clones have been largely overcome by the use of intermittent mist and softwood stem or rhizome cuttings (Kender, 1965, 1966, 1967; Aalders, 1966). The use of seedlings and clonal propagations in establishing matted rows of lowbush blueberries for intensive culture looks very promising (Kender, 1967; Aalders and Hall, 1968). Studies have progressed to the point at which directions for the establishment of commercial lowbush blueberry fields have been offered to the public (Hall, et al., 1971). Improvement of the lowbush by the intercrossing of select clones continues at the Kentville Station of the Canada Department of Agriculture, the University of Laval in Quebec, and the University of Maine.

The lowbush species has contributed genes, through 'Russell,' to the development of over half of the 34 highbush cultivars introduced from the USDA cooperative blueberry breeding programs. The lowbush contribution is represented variously from F_1 hybrids with highbush ('Greenfield') to fourth backcrosses to highbush ('Morrow') types in these cultivars.

Moore (1966) summarized the early generation results of lowbush-highbush crosses up to the F_2 and BC_1 generations from work at Michigan and several eastern United States locations. Noteworthy was the F_1 generation uniformity of intermediate plant height, extreme productivity, a preponderance of early maturity, small to moderate fruit size, dark color, and fair flavor. Fruit firmness of the lowbush species seems to be predominate in lowbush-highbush hybrids.

F_2 and subsequent hybrid generations from lowbush-highbush crosses usually segregate for growth habit and fruit size and color. A lowbush-highbush species hybridization program in Michigan has been carried through five generations (Johnston and Moulton, 1968) with selection for: 1) extreme lowbush segregates with large berries for house or garden border plants; 2) large-fruited halfhigh 0.6-1 m segregates for commerical cultivation in northern Michigan where they would have to overwinter under a snow cover for cold protection; and 3) lower growing 1.5 m highbush segregates which would be easier to harvest than the 3.6-4.4 m tall bushes often found now in commercial fields. Two hybrid clones from the third generation of this program, 'Bluehaven' and 'Northland', have been introduced (Johnston and Moulton, 1967). These cultivars represent first backcross to highbush segregates, and they are lower growing highbush types (1.5 and 1.2 m tall at maturity, respectively). The Michigan hybrid program is continuing with emphasis on securing taller and more upright bushes with concentrated ripening and seasonal spread (J. E. Moulton, personal communication, 1971). Breeding programs based on lowbush-highbush hybridizations have been carried on at West Virginia University and are continuing at the University of Minnesota. In West Virginia a 0.9 m hybrid of *V. pallidum* x 'Concord' (*V. corymbosum* x *V. australe*) was recently named 'Ornablue' (Childs, 1969).

The lowbush species can contribute the following desirable features to a hybrid gene pool: low stature, early fruit maturation season, concentrated ripening, precocity, drought resistance, bud hardiness, productivity, and sweetness. The fine bright blue color should also be inherited, but appears recessive in first generation hybrids. Undesirable lowbush traits include small fruit size, small stature (in some instances), spreading habit, softness of fruit, and low fruit acidity. A recent survey of the fruit qualities of several species (Ballinger et al., unpublished) indicated that enough variability existed within the lowbush species that these undesirable traits could be avoided by careful selection of parent clones.

Highbush blueberries: Credit for the domestication of the highbush blueberry must go to Dr. Frederick Vernon Coville (USDA botanist during the period 1888–1937) and his associates, who recognized the potential of this fruit species early. Excellent reviews dealing with the domestication and subsequent improvement of the highbush blueberry have been written by Coville (1937), Darrow (1960a, 1966), Moore (1965, 1966), and Draper and Scott (1967). The Coville article has the value of a personal account combined with expertise in outlining the origin and evolution of his breeding program. The Darrow 1960a article in particular traces the progress of blueberry breeding with clarity and imagination. This article is beautifully illustrated, and deals effectively with the potential in recombining germ plasm from several subgenera within *Vaccinium*. Moore's 1966 paper includes sections in interspecific hybridization and the highbush blueberry's history of improvement, progress, and present status, as well as highbush-lowbush crosses. The Draper and Scott article gives a concise history of the USDA blueberry breeding program which involves cooperators in several states and the improvement of several species.

Coville began his blueberry domestication work in 1906 (Coville 1910, 1921). During the next four years be established the growth peculiarities and worked out the developmental patterns of the highbush blueberry from seed germination to fruit maturation. Noteworthy peculiarities were the need for acid soil with good drainage, thorough aeration, and permanent but moderate soil moisture. The moisture and aeration needs resulted from a lack of root hairs in the highbush blueberry. Coville also found that the highbush blueberry needed winter chilling to break bud dormancy, and insect pollination of the flowers. He also established that the blueberry could be vegetatively propagated by layering, budding, grafting, or from cuttings. This information provided an unusually sound basis for the breeding work which was to follow.

Coville began selecting wild blueberries for breeding in 1908 (Coville 1910, 1921). His initial selections were 'Brooks' (*V. corymbosum* L.) and 'Russell' (*V. angustifolium* Ait.) from New Hampshire. His first successful controlled pollination was 'Brooks' by 'Russell', made in 1911. During the same year, Miss Elizabeth White of Whitesbog, New Jersey, a blueberry enthusiast, offered to assist Coville in blueberry improvement. She provided land for the growing of thousands of seedlings, and assisted in selecting superior native plants for use in breeding. This fortunate association of Miss White and Dr. Coville continued for many years, and it led to the use of native *V. australe* Small selections in the initial stages of the breeding program and the rapid evaluation of many seedlings over several breeding generations.

The first highbush blueberry cultivars introduced to the public were the first generation hybrids of *V. corymbosum* and *V. australe* 'Pioneer', 'Cabot', and 'Katherine', released in 1920. These clones, and the wild *V. australe* selection 'Rubel', provided the basis for the initiation of a blueberry industry in New Jersey.

As commercial interest in blueberry production increased, Coville's breeding activities also increased. At the time of his death in 1937, a total of 68,000 hybrid seedlings had been fruited and 15 cultivars had been released, the last of which, 'Weymouth' and 'Dixi', were third generation hybrids. By this time, a still small but viable blueberry industry was underway in several American states. An additional 15 cultivars were named in the period 1939–1959 from seed and seedlings left by Dr. Coville at his death. These included the large-sized "Big Seven" cultivars introduced from 1949–1959.

In 1937 Dr. George M. Darrow assumed the blueberry breeding leadership for the USDA. He greatly expanded blueberry breeding in the United States by the development of additional cooperative programs with several state experiment stations and private growers. Using this procedure, blueberry seedlings were grown and evaluated under widely differing soil and climatic conditions. By 1965 experiment stations and private growers in 13 states were cooperating in blueberry improvement (Moore, 1966).

In addition to Coville and Darrow, other breeders who have provided leadership and/or have made important contributions to highbush blueberry improvement in the United States are J. S. Bailey (Massachusetts), R. M. Bailey (Maine), W. H. Childs (West Virginia), A. D. Draper (USDA), H. Dermen (USDA), G. J. Galletta (North Carolina), L. F. Hough (New Jersey), G. Jelenkovic (New Jersey), S. Johnston (Michigan), R. J. Knight (USDA), E. M. Meader (New Hampshire), J. N. Moore (USDA), E. B. Morrow (North Carolina), J. E. Moulton (Michigan). D. H. Scott (USDA), R. H. Sharpe (Florida), W. B. Sherman (Florida), and C. Stushnoff (Minnesota).

Since the beginning of blueberry improvement in the United States, interest has spread to other areas of the world. Blueberry cultivar trials were started at the Kentville Branch Research Station of the Canada Department of Agriculture in 1926 (Eaton, 1949). Breeding work started in 1930, and continues under the direction of Hall and Aalders. Experimental blueberry culture in Europe started as early as 1923 in the Netherlands and 1929 in Germany with trials of American cultivars. England, Denmark, and Austria all had limited pre-World War II trials. Early breeding and selection work were carried on in Germany, Austria, and Denmark. Breeding programs are now underway in Ireland, Italy, Scotland, Finland, and Yugoslavia. A recent report of the International Society for Horticultural Science Working Group, *Blueberry Culture in Europe* (1967), indicates good preliminary success with culture of American and European blueberry species in most countries of Europe, and highly enhanced interest in blueberry culture.

Along with the development of blueberry breeding and selection, much fine research has been reported which has or will contribute to the genetic improvement of the blueberry. These include studies on: 1) chromosome counts, morphology, and meiotic behavior (Longley, 1927; Newcomer, 1941; Darrow et. al., 1944; Rousi, 1966*b;* Stushnoff and Palser, 1969; Jelenkovic and Hough, 1970; Jelenkovic and Draper, 1970; Jelenkovic and Harrington, 1971; Hall and Galletta, 1971); 2) species hybridization (Coville, 1927, Darrow and Camp, 1945; Darrow et al., 1949, 1952, 1954; Moore et al., 1964; Moore, 1966; Rousi,

1966*a*; Sharpe and Sherman, 1971); 3) inheritance and breeding value (Darrow et al., 1939; Johnston, 1942; Goldenberg, 1964; Moore and Scott, 1966; Draper and Scott, 1967, 1969, 1971; Ballinger et al., 1972); 4) pollen, egg, and seed development (Eaton and Jamont, 1966; Stushnoff and Hough, 1968; Stushnoff and Palser, 1969; Edwards et al., 1972; Cockerham, 1972); 5) pollination (Merrill, 1936; Morrow, 1943; Moore, 1964; Knight and Scott, 1964; Marucci, 1966*a*; Shutak and Marucci, 1966; Dorr and Martin, 1966; Brewer and Dobson, 1969*a*, *b*, *c*); 6) seed counts and germination (Darrow and Scott, 1954; Scott and Ink, 1955; Darrow, 1958; Scott and Draper, 1967; Stushnoff and Hough, 1968*b*); 7) fruit size and development (Eaton, 1967; Edwards et al., 1970; Moore et al., 1972); 8) breeding techniques (Morrow et al., 1949, 1954; Aalders and Hall, 1963c; Moore et al., 1964; Rousi, 1966*b*; Constante and Boyce, 1968; Galletta, 1970; Galletta and Mainland, 1971a; Galletta et al., 1971; Sharpe and Sherman, 1971; Quamme et al., 1972); 9) rootstocks (Galletta and Fish, 1971); 10) mechanical harvesting (Galletta and Mainland, 1971*b*); and 11) disease and insect resistance (Demaree and Morrow, 1951; Goldenberg, 1964; Varney and Stretch, 1966; Marucci, 1966*b*; Milholland and Galletta, 1967, 1969; Lockhart and Craig, 1967; Johnson, 1969; Pepin and Toms, 1969; Lockhart, 1970; Milholland, 1970a, 1972; Nelson and Bittenbender, 1971; Cram, 1971; Draper et al., 1972). Darrow (1960*b*) summarized improvement and other related blueberry research over a 50-year period.

Rabbiteye blueberries: The vigorous and tall-growing *V. ashei* (rabbiteye) species is a polymorphic group of hexaploid populations located from southern Georgia and northern Florida along the Gulf Coast west to Alabama. Major wild populations exist along the Satilla River in Georgia (southeastern Georgia or Satilla race), the Suwannee River in Florida (Suwannee or northcentral Florida race), and the Yellow River in south Alabama and Florida (west Florida or northwest Florida race) (Camp, 1945; Brightwell, 1960). This species is purported to have originated through hybridization among the five tetraploid species, *V. australe*, *V. myrsinites*, *V. arkansanum*, *V. fuscatum*, and *V. virgatum*, and thus to possess germ plasm tracing to the four distinctive extant *Vaccinium* diploid species, *atrococcum*, *caesariense*, *darrowi*, and *tenellum*.

The rabbiteye species is considered worthy of improvement and gene transfer to other species because of the following strengths: extreme vigor, longevity, and productivity; tolerance to heat, drought, many fungus troubles, and soil variations; firm, long-keeping, acid fruit with a very small dry pedicel attachment region (scar); low chilling requirement. The species has the following limitations; small to modest fruit size, full color development prior to attaining best edibility, variable consistency of fruit "skin" (tendency to cracking in wet periods), flesh and "seeds" (grittiness), long period until full production is attained, long period from anthesis to fruit maturation (late season), dark skin color correlation with sweetest flavor, cold tenderness, difficulty of propagation, and self-sterility. Many of these weaknesses have already been overcome by the use of superior wild selections as parents in hybridizations followed by further selection.

Rabbiteye blueberry domestication was apparently started by M. A. Sapp of Crestview, Florida, who transplanted selected wild seedlings into his field around 1893 (Moore, 1966). The success of this venture led to a phenomenal increase in Florida of commercial plantings from transplanted wild seedlings. By 1921 Coville felt constrained to warn Florida plant purchasers that high-priced unselected wild seedlings would not fulfill the extravagant claims, and in at least one case the fraudulent photographs, of the nurserymen. Subsequent competition with the improved highbush cultivars led to the demise of the initial Florida rabbiteye industry (Sharpe, 1954).

Many wild rabbiteye blueberry selections were made and evaluated, especially at Tifton, Georgia, during the 1930s and early 1940s. In 1940 G. M. Darrow (USDA) started a cooperative rabbiteye breeding program with O. Woodard (Georgia Agricultural Experiment Station) and E. B. Morrow (North Carolina Agricultural Experiment Station). These programs have continued to the present with Draper (USDA), Brightwell (Georgia), Galletta (North Carolina), and Sharpe and Sherman (Florida) currently producing new rabbiteye hybrids, and Sefick (South Carolina) evaluating USDA seedling progenies. The initial use of the rabbiteye species in Florida was as the hexaploid parent in the 6x by 2x (*V. darrowi*) synthetic low chilling tetraploid development. This difficult heteroploid cross resulted in only five true hybrids from over 7500 cross-pollinations in a four-year period. At the same time (1948–1949), an improvement program was initiated at Florida within the rabbiteye species to obtain better adapted selections. By 1959 a total of 4000 seedlings had been fruited and 23 selections made (Sharpe and Darrow, 1959). The 'Bluegem' cultivar has been recently introduced as a companion cultivar for 'Woodard' (Sharpe and Sherman, 1970).

The very large Georgia-USDA program has been devoted almost entirely to rabbiteye breeding with significant improvements in fruit color, size, texture, and attractiveness over wild progenitors. Present goals also include flavor, crack resistance, limited suckering, leaf disease resistance, uniform fruit ripening, and maturation season extension in both directions (W. T. Brightwell, personal correspondence, 1971). This program has originated eight rabbiteye cultivars: 'Callaway' and 'Coastal' (1950); 'Tifblue' and 'Homebell' (1955); 'Woodard' (1960); and 'Southland', 'Briteblue', and 'Delite' (1969) (Brightwell, 1971a). 'Tifblue' has become the standard commercial hybrid in all of the rabbiteye producing regions of the South. 'Woodard' has fine fruit flavor, texture, and attractiveness, but the bush is not as widely adapted as 'Tifblue', and it is susceptible to powdery mildew, which it transmits to its seedlings.

The USDA-North Carolina program in rabbiteye breeding has been a steady but secondary effort. Two hybrids suitable for home garden culture were released in 1958 ('Garden Blue' and 'Menditoo'). The program has greatly enlarged in recent years; several more recent hybrids from later generations of crossing look promising. The rabbiteye parent germ plasm on hand was enlarged recently by the collection of more native material from Georgia and Florida. At the present time, rabbiteye blueberry development seems to be at about the period of Dr. Coville's death in the corresponding highbush development time scale. There should be some very interesting developments ahead, especially since, in the experience of the several rabbiteye blueberry breeders, the most promising material comes from the intercrossing of hybrids which have diverse parentage including south Georgia and west Florida types.

Modern Breeding Objectives

Characters which are considered important in blueberry improvement have been reviewed by Darrow (1960a), Moore (1965, 1966), Darrow and Scott (1966), and Draper and Scott (1967). Coville's initial breeding aims were principally the fresh fruit characters sweet and excellent flavor, large size, light blue color (bloom), dryness and plumpness at maturity, and the "possession of a foliage surface adequate to the nourishment of a large crop of berries "(Coville, 1921).

As the industry spread to other parts of the United States, bush adaptation to climatic and pest variability became more critical. Examples were the breeding for low chilling requirement in Florida, cane canker resistance in North Carolina, and bud and wood cold hardiness in New Jersey, New England, and Michigan.

With the blueberry promising to become a fruit crop of world-wide interest, there are certain plant, fruit, disease and pest tolerance, and adaptational traits that will need to receive special attention in future breeding programs. These are:

1) broader soil adaptation with less dependence on acid, organic, poorly drained soils;
2) broader climatic adaptation to either warm, long-growing season areas or cool, short-season areas;
3) early bearing onset to reduce time of initial commercial harvest from three to four years to two or three years depending on the species involved;
4) disease and pest resistance to problems such as mummy berry and *Botrytis* fruit rot, *Fusicoccum* and *Botryosphaeria* cankers, *Phytophthora* root rot, many leaf spotting fungi, stunt and red ringspot viruses, the several fruit worms, the bud mite, the sharp nosed leafhopper and the *Xiphonema* nematode;
5) mechanical handling tolerance with respect to harvesting, pruning, reduced cultivation, mulch, or herbicide culture, and increased plant densities;
6) superb fruit flavors including the elongation of the peak flavor period and the capturing of high quality in the processed as well as the fresh product.

For any one locale, objectives usually include a combination of the following traits:

1) plant characters: type, vigor, precocity, productivity, adaptation to mechanization, rooting or propagating ability, season of maturity, ornamental value, disease and pest resistance, cold hardiness, rest period, heat and drought resistance, and upland soil adaptation;
2) fruit characters: size, color, cluster habit,

scar, texture, firmness, flavor, period of fruit development, nutrient content, and processing quality.

Techniques of Breeding

Sources of these techniques are the blueberry improvement articles cited by Moore, and Draper and Scott, the author's techniques, and those of the following blueberry breeders and evaluators: L. E. Aalders and I. V. Hall (Nova Scotia), A. D. Draper (Beltsville, Maryland), H. J. Sefick (South Carolina), W. H. Childs (West Virginia), W. T. Brightwell (Georgia), R. H. Sharpe (Florida), J. E. Moulton (Michigan), and C. Stushnoff (Minnesota).

Biology: Camp (1945), Eck (1966), and Stushnoff and Palser (1969) have summarized the floral and fruit biology of *Vaccinium* species. Members of the genus bear their flower buds usually in the axillary position, but sometimes terminally. The number of flowers per inflorescence bud varies from solitary to a cluster (raceme type) of six to 14 flowers (Galletta and Mainland, unpublished). The individual flowers are perfect and regular (actinomorphic), with a sympetalous four- or five-lobed corolla. The corolla is borne usually in an inverted position, and may vary in shape from cylindrical to bell (campanulate) to urn (urceolate) or even globe-like. The corolla usually encloses the stamens and pistil, but does not in the *Polycodium* subgenus and it is cleft and deeply reflexed in *Oxycoccus* and *Hugeria* subgenera. The stamens number eight to ten, usually twice as many as the corolla lobes, and they are inserted at the base of the corolla forming a circle around the style. The stamen is composed of a filament and an inverted anther with two lobes and four thecae (pollen sacs). Anthers commonly have elongated hollow tubes with a pore at the tip through which the pollen is shed. Anthers of some sections of the genus have two spur-like awns on the filament attachment side of the anther. The pistil has an undifferentiated stigma borne on a single filliform (thin, threadlike) style, which is borne on a three- to ten-celled (commonly four- to five-celled or eight- to ten-celled by false partition) ovary. The calyx tube is adnate to the blueberry ovary, placing it in the inferior position category. The floral whorls in a single flower develop in acropetal sequence.

Following normally regular microsporogenesis, microspore formation is simultaneous (cytokinesis after the second division). Pollen is compound composed of four united grains (usually tetrahedron) of which each is capable of germinating *in vitro*. Each cell of the tetrad pollen grain is two-nucleate at the time of shedding. Shedding pollen tetrads normally fall past the stigma and out of the open corolla without effecting pollination. (Hence insect pollinators are needed in nature.) Pollen germination occurs in one to two hours *in vitro* and the generative nucleus divides in the pollen tube after four to eight hours, depending on the species (Stushnoff and Palser, 1969).

The megaspore mother cell of *Vaccinium* ovules is the micropylar of two enlarged cells in the tip of the nucellus just under the epidermis (Stushnoff and Palser, 1969). Megasporogenesis usually lags behind microsporogenesis. The difference in timing is greatly lessened by growing the plants in the greenhouse. Of the four usually linear megaspores the chalazal one is functional and forms the megagametophyte (embryo sac) according to the Polygonum type pattern. Pollen tubes can grow down to the base of the style in two days, but fertilization does not occur until six to eight days after pollination. Endosperm development precedes zygote development after a normal double fertilization. Zygote development starts at the eight- to ten-celled stage of the endosperm. The mature embryo is linear with an organized radical toward the micropylar end and two relatively short cotyledons at the chalazal end. Stushnoff and Palser conclude that *Vaccinium* embryo development is of the Solanad type.

The *Vaccinium* seed starts from a unitegmic (one integument) anatropous ovule prior to fertilization. In the mature seed, the embryo is surrounded by a starch-rich endosperm. The seed coat (derived primarily from the outer epidermis of the integument) has pitted, thickened, lignified walls. *Vaccinium* fruits bear many small seeds, ranging from eight to 90 with about 20–55% developed (Darrow, 1958; Bell, 1957; Edwards et al., 1972 and Moore et al. 1972).

Pollen collection and storage: Controlled blueberry pollinations may be made in the field, greenhouse, screenhouse, lathhouse, or coldframe. If plants are brought into the greenhouse, the pollen parents are often forced a few days earlier than the seed parents to insure a ready pollen supply. Cut blueberry branches may also be forced for pollen. In areas with very severe winters, branches with tightly closed buds may be brought into a greenhouse and forced with the base of the branch placed in a rooting medium supplied with intermittent mist, or in a jar under mist, until the flowers start to separate. The branch is then placed in a jar of water without mist. The base of the stem is freshly cut to prevent clogging of vascular tissues. The branches should be cut so that some two-year wood is left at the base of the branch.

Many breeders bring in branches bearing flower buds just prior to their flowering in the field. The branches are recut at the base and placed in jars of water. A small amount of fungicide added to the water will retard fungal growth. Water should be changed frequently, and the base of the stem recut with each water change. Weak flower buds may be encouraged to open by snipping the tips off the flower buds.

Fresh pollen can be collected daily from flowers which have just opened and not been visited by insects.

Pollen can be extracted by vibrating or shaking flowers, or by picking the flowers and rolling them gently between the thumb and forefinger onto a flat surface for immediate use, (such as fingertips, thumbnails, spatulas, and microscope slides), or into a container that may be stored (such as petri dishes, a deep-well microscope slide which can be sealed with a cover slip, a glass or plastic vial, or a zero or double zero pharmaceutical gelatin capsule). The capsules have the advantage of easy transport and refrigeration to and in the field and they can be labeled with a very small stringed tag, in which the string is held in place by the capsule lid.

Most breeders prefer to use pollen that is fresh or just a few days old. Pollen to be held a few days may be stored in a cool place in a desiccator or placed in a closed container in a refrigerator at 1.6 to 7.2 C. Pollen held dry in a refrigerator at these temperatures for a year has effected fertilization for the author and for Draper. Stushnoff (personal communication, 1972) has stored pollen at 50% relative humidity at −31.7 C successfully for a year.

Blueberry pollen viability studies were conducted at West Virginia in the early 1940s by Duis and Childs, using highbush and pollen of four other species, subjected to temperatures of 0–20 C and 0–50% relative humidities for 4½ years (Childs, personal communication, 1971). Pollen germination tests at six-month intervals showed that storage conditions of 0–2 C and relative humidities of 12.5–50% retained viability for almost two years for most species, and that V. *myrtilloides* pollen retained about 40% of its original viability after 4½ years stored at 2 C and 37.5% relative humidity.

Emasculation and controlled pollination: Because the position and structure of the blueberry flower do not permit ready self-pollination, and in species where self-incompatibility is widespread (such as lowbush and rabbiteye), some blueberry breeders do not emasculate the flowers in controlled varietal cross-pollinations. This applies especially to greenhouse pollinations performed in screened sections, or at times of the year when insects are not flying. Even outdoor pollinations may be made without emasculation if the flowering branches are covered before and after pollination with paper bags or some similar covering. In self-compatible species (such as highbush), unemasculated flowers may shed pollen onto the pollinating surface and contaminate the desired pollen. In species with very densely packed flower clusters, it is often helpful to tear open the corolla and stamens on one side to expose the stigma and style. Emasculation should be done for inheritance, self- and cross-fertility studies, and so forth.

Emasculation is accomplished by cutting or tearing away the corolla and stamens with small forceps (curved and pointed types are most useful) or small manicuring scissors. On large-flowered species which are quite turgid, the pinching and tearing action of the thumb- and middle fingernails can be rapid and effective. When emasculating, it is necessary not to penetrate the flower for more than a quarter to a third of its depth. A safe method is to grasp two "ribs" of the corolla gently with forceps while holding the calyx of the flower between the thumb and forefinger of the other hand. The forceps is then squeezed together, and rotated toward the side, tearing off the corolla and stamens in two or three motions. Emasculation may sometimes be accomplished in one or two motions by grasping the corolla on each side, squeezing to break the tissues and sliding the corolla off over the style (stripping the corollas). The stamens will then have to be removed. Rapidity should not be sought because the flower bud may be pulled off or the delicate style may be easily broken or bruised.

Transfer of the pollen is effected one to six (preferably two to four) days after emasculation by touching or very lightly brushing the stigmatic surface with a pollen coated fingertip, thumbnail, slide, petri dish, spatula, or short polished glass rod. The short rod is especially effective used with the gelatin capsule or glass vial. Up to 20 or 30 flowers can be pollinated from a well-coated surface before redipping. The pollen transfer tool and the pollinator's hands should be thoroughly washed in 70% or higher alcohol prior to the next pollination or emasculation. If time and an ample pollen source permit, it is always wise to repollinate any stigmas that are still receptive two to three days after the first pollination. Late flower buds can be removed at this time and later.

Field-pollinated plants need to be caged or the appropriate branches bagged prior to pollination to prevent wind or insect contamination. These should be replaced after pollination until the bloom period is over. Labels are attached to the pollinated branch showing the seed and pollen parents, the pollination dates, and the number of flowers pollinated. Set counts can be placed on the same tag later, and it can accompany the ripe fruit at maturity with the number of harvested fruits added to the label.

Fruit from pollinations made anywhere where birds can get to them are vulnerable. It is necessary to rebag the clusters with paper bags, cheesecloth, saran, or small mesh nylon netting. The latter, sewn with a draw string included, has the advantages of not interfering with light or air movement, and catching the overripe berries as they fall. Fruit should be harvested quite mature to insure maximum seed development.

Seed extraction and germination: Accumulated fruit from a single cross is held in polyethylene bags or rigid plastic or glass containers under ordinary refrigeration temperatures until seed extraction time. For special studies and special crosses, seed are removed from the fruit by hand. This usually involves mashing or scooping the fruit pulp onto an absorbent surface (such as a paper towel) and picking out the good seeds after drying. Childs (personal communication, 1971) used to crush the ripe fruits directly into the peat germinating medium without extracting the seeds. For extracting seeds from a quantity of fruit, most blueberry breeders macerate a quantity of fruit in a food blender with a minimum of water. The blender is run from just a few to 30 seconds. Usually a number of short electrical impulses will macerate the fruit and skin without cutting the seeds. The good seeds sink to the bottom. The pulp and skin floats and are decanted several times, and the remaining seeds are usually collected on an absorbent surface (toweling or filter paper) in a funnel. Seed may then be germinated immediately (without letting them dry), rinsed with a fermate solution, and stored moist in sealed plastic bags at 1.6 to 4.4 C (stratified) for two months and subjected to alternating temperature for the last month of stratification; or air dried, refrigerated (1.6–7.2 C) dry in a sealed container for five to six months, and then sown. Dried seed may be refrigerated in paper toweling, in small vials, or in coin envelopes placed in an air-tight container. Adequate germination (criterion was 50% or better germination) has been obtained by the author from seed lots over 20 years old, using the dry refrigerated storage method. Blueberry seed, properly stored, can be held at least ten to 15 years without much loss of viability.

Blueberry seed may be germinated directly after extraction from the fruit if placed in a suitable medium and under proper light and temperature conditions. If dry, seed must be after-ripened (dry) for five to six months at common refrigeration temperatures to overcome dormancy. Seeds stored in refrigerated fruit for one to three months and then extracted and stored dry and cold for an additional three months will also germinate in the normal six- to eight-week period. *V. angustifolium* seeds evidently germinate a bit more rapidly (three to four weeks) than highbush and rabbiteye seeds (Hall et al., 1971).

Scott and Draper (1967) demonstrated that light was necessary for the germination of six-month-old blueberry seed that had been held in dry cold storage at 4.4 C. Germination of uncovered seed under natural light and fluctuating diurnal greenhouse temperatures was essentially complete after six weeks. Comparable seed germinated in the dark for six weeks and then exposed to natural light required an additional seven weeks to catch up in germination percentage to the seeds originally germinated under light. Stushnoff and Hough (1968*b*), working with freshly extracted seed, found that pregermination and germination temperatures were as important or more important than light treatments in promoting germination of open-pollinated seeds from the highbush cultivars 'Bluecrop' and 'Darrow.' However natural daylength was not included in the illumination treatments. There was no germination at any light treatment if seeds were germinated at 23.9 C constant temperature. Fluctuating diurnal greenhouse temperatures of 10 to 32.2 C gave considerably higher germination percentages than a constant germination temperature of 16.7 C. Alternating four-day stratification temperatures of 0.6 C and 21.1 C for periods of 16, 32, or 64 days (two, four, or eight cycles of alternation) gave higher germination percentages under greenhouse conditions than comparable periods of stratification under a continuous temperature of 0.6 C. Exclusion of light, or continuous red or white light, resulted in little or no germination at constant germination temperatures. Under greenhouse conditions germination was highest (42%) under continuous red light, next highest (29–30%) after 24 hours of either red or white light followed by darkness, and lowest (17–20%) with no light or continuous white light. The highest germination (70–98%) was secured after 64 days of alternating temperature stratification, germination in the fluctuating temperatures of the greenhouse, and 24 hours exposure to red or white light or continuous red light. A suggested method for large-scale blueberry seed germination is: 32 days of stratification at alternating four-day periods of 0.6 and 21.1 C, germination in flats covered by two pieces of glass between which a sheet of red cellophane had been placed, in a greenhouse in which temperatures ranged from 10–23.9 C. Some crosses germinated 90% of their seeds in 40 to 50 days using this method. Chemical treatment of blueberry seeds to hasten germination has not been successful.

The germination medium should be acidic and largely organic. Successful germination media used include: acid soil containing at least one third peat; half peat:half sand; fine milled sphagnum moss; peat moss; half sand:half milled sphagnum moss; and Canadian peat. Germination containers include flats or standard, shallow bulb, or azalea clay pots with cooked sand in the bottom and the upper inch containing the germinating medium, or the entire pot containing the medium. Shredded sphagnum moss is either lightly sifted over the seed to a depth of 3 mm, or the seed is left uncovered. Germinating containers must be kept uniformly moist with mist or fine sprays.

Blueberry seed has been successfully germinated in flats in a covered and shaded coldframe until germination is progressing well. The flats are then moved to a greenhouse. Most blueberry breeders germinate seed initially in the greenhouse and grow it there until transplanting size. Usually blueberry seed germination containers receive neither supplementary light nor fertilizer. Continuous fluorescent light 15.2 cm above the germinating seedlings is helpful under poor winter light conditions. Some breeders apply a dilute nutrient solution once a week, or once one to two weeks prior to transplanting the seedlings from the germination container.

Transplanting and growth to field size: Blueberry seedlings may be transplanted from the seed germination container as early as the cotyledons are fully expanded (1 cm tall). Most breeders wait to transplant until most of the seedlings in the seed pot are 2–3 cm tall and bearing two to four true leaves.

Soil media which have been used successfully to grow blueberry seedlings to field size include: 1 sand:1 peat; milled sphagnum; (Canadian) peat; 1 soil:1 sand:1 peat; 2 peat:1 sand; 2 peat:1 sand:1 composted soil; 2 peat:1 soil; Jiffy Mix; 1 garden soil:2 peat:1 sand and 150 g $FeSO_4$, 150 g $MgSO_4$ and 1500 g superphosphate in 436.4 liters (12 imperial bushels); and 2 sand:2 peat:1 soil. Note that the smallest amount of acid organic material in any of these mixes is a third. For best results the organic content of the soil should be greater. Childs (1946) reported a 67.3% increase in the average linear growth of 268 seedlings from 19 progenies when grown on shredded sphagnum as compared with 1 peat:1 sand.

Because of the expense, time, and land use involved in growing blueberry seedlings to maturity for field evaluation, the breeder should grow the seedlings to field size as rapidly and economically as possible, restrict population size where practical, evaluate as many characters as possible in the seedling or juvenile periods, and increase the plant density in field plantings. To accomplish these purposes, blueberry breeders have modified their programs in terms of containers, fertilization, and method of handling the seedlings to field transplanting size.

It can be generalized that young blueberry seedlings should be grown in a high organic acid medium (over a third organic material), and that they should be fed very lightly and infrequently unless they are being forced with supplemental lighting. Supplemental light is necessary if the seedlings are grown over the winter period for spring planting. Seedlings in northern climates should be protected or stored under refrigeration over winter for spring planting. Growing containers which permit the plant to be directly transplanted to the field with a well-developed one- to two-fistful root ball area are much more economical of labor and time and insure best survival. Very close field spacings for evaluating initial fruiting potential are adequate, but it takes at least 45 cm (lowbush) or 61 cm (highbush and rabbiteye) spacing in the row to estimate mature fruiting potential.

Field evaluation and selection: Seedling populations need to be large due to extreme heterozygosity, quantitative inheritance of economic traits, the polyploid condition of the cultivated species, and the very large number of traits which must be considered. Furthermore, the soil requirements are special, often requiring the leasing of privately owned land, and the life cycle is long. In many of the species, one may observe some fruit in the second year in the field. Some culling could be done on the basis of fruit color, size, skin, flesh or seed texture, scar, or firmness. In the highbush types, however, these fruits borne on young plants can be misleading (larger than at maturity, with tighter clusters, poorer scars, etc.). Nevertheless, outstanding young seedlings should be identified for future observation.

In general, each breeding program has certain special emphases which can be observed in young seedlings and selected for or against. These include general vigor and relative freedom from serious pests. The Florida breeding team (Sherman et al., 1973) obtains a first evaluation of very close-planted material (the "fruiting nursery" concept) after 12 months in the field and two years from the cross. Approximately 2% of the seedlings are selected and moved to a wider spacing where they are reevaluated for up to five seasons. The Nova Scotia breeders screen their lowbush seedlings once when they are about 28 months from seed and have been in the field for 15 months. In the USDA program, evaluation is preliminarily in the second year and finally in the third year. Selected individual seedling plants are moved to a row and cuttings are taken from them. If the original seedling plant is still superior after refruiting, it is increased to a row of 25 to 50 plants. Selections promising in this row trial are increased and tested in 0.2 to 0.4 ha blocks. Most other breeders start seedling evaluations in the third or fourth seasons, and continue evaluations for three to five fruiting seasons. This is to insure that all materials have had an opportunity to fruit, that sufficient variation in climate and pest infestation has occurred, that selected materials are consistent in the desirable qualities for which they were chosen, and to insure the selection of seedlings which may have parental but not varietal promise also. This is necessary because the best parents are often rated average for many qualities, and the varietal selections may not transmit the traits for which they were chosen.

Selected seedlings are usually marked with tags upon which a few pertinent facts may be noted. A semi-permanent identification which will serve for several seasons is a bright-colored length of plastic ribbon.

It is well to propagate enough plants of a promising selection to permit the planting of duplicate or triplicate three- to five-plant plots at each of several sites, which represent different soil and climatic conditions. The selections should be interplanted with standard cultivars as checks. Several plants of each selection should be retained in a permanent block for identification verification or for parent source material.

In evaluating selections, each individual selection is scored at bloom and at fruiting time, as well as at appropriate times for disease ratings, propagation trials, processed fruit evaluations, and hardiness screening. Important observations at bloom time are the duration and peak of the flowering period, floral insect attractiveness, pollen fertility, natural set, and self-fruitfulness. At fruiting time, essential observations are the duration of the fruiting period, and at each picking the total yield, size, and fresh fruit quality and firmness. Several times during the season the color, scar, and other notable qualities like mechanical harvesting ability and acidity should be noted.

Initial characterizations of the selections can be made by subjective scoring against a standard for each character, such as the performance of a certain cultivar, a scale of five or ten intervals around a selected arbitrary value or by dividing the range of variation into 10% or 20% classes and assigning a scale number to each class. Of course, traits can also be measured objectively. Scores can be subjected to the usual statistical tests. Scalar standards can be changed as populations improve.

When the superior selections are interplanted with standard cultivars in a statistical design (advanced test), or if the relationships among variables need to be determined, more precise physical and chemical methods should be used. Such traits as yield, color, firmness, acidity, sugars, fruit removal force, size, scar size, bud set, cold endurance, and others can be measured directly and precisely.

Breeding Systems

Hall and Galletta (1971) studied the mitotic morphology of the nine diploid *Cyanococcus* subgenus species recognized by Camp, as well as *V. crassifolium* (*Herpothamnus* subgenus) and *V. macrocarpon* (*Oxycoccus* subgenus). The karyotypes of the 11 species were indistinguishable morphologically. An average idiogram for the diploid *Cyanococcus* species was represented as having two long chromosomes (slightly over 2μ in length), eight medium chromosomes (between 1.5 and 2.0μ) one of which is satellited, and two short chromosomes (under 1.5μ). All of the chromosomes had median or submedian centromeres. The *Cyanococcus* genome had chromosomes averaging 1.4–2.3μ in length, a total average chromosome length of 43.3μ, and an average ratio between the longest and shortest chromosomes of 1.68. The authors concluded that the basic *Vaccinium* genome has evidently not evolved much from the ancestral form, being numerous, small in size, and symmetrical in shape. Apparently the genetic sources of the richly diverse morphological and ecological specialization exhibited by the modern North American diploid blueberry species have been gene mutation and recombination rather than gross chromosomal rearrangements, duplications, or deficiencies.

Basically regular chromosome pairing and disjunction has been reported for many species and cultivars (Longley, 1927; Newcomer, 1941; Rousi, 1966*b*; Stushnoff and Palser, 1969; and Cockerham, 1972). Exceptions were the cultivar 'Coville' and others (see papers by Stushnoff and Palser, 1969; Stushnoff and Hough, 1968a; Jelenkovic and Harrington, 1971; Jelenkovic and Hough, 1970; Rousi, 1966*b*; and Newcomer, 1941), and the hexaploid species *V. constablaei* and *V. ashei* (Longley, 1927; Stushnoff and Palser, 1969) and *V. amoenum* (Cockerham, 1972).

In many of the tetraploid cultivars and species studied, there were true multivalents and/or secondary associations of bivalents present at the first meiotic metaphase. Newcomer and Jelenkovic and Hough reported secondary associations in six tetraploid cultivar meioses. The presence of one to five quadrivalents per cell in polyploids whose metaphase associations were preponderantly bivalent was shown by these authors, as well as Rousi (1966*b*) and Jelenkovic and Harrington (1971). Although the 'Adams' highbush and 'Mich. Lowbush No. 4' both showed normal bivalent pairing, their hybrid showed secondary pairing and lagging of chromosomes during anaphase (Newcomer, 1941).

Observations on bivalent and multivalent form in the cultivar 'Jersey' suggested one to three chiasmata/bivalent, with at least one distal chiasma/bivalent obligatory (Jelenkovic and Harrington, 1971). In view of the smallness of *Vaccinium* chromosomes, one or more chiasmata/bivalent would indicate a fair amount of recombination potential, even though the chiasma frequency is low.

The regularity of chromosome disjunction in polyploid clones is probably best explained by their selection for their high degree of fruitfulness, thus implying, at least, regular megasporogenesis and embryo sac development. Cockerham (1972) found one or two multivalents in about half of the analyzable cells of the hexaploid *V. amoenum*. Cockerham also found that unselected tetraploid species representatives had a higher proportion of stainable pollen than comparable diploid or hexaploid species representatives. All of the species studied had enough unreduced gametes (1.4% average across all species) to account for the origin of polyploid forms over a long period of time. Frequent pollen formation irregularities were early abortions (empty or granular grains), unreduced grains, reduction in spore number, and distorted, partially stained grains.

Species crossability: It has been noted repeatedly in the blueberry literature that *Vaccinium* species of the same chromosome level (homoploid crosses) were completely interfertile whereas species of different chromosome levels (heteroploid crosses) hybridized with great difficulty, if at all. Moore (1966) lists a number of successful homo- and heteroploid crosses, but data on degree of crossability is scarce. Fruit set and seed count per berry data are indicative of crossability, but the number of germinable true hybrid seed per pollination is the best criterion. For example, Sharpe and Sherman (1971) reported five true hybrids from 7500 *V. darrowi* x *V. ashei* cross-pollinations and 31 true hybrids from 1600 *V. darrowi* x tetraploid northern highbush cross-pollinations.

To determine the true crossability of two species one should secure F_2 and BC_1 populations to detect possible hybrid weakness, sterility, or inviability in later generations. Since individuals within species populations are known to vary widely in pollen and egg fertility, it is important which individuals are chosen to represent the parent species. (Perhaps mixed pollen of Species A on several females of Species B and vice-versa would be the best method to determine the crossability of two species.) Reciprocal differences in species crossability need to be studied, especially in heteroploid or wide crosses. The author has found that V. *corymbosum* x V. *ashei* crosses are considerably more successful than the reciprocal crosses. (The V. *ashei* x *corymbosum* crosses fail or yield very little seed.) In tetraploid x hexaploid *Vaccinium* crosses, the cross is often more successful or only successful if the tetraploid is the female parent.

Polyploid forms of *Vaccinium* have been obtained following use of colchicine as a seed treatment (Rousi, 1966*b*; Aalders and Hall, 1963c), or as a treatment of growing shoot tips (Moore et al., 1964; Rousi 1966*b*). Polyploid induction is a valuable tool for transferring genes from diploid to tetraploid species or for restoring partial fertility to sterile hybrids. Successful seed treatments were immersing germinating seedlings in a 0.5% colchicine solution for six hours (Rousi), or for four or seven days in a 0.5% solution into which air had been bubbled at 100 to 200 bubbles per minute (Hall and Aalders). Successful vegetative treatments were five to seven applications at two-day intervals of either growing seedlings or a strong pinched shoot with either 0.25 or 0.5% colchicine, the latter with a surfactant added.

Moore et al. (1964) induced fertility in a previously sterile pentaploid rabbiteye x highbush hybrid. Forcing buds from affected areas by pruning resulted in a sectorial branch bearing 5x tissue on one side and a cytochimera of 5x epidermal and 10x internal tissues on the other side. Flowers produced on this decaploid branch were larger, bore more fertile pollen, and were self-fertile whereas the smaller flowers of adjacent pentaploid branches bore little good pollen and were self-sterile. A decaploid species bridge has been proposed involving crossing the 10x to either 6x or 2x species to produce 8x or 6x breeding lines. These lines when crossed to 4x and 6x hybrids respectively, would yield 6x clones. However, Jelenkovic and Draper (1970) noted that this decaploid subsequently has a low level of fertility in outcrosses to other *Vaccinium* species. Selfing of the decaploid resulted in less than one seedling per pollinated flower. When the decaploid was used as a seed parent, crosses with 6x pollen parents yielded zero to 27 seedlings per 100 pollinations and crosses with 2x pollen parents yielded four to 16 seedlings per 100 pollinations. The reciprocal 6x by 10x crosses yielded from zero to 42 seedlings per 100 pollinations. The decaploid chromosome pairing was almost completely as bivalents, but practically all metaphase I plates had univalent chromosomes present and anaphase disjunction was generally asynchronous. The breeding use of the decaploid as a species bridge is possible, but the low crossability is a formidable obstacle.

It has been previously noted that the lowbush and rabbiteye species are basically self-unfruitful. Stushnoff and associates (1968–1969) have shown abnormal microsporogenesis in the 'Coville' highbush cultivar and in certain V. *constablaei* microsporangia in which nonfunctional pollen tetrads are associated with globules of an unidentified substance. Hall and Aalders dealt with male and female sterility in V. *angustifolium* in a series of four papers from 1961 to 1966. A survey of 1171 clones fell into three categories: 5.2% produced no pollen, 45% yielded less pollen than normal, and 49.8% produced abundant pollen. They concluded that the male sterile factor was widespread in Eastern Canada. Of 21 clones tested for self-compatibility, 12 were completely self-incompatible and nine were partially self-compatible. F_1, F_2, and backcross generations of three male sterile by three male fertile clones demonstrated complex multiple factor inheritance for male sterility. Each of the three male steriles underwent pollen degeneration at a different time. Three instances of female sterility were shown to be associated with abnormal ovule development. Intermediate degrees of female sterility were also reported.

Self-incompatibility must have arisen early in the evolution of *Vaccinium*. Ballington in North Carolina species fertility studies (unpublished) has preliminary information indicating that *Cyanococcus* diploid species V. *atrococcum, caesariense, darrowi, tenellum,* and *vacillans* representatives are largely self-unfruitful.

There has been much speculation on the putative parentage of some of the extant polyploid *Vaccinium* species, as well as whether they are auto or alloployploids, or whether *Vaccinium* species are really "valid species," since many of them intercross. Much more work has to be done to answer these problems and to discern the phylogenetic relationships within the genus. Concerning species parentage, there is no substitute for adjacent garden culture, detailed morphological and cytological comparisons, and crossability studies of the putative parents, intermediate generations, and the putative offspring species. An artificial resynthesis by cross-pollination should also be attempted. Studies like those of Aalders and Hall (1963a) where the inheritance of the major features separating two species is examined, are an example of the type of investigation that would be helpful in determining species relationships. Their designation of *V. brittonii* as a taxonomic form of *V. angustifolium* on the basis of the inheritance of leaf bloom as single semidominant gene is provocative but not conclusive at this point. The question of auto- vs. allopolyploidy is complicated by the fact that the *Vaccinium* genome is relatively undifferentiated, and the putative diploid parents are not entirely unrelated. Add to this the strong bivalent pairing tendencies in colchicine-induced tetraploids and in wide species hybrids, and the most likely conclusion is that our natural polyploid *Vaccinium* species are mostly segmental allopolyploids. The ability of homoploid species within subgenera to intercross indicates that isolating mechanisms are still being established. There must still be some homology or homeology between the *Cyanococcus* and *Euvaccinium* subgenera, between *Cyanococcus* and the whortleberries (Rousi 1966a), and between *Cyanococcus* and *Polycodium*. However, subgenera like *Cyanococcus* and *Oxycoccus* seem to have diverged early.

Inheritance patterns: Blueberry breeders have been hampered by a lack of simply inherited marker genes in the genus. This situation is common in polyploid species (in blueberry, the tetraploid and hexaploid chromosome levels seem to be especially adaptive levels at which to secure climatic and edaphic tolerance, consistent high yields, large fruit size, and mechanical handling ability). Heterozygosity *per se* would appear to be advantageous in promoting high vigor, developing new bush types, and achieving flexibility in adaptation to variable soil, climate, and pest problems. Representative seedling progenies of all of the natural species exhibit a high degree of variability for practically every trait. The traits in which the breeder is interested are related to the normal development and fertility of the plant; it is expected that such important traits would be homeostatically buffered by being numerous and spread across the entire genome, and that they would show quantitative inheritance.

Hall et al. (1964) cited eight putative mutant forms of lowbush blueberry: crinkle leaf, diffuse necrotic leaf spot, bent style, dissected corolla, variegated leaf, puckered necrotic leaf base, pink spotted corolla, and protruding pistil. Studies (Galletta, unpublished) on the morphology and transmission of 30 variegated highbush blueberry seedlings from six progenies over a four-year period suggest six phenotypic variegation patterns and a variety of developmental sequences. Inheritance studies suggest that two to four genes are involved in tetraploid highbush blueberries.

One trait which appears to be simply inherited is albino (white or pink) fruit color. Aalders and Hall (1962a) demonstrated that white fruit is a simple one-gene recessive in the diploid species *V. myrtilloides*. Hall and Aalders (1963) suggested that in tetraploid *V. angustifolium*, white fruit was controlled by two recessive genes, but Draper and Scott (1971) have interpreted the Hall and Aalders data with a single recessive gene tetrasomically inherited with random chromosome pairing. Galletta concluded that pink fruit in tetraploid highbush was most likely a one-gene tetrasomically inherited, recessive genotype (see Ballinger et al., 1972). However, anthocyanin content in normally blue-purple colored fruit genotypes appears to be multigenic.

One example of quantitative inheritance is fruit size inheritance in tetraploid highbush blueberries (Draper and Scott 1969). Since their smallest fruited parent represents an improvement of two to 2½ times the size of wild highbush seedlings, their findings are indicative of segregations in the medium to very large fruit size types. Large by large-fruited crosses yielded progenies with more large-fruited seedlings than large by medium or large by small-fruited crosses. The means of the large by smallest crossed progenies were very close to the mean of the small parent, indicating a dominance of small fruit size alleles over large size alleles. There was no difference in mean progeny yield scores, which led to the conclusion that large fruit size was not linked with low yield. While the overall frequency distribution of the seven test progenies approached normality, individual progeny distributions were skewed toward the smaller fruit sizes. The number of transgressive segregants on the large end of the scale was considerably smaller than those on the small fruit size end of the scale.

Tetrasomic inheritance in the highbush blueberry has been found in the recessive lethal trait, albino seedling (Draper and Scott, 1971). Three duplex genotypes selfed and intercrossed yielded green to albino seedlings in a 43.4:1 ratio (rather than the 35:1 ratio expected with random chromosome segregation). The excess of green seedlings was explained by supposing an increase in homogenic pairing. The excess of green seedlings might also be explained by the nongermination of some of the lethal albino-bearing seeds. Draper and Scott indicate that their tetrasomic genetic data offers partial support of Camp's hypothesis of autotetraploid origin of V. *australe*. Interestingly, the albino seedling trait was first detected in an F_2 population from the interspecific tetraploid hybridization between the highbush and lowbush species. This population segregated 3 green:1 albino.

Methodology: The development of breeding schemes for blueberries is in its infancy. Remarkable progress has been achieved from hybridizations of two general types followed by intensive selection, often for transgressive segregants. The remarkable increase in fruit size over four to five generations of breeding has been secured mainly by the use of phenotypic assortative mating (large-fruited x large fruited crosses). The second type of successful hybridization may be termed "complementary" phenotypic disassortative mating. An example would be the cross of a cane canker susceptible, large-fruited selection with light blue fruit by a canker resistant medium-sized clone with medium to dark blue fruit, followed by stringent selection for canker resistance, large fruit size, and light blue color. To be successful, either hybridization scheme must use as parents clones which do not display marked weaknesses in any traits which are not being actively considered in the selection. And of course, the parents must transmit the desired phenotype to a large proportion of the seedling progeny.

Inbreeding has been a useful method in the breeding of many plant species for increasing homozygosity, obtaining desirable combinations of characteristics, assessing genotypic constitutions, and testing the prepotency of a parent for the transmission of certain characters to its progeny. Many of our named highbush and rabbiteye cultivars may be considered partial inbreds because such parents as 'Rubel', 'Russell', 'Brooks', and 'Ethel' occur in the lineage of both immediate parents, or more than once in the ancestry of one parent.

Severe forms of inbreeding like selfing are hampered in the blueberry because of widespread complete or partial self-incompatibility, and because of reduced seed germination (Draper and Scott, 1971), and later inbreeding depression in the field (Galletta, unpublished). Galletta (1970) reported a marked reduction in fruit set on selfing partially self-compatible or self-incompatible rabbiteye cultivars. Rabbiteye self pollinations resulted in a reduction of morphologically good appearing seeds (663 per 100 pollinations) when compared to outcrossed progenies (1395 good seeds per 100 pollinations). Selfing or moderate inbreeding also reduced the number of vigorous seedlings which could be taken to the field. The number of good appearing seeds is evidently a poor criterion for judging the results of a pollination. The number of good seeds was not correlated with the percent fruit set, the number of seedlings germinated, surviving, or vigorous. There must be a number of abortive or nongerminable embryos within the plump seed category. Full-sib crosses exhibit reasonable vigor, and half-sib crosses are usually as vigorous as outcrosses. Moderate levels of inbreeding like the half-sib cross ($r = 1/4$) may be useful in developing reasonably homozygous parent stocks. Some blueberry genotypes appear to yield large enough progenies for evaluation when selfed, backcrossed, or full-sib crossed.

Breeding procedures may be simplified if a mature plant trait can be shown to be correlated with a trait expressed in the juvenile stage. Morrow et al. (1949) established a significant direct relationship between leaf and fruit glaucescense in rabbiteye blueberries and demonstrated its use. The author has used the method by rouging the least glaucous seedlings in nursery populations of rabbiteye seedlings to eliminate the darkest fruited seedlings prior to transplanting to the fruiting field. However, one observes that high yield, dark fruit color, and high plant vigor are often associated in natural populations. Too stringent a selection for leaf glaucousness may also eliminate the best yielding seedlings. This same glaucousness correlation may be observed in species related to V. *ashei*, such as V. *darrowi*.

Moore et al. (1962) found that the response of highbush blueberry seedlings in the greenhouse to powdery mildew infection was closely correlated to seedling response later under field conditions. However, the correlation, while useful in permitting the discarding of very susceptible seedlings, was not adequate to detect escapes among the seedlings classed as resistant in the greenhouse, nor the seedlings which showed field resistance but greenhouse susceptibility. Sharpe has frequently eliminated high-chilling requiring segregants from the Florida program prior to seedling transplanting by winter exposure of very young seedlings.

An approach to breeding for improvement of specific characters involves the partitioning of the characters into their component parts with appropriate weight assigned to each part, if possible. Galletta and Mainland (1971a) attempted to do this for berry yield. For young plants the size of the bearing area as measured by plant height, width, and number of bearing canes is sufficient to account for 50% of the yield determination by simple correlation analysis. A more complete estimate of yield ($r^2 = .70–.82$) can be secured if the number of flower buds per bush is calculated from the components fruiting canes per plant, fruiting shoots per cane, and flower buds per fruiting shoot. For mature plants, yield is best predicted by using an index which estimates the number of fruits per plant and the fruit size.

It may be possible for a breeder to secure a considerable genetic improvement in a trait by selecting for certain individual components of the trait and not others (e.g., for higher yield through selecting for more fruiting shoots per cane, or increased branching), if the characters do not compensate. Heritability estimates of each trait are desirable to determine the efficacy of this system.

Parent selection: There is no substitute for carefully characterizing all of the phenotypic characters of interest for each selection, in comparison with standard clones, prior to using the selection as a parent. Preferably, this selection evaluation should be carried out over several growing seasons and at several varying locations. If the selection shows any marked weaknesses, it should not be used as a parent. It is very difficult, for example, to overcome the deficiencies introduced into seedling progenies by the use of parents which transmit soft fruit, poor scars, dark fruit, small fruit size, or spreading bush habit.

To be potentially useful as a parent, a clone or seedling should display marked phenotypic superiority for at least two or three traits and be no poorer than average for the other traits under consideration. In the long run it is also wise to evaluate as many traits as possible under the existing local conditions. The more traits considered in selection, the longer it will take to find outstanding germ plasm. But the introductions resulting from a broadly based and conceived breeding program will be all the more valuable and more durable.

Unfortunately, the best phenotypes are not always transmitted by clones which possess them. The best parents in highly heterozygous material are often mediocre for a number of traits. If there prove to be many instances of epistatic and dominance gene interactions in blueberries, the ability to predict progeny performance from parental phenotype would be decreased. However, the large gains that have been obtained in blueberry by phenotypic selection of parents suggests that there may be a significant additive component in at least the inheritance of such traits as fruit size.

Moore (1966) summarized the early character transmission reports of Darrow et al., Johnston, and others for the original highbush and rabbiteye blueberry breeding parents. Moore and Scott (1966) reported the breeding value of the parents used in the USDA breeding program for the northeastern United States for the period 1949–1957. Draper and Scott (1967) made a similar report for the period 1957–1967.

Character transmission ability has been variously reported from the number of times a certain clone has been used as a parent, and the number of selections made from particular crosses, to the frequency distributions of seedlings of different crosses for particular characters. The last method (frequency distribution of seedlings) has the most merit, because comparative population sample statistics can be derived and statistically analysed. Transmission data can then be quantified, and perhaps predicting equations for the potential parental value of each meritorious selection could be derived.

A potential parent should be progeny tested by selfing and crossing to tester clones, but one generation would be lost in this type of study. In a perennial woody crop, one would have to weigh the advantages to be gained from the genetic transmission studies against the delay in using the particular selection as a parent.

Breeding for Specific Plant Characters

Major components of plant type include plant size; general habit (prostrate, rhizomatous, suckering, crown-forming, upright, spreading, etc.); leaf habit (evergreen, semi-persistent, deciduous); leaf size, shape, and color, presence of cuticular waxes and glands; types of buds (flower, leaf, mixed); flower clustering, size, shape, odor, and nectar.

Older breeding objectives have always listed a large, vigorous, upright bush. It seems that this is now a limited objective. There seem to be four interesting commercial plant type possibilities: 1) low-growing rhizomatous matted rows (up to 30 cm tall) or ground covers with large fruit offer an interesting possibility for mechanized culture; 2) intermediate height (waist-high, around 1 m tall) very productive hedgerows of the highbush x lowbush F_1 or the *V. darrowi* or *myrsinites* types. These should be relatively easy to handle mechanically, particularly if the evergreen trait is incorporated and the bush is hedged to place the fruit on the periphery of the bush; 3) "head" high (1.5–2 m tall) bushes in which the many stocky canes would be relatively unitary (unbranched) and bear highbush type fruit in long loose *V. amoenum*-like clusters. Such canes are supple enough to be bent over close to the ground for mechanical detachment, and the unit cane could be mechanically thinned out periodically; and 4) very high (2.25–3 m tall) rabbiteye-type bushes with very firm, large, high quality fruit. This plant is adapted to over-the-row harvesters for which the bushes can be mechanically shaped.

Leaf size and shape are neutral characters as long as the total leaf area is adequate. Thick green leaves are often associated with high vigor. Dark green leaves may be indicative of higher chlorophyll content than light green leaves. Heavily waxed leaves may resist excessive transpiration. Relatively few flower buds per fruiting shoot seems to offer the best handling possibilities. Many fruiting shoots per cane and a moderate number of flowers per bud appear best for later fruit detachment. Large flower size, high nectar and pollen production, and odor are important in determining and attracting the insect pollinators.

Vigor is usually estimated by stem length or stockiness. Actually, some of the most vigorous blueberries the author has seen were rather squat bushes like 'Bluetta', which bore a tremendous crop of leaves and fruit consistently. A more accurate definition of vigor is plant vitality. Its major component is a good balance of growth and development among the major plant organs. Strong stem growth should be accompanied by adequate leaf, bud, and root development. Vigor ratings should be an estimate of the plant's biological efficiency in terms of the production of healthy and plentiful vegetative plant organs sufficient to support a heavy crop. In the future, more vigorous cultivars may have to be selected to compete with weedy, mulched, or increased plant density culture.

Fruitfulness has many components in the blueberry. Unpublished data by Galletta and Mainland suggest that in very young plants the size of the plant body is the major yield component. After two to three years of age the composition of the fruiting cane is much more important. Estimating yield factors then involves the number of fruiting canes per plant, number of fruiting shoots per cane, number of flower buds per shoot, number of flowers per bud, and the natural set of these flowers (an indirect estimate of pollen and egg production and viability). These components will closely estimate the number of fruits per plant. An estimate of the fruit size along with the fruit number is also necessary to predict yield. Many blueberry clones ('Murphy', 'Wolcott', *V. angustifolium* seedlings) have a tendency to differentiate a second crop of flowers after fruit ripening. In North Carolina the last flush of growth is often determinate, whereas the first two or three are indeterminate (apical death). This raises the possibility of having fall-fruiting *Vaccinium* clones as in *Rubus*. *V. arboreum* does fruit naturally on wood produced in the current year, instead of wood produced in the previous year, as in the *Cyanococcus* subgenus.

Earlier development of the plant to the beginning of commercial fruit production is desirable. Such precociousness must not, however, sacrifice later vigor or length of plant life. Selection is essentially for rapid plant body development and early flowering. Since many blueberry seedlings produce flower buds in the field the first season, it should be possible to select precocious individuals which will yield a good crop during the second year without an adverse effect on subsequent productivity or longevity.

Ease of asexual propagation is an essential character in blueberry. There is clonal and species variation in the ability of cuttings to initiate adventitious roots. Many techniques improve rooting of difficult-to-root clones. Stem cutting position, presence of flower buds, juvenile wood selection, proper attention to media, moisture, and shading are all important. Evergreen species root much more readily than many of the deciduous species. This may suggest clonal differences in endogenous auxin and cofactor levels which could be assayed to provide a rooting ability criterion. The most important criterion for successful blueberry rooting appears to be proper timing of taking either softwood or hardwood cuttings with respect to shoot development and maturation.

As agricultural labor becomes more scarce and expensive, the use of mechanization in blueberry production becomes more important. Mechanized harvest of blueberries is now a common commercial practice and mechanized pruning is a distinct possibility. For mechanical harvest of highbush types, canes must be rigid enough to transmit the shaking force well, but supple enough to withstand bending and twisting without breaking. They must also withstand bark "skinning" and be able to rejuvenate shoots readily behind a break. The degree of branching and the arrangement of fruit clusters on the shoots would depend on the number of canes permitted to develop in the chosen training system. For lowbush plants, where a suction or raking action may be used to detach the fruit, the maximum number of fruiting shoots per surface area would be desirable. These shoots would also have to combine strength and flexibility.

Several possibilities exist for making mechanical pruning a reality. Selection for genotypes that do not require extensive pruning and that can be adapted to partial mechanical pruning offers possibilities. Also, there is wide variation in natural plant habit and in plant growth response following pruning from which types adapted to mechanical shaping might be selected.

The best returns are usually obtained for fruit ripening very early or very late in the season. It is difficult to secure high fruit quality in clones ripening at either extreme because vigor is often low or modest in the earliest clones, and both early and late clones ripen under more uncertain (variable) environmental conditions than midseason types. It should be possible to select for a succession of blueberry cultivars maturing over an eight- to 16-week period in any locality, particularly if germ plasm from a number of species is included in the breeding lines.

Uniform ripening for mechanical harvesting purposes is often cited as a breeding aim. However, intense selection for ripening all at one time would have some disadvantages. Some reduction in total yield might occur if the maturation or development of flower buds within the same inflorescence, and in buds of different shoot and growth flush positions, had to occur simultaneously. Also, having the entire crop mature at one time exposes the grower to maximum loss from severe weather and pest infestation conditions. If the maturing period temperatures are 30 C or higher, cultivars like 'Croatan', 'Collins', 'Morrow', and 'Earliblue' will mature over 80% of their crop in a seven-day period. These and other clones (the *V. ashei* cultivar 'Tifblue', for example) will then hold ripened fruit on the plant in reasonably good eating and keeping quality condition until 90% or more can be removed at one harvest. Concentrated ripening may be achieved by securing types in which the blooming period from first to last flower is short, in which the berry ripening period is uniform, or types which can "store" matured fruit on the bush in prime condition.

It should be possible to select blueberries that would thrive under climatic conditions as variable as from those of south Florida to those of the upper Scandinavian peninsula. Species like *V. darrowi* and *V. myrsinites* can be grown in a greenhouse with little or no chilling requirement. Species like *V. erythrocarpum* or *V. constablaei* seem to require in excess of 1500 hours below 7.2 C to break dormancy. A high to moderately high chilling requirement would be useful also in middle latitudes to prevent bud break during intermittent warm periods in late winter and early spring months. For the maximum effectiveness a high chilling requirement would have to be coupled with short fruit development time and high self-fruitfulness. The use of low chilling parent materials would help in extending blueberry culture into subtropical or Mediterranean climates. A high degree of self-fertility and heat and drought resistance would be very useful in low chilling cultivars.

Cold hardiness is critical for areas having severe winters and springs, particularly if the snow cover is not constant, or if highbush types are desired for maximum yields. Hardiness appears to have several components. The obvious one is the maximum amount of cold a particular tissue can stand. Stushnoff (1972) and Quamme et al. (1972) found that the highbush cultivars 'Rancocas' and 'Earliblue' could be hardened to withstand −40 C like native *V. angustifolium*, but that the highbush clones acclimated much more slowly than the native lowbush. They suggested that the breeder might more profitably select for more rapid onset of dormancy than for maximum cold tolerance. Practically speaking, one could select for early fall maturation (bud maturity, leaf coloring, and dropping). One might be able to combine early fall maturity with late blooming resulting from a long rest period requirement to achieve a shorter growing season. Quamme et al. also reported that xylem could tolerate lower temperatures than bark, thus demonstrating tissue specificity for cold tolerance in the blueberry, as known in other deciduous fruit plants. It would be necessary to have hardy bud and flower parts to withstand both severe winter temperatures and spring frosts. Artificial freezing trials of wood and buds of selected clones might prove necessary. Bud hardy cultivars would prove useful throughout the temperate zone.

Adaptation to high summer temperatures and dry soil and atmospheric conditions may be associated with low respiration rate of the plant. The broadly adapted 'Bluecrop' had the lowest respiration rate of six large-fruited highbush cultivars (Coorts et al., 1968). Conversely, the slow growing and maturing 'Coville', which is of limited utility at both the southern and northern cultural latitude extremes, had the highest respiration of the six clones tested.

The restrictive dependence of most blueberry species to acid, high organic matter soils with a high water table must be overcome if blueberries are to expand into wider geographic areas. The traditional imperfectly drained acid sand and peat blueberry soils are limited in area, and future expansion is dependent on the development of germ plasm adaptation. Species such as *V. ashei*, *V. angustifolium*, *V. pallidum*, *V. arboreum*, and *V. stamineum* seem to grow well on either organic or mineral soils. The ability of the root system to effectively extract water and nutrients over a wide range of soil types and reactions is probably the key factor in selecting for types adapted to upland soils. Evidently tap-rooted, rhizomatous, or suckering species have an advantage over strictly crown-forming species in this respect. Galletta and Fish (1971) reported that grafting compatibilities within the *Vaccinium* genus seemed broad, and that the broader soil adaptation and resistance to soil-inhabiting pests of some species could be exploited by using them as rootstocks for scions of species having more limited soil adaptability. A particularly promising combination for the southern United States was highbush (*V. corymbosum*) scions on rabbiteye (*V. ashei*) rootstocks. Rootstock selection seems feasible in several *Vaccinium* species and in certain species hybrids. At present, it seems that a suitable rootstock should possess the following traits: highbush habit, narrow base, limited suckering, stout canes, upright habit, ease of propagation, broad soil variation tolerance, and broad scion compatibility. It is possible that some of the taller rhizomatous or clump-forming species could be crown-grafted by side-grafting techniques to make suitable rootstocks. Such use of rootstocks offer an immediate solution to wider culture of highbush, and for special purpose situations. However, the development of adapted "direct-producers" is the best long-term approach.

Though commercial blueberry culture is only about 50 years old, the crop has become heir to a substantial number of serious diseases and pest. Though cultivar variation in succeptibility has been demonstrated for many diseases, the only continuing breeding program incorporating disease resistance has been that of North Carolina for cane canker resistance.

It is impractical to breed for resistance to a large number of diseases and pests simultaneously because of the polyploid nature of the developed cultivars and the presumed quantitatively inherited nature of resistance. However, resistance to several pests might be sought together if the major genes for resistance were linked. Field tolerance to a large number of organisms may be secured by simple negative selection in seedling and selected clone populations. There is, of course, the later danger that new pests or new strains of old pests will attack previously resistant or tolerant cultivars. This is an additional reason for maintaining a collection block of introduced cultivars and select breeding clones, even though some of them are susceptible to diseases. It is also still wise to plant more than one cultivar in a block to prevent complete loss from an epiphytotic, as well as to insure maximum pollination.

Table 2 lists the maladies of the blueberry which presently are, or could become, serious enough to warrant seeking tolerance or resistance in developing cultivars. Disease and pest resistance may become even more important in the future if certain chemical controls are shown to be environmental hazards.

TABLE 2. Potentially serious diseases and pests of blueberry (Principal literature sources were Marucci, 1966*b*; Nelson, 1966; Varney and Stretch, 1966; and Stretch 1967*a* and 1967*b*.)

Common Name	Causal Organism	Principal Areas of Concern
FUNGI		
Mummy berry	*Monilinia vaccinii-corymbosi* (Reade) Honey; formerly *Sclerotinia vaccinii* Wor.	U.S. and Canada, especially northern areas
Cane (stem) canker	*Botryosphaeria corticis* (Demaree and Wilcox) Arx and Muller	North Carolina and southern U.S.
Fusicoccum canker	*Godronia cassandrae* f. *Vaccinii* (Peck) Grove; conidial stage—*Fusicoccum putrefaciens* Shear	Mass., Mich., Me., British Columbia, New Brunswick, Nova Scotia, and Denmark
Coryneum canker	*Coryneum microstictum* Berk. and Br.	Massachusetts
Botrytis twig and cane blight	*Botrytis cinerea* Ters. ex Fr.	N.J., Alaska, Wash., England, Germany, Ireland, Netherlands
Powdery mildew	*Microsphaera penicillata* var. *vaccinii* (Schw.) Cooke. = *M. alni* var. *vaccinii*	General fungus—all areas
Red leaf	*Exobasidium vaccinii* Wor. (*E. myrtilli* Sieqm. in Europe)	Lowbush plantings in N. Amer. and Sweden
Leaf rust	*Pucciniastrum myrtilli* (Schum.) Arth. and *P. spp.*	Maine, southern and western U.S.
Phytophthora root rot	*Phytophthora cinnamomi* Rands	N.J., Md., N.C. and South
Witches-broom	*Pucciniastrum goeppertianum* (Kuhn) Kleb	Me., south to Pa. and west to Calif.
Stem blight	*Botryosphaeria ribes* (Tode ex Fr.) Gross. and Dug. [*B. dothidea* (Mouq. ex Fr.) Ces. and deNot.]	North Carolina
Septoria leaf spot	*Septoria albopunctata* Cke	Southern U.S., also lowbush areas
Double spot	*Dothichiza caroliniana* Demaree and Wilcox	North Carolina
Anthracnose	*Glomerella cingulata* Spaulding and von Schrenk (*Gloeosporium frustigenum*—imperfect stage) *Gloeosporium minus* Shear	N.J., New England and N.C.
Branch and bush blight	*Crumenula urceolus* (Fries) de Nortaris (probably same organism as *Godronia cassandrae* Peck—see *Fusicoccum* canker)	Scotland
Fruit rotting fungi	*Botrytis, Alternaria* and *Gloeosporium* sp.	General postharvest decay fungi
VIRUS DISEASES		
Stunt	Blueberry stunt virus	General—all areas
Red ringspot	Blueberry red ringspot virus	Eastern U.S.
Necrotic ringspot	Tobacco ringspot virus	Conn., N.Y., Mich., Ill., and N.J.
Shoestring	Blueberry shoestring virus	Mich., N.J., and eastern U.S.
Mosaic	Blueberry mosaic virus	Mich., N.C., N.J., and eastern U.S.
NEMATODES		
Necrotic ring spot vector	*Xiphinema americanum* Cobb	See Necrotic ringspot virus
Root-knot nematode	*Meloidogyne incognita* Kofoid and White	North Carolina
Stubby-root nematode	*Tetylenchus christiei,* also *T. joctus* Thorne	New Jersey
Other nematodes	*Hemicycliophora similis* Thorne and *H.* sp., *Trichodorum* sp.	Mass. and N.J.

TABLE 2—Continued

Common Name	Causal Organism	Principal Areas of Concern
BACTERIA		
Bacterial canker	*Pseudomonas syringae* Van Hall	Oregon
Crown gall	*Agrobacterium tumefaciens* (Smith Town.) Conn or *A. rubi* (Hildebrand) Starr and Weiss	Often a nursery problem-field problem in areas of higher pH (6.8)
INSECTS		
Blueberry maggot	*Rhagoletis pomonella* Walsh.	General—all areas
Cranberry fruit worm	*Mineola vaccinii* Riley	N.J., Mich., N.C. and Mass.
Cherry fruit worm	*Grapholitha packardi* Zell	N.J., Mich., N.C. and Mass.
Plum curculio	*Conotrachelus nenuphar* Herbst.	N.J., Mich., N.C.
Putnam scale	*Aspidiotus ancylus*	New Jersey
Blossom weevil	*Anthonomus musculus* Say.	N.J. and Mass.
Bud mite	*Aceria vaccinii* Keifer	North Carolina
Sharp-nosed leafhopper (Stunt virus vector)	*Scaphytopius magdalensis* Prov.	N.J. and Mass.
Stem borer	*Oberea myops.* Hald.	North Carolina
Black vine borer	*Brachyrhynus sulcatus* Fabricius	Wash. and Canada
Black army cutworm	*Acetebia fennica* Tausch.	Canada and Maine
W-marked cutworm	*Spaelotis clandestina* Harr.	Canada and Maine
Polia cutworm	*Polia purpurissata* Grt.	Canada and Maine
Chain-spotted geometer	*Cingilia catenaria* Drury	Canada and Maine
Blueberry thrips	*Frankliniella vaccinii* Morgan and *Iaeniothrips vacciniophilus* Hood	Canada and Maine
Blueberry flea beetle	*Altica sylvia* May	Maine

The majority of the *Vaccinium* species have a good deal of aesthetic appeal. Often cited are the waxy or glaucous deep green leaf colors. In some of the evergreen phenotypes the glaucousness on leaves and stems is so heavy that the plant has a pleasant bluish-green cast. The flowers are white or pinkish (red or yellow in tropical species), appealingly shaped and clustered with a pleasant, subtle fragrance. Fruits vary from white or pink through red and blue-purples to jet black (red and yellow in many tropical species). Fall leaf coloration is bright red-orange and red or yellow-orange in some clones. Many of the clones have bright yellow and red colored current season wood contrasted with the grays and browns of older wood. The plant habit is variable enough that various species could be used as ground covers, border plants, hedges, area dividers and screens, home garden plantings, or specimen plants. In wildlife control programs *Vaccinium* species are often used to provide food and cover. Certainly an ornamental blueberry planting would also attract birds to the planted area during the fruiting season.

Breeding for Specific Fruit Characters

Large fruit is inherently more appealing to the producer and consumer and is much easier to harvest and handle than small fruit. Large fruit used to bring a premium on the market, and fruit size was the principal grading criterion in marketing highbush fruit. Many workers (e.g., Moore et al., 1972; Brewer and Dobson, 1969*b*) have demonstrated that fruit size is directly related to number of seeds per berry, which is influenced by flowering sequence and pollinator level. There is a large heritable fruit size component, evidenced by cultivar variation, and the prepotency of certain cultivars to transmit large or small size. Large fruit size can be accompanied by fewer seeds in some clones than in others, suggesting a differential hormonal stimulatory contribution per developed seed to fruit ripening (Moore et al., 1972). Seed size also varies among cultivars, and parthenocarpic fruit development has been induced in several cultivars. It is often stated that small fruit is better for pie fillings, for flavor, and for mechanical harvesting. The present lowbush blueberry industry notwithstanding, large berries make equally good or better pies without the necessity of adding gelatinous filler. In natural *Vaccinium* populations, fruit size and flavor vary independently. Small berries harvest mechanically better than large berries only if the small fruit is borne on looser clusters and if it does not adhere tightly to the stem. Experience with various species indicates that an average berry size of 1 g should be the minimum acceptable size for heavy yielding mechanically harvested bushes. An average size of 2 g will be necessary for hand harvested cultivars in the future. There may be some market for 3 g berries, but it does not seem practical to exceed this size because of cluster tightness and fruit shape limitations in cluster-fruited types. There would be no size limitation on singly-borne fruits.

Ballinger et al. (1970, 1972) have indicated that blueberry anthocyanin pigments are located in fruit epidermal and hypodermal cells. Highbush pigment studies were in agreement with previous lowbush studies that the *Vaccinium* anthocyanins are unusually complex, showing all 15 possible combinations of three 3-monoglycoside sugars with five aglycones. A pink (albino) seedling had the same 15 possible pigments as the normally colored 'Croatan' cultivar, but in much reduced quantities. The use of blue coloration as the principal ripeness criterion may not be valid for all cultivars (Ballinger et al., 1972).

The degree of waxy bloom development over the pigmented fruit surface is responsible for the lightness or darkness of the blue color. Light blue berries (possessing a heavy coating of surface bloom) are considered more attractive by the consuming public than the dark blue berries (possessing little, no, or a nonpersistent bloom) for fresh fruit consumption, as well as for the canned and frozen product. Light blue cultivars also retain a much better appearance after mechanical harvesting than dark blue berries (Galletta and Mainland, 1971*b*). The association of high flavor and dark fruit color in some of the early highbush and rabbiteye breeding lines has been largely overcome by selection. It has been suggested that an abundant waxy bloom may aid in the prevention of shriveling of ripe fruit on the plant from water loss.

Very fruitful clones produce at least one inflorescence bud per node. If the fruit set is heavy, fruit from different clusters may overlap, giving the impression of one very large cluster. It is desirable to select for the presence of several clusters per fruiting shoot. The clusters should be loose in arrangement and possess no more than a medium number of berries per cluster, especially where the fruit size is over 2 g. Tight clusters can also be avoided by selecting for longer peduncles and pedicels. Tight clusters can cause poor coverage with spray and dust materials for insect and disease control, incomplete insect visitation for pollination, fruit coloring deficiency at the stem end, misshapen berries, and harvesting difficulty.

The fruit "scar" is the opening in the fruit epidermal wall that remains after the berry is removed from the pedicel. A small, shallow, dry scar is much preferred over a large, deep, moist scar. The scar area is thought to be the principal point of entry for spores of postharvest decay organisms. Larger fruits will normally have larger pedicels than small fruits, hence larger scars. Some clones have fruit which separates from the stem with difficulty, either tearing the skin, or leaving a hole in the stem end of the fruit as part of the fruit vascular system adheres to the stem. Another important trait relating to the scar is dropping ("shattering") of ripe fruit. *V. stamineum* of the *Polycodium* subgenus routinely drops its ripe fruit, an abscission layer seemingly being formed across the pedicel within 2–5 mm of the fruit. *Cyanococcus* subgenus species do not seem to form an abscission layer, and they hold their fruit well after ripening. Selection has always been strongly positive for nondropping cultivars in blueberry improvement programs. Galletta and Mainland (1971*b*) found no relationship between average scar score and the mechanical harvesting potential of 14 highbush blueberry clones.

The initial appeal of *V. corymbosum* and *V. australe* fruits which led to their domestication was their intriguing combination of high flavors and succulence. Contributing to the succulence of highbush *Vaccinium* fruits is the high water content, elastic "skin," relatively uniform parenchymatous fruit tissue, and inconspicuous small-seededness. However, certain highbush clones and species like *V. ashei* have a tougher or less flexible epidermis which splits (cracks) after a heavy rain following a prolonged dry period. Certain clones possess stone cells in the placental area and hard seed coats, which gives the consumer (especially denture wearers) an unpleasant sensation of "grittiness." In improving all blueberry species care should be exercised in selecting berries of uniform juicy consistency with as little skin toughness and seediness as possible.

The overall consistency (firmness) of blueberry fruits varies from mushy and mealy or juicy through soft, medium, and firm categories to quite crisp (firm apple-flesh texture). Ballinger and Kushman (1971) have developed a technique to measure blueberry fruit compression, using an

Instron Universal Testing Machine, which has proven very useful in evaluating firmness. Firmness has always been a selective criterion because firm fruits were known to withstand shipment to market better than soft fruits. It now appears that firm-fruited clones may be better able to resist bruising and subsequent decay than soft-fruited clones, though there is variation in the rate of softening of medium firm and firm clones after mechanical handling.

The flavor of blueberries depends on the pleasurable balance of sweetness, juiciness, acidity, and aroma. In the past, breeders rated the subacid and aromatic berries as best. At present, berries to be consumed fresh should be selected for high balanced soluble solids and acid contents combined with pleasant aroma and texture. Low sugar, low acid berries are too bland for most palates and such fruit does not keep well. High acidity has been demonstrated to inhibit the growth of fruit rotting organisms (Ballinger and Kushman, 1970). Acidity is heritable, and parameters for the selection of fruit of high keeping quality involving soluble solids, pH, and titratable acidity levels, ratios, and products were established by Galletta et al. (1971). Selection might also be made for longer retention of peak flavor after ripening. Processing blueberries require higher acidity than those for fresh consumption because the typical tart blueberry flavor is diluted on mixing with doughs, sugar, and spices.

The length of the period from anthesis to fruit ripening is relatively shorter in the lowbush and highbush species than in the rabbiteye species (Young, 1952; Edwards et al., 1970). The period of fruit development can be characterized by a double sigmoid curve divided into three stages. Stage I is a period of rapid pericarp development characterized by cell division and an accelerated endosperm growth in the shorter-cycled clones. Stage II is characterized by rapid endosperm and embryo development in all clones and little or no mesocarp growth. Stage III is the period of final rapid mesocarp growth due principally to cell enlargement. The duration of Stages II and particularly III is correlated with fruit maturity. Under field conditions temperature variations can considerably advance or retard the length of Stages I and II. Influential factors in determining the length of fruit maturity appear to be chilling requirement, minimum temperature at which flower buds will open, abundance of pollinators, number of developed seeds, time of blossom opening within the cluster, temperature, sunlight, moisture, and inherent development period of the clone under consideration. Within each species there is variation in ripening time from early to late. For consistently producing early and late cultivars, the breeder can select for clones with shorter or longer Stage II and Stage III lengths, and for clones which will blossom, set the majority of their flowers, and develop rapidly under cool, cloudy, or dry conditions.

Compared with other fruits, canned blueberries are an excellent source of iron, a fair source of Vitamin A, about average in protein, fat, carbohydrate, calories, and calcium, and low in phosphorus (Anonymous, in Hodgman et al., 1956). Fresh blueberries are reasonably good sources of Vitamin C, but European work (Roach, Gersons, and Matzner, all 1967) indicates that Vitamin C is lost after cold storage of a few months, and that Vitamin C content measurement in blueberries should include both ascorbic acid and dehydroascorbic acid for accuracy. The common frozen and canned products are often insipid in flavor compared to the fresh product. The breeder thus has an opportunity to select for clones with high and durable vitamin and mineral content and flavor. As the culture of blueberries expands, the breeder may have to select clones that are specifically suited for either processing or fresh consumption. It may be difficult to obtain cultivars with high quality for both uses.

Table 3 lists many of the available and potential parents for specific traits and comments on inheritance patterns where appropriate. Potential parent sources available from European and tropical *Vaccinium* species have been largely omitted from the table due to the lack of published information.

Achievements and Prospects

Blueberry culture is still very young. Wild blueberries began to be canned in Maine in 1866. Dr. Coville initiated his research on domestication of the wild blueberry in 1906 and blueberry improvement through the selection of seedlings resulting from controlled pollinations dates back to 1913. All of the highbush cultivars released through 1969 by the USDA and the cooperating experiment stations trace back to just 13 wild parental selections, (one of *V. corymbosum*, three of *V. angustifolium* and nine of *V. australe*); the introduced

TABLE 3. Summary of prepotent and potential parent sources for certain blueberry traits

Traits	Current Parents	Potential Parents	Remarks
Upright plant habit	Bluecrop, Ivanhoe, Tifblue, Rubel, Adams, Croatan, Jersey.	Darrow, Pemberton, Rancocas, Scammell, Stanley, selected clones of *V. ashei, australe, amoenum, atrococcum, elliottii.*	
Lowbush habit	*V. angustifolium, tenellum, vacillans, pallidum, darrowi, myrtilloides, myrsinites, brittonii.*	*V. myrtillus, uliginosum,* and several of *Euvaccinium* species of Western North America.	May be partially dominant to highbush habit.
Evergreen leaf habit	*V. darrowi* and *V. myrsinites.*	Certain *V. ashei* seedlings and tropical species like *V. meridionale* and *reticulatum,* and *V. ovatum.*	
Plant vigor	Tifblue, Garden Blue, Homebell, Rubel, Jersey, Stanley, Wolcott, Croatan, Pemberton, Berkeley, Bluecrop, Blueray, Ivanhoe, Darrow, Murphy, Scammell.	Delite, *V. ashei,* and selected seedlings from all species.	
Precocity (early bearing)	Most of the lowbush species, especially *V. angustifolium.*	Morrow, Herbert, Bluecrop, Blueray, Collins, Earliblue, *V. atrococcum,* and selected highbush and rabbiteye seedlings.	
Productivity	Bluecrop, Blueray, Herbert, Croatan, Mich. 19-H, Tifblue, Homebell, G-107, Mich. #1, Weymouth, Ivanhoe, Darrow.	*V. ashei* selections, *V. ashei* x *amoenum,* highbush x lowbush, Woodard, Berkeley, Dixi, Pemberton, selected highbush seedlings.	
Mechanical harvesting ability		Wolcott, Morrow, Weymouth, Croatan, Collins, Meader, Tifblue, and selected highbush and rabbiteye seedlings.	
Early ripening season	Selected *V. angustifolium* and *V. atrococcum* seedlings, and Morrow, Weymouth, Earliblue, Croatan, Bluetta, Wolcott, Angola.	Collins, Murphy, Cabot, June, selected highbush seedlings.	Many of these parent sources have serious faults.
Late ripening	Coville, Herbert, Burlington, US-1, Jersey, Lateblue.	Selections of *V. ashei, darrowi, tenellum, virgatum, amoenum,* highbush, Darrow, Herbert, Dixi, Wareham.	Many of these parent sources have serious faults.
Ornamental forms		For ground covers—*V. crassifolium* and rhizomatous lowbush species; for hedges—*V. myrsinites, darrowi* and lowbush alone and hybridized with highbush; for tall hedges—*ashei* and *ashei-amoenum* and *ashei*-highbush hybrids. For specimen plants—selected seedlings from halfhigh and highbush species and species hybrids.	
Large fruit size	US 11-93, E-30, M-23, Berkeley, Herbert, G-80, Darrow.	Blueray, Dixi, Coville, new NC, G and E selections among the highbush; Woodard, Black Giant, Ethel, and other rabbiteye selections.	Small size appears dominant to large size.

TABLE 3—Continued

Traits	Current Parents	Potential Parents	Remarks
Light blue fruit color	Bluecrop, Berkeley, Tifblue, Woodard, Ethel, Stanley, Coville.	Earliblue, Collins, Blueray, Darrow, Ivanhoe, Morrow, selected seedlings of *V. augustifolium, caesariense, australe,* and *darrowi*.	Dark color appears dominant to light color.
Loose fruit cluster		Concord, Burlington, Morrow, Angola, Atlantic, Berkeley, Bluecrop, Collins, Coville, Croatan, Darrow, Jersey, June, Murphy, Pemberton, Rubel, Wolcott, selected seedlings of *V. virgatum, darrowi, amoenum, ashei,* and *australe*.	
Good fruit scar	Many rabbiteye cultivars and selections, Burlington, Wolcott, Ivanhoe, Bluecrop, F-72.	Berkeley, Croatan, Darrow, many of lowbush species, selections of highbush species.	
Uniform fruit texture		Woodard, Menditoo, Collins, Morrow, Murphy, and other selections.	
Fruit cracking resistance		Angola, Atlantic, Berkeley, Bluecrop, Blueray, Burlington, Collins, Coville, Croatan, Darrow, Earliblue, Herbert, Ivanhoe, Jersey, Morrow, Murphy, Stanley, Wolcott, selected species seedlings.	
Non-dropping fruit	Morrow, Earliblue, Collins, Darrow, Coville, Tifblue.	Most of the highbush and rabbiteye cultivars and many species seedlings.	
Fruit firmness	US 11-93, F-72, Ivanhoe, Earliblue.	Most of named highbush cultivars and NC 61-3 and NC 57-31 especially. Tifblue and several rabbiteye cultivars and selections, seedlings of many species, USDA tetraploid species hybrids, *ashei-amoenum* hybrids.	
High fresh fruit flavor	Earliblue, Blueray, Ivanhoe, Herbert, Wareham, F-72, Dixi, Coville, Stanley, Pioneer, Darrow, Rancocas, Tifblue, Garden Blue.	Bluecrop, highbush selections, Woodard, Blau-weiss Goldtraube, *V. australe, ashei, myrsinites,* and *darrowi* selections, Scammell, Delite, also *Euvaccinium* and bilberry species.	High acid berries crossed to low acid berries may yield superior flavored progenies.
Fruit aroma	Earliblue, Blueray, Coville, Herbert, Ivanhoe, Pioneer, Stanley, Wolcott.	Morrow, Angola, Atlantic, Berkeley, Bluecrop, Burlington, Cabot, Collins, Concord, Croatan, Dixi, June, Murphy, Pemberton, Rubel, Scammell.	
Good fruit processing ability		Atlantic (f),* Coville (cf), Berkeley (cfp), Jersey (cfp), Dixi (cp), GN-87 (cfp), Bluecrop (cf), Ivanhoe (f), Pemberton (cfp), Concord (c), Blau-weiss Goldtraube (f), Burlington (cf), Herbert (cf), Pioneer (cf), Stanley (c), Cabot (f), Earliblue (f), June (f), Adams (f), several lowbush species (cfp).	These cultivars are good for different types of processing at varying locations. Suitability varies with season and site. * c = canned or bottled; f = frozen; p = pie

TABLE 3—Continued

Traits	Current Parents	Potential Parents	Remarks
Good fruit keeping quality	Jersey, Tilfblue, most rabbiteye cultivars.	Selected highbush seedlings, Coville, Croatan, Berkeley, Bluecrop, Burlington, V. *ashei, darrowi, australe,* and *corymbosum.*	
Fusicoccum canker resistance		Rubel, Rancocas, Concord, Stanley, Burlington, Berkeley (?).	Varietal susceptibility varies with locale, suggesting fungal racial specialization.
Powdery mildew resistance		Berkeley, Earliblue, Ivanhoe, Rancocas (?), Stanley (?), Harding, Katherine, Dixi (?), June (?), Weymouth (?). Cultivars followed by (?) have been reported both resistant and susceptible by different authors.	Susceptibility is partially dominant to resistance. Greenhouse and field resistance were well correlated.
Mummy berry resistance		Jersey, Rubel, Burlington, Pemberton, Dixi, V. *myrtilloides, uliginosum,* Atlantic, Fraser, Grover, Johnston, Pacific, Collins, Stanley, Bluetta, Darrow, and selected highbush seedlings.	Varieties show differential susceptibility to blight and mummy phases of the disease, also to conidial and ascospore infection; resistance inheritance seems multigenic.
Double spot resistance		Grover, Harding, June, Sam.	All of the older cultivars can be infected. The 4 listed show the most tolerance. No information on newer cultivar susceptibility is available.
Cane canker (*Botryosphaeria*) resistance	Crabbe 4, Murphy, Angola, Croatan, all V. *ashei* cultivars.	Morrow, Bluecrop, V. *ashei,* a number of NC highbush selections, Adams, V. *arboreum, darrowi,* Atlantic, Jersey, Scammell, Pemberton, Rubel, Rancocas, V. *australe, atrococcum, caesariense,* and *elliottii.*	The fungus has specialized into at least 6 races. Infection varies with temperature. Resistance is partially dominant, probably multigenic. Genetic variability is predominantly additive with heritability being .90.
Leaf rust resistance		Dixi, June, Pioneer, Rancocas, Weymouth, also Cabot, Owens, Myers, and Black Giant.	Information on newer cultivar susceptibility not available.
Phytophthora root rot resistance		Most V. *ashei* cultivars and selections, Me-US 32, US 41, Mich. Lowbush #1, V. *atrococcum, angustifolium,* and *darrowi* selections.	Field observations and inoculations indicate no resistance present in highbush cultivars or breeding selections.

TABLE 3—Continued

Traits	Current Parents	Potential Parents	Remarks
Blueberry anthracnose		Morrow (?), Murphy (?).	Milholland (1970*b*) noted cultivar differences in leaf and stem reactions. All of the few inoculated cultivars took the disease; Morrow and Murphy show good field tolerance.
Blueberry stunt resistance		Rancocas (?) (Symptomless carrier).	All cultivated varieties seem to be susceptible.
Stem blight			Milholland (1972) reported 10 highbush cultivars susceptible to 6 *Botryosphaeria dothidea* isolates; 8 rabbiteye cultivars were susceptible to 5 of 6 fungal isolates.
Necrotic ringspot resistance		Berkeley, Bluecrop, Blueray, Herbert, Jersey, Rancocas.	These cultivars become infected but apparently overcome the virus.
Red ringspot virus		Jersey.	Berkeley and Stanley have been successfully infected in some tests but not others; Weymouth, Bluecrop and Jersey show field resistance. Jersey may be a symptomless carrier.
Shoestring resistance		Selected seedlings of *V. angustifolium*.	
Root gall resistance		Jersey, Rubel, Dixi.	
Bacterial canker resistance		Weymouth, June, Rancocas, highbush x lowbush hybrids.	
Bud mite resistance		Burlington, Morrow, Croatan, NC 61-3.	
Black vine borer resistance		Weymouth, Cabot.	
Cold hardiness	Ashworth, *V. angustifolium*, Me 2822, Me 5003, Sebatis.	Rancocas, Earliblue, *V. uliginosum, myrtillus, myrtilloides*, highbush x lowbush hybrids, selected *V. corymbosum* seedlings, *V. brittonii, constablaei*.	
Drought resistance		*V. darrowi, tenellum, ashei, myrsinites, pallidum, vacillans, altomontanum*, and *membranaceum*.	
Short chilling	*V. darrowi, myrsinites*, and *ashei*.	*V. elliottii* and *atrococcum*.	
Long rest	*V. constablaei*.	Selections of *V. simulatum, altomontanum, angustifolium, brittonii, myrtilloides, corymbosum*, and *erythrocorpum*.	

TABLE 3—Continued

Traits	Current Parents	Potential Parents	Remarks
Broad climatic tolerance	Bluecrop, Herbert, Jersey, US 11-93, Tifblue.	Selections from V. *australe, atrococcum, corymbosum,* and *ashei* and possibly V. *vacillans* and *pallidum,* Berkeley (?), Garden Blue.	
Upland soil adaptation	V. *ashei* selections and seedlings.	V. *atrococcum,* Me-US 32, V. *angustifolium, brittonii, elliottii, myrtilloides, myrsinites, darrowi, myrtillus, vacillans, pallidum, tenellum,* and *virgatum.*	

rabbiteye cultivars trace back to four or five wild V. *ashei* parents.

Remarkable increases in fruit size, regular productivity, consistent fruit quality, bush adaptation, and disease resistance have been achieved utilizing a very narrow genetic base and two to six generations of hybridization and selection. The voluminous accompanying research on blueberry physiology, pathology, entomology, and food processing has enabled blueberry culture to become the principal small fruit industry in much of southeastern Canada, in Maine, New Jersey, North Carolina, Michigan, and Indiana, and in other limited areas of the United States. The interest of the Europeans in *Vaccinium,* and the apparent potential success of the American cultivars in other temperate parts of the world, bode well for further improvement of the blueberry.

Further blueberry improvement will have to use germ plasm from throughout the genus in order to incorporate broader tolerance to soil and climatic variability and resistance to more pests into clones which bear superior fruit. Adaptation to mineral soils and continental climates will be the key to widespread utility of blueberries by future generations. Improved flavor after processing and higher Vitamin C content would also increase the popularity of this versatile fruit.

In addition to the American and European *Cyanococcus* (blueberry), *Euvaccinium* (bilberry), and whortleberry species and cultivars of *Vaccinium* mentioned as possible breeding parents in this article, Uphof (1968) lists the following species reputed for their fruit in other parts of the world:

1) V. *andringitrense* Perr., found in Madagascar, is a woody shrub bearing fruit of excellent quality.
2) V. *arbuscula* (A. Gray) Mart., grown in western North America to Alaska, is a small shrub bearing fruit esteemed by natives of southeastern Alaska.
3) V. *berberifolium* (A. Gray) Skottsb. (Ohelo) is a dwarf shrub of the Hawaiian Islands bearing edible fruit.
4) V. *dentatum* J. Sm. (V. *penduliflorum* Gaud., *Metagonia penduliflora* Nutt.) (dentate Ohelo or Ohelo) is a leafy shrub endemic in Oahu, Hawaii, which bears pleasantly acid red fruits.
5) V. *floribundum* H.B.K., known as the Columbian or Andean blueberry or Mortiño, is an Ecuadorean or Peruvian shrub with fruit esteemed by Andean natives.
6) V. *leucanthum* Schlecht., a Mexican shrub, has edible fruit.
7) V. *meridionale* Sw., the bilberry, is a Jamaican shrub producing berries used for jellies, pies, and tarts.
8) V. *myrtoides* (Blume) Miq., a Malayan shrub, provides fruits used as preserves in the Philippines.

It is quite possible that these species can be improved by selection and hybridization to form the basis for blueberry culture in other areas. Or these species may contribute some unusual flavors, disease resistance, climatic adaptation, and nutrient content to the already improved blueberry types. Our American notions of what constitutes a blueberry may have to be modified following hybridizations with exotic species.

The blueberry has a place in reforestation and in wildlife and soil management programs. The genus also has tremendous potential for ornamental purposes. The fruit can be eaten fresh by itself, with additives, as a meat or cereal garnish, or mixed with other fruits. The blueberry produces abundant tasty juice of a rich red-purple color, and the juice can be fermented into fine table or dessert wines, or fortified to make an outstanding brandy or cordial. The blueberry cooks or bakes well to make a variety of outstanding products such as jams, jellies, preserves, compotes, toppings, sauces, pies, cakes, cobblers, and varied pastries. Blueberries can be preserved by bottling, canning, drying, freezing, or dehydrofreezing.

If the future course of *Vaccinium* improvement is as successful as in the past, we can expect that many types of blueberry-like fruits will be available to practically anyone who cares to grow them, and that the versatile *Vaccinium* fruits will surely receive a significant and secure place in the human diet.

Cranberries

Origin and Early Development

The *Oxycoccus* subgenus of the Ericaceous genus *Vaccinium* is composed of slender, creeping, evergreen vines with flowers borne on short upright shoots. The flower corollas are deeply divided and reflexed at maturity. The related *Hugeria* subgenus differs from *Oxycoccus* in being composed of erect deciduous shrubs. The similar *Vitis-Idaea* subgenus differs from *Oxycoccus* in not having a deeply divided corolla and having its flowers borne in small terminal clusters, whereas the *Oxycoccus* flowers are usually borne singly (Camp, 1945).

In addition to the *Cyanococcus* and whortleberry subgenera (species *V. uliginosum* L.) of the genus *Vaccinium*, the *Oxycoccus* subgenus also possesses natural diploid ($2n = 24$, $x = 12$), tetraploid ($2n = 48$), and hexaploid ($2n = 72$) species (Hagerup, 1940; Newcomer, 1941; Camp, 1944; Shultz, 1944; Ahokas, 1971).

Although many authors have used the generic name *Oxycoccus* (Tourn.) Hill rather than *Vaccinium* L. subgenus *Oxycoccus* for the cranberries, Shultz (1944) concluded that the cranberries should remain in the *Vaccinium* genus because 1) there are valid morphological species intermediate between cranberries and blueberries (such as *V. uliginosum* L., *V. Vitis-Idaea* L., *V. ovatum* Pursh, *V. crassifolium* Andr., *V. erythrocarpon* Michx., and *V. Japonicum* Miq.) and "thus there is no major taxonomic character that will consistently distinguish cranberries from all species of blueberries"; 2) the cranberry (*V. macrocarpon* Ait.) was successfully hybridized by Schultz with *V. Vitis-Idaea* and with *V. ovatum*; 3) both cranberries and blueberries have the same basic chromosome number and exhibit a polyploid series from diploid to hexaploid.

Although the cranberry taxonomic equivalents in the literature are confused, there appear to be two presently disjunct diploid species, a large segregating circumboreal tetraploid species which can be subdivided into three major botanical varieties, and a vigorous hexaploid species sporadically appearing in areas bearing predominantly tetraploid species (Porsild, 1938; Hagerup, 1940; Camp, 1944; Ahokas, 1971). The hexaploid form may be sterile or fertile (Hagerup, 1940; Camp, 1944, Ahokas, 1971). A tentative listing of the natural cranberry species, their synonyms, distinguishing characteristics, and range is presented in Table 4.

Camp (1944) considered the large-fruited, southern diploid species *V. macrocarpon* Ait. the most primitive, and suggested that the small-fruited, northern diploid *V. microcarpon* (Turcz.) Hook was derived from *macrocarpon* by a series of plant size reducing mutations, thus enabling the smaller diploid to colonize subarctic regions more effectively. The tetraploid *V. quadripetalum* Gilib. (or *V. oxycoccus* L.) is considered an allotetraploid originating from hybridizations between the two diploid species. The hexaploid *V. hagerupii* (L. & L.) Ahokas is postulated to have arisen as a hybrid between *V. oxycoccus* and *V. microcarpon* (Hagerup, 1940), or as an autoallohexaploid arising from the union of reduced and unreduced gametes from the tetraploid species (Camp, 1944).

Camp further postulated (1944) that the primary divergence of the two diploid species occurred in North America, probably not later than the mid-Tertiary. The tetraploid arose prior to the Pleistocene at a time when the continental masses of Europe, Asia, and North America were more closely connected. The tetraploid was presumed to have originated on what is now the North American segment of Holarctica, and to have

TABLE 4. The natural cranberry species
(Sources: Porsild, 1938; Hagerup, 1940; Camp, 1944; Shultz, 1944; Ahokas, 1971).

A. Flower pedicels pubescent, leaves elliptic, broadest near the middle
- 1.*Vaccinium macrocarpon* Ait. = *Oxycoccus macrocarpus* (Ait.) Pers.
  - a. Chromosome number: diploid (2n = 24, χ = 12)
  - b. Characteristics: stout stems trailing to several meters, leaf arrangement on branch even and symmetrical; leaves (5)6-15(17) mm long, 2-8 mm wide, elliptic-oblong, flat or slightly revolute, apically rounded; flowers 2-6 solitary in axils of basal leaves of uprights, pedicels pubescent with 2 green bracts; fruits 10-20 mm diam., deep pink to red, tart.
  - c. Habitat: open bogs and swamps
  - d. Distribution: endemic to eastern North America; Newfoundland to Minnesota, south to Tenessee and North Carolina—*cultivated forms derived from this species.*
- 2. *Vaccinium oxycoccus* L. var. *oxycoccus* = *O. quadripetalus* Gilib. "typical" material or "main species." = *O. palustris* Pers.
  - a. Chromosome number: tetraploid (2n = 48)
  - b. Characteristics: stems slender, creeping, bark brown or black; leaves 6-10 mm long, 2-5 mm wide, flat or slightly revolute, elliptic, apex pointed; flowers 1-4, inflorescence terminal bearing 2 scaly bracts; fruit 8-12 mm diam., pink to red, often mottled when young.
  - c. Habitat: wet marshy places, full sunlight
  - d. Distribution: North America, Asia, Europe; Nova Scotia to British Columbia, south to Pennsylvania, Wisconsin, and Michigan, mostly in mid-north continental areas such as the Canadian Zone and similar areas in Asia and Europe.
- 3. *Vaccinium oxycoccus* L. var. *microphyllus* Lange = *O. quadripetalus* Gilib. var. *microphyllus* Lange. M. P. Porsild = *O. palustris* Pers. f. *microphylla Lange.*
  - a. Chromosome number: tetraploid (2n = 48)
  - b. Characteristics: diminutive form of *V. oxycoccus*; stems filiform, leaves arranged somewhat unilaterally on stems; leaves 2-5 mm long, 1.5-3 mm wide, strongly revolute, apex rounded; flowers 1 or 2, strictly terminal, pedicels strongly pubescent, style shorter than that of other cranberries, barely exceeding the anthers; fruit less than 10 mm diam., pink, not spotted.
  - c. Habitat: wet mossy bogs, but on the highest and driest sphagnum tufts, not tolerant of shade or standing water.
  - d. Distribution: Denmark, Western Greenland, Labrador, Newfoundland, and Gulf of St. Lawrence.

B. Flower pedicels essentially glabrous, leaves distinctly broader near the base.
- 1. *Vaccinium microcarpon* (Turcz.) Hook = *O. microcarpus* Turcz.
  - a. Chromosome number: diploid (2n = 24)
  - b. Characteristics: stems slender and filiform with red or reddish brown bark, leaf arrangement somewhat unilateral; leaves 2-6 mm long, 1.5-2 mm wide, strongly revolute, ovate when flattened, apex pointed; flowers 1-2(3), determinate, pedicels red, glabrous, bearing a pair of red scaly bracts; fruit 5-7 mm diam., pale pink, insipid.
  - c. Habitat: driest subarctic sphagnum bogs
  - d. Distribution: northermost oxycoccus representative, circumpolar, Scandinavia, Central Alps, Iceland, Asia and northwest North America, Alaska to Hudson Bay, Yukon Territory south in mountains to British Columbia and Alberta.
- 2. *Vaccinium oxycoccus* L. var. *ovalifolium* Michx. = *V. oxycoccus* var. *intermedium* Gray = *Oxycoccus intermedius* Gray = *Oxycoccus ovalifolius* (Michx). n. comb. = *V. quadripetalum* (Gilib.) n. comb.
  - a. Chromosome number: tetraploid (2n= 48)
  - b. Characteristics: stems stouter than *V. microcarpon* with dark brown or black bark: leaves 6-8 mm long, 2-3 mm wide, flat or slightly revolute; flowers 1-8, basal (lateral) inflorescence; fruit 10-12 mm diam. with a bloom.
  - c. Habitat: wet sphagnum bogs, sea level to 1500 m.
  - d. Distribution: Alaska to California, across Canada to Nova Scotia, across northern United States to New England and New Jersey; also in East Asia.

C. Flower pedicels pubescent, leaves ovate to elliptic
- 1. *Vaccinium hagerupii* (L. & L.) Ahokas comb. nov. = *Oxycoccus gigas* Hagerup
  - a. Chromosome number: hexaploid (2n = 72)
  - b. Characteristics: very vigorous vine with vertical shoots, gives appearance of erect shrub; leaves up to 15 mm long and 7 mm wide, flat or slightly revolute, ovate to elliptic; flowers up to 6(?), inflorescence terminal, pedicels pubescent; may be sterile or fruitful, no fruit size given.
  - c. Habitat: bogs where it tops or covers moss tufts, drought and shade tolerant.
  - d. Distribution: isolated areas from within the range of V. oxycoccus, Finland, Denmark, East Prussia, and possibly to be found in Asia and North America.

migrated from there into Europe by way of Northern Asia.

There is also some breeding interest in Europe, Canada, and Minnesota in the lingonberry, cowberry, or foxberry V. *Vitis-Idaea* L., the American representative being the mountain or rock cranberry var. *minus* Lodd., a dwarf diploid form.

The cultural peculiarities of the cranberry are a requirement for acid soils on sites which can be diked and flooded periodically. Flooding is used for winter protection against desiccation, frost protection during the growing season, for facilitating harvest, and for adjusting the ripening season. Sprinkler irrigation is substituted for flooding for freeze and frost protection in Washington and Oregon. The vines can photosynthesize for long periods under water if light is not excluded. The exclusion of light may result in oxygen deficiency. A thin layer of sand is added to the cranberry bog every three to four seasons, basically to provide a new rooting medium for the old vines. The initial setting of a cranberry bog is often, but not necessarily, preceded by sanding to provide initial rooting and growth, to encourage early bearing, to help prevent frost damage, and to help anchor the plant (Dana and Klingbeil, 1966; Doughty, personal communication, 1973).

History of Improvement

Chandler and Demoranville (1958) have given the origin, description, characteristics, and synonyms of the principal cranberry clones in the United States during 1955 to 1957. Most of the cranberry cultivars originated as single vine selections from populations in native bogs, and four of these wild vine selections accounted for 91.5% of the total American production area in 1955–1957:

1) 'Early Black' (1852, Harwich, Mass.) is important in Massachusetts and New Jersey. An early ripening clone with very good color, firmness, and quality, it has resistance to false blossom disease. It offers broad adaptation but is poor in productivity and keeping quality, and is small-fruited.
2) 'Howes' (1843, East Dennis, Mass.) is important in Massachusetts and New Jersey. A late clone, it has medium color and size, excellent keeping quality, and attractive fruit, but moderate productivity.
3) 'McFarlin' (1874, South Carver, Mass.) is the principal cultivar in Washington, Oregon British Columbia, and is important in Wisconsin and Massachusetts. It is late and firm, and has good quality, well-colored, and large fruit. Plants are very productive, frost resistant, and resistant to false blossom. The fruit colors poorly in storage.
4) 'Searles' (1893, Walker, Wis.) is the principal cultivar in Wisconsin and is important in Canada, also. A midseason producer, it offers medium large, firm fruit with excellent productivity. Its fruit colors well in storage and has good fresh color, but only fair quality.

In 1929 the USDA initiated a cooperative cranberry breeding program with the New Jersey and Massachusetts Agricultural Experiment Stations. The Wisconsin Department of Agriculture and Markets and the Wisconsin Cranberry Sales Company joined the cooperative program in 1939 (Chandler et al., 1947).

The initial aims of the program were the origination of high-yielding, high-quality cultivars, with resistance to the false blossom disease. Seventeen clones with some known resistance to the false blossom virus were reciprocally crossed in Wisconsin, and six clones were reciprocally crossed in Massachusetts. An additional cross was made in New Jersey.

From 1934–1946, 12,715 seedlings representing over 34 hybrid progenies were planted in New Jersey, Massachusetts, and Wisconsin. From these plantings 222 seedlings were selected for further evaluation. Initial selection criteria were yield; the appearance traits (berry size, shape, color, and gloss); ripening season; the storage traits (fruit rot and fruit ripening); false blossom, and fresh fruit rot resistance.

Second test criteria of horticulturally outstanding cranberry selections included the additional measures: processing ability for strained sauce (flavor and sauce yield), cocktail (juice), and whole berry sauce, yield per acre, berry size, specific gravity, appearance and keeping quality in cellophane bag packs, and vine resistance to feeding trials by caged blunt-nosed leafhoppers (the vector of the false blossom virus disease).

Six new cultivars have been released from this cooperative cranberry breeding program:

1) 'Beckwith', introduced in 1950 ('McFarlin' x 'Early Black'), is late, large, deep red, productive, very high in flavor, and makes excellent sauce. It is of promise for New Jersey.
2) 'Bergman', introduced in 1961 ('Early Black' x 'Searles'), is midseason, medium large, red, very productive, and processes well as sauce

and juice. It has been suggested for trial in Canada and Massachusetts.

3) 'Franklin', introduced in 1961 ('Early Black' x 'Howes'), is early, medium large, red to very dark red, and not yet sufficiently tried.
4) 'Pilgrim', introduced in 1961 ('Prolific' x 'McFarlin'), is late, very large, purplish-red, and not yet sufficiently tried.
5) 'Stevens', introduced in 1950 ('McFarlin' x 'Potter'), is midseason, large, deep red, very productive, resistant to breakdown or softening, and water harvests well. It is of promise in Wisconsin, Washington, and Canada.
6) 'Wilcox', introduced in 1950 ('Howes' x 'Searles'), is early, medium, deep red, productive, resistant to blunt-nosed leafhoppers, and is of promise in Massachusetts and New Jersey.

In addition, the cooperative cranberry breeding program developed several periclinal chimeral and total tetraploid clones using colchicine on lateral buds of runners. Increased fruit size and color were sought from tetraploids. Intervarietal tetraploid hybrids and tetraploid species hybrids between V. *macrocarpon* and V. *oxycoccus* were made and planted in Wisconsin for evaluation. Tetraploid seedlings were vigorous, but were generally less hardy and grew more slowly than the diploids. The tetraploids set more flower buds than the diploids but the seed count in tetraploid fruit was reduced, thus reducing the fruit size. Tetraploid hybrid fruit production records had evidently not been completed when the cooperative cranberry breeding program terminated (Kust, 1965). Tetraploid fruit production was considerably less than that of diploids in Washington tests (Doughty, personal communication, 1973).

The cooperative cranberry breeding program existed for only one generation of improvement, but it was a farsighted effort involving an entire team of researchers and industry in all of the cranberry growing areas. It demonstrated that many cranberry traits could be improved by hybridization and selection. Selections from this program are still being evaluated by the New Jersey, Wisconsin, Massachusetts, and Washington Agricultural Experiment Stations. Each of these stations has had a modest breeding effort. The Washington station recently released 'Crowley' (Doughty and Garren, 1970) as an earlier, larger, more productive cultivar with a higher pigment content than the standard 'McFarlin.' 'Crowley' was one of 13 second-test selections from 1,200 original seedlings. The Wisconsin station hopes to activate a cranberry breeding program shortly.

Modern Breeding Objectives

Breeding aims for cranberries are outlined by Chandler et al. (1947), Chandler and Demoranville (1958), Dana (unpublished), and Kust (1965). The following list of desirable vine characters includes these sources plus the results of an informal survey taken by the author: medium to coarse vine, rapid root initiation by stem cuttings; minimum runnering; a high density of short, stout uprights which bear fruit at a uniform level above the bog; very high productivity; precocious plant and fruit development; tolerance to water submersion; early fruit maturation season; broad adaptation; fungus, insect, and virus tolerance; vine and bud hardiness; and increased vine vigor.

The cranberry bog is usually established by pressing stem cuttings into the surface of a bog that has been freshly sanded. The cuttings must then quickly establish roots and spread by runners. Vines must be vigorous enough to quickly cover the area, and to compete with weeds in the "solid bed" stand in which they are usually grown. A fine- to medium-leaved vine was desirable, but coarse vines are now permissable with the advent of mechanical and water harvesting. Vines should not runner excessively because upright stem production is then reduced. A high density of flower-producing upright stems per unit area, 200–300 uprights per 30 sq cm (Roberts and Struckmeyer, 1942), is essential to high productivity. A large number of flowers per upright with high pollen and egg fertility and insect attractiveness is also essential to heavy fruit production. Tall uprights have been observed to be less fruitful and more difficult to harvest than short uprights. Fruit borne at a uniform level is less likely to be lost in harvesting operations.

Cranberry vines will bear some fruit in the second and third seasons, but three to five years are necessary for full production. This time could be shortened by appropriate selection for precocious vines. Vines need to be more tolerant to cold and to spring frosts when they are not flooded. When flooded, vines should be able to photosynthesize more effectively under oxygen-deficient conditions. It may not be possible to select vines which will perform well in all of the growing areas; however, vines like 'Stevens' appear to be broadly adapted while 'Searles' does best in Wisconsin, 'McFarlin' on the West Coast, and 'Beckwith' on the East Coast. Early-maturing fruit is much more desirable than late-maturing fruit because the late season has more frost hazards, and early-ripening fruit can be sold to vactioning tourists to provide a fresh market outlet and to stimulate demand. Some delay of the harvest season can be accomplished by appropriately timed winter flood removal, if desired.

The principal vine disorders and pests for which resistance may be developed are: the blunt-nosed leafhopper (*Scleroracus vaccinii* [Van Duzee]), twig blight (*Lophodermium hypophyllum* and *L. oxycocci* [Fr.] Karst.), blossom blight (*Botrytis cinerea* and *Guignardia vaccinii* Shear), fairy rings (*Psilocybe agrariella* Atk. var. *vaccinii*, Charles), red leaf spot and rose bloom (*Exobasidium vaccinii* [Fuckel] Wor.), tip blight (*Monilinia vaccinii-corymbosi* [Reade] Honey), the black-headed fire worm (*Rhopobota naevana* [Hubner]), the black vine weevil (*Brachyrhimus sulcatus* [Fabricus]) and the tip worm (*Dasyneura vaccinii* Smith) (Doughty and Dodge, 1966; Hall et al., 1969; P. E. Marucci and F. Johnson, personal correspondence). Thse severe pests and diseases are now readily controlled by chemicals, and flooding in some instances, but resistance may become more important if agricultural pesticide use becomes more limited.

The following list of desirable fruit characters is compiled from the same sources as the earlier list for vine characters: sweeter and better colored fruit for raw consumption and use in pies; increased aroma, firmness, size, flavor, organic acids, anthocyanin content, glossiness, attractiveness, keeping quality (resistance to fruit worms, physiological breakdown, and fruit rots), pectin content; uniform ripening; ability to color well (ripen) in storage; uniform shape; suitability for juice, sauce, jelly, preserves, and wine manufacture.

Cranberry consumption can be greatly expanded by increases in fresh use and/or in various baked products. To achieve this end, selection should be for a pleasant balance of sweetness combined with high organic acid content, high aroma, deep or bright red color with a high gloss, and a uniform firm ("crunchy") texture in a large-sized berry.

A uniform shape is important in grading the fruit by screening (round or oval berries are preferred). Pointed ends are avoided since they may be bruised easily. Increased size increases yield, simplifies harvesting, and enhances consumer appeal. Increased firmness enables fruit to withstand bruising and rotting and results in better shipping quality. Uniform ripening is an aid to harvesting. Coloring well in storage would permit flexibility in harvesting time. Increased color and flavor and pectin content are very important to the attractiveness, taste appeal, and "gelling" ability of the processed products.

Extended shelf-life can be achieved by selecting for firmer and more acid fruit, and for resistance to infestations by the cranberry fruit worm (*Acrobasis vaccinii* Riley) and certain fruit rotting fungi (*Sporonema oxycocci* Shear, *Guignardia vaccinii* Shear, and *Godronia cassandrae* PK. f. *vaccinii* Groves [*Fusicoccum putrefaciens* Shear]).

The idealized cranberry cultivar has been described by Dana (unpublished) as one that would combine the size of 'Mammoth', the productivity of 'Searles', the keeping quality of 'Stevens', the color of 'Early Black', and the earliness of 'Ben Lear'.

Breeding Techniques

Cranberry cultivars all seem to be self- and cross-fruitful, but the pollen tetrad is heavy and sticky. Bees are effective cranberry pollinators. The cranberry flower hangs upside down; the pistil of the cranberry blossom is equal in length to the ring of stamens when the blossom first opens, but the style expands in length to grow past the anthers in one to three days after opening, exposing the stigma to pollen delivered down the anther tube in response to jarring by visiting insects.

The cranberry flower has been said to be protandrous, but tests by Rigby and Dana (1972) indicated that the stigma was receptive as soon as the flower parts opened. Thus the seeming protandry results from the separated positions of the sex organs as the flower opens. It takes between 24 and 72 hours for the pollen tube to germinate and effect fertilization (Rigby and Dana, 1972). Wind and mechanical jarring of flowers are ineffective in pollinating cranberries; hence, emasculation need not be practiced in many instances (Marucci, 1967, and personal correspondence, 1972), although Doughty (personal correspondence, 1972) indicates that cranberry pollen may be discharged prior to flower opening. Each of the four grains of the pollen tetrad is capable of producing a functional pollen tube (Roberts and Struckmeyer,

1942), and the long blooming period (up to four weeks) helps to insure pollination (Marucci, 1967).

Stem cuttings bearing flower buds may be harvested in November from the field, stored for a few days in a refrigerator, rooted in a mist chamber, and grown in nutrient culture to flower, bear, and mature fruit. Thus it is possible to effect controlled pollinations on plants two months after rooting in a greenhouse. Field pollinations may also be made.

Pollen collection is by tapping the tips of the anther tubes onto finger tips, thumb nails, slides, petri dishes, or into capsules. The blooming periods of early- and late-flowering clones usually overlap so that fresh pollen is transferred immediately to the selected stigma. Pollen may also be stored for a short time in the refrigerator at about 4.4 C.

Emasculation is done at the full pink bud stage with forceps. The flowers may be thinned to one per stem to insure better fruit set. Pollen transfer is effected with a thumb nail or other smooth surface, or a small camel's hair brush. Pollinated flowers are covered with cheese cloth bags or glass bottles. Each pollinated flower will produce from ten to 50 seeds, which will be mature about three months after pollination.

Seed may be extracted from the berry by hand or decanted after maceration with a food blender. Seed germination is good following storage of three to five months in the fruit, cold storage of the extracted seed for six to twelve weeks below 7.2 C, or sown directly following extraction from the fruit (especially if germinated at high temperatures such as 35 C). Treatment with gibberellic acid stimulates germination readily. After-ripened seed germinates well at 20–25 C. Shultz (1944) found that fresh cranberry seed was dormant, and that the dormancy was located in the seed coat, and in the membranous layer of crushed cells between the endosperm and seed coat. Dry dormant seeds also respond to high temperature germination. Light is beneficial to cranberry seed germination. Seed germination can be effected on the following media: nutrient agar, finely ground peat moss, vermiculite, 1 sand:1 peat, or 1 sand:1 peat:1 soil. Germination takes from 11 to 24 days, and the seedlings can be transferred to pots in from as little as two to three weeks or as much as four to five months.

The potting medium is usually a mixture of sand and peat. The seedlings are grown in pot culture until they are vigorous and starting to runner. If grown over winter, supplemental light and high nutrition are necessary. Transferring the potted seedlings to the field is preceded by a "hardening off" period of one to two months at 4.4 to 7.2 C. Without the hardening period, the seedlings are very sensitive to cold injury. Field setting is at 1.5 x 1.5 m or 1.8 x 1.8 m. Some fruit may be observed the year after setting, but a complete evaluation and second testing of the better selections involves ten to 15 years from seed.

Breeding procedures for the foxberry or cowberry (*Vaccinium vitis-idaea* var. *minus* Lodd) appear to be similar to those for the cranberry and for lowbush blueberry (*V. angustifolium* Ait.) (Hall and Beil, 1970).

Breeding Systems

The genome of the cranberries is small-sized and relatively undifferentiated, like that of the blueberries (Hagerup, 1940; Ahokas, 1971). Cranberry species also produce unreduced gametes as a frequent cytological anomaly. Homoploid species interfertility has been demonstrated in the cranberries (Camp, 1944; Chandler et al., 1947), at least at the diploid and the tetraploid levels.

The original hexaploid form described by Hagerup (1940) demonstrated chromosomal sterility in the form of Anaphase I laggards and irregular numbers of spore quartets, varying from one to ten, suggestive of restitution nuclei in some instances and micronuclei in others. Ahokas (1971) described a fertile hexaploid in which he demonstrated interchromosomal fibers during the first meiotic prophase-anaphase and in the mitotic prophase and metaphase. He felt that these fibers permitted homeologous bivalents to exhibit secondary pairing. Colchicine treatment seems to prevent the formation of these fibers or destroys them.

Marucci (1967) indicated that cross-pollination of cranberries results in higher percentage sets, larger berries, and higher seed counts than self-pollination. Kust (1965) speculated that the vigorous, large-fruited vines selected from the wild and clonally propagated may have been interclonal (intraspecific) hybrids. One could thus infer a degree of inbreeding depression if cranberries were systematically inbred for improvement. Conversely, hybridization and selection would be expected to provide the quickest breeding advances, as in other *Vaccinium* species.

Though cranberry improvement has all been accomplished at the diploid level and within one species, species hybridizations, intercrossing natural and artificial polyploid forms, and polyploid with diploid species, might provide types, for example, with the superior fruit of the diploid and the shade and moisture tolerance and late flowering of the hexaploid. However, the fertile hexaploid described by Ahokas (1971) was reproductively isolated from the tetraploid and the small-fruited diploid species. Similarly, crosses of *Oxycoccus* species representatives with *Euvaccinium*, whortleberry, and *V. ovatum* types might provide some very interesting and valuable new cranberry-blueberry hybrids.

While one generation of improvement has produced remarkable advances in fruit yield, size, color, insect and disease resistance, and plant vigor, it is not yet known whether the improved hybrids will transmit their superior characteristics to a high proportion of their offspring. Very little is known about the inheritance patterns of specific cranberry traits.

Achievements and Prospects

The large-fruited cranberry (*V. macrocarpon* Ait.), adapted to lowland acidic soil situations with a ready water supply, is an attractive acid fruit with a high Vitamin C content that can be grown in cooler climates. Culture of clonal selected wild vines is widespread in certain areas of eastern and western North America and in Wisconsin. Cranberry culture has shown experimental promise in several European countries. A cooperative USDA cranberry breeding program was carried on in the period 1929 to 1961, and six first-generation improved hybrids between selected native cultivars were introduced. A seventh hybrid cultivar was recently introduced by the Washington Agricultural Experiment Station. Evaluation of later selections is continuing at several locations. Heritable differences in fruit size, shape, color, appearance, keeping quality, vine type, flavor, processing ability, and resistance to leafhopper infestation were demonstrated during the life of this breeding program. Cranberry improvement, although admittedly long-term and expensive, could provide mankind with another "year-round" source of Vitamin C in an attractive and convenient form.

Literature Cited

The following persons supplied the author with information on the status and development of the cranberry industry and its cultivars: I. S. Cobb, M. N. Dana, I. E. Demoranville, C. C. Doughty, A. D. Draper, P. Eck, I. V. Hall, F. Johnson, P. E. Marucci, A. Y. Shawa, and A. W. Stretch.

Aalders, L. E. 1966. Canada: progress in the selection and culture of lowbush blueberries. *In* W. J. Kender and D. A. Abdalla (eds.). *Proc. N. Amer. Blueberry Workers Conf. Maine Agr. Expt. Sta. Misc. Rpt.* 118. p. 90–92.

Aalders, L. E., and I. V. Hall. 1961. Pollen incompatibility and fruit set in lowbush blueberries. *Can. J. Gen. Cyt.* 3:300–307.

———. 1962*a*. The inheritance of white fruit in the velvet-leaf blueberry, *Vaccinium myrtilloides* Michx. *Can. J. Gen. Cyt.* 4:90–91.

———. 1962*b*. New evidence on the cytotaxonomy of *Vaccinium* species as revealed by stomatal measurements from herbarium specimens. *Nature* 196-(4855):694.

———. 1963*a*. The inheritance and morphological development of male sterility in the common lowbush blueberry, *Vaccinium angustifolium* Ait. *Can. J. Gen. Cyt.* 5:380–383.

———. 1963*b*. The inheritance and taxonomic significance of the "nigrum" factor in the common lowbush blueberry *Vaccinium angustifolium*. *Can. J. Gen. Cyt.* 5:115–118.

———. 1963*c*. Note on aeration of colchicine solution in the treatment of germinating blueberry seeds to induce polyploidy. *Can. J. Plant Sci.* 43:107.

———. 1968. The effect of depth of planting on the survival, yield, and spread of the common lowbush blueberry, *Vaccinium angustifolium* Ait. *HortScience* 3:72–74.

Ahokas, H. 1971. Cytology of hexaploid cranberry with special reference to chromosomal fibers. *Hereditas* 68:123–135.

Anonymous. 1956. Composition and value of foods. *In* C. D. Hodgman, R. C. Weast, and S. M. Selby (eds.). *Handbook of chemistry and physics.* 38th ed. Cleveland: Chemical Rubber Publ. Co. p. 1814–1828.

Bailey, R. M. 1966. Blueberry varieties and breeding—Maine. *In* W. J. Kender and D. A. Abdalla (eds.). *Proc. N. Amer. Blueberry Workers Conf. Maine Agr. Expt. Sta. Misc. Rpt.* 118. p. 88–90.

Ballinger, W. E., and L. J. Kushman. 1970. Relationship of stage of ripeness to composition and keeping quality of highbush blueberries. *J. Amer. Soc. Hort. Sci.* 95:239–242.

———. 1971. Interrelationships of mechanical harvesting, bruising and decay of highbush blueberries. *In* C. M. Mainland and R. P. Rohrbach (eds.). *Proc. 1971 Highbush Blueberry Mechanization Symp.* N.C. State Univ., Raleigh, p. 38–47.

Ballinger, W. E., E. P. Maness, G. J. Galletta, and L. J. Kushman. 1972. Anthocyanins of ripe fruit of a "pink-fruited" hybrid of highbush blueberries, *Vaccinium corymbosum* L. *J. Amer. Soc. Hort. Sci.* 97:381–384.

Ballinger, W. E., E. P. Maness, and L. J. Kushman. 1970. Anthocyanins in ripe fruit of the highbush blueberry, *Vaccinium corymbosum*. *J. Amer. Soc. Hort. Sci.* 95:283–285.

Barker, W. G., I. V. Hall, L. E. Aalders, and G. W. Wood. 1964. The lowbush blueberry industry in Eastern Canada. *Econ. Bot.* 18:357–365.

Bell, H. P. 1957. The development of blueberry seed. *Can J. Bot.* 35:139–153.

Brewer, J. W., and R. C. Dobson. 1969*a*. Seed count and berry size in relation to pollinator level and harvest date for the highbush blueberry, *Vaccinium corymbosum*. *J. Econ. Ent.* 62:1353–1356.

———. 1969*b*. Pollen analysis of two highbush blueberry varieties, *Vaccinium corymbosum*. *J. Amer. Soc. Hort. Sci.* 94:251–252.

———. 1969*c*. Nectar studies of the highbush blueberry, *Vaccinium corymbosum* L. cv. 'Rubel' and 'Jersey.' *HortScience* 4:332–333.

Brightwell, W. T. 1960. Present status of rabbiteye breeding. *In* J. N. Moore and N. F. Childers (eds.) *Blueberry research—fifty years of progress*. New Brunswick: N. J. Agr. Expt. Sta. p. 49–50.

———. 1966. Blueberry varieties and breeding—Georgia. *In* W. J. Kender and D. A. Abdalla (eds.). *Proc. N. Amer. Blueberry Workers Conf. Maine Agr. Expt. Sta. Misc. Rpt.* 118. p. 78–79.

———. 1971*a*. Rabbiteye blueberries. *Ga. Agr. Expt. Sta. Res. Bul.* 100.

———. 1971*b*. Rabbiteye blueberry culture. *Proc. 7th Ann. Mtg. N. Amer. Blueberry Council* p.37–44.

Camp, W. H. 1941. Studies in the Ericales: A review of the North American *Gaylussacieae*: with remarks on the origin and migration of the group. *Bul. Torrey Bot. Club* 68:531–551.

———. 1942*a*. On the structure of populations in the genus *Vaccinium*. *Brittonia* 4:189–204.

———. 1942*b*. A survey of the American species of *Vaccinium*, subgenus *Euvaccinium*. *Brittonia* 4:205–247.

———. 1944. A preliminary consideration of the biosystematy of *Oxycoccus*. *Bul. Torrey Bot. Club* 71:426–437.

———. 1945. The North American blueberries with notes on other groups of *Vacciniaceae*. *Brittonia* 5:203–275.

Camp, W. H., and C. L. Gilly. 1942. Polypetalous forms of *Vaccinium*. *Torreya* 42:168–173.

Chandler, F. B., and I. Demoranville. 1958. Cranberry varieties of North America. *Mass. Agr. Expt. Sta. Bul.* 513.

Chandler, F. B., R. B. Wilcox, H. F. Bain, H. F. Berman, and H. Dermen. 1947. Cranberry breeding investigation of the USDA. *Cranberries* 12(1):6–9 (May); 12(2):6–10 (June).

Chapman, A. W. 1893. *Flora of the southern United States*. 2nd ed. New York: Ivison.

Childs, W. H. 1946. Shredded sphagnum vs. peat and sand as a medium for transplanted blueberry seedlings. *Proc. Amer. Soc. Hort. Sci.* 47:206–208.

———. 1969. 'Ornablue', new blueberry variety. *West Va. Agric. Forestry* 2(4):10–12.

Cockerham, L. E. 1972. Cytological analyses of *Vaccinium* species. M.S. thesis, North Carolina State University.

Constante, J. F., and B. R. Boyce. 1968. Low temperature injury of highbush blueberry shoots at various times of year. *Proc. Amer. Soc. Hort. Sci.* 93:267–272.

Coorts, G. D., J. G. Hapitan, Jr., and V. G. Shutak. 1968. Comparative rate of leaf respiration for six cultivars of highbush blueberry. *HortScience* 3:69–70.

Coville, F. V. 1910. Experiments in blueberry culture. *USDA Bureau of Plant Industry Bul.* 193.

———. 1921. Directions for blueberry culture, 1921. *USDA Bul.* 974.

———. 1927. Blueberry chromosomes. *Science* 66:565–566.

———. 1937. Improving the wild blueberry. In *USDA Yearbook of Agriculture*. 1937:559–574.

Cram, W. T. 1971. The black vine weevil on highbush blueberry. *Can. Agr.* 16(1):36–37.

Dana, M. N., and G. C. Klingbeil. 1966. Cranberry growing in Wisconsin. *Wisc. Agr. Ext. Cir.* 654.

Darrow, G. M. 1958. Seed number in blueberry fruits. *Proc. Amer. Soc. Hort. Sci.* 72:212–215.

———. 1960*a*. Blueberry breeding: past, present, future. *Amer. Hort. Mag.* 39(1):14–33.

———. 1960*b*. History of blueberry research. *In* J. N. Moore and N. F. Childers (eds.). *Blueberry research—fifty years of progress*. New Brunswick: N. J. Agr. Expt. Sta. p. 1–13.

———. 1966. Blueberry research. *In* W. J. Kender and D. A. Abdalla (eds.). *Proc. N. Amer. Blueberry Workers Conf. Maine Agr. Expt. Sta. Misc. Rpt.* 118. p. 3–8.

Darrow, G. M., and W. H. Camp. 1945. *Vaccinium* hybrids and the development of new horticultural material. *Bul. Torrey Bot. Club* 72:1–21.

Darrow, G. M., W. H. Camp, H. E. Fisher, and H. Dermen. 1944. Chromosome numbers in *Vaccinium* and related groups. *Bul. Torrey Bot. Club* 71:498–516.

Darrow, G. M., J. H. Clark, and E. B. Morrow. 1939. The inheritance of certain characters in the cultivated blueberry. *Proc. Amer. Soc. Hort. Sci.* 37:611–616.

Darrow, G. M., H. Dermen, and D. H. Scott. 1949. A tetraploid blueberry from a cross of diploid and hexaploid species. *J. Hered.* 40:304–306.

Darrow, G. M., E. B. Morrow, and D. H. Scott. 1952. An evaluation of interspecific blueberry crosses. *Proc. Amer. Soc. Hort. Sci.* 59:277–282.

Darrow, G. M., and D. H. Scott. 1954. Longevity of

blueberry seed in cold storage. *Proc. Amer. Soc. Hort. Sci.* 63:271.

———. 1966. Varieties and their characteristics. *In* P. Eck and N. F. Childers (eds.). *Blueberry culture.* New Brunswick: Rutgers Univ. Press. p. 94–110.

Darrow, G. M., D. H. Scott, and H. Dermen. 1954. Tetraploid blueberries from hexaploid x diploid species crosses. *Proc. Amer. Soc. Hort. Sci.* 63:266–270.

Demaree, J. B., and E. B. Morrow. 1951. Relative resistance of some blueberry varieties and selections to stem canker in North Carolina. *Plant Dis. Rptr.* 35:136–141.

Door, J., and E. C. Martin. 1966. Pollination studies on the highbush blueberry, *Vaccinium corymbosum* L. *Quart. Bul. Mich. Agr. Expt. Sta.* 48:437–448.

Doughty, C. C., and J. G. Dodge. 1966. Cranberry production in Washington. *Wash. Agr. Ext. Manual* 2619.

Doughty, C. C., and R. Garren, Jr. 1970. 'Crowley', a new, early maturing cranberry variety for Washington and Oregon. *Fruit Var. Hort. Dig.* 24:88–89.

Draper, A. D., and D. H. Scott. 1967. Blueberry breeding program of the USDA and cooperators. *In* Int. Soc. Hort. Sci. Working Group. *Blueberry Culture in Europe Symp.* Venlo, Netherlands. p. 83–94.

———. 1969. Fruit size inheritance in highbush blueberries, *Vaccinium australe* Small. *J. Amer. Soc. Hort. Sci.* 94:417-418.

———. 1971. Inheritance of albino seedling in tetraploid highbush blueberry. *J. Amer. Soc. Hort. Sci.* 96:791–792.

Draper, A. D., D. H. Scott, and L. F. Hough. 1966. Highbush blueberry variety and selection evaluation. *In* W. J. Kender and D. A. Abdalla (eds.). *Proc. N. Amer. Blueberry Workers Conf. Maine Agr. Expt. Sta. Misc. Rpt.* 118. p. 93–96.

Draper, A. D., A. W. Stretch, and D. H. Scott. 1972. Two tetraploid sources of resistance for breeding blueberries resistant to *Phytophthora cinnamomi* Rands. *HortScience* 7:266–268.

Eaton, E. L. 1949. Highbush blueberries. *Can. Dept. Agr. Dom. Exptl. Sta., Kentsville, N.S., Prog. Rpt. 1937–1946.* p. 20–25.

Eaton, G. W. 1967. The relationship between seed number and berry weight in open pollinated highbush blueberries. *HortScience* 2:14–15.

Eaton, G. W., and A. M. Jamont. 1966. Megagametogenesis in *Vaccinium corymbosum* L. *Can. J. Bot.* 44:712–714.

Eck, P. 1966. Botany. *In* P. Eck and N. F. Childers (eds.). *Blueberry culture.* New Brunswick: Rutgers Univ. Press. p. 14–44.

Edwards, T. W., Jr., W. B. Sherman, and R. H. Sharpe. 1970. Fruit development in short and long cycle blueberries. *HortScience* 5:274–275.

———. 1972. Seed development in certain Florida tetraploid and hexaploid blueberries. *HortScience* 7:127–128.

Galletta, G. J. 1960. Breeding for cane canker resistance. *In* J. N. Moore and N. F. Childers (eds.) *Blueberry research—fifty years of progress.* New Brunswick: N. J. Agr. Expt. Sta. p. 51–52.

———. 1966. Blueberry varieties and breeding—North Carolina. *In* W. J. Kender and D. A. Abdalla (eds.). *Proc. N. Amer. Blueberry Workers Conf. Maine Agr. Expt. Sta. Misc. Rpt.* 118. p. 80–85.

———. 1970. Measures of inbreeding potential in cultivated polyploid blueberries. *HortScience* 5:349–350. (Abstr.)

Galletta, G. J., W. E. Ballinger, R. J. Monroe, and L. J. Kushman. 1971. Relationships between fruit acidity and soluble solids levels of highbush blueberry clones and fruit keeping quality. *J. Amer. Soc. Hort. Sci.* 96:758–762.

Galletta, G. J., and A. S. Fish, Jr. 1971. Interspecific blueberry grafting, a way to extend *Vaccinium* culture to different soils. *J. Amer. Soc. Hort. Sci.* 96:-294–298.

Galletta, G. J., and C. M. Mainland. 1971a. Correlations between highbush blueberry yield and other plant measures. *HortScience* 6:293. (Abstr.)

———. 1971*b*. Comparative effects of mechanical harvesting on highbush blueberry varieties. *In* C. M. Mainland and R. P. Rohrbach (eds.). *Proc. 1971 Highbush Blueberry Mechanization Symp.* N. C. State Univ., Raleigh. p. 80–85.

Gersons, L. 1967. Some aspects of the processing of blueberries. *In* Intern. Soc. Hort. Sci. Working Group. *Blueberry Culture in Europe Symp.* Venlo, Netherlands. p. 169–173.

Goldenberg, J. 1964. A study of the disease resistance of plants when resistance is quantitatively inherited (in Spanish). *Rev. de Invest. Agropecuarias, B. Aires: Ser. Pat. Veg.* 1:107–132.

Hagerup, O. 1940. Studies on the significance of polyploidy. 4. Oxycoccus. *Hereditas* 26:399–410.

Hall, I. V., and L. E. Aalders. 1961a. Cytotaxonomy of lowbush blueberries in Eastern Canada. *Amer. J. Bot.* 48:199–201.

———. 1961*b*. Note on male sterility in the common lowbush blueberry, *Vaccinium angustifolium* Ait. *Can. J. Plant Sci.* 41:865.

———. 1963. Two factors inheritance of white fruit in the common lowbush blueberry, *Vaccinium angustifolium* Ait. *Can. J. Gen. Cyt.* 5:371–373.

———. 1968. The botanical composition of two barrens in Nova Scotia. *Naturaliste Canadien* 95:393–396.

Hall, I. V., L. E. Aalders, and L. Jackson. 1971. Establishing superior lowbush blueberry fields. *Can. Dept. Agr. Publ.* 1436.

Hall, I. V., L. E. Aalders, and C. L. Lockhart. 1964. Note on some putative lowbush blueberry mutants. *Can. J. Bot.* 42:122–125.

Hall, I. V., L. E. Aalders, and G. W. Wood. 1966. Female sterility in the common lowbush blueberry, *Vaccinium angustifolium* Ait. *Can. J. Gen. Cyt.* 8:296–299.

Hall, I. V., and C. E. Beil. 1970. Seed germination, pollination and growth of *Vaccinium vitis-idaea* var. *minus* Lodd. *Can. J. Plant Sci.* 50:731–732.

Hall, I. V., L. R. Townsend, C. L. Lockhart, K. A. Harrison, G. W. Wood, and G. T. Morgan. 1969. Growing cranberries. *Can.* Dept. *Agr. Publ.* 1282.

Hall, S. H., and G. J. Galletta. 1971. Comparative chromosome morphology of diploid *Vaccinium* species. *J. Amer. Soc. Hort. Sci.* 96:289–292.

Intern. Soc. Hort. Sci. Working Group. 1967. *Blueberry Culture in Europe Symp. Proc.* Venlo, Netherlands.

Jelenkovic, G., and A. D. Draper. 1970. Fertility and chromosome behavior of a derived decaploid of *Vaccinium*. *J. Amer. Soc. Hort. Sci.* 95:816–820.

Jelenkovic, G., and E. Harrington. 1971. Nonrandom chromosome associations at diplotene and diakinesis in a tetraploid clone of *Vaccinium australe* Small. *Can. J. Gen. Cyt.* 13:270–276.

Jelenkovic, G., and L. F. Hough. 1970. Chromosome associations in the first meiotic division in three tetraploid clones of *Vaccinium corymbosum* L. *Can. J. Gen. Cyt.* 12:316–324.

Johnson, F. 1969. Diseases of blueberries. *In* 1969 Research Progress. *Wash. Agr. Expt. Sta. Bul.* 707. p. 54–55.

Johnston, S. 1942. Observations on the inheritance of horticulturally important characteristics in the highbush blueberry. *Proc. Amer. Soc. Hort. Sci.* 40:352–356.

———. 1966. Blueberry varieties and breeding—Michigan. *In* W. J. Kender and D. A. Abdalla (eds.). *Proc. N. Amer. Blueberry Workers Conf. Maine Agr. Expt. Sta. Misc. Rpt.* 118. p. 87–88.

Johnston, S., and J. E. Moulton. 1967. The 'Bluehaven' and 'Northland' blueberry varieties. *Mich. St. Univ. Quart. Bul.* 50:46–49.

———. 1968. Blueberry distribution, variety evaluation and breeding in Michigan. *Mich. Agr. Expt. Sta. Res. Rpt.* 67.

Kender, W. J. 1965. Some factors affecting the propagation of lowbush blueberries by softwood cuttings. *Proc. Amer. Soc. Hort. Sci.* 86:301–306.

———. 1966. Domesticating the lowbush blueberry. *In* W. J. Kender and D. A. Abdalla (eds.). *Proc. N. Amer. Blueberry Workers Conf. Maine Agr. Expt. Sta. Misc. Rpt.* 118. p. 147–149.

———. 1967. On the domestication of the lowbush blueberry. *Fruit Var. Hort. Dig.* 21:74–76.

Klein, K. 1971. Cranberry cultivation (in German). *Mitteilungen Rebe und Wein, Obstbau und Früicheverwertung* 21(4):322–326. (*Hort. Abstr.* 42 [2]:3297, June 1972.).

Knight, R. J. 1960. Breeding blueberries for horticultural characters. *In* J. N. Moore and N. F. Childers (eds.). *Blueberry research—fifty years of progress.* New Brunswick: N. J. Agr. Expt. Sta. p. 55–56.

Knight, R. J., and D. H. Scott. 1964. Effects of temperatures on self- and cross-pollination and fruiting of four highbush blueberry varieties. *Proc. Amer. Soc. Hort. Sci.* 85:302–306.

Kolupaeva, K. G. 1971. Cranberry productivity in the Kirov province (in Russian). *Rastut. Resursy.* 7:99–103. (*Hort. Abstr.* 41[4]: 8492, Dec. 1971.)

Kust, T. 1965. The need for a cranberry breeding program. *Univ. Wis. Dept. Hort. Mimeo* 627.

Liebster, G. 1971. Cranberry, the cultivated whortleberry, *V. macrocarpon*, a new kind of fruit for Germany (in German). *Erwerbsobstbau* 13:29–32. (*Hort. Abstr.* 41[4]:8493, Dec. 1971.)

Lockhart, C. L. 1970. Leafspot of highbush blueberry caused by *Godronia cassandrae f. vaccinii*. *Can. Plant Dis. Survey* 50:93–94.

Lockhart, C. L., and D. L. Craig. 1967. *Fusicoccum* canker on highbush blueberry in Nova Scotia. *Can. Plant Dis. Survey* 47:17–20.

Longley, A. E. 1927. Chromosomes in *Vaccinium*. *Science* 66:566–568.

Marucci, P. E. 1966*a*. Blueberry pollination. *Amer. Bee J.* 106:250–251.

———. 1966*b*. Insects and their control. *In* P. Eck and N. F. Childers (eds.). *Blueberry culture.* New Brunswick: Rutgers Univ. Press. p. 199–235.

———. 1967. Cranberry pollination. *Amer. Bee. J.* 107:212–213.

Matzner, F. 1967. The contents of ascorbic acid and dehydroascorbic acid in blueberries (In German, English summary). *In* Intern. Soc. Hort. Sci. Working Group. *Blueberry Culture in Europe Symp.* Venlo, Netherlands. p. 174–177.

Merrill, T. A. 1936. Pollination of the highbush blueberry. *Mich. Agr. Expt. Sta. Quart. Bul.* 22:112–116.

Milholland, R. D. 1970*a*. Histology of *Botryosphaeria* canker of susceptible and resistant highbush blueberries. *Phytopathology* 60:70–74.

———. 1970*b*. Effect of leaf exudates on blueberry leaf spots caused by *Gloeosporium minus*. *Phytopathology* 60:635–640.

———. 1972. Histopathology and pathogenicity of *Botryosphaeria dothidea* on blueberry stems. *Phytopathology* 62:658–660.

Milholland, R. D., and G. J. Galletta. 1967. Relative susceptibility of blueberry cultivars to *Phytophthora cinnamomi*. *Plant Dis. Rptr.* 51:998–1001.

———. 1969. Pathogenic variation among isolates of *Botryosphaeria corticis* on blueberry. *Phytopathology* 59:1540–1543.

Moore, J. N. 1964. Duration of receptivity to pollination of flowers of the highbush blueberry and the cultivated strawberry. *Proc. Amer. Soc. Hort. Sci.* 85:295–301.

———. 1965. Improving highbush blueberries by breeding and selection. *Euphytica* 14:39–48.

———. 1966. Breeding. *In* P. Eck and N. F. Childers (eds.). *Blueberry culture.* New Brunswick: Rutgers Univ. Press. p. 45–74.

Moore, J. N., H. H. Bowen, and D. H. Scott. 1962. Response of highbush blueberry varieties, selections, and hybrid progenies to powdery mildew. *Proc. Amer. Soc. Hort. Sci.* 81:274–280.

Moore, J. N., B. D. Reynolds, and G. R. Brown. 1972. Effects of seed number, size and development on fruit size of cultivated blueberries. *HortScience* 7:268–269.

Moore, J. N., and D. H. Scott. 1966. Breeding value of various blueberry varieties and selections for northeastern United States. *Proc. Amer. Soc. Hort. Sci.* 88:331–337.

Moore, J. N., D. H. Scott, and H. Dermen. 1964. Development of a decaploid blueberry by colchicine treatment. *Proc. Amer. Soc. Hort. Sci.* 84:274–279.

Morrow, E. B. 1943. Some effects of cross-pollination vs. self-pollination in the cultivated blueberry. *Proc. Amer. Soc. Hort. Sci.* 42:469–472.

Morrow, E. B., G. M. Darrow, and J. A. Rigney. 1949.

A rating system for the evaluation of horticultural material. *Proc. Amer. Soc. Hort. Sci.* 53:276–280.
Morrow, E. B., G. M. Darrow, and D. H. Scott. 1954. A quick method of cleaning berry seed for breeders. *Proc. Amer. Soc. Hort. Sci.* 63:265.
Nelson, J. W. 1966. Disease problems of highbush blueberries. *In* W. J. Kender and D. A. Abdalla (eds.). *Proc. N. Amer. Blueberry Workers Conf. Maine Agr. Expt. Sta. Misc. Rpt.* 118. p. 64–66.
Nelson, J. W., and H. C. Bittenbender. 1971. Mummy berry disease occurrence in a blueberry selection test planting. *Plant Dis. Rptr.* 55:651–653.
Newcomer, E. H. 1941. Chromosome numbers of some species and varieties of *Vaccinium* and related genera. *Proc. Amer. Soc. Hort. Sci.* 38:468–470.
Norton, J. S. 1969. Cultural practices and mechanization of cranberry and wild blueberry (in three parts). *Cranberries* 33(12):20, 34(1):14, 34(2):20.
Pepin, H. S., and H. N. W. Toms. 1969. Susceptibility of highbush blueberry varieties to *Monilinia vacciniicorymbosi. Phytopathology* 59:1876–1878.
Peterson, B. S., C. E. Cross, and N. Tilden. 1968. The cranberry industry in Massachusetts. *Mass. Dept. Agr. Bul.* 201.
Porsild, A. E. 1938. The cranberry in Canada. *Can. Field Nat.* 52:116–117.
Quamme, H. A., C. Stushnoff, and C. J. Weiser. 1972. Winterhardiness of several blueberry species and cultivars in Minnesota. *HortScience* 7:500–502.
Rigby, B., and M. N. Dana. 1972. Flower opening, pollen shedding, stigma receptivity and pollen tube growth in the cranberry. *HortScience* 7:84–85.
Roach, F. A. 1967. Highbush blueberry production in England. *In* Intern. Soc. Hort. Sci. Working Group. *Blueberry Culture Symp.* Venlo, Netherlands. p. 22–33.
Roberts, R. H., and B. E. Struckmeyer. 1942. Growth and fruiting of the cranberry. *Proc. Amer. Soc. Hort. Sci.* 40:373–379.
Robinson, B. L., and M. L. Fernald. 1908. *Gray's new manual of botany.* 7th ed. New York: American Book Co.
Rousi, A. 1966*a*. The use of North-European *Vaccinium* species in blueberry breeding. *Acta Agriculturae Scandinavica Suppl.* 16:50-54.
———. 1966*b*. Cytological observations on some species and hybrids of *Vaccinium. Zuchter/Gen. Brdg. Res.* 36:352–359.
Scott, D. H., and A. D. Draper. 1967. Light in relation to seed germination of blueberries, strawberries and *Rubus. HortScience* 2:107–108.
Scott, D. H., and D. P. Ink. 1955. Treatments to hasten the emergence of seedlings of blueberry and strawberry. *Proc. Amer. Soc. Hort. Sci.* 66:237–241.
Sharpe, R. H. 1954. Horticultural development of Florida blueberries. *Proc. Fla. St. Hort. Soc.* 66: 188–190.
———. 1966. Blueberry varieties and breeding—Florida. *In* W. J. Kender and D. A. Abdalla (eds.). *Proc. N. Amer. Blueberry Workers Conf. Maine Agr. Expt. Sta. Misc. Rpt.* 118. p. 72–77.
Sharpe, R. H., and G. M. Darrow. 1959. Breeding blueberries for the Florida climate. *Proc. Fla. St. Hort. Soc.* 72:308–311.
———. 1960. Breeding blueberries for the Florida climate. *In* J. N. Moore and N. F. Childers (eds.). *Blueberry research—fifty years of progress.* New Brunswick: N. J. Agr. Expt. Sta. p. 44–48.
Sharpe, R. H., and W. B. Sherman. 1970. Bluegem blueberry. *Fla. Agr. Expt. Sta. Cir.* S209.
———. 1971. Breeding blueberries for low chilling requirement. *HortScience* 6:145–147.
Sherman, W. B., R. H. Sharpe, and J. Janick. 1973. The fruiting nursery: ultrahigh density for evaluation of blueberry and peach seedlings. *HortScience* 8:170–172.
Shultz, J. H. 1944. Some cytotaxonomic and germination studies in the genus *Vaccinium.* Ph.D. dissertation, Washington State Univ.
Shutak, V. G., and P. E. Marucci. 1966. Plant and fruit development. *In* P. Eck and N. F. Childers (eds.). *Blueberry culture.* New Brunswick: Rutgers Univ. Press. p. 179–198.
Sleumer, H. 1941. Studies in the *Vacciniaceae* (in German). *Bot. Jahrb.* (*Sonder-Abdr.*) 71:375–510.
Small, J. K. 1933. *Manual of the southeastern flora.* New York: privately published.
Soczek, Z., and R. Scholz. 1969. Cropping of large-fruited cranberry, *Oxycoccus macrocarpus,* grown on high moor peat in Poland (in Polish, English and Russian summaries). *Prace Inst. Sadown w Skierniewicach* 13:75–80. (*Hort. Abstr.* 41[4]:8497. Dec. 1971.)
Stretch, A. W. 1967*a*. Important fungus diseases of cultivated highbush blueberry in North America. *In* Intern. Soc. Hort. Sci. Working Group. *Blueberry Culture in Europe, Symp.* Venlo, Netherlands. p. 133–143.
———. 1967*b*. Virus diseases of cultivated highbush blueberries in North America. *In* Intern. Soc. Hort. Sci. Working Group. *Blueberry Culture in Europe Symp.* Venlo, Netherlands. p. 144–154.
Stushnoff, C. 1972. Breeding and selection methods for cold hardiness in deciduous fruit crops. *HortScience* 7:10–13.
Stushnoff, C., and L. F. Hough. 1966. Blueberry varieties and breeding—New Jersey. *In* W. J. Kender and D. A. Abdalla (eds.). *Proc. N. Amer. Blueberry Workers Conf. Maine Agr. Expt. Sta. Misc. Rpt.* 118. p. 85–87.
———. 1968*a*. Sporogenesis and gametophyte development in 'Bluecrop' and 'Coville' highbush blueberries. *Proc. Amer. Soc. Hort. Sci.* 93:242–247.
———. 1968*b*. Response of blueberry seed germination to temperature, light, potassium nitrate, and coumarin. *Proc. Amer. Soc. Hort. Sci.* 93:260–266.
Stushnoff, C., and B. F. Palser. 1969. Embryology of five *Vaccinium* taxa including diploid tetraploid and hexaploid species or cultivars. *Phytomorphology* 19: 312–321.
Uphof, J. C. Th. 1968. *Vaccinium. In* J. C. Th. Uphof, *Dictionary of economic plants* (2nd ed.) Lehre, Germany: Cramer. p. 537–538.
Varney, E. H., and A. W. Stretch. 1966. Diseases and

their control. *In* P. Eck and N. F. Childers (eds.). *Blueberry culture*. New Brunswick: Rutgers Univ. Press. p. 236–279.

Whitton, L. 1960. Breeding for winterhardiness. *In* J. N. Moore and N. F. Childers (eds.). *Blueberry research—fifty years of progress*. New Brunswick; N. J. Agr. Expt. Sta. p. 53–54.

———. 1964. The cytotaxonomic status of *Vaccinium angustifolium* Aiton in commercial blueberry fields of Maine. Ph.D. thesis, Cornell Univ.

Young, R. S. 1952. Growth and development of the blueberry fruit (*Vaccinium corymbosum* L. and *V. angustifolium* Ait.) *Proc. Amer. Soc. Hort. Sci.* 59: 167–172.

Currants and Gooseberries

by *Elizabeth Keep*

Although some authors still consider that the gooseberries and currants belong in separate genera, *Grossularia* and *Ribes* respectively, here they will both be grouped under *Ribes*, since, as will be seen later, this is in accord with their crossing relationships. In this wider sense, the genus includes some 150 species, distributed mainly in the temperate regions of the Northern Hemisphere, but also in North Africa and in the Andes.

All species and commercial cultivars are diploid with the basic chromosome number x = 8. PMC meiosis is usually regular (Meurman, 1928; Zielinski, 1953; Goldschmidt, 1964*a*). Some earlier workers considered that evolution in the genus occurred mainly through the accumulation of gene mutations (Darlington, 1929; Sax, 1931; Zielinski, 1953). However, Goldschmidt (1964*a*, 1966) described configurations involving three or four chromosomes in *R. aureum* Pursh., *R. 'rubrum'* and in intersectional hybrids, and suggested that these and the occasional presence of bridges and/or fragments (reported also by Meurman, 1928) indicate that structural chromosome changes may have played a part.

On morphological grounds, the genus has been divided into subgenera and sections or series (Janczewski, 1907; Berger, 1924; Rehder, 1954), with the number of these subdivisions varying considerably according to the taxonomist. An alternative to purely morphological classification, based on fertility relationships, has been proposed by Keep (1962*b*).

The European blackcurrant, *R. nigrum* L., and other black or brown fruited species with resinous glands comprise Berger's Eucoreosma. The subgenus includes the North American species *R. americanum* Mill. (the American blackcurrant); *R. bracteosum* Dougl.; *R. petiolare* Dougl. (the western blackcurrant) and its sister species *R. hudsonianum* Rich. (the northern blackcurrant); and the eastern Asiatic *R. dikuscha* Fisch. and *R. ussuriense* Jancz. Exceptionally for this group, the Californian *R. viburnifolium* A. Gray is evergreen.

The redcurrants are grouped by Berger in the subgenus Ribesia. The commercial cultivars are derived from any one or a combination of three species: *R. rubrum* L. (central and northern Europe and northern Asia); *R. petraeum* Wulf. (from the mountains of western and central Europe), and *R. sativum* Syme (*R. vulgare Jancz.*) (western Europe). Whitecurrants are a color form of the red. The eastern European *R. multiflorum* Kit., the western Chinese *R. longeracemosum* Franch., and *R. triste* Pall. from northern Asia and North America are included in this group. Like their commercial relatives, these species have crystalline glands and, except for *R. longeracemosum*, red fruit.

Species of the gooseberry group (Berger's genus *Grossularia*, subgenus Eugrossularia) are characterized by the presence of nodal spines. The European gooseberry, derived from *R. grossularia* L., with a distribution ranging into the Caucasus and North Africa, and the American gooseberry, *R. hirtellum* Michx., are grouped with the North American *R. niveum* Lindl., *R. oxyacanthoides* L., *R. divaricatum* Dougl., *R. cynosbati* L., *R. leptanthum* Gray, and the western Chinese *R. alpestre* Dcne and *R. stenocarpum* Maxim., and the Japanese *R. grossularioides* Maxim. North American gooseberry cultivars are mostly hybrids between *R. hirtellum* and *R. grossularia.*

Ribes grown as flowering shrubs include the golden currants, or American blackcurrants, native to central and northwestern America. These comprise Berger's subgenus Symphocalyx, characterized by fragrant yellow flowers and yellow or black fruits. The best known species in this group are the very similar *R. aureum* Pursh. and *R. odoratum* Wendl., from which the edible-fruited garden cultivar 'Crandall' is derived.

The large and variable subgenus Calobotrya, subdivided by Berger into three series, includes western North American ornamental currants such as *R. sanguineum* Pursh. (the widely cultivated flowering currant), its sister species *R. glutinosum* Benth., and the waxy squaw currant, *R. cereum* Dougl. Species in this group are characterized by overall glandularity and red or black fruit.

Subgenera of which little use has so far been made by *Ribes* breeders include Berger's North American and northeastern Asiatic Heritiera, comprising dwarf currants (often with procumbent, rooting branches) such as *R. glandulosum* Weber and *R. ambiguum* Maxim.; his North American and northeastern Asiatic Grossularioides (gooseberry-stemmed currants, with nodal spines and currant-like flowers in drooping racemes), e.g. *R. lacustre* (Pers.) Poir., *R. horridum* Rupr.; Berger's primarily western North American gooseberry subgenera Robsonia, Hesperia, and Lobbia, exemplified by *R. speciosum* Pursh., *R. menziesii* Pursh., and *R. pinetorum* Greene, respectively; and the dioecious subgenera Berisia (from Europe and Asia) with erect racemes, e.g. *R. alpinum* L. and *R. orientale* Desf., and Parilla (from South America) with pendulous racemes.

Relationships within and between the blackcurrant, redcurrant, and gooseberry groups and between these and other sections of the genus are discussed later.

The major producers of blackcurrants are Great Britain, Poland, and West Germany, with lesser acreages in Hungary, France, Yugoslavia, Holland, and Bulgaria. About 60,000 ha were grown in Russia in 1972. The crop is scarcely grown in North America as it is an alternate host of white pine blister rust (*Cronartium ribicola* Fisch.). About 75% of the British crop is sold directly to processors, mainly for juice, but also for canning, jam and pie fillings. Much of the European crop is used for juice, for blackcurrant crême and liqueurs, and for converting white wines to "rosé."

TABLE 1. Crop production in Europe (in metric tons)
(after Roach, 1972)

Country	Black-currants, 1969	Red- and white-currants, 1969–71	Goose-berries, 1969
Britain	24,000	1,700	13,000
Poland	18,000	16–25,000	19,000
West Germany	15–20,000[y]	48–54,000[y,x]	23,000[y]
Hungary	4,000	2–3,000	4,000
France	3,500	3,000	2,800
Yugoslavia	3,000	—	—
Bulgaria	1,500	—	—
Holland	1,400	6–8,000	600
Belgium	275	3–6,000	1,000

[z] includes blackcurrants
[y] probably not all sold commercially

The bulk of the European redcurrant crop is grown in Germany and Poland, with lesser acreages in Holland, Belgium, Hungary, and France. Redcurrants are of little commercial significance in Britain. Most of the fruit is utilized in juice production.

Most of the European crop of gooseberries is grown in Germany, Poland, and Britain, with lesser amounts in Hungary, France, Belgium, and Holland. About 19,900 ha are grown in the U.S.S.R. Most of the fruit is sold for processing.

Origin and Early Development

Blackcurrant

The European blackcurrant is a comparative newcomer among the temperate fruit crops having been domesticated in northern Europe within the last 400 years. It was introduced into Britain from the Continent, the first published records of it in England being in seventeenth century herbals. Despite its "stinking and somewhat loathing savour" the blackcurrant was valued for its medicinal properties, hence the name "squinancy berry" from its use in treating quinsy (Hedrick, 1925). Thomas Hitt noted in 1757 that a good wine could be made from blackcurrants and that both fruits and leaves were of use in medicine. To the present day, the fruit is valued for its exceptionally high Vitamin C content.

Lindley's (1831) *Orchard and Kitchen Garden* includes only two blackcurrant cultivars, 'Black Naples' and 'Common Black' (Hatton, 1920). The 1838 *Catalogue of Fruits Cultivated in the Gardens of the Royal Horticultural Society* adds four further cultivars: 'Wild', 'Black Grape' (Ogden's Black'), 'Russian Green', and 'Green Fruited'. Hogg (1875) includes a further cultivar, 'Lee's Prolific Black'. In 1925, Bunyard described 15 cultivars, one of which was green-fruited. He considers that the cultivar 'Baldwin' is the original 'Black Naples', noted for its early leafing out.

It seems likely that the European blackcurrant reached America with, or shortly after, the Pilgrim Fathers in 1629 (probably as seed). 'Black Naples' was added to the fruit catalogue of the American Pomological Society in 1852. In 1925 Hedrick described 61 cultivars of *R. nigrum* of which 40 (including 'Climax', 'Clipper', 'Eclipse', 'Kerry', 'Magnus', 'Saunders', and 'Topsy') had been raised in Canada, most of them by Saunders of Ontario. The remainder were of European origin. At this date only six of the cultivars appear to have been at all widely grown in America: 'Baldwin', 'Boskoop Giant', 'Carter's Champion', 'Lee's Prolific', 'Naples' ('Black Naples'), and 'Saunders'. The European blackcurrant is highly susceptible to the white pine blister rust (*Cronartium ribicola*), and the introduction of this fungus into North America on pine seedlings from Europe in (or before) 1892 (Leppik, 1970) led to severe epiphytotics on pine. As a result, the eradication of currants and gooseberries over large areas was enforced in many states to protect white pine forests. In consequence, blackcurrants are now of little commercial significance in North America.

Seven blackcurrant cultivars derived from two native American species, *R. odoratum* and *R. americanum*, are described by Hedrick (1925); the earliest references quoted are 1832 for 'American Black' (*americanum*) and 1867 for 'Deseret' and 'Missouri Black' (*odoratum*). 'Crandall' (*odoratum*), introduced in 1888, was probably the most widely cultivated of this group, particularly as an ornamental, but like the others of the type, was never grown on a commercial scale for fruit.

Further data on the origin and development of blackcurrants are given by Hatton (1920), Hedrick (1925), Zhukovsky (1950), Bagenal (1955), and Raphael (1958).

Red- and Whitecurrant

According to Bunyard (1917a) the redcurrant was unknown to the Greeks and Romans. It was first mentioned in a German manuscript of the early fifteenth century and the plant was illustrated in the *Mainz herbarius* of 1484. In 1542 *R. vulgare* (*R. sativum*) was the only species grown, but in 1561 Konrad Gesner found a bush of *R. petraeum* near Berne, and this species was introduced into England about 1620. *R. rubrum* was brought into the redcurrant complex at a much later date (Bunyard, 1917a; Thayer, 1923); 'Raby Castle', raised about 1820, is thought to be a pure descendant of *R. rubrum pubescens* Swartz while 'Houghton Castle', introduced about 1820, is *R. rubrum* x *vulgare* (Bunyard, 1917*a* and *b*).

According to Hedrick (1925) the redcurrant was probably first cultivated as a common garden plant in Holland, Denmark and the coastal plains about the Baltic. However, Bunyard (1917*a*) and Thayer (1923) consider that the earliest cultivation was in France and adjoining countries, where the first domesticated redcurrant species, *R. vulgare*, is indigenous, and where early herbals show the currant to have been already established in the first half of the sixteenth century. In Britain, Rea in his *Flora, Ceres and Pomona* of 1665 describes the "greatest dark red Dutch curran" and the "well-tasted" whitecurrant; this is assumed by Hedrick to be an early mention of the 'Red Dutch' and 'White Dutch' varieties.

Mawe, in his dictionary of gardening and botany published in 1778, names six varieties: 'Common Small Red', 'Long-bunched Red', 'Champagne Pale-red', 'Common Small White', 'Large Red Dutch', and 'Large White Dutch'. Hogg (1875) adds a further five cultivars, 'Cherry', 'Knight's Early Red', 'Knight's Large Red','Knight's Sweet Red', 'La Fertile', 'La Hâtive', and 'Raby Castle'.

The redcurrant is said to have been introduced into North America by the earliest English settlers in Massachusetts in the early seventeenth century. Hedrick (1925) considers that the cultivars brought in were probably 'Red Dutch' and 'White Dutch'—varieties which are still being grown today. McMahon (1806) names 'Common Red', 'Large Red', 'Pale White Dutch', 'Large White', and 'White Crystal', and notes that currants may be raised from seed and "improved sorts obtained thereby." In a revised edition of *The Fruits and Fruit Trees of America*, Downing in 1857 described 25 varieties of red- and whitecurrants and it is noteworthy that all of these came from Europe (Hedrick, 1925). In 1868 Lincoln Fay raised 'Fay's Prolific' and this was the beginning of an era of development of American-raised cultivars such as 'Eclipse', 'North Star', 'Pomona', 'Red Cross', 'Ruby', and 'Wilder'.

Further details on the origin and development of red- and whitecurrants are given by Bunyard (1917*a* and *b*), Thayer (1923), Hedrick (1919, 1925), and Bagenal (1955).

Gooseberry

According to Rake (1958) the gooseberry was not mentioned by Greek or Roman writers. The earliest record of gooseberry cultivation in Britain occurs in bills of the fruiterer to Edward I dated 1276–92, in which accounts for gooseberry bushes from France are mentioned. Turner, in his *New Herbal* (1548), noted the superiority of the gooseberries grown in English gardens compared with cultivars in other parts of Europe. Descriptions of red- green-, and "blue"-fruited gooseberries appear in Parkinson's *Paradisi in Sole Paradisus Terrestris* (1629) and in John Rea's *Flora, Ceres and Pomona* (1665). Parkinson also names some yellow and white gooseberries, including 'Amber', 'Hedge-hog', and 'Holland'. By 1778, at least 23 cultivars were available (Mawe, 1778, quoted by Hedrick, 1925).

Hedrick notes that Mawe's descriptions are very brief and that some of his "kinds" are probably group rather than varietal names. With the establishment in Britain of gooseberry clubs giving prizes for the heaviest fruit, the number of cultivars rose rapidly and by 1821 a total of 300 varieties were listed in trade catalogues (Rake, 1958), and ten years later 722 were in existence (Lindley, 1831). According to Hedrick, this number had probably run into four figures by 1925.

Growing of gooseberries on a commercial scale in Britain began at the end of the nineteenth century as a result of the stimulus of the abolition of the sugar tax in 1874. Gooseberries were more widely planted than blackcurrant because they were unaffected by gall (big bud) mite (*Cecidophyopsis ribis* [Westw.]). However the appearance of American gooseberry mildew in 1905 soon drastically reduced gooseberry acreages and the crop has never regained its former popularity (Rake, 1958).

European gooseberries brought to America by the early settlers did not thrive owing to their susceptibility to American gooseberry mildew and unsuitable climatic conditions, and this crop has never been widely grown in America. Before the 1850s, small-fruited, mildew-resistant derivatives of native species, such as 'Pale Red', were grown in gardens (Hedrick, 1925). 'Houghton', a hybrid between a European gooseberry and the native *R. hirtellum*, raised in 1833, and its seedling 'Downing', raised in about 1855, were still the most widely grown varieties in America in 1925, probably because of their wide adaptability, ease of propagation, and mildew resistance. Of the large-fruited European cultivars introduced in the nineteenth century, 'Chautauqua' and 'Whinham's Industry' were the most successful. Despite the breeding of new and improved mildew resistant hybrids between European and American sorts in the late nineteenth and early twentieth centuries, gooseberries continued to lose popularity and at present are scarcely grown as a commercial crop.

Further details on the origin and development of gooseberries are given by Hedrick (1925), Pavlova (1935), Darrow (1937), Zhukovsky (1950), Bagenal (1955), and Rake (1958).

History of Improvement

Until about the end of the nineteenth century, the raising, selection, and introduction of new cultivars of soft fruits were largely haphazard and usually the work of private individuals or nurserymen. In the early part of the twentieth century, as the commercial value of these crops grew, research stations financed by public funds or by growers' cooperatives began to appoint full-time plant breeders. At first, worthwhile advances were often made by simple intercrossing between existing local cultivars. As the science of genetics advanced, and the potential of plant breeding was more fully appreciated, breeding programs deliberately and scientifically planned to meet the more exacting needs of a developing industry were initiated. Present-day programs often make considerable use of foreign cultivars and wild species, and exchange of plant material occurs on an international scale.

Blackcurrant

By the nineteenth century, several cultivars of blackcurrant were available including 'Baldwin', 'Black Naples', and 'Black Grape', all of which are still extant. Indeed, 'Baldwin', though over 150 years old, is the leading cultivar in Britain and is also grown on the Continent. During the nineteenth century, the numbers of named cultivars increased and by the early twentieth century, when Bunyard (1925) described 15 cultivars, there was already a remarkable confusion in blackcurrant nomenclature, several distinct types often having the same name, or the same cultivar having a number of different names (Hatton, 1920). Hatton attributes this to the small range of botanical variation in cultivars and to the fact that blackcurrants are easily raised from seed and breed almost true. He was able to classify 26 cultivars into four main groups of similar (or sometimes synonymous) varieties, the types being 'French Black', 'Boskoop Giant', 'Goliath' (or 'Victoria'), and 'Baldwin'.

Almost all the "newer" established western European cultivars, most of them raised in Britain, are derived from a very restricted range of genetic material, 'Baldwin' and 'Boskoop Giant', particularly, having been repeatedly used as parents (Table 2).

A survey of the inheritance of commercial characters in 20 crosses between western European cultivars convinced Tydeman (1938) that the breeder would have to look outside *R. nigrum* for major improvements. From about this time on, increasing use was made both of Scandinavian and Russian wild forms of *R. nigrum* and of related species wherever blackcurrant breeding was in progress.

In Russia the indigenous *R. nigrum sibiricum* and *R. dikuscha* have been widely used as donors of hardiness and disease resistance in crosses with western European cultivars (Mosolova, 1963; Melehina, 1964; Lihonos and Pavlova, 1969). 'Primorskij Čempion' (*nigrum* x *dikuscha*) was one of the best known cultivars so produced. Selection among wild forms of *R. nigrum sibiricum* extended the northern limits of blackcurrant cultivation in the U.S.S.R. to inside the Arctic Circle (Vitkovskij, 1964*a*). In Europe, where Ameri-

TABLE 2. Origin of some present-day West European blackcurrant cultivars

Cultivar	Parentage	Year		Country of origin
		Raised	Introduced	
Amos Black	Goliath x Baldwin	1927	1951	Britain
Baldwin	Unknown	Before 1820		Unknown
Boskoop Giant	Unknown	c. 1885	1895	Holland
Daniels' September	Poss. bud sport Baldwin	c. 1915	1923	Britain
Goliath	Victoria o.p.	Before 1920		Britain
Malvern Cross	Baldwin x Victoria	1920	1946	Britain
Mendip Cross	Baldwin x Boskoop Giant	1920	1933	Britain
Seabrook's Black	Prob. French Black o.p.	Before 1885	1913	Britain
Silvergieters Zwarte	Boskoop Giant o.p.	1926	1936	Holland
Tor Cross	Baldwin x unknown	1924	1962	Britain
Wellington xxx	Boskoop Giant x Baldwin	1913	1927	Britain
Westwick Choice	Baldwin seedling	1913		Britain

can gooseberry mildew, *Sphaerotheca mors-uvae* (Schw.) Berk., hitherto rare on blackcurrants, reached epidemic proportions in the 1960s, Finnish and Swedish cultivars (*R. nigrum scandinavicum*) such as 'Brödtorp', 'Janslunda', and 'Öjebyn', have been increasingly used as donors of resistance (Knight and Keep, 1966; Tamás, 1966*a*; Anderson, 1967). In Germany and Britain, *R. dikuscha* has provided resistance to leafspot (*Pseudopeziza ribis* [Kleb.]), and *R. nigrum sibiricum* and the gooseberry resistance to gall mite (Bauer, 1955; Knight, 1962*a* and *b*; Wood 1962; Anderson, 1967; North, 1968). In Canada, the rust-resistant varieties 'Crusader', 'Coronet', and 'Consort' were selected from the F_1 of *R. nigrum* x *R. ussuriense* (Hunter, 1950*a*, 1955*a*).

In recent years, high costs of production in western Europe, particularly for hand picking, have resulted in increased emphasis on selecting types suitable for mechanical harvesting. 'Black Reward' (Dutch) and 'Invigo' (German) are modern cultivars specially suited to machine picking and the old Canadian variety 'Magnus' is showing promise for this purpose. To reduce hand-picking costs and improve crop potential, longer strigs with a good "handle" are being introduced from species such as *R. bracteosum* and *R. longeracemosum* (Anon., 1963*b*; Knight and Keep, 1964; Wilson and Jones, 1971) (Fig. 1).

At present (1972) the main cultivars in western Europe are still either representatives of Hatton's (1920) four basic groups (e.g., 'Baldwin', 'Goliath', 'Boskoop Giant'); open-pollinated seedlings from them (e.g., 'Silvergieters Zwarte', 'Westwick Choice'); or else derivatives from intercrossing between these groups (e.g., 'Wellington xxx', 'Amos Black', 'Mendip Cross'). Blackcurrant acreages here are tending to decrease, but the impact of disease-resistant, heavy-cropping cultivars emerging from modern breeding programs and of machine picking may reverse this trend in the next few years. In Poland where acreages are increasing, and the main cultivars so far are 'Blacksmith', 'Cotswold Cross', and 'Mendip Cross', there is a swing towards the hardier compact Scandinavian types such as 'Öjebyn' and 'Brödtorp'.

Red- and Whitecurrant

As with blackcurrants, by the end of the nineteenth century, redcurrant nomenclature was in a state of extreme confusion. Thayer (1923) gives a full list of synonyms obtained from a review of the literature and notes that by recording the synonyms, and the synonyms of the synonyms, of 'Houghton Castle', over half the cultivars he listed including one white cultivar are covered. This confusion is characteristic of a crop commonly raised from seed by amateur horticulturalists.

Bunyard (1917*b*) classified redcurrant cultivars from Britain, the Continent, and America into five groups according to the species they most nearly resembled. The 'Raby Castle' group, derivatives of *R. rubrum pubescens*, include 'Raby Castle' itself, raised before 1860, and 'Houghton Castle' (*rubrum* x *vulgare*) introduced ca. 1820. The numerous Versailles group includes descendants of *R. macrocarpum*, which is probably a large-fruited form of *R. vulgare*. 'Versailles' was raised about 1835 in France; 'Cherry' was imported from Italy in 1840; 'Fay's Prolific' (probably 'Victoria' x 'Cherry') was raised in New York in 1868; and 'Red Cross', 'North Star', 'Perfection', and 'Wilder' are other American-raised cultivars of this group. The group as a whole is characterized by large fruits and a tendency to go "blind." The Gondouin group comprises the derivatives of *R. petraeum* which are recognizable by their stout branches and red campanulate flowers. Cultivars in this group are often resistant to leafspot. 'Gondouin' (*petraeum* x *vulgare*), 'Prince Albert' (which has been grown for some 200 years), and 'Seedless Red' ('Kernlose') are included in this group. The Scotch group, descended from *R. rubrum*, have bell-shaped, more or less red flowers, and include 'Scotch', 'La Constante', and 'Moore's Ruby'. The Dutch group have flowers of the *vulgare* type and include 'Dutch' ('Goliath'), 'Utrecht', and 'Laxton's Perfection' (introduced in 1910).

Redcurrant cultivars introduced since Bunyard's (1917*b*) and Thayer's (1923) reviews include the American 'Red Lake' (introduced 1933), of unknown parentage, which is now the leading cultivar in America and is also grown extensively in Europe; the Canadian 'Stephens No. 9' (introduced 1933); and the two Dutch cultivars 'Jonkheer van Tets' (introduced 1941), an early ripening seedling of 'Fay's Prolific', and 'Maarse's Prominent' ('Jonkheer van Tets' x 'Fay's Prolific'). These cultivars are, with the possible exception of 'Red Lake', in direct line of descent from those raised in the previous century.

A new factor in redcurrant breeding in the mid-twentieth century was the use of the eastern

European *R. multiflorum,* with its late-flowering, dense, many-flowered strigs (Fig. 1) and leafspot resistance. The Dutch 'Rondom', and the German 'Heinemanns Rote Spätlese' and 'Mulka', all backcrosses of *multiflorum* to redcurrant cultivars, are characterized by late flowering and ripening and heavy cropping. 'Rondom' is resistant to leafspot.

In Russia, 'Red Dutch' has proved the only cultivar, of many tested, hardy enough for the far north (Vitkovskij, 1964*b*), and it is being used as a parent in breeding for this area. Large-fruited hardy local forms ('Kandalakša', 'Varzuga') have also been selected.

Gooseberry

By the end of the nineteenth century, white, green, yellow, and red gooseberry cultivars were grown in Europe. Selection before this time was largely for fruit size, and unlike currants, gooseberry cultivars now bore very much larger fruits than the wild species. Many present-day commercial varieties were raised before 1900; 'Careless', 'Leveller', 'Lancashire Lad', 'Whitesmith', 'Whinham's Industry', 'Hönings Früheste', and 'Keepsake' are examples.

Unfortunately, these large-fruited cultivars all proved more or less susceptible to American gooseberry mildew and the main aims of European gooseberry breeders in the twentieth century have been to combine fruit size with mildew and leafspot resistance.

The use of the North American small-fruited species *R. divaricatum* as donor of combined mildew and leafspot resistance was pioneered by Bauer (1938, 1955). Repeated backcrossing to large-fruited European cultivars resulted in the introduction in 1953 of 'Resistenta', 'Perle von Müncheberg', and 'Robustenta'. Later mildew-resistant selections with larger fruits and covering the whole range of seasons were 'Remarka', 'Rokula', 'Risulfa', 'Ristula', and 'Reverta'; 'Remarka' is also resistant to leafspot. Other workers have used Bauer's material as a basis for mildew resistance breeding and are also exploring the donor potentialities of additional North American wild species such as *R. lepthanthum* Gray and *R. watsonianum* Koehne (Knight and Keep, 1966; Knight, 1972).

By 1900 in North America, cultivars such as 'Carrie', 'Josselyn' ('Red Jacket'), 'Oregon', 'Poorman', and 'Downing', all 'Houghton' derivatives, mostly with larger fruits than their parent, were already available, and were still listed by Shoemaker (1955) as being among the leading gooseberry cultivars. 'Como' ('Pearl' x 'Columbus') was introduced in 1922. 'Glenndale', from *R. missouriense* x *R. grossularia,*[3] was introduced in 1932, for its adaptation to the southern limit of gooseberry growing. 'Abundance', 'Perry', and 'Pixwell' ('Oregon' x *R. missouriense*), with few thorns and fruits borne in clusters on long pedicels, were also introduced in 1932. 'Welcome', an open-pollinated seedling of 'Poorman' introduced in 1957, like its parent, has few spines. A major advance in breeding for spinelessness at Ottawa was the use of a spineless form of *R. 'oxyacanthoides'* as donor (Davis, 1926), resulting in the introduction of 'Spinefree' in about 1935 and of 'Captivator' in 1952 (Hunter, 1955*b*). Breeding for spinelessness is being undertaken in Britain, using Hunter's *oxyacanthoides* material and the North American *R. sanguineum* as donors (Anon., 1970).

In Russia, as in North America, most progress in breeding gooseberries has resulted from intercrossing large-fruited European types with North American or Russian wild species or cultivars, the main aims being disease resistance, hardiness, and spinelessness. The following hardy mildew resistant cultivars are from European x American crosses: 'Otličnik', 'Smena', 'Pjatiletka', 'Izumrud', and 'Rekord' (Mosolova, 1956; Sergeeva, 1962; Potapkova, 1969). Further backcrosses to large-fruited cultivars to improve fruit size resulted in 'Russkij', 'Malahit', and 'Plodorodnyj', all resistant to mildew, leafspot, and *Mycosphaerella* (Kuznecov, 1969). Kuminov (1962*a*) describes the cultivars 'Pervenec Minusinska' and 'Muromec', both hardy mildew-resistant derivatives of the native *R. aciculare* Sm. x European cultivars.

In a review of breeding for spinelessness, Kovtun (1962) refers to completely or partially spineless forms occurring in inbreds of 'Houghton', and seedlings of 'Finik'. 'Samburskij 24' ('Houghton' x 'Karelles') sheds the spines from fruiting branches and the cultivars 'Afrikanec' and 'Novosibirskij Velikan', both 'Houghton' derivatives, are almost spineless (Kruglova, 1965; Šitakov, 1967).

Breeding Objectives

The primary objective in any plant breeding program is to produce high quality cultivars which are easier and more profitable to grow. Direct reduction in costs of production can be brought about by reducing the need for expensive hand labor, and by obviating the need for spray control of pests and diseases. Profits can be increased by higher yields of better quality fruit, better adapted to commercial requirements.

Blackcurrant

In blackcurrants in Britain, hand-picking costs represent about 40% of the gross returns, and this figure is probably typical of western Europe generally. Cultivars suited to mechanical harvesting both in fruiting and growth habit, or plants which are easier and quicker to hand pick, are therefore important breeding objectives of almost all west European blackcurrant breeding programs. Spraying to control gall mite, leafspot, and American gooseberry mildew is routine wherever these pests and diseases occur, while sprays to control aphids, blackcurrant leaf midge (*Dasyneura tetensi* Rübs.), capsid (*Lygocoris pabulinus* [L.]), blackcurrant sawfly (*Nematus olfaciens* Benson), blister rust, and botrytis are applied as required. Sources of resistance to all these organisms except capsid have been reported and in all cases resistance breeding is in progress. Reversion, the only known virus of commercial importance, is catered for by breeding for direct resistance and for resistance to the gall mite vector.

The steady loss of popularity of blackcurrants in western Europe is partly due to irregular cropping. This is caused by the early-flowering habit of most cultivars, which often results in loss of crop through spring frost damage or absence of pollinating insects at flowering time. Spring frost resistance or avoidance and an inherently better cropping capacity are both required to make blackcurrants a more reliable commercial proposition, and work to incorporate these characteristics is in progress.

Adaptation to commercial requirements is primarily a matter of suitability for juice production and a high Vitamin C content. The Vitamin C content of many cultivars is known, and selection of parents and progeny for high content is sometimes practiced. In Russia the extreme hardiness of some blackcurrant and related *Ribes* species is being utilized to extend commercial blackcurrant growing to areas too cold for other fruits.

Red- and Whitecurrant

Planned redcurrant breeding is less common and on a smaller scale than that of blackcurrants, although objectives are generally similar. There is rather less emphasis on mechanical harvesting, since hand-picking costs are lower. Resistance to leafspot is a common objective, whereas breeding against American gooseberry mildew on a systematic basis has not yet been reported, although some redcurrant cultivars are susceptible and others are highly resistant and should prove good donors. Deliberate breeding for resistance to the two most serious viruses, spoon leaf and gooseberry veinbanding, has yet to be undertaken.

Aphids, particularly *Cryptomyzus ribis* (L.) and *Hyperomyzus lactucae* (L.), are probably the most serious pests of redcurrants in western Europe. No planned breeding for resistance to these pests appears to be going on. The same applies to the less damaging or less common pests, capsid, gooseberry and blackcurrant sawflies (*Nematus ribesii* [Scop.] and *N. olfaciens*), currant clearwing moth (*Synanthedon* [*Aegeria*] *tipuliformis* [Clerck]), the redcurrant gall-forming mite *Cecidophyopsis selachodon* van Eynd., and the so-called blackcurrant gall mite, which kills redcurrant buds without gall formation.

Redcurrants are generally less liable to spring frost damage than blacks and more regular in cropping so that frost resistance or avoidance is of somewhat lower priority in this crop. Wind resistance, however, is important, as some redcurrant cultivars are liable to branch breakage in high winds. As already described, *R. multiflorum* is being used as donor for increased crop potential.

Gooseberry

Since much of the gooseberry crop is picked when green and hard, suitability for mechanical harvesting should be easier to achieve than with currants. Resistance to leafspot and to American gooseberry mildew continue to be major objectives. Breeding for resistance to dieback, caused by *Botrytis cinerea* Pers. and *Botryosphaeria ribis* Gros. and Dug., is a more or less automatic selection process. Deliberate breeding for resistance to the ubiquitous gooseberry veinbanding virus and to its aphid vectors, *Nasonovia ribisnigri* (Mosley) and *Aphis grossulariae* Kltb., has been reported fairly recently. Another important pest, necessitating spray control, is the gooseberry sawfly *Nematus ribesii*. Differences in susceptibility to this pest occur in species of the gooseberry subgenus, and at least one small-scale resistance breeding project is in operation.

Spinelessness continues to be a major objective.

Breeding Techniques

Floral Structure and Biology

Apart from Berger's dioecious subgenera Berisia and Parilla, *Ribes* are perfect flowered, with racemose inflorescences bearing from two to about 70 flowers. Very rarely, flowers are solitary. The flowers are usually 5-merous, the calyx tube being cylindric to rotate and, like the sepals, usually colored. The petals are generally smaller than the sepals, and sometimes absent. The stamens may be shorter or longer than the sepals. The ovary is inferior, one-celled, and many-ovuled with two more or less connate styles. The fruit is a berry, usually with many seeds.

Pollen morphology of numerous *Ribes* species is described in great detail by Agababyan (1963) and further data are given by Wills (reported by Haskell, 1963, 1966). Pollen fertility in cultivars is usually high, judged on acetocarmine staining; about 95% of fertile grains were seen in seven European blackcurrant, six redcurrant, and five gooseberry cultivars; the only exception was 'Red Dutch' currant, with about 40% sterile grains (Fernqvist, 1961).

The blackcurrant raceme which is borne on year-old and older wood is pendulous and, according to cultivar, often compound with a primary and one or more secondary racemes (Baldini and Pisani, 1961). In Sweden over a four-year period, average numbers of flowers per inflorescence in western European cultivars ranged from 6.6 to 9.6 (Fernqvist, 1961). At East Malling in 1969, total numbers of flowers per node, based on an average of five nodes, ranged from 12.2 for 'Blacksmith' to 27.2 for 'Greens Black', the latter high figure being due to large numbers of fairly short inflorescences.

Broadly, two types of flowers occur; in one the stigma protrudes beyond the anthers, and in the other the stigma is just below the anthers (Hatton, 1920; Baldini and Pisani, 1961). The 'Baldwin' group has 99% of its flowers with anthers and stigmas more or less level, while 'Boskoop Giant' shows 57% of this type and 42% with a protruding stigma. In 'French Black' and 'Goliath', these percentages were 48:50 and 79:20 respectively (Wellington et al., 1921). All groups showed 1-2% of flowers with the style shorter than the stamens, and in all groups the style was shortest in the basal flower of the raceme and tended to lengthen progressively towards the tip. According to Tamás and Porpáczy (1967) numbers of ovules per fruit decreased towards the tip of the raceme.

Blackcurrant pollen is sticky and wind pollination does not occur. Wellington et al. (1921) showed that all types except the 'Baldwin' group, (which, with anthers and stigmas level, is capable of self pollination) rely on insect pollination for cropping.

Numerous workers have investigated flower initiation in blackcurrants, the first primordia being seen from mid-late May (Pisani, 1962; Grbić and Vujanić-Varga, 1968) to late August (Nasr and Wareing, 1961a) and September (Hårdh and Wallden, 1965) according to locality, season, and cultivar studied. In Majkop, flower differentiation began 30 to 36 days earlier than in the same cultivars grown in Leningrad (Vitkovskij et al., 1970). Flower initiation begins in the lower middle nodes and then proceeds acropetally; in southeast England, in 'Wellington xxx', primordia were present in all except the apical bud by mid-August (Thomas and Wilkinson, 1962a). When flower primordia were seen in June and July, all floral organs were differentiated before the winter (Rudloff and Lenz, 1961). The order of differentiation of floral organs in blackcurrant, as in gooseberry, is sepals → stamens → petals → carpels, and microsporogenesis precedes macrosporogenesis (Komar, 1970). Meiosis takes place in spring about a month before the flowers open. In southeast England, flowering occurs in April and May, the exact season varying considerably

A
B
C
G
H

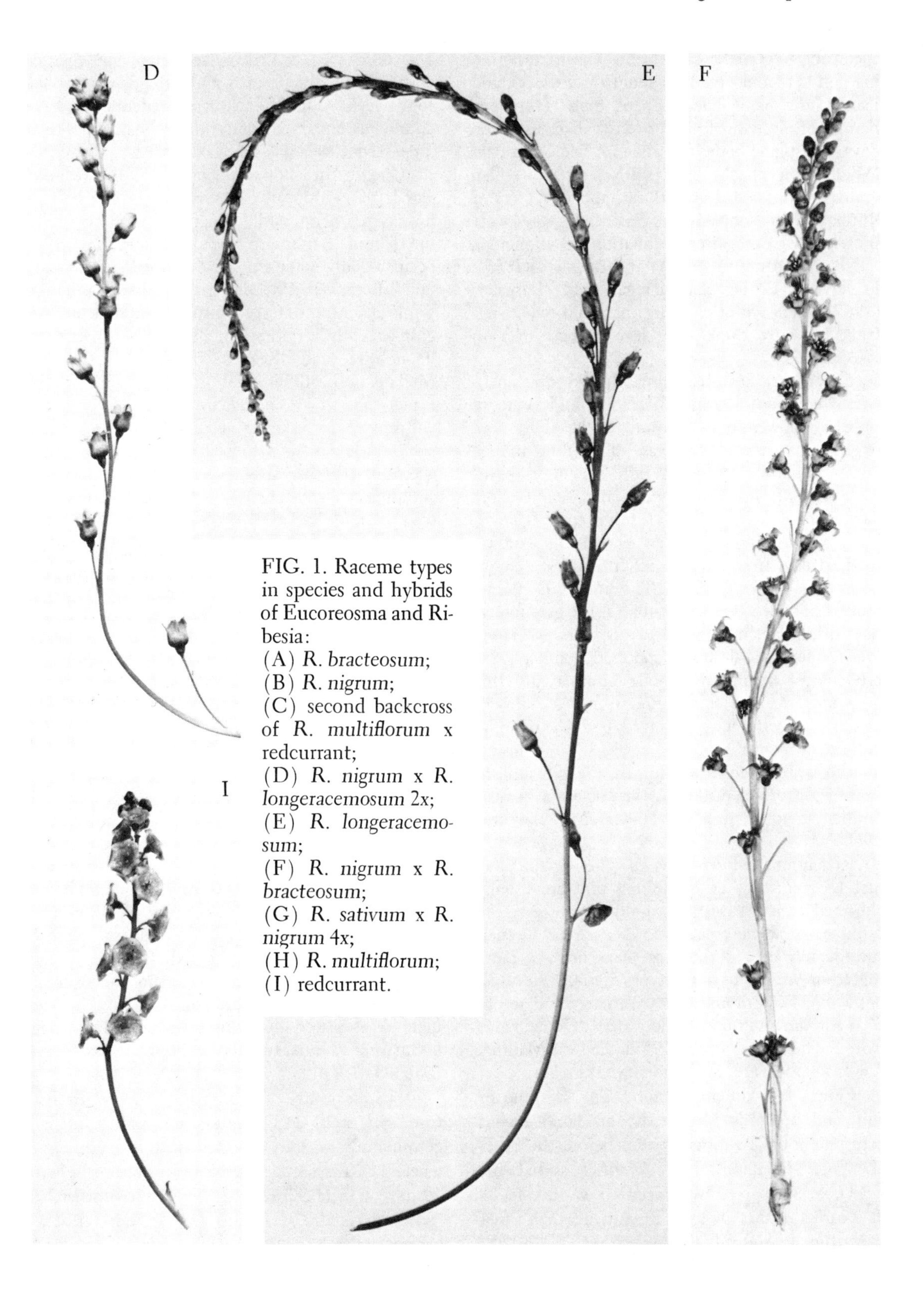

FIG. 1. Raceme types in species and hybrids of Eucoreosma and Ribesia:
(A) *R. bracteosum*;
(B) *R. nigrum*;
(C) second backcross of *R. multiflorum* x redcurrant;
(D) *R. nigrum* x *R. longeracemosum* 2x;
(E) *R. longeracemosum*;
(F) *R. nigrum* x *R. bracteosum*;
(G) *R. sativum* x *R. nigrum* 4x;
(H) *R. multiflorum*;
(I) redcurrant.

according to climatic conditions. The total flowering period of a raceme lasts about three weeks, and usually the basal flower opens first. The open period of individual flowers lasts five to seven days, the stigma being receptive for five or six days (Wellington et al., 1921). Blackcurrant pollen grains are two-celled at anthesis, and in a cross of 'Junnat' x 'Boskoop Giant', fertilized ovules were present two days after pollination (Radionenko, 1970). According to Teaotia and Luckwill (1955), Fernqvist (1961), and Tamás and Porpaćzy (1967), there are often more than 100 ovules per fruit, but only 25–30% or less of these develop into seeds.

Several workers have investigated photoperiodic effects on flower initiation. Nasr and Wareing (1961*b*) showed that exposure to 16 short days (eight hour photoperiod) was sufficient to induce flower initiation, providing fully expanded leaves were present; in decapitated plants, eight short days were sufficient. On returning to long days, many axillary buds expanded, and plants exposed to short days from 21 June set fruit in September (Nasr and Wareing, 1958). Tinklin (Ph.D. thesis, Univ. London, 1968) found that in the greenhouse the critical daylength was 16 to 16½ hours. However, "ripeness-to-flower of shoots depends on the presence of a minimum node number and this over-rides the daylength effect; in 'Wellington xxx' shoots with fewer than 12 nodes cannot be induced to flower" (Tinklin et al., 1970). Flowering can be induced in autumn by a single application of gibberellic acid (GA) at 100 ppm in September, October, and November (Modlibowska, 1960).

Redcurrant inflorescences may be horizontal at first (*R. rubrum*), or pendulous, and are usually clustered at the base of year-old shoots. Average numbers of flowers per inflorescence in Sweden considerably exceed those of blackcurrants, ranging from 9.2 to 13.6 for types derived from *R. petraeum*, *R. rubrum* and *R. sativum* and up to 27.6 for the *multiflorum* derivative 'Heinemanns Rote Spätlese' (Fernqvist, 1961). At East Malling, flower numbers are usually somewhat higher.

Flower bud differentiation begins in June or July and proceeds acropetally, all floral organs being more or less differentiated before the onset of winter (Genevès, 1958; Rudloff and Lenz, 1961). At Majkop, flower initiation was five to 15 days earlier in three redcurrant cultivars than when grown in Leningrad (Vitkovskij et al., 1970). Flowering starts a little after that of gooseberries and before blackcurrants, and extends over about four weeks. On the average, redcurrants have approximately 25 ovules, from which about five large seeds develop (Fernqvist, 1961).

Gooseberries flower on one- and two-year-old wood, the European cultivars having one or two flowers per node, while American types may have up to four. At Alnarp, Sweden, flower initiation occurs about the end of July, two weeks after currants (Fernqvist, 1961), whereas in Estonia, flower primordia were not seen until the middle of August for early varieties and the end of August for 'Houghton' and 'Rekord' (Sarapuu, 1966). In Majkop, flower differentiation began 30 to 36 days earlier than in Leningrad, in the same gooseberry cultivars (Vitkovskij et al., 1970). Meiosis takes place about a month before flowering, which occurs a little earlier than in currants. As in blackcurrants, only 25 to 30% of the 100 or so ovules develop into seeds at Alnarp.

Emasculation and Pollination

Flowers of blackcurrant and gooseberry are emasculated in the bud stage with forceps or by cutting round the calyx tube with a sharp scalpel or with a hook sharpened on both sides, just above the ovary, so removing anthers, petals and sepals in one. With the smaller, crowded redcurrant buds, it is sometimes easier to remove anthers and sepals with fine forceps. In currants, fruit set is usually improved by removing the tip of the inflorescence.

From one to three days after emasculation, pollen is applied to the stigma with a fine paint brush or by touching it with a newly dehisced anther. A paint brush is recommended for crossing redcurrants and for selfing all species, although Wilson (1964*a*) found that branches of blackcurrants which were shaken during flowering set almost as well as those brush-pollinated. In areas where late spring frosts are common, crossing is best done on pot plants in insect-proof glasshouses. In the field, emasculated buds must be protected from pollinating insects by, for example, waxed bags fastened with plastic wire.

Data on storage of *Ribes* pollen are few. In diffuse light, at 20–24 C, blackcurrant pollen lost its germinating capacity by the 10th to 13th day, whereas in dry storage, it proved viable for 25 to 30 days with H_2SO_4, and for 40 days with $CaCl_2$ (Rainčikova, 1967). Adams (1916) found blackcurrant pollen was inviable after 11 weeks dry storage. *In vitro* germination of pollen was reduced from about 75% to less than 2% by spraying open blackcurrant flowers with lime sulphur or thiodan (Anon., 1963a). Fernqvist (1961) and Wills (reported by Haskell, 1962) obtained good germination of pollen of *Ribes* species in 15% sucrose solution.

Attempts to increase the effectiveness of pollination include the use of mixed pollen, of ultraviolet and gamma irradiation and of GA. Russian workers have commonly used pollen mixtures, particularly in crosses between American mildew-resistant gooseberries and European types (Kruglova, 1965). Tolmačev (1940) obtained seed from the cross of 'Crandall' (*R. odoratum*) x gooseberry only when mixtures of pollen were used; pollen of the individual gooseberry cultivars gave no set. Self seed set was induced in the self-incompatible *R. odoratum* and *R. bracteosum* by using pollen UV-irradiated for two or four min at 14,000 ergs/cm^2/sec; similar treatment failed for *R. sanguineum*, *R. multiflorum* and *R. longeracemosum* as did X-irradiation (20 r) of pollen of *R. longeracemosum* (Arasu, 1968). Gamma irradiation of pollen with 100-800 r improved seed set in the cross of blackcurrant x *R. americanum*, and also enhanced the germination rate of the resulting seeds in blackcurrant x redcurrant crosses (Melehina, 1968). Čuvašina (1961, 1962) improved seed set in crosses of blackcurrant x redcurrant and x *R. aureum* by treating stigmas with 0.02–0.03% GA solution before pollination. Fruit set in blackcurrant x whitecurrant crosses and reciprocals was much enhanced by applying 250 mg/liter and 500 mg/liter solution of GA to flowers before pollination (Čuvašina, 1963). Zatykó and Simon (1964) also used GA to reduce fruit drop and so improve seed set in incompatible or inbred cultivars; with inbred plants from 'Boskoop Giant', the number of viable seeds was increased five-fold.

Seed Harvesting, Cleaning, and Germination

Fruits are harvested when ripe and the seeds extracted as soon as possible, before the fruit dries up. At East Malling, swift and efficient currant seed cleaning is achieved by five to ten second treatment in water in an Ato-Mix laboratory blender and subsequent decanting of pulp and water. With gooseberries, seeds are best extracted from the fruit before cleaning in the blender. Morrow et al. (1954) recommend 15 to 45 seconds in a Waring blender for berry fruits.

Providing they are sown soon after extraction, blackcurrant seeds usually germinate in a week or two, although in the absence of prechilling or damage to the seed coat, alternating temperatures are essential for germination (Adam and Wilson, 1967). These workers obtained good germination by alternating treatments of 12 to 16 hours at 24–30 C with 8 to 12 hours at 4–15 C. Higher germination, especially in dried seed, was induced by 12 to 16 weeks prechilling at 4 C. Kronenberg and Hofman (1965) obtained better germination by sowing whole fruits rather than cleaned seed. Redcurrant and gooseberry seeds both require chilling. At East Malling this is done by sowing the seed in seed boxes and placing them in a cold store at about 3 C for three to four months. Alternatively, moist chilling in sand or a similar medium in a refrigerator is equally effective. Some *Ribes* species will germinate at very low temperatures and towards the end of their cold period, seed boxes or containers should be inspected regularly. In areas with severe winters, adequate chilling would be obtained by overwintering the seed boxes in a cold greenhouse.

Germination percentages from reciprocal crosses of blackcurrant cultivars sometimes differ markedly (Wilson and Campbell (*née* Adam), reported by Luckwill, 1969); their data are reproduced below:

Germination percentages

	Pollen parent				
Seed parent	Tor Cross	Roodknop	Baldwin	Amos Black	Means
Tor Cross	13.4	6.7	48.2	60.8	32.3
Roodknop	11.3	0.8	13.9	39.5	16.4
Baldwin	10.1	8.7	27.4	22.8	17.2
Amos Black	52.3	40.7	59.8	50.0	50.7
Means	21.8	14.2	37.3	43.3	

Seedling Growth and Cropping

In southeast England, seedlings potted in late March or early April at about the two-rough-leaf stage and grown with gentle heat if necessary in greenhouse or frames will be large enough to plant in the field about three months later. According to subsequent growth in their first year, these plants will usually bear their first crop in one or two years' time. The period between seed sowing and cropping can be considerably reduced by planting the seedlings out in the greenhouse and giving long day and high temperature treatments; Karnatz (1969) outlines such a method for obtaining fruiting blackcurrant plants one year after seed sowing.

Use of Colchicine and Mutagens

Colchicine has been widely used by *Ribes* breeders, especially in the production of fertile allotetraploids from wide species crosses. Mature plants are highly resistant, repeated applications of fairly concentrated solutions being required. Nilsson (1949) produced one tetraploid branch on a blackcurrant-gooseberry hybrid after colchicine treatment over several years. Similarly, repeated treatments were needed to induce tetraploidy in a hybrid of *R. nigrum* x *R. longeracemosum* (Anon., 1960). Bauer (1955) obtained amphiploids by treating terminal buds with 1% colchicine, while Žironkin (1965) treated axial buds with 0.5% colchicine solution in castor oil.

Treatment of young seedlings with 0.5–1.0% colchicine solution has been more successful. Knight and Keep (1957) obtained 4x blackcurrant-gooseberry hybrids by applying 0.6% colchicine solution three or four times daily for several days to wisps of cotton wool placed on growing points of seedlings at the cotyledon stage. Nilsson (1966) describes colchiploids of a number of species and hybrids obtained by colchicine treatment of germinating seeds and young seedlings. Application of a mixture of 0.8% colchicine and 50 ppm GA produced 27% of mixoploid blackcurrant seedlings, whereas colchicine alone was ineffective (Zatykó and Simon, 1964).

Mutagenic effects of colchicine on *Ribes* have been reported by Vaarama (1949) who found meiotic (but not mitotic) irregularities in blackcurrant plants which remained diploid after colchicine treatment. At M_1 spindles were sometimes split, incompact, or unipolar. Abnormal chromosome behavior included non-orientation of centromeres, and non-congression on the metaphase plate; deletions, translocations, fragments, multivalents, and univalents were seen. Vaarama concluded that colchicine causes permanent changes in the chromosomes, the centromere being chiefly affected. Colchicine-induced morphological mutants of redcurrants were obtained by Zeilinga (Anon., 1963*b*). Meiotic irregularities, including unreduced microspores, were seen in the blackcurrant cultivar 'Pamjat' Micurina' following seven days' exposure to acenaphthene vapor (Čuvašina, 1968).

X-irradiation of blackcurrant material has produced a wide range of mutations. From 1949 to 1953, Bauer (1957) treated a total of 343 dormant cuttings of ten cultivars with 3,000 r and obtained 324 different selections. Increased vigor was shown by 9% of the material; shortened internodes by 4%; longer strigs, 4%; heavier strigs, 9%; larger fruit, 5%; reduced running off, 3%; better flavor, 7%; later ripening, 1%; and moderate resistance to *Cronartium*, 2%. The frequency of mutants in which only positive characters were combined was very small, but represented twice the proportion of outstanding selections obtained from selfing or intercrossing the varieties over the same period. Following x-ray treatment of cuttings and seedlings, Kaplan (1953) recorded variation in leaf color, shape, size, and pose, and in growth habit, internode length (both shorter and longer), inflorescences (longer), yield (increased and decreased), and, in 'Goliath', better flavor. Undesirable changes included greater disease susceptibility and reduced numbers of seeds per fruit. Gröber (1967) reported a significant increase in ascorbic acid content of fruits following irradiation. So far only one mutant resulting from x-irradiation has been introduced into cultivation; this is the cultivar 'Westra' (from 'Westwick Choice') which has a compact habit and is very suited to mechanical harvesting by shaking (Bauer, 1970).

Treatment of dormant and germinating seeds of blackcurrant with x-ray doses of 1,000–3,000 r and gamma doses of 36,000 r resulted in abnormal growth of cotyledons and leaves and nanism; gamma dosage of more than 72,000 r was lethal (Cociu et al., 1966). Irradiation of dormant cuttings with 2,000–3,000 r gamma rays and with 300–900 r fast neutrons induced variability in leaf and shoot characters and bush habit in four cultivars, with 0.7–6.7% of the morphological changes transgressing the specific limits of the blackcurrant (Melehina, 1966). The proportion of these and other mutants was greater in recent hybrids than in old established cultivars. The mutagenic effect of fast neutrons was greater than that of x-rays, and cultivars differed in sensitivity to different exposures of both mutagens.

The results of gamma irradiation and chemical mutagen treatments of blackcurrants are described by Nybom and Bergendal in the Balsgård Fruit Breeding Institute's annual reports for the years

1959–70. More or less homogeneous mutations were isolated by repeated decapitation of irradiated plants, the shoot apex plus three to four leaves being rooted under a mist propagation unit. "Drastic" and "micro-mutations" were found after irradiation and after chemical treatments (report for 1965); chemicals used were EMS (ethyl methane sulphonate) (report for 1961), MNU (nitroso methyl urea), and NG (methyl nitro nitroso guanidine) (report for 1964). Mutations described include early and late leafing, changes in time of flowering, variation in seed set, fruit size, and numbers of fruits per cluster. However, no outstanding mutants were obtained (report for 1969–70).

Breeding Systems

Self-fertility and Self-sterility

Relatively few *Ribes* species have been assessed for self-fertility. The available data are summarized in Table 3, the species being arranged according to Berger's (1924) classification.

No fruit was set by any of the ten species self-pollinated by Offord et al. (1944) and by Keep (unpub.) and no seed was obtained from ten out of 11 selfings made by Arasu (1970); Arasu's exceptional plant (one accession of *R. bracteosum*) set an average of 3.7 seeds per fruit, compared with 25 to 30 from intercrossing different accessions.

Some of the data in Table 3 are based on results from selfing only one accession, and certainly cannot be considered as representative of the species as a whole. Regarded as a random sample of the genus, however, the data suggest that self-sterility (self-incompatibility) predominates in *Ribes* in the wild.

Among blackcurrant cultivars, only two well-authenticated cases of complete self-sterility on a par with that shown by wild *Ribes* species are known. A rogue of 'Invincible Giant Prolific' set no fruit at all on selfing (Luckwill, 1948) and

TABLE 3. Self-compatibility in *Ribes* species

Subgenus	Species	Self-compatibility	Authority
Ribesia	*R. longeracemosum* (1)	incompatible	Arasu, 1970
	R. multiflorum (1)	incompatible	Arasu, 1970
Eucoreosma	*R. bracteosum* (2)	incompatible	Arasu, 1970
	R. dikuscha	compatible	Voluznev, 1966; Keep, unpub.
	R. dikuscha	incompatible	Voluznev, 1968
	R. fontaneum	compatible	Lihonos & Pavlova, 1969
	R. nigrum sibiricum	incompatible	Kuminov, 1965; Voluznev, 1968
Symphocalyx	*R. aureum* (1)	incompatible	Arasu, 1970
	R. odoratum (3)	incompatible	Arasu, 1970
Calobotrya	*R. glutinosum*	incompatible	Offord et al., 1944
	R. nevadense	incompatible	Offord et al., 1944
	R. sanguineum (2)	incompatible	Arasu, 1970
	R. viscosissimum	incompatible	Offord et al., 1944
Grossularioides	*R. lacustre* (1)	incompatible	Keep, unpub.
Hesperia	*R. menziesii* (1)	incompatible	Arasu, 1970
	R. roezlii	incompatible	Offord et al., 1944
Eugrossularia	*R. alpestre* (1)	incompatible	Keep, unpub.
	R. gracile (1)	slightly compatible	Colby, 1927
	R. grossularioides (1)	incompatible	Keep, unpub.
	R. inerme (1)	incompatible	Keep, unpub.
	R. niveum (1)	incompatible	Keep, unpub.
	R. non-scriptum (1)	incompatible	Keep, unpub.

No. in brackets = no. of accessions tested, where known

a form of 'Lee's' also failed to set fruit in the absence of foreign pollen (Ledeboer and Rietsema, 1940). 'Noir de Bourgogne' is highly self-sterile, but does set a few fruits with self pollen (Lantin, 1970).

Wellington et al. (1921) found cultivars in the four main western European blackcurrant groups ('Boskoop Giant', 'French Black', 'Baldwin', and 'Goliath') to be self- and cross-fertile, and Voluznev (1966) refers to highly self-fertile European forms. In Sweden, Fernqvist (1961) also considered four such cultivars ('Goliath', 'Silvergieters Zwarte', 'Wellington x', and 'Wellington xxx') to set almost as well with artificial self-pollination as with open-pollination and artificial cross-pollination. In France, the Dutch cultivar 'Black Reward' proved highly self-fertile (Lantin, 1970). In Britain, natural cross-pollination in the field did not increase the crop of 'Baldwin', 'Cotswold Cross', 'Westwick Choice', 'Wellington xxx', 'Davison's Eight', 'Mendip Cross', or 'Seabrook's Black' (Free, 1968), while cross-pollination by honey bees of 'Mendip Cross' on 'Baldwin' also failed to increase the crop (Hughes, 1966).

In Italy, however, detailed comparisons of percentage fruit set following artificial self- and cross-pollination led Baldini and Pisani (1961) to conclude that although 'Amos Black', 'Goliath', 'Silvergieters Zwarte', and 'Westwick Choice' were highly self-compatible, 'Baldwin Hilltop', 'Boskoop Giant', and 'Wellington xxx', were only moderately so, fruit set being significantly improved by cross-pollination with certain cultivars. 'Roodknop' and particularly 'Mendip Cross' were rather less self-fertile; cross-pollination with all cultivars tried improved fruit set considerably although some pollen parents were more effective than others.

Tamás and Porpáczy (1967) assessed self-fertility by the percentage of fertilized ovules five to seven days after pollination. 'Amos Black' and 'Brödtorp' were highly self-fertile, with about 60% of embryos developing, while 'Mendip Cross' and 'Wellington xxx', were much less self-fertile, with only about 30% of embryos starting to develop. With suitable pollinators, the number of viable embryos could be increased to about 80% in some cultivars, but from 40–95% of these (according to parentage) failed to mature into seeds. In the European group, self-incompatible types were more influenced by the choice of pollen parent than were the self-compatibles. In general, Tamás (1964) considers northern European types to be more self-fertile than the west European, and in the case of 'Brödtorp' this high self-fertility is transmitted to its offspring (Tamás, 1966*b*).

Other workers reporting increased yield in some cultivars following cross-pollination include Voluznev (1948), Neumann (1955), Hofman (1963), and Lantin (1970). The need for bee pollination, relatively greater in cultivars with styles projecting beyond the anthers, is emphasized by Schanderl (1958) and Hughes (1966). Klämbt (1958) considered that with open-pollination, blackcurrants were usually selfed in contrast to redcurrants.

The Canadian rust-resistant cultivars 'Coronet' and 'Crusader' (*R. nigrum* x *R. ussuriense*) are, under most conditions, unable to pollinate themselves or one another. 'Crusader' pollen tends to stick together in clumps and bees do not easily pick it up. The flowers of 'Coronet' do not open widely at the top; bees often reach in from the side without contacting the anthers (Anon., 1950*b*). Tamás (1966*b*) found that some cultivars were totally cross-incompatible with 'Coronet', but when the latter was pollinated by fully compatible types, up to 90% of ovules were fertilized resulting in a 50–100% increase in berry size as compared with open pollination.

Of Russian cultivars, Kuminov (1962*b*) reports that all those with European cultivars or the self-fertile 'Primorskij Čempion' (*R. nigrum* x *R. dikuscha*) in their parentage are highly self-fertile, whereas those bred from *R. nigrum sibiricum* set few or no fruits on selfing. The self-fertile 'Golubka' is a cross of 'Saunders' x 'Primorskij Čempion'; other self-fertile derivatives of *R. dikuscha* are 'Koksa', 'Zoja', 'Družnaja', and 'Dymka' (Potapenko, 1966). Good pollinators for some Siberian cultivars included 'Sinjaja', 'Družnaja' and 'Nočka' (Kuminov, 1962*b*). Selections from crosses of European cultivars with *R. nigrum sibiricum* are basically self-sterile; these include 'Lošickaja', 'Barhatnaja', 'Mečta', 'Belorusskaja Pozdnjaja', and 'Minskaja' (Voluznev, 1968). Cultivars derived from the Siberian wild currant *R. cyathiforme* are also more or less self-sterile (Potapenko, 1966).

Volodina (1964) reported that in the Leningrad district, high yields depended directly on high self-fertility and capacity for self pollination, while Osipov (1968) found that productivity was related to self-fertility and spring frost resistance, self-sterile types and a few partially self-sterile types cropping irregularly. Thirteen self-fertile cultivars recommended by Osipov as the most regular croppers are: 'Altajskaja Desertnaja', 'Černaja Grozd', 'Černaja Lisavenko', 'Družnaja', 'Golubka', 'Koksa', 'Moskovskaja Rannjaja', 'Nina', 'Pobeda', 'Primorskij Čempion', 'Stahanovka Altaja', 'Uspekh', and 'Zoja'.

Hooper (1911) reported that red- and whitecurrants fruited well under self-fertilization and Blake et al. (1913) found 'Wilder', 'Pomona', and 'Red Cross' to be self-fertile. Fruit set in four redcurrant cultivars was approximately 55% following both open and artificial self pollination (Fernqvist, 1961). Seljahudin and Brózik (1967), however, found that self pollination considerably reduced fruit set from the 80–90% obtained from open pollination, with only 'Heinemanns Rote Spätlese' setting more than 60% with self pollen. In contrast, Klämbt (1958) considers 'Heinemanns Rote Spätlese' to be partially self-sterile, together with 'Heros', 'Rote Hollander' ('Red Dutch'), and 'Rote Vierländer' ('Erstling aus Vierlanden'). With open pollination he found redcurrants were usually cross fertilized, in contrast to blackcurrants.

Like red- and blackcurrants, gooseberries are, in general, self-fertile (Hooper, 1911; Colby, 1927; Smith and Bradt, 1965), although several workers obtained considerably better fruit set with open than with self pollination in some cultivars (Loginyčeva, 1958; Fernqvist, 1961; Sarapuu, 1963). Of ten cultivars, 'Zöld Óriás' was the only one with more than 50% set on selfing; open pollination increased fertilization by 30–50% (Seljahudin and Brózik, 1966). Sarapuu (1963) considered 'Otbornyj Leba', 'Houghton', and 'Pervenec Polli' to be wholly self-fertile. Hughes (1962) obtained moderate crops from gooseberries caged to exclude insects, although in these circumstances the berries had few or no seeds and the fruits were rather smaller than usual.

The possibility of introducing self sterility from donor wild species is exemplified by the American cultivar 'Perry' ('Oregon Champion' x *R. missouriense*) which, unlike its sibs 'Pixwell' and 'Abundance', is totally self-sterile (Yeager, 1938).

Pollen Tube Growth

Self sterility in the rogue 'Lee's' and 'Invincible Giant Prolific' was due to failure of fertilization, although pollen tubes reached the ovules (Ledeboer and Rietsema, 1940; Luckwill, 1948). Similarly in the seven self-incompatible species studied by Arasu (1970), fertilization did not normally occur although pollen tubes were not inhibited in the style, and sometimes penetrated into the ovules. In these species the styles are hollow, and Arasu, following Brewbaker, attributed the absence of stylar inhibition to this characteristic. Pollen tube growth in 'Baldwin', 'Wellington xxx', and 'Victoria' was similar following self and cross pollination, but there was some indication that penetration of the ovary was slower and the number of fertilized ovules lower in selfed flowers (Williams and Child, 1963).

Effect of Inbreeding

According to Kronenberg and Hofman (1965), selfed progenies of 'Baldwin', 'Daniels' September', 'Seabrook's Black', and 'Wellington xxx' showed marked inbreeding depression, whereas that of 'Consort' (*R. nigrum* x *R. ussuriense*) did not. In general, inbreeding depression in the S_1 of western European cultivars was characterized by a decrease in self fertility and an increased sensitivity to diseases, particularly to mildew; in S_2 the increase in susceptibility was still more pronounced (Tamás, 1966*a*). Inbred progenies of 'Wellington xxx', particularly, showed increased disease sensitivity.

Tydeman (1930) studied S_1 and S_2 progenies of 'French Black', 'Seabrook's Black', 'Siberian', 'Boskoop Giant', 'Goliath', and 'Baldwin'. In S_1, cropping was poor, and in S_2 crops were negligible or nil. Unfortunately, the conditions under which the seedlings were growing were "somewhat unsatisfactory," the site being exposed on all sides, and Tydeman was unable to say how far poor cropping was due to this rather than to inbreeding depression. However, Spinks (1947) found no heavy-cropping seedlings in selfed progenies of 'Baldwin', 'Boskoop', and 'French [Black]', and concluded from these and other selfed progenies that no outstanding plants were likely to be produced by "selfing" well-known cultivars.

A detailed study on the effects of increasing heterozygosity in blackcurrants was made by Wilson (1970) who selfed and intercrossed 'Amos Black' ('Goliath' x 'Baldwin'), 'Malvern Cross' ('Baldwin' x 'Victoria'), and 'Mendip Cross' and 'Wellington xxx' (both 'Boskoop Giant' x 'Baldwin'). 'Mendip Cross' and 'Wellington xxx' are sibs, as are 'Amos Black' and 'Malvern Cross', if it can be accepted that 'Goliath' and 'Victoria' are synonymous (Todd, 1962). Since 'Baldwin' was a parent common to both groups, there were four half-sib relationships.

Wilson summarized the effects of inbreeding as follows:

> By comparing progenies grouped according to whether they arose from selfing, crossing of sibs or crossing of half-sibs it was possible to investigate the effects of different levels of heterozygosity on character expression. There were no apparent effects on bud-burst, time of flowering, number of flowers per strig, fruit maturity or mildew tolerance. . . . In bush vigour there was a definite progression to higher ratings with increasing heterozygosity. In self progenies there was a shortage of seedlings with long internodes, and the number of seedlings with short or very short internodes increased with increasing homozygosity. Trends in regeneration [assessed on growth produced after cutting seedlings to ground level in spring] were not so consistent, but selfing produced the most seedlings that were weakly regenerative and half-sib crossing produced the most showing strong regeneration. Self progenies contained many more seedlings which carried flowers to the base of the shoot, and . . . there is evidence that sib crosses produced more seedlings flowering on at least three-quarters of the shoot than half-sib crosses. The proportion of seedlings with short stigmas decreased with increasing heterozygosity, and a corresponding trend for long stigmas was also present but less well differentiated. For fruit size and strig length there are again well defined trends in size or length consistent with degree of inbreeding [both decrease with increasing homozygosity]. The leaf-spot susceptibility distributions hint at a possible influence of inbreeding and especially of selfing, but differences are not as large or consistent as for some of the above characters.

In contrast to the effect of inbreeding on western European cultivars, the S_1 progeny of the Finnish 'Brödtorp' showed little loss of vigor at East Malling, and in Scotland, Anderson and Fordyce (reported by Wood, 1960) considered the S_1 of this cultivar to be among the more promising families then under observation.

There appear to be no published data on the effect of inbreeding on redcurrant and gooseberry, but general observations at East Malling have shown there is usually a loss of vigor comparable with that in inbred blackcurrants.

Interspecific Cross Compatibility

Within species, cross compatibility is the rule in *Ribes*, with very rare exceptions. Cross compatibility between species (in the sense that viable hybrids can be raised to maturity) is also usual between the morphologically similar members of the groups which comprise the series (and sometimes subgenera) of Berger (1924), the sections of Rehder (1954), and the subgenera of Keep (1962*b*). Within these groups, all known interspecific hybrids are at least partially fertile. Crosses between species of different sections or subgenera are often less successful, barriers to hybridization being evident at all stages, including initial failure to set fruit, failure of seed germination, death of the young seedlings (as in certain *R. nigrum* x *R. sangineum* and *R. grossularia* x *R. sanguineum* combinations) and failure to flower. Finally, such hybrids as can be raised are usually sterile. In a review paper based on Rehder's classification, Keep (1962*b*) summarizes information on interspecific hybridization: 33 fertile intrasectional hybrids in six of Rehder's 15 sections are known; mature flowering plants have been obtained from intersectional crosses involving the combinations Symphocalyx x Calobotrya, Symphocalyx x Eucoreosma, Calobotrya x Eucoreosma, Eucoreosma x Heritiera, Eucoreosma x Ribesia, Eugrossularia x Robsonia, Euberisia x Eugrossularia, Symphocalyx x Eugrossularia, Calobotrya x Eugrossularia, Eucoreosma x Eugrossularia, and Ribesia x Eugrossularia. Of these, only the combinations Eugrossularia x Robsonia and Calobotrya x Eucoreosma showed any fertility. Further data on interspecific hybridization in *Ribes* have been abstracted and indexed by Knight and Keep (1958*a*) and Knight et al. (1972).

Considering the three commercial groups, hybrids of blackcurrant x redcurrant and of blackcurrant x gooseberry are fairly easily obtained and are often vigorous but invariably sterile. Crosses between redcurrant and gooseberry nearly always fail; the one seedling obtained by Keep (1962*b*) grew poorly and died before flowering, and no mature hybrids are known. The blackcurrant has undoubtedly been more widely used in interspecies crosses than any other *Ribes* species, and flowering plants have been raised from crosses with the following: *aureum**, *bracteosum*, *cereum*, *dikuscha*, *divaricatum**, *glutinosum*, *grossularia**, *hudsonianum*, *longeracemosum**, *niveum**, *odoratum**, *oxyacanthoides**, redcurrant*, *rubrum**, *sanguineum*, *sativum**, *ussuriense*. Species marked with an asterisk form sterile hybrids with *R. nigrum*, the remainder being either fully fertile, or sufficiently so to use as backcross parents.

The gooseberry has also been widely used in interspecific hybridization, and crosses with the following species have produced mature plants: *aciculare, alpestre, alpinum, cynosbati, divaricatum, gracile, hirtellum, inermis, irriguum, missouriense, nigrum, niveum, non-scriptum, odoratum, orientale, oxyacanthoides, pinetorum, rotundifolium, sanguineum, stenocarpum*. Hybrids with *nigrum* are sterile, while the fertility status of crosses with the dioecious species *R. alpinum* and *R. orientale,* and with *R. odoratum,* has not been reported. Crosses with the remaining species produce fully or partially fertile hybrids.

Well-authenticated hybrids of the redcurrant group of species (*rubrum, petraeum, sativum*) are few; sterile hybrids of all three species with *nigrum* are known, as are fertile hybrids of redcurrant x *multiflorum* and of *sativum* x *multiflorum* and x *warscewiczii*.

Cytology of Interspecies Hybrids

Pollen mother cell meiosis has been studied in a number of diploid intra- and intersectional hybrids by, among others, Meurman (1928), Keep (1958), and Goldschmidt (1964a). Meiosis in fertile intrasectional hybrids is usually more or less regular; Meurman found not more than 10% of irregular divisions in five hybrids studied, and Goldschmidt considered PMC meiosis in five such hybrids to be as regular as that of ten species also under investigation. In more or less sterile intersectional hybrids, however, chromosome pairing is much reduced, average bivalent frequency per cell at metaphase I ranging from about 2.0 (*R. nigrum* x *R. aureum,* Goldschmidt [1964a]) to 6.1 (*R. sanguineum* x *R. odoratum* [*R.* '*gordonianum*'], Meurman [1928]). In *R. nigrum* x *R. grossularia,* average bivalent frequency is about 3.4 to 3.8 with a range of 0 to 8, while in *R. nigrum* x redcurrant, average bivalent frequencies are 3.4 to 4.7 with a range of 1 to 8 (Meurman, 1928; Keep, 1958; Goldschmidt, 1964a). On this evidence, homology between blackcurrant and gooseberry chromosomes is approximately as close as that between blackcurrant and redcurrant, suggesting that gooseberry is rightly included in the genus *Ribes*.

Meurman (1928) noticed chromosome fragments in *R.* '*gordonianum*' and in *R.* '*carrierei*' (*R. nigrum* x *R. glutinosum albidum*) and anaphase bridges were "by no means rare." Similarly Goldschmidt (1964a) mentioned the occurrence of bridges and fragments, and both workers disagreed with Darlington's conclusion (1929) that the "striking constancy" of chromosome form and number in *Ribes* indicates that structural chromosome changes have played little part in the evolution of the genus.

In blackcurrant x gooseberry hybrids, only one satellite is evident, although each of the parents has two satellited chromosomes (Keep, 1960). By studying satellite and nucleolar number in allotetraploid hybrids and in backcrosses of these to diploid and tetraploid blackcurrant and gooseberry, Keep (1962a) showed that the blackcurrant satellites almost invariably "disappear" and fail partially or completely to organize nucleoli ("nucleolar suppression") in the presence of the gooseberry genome. This phenomenon has since proved to be fairly common in *Ribes* species hybrids, although the degree of nucleolar suppression, which tends to increase with increasing parental differentiation, is not always as marked as in blackcurrant x gooseberry (Keep, 1971). In three backcross progenies involving *R. nigrum* with *R. cereum, R. glutinosum,* and *R. sanguineum* all segregated for nucleolar suppression, a range from little or no suppression to suppression equal to or exceeding that found in the F_1 parents was observed. No obvious deleterious effects of this phenomenon on vigor or any other plant characteristics were noticed.

Polyploidy

Polyploidy is unknown in *Ribes* in the wild and in cultivation only two cases of natural polyploidy have been reported. Kuz'min (1956) obtained both sterile and fertile F_1 plants from the cross 'Kyzyrgan' redcurrant x 'Davison's Eight' blackcurrant. Subsequent cytological investigations showed the sterile plants (which more closely resembled the seed parent) were 2*x*, while the fertile plants (which were more like the pollen parent) were 3*x* (Žironkin, 1962). In addition, 4*x* and haploid ($2n = 8$) plants were obtained, the former derived from a shoot of one of the sterile F_1 seedlings, and the latter from a branch occurring on a F_2 seedling raised from open pollination of one of the 3*x* F_1 plants.

A 3*x* seedling of the dioecious *R. orientale* was described by Keep (1965), who also noted the occurrence of occasional 4*x* PMCs in the blackcurrant 'Boskoop Giant' and of restitution nuclei in PMCs of the gooseberry 'Broom Girl'. Clearly, the production of triploid seedlings is the result of the functioning of unreduced gametes, and this is apparently not uncommon in *Ribes*. Although a potential for natural polyploidy is present, Nilsson (1959) found fertility of colchiploid *Ribes* species to be much reduced and considered that a comparable reduction in fertility would prevent the establishment of polyploids in the wild.

Colchiploids of species and of intra- and intersectional hybrids are described by Nilsson (1959, 1966). In 4x *R. nigrum*, pollen fertility is high (93%) but fruit and seed set are reduced. Tetraploids will cross with diploids to produce triploids which are also fertile but less so than the tetraploids. Tetraploid *R. grossularia* is coarser and more spiny than the 2x form; pollen fertility is 98%, but female sterility is almost total. In 4x *R. rubrum*, pollen fertility is 80%, but fruit set is poor and number of flowers per raceme is reduced. Tetraploid *R. oxyacanthoides* produces little pollen (but 91% fertile) and sets poorly, although a few C_1 seedlings have been raised. In *R. alpinum*, 4x plants of both sexes have been obtained. Pollen fertility is good, but fertility in pistillate plants is reduced; several C_1 seedlings have been raised. Nilsson attributes the reduced fertility of autopolyploids to slow pollen tube growth, rather than to meiotic irregularities (although multivalent formation is common).

Allotetraploids differ in fertility according to the relationship of the parent species. Tetraploids from fertile intrasectional hybrids show much reduced fertility, but 4x forms from sterile intersectional hybrids usually show good fertility. The 4x *R. nigrum* x *R. grossularia*, which has black fruits intermediate in size between those of its parents, breeds fairly true and can be backcrossed to 4x and 2x forms of both parents (Nilsson, 1959, 1966; Knight, 1962*a*; Keep, 1962*a*). The allotetraploid *R. sativum* x *R. nigrum* is vigorous and fertile with dark red fruits, and can be backcrossed to 4x *R. nigrum* and 4x *R. rubrum*; backcrosses to 2x *R. nigrum* have failed. Other allotetraploids described by Nilsson include *R. nigrum* x *R. sanguineum* (4x F_2s have slightly reduced fertility); *R nigrum* x *R. aureum* (few seeds set); *R. nigrum* x *R. niveum* (vigorous, fertile, immune to gooseberry mildew, has been backcrossed to 4x *R. nigrum*); *R. nigrum* x *R. divaricatum* and *R. nigrum* x *R. oxyacanthoides* (both fertile, resistant to mildew and gall mite).

Cytology of Polyploids

PMC meiosis in auto- and alloploids has been studied by Goldschmidt (1964*b*, 1966). Bivalents, quadrivalents and, to a lesser extent, trivalents were common in colchiploid *R. nigrum*, *R. grossularia*, and *R. oxyacanthoides*. In 4x *R. 'rubrum'* fewer quadrivalents and more univalents occurred. Pairing in the intrasectional alloploid *R. grossularia* x *R. divaricatum* resembled that in the autoploids, with a relatively high proportion of multivalents and an average of only four to six bivalents. In the intersectional 4x hybrids *R. nigrum* x *R. grossularia*, *R. nigrum* x *R. niveum* and *R. sativum* x *R. nigrum*, a much lower proportion of multivalents and an average of from 13 to 14 bivalents occurred.

Mitotic instability has been described by Vaarama (1949) in colchiploid and C_2 *R. nigrum* and by Keep (1965) in second backcrosses of 4x (*R. nigrum* x *R. grossularia*) x 4x *R. nigrum*. Chromosome numbers in root tips of colchiploid *R. nigrum* ranged from four to 32. The diploid number was most frequent but all chromosome numbers divisible by four were more frequent than might have been expected on a random basis; Vaarama concluded that four and not eight is the basic number in *Ribes*. The reduction in chromosome number was caused by the formation of two separate and independent spindles. In five out of 25 second backcross seedlings, root tip chromosome numbers ranged from eight to 32.

Keep considered somatic instability to be one mechanism tending towards the restoration of diploidy in derivatives of *Ribes* colchiploids. Another mechanism is the formation and preferential functioning in crosses with 2x seed parents, of 'haploid' (i.e., 2x) male gametes in 4x blackcurrant x gooseberry hybrids. Two such crosses gave a total of three 2x seedlings, one 3x and one 4x, in comparison with six 3x (or near 3x) and one 4x seedling from reciprocal crosses.

Apomixis

There are few reports of natural apomixis in *Ribes* and this does not appear to be an important reproductive mechanism in the genus.

Since in *Ribes* pollination sometimes occurs in the unopened bud, it is not easy to be certain that seeds set by emasculated and unpollinated flowers are truly apomictic. However, most reports of apomixis concern blackcurrants, whose flowers are comparatively easy to emasculate and also large enough to be inspected fairly effectively for the presence of stray pollen on the stigma.

Zatykó (1962) considered 'Goliath' to be apomictic, as did Seljahudin and Brózik (1967).

Schmidt (1968) reported a limited occurrence of apomixis in this cultivar and in 'Amos Black' and 'Seabrook's Black'. Maternal types occurred in several crosses of 'Baldwin' with other *Ribes* species (Swarbrick, 1959), and progenies from crossing and backcrossing other blackcurrant cultivars with species often include a proportion of pure blackcurrants (Keep, unpub.). Such types are usually attributed to accidental self pollination, but might well be due to apomixis, as suggested by Swarbrick.

Chemical treatments reported to have caused apomictic seed set in blackcurrants include spraying emasculated flowers of 'Amos Black', 'Seabrook's Black', and 'Goliath' with a mixture of 100 ppm indoleacetic acid (IAA) and 100 ppm GA (Zatykó, 1962). Spraying these cultivars and 'Silvergieters Zwarte' and 'Boskoop Giant' with 50 ppm GA alone three days after emasculation also induced apomictic seed set (Zatykó and Simon, 1969). The resulting seedlings were all 2x.

In redcurrants, a full seed set was obtained in 'Fay's Prolific' by spraying emasculated flowers with 20 ppm β-IAA (Zatykó and Simon, 1960). Wills (reported by Haskell [1964]) obtained a 0.5% set in blackcurrants, 5% in redcurrants and 10% in gooseberries by spraying emasculated flowers with a mixture of equal parts of 100 ppm GA + 100 ppm indolebutyric acid. The fruits usually contained one seed only.

Parthenocarpy

Natural parthenocarpy in *Ribes* species is rare; the redcurrant cultivar 'Kernlose' ('Seedless Red'), will set seedless fruit, and Hughes (1962) obtained a moderate crop of fruit with few or no seeds from gooseberries caged to exclude insects. Diploid blackcurrant x gooseberry hybrids sometimes set parthenocarpic fruits, the amount set varying considerably from season to season.

Parthenocarpy has been induced in blackcurrants, redcurrants and *R. aureum* by applying GA, either as a solution at 100 ppm or in lanolin at concentrations of up to 5% (Čuvašina, 1962; Wilson, 1962; Zatykó, 1962). GA at 50 ppm induced 87% parthenocarpic fruit set in *R. 'culverwellii'* (*R. nigrum* x *R. grossularia*) and 31% set in *R. 'gordonianum'* (*R. sanguineum* x *R. odoratum*) (Zatykó, 1963). Zatykó (1962) achieved a higher set in black- and redcurrants by applying a mixture of GA (100 ppm) and β-IAA (100 ppm).

Breeding Methodology

Surveys of Sources of Economic Characters

The planning of a breeding program for any crop necessarily involves a survey of the literature to determine how far characteristics of existing cultivars fall short of requirements, and where sources of improved characters can be found. Some of the literature concerning *Ribes* species and cultivars has already been cited. Abstracts of most of the papers published before 1970 dealing with matters of interest to *Ribes* breeders (e.g., cultivar descriptions, pest and disease resistance, hardiness, mechanical harvesting, processing, breeding behaviour, genetics, cytology, pollen and pollination, etc.) are included in bibliographies published by the Commonwealth Bureau of Horticulture and Plantation Crops (Knight and Keep, 1958a; Knight et al., 1972). Brief descriptions of new cultivars introduced all over the world are included in the annual register of new fruit and nut varieties now published in *HortScience*.

To obtain knowledge of the response of cultivars and species to local conditions, field surveys of varietal and species collections are invaluable, and by leaving the plot unsprayed, useful information on pest and disease resistance can be obtained.

Evaluation of Parents

Methods of evaluating parents in currant and gooseberry breeding are inevitably dominated by three fundamental aspects of these crops—they are perennials, clonally propagated, and basically outbreeding. Evaluation must include both phenotypic and genotypic selection.

Perennial crops in temperate regions experience widely different climatic conditions from year to year and these affect all aspects of growth and development, so it is essential to make observations over several seasons before making final selections. In *Ribes*, this is particularly important in assessing response to late spring frosts, as these occur erratically, but then can drastically reduce the yield of "susceptible" seedlings. The incidence of pests and diseases also varies from season to season and field response is best observed over several years.

Since in clonally propagated crops the ability to produce outstanding individuals, rather than a general high level of combining ability, is required of parents, techniques designed to assess the former are of critical importance. In a series of papers, Wilson and his colleagues reported the results of intercrossing the blackcurrant cultivars 'Baldwin', 'Boskoop Giant', 'Seabrook's Black', and 'Victoria'. These were chosen as representative of the four morphologically distinct groups of Hatton (1920). Six components of yield (Wilson and Adam, 1966), three characters affecting mechanical harvesting (Wilson, 1963), and leafspot and mildew incidence (Wilson et al., 1964) were recorded.

Wilson and Adam (1966) concluded that the majority of the differences between progenies could have been predicted by studying parental phenotypes. In 1970, following a study of the inheritance of 13 characters in progenies derived from selfing and intercrossing four further cultivars, 'Amos Black', 'Malvern Cross', 'Mendip Cross', and 'Wellington xxx', Wilson concluded that the breeding value of these cultivars also could have been predicted fairly accurately from a close study of phenotypes, and that "phenotypic selection of parents should be adequate for the phase of blackcurrant breeding in which new characters or new extremes of character expression are required."

It seems that the time and space available to *Ribes* breeders are normally best devoted to selecting plants with maximum expression of the required characters and intercrossing complementary phenotypes. Where the inheritance of the characters concerned is not known, by using several donors in parallel it is possible to discard lines in which the genetic control proves unsatisfactory. (See the later section on the role of major and minor genes.)

Only Tydeman (1930) appears to have studied the effects of inbreeding beyond the S_1 generation, and his results suggest that probably in S_2 (and by deduction from this, certainly in later inbred generations) inbreeding depression in blackcurrants would be so severe as to preclude selection for agronomic characters. Hence in this crop inbreeding appears to be of limited value in assessing parental quality and in establishing lines pure breeding for any but genetically simple characters. Following his study on the effects of increasing heterozygosity in blackcurrants, Wilson (1970) concluded that the degree of inbreeding had as much if not more influence on the resultant progeny than the actual parental phenotypes, and emphasized the need to take into account the relationships between potential parents, particularly when an extensive backcrossing program is envisaged. Clearly, final crosses to produce commercial types should, as far as possible, be between plants of different origins.

Seedling Evaluation

Details of seedling evaluation are included in a later section and only general aspects will be considered here.

It is obviously desirable to select as early in the life of the seedling as possible to avoid wasting time and energy on useless material. Some workers have selected for response to American gooseberry mildew, white pine blister rust, and leafspot on young seedlings in pots in the greenhouse. Field and greenhouse response to these diseases have generally been similar, although some plants which are mildew resistant in the greenhouse may suffer field attacks, and plants showing field resistance are sometimes slightly attacked under greenhouse conditions (Brauns, 1959; Keep, unpub.). As yet, data on pest resistances are inadequate to recommend full reliance on greenhouse testing; certainly, field and insectary responses to aphids are not always identical (Keep and Briggs, 1971).

Selection for vigor and for developmental abnormalities such as leaf chlorosis and dwarfing can sometimes be done in the seed box or pot stage, and in breeding for spinelessness in gooseberries the really thorny seedlings can be culled as pot plants.

Field Selection

Selection techniques must be adapted to suit the material under observation, and in F_1s of interspecies hybrids and sometimes in first backcrosses, most emphasis must be placed on selecting the particular character or characters for which the progeny was bred. As an example, in breeding for many-flowered strigs, numbers of flowers rather than fertility or fruit size would be the main criterion in the F_1 generation. Again, in resistance breeding, maximum development of resistance is more important in the early generations than fruiting qualities.

In later backcrosses when material is approaching commercial standards, all characters affecting commercial potential must be studied.

Assuming good growth in the year of planting out in the field, field observations for four years are usually adequate for selection of parents and of potential cultivars, providing two good cropping years are included. The first year's crop is often too small to permit adequate judgement of fruit qualities (although Kronenberg and Hofman [1965] found that in blackcurrant poor flower production in the first year of cropping usually indicated a poor cropping potential at maturity also).

Wilson (1970) scored 13 characters in assessing blackcurrant cultivars for breeding value and even then did not include growth habit, running off, and pest resistance. Where so many factors have to be taken into account, a rapid scoring system for each is essential, and a 1–5 scale is usually adequate. With certain characters, such as internode length and numbers of flowers per strig, direct measurements or counts are fairly rapid and can be undertaken for key families.

Preliminary selection for growth habit is usually possible at the end of the second year and initial selection for pest and disease response can be made during the second season. In the third year there is often enough crop to permit selection for flowering season and for fruiting characteristics. Response to late spring frosts can also be scored, given suitable weather conditions, and observations on pest and disease resistance continued. Clonal propagation of really outstanding selections can be undertaken in the third year although in most cases, it is better to await one more season's assessment of cropping and fruit quality. Final selections of parents and of potential commercial types can usually be made in the fourth year, when clonal propagation of most selections is initiated.

Some seedlings may be so outstanding as to be included in replicated cropping trials without passing through a second selection "sieve," but in most cases, further observations on short, clonally propagated rows of selections, in comparison with standard cultivars, are valuable, particularly in judging plants which appeared very similar in the single plant stage.

Interspecific Hybridization

Wild species are being increasingly used in *Ribes* breeding as sources of major improvements. The general wide interspecific "crossability" and the existence of considerable homology between chromosomes of well-differentiated species mean that a very extensive gene pool is available to plant breeders. Breeding methodology in gene transferences is highly influenced by whether the initial hybrid is sterile or fertile; in *Ribes* this usually depends on whether the hybrid is inter- or intrasectional.

Intersectional: Once fertile colchiploids have been obtained from sterile diploid intersectional hybrids, the decision as to whether to backcross to diploid or autotetraploid parent species depends on two main considerations. In backcrossing to the diploid, the first backcross seedlings are usually triploids with approximately two genomes of the recurrent parent and one of the donor. Selective pairing is likely to occur at meiosis in these triploids, and with preferential elimination of unbalanced gametes and zygotes, genetically complex characters (e.g., fruit size) would have less chance of being transferred to the second backcross generation than would characters controlled by one or two genes (e.g., some pest and disease resistances). By making the first backcross to the autotetraploid, first backcross seedlings would be tetraploids with approximately three genomes of the recurrent parent and one of the donor. These tetraploid seedlings would normally be intrinsically more fertile than the triploids and also pairing at meiosis between recurrent and donor genomes is likely to be more complete, so that genetically complex characters would have a better chance of being transferred successfully.

The advantages of backcrossing to the diploid are that a more rapid elimination of the donor genome occurs and fewer generations are needed to restore commercial quality. As will be described in detail later, gall mite resistant, self-fertile blackcurrants have been obtained in the third backcross of 4x blackcurrant x gooseberry to diploid blackcurrants.

If the genetic control of the character concerned is not known at the start of a transference program, it is probably best to backcross to both the diploid and the tetraploid recurrent parent (as was done in the case of gall mite resistance). Once it is certain that the second backcross generation in the diploid line carries the required character, further work on the tetraploid line can be discontinued. For, since autoploid *Ribes* invariably show reduced fertility, it is essential to get back to the diploid level to restore commercial quality, and in tetraploid transferences this would involve at least two further crosses to diploids.

Sometimes the cross of the allotetraploid F_1 with diploid parent species fails. This is so in the case of blackcurrant x redcurrant and blackcurrant x *R. longeracemosum*, although backcrosses to tetraploid blackcurrants have succeeded. In such a situation, if reciprocal crosses of successive 4x backcross selections to the diploid recurrent parent also fail, it is worth sowing open pollinated seed, providing the recurrent parent species is growing nearby.

Since the outcome of attempts to make intersectional gene transferences is less certain than in intrasectional and intraspecific hybridization, it is even more important to have alternative lines of attack on such problems. An example of this is the failure at East Malling to transmit spinelessness from blackcurrant to gooseberry; once it became evident that no worthwhile reduction in spininess would be achieved, efforts were concentrated on alternative lines carried on in parallel, in which *R.* '*oxyacanthoides*' (or *R. hirtellum*) and *R. sanguineum* were donors of spinelessness.

Intrasectional: In fertile intrasectional species crosses, since chromosome pairing will usually be more or less regular, transferences of complex characters are likely to be little more difficult than in intraspecific breeding. The chances of transferring undesirable characters from the donor will be higher than in intersectional transfers, and restoration of recurrent parent phenotypes will be more dependent on stringent selection. An example of a successful intrasectional transfer is the breeding of redcurrants with many flowered racemes, using *R. multiflorum* as donor. Commercial selections (e.g., 'Mulka') have been obtained in the first backcross.

Mutations

Natural mutations ('bud sports') known in *Ribes* include those affecting season of leafing out, flowering, and ripening in blackcurrants (e.g. 'Daniels' September', 'Laleham Beauty', Hughes [1963]) and redcurrants (Knight and Keep, 1958*b*), fruit color in redcurrants (e.g., 'Rosa Sport', Schuppe [1962]) and leaf shape in blackcurrants (Porpáczy et al., 1964). Mutations induced by irradiation and treatment with chemical mutagens have been described in a previous section.

Little use appears to have been made of these mutants in *Ribes* breeding, probably because the improvements so effected are relatively minor and by no means match in "breeding potential" what is available naturally in the genus as a whole. Results to date with blackcurrants suggest that further time and labor spent in inducing and isolating mutants would be largely wasted.

Breeding for Specific Plant Characters

The Role of Major and Minor Genes

When considering morphological and developmental characteristics, major gene control, if available, is usually to be preferred, since a much higher proportion of the progeny will carry the character and selection will be simple. Where pest and disease resistance are involved, the question of relative permanence of the resistance has to be considered. The recent trend towards the use of polygenic "field" resistance is largely the result of experiences with cereals and other annual crops, where the introduction of major genes for resistance to diseases such as the rusts and smuts has soon been matched by the development of new races of the fungus. However, as Knight and Alston (1974) have pointed out, major gene resistance to pests has generally proved to be long lasting, and, in fruit-crops, there is little evidence for rapid breakdown of such resistance to diseases.

Since some degree of inbreeding is required to isolate recessive characters, in perennial outbreeding crops dominant genes are more rapidly and easily transferred and would usually be preferred where both types of control are available.

In a discussion of fruit breeding methodology, Watkins (1971) recommends the use of major economic genes, selected as early as possible in the life of seedlings, so that final selection can be concentrated on polygenically inherited characters. Where no major genes are available, "then quantitative genetic procedures should be considered." Watkins suggests that "if possible, the base populations for such quantitative genetic studies should consist of plants homozygous for all the major economic genes," but points out that the relatively small number of selections homozygous for these genes would limit the use of the quantitative approach in outbreeding fruit crops.

Habit

The ideal fruiting habit in currants and gooseberries is fairly erect and compact so that both hand and machine picking is facilitated and reasonably close spacing is possible. In cultivars of currants and gooseberries there is considerable variation in plant habit and mature plant size, and in general, phenotype is a good indicator of genetic behavior for this character. The erect, compact habit of 'Westra', induced by x-ray treatment of 'Westwick Choice', appears to be heritable (Bauer, 1970), and there is some evidence of major gene components in growth habit in gooseberries (see later).

According to Somorowski (1964) blackcurrant seedlings producing only one shoot in the first year eventually formed rather upright bushes, while those with several shoots formed spreading bushes.

The western European group of blackcurrants are generally at least moderately erect and make medium to large bushes at maturity. Erect growing cultivars include 'Amos Black', 'Baldwin', 'Black Reward', 'Cotswold Cross', 'Goliath', 'Greens Black', 'Invigo', 'Laxton's Giant', 'Laxton's Grape', 'Roodknop', 'Seabrook's Black', and 'Silvergieters Zwarte'. Of this group, 'Goliath' is outstanding as a donor of erect, compact growth having produced a good proportion of seedlings with this habit in crosses with 'Baldwin', 'French Black', 'Seabrook's Black', and 'Boskoop Giant' (Tydeman, 1938). More or less erect cultivars derived from 'Goliath' include 'Westwick Triumph', 'Matchless', 'Invigo', and 'Silgo'. Selfed progenies of blackcurrants includes the dwarfed, local Swedhigh proportion of erect seedlings, as did the F_1 derived from these two cultivars (Spinks, 1947).

The northern European (Scandinavian) group of blackcurrants includes the dwarfed, local Swedish types 'Erkheikki', 'Haparanda', 'Sunderbyn', 'Öjebyn', and the Finnish 'Åström'. In some crosses, 'Öjebyn' transmits its fairly erect habit to a proportion of its progeny. In contrast, the Finnish 'Brödtorp' is notorious for its spreading habit. Anderson and Fordyce (reported by Wood, 1962) describe progenies derived from crossing this cultivar with 'Silvergieters Zwarte' and with 'Janslunda' as very variable in plant habit, the most vigorous and erect seedlings being almost barren and the very spreading, weak-growing seedlings generally cropping well. This variation was less evident in some other 'Brödtorp' crosses, however, and at East Malling, in a selfed progeny of 'Brödtorp', the few erect-growing, poor-cropping seedlings were also the least vigorous (Keep, unpub.).

Species which have proved good donors for growth habit in blackcurrants include *R. sanguineum*, *R. glutinosum*, and *R. bracteosum*, the two latter having produced a good proportion of erect seedlings in backcrosses to 'Brödtorp' (Keep, unpub.). Some *bracteosum* derivatives also inherit a sparsely branched habit, which may prove particularly suitable for mechanical harvesting.

Many of the most commonly grown redcurrant cultivars are markedly erect (e.g., 'Erstling aus Vierlanden', 'Maarse's Prominent', 'Red Lake', 'Jonkheer van Tets') and general observations at East Malling suggest this character is transmitted to a high proportion of the progeny.

Backhouse and Bailey (reported by Crane and Lawrence, 1952), showed that two erect-growing gooseberries, 'Early Green Hairy' and 'Rumbullion', and one spreading type, 'Echo', all bred true for their respective habits on selfing. In reciprocal crosses of 'Echo' x 'Early Green Hairy', the spreading habit was dominant. However two cultivars of intermediate habit, 'May Duke' and 'Pitmaston Greengage', gave ratios of 16:70 and 83:47 for intermediate and spreading habit, respectively, showing that more than one pair of genes are involved. Of the European group, in addition to 'May Duke', 'Whinham's Industry', and 'Whitesmith' are moderately erect, while 'Leveller' is spreading and transmits the character to much of its progeny. Recent mildew-resistant introductions selected for at least moderately erect habit include 'Ristula', 'Rokula', 'Risulfa', and 'Reverta' from Germany, and 'Izumrud', 'Rubin', and 'Smena' from Russia (Potapov, 1961).

Kovalev (1934) cites *R. niveum* as a donor for compact growth habit, and at East Malling, *R. alpestre*, *R. watsonianum*, *R. leptanthum*, and *R. sanguineum* have provided some seedlings of good habit in F_1s or first backcrosses with gooseberry.

Regeneration

In blackcurrants, one system of mechanical harvesting ("destructive") used particularly in Britain, involves cutting off the fruiting wood, or a proportion of it, and mechanically stripping the currants in a stationary harvester. The success of this operation depends on the ability of the plant to regenerate adequately in the following season. The custom of hand pruning to maintain fruit size also relies on adequate replacement growth. Tamás (1961) classified cultivars into three types according to growth habit: 1) vigorous types such as 'Brödtorp', 'Consort', and 'Coronet' which produce numerous new basal and terminal shoots on which most of the inflorescences are borne; 2) cultivars such as 'Kajaana Seminarium', forming no new basal shoots and few new terminal shoots, but able to flower on old wood; 3) intermediate types such as 'Cotswold Cross'. In a study of the inheritance of ability to regenerate, Wilson (1963) found 'Victoria' to be outstanding as a parent, in comparison with 'Baldwin', 'Boskoop Giant', and 'Seabrook's Black', for the production of vigorous seedlings with strong basal growth. In crosses between 'Amos Black', 'Malvern Cross', 'Mendip Cross', and 'Wellington xxx', 'Malvern Cross' was the best parent for seedlings which regenerated strongly, although in the whole population, regeneration was at least adequate in 75% of the seedlings (Wilson, 1970). Selfing greatly decreased the proportion of strongly regenerating seedlings. These results suggest that regeneration would normally be quite adequate in vigorous new selections, particularly those derived from wide crosses.

Ease of Propagation

In commerce, currants and gooseberries are normally propagated by hardwood cuttings taken in the early autumn. For rapid bulking up, other techniques include the use of single node cuttings for blackcurrants (Thomas and Wilkinson, 1962*b*) and softwood cuttings rooted in a mist propagator.

Blackcurrants generally take well from hardwood cuttings, and redcurrants fairly well, but European gooseberries are less easy to propagate in this way. The percentage rooting in the nursery at East Malling (which is run on commercial lines) of three blackcurrant, redcurrant, and gooseberry cultivars over three years is shown in Table 4.

These data suggest there is some variation between cultivars within each group in rooting ability. The data for gooseberries confirm Rake's (1953) statement that nurserymen consider 50% a satisfactory take, but few reach this figure and many obtain only 10%. Another factor which may affect rooting is virus infection. In 'Whitesmith' gooseberry, van der Meer (1965) obtained 49.5% rooting with a healthy clone compared with 23.9% from a clone infected with gooseberry veinbanding virus.

TABLE 4. Percentage rooting of hardwood cuttings of blackcurrants, redcurrants and gooseberries at East Malling

	% of rooted hardwood cuttings			
	Year			
Clone	1	2	3	Average % rooted
Blackcurrant				
Baldwin	80.0	32.4	29.4	47.3
Seabrook's Black	88.0	99.3	66.1	84.5
Wellington xxx	98.0	96.0	58.5	84.2
Redcurrant				
Laxton's No. 1	85.3	96.8	58.7	80.3
Red Lake	60.0	96.0	42.0	66.0
Wilson's Long Bunch	49.3	87.2	56.0	64.2
Gooseberry				
Bedford Red	42.3	44.7	37.5	41.5
Keepsake	69.2	10.9	18.2	32.8
Whinham's Industry	61.3	7.2	34.4	34.3

The ease of rooting of blackcurrants appears to be transmitted to some extent in crosses with both redcurrant and gooseberry. Smirnov (1965) obtained 39–82% of rooted hardwood cuttings from different blackcurrant cultivars but only 2–37% from redcurrants. He reported that most redcurrant x blackcurrant F_1 seedlings were intermediate between their parents in rooting capacity, while blackcurrant x gooseberry hybrids approached blackcurrant in rooting ability. However, in these as in the redcurrant x blackcurrant F_1s, there was considerable variation between seedlings.

The ease with which American gooseberry cultivars (mostly *R. hirtellum* derivatives) can be propagated in comparison with European types was noted by Pavlova (1940). Osipov (1970) also reported that 'Houghton', 'Smena' ('Houghton' x 'Zelenyj Butyločnyj'), 'Russkij' ('Bočenočnyj' x mixed pollen of American cultivars), and 'Jubilejnyj' rooted best out of 15 cultivars tested, while green cuttings of European varieties rooted considerably less well. Other American gooseberry species which transmit ease of rooting to hybrids with *R. grossularia* include *R. divaricatum*, *R. 'succirubrum'*, and *R. oxyacanthoides* (Schmidt, 1952). In backcrossing programs involving American species as donors of disease resistance, screening for rooting ability would certainly seem worthwhile.

Winterhardiness

According to Ravkin (1965, 1966) winter-hardy cultivars of currants and gooseberries are characterized by longer and deeper dormancy, more complete and faster starch hydrolysis, slight dehydrolysis during a prolonged thaw, and the early appearance of large quantities of sucrose in the buds and of raffinose in the shoots. The highest sucrose content was observed in January, in the blackcurrant 'Karel'skaja', the redcurrant 'Red Dutch', and 'Houghton' gooseberry. Tumanov et al. (1970) describe a sharp increase at the end of February and in March in content of gibberellin-like substances, especially GA_3, in the buds and bark of blackcurrant. There was no response to exogenous gibberellins during deep dormancy, and it is suggested gibberellin could be used to indicate the depth of dormancy when selecting frost-resistant forms.

Tamás (1960) studied winterhardiness in currants by direct observations on plants exposed to low temperatures and by comparing the electrolytic resistance of control shoots and those exposed to −25 C for 24 hours. The ratio of electrolytic resistance of untreated shoots to electrolytic resistance of cold-treated shoots (RLF) was found to increase in blackcurrants with increasing frost susceptibility, providing disbudded shoots were used. The RLF values of shoots and the frost susceptibility of buds, however, were not correlated.

Western European blackcurrant cultivars are, in general, considerably less hardy than Scandinavian and Russian varieties and may suffer injury during severe winters (e.g., in Sweden [Bjurman, 1961] and in Holland [Anon., 1957]) although they can be grown in snow-covered areas (Larsson, 1959). Particularly susceptible types are 'Mendip Cross', 'Cotswold Cross', 'Baldwin', and 'Silvergieters Zwarte' while 'Roodknop' and 'Blacksmith' are rather hardier (Anon., 1957; Bystydzienski and Smolarz, 1966). At Balsgård, Sweden, eight cultivars of various origins were found to be decreasingly hardy, according to field observations and/or electrolytic studies, in this order: 'Consort', 'Brödtorp', 'Åström', 'French Black', 'Giant Prolific', 'Silvergieters Zwarte', 'Rosenthal's Black', and 'Wellington xxx' (Tamás, 1960).

Of the Scandinavian types, the dwarf local Swedish cultivars 'Ostersund', 'Erkheikki', 'Haparanda', 'Öjebyn', and 'Sunderbyn' are all very hardy and adapted to northern conditions as are the comparable Finnish 'Brödtorp', Åström', 'Lepaan Musta', and 'Kajaanin Musta' (Nilsson, 1958; Säkö, 1963).

Much of the more recent Russian literature concerns the use of the wild Siberian form of blackcurrant (*R. nigrum sibiricum*) and of *R. dikuscha* as sources of hardiness (and of fruit size and disease resistance). As early as 1938, Andreičenko drew attention to the hardiness of *R. dikuscha* and in 1940 Bologovskaja described the wild Siberian currant (presumably *R. nigrum sibiricum*) as being much hardier and having larger fruit than European cultivars. Particularly hardy selections of *R. nigrum sibiricum* from the wild include forms from the Altaj, Pečora, Varzuga, and Igarka basins, the latter being outstanding (Vitkovskij, 1958). Selected seedlings of these forms were named 'Hibinskaja Rannjaja', 'Pečora', 'Altaj', and 'Lenskaja', and these have extended the northern limits of blackcurrant cultivation in European areas of the USSR to inside the Arctic Circle (Vitkovskij, 1964a).

In breeding for hardiness in western Europe, much use has been made of Scandinavian cultivars such as 'Brödtorp' and Öjebyn' and also of *R. dikuscha* (e.g. Anderson and Fordyce reported by Wood, 1957; Nybom and Bergendal, 1960), although the results of this work do not appear to have been fully evaluated as yet. In Russia, hybridization between western European types and *R. nigrum sibiricum* and *R. dikuscha* has been undertaken on a large scale particularly to improve cropping in the self-sterile *R. nigrum sibiricum* selections. According to Voluznev (1968) *R. dikuscha* is also self sterile (although an accession at East Malling is self fertile) and shares other faults with *R. nigrum sibiricum*, including early flowering, need for high light intensity, growing best at high altitudes, and having brittle wood. By crossing selected forms of *R. nigrum sibiricum* with European cultivars a range of cultivars has been produced which are completely hardy in the Altaj and in many areas of Siberia (Pavlova, 1962), although according to Kuminov (1964) the use of European types in hybridization produces progeny with diminished hardiness. Such hybrids are also basically self sterile (Voluznev, 1968) and Pavlova (1962) recommends further backcrossing. Self-sterile hybrid derivatives of *R. nigrum sibiricum* include the cultivars 'Lošickaja', 'Barhatnaja', 'Mečta', 'Belorusskaja Pozdnjaja' and 'Minskaja'. *R. dikuscha* is even hardier than *R. nigrum sibiri-*

cum and was considered by Bočkarnikova (1964) to be the most valuable of four native Siberian species as donor of hardiness, earliness, yield, and disease resistance. Derivatives of this species are usually self-fertile. 'Primorskij Čempion' (*R. dikuscha* x 'Lee's Prolific'), which is self-fertile, very hardy, and early ripening, has been much used in further backcrossing, giving rise to such self-fertile cultivars as 'Stahanovka Altaja' ('Goliath' x 'Primorskij Čempion') with fair frost resistance, 'Golubka' ('Saunders' x 'Primorskij Čempion') with great winterhardiness, and 'Zoja' ('Primorskij Čempion x a wild Siberian form). Kuminov (1965) mentions that hybrids of 'Primorskij Čempion' with certain Siberian cultivars had larger fruit and double or treble the yield of the parents. Kuminov also gives the proportion of valuable selections arising from crosses between European x Siberian types (whether *R. dikuscha* or *R. nigrum sibiricum* derivatives is not clear). At Krasnojarsk, of 7604 hybrids of this origin, 27 (including 'Krasnojarskaja Desertnaja' and 'Kollectivnaja') were frost resistant and had large fruits, while at Minusinsk, such crosses yielded 20–50% of hardy types.

Other hardy species of the Eucoreosma which do not yet appear to have been used as donors include *R. nigrum pauciflorum* (Zhukovsky, 1965) and *R. procumbens;* the latter species will survive temperatures of −60 C (Kovalev, 1936).

Rehder (1954) considers that the three main progenitor species of redcurrant cultivars, *R. rubrum, R. sativum,* and *R. petraeum* will grow satisfactorily in his climatic Zones III, IV, and V respectively. The average annual minimum temperatures of these zones are: Zone III, −35 to −20 C; Zone IV, −20 to −10 C; Zone V, −10 to −5 C. In general, the little available information on hardiness in redcurrant cultivars suggests derivatives of the three species tend to inherit something of the parental grade of hardiness. Thus, in Sweden from 1943–1957 'Red Cross' and 'Versailles' (*R. sativum macrocarpum*) showed most winter injury, 'Earliest of Fourlands' ('Erstling aus Vierlanden', *R. petraeum*) was intermediate, and 'Red Dutch' and 'Houghton Castle' (probably *R. petraeum* x *R. rubrum* and *R. rubrum* x *R. vulgare*) were undamaged (Bergelin and Sahlström, 1958). Macoun and Davis (1920) considered 'Red Dutch' and 'Raby Castle' (*R. rubrum pubescens*) to be very hardy, and Vitkovskij (1958) found 'Red Dutch' to be the only European cultivar sufficiently hardy to be grown in the Kola peninsula. This cultivar was uninjured following a snowless winter with an average December air temperature of −28 C, while 'Komovaja Markina' and 'Houghton Castle' were the hardiest of five cultivars which had little frost injury (Barsukov and Savvina, 1969). Other cultivars described in the literature as hardy include 'Red Lake', 'Victoria', and 'White Grape' (Macoun and Davis, 1920; Klingbeil et al., 1965).

In breeding for hardiness, Vitkovskij (1964*b*, 1969) recommends using the local forms 'Kandalakša' and 'Varzuga' and *R.* [*petraeum*] *atropurpureum;* seedlings of *R. vulgare* and its hybrids with *R.* [*rubrum*] *pubescens* were not hardy enough for the Kola peninsula, whereas local *rubrum* forms ripened early and yielded well. *R. petraeum* proved unsuitable. The most hardy species of all in the redcurrant group is *R. triste* according to Kovalev (1936), but this does not appear to have been used in breeding.

Hardiness in gooseberry cultivars to some extent parallels that of blackcurrants, with western European types being considerably less hardy than Scandinavian and Russian varieties.

In Sweden, only small-fruited introductions from the wild such as 'Scania' are completely hardy in Norrland (Nilsson, 1940), while at Öjebyn, just south of the arctic circle, of 50 Swedish and foreign cultivars the Finnish 'Hinnonmäkis Gula', 'Hankkijas Delikatess', and 'Finland 1' were superior in hardiness, yield, mildew resistance, and fruit quality (Larsson, 1961). Other hardy cultivars of this group are 'Dr. Törnmarck', 'Pellervo', and 'Packalén'.

Breeding for hardiness has almost invariably involved crossing large-fruited *R. grossularia* derivatives with local, hardy, small-fruited species. American *grossularia* x *hirtellum* derivatives such as 'Houghton', 'Downing', 'Silvia', and 'Poorman' are very hardy, and 'Houghton', particularly, has been much used in Russia as donor of hardiness in crosses with European types. Hardy derivatives of 'Houghton' include 'Smena', 'Pjatiletka', 'Izumrud', and 'Rekord' (Sergeeva, 1962; Potapkova, 1969). Other North American wild species used as donors of hardiness include *R. divaricatum* from which 'Černyš' (Lancer x *R. divaricatum*) was derived; *R. 'arcuatum'* (*R. hirtellum* x *R. missouriense*) of which 'Isabella', 'Doškol'nik', and 'Višnevyj' are derivatives; and *R. rotundifolium* from which 'Skaidrais Ūdens', 'Varonis', and 'Avēnite' were derived (Mosolova, 1956). In addition, Romanovskaja (1955) mentions the use of *R. cynosbati* and *R. gracile*, and of the Chinese *R. stenocarpum* in crosses with European cultivars.

Of species native to Russia, *R. aciculare*, reputed to be the hardiest species of Eugrossularia (Bologovskaja, 1940), has been crossed and backcrossed with large-fruited types to produce 'Pervenec Minusinska' and 'Muromec' and selections have been made in F_2 and F_3 progenies of this origin (Kuminov, 1962*a*). Another species outstanding for hardiness is the Far Eastern *R. burejense* which survives temperatures of −50 C but does not yet appear to have been used in breeding for this character.

Spinelessness

Within the gooseberry group of species a whole range of spininess occurs from the large, stout spines of *R. stenocarpum* through the rather smaller, less numerous spines of *R.* grossularia to the occasional, small single spines of *R. hirtellum* and *R. rotundifolium* (Fig. 2). Cultivars of *R. grossularia* differ somewhat in size and number of spines, and there are forms with very few spines, rather lacking in vigor, such as *R. grossularia inermis* and the old French 'Souvenir de Billard', 'Belle de Meaux', and 'Edouard Lefort'. The American cultivars derived from hybridizing *R. grossularia* and *R. hirtellum* are usually less spiny than European types.

'Belle de Meaux', 'Edouard Lefort', and *R. grossularia inermis* bred true for reduced spines on selfing, and in crosses of 'Edouard Lefort' and *R. grossularia inermis* with spiny forms such as 'Crown Bob' and 'Whinham's Industry', all the F_1 progeny had numerous prickles (Backhouse and Bailey, reported by Crane and Lawrence, 1952). This suggests that spinelessness is recessive in *R. grossularia.* However, Lorenz (1929) pointed out that the existence of spineless gooseberries and of various slightly spiny forms implies that spinelessness is controlled by a series of similar genes. More recent studies have confirmed the genetic complexity of the character, Hunter (1950*b*) considering that spinelessness as derived from *R. 'oxyacanthoides'*, is recessive and controlled by several pairs of genes. Hunter based his conclusions on a progeny of the cultivar 'Spinefree'. According to Davis (1926) this cultivar was derived from crossing a thornless segregate from the F_2 of (thornless *R. 'oxyacanthoides'* x 'Victoria' [*R. grossularia*]) with 'Mabel' (a Canadian cultivar).

FIG. 2. Spines in species of Eugrossularia; from left, *R. 'oxyacanthoides'*, *R. grossularia* (Whitesmith), *R. divaricatum*, *R. stenocarpum*.

Since both Berger (1924) and Rehder (1954) agree that *R. oxyacanthoides* is spiny, and is often confused with the relatively spineless *R. hirtellum*, it seems more probable that Davis' thornless *R. 'oxyacanthoides'* was actually *R. hirtellum*, and on this assumption the breeding behavior of 'Spinefree' can be compared with that of other *R. hirtellum* derivatives (such as 'Houghton') in respect of spinelessness.

By crossing 'Spinefree' with 'Clark' (probably European x American), Hunter obtained a ratio of 64 spiny:37 more or less spineless seedlings, from which the almost spineless 'Captivator', '0–271', '0–273', '0–274', and '0–275' were selected. The seedling '0–261' was selected from the reciprocal cross. Although 'Captivator' and its sibs cropped well in Canada (Hunter, 1955*b*; Spangelo et al., 1970) at East Malling the fruit proved too small for the English market and cropping was erratic and usually poor.

In Britain, further backcrosses were made of both 'Captivator' and '0–261' to large-fruited English cultivars or selections, and four F_2 progenies were raised. Progenies were graded for degree of spininess when at least two years old, as in the first year there is considerable spine development in seedlings later found to be almost or completely spineless. The four spininess grades were: 1, spineless; 2, occasional spines; 3, moderately spiny; 4, full *grossularia* spininess. Results are summarized in Table 5.

TABLE 5. Segregations for spinelessness in *R. 'oxyacanthoides'* derivatives

Family	Parentage	Donor parent grade	No. of seedlings of spininess grade 4	3	2	1	% grade 1+2
B 23	Captivator x Lancashire Lad	2	4	7	9	1	47.6
B 225	B 23/18 S	3	13	0	0	5	27.8
B 223	B 23/9 S	2	1	2	1	15	84.2
B 224	B 23/10 S	2	11	0	2	9	50.0
B 226	B 23/19 S	2	3	0	0	17	85.0
B 227	B 23/10 x Broom Girl	2	33 (4+3)		0	0	0
B 229	B 23/19 x Broom Girl	2	31	14	2	1	6.3
B 339	0-261 x Leveller S_1	2	33	9	9	38	52.8
B 340	0-261 x Keepsake	2	69	14	8	4	12.6

These results agree with Hunter's conclusions that genetic control of spinelessness is complex. The differing results in percentage of spinelessness in Families B 339 and B 340 show that European cultivars may differ genotypically; phenotypically, 'Keepsake' has more and larger spines than the 'Leveller' S_1 seedling parent of Family B 339.

Fourteen of the 21 seedlings in Family B 23 had more or less contabescent anthers and produced very little pollen. Cropping was invariably poor, as it was also in the F_2 families in B 227, B 229, and in nearly all seedlings of Families B 339 and B 340. One seedling in Family B 339, with medium-large fruits which cropped fairly well and proved to be partially self-fertile and mildew resistant, has been used in further backcrossing. Two selections, B 942/9 and B 942/10, from an open-pollinated progeny of B 226, also crop fairly well, propagate well, are mildew resistant, self-fertile, and have medium-large fruits and only occasional spines. These selections have been propagated for trial and further backcrosses have been made from them to improve plant habit and fruit size.

Popova (1967) considered 'Captivator', '0–271', and '0–274' to be promising initial material for breeding spineless varieties, being resistant to *Sphaerotheca*, as winter-hardy as 'Smena' ('Houghton' x 'Zelenyj Butyločnyj'), and partially self-fertile. Open pollination of 'Captivator', '0–271', and '0–274' resulted in 47.8, 24.0, and 49.0% fruit set respectively, while 'Smena' set 49.6%. Percentages of spineless plants in the resulting progenies were 7.5, 12.5, 21.4, and 5.6. The corresponding sets and percentages of spineless seedlings following selfing were 46.1, 68.8, 78.4, and 45.8 and 11.1, 56.6, 65.4, and 8.3, respectively.

Reference has already been made to the use of 'Houghton' in breeding for spinelessness in Russia. In North America Colby (1934) considered 'Poorman' (probably 'Houghton' x 'Downing') to be a good parent for combined fruit size and spinelessness, and the American 'Welcome', which is almost spineless, yields well, and is more or less disease-free, is a fairly recent open-pollinated derivative of this cultivar. 'Poorman' has also been used in crosses with European cultivars in Russia (Kruglova, 1965), its progenies being less spiny than those derived from 'Josselyn'.

Other species which have been used as donors of relative spinelessness include *R. inerme*, *R. cynosbati*, and *R. 'robustum'* (Sergeeva, 1966a), while 'Pixwell' was derived from 'Oregon Champion' x *R. missouriense* (Brooks and Olmo, 1947).

F_1 hybrids between gooseberry species and spineless species from other subgenera are almost invariably totally spineless at maturity. Knight and Keep (1966) attempted to transfer spinelessness from blackcurrant to gooseberry via the allotetraploid, and in the second backcross to diploid gooseberry six out of 75 seedlings were more or less spineless in their first year but all subsequently developed spines. Although these spines were less numerous and smaller than those of European gooseberry cultivars, the reduction in spininess was less than that achieved in *R. 'oxyacanthoides'* derivatives and the project was abandoned. At present, attempts are being made at East Malling to transfer spinelessness to the gooseberry from *R. sanguineum*. The 2x F_1 hybrid is partially fertile and a proportion of totally spineless seedlings has been obtained in the first backcross generation.

Wind Resistance

According to Bunyard (1921) all cultivars of the 'Versailles' group have stout shoots which are easily broken off by wind or overcropping. Cultivars still grown today which are liable to wind damage include 'Fay's Prolific', 'Versailles', 'Laxton's Perfection', 'Laxton's No. 1', and 'Heinemanns Rote Spätlese'. Types reported as wind resistant include 'Erstling aus Vierlanden', 'Ayrshire Queen', 'New Red Dutch', 'Victoria' ('Wilson's Long Bunch') (Wood, 1964), 'Jonkheer van Tets', 'Red Lake', and 'Prince Albert'. Obviously, in breeding red- and whitecurrants, care should be taken to avoid intercrossing parents both of which are liable to wind damage, and wind resistance should be a selection criterion.

Flowering Season and Spring Frost Response

The flowering season in currants and gooseberries (late March-early May in southeast England) is so early that late spring frost damage is a constant hazard, particularly for early-flowering cultivars such as 'Wellington xxx', 'Westwick Choice', 'Brödtorp', and 'Baldwin' blackcurrants, 'Fay's Prolific' redcurrant, and 'Hönings Früheste' and 'Mauks Frühe Rote' gooseberries. Cold, windy weather during blossoming may also cause a scarcity of pollinating insects, the result being that currants and gooseberries, and particularly blackcurrants, tend to be unreliable croppers and a poor commercial proposition. Obviously, breeding for regularity of cropping, either through incorporating direct frost resistance or frost avoidance mechanisms, should have high priority in *Ribes* breeding programs.

The cultivars 'Consort', 'Roodknop, and 'Amos Black', all well known for their relative spring frost resistance, (e.g., Wassenaar, 1958), cover a range in flowering season from rather early, through midseason, to late, respectively. This suggests that factors other than flowering season may be involved in frost resistance. Little work has yet been done to discover if direct spring frost resistance occurs in the genus. Modlibowska and Ruxton (1953) subjected a range of western European blackcurrant cultivars to controlled low temperatures (−3.0 C to −4.2 C) for about two hours, either as cut racemes with both open flowers and buds, or as pot plants. Results showed no marked resistance in any variety tested. Tests with cut racemes, to demonstrate any basic resistance, showed a tendency for 'Seabrook's Black' and 'Cotswold Cross' to be less damaged and 'Mendip Cross' and 'Goliath' to be more damaged than other cultivars tested. Results with bushes in pots showed larger and more consistent differences, with the late-flowering 'Seabrook's Black' and 'Amos Black' (which was not included in the raceme test), being least damaged, and 'Wellington xxx' and 'Boskoop Giant' most damaged. Open flowers of 'Seabrook's Black' were more resistant to frost than those of the other cultivars tested, but in general, buds were more damaged than open flowers. This indicates that the blackcurrant increases in frost susceptibility up to the "grape" stage, and then, as the flowers open, becomes more resistant. Thus, while in most years the later-flowering cultivars (of which 'Amos Black' is an outstanding example) may experience less severe frosts at blossom times and so suffer less frost damage than the earlier-flowering ones, in certain years frosts may occur at periods when earlier-flowering cultivars have passed into the more resistant stage, and later-flowering ones are more susceptible. In spite of this, until sources of inherent spring frost resistance are located, breeding for frost escape mechanisms, such as really late flowering, seems the most promising line of attack.

There is some evidence of a major gene component in control of flowering season in both black- and redcurrants. The blackcurrant cultivars 'Daniels' September' and 'Laleham Beauty' both produce shoots of two types, the first type leafing out and flowering early and ripening fruit with 'Baldwin', the second being later in leafing and flowering, and ripening fruit a fortnight after 'Baldwin'. Hughes (1963) considered these cultivars to be chimeras, and was unable to fix either type by propagation of selected material, although single bud propagation was fairly successful in fixing late types. 'Daniels' September' is reputed to be a bud sport of 'Baldwin', but in the light of morphological and other differences, Hughes concluded this was unlikely. Whatever the origin of these two cultivars, it seems certain that their chimeral nature is the result of a major gene mutation in somatic tissue.

Although there is not an exact correlation between season of bud burst (or leafing-out) and season of flowering, there is inevitably a strong tendency for early leafing to be associated with early flowering and late leafing to be associated with late flowering, so that major genes controlling season of bud burst would also have an effect on flowering season. On evidence from progenies derived from intercrossing 'French Black', 'Seabrook's Black', 'Goliath', 'Boskoop Giant', and 'Baldwin', Tydeman (1938) postulated that season of leafing-out in blackcurrants is controlled by two additive genes, late-leafing types being the double recessive, midseason types having one dominant allele, and early types both dominant alleles. Wilson (1970) also obtained evidence of major gene segregation for this character, in progenies from selfing and intercrossing 'Amos Black', 'Mendip Cross', 'Malvern Cross', and 'Wellington xxx'.

In redcurrants, an early-leafing, early-flowering, and early-ripening bud sport of a seedling selection, attributed to a mutant gene, was described by Knight and Keep (1958*b*). In 1957, the sport was in full bloom on 14 March, more than four weeks before the normal type. The difference in season of ripening was considerably less—about ten days in 1957—as fruit from early flowers took longer to ripen owing to colder weather. Fruit size, seed number, and set were reduced in the sport, probably due to poor pollination conditions so early in the year. A similar reduction in size and seed number in fruit from the early branches of 'Daniels' September' and 'Laleham Beauty' blackcurrants was noticed by Hughes (1963).

The flowering season of a range of European blackcurrant cultivars is given by Todd (1962) and dates of first open flower, full flower, and end of flowering for 38 cultivars during the period 1960–69 are included by Anonymous (1973). Late-flowering cultivars include 'Mite Free', 'Kippen's Seedling', 'Roodknop', 'Amos Black', 'Goliath; 'Noir de Bourgogne', 'Seabrook's Black', 'French Black', and 'Victoria'. The recent introductions 'Black Reward' and 'Invigo' are also late-flowering. 'Brödtorp', 'Wellington xxx' and 'Baldwin' are among the earliest to flower. *R. nigrum sibiricum* and *R. dikuscha*, both used so extensively in breeding for winterhardiness, are both early-flowering and somewhat spring frost susceptible.

Data on the inheritance of flowering season in blackcurrants are given by Spinks (1947), Kronenberg and Hofman (1965) and Wilson (1964*b*, 1970). Cultivars producing a fairly high proportion of late-flowering seedlings in their progenies include 'Amos Black', 'Victoria', 'French Black', 'Seabrook's Black', and 'Roodknop'. The early-flowering 'Baldwin' is outstanding for the production of early-flowering seedlings. From intercrosses and selfs of 'Amos Black', 'Malvern Cross' (early flowering), 'Wellington xxx', and 'Mendip Cross' (midseason), Wilson (1970) obtained a total of 10% of late-midseason to late-flowering seedlings.

Data on the flowering season of redcurrants are given by Hedrick (1915) and Anonymous (1973), the latter tabulating dates of first flower, full flower, and end of flowering for 36 cultivars during the period 1961–68. Late-flowering cultivars include 'Prince Albert' which is frost escaping (de Bruyne et al., 1958), 'Heinemanns Rote Spätlese' (also frost escaping), 'Stanza', 'Kernlose', 'Wilson's Long Bunch', 'Moore's Ruby', and 'Raby Castle'; while 'Jonkheer van Tets', 'Fay's Prolific', and 'Cherry' are early-flowering. There are few published data on the inheritance of season of flowering in redcurrants, but it can be assumed that phenotype is generally a good indicator of genotype for the character, as in blackcurrants.

There is a range in flowering season in gooseberry cultivars as in currants, late types being 'Houghton', 'Lancer', 'Lord Derby', 'Trumpeter', 'Broom Girl', 'Captivator', 'Resistenta', and 'Robustenta'. Early-flowering types include 'Gem', 'Langley Gage', and 'Telegraph' (Hedrick, 1915; Anon., 1973).

Accessions of some species at East Malling are exceptionally late-flowering, including *R. bracteosum*, *R. petraeum*, and *R. divaricatum*. *R. bracteosum* has been used as a donor of this character in blackcurrant breeding, late-flowering selections being obtained in first and second backcrosses to the early flowering 'Brödtorp' and 'Wellington xxx', respectively (Knight and Keep, unpub.). Derivatives of *R. petraeum* (e.g., 'Prince Albert') and of *R. divaricatum* (e.g., 'Resistenta', 'Robustenta') also appear to inherit late flowering. Other late-flowering species include *R. manshuricum* and *R. multiflorum*, and it seems probable there are many more species of good potential for breeding frost-escaping cultivars; further work on these lines would be well worthwhile.

Another possible line of attack on the spring frost problem, which has apparently not yet been investigated, would be to breed for relatively early leaf development combined with late flowering. Although leafing-out and flowering are inevitably correlated to some extent, there is some variation between cultivars and species in relative rates of development of vegetative and floral organs, and well-developed leaves at the time of flowering should help to protect the inflorescence from frost damage.

Leaf Characters

Tydeman (1930) showed that 'Seabrook's Black' and 'Siberian' are heterozygous for two genes, *S* and *R*, controlling leaf shape, while 'French Black' is heterozygous for *R*. Leaf type *Sr* is more attenuated and pointed than the parental (*SR*) type, the serrations being markedly more acute, and the extremity of the terminal lobe being frequently surmounted by a filamentous extension of the midrib. Leaf type *sR* is almost rounded, the terminal and lateral lobes being much suppressed, but the exact form varies considerably among leaves even on the same plant. Leaf type *sr* is dark green, and in the most extreme form is almost round and adherent along the lower margin for a short distance round the petiole, giving the whole leaf the appearance of a hollow cone.

Leaf types are associated with vigor, *SR* and *Sr* seedlings being almost as vigorous as the parents; *sR* types being rather less vigorous with whippy, slender growth; and *sr* types often markedly dwarfed. Raceme lengths of *SR* and *Sr* seedlings are also appreciably longer than those of *sR* and *sr* types.

Tydeman reported that natural infestations in the field and results from artificial inoculations with gall mites suggested the parent ('Seabrook's Black') and *sR* and *Sr* type seedlings were mite resistant.

Sulphur Response

Sulphur sprays to control gall mite on blackcurrants and American gooseberry mildew on gooseberries have now been largely superseded by newer pesticides and fungicides, but it is of theoretical interest, and might become of practical importance again, that both crops, and particularly the gooseberry, contain types markedly susceptible to sulphur applications.

In blackcurrants, 'Goliath', 'Davison's Eight', 'Wellington xxx', 'Monarch', 'Victoria', and 'Edina' are sulphur shy, while Shipton (1969) found 'Wellington xxx', 'Laxton's Giant', 'Boskoop Giant', 'Seabrook's Black', and 'Cotswold Cross' to be adversely affected by lime sulphur.

Of 100 gooseberry cultivars tested, Kock (1914 *a* and *b*) found 56 including 'Leveller' to be sulphur shy. Present-day cultivars known to be susceptible to sulphur damage include 'Cousen's Seedling', 'Early Sulphur', 'Golden Drop', 'Grüne Flaschenbeere', 'Hönings Früheste', 'Leveller', 'Yellow Rough', and sometimes, 'Careless'. Resistant types include 'Whinham's Industry', 'May Duke', 'Gunners Seedling', 'Grüne Kugel', 'Lancer', 'Lady Delamere', 'Roaring Lion', and 'Whitesmith'.

In a study of the inheritance of sulphur susceptibility in gooseberry, Crane and Lawrence (1952) found that 'Whinham's Industry' and 'May Duke' bred true for resistance while selfed progenies of 'Early Green Hairy', 'Golden Drop', 'Langley Gage', 'Ostrich', and 'Pitmaston Greengage' segregated into resistant ('non-leaf-droppers') and susceptible ('leaf-droppers') in ratios approximating 3 or 2:1.

Disease Resistance

Leafspot: Leafspot (*Pseudopeziza ribis* Kleb.) attacks mature leaves of red- and blackcurrants, and gooseberries, from early summer onwards, resulting in premature defoliation, loss of fruit quality, and reduced cropping in the following year. Control of this pathogen is obligatory if plantations are to remain economic, so that resistance breeding should take high priority.

Leafspot occurs wherever *Ribes* host plants are grown and has been reported from all over Europe, New Zealand, Australia, and North America (Blodgett, 1936). The fungus can be cultured on nutrient agar, and detailed studies by Blodgett showed that the optimal temperature for conidial germination was about 20 C, for ascospores about 12 C, for conidial production 20–24 C, and for mycelial growth 20 C. Optimal pH for growth in culture was 5.4–7.0. Blodgett inoculated plants by spraying lower leaf surfaces with conidial suspensions or by discharging ascospores on to the leaf surface. The plants were then placed in a moist chamber for 12 or more hours. Plants maintained at higher temperatures prior to inoculation were more susceptible than those held at lower temperatures. Inoculations of excised leaves in moist dishes were as reliable as those on pot plants. Blodgett found that diseased leaves maintained over winter at 8–12 C would form ascospores for spring inoculations, and viable ascospores were obtained from gooseberry leaves collected in May 1933 and stored at 4 C until April 1934.

Klebahn (1906) described three morphologically distinct races from blackcurrant, redcurrant, and gooseberry, respectively; *R. alpinum* was resistant to conidia of all three. Eleven isolates from North America and Europe studied by Blodgett (1936) differed in several morphological and physiological characters and showed different levels of pathogenicity. Spores from isolates from currants were more pathogenic to currants than to gooseberries and vice versa. In Germany, Schmidt (1948) also isolated strains of *Pseudopeziza*, one attacking redcurrants, the other gooseberries; *R. alpinum* proved resistant. Zakhryapina (1959) confirmed that in Russia three morphologically distinct races occurred on red- and blackcurrants and gooseberries respectively, each being most pathogenic to its main host but capable of infecting the other two. He also described two further races on *R. aureum* and *R. alpinum*, and found evidence of several other races on the basis of differential host ranges. *R. alpinum*, severely attacked by its own strain, was resistant to all other races tested except one found on wild *R. rubrum*. Ascospores from black- and redcurrant and gooseberry had a wider host range than conidia. In America, Schoeneweiss (1966) described a strain specific to *R. alpinum*, which differed from that on gooseberry. In Hungary, leafspot was recorded on gooseberry for the first time in 1951. This form of the disease was considered to differ from that on redcurrants, and neighboring bushes of redcurrant, blackcurrant, *R. aureum*, and *R. alpinum* were unaffected (Kaszonyi, 1951).

Of the western European group of blackcurrant cultivars, none of the widely grown commercial types show really high resistance (although differences in susceptibility occur) and all need spraying to control the disease. The 'Baldwin' group has generally been found the most susceptible, while 'Boskoop Giant', 'Goliath', 'Laxton's Raven', 'Mendip Cross', 'Seabrook's Black', 'Victoria', and 'Westwick Triumph' have usually proved least susceptible. The little- grown Dutch 'Anger van Oeffelt' has shown resistance in some areas, although at East Malling it proved moderately susceptible. Of the Scandinavian group, 'Brödtorp', 'Gerby', and 'Öjebyn' show fairly good resistance. Russian cultivars reported as resistant include 'Barhatnaja', 'Belorusskaja Pozdnjaja', 'Izbrannaja', 'Junnat', 'Lošickaja', 'Metčpa', 'Minskaja', 'Nahodka', 'Neosypajuščajasja', 'Podmoskovnaja', 'Sopernik', and 'Uspekh' (Anon. 1961*a*; Žitneva et al., 1961; Voluznev, 1963). The Hungarian 'Victorie' is seldom attacked (Bordeianu et al., 1969).

In intercrosses of 'Baldwin', 'Boskoop Giant', 'Seabrook's Black', and 'Victoria', susceptibility appeared to be polygenic (Wilson et al., 1964). Two types of resistance occurred in these progenies: a resistance to initial infection, occurring in some 'Seabrook's Black' progenies, and a resistance to infection spread found in 'Victoria' progenies. In Scotland, high resistance both to leafspot and to American gooseberry mildew occurred in progenies of 'Anger van Oeffelt' x 'Stahanovka Altaja', 'Invincible Giant Prolific' x 'Öjebyn', and ('Consort' x 'Magnus') x ('Brödtorp' x 'Janslunda') (Anderson, 1969). Wilson has selected resistant seedlings in progenies of 'Baldwin' and 'Victoria' x 'Brödtorp' (Luckwill, 1966, 1967).

The resistance of Russian cultivars is derived mainly from *R. nigrum sibiricum* or *R. dikuscha* (Pavlova, 1955; Melehina, 1964; Kuminov, 1965; Mosolova and Volodina, 1969). *R. dikuscha* is being used as donor of leafspot resistance in western Europe also. Field resistant derivatives of this species showed restricted, non-sporulating lesions only and segregations in F_1, F_2, and first backcross progenies were consistent with the hypothesis that *R. dikuscha* is heterozygous for two dominant complementary genes Pr_1 and Pr_2, some *R. nigrum* cultivars being heterozygous for one or other of these genes (Anderson, 1972). All the F_1 hybrids were sub-fertile, the resistant seedlings being generally inferior to the susceptible in this respect (Anderson, 1967).

Other species which make viable hybrids with blackcurrant and which appear to be highly resistant to leafspot at East Malling include *R. americanum* (three accessions) and *R. aureum* (one accession) (both susceptible to the blackcurrant race in Russia [Zakhryapina, 1959]), *R. ciliatum* (one assession), *R. glutinosum* (two accessions), *R. nevadense* (one accession), *R. odoratum* (two accessions), and *R. sanguineum* (four accessions). The accessions of *R. ciliatum* and *R. sanguineum* appear to be immune. F_1 hybrids of blackcurrant x *R. glutinosum* and x *R. sanguineum* also show high resistance (but not immunity) and backcross progenies are now under selection.

Of the older redcurrant cultivars, those derived from *R. petraeum* usually show fairly high resistance to leafspot although none are immune. Resistant cultivars of this origin include 'Erstling aus Vierlanden', 'Prince Albert', 'Red Dutch', and 'Viking'. The most susceptible cultivars are those derived from *R. sativum* and include 'Fay's Prolific', 'Heros', 'Laxton's No. 1', 'Laxton's Perfection', 'Maarse's Prominent', 'Red Versailles', and

'White Versailles'. *R. rubrum* derivatives are often intermediate in resistance. The more recent *multiflorum* derivatives, 'Heinemanns Rote Spätlese' and 'Rondom', also show resistance, and Op't Hoog (1964) classifies a range of cultivars in decreasing order of resistance as follows: 'Duitse Zure' (= 'Prince Albert'), 'Erstling aus Vierlanden', 'Rondom', 'Heinemanns Rote Spätlese', 'Jonkheer van Tets', 'Fay's Prolific', 'Laxton's No. 1', and 'Maarse's Prominent'.

In Russia, wild forms of *R. petraeum* (including subsp. *atropurpureum* and *altissimum*), *R. rubrum*, *R.* [*rubrum*] *pubescens*, and *R. warscewiczii* are resistant (Anon., 1934; Bologovskaja, 1940; Zhukovsky, 1950). Other redcurrant species in the collection at East Malling which have proved resistant include *R. longeracemosum*, *R. moupinense*, and *R. multiflorum*.

In the light of present knowledge, *R. petraeum* and *R. multiflorum* appear to be the most satisfactory donors of leafspot resistance in redcurrant breeding.

As with blackcurrants, all widely grown western European *R. grossularia* cultivars are more or less susceptible to leafspot, often necessitating spray control, although the disease is usually less severe than on red- and blackcurrants. According to Wormald (1928) 'Careless' and 'Freedom' are highly susceptible, while 'Greengage', 'Gunner', 'White Lion', and 'Whitesmith' are rather less so.

Of the American cultivars, the following have been described as more or less resistant: 'Carrie', 'Como', 'Rideau', and 'Transparent' (Colby, 1935; Anderson, 1956). More susceptible cultivars include 'Columbus', 'Downing', 'Glenndale', 'Houghton', 'Josselyn', 'Oregon Champion', 'Pearl', 'Poorman', and 'Portage'. 'Como', 'Rideau', and 'Transparent', particularly the latter, were the best parents for resistance, although none of the seedlings were immune (Colby, 1935; Anderson, 1956). At East Malling the Canadian 'Captivator' and the American 'Pixwell' show fairly high resistance.

In Russia, cultivars described as resistant include 'Černyj Negus', 'Malahit', 'Pioner', 'Plodorodnyj', and 'Russkij' (Dement'eva, 1953). Newer resistant Russian cultivars include 'Belorusskij', 'Izjumnyj', 'Jarovoj', 'Krasavec Lošicy', 'Ogonek', and 'Ščedryj' (Anon., 1965c).

Bauer (1938, 1955) has used the small-fruited North American *R. divaricatum* as donor of combined mildew and leafspot resistance. First introductions following backcrossing to large-fruited *R. grossularia* cultivars were 'Resistenta' and 'Robustenta', both late-ripening and with medium-sized green fruits. A more recent leafspot and mildew resistant introduction of this origin is the early-maturing 'Remarka' with large red fruits. Work on these lines based on Bauer's early material is continuing at East Malling, where in addition some selections derived from backcrossing 'Captivator' to large-fruited English cultivars have been selected for moderate resistance.

Gooseberry species showing strong resistance to leafspot at East Malling include *R. niveum* (two accessions), *R. gracile* (one accession), *R. rotundifolium* (one accession), *R. oxyacanthoides* (two accessions), *R. irriguum* (one accession), *R. non-scriptum* (one accession), *R. cynosbati* (two accessions), *R. burejense* (one accession), and *R. watsonianum* (one accession). Selections in F_1 progenies of *R. grossularia* x *R. leptanthum*, x *R. stenocarpum* and x *R. watsonianum* showed high resistance to leafspot and further backcrossing is being undertaken to improve fruit size and quality.

Septoria leafspot: In Europe and North America, this disease (*Mycosphaerella ribis* [Fuckel] Kleb. [*Septoria ribis* Desm.]) occurs occasionally on both currants and gooseberries but it is considerably less common and much less damaging than *Pseudopeziza*. It is generally distributed on wild species in North America (Anderson, 1956). In New Zealand, it is the most serious disease of blackcurrants (Raphael, 1958). The disease overwinters on fallen leaves, and infection is initiated in the spring by ascospores.

Blackcurrants described as resistant in the literature include a seedling derivative of *R. dikuscha*, 'Junnat' ('Ščedra' x 'Goliath'), and 'Victorie' ('Rosenthals Schwarze' x 'Pamjat' Mičurina').

Resistant gooseberry cultivars include the *R. divaricatum* derivatives 'Resistenta', 'Robustenta', 'Perle von Müncheberg', 'Rezistent de Cluj', and 'Cluj V/3'. Russian cultivars reported as resistant include 'Malahit', 'Pioner', 'Plodorodnyj', and 'Russkij' (Kuznecov, 1969).

American gooseberry mildew: This disease, *Sphaerotheca mors-uvae* (Schw.) Berk., attacks young shoots and leaves of red- and blackcurrants and, particularly, gooseberries and causes death of growing points, stunting, and distortion. Gooseberry fruits are also attacked, making them unsaleable. Pruning helps to control this disease but fungicide sprays are often necessary, and breeding for resistance to mildew should be given high priority.

Until 1900, when American gooseberry mildew was first recorded on gooseberries in Ireland, the disease was confined to North America (Salmon, 1900–01). From Ireland the mildew spread rapidly to England and the rest of Europe, where it is now endemic. Although a few cases of mildew on red- and blackcurrants were reported in the early part of this century, the disease did not reach epidemic proportions on blackcurrants until the 1960s. In Britain this occurred more or less concurrently with changes in hitherto standard cultural and spray practices, notably the abandonment of sulphur sprays to control gall mite.

The life history in Britain of American gooseberry mildew on gooseberries and blackcurrants appears to differ slightly in that the fungus overwinters primarily as mycelium in the buds of gooseberries, whereas in blackcurrants ascospores initiate the spring infections (Merriman and Wheeler, 1968). Since *S. mors-uvae* cannot be cultured, for greenhouse tests on seedlings Bauer (1938) maintained inoculum through the autumn on plants cut back to induce new shoot growth, and through the winter either on *R. alpinum* seedlings or on plants forced into growth following low temperature treatment. Bauer sprayed test seedlings with spore suspensions then kept plants at 18–22 C for 24 hours in a saturated atmosphere for spore germination; relative humidity was then reduced to 80–90%. At East Malling, dusting shoot tips and young leaves repeatedly with conidia at rather lower humidities has proved effective. Brauns (1959) reports that some selections showing field resistance were slightly attacked in the greenhouse, and vice versa; similar results have occasionally been obtained at East Malling. Since the physiological and developmental state of the plant has a considerable effect on its mildew response (Bauer, 1953), the soft growth induced by greenhouse conditions may sometimes partially overcome genetic resistance; in types which are greenhouse resistant but field susceptible, it is possible that as yet unidentified strains may play their part.

There is some evidence of strain differentiation in *S. mors-uvae*. Salmon (1907) distinguished conidia on redcurrant from those on gooseberry by differences in shape and other characters, and Bouwens (1927) failed to cross infect red- and blackcurrant from gooseberry. Bauer (1953) described physiologic races following microspore inoculation of unspecified host species. However, Jordan (1968) was able to infect blackcurrants with conidia from gooseberry quite readily, and Keep (unpub.) also achieved this and the reciprocal cross infection.

Western European blackcurrant cultivars commonly grown in Britain are all more or less mildew susceptible, although 'Mendip Cross', 'Laxton's Giant', and 'Seabrook's Black' appear to be less seriously attacked in some areas. Of the Scandinavian group, 'Åström', 'Brödtorp', 'Gerby', 'Kajaanin Musta', 'Lepaan Musta', and 'Öjebyn' show moderate to fairly high resistance. Russian cultivars reported to show resistance in Scotland include 'Golubka', 'Gornoaltajskaja', and 'Ukraina' (Anderson, 1967).

Western European cultivars usually breed true for mildew susceptibility, although Wood and Anderson obtained a few resistant seedlings in the S_1 of 'Laxton's Giant' (Wood, 1966). Wilson (1964*b*, 1970) showed that eight cultivars of this group differed in the level of susceptibility transmitted to their progeny, 'Malvern Cross' ('Baldwin' x 'Victoria') and 'Victoria' being outstanding for transmitting high susceptibility, and 'Wellington xxx' and 'Mendip Cross' (both 'Boskoop Giant' x 'Baldwin') transmitting markedly lower levels of susceptibility. Genotypic differences between cultivars of this group are also demonstrated by differential behavior in crosses with resistant Scandinavian cultivars as discussed below.

Of the Scandinavian group, 'Brödtorp' and 'Lepaan Musta' are heterozygous for a dominant resistance gene *M* conferring moderate resistance (Rousi, 1966). Rousi suggests this gene is probably widespread in the wild north European *R. nigrum*. 'Goliath', 'Silvergieters Zwarte', 'Invincible Giant Prolific', 'Stahanovka Altaja', and 'Vystavočnaja', but not 'Baldwin' or 'Wellington xxx', carry factors for susceptibility epistatic to *M*, resulting in ratios approximating to 1:3 and 1:1 in crosses with *Mm* and *MM* plants respectively (Rousi, 1966; Anderson, 1967).

In western Europe, considerable use has been made of Scandinavian cultivars such as 'Brödtorp' and 'Öjebyn' as donors of resistance. Anderson (1969) obtained outstanding resistance in progenies of 'Invincible Giant Prolific' x Öjebyn' and ('Consort' x 'Magnus') x ('Brödtorp' x 'Janslunda'). Wilson and Adam reported high field resistance to both mildew and leafspot in several F_1 seedlings of 'Victoria' x 'Brödtorp' (Luckwill, 1967) and the recently released 'Blackdown' ('Baldwin' x 'Brödtorp') is field resistant. Other cultivars used as donors include the Russian 'Golubka', 'Gornoaltajskaja', 'Vystavočnaja', and 'Doč Altaja' (Wood, 1966). However, resistant seedlings derived from these cultivars succumbed on retest; a similar variability in response to mildew has been noticed at East Malling in derivatives of Scandinavian cultivars, resistant plants sometimes showing sizeable sporulating lesions on the undersides of leaves.

Alternative donors of resistance include *R. dikuscha* (reported to be heterozygous for susceptibility by Anderson [1967]), *R. glutinosum* and *R. sanguineum*. Second and first backcrosses from the two latter species, respectively, have included seedlings showing high resistance, and segregations suggest dominant major gene control (Keep, unpub.).

Keep (1970) summarized data on the response of 59 *Ribes* species to mildew in the field. Several species which make fertile hybrids with *R. nigrum* have not yet been used as donors and might repay investigation. In crosses between *R. nigrum* and members of the gooseberry group, high resistance or immunity (*R. nigrum* x *R. niveum*) has been reported where the gooseberry parent was resistant (Nilsson, 1966), and resistant seedlings have been selected in a first backcross of (*R. nigrum* x *R. divaricatum*) 4x x *R. nigrum* 4x (Anonymous, 1968*b*). However, transferences via an allotetraploid take considerably longer, and the alternative lines described above are to be preferred at present.

Little information is available on the incidence and inheritance of mildew response in red- and whitecurrants. Cultivars reported as susceptible in the literature include 'Prince Albert', 'Raby Castle', and 'Victoria' (= 'Wilson's Long Bunch'), while 'Viking' has proved susceptible at East Malling. Field resistant cultivars at East Malling include 'Fay's Prolific', 'Houghton Castle', 'Jonkheer van Tets', 'Minnesota', 'Red Lake', 'Rondom', 'Versailles' (red), 'Wentworth Leviathan' (white), and 'White Transparent'. Of the progenitor species, certain accessions of *R. rubrum*, *R.* [*petraeum*] *atropurpureum* and *R. sativum* have been reported as more or less susceptible, while other accessions of all these species and also of *R. longeracemosum*, *R. multiflorum*, and *R. warscewiczii* have proved field resistant (Vukovits, 1961; Keep, 1970).

Segregations for resistance in backcrosses and F_2s of backcrosses of *R. 'koehneanum'* (*R. sativum* x *R. multiflorum*) x redcurrant cultivars show that resistance is dominant and suggest that one or more major genes are involved (Keep, unpub.); from these data 'Jonkheer van Tets', La Constante', 'Red Lake', 'Rondom', and also *R. 'koehneanum'* appear to be heterozygous.

There is no strong resistance to mildew amongst western European gooseberry cultivars, although there are variations in susceptibility, 'Keepsake', 'Leveller', and 'Whinham's Industry' being particularly susceptible while 'Lancashire Lad' is fairly resistant.

Many cultivars grown in Scandinavia are both hardy and resistant, including the small-fruited Swedish 'Scania' and 'Dr. Törnmarck' and the Finnish 'Hinnonmäen Keltainen', 'Hankkijan Herkku', 'Hankkijas Delikatess', 'Hinnonmäkis Gula', and 'Pellervo'.

North American cultivars are nearly all mildew resistant, deriving their resistance from the native *R. hirtellum*. According to Dement'eva (1958), resistant cultivars such as 'Houghton' have a higher peroxidase and lower catalase activity in the leaves, and their fruits have more acid and less sugar. The stomata of resistant types are less widely open, and they have a lower transpiration rate and a thicker cuticle.

In breeding for resistance, much use has been made of American cultivars as donors, particularly in Russia. Crossing and backcrossing to large-fruited European types is the standard method. Progenies of 'Houghton' x European types include relatively resistant seedlings with good yields but small fruits; in backcross generations fruit size increases but the proportion of resistant seedlings decreases (Kruglova, 1965; Sergeeva, 1966*b*). Hybrids of 'Josselyn' have larger fruits than 'Houghton' derivatives but are very spiny, while those of 'Poorman' are less spiny but yield less well (Kruglova, 1965). Resistant selections derived from 'Houghton' include 'Izumrud', 'Rekord', and 'Smena'. Other resistant Russian cultivars derived from American types include 'Russkij', 'Malahit', and 'Plodorodnyj' (Sergeeva, 1954). According to Sergeeva, 'Russkij' and 'Smena' are among the most popular cultivars, 'Finik' being the only one of the old susceptible types still grown.

The use of *R. divaricatum* in Germany as donor of mildew resistance resulting in the introduction of the resistant cultivars 'Remarka', 'Rokula', 'Risulfa', 'Ristula', and 'Reverta' has been described in an earlier section. Latterly, material of similar origin at East Malling, previously highly resistant, has shown small mildew lesions on both plant and fruit, and other potential donors are being investigated as an insurance.

The Canadian 'Captivator' and its sibs have been used as donors of combined spinelessness and mildew resistance in several breeding programs (Anon., 1962; Popova, 1967). In F_2s of first backcross derivatives of 'Captivator', ratios of resistant to susceptible varied from approximately 3:1 to 1:3; in a second backcross the ratio was 1:1. In two first backcrosses to '0–261', ratios of about 1:10 and 1:4 were obtained (Knight and Keep, unpub.). Some of these selections showed high resistance, while others were slightly attacked.

Apart from *R. hirtellum* and *R. divaricatum*, a whole group of North American gooseberry species show high resistance to mildew, including *R. cynosbati*, *R. irriguum*, *R. niveum*, and *R. oxyacanthoides*, etc. (data summarized by Keep, 1970). Resistant Asiatic species include *R. aciculare* and *R. stenocarpum*. In Russia F_1 progenies between European cultivars and four resistant American species (and two resistant hybrids) all included resistant seedlings, the resistance varying according to the species and the cultivar used (Pavlova, 1940). In contrast, progenies of the Asiatic *R. aciculare* and *R. stenocarpum* x European types were entirely susceptible. These results have been largely confirmed at East Malling where a total of eight out of nine North American species contributed resistance to F_1 seedlings (the exception was *R. pinetorum*). In the F_1 of 'Leveller' x *R. stenocarpum*, however, only five out of 35 seedlings were fairly resistant. Donors of really high resistance included *R. leptanthum*, *R. niveum*, and *R. watsonianum*.

The resistance of the North American species is obviously dominant except in the case of *R. pinetorum*, and although Brauns (1959) considered resistance of *divaricatum* origin to be complex, sharp segregations into resistant and susceptible classes in some of the F_1 progenies studied at East Malling suggested relatively simple major gene control.

European gooseberry mildew: Attacking gooseberries and occassionally black- and redcurrants, *Microsphaera grossulariae* (Wallr.) *Lév.* occurs mostly on the upper leaf surfaces of overcrowded or shaded bushes and only rarely on the berries. It usually causes little economic damage, although severe attacks may result in early defoliation. The Romanian 'Galbene Mari', 'Rezistent de Cluj', and 'Cluj V/3', the two latter being first backcross derivatives of *R. divaricatum*, combine resistance to both American and European gooseberry mildew (Bordeianu et al., 1969).

White pine blister rust: The main economic importance of white pine blister rust (*Cronartium ribicola* Fisch.) lies in its effect on the five-needled pine species on which the aecial stage occurs. The uredial and telial stages occur on *Ribes* species only in areas where pines are present, and little harm is caused to *Ribes* hosts except occasionally when early severe infection causes premature leaf drop.

The disease originated in Asia and was first described on *Ribes* in western Europe in the mid-nineteenth century. It was reported in the United States on *R. aureum* in 1892 (Leppik, 1970) and subsequently spread rapidly both on pines and *Ribes* species and cultivars, resulting in the widespread banning of currant and gooseberry growing wherever five-needled pines were an important crop.

According to Pierson and Buchanan (1938), on inoculated plants of *R. petiolare* Dougl. the younger leaves were most susceptible to infection by both aecidiospores and uredospores. Uredospores developed on several species at day temperatures of 16–28 C and night temperatures of 2–20 C (van Arsdel et al., 1956). Two weeks of suitable temperatures were needed for fertile teleutospore production, which was inhibited by 20 C at night and 35 C by day.

There is little information about strain differentiation in this pathogen. Anderson and French (1955) found that rust collections from *Pinus strobus* and *P. monticola* produced symptoms on leaves of clonal *R. hirtellum* differing from those caused by collections from *P. lambertiana*. Kovaleva and Natal'ina (1968) found gooseberry to be resistant to spores from blackcurrants and described morphological differences in spores from the two hosts.

R. nigrum is among the most rust-susceptible species and blackcurrant cultivars were probably the main agents in the rapid spread of the disease throughout the United States. Western European cultivars are apparently all more or less susceptible wherever grown. Cultivars which are highly susceptible in Holland and/or France include 'Baldwin', 'Mendip Cross', and 'Wellington xxx'; moderately susceptible cultivars include 'Silvergieters Zwarte', 'Boskoop Giant', and 'Amos Black'; cultivars suffering slight to moderate attacks include 'Roodknop' and 'Noir de Bourgogne'. 'Cotswold Cross', described as highly susceptible in Holland, is little attacked in France, while 'Davisons's Eight', reported as resistant in Russia, suffers moderate to severe attacks in Holland (Anon., 1957; Decourtye and Lantin, 1965; Šerengovyj, 1969).

In Russia some cultivars derived from *R. nigrum sibiricum* are resistant to both rust and leafspot (Voluznev, 1966) and in Poland 15.2% of open-pollinated seedlings raised from hybrids of *R. nigrum europeaum* x *R. nigrum sibiricum* were field resistant to both diseases; named cultivars of this origin include 'Bzura', 'Dunajec', 'Łoda', 'Ner', 'Odra', 'Warta', and 'Wisła' (Somorowski, 1964).

R. dikuscha appears to vary in its response to rust from immune (Evréinoff, 1942) to susceptible (Guseva, 1964); an F_1 derivative, 'Primorskij Čempion', is described as resistant by Šerengovyj (1969).

The Canadian cultivars 'Crusader', 'Coronet', and 'Consort' ('Kerry' x *R. ussuriense*) all carry a dominant gene for immunity derived from *R. ussuriense* (Hunter, 1950a, 1955a). This gene was allocated the symbol *Cr* by Knight et al. (1972). The Swedish seedling 'Ri 1800' ('Consort' x 'Kajaanin Musta') combines resistance to rust and leafspot (Tamás, 1968), and work is in progress in Britain to combine the rust immunity of 'Consort' with gall mite and American gooseberry mildew resistance (Knight and Keep, unpub.).

The response of redcurrant cultivars to rust varies according to their origin, *R. petraeum* derivatives often showing immunity, *R. rubrum* derivatives having moderate to high resistance, and *R. sativum* types usually being very susceptible (Hahn, 1943). Cultivars described as more or less immune, or highly resistant, include 'Birnförmige Weisse', 'Erstling aus Vierlanden', 'Eyath Nova', 'Franco-German', 'Gögingers Birnförmige', 'Heros', 'Hochrote Frühe', 'Holland Redpath', 'London Market', 'Red Dutch', 'Rivers', 'Rote Kernlose', 'Simcoe King', 'Victoria', 'Viking', and 'Weisse Burgdorfer' (von Tubeuf, 1933; Kotte, 1958).

Although 'Viking' has frequently been described as immune, young leaves may be infected but the hyphae soon die, leaving minute necrotic flecks (Anderson, 1939). The resistance of this cultivar is dominant and in crosses with 'Cascade', 'Red Lake', and 'Stephens No. 9', all F_1 seedlings sampled were resistant (Anon., 1950a). Immune seedlings also occurred in progenies of 'Erstling aus Vierlanden', 'Gögingers Birnförmige', 'Hochrote Frühe', 'Houghton Castle', and 'Rote Kernlose' (von Tubeuf, 1933).

Very little information is available on the response of gooseberry cultivars to rust. On a 0–10 scale where 10 represented full susceptibility, Schellenberg (1923) graded *R. grossularia* in the range from 2 to 8, and commented on the variability of response in gooseberries in comparison with that of blackcurrants (graded 10). Darrow (1937) describes gooseberry cultivars in general as being very resistant and Kimmey (1938) and Kotte (1958) considered they showed some resistance.

There are numerous reports of the response of *Ribes* wild species to blister rust, the majority concerning North American species and published in the first half of this century. Most species are susceptible, and only potential donors of resistance will be mentioned here. Strong resistance or immunity has been shown by individual plants of *R. viburnifolium* (Anderson and French, 1955), *R. glutinosum* (Mielke, 1938), *R. aciculare*, *R. cynosbati*, *R. diacanthum*, *R. giraldii*, *R. glaciale*, *R. orientale*, *R. pinetorum*, and *R. stenocarpum* (although seedlings of *R. stenocarpum* were susceptible) (von Tubeuf, 1933). Varying degrees of resistance have been found in *R. americanum*, *R. innominatum*, *R. leptanthum*, *R. luridum* (von Tubeuf, 1933), *R. hallii*, and *R. cereum* (Kimmey, 1935, 1938). Hahn (1939) found a staminate clone of *R. alpinum* to be immune, while a pistillate clone was susceptible, in agreement with an earlier report of a sex-linked differential response.

Cane blight: According to Anderson (1956) this disease, *Botryosphaeria ribis* Gros. and Dug., is found only in the United States where it occurs on currants and gooseberries and other hosts. Attacked branches or whole bushes suddenly wilt and die during the summer, particularly just before the fruit ripens. The disease is now of minor importance since it can be controlled by cutting out and burning diseased branches.

The disease has been reported as most common

on blackcurrants and 'Fay's Prolific' (Adams, 1923) and on redcurrants (Anderson, 1956). Hildebrand and Weber (1944) described a number of strains differing in pathogenicity. On the basis of results following repeated inoculations with one highly virulent culture, they describe the following cultivars as immune: 'Crandall', 'Everybody', 'Fay's Prolific', 'Fox New', 'Knight's Sun Red', 'La Constante', 'Littlecroft', 'Minnesota No. 77', 'Perfection', 'Prince Albert', 'Red Dutch', and 'White Dutch'. 'Stephens No. 9', 'White Grape', and 'Pomona' were nearly immune. Adams' earlier report that 'Fay's Prolific' was commonly attacked contrasts with the findings of Hildebrand and Weber, again suggesting the occurrence of strains of the fungus (or possibly varietal misidentification).

Die-back and fruit rot: Dieback due to the grey mold fungus *Botrytis cinerea* Pers. is fairly common on European gooseberries, causing the death of branches or whole bushes. Rake (1966) showed the main site of entry was through fresh pruning wounds, winter inoculations being most successful and often resulting in the death of four-year-old bushes in one to two months. The disease is much less common on established currant bushes.

Of European gooseberry cultivars, 'Whinham's Industry' and 'Keepsake' are particularly susceptible, while in Russia, Dement'eva (1953) reported high resistance in 'Carrie', 'Černyj Negus', 'Houghton', 'Mičurinskij', 'Mysovskij 17', 'Mysovskij 37', and 'Pjatiletka'.

In gooseberry breeding, selection against excessive susceptibility to dieback is automatic, but as yet there are no reports of deliberate breeding for resistance. It is possible that further observations on cultivars and wild gooseberry species would reveal promising donors of resistance, and planned breeding to incorporate this into improved types would be well worthwhile.

Fruits of both currants and gooseberries are sometimes attacked by grey mold, the dead calyx or a wound providing a point of entry. Firm fruits with a tough skin would inevitably be less liable to damage and therefore to fruit rot. Intrinsic differences in susceptibility are suggested by the work of Wilson (1966) who sprayed blackcurrant fruits with a spore suspension and graded cultivars in order of increasing susceptibility to fruit rot as follows: 'Gerby', 'Invincible Giant Prolific', 'Kajaanin Musta', 'Consort', 'Brödtorp', 'Tinker', 'Bogatyr', 'Merveille de la Gironde', 'Baldwin', 'Ahornblättrige', and 'Laxton's Giant'. Redcurrants differ considerably in susceptibility to botrytis of the fruit and Kronenberg (unpub.) has graded cultivars for field response in Holland as follows: Very susceptible—'Erstling aus Vierlanden', 'Gondouin', 'Laxton's No. 1', 'Laxton's Perfection', 'Moore's Ruby', 'Prince Albert', 'Raby Castle', 'Scotch'; least susceptible—'Fay's Prolific', 'Hoornse Rode', 'Maarse's Prominent', 'Palandts Seedling', 'Stern des Nordens', 'Victoria'.

Reversion virus: The only virus of major economic importance on blackcurrants, reversion causes an estimated crop loss in Britain of at least 20% (Thresh, 1966). It is transmitted by the gall mite *C. ribis* (see under "Pest resistance"). The disease was first described in Holland and Britain between 1904 and 1912, and is still rare in New Zealand, Australia, and Canada (Thresh, 1970).

Thresh described symptoms of reversion in blackcurrants, method of graft transmission, life cycle of the gall mite vector, natural spread, and the use of heat therapy. Patch grafts are used in transmitting or testing for the presence of reversion, 'Baldwin' being recommended as an indicator variety; symptoms show up in the year following grafting. Shoot tips taken from reverted 'Baldwin' heat-treated for 20 to 30 days at 34 C were still free from virus after two years (Campbell, 1965). Thresh (1970, 1971) recognizes strains of reversion in blackcurrants, varying in virulence and symptom expression, including in particular a strain which occurs mainly in Russia and Scandinavia, causing malformation of flowers.

All western European cultivars are more or less susceptible, virulent strains reducing crops to almost nil (Thresh, 1971), but Tiits (1964, 1970) reported that some wild forms of *R. nigrum* and 'Leopoldorina' were tolerant, and some material of *R. nigrum sibiricum* was resistant to the movement of the pathogen in the plant owing to hypersensitivity reactions. Some Russian cultivars derived mostly from *R. nigrum sibiricum* or *R. dikuscha* are reported as resistant. These include 'Blestjaščaja', 'Golubka', 'Novost', 'Pamjat' Mičurina', 'Podmoskovnaja', 'Sejanec Černyj', 'Stahanovka Altaja', and 'Uspekh' (Šaumjan, 1964; Potapov and Grinenko, 1968). 'Rus', 'Narjadnaja', and an accession of *R. ussuriense* were resistant to the gall mite vector and failed to develop galled buds or reversion symptoms in a replicated infection plot, and 'Koksa', although more or less susceptible to the mite, also remained free from reversion (Anderson, 1971a). More recent work has confirmed that 'Rus' is reversion resistant, but has shown that 'Narjadnaja' and *R. ussuriense* are susceptible (Anderson, unpub.).

In breeding for reversion resistance, Anderson (1971a, 1971b) has used 'Rus' as donor, and obtained a proportion of resistant plants in its progenies. Natural infections of reversion in other cultivars and species have been detected only in *R. bracteosum*, *R. rubrum pubescens*, and *R.* 'carrierei' (*R. nigrum* x *R. glutinosum*) (Thresh, 1970). Although neither Thresh (Anon., 1966) nor Tiits (1970) was able to infect *R. grossularia* with reversion by graft inoculation, a hybrid of *R. nigrum* x *R. grossularia* was infected in this way (Anon., 1966), suggesting that immunity in the gooseberry is recessive. Other species reported as immune by Tiits include *R. vulgare* Lam., *R. rubrum*, *R. petraeum*, *R. multiflorum*, and some individuals of *R. aureum* and *R. hirtellum*. In contrast to Tiits' findings with redcurrant species, Thresh (Anon. 1968a,) succeeded in transmitting the virus to several redcurrant cultivars.

Tolerance to infection was shown by some plants of *R. aureum*, *R. cereum*, *R. 'fuscescens'*, *R. 'gordonianum'* (*R. sanguineum* x *R. odoratum*), *R. longeracemosum*, and *R. niveum* (Anon., 1968a; Tiits, 1970). Species considered resistant to movement of the pathogen within the plant owing to hypersensitivity reactions were *R. fasciculatum* Sieb. and Zucc., and *R. sanguineum* hort. (Tiits, 1970).

Vein banding virus: According to Posnette (1970), gooseberry vein banding disease was first noted in Czechoslovakia in 1930. It is now widespread on gooseberries in Europe, and a disease causing similar symptoms is common on redcurrants, although it is not yet absolutely certain that the diseases are identical (Posnette, 1970; van der Meer, 1970). Transient vein banding symptoms sometimes occur on blackcurrants in the spring foliage that subtends flowers, but the disease appears to be of little importance in this crop.

All the main aphid species occurring on currants and gooseberries have proved capable of transmitting the vein banding virus from either redcurrant or gooseberry or both (van der Meer, 1962; Karl and Kleinhempel, 1969). The virus from gooseberry is semipersistent in *Hyperomyzus pallidus* and *Nasonovia ribisnigri*, the aphids requiring an acquisition access period of at least one hour and an inoculation period of at least 30 minutes (Posnette, 1964). The virus is also readily transmitted from gooseberry to gooseberry and redcurrant to redcurrant by chip budding or patch grafting, but interspecific graft transmission is less reliable. Neither aphid- nor graft-transmission tests suggest the occurrence of well-differentiated strains (Karl and Kleinhempel, 1969).

Kleinhempel (1970) reported a 51% depression in growth of redcurrant seedlings one year after infection with gooseberry vein banding virus. All common redcurrant cultivars are susceptible to vein banding, 'Fay's Prolific', 'Jonkheer van Tets', 'Prince Albert', and 'Red Lake' being sensitive, while 'Rondom' is tolerant (van der Meer, 1970). Susceptible redcurrant species and hybrids include *R. 'holosericeum'* (*R. petraeum* x *R. rubrum*), *R. 'Koehneanum'* (*R. multiflorum* x *R. sativum*), *R. longeracemosum*, and *R. multiflorum* (Posnette, 1970). There appear to be no data on potential donors of resistance in this group, and no planned resistance breeding is in progress.

Western European gooseberry cultivars are all more or less susceptible to vein banding, 'Leveller' (and its seedlings) and 'Lancashire Lad' being particularly sensitive and 'Careless' and 'Crown Bob' among the more tolerant (Posnette, 1970). Of 123 cultivars studied for eight years in Denmark, the American 'Mountain Seedling', 'Oakmere', and an unnamed accession were the only types showing no symptoms, although the severity of symptoms varied in the other cultivars (Thomsen, 1970). The disease is so widespread that uninfected clones of some cultivars have proved difficult to locate (Posnette, 1954), but healthy clones of 'Careless' have been established by meristem culture (Anon., 1968a). Heat therapy has so far failed as gooseberries are particularly sensitive to high temperatures (Thomsen, 1970).

Severe attacks of vein banding cause stunting and have a deleterious effect on rooting of hardwood cuttings, so that the introduction of resistant cultivars would be of considerable benefit to gooseberry growers.

Little is known about wild species as potential donors of resistance, although vein banding symptoms have been seen on *R. 'robustum'* (*R. niveum* x *R. inerme*) and *R. 'rusticum'* (*R. grossularia* x *R. hirtellum*) (Posnette, 1970). In Britain, ten seedlings of 'Green Ocean' x *R. divaricatum* have remained free of symptoms for seven years although inoculated with viruliferous aphids and subsequently graft inoculated (Adams and Keep, unpub.). Back grafts from the F_1 seedlings failed to produce symptoms in gooseberry indicators. A backcrossing program is planned to restore fruit size and quality.

Other viruses: A number of other viruses occur sporadically in *Ribes* but all of them are of relatively minor importance at present. Further information on host range, symptoms and transmission is given by Frazier (1970).

Pest Resistance

Blackcurrant gall mite: Identified by Westwood (1869) in Britain, *Cecidophyopsis* (*Eriophyes, Phytoptus*) *ribis* (Westw.) now occurs throughout the northern temperate zone where blackcurrants are grown; it has not yet been reported from Tasmania (Raphael, 1958). Apart from the serious economic consequences of reversion virus (of which the mite is the sole vector [Massee, 1952] and which almost invariably follows the mite infestations), attacked buds fail to grow out and crop loss from this cause alone can be considerable if the mite is not controlled.

Smith (1959, 1960) described the life cycle of the mite in detail. The majority of mites emerge from galled buds from April until late June, and enter buds on new shoots shortly afterwards. The mites feed inside the buds, and reproduce throughout the summer and autumn and again from January to April, causing the buds to swell into the familiar "big bud." Collingwood and Brock (1959) observed average peak populations of 4,000 and 30,000 individuals per galled bud in October and late March respectively.

Hardwood cuttings can be freed of mites by immersing in warm water at 40–47.5 C for from five to 40 minutes, the time decreasing with increasing temperature (Thresh, 1964a). Unless carried out in early autumn, this treatment may cause precocious shoot growth which is susceptible to frost damage.

Taylor (1914) suggested an association between plant morphology and mite resistance, attributing the inability of the mite to invade the gooseberry bud to the greater overlapping of scale leaves, the denser fringe of hairs at their margins and the general tightness of the buds compared with blackcurrant. Thresh (1964*b*) found that relative resistance to the mite was correlated with the hairiness of stipules, petiole base, young leaves, and stems. 'Seabrook's Black', the most hairy of eight blackcurrants studied, was the most resistant, and 'Goliath', the least hairy, was the most susceptible.

Suncova (1958) attributes the mite resistance of Altaj blackcurrant cultivars and of 'Laxton' to high osmotic pressure of the cell sap; the higher the pressure, the greater the resistance.

In screening for resistance, Smith (1962) indicated the difficulties of handling mites, which are soft skinned, delicate, and rapidly desiccated, and have not yet been cultured outside the environment of blackcurrant buds. He successfully transferred mites from galled buds individually or in groups, on a single hair, and in larger numbers by pinning halves of galled buds to new shoots. Some plants on which a single mite had been placed subsequently developed reversion symptoms.

For large-scale rapid inoculation of seedling populations, Knight et al. (unpub.) stood mite-infested shoots at intervals among seedlings in pots in the greenhouse, maintaining a high humidity and shading against strong sunlight. Autumn inoculations of rapidly growing seedlings with suspensions of mites in water have also proved successful as judged by subsequent development of big buds on treated and control seedlings planted the following spring in a mite infection plot.

Thresh (1966) has shown that infection with reversion virus increases susceptibility of blackcurrants to the mite, possibly through the reduction in hairiness following virus infection. In screening seedlings for resistance, therefore, greenhouse inoculations are best followed by planting in a field infection plot in which alternate rows (or perhaps one in every four or five rows) consist of heavily infested blackcurrants. Under these conditions at East Malling, occasional seedlings in segregating populations have developed galled buds for the first time in their fourth year in the infection plot. Although such plants have a measure of resistance, in testing for high resistance or immunity, a prolonged period of exposure to mite infestation is obviously required.

Eriophyid mites have been reported fairly frequently on redcurrant, gooseberry and other *Ribes* species, but until recently gall formation has been described only on blackcurrants. In the past few years, a gall-forming mite has been reported on redcurrants at Perleberg in Germany and in northern Holland (van de Vrie, 1956; Behrens, 1964; van Eyndhoven, 1967). Cross transferences between red- and blackcurrant gall-forming mites showed that two different types ("soort") were involved, neither mite being able to maintain itself for long, nor to reproduce on the other host (van de Vrie, 1958, 1959). Although van de Vrie could find no morphological differences between the two gall mites, van Eyndhoven (1967) considered them sufficiently unlike to warrant specific status for the redcurrant type, which he named *C. selachodon*. Briggs (Anon., 1965*d*) transferred gall mites from blackcurrant to gooseberry ('Keepsake') and again the mites failed to maintain themselves, none being found two weeks after inoculation.

Recently, a non-gall-forming Eryophyid mite has been found at East Malling living on certain blackcurrant cultivars, notably 'Brödtorp' and other Scandinavian types, and on a gall mite-immune seedling selection, 'B655/56'. Similar mites have been found living and reproducing on redcurrant and gooseberry cultivars and wild species. It seems probable that earlier reports of 'gall' mites causing death but not galling of redcurrant buds and living on bud scales of gooseberries (e.g., Taylor, 1914; Massee, 1928) refer to this type of mite and not to the common gall mite of blackcurrant.

Clearly, there are at least three races (or possibly species) of Erophyid mites occurring on *Ribes* cultivars, two causing galled buds on blackcurrants and redcurrants respectively, and a third type which causes necrosis and sometimes death of buds, but not gall formation. The relationships between the non-gall-forming mites which occur on redcurrant, gooseberry, and blackcurrant are not yet clear, but investigations on host range and virus vector relationships of these mites have been started in Britain (Briggs et al., unpub.).

Widely grown western European blackcurrant cultivars are all more or less susceptible to *C. ribis*, although the French group, including 'Seabrook's Black', 'Mite Free', 'Resister', and 'Siberian' shows a measure of resistance (Lees, 1918; Hatton, 1920; Tydeman, 1930). Proeseler (1967) graded 67 cultivars for severity of attack in the field in Germany. No galled buds occurred on 'Kippen's Seedling' or 'Taylor's Seedling', while 'Coronation', 'Merveille de la Gironde', and 'Nulli Secundus' were only slightly attacked.

A number of Russian cultivars are reputedly resistant, Pavlova (1964) reporting that newer resistant types from Siberia and the Altaj were all derivatives of *R. nigrum sibiricum*. Work in Scotland has confirmed the resistance of the *R. nigrum sibiricum* derivatives 'Rus' and 'Narjadnaja' (North, 1970; Anderson, 1971a). Mites can infest but not survive in buds of these cultivars, and *R. ussuriense* responds similarly to mite infestations.

Anderson (1971a) showed that resistance to galling in 'Rus' and 'Narjadnaja' is due to a single dominant gene *P*. The occurrence of reverted, non-galled plants showed that this type of resistance permitted mites to feed long enough to transmit the virus.

Other potential donors of resistance in blackcurrants include *R. nigrum pauciflorum* and wild forms of *R. nigrum europaeum* (Pavlova, 1964).

Knight and his coworkers (1974) selected gooseberry as donor for combined gall mite and reversion resistance. As already indicated, the resistance to reversion in the gooseberry proved to be recessive, but the F_1 allotetraploid (Fig. 5) was highly resistant if not immune to the gall mite. First backcross plants from crossing to both 2x and 4x blackcurrants were also highly resistant. In the second backcross diploid line almost 2% of plants, including 'B655/56', were resistant; in the tetraploid line, over 50%. The majority of the plants in the diploid line were morphologically closely similar to blackcurrants and all except one of those sampled for cytological examination were diploid. Third backcrosses at both levels also segregated for resistance, 1:1 ratios demonstrating the presence of a major gene, *Ce*, governing resistance (Knight et al., 1974). Since resistance had been successfully transferred at the diploid level, work on the tetraploid line was discontinued. A third backcross diploid progeny is now under selection for commercial qualities, and other third and fourth backcrosses are under test. Preliminary observations suggest that mite resistant blackcurrants of near commercial quality will be obtained in the third backcross.

Aphids: Börner (1952) describes 12 aphid species living on *Ribes*, but only six or seven of these are sufficiently widespread on currants and gooseberries to be of much commercial significance. The main species are vectors of vein banding virus, and severe infestations also cause stunting and leaf distortion. Spray control is essential.

Aphis grossulariae Kltb. commonly occurs on redcurrant and gooseberries, while *A. schneideri*

Born. is a less common pest of red- and blackcurrants. Both species remain on currants or gooseberries all year.

Hyperomyzus lactucae (L.) is the most common aphid on blackcurrants and is occasionally found on redcurrants, while *H. pallidus* (H.R.L.) occurs, less commonly, on gooseberries. Both species produce phytotoxic substances which cause vein banding type symptoms on leaves. These species overwinter as eggs on *Ribes* plants, and after a few generations of apterae in the spring, alatae develop and migrate to herbaceous hosts. The aphids return to currants and gooseberries in late autumn.

Cryptomyzus ribis L. commonly causes "blistering" of redcurrant leaves, and sometimes occurs on blackcurrants. *C. galeopsidis* Kltb. is found on mature leaves of both black- and redcurrants causing few marked symptoms. There are non-migrating forms of both these species which remain on *Ribes* throughout the year (Hille Ris Lambers, 1953).

Nasonovia ribisnigri (Mosley) is the most common gooseberry aphid, and is occasionally found on black- and redcurrant. The summer generations are a serious pest of lettuce.

There are few reports of cultivars resistant to the aphids commonly found on other members of their group, although as has been indicated, some aphid species occur only, or primarily, on one or two of the three *Ribes* crops. Hunter (1950c) described the redcurrant 'Viking' as showing some aphid resistance (aphid species unspecified) and obtained a markedly resistant seedling in the F_1 of 'Viking' x 'Stephens 9'. Three selections from 'Crandall', 'Uzbekistanskaja Krupnoplodnaja', 'Plotnomjasaja', and 'Uzbekistanskaja Sladkaja' were resistant to *C. ribis* (Jagudina, 1968).

In a search for sources of strong resistance as a preliminary to resistance breeding, Keep and Briggs (1971) surveyed a total of 95 plants comprising accessions of 56 species, for the presence of *H. lactucae, H. pallidus, N. ribisnigri, A. grossulariae, A. schneideri,* and *C. ribis.* So many species appeared field resistant that insectary tests on cut shoots were undertaken with the two most serious aphid pests of blackcurrant and gooseberry, *H. lactucae* and *N. ribisnigri,* on plants considered suitable as donors from an agronomic point of view. Response to *A. grossulariae* was checked by sleeve inoculations in the field. It was assumed that such tests would eliminate types whose field resistance depended on escape mechanisms likely to be vitiated in the course of backcrossing. Two young shoots from each test species were maintained in water in an insectary and inoculated with two or three second generation adult apterae. Missing apterae were replaced after 24 hours and the total number of aphids present were counted 24 hours later. Shoots on which an average of less than one nymph per added adult had been produced over the 48 hour period were classed as resistant. Results for the more abundant aphid species are summarized in Table 6.

TABLE 6. Response of *Ribes* species to aphids

	Field observations		Insectary tests on field res. plants	
Aphid species	No. of plants examined	No. field resistant	No. tested	No. resistant
Hyperomyzus lactucae	95 (56)	82 (51)	63 (40)	32 (22)
H. pallidus	95 (56)	83 (50)	—	—
Nasonovia ribisnigri	95 (56)	57 (37)	15 (12)	4 (4)
Aphis grossulariae	95 (56)	64 (46)	17+ (13)	15+ (12)

+ Tested by sleeve inoculations in the field.
Nos. in brackets = no. of species represented.

In selecting donors of resistance, plants were chosen which had proved highly resistant or immune both in field and insectary tests, and which were known, or thought likely, to produce more or less fertile F_1s with the recurrent parent (Keep and Briggs, 1971).

In testing progenies for resistance, young seedlings in pots were inoculated with up to five apterous adults and the numbers of nymphs recorded. Plants from which at least five adults "disappeared" and on which few or no nymphs were produced were classed as resistant.

For breeding blackcurrants resistant to *H. lactucae*, species in Berger's section Calobotrya were selected. All accessions of species in this section examined (*R. cereum, R. ciliatum* H. and B., *R. glutinosum, R. malvaceum* Sm., *R. nevadense* Kellogg, *R. sanguineum, R. viscosissimum* Pursh.) were highly field resistant or immune.

Preliminary results from F_1 and first backcross progenies suggest that resistance in *R. sanguineum, R. glutinosum,* and *R. cereum* is recessive. One out of five F_2 progenies from selfing susceptible first backcross derivatives of *R. cereum* contained three out of 50 seedlings showing moderate or high resistance in the insectary; the other four progenies were all susceptible.

In research on *N. ribisnigri*, a field immune accession of the Californian *R. roezlii* produced one F_1 seedling when crossed with 'Leveller'. This seedling proved fully susceptible in the insectary, again suggesting resistance may be recessive.

Segregations for high resistance to *A. grossulariae* occurred in an F_1 progeny of the Far Eastern *R. alpestre* crossed with *R. grossularia*. A ratio of 35 resistant to 14 susceptible, determined by natural infestations supplemented with sleeve inoculations in the field, was obtained; 31 weak dwarfs were omitted from the tests. Two backcross progenies have been raised.

Preliminary results in breeding for resistance to *Ribes* aphids suggest that it would be well worthwhile investigating additional species. Dominant major gene resistance to *H. lactucae* and *N. ribisnigri* may be available in the numerous potential donors, and breeding for resistance to *C. ribis* in redcurrants and to *H. pallidus* in gooseberries would be of commercial benefit.

Sawflies: Four species of sawflies occur on *Ribes*, three of them being sufficiently prevalent to be of commercial significance. These are the common gooseberry sawfly *Nematus ribesii* (Scop.), the pale spotted gooseberry sawfly *N. leucotrochus* Hart., and the blackcurrant sawfly *N. olfaciens* Benson. *N. ribesii*, which usually has three generations a year, is the most common pest, occurring mainly on gooseberries and less commonly on redcurrants. *N. leucotrochus*, with the same host range, has only one generation a year. *N. olfaciens* which was first described in Britain by Benson (1953) has two or more generations a year, and attacks black- and redcurrants. If unchecked, heavy infestations of sawfly larvae may result in complete defoliation.

Records of natural sawfly damage at East Malling, where gooseberry cultivars, and to a lesser extent redcurrants, are regularly and heavily infested with *N. ribesii*, suggest that strong resistance to this species is unlikely to be found within cultivars or species of the gooseberry group.

In the dioecious subgenera, no natural damage has been seen over a period of six or more years on any of 12 species, although slight feeding symptoms occurred on some plants when larvae were enclosed on shoots in nylon sleeves. Similarly, two accessions of *R. roezlii* Reg. and one of *R. menziesii* (both included by Berger in the subgenus Hesperia), have shown no natural infestations over the same period, and few or no feeding symptoms were seen when larvae were sleeved on *R. menziesii* and on one of the accessions of *R. roezlii*. Six species in Berger's subgenus Calobotrya, including *R. sanguineum*, have also proved field immune.

Hybrids between gooseberry (*R. grossularia*) cultivars and field resistant accessions of *R. nigrum*, *R. sanguineum*, *R. roezlii*, and *R. orientale* have been raised at East Malling. Several F_1s of *R. nigrum* x *R. grossularia* have suffered slight to moderate natural infestation but one accession of *R. grossularia* x *R. sanguineum* has shown no field damage over six years, although sleeved larvae caused slight damage. A first backcross progeny from this hybrid is under observation. Hybrids with *R. roezlii* and *R. orientale* have yet to be tested in the field.

Little is known about the host range of *N. olfaciens*, but since it commonly occurs on currants but not gooseberries, donors might be found in the gooseberry group. Species in the Calobotrya subgenus, which appear to be highly resistant to *N. ribesii*, might also repay investigation.

Leaf curling midge: This blackcurrant pest, *Dasyneura tetensi* Rübs., was first described in Germany in 1891 and was first recorded in England in 1917 (Greenslade, 1941). Eggs are laid in early May on shoot tips of blackcurrants where larval feeding causes scarring, distortion, and curling of the young leaves. There are three or four generations during the summer (Greenslade, 1941). Established fruiting plantations are little affected by midge attacks, but heavy infestations in the nursery may prevent cuttings from growing and hinder identification when roguing and inspecting for certification.

Although there are no reports of strong resistance in blackcurrant cultivars, 'Goliath' was significantly more susceptible in a replicated trial than 'Seabrook's Black', 'Baldwin', and 'Boskoop Giant'; 'Boskoop Giant' was the least damaged of these four cultivars, and 'Davison's Eight', growing alongside, was slightly less damaged than 'Boskoop Giant' (Greenslade, 1941). According to Stenseth (1966), 'Brödtorp' is less susceptible than other cultivars tested, while 'Bang Up' is significantly less susceptible than 'Wellington xxx' and 'Boskoop Giant'.

The midge does not attack redcurrants and gooseberries and in 1963, when all blackcurrant cultivars in the type collection at East Malling were attacked, as well as one accession of *R. dikuscha*, all dioecious species including *R. alpinum* and *R. orientale* remained unharmed, as did members of the subgenera Calobotrya (*R. cereum*, *R. ciliatum*, *R. glutinosum*, *R. nevadense*, *R. sanguineum*) and Symphocalyx (*R. aureum*, *R. odoratum*) (Keep, unpub.). Anderson (1967) reported that *R. ussuriense* was resistant but its derivatives 'Consort', 'Coronet', and 'Crusader' were susceptible, as were 190 seedlings of 'Consort'. Other evidence suggesting midge resistance is recessive is the susceptibility of a number of F_1 hybrids of blackcurrant with various species of which certain accessions have proved resistant. These susceptible hybrids include *R. 'carrierei'* (*R. glutinosum albidum* x *R. nigrum*), 'B22' (*R. nigrum* x *R. longeracemosum*), and *R. 'culverwellii'* and 'B21/13' (*R. nigrum* x *R. grossularia*) (Keep, unpub.).

In breeding for resistance to this pest, further work with *R. ussuriense* and species of the subgenera Calobotrya and Symphocalyx as donors should prove a promising line of attack.

Bud eelworm: The blackcurrant bud eelworm, *Aphelenchoides ritzemabosi* (Schwartz) Steiner & Buhrer, has been reported on 'Westwick Choice', 'Daniels' September', 'Baldwin', 'Malvern Cross', and 'Amos Black'. It is of commercial importance in Britain only on the first two cultivars in certain localities. The eelworms live in the buds, many of which die following severe infestations.

In New Zealand, 'Daniels' September', 'Cotswold Cross', and 'Goliath' are very susceptible, while 'Magnus' shows satisfactory resistance or tolerance (W. R. Boyce, personal communication). Red- and whitecurrants are also attacked in New Zealand, but no eelworm has been seen on gooseberry at the Levin Horticultural Research Centre (Boyce, unpub.).

Bōys (reported by Knight, 1972) found no eelworms at all in buds of the two gall mite resistant gooseberry derivatives 'B655/56' and 'B439/1', which were growing in a gall mite infection plot; on closely adjacent blackcurrant infector plants, however, from 2.4–16.1% of non-galled buds and from 9.3–50.0% of galled buds per bush were infested with *A. ritzemabosi*. Clearly, galling renders buds more susceptible to eelworm infestation, but the total absence of eelworms from 'B655/56' and 'B439/1', shows that these are promising donors, The origin of the resistance of these two seedlings is not certain as the response of their blackcurrant parents to eelworm is not yet known.

Breeding for Specific Fruit Characters

Yield

The cropping capacity of a cultivar depends both on yield potential and on ability to develop this potential to the fullest. Factors contributing to yield potential include amount of fruiting wood (or "yielding surface"), number of fruiting nodes per unit length of branch, number of flowers per node, and fruit size. Actual crop yield depends on the proportion of flowers which set, and the absence of subsequent running-off. The proportion of flowers setting depends on amount of spring frost damage, and relative ability to set with self pollen even in the absence of pollinating insects. These factors may be more important in final crop yield than intrinsic crop potential.

Blackcurrant: Tamás (1961) and Wilson and Adam (1967) studied the phenotypes of a range of blackcurrant cultivars from the point of view of crop potential. In Sweden, Tamás measured the yield components' yielding surface (or total shoot length per individual, capable of forming flower buds), yielding ability (number of berries per shoot meter), and average weight per berry for eight cultivars. He found 'Coronet' to be outstanding for yielding surface, followed by 'Consort', 'Wellington xxx', and 'Brödtorp', while 'Rosenthals Langtraubige', 'Cotswold Cross', and 'Kajaana Seminarium' produced much less fruiting wood. 'Brödtorp' was outstanding for numbers of fruits per unit length of shoot, followed by

'Kajaana Seminarium'; 'Cotswold Cross' and 'Wellington xxx' produced fewest fruits. For fruit size, 'Coronet' was best, followed by 'Silvergieters Zwarte' and 'Wellington xxx'. 'Brödtorp' and 'Cotswold Cross' had the smallest fruit.

In Britain, out of 14 English cultivars, 'Cotswold Cross' produced most wood per bush, followed by 'Malvern Cross', 'Tinker', 'Blacksmith', and 'Baldwin'; 'Westwick Choice', 'Laxton's Giant', and 'Amos Black' produced least (Wilson and Adam, 1967). 'Greens Black' was outstanding for number of buds per shoot, a high proportion of these being fruit buds. 'Wellington xxx' and 'Laxton's Giant' had fewest buds per shoot but most of these were fruit buds. In contrast, 'Boskoop Giant', 'Tinker', 'Westwick Choice', 'Seabrook's Black', and 'Mendip Cross' combined a relatively low total number of buds with a low proportion of flower buds. Wilson and Adam estimated mean number of fruit buds per bush and again found 'Cotswold Cross' to be outstanding, followed by 'Malvern Cross', 'Greens Black', 'Baldwin', and 'Wellington xxx', with 'Westwick Choice' and 'Boskoop Giant' at the other end of the scale. Unfortunately, both 'Cotswold Cross' and 'Malvern Cross' have proved unsatisfactory owing to their tendency to biennial cropping. 'Greens Black', with only a moderate number of shoots of only medium length, had a large number of buds, a high proportion of which were fruit buds. At East Malling this cultivar also produces outstandingly high numbers of flowers per node. Other cultivars recommended by Wilson and Adam for crop potential are 'Wellington xxx' and 'Baldwin'.

Studies on breeding behavior of western European cultivars for yield components by Tydeman (1930, 1938), Spinks (1947), Wilson (1964*b*), Kronenberg and Hofman (1965), Wilson and Adam (1966), and Wilson (1970) showed that, in general, phenotype is a good indicator of genotype, and that self progenies are unlikely to produce plants of commercial merit.

Tydeman and Spinks both found 'Boskoop Giant' to be a good parent for high flower number per raceme, 'Baldwin' was best for fruit set, and 'Boskoop Giant', 'Goliath', and 'Victoria' were the best donors for fruit size. Tydeman commented on the limited range of variation derived from intercrossing 'French Black', 'Seabrook's Black', 'Boskoop Giant', 'Goliath', and 'Baldwin', having found no seedlings with racemes longer than those of 'Boskoop Giant' and none with fruits larger than those of 'Goliath', while long racemes and large fruits were combined in only 1% of all seedlings. He concluded that for marked advances in these two characters, breeders would have to look outside *R. nigrum*.

The later studies of Wilson, Wilson and Adam, and Kronenberg and Hofman confirmed and extended Tydeman's and Spinks' findings: 'Baldwin', the most widely grown cultivar, was proved the best parent overall.

In the 12 intercross families from 'Baldwin', 'Boskoop Giant', 'Seabrook's Black', and 'Victoria', the mean numbers of flowers per node varied from five in some seedlings to 17 in others, the mode varying with family between nine, ten, and eleven (Wilson and Adam, 1966). In progeny tests of 'Amos Black', 'Mendip Cross', and 'Wellington xxx', the latter two were the best parents for high flower count, while 'Amos Black', as expected from its phenotype, was the worst (Wilson, 1970). Kronenberg and Hofman (1965) crossed the Canadian 'Consort' with 'Blacksmith', 'Daniels' September', 'Seabrook's Black', and 'Wellington xxx' and found 'Blacksmith' and 'Wellington xxx' were the best donors of high flower number per raceme.

The best parents for short internodes were 'Baldwin', 'Seabrook's Black', 'Malvern Cross', and 'Amos Black', the combination 'Baldwin' x 'Seabrook's Black' producing 41% of seedlings élite for this character (Wilson and Adam, 1966). The distribution for the whole population was skewed towards long internodes.

For high proportion of fruiting nodes, 'Amos Black' and 'Wellington xxx' were better than 'Mendip Cross' and 'Malvern Cross', as expected from their phenotypes (Wilson, 1970). Almost half the seedlings in the whole population had flowers on about three-quarters of the shoot, a further 14% flowering to the base.

Wilson (1970) showed that 'Wellington xxx' was much the best parent (out of the four studied) for producing seedlings having flowers with short styles, while 'Amos Black' was outstanding for the production of seedlings with stigmas protruding beyond the anthers.

The percentages of seedlings élite for various combinations of the four yield components (fruits per node, vigor, internode length, and fruit size) are tabulated by Wilson and Adam (1966). 'Baldwin' was the best parent for all four characters combined, although only 1.3% of its seedlings were élite for all four.

As originally suggested by Tydeman, marked advances in cropping potential are likely to be achieved only by going outside *R. nigrum* (or certainly outside the western European *nigrum* group) for characters such as very long strigs with many flowers, and much larger fruits (see later).

Red- and whitecurrant: The main components of yield in red- and whitecurrants are similar in some respects to those of blackcurrants, important factors being yielding surface, numbers of flowers per raceme, fruit size, and fruit set. Avoidance of frost and wind damage are also important in this crop.

Of the older cultivars, 'Houghton Castle', 'Laxton's No. 1', and 'Red Dutch' are usually heavy cropping wherever grown. Of the newer "traditional" redcurrants, 'Jonkheer van Tets', 'Erstling aus Vierlanden', and 'Red Lake' crop well. The recently introduced *R. multiflorum* derivative 'Rondom' is exceptionally heavy cropping and 'Heinemanns Rote Spätlese', of similar origin, is also outstanding for this character.

Heavy cropping in some *R. multiflorum* seedlings (e.g., 'Heinemanns Rote Spätlese') is at least partly due to high numbers of fruits per strig, inherited from *R. multiflorum*. At East Malling, some backcross derivatives of this species have been selected in addition for a more even distribution of crop, part of it being borne on year-old wood, unlike that of most redcurrant cultivars. This represents a large increase in fruiting surface and seems a promising line in breeding for yield. Kuz'min and Čuvašina (1960) refer to heavy cropping derivatives obtained from open-pollinated seed of a triploid hybrid from 'Kyzyrgan' (*R. rubrum* x *R. petraeum*) x 'Davison's Eight'. These seedlings had their crop evenly distributed along the branch on long-lived fruit spurs. It is possible that selection for cropping on year-old wood, derived from blackcurrant, would be possible in this material also.

Gooseberry: In gooseberries, the bulk of the crop is picked underripe either for processing or for sale to housewives. For the latter trade, cultivars whose fruits reach marketable size early in the season (late May–early June) are most profitable. Rake (1966) found that of seven cultivars studied, 'May Duke', closely followed by 'Keepsake', had the heaviest berries at the first two pickings on 29 May and 5 June. At the third picking (12 June) and subsequently, 'Keepsake' had the heaviest berries. 'Careless' had the highest yield at each picking date, but fruit size was rather low at the first pick.

Other English cultivars outstanding for high yields include 'Green Gem', 'Whitesmith', 'Lancashire Lad', and 'Whinham's Industry'. Of American cultivars, 'Houghton', 'Poorman', 'Downing', and 'Como' crop well, while in Russia, 'Smena', 'Russkij', 'Izumrud', 'Pioner', and 'Malahit' give high yields (Nitočkina, 1967; Žukovskaja, 1967).

Schmidt (1952) obtained heavy cropping selections in the second generation of crossing gooseberry cultivars with *R. divaricatum*, *R. 'succirubrum'*, and *R. oxyacanthoides*, and the high yielding, regular cropping Rumanian 'Rezistent de Cluj' is a selection in the first backcross of *R. divaricatum* to gooseberry (Bordeianu et al., 1969). Schmidt's selections usually had two or three fruits per strig, a character which in gooseberries is sometimes associated with an undesirable variability in fruit size. Another very productive gooseberry hybrid is the American 'Abundance' (*R. missouriense* x 'Oregon Champion' (Yeager and Latzke, 1933).

Size

Two interesting aspects of fruit size in *Ribes* cultivars are the contrast between the very large fruits of the European gooseberry in comparison with those of the rest of the genus, and the close correlation between seed content and fruit size in all three crops.

Rootsi (1967) tabulated average individual fruit weights, over a three- or four-year period, of 33 gooseberry cultivars, 14 red- and whitecurrants, and three blackcurrants. Weights of gooseberry fruits ranged from 11.43 g to 1.61 g according to cultivar, those of red- and whitecurrants from 0.88 to 0.33 g and of blackcurrants from 0.98 to 0.73 g. The upper end of the gooseberry range comprised European cultivars, the lower end American types.

Correlations between seed number and fruit size have been established for blackcurrants by Teaotia and Luckwill (1955), Lenz (1960), (reported by Kronenberg, 1964), Tamás (1963*b*) and Hårdh and Wallden (1965); for redcurrants by Lenz (Kronenberg, 1964); and for gooseberries by Wills (reported by Haskell, 1963). A relationship between self-fertility status and variability in fruit size was described in blackcurrants by Tamás. For the self-fertile 'Brödtorp' and 'Kajaanin Musta', the distribution of berry weight after open pollination was a normal curve; for the self-sterile 'Coronet', the curve was skew and flattened; and for the intermediate types 'Boskoop Giant' and 'Wellington xxx', the curve also was intermediate. Berry weight varied between, rather than within, strigs in 'Brödtorp' and 'Kajaanin Musta', varied irregularly within strigs in 'Coronet', and diminished regularly from strig base to tip for the intermediate types.

No major gene components of intrinsic fruit size have been described for *Ribes*, the genetic control in both currants and gooseberries being apparently multifactorial. Since seed number and therefore fruit size is affected by self-fertility, combined breeding for high self-fertility and for intrinsic fruit size are likely to prove most effective.

The use of the large-fruited western European cultivars 'Boskoop Giant', 'Goliath', and 'Victoria' in breeding for fruit size was described in relation to other components of yield in a previous part of this section. Very large-fruited selections have been obtained by Anderson in a progeny of ('Consort' x 'Janslunda') x ('Magnus' x 'Brödtorp'). For even more marked advances in fruit size, large-fruited forms of *R. nigrum* or of other wild species such as *R. aureum* seem the most promising donors. Zhukovsky (1965) described wild *R. nigrum* seedlings with fruits of 2.5 cm in diameter and Pavlova (1962) mentions very large-fruited forms of *R. nigrum sibiricum*. Such native forms are already being used in breeding for large fruits and high yields in Russia, usually in crosses and backcrosses with western European cultivars (Pavlova, 1963; Melehina, 1964). Kuminov (1964, 1965) describes the large-fruited 'Krasnojarskaja Desertnaja', 'Kollektivnaja', 'Diploma', 'Hasanovic', 'Hakaska', 'Tagarsk', and 'Minusinska', all derivatives of *R. nigrum sibiricum*. 'Lee's Prolific' was among the most useful European cultivars for crossing with Siberian currants, and 'Goliath' has also been used.

Kronenberg (1964) classified 11 redcurrant cultivars for fruit size, 'Laxton's Perfection', 'Fay's Prolific', 'Jonkheer van Tets', and 'Versailles' having the largest fruits. In Germany, 'Heros', 'Rosa Sudmark', 'Versailles', 'Rote Kirsch', and the whitecurrant 'Weisse aus Juterbog' all had larger fruits than 'Fay's Prolific' (Rootsi, 1967).

According to Gruber (1934), neither raceme length nor number of flowers is correlated with fruit size. Kronenberg (1964) suggests redcurrants may differ genetically in numbers of seeds necessary to make a good-sized berry, 'Rondom' and 'Heinemanns Rote Spätlese' having an average of four seeds per berry of 0.7 g, while 'Erstling aus Vierlanden' averaged seven seeds in a berry of 0.5 g. Reduction in seed number is very desirable in redcurrants since the seeds are generally large and hard; a potential donor (as pollen parent) is the more or less seedless 'Kernlose'.

In breeding for fruit size, cultivars of the *R. macrocarpum* group, which includes 'Fay's Prolific', 'Cherry', and 'Versailles', have been most successful as donors. 'Fay's Prolific', particularly, has been much used as a parent; recent large-fruited derivatives include 'Jonkheer van Tets' and 'Maarse's Prominent'.

Little is known about the fruit size breeding potential of wild redcurrant species. Kovalev (1936) and Vitkovskij (1969) recommend *R.* [*petraeum*] *atropurpureum*, native in eastern Russia, for its exceptionally large fruits. Further studies on wild forms of *R. petraeum*, *R. rubrum* and *R. sativum* might well prove worthwhile.

Of 33 gooseberry cultivars, Rootsi (1967) found the European 'Grüne Edelstein', 'Grüne Flaschenbeere', 'Dan's Mistake', 'Keepsake', and 'Maurer's Seedling' to have the largest fruits, while the American 'Houghton' followed by 'Rauche Rote' and 'Red Jacket' (= 'Josselyn') had the smallest. Cultivars with fruits listed as very large by Bunyard (1925) include 'Careless', 'Broom Girl', 'Green Ocean', 'Gunner', 'Lord Derby', 'Telegraph', and 'Thumper', while 'Antagonist' is called "enormous."

There seems no reason to look outside the *R. grossularia* cultivars in breeding large-fruited gooseberries. Wild gooseberry species all have very much smaller fruits and when used in transference programs in Europe, North America, and Russia, have invariably been crossed and backcrossed with European gooseberries. Gruber (1934) emphasized that such programs involve repeated backcrossing to large-fruited cultivars to restore fruit size. Bauer's mildew resistant 'Resistenta' and 'Robustenta', derived from backcrossing *R. divaricatum* to large-fruited cultivars, have only medium-sized fruits. Two further backcrosses to large-fruited types at East Malling have increased size somewhat, but still not quite to the standard of the backcross parents.

Strig Length in Currants

Differences in strig length (used here to imply numbers of flowers per raceme also), are not large in blackcurrant cultivars and the relative ineffectiveness of selecting for this character within the *R. nigrum* group has already been discussed under "Yield."

Very much longer strigs with many more flowers are common in the redcurrant group and in a few species of the blackcurrant section. Species in both groups are at present being used as donors in blackcurrant breeding programs.

At East Malling two accessions of the North American species *R. bracteosum*, both of which are self-sterile, have up to 60 to 70 flowers per raceme (Fig. 1). An F_1 hybrid with *R. nigrum* has long been known under the name *R. 'fuscescens'*, and is fairly fertile at the diploid level. Five-year-old F_1 seedlings from crossing an accession of *R. bracteosum* with 'Boskoop Giant' and 'Baldwin' had maximum flower numbers per strig ranging from 25 to 43 and 23 to 29, respectively. Selected seedlings of the same age from the first backcross of the original *R. 'fuscescens'* as male with 'Boskoop Giant', 'Baldwin', and 'Brödtorp' had maxima ranging from 11 to 30, 9 to 20, and 10 to 28 respectively. Multiple strigs per node were common in the backcrosses to 'Baldwin' and 'Brödtorp', and rather less so in the 'Boskoop Giant' progenies. Flower number and number of racemes per node varied considerably from season to season in some seedlings. In second backcrosses to 'Wellington xxx' and 'Boskoop Giant', seedlings with up to 25 flowers per strig were selected. An F_2 progeny of a first backcross seedling showed little improvement over the parent in flower number, but some seedlings in two families from sibbing first backcross selections exceeded both parents in flower number.

FIG. 3. Strigs of a first backcross of *R. bracteosum* x blackcurrant (left) and of 'Amos Black'.

In the first backcrosses, fertility as judged on fruit set and relative variability of fruit size within the plant was much improved over the F_1, particularly in the 'Brödtorp' progeny. One plant from this progeny, 'B674/108', (now named 'Jet') was selected for replicated cropping trial. Like most of the first backcross seedlings, fruit size in this selection was only medium and the fruit ripened late, but in this case other qualities included remarkably good setting, easy hand-picking, even-sized fruit, very good hanging, late flowering for frost avoidance, and a dry scar where the fruit separated from its stalk (Fig. 3).

In the second backcrosses, eight seedlings were selected for further trial. In general, in these seedlings fruit size was rather larger, but less even than in the 'Brödtorp' first backcross progeny, suggesting a somewhat lower level of self-fertility (although all selections tested set well with self pollen). In the 'Boskoop Giant' progeny, several selections were exceptionally late flowering and ripening and in both families some seedlings had a good upright habit of growth.

Further backcrosses to improve fruit size and fertility and incorporate pest and disease resistances have been made to cultivars and seedling selections.

R. longeracemosum, a redcurrant indigenous in western China, is remarkable for its extremely long racemes which in some seasons carry up to about 75 well-spaced flowers (Fig. 1). Tydeman crossed this species with blackcurrant at East Malling in 1930, but it was not until 1958 that colchicine treatment produced a fertile allotetraploid branch on the original diploid plant. Backcrossing at the diploid level failed but in first and second backcrosses to colchiploid blackcurrants, selections with long strigs carrying up to 20 and 15 flowers respectively, were obtained. The fertility of these hybrids is low and the relatively minor improvement in numbers of flowers per strig (possibly partly due to tetraploidy) has resulted in considerably less emphasis being placed on this line of work.

The method of transference and results obtained from the use of redcurrant as donor of strig qualities for the blackcurrant are similar to those from *R. longeracemosum*. In this case fertility in the first backcross to 4x *R. nigrum* is very low and further backcrossing has so far failed.

Attempts at East Malling to use *R. multiflorum* (Fig. 1) as donor of strig characters for the blackcurrant either directly or by using redcurrant as a bridging species have so far failed. Although seed has been obtained from the cross of *R. nigrum* x *R. multiflorum* and *R. nigrum* x (*R. multiflorum* x redcurrant[3]), it has invariably failed to germinate.

As a donor of many-flowered strigs for the redcurant, *R. multiflorum* has proved much more successful, the cultivars 'Heinemanns Rote Spätlese' and 'Mulka' being backcross derivatives. At East Malling attempts to breed for still further increases in numbers of flowers per strig have met with considerable success, seedlings with up to 48 flowers having been selected in first and second backcrosses of *R.* '*koehneanum*' (redcurrant x *R. multiflorum*) to 'Red Lake'. One of the second backcross seedlings, 'B608/3', has medium-large fruits, up to 32 flowers per strig, sets well, and crops heavily and regularly, and is at present under observation at the National Fruit Trials, Brogdale, for commercial qualities. In a third backcross to 'Houghton Castle', to introduce greater leafspot resistance, a seedling with up to 35 flowers per strig has been selected.

Inbreeding resulted in even higher flower numbers, up to 60 in F_2s of first backcross selections and up to 58 in F_2s of second backcross selections.

Although sufficient data were not recorded to make possible exact determination of the inheritance of the many-flowered *multiflorum* strig type, the high number of flowers maintained over three generations of backcrossing to redcurrant suggests that a gene, or a very few genes, of relatively large effect are involved.

In 1968, a small F_1 progeny of *R. longeracemosum* x *R. multiflorum* was raised at East Malling. These seedlings flowered for the first time in 1971 or 1972, and one seedling in particular has spectacular strigs with up to 61 fairly closely spaced flowers (Fig. 4). Crosses with a range of redcurrant cultivars are planned.

Absence of Running Off

Running off in currants is the precocious falling of young developing fruits following fertilization. Berries towards the tip of the raceme, which usually have fewer seeds, are most likely to be affected. According to Teaotia and Luckwill (1955) running off in blackcurrants reaches its peak about three weeks after bloom. Wellington and his coworkers (1921) obtained much better sets in 'Boskoop Giant' with hand as opposed to open pollination, and found few or no developing ovules in "run off" fruit. They concluded that poor pollination or fertilization was the cause of the trouble. However, even with optimal (hand) pollination, Teaotia and Luckwill found that much running off occurred, the dropped berries having fewer seeds because of the abortion of fertilized ovules at various stages of development. They concluded, "The crux of the 'running off' problem would. . . seem to lie in the factors responsible for the relatively low seed content towards the apex of the truss." Klämbt (1958) suggested that slower seed development after self-pollination is primarily responsible for running off, while Rudloff and Lenz (1960) considered a contributory factor to be an "unfavorable" position of the inflorescences in the shoot system.

Cultivars of the 'Boskoop Giant' and 'French Black' groups are more liable to running off than those of the 'Baldwin' and 'Goliath' groups (Wellington et al., 1921). In detailed studies of running off and pre-harvest drop, Teaotia and Luckwill (1955) found the following percentages of total fruit drop per raceme: 'Seabrook's Black', 52.0; 'Westwick Choice', 49.5; 'Baldwin', 36.7; 'Mendip Cross', 35.9; 'Cotswold Cross', 31.4. Preharvest drop in 'Baldwin' accounted for 12.9% of fruits; in the other four cultivars it ranged from 3.5–6.1%.

Redcurrants are generally rather less susceptible to running off than blackcurrants, but there are considerable varietal differences in this crop also. Op't Hoog (1964) classified 12 redcurrant cultivars and seedlings for resistance to running off,

FIG. 4. Four-year-old F_1 of *R. longeracemosum* x *R. multiflorum*.

named cultivars showing decreasing resistance in this order: 'Rondom', 'Heinemanns Rote Spätlese', 'Erstling aus Vierlanden', 'Duitse Zure' (= 'Prince Albert'), 'Fay's Prolific', 'Jonkheer van Tets', 'Laxton's No. 1', 'Maarse's Prominent'.

Planned breeding for resistance to running off has not been described in the literature, but inevitably in any currant breeding program the use of heavy cropping parents and the selection of high yielding seedlings will automatically tend to eliminate types particularly susceptible to this trouble. Breeding and selection for high self fertility, short styles, and spring frost avoidance would all be likely to contribute to good initial fruit setting and reduction in subsequent fruit drop.

Toughness of Skin and Firmness of Fruit

Toughness of skin and fruit firmness in currants, which are associated with resistance to mechanical damage, will become more important as mechanical harvesting increases. A few data are available on varietal phenotype and breeding behavior for these characters in currants.

English cultivars known to travel well, due to toughness of skin or firmness of berry or both, include 'Amos Black' (fairly tough skin), 'Baldwin' (tough), 'Blacksmith' (rather thin and tender), 'Cotswold Cross' (thick) and 'Seabrook's Black' (medium thick and tough) (Anon., 1965*a*). 'Boskoop Giant' and 'Raven' have thin, tender skins, while that of 'Mendip Cross' is thin but fairly tough and 'Wellington xxx' is moderately tough. 'Tor Cross' has a tougher skin and travels better than other early cultivars.

Wilson (1963) graded four blackcurrant cultivars in order of decreasing skin toughness as follows: 'Baldwin', 'Seabrook's Black', 'Boskoop Giant', 'Victoria'. On intercrossing these cultivars, no large inter-progeny differences were found for skin texture, although 'Baldwin' had rather fewer seedlings with tender skins and 'Boskoop Giant' rather fewer with tough skins. 'Baldwin' x 'Seabrook's Black' was the best combination for toughness. Percentages of seedlings with skin classed as tough or very tough ranged from 11% for 'Boskoop Giant' to 15% for 'Baldwin' and 'Seabrook's Black' (Wilson, 1964*b*).

Tydeman (1930, 1938) also gave data on skin

toughness in progenies derived from selfing and intercrossing 'Baldwin', 'Goliath', 'Boskoop Giant', 'French Black', and 'Seabrook's Black'. Selfs of 'Baldwin' had mostly medium or tough skins, of 'Goliath' mostly medium, and of 'Boskoop Giant', mostly medium or, rarely, tender.

Of the more recently introduced blackcurrant cultivars, 'Invigo' and 'Black Reward' have very firm fruit, while 'Lissil', 'Meitgo', and 'Wassil' are classed as firm (Kronenberg, 1965; Dorsman et al., 1969).

Kronenberg (1964) measured fruit firmness in 18 redcurrant cultivars with a penetrometer. The six firmest types were 'Gondouin', 'Rode Komeet', 'Rondom', 'Heinemanns Rote Spätlese', 'Prince Albert', and 'Erstling aus Vierlanden' (in descending order of firmness), and the five least firm were 'Laxton's Perfection', 'Laxton's No. 1', 'Utrecht', 'New Red Dutch', and 'Jonkheer van Tets' (in ascending order of firmness).

Ripening Season

Early-ripening blackcurrant cultivars include 'Boskoop Giant', 'Brödtorp', 'Greens Black', 'Laxton's Giant', 'Mendip Cross', 'Rosenthals Schwarze', 'Tor Cross', and the Russian 'Nahodka', 'Vystavočnaja', and 'Pamjat' Mičurina', the latter being particularly early. Midseason types include 'Blacksmith', 'Consort', 'Coronet', 'Cotswold Cross', 'Davison's Eight', 'French Black', 'Goliath', 'Noir de Bourgogne', 'Roodknop', 'Seabrook's Black', 'Silvergieters Zwarte', 'Wellington xxx', and 'Westwick Triumph'. Late cultivars include 'Baldwin', 'Black Reward', 'Invigo', 'Malvern Cross', 'Westwick Choice', and the very late 'Amos Black' and 'Daniels' September'.

Data on the relationship between season of ripening, length of maturation period, and season of flowering are included in Table 7. Data for ripening season and maturation period are taken from papers by Wilson (1958) and Lantin (1967), and for order of flowering from Anonymous (1973). Although the different environments of the three localities where records were made (Long Ashton, southwest England; Angers, central France; Brogdale, southeast England) obviously affect the three variables, the order of ripening, relative lengths of maturation periods, and order of flowering are probably more or less constant, so that useful comparisons can be made.

These data suggest, as Wilson (1958) indicated, that time of flowering is not a good indicator of ripening season, some early-flowering types ripening late (e.g., 'Baldwin'), and some relatively late-flowering types ripening early (e.g., 'Boskoop Giant'). In general, late-ripening plants require a longer maturation period than early cultivars. The data suggest that in breeding for late flowering, by choosing parents with a short maturation period it should be possible to select late-flowering types with a range of ripening seasons. In this connection, Lihonos and Pavlova (1969) describe forms of *R. nigrum sibiricum* with an extremely short growing season.

Phenotype appears to be a good indicator of genotype for ripening season in blackcurrants, as shown by Tydeman (1930, 1938), Spinks (1947), Wilson (1964*b*, 1970) and Kronenberg and Hof-

TABLE 7. Season of ripening, maturation period and flowering season in blackcurrant cultivars

Long Ashton		Brogdale	Angers		Brogdale
Order of ripening	Maturation period (days)	Order of flowering	Order of ripening	Maturation period (days)	Order of flowering
Boskoop Giant	79	5	Boskoop Giant	74	5
Mendip Cross	84	=3	Rosenthals Langtraubige	77	
Seabrook's Black	82		Noir de Bourgogne	77	
Cotswold Cross	91	=1	Consort	81	6
Malvern Cross	96	2	Merveille de la Gironde	80	
Baldwin	96	=1	Silvergieters Zwarte	79	=7
Westwick Choice	95	=3	Cotswold Cross	85	=1
			Wellington xxx	87	2
			Davison's Eight	87	4
			Goliath	83	=7
			Baldwin	92	=1
			Amos Black	97	8

man (1965). The best parents for lateness were 'Baldwin', 'Daniels' September', and 'Amos Black', while 'Mendip Cross' and 'Consort' gave relatively high percentages of early seedlings. Selfed progenies of 'Amos Black' and 'Malvern Cross' contained no early ripening types and 74% and 18% of very late ripening seedlings, respectively, while 'Mendip Cross' selfed gave 78% of early seedlings and none classed as very late (Wilson, 1970). Tydeman (1938) obtained the highest percentages of early seedlings from crosses of 'French Black' x 'Boskoop Giant', 'Boskoop Giant' x 'Goliath', and 'Boskoop Giant' x 'Baldwin', and the most late seedlings from 'Seabrook's Black' x 'Goliath' and 'Goliath' x 'Baldwin'.

Exceptionally late ripening selections have been obtained at East Malling in first and second backcrosses of *R. bracteosum* to blackcurrant. Other species of potential use in extending season of ripening in currants (both red and black) include the eastern Siberian *R. palczewskii* (Jan.) Pojar and *R. manshuricum* Komar., the former ripening in June, and the latter in September or October (Kovalev, 1936).

Redcurrant cultivars of the *petraeum* group, such as 'Gondouin', 'Prince Albert', and 'Kernlose', are characterized by late leafing and late ripening. Similarly, the more modern cultivars derived from *R. multiflorum*, 'Heinemanns Rote Spätlese', 'Mulka', and 'Rondom', are all late maturing. In the other redcurrant groups, season is more variable. Early-ripening types include 'Erstling aus Vierlanden', 'Fay's Prolific', 'Jonkheer van Tets', 'Laxton's No. 1', 'Maarse's Prominent', and 'Scotch'. The midseason group includes 'Houghton Castle', 'Laxton's Perfection', 'Red Lake', and 'Versailles', while 'Fertility', 'La Constante', 'Raby Castle', and 'Wilson's Long Bunch' are late or very late.

As in blackcurrants, late-ripening types tend to need longer to mature, average maturation periods for eight cultivars, arranged in order of ripening in Holland, being as follows (Kronenberg, 1964): 'Jonkheer van Tets', 70 days; 'Maarse's Prominent', 77; 'Erstling aus Vierlanden', 74; 'Fay's Prolific', 78; 'Correction', 72; 'Rondom', 81; 'Prince Albert', 80; 'Heinemanns Rote Spätlese', 98.

Most of the gooseberry crop is picked underripe for processing so that season of maturity is of less general significance than in currants.

Early types include 'Bedford Red', 'Broom Girl', 'Early Sulphur', 'Gautry's Earliest', 'Hönings Früheste', 'Mauks Frühe Rote', 'May Duke', 'Remarka', 'Reverta', 'Risulfa', 'Rokula', the American 'Abundance' and 'Josselyn', and the Russian 'Lučistyj' and 'Solnečnyj'. Midseason cultivars include 'Achilles', 'Careless', 'Crown Bob', 'Grüne Flaschenbeere', 'Grüne Kugel', 'Howard's Lancer', 'Keepsake', 'Lancashire Lad', 'Leveller', 'Rixantha', 'Whinham's Industry', 'Whitesmith', the North American 'Captivator', 'Chautauqua', 'Clark', 'Glenashton', and 'Poorman', and the Russian 'Smena' and 'Pjatiletka'. Late or very late types include 'Green Gem', 'Lady Delamere', 'Perle von Müncheberg', 'Resistenta', 'Robustenta', 'Rochus', 'Rote Preis typ Goliath', the American 'Fredonia' and 'Houghton', and the Russian 'Izumrud' and 'Rekord'.

Color

Green- and yellow- or white-fruited forms of *R. nigrum* have been known under the names *R. nigrum chlorocarpum* Spaeth and *R. nigrum xanthocarpum* Spaeth since before 1838 and 1827, respectively. In the wild, the fruit color of *R. nigrum sibiricum* ranges from black through dark violet, brown, and red to green. The pigments in fruits of blackcurrant cultivars comprise two cyanidin and two delphinidin derivatives, these being absent in "white"-fruited types (Harborne and Hall, 1964; Anon., 1969).

Keep and Knight (1970) postulated major gene (*rb*) control of yellow-green fruit color in derivatives of a Finnish blackcurrant, with a linked lethal, *l*, to account for the absence or deficit of yellow-fruited seedlings in six out of ten small F_2 progenies. However, the occurrence of both whitish-green and yellowish-green blackcurrants and the various color forms of *R. nigrum sibiricum* show that additional color genes must occur in this group.

In blackcurrant-gooseberry hybrids, fruit color is invariably more or less black, regardless of the color of the gooseberry parent. Blackcurrant-redcurrant hybrids have dark red fruits.

In *R. odoratum*, Baird of North Dakota showed that black fruit color is dominant to red and yellow (quoted by Darrow, 1937).

In red- and whitecurrants, fruit color ranges from very dark to light red, through pink to yellowish and "off-white." The pigments in redcurrant fruits are cyanidin derivatives, the number of these varying according to the origin of the cultivar (Harborne and Hall, 1964). Nybom (1968)

FIG. 5. Branch of *R. nigrum* x *R. grossularia*.

showed that the *sativum* and *rubrum* groups had four pigments, and the *petraeum* type, six.

Keep and Knight (1970) showed 'Minnesota' and 'Red Lake' to be heterozygous for a single dominant gene, *Rc*, controlling red fruit color. 'Houghton Castle' and 'Fay's Prolific' are homozygous. In F_2s of *R. 'koehneanum'* (*R. sativum* x *R. multiflorum*) x 'Red Lake' there was an association between high numbers of flowers per strig and red fruit color.

The selfed progeny of a pink-fruited form of *R. rubrum* (36 plants) were all white-fruited; this was interpreted as showing that the parent plant was a periclinal chimera with the pigment confined to the outer cell layer (Anon., 1938).

Keep and Knight (1970) suggest that, as in blackcurrants, there must be color modifying genes in addition to *Rc* to account for the range in redcurrant fruit color.

The color of gooseberry fruits ranges from dark and light red through various shades of green to yellow and almost white. Unlike currants, cultivars covering the whole range of fruit color are grown commercially.

Yeager and Latzke (1933) considered red to be dominant to green and Crane and Lawrence (1952) and Keep and Knight (1970) showed the following red-fruited cultivars (all those tested) to be heterozygous for a single basic gene (assigned the symbol *Rg* by Keep and Knight) controlling red fruit color: 'Bedford Red', 'Champagne Red', 'Crown Bob', 'Echo', 'Gautry's Earliest', 'Lion's Provider', 'May Duke', and 'Whinham's Industry'. 'May Duke' showed a significant deficit of non-red seedlings on selfing, suggesting this cultivar may carry an additional color gene. Green cultivars true breeding for green of various shades on selfing were 'Early Green Hairy' and 'Lancer' while the yellow 'Leveller' and 'Broom Girl' bred true for yellow. Other green and yellow types which segregated for colors in the yellow-green spectrum were 'Bedford Yellow' and 'Golden Ball' (yellow) and 'Bellona', 'Keepsake', 'Langley Gage', 'Shiner', and 'Whitesmith' (green).

'Golden Drop' (yellow) and 'Pitmaston Greengage' (green) gave ratios of 4:12 and 1:58 red to non-red fruits respectively. To explain this, Crane and Lawrence postulated the presence of an inhibtor, *I*, capable of suppressing red. The red cultivar 'Ostrich' unexpectedly gave only non-red seedlings on selfing, from which Crane and Lawrence concluded this cultivar is a periclinal chimera, the inner tissues being genetically non-red.

Vitamin C Content

The antiscorbutic properties of the blackcurrant were well established by 1820 (Bagenal, 1955) and this crop is still valued for its exceptionally high Vitamin C content. Values of over 200 mg per 100 g fruit are not uncommon in blackcurrant cultivars, in contrast to commercial redcurrants, where the few data available suggest contents of 30–40 mg % are the norm (Zubeckis, 1962). Voščilko (1969) reports that wild forms of *R. nigrum sibiricum* contain 200 to 800 mg % of Vitamin C, while a selection of wild *R.* [*petraeum*] *atropurpureum* contained 90 mg %. In *R. aureum*, 15 accessions showed a range of 32–58 mg % (Gomoljako, 1941).

In a study of the Vitamin C contents of blackcurrant fruits from various localities during 1963–7, Nilsson (1969) detected significant interactions for year x cultivar, cultivar x locality, and year x locality x cultivar. Thus, in considering the numerous data on Vitamin C content, direct comparisons of actual values obtained by different workers are probably of less significance than the relative order in which the cultivars are placed.

Todd (1962) tabulated Vitamin C values (supplied by Messrs. Beecham Foods) for 21 western European cultivars. These have been arranged in descending order in Table 8.

TABLE 8. Vitamin C ranges in European blackcurrant cultivars (after Todd, 1962)

Cultivar	Vitamin C range (mg per 100 g)
Laxton's Giant	181–248
Baldwin	178–236
Daniels' September	182–235
Boskoop Giant	199–234
Raven	195–227
Laxton's Tinker	190–221
Laxton's Grape	185–214
Blacksmith	204–212
Westwick Choice	199–211
Wellington xxx	187–209
Westwick Triumph	161–181
Seabrook's Black	157–179
Davison's Eight	166–174
Supreme	148–168
Mendip Cross	141–168
Malvern Cross	118–144
Amos Black	136–143
Cotswold Cross	117–125
Victoria	111–120

Other western European cultivars reported as having high Vitamin C contents include 'Roodknop' (Dorsman et al., 1969), 'Noir de Bourgogne' (Lorrain, 1962), and 'Akkermans Bes' (Kronenberg and Doesburg, 1955), while 'Goliath' has a relatively low content (Kronenberg and Doesburg, 1955). In France, the Canadian 'Consort' exceeded 'Baldwin' in Vitamin C content (Decourtye and Lantin, 1965).

Scandinavian cultivars tend in general to have rather less Vitamin C than the western European groups. Kuusi (1965) classes 12 cultivars in descending order as follows: 'Boskoop Giant', 'Roodknop', 'Westwick Choice', 'Wellington xxx', 'Gerby', 'Wellington x', 'Åström', 'Brödtorp', 'Goliath', 'Silvergieters Zwarte', 'Janslunda', and 'Black of Lepaa'. In Sweden, 'Öjebyn' contained 130 mg % in comparison with 170 mg % for 'Boskoop Giant' (Larsson, 1959).

According to Samorodova-Bianki (1969), of 19 cultivars derived from *R. nigrum*, *R. nigrum sibiricum*, *R. dikuscha*, or hybrids between them, Vitamin C content was highest in the European group. Russian cultivars with relatively high Vitamin C contents include 'Altajskaja Desertnaja', 'Golubka', 'Nahodka', 'Novost', 'Pobyeda', and 'Uspekh' (Žitneva et al., 1961; Rilškis, 1964).

The few data on stability of Vitamin C in fresh and processed fruit material suggests that this varies considerably with cultivar. Vestrheim (1965) described a general decrease in ascorbic acid during ripening, ranging in five cultivars investigated in 1963 and 1964 from 1.97% per day for 'Wellington xxx' to 0.67% per day for 'Brödtorp'. The Vitamin C content of juice of 'Consort', 'Black Reward', and 'Goliath' decreased in six months from 141 to 91 mg, 134 to 93 mg, and 94 to 34 mg respectively (Gersons, 1966*a*). Tamás (1968) found that the stability of Vitamin C content in seedlings of *R. nigrum* was considerably less than in *R. nigrum* x *R. dikuscha* progenies. Individual variation in both the content and stability of ascorbic acid was greatest in *R. nigrum* x *R. ussuriense* progenies.

Apart from Tamás' work, detailed investigations of the inheritance of Vitamin C content do not appear to have been undertaken, but from published data on parents and offspring it can be inferred that the range in values for the progeny usually lie somewhere between those of the two parents. Thus, as seen in Table 8, the Vitamin C contents of the 'Baldwin' x 'Boskoop Giant' selections, 'Laxton's Giant', 'Laxton's Grape', 'Wellington xxx', and 'Mendip Cross', are 181–248, 185–214, 187–209, and 141–168 mg % respectively, while parental values are 178–236 and 199–234. The values for 'Malvern Cross' and 'Cotswold Cross' (both 'Baldwin' x 'Victoria') are 117–125 and 118–144 mg %, nearer to that of 'Victoria' (111–120 mg %) than 'Baldwin'. In breeding for high Vitamin C content, intercrossing parents with high values and deliberate selection for high expression in the progeny should prove successful.

Dessert Quality

Dessert quality in currants and gooseberries comprises a combination of flavor, size, and appearance, a skin that is not too tough, and, in redcurrants, an absence of excessive seediness.

Blackcurrants are not usually considered a dessert fruit to be eaten raw, as the flavor of most cultivars is too strong for many palates. 'Brödtorp' and 'Goliath' both have a mild, sweet flavor which is widely acceptable, while 'Silvergieters Zwarte' is also considered to be well flavored. At East Malling, several yellow-fruited selections of Scandinavian origin were pleasantly sweet and almost totally lacking in the astringency characteristic of many blackcurrants.

Redcurrants are grown primarily for processing and are not widely popular as dessert fruit. In general, large-fruited cultivars of the 'Versailles' group are the best dessert cultivars. Kronenberg (1964) evaluated the flavor of redcurrant cultivars in relation to their sugar/acid balance and aroma, and considered 'Laxton's No. 1' to be best, followed by 'Laxton's Perfection' and 'Hoornse Rode'. 'Prince Albert', 'Rondom', and 'Heinemanns Rote Spätlese' were very acid. Aroma was good in 'Fay's Prolific', 'Versailles', 'Laxton's No. 1', 'Laxton's Perfection', and 'Correction' and in the small-fruited 'Hoornse Rode' and 'Hoornse Geelsteel'. The aroma of 'Jonkheer van Tets', 'Rondom', 'Maarse's Prominent', and 'Erstling aus Vierlanden' was rather weak. The well-flavored cultivars contained proportionally more citric acid, little tartaric acid, and still less malic acid, and Kronenberg quotes Jordan et al. (1957) who found that citric acid is agreeable to the taste but malic acid is not. Of the cultivars examined by Kronenberg, 'Heinemanns Rote Spätlese' and 'Erstling aus Vierlanden' were the most "seedy," seed weight in these cultivars comprising about 7% of total berry weight in comparison with 4–5% for 'Laxton's Perfection', 'Rondom', 'Prince Albert', and 'Correction'.

Gooseberries are generally more popular than currants for dessert and a small proportion of the crop is sold ripe for this purpose. In Britain 'Leveller', which has very large, yellow, well-flavored fruit, is grown only for the dessert trade. In Germany the early 'Hönings Früheste' serves the same purpose. Red cultivars such as 'Lancashire Lad', 'May Duke', and 'Whinham's Industry' are sometimes allowed to ripen for dessert purposes, as are whitish or pale green cultivars such as 'Careless', 'Howard's Lancer', 'Keepsake', 'Whitesmith', and 'White Lion'. 'Whinham's Industry', 'Whitesmith', and 'Howard's Lancer' are generally considered to have a very good flavor. Of the older smaller-fruited cultivars, Rake (1958, 1963) recommends the following for their excellent flavor: 'Glenton Green', 'Golden Drop', 'Green Walnut', 'Ironmonger', 'Langley Gage', 'Pitmaston Greengage', 'Red Champagne', 'Roseberry', 'Scotch Red Rough', 'Yellow Champagne', and 'Warrington'.

At present there seems to be little call for breeding new gooseberries specifically for dessert purposes. General purpose, disease and pest resistant, spineless types would be more in line with modern requirements.

Processing

Most of the blackcurrant crop is used to produce juice, but a proportion goes for liqueur making, for jamming and canning, for pie fillings, and for the manufacture of sweets and pastilles. Most commonly grown western European cultivars are acceptable for all these purposes. For juice production, color intensity and quality as well as flavor are important. In Holland, French types are considered best for color intensity, while 'Black Reward' and 'Roodknop' are thought unsatisfactory owing to the brownish color of their juice (Gersons, 1966*b*; Dorsman et al., 1969). 'Consort' is not recommended in Holland owing to the poor flavor of the juice (Kronenberg et al., 1961), although in Sweden this is considered good and full-bodied (Tamás, 1963*a*).

Since there appear to be only minor differences between blackcurrant cultivars in suitability for processing, apart from juice production, breeding and selection for such characters as regular heavy cropping, ease of harvesting, pest and disease resistance, and good plant habit should take precedence. The majority of selections should prove suitable for processing, the very few that are not being satisfactorily eliminated during cropping trials prior to release.

Small quantities of whitecurrants are used in winemaking, but apart from this, whitecurrants are not in demand for processing. Redcurrants are used primarily for juice production but also for canning, freezing, and for jams and jellies. As with blackcurrants, most cultivars now grown appear to be more or less suitable for processing, particularly the acid types not favored for dessert. Although the juice of 'Rondom' is a poor color, this cultivar is among the best for freezing and canning (Anon., 1965*b*). 'Prince Albert' is particularly good for processing, especially for jamming, while 'Jonkheer van Tets' and 'Fay's Prolific' are dual-purpose varieties acceptable both for dessert and processing.

In breeding redcurrants, programs aimed at producing dual-purpose types combining the cropping capacity of *R. multiflorum* derivatives with the fruit size and dessert quality of the 'Versailles' group would seem to hold most promise.

The bulk of the gooseberry crop is grown for jamming and canning. For canning, the fruits are picked when hard and green; for jamming, just before they begin to show color. The main cultivars grown for processing include 'Careless', 'Gelbe Triumphbeere', 'Howard's Lancer', 'Keepsake' (= 'Berry's Early Kent'), 'Lady Delamere', 'Lancashire Lad', 'Whinham's Industry', and 'Whitesmith'.

At East Malling, the gooseberry breeding program includes lines designed to introduce American gooseberry mildew resistance and spinelessness into types suitable for processing. The mildew resistant cultivars recently introduced in Germany by the Max Planck Institute have yet to be evaluated for processing qualities (see under "Disease Resistance").

Handpicking

Currant cultivars, particularly blackcurrants, differ considerably in the ease and speed with which they can be picked by hand. In 1934 Swarbrick and Thompson estimated that picking costs for blackcurrants increased in this order: 'Boskoop Giant', 'Baldwin', 'Davison's Eight', 'Edina', 'French Black', and 'Taylor'. The picking costs for 'Taylor' were just over double those for 'Boskoop Giant'. More recently Sorge (1961) gives picking rates for 'Goliath', 'Rosenthals', and 'Silvergieters' as 3.4, 5.5, and 6.1 kg/hr respectively. Brandis (1967) gives picking times in min/kg for 23 cultivars. The most rapidly picked were 'Invincible Giant [Prolific]', 'Westwick Choice' and 'Boskoop Giant'; the least rapidly, 'Holger Danske', 'Mendip Cross', and 'Wellington xxx'. Berning (1966) classifies blackcurrant cultivars for ease of picking as follows: good, 'Daniels' September'; fairly good, 'Cotswold Cross', 'Wellington xxx', Westwick Choice', 'Westwick Triumph'; rather poor, 'Amos Black', 'Goliath'; poor, 'Lees Schwarze' (Baldwin type), 'French Black', 'Laxton's Standard'; very poor, 'Hatton Black'.

In breeding for cheap and easy hand picking, the main objectives should be long strigs hanging well clear of the leaves, a good picking handle, and large fruit. The use of long-strigged cultivars such as 'Boskoop Giant' and of species such as *R. bracteosum*, with a good picking handle and very many flowered strigs, should produce selections which can be picked much more rapidly. The long-strigged *R. bracteosum* derivative, 'B674/108' (= 'Jet'), described previously (under "strig length in currants") is picked in approximately half the time needed for the same weight of 'Amos Black', although the fruits are considerably smaller.

Redcurrants in general have strigs better adapted to handpicking than those of blackcurrants, but in some cultivars overcrowding of strigs makes picking difficult. Wilking and Hardenberg (1965) tabulate output in kg/hr for five redcurrant cultivars as follows: 'Heinemanns Rote Spätlese', 14.7; 'Red Lake', 11.9; 'Rondom', 13.5; 'Red Dutch' (= 'Rote Holländische'), 7.4; 'Erstling aus Vierlanden' (= 'Rote Vierlanden'), 5.6. Kronenberg (1966) comments that the strig of 'Erstling aus Vierlanden' is rather tough, making picking difficult. Berning (1966) considers 'Rondom' and 'Jonhkeer van Tets' to be easy to pick, 'Laxton's No. 1' and 'Mulka' to be fairly easy, and 'Houghton Castle' to be difficult. Op't Hoog (1964) classifies in order of increasing difficulty of handpicking as follows: 'Rondom', 'Heinemanns Rote Spätlese', 'Erstling aus Vierlanden', 'Fay's Prolific', 'Jonkheer van Tets', 'Maarse's Prominent', 'Laxton's No. 1', and 'Duitse Zure' (= 'Prince Albert').

Breeding for longer strigs more evenly distributed over the plant seems the best approach in breeding redcurrants for easier handpicking.

Mechanical Harvesting

According to Christensen (1966) the first report on the mechanical harvesting of *Ribes* crops appeared as recently as 1961. Although some forms of mechanical harvesting machines have now been in commercial use for several years, others are still in the developmental stage. So far, results with existing cultivars in different areas have not always been consistent, probably partly because of the different types of machines used, and also because of the varying degrees of ripeness at the time of harvesting; several workers have found it necessary to classify their material for state of maturity at the time of the investigation.

At present, there are two methods for mechanically harvesting *Ribes* crops, the pruning system and the shaking system. In the pruning system, suitable only for blackcurrants, the fruit is mechanically stripped from cut branches in a stationary harvester. In the shaking system, which has been tried out on all *Ribes* crops, the branches are vibrated mechanically until the fruits drop off.

For the pruning system, heavy-cropping blackcurrants which produce plenty of vigorous new basal growths are required. Since the mass of new shoots developing are very liable to mildew infection, resistance to this disease is particularly desirable. The fruits should be even ripening, fairly firmly attached (so that they are not shed on the way to the machine), should have a tough skin, and should be firm and probably fairly small, since hard pruning will increase fruit size and overlarge fruits tend to be soft.

For the shaking system, heavy-cropping plants with firm, even-ripening fruits with a tough skin which shake off fairly readily are required. If the fruits detach from their pedicels, there should be a dry scar at the point of attachment. Leaves should be small and hard to reduce leaf loss to a minimum, since cropping is reduced in the following year proportionally to amount of leaf lost. Blackcurrants which produce relatively few long, erect, sparsely branched shoots are preferred. Bauer (1969) advocates growing blackcurrants on a "leg" for shaker harvesting. At present, the shaking system appears to be gaining ground for blackcurrants, and breeding cultivars suited to this rather than the pruning system should be given priority.

According to Neumann (1965), the correlation between high yields and labor productivity is higher for mechanical than for hand picking. She considers cultivars with short strigs and large fruits to be more satisfactory, and Christensen (1966) also recommends cultivars with short strigs. Kronenberg (1966), however, found longer strigs were also suitable. Types which shed their fruit fairly easily include 'Invigo', 'Black Reward', and 'Roodknop' (Kronenberg, 1965, 1966), 'Topsy' (which has large, firm, even-ripening fruit [Ricketson, 1969], and 'Bang Up' and 'Goliath' (Bogdański et al., 1957). Of 21 cultivars tested by Berning (1967), 'Invigo' was the only one completely free from mechanical damage. This cultivar could be harvested over a ten-day period. Of seven other cultivars which were slightly injured, only one had a four-day harvesting period, the others even less. 'Wellington xxx', 'Rosenthals Langtraubige', 'Wassil', 'Mendip Cross', and 'Roodknop' were severely damaged.

When shaken, single berries without stalks, or single fruits with stalks attached, or whole strigs may drop. Berning (1967) emphasizes the importance of firm attachment of berries to their individual stalks, since the escape of juice from torn fruit makes cleaning by air jets difficult and favors the incidence of pests and molds. Kronenberg (1966) assessed proportions by weight of single sound fruit, and of berries on strigs, for eight cultivars. 'Black Reward' (= 'M20') was outstanding with 86% of single sound fruits, followed by 'Wellington xxx' and 'Goliath'. These cultivars had 13, 18, and 24% of their berries on strigs respectively, while 'Daniels' September' ('Tinker' type), 'Cotswold Cross', 'Baldwin', and 'Roodknop', with 38–58% of single sound berries, had 58, 41, 40 and 39% of their berries on strigs, respectively. Percentage of damaged and rotten fruit ranged from 1 for 'Black Reward' and 'Cotswold Cross' to 8 for 'Wellington xxx'.

Wilson (1963) assessed the breeding behaviour of 'Victoria', 'Boskoop Giant', 'Seabrook's Black', and 'Baldwin' with respect to mechanical harvesting qualities. 'Seabrook's Black' produced the most seedlings which shed their fruit readily on shaking, while 'Baldwin' progenies contained the fewest seedlings with tender skins. 'Victoria' was the best parent for good regeneration from the base of the plant.

Knight and Keep (1966) are using a number of species such as *R. bracteosum*, *R. longeracemosum*, and more recently *R. sanguineum* and *R. glutinosum*, as donors of various characters. Such material will produce a range of strig types from which selection of those best adapted to mechanical harvesters, as they evolve, will be possible. In first and second backcrosses from *R. bracteosum*, erect, sparsely branched selections have been obtained with fruits showing, variously, very good hanging, tough skins, and ready separation from the strig with a dry scar.

Kronenberg (1966) considered the tender skins of most redcurrant cultivars to be the main drawback to mechanical harvesting, and considers those which shed whole strigs to be most promising. 'Rondom' dropped 90% of its fruit (by weight) as whole strigs and the 10% of single fruits were completely undamaged. Other cultivars tested showed a range of from 15–63% of damaged and rotten single fruits; 'Laxton's No. 1', followed by 'Versailles' and 'Erstling aus Vierlanden', were worst for this characteristic. In contrast to Kronenberg's results and those of Berning (1967), Wilking (1967) and Blommers et al. (1969) found that the strigs of redcurrants separated more readily, and the fruits were less damaged, than those of blackcurrant. Winkler (1967) reported that 'Erstling aus Vierlanden' and 'Holländische' (probably 'Red Dutch') were unsatisfactory for mechanical harvesting because their branches were too inflexible.

In breeding redcurrants suitable for mechanical harvesting, the few data available suggest that the objective should be a combination of ready detachment of whole strigs with firmness of fruit.

Very few published data on mechanical harvesting of gooseberries are available. Winkler (1967) considered this crop to be satisfactory for machine harvesting, and Bradt et al. (1968) describe 'Captivator', 'Fredonia', 'Glenashton', and 'Silvia' as being at least moderately easy to shake off. On the other hand, 'Abundance' and 'Ross' (= 'Davidson') are unsatisfactory for harvesting by machine.

Achievements and Prospects

Over the past few decades, considerable advances have been made in basic knowledge of the genus *Ribes* as a whole, both in matters of direct interest to plant breeders, such as breeding behavior, genetics, and cytology, and in related fields such as biochemistry and physiology.

In the last few years, it has been shown how, by day length, temperature, and gibberellin treatments of blackcurrants, fruit can be obtained from seedlings one year after seed sowing (Karnatz, 1969), and in established plants, flower initiation and flowering can be induced more or less at will (Modlibowska, 1960; Nasr and Wareing, 1961*b*; Tinklin et al., 1970). Such techniques might well prove invaluable in shortening the early stages of gene transference programs, particularly in resistance breeding, when agronomic characters are of secondary importance.

Recent work on self incompatibility (e.g., Arasu, 1970) has shown that the majority of wild species are obligate outbreeders, and this and the inbreeding depression commonly occurring in self progenies of cultivars suggests that their self-compatibility is of recent origin. The effect of increasing homozygosity on agronomic characters in blackcurrants has been clearly defined by Wilson, and Spinks and other workers have emphasized the need for hybrid vigor in breeding commercial cultivars.

Work in Britain, Sweden, and Holland has confirmed that in general phenotype reflects genotype and at the present stage of evolution of *Ribes* crop plants, rapid techniques designed to select outstanding individuals rather than attempts to assess performance of progenies as a whole are appropriate.

Recent findings from intercrossing standard western European blackcurrant cultivars have reemphasized the need to introduce new "blood" into the gene pool, and this is occurring on an increasing scale, both through interspecific hybridization and through intercrossing of cultivars and wild selections of widely different origins.

A comprehensive review of interspecific hybridization in the genus was published by Keep in 1962. Recent work by plant breeders and others has led to greater knowledge of the potential of species crosses, and cytological investigations by several workers, notably Goldschmidt, have suggested that there is sufficient chromosome homology in most cases to permit gene transferences even in wide crosses.

Techniques for the induction of polyploidy by colchicine treatment of young seedlings have been perfected and a number of colchiploid species and hybrids have been raised in several countries (e.g., Nilsson, 1959, 1966). Fertile allotetraploids are being used in gene transference programs, notably for the introduction of gall mite resistance from gooseberry into diploid blackcurrants by Knight et al.

At the diploid level also, wild species have been increasingly used as donors, sometimes to introduce characters transgressing the natural limits of cultivars. In blackcurrants, these donors include *R. aureum* for fruit size, *R. bracteosum* for strig length and mechanical harvesting properties, *R. cereum* for aphid resistance, *R. dikuscha* for hardiness and for leafspot and American gooseberry mildew resistance, *R. glutinosum* and *R. sanguineum* for resistance to mildew, leafspot, and aphids, and *R. ussuriense* for resistance to blister rust, reversion, and gall mite. *R. ussuriense* has been shown to carry a dominant gene *Cr* controlling resistance to blister rust. In redcurrants, the main donor species has been *R. multiflorum* for strig length and yield. Donors for gooseberries include *R. aciculare* for hardiness, *R. alpestre* for aphid (*A. grossulariae*) resistance, *R. divaricatum* for hardiness, leafspot, and mildew resistance, *R. leptanthum* and *R. watsonianum* for leafspot and mildew resistance, *R. roezlii* for aphid (*N. ribisnigri*) resistance, and *R. sanguineum* for spinelessness, and possibly also for sawfly and aphid (*N. ribisnigri*) resistance.

A great increase has also occurred in hybridization between selections from geographic races of wild relatives of crop plants and cultivars of widely different origin, mainly in blackcurrant breeding. In Russia, selected forms of *R. nigrum sibiricum* have been used as donors of hardiness, fruit size, leafspot, mildew, reversion, and gall mite resistance, and hybrids and backcrosses of these with western European cultivars have been introduced. In Europe, Russian and Scandinavian blackcurrant cultivars have been crossed with European types with similar objectives in view. In the course of this work, the dominant genes *M* and *P* controlling resistance to American gooseberry mildew and gall mite, respectively, have been isolated.

As a result of this hybridization, the gene pool now available to plant breeders is very much broader than in the early part of the century and the benefits of this are only just beginning to be felt commercially.

Recently introduced blackcurrant cultivars in Europe include the Dutch 'Black Reward' and the German 'Invigo', both reported to be heavy cropping, late flowering for frost avoidance and well suited to machine picking, having firm fruits which shake off readily and an erect growth habit. The British 'Blackdown' is heavy cropping and derives its mildew resistance from 'Brödtorp'. The very erect German 'Westra' is unique in being the first x-ray induced mutant in *Ribes* which has been introduced into commerce.

In England promising seedlings at a late stage of selection include first and second backcross derivatives of *R. bracteosum*, some of which combine late flowering, long strigs both for ease of handpicking and for crop potential, good hanging, ready separation of fruits from stalks leaving a dry scar suggesting mechanical harvesting potential, moderate mildew resistance, and an erect habit. Outstandingly large-fruited, somewhat mildew and leafspot resistant, seedlings of good crop potential ('93/16', '93/20', and '93/28') have been bred in Scotland from a combination of Canadian ('Consort', 'Magnus') and Scandinavian cultivars ('Brödtorp', 'Janslunda') (Anderson, unpub.).

Blackcurrant seedlings at an earlier stage of selection in England include third and fourth backcross derivatives of gooseberry, carrying gall mite and mildew resistance, highly mildew resistant first and second backcross derivatives of *R. sanguineum* and *R. glutinosum*, and second backcross derivatives of *R. cereum* of very high crop potential. In Scotland leafspot resistant derivatives of *R. dikuscha* carrying the genes Pr_1 and Pr_2 and gall mite and reversion resistant types bred from Russian *R. nigrum sibiricum* cultivars are under selection.

It can be anticipated that future blackcurrant cultivars will show considerable improvement over those now in commerce with respect to adaptation to mechanical harvesting, regularity and amount of crop, and in pest and disease resistance.

Far more effort is devoted to blackcurrant than to redcurrant breeding, and this is reflected both in the output of new redcurrant cultivars and in the breeding material available for future work. The most interesting development in this field has been the introduction of heavy-cropping derivatives of *R. multiflorum*. 'Rondom' and 'Heinemanns Rote Spätlese' are leafspot resistant, midseason and late flowering respectively, have firm fruit, and a report from Holland (Kronenberg, 1966) suggests that 'Rondom' is suitable for mechanical harvesting, unlike most redcurrant cultivars. 'Mulka' is another late-flowering cultivar of this origin, introduced in Germany in 1964. Second and third backcrosses from *R. multiflorum* are under selection in England; some of these have long strigs and are of better dessert quality than the named derivatives of this species.

A potential source of greater variability and heavy cropping for the redcurrant is the F_1 progeny of *R. longeracemosum* x *R. multiflorum* described under "Strig length in currants."

Of recently introduced gooseberry cultivars, Bauer's range of mildew resistant *R. divaricatum* derivatives, 'Remarka', 'Rokula', 'Risulfa', 'Ristula', and 'Reverta' represent a distinct advance on his earlier 'Resistenta' and 'Robustenta' in respect of fruit size, range of season, and fruit color. 'Remarka' is also resistant to leafspot. The Rumanian 'Rezistent de Cluj' and 'Cluj V/3', which are first backcrosses from *R. divaricatum*, combine resistance to American and European gooseberry mildew with hardiness and drought resistance.

Gene list (based on Knight et al., 1972)

Symbol suggested	Gene effect	Authority	Species
Ce	resistance to *Cecidophyopsis ribis*	Knight et al., 1974 (in press)	*grossularia*
Cr	resistance to *Cronartium ribicola*	Hunter, 1950a	*ussuriense*
l	lethal linked with *rb*	Keep & Knight, 1970	*nigrum*
M	resistance to *Sphaerotheca mors-uvae*	Rousi, 1966	*nigrum*
P	resistance to *Cecidophyopsis ribis*	Anderson, 1971a	*nigrum sibiricum*
Pr_1 } Pr_2 }	complementary genes for resistance to *Pseudopeziza ribis*	Anderson, 1972	*dikuscha*
rb	greenish fruit	Keep & Knight, 1970	*nigrum*
rc	white fruit	Keep & Knight, 1970	redcurrant
rg	non-red fruit	Keep & Knight, 1970	*grossularia*
r } *s* }	leaf shape	Tydeman, 1930	*nigrum*
Sph	resistance to *Sphaerotheca mors-uvae*	Keep, 1974	*oxyacanthoides*

Gooseberry seedlings at a late stage of selection in England include large-fruited mildew resistant late backcross derivatives of *R. divaricatum* and more or less spineless mildew resistant derivatives

Bud sports

Nature	'Parent' cultivar	Authority
Blackcurrant		
High flower number, regular stamen number, leaves less deeply divided, flavor less 'foxy'	Juharlevelü	Porpáczy et al., 1960
Early leafing, flowering, and ripening	Daniels September 'Laleham Beauty	Hughes, 1963
Redcurrant		
Fruit color, red→pink ('Rosa Sport')	Heros	Schuppe, 1962
Early leafing, flowering, and ripening	seedling	Knight & Keep, 1958*b*

of *R.* '*oxyacanthoides*'. Progenies at an earlier stage of selection include a first backcross from *R. sanguineum* which segregated for complete spinelessness and should carry resistance to the aphid *N. ribisnigri* and possibly to the sawfly *N. ribesii*, first backcrosses from *R. alpestre* which should provide dominant resistance to *A. grossulariae*, and F_1s of gooseberry x *R. roezlii*, x *R. leptanthum* and x *R. watsonianum*, the former bred for resistance to *N. ribisnigri*, the two latter for combined high resistance to leafspot and American gooseberry mildew.

It seems probable that gooseberry cultivars of the fairly near future will combine fruit size, spinelessness, leafspot and mildew resistance, but pest resistances will take rather longer to incorporate into cultivars of commercial quality.

Literature Cited

The first few pages of this chapter were written in collaboration with the late Dr. R. L. Knight. Much of the rest of the chapter has been deeply influenced by wide ranging discussions covering all aspects of plant breeding held while we were colleagues for many years in the soft fruit breeding team at East Malling. It is a great pleasure to acknowledge my indebtedness to him for freely sharing his wide knowledge and wisdom gained during a life time devoted to plant breeding.

Adam, J., and Wilson, D. 1967. Factors affecting the germination of blackcurrant seed. *Rpt. Long Ashton Res. Stn. 1966* p. 96–103.

Adams, J. 1916. On the germination of the pollen grains of apple and other fruit trees. *Bot. Gaz.* 61: 131–147.

Adams, J. F. 1923. Diseases of small fruits. *In* Diseases of fruit and nut crops in the United States in 1922. *Plant Dis. Bul., Suppl.* 28: 357–375.

Agababyan, V. Sh. 1963. A study in pollen morphology of the genus *Ribes* L. (in Russian). *Izv. Akad. Nauk Est. SSR, Ser. Biol.* 16(4): 93–98.

Anderson, H. W. 1956. *Diseases of fruit crops*. New York: McGraw-Hill.

Anderson, M. M. (Reported by North, C. 1967). Plant breeding: Blackcurrant. *Rpt. Scot. Hort. Res. Inst. 1966* p. 40–42.

———. 1969. Resistance to leaf spot and American gooseberry mildew. *Rpt. Scot. Hort. Res. Inst. 1968* p. 39.

———. 1971*a*. Resistance to gall mite (*Phytoptus ribis* Nal.) in the Eucoreosma section of *Ribes*. *Euphytica* 20: 422–426.

———. 1971*b*. Gall mite and reversion virus. *Rpt. Scot. Hort. Res. Inst. 1970* p. 37.

———. 1972. Resistance to blackcurrant leaf spot (*Pseudopeziza ribis*) in crosses between *Ribes dikuscha* and *R. nigrum*. *Euphytica* 21: 510–517.

Anderson, O. C. 1939. A cytological study of resistance of Viking currant to infection by *Cronartium ribicola*. *Phytopathology* 29: 26–40.

Anderson, R. L., and French, D. W. 1955. Evidence of races of *Cronartium ribicola* on *Ribes*. *For. Sci.* 1: 38–39.

Andreičenko, D. A. 1938. Let us utilize the many wild bush fruits of eastern Siberia (in Russian). *Plodoov. Khoz.* 3: 56–62.

Anonymous. 1934. The plant resources of the world as initial material in plant breeding. Botanic-ecological and economic characteristics. No. 5. Fruits and small fruits and their wild relatives (in Russian). *Lenin Acad. Agr. Sci., Inst. Pl. Ind.* Leningrad.

———. 1938. Pomology department. *Rpt. John Innes Hort. Instn. 1937* p. 12–15.

———. 1950*a*. *Prog. Rpt. 1934–1948 Div. Hort. Cent. Expt. Fm, Ottawa.*

———. 1950*b*. *Rpt. Minister Agric. Can.*

———. 1957. Blackcurrant trials; results 1948 to 1956 (in Dutch). *Tuinbouwber., Groningen* 12(48 & 49): 198–199.

———. 1960. Pomology. *Rpt. E. Malling Res. Stn. 1959* p. 6–11.

———. 1961*a*. The blackcurrant Nahodka (in Russian). *Sadovodstvo* 4: 32.

———. 1961*b*. Pomology. *Rpt. E. Malling Res. Stn. 1960* p. 7–12.

———. 1962. Pomology. *Rpt. E. Malling Res. Stn. 1961* p. 7–13.

———. 1963*a*. Summary of research, 1962. *Rpt. Long Ashton Res. Stn. 1962* p. 13–42.

———. 1963*b*. Artificial mutations and Breeding red- and blackcurrants (in Dutch). *Jversl. tuinbouwk. Onderz.*: 68.

———. 1965*a*. Bush fruits. *Bul. Minist. Agric. Fish. Fd, London* 4, 8th edn.

———. 1965*b*. Varietal trials with redcurrants (in Dutch). *Tidsskr. PlAvl* 68: 903–906.

———. 1965*c*. Anatolij Grigor'evič Voluznev (in Russian). *Sadovodstvo* 4: 62.

———. 1965*d*. Entomology. *Rpt. E. Malling Res. Stn. 1964* p. 34–37.

———. 1966. Plant pathology. *Rpt. E. Malling Res. Stn. 1965* p. 40–49.

———. 1968*a*. Plant pathology. *Rpt. E. Malling Res. Stn. 1967* p. 34–44.

———. 1968*b*. Fruit breeding. *Rpt. E. Malling Res. Stn. 1967* p. 24–27.

———. 1969. Breeding currants, gooseberries and raspberries. *Prog. Rpt. Fruit Breed. Stn. Angers 1965–66* p. 44–47.

———. 1970. Fruit breeding. *Rpt. E. Malling Res. Stn. 1969* p. 37–38.

———. 1973. Flowering periods of tree and bush fruits. *Tech. Bul. Minist. Agr. Fish. Fd.* 26.

Arasu, N. T. 1968. Overcoming self-incompatibility by irradiation. *Rpt. E. Malling Res. Stn. 1967* p. 109–112.

———. 1970. Self incompatibility in *Ribes*. *Euphytica* 19: 373–378.

van Arsdel, E. P., Riker, A. J., and Patton, R. F. 1956. The effects of temperature and moisture on the spread of white pine blister rust. *Phytopathology* 46: 307–318.

Bagenal, N. B. 1955. History and development of the cultivated fruits. 6. *Ass. Agr. Rev.* 27: 9–20.

Baldini, E., and Pisani, P. L. 1961. Research on the biology of flowering and fruiting in blackcurrants (in Italian). *Riv. Ortoflorofruttic. ital.* 45: 619–639.

Barsukov, N. I., and Savvina, I. V. 1969. Redcurrants in Siberia (in Russian). *Sadovodstvo* 5:24.

Bauer, R. 1938. The method of mass infection in the breeding of varieties resistant to mildew and leaf fall in the genus *Ribes* (in German). *Forschungsdienst* 6: 575–584.

———. 1953. Immunity and resistance to *Sphaerotheca mors-uvae* (Schw.) Berk. in *Ribes* (in German). *Proc. 7th Int. Bot. Congr., Stockholm 1950* p. 701 (Abstr.).

———. 1955. Resistance problems in the genus *Ribes* and possibilities of their solution by making intra- and inter-sectional crosses (in German). *Rpt. 14th Int. Hort. Congr., Scheveningen* p. 685–696.

———. 1957. The induction of vegetative mutations in *Ribes nigrum*. *Hereditas* 43: 323–337.

———. 1969. A "stem form" for complete mechanization of culture and harvest of blackcurrant (in German). *Erwerbsobstbau* 11: 148–151.

Bauer, [R.] 1970. Working group for breeding research. Max Planck Institute of Breeding Research (Erwin Baur Institute), Köln-Vogelsang (in German). *Naturwissenschaften* 57: 666.

Behrens, E. 1964. The biology and ecology of the currant gall mite *Eriophyes ribis* Nal. and its control in the Perleberg currant plantation (in German). *Wiss. Z. Univ. Rostock, Mathematisch-Naturwissenschaftliche Reihe* 13: 279–288.

Benson, R. B. 1953. A new British *Nematus* (Hym., Tenthredinidae) attacking blackcurrant. *Ent. Mon. Mag.* 89: 60–63.

Bergelin, E., and Sahlström, H. 1958. Varietal trials with redcurrants 1943–1957 (in Swedish). *Meddn St. TrädgFörs. Malmö* 116: 1–18, and *Fors. Forsk.* 10: 96–97.

Berger, A. 1924. A taxonomic review of currants and gooseberries. *Bul. N.Y. St. Agr. Expt. Stn.* 109.

Berning, A. 1966. Testing varietal characters in redcurrants and blackcurrants in relation to mechanical harvesting (in German). *Erwerbsobstbau* 8: 66–69.

———. 1967. Further testing of black and redcurrant varieties for suitability for mechanical harvesting by shaking (in German). *Erwerbsobstbau* 9: 106–108.

Bjurman, B. 1961. Variety trials with blackcurrants in south and central Sweden 1948–1960 (in Swedish). *Meddn St. TrädgFörs. Alnarp* 137: 1–14.

Blake, M. A., Farley, A. J., and Connors, C. H. 1913. Report of the horticulturist. *Rpt. N.J. Agr. Expt. Stn.* p. 89–173.

Blodgett, E. C. 1936. The anthracnose of currant and gooseberry caused by *Pseudopeziza ribis*. *Phytopathology* 26: 115–152.

Blommers, J. and van Oosten, A. A. 1969. Mechanical harvesting (in Dutch). *Jversl. Proefstn. Fruitteelt Wilhelminadorp* 1969: 63–67.

Bočkarnikova, N. M. 1964. *Ribes dikuscha* Fisch., a valuable blackcurrant species in the Far East (in Russian). *Tr. prikl. Bot. Genet. Selek.* 36(3): 25–39.

Bogdański, K., Zalewski, W., and Bogdańska, H. 1957. Study of the ascorbic acid content, colour and other compounds in blackcurrant varieties, *Ribes nigrum* (in Russian). *Roczn Nauk roln.* 73A: 123–143.

Bologovskaja, R. P. 1940. Breeding small fruits in Siberia and the Far East (in Russian). *Vest sel'.-khoz. Nauki, Mosk.* 4: 64–68.

Bordeianu, T., Constantinescu, N., and Stefan, N. (eds). 1969. Local varieties and promising hybrids 8 (in Romanian). *Editura Academiei Republicii Populare Romîne.*

Börner, C. 1952. Europae centralis Aphides. The aphids of central Europe. Names, synonyms, host plants, and life cycles (in German). *Schr. Landesarbeits. Heilpflanz.* 4 (*Mitt. thüring. bot. Ges. Suppl.* 3).

Bouwens, H. 1927. Further investigations in the Erysiphaceae (in German). *Meded. phytopath. Lab. Willie Commelin Scholten* 10: 3–31.

Bradt, O. A., et al. 1968. Fruit varieties. *Pub. Ont. Dept. Agr.* 430.

Brandis, A. 1967. Yield and picking performance of blackcurrants and redcurrants (in German). *Erwerbsobstbau* 9: 113–115.

Brauns, M. 1959. A contribution to the breeding of mildew resistant gooseberries. 1. Investigations on the inheritance of resistance and the possibilities of using early selection (in German). *Züchter* 29: 51–57.

Brooks, R. M., and Olmo, H. P. 1947. Register of new fruit and nut varieties. List 3. *Proc. Amer. Soc. Hort. Sci.* 50: 426–442.

De Bruyne, A. S., Kronenberg, H. G., and de Mos, D., et al. 1958. *9th List of Fruit Varieties* (in Dutch). Wageningen, The Netherlands. p. 73–75, 111–116, 117–121.

Bunyard, E. A. 1917*a*. The history and development of the red currant. *J. Roy Hort. Soc.* 42: 260–270.

———. 1917*b*. A revision of the red currants. *Gard. Chron.* 62: 205–206, 217, 232, 237.

———. 1921. A revision of the red currants. *J. Pomol.* 2: 38–55.

———. 1925. *A handbook of hardy fruits. Stone and bush fruits, nuts, etc.* London: John Murray.

Bystydzieński, W., and Smolarz, K. 1966. Results of yield trials with 8 blackcurrant varieties in the Silesia region (in Polish). *Pr. Inst. Sadow. Skierniew.* 10: 239–246.

Campbell, A. I. 1965. The inactivation of blackcurrant reversion virus by heat therapy. *Rpt. Long Ashton Res. Stn. 1964* p. 89–92.

Christensen, J. V. 1966. Mechanical harvesting of berries. Prospects and problems. *Proc. Balsgård Fruit Brdg Symp., Fjälkestad 1964*: p. 90–92.

Cociu, V., et al. 1966. Influence of mutagens on the hybrid progeny of tree fruits, strawberries and bush fruits (in Romanian). *Lucr. Stiint. Inst. Cerc. Hortivitic.* 8:329–336.

Colby, A. S. 1927. Notes on self-fertility of some gooseberry varieties. *Proc. Amer. Soc. Hort. Sci.* 23: 138–140.

———. 1934. Size inheritance in gooseberry fruits. *Proc. Amer. Soc. Hort. Sci.* 30: 105–107.

———. 1935. Inheritance of gooseberry leaf infection. *Proc. Amer. Soc. Hort. Sci.* 32: 397–399.

Collingwood, C. A., and Brock, A. M. 1959. Ecology of the blackcurrant gall mite (*Phytoptus ribis* Nal.). *J. Hort. Sci.* 34: 176–182.

Crane, M. B., and Lawrence, W. J. C. 1952. *The genetics of garden plants.* 4th ed. London: Macmillan.

Čuvašina, N. P. 1961. The effect of gibberellin on crossability between distantly related plants (in Russian). *Tr. Cent. Genet. Lab. Mičurina* 7: 183–189.

———. 1962. The effect of gibberellic acid on crossability of different species in the genus *Ribes* L. (in Russian). *Rpt. Soviet Scientists 16th Int. Hort. Congr., Moscow*: 123–128.

———. 1963. Overcoming incompatibility in distant hybridization in fruit crops by means of gibberellin. *In* Gibberellins and their effect on plants (in Russian). *USSR Acad. Sci., Moscow*: 202–206.

———. 1968. The action of acenaphthene on microsporogenesis in black currant. In *Plant breeding by mutation* (in Russian). Moscow: Nauka. (from *Referat. Zh.* 1969: 283–284).

Darlington, C. D. 1929. A comparative study of the chromosome complement in *Ribes*. *Genetica* 11: 267–269.

Darrow, G. M. 1937. Improvement of currants and gooseberries. In *USDA Yearbook of Agriculture* 1937: 534–544.

Davis, M. B. 1926. Gooseberry breeding. *Rpt. Dept. Agr. Can. Hort. 1925* p. 11–14.

Decourtye, L., and Lantin, B. 1965. Improvements in cultural practices in blackcurrants (in French). *Pomol. fr.* 7: 103–104, 107–108, 113–114, 117–118, 121.

Dement'eva, M. I. 1953. Resistance of gooseberry to a group of diseases (in Russian). *Proc. Lenin Acad. Agr. Sci.* 6: 32–34.

———. 1958. Biochemical and physiological factors governing resistance of gooseberry to mildew (in Russian). *Izv. timiryazev. sel'.-khoz. Akad.* 5: 149–160.

Dorsman, C., et al. (eds.). 1969. *14th List of Fruit Varieties* (in Dutch). Wageningen, The Netherlands. p. 51, 81–86.

Evréinoff, V. A. 1942. The blue currant, *Ribes dikuscha* Fisch. (in French). *Revue hort.* 28: 46–47.

van Eyndhoven, G. L. 1967. The redcurrant gall mite, *Cecidophyopsis selachodon* n. sp. *Ent. Ber. Amst.* 27: 149–151.

Fernqvist, I. 1961. Investigations on floral biology in blackcurrants, redcurrants and gooseberries (in Swedish). *Kgl. LantbrAkad. Tidsk.* 100: 357–397.

Frazier, N. W. (ed.). 1970. *Virus diseases of small fruits and grapevines*. Berkeley: Univ. of Cal. Division of Agricultural Sciences. p. 75–104.
Free, J. B. 1968. The pollination of blackcurrants. *J. Hort. Sci.* 43: 69–73.
Genevès, L. 1958. On the principal stages of the development of flowers of *Ribes rubrum* L. (Grossulariaceae) (in French). *C.r. Hebd. Séanc. Acad. Sci., Paris* 247: 2175–2178.
Gersons, L. 1966*a*. Quality research on processed horticultural produce: Quality standards for processed vegetable and fruit products. *Rpt. Inst. Res. Stor. Process. Hort. Prod. Wageningen 1965*. p. 68–71.
———. 1966*b*. Assessing the color of blackcurrant juice (in Dutch). *Bul. Inst. Res. Stor. Process. Hort. Prod. Wageningen* 48.
Goldschmidt, E. 1964*a*. Cytological studies on diploid species and interspecific hybrids of the genus *Ribes* L. *Hereditas* 51: 146–186.
———. 1964*b*. Cytological studies on tetraploid plants of the genus *Ribes* L. *Hereditas* 52: 139–150.
———. 1966. Cytological investigations in the genus *Ribes*. *Proc. Balsgård Fruit Brdg. Symp., Fjälkestad 1964*: 205–208.
Gomoljako, L. J. 1941. A new source of carotene (Provitamin A) (in Russian). *Dokl. Acad. Sci. U.S.S.R.* 32: 142–143.
Grbrić, O., and Vujanić-Varga, D. 1968. The time and course of flower bud differentiation in blackcurrants (in Hungarian). *Zborn. Inst. Vinogr. Voć. Sremski Karlovic* 1(1): 73–78.
Greenslade, R. M. 1941. The black currant leaf midge (*Dasyneura tetensi* (Rübs). *Rpt. E. Malling Res. Stn. 1940* p. 66–71.
Gröber, K. 1967. Induced mutations and their utilization. Erwin Baur memorial lectures, 4, 1966. Some results of mutation experiments in apples and blackcurrants (in German). *Abh. dt. Akad. Wiss. Berl.: Klasse Med.* 2: 377–382.
Gruber, F. 1934. Studies in variation statistics of some economically important characters in small bush fruits. Preliminary communication (in German). *Züchter* 6: 294–296.
Guseva, A. N. 1964. Blister rust on Siberian cedar pine in southern Yakutia (in Russian). *Les. Hoz.* 17(11): 51–52.
Hahn, G. G. 1939. Immunity of a staminate clone of *Ribes alpinum* from *Cronartium ribicola*. *Phytopathology* 29: 981–986.
———. 1943. Blister rust relations of cultivated species of red currants. *Phytopathology* 33: 341–353.
Harborne, J. B., and Hall, E. 1964. Plant polyphenols. 13. The systematic distribution and origin of anthocyanins containing branched trisaccharides. *Phytochemistry* 3: 453–463.
Hårdh, J. E., and Wallden, J. 1965. Flower formation and fruit growth in blackcurrants (in Finnish). *Maataloust. Aikakausk.* 37: 61–75.
Haskell, G. 1962. Genetics. *Rpt. Scot. Hort. Res. Inst. 1961–2* p. 54–64.
———. 1963. Genetics. *Rpt. Scot. Hort. Res. Inst. 1962–3* p. 56–67.
———. 1964. Genetics. *Rpt. Scot. Hort. Res. Inst. 1963–4* p. 56–62.
Haskell, G. M. L. 1966. Genetics. *Rpt. Scot. Hort. Res. Inst. 1964 and 1965* p. 39–41.
Hatton, R. G. 1920. Black currant varieties. A method of classification. *J. Pomol.* 1: 65–80, 145–154.
Hedrick, U. P. 1915. The blooming season of hardy fruits. *Bul. N.Y. St. Agr. Expt. Stn.* 407: 386–389.
———. (ed.). 1919. Sturtevant's notes on edible plants. *Rpt. N.Y. Agr. Expt. Stn.* 27.
———. 1925. The small fruits of New York. *Rpt. N.Y. St. Agr. Expt. Stn.* 33: 243–354.
Hildebrand, E. M., and Weber, P. V. 1944. Varietal susceptibility of currants to the cane blight organism, and to currant mosaic virus. *Plant Dis. Rptr.* 28: 1031–1035.
Hille Ris Lambers, D. 1953. Contributions to a monograph of the Aphididae of Europe. 5. *Temminckia* 9: 1–176.
Hofman, K. 1963. Fruit set in a number of blackcurrants (in Dutch). *Fruitteelt* 53: 334–335.
Hogg, R. 1875. *The fruit manual*. (4th ed.). London: Fleet Street.
Hooper, C. H. 1911. Experiments in the pollination of our hardy fruits. *Agr. Stud. Gaz.*, Wye N.S. 15: 110–113.
Hughes, H. M. 1962. Preliminary studies on the insect pollination of soft fruits. *Expl. Hort.* 6: 44.
———. 1963. A study of two blackcurrant chimaeras. *J. Hort. Sci.* 38: 286–296.
———. 1966. Investigations on the pollination of blackcurrant var. Baldwin. *Expl. Hort.* 14: 13–17.
Hunter, A. W. S. 1950a. Small fruits: Black currants. *Prog. Rpt. Cent. Expt. Fm, Ottawa, 1934–48* p. 26–29.
———. 1950b. Small fruits: Gooseberries. *Prog. Rpt. Cent. Expt. Fm, Ottawa, 1934–48* p. 29–30.
———. 1950c. Small fruits: Red currants. *Prog. Rpt. Cent. Expt. Fm, Ottawa, 1934–48* p. 29.
———. 1955*a*. Black currants. *Prog. Rpt. Cent. Expt. Fm, Ottawa, 1949–53* p. 28–29.
———. 1955*b*. Gooseberries. *Prog. Rpt. Cent. Expt. Fm, Ottawa, 1949–53* p. 28.
Jagudina, S. 1968. New varieties of berry crops (in Russian). *Sadovodstvo* 7: 29–30.
Janczewski, E. de 1907. Monograph of the currants *Ribes* L. (in French). *Mem. Soc. Phys. Hist. nat. Genève* 35: 199–517.
Jordan, C., Korte, F., and von Sengbusch, R. 1957. Paper chromatographic estimation of certain acids and sugars as a basis for the selection of good flavour in top and soft fruits and vegetables (in German). *Züchter* 27: 69–76.
Jordan, V. W. L. 1968. The life history and epidemiology of American gooseberry mildew on blackcurrants. *Ann. Appl. Biol.* 61: 399–406.
Kaplan, R. W. 1953. On the potentialities of mutation induction in plant breeding (in German). *Z. Pflanzenzücht.* 32(2): 121–131.
Karl, E., and Kleinhempel, H. 1969. Experiments on aphid- and graft-transmission of various isolates of veinbanding virus of gooseberry and currant (in Hungarian). *Acta Phytopath. Acad. Sci. Hung.* 4: 19–28.
Karnatz, A. 1969. Raising *Ribes nigrum* seedlings under

cover (in German). *Mitt. Klosterneuburg* 19: 319–321.
Kaszonyi, S. 1951. *Gloeosporium ribis* (Lib.) Mont. & Desm. f. sp. *grossulariae* Kleb. A disease of gooseberry caused by *Gloeosporium* in Hungary (in Hungarian). *Agrartud. Egyet. erdömern. Kar. Évk.* 2 (*Agrartud, egy.* 15[1]): 205–210.
Keep, E. 1958. Cytological notes. *Rpt. E. Malling Res. Stn. 1957* p. 75–78.
———. 1960. Amphiplasty in *Ribes. Nature, Lond.* 188: 339.
———. 1962*a*. Satellite and nucleolar number in hybrids between *Ribes nigrum* and *R. grossularia* and in their backcrosses. *Can. J. Gen. Cyt.* 4: 206–218.
———. 1962*b*. Interspecific hybridization in *Ribes. Genetica* 33: 1–23.
———. 1965. Cytological notes 2. *Ribes. Rpt. E. Malling Res. Stn. 1964* p. 104–107.
———. 1970. Response of *Ribes* species to American gooseberry mildew, *Sphaerotheca mors-uvae* (Schw.) Berk. *Rpt. E. Malling Res. Stn. 1969* p. 133–137.
Keep, E., and Knight, R. L. 1970. Inheritance of fruit colour in currants and gooseberries. *Rpt. E. Malling Res. Stn. 1969* p. 139–142.
Keep, E. 1971. Nucleolar suppression, its inheritance and association with taxonomy and sex in the genus *Ribes. Heredity, London* 26: 443–452.
Keep, E., and Briggs, J. B. 1971. A survey of *Ribes* species for aphid resistance. *Ann. Appl. Biol.* 68: 23–30.
Keep, E. 1974. Breeding for resistance to American gooseberry mildew, *Sphaerotheca mors-uvae*, in the gooseberry (*Ribes grossularia*). *Ann Appl. Biol.* 76: 131–135.
Kimmey, J. W. 1935. Susceptibility of principal *Ribes* of southern Oregon to white-pine blister rust. *J. For.* 33: 52–56.
———. 1938. Susceptibility of *Ribes* to *Cronartium ribicola* in the West. *J. For.* 36: 312–320.
Klämbt, H.-D. 1958. Studies on pollination behaviour in black- and red-currants (in German). *Gartenbauwissenschaft* 23: 9–28.
Klebahn, H. 1906. Studies on certain Fungi Imperfecti and the associated Ascomycetous forms. 3. *Gloeosporium ribis* Mont. et Desm. (in German). *Z. Pflkrankh.* 16: 65–83.
Kleinhempel, H. 1970. The distribution and damaging effects of virus diseases of currants and gooseberries (in German). *Arch. Gartenb.* 18: 319–325.
Klingbeil, G. C., Wade, E. K., and Libby, J. L. 1965. Currants and gooseberries. *Cir. Wis. Univ. Agr. Ext. Serv.* 644.
Knight, R. L., and Keep, E. 1957. Fertile blackcurrant-gooseberry hybrids. *Rpt. E. Malling Res. Stn. 1956* p. 73–74.
———. 1958*a*. Abstract bibliography of fruit breeding and genetics to 1955. *Rubus* and *Ribes*—A survey. *Tech. Commun. Commonw. Bur. Hort. Plantn. Crops* 25.
———. 1958*b*. An early sport of a redcurrant. *Rpt. E. Malling Res. Stn. 1957* p. 74.
Knight, R. L. 1962*a*. Fruit Breeding. *J. R. Hort. Soc.* 87: 103–113.
———. 1962*b*. Heritable resistance to pests and diseases in fruit crops. *Proc. 16th Int. Hort. Congr., Brussels* 3: 99–104.
Knight, R. L., and Keep, E. 1964. Soft fruit breeding. *Rpt. E. Malling Res. Stn. 1963* p. 158–160.
———. 1966. Breeding new soft fruits, p. 98–111. *In* P. M. Synge and E. Napier, eds., *Fruit—present and future*. London: Royal Horticultural Society.
Knight, R. L. 1972. Fruit breeding. *Rpt. E. Malling Res. Stn. 1971* p. 107–109.
Knight, R. L., Parker, J. H., and Keep, E. 1972. Abstract bibliography of fruit breeding and genetics 1956–1969. *Rubus* and *Ribes. Tech. Commun. Commonw. Bur. Hort. Plantn. Crops* 32.
Knight, R. L., and Alston, F. H. 1974. Pest resistance in plant breeding. In *Biology of pest and disease control*. Eds. D. Price Jones, M. E. Solomon. Oxford: Blackwell p. 87–96.
Knight, R. L., Keep, E., Briggs, J. B., and Parker, J. H. 1974. Transference of resistance to blackcurrant gall mite, *Cecidophyopsis ribis*, from gooseberry to blackcurrant. *Ann. Appl. Biol.* 76: 123–130.
Kock, G. 1914*a*. Varietal resistance of the gooseberry against mildew and the effects of sulphur treatment therefore (in German). *Z. landw. VersWes. Öst.* 17: 634–637.
———. 1914*b*. The resistance of various gooseberry varieties against North American gooseberry mildew and their behaviour on treatment with sulphur (in German). *K. K. Pflanzensch., Vienna.*
Komar, G. A. 1970. The development of flower and inflorescence in some representatives of the family Grossulariaceae (in Russian). *Bot. Zh. SSSR* 55: 954–971.
Kotte, W. 1958. *Diseases and pests of fruit* (in German). (3rd ed.). Berlin: Paul Parey.
Kovalev, N. V. 1934. Scientific fruit culture in the U.S.S.R. in the new era. Summary of recent scientific investigations (in Russian). *Tr. prikl. Bot. Genet. Selek., Ser. A.* 10: 67–92.
———. 1936. The plant resources of fruit trees in the far east and their significance (in Russian). *Tr. prikl. Bot. Genet. Selek., Ser. A* 18: 21–38.
Kovaleva, E. S., and Natal'ina, O. B. 1968. Study of specialization in the causal organism of rust—the fungus *Cronartium ribicola* Dietr. (in Russian). *Tr. saratov. sel'.-khoz. Inst.* 17(1): 253–256.
Kovtun, I. M. 1962. The efficacy of different methods of developing a spineless gooseberry (in Russian). *Nauč. Trud. ukrain. nauchno-issled. Inst. Sadov.* No. 39: 23–24.
Kronenberg, H. G., and Doesburg, J. J. 1955. The vitamin C content of black currant varieties requires more attention! (in Dutch). *Fruitteelt* 45: 869.
Kronenberg, H. G., and Gersons, L. 1961. Can the blackcurrant cultivar 'Consort' be recommended? (in Dutch). *Fruitteelt* 51: 1305.
Kronenberg, H. G. 1964. Some varietal differences in redcurrant. *Hort. Res.* 3: 72–78.
———. 1965. New German currant and raspberry varieties (in Dutch). *Fruitteelt* 55: 1397.
Kronenberg, H. G., and Gersons, L. 1965. The redcurrant cultivar 'Rondom' (in Dutch). *Fruitteelt* 55: 386.
Kronenberg, H. G., and Hofman, K. 1965. Research on

some characters in blackcurrant progenies. *Euphytica* 14: 23–35.
Kronenberg, H. G. 1966. Preliminary experiences with mechanical harvesting of some varieties of black and redcurrant. *Proc. Balsgård Fruit Brdg. Symp.,Fjälkestad 1964* p. 103–105.
Kruglova, A. P. 1965. Breeding gooseberry varieties resistant to *Sphaerotheca* under the conditions of the Saratov area of the Volga Basin (in Russian). *(Proc.) 4th All-Union Conf. on Immunity in Agr. Plants, Vines & Fruit Crops, Kišinev*: 223–226.
Kuminov, E. P. 1962a. The wild Altaj gooseberry and its hybrid progeny (in Russian). *Sadovodstvo* 1: 16–17.
———. 1962b. Self-fertility and cross-fertility of Siberian currant varieties (in Russian). *Sel'sk. Hozjajstv. Sibir.* 12: 57–59.
———. 1964. *Ribes nigrum* var. *sibiricum* as initial breeding material (in Russian). *Tr. krasnojarsk nauč.-issled. Inst. sel'sk. Hoz.* 2: 115–119.
———. 1965. Siberian blackcurrant—for breeding (in Russian). *Sadovodstvo* 8: 27–28.
Kuusi, T. 1965. The most important quality criteria of some home-grown blackcurrant varieties. 1. Ascorbic acid. *Maataloust. Aikakausk.* 37: 264–281.
Kuz'min, A. Ja. 1956. New forms of currant (in Yugoslavian). *Priroda. Zagr.* No. 10: 94–95.
Kuz'min, A. Ja., and Čuvašina, N. P. 1960. Distant hybridization in the gooseberry family (in Yugoslavian). *Otdalennaya Gibrid. Rast. Zhivotnykh:* 113–126. NLL Translation RTS 5720.
Kuznecov, V. F. 1969. Resistance to fungal diseases in gooseberry (in Russian). *Sadovodstvo* 3: 27.
Lantin, B. 1967. A study of some blackcurrant cultivars (in French). *Revue hort.* No. 2.277: 1.303–1.308.
———. 1970. Importance of cross pollination in the blackcurrant (in French). *Pomol. fr.* No. 8: 237–243.
Larsson, G. 1959. Blackcurrant variety trials in Norrland, 1944–1958 (in Swedish). *Meddn St. Trädg-Förs. Malmö* 122: 30 pp.
———. 1961. Varietal trials of gooseberry at the Öjebyn Experimental Station (in Swedish). *Sver. pomol. För. Årsskr.* 62: 87–97.
Ledeboer, M., and Rietsema, I. 1940. Unfruitfulness in blackcurrants. *J. Pomol.* 18: 177–181.
Lees, A. H. 1918. "Reversion" and resistance to "big bud" in blackcurrants. *Ann. Appl. Biol.* 5: 11–27.
Leppik, E. E. 1970. Gene centers of plants as sources of disease resistance. *Ann. Rev. Phytopath.* 8: 323–344.
Lihonos, F. D., and Pavlova, N. M. 1969. Fruit crops (in Russian). *Tr. Prikl. Bot. Genet. Selek.* 41(1): 264–284.
Lindley, G. 1831. *In* J. Lindley, ed. *A guide to the orchard and kitchen garden.* London.
Loginyčeva, A. G. 1958. The self-fertility and cross-pollinating ability of some varieties of blackcurrant, gooseberry, and raspberry (in Romanian). *Agrobiologiya* 6: 125–126.
Lorenz, P. 1929. Hybridization in the genus *Ribes* (in German). *Züchter* 1: 66–68.
Lorrain, R. 1962. The blackcurrant (in French). *Bul. hort., Liège* 17: 144–147.
Luckwill, L. C. 1948. A note on the unfruitfulness of a rogue strain of the blackcurrant variety 'Invincible Giant Prolific.' *Rpt. Long Ashton Res. Stn.* p. 22–25.
———. 1966. Pomology and plant breeding. *Rpt. Long Ashton Res. Stn. 1965* p. 22–29.
———. 1967. Pomology and plant breeding. *Rpt. Long Ashton Res. Stn. 1966* p. 20–27.
———. 1969. Pomology and plant breeding. *Rpt. Long Ashton Res. Stn. 1968* p. 14–24.
Macoun, W. T., and Davis, M. B. 1920. Bush fruits and their cultivation in Canada. *Bul. Can. Dept. Agr.* 94.
Massee, A. M. 1928. The blackcurrant gall mite on redcurrants. *Rpt. E. Malling Res. Stn. 1926 & 1927, 2, Suppl.* p. 151–152.
———. 1952. Transmission of reversion of black currants. *Rpt. E. Malling Res. Stn. 1951* p. 162–165.
McMahon, B. 1806. *The American gardener's calendar.* Philadelphia.
van der Meer, F. A. 1962. Virus diseases of currants (in Dutch). *Jversl. Inst. plziektenk. Onderz.*: 91–92.
———. 1965. Veinbanding mosaic in gooseberries (in Dutch). *Fruitteelt* 55: 245–246.
———. 1970. Red currant vein banding. *In* N. W. Frazier, ed., *Virus diseases of small fruits and grapevines,* p. 91–92. Berkeley: Univ. of Cal., Division of Agricultural Sciences.
Melehina, A. A. 1964. Varieties and the breeding of blackcurrants (in Russian). *Sadovodstvo* 12: 33.
———. 1966. Variability of blackcurrant under the influence of ionizing radiation (in Russian). *Latv. PSR Zinãt. Akad. Vest.* 10: 83–91.
———. 1968. Effect of γ irradiation on the viability and fertilizing ability of pollen in intraspecific crosses of black currants. *In* Crop plants in the national economy. 4. (in Russian). Zinātne, Riga 4 (from *Referat. Zh.* 1970: 145–155.)
Merriman, P. R., and Wheeler, B. E. J. 1968. Overwintering of *Sphaerotheca mors-uvae* on blackcurrant and gooseberry. *Ann. Appl. Biol.* 61: 387–397.
Meurman, O. 1928. Cytological studies in the genus *Ribes* L. *Hereditas* 11: 289–356.
Mielke, J. L. 1938. Spread of blister rust to sugar pine in Oregon and California. *J. For.* 36: 695–701.
Modlibowska, I., and Ruxton, J. P. 1953. Preliminary studies of spring frost resistance of black currant varieties. *Rpt. E. Malling Res. Stn. 1952* p. 67–72.
Modlibowska, I. 1960. Breaking the rest period in blackcurrants with gibberellic acid and low temperature. *Ann. Appl. Biol.* 48: 811–816.
Morrow, E. B., Darrow, G. M., and Scott, D. H. 1954. A quick method of cleaning berry seed for breeders. *Proc. Amer. Soc. Hort. Sci.* 63: 265.
Mosolova, A. V. 1956. New varieties of gooseberry bred by the Institute of Plant Industry (in Russian). *Byull. vses. Inst. Rasteniev.* 2: 39–40.
———. 1963. Siberian plant breeders in Leningrad province (in Russian). *Sadovodstvo* 3: 33–35.
Mosolova, A. V., and Volodina, E. V. 1969. Anthracnose-resistant varieties of currant in the collection of the All-Union Institute of Plant Breeding (in Russian). *Tr. Prikl. Bot. Genet. Selek.* 40(3): 140–145.
Nasr, T. [A. A.], and Wareing, P. F. 1958. Photo-

periodic induction of flowering in blackcurrant. *Nature, London* 182: 269.

———.1961*a*. Studies on flower initiation in blackcurrant. 1. Some internal factors affecting flowering. *J. Hort. Sci.* 36: 1–10.

———. 1961*b*. Studies on flower initiation in blackcurrant. 2. Photoperiodic induction of flowering. *J. Hort. Sci.* 36: 11–17.

Neumann, U. 1955. The importance of fertilization conditions and cultural measures for premature fruit drop in black currants (in German). *Arch. Gartenb.* 3: 339–354.

———. 1965. Yield increases and mechanization of harvesting as requirements for large-scale culture of blackcurrants (in German). *Obstbau. Berl.* 5: 21–23.

Nilsson, F. 1940. Some discoveries and impressions of horticulture in Norrland (in Swedish). *Fruktodlaren* No. 1: 16–19.

———. 1949. Polyploids in *Ribes, Fragaria, Raphanus* and *Lactuca*. *Proc. 8th Int. Congr. Genet., Stockholm (Suppl. Hereditas)*: 34–35.

———. 1958. *Cultivation of top and berry fruits* (in Swedish). Stockholm: Saxon & Lindströms Forlags Tryckeri.

———. 1959. Polyploidy in the genus *Ribes*. *Gen. Agr.* 11: 225–242.

———. 1966. Cytogenetic studies in *Ribes*. *Proc. Balsgård Fruit Brdg. Symp., Fjälkestad, 1964* p. 197–204.

———. 1969. Ascorbic acid in blackcurrants. *LantbrHögsk. Annlr* 35(1): 43–59.

Nitočkina, A. P. 1967. Testing gooseberry varieties (in Russian). *Sadovodstvo* 5: 33–35.

North, C. 1968. Plant breeding. *Rpt. Scot. Hort. Res. Inst. 1967* p. 37–45.

———. 1970. Plant breeding. *Rpt. Scot. Hort. Res. Inst. 1969* p. 38.

Nybom, N., and Bergendal, P.-O. 1960. Pome fruits and bush fruits (in Swedish). *Rpt. Balsgård Fruit Brdg. Inst. 1959*: 7–13.

Nybom, N. 1968. Theoretical investigations (in Swedish). *Rpt. Balsgård Fruit Brdg. Inst. 1967*: 13–14.

Offord, H. R., Quick, C. R., and Moss, V. D. 1944. Self-incompatibility in several species of *Ribes* in the western states. *J. Agr. Res.* 68: 65–71.

Op't Hoog. G. T. 1964. All about redcurrant varieties (in Dutch). *Groent. en Fruit* 21: 873–877.

Osipov, Ju. V. 1970. Types of green gooseberry cuttings and their rooting ability, p. 216–32. In *Breeding, variety studies and cultural practices in top and small fruit crops* (in Russian). Minist. Agric. R.S.F.S.R., Orel.

Osipov, K. V. 1968. Regular annual yield of currants (in Russian). *Sadovodstvo* 8: 41.

Parkinson, J. 1629. *Paradisi in sole paradisus terrestris*. London.

Pavlova, N. M. 1935. The gooseberry (in Russian). *Lenin. Acad. Agr. Sci., Inst. Pl. Ind., Sci. Pop. Ser.* 53.

———. 1940. Initial material for the breeding and cultivation of bush fruits (in Russian). *Vestn. sots. Rasteniev.* 5: 33–46.

———. 1955. Breeding black currants in the U.S.S.R. (in Russian). *Agrobiologiya* 4: 264–271.

Pawlowa (Pavlova), N. M. 1962. Breeding currants for winterhardiness in the USSR (in Russian). *Proc. 16th Int. Hort. Congr., Brussels* 1: 178–180. (Abstr.).

———. 1963. Breeding blackcurrants for winterhardiness (in Russian). *Vest. sel'.-khoz. Nauki, Mosk.* 8(2): 119–121.

———. 1964. Possibilities of breeding blackcurrant varieties resistant to the bud gall mite (in Russian). *Tr. prikl. Bot. Gen. Selek.* 36(3): 94–102.

Pierson, R. K., and Buchanan, T. S. 1938. Age of susceptibility of *Ribes petiolare* leaves to infection by aeciospores and urediospores of *Cronartium ribicola*. *Phytopathology* 28: 709–715.

Pisani, P. L. 1962. Flower-bud differentiation in blackcurrants (in Italian). *Riv. Ortoflorofruttic. ital.* 46: 3–9.

Popova, I. 1967. Spineless gooseberry (in Russian). *Sadovodstvo* 12: 28.

Porpáczy, A. et al. 1960. Study of the bud mutations of fruit trees and bushes. 1. (in German). *Kisérl. Közl.* 53C: 3–17.

Porpáczy, A., Garay, A. S., and Garay, M. 1964. Comparative physiological and biochemical investigations on the formation of yield in different blackcurrant cultivars (in German). *TagBer. dt. Akad. Landw-Wiss. Berl.* 65: 81–85.

Posnette, A. F. 1964. Transmission studies of gooseberry vein-banding virus. *Rpt. E. Malling Res. Stn. 1963* p. 110–112.

———. 1970. Gooseberry vein banding. *In* N. W. Frazier, ed., *Virus diseases of small fruits and grapevines*, p. 79–81. Berkeley: Univ. Cal., Division of Agricultural Sciences.

Potapenko, A. A. 1966. The choice of parental pairs in breeding self-fertile varieties of blackcurrants (in Russian). *Nauč. Tr. omsk. sel'skohoz. Inst.* 64(1): 104–108.

Potapkova, L. A. 1969. On the Vladimir variety plot (in Russian). *Sadovodstvo* 8: 33–34.

Potapov, S. 1961. Improving the efficiency of labor in picking gooseberries (in Russian). *Sadovodstvo* 8: 17–19.

Potapov, S., and Grinenko, A. 1968. Reversion in blackcurrants, and problems of breeding (in Russian). *Vest. sel'.-khoz. Nauki. Mosk.* 13(2): 77–80.

Proeseler, G. 1967. Occurrence of *Cecidophyes ribis* Nal. in blackcurrants in the DDR (in German). *Obstbau, Berl.* 7: 106–108.

Radionenko, A. Ja. 1970. The fertilization process in *Ribes nigrum* (in Russian). *Bot. Zh. SSSR*, 55: 807–814.

Rainčikova, G. P. 1967. Germination of blackcurrant pollen and conditions influencing its viability. p. 232–37 in *Fruit and berry crops* (in Russian). Minsk: Urožaj.

Rake, B. A. 1953. The propagation of gooseberries. 1. *Rpt. Long Ashton Res. Stn.* p. 79–88.

———. 1958. The history of gooseberries in England. *Fruit Yb.* 10: 84–87.

———. 1963. Gooseberry varieties for flavour. *Gdnrs. Chron.* 154: 212–213.

———. 1966. Some effects of picking dates on gooseberry returns. *Comml Grow.* 3674: 1091.

Raphael, T. D. 1958. Blackcurrants in Tasmania. *Tasm. J. Agr.* 29: 5–12.

Ravkin, A. S. 1965. Content of sugars and free amino-acids in the autumn-winter period in relation to winterhardiness of gooseberry and currant varieties (in Russian). *Sb. Rab. mold. Uč. nauč-issled. zon. Inst. Sadovod. nečernozem. Pol.* p. 252–257.

———. 1966. The winterhardiness of currants and gooseberries in the central Nonchernozem area. p. 361–77 *in Breeding and varietal study of fruit and berry crops in the Nonchernozem Zone* (in Russian). Moscow: Kolos.

Rea, J. 1665. *Flora, Ceres & Pomona.*

Rehder, A. 1954. *Manual of cultivated trees and shrubs.* 2nd ed. New York: Macmillan.

Ricketson, C. L. 1969. Small fruits for the home gardens. (rev. ed.) *Publ. Ont. Dept. Agr.* 475.

Riliškis, A. I. 1964. Altaj varieties of blackcurrant in Lithuania (in Russian). *Sadovodstvo* 8: 32.

Roach, F. A. 1972. Prospects for the future U.K. production of fruit crops and the challenge of the E.E.C. Norwich Conf., March 1972.

Romanovskaja, O. I. 1955. Mičurinist horticulturists in the Latvian S.S.R. (in Russian). *Agrobiologiya* 4: 350–352.

Rootsi, N. 1967. Some fruit characteristics of cultivated small fruit varieties (in German). *Gartenbauwissenschaft* 32: 459–474.

Rousi, A. 1966. A probable case of monogenically determined resistance to American gooseberry mildew in blackcurrant. *Ann. Agr. Fenn.* 5: 256–259.

Rudloff, C. F., and Lenz, F. 1960. The potential fertility of currant flowers (in German). *Erwerbsobstbau* 2: 214–217.

———. 1961. Flower differentiation in currants (in German). *Erwerbsobstbau* 3: 36–38.

Säkö, J. 1963. The cropping of certain Finnish blackcurrant varieties (in Finnish). *Maatalous Koetoim.* 17: 168–175.

Salmon, E. S. 1900–1. The gooseberry mildew (*Sphaerotheca mors-uvae* [Schwein.] Berk. & Curt.). *J. R. Hort. Soc.* 25: 139–142.

——. 1907. The American gooseberry mildew attacking the redcurrant. *Gdnrs. Chron.* 42: 26.

Samorodova-Bianki, G. B. 1969. Biologically active substances in blackcurrant under the conditions of the Leningrad district (in Russian). *Tr. prikl. Bot. Gen. Selek.* 40(3): 146–154.

Sarapuu, È. 1963. Yield formation in the common gooseberry varieties in the Estonian SSR (in Russian). *Tead. tööde kogum. Eesti Maaviljel. Maararand. Tead. Uurim. Inst.* 3: 84–90.

———. 1966. Yield formation of the common gooseberry varieties in Estonia (in Russian). *Tartu Ülik. Toim.* 185: 577–583.

Šaumjan, K. V. 1964. Resistance of blackcurrants to the gall mite and reversion (in Russian). *Sadovodstvo* 4: 40–41.

Sax, K. 1931. Chromosome numbers in the ligneous Saxifragaceae. *J. Arnold Arbor.* 12: 198–206.

Schanderl, H. 1958. Pollination in blackcurrants (in German). *Flüssiges Obst.* 25(9): 46.

Schellenberg, H. C. 1923. The susceptibility of *Ribes* species to white pine blister rust (in German). *Schweiz. Z. Forstw.* 74: 25–50.

Schmidt, H. 1968. Apomixis in fruit trees (in German). *Erwerbsobstbau* 10: 183–186.

Schmidt, M. 1948. Achievements and aims in fruit breeding (in German). *Züchter* 19: 135–153.

———. 1952. Crosses between currant and gooseberry species (in German). *Dt. Baumsch.* 10: 280–283.

Schoeneweiss, D. F. 1966. Anthracnose of Alpine currant. *Plant Dis. Rptr.* 50: 196–200.

Schuppe, E. 1962. Results of varietal trials with currants (in German). *Erwerbsobstbau* 4: 166–169.

Seljahudin, A., and Brózik, S. 1966. Fertilization conditions in berry producing fruit varieties. 2. Strawberry-gooseberry. *Acta agron. hung.* 15: 187–198.

———. 1967. Fertilization conditions of berry fruit varieties. 3. Raspberry, black-, redcurrant. *Acta agron. hung.* 16: 63–74.

Šerengovyj, P. Z. 1969. Blister rust of currant (in Hungarian). *Zashch. Rast. Vredit.* 14(9): 40–41.

Sergeeva, K. D. 1954. New gooseberry varieties (in Russian). *Priroda, Mosk.* 9: 100–102.

———. 1962. Breeding *Sphaerotheca*-resistant gooseberry varieties by geographically distant crossing (in Russian). *Rpt. Soviet Scientists 16th Int. Hort. Congr., Moscow* p. 106–115.

———. 1966*a*. Breeding gooseberries without thorns (in Russian). *Rpt. Soviet Scientists 17th Int. Hort. Congr., Moscow* p. 250–257.

———. 1966*b*. Breeding gooseberry varieties resistant to *Sphaerotheca* by the methods of wide crossing and directed training (in Russian). *(Proc.) 4th All-Union Conf. Immunity Agr. Plants, Vines & Fruit Crops, Kišinev* p. 225–230.

Shipton, P. J. 1969. Observations on the phytotoxicity and other effects of fungicides used on blackcurrants. *Plant Path.* 18: 133–137.

Shoemaker, J. S. 1955. *Small fruit culture.* New York: McGraw-Hill.

Šitakov, I. I. 1967. Spineless forms of gooseberry (in Russian). *Sadovodstvo* 8: 33.

Smirnov, A. L. 1965. An important varietal property (in Russian). *Sadovodstvo* 10: 26–27.

Smith, B. D. 1959. The behaviour of the blackcurrant gall mite (*Phytoptus ribis* Nal.) during the free-living phase of its life cycle. *Rpt. Long Ashton Res. Stn.* p. 130–136.

———. 1960. Population studies of the blackcurrant gall mite (*Phytoptus ribis* Nal.) *Rpt. Long Ashton Res. Stn.* p. 120–124.

———. 1962. Experiments in the transfer of the blackcurrant gall mite (*Phytoptus ribis* Nal.) and of reversion. *Rpt. Long Ashton Res. Stn. 1961* p. 170–172.

Smith, M. V., and Bradt, O. A. 1965. Fruit pollination. *Publ. Ont. Dept. Agr.* 172.

Somorowski, K. 1964. Preliminary results on the breeding of blackcurrant (in Polish). *Pr. Inst. Sadow. Skierniew.* 8: 3–19.

Sorge, P. 1961. Yield, frost susceptibility and ease of picking of blackcurrants (in Dutch). *Neue dt. Obstb.* 7: 102–103, 106.

Spangelo, L. P. S., et al. 1970. Combining ability analysis and interrelationships between thorniness and

yield traits in gooseberry. *Can. J. Plant Sci.* 50: 439–444.
Spinks, G. T. 1947. Black currant breeding at Long Ashton. *Rpt. Long Ashton Res. Stn.* p. 35–43.
Stenseth, C. 1966. The blackcurrant leaf midge, *Dasyneura tetensi Rübs.* Investigations on the biology, control measures and the effect of attack on growth and yield (in Swedish). *Forsk. Fors. Landbr.* 17: 241–258.
Suncova, M. P. 1958. The effect of osmotic pressure of the cell sap on susceptibility to bud mite injury of blackcurrants (in Russian). *Zap. leningr. sel'.-khoz. Inst.* 11: 171–178.
Swarbrick, T. 1959. Genetics. *Rpt. Scot. Hort. Res. Inst. 1958–59* p. 34–35.
Tamás, P. 1960. Research on the characterization of winter resistance of black- and redcurrant (in German). *Züchter* 30: 242–247.
———. 1961. Yield analysis in blackcurrants (in Swedish). *Sver. pomol. För. Årsskr.* 62: 117–126.
———. 1963*a*. Bush fruits (in Swedish). *Rpt. Balsgård Fruit Brdg. Inst.* 1962 p. 23–25.
———. 1963*b*. The interrelations between fertility and berry size in blackcurrant (in German). *Züchter* 33: 302–306.
———. 1964. Bush fruits (in Swedish). *Rpt. Balsgård Fruit Brdg. Inst. 1963* p. 25–28.
———. 1966*a*. Bush fruits (in Swedish). *Rpt. Balsgård Fruit Brdg. Inst. 1965* p. 26–29.
———. 1966*b*. Bush fruit breeding at Balsgård. *Proc. Balsgård Fruit Brdg. Symp., Fjälkestad* 1964 p. 119–128.
Tamás, P., and Porpáczy, A. (Jr) 1967. Some physiological and breeding problems in the fertilization of the genus *Ribes*. 1. Variability in compatibility in blackcurrants (in German). *Züchter* 37: 232–238.
Tamás, P. 1968. Bush fruits (in Swedish). *Rpt. Balsgård Fruit Brdg. Inst. 1967* p. 21–26.
Taylor, A. M. 1914. *Eriophyes ribis* (Nal.) on *Ribes grossularia*. *J. Agr. Sci., Camb.* 6: 129–135.
Teaotia, S. S., and Luckwill, L. C. 1955. Fruit drop in black currants: 1. Factors affecting "running off." *Rpt. Long Ashton Res. Stn.* p. 64–74.
Thayer, P. 1923. The red and white currants. *Bul. Ohio Agr. Expt. Stn.* 371: 309–394.
Thomas, G. G., and Wilkinson, E. H. 1962*a*. Vegetative growth and flower bud initiation of the blackcurrant in the field. *Proc. 16th Int. Hort. Congr., Brussels* 3: 363–369.
———. 1962*b*. Propagation of blackcurrants from single-bud cuttings. *J. Hort. Sci.* 37: 115–123.
Thomsen, A. 1970. Vein banding disease of gooseberry (in Dutch). *Tidsskr. PlAvl* 74: 313–317.
Thresh, J. M. 1964*a*. Warm water treatments to eliminate the gall mite *Phytoptus ribis* Nal. from blackcurrant cuttings. *Rpt. E. Malling Res. Stn. 1963* p. 131–132.
———. 1964*b*. Increased susceptibility to the mite vector (*Phytoptus ribis* Nal.) caused by infection with blackcurrant reversion virus. *Nature, Lond.* 202: 1028.
———. 1966. Virus diseases of blackcurrant. *Rpt. E. Malling Res. Stn. 1965* p. 158–163.
———. 1967. Virus diseases of redcurrant. *Rpt. E. Malling Res. Stn. 1966* p. 146–152.
———. 1970. Reversion of blackcurrant. p. 82–84 *In* N.W. Frazier, ed., *Virus diseases of small fruits and grapevines*. Berkeley: Univ. Cal., Division of Agricultural Sciences.
———. 1971. Some effects of reversion virus on the growth and cropping of black currants. *J. Hort. Sci.* 46: 499–509.
Tiits, A. 1964. Some observations on the blackcurrant reversion disease transmitted by grafting (in Russian). *Izv. Akad. Nauk. eston. SSR. Ser. biol.* 13: 267–271.
———. 1970. Studies on the etiology and pathology of the blackcurrant reversion. 3. Further information on the reactions of currant species to the pathogen of reversion (in Russian). *Izv. Akad. Nauk. eston. SSR. Ser. biol.* 19: 183–186.
Tinklin, I. G., Wilkinson, E. H., and Schwabe, W. W. 1970. Factors affecting flower initiation in the blackcurrant (*Ribes nigrum* [L.]) *J. Hort. Sci.* 45:275–282.
Todd, J. C. 1962. Blackcurrant varieties: their classification and identification. *Tech. Bul. Minist. Agr. Fish. Fd* 11.
Tolmačev, I. A. 1940. An experiment on overcoming incompatibility (in Hungarian). *Jarovizacija* No. 5 (32): 125–126.
Tubeuf, C. V. 1933. Studies on symbiosis and tendency to parasitic infection and on the inheritance of pathological characters in our woody plants. 4. The tendency of the five-needled species of pine, on the one hand, and of the various genera, species, hybrids, and horticultural forms of *Ribes*, on the other, to attack by *Cronartium ribicola* (in German). *Z. PflKrankh.* 43: 433–471.
Tumanov, I. I., Kuzina, G. V., and Karnikova, L. D. 1970. The effect of gibberellins on dormancy and frost resistance in plants (in Russian). *Fiziologiya Rast.* 17: 885–895.
Turner, [?]. 1548. *New herbal.*
Tydeman, H. M. 1930. Some results of experiments in breeding blackcurrants. 1. The self-pollinated families. *J. Pomol.* 8: 106–128.
———. 1938. Some results of experiments in breeding black currants. 2. First crosses between the main varieties. *J. Pomol.* 16: 224–250.
Vaarama, A. 1949. Spindle abnormalities and variation in chromosome number in *Ribes nigrum*. *Hereditas* 35: 136–162.
Vestrheim, S. 1965. Ascorbic acid in blackcurrants (in Norwegian). *Meld. Norg. LandbrHøgsk.* 44(18).
Vitkovskij, V. L. 1958. Soft fruits in the Kola Peninsula (in Russian). *Sad Ogorod* 6: 54–57.
———. 1964*a*. Small fruit breeding under conditions of the Far North (in Russian). *Tr. Prikl. Bot. Genet. Selek.* 36(3): 149–157.
———. 1964*b*. A phenological appraisal of varieties and forms of redcurrant of different origin under the conditions of the Kola Peninsula (in Russian). *Dokl. fenol. Komis. geogr. Obšč. SSRS* 1(2): 38–49.
———. 1969. Cultivars and forms of redcurrant of diverse provenance under the conditions of the Kola

Peninsula (in Russian). *T. prikl. Bot. Genet. Selek.* 40(3): 155–162.

Vitkovskij, V. L., Lazareva, A. G., and Čuvašina, N. P. 1970. Effect of the latitude of the place of cultivation on the formation of flowers in *Ribes* (in Russian). *Byull. vses. Inst. Rasten. Vavilov* 16: 65–69.

Volodina, E. V. 1964. Self-fertility and yield of the blackcurrant (in Russian). *Vest. sel'.-khoz. Nauki Mosk.* 9(7): 96–100.

Voluznev, A. G. 1948. On self fertility of black currants (in Russian). *Sad Ogorod* 8: 33–34.

———. 1963. Breeding blackcurrants in White Russia (in Russian). *Sadovodstvo* 2: 22–23.

———. 1966. Breeding blackcurrants for a high degree of self-fertility (in Russian). *Vestsi Akad. Navuk BSSR: Ser. sel'skagasp. Nauk* 3: 53–59.

———. 1968. Utilization of the Siberian species *Ribes nigrum* ssp. *sibiricum* Pav. and *R. dikuscha* Fisch. in developing highly self-fertile varieties of blackcurrant in Belorussia, p. 88–93 *in* Papers of a scientific conference on problems of genetics, breeding and seed production in plants. Section top and small fruits and ornamental crops (in Russian). Gorki.

Voščilko, M. E. 1969. Some biological characteristics and morphological features of selected forms of wild berry fruits from the Salair ridge (in Russian). Nauka, Novosibirsk (From *Referat Zh.* 1969: 219–222).

van de Vrie, M. 1956. Redcurrant. Big bud mite, *Eriophyes ribis* Nal. (in Dutch). *Jversl. Inst. plziektenk. Onderz.*: 39–40.

———. 1958. Big bud mite, *Eriophyes ribis* Nal. (in Dutch). *Jversl. Inst. plziektenk. Onderz:* 43–44.

———. 1959. Acarological studies (in Dutch). *Jversl. Proefstn Fruitteelt volle Grond, 1958*: 53–77.

Vukovits, G. 1961. American gooseberry mildew on blackcurrants (in German). *Pflanzenarzt* 14: 105–106.

Wassenaar, L. M. 1958. Experiments and varietal trials with blackcurrants (in Dutch). *Fruitteelt* 48: 1043–1044, 1047.

Watkins, R. 1971. Fruit breeding methodology. Major gene and quantitative genetics. *Proc. Angers Fruit Brdg. Symp., 1970* p. 251–263.

Wellington, R., Hatton, R. G., and Amos, J. 1921. The "running off" of blackcurrants. *J. Pomol.* 2: 160–198.

W[estwood], I. O. 1869. Currant bud disease. *Gdnrs' Chron.* 32: 841.

Wilking, E., and von Hardenberg, D.-W. 1965. Harvesting bush fruits (in German). *Erwerbsobstbau* 7: 108–111.

Wilking, E. 1967. Second report on time studies on the mechanical harvesting of blackcurrants and redcurrants (in German). *Erwerbsobstbau* 9: 109–112.

Williams, R. R., and Child, R. D. 1963. Some preliminary observations on the development of self- and cross-pollinated flowers of blackcurrants. *Rpt. Long Ashton Res. Stn. 1962* p. 59–64.

Wilson, A. R. 1966. Grey mould of soft fruit: varietal susceptibility. *Rpt. Scot. Hort. Res. Inst. 1964 & 1965* p. 56.

Wilson, D. 1958. Prediction of harvest date for blackcurrants. *Rpt. Long Ashton Res. Stn.* p. 78–81.

———. 1962. Induced parthenocarpy in blackcurrants. *Rpt. Long Ashton Res. Stn. 1961* p. 58–60.

———. 1963. Some aspects of breeding blackcurrant varieties adapted to mechanical harvesting. *Rpt. Long Ashton Res. Stn. 1962* p. 55–59.

———. 1964*a*. Cross-pollination can be hard to achieve. *Grower* 61: 159–160.

———. 1964*b*. A summary of the breeding behaviour of four blackcurrant varieties. *Euphytica* 13: 153–156.

Wilson, D., Corke, A. T. K., and Jordan, V. W. L. 1964. The incidence of leaf spot and mildew on blackcurrant seedlings. *Rpt. Long Ashton Res. Stn. 1963* p. 74–78.

Wilson, D., and Adam, J. 1966. The inheritance of some yield components in blackcurrant seedlings. *J. Hort. Sci.* 41: 65–72.

———. 1967. A comparative study of vegetative growth and flower bud differentiation in blackcurrant varieties. *Rpt. Long Ashton Res. Stn. 1966* p. 104–111.

Wilson, D. 1970. Blackcurrant breeding: A progeny test of four cultivars and a study of inbreeding effects. *J. Hort. Sci.* 45: 239–247.

Wilson, D., and Jones, R. P. 1971. Fruit breeding. Black currant. *Rpt. Long Ashton Res. Stn. 1970* p. 18–19.

Winkler, W. 1967. Results of mechanical harvesting of bush berry fruits (in German). *Obstbau*, Berlin 7: 119–122.

Wood, C. A. 1957. Pomology: Plant Breeding. *Rpt. Scot. Hort. Res. Inst. 1956–57* p. 11–12.

———. 1960. Pomology. *Rpt. Scot. Hort. Res. Inst. 1959–60* p. 13–23.

———. 1962. Pomology. *Rpt. Scot. Hort. Res. Inst. 1961–62* p. 13–28.

———. 1964. Pomology. *Rpt. Scot. Hort. Res. Inst. 1963–64* p. 19–30.

———. 1966. Pomology. *Rpt. Scot. Hort. Res. Inst. 1964 & 1965* p. 15–24.

Wormald, H. 1928. Notes on plant diseases in 1926. *Rpt. E. Malling Res. Stn. 1926 & 1927*. 2. Suppl. p. 75–88.

Yeager, A. F., and Latzke, E. 1933. Gooseberries, varieties, breeding, culture and use. *Bul. N. D. Agr. Expt. Stn. 267*.

Yeager, A. F. 1938. Pollination studies with North Dakota fruits. *Proc. Amer. Soc. Hort. Sci.* 35: 12–13.

Zakhryapina, T. D. 1959. Differentiation of the pathogen of anthracnose of currant and gooseberry (in Russian). *Bot. Zh.* 54: 836–843.

Zatykó, J. [M.], and Simon, I. 1960. Induced apomixis in the genus *Ribes* (in German). *Kísérl. Közl.* 53C(3): 19–25.

Zatykó, J. M. 1962. Parthenocarpy and apomixis in the *Ribes* genus induced by gibberellic acid. *Naturwissenschaften* 49: 212–213, and *Acta Biol. Acad. Sci. Hung.* 13, Suppl. 5: 60 (1963).

———. 1963. Parthenocarpy induced in sterile *Ribes* species by gibberellic acid. *Naturwissenschaften* 50: 230–231.

Zatykó, J. M., and Simon, I. 1964. Possible uses of gibberellic acid in fruit breeding (in German). Z. *Pflanzenzücht.* 52:262–272.

———. 1969. The use of gibberellic acid in fruit breeding, p. 99–103 *in* The application of the most recent findings in breeding research to fruit breeding (in German). *TagBer. dt. Akad. LandwWiss., Berl.* 96.

Zhukovsky, P. M. 1950. *Cultivated plants and their wild relatives* (in Russian). Moscow: State Publishing House Soviet Science. Abridged translation by P. S. Hudson, Commonw. Agr. Bur., 1962.

———. 1965. Main gene centres of cultivated plants and their wild relatives within the territory of the U.S.S.R. *Euphytica* 14: 177–188.

Zielinski, Q. B. 1953. Chromosome numbers and meiotic studies in *Ribes. Bot. Gaz.* 114: 265–274.

Žironkin, I. M. 1962. Polyploid forms in the hybrid family of the red Kyzyrgan currant (hybrid of the series *rubra* x *petraea*) x the blackcurrant Davison's Eight, *Ribes nigrum* L. (in Russian). *Tr. mosk. Obšč. ispyt. Prir.* 5:313–321.

———. 1965. Amphidiploidy in currants, p. 271–73 in *Polyploidy and breeding* (in Russian). Moskva-Leningrad: Nauka.

Žitneva, P. I., Šaumjan, K. V., and Bronštein, E. V. 1961. New varieties of blackcurrant (in Russian). *Dokl. mosk. sel'.-khoz. Akad. K.A. Timiryazeva* 62: 303–313.

Zubeckis, E. 1962. Ascorbic acid content of fruit grown at Vineland, Ontario. *Rpt. Hort. Expt. Stn. Prod. Lab. Vineland* p. 90–96.

Žukovskaja, A. A. 1967. Gooseberries at the varietal field station (in Russian). *Sadovodstvo* 5: 35–36.

Minor Temperate Fruits

by George M. Darrow

There is perhaps no better introduction to be made for this chapter than the one to a similar article written in the 1937 United States Department of Agriculture *Yearbook of Agriculture* (Darrow and Yerkes, 1937):

> Opportunities for the development of new and improved plants by breeding are by no means limited to those now grown in home or commercial gardens. All our present cultivated plants, it must be remembered, have been derived from wild plants. Those that were most outstandingly useful or more readily adaptable to cultivation, man took from forest and field and grew in his own dooryard. Others he left in their wild state, for one reason or another though he continued to use their products. One of the plants left wild until very recently was the blueberry. The work of the late Frederick V. Coville, described elsewhere in this Yearbook, shows how modern knowledge and modern technique, applied to suitable wild material, can change and improve it enormously for human uses. Not all neglected wild plants, undoubtedly, would produce such splendid results as the blueberry under Coville's handling, but the achievement suggests that there is a wealth of material not yet touched, awaiting merely the right imagination and the right opportunity for the breeder to transform it in greater or lesser degree.
>
> Some of the native wild plants and introductions from foreign countries need only careful selection of superior strains to increase their usefulness. In other cases a planned program of breeding is necessary, including crosses with types already in cultivation. It must be recognized that hybrids between distinct species are made with considerable difficulty in most cases, and only rarely are they directly valuable in a horticultural way. However, occasionally valuable things do come from distant crosses, and the only way to find them is to make the attempt.

The work of F. V. Coville has been continued and expanded by the USDA, by workers of state experiment stations, and by private individuals, and now the blueberry is an important fruit crop; all this occurred during my lifetime. A beginning has now been made with other little-known foreign and native fruits. Named cultivars of some of these are now available. As such cultivars are more widely available and more widely grown, further improvements in plant and fruit will suggest themselves and new fruits will come to our gardens and farms and finally into our markets. And as we develop these new fruits and solve problems in their breeding, ways to develop still other fruiting plants into desirable garden and crop plants will suggest themselves. The increasing world population makes the exploration of new food plants especially important. In this chapter the discussions are designed to indicate some of the problems and potentials of some lesser-known fruits.

In addition to fruits discussed in this chapter many more could be added. Several are related to major fruits and should be considered in relation to the improvement of such crops. Among these are American plum (*Prunus americana* Marsh.), black plum (*Prunus nigra* Ait.), beach plum (*Prunus maritima* Marsh.), choke cherry (*Prunus virginiana* L.), pin cherry (*Prunus pennsylvanica*

L.), western sand cherry (*Prunus Besseyi* Bailey), sand cherry (*Prunus pumila* L.), and the many native *Rubus, Ribes,* and *Viburnum* species as described by Yeager et al. (1935) for North Dakota and other fruits in other states. The opportunity for improvement widens as basic information on relationships, chromosome numbers, and cytogenetics increase.

Native American Persimmon

The persimmon (*Diospyros virginiana* L.) belongs to the ebony family, *Ebenaceae,* which has six genera and nearly 300 species, mostly tropical. There are about 200 species of the genus *Diospyros*. Two are edible: *D. kaki* L. of China and Korea from which the large-fruited garden and orchard cultivars come, and *D. virginiana* L. of the United States, which is smaller-fruited and sweeter than *D. kaki*. One other species, *D. Texana* Scheede, a small tree with small black fruit, is native to Texas and New Mexico.

The American persimmon is native from Connecticut to Florida and west to Kansas and Texas. The trees grow to 10 to 15 m in the open but to 25 m or more in the forest. The wood is hard and heavy. The species is dioecious, although some "male" trees have a few flowers with pistils and bear a few fruits and some "female" trees may have some staminate flowers. The pistillate flowers are borne singly and the staminate are generally in threes (Fletcher, 1942). The flowers are greenish-yellow and blossom in late spring after danger of spring frost is past. The pollen is very light and though generally carried by bees it is also wind blown. The time of fruit ripening varies greatly—from August 1 (midsummer in the deep South) to late autumn, even to December. In general those ripening before frost are considered to have the best flavor. Frost injures the fruit quality but the fruit of late cultivars continues to ripen on warm days through the fall and into early winter. The fruits are very astringent when green but the green color and the astringency disappears as the fruit ripens and softens (Griffith 1971, 1972). Some never ripen. Griffith (1972) also demonstrated that the American persimmon could be ripened and the astringency removed by treatment with ethylene gas. Apples may be used as a source of ethylene for the *kaki* cultivars, but the American persimmon needs a greater gas concentration than does the Oriental to remove all the astringency. The fruits are 1½ to 5 cm diameter and the more desirable are bright yellow to red when ripe. McDaniel (personal communication) considers the fruit ideally adapted to freeze preservation that will keep for years. Seeds vary in size and shape on different trees. A full set of seed is eight and fruits with fewer seeds are usually somewhat smaller. Seedless fruits are usually small but some may approach seeded fruits in size. For germination, seed should not be allowed to dry and should be given a moist chilling period of 90 days before planting (McDaniel 1970*a*).

Selected trees to be propagated may be grafted or budded to seedlings when leaves start and should begin bearing in three to four years. Top-worked old trees bear in two to three years. Selected seedlings may be readily propagated by root cuttings 15 to 20 cm long with both ends sealed. Seedlings are difficult to transplant (Gerardi, 1962) because of the long tap root and only few fibrous roots which are easily broken, and tops should be cut back severely on transplants. Trees should be planted 5 to 7 m apart and should be kept low-headed for ease in picking. Gerardi (1962) considers the persimmon easiest to graft and bud in the fall or spring using a T bud. Kowa (1969) suggests using side-bark grafts on stocks 2½ to 4 cm in diameter. Water sprouts and scion wood from the tree top are used for grafting and grafts are enclosed in a polyethylene bag with a brown paper bag over the plastic bag for one to two weeks; then the plastic bag is removed but the paper bag is left for a few days longer. A brace is tied to stock and scion to prevent the wind blowing the new growth out; another method is to cover the stock and a considerable portion of the scion with grafting compound. This eliminates desiccation and a thick coating of grafting compound gives some support to the scion, eliminating the need for bracing. Griffith (personal communication) considers that there is little need for spraying. The twig girdler can be damaging at times and the principal disease, persimmon wilt (caused by *Cephalosporium diospyri* Crandall), occurs from North Carolina, Tennessee, and Arkansas south, but is not spreading rapidly.

Cytology

Baldwin and Culp (1941) found two chromosome races of *D. virginiana* with 60 and 90 chromosomes. They reported that D. *Texana* was a diploid with 2n = 30. Thus the two species *D. lotus* and *D. Texana* are diploid (2n = 30), *D. virginiana* has both tetraploid and hexaploid forms (2n = 60 and 2n = 90), and D. *kaki* is a hexaploid (2n = 90). On a map showing the location of the source of their material, all places west of the Mississippi River and north of the Ohio River including also Mississippi, Virginia, Delaware, and Pennsylvania, had 90 chromosome *D. virginiana.* In the southeast only the 60-chromosome trees were found, except in Georgia, where there were both. McDaniel (1970*b*, 1971, 1972) felt that the evidence indicated that many of the best of the present American cultivars belong to the 90-chromosome group of American persimmons, the same number as *D. kaki* L. of China. 'Penland' from North Carolina and 'Mood Indigo', a seedling in Illinois but of the southeastern race, are in the 60-chromosome group. He includes the 'William', a male, in the probable 90-chromosome group. Mature fruits from the cross *virginiana* x *kaki* had good seed but *kaki* x *virginiana* seed were imperfect.

McDaniel (1972) reported setting fruit and seed of 'Garretson' with pollen from a staminate branch found on a 'Garretson' tree. Seed from this self-pollination produced good trees. He also reported that E. M. Meader of Rochester, N. H., has a seedling of 'Garretson' isolated from all others which bears well and has mostly seedless fruit (McDaniel 1971).

Breeding Objectives

Gibson (1962) pointed out serious limitations of the native persimmon for commercial growers: 1) the tree is too tall; 2) it is self sterile (self-fruitful cultivars are important); and 3) the fruits are too small. Though fruits of commercial cultivars of the Oriental persimmon are large, fruits of some of its seedlings are tiny (Gibson, 1970), and the species lacks winterhardiness. A few cultivars ('Fuju', 'Hachiya') of the Oriental persimmon may be grown as far north as Maryland.

McDaniel (1970*b*) noted that the best cultivars for Illinois originated in Illinois or Indiana or trace to 'Early Golden' of Illinois origin as an ancestor. He listed the following qualities as important: 1) a pleasant flavor with no astringency when ripe; 2) fruit 42 g or more in size; 3) bright yellow, orange, red, or red-blush fruit with uniform flesh color; 4) small, few, or no seeds as in 'Florence'; 5) a small adhesive calyx; 6) non-bursting when falling to the ground or, preferably, non-dropping; 7) earliness (those maturing before frost are likely to have best flavor and can be grown and still ripen farther north); 8) a flat or oblate shape, which should pack best; 9) reasonably heavy cropping but not so heavy as to start alternate bearing or induce excessive limb breakage; 10) a spreading tree with stronger than average branches as in 'Early Golden' and its offspring; 11) monoecious, self-fertile trees for better pollinization. Other objectives might be added to this list including easy abscission of fruit with calyx attached, and much firmer fruit for storage and marketing.

McDaniel (1970a) listed for Illinois as the best cultivars 'John Rick', 'Garretson', 'Killen', 'Early Golden', 'Craags', 'Beavers', 'Wabash', and 'Florence'. He listed 'Williams' as the best male for pollination. 'Garretson', 'John Rick', 'Hicks', and 'Juhl' are recommended for northern areas (Ourecky, 1972).

Elderberry

The elder or elderberry is the fruit of species of the genus *Sambucus* (Caprifoliaceae) but the names are applied chiefly to three species: 1) *Sambucus nigra* L., the European elder, native to Europe, northwest Africa, and western Asia; 2) the American elderberry, *Sambucus canadensis* Hesse, native to eastern North America from Nova Scotia to Manitoba and south to Florida and Texas with larger, sweeter, better-flavored fruit and more productive plants than S. *nigra*; and 3) the blueberry elder, *Sambucus cerulea* Raf., native from British Columbia to California and east to Montana and Utah with larger fruit than *S. canadensis* but smaller clusters. There are, however, about 20 species native to the temperate and subtropical climates of the Northern Hemisphere. While the above three are useful for their fruit, *nigra, canadensis,* and other species also have me-

FIG. 1. Fruiting clusters of the elderberry (*Sambucus canadensis*).

dicinal properties. *S. nigra* without stolons, and *S. canadensis* spreading by stolons, are large coarse shrubs while *S. cerulea* is still larger and may grow to 15 m. The entire plant of *S. canadensis* or a closely related species is used by the Indians of Central America for soil enrichment, being cut up and buried in trenches where crops are to be planted. These three as well as many of the other species are also prized as landscape plants, especially for their quick growth. They may be planted in the spring and by autumn may be 2 m or more high. The large flower clusters of May, June, and July are followed by purplish black (*canadensis*), blue-black (*cerulea*), or black (*nigra*) fruit. The scarlet-fruited species *S. racemosa* L. (Europe) and *S. pubens* Michx. (North America) and the salmon-red *S. Schweriniana* Rehd. of western China have less showy flowers but very handsome although rather tasteless or objectionable tasting fruit. Various landscape forms of *S. canadensis* have been selected: *nana* (dwarf); *maxima* Schwer. (vigorous with immense flower and fruit clusters); *acutiloba* E & B (pinnate-leafed); *aurea* Cowell (golden yellow-leafed with cherry-red berries); *Chlorocarpa* Rehd. (greenish fruit and pale yellow foliage). Similar forms of *nigra* and the red-berried *racemosa* and *pubens* are known. Also hybrids between most species are known.

History and Use

All early improvement of the elderberry was by selection of superior wild plants. Ritter and McKee (1964) reviewed the history of improved cultivars beginning with one called 'Improved Elderberry' in 1890 in Brewen, Ohio. Introductions of two of the Adams selections in New York were made about 1920 and were known as 'Adams # 1' and 'Adams #2'. The New York Fruit Testing Association introduced one of these ('Adams #2') as 'Adams'. Both the 'Adams #1' and '#2' are large in plant and fruit and have large fruit clusters. 'Superb' of the *cerulea* species was introduced by Luther Burbank in 1921 and was said to blossom and fruit all summer. To date, at least 24 cultivars have been named. All cultivars of the elderberry are partially self-fruitful (Way 1967), but crops will be light unless cross-pollination is provided.

The fruit of S. *canadensis* was used by the American Indians long before Europeans came to the United States. The early settlers also used the elder and its use has continued to the present, chiefly from wild plants. Estimates are that in the 1960s, 2000 to 2500 tons were harvested annually from the wild in northwestern Pennsylvania, Ohio, and New York. Many test plantings up to several acres have been made, but the chief source of the fruit is still from wild plants. The fruit with its unique flavor is used for jam, jelly, pies, juice, and wines. Its Vitamin C content is reported by Andross (1941) as 25–30 mg/100 g. Harvesting and processing has been mechanized to some extent.

Cytology and Breeding

The basic chromosome number is 18 ($2n = 36$). Ourecky (1970) lists *canadensis, ebulus* L., *nigra* L., *simpsonii* Rehd., and *williamsii* Hance with $2n = 36$, and *buergeriana* Nakai, *miquelli* Nakai, *kamschatica* Wolf, *melanacarpa* Gray, *sieboldiana* Blume, *siberica* Nakai, *mexicana* Presl., *callicarpa* Green, *racemosa* var. *aborescens, cerulea* Raf., and *glauca* with $2n = 38$. Hounsell (1968) lists *sieboldiana* Graebn. and *racemosa* var. *arborescens* as having $2n = 38$ and a hybrid of *canadensis* x *pubens*, $2n = 37$. One plant of *pubens* had $2n = 42$.

Slate (1955) at Geneva, New York, crossed the 'Adams' with *cerulea* from Oregon but the resulting seedlings were not promising enough to continue. The population raised by Slate were very uniform. Work at the New York Experiment Station was continued by Way who crossed both 'Adams #1' and 'Adams #2' with 'Ezyoff', a selection made at Ithaca, N.Y. (Way, 1964). Out of 1900 seedlings, 14 were first selected. Of these, 'York', introduced in 1964, is considered the best. The bushes of 'York' are larger and more productive than most cultivars and the berries are also larger. Eaton et al. (1959) at Kentville, Nova Scotia, obtained crosses of *canadensis* ($2n = 36$) x *pubens* ($2n = 36$) and over 1000 seeds germinated. Plants set in 1954 flowered in 1955 and for three succeeding years. The hybrids bore moderate crops although none produced seeds with embryos. Colchicine treatment failed to produce balanced tetraploids. They also raised open-pollinated seedlings, tested selections, and named several cultivars. Of these, 'Scotia', an open-pollinated seedling of the 'Adams', was released in 1959 and ranks high in flavor. It had a lower acid and sugar content than other cultivars tested at the Pennsylvania Agricultural Experiment Station. Eaton also named the 'Johns', a selection of a century earlier in Waterloo County, Ontario. It is about two weeks earlier than 'Adams', its clusters and berries are large, and the plant is very vigorous. Thus, at present, 'Johns', 'Scotia', and 'Adams' are considered superior under most conditions in the northern regions. Craig (1966) lists the 'Kent', 'Nova', 'Scotia', and 'Victoria' for eastern Canada.

Breeding objectives should include: 1) large berry size; 2) firmer berry texture; 3) large cluster size; 4) small seed; 5) self-fruitfulness; 6) increased productivity, which is reflected in number and size of cymes and berry size; 7) vigor and stronger canes to support heavy crops; 8) uniformity of ripening, both within clusters and between clusters of a cultivar; 9) glossy, dark, attractive berry color with selection against muddy color; 10) better quality, especially high flavor and low astringency; 11) resistance to shelling before harvest and during harvest; 12) resistance to fungus diseases; 13) immunity or high tolerance to virus diseases since a serious virus problem exists; 14) wider adaptation to different climatic conditions. An additional objective might be the development of genotypes with pendulous clusters since bird depredation is greater on erect clustered types.

Larger-fruited cultivars, better handling methods, and mechanical separation of seeds have largely overcome processing limitations so that elderberry jam of superior quality is now obtainable in many markets throughout the year.

Serviceberry, Juneberry, or Saskatoon

Many local names in addition to those above are used for this fruit, including shadbush, sarvisberry, sugar pear, and many others. This fruit is from species of the genus *Amelanchier* of the rose family and, botanically speaking, the fruit is a pome, not a berry. Though fruit of all species is edible, those of chief interest for their fruit are A *alnifolia* Nutt., A. *spicata* (Lam.) K. Koch

FIG. 2. Native wild Juneberries (*Amelanchier* sp.).

(includes *stoloniferi* Wieg. and *humilis* Wieg.), A. *oblongifolia* T & S (Roemer), and A. *canadensis* (L) Medic. Those most highly prized for landscape use are A. *laevis* Wieg., a graceful, very early flowering, slender tree to 13 m, and A. *grandiflora* Rehd., large-flowered, tetraploid, and not productive of fruit. A. *cusickii* Fern. of Washington and Oregon to Montana and Utah with large fruit and flowers has been largely neglected (Rehder, 1949).

The genus is circumpolar and consists of about 25 species, mostly native to North America. Extensive studies of the classification and characteristics have been made by Wiegand (1912), Fernald (1950), Jones (1946), and Miller and Stushnoff (1971). McDaniel (1962) and Miller and Stushnoff (1971) also evaluated their possible value for breeding. Many species are very hardy and selected cultivars can be grown in fruit gardens in the far north. *Amelanchier* was one of the fruits most used by the American Indians (in the making of pemmican) and by the early settlers of North America but it is also widely used in Europe and northern Asia. In contrast to the blueberry it is adapted to a wide range of soils from acid to alkaline and is far hardier; one or more species are found in every state in the United States and every province of Canada (Miller and Stushnoff, 1971). Some species are found along stream banks and in bog areas, while in the West, A. *alnifolia* is found in very dry locations, bearing heavy crops of large berries.

Nearly all species are used as landscape plants because of their very early flowering, especially A. *laevis*. Several are shrub species 1 to 3 m in height, suitable as landscape plants, and may bloom the second year in the nursery. Selections of the shrub species, A. *alnifolia* and A. *oblongifolia*, are the chief ones grown for their fruit. No sizeable plantings for fruit for market are known although in 1919 small plantings were grown near Atlanta, Georgia, and the fruit was marketed as currants. Large bushes 5 m or more high and 8 m across were not considered as profitable as the small 2 m cultivars grown like erect-stemmed blackberries on contours.

Breeding

The flowers of the species used for their fruit open before their leaves and before the leaves of most other trees appear. They are remarkably frost hardy. D. N. Madison (personal communication) noted that the flowers of the species he knew have 20 stamens, 15 of which are around the outer edge of the ovary, and the other five at the base of the style. He also states that there is an overlap in flowering time for all species and cultivars that he has observed. The flowers on many plants are self-sterile but are self-fertile on 'Smoky', 'Pembina', and 'Success'. Emasculation in the bud stage and bagging after pollination are standard practices. The fruit of most species ripens in about six weeks from flowering and in season with strawberries. The seed requires chilling (at 1 to 5 C) while moist for about three months for best germination. Seedlings bloom in about three years.

The first notable cultivar was the 'Success' (see Promising New Fruits, *USDA Yearbook*, 1888), introduced by H. E. Van Deman of Kansas about 1878, and said to have been grown in Illinois from seeds obtained in Pennsylvania. It grows about 2 m high and into a clump about 1 to 1½ m across and has large racemes of berries that resemble blueberries. It has been referred to *A. oblongifolia*, a species native to lowlands from southern Maine to South Carolina. Birds seem fonder of its fruit than any other fruit. Beginning about 1937, Macoun (Harris, 1964) started improvement work at the Canadian Research Station at Beaverlodge, Northern Alberta, and the Brooks Provincial Station of Southern Alberta.

Four cultivars have been named at the research station at Beaverlodge, Alberta: 'Ataglow', which is self-fertile, white-fruited, ornamental, and grows to 6 m in height; 'Frostburg', which is large-fruited; 'Pembina', which is full-flavored with a long cluster, and which grows to 3½ m in height; 'Smoky', an unusually sweet and highly flavored cultivar which grows to 2–2½ m (Harris, 1966). One selection, 'Carloss', has been made by Erskine of Alberta, Canada (Madison, 1965). 'Northline' is a recent cultivar similar to 'Pembina' introduced by the Beaverlodge Nursery of Alberta. 'Indian' and 'Shannon' are new and very productive with large fruit (Ourecky, 1972). The Minnesota Experiment Station at Excelsior, Minnesota, is testing improved cultivars for fruit.

Records at the Beaverlodge Station, Alberta, Canada, indicate a possible yield of 8 liters per bush or almost 14 mt/ha (30,000 lb./acre). Under favorable conditions, all berries can be harvested at one or two pickings. No mechanical harvesting is known, but trials indicate that it may be possible to harvest all fruit at one picking by use of a shaker device. The fruit is used fresh, in pies and puddings, for wine, and other uses.

Cytology

The chromosome numbers of *Amelanchier* are 2n = 34 and 68. The following species were examined and reported by Cruise (1964), Moffett (1931), and Sax (1931):

2n = 34	2n = 68
asiatica (Sax, 1931)	*arborea* (Cruise, 1964)
humilis (Sax, 1931)	*bartramiania* (Love and Love, 1948)
oblongifolia (Sax, 1931)	*canadensis* (Moffett, 1931)
sanguinea (Sax, 1931)	*grandiflora* (Sax, 1931)
spicata (Sax, 1931)	*laevis* (Moffett, 1931)
stolonifera (Sax, 1931)	*ovalis* (rotundiflora) (Moffett, 1931)
ovalis (Favarger and Corrivon, 1967)	*spicata* (Sax, 1931)
	stolonifera (Moffett, 1931)
	sanguinea (Moffett, 1931)

Breeding Objectives

Improvements that would make this fruit much more useful are: 1) later flowering to escape frost where severe frosts occur and later ripening cultivars to extend the season. *A. sanguinea* of America an *A. asiatica* of northern Asia blossom early, ten to 14 days after *A. laevis*, but ripen two or more months later than the named cultivars. 2) Higher yields by means of long clusters like those of 'Pembina', larger fruit, and increased flower bud set. *A. cusickii* has long clusters, large fruits, and large flowers. 3) Increased number of fruits per cluster. 4) Smaller seed. 5) Greater firmness. 6) Increased self-fruitfulness. 7) Easier separation of the fruit from the stem and uniformity of ripening within a plant for mechanical harvesting. 8) Though 'Smoky' is highly flavored, other flavors and more acidity by selection among hybrid seedlings are desired. 9) Larger, richer pink flowers and brilliant fall foliage for dual purpose use for fruit and landscaping. Other characters may be needed for different regions of North America such as greater fire-blight resistance and leaf-disease resistance, lower plants for simpler pruning and harvesting practices, and methods for bird control. Wright (1969) of Saskatoon, Saskatchewan, Canada, has suggested the need for a suitable understock to eliminate the problem of loss of plants by root or crown weakness. He has suggested trials of the numerous hybrids of saskatoon x mountain ash for this purpose.

Papaw

Papaw or pawpaw (*Asimina triloba* (L.) Dun.) is the only temperate climate fruit of the *Annonaceae*, a family which includes such fine tropical fruits as cherimoya and custard apple; and papaw is the only tree species (to 12 m) in the genus *Asimina* which includes eight to ten species, the fruit of all being edible "with a taste suggesting a sweet avocado." The papaw is native in southern Ontario and the eastern United States except New England, west from New York to Wisconsin, southern Iowa and eastern Nebraska, south through east Texas to the Gulf of Mexico, and east to northern Florida. A second species, *A. obovata* (Wild) Nash (or *A. grandiflora* Bartr.), is a small shrub to 2 to 3 m, (rarely 4.5 m), grown for its flower but also bearing large fruit which is rarely eaten. Due to its fruit shape and the peculiar arrangement of the fruit in the clusters, the papaw is often referred to as the West Virginia or Kentucky or Indiana banana (or whatever state it happens to grow in).

Kral (1960) separated what had generally been regarded as one genus into two genera, *Asimina* and *Deeringothamnus*, the latter genus with only two species and native to Florida having six to 12 petals instead of just six, and 25 or fewer stamens instead of 30 or more.

In general the fruit of all Florida species is edible but not pleasant (Kral, 1960). Of the *Asimina* species, *parviflora* (Michx.) Dun., a shrub or low tree, is similar to the papaw but is restricted to the South (Florida to Texas to southeastern Virginia and Tennessee). Its fruit is only 3 to 6 cm long. It suggests a poorly grown *triloba* and is found in similar but more coastal areas. *A. incarna* Bartr. (or *A. speciosa* Nash) is a low shrub to 1.5 m in southeastern Georgia and northeastern Florida with large fragrant white flowers and delicious small fruits. *A. obovata* is a larger shrub of southeastern, northeastern, and northcentral Florida with large (6 to 10 cm) fragrant white flowers, the showiest of the genus, and with small fruit (5–9 cm). *A. reticulata* Shuttlew, a plant (1.5 m) of poorly drained sands of Florida, has fruits of 4–7 cm. *A. tetramera* Small is a 1–3 m shrub of the coastal dunes of east Florida, flowering May to August. Its flowers are always four-merous. *A. pygmaea* (Bartr.) is very dwarf, 20–30 cm, with maroon flowers and fruits 3–4 cm, from central Florida to southeastern Georgia. *A. longifolia* Kral, (1–1.5 m) has fragrant white flowers, 4–10 cm fruit, and is found in northeastern Florida and southeastern Georgia.

Bowden (1948) found all *Asimina* except *pygmaea* to be $2n = 18$. The chromosomes are small and clump so that counts are difficult. Zimmerman (1941) crossed *triloba* with *longifolia* and *obovata*. Kral (1960) notes supposed hybrids of *triloba* with *parviflora*, and recognized natural hybrids between *longifolia* and *pygmaea*, *pygmaea* and *reticulata*, *reticulata* and *obovata*, *pygmaea* and *obovata*, and *speciosa* and *longifolia*, and describes other supposed hybrids. He noted that in fruit and seeds the hybrids were as prolific as the parent species. Hybrids between species were common in *Asimina* either as intermediates or as backcrosses.

The papaw flowers open about with the apple, are dark maroon and are 4 to 6 cm in diameter. Cross pollination is necessary for good fruit set. The fruits, maturing from about September 1 to the first week in October in Maryland, are the largest native fruit of the United States (8 to 13 cm). As they ripen they fall to the ground, but can be picked from the tree as they turn from green to yellowish-green and are edible after a few days when they soften. They can be kept up to six months if cold stored while still firm (Bartholmew, 1962). Green- or white-fleshed late-ripening seedlings have an offensive odor. A yield of 48 kg from a single tree has been reported; good trees should bear 50 to 100 fruits each. No orchard yield records are known.

The papaw is eaten when of a custard consistency. It has a characteristic flavor and aroma and is used in making preserves, pies, ice cream, cookies, cakes, and other items. Cold, even above 0 C, was reported to blacken the fruit and spoil its flavor. The Vitamin C content has been reported as 35 mg/100 g; it is said to have a much higher protein content than the banana, and contains 959 Calories/kg.

All fruits are many-seeded. Seeds are large, about 2.5 cm long, flat, and very dark brown. Seeds are difficult and very slow to germinate (normally not until midsummer), and seedlings are difficult to transplant except when quite small. Seeds which seem to require a cold rest period have been germinated if never allowed to dry and if held cold for 90 days. Dabb (1970) planted stratified seed in cans in February, placing them on the shady side of a building. The seedlings emerged in July. Davis (1962) suggests planting the seed in the fall in well-rotted sawdust in cans and over-wintering in a cold frame. They begin bearing at four to six years of age. Whip-grafting in the fat-bud stage is suggested in propagating superior seedlings. Grafts are said to take with 80% success if made early when the buds of the stock are swelling. A few suckers grow from stolons but sucker plants are slow to develop a root system and are difficult to transplant. Separation of the suckers by cutting the stolons a year ahead may force the development of a root system. Root cuttings taken about the end of the frost period are said to develop into good plants. At present two cultivars are known: 'Davis' and 'Overleese'. Both are yellow-fleshed and considered excellent in quality.

To stimulate the search for good papaws, in July 1916 the *Journal of Heredity* offered $50 prizes for photos of the largest tree and for samples of the best fruit (Anon., 1916, 1917). Seventy-five samples of fruit were received between August 18 and October 28. The best selection, 'Ketter', was from a Mrs. Ketter of southern Ohio, but good fruit came also from Kansas, Maryland, Indiana, and Missouri. Nine named cultivars were sent in from Illinois but these originated in Arkansas, Virginia, and Ohio, as well as Illinois. The flesh of the best was light to rich yellow and there was great variation in fruit and some variation in seed size. The largest tree was photographed in Indiana and was 1–1½ m in diameter at 1 m above the ground and was 8 m high. Although the papaw tree is relatively short-lived, trees over 50 years old are known. G. A. Zimmerman (1941) of Harrisburg, Pa., collected and grew all available cultivars. He also grew seedlings from the best cultivars and of other species from Florida and Georgia (Flory, 1958). Seedlings of the best selections ranged from too mild to too highly flavored. Zimmerman considered an early ripening seedling of the 'Ketter' called 'Fairchild' to be the best, but the original 'Ketter' was a close second. Several, late to very late, were listed. One, 'Martin', was considered to be somewhat cold resistant in fruit.

Possible objectives in breeding for better papaws would include 1) greater productivity, 2) large fruit size, 3) early ripening, 4) retention of ripe fruit on tree until harvest, 5) mild flavored fruit, 6) reduced seed size and number, and 7) attractive ripe fruit color and lack of browning at maturity.

Serious limitations of the papaw as a cultivated fruit suggest studies also of 1) slowness of seed to germinate and of young trees to grow, 2) difficulty in transplanting, 3) lack of self-fruitful cultivars, 4) slow development of the root system, possible harmful or lack of beneficial fungi of their roots, 5) temperatures needed for root growth, and 6) ripe fruit storage conditions.

American Cranberrybush

The American cranberrybush (*Viburnum trilobum* Marsh.), also called "highbush cranberry" and "pembina", is in the family *Caprifoliaceae*. A close relative of the elderberry, it grows to about the same height and in similar clumps. It is hardy even across the northern United States. It is often confused with the European *V. opulus* L. which is widely used as an ornamental but is readily distinguished by its intensely bitter fruit in contrast to the clear acid fruit of *V. trilobum*. The latter

FIG. 3. Flowering branch of the American cranberrybush (*Viburnum trilobum*). The outer large sterile flowers surround the small fertile flowers; however, only 5 to 10% of the fertile flowers set.

has leaves quite free from the aphid, which distorts the leaves severely and lessens the beauty of *V. opulus*. *V. trilobum* is a widely used ornamental, beautiful in flower and fruit, with rich green summer foliage which becomes highly colored in the fall.

Egolf (1962) states that there are about 250 species of *Viburnum*, nearly all natives of the cooler regions of the Northern Hemisphere. Rehder (1949) recognized nine sections of *Viburnum* and placed *V. trilobum* in section nine, (OPULUS DC.), all species of which may be expected to interbreed and all of which have a 2n =18 as their chromosome number. *V. edule* (Michx.) Raf., *V. kansuene* Batal., *V. orientale* Pall., *V. Sargenti* Koehne, *V. opulus* 'Notcutts', and *V. flavum* Rehd. are among suggested parents to cross with *V. trilobum*.

The fruit of the native American cranberrybush is highly prized as a source of jelly in sections of northern United States and Canada. The jelly is as rich-colored and as high in pectin as that made from cranberries and currants. A limitation is the disagreeable odor when the fruit is heated in jelly-making, unless the berries are harvested while still firm. The fruit resembles the cranberry but it is borne on a high bush and in clusters like the elderberry.

In the spring of 1921, the United States Bureau of Plant Industry took over for ten years a planting of the cranberrybush established by A. E. Morgan at East Lee, Mass. The plants were the best obtainable selections in a personal survey by Morgan (a former president of Antioch College) in the wilds of New York and New England and including observations on it even in Manitoba and Saskatchewan. Fruit was also obtained from Alaska and Newfoundland. After a study of the selections at East Lee and of plants in the wild at various regions, three were named, propagated, and introduced: 'Wentworth', 'Hahs', and 'Andrews'. Analyses for acid and pectin and tests for jelly by C. A. Magoon (see Darrow, 1923, 1924) indicated that these three were superior to the other selections. They also covered a long season, 'Wentworth' being early, 'Hahs' midseason, and 'Andrews' late. 'Wentworth' has proven quite out-

FIG. 4. Fruiting cluster of the American cranberrybush or highbush cranberry.

standing. These cultivars are established at a number of arboreta as well as at other places and in nurseries. The 'Manitou', selected at the south end of Lake Manitoba, Canada, for its large fruit, was named by Canada Experimental Farm, Morden, Manitoba, in 1947. Meader (1965) selected a plant at West Acton, Maine, that has been named 'Phillips' and is said to be free of objectionable flavors with jelly equal to red currant jelly. It has been tested at the New Hampshire Experiment Station at Durham.

Suggested breeding objectives would be more erect plant habit to hold fruit clusters off ground, bright attractive red fruit color, resistance to berry shelling upon maturity, and high flavor, including reduced astringency.

Some suggested breeding approaches are: 1) establishing plantings of all available species for study of their qualities, 2) intercrossing species of section nine (OPULUS), especially *orientale*, to obtain hybrid vigor and wider adaptation especially in the southern regions, 3) selfing the first generation crosses with V. *opulus* to determine if segregation of nonbitter-fruited plants appear.

Interspecific hybridization of V. *trilobum* having a chromosome number of 2n = 18 with ornamental species having a chromosome number of 2n = 16 has not been successful.

Tara or Wild Fig

There are about 25 species of *Actinidia*, vigorous climbing plants chiefly of east Asia from Saghalin south to Java and to the Himalayas. The subtropical A. *chinensis* Planch, with a deciduous vine hardy to Washington, D.C., and a fruit called "kiwi," about the size of a hen's egg, is just beginning to be grown in California but is commercially important in New Zealand and Australia. The tara or wild fig, A. *arguta* Miq., and A. *Kolomikta* (Rupr.) Maxim are much hardier but have smaller fruit. A. *arguta* is hardy into Canada and A. *Kolomikta* is still hardier. Seven species have been cultivated as ornamentals because of their beautiful foliage which is unusually resistant to diseases and insects. The sprays of white bell-shaped flowers of A. *arguta* are particularly lovely. The base chromosome number of the genus *Actinidia* is 58, A. *polygama* Miq. having 58 and 116, A. *Kolomikta* 112, A. *arguta* 116, A. *chinensis* 116 and 160, and Fairchilds' hybrid of *arguta* x *chinensis* 132 (Bolkhovskikh et al., 1969).

The plants of A. *arguta* are usually dioecious but sometimes have both sexes on the same plant. Opinions vary on the quality of the fruit from the "best fruit in Japan" to "compares with mulberry." Michurin named the following five cultivars: 'Ananasaia Michurin', 'Clara Zetkin', 'Pozdniaia' (late), 'Raniaia' (early), 'Urezhainaia' (high yielding) (Darrow and Yerkes, 1937). The fruit ripens in September and October, varies from oblong to globose, is the size of a large sweet cherry, is green in color, has flesh the texture of a ripe fig, and is acid until ripe, then becomes sweet. It is said to be a laxative and is used fresh and for sauce. Its Vitamin C content has been reported as about that of A. *chinensis*, the kiwi (250–380 mg/100) or

FIG. 5. A heavy crop of the tara or wild fig (*Actinidia arguta*) at Glenn Dale, Md.

ten times that of a lemon (Anon., 1949). No natural hybrids with A. *Kolomikta* or A. *polygama* have been reported (Shaskkim, 1938). Fairchild (1927) reported uproductive hybrids with A. *chinensis*.

Selections of high-producing and hermaphrodite-flowered vines of A. *arguta* bearing high-flavored fruit are needed. Fairchild (1913) discussed the great variation of the fruit of both A. *chinensis* and A. *arguta*. Selections are needed with a shorter growing season requirement for successful culture in northern areas. Studies on climatic adaptation, number of male vines to female vines for best production, time of fruit maturing, keeping quality, uses of fruit, and laxative qualities are other objectives. The extraordinarily long keeping quality of A. *chinensis* (four to 12 months) suggests a study of the keeping quality of cultivars of A. *arguta*.

Mountain Ash

There are about 80 *Sorbus* species of North America, Europe, and Asia, mostly with bitter fruit, although some are sweet. They are widely used in northern regions as ornamentals for the beauty of their trees and their brilliant fruit clusters. They are also extensively used in northern Europe for windbreaks; the 'Seljeron' (S. *intermedia* Pers.) is preferred in Denmark for this use (Lauritsen, personal communication). Petrov (1957) describes several species of *Sorbus* and their cultivars native to the Soviet Union. Two species are shown with a distribution above the Arctic Circle and a third species almost as far north. *Sorbus* is widely cultivated in northern regions of the Soviet Union where it is often referred to as "northern grape." Pieniazek (personal communication) states that *Sorbus aucuparia* var. *moravica* Zengerl. is grown to some extent by amateurs in Poland, Czechoslovakia, and East Germany, the fruits being used as compote. In Poland the fruits of S. *aucuparia* L. are sugar-coated as a confectionary.

Bolkhovskikh et al. (1969) have reported chromosome numbers of 11 species as $2n = 34$, six species as $2n = 51$, and 14 species as $2n = 68$.

Three intergenetic hybrids with *Amelanchier* were grown at the Arnold Arboretum and called *Amelosorbus* (Sax and Sax, 1947). Hybrids with *Aronia* are called *Sorbaronia;* one, S. *Dippelii* (Zab.) Schneid., is fertile. *Sorbopyrus auricularis* (Knoop) Schneid. is *Sorbus Aria* (L.) *Crantz* x *Pyrus communis* L. and is a triploid. Its seedlings are variable.

Hunter (1969) reported having in Michigan two species with edible fruit: S. *aucuparia edulis* Dieck., used for preserves with an agreeable acid flavor, and having larger fruit and leaves than the ordinary species; and S. *mougeottii* Soy. Willem and Godr., from the mountains of Central Europe but with less desirable fruit. Meader at Rochester, N. H., has the red Korean mountain ash (S. *commixta* Hedl.) which shows much greater vigor in that latitude than any mountain ash I have seen. It is native north to Sakhalin Island.

Trofimov (1959) refers to large-fruited selections in Vladimir and Ivanov provinces of the Soviet Union. One selection had a sugar content of 13.1 vs. 4.28% in the common wild *Sorbus*. Petrova (1937) gives the Vitamin C content of Michurin's cultivars of the mountain ash as up to 20 mg/100 g. Flemion (1931) obtained 71 to 76% seed germination after four months storage at 1 C. After a year of storage at room temperature and storage of four months at 1 C, germination of 90 to 100% was obtained. Among obvious objectives in breeding are selection for larger fruit size, higher sugar content, and better adaptation to more southern regions.

Cornelian Cherry

Cornus mas L., a shrub or small tree (to 8 m) of the dogwood family and genus, native to western Asia and eastern Europe, is grown in the United States for its ornamental value but in western Asia as an orchard fruit to some extent (Kovleva, 1950). It is attractive with its flowers that open in March in Maryland considerably before its leaves and before most other flowers.

The buds and flowers are extremely hardy and are not injured by temperatures in northeastern United States. They are readily forced into bloom in January. The present cultivars may not be fully self-fertile, for trees with another kind grafted onto them for cross-pollination have been reported as becoming more productive. The brilliant deep red fruit matures in August in Maryland. Although the fruit of the common ornamental cultivars is too acid for use, much larger and good flavored fruits, rich in sugar with a high Vitamin C content, are reported in the Crimea and Caucasus areas of south Russia. These are not known to be available in the United States as yet. Ascorbic acid (Vitamin C) content of *C. mas* is 97.4 to 120.4 mg/100 g, over twice that of the orange.

The base chromosome numbers of *Cornus* are $n = 10$ and $n = 11$. The chromosome number of *C. mas* is given as $2n = 18$ and 27 by D'Amato (1946) and by Ferguson (1966); that of *C. officinali* Sieb. and Zucc. of Japan and Korea as $2n = 18$; that of other *Cornus* species as 20 or 22; and of one, the bunch berry (*C. canadensis* L.), as $2n = 44$ (Dermen 1932).

Obvious breeding objectives would be increased fruit size, much smaller seed size, less acid fruits, more uniform ripening of fruit within a tree, retention of fruit until fully ripe, and more productive trees. Hybrids with other species of *Cornus* might give new plant and fruit characteristics.

Buffalo Berry, Barberry, and Mahonia

The buffalo berry, *Shepherdia argentea* Nutt., also called rabbit berry and Nebraska currant, is a thorny spreading shrub (to 6 m high), native to the dry northwestern Great Plains from Minnesota to Saskatchewan, Kansas, and Nevada. It is extremely hardy and drought tolerant. It blossoms very early and its flowers are frost resistant. Staminate and pistillate flowers are borne on separate plants, with both sexes being required for fruit production. The plants are very productive of small (4 to 6 mm) scarlet berries that are gathered from the last of July until late fall from the wild and made into jelly. Its ascorbic acid (Vitamin C) content was determined as over 150 mg/100 g. The juice has a disagreeable odor and should not be stored in a refrigerator with other foods. Because of the thorniness of the plants, harvesting is difficult. Propagation of selections of thornless "male" and "female" plants of the buffalo berry, a study of hybrids with *S. canadensis* to determine if productive perfect-flowered and juicy-fruited ones are possible, are first steps to be taken in improvement. No intergenetic hybrids with the larger fruited species of the related genus *Elaeagnus* are known; both *E. multiflora* Thunb. and *E. pugens* Thunb. are commonly grown and prized as ornamental shrubs of North America.

The barberries (*Berberis* spp.) and mahonias (*Mahonia* spp.) are useful in making jelly; fruits of most species are small but grow in clusters so that they are widely used for this purpose. In many areas the barberries are exterminated because many species are alternate hosts for *Puccinia graminis*, causing stem rust of wheat. One species, *Mahonia Swaseyi* (Buchl.) Fedde, with larger fruits sometimes over 1 cm in diameter, is a native of the semi-desert of west Texas from about Austin to the Rio Grande river (Ramsey, 1926). The plants stand extreme changes in temperature and moisture. It differs from the species *M. trifoliolata* (Moric) Fedde with which it grows in having more than three leaflets, five to 11, and even up to 17; in its later ripening, June instead of April and May; and in its large (twice the size) fruit and its somewhat more erect bushes. It is easily harvested by jarring the bushes or beating the cut branches on sheets on the ground. No breeding has been reported. It is also reported to be a host of the wheat rust fungus. Planned breeding would need to consider resistance to wheat rust, larger size of fruit, and adaptation of the plant to climate.

Che

Che, also known as cudrang and silkworm thorn (Winters, 1972), is the tree from which the Chinese harvest leaves to feed silkworms when the supply of mulberry leaves fails. *Cudrania tricuspidata* (Carr.) Bar. is a small thorny tree of the mulberry, fig, and Osageorange family (Moraceae) and is about as hardy and as large (to 10 m high) as the Osageorange. It has flowers somewhat like those of the mulberry but they appear in early June when mulberries are ripening. The fruits are 2½ to 4 cm in diameter and mature in October and November in Maryland. They turn a dull maroon color about October 1 but while still firm they are quite tasteless. When they become soft ripe like a ripe persimmon they may be quite delicious (Darrow, 1970).

Although the Che was introduced into England in 1872 and has been known in Europe for about 100 years, no reports on its fruit production and use are known. Forbes (1883) tells of a Dr. Hance first describing it from Shantung Province in north China and later from near Shanghai in Kiangson Province. Forbes states that the Che is common at Chefoo and Shanghai and "grows everywhere in the North."

One hybrid with *Maclura pomifera* Schneid, the Osageorange, has been reported in France but the supposed hybrid seen in the National Arboretum (Washington, D. C.) appears to be only a thornless *Maclura* with thorny suckers. However, the original hybrids were said to resemble the *Cudrania* (pollen) parent much more closely than the *Maclura* (seed) parent.

FIG. 6. Fruiting branches of the Che (*Cudrania tricuspidata*).

The Blandy Experimental Farm at the University of Virginia at Boyce planted a staminate and a pistillate tree in 1935. The "female" tree has fruited since 1942; the "male" tree bears a few small fruits, and thus tends to be an intersex (Darrow, 1972). Both trees are thornless but the suckers from the roots and all those from the base of the trunks are thorny. My tree propagated from the above female tree bore heavily in its fifth year at Glenn Dale, Maryland, on the one-third of the tree that was thornless in 1969 and again in 1970; the thorny part bore no fruit. I then pruned off the thorny part; with the thorny part cut off, the tree bore only two fruits in 1971 though the thornless branches blossomed heavily, as before. The thorny part was allowed to regrow and in 1972 it bore again a very heavy crop on the thornless part. Evidently a few unseen staminate flowers were borne by the thorny branches.

The chromosome number for *C. tricuspidata* is given as 56 by Sinoto (1928, 1929), and that of *C. Javanensis* Trec. as 28 by LeCog (1963). It is considered a tetraploid. Among related genera *Ficus* has 26 and *Morus* 28 as basic 2n chromosome numbers. *Maclura pomifera* was reported to have 56 chromosomes by Das and Sikdar (1968). Forbes (1883) described a species (variety) *polymorpha* under *tricuspidata* which often had thornless branches and which may be our plant.

To obtain thornless productive trees, male branches may be grafted onto the female, or possibly trees may be found or bred that produce flowers of both sexes. Surveys for trees with large and flavorful fruit in China are suggested. Earlier ripening cultivars that separate readily from the branches should be sought. No pests of the tree have been noted. Its ease of growing, heavy annual production, and the wide adaptation of its relative, Osageorange, suggest its possible value as a garden fruit. The fruit is said to have laxative properties (Forbes, 1883).

Bush Honeysuckle

Though fruit of most honeysuckle species is insipid or objectionable in flavor, that of a few northern species is edible. In eastern Siberia and Tibet, *Lonicera coerulea edulis* Reg. is eaten and made into preserves. The berries are quite tart. In Canada, Simonet (1971) reported that two cultivars, 'George Bugnet' and 'Marie Bugnet', have been selected for their milder taste. The 1½ m plants have stiff erect branches and can be grown on soil too high in lime for blueberries. The berries are very early, very juicy, dark blue, and have many small seeds. They mature from late May to early August in Alberta, Canada, earlier than blueberries. They remain on the bushes in good condition for many weeks.

In the U.S.S.R., several stations are reported as conducting breeding work and cultivation studies with *Lonicera*. Six species are mentioned as being worthy of attention there (Zajcev, 1969). Larger size, firmer fruit, and fewer and smaller seeds through selection and breeding are suggested objectives. Their long keeping in good condition after ripening on the bushes suggests selection for this quality for mechanical harvesting. The base diploid chromosome number of *Lonicera* is 2n=18. The following 2n chromosomes counts have been recorded for *Lonicera:*

chromosome no. (2n)=	16	18	24	36	45	54
number reports in literature	2	68	1	4	1	2

The species *L. villosa* (Michx.) R. & S. to which American edible-fruited species are referred is closely related to *L. coerulea edulis,* the edible honeysuckle of Siberia.

Literature Cited

Andross, M. 1941. Vitamin C content of wild fruit products. *Analyst* 66:358–362.

Anonymous. 1916. Where are the best papaws? *J. Hered.* 7:291–299.

———. 1917. The best papaws. *J. Hered.* 8:22–33.

———. 1949. Notes on varieties of actinidias. *Rev. Hort.* 131:155–158.

Baldwin, J. T., and R. Culp. 1941. Polyploidy in *Diospyros virginiana* L. *Amer. J. Bot.* 28:942–944.

Bartholmew, E. A. 1962. Possibilities of the papaw. *No. Nut Growers Assn. Ann. Rpt.* 53:71–74.

Bolkhovakikh, Z., V. Grif, T. Matvejeva, and D. Zakharyeva. 1969. *Chromosome numbers of flowering plants* (in Russian). Academy of Sciences of the U.S.S.R. Botanical Institute of V. L. Kamarov 1–926.

Bowden, W. M. 1948. Triploid mutant among diploid seedling populations of *Asimina triloba. Bul. Torrey Bot. Club* 76:1–6.

Craig, D. L. 1966. Elderberry culture in eastern Canada. *Can. Dept. Agr. Publ.* 1280.

Cruise, J. E. 1964. Studies of natural hybrids in *Amelanchier. Can. J. Bot.* 42:651–663.

Dabb, C. H. 1970. Pomegranates, jujubes and papaws. *Pomona* 3:150.

D'Amato, F. 1946. Cyto-embryological observations in *Cornus mas* L. with particular regard to sterility of triploid biotypes (in Italian). *Bot. Ital.* 83:170–209.

Darrow, G. M. 1923. *Viburnum americanum* as a garden fruit. *Proc. Amer. Soc. Hort. Sci.* 21:44–54.

———. 1924. The American cranberry bush. *J. Hered.* 15:243–253.

———. 1970. Che, a little known fruit: *Cudrania tricuspidata. Pomona* 3:85, 91.

———. 1972. More about the Che—*Cudrania tricuspidata. Pomona* 5:8, 10.

Darrow, G. M., and G. E. Yerkes. 1937. Some unusual opportunities in plant breeding. In *USDA Yearbook of Agriculture.* 1937:545–558.

Das, B. C., and Sikdar, A. K. 1968. Karyotype studies on *Maclura pomifera* (Raf.) Schenid. *Science & Culture* 34:468.

Davis, C. 1962. Graftable papaw seedlings in one year. *Nutshell* 9:16.

Dermen, H. 1932. Cytological studies of *Cornus. J. Arnold Arboretum* 131:410–416.

Eaton, E. L., L. E. Aalders, and I. V. Hall. 1959. Hybrids of an interspecific cross of elder. *Proc. Amer. Soc. Hort. Sci.* 74:145–146.

Egolf, D. R. 1962. A cytological study of the genus *Viburnum. J. Arnold Arboretum* 43:132–172.

Fairchild, D. 1913. Some Asiatic Actinidias. *U.S. Plant Ind. Cir.* 110:7–12.

———. 1927. The fascination of making a plant hybrid. *J. Hered.* 18:49–60.

Favarger, C., and P. Correvon. 1967. Evidence for chromosome races in *Amelanchier* (*ovalis*). *Inst. Bot. Univ. Meuchatel, Bul. Nat. Sci.* 90:215–218.

Ferguson, I. K. 1966. The *Cornacea* in the southwestern United States. *J. Arnold Arboretum* 41:106–116.

Fernald, M. L. 1950. *Gray's manual of botany*. 8th ed. New York: American Book Co.
Flemion, F. 1931. After-ripening, germination and vitality of seeds of *Sorbus aucuparia* L. *Contr. Boyce Thompson Inst.* 3:413–440.
Fletcher, W. F. 1942. The native persimmon. *USDA Farmers' Bul.* 685.
Flory, W. S., Jr. 1958. Species and hybrids of *Asimina* in the northern Shenandoah Valley of Virginia. *No. Nut Growers Assn. Rpt.* 49:73–75.
Forbes, F. B. 1883. *Cudrania triloba* Hance and its uses in China. *J. Bot.* 21:145–149.
Gerardi, L. J. 1962. Persimmon varieties and propagation. *No. Nut Growers Assn. Ann. Rpt.* 53:25–26.
Gibson, M. D. 1962. Domesticating the American persimmon. *No. Nut Growers Assn. Rpt.* 53:26–28.
———. 1970. What is a persimmon? *Pomona* 3:89–91.
Griffith, E. 1971. Fallacies about the persimmons. *Pomona* 4:5–7.
———. 1972. Removing astringency from American persimmons. *Pomona* 5:85–87.
Harris, R. E. 1964. The Saskatoon berry. *Can. Dept. Agr. Bul.* 1246.
Hounsell, R. W. 1968. Cytological studies in *Sambucus*. *Can. J. Gen. Cyt.* 10:235–247.
Hunter, I. R. 1969. Fruit of the mountain ash. *Pomona* 2:37.
Jones, G. N. 1946. American species of *Amelanchier*. *Ill. Biol. Memo.* 20:1–126.
Kovleva, T. N. 1950. Cornelian cherry growing in U.S.S.R. (in Russian). *Sad i Ogorod* 1:31–33.
Kowa, C. 1969. Topworking the native persimmon. *Pomona* 2:21.
Kral, R. 1960. A revision of *Asimina* and *Deeringothamnus* (*annonaceae*). *Brittonia* 12:233–278.
LeCog, C. 1963. Contribution to the study of the cytotaxonomy of the *Moraceaes* and the Urticaceaes (in French). *Rev. Gen. Bot.* 70:385–420.
Madison, D. N. 1965. *Amelanchier:* three seasons of beauty. *No. Nut Growers Assn. Ann. Rept.* 59:133–135.
McDaniel, J. C. 1962. A quick survey of serviceberries—juneberries. *Nutshell* 9:17–19.
———. 1970*a*. Native persimmon. *Pomona* 3:27.
———. 1970*b*. Persimmons in Illinois. *Pomona* 3:137–140.
———. 1971. Persimmon breeding source material. *Pomona* 4:45–46.
———. 1972. Persimmon breeding notes. *Pomona* 5:123–124.
Meader, E. M. 1965. A promising selection of American highbush cranberry. *Nutshell* 12:9.
Miller, W. S., and C. Stushnoff. 1971. A description of *Amelanchier* species in regard to cultivar development. *Fruit Var. Hort. Dig.* 25:3–10.
Moffett, A. A. 1931. A preliminary account of chromosome behavior in the Pomoideae. *J. Pomology* 92:100–110.
Ourecky, D. K. 1970. Chromosome morphology in the genus *Sambucus*. *Amer. J. Bot.* 57:239–244.
———. 1972. Minor fruits in New York State. *N. Y. Infor. Bul.* 11.
Petrov, E. M. 1957. *Riabina* (*Sorbus*). Moscow. p. 150.
Petrova, N. E. 1937. Anti-scorbutic vitamin content in certain varieties of Michurin fruits. *Problems of Vitamins* 2:1–13.
Ramsey, F. T. 1926. The Swazey barberry. *J. Hered.* 12:426–427.
Rehder, A. 1949. *Manual of cultivated trees and shrubs*. 2nd ed., New York: MacMillan Co.
Ritter, C. M., and G. W. McKee. 1964. The elderberry. *Pa. Agri. Expt. Sta. Bul.* 709.
Sax, H. J., and K. Sax. 1947. The cytogenetics of genetic hybrids of *Amelosorbus* *J. Arnold Arboretum* 28:137–140.
Sax, K. 1931. The origin and relationships of the Pomoideae. *J. Arnold Arboretum* 12:3–22.
Shaskkim, I. N. 1938. Notes on *Actinidia* (in Russian). *Sci. Fruitgrowing. Mitchurinsk* 5:49–51.
Simonet, R. 1971. Notes on results. (Breeding honeysuckle.) *Pomona* 4:112.
Sinoto, Y. 1928. On the chromosome number and the unequal pairing of chromosomes in some dioecious plants. *Proc. Imp. Acad. Japan* 44:175–176.
———. 1929. Chromosome studies in some dioecious plants with special reference to the allosomes. *Cytologia* 12:109–141.
Slate, G. L. 1955. Minor fruits. *Nat. Hort. Mag.* 34:139–149.
Trofimov, T. 1959. Some species of *Sorbus* (in Russian). *Sad i Ogorod* 12:48–49.
Way, R. D. 1964. Elderberry varieties and cultural practices. *N.Y. St. Hort. Soc. Proc.* 110:233–236.
———. 1967. Elderberry growing in New York State. *Cornell Ext. Bul.* 1177.
Wiegand, K. M. 1912. The genus *Amelanchier* in eastern North America. *Rhodora* 14:116–164.
Winters, H. F. 1972. Introduction of cudrang or silkworm thorn into the United States. *Amer. Hort.* 51:6–8.
Wright, P. H. 1969. The Saskatoon bush for fruit and ornament. *Pomona* 2:32–33.
Yeager, A. F., E. Latzke, and D. Berrigan. 1935. The native fruits of North Dakota and their use. *No. Dak. Agr. Expt. Sta. Bul.* 281.
Zajcev (Zaitoev), G. N. 1969. Honeysuckle with edible fruit (in Russian). *Priklad Bot. Genet. Seleke* 40:183–189.
Zimmerman, G. A. 1941. Hybrids of the American papaw. *J. Hered.* 32:83–91.

Peaches

by Claron O. Hesse

Few fruit tree species have spread so rapidly and become adapted to so many climatic situations as has the peach, *Prunus persica* (L.) Batsch. The popularity of the peach and its smooth-skin mutant, the nectarine, is closely related to the wide and general appeal of the fresh fruit, its ease of use, and its utility in the production of a variety of preserved products.

The peach is primarily a tree of the temperate zones. Important centers of commercial production usually lie between latitudes 30° and 45° N and S. At higher latitudes minimum winter temperatures and spring frosts are the usual limiting factors. The peach flower is bud hardy to about −23 to −26 C. Its northern range is extended where minimum temperatures are ameliorated by large bodies of water, as western continental seaboards, the Great Lakes in the United States and Canada, and the Caspian and Black seas of Eurasia. Midcontinental areas, characterized by lower winter temperatures and usually more severe spring frost hazards, are seldom centers of production.

Most peach cultivars require from 500 to 1,000 or more hours of cold below 7.2 C to foliate and bloom normally in the spring. At low latitudes the winter rest requirement of this deciduous species is not met. Foliation, bloom, and production become erratic. Cultivars with less than 100 hours of chilling requirement are known. Also strains of the peach are grown throughout the subtropical and tropical zones, especially at higher elevations, where the heat of low-elevation tropics is ameliorated. Usually the quality is poor.

The peach attains best quality in areas where the summer temperatures are warm to hot. Many cultivars become more astringent when grown under cool summer temperatures. Some, as the low acid Honey and flat- or saucer-shaped Peen-to forms from southern China, are naturally adapted to the southern limits of peach production. As a native or naturalized species, the peach requires considerable rainfall spread over the summer period. Irrigation has therefore opened up wide areas to peach culture under highly favorable conditions. The absence of summer rains reduces disease and pest hazards considerably.

The climates of the eastern and western seaboards of North America exemplify in high degree the range of climates under which the peach is highly successful. Areas from New Jersey to the Florida border represent a region of intensive and successful peach culture. On the west coast the peach is produced in quantity from central California to British Columbia. The latitudes limiting successful culture are somewhat higher on the west coast due to the moderating influence of the west to east flow of air from the Pacific Ocean.

Table 1 gives world peach production for those nations reporting such statistics to the Food and Agricultural Organization of the United Nations. The United States is shown to be the leading peach-producing country of the world. This is due

TABLE 1. World peach production, 1970, by specified areas and leading countries, except Russia, China, and southern Asian countries (from *FAO Production Yearbook*, vol. 25, 1971)

Area	1000 mt	Leading countries
Europe	2442	Italy (1128), France (497), Bulgaria (167), Spain (165)
North America	1555	U.S.A. (1426), Mexico (81), Canada (48)
Latin America	469	Argentina (235), Brazil (120), Chile (40), Peru (33)
Near East	153	Turkey (112)
Far East	357	Japan (279)
Africa	200	South Africa (155), Morocco (14)
Oceana	145	Australia (116)
Total	5321	

to its size, to several geographic areas favorable to peach production, to the popularity of the peach in the United States, and to advanced technology for shipping and preserving. The importance of the latter is reflected in the fact that about 45% of the national production consists of canning clingstone cultivars, practically all of which are processed.

The European nations bordering the Mediterranean basin and their neighbors constitute a major production area. Italy is the leading peach-producing nation of this region, and approaches the United States in total production. Again, large markets are nearby and transport and processing techniques are highly developed.

It is unfortunate that production figures are not available for Russia, China, and their southern neighbors on the Asian continent. Afghanistan, Pakistan, and India undoubtedly produce peaches in significant amounts. Certainly China and southern Russia are important producing areas.

In the Southern Hemisphere, Argentina, Brazil, South Africa, and Australia are the leading producing nations. The South American nations rely largely on their own and other South American markets to dispose of their fruit. South Africa and Australia produce for export. Australia, like California in the United States, is a major center for the production of canning clingstone cultivars.

The utilized production of the 1971 peach crop in the United States, in thousands of metric tons, was: California, 817; South Carolina, 141; Georgia, 58; New Jersey, 61; Pennsylvania, 51; Michigan, 40; Washington, 20; Virginia, 18; North Carolina, 17; Ohio, 14; West Virginia, 13; Illinois, Maryland, and Colorado, 11 each; Missouri, 10; New York, 9; and other, 68, with a national total of 1370. Nectarine production, also in thousands of metric tons, ranged in California from 60 in 1970 to 80 in 1972. All other states produced less than one, with Florida, Pennsylvania, Washington, and New Jersey as the principal producers (USDA Econ. Res. Serv. TFS 186, Feb. 1973).

California thus produces about 60% of the peaches in the United States, and nearly 100% of the canning clingstone cultivars. The latter constitute the bulk of the California production. The state produces about 13% of the dessert peaches, about the same as South Carolina, which is the second most important state in peach production. Nectarine production in the United States is mostly in California, with only minor amounts being produced in other states. Approximately 400 ha have been planted with nectarines in Florida in recent years. Production can be expected to increase rapidly.

Tremendous variability exists in the types of peaches grown throughout the world. Variability is undoubtedly greatest in producing regions which cater primarily to their own local markets. Standardization is encouraged, or becomes channeled, when transport and/or processing qualities for distant markets become more important than diversity and flavor quality. Nevertheless, it is this variability with which we are primarily concerned, and particularly with its measurement, mode of heritable transmission, and methods for exploiting it. New improved cultivars are produced at an accelerating rate, particularly where the peach is an important factor in international or interstate commerce. Continued improvement, extension of range, and control of pests and diseases all require greater understanding of the genetic mechanisms involved. We will, therefore, be concerned with the breeding behavior of the species, and only incidentally with cultivars *per se*.

Extensive cultivar listings and descriptions are given by Hedrick (1917), Morettini, et al. (1962), Olmo (1965), Brooks and Olmo (1972), and Savage and Prince (1972). Information about cultivar trials and new cultivars originating in the socialist countries of eastern Europe and western Asia appear regularly in *Plant Breeding Abstracts*.

Apparent in the extensive review of important cultivars in many of the temperate and subtropical zone nations (Olmo, 1965) is the reliance upon cultivars originated in the United States. Even in Italy, long the center of European peach production, seven of the ten leading cultivars were of United States origin as of 1965.

Good reviews of peach genetics are those of Cullinan (1937), Bailey and French (1949), Weinberger (1955), Monet (1965), Pisani (1965), and Lalatta (1956). This discussion is based primarily upon the reports of investigations conducted in North America.

Origin and Early Development

Hedrick (1917) thoroughly marshalled the evidence acknowledging China to be the native home of the peach. According to Evreinoff (1957), other *Prunus* species usually referred to as peaches are also native to China. The history of the peach as a garden crop in China far antedates any claim that can be substantiated for other geographic areas.

Frank N. Meyer (in Hedrick, 1917), a USDA plant introduction explorer, travelled in China early in the present century looking for valuable plants. He found peaches growing semi-cultivated and in apparently wild stands in many of the provinces of central China. His descriptions indicate that all of the discrete variations known today existed then, either in wild or garden strains. White-, yellow-, and red-fleshed forms were noted, for example. Among apparently wild forms his descriptions mention clouds of pink-flowered peaches on hillsides, which would indicate that some, if not most, of the wild peach stands consisted of showy-flowered types. Freestones and clingstones were mentioned; he describes wild stands as freestones. The Honey and Peen-to types were described much later by Fortune, an English explorer.

Turkestan and Chinese Turkestan, if not included in the original range of the wild peach, were undoubtedly an early extension of its range—almost certainly long before the peach was known to Western civilizations. In these areas early European visitors and settlers described all types of peaches. Nectarines appear to have been much more common than in China proper if the space given to their description indicates their relative importance, and may have originated here. As noted by Meyer, nectarines were less esteemed than the peach in China and may have been introduced from these western locales.

One of Vavilov's principles is that a wild species shows its greatest variability at or near its center of origin. This great variability of the peach in China is quite apparent in the evidence given by Hedrick (1917). That the peach came to the Mediterranean basin from Persia was reiterated by early Greek and Roman writers, and is reflected in the name they gave to the peach. It appears to have been introduced to Greek culture between 400 and 300 B.C., and to the Romans shortly after the beginning of Christianity, as noted by Pliny (see Hedrick, 1917).

The Romans spread the peach throughout their realm, although earlier migrations may have brought the peach to northern Africa, and to Spain by the Moors. But its spread through the European Mediterranean countries was mainly by the Romans. They undoubtedly brought the peach to present-day France, and may have carried it as far as England. In the latter case, the introductions did not outlive the Roman stay in England, to which peaches were later brought primarily from France. Through these introductions the peach may have become wild in regions best adapted to it, but remained a cultivated exotic in other regions.

Although grafting and budding were skills known to the Greeks, it appears that early peach culture relied almost solely upon growing of seedlings. Seed would have been the simplest and easiest method of introduction of peaches to north Africa, Spain, France, and England from the eastern Mediterranean countries. Successive generations of seedlings grown in such new locations would have provided the raw material for selection to modify adaptation to the local climates encountered, and for the ultimate selection of types differing substantially in their general characteristics. The earliest described cultivars of France and England, for example, were often white-, rather soft-fleshed types, whereas those of Spain were described as firm and yellow-fleshed.

The era of exploration and colonization in the sixteenth and seventeenth centuries was accompanied by distribution of plants from the home countries. It was quite characteristic of the Spanish, for example, to quickly export their homeland crop plants to the limits of their conquered lands. The peach serves as an example. Seed could be readily transported to new lands. It was introduced to continental America both via the Spanish conquest of Mexico, and into Florida as early as 1565 with the founding of St. Augustine. The Portuguese apparently introduced the peach to the east coast of South America at an early date also. It is tempting to speculate that if these lands had been settled by the English the peach might have failed. Instead it became wild, especially in Mexico and the South and Southwest of present-day United States. The Indians of America also appreciated the peach, and spread it throughout the South, Southeast, and East practically to the limits of present-day culture. Remnants of these early movements still remain in so-called Tennessee naturals (rootstocks) and the garden plots of the Havasu Indians of Arizona. Early settlers of the North American continent noted and used the fruit of such stands.

Another factor in the European development of the peach which was to eventually affect peach culture in the United States closely followed the industrial revolution of the sixteenth century. This was the formation of a prosperous class whose interest in gardening reached unprecedented heights. A great expansion of fruit culture of all kinds occurred, as is evident from the descriptions of famous estate gardens of the time. In England, with a climate inhospitable to the peach, trees were grown against walls, in fruit houses, or even in the open. Prized forms were quickly propagated and distributed. Such notable professionals as John Rivers hybridized various tree fruits, and introduced superior clones from France and elsewhere. A great multiplication of cultivars occurred. Some of these were eventually introduced to the colonies. As the European gardeners were skilled in the arts of graftage and propagation, the cultivars described may be assumed to be clonal. However, earliest descriptions also may have been of seedling "types." There was unquestionably a good deal of confusion in nomenclature. Nevertheless, both the number and variety of available cultivars increased apace.

In the American colonies these two sources of plant materials—the indigenous seedlings via Mexico and the Southeast, and the cultivars or seed sources brought from England—met and provided two basic sources of peach material for further advances through selection.

The third source of importance to the production of the present-day United States peach cultivars was a single introduction—that of 'Chinese Cling' from China via England.

Up to the time of the American Revolution, North American peach culture relied almost solely on seedling stands. These were extensive, but of lowly estate. The product of the orchard was more likely to end up as food for hogs or in the distillery than as a product for the table. It was an era of overproduction of a perishable product.

Budded trees were first offered for sale in North America just before the Revolution (although amateur connoisseurs had long exchanged budwood and scionwood) by the nursery of Robert Prince of Flushing Landing, Long Island, New York. Following the Revolution, the nursery of John Kenrick, of Brighton and Newton, Massachusetts, also offered "inoculated" (budded) trees in the 1790s (Hedrick, 1950).

Following the Revolution, the peach was utilized by settlers of the heartland of the North American continent. In the first round of planting by settlers, seedlings were often the sole stock, but budded trees soon became common. New cultivars originated in these seedling plantings—a form of frontier-day selection. But the measure of quality was changing too. No longer was local use the criterion of excellence, but how well they would carry to the nearest settlement or town for sale. Included among such selections of local origin were the prominent or popular cultivars brought from the eastern seaboard.

Between the Revolution and the Civil War, a number of cultivars came to the fore, usually seedlings of unknown origin. Some of the better were 'Early Crawford' and 'Late Crawford' and 'Oldmixon Cling'. However, it was not until after the Civil War that many of the genotypes originated upon which modern cultivars are based.

'Chinese Cling' was introduced to the United States by Charles Downing in 1850, from England. It had been discovered in China by Charles Fortune, an English plant explorer, a few years previously. Downing sent trees of this new introduction to Henry Lyons in South Carolina, who first fruited the peach. Following the Civil War,

Samuel H. Rumph of Marshallville, Georgia, fruited seedlings of 'Chinese Cling'. 'Belle of Georgia' ('Belle') and 'Elberta' came from this planting. Although only speculation is possible regarding the pollen parents of these two cultivars ('Chinese Cling' was pollen sterile according to Connors [1927]), their practically simultaneous discovery and introduction is notable. Both became leading commercial clones. Also important, 'Hiley' (a seedling of 'Belle') and 'J. H. Hale' (supposedly a seedling of 'Elberta') were found only a few years later.

As practically all of the early hybridization which was to follow a few years later involved one or more of these cultivars, and as most modern cultivars can be traced back to lines which included one or more of them, the importance of this small group of cultivars cannot be overestimated. Among the most successful cultivars grown today, many, if not most, trace back to 'J. H. Hale', and hence through 'Elberta' or 'Belle' to 'Chinese Cling'.

From our viewpoint today, the most remarkable fact concerning this lineage is that most of our present-day cultivars are derived from a rather narrow base, and hence are genetically restricted. Much of the variation in the peach available from China and the early introductions into this country may have been lost. A few unrelated cultivars, such as 'Champion', 'Alexander', 'Admiral Dewey', 'Greensboro', and 'Carmen' were also used extensively in early breeding programs.

Recent Improvement in North America

Many individuals contributed to the growing list of peach cultivars through selection, growing seedlings of prominant clones, or even hybridizing during the eighteenth and nineteenth centuries (Cullinan, 1937). The modern era of peach breeding in the United States can be marked by the initiation of sustained programs by several of the state agricultural experiment stations in the early decades of the twentieth century. These programs were designed to develop cultivars equal to the demands of growing distant markets, and encouraged by the then still recent rediscovery of Mendel's principles of heredity.

Among the early, and mostly continuing, state programs were those of New York (recently R. C. Lamb and J. P. Watson), Iowa (S. A. Beach), Illinois (C. A. Crandall, J. B. Mowry), California (E. B. Babcock and later J. W. Lesley, W. E. Lammerts, G. L. Philp and L. D. Davis), New Jersey (M. A. Blake and later L. F. Hough and C. Bailey), Massachusetts (J. S. Bailey and A. P. French), and the USDA (F. P. Cullinan, W. F. Wight, J. H. Weinberger, and later L. Havis, H. W. Fogle, T. Toyama, and V. E. Prince). Michigan (R. A. Andersen) began a very successful program in 1924 under the direction of Stanley Johnston. Other state programs of note are those of Virginia (G. Oberle), Maryland (I. C. Haut), Arkansas (J. N. Moore, R. C. Rom), North Carolina (G. W. Schneider, F. E. Correll, and C. N. Clayton), and Louisiana (W. L. Hawthorne and J. C. Taylor). Texas (H. H. Bowen) has been more or less active in peach breeding since 1935. Florida (R. H. Sharpe, W. B. Sherman) began an active breeding program in 1951. Other state programs are those of Missouri (K. W. Hanson and A. D. Hibbard), Oklahoma (H. A. Hinricks), Georgia (F. E. Johnstone), South Carolina (W. S. Jordan), Mississippi (J. P. Overcash), and Kentucky (H. C. Mohr). For more detailed information on these programs see Cullinan (1937), Havis et al. (1947), and Olmo (1965).

The Canadian stations at Vineland and Harrow (O. A. Bradt, R. C. Layne) have maintained a breeding program since 1914. In Europe, the efforts of A. Morettini in Italy have been particularly productive, and peach breeding is also conducted in other European countries. In the Southern Hemisphere, South Africa, Australia, and some of the South American nations have undertaken work in this field. Russia has a strong breeding program with many tree fruits, including the peach (Weaver, 1968). Cullinan (1937) and Olmo (1965) give greater detail concerning some of these ongoing efforts.

Following the passage of the United States Plant Patent Law in 1930, private peach and nectarine breeding endeavors developed. The three most notable and continuing programs have been those of Grant Merrill since about 1932, F. W. Anderson since the early 1930s, and the Armstrong Nursery Company, begun just slightly later. Collectively these programs are responsible for nearly 150 patents on peach and nectarine cultivars. For the contributions of private individuals in earlier years see Cullinan (1937).

Of the earlier programs in the United States and Canada, those of the Massachusetts, New Jersey, and the USDA were the most productive in elucidating inheritance of qualitative characters (see Table 4). The New Jersey program has given considerable attention to modern technology of breeding. As might be expected, the northern and north central state programs have given us the most information relating to tree and bud hardiness, just as the early investigations in California at Riverside and Los Angeles, and of the USDA in Georgia emphasized low chilling.

Much of this earlier effort soon centered upon cultivar development, however. New Jersey, Michigan, Virginia, and the USDA programs have been particularly prolific in this regard, as have the Vineland and Harrow stations in Canada. The USDA and the California station have released several canning clingstone cultivars. The productivity of the private programs has been notable, not only for the number of cultivars developed, but also for the prominence many have attained. Several of Merrill's peach cultivars are grown in California and France. Anderson completely revolutionized the nectarine industry of California, and contributed to the development of nectarine growing in other states and nations. The Armstrong Nursery Company program, originally developed to supply improved cultivars for southern California home gardening, is now one of the leading sources of new early maturing peach and nectarine cultivars for California's Central Valley peach districts.

As a result of this intensive effort, cultivars of North American origin have come to occupy leading positions in many foreign countries (Olmo, 1965) and are being used in intensive breeding efforts in foreign lands.

Cultivar development has reached an advanced stage. Opportunities for continued improvement exist, but further improvement of the peach will necessarily rely more and more on subtler quantitative traits. At the extremes of range of commercial production, opportunities for continued improvement are greater. Thus, the extension of commercial culture into the warmer winter areas, as northern and central Florida, the Gulf States, and northern Mexico requires improved cultivars which will probably be rather easily obtained. The opportunities in these areas are of course related to an early season of maturity, which cannot be equalled in more northerly locations. At the northern extreme, a more difficult problem needs solution, namely, acceptable bud- and winterhardiness.

Modern Breeding Objectives

Recent investigations in peach and nectarine breeding are less concerned with the inheritance of qualitative characters and more with an understanding of the transmission of quantitative traits. The manipulation of major genes to yield wanted types of cultivars, as dessert freestones, canning clingstones, or nectarines is well understood. An understanding of the genetic systems controlling quantitative traits, as exemplified by climatic adaptations and quality traits, and the development of efficient breeding systems to maximize transmission of favorable variations is now of greater interest and importance.

Extension of season of maturity remains an important objective in many breeding programs. Two factors account for this: 1) existing cultivars are deficient in desirable traits at the extremes of the season, and 2) market opportunities are greatest at these times. Most emphasis is being given to earlier maturity.

Consumption of fresh dessert fruit continues to decrease, and a greater proportion of the total production is processed. Therefore, greater emphasis is being given to processing characteristics and performance. Firmness of flesh, freedom from loose fiber, fine texture, attractive color, and non-browning of the flesh are all important traits needing improvement for processing outlets for freestone peaches.

Cultural, transport, and marketing requirements have been emphasized inordinately in the past. Improvement in flavor quality, for both dessert and processed use, has become an important objective in most current breeding programs.

Changing cultural practices, especially increasing mechanization, will require modification of tree structure and fruiting response. These changes have been recognized and investigations begun. Control of tree vigor, both as an aid to mechanization and to assist in lowering costs of manual labor in pruning, thinning, and harvesting, are a part of these developing changes. We know too little of the relation between tree size, precocity, and the potential for high-density plantings with the peach and nectarine to understand the variables to be controlled. Can the brachytic dwarf gene or the bushy genes (see Table 4) be used to advantage in cultivars designed for high density plantings?

Genetic control of some disease and pest problems are being more intensively studied at the present time.

Some of the research needed to answer these problems will be done by horticulturists, physiologists, and pathologists, but the breeder should have available appropriate materials for discriminating tests.

A broadening of the genetic base for cultivar development is being encouraged. The main thrust in this regard is evident at the northern and southern extremes of commercial peach production, where reliance upon distinctive gene sources was required to meet the limiting conditions existing. We note elsewhere the severely restricted base for modern cultivar improvement in the United States, and the tendency for breeding programs to become rather highly inbred. Attractive new variants are needed for diversity in marketing opportunities. Unique flavor and nutritive factors will be valuable assets.

Within the past decade or two many institutional programs, not only in breeding but in all agricultural research, have become more restricted because of economic limitations. This trend is more liable to persist and intensify than to be reversed. In response peach and nectarine breeding programs may have to become much more selective in their objectives, especially regarding cultivar improvement. More attention to basic problems may better generate support. Cultivar development may come to rely more upon the non-institutional peach breeders.

Breeding Techniques

The peach belongs to the family *Rosaceae*, subfamily *Prunoidea*, genus *Prunus* L., and subgenus *Amygdalus*. All commercial cultivars belong to *P. persica* (L.) Batsch. Other peach species are *P. davidiana* (Carr.) Franch., *P. mira* Koehne, *P. ferghanensis* (Kost. et Rjab) Kov. et Kost., and *P. kansuensis* Rehd. All are native to China. *P. andersonii* Gray is the lone representative found in the United States.

All commercial clones grown for their fruits belong to *P. Persica*, *P. davidiana*, *P. ferghanensis*, *P. kansuensis*, *P. andersoni*, and *P. mira* are not described, as no cultivars are referred to these species, although they are of some interest for potential rootstock development. Other peach species are as yet unimportant and uninvestigated, at least in western horticulture.

The flowers of *P. persica* are commonly borne on one-year-old shoots, with very few spurs forming. One or two flower buds form laterally to the vegetative buds but some cultivars may form up to six or more flower buds per node. Usually the more basal and terminal nodes are less floriferous, or not at all.

The flowers are mostly one per bud; anthesis is with or before leafing. The flowers are perfect, complete, perigynous, usually with a single pistil. The gynoecium is superior, and the single carpel bears two anatropous ovules, one of which normally aborts following fertilization. The fruit is normally one-seeded. The style is elongated, ending in a small, capitate stigma, which becomes receptive at bloom. The androecium is borne on the fused corolla, just below the five sepals and five petals, which are arranged alternately. Stamens number 30 or more. The filaments are long and thin, bearing the four-loculed anthers which are reddish-yellow to yellow. The corolla cup is marked by the presence of nectaries, which vary in color from greenish through yellow to orange.

Microsporogenesis begins in December (Knowlton 1924; Draczynski, 1958), and meiosis occurs about the time of bud-swell in the spring. Fertilization normally occurs within 24 to 48 hours of pollination. The megagametophyte is of the common eight nucleate, seven-celled type. Following fertilization the endosperm becomes and remains multinucleate for some time, becoming cellular near the time of pit-hardening.

The fruit is a typical drupe, with a thin epidermis, a fleshy mesocarp of considerable thickness, and a stony endocarp which may or may not become free from the mesocarp. The seed is cotyledenous.

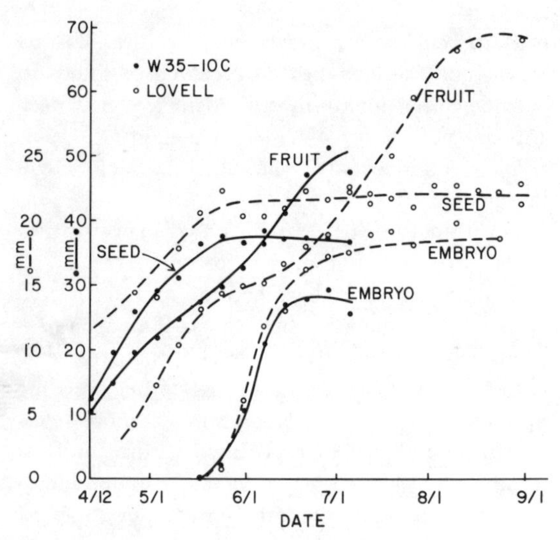

FIG. 1. Growth curves of fruit (suture diameter), seed (length), and embryo (length) of early ('W35-10C') and late maturing ('Lovell') peaches.

Fruit Development

After fertilization, embryo development is very slow for some weeks. First division of the fertilized egg may occur within 24 to 48 hours, and a small, microscopic proembryo usually forms within a few days. Thereafter, for some time, embryo development is very slow.

The peach fruit undergoes characteristic growth. First, growth is rather rapid, slows to a low rate, and then undergoes a second period of rapid growth, which ends at maturity (Fig. 1). The period of slow growth occurs when the fruit is approximately half its final size if thinned to normal crop loads. The endocarp is hardened before the final swell and does not enlarge during the final growth stage.

The first period of rapid growth is nearly equal in duration for all cultivars. The timing of the two phases of rapid growth differ tremendously. In early-maturing cultivars the period of reduced growth is barely noticeable, whereas in late ones it may extend for several weeks.

At about the same time the fruit growth slows following its initial rapid expansion, the endosperm and embryo undergo a period of extremely rapid growth; the endosperm becomes cellular and invades and absorbs the nucellar tissues, while the embryo is at the same time rapidly absorbing the endosperm. In a relatively short time, the nucellus and endosperm are absorbed, and the embryo fills the testas. Just at this time the percent dry weight of the embryo is at a minimum; thereafter its growth is by accumulation of solids, both organic and inorganic. At maturity the seed stores approximately 50% oils.

The seed of early cultivars are incompletely developed at the time of fruit maturity. Indeed, in some fruits of 'Springtime', under California conditions, the embryo may still be microscopic in size, with both nucellar and endosperm tissue abundant, although it is generally visible, and may be approximately 20% as long as the testas (PF_1 = 20, Hesse and Kester, 1955). Slightly later cultivars have embryos which fill the testas, but are still physiologically quite immature; i.e., they contain little stored reserves. All such seed are ungerminable unless special methods are used.

Pollen Collection

Pollen is collected from flowers well advanced, but not open. Nonshowy flowers may have the tips of the anthers exposed before anthesis. Unless pollen is evident, such flowers are suitable for collection. It is not important if some of the anthers of the showy-flowered types have shed before opening if the flowers are gathered before the petals separate. A double handful of flowers in the proper stage will supply all the pollen needed for a few thousand pollinations, even if the clone does not supply large amounts of pollen.

Flowers are collected in small paper bags, using a new one for each collection. Moisture will condense rapidly in film bags and may ruin the pollen. King (1955) has described a vacuum apparatus for gathering anthers and/or pollen.

Separation of the pollen from the flowers is most easily accomplished by rubbing them lightly over a piece of hardware cloth—4 to 6 mm mesh (Barrett and Arisumi, 1952). Excess pressure may result in bruised anthers which will not dehisce well. For small numbers of flowers, cutting or tearing the flower open vertically, rolling it open between the fingers, and raking the anthers off with a small comb or the back of a scissors blade will insure the greatest yield of anthers. It is quicker to collect more flowers and use the screen if the flowers are available. It is best to make the separation as soon as possible, while the flowers are still crisp and turgid. Anthers that cannot be separated from the flowers within a few hours of collection should be stored in their collecting bags in a refrigerator at 2 to 4 C until separated.

Screen separation results in debris mixed with the anthers. The debris can be separated from the anthers with a colander of proper mesh size. Clean all equipment with 70% alcohol before using on another pollen. Sift the anthers onto unglazed, permeable paper, such as newspaper or paper towelling, for drying. Place in a warm, but not hot area—the ambient temperature of the laboratory is fine. Do not expose to the sun. Dehiscence and air-drying is usually complete in 12 to 24 hours. Drying may be assisted by very gentle heat if necessary (King, 1955). We prefer the use of permeable paper to petri dishes or other solid surfaces because it assists in the removal of moisture, but good results can be obtained by drying in petri dishes or on glass plates if the pollen is not piled.

After drying, the anthers and pollen are collected in a convenient container. Glass shell-vials of convenient size are satisfactory. They are loosely stoppered with non-absorbent cotton. If corks are used for stoppers *and* pollination, loosen them during storage periods to prevent moisture condensation. Some prefer larger containers, as small shallow tin cans with friction lids. The latter are especially convenient when a brush is used for pollination.

For use in the same season, no special precautions are necessary to prevent loss of viability. Hold at ambient temperatures and do not expose to direct sunlight. Pollen can be air-mailed to any location without serious loss of viability.

When it is necessary to use year-old pollen, as in the cross early blooming x late blooming, pollen should be stored in a deep freeze at −30 C (Griggs et al., 1953). Under these conditions, pollen will retain its viability for at least two years. If a deep freeze is not available, storage at 0–2 C at 25% relative humidity will preserve its viability for one to two years (King and Hesse, 1938). Following low-temperature storage, it is best to remove only as much pollen as will be used the same day.

Pollen Germination

It is usually unnecessary to check pollen viability if the pollen has been correctly gathered and prepared. Pollen stored for long periods should have germinability checked. Two methods are generally used: the hanging drop and agar plates. For the hanging drop method use either depression slides or Van Teigham cells. The Van Teigham cell may be fixed to an ordinary slide with melted paraffin. A drop of 12 to 15% sucrose solution is placed on a coverslip, the pollen dusted on the drop with a teasing needle or glass rod, the coverslip quickly inverted and placed on the cell, which has been prepared by placing a film of vaseline on the upper edge. A drop of the sugar solution should be placed in the bottom of the cell to prevent a change in the culture solution concentration.

Agar plates are prepared by making up a 12 to 15% sucrose solution in 1 to 2% agar-agar (or Difco Agar), which is poured into petri dishes and sterilized. Six to eight pollens may be tested simultaneously in a 100mm petri dish by dusting pollen on the agar near the rim, at equally spaced intervals around the dish.

Germination of good pollen will begin very quickly at room temperatures. Gross viability can be observed within three or four hours, but eight to 12 hour cultures are usually used for counts. Two to several fields of 100 to 200 grains are usually counted, depending upon the accuracy required. Good, viable pollen will often burst before germination or after the tube has grown for a short distance. Such pollen should be counted as germinated.

Germination of freshly-prepared pollen will vary greatly, but good peach pollen should usually show 50 to 85% of the grains with strong tubes. In 24 hours strong cultures should show tubes many times longer than the diameter of the pollen grain. As pollen ages, both germination and length of tubes decrease. If the pollen seems weak the presence of some strong tubes will indicate that the pollen is good enough to secure at least moderate fruit set, even though the germination percentage is low.

The sucrose source is vital. Unless reagent grade sucrose is available use cane sugar. Beet sugar *may* greatly reduce or even completely inhibit pollen germination.

Vital-staining reactions may not correlate well with *in vivo* or *in vitro* germination or ability to effect fertilization, especially with stored pollen. But acetocarmine, aceticoreicin, or vital stains such as tetrazolium salts (Oberle and Watson, 1953), may be used to estimate percentage of good pollen.

Emasculation

Stamens of peach flowers are attached distally on the corolla, and can be removed by cutting through the floral tube below the point of attachment. This is easily accomplished by grasping the flower at one side between the thumb nail and the nail of the third finger. Pinching the tube breaks the tissue, which is then torn loose by pulling, usually with a slight side-twist. For those who do not find hand emasculation convenient, emasculating tools (Barrett and Arisumi, 1952; Jones and Thompson, 1955) are useful. The adept find the fingers are quicker than the tool. A uniform bloom of flowers in the proper stage can be emasculated at the rate of 750 to 1000 per hour, or faster.

Emasculation is best done as the flowers near anthesis. The anther tips and even the stigma of non-showy flowers may be exerted before any pollen is shed. Such flowers are not visited by pollinating insects and may safely be used in crossing. In dry, hot, or windy weather anthers of showy flowers may dehisce before the petals separate. If facultative cleistogamy is suspected several flowers should be carefully inspected to determine the largest "balloon" stage that can be used safely. To minimize accidental pollination by wind- (Branscheidt, 1933) or gravity-borne pollen, emasculate branches from the top down as far as needed to secure the desired number of flowers. Use branches on the windward side of the tree. When the bloom is extended, as in mild-winter areas, only a small portion of the bloom may be in the proper condition at one time. Emasculate the chosen branch in two or three stages, at one- or two-day intervals.

Check the branches a week to ten days after pollination and remove any buds missed earlier.

Protection of Bloom

The need to protect emasculated and pollinated pistils from chance pollination is debatable for crosses made specifically for cultivar improvement. Crosses designed to give genetical information may justify protection against accidental contamination. There are three sources of accidental self- or cross-pollination of emasculated pistils: 1) wind-blown or gravity movement of pollen, 2) visitations of bees or other pollinating insects to emasculated flowers, and 3) facultative cleistogamy preceding emasculation.

Prunus pollens are relatively heavy, so wind pollination is limited. Also, most of the pollen is collected by bees, with relatively little remaining to be blown about. Gravity (with wind) may distribute some pollen, particularly from upper branches to lower parts of a tree (Singh, R., unpublished). Facultative cleistogamy is more common with showy-flowered types, and may prove to be an important source of contamination.

Effective protection may be provided with a variety of coverings. With good bloom, 40 x 25 x 15 cm paper bags will cover approximately 100 to 150 flowers. If inclement weather is expected during the bloom period paper bags can be protected with polyethylene bags, well vented with several holes. Cheesecloth sleeves have the advantage of covering larger units, up to whole scaffold branches, but will not prevent wind- or gravity-pollen movement. Frames, covered with cheesecloth or other screening, or solid films are sometimes used, especially in locations where the weather is likely to be mild during the bloom period. Tight enclosures of solid films usually require ventilation to avoid excessive moisture condensation. In areas where storms are expected to occur during the bloom period, cages must be strong and well anchored. Limited usefulness during a single season makes these expensive.

Fogle and Dermen (1969) calculated 14% chance pollination in a closely-planted seedling block. Much of the chance contamination may be identified in seedling or fruiting stage by use of genetic markers when simply inherited characters are identified in the parents. This procedure is less expensive than providing assured isolation. The amount of residual, undetectable contamination is usually too low to be of importance in crosses made for cultivar improvement.

The most useful instance of assuring protection against accidental contamination is in self-pollination. Protection of unemasculated bloom by any of the methods mentioned serves to keep out bees, which are the main source of contamination in this instance. Even so, with considerable evidence that cross-pollination will normally be less than 5% under *open pollination*, even in mixed stands, the greater efficiency of using open-pollination should not be ignored, especially if genetic information is not the critical goal.

One helpful use of whole-tree cages would be their use with a colony of bees and bouquets when large amounts of seed are needed from a pollen sterile female, such as 'J. H. Hale'. Erecting the cage may be less costly, and time saving, as compared to emasculation and pollination, which would be required to discourage visits by pollinating insects.

Pollination

Pollination is easily accomplished by a variety of methods: a camel's hair brush, the rubber tip of a pencil, a glass rod, or a finger. All readily show the pollen load. After picking up the pollen, the stigmatic surface is touched; pollen transfer is usually visually evident. Pollen remaining on the applicator after pollination is killed by rinsing in 70% alcohol.

If a record of set is desired, the pistils pollinated should be counted, but pollination (or counting) should be delayed one day or more after emasculation so that injured pistils will be evident.

The presence of stigmatic fluid indicates receptivity. However, pollination immediately after emasculation is effective. Pistils are usually functional for from four to seven days following emasculation. Cool, humid weather favors longer receptive time but hot drying winds and strong sunlight can dessicate pistils. Experience has forcefully emphasized the futility of pollinating when day-time temperatures do not exceed 12–15 C for appreciable periods—about the same conditions necessary for bee flight. Though the temperature may rise the day following pollination, sets obtained are much reduced or nil. Sets considered satisfactory will range from a usual 10% to 40%, or rarely higher. Tests have shown that thinning flowers does not appreciably increase set.

Seed Handling

Seed from fruit ripening two to three weeks before 'Elberta' are normally capable of germination following ordinary after-ripening. The fruit should be ripe when harvested. The flesh should be removed soon after harvest and should not be allowed to rot or ferment (Haut and Gardner, 1934).

Seeds to be stored for later stratification should be air-dried and held under cool, dry conditions or at 0 to 5 C if to be held for many months. Such dry seed will retain its viability for two or three years or more. If seed out of the endocarp is to be stratified it can be removed before drying or storing. The easiest method to remove the seed from the endocarp is by applying pressure in the dorsal-ventral axis with a vise.

Stratification of peach pits (Hartmann and Kester, 1959) normally takes from 90 to 120 days, depending upon season and the inherent rest requirement of the seed parent.

Following stratification, the seed may be planted in a nursery, or in flats or pots in the greenhouse. Removing the endocarp, if it has not split naturally, may increase germination. Sterilized sand is a better media than soil as it eliminates most fungus problems.

Seed to be used for some shortened breeding-cycle methods or for embryo culture should never be allowed to dry.

Stratification and Germination

Hybrid seed are produced at considerable expense and effort. A high germination is desirable, and may be critical to avoid possible bias in genetic studies (Lesley, 1939). Thus, most peach breeders use a modified stratification method. A simple one is described:

About 25 seeds are placed in a 250 ml Erlenmeyer flask, to cover the flask bottom in a single layer, and soaked overnight in water to which has been added a nonphytotoxic fungistat (e.g., Seresan 1 g/liter). The next day the water is decanted until that remaining does not cover the seed; the fungistat slurry is left to maintain sterility. The flask is stoppered with non-absorbent cotton, film, or foil, and placed in the cold room at 2–4 C. Water is added as necessary. When the rest requirement is met germination starts. Although this method avoids the necessity of using aseptic techniques, as in the method of Gilmore (1950), we recommend that sterilized glassware be used.

At the temperature recommended, the seed will start germination in the flask. If all the seed germinate at one time, they can be removed for planting when the radicles are .5 to 1 cm long. An advantage of the method is that if germination is erratic those seeds germinating first can be removed, and the remainder returned to the cold room for additional stratification.

If stratification is timed to give germination in the spring, the seed may be planted directly in the nursery row. Place the radicle down at a depth of about 5 cm. To obtain more assured emergence and better early growth, plant the seed in sterile sand or potting mixture in flats or individual containers, and grow the seedlings to the five- to seven-leaf stage in a protected place, as a glasshouse, plastic house, or cold frame. They can then be transplanted to the nursery or directly to the orchard site.

If there is any question regarding the germinability of the seed, it is best to remove the seed coats, and plant the seed in sterile sand, with about half of the cotyledons exposed to continuous light.

Shortened Breeding Cycles

Because of the long generation life of trees there has been much interest in shortening the breeding cycle of peaches. Significant shortening may require considerable expenditure of time and effort in early care, but some of the methods are efficient over the long run. Following normal seed stratification, a nursery year, and transplanting to an orchard site, the normal cycle to first significant fruiting is five years. The following methods shorten this period.

Several peach breeders use a system which reduces this cycle by one or two years, largely through the elimination of the nursery operation. It is useful with seeds with a normal potential for germination after drying. The seed are stratified (as above) and germinated in flats or individual containers somewhat earlier than would be normal if the next step were to be nursery planting. The resulting seedlings are then planted in the orchard site, by bare-rooting them if grown in flats at about the five- to seven-leaf stage. They are irrigated immediately after planting. Bare-rooted seedlings may lose their leaves, but nearly all recover. Planted at 1 to 1.5 m apart in the row, such plants will normally fruit following their first or second year in the orchard, or two to three years from making the cross.

A modification of this system, which is applicable to seeds whose embryos are from about the length of the testas at fruit maturity through midsummer varieties, has been described by Taylor (1957). The seed are removed immediately upon harvest, the testas removed and, without drying, are planted as described above with about half of the cotyledonus exposed. For seed from later-maturing cultivars it may be necessary to soak the seed overnight in water to aid in removing the testas. The planted seed are placed under mist, and, especially for seed from earlier-maturing mother plants, continuous light is provided. If the plants are forced by continuous light, occasional feeding with half-strength Hoagland's solution or other nutrient mixtures and application of bottom heat, the seedling will reach considerable height before rosetting. Rosetting eventually occurs with unstratified seed.

Resulting seedlings can be handled in either of two ways: 1) buds can be taken and placed in shoots of prepared rootstocks, or 2) the seedlings can be planted in a nursery or in the orchard site where they will fruit. If the first alternative is followed, the rootstocks are usually orchard trees which have been dehorned the previous dormant period to provide shoots for budding, or nursery seedlings. Several seedlings may be budded on shoots of a single dehorned tree. Budded nursery seedlings would normally be treated as dormant budded trees, and planted out in the winter following budding with no temporal advantage. Both methods require a great deal of effort, may not conserve space, and under option 1) the danger of transmitting virus to the new seedling is very real—as has been experienced! Nevertheless, using option 1, it is possible to secure fruiting in the second season from crossing from seed of early-maturing mothers, albeit restricted. Option 2 is very similar to that described earlier, but will accommodate seed from mother plants which are somewhat earlier-maturing.

Taylor (1957) noted that seed taken from late-maturing cultivars failed to germinate when planted immediately. Apparently seed of late-maturing cultivars have entered the rest period, which prevents germination without recourse to stratification.

Embryo Culture

The objective in embryo culture is to germinate seed too young to respond to the methods described above. Peach breeders have been vitally interested in embryo culture to germinate seed from early-maturing cultivars, so that both parents of a cross could be selected for early maturity.

Tukey (1933, 1934) first described embryo culture of *Prunus* seeds. Embryo culture involves the excision of embryos from the seed, after which they are cultured on .67% agar containing 2 to 4% sucrose, with or without nutrient salts and growth regulators. All manipulations are done under aseptic conditions. The fruit is surface sterilized with 5% phenol or sodium hypochlorite (NaOCl), the endocarp cracked, and the seed removed. The seed may be surface sterilized with 1% NaOCl for three to five minutes, or simply removed from the cracked endocarp with sterile tweezers and laid between layers of cheesecloth dampened with either of the sterilizing agents. Embryo excision is usually carried out in a transfer room, but a laboratory, adequately enclosed to exclude drafts and dust, will serve about as well.

The table top is usually washed with a mild disinfectant. A scalpel and tweezers are the only tools needed. The embryos are transferred to the agar culture tubes or bottles with a long forceps, and the tube plugged or stoppered. Bottles with screw caps are sometimes used. An autoclave is usually necessary for preparation of the culture media. Implements can be surface sterilized with 70% alcohol and flamed.

Tukey (1934) allowed the cultures to germinate immediately. Brooks and Hough (1958) stored the cultured embryos for two to four weeks at 4 C before allowing them to germinate. This vernalization treatment increased the percentage of germination and resulted in stronger growth. Lesley and Bonner (1952) stored the fruit for seven weeks at 2 C before culturing the embryos. This type of vernalization appears to be as effective as the method used by Brooks and Hough.

Immature embryos definitely require a source of energy, usually supplied by 2% sucrose, but later maturing embryos are inhibited by sucrose (Hesse and Kester, 1955).

A characteristic of all excised peach embryos is that germination starts immediately. There is no further development of the embryo regardless of treatment. Growth regulators and special nutrients, as amino acids, do not seem to enhance germination (Lesley and Bonner, 1952).

Most investigators have based embryo development upon days from full bloom until the fruit is picked, whether ripe or immature. Reports of successful embryo culture experiments indicate that cultivars ripening at about 70 to 75 days from full bloom may be successfully cultured. Because of variations in climatic conditions where the trees are grown, there may be marked differences in the time necessary for comparable embryo development. Also, the embryos of all cultivars do not reach the same stage of development following a given number of days from full bloom, even in one locality. Thus Brooks and Hough say, "in comparison with 'Mayflower' and some other early-ripening selections from some other progenies, the progenies of 'Jerseyland' x 'Mayflower' and 'Jerseyland' x N. J. 156 have precocious embryo development."

Hesse and Kester (1955) developed an index designated as PF_1, (embryo length/seed length) to measure comparative embryo development. Their results indicated that embryos with a PF_1 of approximately .70 were the least developed that could be successfully cultured. This is in agreement with all such values that can be deduced from evidence in the literature.

Embryos cultured without vernalization soon rosette following germination. According to Brooks and Hough (1958) and Lesley and Bonner (1952), those cultured following vernalization also rosette, but not as soon, and following more vigorous growth.

Rosetting can be partially overcome by placing the cultures or potted plants at 4 C for two or three weeks. If the germinated embryos have been removed from the culture tubes and planted, the rosetted terminal may be pinched out. The lateral bud forced by pinching will be normal. Gibberellin sprays (100 ppm) have been effective in overcoming rosetting of sweet cherries (Fogle, 1959).

Transfer of seedlings from the culture tubes to a soil substrate usually results in substantial loss of seedlings. A sterile potting mixture should be used, and the seedlings protected from drying, perhaps with intermittent mist. A gradual hardening will accustom them to normal glasshouse conditions, from which they can be transferred to the outside.

The difference in seed development from embryos that can be germinated directly in sand, and those that require embryo culture, is very small in terms of days difference in maturity of seed clones. In addition, by discovering those items that may be "precocious in development," as indicated by Brooks and Hough, greater gains can be made than in using less-developed seed parents. In summary, no embryo culture technique has given unequivocal, practical levels of germination of embryos too small to be handled by simpler methods. The method has not been as extensively and successfully used in developing hybrid population as was expected.

Further investigations need to be made using more modern culture techniques. Some of the important questions needing answers are: 1) can the embryo be stimulated to further development in culture, rather than immediate germination; 2) can growth regulators be used to stimulate radicle and plumule development sequentially; 3) will subjecting fruit on the tree to higher temperatures stimulate embryo development more than fruit development, thereby yielding embryos of greater development at fruit maturity, and 4) can the period of fruit development be delayed without slowing embryo development proportionally, also resulting in embryos more advanced at fruit maturity.

The third suggestion is inferentially supported by the observations of Sharpe and Sherman (personal communication) who have found in some years *normally germinable* embryos in items ripening in 70 to 90 days from full bloom. 'Springtime' embryos from New Jersey appeared to be more fully developed at fruit maturity than those of the same cultivar in California. Both Florida and New Jersey have higher average temperatures between bloom and maturity of these early-ripening cultivars than is the case in California.

Planting Systems

The best measure of efficiency of land use for evaluating seedling populations is the area occupied by the seedling tree under conditions which permit adequate, normal fruiting response multiplied by the time required to complete evaluation. A common method is to plant close in rows, usually at 1 to 2 m, with the rows from 3 to 4.5 m apart, depending upon available cultivation equipment to be used in land management. These spacings utilize 3 to 9 sq m per tree. This can be halved by planting two trees per location, and treating each tree as a separate scaffold of a very low-headed tree. As the latter complicates record-taking, single tree plantings are preferred. Some peach breeders plant their seedlings at very close distances—30 to 50 cm in rows. This probably delays fruiting of some of the items, but as fast as the more precocious do fruit and are discarded, the dense planting is quickly thinned. Under this method a seedling can be grown on as little as 1 sq m, although the average length of time needed for selection may be lengthened, and the spatial advantage reduced.

Sherman et al. (1973) have described a high-density planting system described as the "fruiting nursery." Seedlings with five to seven leaves, germinated and grown in flats in sterile soil in a greenhouse, were set at 13 cm in rows 1 m apart. The density was 7.7 plants/sq m. Intensive care was given both before and after planting; pre-planting soil fumigation, post-planting shading for two weeks, frequent irrigation during the growing season, approximately monthly fertilization, and adequate disease and pest control. Under the conditions prevailing in Florida, and because the seed cultivars were early maturing, germination of the seed and growth to planting height were attained in August or September. At that time six to eight weeks of additional growing weather occurs in Florida. More than 50% of the seedlings fruited in the second year in the fruiting nursery. In more temperate climates, and with mid-season and late-maturing seed cultivars, planting of the fruiting nursery would probably be delayed until the spring following hybridization.

Selection in the fruiting nursery emphasized a few more obvious characters and traits. Final evaluation was accomplished with propagules planted at more normal distance. Thus, while complete evaluation of selections may be delayed, some 95% to 98% of the seedlings could be discarded in the fruiting nursery. We can see but two objections to the use of this method. First, if sufficient fruit for processing evaluation is required, the density of even 5% of the original planting will be too close for the development of adequate bearing surface. Second, if quantitative genetic studies are contemplated the density will not allow normal growth and development for many of the traits which might be under study. For cultivar development only, this method appears highly efficient.

Topworking has been favored at times, and is essential to the quickened breeding cycle described by Taylor (1957). In spite of somewhat quicker fruiting—usually plentiful in the second leaf from grafting—it involves added work and expense in the care of the grafts. It also has the serious fault of transmitting any virus disorder the rootstock may have acquired.

Selection

Selection of individuals to be continued as candidates for introduction as new cultivars is largely subjective, based on the experience of the plant breeder and the characteristics of established cultivars against which the new selection must compete. If there is any rule that can be given it is, "when in doubt, discard!" Too much marginal material is often carried in breeding programs. Advancing in fruit breeding is a step-wise process; unless selections are firmly believed to be advances over already available material they serve little purpose, either as potential varieties or as parental material.

If selection is to be based on quantitative traits, adequate studies of trait expression are essential. No one individual will express in high degree all of the quantitative traits that characterize an improved cultivar, so some compromise is usually necessary. Unfortunately, sufficient knowledge of the breeding behavior of quantitative characters in the peach are not yet well enough understood to suggest rational methodology. A more thorough evaluation of correlations among such traits, combined with heritability studies, will be essential to knowledgeable planning. On the basis of the meager heritability studies now available, phenotype appears to be the best measure of the genotype. Hence, selection of one to a few individuals combining appropriate extremes of trait expression offer the best immediate solution to the selection of parental material for further genetic advance, even if not equal to commercial standards.

Selection for specific characters may entail still other criteria. Studies of disease resistance, for example, require adequate evaluation of host-pathogen relationships. Reliance upon natural epidemics may be both time-consuming and/or misleading. Thus appropriate discriminating tests must be devised and evaluated. Once adequate testing methods are developed the identification of suitable parental material is a first requisite, followed by hybridization and evaluation of populations.

Examples of investigations in some of these areas will be discussed in later sections.

Testing

Introduction of new cultivars is usually preceded by some kind of so-called testing procedure. For tree fruits, such tests are often superficial. It is not economic for either private or institutional plant breeders to conduct replicated tests that will insure, within the usual statistical probability limits, that a new clone will actually be superior to one already available either in productivity, longevity, handling and transport characteristics, processing characteristics, or finally, market acceptance. This is true even for programs designed to serve rather restricted areas of production.

New cultivars are accepted by commercial growers because of their promise of profitable production, either through exceptional productivity, adaptibility, or premium returns. Fortunately commercial growers are eager to make experimental plantings of promising selections. These can be very informative, for they can be authorized in such a manner as to test candidates under differing management and soil conditions, and sometimes even in rather markedly different climatic situations. While uncontrolled, in the statistical sense, they have proven quite adequate in a practical manner. The experience and knowledge of the plant breeder, the grower, and all others in the chain from production to marketing are brought to bear. Usually introductions are made following such tests, and such tests have proven to be efficient in terms of time and investment by all parties concerned. The plant breeder, because of his expertise and familiarity with possible weaknesses of the candidate clone as well as those of competing established cultivars, should make the final decision to introduce or not.

Record Collection for Seedling Evaluation

Selection of potential cultivars involves evaluation of many traits taken together, and is usually highly subjective. Most peach breeders take some notes on hybrid populations, or at least upon that part of such populations that meet minimal subjective requirements. Blake (1944) and Weinberger (personal communication), in populations designed for cultivar improvement, score only seedlings to be retained for further characterization. Data concerning such selected portions of a population will not serve for genetic studies, as they will constitute a biased sample.

The degree of record-taking varies from none, to simple notes on a few traits or characters, to fairly extensive descriptions. A common type of record is a ranking system whereby the several characters or traits the peach breeder feels are important are ranked. Often a 1 (poorest) to 10 (best) scale is used with an acceptable standard ranked 5 or 6.

Factors of importance are usually fruit size, fruit appearance and uniformity, freedom from protruding sutures or apices, firmness, ground color at maturity, flesh color, amount of red at the pit cavity or in the flesh, pubescence, etc. Measurement of some of these traits could also be made more or less objective.

Time of bloom, leafing, and maturity are often treated somewhat differently. Leaf and bloom notes usually are taken on a calendar date basis. Time of maturity may also be treated in this fashion, or in comparison by days, half weeks, or weeks before or after selected reference cultivars. Usually calendar dates are used, and transformations later made to other intervals. Commonly weeks before or after 'Elberta', a universal standard, are used in grouping data, or in reporting. 'Redhaven' is also used as the standard. French (1951) stated that the weekly interval gave the most consistent picture of several intervals tested, although he reported time of maturity in weekly intervals from a generalized full bloom date. Bailey and Hough (1959) found a half-week interval to best characterize populations ripening from about 2½ to 12 weeks before 'Elberta'.

Ranked or scaled records are basically descriptive. They serve as a basis for comparative evaluation, summation at the season's end, and in performance recall in following seasons. Rankings of this sort often have little genetical significance. Problems of scaling, difficulty of exact reproduction, seasonal variation, and, if recorded only for favored selections, bias tend to reduce their discriminant value. If truly discriminant they may be valuable, but only if notes are taken on all, or a suitably-sized random selection, of the population. Some of the problems encountered in the use of such data are amply exemplified by the reports of French (1951) mentioned later.

Many fruit breeders use printed cards for record-taking. The characters and traits to be measured or ranked are printed thereon, to be checked with minimal time and effort. Special mark-sensitive or punched cards for use with sorting implements or machines have been described by Fogle (1961). In the accumulation of large amounts of data, as might be required for quantitative population studies, such cards save most of the considerable time required to prepare the data for computer analysis.

For descriptive evaluation, then, the peach breeder may devise a system of note-taking suitable to his needs. For this purpose the system adopted should be relatively simple, quick, and capable of giving him comparative information in its interpretation. Actual selection after two or more seasons of observation on selected individuals will very likely eventually be highly subjective.

If the program objective is to develop fruit for processing, adequate facilities for these operations must be available. Commercial firms will often cooperate by processing a few advanced selections, but normally many more than are considered potential cultivars need to be processed to evaluate variability, uniqueness for specific characters, effects of maturity on processed quality, and other such factors.

Mensuration for Inheritance Studies

Measurement should be made as objective as possible for inheritance studies. It is seldom accomplished because of the time and effort needed to accumulate such data on sufficiently large samples of hybrid populations. If the character has to do with the fruit, it usually must be gathered just when the program leader is involved with the selection processes. Nevertheless, genetical studies most often require just such objective measurements if valid conclusions are to be drawn.

Two general types of data may be gathered: enumeration data on discrete characters, and measurements on quantitative, continuous traits. For unit characters, such as foliar gland type, or fruit flesh color, a single scoring that is both quick and easy to make is all that is involved. Seasonal variability generally is insignificant for such discrete characteristics. As ratios are normally tested by Chi-square statistics, both small and large populations can contribute to the analysis. In fact it is desirable to have several segregating populations available for the application of homogeneity tests, which can aid in detecting variant ratios much better than occasional aberrant ratios which may be due to chance.

Relatively few reports have treated dihybrid or trihybrid ratios. The reports of Bailey and French (1949) in detecting linkage in the peach and of Monet (1967) and Lammerts (1945) in establishing independent inheritance of two or more traits are exceptions. If available in the data, such analyses may be helpful. Dihybrid ratios are essential to detection of linkage between discrete characters. If linkage, particularly close linkage, is evident much larger progeny numbers will be needed to adequately analyze the data.

Phenological data, as mentioned in the previous section, is often taken on a calendar day basis. It may later be transformed to other convenient intervals. A good example is the report of Bailey and Hough (1959) on time of maturity, which was based on a half-week interval. We have found a convenient method is to give March 1 the arbitrary date of 11, and all other temporal data are then transformed to fit. In our location this avoids negative numbers for bloom data to a large extent, and the resulting figures can be handled as simple numbers.

For the particular purpose of evaluating chilling requirement Lesley (1944) and Lammerts (1945) devised systems for recording phenological data which are of interest. Lesley observed seven classes of reaction to mild, "test" winters. Each member of his hybrid populations was placed in one of these seven classes. The weakness of this method was the need for categorization following test winters only. Following colder winters he could detect only four or five classes. Lammerts depended upon a proportionality between the reaction of a standard clone, 'Lukens Honey', which showed low year-to-year variability in time of leafing, and other cultivars which had been characterized in relation to 'Lukens Honey' for two or more seasons. He could then place an unknown in its relative chilling requirement niche by comparison following any type of winter common to his area.

The many continuously varying quantitative traits the peach breeder would like to measure in a genetically meaningful manner present the greatest problems. Broad classes may be established and categorization made. This was the method used by Williams and Brown (1956) to study associated character and trait expression believed to have a genetic base, (see later). French (1951) attempted to determine the mode of inheritance of several quantitative traits, (fruit pubescence, flesh stringiness, flesh coarseness, stone size, juiciness, skin thickness, and skin toughness) by categorization into three to six classes. Fortunately he categorized these quantitative characters over two years, for all individuals in at least two large populations. The correlations between assigned categories for individuals in the two years, though significant for pubescence, stone size, and juiciness, were so low (range + .176 to + .384) that it is evident that little reliance can be placed upon the observations. Skin thickness and skin toughness gave non-significant correlation coefficients for the two years. Environmental effects and measurement error dominated the categorization. Flesh coarseness and flesh stringiness categorizations made in the two years were tested by contingency Chi-square statistics. For flesh toughness categorization in the two years were independent of each other. While flesh coarseness did reveal year-to-year association, French states: "[the data] reveal too many individuals of inconsistent record to justify further speculation on the genetic nature of this character." All of these quantitative factors were deemed to be polygenically controlled. Nevertheless, on the basis of distribution data, averages, and their standard deviations, French drew some conclusions regarding inheritance of some of these quantitative traits. Exactly the same difficulty is suspected to have resulted in the low heritabilities estimated by Hansche et al. (1972) for such ranked data.

The general method and theory used to estimate such discriminant parameters for continuously varying quantitative traits are discussed by Schultz (1955) and Schultz and Schneider (1955).

Randomization

Discontinuous characters seldom present problems in categorization. Randomization is therefore not required for accurate assessment. Occasionally for a character like pollen sterility readings over two years may be required to identify surely the phenotype. But for quantitative characters which are to be analyzed statistically, randomization is very desirable. That randomization of populations may complicate record-gathering is evident, but—as reference to any statistical text will emphasize—severe biases may be introduced by failing to randomize. Complete randomization of populations planted in any one year is best, but some simplification can be obtained by dividing populations into convenient sized groups, as five or ten individuals, and randomizing the planting location for these groups. Records over two years are also needed to assess seasonal variability and year-trait interaction.

Data recorded on the "first 50" or "first 100" individuals of a large hybrid population are not random, and may lead to considerable bias in the data, just as was suggested for gathering of quantitative data on selected individuals.

Cytology

The chromosome number of the peach is 2n = 16, x = 8 (Knowlton, 1924; Kobel, 1927; Darlington, 1928). Though other subgenera of *Prunus, Euprunus* and *Cerasus*, exhibit naturally occurring euploid series (2n = 16, 32, 48, and even higher numbers) exceptions to 2n = 16 are rare in the subgenus *Amygdalus*, and have been of no importance in the development of cultivars.

Dermen and Scott (1939) reported a triploid peach, *P. persica*, among temperature-treated seedlings. Dermen has also (1941, 1947*a*, 1947*b*) obtained triploid and tetraploid forms of standard clones by colchicine treatment. These forms have apparently not been investigated cytologically, except to confirm chromosome number either by direct observation or by inference from nuclear and cell size. These forms are highly sterile, as

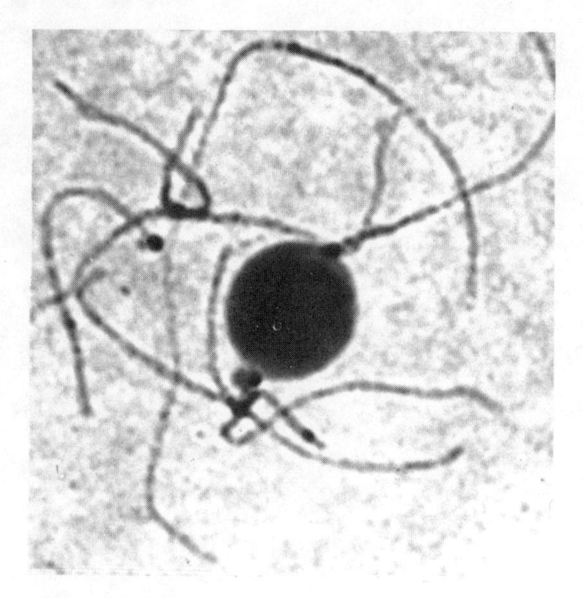

FIG. 2. The eight chromosome bivalents at pachytene of meiosis in *Prunus persica* 'Raritan Rose'. (Photograph courtesy of G. Jelenkovic and E. Harrington, 1972).

would be expected of triploids and of autotetraploids derived from basic diploids.

Pratassenja (1939) and Pratassenja and Trubitzana (1938) produced triploid peach seedlings by pollinating diploids with giant pollen grains obtained by selection under the microscope or by sieving methods. Dermen (1938) showed that diploid clones differed considerably in their tendency to form giant grains, some producing as many as 25% such grains.

Jelenkovic and Harrington (1972) have recently described the peach karyotype from the meiotic pachytene configurations of three clones (Fig. 2). Table 2 shows the pertinent data for the eight chromosomes. Besides length and arm rations, they describe the heterochromatin as proximal, and confined to small chromomeres. Chromomere size, number, and distribution between the two arms aids in distinguishing between chromosomes. Chromosome 5 is distinguished by a proximal knob, and in 'Raritan Rose' a second knob was present distally on the shorter arm on one homolog. Chromosomes 6 and 7 were both nucleolus organizing chromosomes. A variable number of nuclear bodies, described as micronucleoli-like bodies, were observed. Their number was variable, and possibly affected by environmental conditions. These bodies disappeared by MI. No genetic influence was suggested.

Meiosis in pollen sterile cultivars apparently proceeds normally until the tetrad stage (Knowlton, 1924) after which pollen grains do not form. The tetrads eventually shrivel or are resorbed. This failure may not be absolute, and under some environmental conditions, a very small amount of normal pollen may be formed by sterile genotypes.

Monoploid peaches have been obtained from seedling populations (Hesse, 1971). Toyama (personal communication) has estimated that approximately 0.1% of peach seeds will yield monoploids. The mechanism of their production has not been investigated, but Hesse suggested that they are of merogenous origin.

The eight chromosomes were randomly distributed at metaphase I. An appreciable percentage of the metaphase II spindles oriented in a parallel manner. In those cells the products of the very regular mitotic second division, came to lie closely adjacent at telophase II. Reconstituted nuclei formed which regularly contained the whole genome. As a result, dyad spores formed which were presumably isogenic since synapsis was too low to have been a factor in the genome of such dyads. The pollen grains formed were shown to be viable and capable of fertilizing bloom. This same process may account for the fairly high percentage of "giant" pollen grains found in diploid peaches by Dermen (1938) and Pratassenja and Trubitzana (1938), as well as other *Prunus* species. Toyama (personal communication) has doubled such monoploids by colchicine treatment. These should be isogenic and fully fertile, and hence of tremendous value in quantitative genetic studies.

TABLE 2. Pachytene chromosome lengths and arm lengths ratios of the peach, *P. persica* (from Jelenkovic and Harrington, 1972)

Chromosome	Total length in μ	Arm length ratio
1	65.5 ± 1.59	2.0 ± 0.48
2	45.7 ± 1.39	1.0 ± 0.47
3	41.7 ± 1.45	2.2 ± 0.63
4	38.4 ± 1.23	1.3 ± 0.17
5	36.1 ± 0.88	2.7 ± 0.68
6	34.9 ± 1.10	5.4 ± 1.25
7	33.3 ± 0.86	6.0 ± 1.58
8	31.2 ± 0.86	2.7 ± 0.74

Interspecific Hybridization

Table 3 indicates the large number of interspecific hybrids involving the peach. Crosses between the so-called peach species, *P. persica*, *P. davidiana*, *P. mira*, and *P. kansuensis* have been reported repeatedly. That between *P. persica* and *P. davidiana* is involved in the development of nematode resistant rootstocks. Hybrids between *P. persica* x *P. amygdalus* or the reciprocal have also been widely tested as rootstocks because of their vigor and compatibility with either peach or almond.

Meader and Blake (1939, 1940) have reported on F_1 and F_2 populations of *P. persica* x *P. kansuensis*. The F_1 plants were vigorous and somewhat intermediate. F_2s showed various segregations for some of the distinguishing characters, which are discussed in the section on inheritance.

Armstrong (1957) studied hybrid populations of *P. persica* x *P. amygdalus*, including F_1s, F_2s, and BCs, to both parental species. First generation hybrids are notably fertile in this combination, as they also appeared to be in the crosses between *P. persica* x *P. kansuensis* reported elsewhere. However, F_2s were very variable, and were characterized by many morphological abnormalities leading to sterility. This behavior is in strong contrast to the other wide crosses involving *P. persica*, in which such descriptions as are available usually comment upon morphological disorientation and/or sterility of the F_1 hybrid itself. The explanation is probably that *P. persica* and *P. amygdalus* (and *P. kansuensis*) are so nearly equivalent genomically that physiological functions are quite normal—in fact, hybrid vigor is apparent. Pairing at meiosis must also be quite regular or considerable megaspore failure would result. But in the F_2s, abnormalities appear due to chromosomal segments no longer containing all of the alleles needed for normal development. Backcrosses to either parent minimize these deficiencies, and it seems probable that with two or three generations of backcrossing, fertility would be completely restored. Because of this, it has been of interest to attempt to transfer the self-fertility gene of the peach into the almond genome to obviate the necessity for interplanting almond cultivars for commercial production.

TABLE 3. Interspecific hybrids with *P. persica* (from Knight, 1969)

P. amygdalus Batsch	x *P. persica* (L.) Batsch
	x (*P. amygdalus* x *P. davidiana* (Carr.) Franch.)
P. armeniaca L.	x *P. davidiana*
	x *P. persica*
P. besseyi Bailey	x *P. persica*
P. cerasus L.	x *P. persica*
P. hortulana Bailey	x *P. persica*
P. nana Stokes	x *P. persica*
P. persica	x *P. amygdalus*
	x *P. davidiana*
	x *P. cerasifera var. divaricata* (Ledeb.) Bailey
	x cherry (sps?)
	x *P. kansuensis* Rehd.
	x *P. mira* Koehne
	x *P. nana*
	x *p. besseyi* Bailey
	x *P. salicina* Lindl.
	x *P. spinosa* L.
P. salicina	x *P. persica*
P. spinosa	x *P. persica*
P. tenella Batsch	x *P. davidiana*
	x *P. persica*

Russian investigators have been particularly interested in interspecific hybridization among species of *Prunus* in order to develop extremely winter-hardy types with desirable fruiting characteristics. Examples of such hybrids with the peach are those with *P. besseyi* Bailey, *P. nana* Stokes (= *P. tenella* Batsch), *P. davidiana*, and *P. spinosa* L. The peach has also been crossed with unspecified cherry, apricot, and plum species.

The number and repetition of the reports strongly suggests that all of the hybrids listed probably have been obtained. In wide crosses, as between peach and cherry or plum, matrocliny and patrocliny is often indicated. This can indicate that unfertilized gametes were induced to develop under the stimulation of interspecific crossing, and hence the reported hybrids, or some of them, may have not been truly hybrid.

Jakovlev (1946) investigated several purported hybrids between *P. spinosa* ($2n = 32$) x *P. persica* ($2n = 16$). Among the progeny he found ploidy levels of $2n = 32$, 24, and one plant with 16. The seedlings with $2n = 32$ resembled *P. spinosa*. Those with $2n = 24$ were variable, some resembling one parent, some the other, and some still other *Prunus* species. The individual with $2n = 16$ was considered to be a true hybrid, and not a haploid or a product of androgenesis.

One wishes for more cytological studies on these peach hybrids. They could afford considerable information on the comparative morphology of the *Prunus* karyotype, assist in resolving taxonomic problems, and give some insight as to the potential of distant hybrids in *Prunus*. To now, however, interspecific hybridization has been unimportant to the development of new peach cultivars.

Breeding Systems

The peach is said to be a highly heterozygous species. This condition is substantiated by the polymorphy observed for many characters, and by the fact that the peach is usually considered to be a naturally out-crossing species. The latter conclusion has not been demonstrated, although one certainly can assume at least a low to moderate percentage of outcrossing, sufficient to maintain heterozygosity. This heterozygosity is considered desirable. It is the basis for variation among seedlings, and hence the foundation for selection.

Heterozygosity is sustained by two methods: 1) crossing unrelated individuals, sometimes followed by self-pollination of selected F_1s to secure extreme segregants, and 2) selection for heterozygous polygenic systems which yield desirable forms. The expression of traits and characteristics afforded by this second system may be related to homeostatic balance—or "fitness"—as measured for cultural and market requirements.

Outcrossing

Several factors encourage crossing between unrelated parents. The earlier sustained efforts in peach breeding by the Massachusetts and New Jersey Agricultural Experiment Stations exemplify some of these factors. The immediate goal was elucidation of the method of inheritance of the several polymorphic traits readily observable in the cultivars of that time—foliar gland types, flower size and types, fruit flesh color, etc. Selection of parents in these initial efforts relied on clones which were more or less unrelated, but which exhibited the contrasting characters to be studied. Later breeding programs undoubtedly also have relied upon selection of relatively unrelated parents for much the same reason, although selection of parental forms may have been upon particular quantitative traits which, if combined, would be deemed favorable for ultimate commercial exploitation. For example, 'Early Crawford' was often used because of its superior flavor quality.

A second example of conscious outcrossing to incorporate desirable qualities was the program of Professor L. D. Davis, University of California, Davis, to incorporate higher processed fruit flavor in canning clingstone cultivars. Initial rounds of crossing involved hybridization of identified sources of "quality" with so-called conventional canning clingstones. Later, selections of improved quality derived from several such hybrid sources, still unrelated, were combined by further hybridizing. That crosses between related selections were made later does not detract from the fact that the initial crosses were made between unrelated forms.

Other similar examples can be pointed to in the several breeding programs in southern locations designed to improve clones adapted to mild winter conditions. All have hybridized low chilling requirement types, as Honey and Peen-to or lines derived from them, crossed with more standard conventional cultivars (for example, Sharpe and Aitkin, 1971).

Finally, the development of disease or pest resistant cultivars or seed sources, as exemplified by the development of seed cultivars for root-knot nematode resistance, has relied upon crossing of unrelated parents. This has been done to either determine the mode of inheritance or to combine greater adaptability with such resistance (Weinberger et al., 1943; Sharpe et al., 1970). Some of these efforts have involved interspecific hybridization.

Possibly no peach breeder has given serious consideration to manipulation of the polygenic system. If we assume that commercially successful cultivars probably require well-balanced polygenic systems, outcrossing would be considered regressive in that it would tend to destroy such balance. But if desirable polygenic systems involve appreciable heterozygosity it will be maintained in the very act of selection. This will be true whether from an outcrossing or inbreeding system.

Thus, most peach breeding programs involving outcrossing initially resort often to some form of inbreeding, usually through the use of related selections for continued advance.

Inbreeding

Inbreeding is attained most quickly and in highest degree by continued selfing, and many peach breeding programs have relied heavily upon self pollination following original crosses between more or less unrelated parents. The pedigrees of many cultivars exemplify this approach; for example, we can cite the successful program of Stanley Johnston at South Haven, Michigan. The new Canadian introduction 'Harbelle' originated as a selfed seedling of 'Sunhaven', one of Johnston's already highly-inbred introductions.

A less obvious route to significant inbreeding has involved the use of previous selections in the continuing steps of cultivar improvement within programs. Such crosses may involve sibs, half-sibs, or individuals of other similar close relationship.

We have already noted the historical importance of 'Chinese Cling'. This clone appears to be a probable grandparent of 'Elberta', which is supposed to be the parent of 'J. H. Hale', a cultivar much used by earlier-day peach breeders. Other old cultivars which appear and reappear in early pedigrees are 'Early Crawford', 'Oldmixon', 'Admiral Dewey' and 'St. John'—either directly or in clones derived from them. Thus, unconsciously if not by design, much of the cultivar improvement carried on in the United States in this century traces back to a surprisingly few clones, and hence to a limited gene pool.

Lesley (1957) inbred peach lines for up to F_7. The F_1s were wide crosses, between standard clones and low-chilling individuals of Peen-to derivation. In F_4 to F_6 the inbred lines showed evidence for considerable homozygosity, but a large residuum of variability was still apparent, as perusal of his report indicates. In fact, some lines were still segregating for simple Mendelian factors. Several unique characters were described, as willowy growth, wavy-nonshowy petals, and rough-bark conditions not believed to be virus disorders.

Two of nine lines showed marked inbreeding depression, but two were quite vigorous. Crosses between the inbred lines at F_5 and F_6 often resulted in some heterosis, but it was not marked. As inbreeding depression is more characteristic of plants normally exhibiting a high degree of outcrossing, the lack of marked inbreeding depression in the peach may reflect its tendency to a high percentage of selfing. A corollary is that marked heterosis may be difficult to obtain in peach breeding.

Because of the apparent homozygosity within lines, Lesley proposed that new cultivars might be developed by crossing between such highly inbred selections. The seedlings of such crosses would presumably be so nearly homogeneous that variability among them would be insignificant. But when one considers the degree of inbreeding evident in some more recently introduced cultivars, as for 'Harbelle' noted above, the case of isogenicity still needs firm proof. One of the problems connected with Lesley's suggestion for the development of cultivars in this manner is the number of inbred lines needed to supply the seasonal and type diversity required by the peach industry. Four such lines would yield six cultivars *if* all showed equal and desirable specific combining ability. Diversity could be generated only if additional inbred lines were developed. The proposal is nevertheless intriguing.

Other Systems

Peach breeding is based on the selection of the individual vegetatively-propagated plant which is usually heterozygous. Therefore, the use of breeding systems other than those outlined above has been rare. A unique example of recurrent backcrossing is being followed by F. W. Anderson (personal communication) in the development of commercially-acceptable clones of brachytic dwarf nectarines. He has found it necessary to backcross selected brachytic dwarfs to high quality standard cultivars every second or third generation to appreciably improve fruit quality factors, particularly flavor.

Similar situations might well apply to future efforts to incorporate disease or pest resistance into acceptable commercial varieties, especially if the source of resistance is discovered in stock bearing fruit of totally unacceptable quality. Occasional backcrossing of derived forms which still exhibit a high degree of the sought resistance to susceptible quality cultivars may prove essential.

Breeding for Specific Characters

Characters Under Simple Allelic Control

Table 4 lists simply inherited characters described in the peach. Only one or two loci are involved. Some of the characters have been used as a basis for subspecies classification, as for example the S-genotype, referred to the subspecies *P. persica* var. *platycarpa* Dcnse. Most of the loci listed show independent inheritance (see linkage), and are readily categorized. But apparent confusions occur, and for this reason some of the characters are discussed briefly below.

The anthocyanless *anan* character described by Monet (1967) is obviously not identical with *ww* described by Lammerts (1945), for Monet pictures colored flowers among *anan* segregants. Monet clearly shows that *anan* is inherited independently of *Y/y*, *F/f*, and *Sh/sh*.

Brachytic dwarf, *dwdw*, first described in Mendelian terms by Lammerts (1945), is apparently affected by modifying genes, as a considerable range in plant height can be obtained.

The leaf color locus *Gr/gr* described by Blake (1937) is incompletely dominant. The heterozygotes usually can be categorized throughout the foliage season, and most certainly after the spring growth flush.

The status of foliar glands has been long in dispute. Some authors (Gregory, 1915; Hedrick, 1917) describe four gland classes: reniform, mixed (reniform and globose), globose, and eglandular. In our experience any plant exhibiting reniform glands should be so classed, even though some of the glands appear less characteristic. Gland size and number may be affected by developmental and seasonal conditions, by leaf position, tree vigor, etc. The character should be read on mid-shoot leaves of vigorous first-emerging shoots in the spring before the foliar glands dry and/or dehisce. The *E/e* locus, particularly the heterozygote, may interact with the *Wa/wa* locus to yield abnormal gland types and leaf margins (Bailey and French, 1933, 1949).

Ps/ps conditioning pollen fertility may not be absolute in its phenotypic expression. Knowlton (1924) noted irregularities in tetrad formation in 'J. H. Hale' and subsequent abortion of the microgometophytes. Some pollen occasionally may form in *psps* genotypes, although rarely. Observation over two or more seasons will normally establish the genotype of doubtful phenotypes. Pollen-sterile individuals which form a little normal-appearing pollen in some years, are never prolific pollen producers. In truly doubtful cases, only test crosses will reveal the actual genotype. It is not true that pollen sterility and yellow anthers are always associated.

The *Sh/sh* locus is of historical and genetic interest. The early reports on flower size emphasized three classes—large, medium, and small (Connors, 1920, 1923, 1929; Blake, 1931, 1940; Bailey and French, 1939). But it was Bailey and French (1941) who defined the "showy" and "nonshowy" flower types, later verified by Weinberger (1944). Hedrick published *The Peaches of New York* without this insight, so it is almost impossible, except in illustrated or clearly extreme examples, to determine the floral phenotype of many of the cultivars as described. This misunderstanding exists to the present, for in many plant patent descriptions the flower size is designated only by the word "medium." Further, to date I have found no reference to the fact (checked in verbal communications with other peach breeders) that in cultivars grown for commercial fruit production showy flowers are pink, whereas non-showy signifies red flowers. Lammerts (1945) described petal size-determining, incompletely dominant alleles, *L/l*, for the showy-flowered class. If these act independently of the *Sh/sh* locus, and if the simplest explanation is applied, at least six size classes should be discernable.

Lammerts (1945), working with showy forms, presented data indicating that flower color in *shsh* phenotypes is inherited independently of flower size. Williams (1972) has confirmed that in progeny derived from commercial x ornamental forms of the *shsh* genotype, flower color and doubleness are also independently inherited and he recovered red-colored *shsh* D_1D_1 phenotypes. 'Helen Borcher's is a *Sh*–pink-flowered phenotype. Thus, while available data indicate the loci for flower color, flower type, and doubleness are independently inherited, the association between flower type and color in commercial clones remains unexplained.

The *W/w* locus affecting flower pigmentation (Lammerts, 1945) was completely normal in its segregation pattern. Few commercial cultivars are

TABLE 4. Simply inherited characters of the peach

Character	Symbol	Reference[z]	Remarks
TREE			
With anthocyanins/anthocyanless	An/an	Monet, 1967	
Tall, normal/brachytic dwarf	Dw/dw	Lammerts, 1945	
Tall, normal/bushy	Bu_1/bu_1 Bu_2/bu_2	Lammerts, 1945	Bu_1 and Bu_2 are duplicate, independent factors.
Normal/albino	C/c	Bailey & French, 1941	
Resistant to *M. incognita*/susceptible	(Mi/mi)[y] (Mj_1/mj_1)[y]	Weinberger, et al, 1943; Sharpe, et al, 1970	
Resistant to *M. javanica*/susceptible	(Mj_2/mj_2)		Duplicate factors proposed but not fully established.
FOLIAGE			
Red leaf/green leaf	Gr/gr	Blake, 1937	Incomplete dominance
Smooth leaf margin/wavy leaf margin	Wa/wa	Scott & Cullinan, 1942	Recessive gives low tree vigor.
Glandular foliage/eglandular foliage	E/e	Connors, 1922; Blake & Connors, 1936	EE = reniform Ee = globose
FLOWER			
Pollen fertile/pollen sterile	Ps/ps	Connors, 1926; Blake & Connors, 1936; Scott & Weinberger, 1944	
Nonshowy/showy	Sh/sh	Bailey & French, 1942; Weinberger, 1944; Lammerts, 1945	
Large showy flowers/small showy flowers	L/l	Lammerts, 1945	Incomplete dominance
Colored/white	W/w	Lammerts, 1945	w^v = proposed third allele for variegated flower color, white is continuous phase.
Pink/red	R/r	Lammerts, 1945	
Dark pink/light pink	P/p	Lammerts, 1945	
Single/double	D_1/d_1	Lammerts, 1945	
Fewer extra petals/more extra petals	Dm_1/dm_1 Dm_2/dm_2	Lammerts, 1945	Dm_1 and Dm_2 are independent and additive
FRUIT			
White flesh/yellow flesh	Y/y	Connors, 1920	
Pubescent skin/glabrous	G/g	Blake, 1932	
Freestone/clingstone	F/f	Bailey & French, 1941, 1949	
Soft melting flesh/firm melting	St/st	Bailey & French, 1941, 1949	
Melting flesh/nonmelting flesh	M/m	Bailey & French, 1941, 1949	
Saucer shape/non-saucer	S/s	Lesley, 1939	

[z] Reference to earliest and major clarifying reports only are noted. Many later confirming reports have been published for some loci.
[y] Symbols suggested to identify genes; not proposed in reference given.

ww; e.g., 'La Niege' is *wwSh*- and 'Summer Snow' is *wwshsh*. The *ww* genotype appears to have a systemic effect, at least the fruit are anthocyaninless. Lammerts also proposed a third allele at this locus, w^v, to which he attributed the variegated flower color pattern found in the ornamental clone 'Peppermint Stick'. However, this was neither proved, nor its action exemplified in segregating populations.

The D_1/d_1, Dm_1/dm_1, and Dm_2/dm_2 loci controlling extra petal number was described by Lammerts (1945). Doubleness is not due to petaloidy. Inheritance of petal number has not been finally resolved; cultivars such as 'Helen Borcher', with 60 to 70 extra petals, remain unanalyzed and the factors affecting petaloidy in D_1/d_1 or Dm_1/dm_1 and Dm_2/dm_2 phenotypes have not been elucidated.

The Y/y locus deserves further study. Heterozygotes may be determinable in the fruit and are in the flower nectaries. Paper-white vs. cream-colored white-fleshed peaches are commonly notable.

Fruit maturity may play a part in phenotypic characterization.

The *G/g* "locus" (peaches homozygous gg are known as nectarines) deserves added study. Its pattern of inheritance is clear-cut. But Oberle and Nicholson (1953) and others have noted several associated (pleiotropic?) effects in nectarine cultivars and mutants. The fruit is smaller (Williams and Brown, 1956; Oberle and Nicholson, 1953). Soluble solids of nectarines, on the average, are considerably higher than for either the clone from which a mutant form was derived (Oberle and Nicholson, 1953) or sibling peaches in segregating populations (Hesse, unpublished). Nectarines exhibit different reactions to specific insect attacks and fungal infections than do peaches. The nectarine obviously arose as a mutation of peach. But no back mutation to peach has been found unless the nectarine was itself a "bud sport" of a peach. The absence of back mutation is a critical distinction between a point mutation and a deficiency. No such mutations have been observed in millions of trees of non-sported nectarines; i.e., nectarines of sexual origin. Unfortunately, cytological evidence for a deficiency is lacking. Thus, the hypothesis that nectarines are phenotypes for a minor, non-lethal deficiency, which is further characterized by including closely-linked genes or polygenes having effect on fruit size and resistance to certain diseases and pests, must be considered.

Most of the evidence supports the conclusion that flesh adhesion is determined by a single gene pair *F/f*, (Bailey and French, 1933) with complete dominance. However, among the various reports, particularly those concerned with dessert types, irregular segregations have been reported (Connors, 1920, 1923, 1929; Blake, 1940; Weinberger, 1944). Undoubtedly some of the confusion is due to the difficulty of categorizing phenotypes of early-maturing dessert types, a factor noted by Weinberger (1944). He presented his data for flesh adhesion under three categories: freestone, semi-cling, and cling. If one recalculates his data by grouping the semi-clings with the freestones, non-significant Chi-square values are obtained for 14 of 17 segregating crosses for which data are presented. A homogeniety test of 16 of the 17 crosses indicates that two of the aberrant ratios may be considered chance deviations, or perhaps due to a few errors in recording phenotypes. Only one of the 17 crosses, a 'Veteran' x self population, seems completely aberrant. It is noted that the excess individuals are in the clingstone class. Only test crosses of these individuals could determine if any are genotypically freestones.

The *S/s* locus determining the saucer-shaped or Peen-to fruit shape, described by Lesley (1939) is notable for the low germinability of *S-* seed. Because of this he suggested that seed germination might affect genetic ratios observed among viable progeny.

The allelic symbols (*Mi/mi*) and (Mj_1/mj_1), (Mj_2/mj_2) are proposed for the loci determining root-knot nematode resistance. While Weinberger et al. (1943) established the dominance for resistance to *Meloidogyne incognita*, no symbol was suggested and the mode of inheritance was not elucidated. Similarly, only a generalized scheme was presented by Sharpe et al. (1970) for the genes controlling resistance to *M. javanica*. The symbols suggested here indicate the specific organism to which the gene applies, and the duplicate nature of the genes controlling resistance to *M. javanica*. Biological strains of the organisms may be found which will require further modification of these symbols.

Within this group of loci, independent inheritance has been noted for all but *F/f*, *M/m*, *St/st*, and *C/c* (see below). Monet gave conclusive evidence for the independent inheritance of *An/an*, *Sh/sh* and *M/m*, and *Sh/sh*, *M/m* and *Y/y*. Lammerts (1945) states that the color and tree habit genes he reported are independently inherited, including the duplicate loci for extra petals, Dw_1 and Dw_2, and bushy tree habit, Bu_1/bu_1 and Bu_2/bu_2.

Finally, it is noted that from time to time authors have suggested, on observational evidence only, simple inheritance for a number of observed characters or traits without supplying data on which appropriate tests could be made such as by Lesley (1957) for pendulous growth and Willowleaf. These are not included in Table 4.

Multigenic Inheritance

Many traits are phenotypically variable, as red fruit skin color, or do not conform to simple genetic explanation, as fruit shape. Some such traits, as originally described, appeared to be relatively simply inherited, but later investigations have confirmed their complexity. They also tend to show marked interaction with environmental conditions. None of these have been adequately analyzed genetically, but several suggest interaction of a few genes, or main gene effects subject to modifying genes, gene interaction, and other such genetic phenomena. Some may prove to be subject to polygenic control, as will be discussed later. Bailey and French (1949) give a table similar to Table 4 for most of the traits discussed below. Comment on some traits, as time of maturity, chilling requirement, etc., are reserved for more complete discussion in a later section.

Connors (1923) reported that low fruit acidity was dominant to high acidity. This appears to be the case when extreme phenotypes are crossed, as in crosses involving honey-type peaches. 'Babcock', derived from a Peen-to hybrid shows the low acid character. Ryugo and Davis (1958) compared high and low acid clones from crosses of standard cultivars. Their results indicated that acidity in the fruit shows a continuous range of values, as did the investigations of Blake and Davidson (1941). Unquestionably, acidity is also modified by other genes, regardless of any major gene effects. Yoshida (1970) concluded that there was sufficient diversity in acidity among cultivars to be effectively used in breeding and selection. From a study of 60 hybrid populations he concluded that the sweet (low acid) type of peach resulted from cumulative dominant genes.

Connors (1923) reported that upright tree habit was incompletely dominant to a spreading tree habit. Little has been noted regarding this trait in recent work. It is possibly under multigenic control, but modern cultivars used in breeding programs appear to be relatively homogeneous for an intermediate type.

Blake (1932) noted differences in sucker growth between crosses, and concluded that vigorous sucker growth was dominant. As with tree shape, no further reports have been made to verify this analysis. The result is suspect because it was based mainly on an interspecific hybrid population.

Fruit bud set undoubtedly has a genetic component (Blake, 1933*a*, 1933*b*; Blake and Edgerton, 1946; Weinberger, 1944). Meader and Blake (1940) attempted to explain bud set on the basis of a simple gene effect with incomplete dominance in an F_2 progeny of *P. persica* x *P. kansuensis*. But no conclusive evidence has been put forward to support this or any other genetic explanation (see Weinberger, 1944).

P. kansuensis has glabrous fruit bud scales. In F_1 and F_2 progeny of *P. persica* x *P. kansuensis*, Meader and Blake (1939) suggest that the difference is due to a single pair of alleles, with incomplete dominance. One must be wary of genetic data from interspecific hybrids, as segregation may reflect dosage effects acting at the chromosomal level.

Connors (1923) and Blake (1934, 1940) decided that small fruit size was dominant to large fruit size. This may have been due to observations on crosses involving wide differences in inherent fruit size, as 'J. H. Hale' x 'Mexican Honey'. Size reflects volume, which is the product of height x width x depth. Genes controlling these linear parameters often have multiplicative rather than additive effects on size. Such data must be transformed (a logarithmic transformation is usually proper) before statistical analysis. Then the apparent dominance of small size is often shown to be invalid, but due to additive gene effects. In simpler terms, dominance of small size of fruit may be an optical illusion. Weinberger (1955) stated that the data "do not support the idea of large-fruit-size recessiveness. Fruit size appears to be controlled by multiple genes without complete dominance." Hansche et al. (1972) concluded that fruit size is probably an expression of additive gene effects.

Excluding the simply inherited saucer fruit shape, Blake (1934) and Blake and Connors (1936) concluded that all oval forms were dominant over round fruit shape. Later data (Blake, 1940) suggest a more complex pattern of inheritance. Weinberger (1944) stated that in the crosses he classified round form was predominant. This character is difficult to characterize, interacts with climate, and undoubtedly is controlled by multiple gene action.

Blake and Connors (1936) reported that heavy pubescence on the fruit was dominant to light pubescence, but later Blake (1940) presented data indicating that inheritance for this character was more complex. Weinberger (1944) points out that all gradations of pubescence may be found, and that his data did not suggest that heavy pubescence was dominant to light fruit pubescence. French (1951) studied fruit pubescence in self-pollinated and hybrid progeny. By means of a contingency table he secured very low though significant correlations between his ranked measurement of pubescence over two years in two large self-pollinated populations of 'Champion' and 'Belle'. He estimated that less than 5% of the variation in progeny was traceable to hereditary differences among them, and hence drew no conclusions about the method of inheritance of this trait. Multiple gene action is probably involved.

Red skin color contributes greatly to the attractiveness of the fruit. Blake (1932, 1938, 1940) and Weinberger (1944) have reported on the expression of this character, which is also difficult to categorize and interacts with environmental factors. Neither of these investigators could draw firm conclusions regarding the inheritance of this trait. We must conclude that it is controlled by multiple gene action.

Fruit flesh color involves the production of both carotenoids and anthocyanins. The Y allele inhibits carotenoid synthesis (see Table 4), but the presence or absence of anthocyanins in the fruit flesh or at the stone cavity is inherited in a complex pattern. Blake (1932) reported that red color around the pit was dominant over its absence, but gave no data. If true, this character is undoubtedly conditioned by modifying factors as degrees of redness at the cavity are readily observed. Classes which would normally be called free of red at the pit often have a light pinkish cast, but some are completely free of all anthocyanins. There may be a relatively simple major gene effect, but if so it is undoubtedly modified by additional gene action.

"Blood flesh" was reported by Blake (1932) to be dominant in an F_1 cross, but the F_2 gave equivocal segregation (Blake, 1937). This trait is apparently associated with time of maturity, early-ripening individuals expressing the character more fully, and is also environmentally conditioned. Expressitivity is greatest under low temperatures. This trait, like red color at the cavity, needs further study.

French (1951) studied the behavior of several complex traits in peach hybrid progeny: pubescence, flesh stringiness and coarseness, stone size, juiciness, skin thickness and toughness. Studies of these characters were based on contingency tables or distributions with from three to six classes depending upon the trait studied. For two large selfed populations the expression of the character over two successive years was correlated or tested by contingency Chi-square statistics. In most cases the correlation between classification in the two years was either significant at a very low level or non-significant. Thus, individuals were characterized with difficulty. Nevertheless, he drew conclusions regarding the tendency of some cultivars to transmit extremes of trait expression. French's findings for pubescence were reported earlier. Stringiness of the flesh was considered to be recessive to non-stringy on the basis of its appearance in hybrids between two non-stringy cultivars. A population of 'Champion' x self seedlings was noted to give more stringy individuals than selfed or hybrid populations of the three other cultivars used. Flesh coarseness gave similar results, except in this case 'Belle' appeared to transmit coarseness more than the other cultivars. French considered that a genetic basis for the juiciness trait was detected on the basis of distribution means and their standard deviations. 'Champion' tended to transmit this trait to more of its seedlings than did 'Belle'. Other differences were noted, such as that "the juiciness of 'Elberta' expresses phenotypic dominance over that of 'Belle'." But basically, the trait was mediated by polygenes.

The same conclusion was reached for the genetic control of stone size. However, he suggests that there was complete phenotypic dominance of 'Belle' (relatively large stone size) over 'Champion' (small stone size). 'Elberta' progeny had the largest average stone size.

Skin thickness was also very subject to environmental influences. Nevertheless, on the basis of the distribution frequencies observed, French concluded that 'Belle' offspring had thinner skin than offspring of 'Elberta', with 'Champion' progeny intermediate. The same conclusion was drawn for skin toughness, which is not surprising for the two skin traits were rather highly correlated, approximately + .55.

The major importance of environment upon these kinds of traits in the peach is particularly well demonstrated in these studies. They emphasize the need to estimate both environmental and genetic variability, so that only the proportion of the variability assignable to the genetic component is considered. Such traits could probably be more satisfactorily studied in crosses between cultivars or individuals exhibiting extreme phenotypes, even though the probability of securing valuable selections from such crosses would be essentially nil. At least it would maximize the genetic component, and make its statistical estimation more satisfactory. Cultivars are seldom good parents for this type of study simply because the requirements of commercial usage tend to channelize acceptable variation for such traits into narrow limits. F_1s, F_2s or further generations, and BCs based on crosses of extreme phenotypes, all scored or measured over two or more seasons, would be required to yield convincing data.

Inheritance of Some Complex Traits

Many characteristics of peach cultivars are obviously combinations of more or less inseparable components. Some of these traits are considered at this time.

Hardiness: Recurrent crop losses to low temperatures emphasize the need for hardier cultivars. The importance of these losses is reflected in the tremendous amount of research applied to hardiness problems—much too voluminous to be treated here. As Stushnoff (1972) has pointed out, observational and intuitive reports provide a basis of value for the plant breeder, but directed breeding for hardiness based upon the occurrence of "test" winters is both inefficient and may lead to conflicting conclusions. Stushnoff summarized the complexity of the problem as follows: "The components of acclimation and deacclimation can be broken down into the following: a) time of development of cold tolerance, b) rate of development of cold tolerance, c) intensity of tolerance developed, d) retention of cold tolerance, e) onset of loss of tolerance, f) rate of loss of tolerance, g) ability to regain tolerance."

Besides these components of hardiness the interaction of the plant with its environment is well known. Cullinan and Weinberger (1934), Mowry (1964*a*, *b*), Oberle (1957), Blake (1936), and others have reviewed the several environmental, developmental, and cultural conditions that may affect resistance to winter cold. Chaplin (1948) has noted differences in cold intensity needed to kill specific tissues of the peach tree at different times of the winter period, from pre-leaf-fall to bloom.

The ability of the peach tree to tolerate low temperatures without serious or fatal damage generally has not been of great concern. Campbell (1948) observed 32 cultivars after they had experienced −35.5 C on the night of January 3, 1946. All flower buds were killed, but most of the cultivars showed but medium wood injury, and 'Elberta' and 'Gage Elberta' but light injury. As 'Elberta' is usually considered to be less flower-bud-hardy than many of the other cultivars listed, it is evident that absolute hardiness of the whole plant is of less importance than is cold tolerance of the flower buds. Lantz and Maney (1941) report hardiness to early fall low temperatures in hybrids of *P. persica* x *P. davidiana*, but the temperatures were too high to be compared to the results observed by Campbell.

The most important limiting factor preventing peach production in more mid-continental and northern climates is lack of flower-bud hardiness, either to minimal winter temperatures during the dormant period, or to frost conditions from bud-swell to full bloom or shortly thereafter.

A number of methods has been developed to determine hardiness of peach flower buds. Perhaps the most effective has been the use of devices which lower bud temperatures of excised shoots at controlled rates (Cullinan and Weinberger, 1934; Meader and Blake, 1943; Proebsting and Fogle, 1956; Scott and Spangelo, 1964; and Weaver and Jackson, 1969). A number of other physical and chemical tests have been devised to determine killing temperatures (Siminovich et al., 1964; Van der Driessche, 1969; and see Weaver et al., 1969, and Weiser, 1970). These sophisticated methods for evaluation of freezing injury have, for the most part, been used but little with peaches, and not at all in the evaluation of hybrid populations.

Though a number of controlled freezing tests have been conducted on numerous cultivars to identify resistant clones, much of our information regarding cultivars and hybrid populations is derived from reports on bud-kill following natural freezes. Several reports (Lantz, 1948; Armstrong, 1950; Joley and Bradford, 1961; and Lamb and Watson, 1964) report on the bud-kill experienced by various cultivars in such test winters. The overall observations appear reliable in identifying cultivars of exceptional bud hardiness. 'Chili' has long been noted for its hardiness on observational grounds. Meader and Blake (1943) reported 'Cumberland', 'Belle', and 'Golden Jubilee' to be relatively bud-hardy among the varieties tested. Blake (1934) reported on the hardiness of 157 cultivars after test winters in New Jersey. Among 91 cultivars Mowry (1964a) noted 'Belle', 'Calora', 'Comanche', and 'Sunhaven' as producing full crops following the freeze conditions in the year his evaluations were made, following minima of −23.3 C on December 23, 1960; −25.0 C on January 28, 1961; and −7.8 C on April 2, 1961, when nearly all of the cultivars were in bloom.

By means of correlation coefficients Mowry was able to demonstrate differences in flower bud mortality among the 91 cultivars at each period of low temperature, but this also correlated with the chilling requirement of the cultivar. Thus, clones with high chilling requirements tended to have the least bud-kill. The crop following this series of low temperatures was, of course, determined by cumulative effects. Identification of hardiness was therefore not direct, and individual cultivars could only be evaluated on the basis of resulting bloom and crop.

Weaver and Jackson (1969) determined the bud-kill of several cultivars following controlled rate of reduction of flower bud temperature of samples taken on March 18 and 19, 1968. They found marked differences, ranging from 23.5% bud-kill in the most resistant cultivar 'Harrow Blood' to 89.2% bud-kill in 'Elberta' at −20 C, when the most effective rate of temperature decrease of 4.5°/hr was used. As −20 C is generally lower than that reported as a more or less common threshold for significant injury to flower buds so well advanced in the season, the authors interpreted the results to be more discriminant than other tests, or for results following test winters. The cultivar 'Envoy', which showed the highest bud-kill under natural conditions and was among the most sensitive to cold under their test conditions, produced approximately 75% of a full crop under the conditions observed by Mowry. This seems to emphasize the need for caution in developing criteria for maximum hardiness at any one stage of the winter period.

Oberle (1957) reported on tolerance to cold during the bloom period of cultivars maintained in a collection at Blacksburg, Virginia. During each of the years 1950, 1953, and 1955, severe frosts occurred during or after the bloom period. In the 1955 season all cultivars arrived at the late swell to bloom stages without dimunition through winter killing. Temperatures fell rapidly, and appeared to vary about ±1.7 C from −12.2 C in the vicinity of the orchard. Evaluation was based on the amount of thinning required to bring the resulting set to average commercial levels, and the yield obtained at harvest. On this basis, cultivars such as 'Jerseyland', 'Vedette', 'Ambergem', 'Cavalier' nectarine, 'Veteran', 'Belle', 'Shippers Red Late', and 'Rio Oso Gem' appeared most tolerant of these severe frost conditions.

Seedling populations were also observed by Oberle following the 1954 and 1955 frosts (1957). Thirty-five hybrid populations were ranked on a five-point scale (1 = full crop, 5 = no fruit) in 1955. In 1954 and 1955 the number of seedlings yielding full crops were recorded for an additional 29 populations. The distributions observed from the 35 hybrid populations, and the proportion of seedlings yielding full crops following the 1954 and 1955 frosts demonstrated the influence of a hardy parent in transmitting blossom bud hardiness to its seedlings. 'Cavalier' nectarine and 'Veteran' and VPI 13 peaches transmitted budhardiness: 'Redskin' transmitted bud tenderness.

Mowry (1964*b*) also reported on the inheritance of hardiness among seedling populations following five test winters between 1957 and 1964. In these test winters minimum temperatures varied, but always included minimums sufficiently low to cause some bud-kill. The lowest minimum was that described earlier, upon which the correlation studies were made. Each individual of a progeny was assigned to its appropriate 10% interval bud-kill class, yielding a distribution curve for each population. Mowry concluded that most progenies exhibited a normal distribution for hardiness with little evidence of discrete classes. All progenies exhibited some degree of transgressive segregation, as compared to the parental means. He concluded that dormant bud hardiness is inherited on a quantitative basis involving an undetermined number of genes with small additive effects. The greatest positive skewness observed, indicating greater average resistance to cold injury, was in a 'Redskin' x 'Ambergem' population. 'Ambergem' was noted as a hardy cultivar by Oberle, but 'Redskin' was among the less tolerant cultivars in the same observations. Of Mowry's selfed populations, only the 'Redskin' population was significantly skewed in the direction of greater hardiness, as were a third of the crossed progeny involving 'Redskin'. In each case the other parent in the combination was rated hardier than 'Redskin' itself. Of the normally distributed progeny involving 'Redskin', the other parents were less hardy than 'Redskin'.

Other cultivars which gave positively skewed progeny were some of the crosses of 'Boone County', 'Blake', 'Ranger', and 'Redhaven'—all clones which have been noted by others as being moderately or highly resistant to winter bud-kill under conditions of "test" winters. Thus, Mowry concluded that 'Redskin', a cultivar not particularly noted for its own hardiness, was prepotent "for transmitting bud hardiness." He concludes, "In the absence of definite information concerning the prepotency and combining ability of specific parents, a reasonable choice of parents for transmitting dormant peach flower bud hardiness can be based upon the mean bud hardiness of the parents used."

A more rational approach to the problem will require much new research. Stushnoff (1972) summarizes the current status well in the following remarks:

> Assuming that the breeder has identified one specific aspect of cold hardiness as a prime objective of his total program, the sequence of events to be considered are as follows: a) identify the primary and secondary problems and specific tissues involved; b) develop data on seasonal temperature tolerance response and relate to primary and secondary hardiness problems; c) develop an appropriate preconditioning regime; d) select appropriate artificial freezing technique, i.e., equipment, freezing rate, duration, thawing, etc.; e) determine most appropriate viability estimation technique; f) develop breeding systems on the basis of selection intensity, mating systems, source population improvement and juvenile selection procedures incorporating cultivars, selection, seedlings and unique germplasm.

Such a program seems beyond the capacity of most ongoing peach breeding programs designed primarily for the improvement of cultivars, despite the importance of hardiness in marginal producing areas.

Chilling requirement: The extension of peach and nectarine production into areas characterized by mild winters has been marked by notable successes from several fruit breeding programs. Most cultivars grown in the more important production regions of the United States and Canada require from 750 to 1,000 hours of chilling temperature below 7.2 C during the dormant period for them to bloom and foliate normally. Cultivars requiring but 200 to 600 hours chilling for normal spring development are required in many of the milder winter areas, as Florida, southern California, and the Gulf Coast states. Outside the United States, clones with even lower chilling requirements could be used in many areas.

The genes for low chilling have been derived primarily from peaches of south China origin. The Honey and Peen-to types have been most used.

Weinberger (1950) published on the chilling requirements of some 83 cultivars growing at Fort Valley, Georgia. Most of these were in the high chilling requirement range. Chilling requirement is sometimes stated in descriptions furnished by the *Register of New Fruit and Nut Varieties* (Brooks and Olmo, 1972), especially if low chilling is a trait of the cultivar. Descriptions of new cultivars from the southern states usually specify the chilling requirement.

As noted by Weinberger, and long recognized by those familiar with the response of peaches and nectarines to insufficient chilling, flower and foliage buds may have different chilling requirements. When different, the foliage buds nearly always have the greater requirement. For example, 'Elberta' flower buds require 850 hours of chilling, while the vegetative buds require 950 hours.

Estimation of chilling requirement has usually been done under conditions of test winters. As has been mentioned for other such complex traits, such types of measurement are highly inefficient, but more accurate determinations are usually both cumbersome and time consuming. A method commonly used to determine the chilling requirement of cultivars is to remove flowering shoots at intervals throughout the winter, usually after some given number of chilling hours have accumulated, and to force them in a warm glasshouse. The rate of emergence of flower and vegetative buds, and the percentage of vegetative buds that break readily is determined. If flowers and a high percentage of the vegetative buds break in seven to ten days, the chilling requirement is considered to have been satisfied. The work involved in testing numerous seedlings in this manner has been prohibitive.

Another common method has been to compare new cultivars or seedlings with the performance of cultivars of known chilling requirement. By proper selection of cultivars with different requirements, the unknown can be rather accurately assessed. Phenological data collected in test winters will provide such information. To more accurately assess chilling requirement under varying annual conditions, Lammerts (1941) has used an index derived from the difference between a given cultivar and 'Luken's Honey'. Seasonal differences were adjusted on the basis of the difference from 'Luken's Honey' in years of adequate chilling as compared to the test year. Lesley (1944) graded cultivars and seedlings on a seven-class scale on the basis of leafing response in test winters. In colder winters he was able to distinguish fewer classes.

Lammerts (1943) noted that in young seedling peach trees those with low chilling requirement were less influenced by short day treatment and lower growing temperatures than were those with higher chilling requirement. He suggested that, "Should this correlation prove general it would seem that hybrids could be accurately classified in the seedling stage with respect to their chilling requirements when mature trees." This suggestion has not been tested.

Chilling requirement is usually estimated on the basis of vegetative bud development. This has practical implications, for under marginal conditions trees may flower and set rather normally without sufficient foliation to develop the crop.

Breeding for low chilling requirement has been largely empirical, but Lesley (1944) and Lammerts (1945) have reported on segregation in hybrid populations. Lesley constructed distribution frequency charts, on the basis of his seven classes, following test winters.

Most of the hybrid populations displayed normal distributions, centering around the class of the mean of the parents used. In some crosses evidence of slightly skewed distributions was obtained, indicating the possible presence of one or a few genes with major gene effects. In all crosses variability was great, usually extending through most, if not all, of the seven classes. He therefore concluded that chilling requirement is based upon "the presence of multiple genes having a cumulative and more or less similar effect on the phenotype, and the absence of dominance." He also concluded that all of the cultivars or selections used in the crosses were highly heterozygous for the multiple genes.

Several inbred lines had been developed. Among these F_2 and F_3 populations still showed considerable variability. The few F_4 and F_5 lines were more nearly uniform, but the few numbers of individuals make conclusions tentative. He notes the occurrence of a 'Sims' bud sport limb which foliated considerably earlier than the cultivar—enough so to change its grade by at least one class. Such evidence would point to the presence of a gene with relatively large effect on chilling requirement.

Lammerts (1945) also graded several seedling populations for reaction to mild winters. In his populations he recognized, following a "test" winter, a class he designated "evergreen." These held much of their foliage through the winter. The number of segregants for this character suggested that it was controlled as a simple recessive character. It could be observed only following nearly frostless winters. Modifying factors were also involved, according to Lammerts, so that low chilling requirement, as with Lesley, was considered to depend upon multigenic control.

The results of these two investigations appear to be similar to those described for other complex quantitative traits. Multi- or polygenic control seems to predominate, but there is some evidence for the presence of one or a few genes having relatively major effects. Interaction with season makes analysis difficult. The development of methods to more accurately characterize chilling requirement (Weinberger, 1961) would undoubtedly improve the efficiency of genetic characterization for the trait. However, for all practical purposes, observational and comparative data appear to serve the plant breeder well in the development of new cultivars.

Time of bloom: In any growing area losses to spring frosts often occur during the bloom period. Hence, later bloom is desirable. In spite of its importance we know little about the genetic components relating to this trait. Time of bloom is conditioned by chilling requirement in mild winter climates. Though the order of bloom of cultivars is constant from season to season, it is almost always a reflection of chilling requirement (Weinberger, 1944). In cold winter climates time of bloom may reflect growing temperatures. Seasonal variability in expression of this trait is also great, so that comparisons are at best but suggestive. Within a season, weather sequences may materially affect the timing between cultivars.

Connors (1920) presented distributions of first bloom for several hybrid populations. The distributions noted were, for the most part, similar to those for such traits as bud hardiness or time of maturity. Little was then known of the chilling requirement for peaches, but an inspection of Connors' distributions indicate the importance of this factor. The most notable abnormality in his distribution curves was for those which must have involved pollen sterility. In these populations, as Connors notes (1922), "'Elberta' and 'Belle' selfed gave seedlings blossoming as much as a week after the parents, but a number of these late blossoming seedlings proved to be sterile."

Yarnell (1945) working in the mild winter conditions of central Texas, considered that time of bloom of peaches was dependent upon their requirements for both heat and cold. By identifying adaptability of cultivars and seedlings to mild climates on the basis of three rather broad groups, he was able to identify three seedlings which showed high adaptability but late bloom, whereas adapted parental cultivars bloomed much earlier. He therefore concluded that, "It should be possible to combine factors for adaptability to a mild climate and for considerable heat for blooming with factors for other desirable characteristics."

It appears that the genetic components for absolute time of bloom have not been fully identified. Due to interaction among the variables of cold and heat requirements for normal bloom, seasonal effects, and associated trait expression conditioned by other genes, time of bloom is extremely complex even though cultivars maintain a rather ordered progression of bloom in any given locality. It will be necessary to partition the variability found in hybrid populations among the components known or suspected to affect time of bloom to gain an estimate of the actual genetic component regulating time of bloom, all other factors being equal. Until then, selection for favorable time of bloom will be at best an empirically arrived-at compromise among the several variables known to have an effect on this trait.

Time of maturity: Possibly more observations have been made on this trait than on any other complex investigated by peach breeders. The commercial value of extreme cultivars, particularly very early maturing ones, has given emphasis to such studies.

Much of the data presented by various investigators can be well exemplified by those of French (1951), Weinberger (1944), Hansche et al. (1972), and Bailey and Hough (1959). Therefore, these particular reports are discussed in connection with the theories which have been put forward regarding the inheritance of this trait. Weinberger presented only data on distribution of time of maturity in populations, either from self- or cross pollination. French presented distributional information on populations resulting from self-pollination and from hybridization. He also analyzed such populations on the basis of their average time of maturity, with their standard deviations as a measure of variability. Bailey and Hough illustrated progeny distributions and presented their data as ratio tables, with Chi-square analyses. Hansche et al. discussed time of maturity on the basis of heritability.

The measurement of time of maturity, though perhaps not as difficult as measuring hardiness or disease resistance, presents some problems, usually associated with the marked interaction between season and time of maturity well recognized by all fruit growers. Normally cultivars, and therefore presumably seedlings, will retain their order of maturity. Only interaction with other decisive factors, as chilling requirement, can upset this sequential order. However, some seasons are earlier than others, some seasons are shorter or longer than average from the time of maturity of the earliest to latest cultivar involved in the observations, and occasionally, as indicated, some cultivars or seedlings may change in order of maturity. French measured time of maturity as weeks from full bloom, the latter being a generalized date—not that specific for each cultivar or seedling involved in his studies. Under proper conditions, as were probably the case in his studies, time of bloom can be very short over all cultivars, and not change the resulting estimate of time from bloom to ripeness greatly.

Weinberger, and Bailey and Hough used a time period, based on earlier or later ripening than 'Elberta' as their measurement criterion. Bailey and Hough used a half-week interval, Weinberger a weekly interval. Hansche et al. used a seasonally corrected day unit. It need only be mentioned here that a scaling effect may be involved. Thus, at the early end of the peach season, any stated interval may have a more meaningful impact on time of maturity than the same interval at the end of the season. No one has investigated this possibility, but it seems inevitable. Thus, increasing or decreasing the time of maturity by one or two days at the early end, 70 to 80 days from bloom, would represent a greater percentage change, and probably an even more significant physiological change, than a similar change in maturity time after 200 days from bloom.

French based his conclusions on eight self-pollinated populations and nine cross-pollinated populations, the latter including seven of the cultivars used in the selfed populations, but not necessarily among themselves. He analyzed distributions for years which did not differ greatly in time of bloom or length of time to maturity. His environmental control, therefore, was largely, though fortuitously, of the seasonal conditions. In comparing observations taken in successive years, the mean ripening time of but one of seven available comparisons demonstrated a significant difference. The time of maturity of the individuals of two large F_2 populations in two years (1938 and 1939) correlated r = .85 and .68. While these values are relatively high for correlation coefficients, they indicate that but 72 and 46% of the ripening sequence variability observed was associated in the two years. Using an interval of one week, a slight discrepancy in assigning ripening dates, or a slight seasonal discrepancy, could have just such a marked effect on the distributions observed. Nevertheless the consistency of French's results testifies to their usefulness in drawing conclusions as to the genetic basis for this complex trait. Also, it should be pointed out that these observations over two-year periods give much greater validity to the distributions observed than do one-year observations of such complex, quantitative traits.

French then reported distributions for time of maturity of self- and cross-pollinated populations which showed either approximately normal, bimodal, or skewed distributions. Distributions that were normal showed high or low variability. The data were characterized further by their population averages and standard deviations.

The level of variability encountered in selfed progeny appeared to be transmitted, suggesting quantitative inheritance. That is, cultivars whose selfed progeny exhibited high variability, as measured by large standard deviations, also gave hybrid populations with high variability when crossed with other cultivars.

The bimodal and skewed distributions observed in some crosses suggested major genes for both earliness, as in 'Champion' crosses, or lateness, as from 'Gold Drop' crosses.

Weinberger's results were accumulated over five years. He did not speculate on the mode of inheritance of this trait, but commented upon the difference in variability exhibited in different crosses, much as outlined by French. An examination of the progeny results presented by Weinberger closely parallels those tabulated by French.

Bailey and Hough constructed distribution charts for five hybrid progeny ripening from three to eleven weeks before 'Elberta'. To minimize the seasonal difference in ripening time, the results obtained in a single season were used in the analyses, 1954 for four of the progeny, and 1953 for one. The most notable characteristic of the frequency graphs was that three exhibited bimodal distributions, one a slightly negative (later) skewness, and the fifth a broad scattering of ripening times with no marked mode. The authors present a theoretical scheme to account for the results. It involved nine "major or dominant" and ten modifying genes. Tentative genotypes involving these 19 loci were assigned to each of the six parents involved in the five hybrid populations. This most unusual analysis involves sufficient loci to meet the usual multigenic explanation for such quantitative traits, or enough to meet French's suggested polygenic hypothesis, with some major gene effects. Bailey and Hough assign dominance, epistasis, interaction, and linkage effects to specific genes of their scheme. It remains a highly theoretical, but intriguing, explanation.

Hansche et al. reported an analysis for several quantitative traits of the peach on the basis of heritabilities derived from midparent-offspring correlations. Ripe dates for individuals were on a daily interval basis, counting from March 1 as day 11. Measurements taken in two years were considered, and the seasonal variation, which was considerable, was adjusted through statistical techniques. By this technique data on parents and offspring over a period of eight years could be used. An heritability estimate for ripe date of .84, the highest among the several quantitative traits studied, was found. This is extremely high in comparisons with heritabilities normally reported for such traits in agronomic crops, for example. The conclusion was that for ripe date, as for other traits showing relatively high heritabilities, gene action is largely additive.

But there is some evidence that major genes affect time of maturity. Many spontaneously occurring "sports" of cultivars vary five to 14 days in time of maturity from their "mother" clone. It is usually assumed that such sports are point mutations.

Though the bulk of the evidence favors additive polygenic control, there is evidence for genes having relatively large effects. But as abnormal distributions tend to be most pronounced when early and late maturing parents are used, some reservations relating to scaling effects in measurement of time of maturity may be appropriate. Bimodal distributions from selfed populations, as for Weinberger's 'Cumberland' self population, from crosses, and from skewed distributions, also offer concrete evidence for some heritable effects due to genes having relatively large effects. Withall, in selecting parents to yield populations whose average ripening time will be at a chosen period in the maturity sequence, the average of the parental ripening times, as suggested by Connors in 1923, still remains the best indicator on which to base such a choice.

Associated Characteristics

A number of simply inherited characters are associated with other observable traits. Associated conditions can be referred to four types: 1) systemic (in which a single locus controls some character in all plant parts where it might be expressed even though the locus was described for and refers to but one of these parts); 2) correlated (a statistical measure of the tendency for quantitative traits to vary together); 3) pleiotropic (effects, usually quantitative or developmental, associated with the allelic state at a locus, but which are completely unrelated to the major gene expression); and 4) linkage (caused by major genes being located on the same chromosome at less than independent cross-over distances).

Observed associations are: fruit flesh color and floral nectary color (Hedrick, 1917); fruit flesh color and foliage color (Connors, 1920); fruit flesh color and the color of senescent leaves (HofMann, 1940); pollen sterility and late blooming (Connors, 1923); saucer shape, early maturity, light fruit weight and poor seed germination (Lesley, 1939); eglandular foliage and susceptibility to mildew, and foliar glands and crenate leaf margins (Connors, 1920). Sherman et al. (1972) reported two foliage characters associated with early maturing cultivars and seedlings. One is a solid red color which is expressed only on old leaves in late fall, and the second is a variegated red pattern which is observable on senescent yellow leaves at any time during the season. Particular cultivars may express either or both of these characteristics. The characters are associated with early fruit maturity. Among the several cultivars examined, 'Redhaven' was the latest maturing to show one of the characters.

The fruit flesh-foliage-flower nectary associations are all systemic effects of a single gene, Y, which inhibits carotenoid production. The association between pollen sterility and late bloom is possibly one of the most clear-cut cases of pleiotropy found in the peach. Pollen sterile blossoms lack normal turgor and hence the physiological changes conditioning anthesis are delayed in development. The association noted by Lesley is undoubtedly a systemic effect associated with shape: flat fruit are of less mass than round or elongated fruit of the same diameter. The association between the presence of glands and mildew susceptibility is not so easily explained. Hesse and Griggs (1950) have shown that eglandular leaves have a distinctive "wetability" pattern, which undoubtedly affects the microclimate at the leaf surface. Even in glandular leaves, the intermediate condition of globose glanded clones was evident, and correlated well with their known susceptibility to mildew. This may be a pleiotropic response. *EE* and *Ee* crenate leaf margins have a mini-gland at each serration, which blunts the tip—hence, they are classified as crenate. This is evident in the serrate margins of weak basal leaves of shoots, which may not have glands even though the variety is clearly reniform or globose. The association is clearly systemic. The nature of the association described by Sherman et al. remains unexplained. Until the genetic basis for these foliage patterns is discovered it is impossible to assign this to any of the four association types.

Systemic associations may be more revealing than the difference for which the gene was originally described. In the case of the association between fruit flesh color and floral nectary color, the heterozygote Y/y may be distinguished by the nectary color (Connors, 1923), though it has not been shown to be possible in the fruit.

Linkage: Bailey and French (1933) reported the first case of true linkage in the peach. This involved the factors for clingstone and non-melting flesh, *ffmm*. Later (1941) they reported that the genes *F* (freestone), *M* (melting flesh), c (albino), and *St* (soft-melting flesh) are linked. Because of the limitations on segregation imposed by the alleles c (lethal) and the effect of *mm* on the expression of *F*, a rather elaborate analysis was required to determine the cross-over values (1949). The values determined indicate that linkage between these four genes is either:

F	M	C	St
0	15	50	60

or

F	M	St	C
0	15	45	55

The dominance relations between the gene pairs regulating flesh texture is $St > M > m$. In addition, there is interaction between the allele for freestone *F* and *mm*. *F* is either epistatic to *mm*, giving an *St* or *M* expression, or the combination *F-mm* is lethal. Only Blake (1937) has reported discovery of a non-melting freestone individual. The relationship between the *F/f*, *St/st*, and *M/m* loci certainly needs further investigation.

Bailey and Hough (1959) suggest very close linkage between one of their proposed factors, *D*, for early maturity and blood flesh. Also, a crossover value of 20% was suggested between their factors B_2 and *D*. The difficulty in this instance is that the genes for earliness are hypothetical and not proven by test crosses.

Correlation: True correlation is a statistical concept, which measures the association between two variables. Correlation studies with peaches have been few.

Mowry (1964a) studied hardiness of peach buds and opening flowers, using correlation techniques. Killing temperatures were experienced in late December, late January, and early April as the flowers were opening. In comparing cultivars, there was a correlation of 0.42 between hardiness in December and proportion of open flowers in April, a measure of freezing kill. At the time of the January freeze, this correlation was essentially 0. A significant correlation coefficient of −0.36 was found between the rated chilling requirement of the cultivars concerned and the proportion of flowers open on April 2. These results were interpreted to indicate that cultivars with high chilling requirements had less bud mortality, as measured in April, than those with less chilling requirement. The lack of correlation following the January freeze was because all cultivars were then out of the rest, and hence had lost the extreme hardiness conferred by the rest condition.

Phenotypic correlations were obtained for the association between 29 traits considered in the heritability studies of Hansche et al. (1972). Correlations found for which r ≥ .30 are shown in Table 5. These correlations were based on over 200 families and involved more than 2000 individuals in most instances, which accounts for the low standard errors.

Correlations between tree performance temporal traits or between alternative fruit measurements are high, as expected. Inspection of the table will show some correlations of considerable magnitude for which no simple explanation is evident, as the r = .80 between amount of bloom and breadth of the fruit base. Correlations between factors of appearance and quality are generally rather high, but they indicate that the rating of seedlings was based more on appearance (r =

TABLE 5. Phenotypic correlations among traits of peaches. Only r ≥ ±.30 shown (Standard errors of r values = ±.00+ to ±.06, most ±.02 to ±.04)

Trait	First bloom	First leaf	% bloom	Ripe date	Fruit length	Fruit cheek diam.	Fruit suture diam.	% drop	Fruit shape	Blush	Flavor	Appear-ance	Qual-ity	Rat-ing
Full bloom	.83	.61												
First leaf	.52													
% drop			.33											
Base			.80						−.32					
Fruit length				.41										
Fruit cheek diameter				.50	.88									
Fruit suture diameter				.44	.91	.96								
Blush				−.40				.30						
Acidity				.47										
Soluble solids					−.32		−.30							
Sweetness											.56		.59	.36
Appearance								.36		.44			.37	
Quality											.61			
Rating								.31		.30	.39	.67	.56	

Bloom ripe dates = calendar days from March 1 = day 11; measurement data in mm; soluble solids, refractometer reading; all others on a ranked scale: 1–5 for base, blush, acidity, 1–10 for flavor and appearance traits. (see Hansche, et al. 1972 for more precise definitions of traits).

TABLE 6. Heritability estimates for several traits in the peach (from Hansche et al., 1972)

Trait	h^2
Full bloom date	0.39
Amount of bloom	0.38
Ripe date	0.84
Crop	0.08
Fruit length	0.31
Fruit cheek diameter	0.26
Fruit suture diameter	0.29
Firmness of fruit	0.13
Fruit acidity	0.19
Fruit soluble solids	0.01

.67), whereas quality was judged primarily on sweetness (r = .59) and flavor (r = .61).

For highly correlated traits, as fruit axes measurements, measurement of one axis yields essentially the same information as obtained from multiple measurement. Such estimates also can aid the breeder by pointing to conflicting trends in relation to selection. Thus, the data indicate that quality is rather highly correlated with sweetness. On the other hand, soluble solids were negatively correlated with two of the three fruit size measurements. Thus, in selecting for high soluble solids one would need to take caution that fruit size was not unduly decreased.

Heritability: Heritability can be considered as that portion of observed variability due to heredity. Its estimation requires the partition of observed variability between gene effects and environmental effects. Traits of high heritability are subject to large genetic gain, under selection, per generation; those with low heritability may not be capable of significant advance through selection.

Such studies, common in agronomic crops, are sparse for tree crops. Hansche et al. (1972) have estimated heritabilities for several peach traits. The heritability estimates, h^2, are given for several of the traits studied in specific peach populations grown at Davis, California (Table 6).

Ripe date, adjusted for year affects, had an extremely high heritability of .84. This indicated that the population contained a great deal of additive genetic variability for this trait. Under high selection pressure, genetic advance in a chosen direction would be rapid.

Such characters as full bloom date, amount of bloom, and fruit suture diameter also had large heritabilities, whereas crop, fruit acidity, and fruit firmness and soluble solids had low heritabilities. It was noted that most of the latter characters were based on either crude estimates (ranks) or inadequate data, as for soluble solids. For some estimates, as crop (productivity), tree age might have affected the results. As pointed out by Hansche, some traits which would for other reasons seem to be genetically controlled, as soluble solids, may yield very low heritability estimates if not adequately measured.

It is therefore suggested that investigations of heritability in peaches be preceded by evaluation of measurements applied to characterize traits, and that only those that are adequate be used. We do not imply that adequately measured quantitative characters will invariably yield high heritability estimates. Rather, that good measurements are necessary for reliable estimates of heritability.

One of the major problems in securing adequate heritability estimates is estimation of seasonal effects on such traits as ripe date. Hansche et al. give references to statistically valid methods to minimize this effect.

The multiple- or polygene controlled traits accounting for continuous variation cannot be under such close breeding control as can genes for discrete characters. In advanced breeding programs, parental differences and variability in hybrid progeny may largely reside in multiple gene systems. Such genes appear to be largely additive in their effect. Therefore, selection of parents exhibiting a trait in the highest degree will best insure progeny with high average expression of that desired trait.

Polygenic systems: Polygenes are genes whose effects are too slight to be identified individually but which, through similar and supplementary (additive) action, can have important effects on total variability. Because of their number and slight effects, they are difficult to study. But some of the loci involved in the polygenic system may lie close to main genes, and hence be tightly linked to them. This association may be tested, though not proved. Part of the relationships discussed as associated characters are probably due to the action of such systems.

Williams and Brown (1956) categorized all cultivars described in two large pomologies and in the then published lists of the *Register of New*

TABLE 7. Distribution of phenotypes among named peach cultivars (Adapted from Williams and Brown, 1956)

Genotype[z]	Observed (n = 363)	Expected[y]
E-Sh-	215	214.5
E-shsh	128	128.5
eeSh-	12	12.5
eeshsh	8	7.5
		$X^2 = 0.0563$; df = 3
		P = >.99
Y-Sh-	132	151
Y-shsh	109	90
yySh-	95	76
yyshsh	27	46
		$X^2 = 18.99$; df = 3
		P = 0.001

[z] E- = *glandular,* ee = *eglandular foliage*
Sh- = *non-showy,* shsh = *showy flower type*
Y- = *white fleshed,* yy = *yellow fleshed fruit*

[y] Expected values calculated on basis of independent assortment of single gene frequencies observed; e.g., expected E-Sh- = all Sh- (227) x all E- (343)/total (363).

Fruit and Nut Varieties (see in Brooks and Olmo, 1972) according to phenotype controlled by certain major genes (*Sh/sh, G/g, Y/y, E/e,*—see Table 4) for order of bloom and maturity. The concept tested was that the simple gene effects should be randomly distributed (each according to its own overall frequency) among either cultivars exhibiting other allelic phenotypes or for the quantitative trait "time of maturity." There was no a *priori* reason why early maturing cultivars should tend to have white-fleshed fruits and/or showy flowers, or that yellow-fleshed clones should have nonshowy flowers. These presumably were not determinative characters in the naming of commercial cultivars grown for the supposed superiority of their fruit. Thus, if associations could be shown, they must have some deeper, more fundamental meaning.

Tables 7–9, modified from their report, are examples of the distributions they found. First they demonstrated that time of maturation was not randomly distributed in relation to time of bloom. Early blooming cultivars tended to ripen earlier than late blooming ones. They attributed this to the simple fact that early blooming cultivars had a longer time to ripen their fruit, and hence would ripen earlier, on the average. This would be a correlated effect. In Table 7 observed and expected frequencies for the association between glandular condition and flower type is shown to be randomly associated, but white-fleshed, showy-flowered and yellow-fleshed, nonshowy-flowered appear together in significantly greater frequency than would be expected by chance. A similar nonrandom association between flesh color and foliar gland type (not illustrated) was found, the white-eglandular (Y-ee) class being particularly deficient among named cultivars.

Table 8 shows the relation between the flesh color and flower type and time of maturation. Of particular note is the excess of white-fleshed and showy-flowered cultivars among the earliest maturing group. To a lesser degree, other similar discrepancies are detectable. Nonrandom distribution of phenotypes among maturation classes was also shown for flower type, flesh color, and in the case of foliar gland type, for the three genotypes. The joint distribution of pairs of these phenotypes was also shown to be distributed in a nonrandom fashion.

Table 9 demonstrates that the gene frequency for nectarines (gg) differs among the maturity groups, being greater in the middle maturity range, and lesser at the extremes. Finally, they report that the average weight of the peach cultivars was approximately twice that of the nectarines.

All of these differences, except the correlation between early maturity and early blooming, were attributed to *pleiotropism*. Thus, association be-

TABLE 8. Distribution of phenotypes for flesh color and flower type in four maturity classes (Adapted from Williams and Brown, 1956)

Order of maturation[z]	Genotype[y]			
	Y-Sh-	Y-shsh	yySh-	yyshsh
1–100	20	65	10	5
101–200	43	25	28	4
201–300	43	15	32	10
301–363	40	8	41	11
Expected[x]	41.6	23.15	22.65	12.6
$X^2 = 111.16$ P = 0.001				

[z] Sequence of maturity of observed cultivars from earliest to latest.

[y] Y- = *white fleshed,* yy = *yellow fleshed fruit*
Sh- = *non-showy,* shsh = *showy flower types*

[x] See footnote [y] Table 7 for calculation of expected distribution.

TABLE 9. Distribution of nectarines in four maturity classes, and the gene frequency for g in each maturity class (Adapted from Williams and Brown, 1956)

Order of maturation[z]	Genotype[y]		Gene frequency for g
	G—	gg	
1–100	97	3	0.17
101–200	85	15	0.39
201–300	88	12	0.35
301–363	98	2	0.14
Mean weight (g)	163.75	87.16	
	$t^x = 4.64$	P = .001	

[z] Sequence of maturity of cultivars: earliest to latest.
[y] G- = *peach*, gg = *nectarine*
[x] Significance of weight difference

tween superiority (as reflected by the fact that a selection had been given a name) of certain combinations of simply inherited genes inferred that peliotropic effects contributed to those often subtle differences which contribute to superiority. The conclusion reached by Williams and Brown does not exclude the possibility of linked polygenic systems accounting for the superiority of particular discrete phenotypes. If it can be shown that similar effects for discrete, simply inherited genes occur among individuals for which no superiority can be claimed, the case for pleiotropism is reduced. For this purpose, some quantitative character which is easily and accurately measurable is required. Hesse (unpublished data) explored this possibility by classifying mono- and dihybrid segregating populations for date of full bloom, according to segregating phenotypes or genotypes for five simply inherited characters.

The average date for full bloom for each phenotype within all segregating populations was calculated. Seven phenotypic comparisons were available: reniform vs. globose, globose vs. eglandular, and reniform vs. eglandular foliage; nonshowy vs. showy flower types; pollen-fertile vs. pollen-sterile; peach vs. nectarine; and freestone vs. clingstone flesh adhesion (Table 10, column 1). Positive values indicate that the dominant phenotype averaged earlier in day of full bloom than the recessive, and negative values that the dominant phenotype was later.

Such differences were calculated first for monohybrid segregations only, whether or not a particular population was segregating at more than one locus. The results from these monohybrid segregations are shown in Table 10, column 2. But two differences were significant: i.e., nonshowy-flowered phenotypes bloomed later, on the average, than did showy-flowered phenotypes, whereas pollen-fertile individuals bloomed earlier than pollen-sterile segregants.

It was obvious that many of the differences observed were confounded because the populations were segregating for two or more allelic states that might either reinforce or oppose the differences being tested. Therefore, a similar study was made of populations segregating for two simply inherited, and independently assorted, allelic states. The average bloom dates for dihybrid phenotypes, as pollen-sterile, freestone vs. pollen-fertile, clingstone, etc., were calculated. The differences found were treated statistically, as for the monohybrid segregations. The results are given in Table 10, column 3. From the dihybrid differences for single gene effects all but the G/g alleles showed a significant effect on time of bloom.

Earlier it was suggested that the later bloom of pollen-sterile individuals observed by Connors (1923) was a pleiotropic effect. The delay in bloom of *psps* genotypes is substantiated in this study.

Only Williams and Brown (1956) have suggested that pleiotropic effects are widely distributed among main gene effects. In the present instance four of five genes are shown to have significant effect on time of full bloom, only one of which we consider to be pleiotropic. By definition

TABLE 10. Average difference in full bloom of phenotypes or genotypes from mono- and dihybrid segregating populations

Genotypic comparison	Difference in days: From monohybrid segregations	From dihybrid segregations
EE/Ee[z]	−0.416	−0.811**
Ee/ee	−1.245	−0.783**
EE/ee	−1.767	−2.182**
Sh/shsh	−0.916**	−0.863**
Ps/psps	+1.273**	+1.441**
G/gg	−0.287	−0.191
F/ff	+1.177	+1.509**

** Significant at P = .01
[z] See Table 4 for phenotypes.

polygenes are widely distributed throughout the genome. Critical investigations with other organisms (see Mather and Jinks, 1971) show that polygenes tend to become associated with and closely linked to main genes as a part of overall "fitness." Therefore, just such effects as we found might well be expected of linked polygenic systems.

It is therefore concluded that such linked polygenic systems exist in the peach, and that they are the primary cause of differences such as those demonstrated by Williams and Brown (1956) and Hesse (unpublished). As a corollary it is postulated that they are associated with overall "fitness" of superior individuals, and hence selection will favor their future continuation and intensification. It is not necessary that "intensification" be synonymous with homozygosity—indeed, heterozygous linked polygenic systems are just as likely to occur.

To test this idea of "fitness," segregating peach populations were scanned for individuals which would result in extreme effect on time of full bloom, as *eeshshPs—F-* individuals, for early bloom time as compared to individuals which would yield more or less average times of full bloom. The examination was confined to populations showing trihybrid segregation to maximize the effect, thus the number of extreme individuals was small. But the survey clearly indicated that such extreme segregants exhibited both lighter bloom and lower productivity than phenotypes or genotypes that would show essentially average time of full bloom. It would be difficult to explain this phenomenon on the basis of pleiotropism.

It is therefore concluded that most trait expressions referred to pleiotropism are probably expressions of linked polygenic systems. Many, if not most, may be due to polygenic systems which, through selection for "fitness," are heterozygous. This may account for some of the heterozygosity observed in many rather highly-inbred peach breeding programs.

Bud-Sports and Chimeras

Mutations, commonly called bud-sports, are found frequently in peach. Most of these behave as simple one-step genetic changes in some major gene or quantitative character, such as showy flowers on an 'Elberta' tree (Weinberger, 1955) or an early-ripening form of a well-known cultivar, as 'Fisher'.

Rather remarkable are the several mutations' shifting time of ripening, five to 14 days earlier or later than the original variety. This suggests main gene effects controlling the time-of-ripening period. But I know of no breeding tests to confirm this.

Bud-sports of the kind just mentioned appear to involve the whole growing point of the mutant limb(s) and are stable when propagated. Other mutant forms, less stable, have long puzzled observers. Mutants were observed which would not transmit in breeding tests (Fogle and Dermen, 1969), and which reverted readily, or otherwise appeared unstable.

Chimeras, plants with genetically differing tissues, have long been known. Through a series of elegant reports, Dermen (1941, 1947*a*, 1953, 1954, 1956) has elucidated these chimeral mutants through careful analyses of cytochimeral mutants (those whose histogenic layers, see below, are distinguished by ploidy differences). It is noted that ploidy may affect growing rates, hence the observations of Dermen should be accepted with this reservation in mind. The peach growing point, in common with most woody plant meristems, involves three histogenic layers. Dermen designated these as LI, a single-layered epidermal layer; LII, a one- or two-cell thick layer just below the epidermal layer; and LIII, that part of the meristem below the LI and LII, and usually not regularly arranged. In normal development the LI histogenic layer gives rise to epidermal tissue only; the LII layer to subepidermal tissues, but more importantly to both the male and female sporogenous tissues. The LIII derivatives form the remainder of the shoot.

As mutations are rare nuclear events, never will all of the cells in the multi-cellular meristem—or even one histogenic layer—be derived from the original mutated cell. The usual organization of chimeras in woody shoots is therefore mericlinal, i.e., a portion of a particular histogenic layer is genetically different. In the latter case, a bud arising wholely in the mutated sector will be periclinal (a hand-in-glove arrangement of normal and mutated tissue), and it is probably this derived shoot that is first seen as a bud-sport.

This periclinal chimera is usually regularly reproduced by the growing point, and hence is stable. However, in derived organs, such as leaves and fruits, Fogle and Dermen (1969) have demonstrated that one histogenic layer may "invade" another, so that the relation to the histogenic meristem organization is much less orderly.

Unless the mutation is found in the LII layer it will have no significance in hybridization.

Shamel et al. (1932) found a mutant of the yellow-fleshed 'Sims' in which the flesh was white except for a thin yellow sector from the epidermis to the endocarp at the suture. Over a hundred seedlings resulting from self-pollination of this mutant were all completely white-fleshed (Hesse, unpublished). It was concluded that the LI layer was not involved in the mutation, and hence expressed the original yellow-flesh character, but that the LII layer was mutant, and hence transmitted the mutant character. 'Kirkman Gem', a late-ripening mutant of 'Rio Oso Gem' is characterized by premature ripening of the flesh along the suture, even to the endocarp. Again, the LI layer was not involved in the mutation. A relatively (more or less) fuzzless mutant of 'J. H. Hale' was investigated by Dermen (1956). He suggested that the LII layer had mutated to nectarine, with reversed dominance, and explained the short, relatively sparse pubescence of sectors of the fruit as due to a diffusable inhibitor which partially suppressed hair formation in the adjacent epidermal layers. Parenthetically, samples of seed of a similar but stable mutant, 'Royal Fay', yielded nectarines and peaches in varying proportions—perhaps because of invasion effects involving the LII and LIII histogenic layers. Dermen did not consider the potential results of somatic crossovers on histogenic layer expression.

Bud-sports have been a relatively prolific source of new cultivars. They can supply the peach breeder with distinctly different phenotypes which differ minimally genotypically. But as the discussion of chimeras demonstrates, the exact nature of the mutant form must be determined to be suitable for breeding usefulness.

Mutation Breeding

Induction of mutations, usually by X- or gamma-ray irradiation, has been termed mutation breeding. The object is to increase mutation rate over that observed naturally. Among the increased number it is also possible that mutant forms not yet observed in nature will be found.

Mutation is induced by exposing sporogenous or meristematic tissues to the mutagen. The tissue may be pollen or vegetative growing points. In peaches only radiation of growing points has been considered because of the necessity of producing second and third generation populations to detect the usually recessive mutants that might be produced in pollen.

The chance of securing mutants depends upon several factors—dosage, the tissue irradiated, and its physiological condition.

Hough and Weaver (1959) exposed year-old peach trees of several cultivars to chronic irradiation in the gamma field at the Brookhaven National Laboratory, New York. They describe mutant forms discovered, which included early- and late-ripening forms of 'Fairhaven' and 'Elberta', and a clingstone mutant of 'Brackett'. In the same mutant forms, changes in flesh firmness were observed. A particularly firm-fleshed variant of 'Elberta' was noted. One of the 'Fairhaven' mutants produced much more anthocyanin pigment(s) in the fruit flesh. The presence of multiple changes in several of these mutants suggests that some chromosomal modification may also have occurred, as point mutations involving discrete loci would not be expected to show multiple effects. These mutants, found among trees irradiated at the rate of 10 to 60 r per day for periods of eight or 20 months, were obtained from total radiation doses of 2,450 to 36,300 roentgens. Mutants observed on these irradiated trees appeared to be mericlinal chimeras; stability under propagation was not reported.

Hough et al. (1964) reported breaking the linkage between "blood" flesh color and early ripening with X-irradiation in dose ranges of 1000 r to 4000 r.

Though mutation breeding has not produced outstanding parental materials or cultivars for peach breeders, it must still be considered an important tool for the plant breeder. Notable results have been secured with other tree fruit species, as apples (color mutants), sweet cherries (self fertility, compact growth habit), and grapes (looser bunch development). A yet unproven use has been the induction of disease resistance in peaches, mentioned later (DeVay et al., 1965).

Much of the mutation breeding has been done by rather massive doses of the mutagen to vegetative tissues. Two aspects which seem not to have been studied adequately are the production of "micromutants" with lower dosages, and irradiation of pollen. The former may produce favorable, though slight, intensification of desirable traits which would be of value *per se* or in further breeding. Pollen irradiation requires the growing of second and third generation progenies to uncover recessive mutants, but may prove more useful in the long run in the development of useful variations.

Selection of Parents

The reports from the initial breeding programs at Massachusetts and New Jersey often included remarks on the "potency" of certain cultivars in contributing specific characteristics to their seedlings. Blake (1941) was particularly conscious of this phenomenon, best exemplified by his report (1932) which describes 'J. H. Hale' as a collection of recessive traits. Today we would suspect that such cultivars were relatively homozygous for multiple and polygenic systems controlling the traits. In combination with parents more heterozygous for genes affecting the same traits, the average phenotype would tend to resemble the noted parent. Under continued hybridization, using derived seedlings, such multiple gene systems soon became more heterozygous. Today such so-called "potency" is seldom mentioned. Cultivars which tend to contribute to some quantitative trait, as fruit size, are usually themselves of large size, as 'J. H. Hale' (Weinberger, 1944). Mowry (1964*a*) concludes that cultivars known to be hardy best transmit hardiness to offspring. Hansche et al. (1972) thus concludes that, "it appears that selection of parents on the basis of their own performance, and their mating *inter se*, should produce rapid gains in these (stone fruits) crops."

Progeny Sizes

The number of seedling required to yield a new cultivar may be quite different from the number required to secure valid genetic information.

For the former, assuming populations among which economic traits can be expected to be expressed at a reasonably high level in most of the progeny, from 2,500 to 5,000 seedlings will probably be needed to assure production of a selection worthy of naming. The higher number would be required to replace well established mid-season cultivars in major producing areas. On the other hand, if the peach breeder could find a combination which would yield seedlings ripening ten days earlier than known cultivars, a few hundred seedlings only would be needed to select one to meet minimal cultural requirements and still find an assured market.

If data are to be gathered for genetic studies, the particular system to be investigated will determine the number of progeny needed. Discrete gene differences under one- or two-gene pair control require relatively few individuals. If contrasting homozygous parental forms can be selected for the factor(s), only a single F_1 individual is needed, and only sufficient F_2 or BC individuals to meet the requirements of the Chi-square test, usually considered to be five individuals in the smallest class. Studies involving linkage tests will require larger populations. Estimates of variability for individual quantitative characters can usually be obtained from 30 to 50 individuals.

Because most cultivars or hybrid parental material are somewhat heterozygous, crosses between such plants will exhibit considerable variation. The hybrid population size should be sufficient to adequately sample this variation. For parents which vary in many loci the number of individuals required to fully explore all genotypes is prohibitive; e.g., parents differing in but ten allele pairs over 59,000 genotypes are possible in the F_2 generation. Therefore, while it is desirable to grow large populations one should remember that improvement necessarily will be a step-wise process. If trait expression is normally distributed, variation may be reasonably well sampled with populations between 100 to 200 seedlings, although larger populations would be desirable if economically feasible. Within the sample population individuals significantly removed from the population mean will be found and used as the base for future improvement. If the parents exhibit *contrasting* quantitative traits, the first cross between them may prove relatively uniform, and lesser numbers will provide the basis for selection for the development of subsequent more variable populations.

Heritability studies require numerous families preferably 100 or more, rather than large numbers within a family.

The efficiency of genetic studies can usually be improved by careful attention to discriminant measurement of the trait under consideration.

Disease and Pest Resistance

Only recently have peach breeders emphasized breeding for disease and pest resistance. This has been understandable in view of the success of pathologists and entomologists in developing adequate controls for many common disorders. But such controls involve millions of dollars expense to the peach grower through investment in equipment to apply controls and in the actual cost of materials. In addition, the current concern with the impact of exotic chemicals upon human and animal ecology is reducing the type and number of miracle compounds available for disease and pest control.

It seems inevitable that, from the viewpoints of economics and ecology, genetic control of disease and pest problems will become more important as time progresses. Fortunately a few investigations point the way to such genetic control, and—at the present stage of development—delineate some of the problems involved. None will be simple in their final resolution. Indeed, it is doubtful if genetic control will ever become the answer to all such problems. But insofar as practical, the peach breeder must be concerned.

Peach canker: One of the most thorough and complete investigations of disease resistance in the peach has been done by the Canadian research group (Weaver, 1963*a*, 1963*b*; Weaver and Jackson, 1963; and Wensley, 1964).

The etiology of peach canker has been elucidated by Wensley. He demonstrated that two organisms were primarily involved in this disease complex in Canadian orchards, *Valsa cincta* Fr. and *V. leucostoma* (Pers.) Fr. Other organisms may be involved secondarily, and in more southern localities may be relatively more important than in Canada. However, all of the organisms appear to be associated with the canker complex. Both *Valsa* species appear to be weak pathogens. Entry is gained through mechanical and disease or pest injuries, but the primary court of entry appears to be through leaf scars at the time of defoliation. Wensley also demonstrated that *V. leucostoma* was the more active pathogen at temperatures above 15.6 C, whereas *V. cincta* was more active below 10 C.

Weaver (1963a) determined that cultivars varied in their rate of defoliation in the fall. Cultivars could be characterized by rapid, medium, and slow rate of fall. He further correlated ($r = -.649$) rate of fall with susceptibility to canker. The more rapid the rate of leaf fall, the more resistant the cultivar. This association appears to be based on defoliation occurring at somewhat higher temperatures and under drier conditions.

Weaver (1963*a*) identified resistant and susceptible cultivars in stands of different ages to obtain the correlation cited. He (1963*b*) further found that rapid rate of leaf fall appears to be inherited as a simple recessive.

Weaver and Jackson (1963) studied resistant and susceptible clones biochemically. This study pointed to an association of rapid leaf fall with a sudden release of bound tryptophan in 'Elberta'. In this cultivar transformation of tryptophan resulted in high activity levels for 5-hydroxytryptamine and 5-hydroxyindolacetic acid. 'Dixired', a susceptible cultivar, showed no evidence of increased activity levels of these compounds, but released tryptophan was evidently quickly transformed to kynurenic and xanthurenic acids. The authors postulate that this difference in chemical pathways controls leaf abscission rate, and via this indirect route the reaction to the canker complex. Further, Wensley (1970) demonstrated that when resistant and susceptible cultivars were artificially inoculated with the causal organisms under controlled conditions no reaction difference could be noted. Therefore, it is not the chemical difference that mediates resistance, but an "escape" mechanism due to early, rapid defoliation.

Cultivars derived from 'Elberta' parentage were generally characterized as resistant. Cultivars from other generic sources proved susceptible.

This unique example of understanding disease resistance infers that seedlings can be selected on the basis of their chemical content at the time of natural defoliation. It will be interesting to follow continued research through hybrid populations.

Peach leaf curl: Resistance to peach leaf curl, *Taphrina deformans* (Berk.) Tul. was noted by Rivers (1906). Eglandular leaves were reported to be less affected than glandular ones. Ackerman (1953) surveyed many of the cultivars and seedlings growing in the Plant Introduction Gardens, Chico, California, in 1952, under conditions of severe natural infection. Contingency tables were developed from Ackerman's data, using only cultivars and clonal selections (Table 11). Data in Table 11, for peaches only, indicate the observation of Rivers to be sound. Globose glanded cultivars seem to be intermediate and somewhat more variable in their reaction than do reniform-glanded cultivars. Eglandular cultivars were not free of the disorder, but were heavily weighted toward the low end of the distribution classes noted.

When a similar distribution was calculated for nectarines only, the results were contrastingly different (Table 11). Only an excess of cultivars in the eglandular 0,0/1 class was significantly different, and in view of the small number of items rated, considerable doubt may be raised as to this significance. It appears that the G/g locus has an

TABLE 11. The distribution of seedlings and cultivars of peaches and nectarines by foliar gland type into six classes of resistance or susceptibility to peach leaf curl (adapted from Ackerman, 1953)

Foliar gland type	Peach leaf curl class 0,0/1	1	2	3	4	5	Total
			Peaches[z]				
Reniform	4 (15.7)	80 (90.3)	72 (72.5)	57 (54.8)	47 (37.0)	38 (27.7)	298
Globose	16 (5.0)	29 (29.1)	27 (23.4)	18 (17.6)	5 (11.9)	1 (9.0)	96
Eglandular	2 (1.3)	18 (7.6)	3 (6.1)	2 (4.6)	0 (3.1)	0 (2.3)	25
Total	22	127	102	77	52	39	419
			Nectarines[z]				
Reniform	2 (3.9)	19 (21.8)	17 (14.0)	6 (4.7)	2 (1.5)	0 (0.0)	46
Globose	0 (0.7)	7 (3.8)	1 (2.5)	0 (0.8)	0 (0.3)	0 (0.0)	8
Eglandular	3 (0.4)	2 (2.4)	0 (1.5)	0 (0.5)	0 (0.2)	0 (0.0)	5
Total	5	28	18	6	2	0	59

[z] Classes based on symptoms: 0 = none; 0/1 = inconclusive; 1 = up to 20% of all leaves showing some symptoms; 2 = 21 to 40%; 3 = 41 to 60%; 4 = 61 to 80%; 5 = 81 to 100%.
[y] Data in ( ) are expected assuming no influence of foliar gland type of peach leaf curl reaction.

effect on the expression of peach leaf curl in peaches, and that nectarines are less liable to suffer severe infestation, as shown by the absence of any individuals in class 5, regardless of gland class.

No genetic basis for resistance is available. Lack of effort to develop peach leaf curl resistant cultivars is probably associated with the ease with which the disease is controlled by chemical sprays. No commonly grown cultivars were listed among the more resistant classes. 'Southhaven' was among those listed in class 1, along with several domestic and imported nectarine cultivars. Tavdumadze (1971) noted "relative resistance" in several Russian cultivars, including one nectarine.

The evidence from these contingency tables suggesting the presence of a polygenic system of control for the trait has been discussed in the section on *Polygenes.*

Crown gall: The crown gall organism, *Agrobacterium tumafaciens* (S. & T.) Conn. attacks peach through root and crown injuries, and may seriously debilitate, or even kill, the susceptible cultivar. Little has been done to discover resistance to this disorder. However, DeVay et al. (1965) have attempted to find resistance among irradiated seeds. Following inoculation, seedlings appearing to carry some factor for resistance have been propagated and crossed. No results of these experiments have been published. It may turn out that suspected resistance following irradiation will have no genetic component.

Peach tree borer: The peach tree borer, *Sanninoidea exitiosa* (Say) may cause serious injury to peach trees. Weaver and Boyce (1965) obtained data relating to levels of resistance to this organism in their investigations in Canada. The research reported is based on seedling populations, rather than cultivars. They were able to show a wide range in susceptibility among seedling populations, from 8% for seedlings of 'Goldray' x self to over 81% in an F_2 population of 'J. H. Hale' x 'Vedette'. A complicating factor was that many individuals in the populations studied were winter killed. A negative correlation coefficient ($r = -.555$) was determined between winter killing and seedling infestation. There were populations that were exceptions to this association. From these exceptional populations the authors identified 'Goldray', 'Elberta', 'Dixired', and 'Golden Jubilee' as possible sources for borer resistance.

Other diseases and pests: Peach and nectarine cultivars are attacked by numerous disease organisms and pests in various parts of the world. Some, like the brown rots, appear to be of world-wide distribution; others are limited to given continents or climatic zones. For example, bacterial leaf spot of peach is practically unknown in California, probably because of the long, dry summer.

For the diseases and pests listed below no definitive research has been reported on the genetics of resistance or immunity. Cultivars have been observed to be more or less resistant, presumably under conditions of annual endemic infection, or under epiphytotic conditions. Thus we may believe that a genetic component is involved, and that more or less resistant cultivars will be developed as these diseases come under more rigorous observation and manipulation.

Wide-spread and common disorders are:

Bacterial Leaf Spot	*Xanthomonas pruni* (E. F. Sm.) Dows.
Peach Leaf Curl	*Taphrina deformans* (Berk.) Tul.
Brown Rot	*Monilinia laxa* (Aderh. & Ruhl.) Honey
Brown Rot	*Monilinia fructicola* (Wint.) Honey
Peach Blight	*Coryneum carpophilum* (Lev.) Jauch.
Peach Twig Blight	*Glomerella cinqulata* (Stou.) Spauld & Schrenk
Crown Rot	*Phytophthora cactorum* (Leb. & Cohn.) Schroet.
Crown Gall	*Agrobacterium tumafaciens* (S. & T.) Conn
Peach Mildew	*Shaerotheca pannosa* (Wallr.) Lev.
Aphids	unspecified

Table 12 shows a typical evaluation of peach cultivars for resistance or susceptibility to three diseases and to aphids. No one cultivar is relatively resistant to the four organisms listed, which suggests that multiple disease resistance will be very difficult to attain.

In some instances there is empirical evidence that conditions favoring resistance to a given organism may be associated with susceptibility to another. Thus, eglandular foliage is associated with tolerance to peach leaf curl, but with susceptibility to peach mildew.

In the descriptions of new cultivars presented by Brooks and Olmo (1972) many of the newer releases from the eastern breeding programs are noted to be resistant to bacterial leaf spot. Sharpe and Aitken (1971) note that the nectarine cultivars released from the Florida breeding program are tolerant of brown rot, and several of the Nectared series of nectarines from the New Jersey program are also stated to be tolerant of this disease. Tolerance to or ability to escape infection from the brown rot organism has been an important objective sought in the nectarine and peach breeding program of the Virginia Agricultural Experiment Station and progress has resulted from these efforts (Oberle, personal communication). This is in contrast to many of the cultivars of California origin, or older cultivars, which are usually considered to be extremely susceptible. The canning clingstone peach cultivars of California are also noted for their susceptibility to brown rot.

Resistance to some organisms may be found in related species. Thus, Taz (1958, in Knight, 1969) observed resistance to peach mildew in *P. ferganensis*, and Evreinoff (1939, in Knight, 1969) states that *Amygdalus petunnikowii* Litw. is immune to peach leaf curl and to brown rot of almond, *Monilinia cinerea*.

The elucidation of the genetics of resistance or immunity to these common peach diseases will require intensive programs which evaluate the genetic components of such resistance, which take into account the variability in virulence of the organism, and the host-pathogen interactions. Some may prove to be correlated with specific biochemical compounds, as the association between low cytokinins and root-knot nematode resistance

TABLE 12. An example of evaluation of peach cultivars for resistance to certain diseases and to aphids (adapted from Mihăescu, 1959)

Cultivar	Peach leaf curl	Peach blight	Brown rot	Aphid
Mayflower	good	moderate	moderate	v. slight
Sneed	slight	v. slight	good	mod. res.
Amsden	good	moderate	moderate	mod. res.
Sunbeam	good	slight	v. good	mod. res.
Red Bird	good	slight	v. good	good
Greensboro	slight	moderate	good	moderate
Carmen	slight	good	good	good
Peen-to	slight	moderate	v. slight	moderate
Orange Cling	moderate	good	v. good	moderate
Elberta	v. slight	good	v. good	good
J. H. Hale	v. slight	slight	moderate	v. good

found by Kochba and Samish (1971, 1972). Others may prove to be tolerant because of an "escape" reaction, such as that demonstrated by the Canadian investigations (Weaver, 1963*a*, 1963*b*; Weaver and Jackson, 1963; and Wensley, 1964, 1970) on resistance to the *Valsa* complex.

Virus disorders: The virus and virus-like disorders affecting peaches have not been investigated genetically. Symptom expression may vary among cultivars, but immunity to these diseases is not known among peaches. The peach breeder must be aware of these disorders, however, and all breeding stocks should be kept free of them. Some, as those of the ring spot complex, have been shown to be seed transmitted (Wagnon et al., 1960) and the virus may pass to the seed from the pollen (Nyland, personal communication).

Rootstock Breeding

Peach seedlings are the most usual rootstock for peaches. This discussion is therefore limited to the use of peach and peach interspecific hybrids as rootstocks for the species. The utilization of plums and other stocks for peaches, though appropriate to discussions of rootstocks for *Prunus*, have not involved peach breeding.

Efforts to improve peach rootstocks by breeding have been relatively few. Havis et al. (1946) suggested that seedlings of several cultivars, primarily those ripening late enough to yield readily-germinable embryos and vigorously-growing seedlings, would serve about equally well. Tukey (1944) and Gleason and O'Rourke (1952) found merit in the "naturals"—seed from wild stands of early-day introductions of peaches to the North American continent.

Some of the main problems associated with the use of peach rootstocks have been various soil-borne diseases and pests, particularly root knot nematodes. Susceptibility to "wet feet," caused by *Phytophora* sp. (see Mircetich and Keil, 1970), is one of the most serious peach rootstock diseases in many peach growing areas. For a few of these troubles research has indicated that improvement through breeding is possible.

Vigor control: The development of smaller trees to lower harvest costs and aid in the establishment of high density peach orchards by use of dwarfing rootstocks has been attempted. Layne (1971) of the Canadian Department of Agriculture Experiment Station, Harrow, Ontario, has recently reported on seed lines 'Harrow Blood' and 'Siberian C'. The former is reputed to dwarf standard peach clones by 20% and the latter by 10 to 15% when used as rootstocks. The factors involved in vigor control have not been determined, nor has the breeding behavior of this valuable trait been investigated. Both 'Harrow Blood' and 'Siberian C' seedlings are indicated to be winter-hardy, a prime consideration in this southern Canadian location. 'Siberian C', in addition, is reported to promote early defoliation, precocious bearing, and increased bud hardiness of scion cultivars.

The use of species hybrids to produce clonal stocks have been investigated by several researchers. Bernhard (1949) produced *P. persica* x *P. amygdalus* hybrids, and selected among these on the basis of ease of propagation and other desirable characteristics. Many show hybrid vigor, and are probably more suited as rootstocks for almonds because of the vigor imparted to the scion cultivar, but are also of value because of their superior performance on poor soils and resistance to chlorosis, lime-induced iron deficiency (Souty et al., 1955). Two selected clones, easily propagated by softwood cuttings, are GF 677 and GF 557 (Bernhard, 1965). A third generation selection, G_3 S 677-9, is multiplied by seed, but the seedlings are somewhat variable, and require roguing. Kester and Hansen (1966) and Kester et al. (1970) conduct a similar program in California, and have reported incorporating root knot nematode resistance into such hybrids by the use of resistant or immune peaches as parents.

Vegetative propagation (Kester and Sartori, 1966) is necessary in the use of selected F_1 hybrids at the present stage of development because segregating F_2 populations are extremely variable in all characteristics, and many are weak or otherwise abnormal (Kester and Hansen, 1966). Peach x almond hybrids can be produced directly, however, and as lines with desired combining ability are developed (Kester et al., 1970) and grown under conditions to yield nearly all hybrid seedlings, direct production will be feasible. Riggoti (1942) reported on the promise of selected *P. persica* x *P. davidiana* hybrids as rootstocks for peaches. A selection, Riggoti No. 2, is seed multiplied. The seedlings are relatively uniform and impart vigor to the scion variety (Bernhard, 1965).

Bernhard (1965) reported that the F_1 hybrid between *P. persica* x *P. kansuensis* shows great hybrid vigor and imparts this vigor to the scion cultivars grafted to it. The results were preliminary.

Current changes in peach culture in the United States appear to favor the development of semi-dwarfing rootstocks, but the hybrid vigor of the interspecific stocks just mentioned may prove useful in poor sites. It appears likely that, through selection, peach rootstocks with many desirable traits can be developed. Related to the production of rootstocks within the peach species is the use of other *Prunus* species, either directly or through interspecific hybridization to incorporate still other desirable characteristics, including vigor control. Hansen (unpublished) has found seedlings of 'Wayland' plum, *P. hortulana* Bailey, to be semi-dwarfing rootstocks for peach cultivars. A. N. Roberts (Oregon State University) selected a *P. subcordata* Benth. clone, 'Klamath 1' which, used as an interstock between peach roots or *P. americana* Marsh roots and peach cultivar tops, gave productive trees. These may prove more suitable than the various clones of *P. domestica* L., which have generally been considered unsuitable under United States conditions.

Root-knot nematode resistance: The discovery by Hutchins (1936) of resistance to root-knot nematodes in seedlings of peaches introduced from Asia, particularly Shalil and Yunnan introductions, was the beginning of investigation in this area. Sharpe et al. (1970) have reviewed the earlier investigations. Chitwood (1949) revised the taxonomy of the root-knot nematode group; six species were recognized, all assigned to the genus *Meloidogyne* in place of the single species *Heterodera marioni* (Cornu) Goodey. It then was soon recognized that of the six *Meloidogyne* species, *M. incognita* (Kofoid and White) Chitwood and *M. javanica* (Treub) Chitwood were the primary species infecting susceptible peach rootstocks. Shalil and Yunnan seedlings were found to be resistant to *M. incognita* but susceptible to *M. javanica*, which explained conflicting observations regarding their resistance.

Weinberger et al. (1943) had crossed cultivars whose seedlings were known to be susceptible to nematodes with Shalil and Yunnan, and described the dominance of resistance. Later Lownsberry et al. (1959) determined that resistance to *M. incognita* is controlled by a single dominant allele.

Identification of other sources of resistance involved the testing of 'Okinawa' seedlings and F_2 or other hybrids of *P. persica* x *P. davidiana*, the former by Sharpe and the latter by Havis. The latter hybrid was made at Chico at the request of Havis, and involved an item, Chico 11, a seedling of 'Shau Tai' from China which Long and Whitehouse (1943) had identified as being resistant to nematodes. 'Nemaguard' is also an apparent *P. persica* x *P. davidiana* hybrid according to Weinberger (see Brooks and Olmo, 1972).

A complicating factor in nematode resistance research has been the recognition of two levels of resistance: 1) so-called field resistance, in which the roots are attacked and galled, under test conditions, but the nematode does not complete its life cycle; and 2) the so-called immune reaction, in which no galls are found. 'Nemaguard' seedlings segregate for field resistance and immunity against *M. incognita*, and show field resistance to *M. javanica*. Field resistance is quite adequate for commercial use, for the nematode population does not reproduce.

Sharpe et al. (1970) in their summary of the investigations, verified that resistance to *M. incognita* was simply inherited as a dominant factor, and suggested that resistance to *M. javanica* was based on duplicate, independent dominant factors. The latter conclusion remains to be verified. Kochba and Samish (1971, 1972) have associated low endogenous cytokinin levels with resistance.

Achievements and Prospects

Peach and nectarine breeding has unquestionably been one of the brightest facets in all tree fruit breeding achievements. The cultivar picture has changed not once, but several times in some production areas. New producing regions have come to the fore through the efforts of peach and nectarine breeders. Industries have been revitalized. Some disease problems have been alleviated, if not conquered. Cultivars for processing have been developed for regions not commonly recognized for such production. The consumer has been given a better product at lower cost. The genetics of the peach have been more thoroughly explored than any other of the deciduous fruit tree species. All of these lines of development are continuing, but with greater emphasis on the fundamental problems involved in such interrelated traits as climatic adaptation, tree and fruit disease and pest resistance, fruit quality and processing, and cultural management and production. Nevertheless, we cannot be even moderately complacent, for economic and social changes are certainly challenging old methods, old criteria, and old objectives.

Spangelo (1968) has recently cited these challenges to fruit breeders and inferred deficiencies in many current institutional plant breeding programs, among which peach and nectarine breeding are representative. Certainly many long term efforts are in a period of reevaluation regarding objectives and methodology. Criticism is probably largely based on the supposed cost of programs which appear to be largely geared to the development of improved cultivars. Concurrently, administrators, even though they may appreciate the products of cultivar improvement, are tending to measure research in terms of input vs. output, and feel that cultivar development *per se* is not currently serving the best interests of a more and more complex and interrelated society. Most long-term breeding programs are expensive in terms of personnel and land resources. In some circles even the appropriateness of government institutional plant breeding, which has as its main objective the development of new cultivars for utilization by a restricted group of growers, is being questioned. All of these forces require that fruit breeders reevaluate their objectives and methods. Nevertheless, the great utility of improved plant resources has never been seriously questioned, so it appears that a reorientation is more likely than discontinuance of programs.

Objectives will need to be thoroughly examined, and changes made in line with newly established priorities. Most of these will require greater attention to quantitative inheritance. Such factors as winterhardiness, low chilling requirement, tree vigor control, quality enhancement, disease and pest control, adaptation to mechanization, and processing requirements readily come to mind. For example, Stushnoff (1972) has listed some six aspects important in understanding winterhardiness. Each must be investigated in depth and the role of genetic control for each needs to be measured and integrated into a whole. The most efficient method of breeding under such a complex system needs study.

Quantitative traits appear, from more recent investigations, to be mostly multi- or polygenically controlled. Methods of handling such genetic systems in the peach have not been investigated in number or detail. Much greater attention will be given to such studies.

Estimates of the relative importance of genetic and environmental variation needs much study. Accumulation of masses of genetically insignificant data can no longer be tolerated.

Intensive attention to particular facets of peach genetics may require in-depth studies based on crosses which are not necessarily appropriate to cultivar improvement. Once an understanding of a trait such as resistance to a disease is well understood, then an appropriate breeding system to incorporate that factor with superior cultivar performance can be intelligently designed.

Limited objectives will be more prominent. For example, quality factors may be determined through chromatographic studies. Then advances will be made by studying the qualitative and quantitative inheritance of compounds rather than rating systems.

Spangelo (1968) emphasized the need for greater control of the evaluation processes. This may or may not be feasible, but certainly is a desirable objective.

The trend toward mechanization of orchard operations will certainly call for a reevaluation of tree structure, factors of precocity, and economics of production. Rootstock control of tree vigor will become more important, but few have tackled the problem of usefulness of dwarf and semi-dwarf, genetically controlled tree size in relation to these problems. Both approaches need more research, and it will necessarily be in cooperation with horticulturists and economists.

The role of the rootstock in nutrition has not been seriously investigated. We have reported on the observed improvement of lime-induced chlorosis through the use of peach x almond hybrid stocks. But many other nutritional problems may be alleviated through the use of improved rootstocks. This area of investigation will require cooperation with other specialists.

The means of increasing our germplasm resources will require cooperation among experiment stations on a regional and national basis. Most experiment stations increasingly find the expense of maintenance of large plant collections too great. Exploration and introduction must be supplemented before some of the sources of variability themselves become restricted or lost.

More efficient and effective means of producing and evaluating hybrid populations certainly must be tested. Much of the expense of peach breeding involves the maintenance of large orchards. The simple expediency of ruthless elimination is not in itself sufficient.

In conclusion, researchers must continually be aware of the needs of society, and recognize and relate to social and economic trends and restrictions.

Literature Cited

Ackerman, William L. 1953. *The evaluation of peach leaf curl in foreign and domestic peaches and nectarines grown at the U. S. Plant Introduction Garden, Chico, Calif.* Div. of Plant Exploration and Introduction, Bur. of Plt. Ind., Soils, and Agric. Eng., USDA. (Mimeo.) p.1–31.

Armstrong, David L. 1957. Cytogenetic study of some derivations of the F_1 hybrid *Prunus amygdalus* x *P. persica*. Ph.D. dissertation, University of California, Davis.

Armstrong, W. D. 1950. Peach fruit bud survival at Princeton, Kentucky, after 8 degrees below zero on November 25, 1950. *Fruit Var. Hort. Dig.* 5:98–99.

Bailey, Catherine H., and L. Frederic Hough. 1959. An hypothesis for the inheritance of season of ripening in progenies from certain early ripening peach varieties and selections. *Proc. Amer. Soc. Hort. Sci.* 73:125–133.

Bailey, J. S., and A. P. French. 1933. The inheritance of certain characters in the peach. *Proc. Amer. Soc. Hort. Sci.* 29:127–130.

———. 1939. The genetic composition of peaches. *Mass. Agr. Expt. Sta. Bul.* (*Ann. Rpt.* 1938):86.

———. 1941. The genetic composition of peaches. *Mass. Agr. Expt. Sta. Bul.* 378 (*Ann. Rpt.* 1940): 91.

———. 1942. The inheritance of blossom type and blossom size in the peach. *Proc. Amer. Soc. Hort. Sci.* 40:248–250.

———. 1949. The inheritance of certain fruit and foliage characters in the peach. *Mass. Agr. Expt. Sta. Bul.* 452.

Barrett, Herbert C., and Toru Arisumi. 1952. Methods of pollen collection, emasculation, and pollination in fruit breeding. *Proc. Amer. Soc. Hort. Sci.* 59: 259–262.

Bernhard, R. 1949. The peach-almond and its utilization (in French). *Rev. Hort.* (Paris) 121:97–101.

———. 1965. The propagation of peach varieties, and of useful rootstocks (in French). *Proc. Int. Peach Congr.* (Verona.) p.365–378.

Blake, M. A. 1931. Flower types developed by peach seedlings. *N. J. Agr. Expt. Sta. Ann. Rpt.* 52:258–259.

———. 1932. The J. H. Hale as a parent in peach crosses. *Proc. Amer. Soc. Hort. Sci.* 29:131-136.

———. 1933a. New Jersey standard for classifying the set of fruit buds upon peaches. *N. J. Agr. Expt. Sta. Cir.* 271.

———. 1933b. Classification of 135 varieties of peaches and nectarines on basis of fruit bud set at New Brunswick. *N. J. Agr. Expt. Sta. Cir.* 274.

———. 1934. Relative hardiness of 157 varieties of peaches and nectarines in 1933 and of 14 varieties in 1934 at New Brunswick, N. J. *N. J. Agr. Expt. Cir.* 303.

———. 1936. Types of varietal hardiness in the peach. *Proc. Amer. Soc. Hort. Sci.* 33:240–244.

———. 1937. Progress in peach breeding. *Proc. Amer. Soc. Hort. Sci.* 35:49–53.

———. 1940. Some results of crosses of early ripening varieties of peaches. *Proc. Amer. Soc. Hort. Sci.* 37:232–241.

———. 1941. An acquaintance with peach varietal types is essential in peach breeding to secure improved varieties. *Proc. Amer. Soc. Hort. Sci.* 38: 144–147.

———. 1944. Some methods used in breeding peaches in New Jersey. *Proc. Amer. Soc. Hort. Sci.* 45:220–224.

Blake, M. A., and C. H. Connors. 1936. Early results of peach breeding in New Jersey. *N. J. Agr. Expt. Sta. Bul.* 599.

Blake, M. A., and O. W. Davidson. 1941. Some results of acidity and catechol tannin studies of peach fruits. *Proc. Amer. Soc. Hort. Sci.* 39:201–204.

Blake, M. A., and L. J. Edgerton. 1946. Standards for classifying peach characters. Their use in identifying New Jersey varieties. *N. J. Agr. Expt. Sta. Bul.* 728.

Branscheidt, P. 1933. On the question of varietal descriptions and fertility relationships in the peach (in German). *Gartenbauwiss.* 8:45–76.

Brooks, H. J., and L. F. Hough. 1958. Vernalization studies with peach embryos. *Proc. Amer. Soc. Hort. Sci.* 71:95–102.

Brooks, Reid M., and H. P. Olmo. 1972. *Register of new fruit and nut varieties.* 2nd ed. Berkeley: Univ. of Calif. Press.

Campbell, R. W. 1948. More than thirty peach varieties survived −32 F. *Proc. Amer. Soc. Hort. Sci.* 52:117–120.

Chaplin, C. E. 1948. Some artificial freezing tests of peach fruit buds. *Proc. Amer. Soc. Hort. Sci.* 52: 121–129.

Chitwood, B. G. 1949. Root-knot nematodes. 1. A revision of the genus Meloidogyne. *Proc. Helm. Soc. Wash.* 16:90–104.

Connors, C. H. 1920. Some notes on the inheritance of unit characters in the peach. *Proc. Amer. Soc. Hort. Sci.* 16:24–36.

———. 1922. Inheritance of foliar glands of the peach. *Proc. Amer. Soc. Hort. Sci.* 18:20–26.

———. 1923. Peach breeding. A summary of results. *Proc. Amer. Soc. Hort. Sci.* 19:108–115.

———. 1926. The sterility of J. H. Hale. *N. J. Agr. Expt. Sta. Ann. Rpt.* (1925) 46:90–91.

———. 1927. Sterility in peaches. *Mem. Hort. Soc. N. Y.* 3:215–222.

———. 1929. Further notes on peach breeding. *Proc. Amer. Soc. Hort. Sci.* 25:125–128.

Cullinan, F. P. 1937. Improvement of stone fruits. In *USDA Yearbook of Agriculture.* 1937:665–702.

Cullinan, F. P., and J. H. Weinberger. 1934. Studies on resistance of peach buds to injury at low temperatures. *Proc. Amer. Soc. Hort. Sci.* 32:244–251.

Darlington, C. D. 1928. Studies in *Prunus*, 1 and 2. *J. Genet.* 19:213–256.

Dermen, H. 1938. Detection of polyploidy by pollen-grain size; investigations with peaches and apricots. *Proc. Amer. Soc. Hort. Sci.* 35:96–103.

———. 1941. Simple and complex periclinal tetraploidy in peaches induced by colchicine. *Proc. Amer. Soc. Hort. Sci.* 38:141.

———. 1947a. Histogenesis of some bud sports and variegations. *Proc. Amer. Soc. Hort. Sci.* 50:51–73.

———. 1947b. Inducing polyploidy in peach varieties. *J. Hered.* 38:77–82.

———. 1953. Periclinal cytochimeras and origins of tissues in stem and leaf of peach. *Amer. J. Bot.* 40:154–168.

———. 1954. Histogenetic factors in color and nectarine sports of peach. *Genetics* 39:964.

———. 1956. Histogenic factors in color and fuzzless peach sports. *J. Hered.* 47:64–76.

Dermen, H., and D. H. Scott. 1939. A note on natural and colchicine-induced polyploidy in peaches. *Proc. Amer. Soc. Hort. Sci.* 36:299.

DeVay, J. E., G. Nyland, W. H. English, F. J. Schick and G. D. Barbe. 1965. Effects of thermal neutron irradiation on the frequency of crown gall and bacterial canker resistance in seedlings of *Prunus* rootstocks. *Rad. Bot.* 5:197–204.

Draczynski, M. 1958. The course of pollen differentiation in time in almond, peach, and apricot and the influence of bud temperatures on these processes (in German). *Gartenbauwiss.* 23:327–341.

Evreinoff, V. A. 1957. Notes on the ancestors of the cultivated peaches (in French). *Ann. Ec. Nutr. Sup. Agron.* (Toulouse) 5:61–68.

Fogle, H. W. 1959. Effects of duration of after-ripening, gibberellin, and other pretreatments on sweet cherry germination and seedling growth. *Proc. Amer. Soc. Hort. Sci.* 72:129–133.

Fogle, H. W., and Eric Barnard. 1961. Punched cards as aids in evaluating seedlings in the field. *Fruit Var. Hort. Dig.* 15:47–50.

Fogle, H. W., and Haig Dermen. 1969. Genetic and chimeral constitution of three leaf variegations in the peach. *J. Hered.* 60:323–328.

French, A. P. 1951. The peach. Inheritance of time of ripening and other economic characters. *Mass. Agr. Expt. Sta. Bul.* 462.

Gilmore, A. E. 1950. A technique for embryo culture of peaches. *Hilgardia* 20:147–170.

Gleason, B. L., and F. L. O'Rourke. 1952. Tests on peach rootstocks. *Amer. Nurseryman* 96:10, 45–47.

Gregory, C. T. 1915. The taxonomic value and structure of the peach leaf glands. *N. Y. (Cornell) Agr. Expt. Sta. Bul.* 365.

Griggs, W. H., George H. Vansell, and Ben T. Iwakiri. 1953. The storage of hand collected and bee collected pollen in a home freezer. *Proc. Amer. Soc. Hort. Sci.* 62:304–305.

Hansche, P. E., C. O. Hesse, and V. Beres. 1972. Estimates of genetic and environmental effects on several traits in peach. *J. Amer. Soc. Hort. Sci.* 97:76–79.

Hartmann, Hudson T., and Dale E. Kester. 1959. *Plant propagation.* p.89–116. Englewood Cliffs, N. J.: Prentice-Hall.

Haut, I. C., and F. E. Gardner. 1934. The influence of pulp disintegration upon the viability of peach seeds. *Proc. Amer. Soc. Hort. Sci.* 32:323–327.

Havis, Leon, P. C. Marth, and F. E. Gardner. 1946. Orchard performance of peach variety seedlings as rootstocks for peaches. *Proc. Amer. Soc. Hort. Sci.* 48:115–120.

Havis, Leon, J. H. Weinberger, and C. O. Hesse. 1947. Better peaches are coming. In *USDA Yearbook of Agriculture.* 1943–1947:304–311.

Hedrick, U. P. 1917. The peaches of New York. *Rpt. N. Y. Agric. Expt. Sta.* 1916.

———. 1950. *A history of horticulture in America to 1860.* New York: Oxford Univ. Press.

Hesse, Claron O. 1971. Monoploid peaches, *Prunus persica* Batsch: Description and meiotic analysis. *J. Amer. Soc. Hort. Sci.* 96:326–330.

Hesse, Claron O., and W. H. Griggs. 1950. The effect of gland type on the wettability and water retention of peach leaves. *Proc. Amer. Soc. Hort. Sci.* 56:173–180.

Hesse, Claron O., and D. E. Kester. 1955. Germination of embryos of *Prunus* related to degree of embryo development. *Proc. Amer. Soc. Hort. Sci.* 65:251–264.

Hough, L. F., J. N. Moore, and C. H. Bailey. 1964. Irradiation as an aid in fruit variety improvement. 2. Methods for acute irradiation of vegetative growing points of the peach, *Prunus persica* (L.) Batsch p.679–686. In *The use of induced mutations in plant breeding.* New York: Pergamon Press.

Hough, L. F., and Gerald M. Weaver. 1959. Irradiation as an aid in fruit variety improvement. 1. Mutations in the peach. *J. Hered.* 50:59–62.

HofMann, F. W. 1940. Some foliar characters for peach breeding. *Va. Acad. Sci. Proc.* 1940:208–209.

Hutchins, Lee M. 1936. Nematode resistant peach rootstocks of superior vigor. *Proc. Amer. Soc. Hort. Sci.* 34:330–338.

Jakovlev, P. N. 1946. Experimental facts from biology

concerning the development of fruit trees (in Russian). *Agrobiologija,* No. 4:57–64. (in *Plant Brdg. Abstr.* 19, p.149).
Jelenkovic, G., and E. Harrington. 1972. Morphology of the pachytene chromosomes in *Prunus persica. Can. J. Gen. Cyt.* 14:317–324.
Joley, L. E., and F. C. Bradford. 1961. Variation in blossom hardiness within a hardy group of peaches. *Proc. Amer. Soc. Hort. Sci.* 43:79–83.
Jones, Robert W., and Lester A. Thompson. 1955. Instruments for emascullating flowers of stone fruits. *Proc. Amer. Soc. Hort. Sci.* 65:279–282.
Kester, Dale E., and Carl J. Hansen. 1966. Rootstock potentialities of F_1 hybrids between peach (*Prunus persica* L.) and almond (*Prunus amygdalus* Batsch). *Proc. Amer. Soc. Hort. Sci.* 89:100–109.
Kester, Dale E., Carl J. Hansen, and B. F. Lownsberry. 1970. Selection of F_1 hybrids of peach and almond resistant and immune to root knot nematodes. *HortScience* 5:349 (Abstr.).
Kester, Dale E., and Elvino Sartori. 1966. Rootings of cuttings in populations of peach (*Prunus persica* L.), almond (*Prunus amygdalus* Batsch) and their F_1 hybrid. *Proc. Amer. Soc. Hort. Sci.* 88:219–223.
King, J. R. 1955. The rapid collection of pollen. *Proc. Amer. Soc. Hort. Sci.* 66:155–156.
King, J. R., and C. O. Hesse. 1938. Pollen longevity studies with deciduous fruits. *Proc. Amer. Soc. Hort. Sci.* 36:310–313.
Knight, R. L. 1969. *Abstract bibliography of fruit breeding and genetics to 1965. Prunus.* London: Eastern Press.
Knowlton, H. E. 1924. Pollen abortion in the peach. *Proc. Amer. Soc. Hort. Sci.* 21:67–69.
Kobel, F. 1927. Cytological investigations in Prunoideae and Pomoideae (in German). Archiv. der Julius Klaus-Stittung uer Vererbungsforschung Socialanthropologie und Rassenhygiene. Band 3, Seite 1–84.
Kochba, J., and R. M. Samish. 1971. Effect of kinetin and 1-naphthylacetic acid on root knot nematodes in resistant and susceptible peach rootstocks. *J. Amer. Soc. Hort. Sci.* 96:458–461.
———. 1972. Level of endogenous cytokinins and auxin in roots of nematode resistant and susceptible rootstocks. *J. Amer. Soc. Hort. Sci.* 97:115–119.
Lalatta, F. 1956. Present knowledge on the transmission of hereditary characters in the peach (in Italian). *Riv. d'lla Ortoflorofruttic. Ital.* 40(1–2):255–270.
Lamb, R. C., and J. P. Watson. 1964. Peach variety and blossom bud hardiness. *Fruit Var. & Hort. Dig.* 18:45–47.
Lammerts, W. E. 1941. An evaluation of peach and nectarine varieties in terms of winter chilling requirements and breeding possibilities. *Proc. Amer. Soc. Hort. Sci.* 39:205–211.
———. 1943. Effect of photoperiod and temperature on growth of embryo-cultured peach seedlings. *Amer. J. Bot.* 30:707–711.
———. 1945. The breeding of ornamental edible peaches for mild climates. I. Inheritance of tree and flower characters. *Amer. J. Bot.* 32:53–61.
Lantz, H. L. 1948. Improved varieties of peaches with special reference to hardiness. *Iowa Agr. Expt. Sta. Rpt.* 1948 p.277–278.
Lantz, H. L., and T. J. Maney. 1941. Peach breeding: A study of inheritance in some cross-bred seedlings. *Proc. Amer. Soc. Hort. Sci.* 38:184–186.
Layne, R. E. C. 1971. Peach rootstock research at Harrow. *Can. Agr.* 16(3)20–21.
Lesley, J. W. 1939. A genetic study of saucer fruit shape and other characters in the peach. *Proc. Amer. Soc. Hort. Sci.* 38:218–222.
———. 1944. Peach breeding in relation to winter chilling requirements. *Proc. Amer. Soc. Hort. Sci.* 45:243–250.
———. 1957. A genetic study of inbreeding and of crossing inbred lines in peaches. *Proc. Amer. Soc. Hort. Sci.* 70:93–103.
Lesley, J. W., and J. Bonner. 1952. The development of normal peach seedlings from seeds of early-maturing varieties. *Proc. Amer. Soc. Hort. Sci.* 60: 238–242.
Long, J. C., and W. E. Whitehouse. 1943. Variations in root knot nematode infection of various lines of peach progenies at Chico, California. *Proc. Amer. Soc. Hort. Sci.* 43:119–123.
Lownsberry, B. F., E. F. Serr, and C. J. Hansen. 1959. Root knot nematode on peach and root-lesion nematode on walnut cause serious problems for California orchardists. *Calif. Agr.* 13(9):19–20.
Mather, Kenneth, and John L. Jinks. 1971. *Biometrical genetics.* Ithaca, N. Y.: Cornell Univ. Press. p.7–13.
Meader, E. M., and M. A. Blake. 1939. Some plant characteristics of the progeny of *Prunus persica* and *Prunus kansuensis* crosses. *Proc. Amer. Soc. Hort. Sci.* 36:287–291.
———. 1940. Some plant characteristics of the second generation of *Prunus persica* and *Prunus kansuensis* crosses. *Proc. Amer. Soc. Hort. Sci.* 37:223–231.
———. 1943. Seasonal trends of fruit bud hardiness in peaches. *Proc. Amer. Soc. Hort. Sci.* 43:91–98.
Mihăescu, G. 1959. The behavior of some peach varieties in the vicinity of the town of Bucharest (in Rumanian). *Grad. Via. Liv.* 8:33–39.
Mircetich, Sreko M., and H. L. Keil. 1970. *Phytophthora cinnamomi* root rot and stem canker of peach trees. *Phytopathology* 60:1376–1382.
Monet, R. 1965. Simply determined genetic characters in *Prunus persica* Stokes (in French). *Ann. Amelior. Plantes* 15:99–106.
———. 1967. A contribution to the genetics of peaches (in French). *Ann. Amelior. Plantes* 17:5–11.
Morettini, A. E., E. Baldini, F. Scaramuzzi, G. Bargioni, and P. L. Pisani. 1962. *Monografia delle principol: cultivar di pesco.* Firenzi, Italy.
Mowry, James B. 1964*a*. Seasonal variation in cold hardiness of flower buds on 91 peach varieties. *Proc. Amer. Soc. Hort. Sci.* 85:118–127.
———. 1964*b*. Inheritance of cold hardiness of dormant peach flower buds. *Proc. Amer. Soc. Hort. Sci.* 85:128–133.
Oberle, George D. 1957. Breeding peaches and nectarines resistant to spring frosts. *Proc. Amer. Soc. Hort. Sci.* 70:85–92.
Oberle, George D., and J. O. Nicholson. 1953. Implica-

tions suggested by a peach to nectarine sport. *Proc. Amer. Soc. Hort. Sci.* 62:323–326.

Oberle, George D., and Richard Watson. 1953. The use of 2,3,5-Triphenyl tetrazolium chloride in viability tests of fruit pollens. *Proc. Amer. Soc. Hort. Sci.* 61:299–303.

Olmo, H. P. 1965. Peach and nectarine varieties and new variety trends. *Proc. Int. Peach Cong.* (Verona) p. 37–79.

Pisani, P. L. 1965. The contribution of genetics to the development of new peach varieties (in Italian). *Proc. Int. Peach Cong.* (Verona) p. 112–144.

Pratassenja, G. D. 1939. Production of polyploid plants. Haploid and triploids in *Prunus persica. Comp. Rend. Acad. Sci. (Doklady) URSS* 22:348–351.

Pratassenja, G. D., and E. M. Trubitzana. 1938. Production of polyploid plants. A triploid *Prunus persica. Comp. Rend. Acad. (Doklady) URSS* 19: 531–533.

Proebsting, E. L., Jr., and Harold W. Fogle. 1956. An apparatus and method of analysis for studying fruit bud hardiness. *Proc. Amer. Soc. Hort. Sci.* 68:6–14.

Riggoti, R. 1942. A new peach stock? *Prunus persica* x *P. davidiana* (in Italian). *Ital. Agr.* 21:664–667.

Rivers, H. S. 1906. The cross-breeding of peaches and nectarines. *3rd Int. Conf. Gen.* (London). p. 463–467.

Ryugo, K., and L. D. Davis. 1958. Seasonal changes in acid content of fruits and leaves of selected peach and nectarine clones. *Proc. Amer. Soc. Hort. Sci.* 72:106–112.

Savage, E. F., and V. E. Prince. 1972. Performance of peach cultivars in Georgia. *Ga. Agr. Expt. Sta. Res. Bul.* 114.

Schultz, E. F., Jr. 1955. Optimum allocation of experimental material with illustrative example in estimating fruit quality. *Proc. Amer. Soc. Hort. Sci.* 66: 421–433.

Schultz, E. F., Jr., and G. W. Schneider. 1955. Sample size necessary to estimate size and quality of fruit, growth of trees, and percent fruit set of apples and peaches. *Proc. Amer. Soc. Hort. Sci.* 66:36–44.

Scott, D. H., and F. P. Cullinan. 1942. The inheritance of wavy-leaf character in the peach. *J. Hered.* 33: 293–295.

Scott, D. H., and J. H. Weinberger. 1944. Inheritance of pollen sterility in some peach varieties. *Proc. Amer. Soc. Hort. Sci.* 45:229–232.

Scott, K. R., and L. P. S. Spangelo. 1964. Portable low temperature chamber for winter hardiness testing of fruit trees. *Proc. Amer. Soc. Hort. Sci.* 84:131–136.

Shamel, A. D., C. S. Pomeroy, and F. N. Harmon. 1932. Bud variation in peaches. *USDA Cir.* 212.

Sharpe, R. H., and J. B. Aitken. 1971. Progress of the nectarine. *Proc. Fla. St. Hort. Soc.* 84:338–345.

Sharpe, R. H., C. O. Hesse, B. F. Lownsberry, V. G. Perry, and C. J. Hansen. 1970. Breeding peaches for root knot nematode resistance. *J. Amer. Soc. Hort. Sci.* 94:209–212.

Sherman, W. B., R. H. Sharpe, and Jules Janick. 1973. The fruiting nursery: Ultrahigh density for evaluation of blueberry and peach seedlings. *HortScience* 8:170–172.

Sherman, W. B., R. H. Sharpe, and V. E. Prince. 1972. Two red leaf characters associated with early ripening peaches. *HortScience* 7:502–503.

Siminovich, D., H. Therrien, G. Feller, and F. B. Rheaume. 1964. The quantitative estimation of frost injury and resistance in black locust, alfalfa, and wheat tissue by determination of amino-acids and other ninhydrin-reacting substances released after thawing. *Can. J. Bot.* 42:637–649.

Souty, J., R. Bernhard, P. Rémy, G. Sanfourche, and M. Thomas. 1955. Work carried out by the fruit tree section during the period 1938–1953 (in French). Pont de la Maye, Girond. *Inst. natn. Rech. agron.*, (Ser. B) 5. p. 121–236.

Spangelo, Lloyd P. S. 1968. The challenging situation in developing superior fruit cultivars. *HortScience* 3:252–253.

Stushnoff, C. 1972. Breeding and selection methods for cold hardiness in deciduous fruit crops. *HortScience* 7:10–13.

Tavdumadze, K. T. 1971. Resistance of peach varieties to leaf curl (original in Russian). *Plant Brdg. Abstr.* 41:3708.

Taylor, J. W. 1957. Growth of non-stratified peach embryos. *Proc. Amer. Soc. Hort. Sci.* 69:148–151.

Tukey, H. B. 1933. Artificial culture of sweet cherry embryos. *J. Hered.* 24:7–12.

———. 1934. Artificial culture methods for isolated embryos of deciduous fruits. *Proc. Amer. Soc. Hort. Sci.* 32:313–322.

———. 1944. Variations in type and germinability of commercial lots of peach seed used by the nursery trade. *Proc. Amer. Soc. Hort. Sci.* 45:203–210.

Van der Driessche. 1969. Measurement of frost hardiness in two-year-old Douglas fir seedlings. *Can. J. Plant Sci.* 48:37–47.

Wagnon, H. Keith, J. A. Taylor, H. E. Williams, and J. H. Weinberger. 1960. Observations on the passage of peach necrotic leaf spot and peach ring spot viruses through peach and nectarine seeds and their effects on the resulting seedlings. *Plant. Dis. Rptr.* 44:117–119.

Weaver, G. M. 1963*a*. A relationship between the rate of leaf abscission and perennial canker in peach varieties. *Can. J. Plant Sci.* 43:365–369.

———. 1963*b*. The biochemistry of resistance to perennial cankers in peach. *Can. J. Gen. Cyt.* 5:109.

———. 1968. Gene resources and fruit research in the Soviet bloc countries. *HortScience* 3:260–262.

Weaver, G. M., and H. R. Boyce. 1965. Preliminary evidence of host resistance to the peach tree borer, *Sanninoidea exitiosa. Can. J. Plant Sci.* 45:293–294.

Weaver, G. M., and H. O. Jackson. 1963. Genetic differences in leaf abscission and the activity of naturally occurring growth regulators in peach. *Can. J. Bot.* 41:1405–1418.

———. 1969. Assessment of winterhardiness in peach by a liquid nitrogen system. *Can. J. Plant Sci.* 49: 459–463.

Weaver, G. M., H. O. Jackson, and F. D. Stroud. 1968. Assessment of winterhardiness in peach cultivars by electrical impedance, scion diameter and artificial freezing methods. *Can. J. Plant. Sci.* 48:37–47.

Weinberger, J. H. 1944. Characteristics of the progeny of certain peach varieties. *Proc. Amer. Soc. Hort. Sci.* 45:233–238.

———. 1950. Chilling requirements of peach varieties. *Proc. Amer. Soc. Hort. Sci.* 56:122–128.

———. 1955. Peaches, apricots, and almonds. *Handbook der Pflanzenzrechtung* 6:624–636.

———. 1961. Some temperature relations in natural breaking of the rest of peach flower buds in the San Joaquin Valley, California. *Proc. Amer. Soc. Hort. Sci.* 91:84–89.

Weinberger, J. H., P. C. Marth, and D. H. Scott. 1943. Inheritance study of root knot nematode resistance in certain peach varieties. *Proc. Amer. Soc. Hort. Sci.* 42:321–325.

Weiser, C. J. 1970. Cold resistance and acclimation in woody plants. *HortScience* 5:402–410.

Wensley, R. N. 1964. Occurrence and pathogenicity of Valsa (Cytospora) species and other fungi associated with peach canker in southern Ontario. *Can. J. Bot.* 42:841–857.

———. 1970. Innate resistance of peach to perennial canker. *Can. J. Plant Sci.* 50:339–343.

Williams, Jacob A. 1972. Occurrence, identification, and inheritance of anthocyanins of peach flower petals in intraspecific and interspecific crosses. Ph.D. dissertation, University of California, Davis.

Williams, Watkin, and H. G. Brown. 1956. Genetic response to selection in cultivated plants; gene frequencies in varieties of *Prunus persica*. *Proc. Roy. Soc. B.* 145:337–347.

Yarnell, S. H. 1945. Temperature as a factor in breeding peaches for a mild climate. *Proc. Amer. Soc. Hort. Sci.* 44:239–242.

Yoshida, M. 1970. Genetical studies on the fruit quality of peach varieties. 1. Acidity (in Japanese). *Bul. Hort. Res. Stat.* (Hiratsuka) No. 9:15.

Plums

by *John H. Weinberger*

Plums are widely grown throughout the world. Europe produces about four times as many plums and prunes as the United States, according to *Agricultural Statistics* (USDA, 1971). Yugoslavia leads in production in these statistics with approximately a million metric tons annually, West Germany is second with 600 mt, and the United States third with a little over 500 mt. France, Italy, Austria, United Kingdom, Spain, Turkey, Argentina, and Japan also produce significant quantities of plums.

In the United States the plum crop is the fourth leading tree fruit crop in total production. The 1964 Census of Agriculture lists cultivated plum trees in every state except Alaska.

Early United States settlers found native plums, which were used by the Indians for food, growing in most parts of the country. The settlers brought with them the European plums (*Prunus domestica* L.) which were far superior in fruit characters to the native plums. These became well established except in the South and in the colder regions of the country. About 1870 the Japanese plums (*Prunus salicina* Lindl.) were introduced, and soon became popular because of their size, productivity, and shipping qualities.

About one-fourth of the plum crop in the United States is used for fresh consumption. With the exception of a small quantity used for canning, most of the remaining crop is dried to make prunes. A prune is a firm-fleshed plum with a high sugar content which can be dried whole without fermenting at the pit. All prune cultivars belong to the species *P. domestica*. Fruits of several prune cultivars are shipped fresh as well as dried. Non-prune type plums may be dried when halved, and the product is known as dried plums.

Some plums are grown for their ornamental flowers and others for colored or variegated foliage.

Origin and Early Development

Plums belong to the genus *Prunus* and include many species. Hedrick (1911) and Bailey (1935) describe a large number of plum species. The ones which have been most useful to horticulturists are listed in Table 1.

Plums resemble other common fruits in the genus—e.g., peaches, nectarines, apricots, cherries, and almonds—but differ from these fruits, except the cherry, in that the fruits are borne on stems. Peaches, nectarines, apricots, and almonds have short peduncles. Plums differ from cherries by having a flattened pit, usually longer than broad; cher-

TABLE 1. Horticulturally important plum (*Prunus*) species.

Species	Chromosome no. (2n)	Location	Characteristics of value in breeding
alleghaniensis Porter	16	North central USA	Fruit, ornamental
americana Marsh.	16	Eastern USA	Winterhardiness, fruit
angustifolia Marsh.	16	Southeastern USA	Earliness, disease resistance
cerasifera Ehrh.	16	S.E. Europe, S.W. Asia	Rootstock, fruit
cocomilia Ten.	16	Italy	Fruit
domestica L.	48	Western Asia	Fruit, productivity
hortulana Bailey	16	Central USA	Fruit
insititia L.	48	E. Europe, W. Asia	Fruit
maritima Marsh.	16	Northeastern USA	Ornamental, fruit
mexicana Wats.	16	Central USA, Mexico	Fruit
munsoniana Wight & Hedr.	16	Central USA	Fruit, disease resistance
nigra Ait.	16	Canada, N. USA	Winterhardiness, fruit
rivularis Scheele	16	Texas	Disease resistance
salicina Lindl.	16	China	Fruit (size and firmness)
simoni Carr	16	China	Fruit
spinosa L.	32	Europe to Siberia	Fruit, *domestica* parent
subcordata Benth.	16	California, Oregon	Fruit, dwarfing rootstock
umbellata Ell.	16	South Carolina & Florida	Fruit

ries have globular pits. Plums inter-hybridize with apricots (King 1940) and with cherries. Plums inter-graft with other stone fruits, with varying degrees of success.

P. domestica is the most important source of commercial cultivars. The species is said to be native to western Asia (Bailey 1935) but was not domesticated until about 2,000 years ago. There is a question whether the species ever existed in the wild. *P. domestica* plums are hexaploids ($2n = 48$, $x = 8$) whereas most plum species are diploids. Crane and Lawrence (1952) suggest that *P. domestica* originated as a hybrid between *P. cerasifera* Ehrh., a diploid, and *P. spinosa* L., a tetraploid. The resulting triploid would have been sterile but if its chromosomes doubled, a fertile hexaploid could have been produced similar to *P. domestica*. They point out that in *P. cerasifera* the fruit has a yellow ground color and red anthocyanin. *P. spinosa* fruits have a green ground color and blue anthocyanin. The variation in these characters is limited. In *P. domestica* fruits have both yellow and green ground colors, and also both red and blue skin colors with an unlimited range in variations. The size, shape and flavor of the fruit support the above origin also.

Werneck and Bertsch (1959) made a comprehensive study of plums in the upper Rhine and Danube region and concluded that *P. domestica* is indigenous to middle Europe. The natural and spontaneous origin of hexaploid plums occurred before the Neolithic Age. Endlich and Murawski (1962) crossed *P. cerasifera* and *P. spinosa*. The F_1 were nearly sterile. Of 26 open pollinated F_2 seedlings, 15 seedlings had $2n = 48$, 7 seedlings $2n = 40$, and 4 seedlings had $2n = 32$. Some of the hexaploids closely resembled *P. domestica*.

P. salicina is of Chinese origin and was brought to this country about a hundred years ago. Luther Burbank hybridized it with *P. cerasifera*, *P. simonii* Carr, *P. americana* Marsh. and other species to produce important cultivars (Howard 1945).

The myrobalan or cherry plum, *P. cerasifera*, is native to southeastern Europe or southwestern Asia. The species is the source of many cultivars, but its seedlings are useful principally as plum rootstocks.

Native American species are the source of hundreds of plum cultivars. *P. americana*, found in the eastern two-thirds of the United States, has greater winterhardiness than *P. domestica*. The tree and fruit are small, and the fruit has a tough skin. The Chickasaw plum, *P. angustifolia* Marsh., the wild goose plum, *P. munsoniana* Wight and Hedr., and *P. hortulana* Bailey are native to the southeastern and southcentral United States. They have furnished cultivars adapted to those regions. Other useful native plums include the Pacific plum, *P. subcordata* Benth. (Roberts and Hammers 1951), the small beach plum, *P. maritima* Marsh., and the Canadian plum, *P. nigra* Ait.

The Damson plum, *P. insititia* L., is native to

eastern Europe or western Asia and is considered to be older than *P. domestica*. Damson plum pits have been found in ancient ruins. The apricot plum, *P. simonii* Carr., is native to China. It is represented in a number of shipping plum cultivars.

Individuals of the various species range in size from trees to shrubs, in vigor from vigorous to weak, and in shape from upright to spreading. The foliage may be green or red. Some species survive in cold climates, and others are adapted to warm climates. The exterior color of fruit may be green, black, purple, blue, red, or yellow. The color of the flesh may be green, yellow, amber, or red. Some fruits are soft and some are very firm. They vary in size from large to small and in shape from ovate to oblate. Some fruits are very high in sugar; others are quite acid and low in sugar. Some fruits are high in tannin. Some plums are resistant to diseases and others susceptible. The breeder has a wide selection in his choice of parental material. The hexaploid *P. domestica* plums are usually not intercompatible with diploid *salicina* plums. Griggs (1953) however notes that the *P. domestica* cultivar 'Tragedy' is a moderately effective pollenizer for several *P. salicina* cultivars but does not set fruit when pollinated by them. The difference in ploidy makes it difficult to transmit the usually high quality of *P. domestica* plums to the more common diploid species.

History of Improvement

Improvement of plums in the United States was begun by early private breeders using native American species. Cullinan (1937) presents a comprehensive review of the early progress. The first attempts involved selection and domestication of wild seedlings to secure more desirable fruits. About 1850, private breeders began to improve on the selection method by cross-pollinations within single species or between different species. H. A. Terry of Crescent, Iowa, started plum breeding about 1860 and named over 50 cultivars, producing more new cultivars than any other private breeder (Cullinan 1937). J. W. Kerr, Denton, Md., began breeding about 1870 and developed several cultivars from native species.

Wight (1915) lists over 600 named cultivars derived from native American species. They were originated mostly in the Mississippi Valley, Texas, and the southeastern United States where *P. domestica* cultivars were not adapted. *P. americana* provided the most cultivars through breeders in Iowa, Minnesota, and South Dakota. In Texas, the native *P. angustifolia* was the leading species in producing new cultivars. A few were hybrids between species but most of the cultivars when introduced were not more than two generations removed from their wild origin.

The best known private plum breeder was Luther Burbank. He imported *P. salicina* seedlings from Japan in 1884 and 1885. Several selections from them were named and introduced, including 'Burbank', 'Abundance', and 'Satsuma'. Later he made crosses between *P. salicina* and other species to produce many commercially desirable cultivars. His 'Santa Rosa' cultivar which contains a mixture of *P. salicina, P. simonii* and *P. americana* (Howard 1945) has long been the leading shipping plum in this country. Very few of our important shipping cultivars of diploid origin are derived from a single species.

Improvement in plum cultivars by breeding was started by agricultural experiment stations in Minnesota in 1889, in New York in 1893, and in South Dakota in 1895. As with private breeders, open-pollinated seedlings of native species were grown first, and from them selections were made. Cross pollinations with other species were made later, largely to improve flavor, hardiness, and productivity. Agricultural experiment stations in California and Iowa and U. S. Department of Agriculture stations in California and North Dakota undertook plum breeding about 1930. Florida, Texas, and Missouri state experiment stations initiated work more recently. Dozens of new cultivars have been developed. Most have only regional adaptation.

Plum breeding work is also in progress in a number of foreign countries including Canada, United Kingdom, France, Spain, Italy, West Germany, South Africa, Sweden, Bulgaria, Yugoslavia, Russia, Hungary, Romania, and Pakistan (Mao 1961, 1965).

Modern Breeding Objectives

The principal objective in a plum breeding program is the development of a desirable plum which can be grown successfully in a particular locality and which can be marketed profitably. A desirable fruit must have an attractive appearance, adequate size and firmness, and acceptable flavor and texture. To be grown successfully, the trees must be productive, and must be resistant or tolerant to problems which affect them. This includes hardiness in northern regions, low chilling requirements for buds in southern regions, and resistance to diseases and physiological problems. Successful marketing involves orchard location, proximity and type of markets, and the fruit's intended usage—shipping, canning, drying, or processing.

United States

A shipping plum cultivar is needed for each week of a ripening season which may continue as long as seven months. The season may start in April in Florida and end in October in California. A succession of cultivars provides the markets and consumers with a steady supply of fresh plums. The cultivars should be similar in character to avoid a drastic interruption in the quality or type of plum available.

More specifically, in Washington State the object is to replace early 'Italian Prune' clones with early-ripening cultivars having fewer physiological weaknesses. The leading cultivar, 'Richard's Early Italian', has problems with leaf curl, premature fruit drop, poor color, softening, internal browning, and poor quality.

In Florida the main plum breeding objectives are to develop desirable plums with low chilling requirements and canker resistance (*Pseudomonas syringae* Van Hall) (Sherman and Sharpe 1970).

At the New York (Geneva) Agricultural Experiment Station, the principal aim is to develop improved cultivars to extend the season for 'Stanley' type prunes that are suitable for baby food processing. The latter is the main outlet for prunes in the state. Specific characters of interest are increased hardiness, better quality, reliable production, relative freedom from disease, and self-fruitfulness.

At the USDA Station at Byron, Georgia, the objectives are to develop a succession of cultivars adapted to growing conditions in the southeastern part of the United States. The principal problems are canker resistance and bacterial leaf spot resistance (*Xanthomonas pruni* [E. F. Sm.] Dows). In California, plum breeding programs of the Agricultural Experiment Station and the USDA Station at Fresno are developing a succession of *P. salicina* cultivars adapted to prevailing climatic conditions. Each present cultivar has distinctive faults of either fruit or tree.

Europe

Plum breeding objectives in Sweden are to develop early-ripening market type fruits and trees having winterhardiness.

In southcentral Europe and the U.S.S.R. much plum breeding work is reported for the development of cultivars with improved fruit characters, resistance to disease and greater winterhardiness (Enikeev 1968). *P. domestica* plums for dessert, drying, and brandy are the main interests. In the U.S.S.R. rather wide crosses have been made to improve upon productivity, hardiness, and fruit characters (Eremin 1971). *P. salicina* has been used in some crosses (Rjabov and Kostina 1957). Mišić (1964) describes the plum breeding work in Yugoslavia aimed at developing frost resistant and disease resistant cultivars.

The experiment station at Pont-de-la-Maye, France, has a plum breeding program principally with *P. domestica* to develop a series of drying prunes and a series of dessert plums ripening in succession, along with basic cytological studies. Rootstocks are being bred also for resistance to winter asphyxiation of the roots and for less sensitivity to chlorine excess, which is a condition induced by calcareous soils.

South Africa

In South Africa, where winters are warm, the chilling requirements to break the rest period of buds are a problem. The principal need is for plum cultivars suitable for export and prunes for drying which are adapted to the climate of South Africa.

There appears to be no reason why these objectives cannot be met. Parental material having the desired characters is available. It remains only for the breeder to select the proper parents and recombine the needed characters in individuals.

Breeding Techniques

The technique used in plum breeding is similar to that used for other deciduous fruits. It involves pollen collection, emasculation and pollination of flowers, seed collection and germination, and an evaluation study of the progeny.

Plum flowers are small but showy. Initially all reproductive parts are enclosed in the five petals. The stamens usually number 20 to 30 and are attached along with petals and sepals to the rim of the calyx cup. When fully open the stamens are about the same length as the petals. The single pistil is attached to the base of the calyx cup. The stigma is positioned only slightly beyond the anthers at full bloom.

Pollen may be collected by gathering unopened blossoms just before the petals separate. They are placed in a wire screen sieve. Maceration of the flowers forces the anthers through the screen to be collected underneath. The anthers are dried overnight at room temperature or with slight extra heat. Pollen is easily shed from the dried anthers by manipulation with a camel's hair brush.

Pollen germinability can be readily determined by staining. Of 12 tetrazolium salts tried, Norton (1966) found 3(4,5-dimethyl thiazolyl 1-2) 2,5-diphenyl tetrazolium bromide (MTT) was best for staining plum pollen to determine its viability. He obtained a correlation of +.99 between per cent stainable pollen and viability in *P. salicina* cultivars and hybrids.

Plum pollen can be successfully stored from one year to the next. King and Hesse (1939) stored *P. domestica* pollen for 550 days at 2 C and 25% humidity, and secured high percentages of germination. Nebel (1940) successfully germinated *P. domestica* pollen after 4½ years storage at 2 to 8 C and 50% humidity.

Emasculation of the flowers may be performed by grasping the calyx cup with the fingernails and tearing away the corolla. The stamens are attached to the rim and are removed with it. The bare pistil and part of the calyx cup remain.

Various tools have been substituted for the use of fingernails in emasculation. One of the simplest is made of a 10-cm section of a flat piece of steel such as a hacksaw blade. The teeth are ground off and a small notch is made in one end with an emery wheel. The notched end of the blade and the inner edges of the notch are sharpened for cutting the blossom. The sharpened blade is maneuvered to cut off the distal end of the calyx cup after placing a finger on the opposite side of the flower to brace it. The petals and stamen parts of the flower are easily flipped off, still attached to the cup.

Cultivars of plums which are self-unfruitful need not be emasculated before pollination. Using them as seed parents reduces the work of making crosses and improves the chances of obtaining a set of fruit. Some "self-unfruitful" cultivars will set a small percentage of blossoms with their own pollen. For breeding purposes this is negligible but for cytological studies emasculation is necessary.

Pollen may be applied to the stigmas with a camel's hair brush. Pollination may be done immediately after emasculation. Some workers apply pollen a second time a few days later. Unemasculated flowers are pollinated as soon as open. Two or three applications of pollen at intervals of several days as the flowers open are necessary to cover the period of bloom of a tree. The flowers should be protected from insect visitation by covering them with a paper bag or cheesecloth cover. More frequently an entire tree is enclosed in a wooden frame covered with cheesecloth. A polyethylene cover on top facilitates pollination in rainy weather. Unemasculated flowers that have been hand pollinated must always be covered for protection from insects. Emasculated flowers are seldom visited by insects. Protecting the flowers often pays in improved set of fruit and in protection from frost injury.

Individual flowers or branches on which a particular pollen is used must be labeled in a manner to last until the fruit is harvested.

Plum fruits are harvested when mature and the pit is removed from the flesh. Seed of early ripening cultivars usually give a very low percentage germination. It is advisable to culture these seed on sterile nutrient agar after removing the endocarp and integuments (Tukey 1935), (Hesse and Kester 1955). After-ripening of freshly cultured seed is not necessary but may be of some help.

A mixture of peat and sand may be substituted for the agar medium. The integuments are removed and the seed is planted partially exposed in greenhouse flats without sterile conditions. The cotyledons soon develop chlorophyll, after which they are quite resistant to infection.

Seed of midseason and later-ripening cultivars usually give satisfactory germination if they are not allowed to dry out. The rehydration of seed which has been dried for storage introduces bacterial and fungal infections which often destroy the seed. Fewer opportunities for infection occur if the seed are cracked immediately after harvest and placed with integuments intact in damp peat moss in a polyethylene bag in a refrigerator at 4 to 5 C. In two or three months the seed will start to germinate. Peat moss inhibits mold formation and keeps the seed moist. Application of a mild fungicide may be helpful also.

Lin and Boe (1972) in studies on dormancy in 'Italian Prune' seeds (*P. domestica*) make the point that dormancy is associated with entire seed: testa, cotyledon, and embryo. Stratification overcame dormancy controlled by the testa, while applications of gibberellic acid and N-6-benzyladenine overcame dormancy in the embryo and cotyledons. The chemicals were not effective if applied when the testa were intact.

At the first signs of germination, the seed should be removed from the refrigerator and planted. A greenhouse is necessary since planting time will be fall or early winter. A mixture of sand and peat makes a satisfactory planting medium and reduces danger of damping-off. Occasional applications of a complete nutrient solution will prevent nutrient deficiencies. A little soil may be included in the planting medium for the same purpose but use of soil might necessitate fungicidal or sterilization treatments.

The young seedlings may be kept in an active condition through the winter if purposely retarded by cool temperatures. An alternative is to harden off the seedlings while small and force them into defoliation and dormancy. Chilling is then needed to break the rest period of the buds. When freezing danger is past, the seedlings, whether vegetative or dormant, are transplanted to seedling test blocks to grow to maturity. It is not necessary to grow them for a year in a nursery. The seedlings may be planted about 1 m apart in rows 3–4 m apart. This allows more than 3,000 seedlings/ha. If trees to be discarded are removed promptly, crowding does not result.

The first fruit are borne on seedlings the third, or sometimes fourth, year after planting. Observations are made on fruit and tree characters. Undesirable seedlings are eliminated, and the best material is saved for further observations and testing. It is not unusual to have eliminated 97 to 99% of the seedlings by the time they are five years old.

One of the leading reasons for elimination of seedlings is small size of fruit. Plums are naturally marginal in this respect. Other causes may be lack of fruit attractiveness, disease susceptibility, lack of productivity, tendency to develop flesh cracks, dry or coarse flesh textures, odd-shaped fruit, flesh or skin astringency, softness of flesh, and poor quality.

If a seedling appears promising when it first fruits, it is propagated by budding or grafting for a second test. Continued good performance warrants additional testing. When the seedling has proved itself to be better than any cultivar ripening at the same season, it can be considered a candidate for naming and introduction. A minimum of ten years usually elapses between the pollination of the flower and the introduction of the new cultivar.

Breeding Systems

Since the two leading groups of commercial plums have different chromosome numbers, inter-hybridization between them often gives poor results. The European plums (*P. domestica*) are hexaploid ($2n = 48, x = 8$) as are *P. insititia* plums. *P. spinosa* plums are tetraploids ($2n = 32$). The Japanese plums (*P. salicina*) are diploid ($2n = 16$) as are *P. cerasifera*, *P. americana* and others.

Where both parents have the same number of chromosomes, interspecific hybridization is generally successful. Where different degrees of ploidy are involved, fruit may be produced but the seeds are rarely viable (Wellington 1927). Recent workers have made many interspecific crosses and some have determined chromosome numbers of progeny.

Olden (1965) made cross pollinations of many combinations of various degrees of ploidy. Regardless of the ploidy status the success of a cross seemed to depend on using a seed parent with a pistil shorter than that of the pollen parent. In reciprocal crosses between *P. cerasifera* x *P. nigra* fruit was produced only when the shorter-styled *P. cerasifera* was used as the seed parent.

In crosses between diploids and tetraploids, Olden found *P. spinosa* cross-compatible with most diploid plums. In crosses between diploids and hexaploids, diploid females produced more fruit than hexaploid females, but the seed had low viability. Unreduced female gametes frequently united with *P. domestica* male gametes and produced offspring with odd ploidy. Hexaploid *P. domestica* females which were self-fruitful were more compatible with diploids than those which were self-unfruitful. The progeny were mostly intermediate in chromosome numbers. Diploids crossed with hexaploids produced tetraploid, pentaploid, hexaploid, heptaploid, and octoploid progeny. Unreduced gametes could have yielded the higher combinations. Octoploids probably resulted from a doubling of tetraploid zygotes.

Olden obtained progeny of diploids x triploids, and triploids x hexaploids, having various ploidies. Open-pollinated triploid, tetraploid, pentaploid, and octoploid seedlings produced fruit and seedlings also. In his studies pollen quality varied considerably. Large percentages of pollen of various progeny were in the large, small, or empty classes rather than normal in size and soundness. The larger size grains were of particular interest because of possibly extra ploidy.

Roy (1939) pollinated *P. divaricata* Led. with *P. domestica* and secured a 6% set of fruits. In the reciprocal cross 15% reached maturity. He found a diploid pollen tube grew more rapidly in a hexaploid style than a hexaploid pollen tube in a diploid style.

Cultivars within the same species are not always compatible with each other. Self-unfruitfulness is common in plums. Pollen sterility occurs occasionally. In a comprehensive study of pollen of plum species and hybrids, Flory and Tomes (1943) found variations from flower to flower in percentages of normal pollen, but little variation from year to year. Pollen from trees of a single species averaged 46% germinability, and from interspecific hybrids only 12%.

Alderman and Weir (1951) in a pollination study of plum hybrids concluded that apomixis is a common occurrence in interspecific hybrids. They found also a great deal of incompatibility among hybrid diploid cultivars they had developed.

Salesses (1970) studied four triploid hybrids and found that three of them showed the phenomenon of cytomixis. He points out the possibility of producing hexaploid plums through nonreduced gametes of triploid plums.

Most of the *P. salicina* cultivars are self-unfruitful (Griggs 1953). Some, like 'Nubiana', are partially self-fruitful, and occasionally some, like 'Santa Rosa', will set full crops of fruit with their own pollen. There is a high degree of inter-incompatibility also, with many combinations such as 'Kelsey' and 'Gaviota' setting only poor or fair crops of fruit. *P. domestica* cultivars perform similarly except that there is apparently no cross-incompatibility among important *P. domestica* cultivars. French prune is self-fruitful (Schmidt 1954).

Incompatibility in plums is a genetic character (Crane and Lawrence 1929). Multiple allelomorphs are believed responsible for self-incompatibility and intersterility of individuals with the same genetic constitution. In *P. domestica* incompatibility is not that simple, for reciprocal crosses are not always compatible. Different degrees of compatibility occur also. Cultivars which carry the same factors for incompatibility are reported to have their pollen tube growth in the stylar tissue inhibited or delayed, so that the gametes cannot reach the embryo sac in time to fertilize the embryo (Afify 1933) (Thiele and Strydom 1964). This applies to self-unfruitful cultivars also. Roy (1939) found a portion of pollen tubes of compatible *P. domestica* cultivars may swell up at their ends and cease growth in the style without effecting fertilization. He concluded pollen of a single cultivar may differ in genotype. The physiological basis of incompatibility in higher plants, according to Crane and Lawrence (1952), is the relation between the male gametophyte and female sporophyte tissue.

The mode of transmission of a few *P. salicina* plum characters has been worked out by Weinberger and Thompson (1962). Time of ripening is quantitatively inherited. The mode of the distribution pattern of ripening dates of progeny was close to the average ripening date of the parents with some individuals ripening earlier or later than either parent.

Size of fruit is quantitatively inherited. When both parents had large-sized fruit approaching the extreme in size, the progeny fruit averaged smaller in size than that of the parents.

Shape of fruit in *P. salicina* is controlled by

multiple factors with neither round nor ovate shape dominant.

Yellow skin color in *P. salicina* is a single gene recessive to black, red, or purple skin color. The latter colors are quantitatively inherited.

Red flesh color in *P. salicina* is dominant over yellow, and a single factor is involved. The intensity of the red color is controlled by multiple genes. Hurter (1962) also found red flesh color dominant over yellow, and the inheritance was monofactorial.

The freestone character is apparently recessive, as occasional seedlings with freestone fruits were found in progeny from clingstone parents. The maturity of the fruit and the firmness of the flesh affected the degree of clinginess. Some plums were "air-free."

In *P. domestica* oval fruit shape was reported to be dominant over round fruit shape; yellow skin color recessive to red, purple and black; thick bloom on fruit dominant over thin bloom; and freestone recessive to cling (Wellington 1927). Crane and Lawrence (1952) reported that purple color of leaves and fruit in *P. pissardi* was controlled by a single pair of genes with heterozygous individuals having intermediate color intensity. In *P. domestica* spreading character of growth appears to be recessive. Hairiness of growth was dominant to sub-glabrous.

Olden (1965) found hairiness of leaves and short stamens to be dominant in *P. domestica*.

In a study of inheritance of winterhardiness of twigs, wood, and flower buds, Dorsey and Bushnell (1925) found when both parents were hardy, as in *P. nigra* and *P. americana* seedlings from the area, the progeny were hardy. When both parents were tender as in *P. salicina* seedlings, the progeny were tender. When a hardy and a tender parent were used, segregation occurred into hardy and tender classes with few intermediates. The progeny seemed more like the pistillate parent. *P. nigra* transmitted greater hardiness than *P. americana* seedlings.

Alderman and Angelo (1935) summarized the relative value of certain cultivars as parents in Minnesota. The *P. salicina* 'Burbank' was outstanding. 'Santa Rosa' and 'Shiro' also gave good progeny.

The selection of parents in breeding new cultivars is most important. They should be strong in the particular characters desired in the progeny. One parent is often chosen because it has a certain character in which the other parent may be deficient, while the second parent may have traits the first parent lacks. The whole purpose in selecting parents is to combine the desired traits into a single individual. Small-fruited parents seldom give large-fruited progeny; thus at least one parent should be large-fruited. Parents should be selected similarly for time of maturity, firmness of flesh, flavor, or other characters. Progeny will tend to be intermediate in various characters which are quantitatively inherited, with occasional seedlings possessing certain characters beyond the range of either parent.

The selected parents must be compatible in order to produce satisfactory quotas of seed. Flory (1947) gives effective pollenizers for cultivars derived from species useful in breeding plums for the southwestern United States: 1) *Hortulana:* other *hortulana*, or *munsoniana* cultivars; 2) *Munsoniana: hortulana* or *salicina* cultivars; 3) *Salicina:* most other diploid cultivars; 4) Asiatic: American species hybrids—native *angustifolia, munsoniana*, and often *salicina* cultivars.

Since plums are highly heterozygous and most characters are controlled by multiple factors, a great deal of variability will occur in any progeny. To secure an individual which will possess most of the desired traits, large progenies are necessary. Individual plum breeding projects of public agencies develop from a few hundred to several thousand seedlings from controlled pollinations each year. The number of seedlings which must be grown to develop a good plum cultivar may average into the thousands, perhaps as many as 5,000. Large populations are necessary in order to have a reasonable chance of attaining success.

Breeding for Specific Characters

Fruit Characters

The 12 most important fruit characters of interest to breeders are: time of maturity, size, shape, amount of waxy bloom on fruit, amount of skin coloring with red coloring preferred, attractiveness of ground color, color of flesh, firmness, freeness of pit, texture, quality, and resistance to disease. These are not listed in order of importance but rather in sequence in which observations are usually made. Attractiveness is perhaps the most important feature, for a fruit must have consumer appeal to be successful in the markets. It must also be firm enough to arrive in markets in good condition, and must have adequate quality to assure repeat sales.

Fruit with the best flavor and quality is found in *P. domestica* cultivars. *P. salicina* hybrids have size, color, and attractiveness. Some have exceptional firmness and keeping quality. *P. cerasifera* plums transmit earliness. They produce progeny which are quite variable in hardiness, fruit form, and other characters even when selfed (Murawski 1959). *P. americana*, native to the northern states, *P. nigra* native to Canada, and *P. ussuriensis* carry factors for winterhardiness. Tough skin to improve shipping quality is often carried in *P. americana* plums. It may be necessary to raise the ploidy level of breeding material to combine the best traits of different species. Results indicate that fertility at the tetraploid level is adequate. Polyploids artificially induced with colchicine are being used to accomplish fertile combinations. Hurter and Van Tonder (1963) induced polyploidy in 28 *P. salicina* seedlings in which 9 were complete tetraploids suitable for breeding and 19 were chimeras.

Griggs and Thompson (1964) examined dividing cells of young leaves of the 'Star Rosa' cultivar, a spontaneous mutation, and found that of the 85 cells studied, 71 were tetraploid and 14 diploid. The mutation was apparently chimeral in nature. The fruit is larger and earlier-maturing than the parent.

Brown rot disease (*Monilinia fructicola* [Wint.] Honey) resistance in progeny is more difficult to attain, for parental differences in resistance are slight. The industry problem is being helped by new handling methods for fruit, new fungicides, and new insecticides to control insect carriers of brown rot inoculum.

Plum breeding work is best accomplished in the regions where the plums are to be grown. Each region has its own climatic distinctions in which a new cultivar must be tested to prove its adaptation. High summer temperatures in the San Joaquin Valley of California, for example, can cause internal browning of flesh of some *P. domestica* cultivars. Testing of potential cultivars is more or less a local problem.

In the northwestern United States a large part of the plum crop is canned. The 'Italian Prune' and early-ripening 'Italian Prune' type plums which have many problems are the main cultivars. Approximately half of the crop is shipped fresh and half is canned. Early plums with high quality are needed to extend the season of harvesting and processing.

About two-thirds of the commercial plum trees in the United States are of the 'French Prune' cultivar. It was introduced to this country in 1854. At present the cultivar has no serious competitor. Some efforts have been made to develop a satisfactory prune with larger size, with little success thus far. The opportunities for improving shipping plums are much greater.

Tree Characters

Canker diseases of various types greatly affect the longevity of plum trees in many regions but particularly in the southeastern United States. Parent material is available which carries some resistance or tolerance to canker diseases (Norton 1967). Red-fleshed *P. salicina* cultivars seem better adapted to the southeast than yellow-fleshed cultivars.

Resistance to bacterial leaf spot is needed for areas from Texas to the Atlantic Coast, where the disease is prevalent, as well as in other parts of the world. In rating 45 cultivars for resistance to bacterial leaf spot, Flory (1941) found 20 which were free of cankers. Most resistant cultivars were derived from native American species, with 'Damson', 'Green Gage', and 'Simon' included also. Popenoe (1959) in Alabama found that the highest degree of resistance to bacterial leaf spot on leaves and fruit as well as stem cankers was in cultivars derived from native species as 'Bruce', 'Compass', 'Munson', and 'Opata'. *P. domestica* cultivars were most susceptible to leaf and fruit infection but no stem cankers developed. In *P. salicina* cultivars no stem cankers were found on 'Burbank', 'Elephant Heart', 'Satsuma', and 'Kelsey' while others were badly damaged. 'Burbank', 'Mariposa', and 'Santa Rosa' had the greatest resistance to leaf and fruit spot.

A spreading type of tree is easier to manage in the orchard than an upright growing tree, and it is desirable from that standpoint to use spreading types in breeding programs. Popenoe (1959) noted that resistance to stem cankers in *P. salicina* cultivars is associated with the spreading growth tree character.

Some *P. salicina* cultivars like 'Queen Ann' are affected with a shot-hole condition of the foliage which resembles bacterial infection. It is a genetic defect, which can be overcome by breeding and selection (Weinberger and Thompson 1962).

Tree productivity is essential for successful cultivars. This can be associated with tree vigor, hardiness or other characteristics. Since many plum cultivars are self-unfruitful, compatibility with other cultivars can strongly influence productivity. Its lack has adversely affected the popularity of cultivars derived from native American species in Minnesota (Anderson and Weir 1967). Late blossoming would be advantageous for a late blooming selection would have greater productivity in an area subject to spring frosts. The chilling requirement to break the rest period of buds would affect production also in warm regions. In cold regions bud hardiness in winter readily limits production.

Rootstocks

The performance of plum trees is affected by the rootstocks on which they are grown. Myrobalan and marianna plums are the leading plum rootstocks, except that 'Nemaguard' peach rootstock is commonly used for *P. salicina* cultivars in California. Marianna is thought to have originated in the cross *P. cerasifera* x *P. munsoniana* (Day and Tufts 1922). Both myrobalan and marianna seedlings vary in their suitability for rootstocks. Selections have been made in these two rootstocks which are then propagated vegetatively to improve their vigor, longevity, hardiness, nematode resistance, dwarfing characteristics, tolerance to wet soil conditions, and drouth resistance. This work is still being carried on, as well as attempts to improve rootstocks by hybridization. Hurter (1969) has noted that progeny of the self-sterile diploid marianna contain a great deal of polyploidy. In France attempts are being made to utilize or adapt plum rootstocks for peach in wet soil conditions. Sloane (1971) reports that 'Victoria Myrobalan' rootstock is highly resistant to high salt concentrations.

Mutation breeding in plums has not proven successful in improving cultivars. Irradiation of seeds and buds has been of little help in obtaining better fruit characters. However, DeVay et al. (1964) after irradiating 3,000 seeds of *P. cerasifera* secured a higher frequency of crown gall (*Agrobacterium tumefaciens* [E. F. Sm. & Towns.] Conn) resistant seedlings from treated seed than in the controls, and a lower frequency of bacterial canker resistance.

Achievements and Prospects

The importance of breeding to the development of the shipping plum industry cannot be overemphasized. The ten leading plum cultivars in fruit shipments in California in 1971 had their origin in breeding programs. 'Laroda', 'Queen Ann', 'Nubiana', and 'Burmosa' were developed in a University of California—USDA project by Dr. C. O. Hesse. 'Beauty', 'El Dorado', and 'Santa Rosa' were originated by Luther Burbank. 'President' came from Thomas River's work in England. 'Casselman' and 'Late Santa Rosa' are natural mutations of Burbank's 'Santa Rosa'. All are *P. salicina* derivatives except 'President' which is of *P. domestica* origin.

The New York Agricultural Experiment Station introduced the 'Stanley' plum which has earned a place for itself in this country wherever *domestica* plums are grown. More recent introductions are the 'Iroquois', 'Oneida', and 'Mohawk' cultivars (Watson 1966). All are of *P. domestica* origin. 'Iroquois' and 'Mohawk' are 'Italian Prune' x 'Hall'. 'Oneida' is 'Albion' x 'Italian Prune'.

More recently the USDA has introduced three *P. salicina* cultivars developed at Fresno, California: 'Frontier', 'Friar', and 'Queen Rosa'. All are being well accepted by the plum industry in California.

The Vineland Experiment Station in Ontario, Canada introduced three new plum cultivars of *P. domestica* origin in 1967. 'Valor' and 'Verity' are 'Imperial Epineuse' x 'Grand Duke' seedlings. 'Vision' is a 'Pacific' x 'Albion' seedling.

Several private breeders and nurseries in California are actively engaged in developing plum cultivars. 'Red Beaut' is an early ripening *P. salicina* plum originated by Fred Anderson, which has been heavily planted. Several other of his *P. salicina* cultivars have been named and are in production, including 'Amazon', 'Andy's Pride', 'Ebony', 'Grandora', and 'Grand Rosa'. John Garabedian has developed the *P. salicina* cultivars 'Angeleno', 'Bee-Gee', 'Black Queen', 'Early Gar Rosa', 'Fire Queen', and 'Fresno Rosa'. The 'Ozark Premier' plum is a cross of 'Burbank' x 'Methley' developed at the Mountain Grove Experiment Station, Missouri. It seems better adapted to southeastern United States conditions than many other *P. salicina* cultivars (Shepard 1948).

Eleven plums developed in the *P. domestica* breeding program in Romania are described in Pomologia (Blaja et al. 1969). Other promising Romanian hybrids are listed by Cociu et al. (1968).

Three new *P. salicina* selections of promise are described by Hurter et al. (1970) as evidence of progress made in the South Africa breeding program. Nineteen plum varieties bred at the central Asiatic branch of the Institute of Plant Industry are described by Kovalev (1960). Some are *P. salicina* hybrids.

The number of plum trees in the United States has declined drastically in the last 50 years. This decline probably accounts for the present low level of interest in plum cultivar improvement. Only Arizona and Michigan have shown a sharp increase in plum tree plantings in recent years. California plum acreage has changed little in total amount but the *P. salicina* cultivars being grown are changing in accordance with their relative value and popularity. In other parts of the United States particularly with *P. domestica* cultivars, the expense of handling the small fruit and the disease problems associated with plum production in humid climates have had adverse effects on new plum plantings.

State agricultural experiment stations in California, Florida, Missouri, New York, and Texas; the Canadian Station at Vineland; and the USDA at four locations—Fresno, California; Beltsville, Maryland; Byron, Georgia; and Prosser, Washington—report promising seedlings under test in 1971. Further improvements in cultivars may be expected which will help the plum industry. The literature from other plum-producing countries throughout the world also promises further improvements in plums.

Literature Cited

Afify, A. 1933. Pollen tube growth in diploid and polyploid fruits. *J. Pomol. Hort. Sci.* 11:113–119.

Alderman, W. H., and E. Angelo. 1935. An analysis of the breeding value of certain plum varieties. *Proc. Amer. Soc. Hort. Sci.* 32:351–356.

Alderman, W. H., and T. S. Weir. 1951. Pollination studies with stone fruits. *Minn. Agr. Expt. Sta. Tech. Bul.* 198.

Andersen, E. T., and T. S. Weir. 1967. *Prunus* hybrids, selections and cultivars at the University of Minnesota Fruit Breeding Farm. *Minn. Agr. Expt. Sta. Tech. Bul.* 252.

Bailey, L. H. 1935. *The standard cyclopedia of horticulture.* New York: Macmillan.

Blaja, D. et al. 1969. Varieties and hybrids of plums (in Romanian, French summaries). *Pomologia* 8:175–211.

Cociu, V., C. Radulescu and I. Bodi. 1968. The plum assortment in Romania and prospects of its improvement (in Romanian). *Rev. Hort. Viticult. Bucuresti.* 17:20–25. (*Plant Brdg. Abstr.* 38:6814. 1968.)

Crane, M. B. and W. J. Lawrence. 1929. Genetical and cytological aspects of incompatibility and sterility in cultivated fruits. *J. Pomol. and Hort. Sci.* 7:276–301.

———. 1952. *The genetics of garden plants.* 4th ed. London: Macmillan.

Cullinan, F. P. 1937. Improvement of stone fruits. p. 703–723. In *USDA Yearbook of Agriculture.*

Day, L. H. and W. P. Tufts. 1922. Nematode-resistant rootstocks for deciduous fruit trees. *Calif. Agr. Expt. Sta. Cir.* 359.

DeVay, J. E., G. Nyland, W. H. English, and F. J. Schick. 1964. Frequency of crown gall and bacterial canker-resistant seedlings from seed of *Prunus* rootstocks irradiated with thermal neutrons. *Phytopathology.* 54:891. (Abstr.)

Dorsey, M. J. and J. Bushnell. 1925. Plum investigation II. The inheritance of hardiness. *Minn. Agr. Expt. Sta. Tech. Bul.* 32.

Endlich, Von J. and H. Murawski. 1962. Contributions to breeding research on plums. III. Investigations on interspecific hybrids of *P. spinosa* (in German). *Zuchter* 32:121–133.

Enikeev, H. K. 1968. Breeding new plum cultivars for the central areas of the Nonchernozem zone (in Russian). *Vestn. sel'skohoz. Nauk.* 8:54–58. (*Plant Brdg. Abstr.* 39:1239. 1969.)

Eremin, G. V. 1971. Aspects of the remote hybridization of plums (in Russian). *Sel'skohoz. Biol.* (*Agric. Biol.*) 6:41–46. (*Plant Brdg. Abstr.* 41:753. 1971.)

Flory, W. S. 1941. Varietal rating of plums with reference to canker resistance. *Texas Agr. Expt. Sta. Prog. Rpt.* 753. (Mimeo.)

———. 1947. Crossing relationships among hybrid and

specific plum varieties, and among the several *Prunus* species which are involved. *Amer. J. Bot.* 34:330–335.

Flory, W. S., and M. L. Tomas. 1943. Studies of plum pollen, its appearance and germination. *J. Agr. Res.* 67:337–358.

Griggs, W. H. 1953. Pollination requirements of fruits and nuts. *Calif. Agr. Expt. Sta. Cir.* 424.

Griggs, W. H., and Thompson, M. M. 1964. Star Rosa—a spontaneous tetraploid Japanese plum. *Proc. Amer. Soc. Hort. Sci.* 84:117–122.

Hedrick, U. P. 1911. The plums of New York. *New York (Geneva) Agr. Expt. Sta. Rpt.* 1910.

Hesse, C. O. and D. E. Kester. 1955. Germination of embryos of *Prunus* related to degree of embryo development and method of handling. *Proc. Amer. Soc. Hort. Sci.* 65:251–264.

Howard, W. L. 1945. Luther Burbank's plant contributions. *Calif. Agr. Expt. Sta. Bul.* 691.

Hurter, N. 1962. Inheritance of flesh color in the fruit of the Japanese plum *Prunus salicina*. *So. Afr. J. Agr. Sci.* 5:673–674.

———. 1969. Cytogenetic studies on the marianna plum rootstock. 1. Pollen sterility, pollination studies and ploidal range in F_1 progeny. *Agroplantae* 1:113–120.

Hurter, N., and M. J. Van Tonder. 1963. Colchicine induction of autotetraploidy in the Japanese plum, *Prunus salicina*. *S. Afr. J. Agric. Sci.* 6:403–410.

Hurter, N., F. N. Matthee, M. J. Van Tonder, and T. R. Visagie. 1970. Promising plum selections and their final evaluation. *Decid. Fruit Grower* 20:290–292.

King, J. R. 1940. Cytological studies in some varieties frequently considered as hybrids between the plum and the apricot. *Proc. Amer. Soc. Hort. Sci.* 37: 215–217.

King, J. R., and C. O. Hesse. 1939. Pollen longevity studies with deciduous fruits. *Proc. Amer. Soc. Hort. Sci.* 36:310–313.

Kovalev, N. V. 1960. New plum and sweet-cherry cultivars for Uzbekistan (in Russian). *Trud. nauc-issled, Inst. Sadavodstv., Vinogradarstv. Vinodel. uzbekistan. Akadisel'skohozjajstv. Nauk.* 24:27–33. (*Plant Brdg. Abstr.* 31:5324. 1961.)

Lin, C. F. and A. A. Boe. 1972. Effects of some endogenous and exogenous growth regulators on plum seed dormancy. *J. Amer. Soc. Hort. Sci.* 97:41–44.

Mao, Y. T. 1961. *World list of plant breeders.* Food and Agr. Org. of the United Nations.

———. 1965. *World list of plant breeders.* Supplement No. 1. Food and Agr. Org. of the United Nations.

Mišić, P. D. 1964. Fruit breeding in Yugoslavia—problems and research (in Yugoslavian). *Hort. Res.* 4: 49–59. (*Plant Brdg. Abstr.* 35:2749. 1965.)

Murawski, H. 1959. Contributions to breeding research on plums. II. Further investigations on the breeding value of seedlings (in German). *Zuchter* 29:21–36.

Nebel, B. R. 1940. Longevity of pollen in apple, pear, plum, peach, apricot and sour cherry. *Proc. Amer. Soc. Hort. Sci.* 37:130–132.

Norton, J. D. 1966. Testing of plum pollen viability with tetrazolium salts. *Proc. Amer. Soc. Hort. Sci.* 89:132–134.

———. 1967. Resistance to bacterial canker in plums. *Proc. Assoc. So. Agric. Workers 64th Ann. Conv. New Orleans* 1967:227–228.

Olden, E. J. 1965. Interspecific plum crosses. Balsgord Fruit Breeding Institute, Fjalkestad, Sweden. *Res. Rpt.* 1.

Popenoe, J. 1959. Relation of heredity to incidence of bacterial spot on plum varieties in Alabama. *Proc. Assoc. So. Agric. Workers 56th Ann. Conv. Memphis.* 1959:176–177.

Rjabov, I. N. and K. F. Kostina. 1957. The achievements in breeding stone fruit trees at the Nikita Botanical Gardens (in Russian). Agrobiologija No. 5:41–54. (*Plant Brdg. Abstr.* 28:4632. 1958.)

Roberts, A. N. and L. A. Hammers. 1951. The native Pacific plum in Oregon. *Ore. Agr. Expt. Sta. Bul.* 502.

Roy, B. 1939. Studies on pollen tube growth in *Prunus*. *J. Pomol.* 16:320–328.

Salesses, G. 1970. The phenomenon of cytomixis in triploid hybrids of *Prunus*. Possible genetic consequences (in French, English summary). *Ann. Amelior Pl.* 20:383–388.

Schmidt, M. 1954. Contribution of research on the breeding of plums. I. On the progeny of a selfed form of *P. insititia* (in German). *Zuchter* 24:157–161.

Shepard, P. H. 1948. New fruit varieties originated and introduced by the Missouri State Fruit Experiment Station, Mountain Grove, Missouri. *Missouri State Fruit Expt. Sta. Bul.* 33.

Sherman, W. B. and R. H. Sharpe. 1970. Breeding plums in Florida. *Fruit Var. Hort. Dig.* 24:3–4.

Sloane, R. T. 1971. A salt resistant rootstock for apricot and Japanese plum. Salinity Symposium, Mildena, Victoria. Canberra, Australia 6(3)3–6(3)5.

Thiele, I. and D. K. Strydom. 1964. Incompatibility studies in some Japanese plum cultivars (*Pr. salicina* Lindl.) grown in South Africa. *S. Afr. J. Agr. Sci.* 7:165–168.

Tukey, H. B. 1935. Artificial culture methods for isolated embryos of deciduous fruits. *Proc. Amer. Soc. Hort. Sci.* 32:313–322.

U. S. Department of Agriculture. 1971. *Agricultural Statistics.* Washington: Government Printing Office. p. 248.

Watson, J. 1966. Iroquois, Oneida, and Mohawk plums named. *New York (Geneva) Agr. Expt. Sta. Farm Res.* 32(2):10–11.

Weinberger, J. H. and L. A. Thompson. 1962. Inheritance of certain fruit and leaf characters in Japanese plums. *Proc. Amer. Soc. Hort. Sci.* 81:172–179.

Wellington, R. 1927. An experiment in breeding plums. *New York (Geneva) Agr. Expt. Sta. Tech. Bul.* 127.

Werneck, H. L. and K. Bertsch. 1959. On the prehistory and early history of plums in the upper Rhine and Danube regions (in German). *Agnew Botany* 33:19–33.

Wight, W. F. 1915. The varieties of plums derived from native American species. *USDA Bul.* 172.

Cherries

by Harold W. Fogle

It is apparent that cherries have been cultivated since the dawn of civilization (Marshall, 1954), but efforts to improve them date back only to about the sixteenth century. Cherries are popular fruits in all temperate regions of the world and most of these regions have significant improvement programs.

The cherries of commerce are not indigenous to North America, nor are they well adapted to most regions of the continent. Therefore, with a few exceptions, interest in their improvement has been localized to the few areas particularly suitable to culture of sweet and sour (tart, acid, or red) cherries. Relatively few institutional breeding programs have been initiated and these usually have had very specific local objectives.

On the other hand, there has been great interest in cherries for home orchards in all areas. Numerous selections of the commercial types and of native cherries have been made by growers and amateur horticulturists and some of these continue to be important commercial cultivars.

Origin and Early Development

Cherries have a basic chromosome number of x=8. Sweet cherries, *Prunus avium* L., are usually diploid (2n=16), but occasionally triploid and tetraploid (2n=24 and 32) forms are found. Sour cherries, *P. cerasus* L., are tetraploid (2n=32). These two are most important commercially of the many cherry species. Duke cherries, *P. gondouini* (Poit. & Turp.) Rehd., also are tetraploid and are presumed to arise from pollination of *P. cerasus* by unreduced (2n) pollen of *P. avium*. Tree and fruit types are intermediate between the species.

The mazzard rootstocks, preferred in most areas for sweet cherries, are seedlings of *P. avium* which escaped from cultivation. The St. Lucie cherry rootstocks are seedlings of *P. mahaleb* L. (2n=16) which adapt well to irrigated, slightly alkaline soils. *P. fontanesiana* (Spach) Schneid. apparently is a hybrid of mazzard and mahaleb.

The Nanking or Hansen Bush cherry, *P. tomentosa* Thunb. (2n = 16), adapts to areas too cold and arid for sweet or sour cherries and is a favored windbreak and hedge species in the severe climate of the Great Plains. It hybridizes with some plum species and has been used to a limited extent in breeding programs of the northern Great Plains and northern Europe. Its principal uses have been as an ornamental shrub and as a dwarfing rootstock for peach and other *Prunus* species.

Prunus fruticosa Pall. (2n = 32), the ground cherry, is considered the probable parent species of both *P. avium* and *P. cerasus*. Crosses in Sweden between *P. fruticosa* and either diploid or tetraploid sweet cherries have given progenies resembling sour cherries (Olden and Nybom, 1968). This species has been tested as a dwarfing rootstock for sweet cherries but lacks adequate anchorage and suckers profusely (Toyama et al. 1964).

All of the preceding cherry species were introduced into North America. All were apparently native to the Caspian-Black Sea regions (De Candolle, 1890) and to parts of Asia as far east as northern India and China (Hedrick, 1915). Men and birds spread seeds of these cherry species to all of continental Europe. Early settlers brought seeds to North America and pioneers moved the cherries westward with them.

Numerous species of cherries are indigenous throughout North America (Wight, 1915) and some of these have been utilized in breeding for cherries and cherry-plum hybrids for severe climates. Among these are the sandcherries (*P. besseyi* Bailey and *P. pumila* L.), wild red cherry (*P. pensylvanica* L. f.), wild black cherry (*P. serotina* Ehrh.) and chokecherry (*P. virginiana* L.). The sandcherries are diploid and the latter two species are tetraploid. The ploidy of *P. pensylvanica* apparently has not been determined, but this species crosses readily with the diploid ones.

Cherries are sometimes classified into various subgroupings on the basis of flesh color and flesh firmness (Hedrick, 1915; Grubb, 1949). Flesh color ranges from almost colorless through shades of yellow and red to a dark, purplish-red, almost black. This range of color divides roughly into light-fleshed and dark-fleshed types with the latter dominant. In sour cherries, cultivars are classified into morello or griotte (dark-fleshed with red juice) and amarelle or Kentish (light-fleshed with colorless juice) types. Sweet cherry cultivars are classified roughly into bigarreau (firm-fleshed) types and guigne (gean or heart) types with soft, tender flesh. Intermediate types are difficult to classify into these groups.

History of Improvement

The cherry apparently was first cultivated in Greece (Hedrick, 1915; Marshall, 1954). It had been described in 300 B.C. and probably had been grown, primarily for its wood rather than its fruit, several centuries earlier.

Little is known about the development of cherries before the sixteenth century, and less than a score of cultivars can be traced to this early period. Apparently most European countries cultivated cherries, with Germany as the center of cultivation. However, significant new cherries were not introduced in Germany until the eighteenth century. The Romans took cherries to England in the first century, but fruit growing essentially disappeared there between the ninth and fifteenth centuries. By 1600, several cultivars including the 'Yellow Spanish' sweet cherry had been described.

Early colonists brought cherries to North America. Until about 1767, propagation of cherry trees was mostly from seeds. Some 20 cultivars, propagated by budding, were offered by at least one nursery in New York by that date. Cherry seeds and trees were planted by pioneers as the West was homesteaded.

The salient date as far as cherry improvement in America is concerned is 1847, when the Lewelling brothers arrived in Milwaukie, Oregon, after their oxcart journey from Iowa with about 300 propagated cherry trees. One brother became interested in growing cherry seeds first for pollenizers and then for selection of promising cultivars. From these seedlings, Seth Lewelling selected 'Bing', still the leading sweet cherry cultivar in North America; 'Lambert', second in importance; and 'Republican', a leading pollenizing cultivar.

About the same time, Professor J. P. Kirkland of Cleveland, Ohio, originated some 30 soft-fleshed sweet cherry cultivars of high quality and adapted to eastern United States growing conditions. About 50 years later, Luther Burbank, a private fruit breeder, made initial crosses from which he named the 'Burbank' sweet cherry in 1911 and three other sweet cherries later. The first institutional sweet cherry breeding program in North America was started in 1911 at the New York Agricultural Experiment Station at Geneva. However, breeding of sour and native species started earlier. Greater detail of the history of cherry improvement is given by Hedrick (1915), Cullinan (1937), Grubb (1949), and Marshall (1954).

Sweet Cherries

Cultivars derived by selection of chance seedlings or mutations are not included in this compilation. These and the cultivars derived from breeding programs are described in the *Register of New Fruit and Nut Varieties*, 2nd edition (Brooks and Olmo, 1972), with supplemental annual lists in *HortScience* or in the *Fruit Varieties and Horticultural Digest* (since 1971, *Fruit Varieties Journal*), a quarterly of the American Pomological Society. Following is a compilation of salient cultivars introduced from institutional and private improvement programs:

Institution or private breeder	Initial crosses	Releases: Cultivar	Releases: Date	Pertinent reference or (breeder)
Luther Burbank (private)	about 1900	Burbank	1911	Hedrick, 1915
		Abundance	1912	
		Giant	1914	
		Honey Heart	1934	Brooks & Olmo, 1972
New York Agr. Expt. Sta.	1911	Seneca	1924	Wellington & Lamb, 1950
		Gil Peck	1936	
		Sodus	1938	
		Hudson	1964	Way, 1968a
		Ulster	1964	
Vineland (Ontario) Hort. Expt. Sta.	1915	Victor	1925	Dickson, 1961
		Velvet	1937	
		Vernon	1937	
		Venus	1959	
		Vic	1959	
		Vista	1959	
		Valera	1967	Tehrani & Kerr, 1968
		Vega	1967	
Summerland (British Columbia) Res. Sta.	1924	Sparkle	1944	
		Van	1944	Mann & Keane, undated
		Star	1949	
		Sam	1953	
		Sue	1954	
		Compact Lambert	1964	Lapins, 1965
		Stella	1968	Lapins, 1970
		Salmo	1970	
Idaho Agr. Expt. Sta.	1934	Lamida	1946	Verner, 1946
		Ebony	1946	
		Spalding	1946	
California Agr. Expt. Sta.	1935	Lambush	1945	Brooks, 1954
		Bada	1964	Brooks & Griggs, 1964
		Berryessa	1964	
		Jubilee	1964	
		Larian	1964	
		Mona	1964	
Washington Agr. Expt. Sta. and USDA cooperating	1950	Rainier	1960	Anony., 1960 (Fogle)
		Chinook	1960	
Oregon Agr. Expt. Sta.		Corum	1961	Zielinski, 1961
		Macmar	1961	Zielinski, et al., 1959
Fred Anderson (private), Calif.		Bingandy	1963	Brooks & Olmo, 1972
John Innes Inst., England		Merton Bigarreau	1947	(Crane)
		Merton Bounty	1947	
		Merton Favourite	1947	
		Merton Glory	1947	
		Merton Heart	1947	
		Merton Premier	1947	
		Merton Reward	1958	

Institution or private breeder	Initial crosses	Releases		Pertinent reference or (breeder)
		Cultivar	Date	
		Merton Crane	1961	Matthews, 1968
		Merton Late	1961	
		Merton Marvel	1961	
Max Planck Inst., Germany		Primavera	1960	Zwintzscher, 1960
		Secunda	1963	Zwintzscher, 1963
Fruit Res. Sta., Jork, Germany		Alma		(von Vohl)
		Annabelle		
		Barbara		
		Bianca		
		Rebekka		
		Rube		
		Valeska		
Wadenswil, Switzerland		Alfa		Schaer, 1968
		Beta		
Verona, Italy		Vittoria		Bargioni, 1970
Mitschurinsk, Russia	about 1930	Valerij Tschkalow	1967	Shukow & Charitonowa, 1968
Melitopol Fruit Res. Sta., Russia		Dneprovka		Oratovskiy, 1956
		Hryashchevataya		
		Kolektivnoya		
		Liza Chaykina		
		Melitopolskaya chernaya		
		Melitopolskaya pozdnaya		
		Melitopolskaya rannaya		
		Melitopolskaya rozovaya		
		Plodorodnaya		
		Priusadebnaya		
		Skorospelka		
		Smuglyanka		
		Yukzhnoukrainskaya		
Moscow, Russia		Kiyevskaya 2	1956	Duka, 1956 (Rodionov)
		Kiyevskaya 5		
		Kiyevskaya 11		
Pitesti, Marcineni, Romania		Negre timpurii	1971	Cociu, 1971
		Uriase di Bistrita		
New South Wales Res. Sta., Australia		Rons	1928	
		Ronson	1955	
		Rival	1955	
		Regina	1955	

The Balsgard Experiment Station in Sweden initiated cherry research in 1949. No cultivars have become available for testing from this program, but promising selections have been noted (Olden, 1959). Considerable inheritance data have come from this program.

At the Institute of Horticultural Plant Breeding at Wageningen, Netherlands, a cherry breeding program was initiated about 1952. New cultivars have not become available, but progress in increasing resistance to bacterial canker has been made.

Several other European programs (particularly Polish, Yugoslavian, Bulgarian, and Romanian) and Asian programs are of particular interest because of the availability to them of germplasm from the areas where sweet cherries originated.

Sour Cherries

New sour cherry cultivars have been introduced from only three North American breeding programs. The Minnesota Agricultural Experiment Station initiated cherry and cherry-plum hybrid breeding programs about 1908 (Alderman, 1926). In 1925 'Nicollet', (*P. avium* x *P. pensylvanica*) x

P. besseyi, but closely approaching *P. cerasus* type, was introduced. In 1950 'Northstar', a morello type of sour cherry, and in 1952 'Meteor', an amarelle type, were introduced. Both were dwarfish in growth habit; the former has attracted interest as a dwarfing interstock for sweet cherries.

In 1937 the Dominion Experimental Station at Morden, Manitoba, Canada, introduced 'Coronation', an open-pollinated seedling of 'Shubianka'.

In 1952 the United States Horticultural Field Station at Cheyenne, Wyoming, introduced 'Dwarfrich', an open-pollinated seedling of 'Vladimir'.

Although limited selection for improved sour cherry types was carried out at the United States Northern Great Plains Research Center at Mandan, North Dakota, and at several state experiment stations, no cultivar was introduced.

The Michigan Agricultural Experiment Station and the Horticultural Experiment Station at Vineland, Ontario, have initiated extensive breeding programs and are utilizing material collected from the supposedly indigenous regions for *P. cerasus* in eastern Europe. The Michigan station has an active program of 'Montmorency' strain selection, and some United States nurseries have spur-type mutations of 'Montmorency' under test.

Most of the new introductions of sour cherries have been mutations or "strain" selections of the 'Montmorency' cultivar. At least ten such "strains" were introduced between 1925 and 1956 (Brooks and Olmo, 1972). These primarily have been selections by growers.

From a Russian program (Shukow and Charitonowa, 1968) started about 1932, the cultivars 'Shukawskaja', 'Standard Urale', 'Schtschedraja', and 'Uralskaja Rubinowaja' were named.

From the breeding program at the Max Planck Institute in Germany (Zwintzscher, 1968), the cultivars 'Nabella', 'Cerella', and 'Successa' were introduced. Hardiness and brown rot (*Monolinia*) resistance are major objectives of this program.

Many other European stations have sour cherry breeding programs in various stages of development. In Denmark, several "brown" cherry cultivars have been developed for use in liqueurs. Several morello-type cultivars are being patented from the Wageningen program in the Netherlands.

Duke Cherries

The Duke cherries named in North America have been chance seedlings. 'Krassa Severa', imported from the Michurin Horticultural Institute, Koslov, Russia, apparently resulted from a controlled cross of sweet and sour cherries. 'Mailot' was derived from a Duke cherry x sour cherry cross at the Max Planck Institute in Germany (Zwintzscher, 1968).

Sand Cherries

The western (*Prunus besseyi*) and eastern (*P. pumila*) sandcherries have been utilized in the northcentral and Great Plains states of the United States and the prairie provinces of Canada for selection of improved cultivars and for hybridizations with other *Prunus* species, mostly plum. At least seven improved sandcherry clones were introduced by the Dominion Experimental Station at Morden, Manitoba, starting about 1896. The South Dakota station named at least seven clones and released six numbered selections for trial. Two additional clones, 'Brooks' in 1934 and 'Alace' in 1942, were introduced by the Horticultural Experiment Station at Brooks, Alberta. Considerable selection for desirable types was done by the North Dakota and Minnesota Agricultural Experiment Stations and by the United States Department of Agriculture at Mandan and Cheyenne. None of these selections was named.

Cherry Plums

In nearly all of the Great Plain states and provinces of North America crosses were made to incorporate the hardiness and drought resistance of sandcherries with various plums and other *Prunus* species. Professor N. E. Hansen of the South Dakota Agricultural Experiment Station probably was the prime motivator of this interest through his extensive collection of native and foreign germplasm. A partial list of introductions from hybridizations he started in 1895 follows:

	Cultivar	Date
Sandcherry x *P. americana*	Sansota	1907
	Owanki	1908
	Cheresota	1911
(Sandcherry x *P. munsoniana*) x *P. salicina*	Opata	1911
Sandcherry x *P. salicina*	Etopa	1908
	Okiyi	1908
	Sapa	1908
	Enopa	1911
	Eyami	1911
	Ezoptan	1911
	Wachampa	1911
	Oka	1924
	Honey Dew	1950

	Cultivar	Date
Sandcherry x P. simonii	Tokeya	1908
Sandcherry x apricot (*P. armeniaca*)	Yuksa	1908
Sandcherry x peach (*P. persica*)	Kamdesa	1908

Other experiment stations involved in sandcherry crosses and the cultivars introduced follow:

	Cultivars	Date	Probable Cross
Minnesota Agr. Expt. Sta.	Zumbra	1920	*P. besseyi* x *P. salicina*
	St. Anthony	1923	*P. besseyi* x *P. salicina*
	Deep Purple	1965	Sioux x Elephant Heart
North Dakota Agr. Expt. Sta.	Cooper	1928	Sandcherry x *P. hortulana*
Dominion Expt. Sta., Morden, Manitoba	Mordena	1930	Sandcherry x *P. hortulana*
	Dura	1940	Sandcherry x *P. salicina*
	Manor	1945	Sandcherry x *P. salicina*
Univ. of Saskatchewan, Saskatoon	Alpha	1960	Sandcherry x *P. salicina*
	Delta	1960	Sandcherry x *P. salicina*
	Epsilon	1960	Sandcherry x *P. salicina*
	Gamma	1960	Sandcherry x *P. salicina*
	Kappa	1960	Sandcherry x *P. salicina*
	Omega	1960	Sandcherry x *P. salicina*
	Sigma	1960	Sandcherry x *P. salicina*
	Zeta	1960	Sandcherry x *P. salicina*
Central Expt. Farm, Ottawa	Algoma	1937	Sandcherry x *P. salicina*
U. S. Northern Great Plains Res. Center, Mandan, N.D.	Hiawatha	1957	Sandcherry x *P. salicina*
	Sacagawea	1957	Sandcherry x *P. salicina*

Nanking (Hansen Bush) Cherry

Selection in *Prunus tomentosa* led to introduction of 'Drilea' by the Morden station in 1938 and of 'Orient' by the Minnesota station in 1949. Several superior selections of this species are available from domestic and plant introduction sources. Professor Hansen of the South Dakota Agricultural Experiment Station was largely responsible for importing and demonstrating the adaptability of this exotic species.

A Canadian nursery introduced 'Eileen', a hybrid between this species and sandcherry.

Other Cherries

Prunus pensylvanica was utilized by Minnesota fruit breeders in the production of the 'Nicollet' and 'Zumbra' cultivars. The former approaches sour cherry type and the latter cherry plum type, each resulting from the cross (*P. avium* x *P. pensylvanica*) x *P. besseyi*. A Canadian nursery crossed this species with sour cherry to produce 'Dropmore', which does not have edible fruit, but has some potential as a rootstock.

Improvement of St. Lucie (*P. mahaleb*) seed sources has been made through selection for fruit type and elimination of diseases. The Mahaleb 900 rootstock introduced in 1946 by the Washington Agricultural Experiment Station is a mixture of similar clones. Turkish mahaleb rootstocks were selected in Washington State from plant introduction accessions. INRA Sainte Lucie 64 originated in France. At Ege University in Turkey, Dr. Mehret Dokuzoguz is selecting for improved compatibility of *P. mahaleb* with sweet cherry.

Mazzard rootstock seed sources have progressed from unselected pollenizer or uncultivated trees to sources with better tree types which are regularly indexed for disease-freedom, and selected for high germination.

At Biriulov, Russia, extensive hybridization within cherry species is utilized in a search for greater hardiness, dwarfness, and adaptability to mechanical harvesting. *P. fruiticosa*, *P. besseyi* and *P. tomentosa*, as well as sweet x sour cherry crosses, are used.

Garner at East Malling Research Station, England, is testing crosses of several cherry species and genetic dwarfs as rootstocks and interstocks for sweet cherries.

Modern Breeding Objectives

Breeding objectives differ with the producing area. Climatic factors, chiefly amount and distribution of rainfall, excessive heat, occurrence of frosts and freezes, diurnal temperature fluctuation, and similar factors determine to a large extent the type of cherry which will succeed in a specific area. In the United States west of the Rocky Mountains, the firm-fleshed (bigarreau) sweet cherries succeed best in the low rainfall, low humidity, widely fluctuating day and night temperatures, and near-neutral soils, providing moisture requirements can be met by irrigation and temperatures remain in a moderate range. In the eastern United States where rainfall can be expected during harvest, softer-fleshed sweet cherries are not cracked so severely by rains. Usually, proximity to large bodies of water, such as the Great Lakes, controls the minimum temperatures sufficiently to avoid winter damage to trees during most years. Often sour cherries succeed in climates too severe for sweet cherries. In the midwestern and Great Plains areas of the United States, rainfall often is inadequate and winter temperatures too severe even for sour cherries. There only the native species and their derivatives will survive. Roughly, the growing areas in the United States divide into three areas: the western sector, where sweet cherries predominate; the prairie or plains states where native species and their hybrids predominate; and eastern areas where soft-fleshed sweet cherries grow well adjacent to the Great Lakes or in the low foothills. Sour cherries thrive in all but the extreme northern and southern parts of the eastern United States. Most cherries do not grow well under southeastern conditions.

Specific breeding objectives are sought in the various species.

Sweet Cherries

Resistance to rain cracking is an objective in all producing areas. Even the semiarid interior valleys of the western United States tend to receive occasional damaging rains during the harvesting period. Tucker (1934) and many other investigators showed cultivar differences in cracking. A positive correlation exists between firm flesh and susceptibility to cracking by rain. Thus the bigarreau cherries generally are not planted in areas of heavy natural rainfall. Some of the firm-fleshed cherries such as 'Van' appear to have greater field resistance to cracking than does 'Bing'; however, stages of ripening in relation to the timing of the rain may vary among cultivars. 'Lambert' fruits tend to crack in greener stages than do 'Bing' fruits. Retention of warm rain on the fruit or rain followed by a sunny period is most conducive to cracking.

Flesh firmness is a necessity in western areas from which cherries are shipped to eastern markets. Thus, the bigarreau types are used despite the rain-cracking hazard. Increased firmness of flesh would increase the competitiveness of eastern-grown cherries because of the shipper and consumer preference for 'Bing' types.

Large fruit size commands a premium on the fresh market. Areas such as the Pacific Northwest, capable of producing cherries averaging over 26mm, enjoy better fresh market acceptance and usually a price advantage.

Dark skin and flesh color is preferred for the fresh market. Usually this color is described as "mahogany." Several cultivars very similar to 'Bing' in other characteristics have been discarded by the industry because of amber or light red flesh, which often is associated with immaturity of 'Bing' fruits. On the other hand, processors prefer 'Napoleon' type cherries for brining because of ease of bleaching.

Many factors enter into the term quality. The term often is used to mean merely attractiveness, large size, or proper maturity. These contribute to an overall impression, which should also include taste factors, firmness, and texture. High soluble solids (mostly sugars), a pleasing soluble solids: acid ratio which permits sensing of both sweetness and acidity (rather than blandness), and intermediate astringency give the best combination of taste components. Soluble solids and acids may be measured, but most quality factors are subjective.

Attractiveness consists of the amount of blush color and undercolor and of the brightness and glossiness of the colors. Any dullness is equated with overmaturity in the fresh market. Persistence of the greenness of pedicels during marketing also is used to denote prime condition of fruit.

Productiveness is a requirement in all areas. However, adequate thinning by mechanical or chemical methods is not presently feasible, so extremely high productiveness may be undesirable because of fruit size reduction. Smaller fruit sizes are unprofitable. A moderately productive but otherwise well adapted cultivar may be preferable to a very prolific one.

Resistance to numerous diseases and insects is desired. Present levels of resistance are relatively low. Mildew, bacterial canker, verticillium wilt, and several virus or virus-like diseases (such as X-disease or sour cherry yellows) cause losses, sometimes relatively unnoticed, in some areas. Resistance to the cherry fruit flies, the lesser peach tree borer, mites, and aphids would markedly reduce the production costs and labor needs.

Presently, the fresh market is based on the cultivar 'Bing' as it is grown in western areas. Marketing occurs within a few weeks, spread chiefly by shipments from different producing areas. In each area, the harvest lasts for only two or three weeks. There is need for a succession of similar cultivars in each area to supply a demand over a longer total period and to permit more orderly marketing.

Lack of flower bud and wood hardiness contributes to wide fluctuations in annual production. Fully dormant, sweet cherry trees withstand almost as much cold as most apple trees. However, cherries are susceptible to early fall freezes before hardening has been completed.

Self-compatibility in all cultivars would eliminate the present necessity for pollinizer trees, which until recently produced fruit less desirable for the market than fruit from the main cultivar. Pollinizers now comprise about 10% of sweet cherry orchard trees.

Adaptability to more extensive areas would be desirable. Sweet cherry cultivars generally are fairly specific as to environmental requirements.

Adaptability of cultivars to canning and other processing uses in addition to fresh market adaptability is desired.

Sour Cherries

Because nearly all sour cherries are processed, fruit size, attractiveness of the fresh fruit, and other consumer appeal factors are not so essential as in sweet cherries. Tree productivity assumes greater importance. As long as fruit size is large enough that the ratio of flesh to stone is not greatly reduced, the potential to set and mature heavy crops annually is a desirable tree characteristic. Improvement in pitting efficiency—greater freeness of flesh from stones and reduced bleeding of juice—is needed.

Adaptability to mechanical harvesting is becoming more important. Over 85% of the Michigan sour cherry crop, which comprises 50 to 70% of the total United States production, is now harvested mechanically. Fruit firmness and high resistance to bruising are essential to mechanical handling. Evenness of ripening in individual fruits and among fruits is necessary for uniform quality cherries. Uniform formation of abscission layers in response to chemical looseners reduces the shaking force necessary and thereby reduces bruising.

Disease and insect resistance need to be increased. Reduction in productiveness by the sour cherry yellows disease can be partially overcome by gibberellic acid applications (Parker et al, 1964), but genetic resistance appears to offer the best eventual solution.

Improved hardiness of flower buds and of wood would expand the adaptability of sour cherries to areas with more severe climates. Later blossoming would reduce loss of crops from frost damage.

Extended harvest period by means of new cultivars or with chemicals would increase the efficiency of available labor and machinery.

Duke Cherries and Others

Little interest has been shown in improving Duke cherry characteristics, probably because of the lack of commercial handling. Improvements could be made in fruit firmness, in soluble solids, in spread of ripening season, in productiveness, in hardiness, and in disease resistance.

In other cherry species, increased fruit size, reduced astringency and acidity, greater hardiness of wood and flower buds, drought-resistance, and improved resistance to diseases and insects are major objectives.

Techniques of Breeding

Breeding techniques applicable to cherries are similar to those used with other stone fruits and most deciduous fruits.

Cherry flowers usually are white-petaled, single, perfect blossoms with glabrous pedicels up to 3 cm long. They arise in groups of two to five from cluster buds,and form shortened corymbs. Double, rose-like, white blossoms also occur. Usually each blossom contains 30 to 36 stamens, attached along with the petals and sepals to the rim of the calyx cup. The glabrous pistil, usually single, is attached to the base of the calyx cup and often is shorter than the stamens. The stigma is positioned below or even with the anthers at full bloom.

Because of self-incompatibility and some cross incompatiblity in sweet cherries, emasculation is not essential in crossing most cultivars. The incompatibility is based on the failure of pollen tubes of genetic constitution like that of the style to penetrate the style at a rapid enough rate to effect fertilization (Crane and Lawrence, 1929; Crane and Brown, 1937; Ascher, 1966). Only 1% or fewer of self-pollinated blossoms will set fruit and any compatible pollen will penetrate the style more rapidly than self pollen. However, it is mandatory that open-pollination by compatible cultivars be prevented by isolation or by caging to prevent bee visitations (Way, 1968*b*).

Other cherries must be emasculated to ensure controlled crossing. This is accomplished by cutting or tearing away the calyx cup by pinching with fingernails, notched scissors, spring steel pinchers, or similar devices. The outer portion of the calyx cup to which the stamens are attached is excised and the pistil is exposed. Care is required to prevent damage to the pistil such as removal of the stigma or bruising of the style. Bees seldom visit emasculated blossoms.

Crossing may be done in the greenhouse at normal blossoming time or blossoming may be forced earlier. Maintaining bearing trees in the greenhouse is difficult; therefore crossing usually is done in the orchard. Caging the tree or branch before the blossoms open is necessary if blossoms

FIG. 1. 'Bing' sweet cherry tree covered with tobacco seedbed cloth to protect ripening fruits from birds. The frame was also covered during blossoming to exclude bees. From this tree at the Irrigated Agriculture Research and Extension Center at Prosser, Washington, came the fruits which produced the 'Rainier' and 'Chinook' cultivars.

are to be emasculated, and desirable for protection against wind and frost, as well as to exclude bees. Well-braced wood frame cages, covered with tobacco seedbed covering or similar cheesecloth or with nylon screening, may be used to cover whole trees (Fig. 1). Limb cages or sleeves covered with the same materials may also be used (Fig. 2). Bracing must be adequate to withstand severe winds and the cages must be accessible for pollination.

Pollen should be collected from unopened flower buds, preferably just before the petals separate. The buds may be forced in water in the greenhouse or warm room. The opened flowers are rubbed across a fine wire screen sieve and the anthers are collected underneath. The anthers are dried for at least 12 hours at 22 C or slightly higher. The dried anthers shed pollen readily, but some crushing increases the yield of pollen. The pollen is stored in stoppered vials and refrigerated until used. For storage over long periods, freezing or freeze-drying delays loss of viability.

FIG. 2. Cheesecloth pollination sleeves on sweet cherry selections at Prosser. Thin plywood discs at either end are connected by laths. The discs are hinged in halves to permit clamping around main limbs.

Pollen is applied to the bare stigma (or complete flower, as the case may be), with a camel's hair brush immediately after emasculation or up to two or three days after emasculation. Usually not more than a day should elapse between emasculation and pollination. Eaton (1959, 1962) showed this varied between cultivars used as seed parents. 'Schmidt' required immediate pollination as the flowers opened; 'Windsor,' 'Bing,' and others did not. To conserve pollen, a glass tube with rounded tip or pencil eraser may be used rather than the brush. Successive applications of pollen may be necessary if all blossoms are not open or cannot be emasculated at once.

Applying pollen by dusting or other nonselective methods usually is wasteful of pollen and often ineffective.

The fruits are harvested at about normal market maturity. Because cherries are so attractive to birds, recaging the tree is desirable to prevent loss of valuable crosses. Evidently the seeds are fully developed ahead of the fruits, so harvesting slightly immature fruit will not adversely affect viability of seeds. The pits should be removed immediately after harvest before any fermentation occurs. The pits should be washed and scrubbed thoroughly to remove all flesh.

Seeds from early-maturing cultivars may germinate poorly. Embryos tend to abort under usual afterripening procedures. Culturing embryos of early-maturing cultivars on nutrient agar increases germination of sweet cherry (Tukey, 1933; and Olden, 1959) and 'Montmorency' sour cherry (Havis and Gilkeson, 1949). Zagaja (1966) showed increased germination of cultured embryos and improved growth of seedlings from a cold treatment of sweet and sour cherry embryos.

Seeds from later-ripening cultivars are afterripened by conventional methods. This process involves a moist, cold treatment for three or more months. Germination of seeds of sweet cherry crosses increased up through six months of afterripening, although gibberellin soaks appear to substitute for the latter half of this cold requirement (Fogle, 1958*b*; Fogle and McCrory, 1960). Pillay et al. (1965) obtained similar results for mazzard rootstock seeds. However, Proctor and Dennis (1968) could find no measurable change in extractable gibberellins between afterripened and nonafterripened seeds. Afterripening proceeds best under uniform moisture conditions at about 0 to 5 C. Storage at the lower portion of this range in the early stages and for the last month or two in the upper portion sometimes gives better germination than a uniform temperature throughout. Removal of the seed coat increases germination at any stage of the afterripening process.

Spaghnum moss makes an excellent medium because of its moisture retention without exclusion of air. A fungicide solution should be used to soak up the spaghnum. Peat moss or mixtures of sand and peat moss also make satisfactory media. Storage may be in synthetic screen or polyethylene bags. The pits should be cracked before or during afterripening to give maximum germination. Commercial walnut crackers may be used.

After the afterripening requirement is satisfied, as indicated by substantial germination in the medium, the seeds should be planted shallowly in a light sandy soil mixture, vermiculite, or similar material in a warm greenhouse. Planting in individual plant bands simplifies the transplanting procedures. The seedlings can be managed for about two months in the greenhouse, until outside soil temperature is favorable and the danger of frost is past. Seedlings which rosette or set terminal buds may be forced by gibberellin sprays (Fogle, 1958*b*). Generally direct seeding to the soil in autumn is not advisable because of reduced germination.

Transplanting of seedlings to the field evaluation block should be timed to avoid late frosts and early hot periods. Keeping the soil around the roots by use of plant bands or clay or plastic pots minimizes the transplanting shock. Water should be added during or after planting to give good contact of the ball of roots with the field soil. Generally, the planting distance should be about 1½m in the row and 4 to 6m between rows, depending on cultivating equipment available.

On sweet cherries, first fruits will be borne on a small percentage of the seedlings during the fourth growing season, and thorough selection can be done during the fifth season. A higher percentage of fruiting can be forced in the fourth season by girdling a main leader of each seedling in mid-May of the third season (Fogle, 1955, Fig. 3). Sour cherries usually bear one season earlier than sweet cherries, and other cherry species even earlier.

Initial selection will depend on the specific objectives of the program. Although some albino or diseased seedlings will eliminate themselves in the greenhouse flats, most are not rogued until fruiting occurs. Elimination of undesirable tree types, trees showing disease or nutrient deficiency symptoms, trees lacking winterhardiness, genetic dwarfs, or those with other objectionable characteristics may be done before fruiting.

Insufficient fruit size probably causes the elimination of more seedlings than any other characteristic. The first crop expresses potential size well. Use of an average diameter of 3 cm as a selective criterion would eliminate perhaps 90% of the seedlings even in areas favorable for the sweet cherry. Large fruit size is essential for the fresh fruit industry. Firmness of flesh is a second requirement, particularly of the West Coast industry. Attrac-

FIG. 3. Fruit on a girdled leader of a sweet cherry seedling in the fourth growing season. The girdle (arrow) was made the previous May. Compare this leader with the parallel one of the same seedling (right). Leon Harris is pictured with the cross-controlled seedlings of the USDA-Washington State program at Prosser.

tiveness—mahogany color and glossiness—such as that of the 'Bing' cultivar is also demanded by the industry. Therefore at least threshold minima for these three characters must be met before quality or other characters can be considered. Size and firmness can be measured and attractiveness determined fairly objectively by comparison to color standards. Most other evaluations are subjective. The evaluator is subject to taste fatigue. Thus in the relatively short ripening season available, the breeder is definitely limited in the number of seedlings he can evaluate.

Selections are made in two categories: 1) those considered to have definite potential as cultivars, which may be marked by colored plastic tape or other appropriate marker, for subsequent propagation of second-test trees and 2) those of possible cultivar status but with one or more questionable characters, or of genetic or parental interest. The latter usually are not propagated but are marked by contrasting colored plastic tape or other appropriate markers. Trees to be rogued are marked at time of fruit evaluation by breaking a main terminal into the center of the seedling from alternate row middles. Original selections are rescreened each season for three to seven years.

Propagated selections are second-tested against standard cultivars. Most promising ones are distributed after two to four crops on the second-test trees to experiment stations for adaptation tests under non-propagation agreements. Grower-cooperator tests, also under non-propagation agreements, are utilized to get commercial volume handling of two or more promising selections in a ripening period. When a selection survives these tests and appears to have cultivar potential, it is named and scionwood is released to nurserymen and growers. A minimum of ten years, and often many more years, is required between the controlled cross and the introduction of a cultivar.

Numerous mutations have occurred in nature in cherries. Mutants for fruit color, tree type, and season of ripening are most noticeable of these. Both the 'Bing' and 'Lambert' cultivars have mutated to light skin and flesh color. These mutants have 'Napoleon' type fruit except for a dark suture line. 'Rainbow Stripe' is such a mutation of the 'Lambert' cultivar.

The spur type of tree is characterized by 'Compact Lambert', an induced mutation of 'Lambert' (Lapins,1963). Likewise self-fertility in sweet cherries was induced by irradiation first in England (Lewis and Crowe, 1954) and later in Sweden (Olden, 1959). Way (1968a) reported apparently self-fertile trees from crosses in 1955 made with pollen treated by x-rays or thermal neutrons crossed on normally incompatible female parents. Shukow and Charitonowa (1968) reported improved pollen production and increased germination of seeds from gamma irradiation.

Breeding Systems

Comparatively, cherry breeding is in an early stage. The genetic variability of the various cherry species has not been fully explored. Hence, good progress is still being made in recombining characters by hybridizing standard cultivars. Selection for the most part has been for simply inherited or readily measurable characters. As the programs have progressed, however, some of the quantitatively inherited characters, such as disease resistance, hardiness, and quality, have become main objectives. These require more sophisticated breeding systems.

Apparently, no sweet cherry program has pursued an inbreeding system. This is understandable because of the universal self-incompatibility of cultivars until English researchers made available three self-fertile selections in 1957. The fertility factor has been transferred to the cultivar 'Stella', introduced by the Summerland Experiment Station in British Columbia, Canada (Lapins, 1970). As self-fertility becomes a component of standard cultivars, the possibilities of concentrating homozygous factors through inbreeding and recombination of inbred lines may be explored. Inbreeding is being used in the sour cherry breeding program at the Max Planck Institute in Germany (Zwintzscher, 1970).

Considerable interspecific hybridization has been done in the development of cherry-plum hybrids. Little difficulty has been encountered in hybridizing certain cherry species with plums, apricots, and even peaches. These crosses have made possible growing cherry-like fruits in the severe conditions of the prairies. Further recombination of these species offers additional advances.

Doubling the chromosome number of sweet cherries and other species for hybridization with sour cherries makes possible recombinations which are not possible within *P. avium* alone. Colchicine has been used in the programs in Sweden and the Netherlands (Gerritson,1956). Some crosses of sweet and sour cherries have occurred naturally through fertilization of sour cherry by unreduced pollen of sweet cherry. Olden (1959) also reports use of colchicine to produce unreduced (giant) pollen grains in sweet cherry for crossing with sour cherry.

The self-fertility gene in sweet cherry also makes possible the use of backcrossing as a breeding system in all cherry species. Usually backcrossing is best adapted to recombining one or possibly two desired characters with an otherwise desirable cultivar. Successive backcrossing to the desirable cultivar and selection for the desired character are used until the recombination is completed.

Induction of mutations by use of various sources of irradiation has been useful in producing self-fertility and spur-type growth habit in sweet cherries. Despite the preponderance of undesirable progeny from irradiated material, this method offers the possibility of destroying or mutating deleterious genes controlling disease susceptibility or similar characters without other significant changes in a clone.

Selection of parents to transmit specific characters did not receive much attention in early breeding programs. As interest increased in the mode of inheritance of specific characters however, specific selections were studied for their ability to transmit desired traits. This required detailed notes on the parents and the progenies to delineate the crosses and the specific parents which transmit higher frequencies of desired characters to their progeny. For this, manual and machine card systems have been developed to measure or rate numerous characters. These range from circling the appropriate term or rating on a printed card to field-punching standard cards which can be analyzed directly by machine (Fogle and Barnard, 1961*b*).

The combining ability of a clone may be compared with that of others by an analysis of progeny from crosses with tester lines. Ideally these tester lines should be relatively homozygous. Development of self-fertile sweet cherries now makes this type of analysis possible. Statistical methods are available (Hansche, et al., 1966) for estimating environmental and genetic variability and the heritability of specific characters in the sweet cherry. These authors estimate high heritability of full bloom date, fruit maturity date, and fruit firmness, and show high positive correlations between early blossoming, early ripening, and firm fruit.

Size of progenies needed to assure recovery of desired phenotypes is dependent on dominance, the number of genes involved, linkages, and often unknown factors which affect the ease of transmission of specific characters. Small progenies are adequate for simply inherited characters such as fruit color. Larger sized white-fleshed fruits come from crosses of two heterozygous dark-fleshed cultivars than from crosses involving light-fleshed ones (Fogle, 1961a) Segregation approximated 3 dark: 1 light-fleshed. Hence a small progeny should give the desired selection. Most characters however, are not simply inherited and other considerations are encountered. Fruit size, for example, tends to be closer to that of the smaller-fruited parent. Hence, in crosses involving a small-fruited parent, very large progenies would be needed to assure a large-fruited selection and, in fact, several generations might be needed to regain the desired fruit size. Arbitrarily, a progeny of about 200 per cross is used until analysis indicates this is too small or too large for the specific objectives sought.

Breeding for Specific Characters

Many vegetative characters are of importance in cherry breeding. Tree type affects orchard management, for example. Very upright growing trees are difficult to keep within convenient picking height without sacrificing bearing surface. Susceptibility to diseases and insects, to fruit cracking, and to production of fruit doubles (because of extreme heat during bud initiation) can limit cultivar adaptability. Many of the desirable tree characters appear to be quantitatively inherited and the source of resistance or other desired characters often is at a low level.

Plant Characters

Some seedling characters permit early roguing to conserve field space. A seedling lethal factor, called albinism or chlorophyll-deficiency, shows single factor inheritance in most crosses (Fogle, 1961a; Kerr, 1963), but more complex inheritance was found in a few crosses in Sweden (Olden, 1959).

Kerr (1963) reported two other apparently genetic abnormalities, crinkle and variegation. Each was controlled by one principal recessive gene but modifiers were probable. Crinkle is thought to be chimeral in origin and there is evidence that it is a seedling lethal when the principal gene is homozygous recessive. Pratt et al. (1968) reported occurrence of normal foliage on irradiated sweet cherry trees affected by crinkle.

Stem color at the seedling emergence stage may be useful to indicate anthocyanin-production capabilities of the mature tree. Progenies studied in Washington gave inconclusive results (Fogle, 1961a) because all seedlings develop red stem color a few days after emergence and counts made included some of the latter type. In Sweden, Olden (1966) found that seedlings lacking anthocyanin from crosses of yellow-fruited cultivars x yellow-red ones mainly produced yellow fruits.

Dwarfness is recessive and is controlled by relatively few factors. Dickson (1961) and Fogle (1961a) reported extremely dwarfed trees in sweet cherry progenies. The latter found up to 17% dwarfs and no apparent intermediate tree sizes and suggested control by duplicate factors. Matthews (1970) found intermediate sizes and suggested control by a 2-gene recessive suppressor system, which explained the segregation ratios observed. Closely linked to dwarfness is a rugose leaf character. Only rarely was this rugosity found on a standard sized seedling (Fogle, 1961a; Olden, 1959). Matthews (1970) found rugose leaves invariably in his progenies. The rugose leaf type may be useful for roguing dwarf types in sweet cherries. In sour cherries 'Northstar' and 'Meteor' produce intermediate-sized trees.

Precosity of fruiting appears to be quantitatively inherited but in cherries some clones are prepotent for this character. 'Van' transmitted early fruiting, i.e., fruiting in the fourth growing season, to 73% of its progeny. Twice as many 'Van' seedlings produced 60% or more of a full crop than did seedlings of any other parent used (Fogle, 1961a).

Vigor appeared to be controlled by relatively few factors in progenies grown in Washington (Fogle, 1961a).

Self-fertility was shown (Way, 1966) to be an allele in the incompatability series described by Crane and Lawrence (1929).

Winterhardiness inheritance in 'Bing' crosses was studied by Tehrani and Kerr (1968). In combinations of 'Bing' with two other tender cultivars, 11 and 19% of the progenies survived a test winter in the nursery. Crosses of 'Bing' with a hardy cultivar gave 62% survival. This suggests possible dominance of hardiness and few controlling factors. Russian scientists (Enikejew, 1968) are crossing *P. fruticosa* with *P. cerasus* for increased hardiness.

Fasciation, caused by failure of lateral buds to separate from the main stem, may be controlled by a single factor (Fogle, 1961a). The condition occurs on only a few branches of a tree and is sometimes found on commercial cultivars. It often does not recur if pruned. Hence, a highly mutable gene may control the weakness or highly specific conditions may be necessary for its development.

A bark canker condition likewise appeared to be controlled by a single factor (Fogle, 1961a). The cankers were not typical of those caused by specific pathogens but appeared to be a genetic weakness.

Incompatibility of the 'Van' cultivar with *P. mahaleb* rootstocks has been noted, but inheritance of this interaction has not been studied.

Fruit Characters

Crane and Lawrence (1934) first showed an apparently monohybrid ratio for flesh color, with red flesh dominant over yellow. Lamb (1953) confirmed this type of inheritance in New York cherry crosses. Fogle (1958a) suggested from his data that this single factor, or a closely linked one, also controlled juice color and is the epistatic factor controlling skin color. A second factor, however, was necessary to explain the segregation ratios obtained for skin color. Dominance appeared incomplete for both factors, giving the illusion of quantitative segregation.

Fruit shape in sweet cherry apparently is controlled by a major factor with heart shape dominant to oblate shape (Fogle, 1961a). However, the occurrence of round, long heart, and pointed shapes suggested other modifying factors.

Fruit size was considered by Lamb (1953) to be controlled by multiple factors in New York crosses. Fogle (1961a) concluded that a major factor explained most of the segregation obtained in crosses in Washington. Because of the three-dimensional growth of the fruits, progenies tend to have a mean fruit size closer to a small-fruited parent than to a large-fruited one.

Fruit texture is considered to be quantitatively controlled (Lamb, 1953, and Fogle, 1961a). Since texture and firmness are closely allied, these two factors have not always been considered separately. Fogle (1961a) considered that a major factor, with firmness partially dominant, is involved in the latter. Lamb (1953) pointed out a linkage between good texture (actually firmness) and late maturity. Tehrani and Kerr (1968) considered this linkage, which they estimated at 12.5 crossover units, to be between a single dominant gene for each character.

Season of maturity was considered to be quantitatively controlled (Lamb, 1953). The range of maturity dates was not so extensive as was expected in each progeny in Washington crosses (Fogle, 1961a) suggesting a major factor. The only very early parent used transmitted earliness to 17% of its offspring. In Canada,Tehrani and Kerr (1968) explain segregation for early maturity obtained in a 'Hedelfingen' x 'Black Tartarian' cross as being controlled by a single dominant gene linked with 12.5 crossover units to fruit firmness, also controlled by a single dominant gene. Vondracek (1968) in Czechoslovakia also noted this correlation and suggested use of it and the correlation between season of maturity and fruit size in breeding programs.

Freeness of the pit, fruit attractiveness, and flavor each appeared to be quantitatively inherited. Certain parents were shown to be superior in transmitting all three characteristics (Lamb, 1953, and Fogle, 1961a).

Resistance to rain cracking appears to be closely linked with softness of flesh. However, certain firm-fleshed cultivars, e.g., 'Van' and 'Hedelfingen', show some field resistance to rain cracking.

Cultivars with highly astringent fruits tend to transmit this flavor component to their progenies, suggesting dominance. Acidity is difficult to separate from astringency but likewise seems to be transmitted readily.

Stem length and thickness, which seem to be closely linked, may be simply inherited. There is a suggestion of dominance of short stem length in crosses involving 'Van'.

In sweet cherry crosses, seedlings have been noted which are more uniform in ripening, and some which abscise readily. These characters can be utilized in breeding sweet cherries for mechanical harvesting.

Disease Resistance

Relatively low levels of resistance to most of the important diseases are found in commercial cultivars. Interspecific hybridization with the native plums undoubtedly added disease resistance to the cherry-plums. It may be necessary to go outside the species to obtain adequate levels of resistance in sweet and sour cherries.

Brown rot now can be controlled adequately by routine fungicide sprays. However, as pesticides are more closely scrutinized, the eventual solution to this problem will be resistant cultivars. Olden (1959) noted the variability in sensitivity to the brown rot organism in sour cherry cultivars and the breeding possibilities. Brown rot resistance in sour cherries also is a major objective in a German breeding program (Zwintzcher, 1968).

Mildew also can be controlled by fungicide sprays. It can become a serious problem in breeding plots where successive plantings of seedlings are grown at close spacings. Higher levels of resistance need to be sought.

In a Russian breeding program, Enikejew (1968) reported that crosses between sour cherry cultivars gave progenies more resistance to cherry leaf spot (*Coccomyces hiemalis*, Higgins) than did crosses of sour cherries and *P. fruticosa*. Borecki and Masternak (unpublished) in Poland suggest polygenic control of leaf spot resistance. They demonstrated that higher percentages of resistant seedlings occur when one parent is resistant.

Resistance to gummosis (bacterial canker, *Pseudomonas* spp.) in England (Matthews, 1968) and Sweden (Olden, 1959) is considered to be quantitatively inherited. Resistant selections have been combined with self-fertile, high quality selections in England.

Some differences in resistance to wilt caused by *Verticillium albo-atrum* (Reincke and Berth) have been noted and these need to be utilized to permit planting sweet cherries after wilt-susceptible crops such as potatoes, tomatoes, and melons.

Several of the virus and viruslike diseases are debilitating to cherry trees but most do not kill the trees. Usually these trees need not be rogued until they become unprofitable. X-disease is an exception. In certain areas of Utah, usual sanitation methods do not control the disease. Even top-working individual leaders on *P. mahaleb* frame-work trees fail to prevent spread of the causal agent. In most areas, a diseased leader can be removed without loss of the remainder of the tree on *P. mahaleb* because of a supersensitive reaction which prevents spread of virus through mahaleb to other leaders. Even this fails in some producing areas. Two parents with relatively good resistance to X-disease have been utilized in cooperative studies by the Utah Agricultural Experiment Station and the USDA to produce several selections, now in advanced testing stages (Wadley, 1970). The introduction 'Salmo' by the Summerland Research Station is reported to have resistance to "little cherry" disease. New York tests (Gilmer and Way, 1961) show pollen transmission of necrotic ringspot virus in sour cherries. Other studies (Way and Gilmer, 1963) showed reduction of fruit set from use of pollen from trees with sour cherry yellows.

Successive indexing of parents and selections for viruslike diseases is practiced in most breeding programs. Sources of greater resistance need to be found and incorporated into these programs, however. Some viruses, such as green ring mottle, are symptomless in commercial cherry species but are widespread in available cultivars (Fridlund, 1963).

The observant fruit breeder undoubtedly selects, knowingly or otherwise, for better than average resistance to adverse factors such as susceptibility to low winter temperature and to diseases and insects. These factors detract from some of the major objectives, such as yield, quality, or tree longevity. However, specific selection of parents with greater innate resistance and prepotency within the species or from compatible related species is necessary to increase the resistance of sweet cherries, for example, to verticillium wilt, bacterial canker, brown rot, or mildew in the scion cultivar and to the root rot organisms in rootstocks.

Rootstocks

Improved rootstocks have come primarily from selection within the *P. avium* and *P. mahaleb* species. Much of this improvement has come through selection of clones free of harmful viruses and other diseases and maintenance of disease-free seed source trees. Likewise, stocks have been selected for compatibility with commercial cultivars. Vigorous and more upright-growing mahaleb sources have been introduced from Turkey and Russia.

The mazzard and mahaleb rootstocks have been hybridized in an effort to remove some of the limitations of both rootstocks. However, this rootstock has not been widely tested.

Efforts have been made to reduce the size of mature sweet cherry trees. Some *P. cerasus* clones have been utilized for this purpose. 'Montmorency' sour cherry interstocks between the mahaleb rootstock and the scion cultivar induce precocity of bearing and help control ultimate tree size. 'Northstar' has been used as the interstock where even more dwarfing is desired. Genetic dwarfs have been noted in most cherry breeding programs and these are now being investigated as possible dwarfing rootstocks at the John Innes Institute in England and in Sweden (Olden, 1966). In the Netherlands, other *Prunus* species are being screened for possible use as sweet cherry rootstocks. *P. incisa* Thunb. and *P nipponica* var. *kurilensis* (Miy.) Wilson show most promise. In Belgium, some clones of ornamental cherries are being tested as rootstocks by Monin (1966).

'Stockton Morello' seedlings and suckers are sometimes used as rootstocks for sweet cherries where the soil remains too moist for standard stocks.

Prunus fruticosa was not a satisfactory rootstock in Washington State tests (Toyama et al., 1964). Way (1968a) reports testing of 47 selected *P. fruticosa* x *P. avium* hybrids as interstocks or rootstocks.

Achievements and Prospects

The impact of the sweet cherry improvement programs on commercial orcharding has just begun. 'Bing' and other leading cultivars probably will remain the major ones for many years. However, the necessary spaces for pollinizer trees are no longer planted to mazzards or unmarketable small-fruited cultivars. Several of the new cultivars combine effective pollinization of the main crop cultivar and desirable marketing characteristics as well. 'Rainier' produces larger and firmer fruits than 'Napoleon', the standard brining cherry, for example. The potential of cherry breeding is only beginning to be realized. Some progress has been made in increasing resistance to rain cracking, but usually at the expense of fruit marketing qualities. The maturity season has been lengthened, both earlier and later, but equivalent quality, firmness, and fruit size is not available over the entire period. The 'Van' cultivar shows markedly better tree hardiness than 'Bing' but hardiness needs to be increased and incorporated into clones that extend over the entire ripening sequence. The induction of the spur-type sweet cherry tree, 'Compact Lambert', in Canada promises a more manageable semidwarf tree size for the future. The potential of increased disease and insect resistance is largely unexplored. Self-fertility of sweet cherry was induced in England and has been incorporated into a promising Canadian cultivar, 'Stella'. This fertility can be transferred by hybridization to forthcoming cultivars.

The impact of sour cherry improvement on the industry to date has been based on strain selection (Gardner and Toenjes, 1948). However, the breeding programs in the upper Midwest have extended the range of this fruit to more severe climates than before possible. As germplasm is incorporated from the parts of the world where sour cherries are indigenous, great progress is expected. The fruit size and quality, productiveness, and tree vigor of 'Montmorency' will be combined with this hardiness.

Growing cherry-like fruits in the prairie states and provinces of the United States and Canada was made possible by the upper Midwest and Canadian breeding programs. Continued use will be made of species hybrids to combine the disease resistance, drought resistance, and hardiness of native species with commercially desirable fruit and tree characteristics.

The need for improved rootstocks was not fully realized in the early phases of cherry breeding programs. The need for compatibility of the rootstock with specific scion clones has been considered, however, and will continue to be. Resistance to verticillium and to root rot organisms, tolerance to excessive or deficient moisture, and superior ability to take up and transport nutrients will be sought. The improvements to date have been primarily clonal selections. Present programs have improved rootstocks as important objectives and progress can be expected.

Literature Cited

Alderman, W. H. 1926. New fruits produced at the University of Minnesota Fruit Breeding Farm. *Minn. Agr. Expt. Sta. Bul.* 230.

Anonymous. 1960. Chinook and Rainier, new cherry varieties. *Wash. Agr. Expt. Sta. Circ.* 375.

Ascher, P. D. 1966. A gene action model to explain gametophytic self-incompatibility. *Euphytica* 15: 179–183.

Bargioni, G. 1970. 'Vittoria' a new sweet cherry cultivar (in Italian). *Revista della Orto. Italiana* 63:12.

Brooks, R. M. 1954. Cherry breeding at the California station. *Fruit Var. Hort. Dig.* 9:30–31.

Brooks, R. M. and W. H. Griggs. 1964. Five new sweet cherry varieties. *Calif. Agr. Expt. Sta. Bul.* 806.

Brooks, R. M. and H. P. Olmo. 1972. *Register of new fruit and nut varieties.* 2nd ed. Berkeley: Univ. of Calif. Press.

Cociu, V. 1971. The assortment of fruit trees (in Romanian). *Capsuni di Arbusti Fruct. Bucuresti.*

Crane, M. B. and A. G. Brown. 1937. Incompatibility and sterility in the sweet cherry, *Prunus avium* L. *J. Pom. Hort. Sci.* 15:86–116.

Crane, M. B. and W. J. C. Lawrence. 1929. Genetical and cytological aspects of incompatibility and sterility in cultivated fruits. *J. Pom. Hort. Sci.* 7:276–301.

———. 1934. *The genetics of garden plants.* London: MacMillan.

Cullinan, F. P. 1937. Improvement of stone fruits: cherries. *USDA Yearbook of Agriculture* 1937: 724–737.

De Candolle, A. 1890. *Origin of cultivated plants.* New York: Appleton.

Dickson, G. H. 1961. Cherry breeding. *Rpt. Ont. Hort. Expt. Sta. & Prod. Lab.* 1959–60:53–58.

Duka, S. H. 1956. *Selection of stone fruits* (in Russian). Moscow: Selhozgiz. 274–277.

Eaton, G. W. 1959. A study of the megagametophyte in *P. avium* and its relation to fruit setting. *Can. J. Plant Sci.* 39:466–476.

———. 1962. Further studies on sweet cherry embryo sacs in relation to fruit setting. *Rpt. Ont. Hort. Expt. Sta. and Prod. Lab.* 1962–63:26–38.

Enikejew, H. K. 1968. Commercial value and biological characteristics of hybrids originating from crosses between *Prunus cerasus, P. fruticosa* and *P. avium.* (in German; English summary). *Proc. Sympos. on cherries and cherry growing (Bonn)* 138–145.

Fogle, H. W. 1955. Girdling sweet cherry seedlings for early fruit production. *Fruit Var. Hort. Dig.* 10:55;56.

———. 1958a. Inheritance of fruit color in sweet cherries (*P. avium*). *J. Hered.* 49:294–298.

———. 1958*b*. Effects of duration of after-ripening, gibberellin and other pretreatments on sweet cherry germination and seedling growth. *Proc. Amer. Soc. Hort. Sci.* 72:129–133.

Fogle, H. W. and C. S. McCrory. 1960. Effects of cracking, after-ripening and gibberellin on germination of Lambert cherry seeds. *Proc. Amer. Soc. Hort. Sci.* 76:134–138.

Fogle, H. W. 1961*a*. Inheritance of some fruit and tree characteristics in sweet cherry crosses. *Proc. Amer. Soc. Hort. Sci.* 78:76–85.

Fogle, H. W. and E. Barnard. 1961*b*. Punched cards as aids in evaluating seedlings in the field. *Fruit Var. Hort. Dig.* 15:47–50.

Fridlund, P. R. 1963. Prevalence of the green ring mottle virus in Washington sweet cherry orchard trees and in foreign cherries introduced prior to 1953. *Plant Dis. Rptr.* 47:345–347.

Gardner, V. R. and W. Toenjes. 1948. Strain differences in the Montmorency cherry. *Mich. Agr. Expt. Sta. Quart. Bul.* 31:83–90.

Gerritsen, C. J. 1956. Improvement of the cherry varieties used in the Netherlands. *Euphytica* 5:101–116.

Gilmer, R. D. and R. D. Way. 1961. Pollen transmission of necrotic ringspot and prune dwarf viruses in sour cherry. *Tids. for Planteavl.* 65:111–117.

Grubb, W. H. 1949. *Cherries*. London: Crosby Lockwood.

Hansche, P. E., V. Beres, and R. M. Brooks. 1966. Heritability and genetic correlation in the sweet cherry. *Proc. Amer. Soc. Hort. Sci.* 88:173–183.

Havis, A. L. and A. L. Gilkeson. 1949. Starting seedlings of Montmorency cherry. *Proc. Amer. Soc. Hort. Sci.* 53:216–218.

Hedrick, U. P. 1915. *The cherries of New York*. Albany, N. Y.: J. B. Lyon.

Kerr, E. A. 1963. Inheritance of crinkle, variegation and albinism in sweet cherry. *Can. J. Bot.* 41:1295–1404.

Lamb, R. C. 1953. Notes on the inheritance of some characters in the sweet cherry *Prunus avium*. *Proc. Amer. Soc. Hort. Sci.* 61:293–298.

Lapins, K. 1963. Note on compact mutants of Lambert cherry produced by ionizing radiation. *Can. J. Plant Sci.* 43:424–425.

———. 1965. The Lambert Compact cherry. *Fruit Var. Hort. Dig.* 19:23–24.

———. 1970. The Stella cherry. *Fruit Var. Hort. Dig.* 24:19–20.

Lewis, D. and L. K. Crowe. 1954. The induction of self-fertility in tree fruits. *J. Hort. Sci.* 29:220–225.

Mann, A. J. and F. W. L. Keane. Undated. *New fruits from Summerland, British Columbia*. Can. Dept. Agric. (Unnumbered).

Marshall, R. E. 1954. Cherries and cherry products. In *Economic crops*, vol. 5. New York: Interscience Publ.

Matthews, P. 1968. Breeding for resistance to bacterial canker in the sweet cherry. *Proc. Symp. on Cherries and Cherry Growing (Bonn)* 153–164.

———. 1970. The genetics and exploitation of dwarf seedlings in the sweet cherry. *Proc. Angers Fruit Brdg. Symp.* 319–335.

Monin, A. 1966. Summary of the main features of fruit breeding activities at Grand Manil. *Proc. Balsgard Fruit Brdg. Symp.* 100–102.

Olden, E. J. 1959. Cherry breeding at Balsgard. *Sv. Pom. Foren. Arrskr.* 48:47–60. Stockholm.

Olden, E. 1966. Current problems and trends in stone fruit breeding. *Proc. Balsgard Fr. Breed. Symp.* 147–153.

Olden, E. J. and N. Nybom. 1968. On the origin of *Prunus cerasus* L. *Hereditas* 59:327–345.

Oratovskiy, M. T. 1956. Sweet cherry selection in southern Ukraine (in Russian). 262–273. In *Selection of stone fruits*. Moscow: Selhozgiz.

Parker, K. G., L. J. Edgerton, and K. D. Hickey. 1964. Gibberellin treatment for yellows-infected sour cherry trees. N. Y. Agr. Expt. Sta. *Farm Res.* 29:8–9.

Pillay, D. T. N., K. D. Brase, and L. J. Edgerton. 1965. Effects of pre-treatments, temperature and duration of after-ripening on germination of mazzard and mahaleb cherry seeds. *Proc. Amer. Soc. Hort. Sci.* 86: 102–107.

Pratt, C., R. M. Gilmer, and R. D. Way. 1968. Occurrence of normal foliage in irradiated sweet cherries with crinkle. *Plant Dis. Rptr.* 52:268–271.

Proctor, J. T. A. and F. G. Dennis. 1968. Gibberellin-like substances in after-ripening seeds of *Prunus avium* L. and their possible role in dormancy. *Proc. Amer. Soc. Hort. Sci.* 93:110–114.

Schaer, E. 1968. The new cherry species 'Alfa' and 'Beta' (in German). *Zeitschr. F. Obst. Weinbau* 104:670–674.

Shukow, O. S. and J. N. Charitonowa. 1968. Breeding and genetic investigations on cherries at the "I. W. Mitschurin" central genetic laboratory (in German, English summary). *Proc. Symp. on Cherries and Cherry Growing (Bonn)* 105–115.

Tehrani, G. and E. A. Kerr. 1968. Fifty years of cherry breeding at Horticultural Research Institute in Ontario, Vineland, Canada (German summary). *Proc. Symp. on Cherries and Cherry Growing (Bonn)* 114–120.

Toyama, T. K., H. W. Fogle, E. L. Proebsting, E. C. Blodgett, and M. D. Aichele. 1964. Three-year production records from a cherry rootstock study. *Proc. Wash. State. Hort. Assn.* 60:119–120.

Tucker, L. R. 1934. A varietal study of the susceptibility of sweet cherries to cracking. *Idaho. Agr. Expt. Sta. Bul.* 211.

Tukey, H. B. 1933. Artificial culture of sweet cherry embryos. *J. Hered.* 24:7.

Verner, L. 1946. The Lamida, Ebony and Spalding sweet cherries. *Idaho Agr. Expt. Sta. Circ.* 109.

Vondracek, J. 1968. Correlations between phenophases in sweet cherry may be useful in sweet cherry breeding (in German; English summary), *Proc. Symp. on Cherries and Cherry Growing (Bonn)* 146–152.

Wadley, B. N. 1970. Developments in disease resistant sweet cherries. *Proc. Utah State Hort. Soc.* 1970: 46–48.

Way, R. D. 1966. Identification of sterility genes in sweet cherry cultivars. *Proc. 17th Int. Hort. Cong.* 1–145.

———. 1968*a*. Breeding for superior cherry cultivars in

New York State (German summary). *Sympos. on cherries and cherry growing (Bonn)* 121–137.
———. 1968*b*. Pollen incompatibility groups of sweet cherry clones. *Proc. Amer. Soc. Hort. Sci.* 92:119–123.
Way, R. D. and R. M. Gilmer. 1963. Reductions in fruit sets on cherry trees pollinated with pollen from trees with sour cherry yellows. *Phytopathology* 53: 399–401.
Wellington, R. and R. C. Lamb. 1950. Sweet cherry breeding. *Proc. Amer. Soc. Hort. Sci.* 55:263–264.
Wight, W. F. 1915. Native American species of *Prunus*. *USDA Bul.* 179.
Zagaja, S. W. 1966. Growth of seedlings from immature fruit tree embryos. *Proc. Balsgard Fr. Breed. Symp.* 181–183.
Zielinski, Q. B. 1961. The Corum sweet cherry. *Ore. Agr. Expt. Sta. Circ. Info.* 609.
Zielinski, Q. B., W. A. Sistrunk, and W. M. Mellenthin. 1959. Sweet cherries for Oregon. *Ore. Agr. Expt. Sta. Bul.* 570.
Zwintzscher, M. 1960. Cherry breeding (in German). *Gartenbauniss* 25:151–161.
———. 1963. The sweet cherry 'Sekunda' (in German). *Der Erwerbs.* 5:101–103.
———. 1968. Classifications in a sour cherry collection that serves as parental material for breeding (In German; English summary). *Proc. Symp. on Cherries and Cherry Growing (Bonn)* 99–104.
———. 1970. Inbreeding as a useful method in fruit breeding. *Proc. 18th Int. Hort. Cong.* 1:21. (Abstr.).

Apricots

by Catherine H. Bailey and L. Fredric Hough

The apricot is considered by many to be one of the most delectable of tree fruits. It has been appreciated and grown with a minimum of cultural care for millenia on the mountain slopes in the temperate regions of central Asia and China. The Soviet Union and the eastern European countries have an estimated annual production of 650,000 mt of apricots (Pieniazek, 1966) which is almost as much as that of the rest of the world—775,000 mt (USDA, 1972). Production information from China is unavailable. In some regions in Asia apricots used for edible seed and for seed oil are more important than apricots grown for fruit.

Although the apricot is a very desirable fruit, the history of apricot culture indicates that apricots are severely restricted in their ecological adaptation. Consequently, although apricots are widely spread geographically, they have not become pomologically important except in areas with very special ecological conditions.

The apricot is often considered to be a drought-resistant species since apricot trees will thrive in areas with low atmospheric humidity, but they are sensitive to a lack of soil moisture. At the present time, in northwestern China where apricots are indigenous, trees grow in solid stands as forests on the dry mountain slopes where other fruits do not grow (Holub, 1958). Where other fruits can be grown apricots are less successful. Favorable areas for apricots in China are found between 35° and 43° N latitude, at altitudes of 700–1500 m where rainfall is not more than about 500 mm per year, and where minimum temperatures in the colder areas do not go below −33 C. Such regions have a continental climate without fluctuating winter temperatures and with warm summers. Similarly in central Asia and Asia Minor apricots are grown at altitudes of 1000–2500 m. Winter temperatures can go as low as −40 C in the most northern area of apricot production in central Asia (Kostina, 1969).

Within extensive collections of apricots from many geographic areas such as Kostina (1969, 1972) has studied, a wide range of phenotypes and genotypes can be recognized, especially with regard to ecological adaptation. With this gene pool available, and with coordinated, ambitious, and persistent breeding programs, certainly cultivars can be developed that will make it possible to grow this high quality fruit in many more areas where temperate fruits are grown.

Origin and Early Development

The apricot belongs to the family *Rosaceae*, subfamily *Prunoidea*, genus *Prunus* L., and sub-genus Prunophora. Most cultivated apricots belong to the species *Prunus armeniaca* L. (*Armeniaca vulgaris* Lam.). Other closely-related species are *P. brigantiaca* Vill., the Briançon apricot from the French Alps; *P. ansu* Komar; *P. mume* Sieb. and Zucc., the Japanese apricot; *P. sibirica* L.; *P. mandshurica* (Maxim.) Koehne; and *P. dasycarpa* Ehrh., the black apricot. This last species is made up of natural hybrids occurring between *P. armeniaca* L. and *P. cerasifera* Ehrh. (Kostina and Riabov, 1959).

Kostina, who includes all species of apricots in the genus *Armeniaca* Juss. while we prefer to include them in the larger genus *Prunus* L., has described the distribution of apricots (1936) as follows:

> A detailed study of literature on the apricot and of herbarium collections made by travelers and botanists has shown that the general area of distribution of wild species of the genus *Armeniaca* Juss. is located in the temperate zone of Asia between 133° and 70° E. and between 52° and 30° N. This area extends from east to west in irregular fashion, passing from Northern Korea and the South Ussurian Region through Manchuria and North China. Following the general course of the mountain ranges, it forms two branches, skirting the great Asiatic deserts on the north and on the south. Then comes a break in the southern branch, which recommences with the Tian-Shan mountain range and extends almost to the limits of its westernmost spurs. The northern branch of the area, starting from the Far Eastern Region and Manchuria, also extends westward, crossing the spurs of the Khingan Mountains, northeastern Mongolia, and the Nerchinsk and Chita Districts of the Trans-Baikal Region, as far as the Selenga Valley. It is separated from the Tian-Shan area by the Altai . . . Mountains.

Cultivated Apricots

Vavilov (1951) includes cultivated apricots in three of his centers of origin. In these centers the regions important for apricots are: 1) the Chinese Center, the mountainous regions of northeastern, central, and western China as far as Kansu Province, and northeastern Tibet; 2) the Central Asian Center, the mountainous area extending from the Tien-Shan south, including the Hindu Kush, to Kashmir; and 3) the Near-Eastern Center, the mountainous continuum that extends from northeastern Iran to the Caucasus and Central Turkey. Vavilov indicated, though, that for apricots, the Near-Eastern Center is a secondary center of origin of cultivated forms.

Vavilov (1926) called attention to the importance of mountains for the manifestation of varietal diversity in cultivated plants. This seems to be especially true in the case of apricots. The association of both the wild species and the older cultivated forms of apricots with mountains is apparent from a study of the relief map of Asia (Fig. 1), which is from page 338 of the *Life Pictorial Atlas of the World*, copyright Rand McNally and Company, R. L., 73-GP-1.

Apricots apparently moved from central Asia through Iran into the Transcaucasian area and westward (Blaha et al., 1966; Goor and Nurock, 1968). Such movement must have occurred as a part of the military, economic, and cultural exchange that followed Alexander of Macedon's penetrations into Turkestan as far as the Fergana Valley during the fourth century B.C. The further movement of apricots westward into Europe seems to have occurred in two steps. Apricots came to be known in Greece and Italy as an aftermath of the Roman-Persian wars during the first century B.C. The specific or generic name, *Armeniaca*, certainly suggests that the apricot was first distributed in Italy and Greece by Armenian traders. It was many years later before apricots were cultivated in the other southern European countries.

From 1928 to 1938, Kostina and others made an effort to collect apricots from all geographical regions (Kostina, 1969). Living collections of seven species and 600 forms and cultivars were established at the Nikitsky Botanic Garden, Yalta, and at the Central Asian Experimental Station of the Institute of Plant Industry, Tashkent. On the basis of the morphological characters, and the pomological and biological responses of these apricots as they performed in these experimental orchards, Dr. Kostina concluded that most of the cultivated apricots of the world belong to one species, *P. armeniaca*. Wild forms still exist in some places in the forests on the slopes of the western, central, and eastern Tien-Shan; in eastern Tibet; and in the forests from the Tsinling Shan range up to the mountains north of Peking.

ROMANIA
CRIMEA
Yalta
Black Sea
TURKEY
DAGESTAN
CAUCASUS MTS.
Caspian Sea
Aral Sea
HORZEM
KOPET-DAG
IRAN
Tashkent
Samarkand
Shahrisyabz
ZERAVSHAN
HINDU KUSH
KASHMIR
Alma Ata
Fergana
L. Balkash
ZAILIJ MTS.
DZHUNGAR MTS.
TIEN SHAN
SINKIANG
ALTAI MTS.
USSR
MONGOLIA
GOBI DESERT
Selenge River
L. Baikal
Chita
KHINGAN MTS.
Amur River
Peking
Yellow River
Lanchou
TSINLING SHAN
CHINA
© Rand McNally & Company, R.L. 73-GP-1

After extended studies of these collections Kostina (1969) distinguishes four major eco-geographical groups and 13 regional subgroups within the common species *P. armeniaca*:

1) Central Asian:	Fergana Upper Zeravshan Samarkand Shahrisyabz Horezm Kopet-dag
2) Irano-Caucasian:	Irano-Caucasian Dagestan
3) European:	West European East European North European (*zerdel* or Ukrainian type)
4) Dzhungar-Zailij:	Dzhungar Zailij.

It is interesting to note how faithfully Kostina's careful study of this extensive collection supports Vavilov's thesis which presumes that original botanical species will become further differentiated into eco-geographical groups during cultivation. Wherever she made an extensive collection, especially in the central Asian region, she ultimately regrouped the cultivars on the basis of collection loci. This dramatizes the ecological, geographical (Fig. 1), and social isolation of these sites during past millenia when the cultivars were being developed.

Descriptions of the four major groups together with consideration of their ecological adaptation should be very useful to fruit breeders (Kostina, 1969, 1972).

The Central Asian Group is the oldest and the richest in diversity of forms. It includes the local apricots from central Asia, Sinkiang (China), Afghanistan, Baluchistan, Pakistan, and northern India. The trees are vigorous and long-lived, forming dense, small-branched crowns. The trees have a long rest period, resisting fluctuating temperatures in late winter. They are late blooming, needing a high accumulation of warm temperatures in the spring for bud break.

The trees have comparatively high resistance to a dry atmosphere but are sensitive to a lack of soil moisture. Most are self unfruitful. The fruits have high sugar content, are small to medium sized, and have a sweet kernel (edible, without much bitterness). They are used fresh or for drying—often drying on the trees. Cultivars ripen from May to September. The susceptibility of the trees to fungus diseases, especially brown rot (*Sclerotinia* sp.) and shot hole (*Coryneum beijerinckii* Oud.), limits their planting in regions of high humidity.

The Irano-Caucasian Group includes the local selections of Armenia, Georgia, Azerbaijan, Dagestan, Iran, Syria, Turkey, North Africa, and, in part, Spain and Italy. The trees are not as vigorous or as long-lived as those of the Central Asian Group. They have thicker branches and shoots and larger, shiny leaves. They are less winter-hardy and begin growth earlier in the spring, having a lower chilling requirement. Most cultivars are self unfruitful. The fruit is larger but not as variable as in the Central Asian Group. The kernels are sweet. The larger types are used fresh or for canning, while the smaller sorts with higher sugar content are used for drying. The flesh is often white or light colored.

We believe that if Dr. Kostina had had access to a larger collection of apricots from North Africa, Tunisia, and the Valley of the Oases south of the Atlas Mountains, for example, she would have included a third (North African) subgroup adapted to a much warmer climate within the Irano-Caucasian Group.

The European Group is the youngest in origin and the least variable. It originated from a relatively few forms which were brought to Europe from Armenia, Iran, and the Arab countries during the past 2000 years. It has been assumed (Kostina, 1936; Löschnig and Passecker, 1954; Blaha et al. 1966; Goor and Nurock, 1968) that the English word *apricot* was derived from *praecocia*, the Latin word for apricot. This suggests that the Romans, and later the English, knew only early-ripening forms of apricots.

The commercial cultivars of North America, South Africa, and Australia also belong to the European Group. The trees of this group come into bearing earlier then those of the Central Asian and Irano-Caucasian groups. The trees of the European Group are not as vigorous, have a shorter period of dormancy and a still faster tempo of bud break. Although they have a shorter chilling requirement, some cultivars, especially of the North European subgroup, can withstand greater cold in the dormant period than can most cultivars in the first two groups. The types are mostly self-fertile. The fruit of all ripen within about one month, and the flesh is yellow or orange, with less sugar and higher acidity than the cultivars of the Central Asian and Irano-Caucasian groups. They have a characteristic apricot aroma. The flesh is firmer and drier than the flesh of the previous two groups. ("Mealy" is the best translation that we can make of the Russian word *muchnistij* that Kostina uses to describe the flesh of this group.) Most have bitter kernels. As a whole, the European Group has greater resistance to fungus diseases, especially brown rot and shot hole.

The lack of variation in the European Group may certainly be inferred from the statement of Dr. F. Nyujto, Kerteszeti Kutato Intezet, Cegled, Hungary (personal communication) that, in the apricot material available to him in Hungary, he could make more rapid improvement by clonal selection than by hybridization and selection in the progenies. Most American breeders would also admit to the lack of variation in the apricot material known to them.

The Dzhungar-Zailij Group is the most primitive. It includes local selections from the Panfilov (Dzharskent), Taldy-Kurgan, and Alma Alta regions in Kazakhstan, as well as the Ining (Kuldja) region of Sinkiang. These apricots have increased winter hardiness (temperatures often reach −30 C). They are mostly small-fruited types, intermediate between the semi-cultivated seedlings of the Central Asian Group and of the northern or Ukrainian subgroup of the European Group.

These descriptions of the characteristics of the apricots in each of the four groups show the differences in the predominating types of trees and fruit in each group. Actually there are a number of intermediate, transitional forms within most of the groups and subgroups.

We believe that there must also be a North Chinese Group of apricot cultivars with four subgroups and an East Chinese Group:

5) North Chinese	Northeastern Tibet North China Southeastern Manchuria Khingan Mountains
6) East Chinese	

This proposed grouping is based on the distribution of the mountains of this region (Fig. 1) together with an extension of Kostina's observation of the development of groups of similar cultivars in eco-geographically isolated areas. Most of the apricots of the Southeastern Manchurian subgroup would be included within the species *P. mandshurica*, most of those of the Khingan Mountains subgroup would be included in *P. sibirica*, while the East Chinese Group would be included within the species *P. ansu*. The North China subgroup was surely richer in forms one or two centuries ago before the population pressure caused the agriculture of this region to concentrate on annual crops for basic food production and destroy most of the trees which could have provided the protective and aesthetic food qualities that come from fine fruits.

Related Species

Prunus brigantiaca, sometimes called alpine plum, obviously is derived from *P. armeniaca*. *P. brigantiaca* is a shrub or small tree that is established in the foothills of the French Alps. The fruits are small and smooth.

Prunus ansu is the cultivated apricot of humid eastern China and Japan. Many authors consider *P. ansu* to belong to *P. armeniaca*. However since the cultivars are adapted to more humid growing conditions, grouping them in a separate species has meaning for pomologists as well as for botanists.

Prunus mume, the Japanese apricot, grows in Japan and eastern China in warmer and more humid regions than *P. armeniaca*. It is more disease resistant. It is usually grown as an ornamental for its bloom which can be single or double, with white, pink, or bright red petals. The plant may vary from a shrub to a tree similar to the common apricot. The fruits are clingstone and usually somewhat smaller than common apricots. The fresh fruit is inedible but the fruit is used pickled.

Prunus sibirica and *P. mandschurica* are very cold resistant species withstanding −40 to −50 C when fully dormant. *P. mandshurica* occurs in the mountains of eastern Manchuria from North Korea northeastward almost to the confluence of the Ussuri and Amur rivers. The eastern range of *P. sibirica* largely overlaps the distribution of *P. mandshurica*. *P. sibirica* extends westward across Manchuria and continues in the mountains north and south of the Gobi Desert. On the north it occurs as far west as the Selenge River. On the south *P. sibirica* extends through the mountains of Inner Mongolia and northern China as far west as the northern loop of the Yellow River. Thus, both species are grown in cold, northern climates with cold winters and cold, late springs. When planted in warmer areas, they bloom early since they have a short rest period and need only a small amount of warmth in the spring to flower. These two species are the basis of the highly cold resistant cultivated forms of Mongolia, northern Manchuria, and eastern Siberia. These species were also the source of cold hardiness for the breeding work of I. V. Michurin. *P. mandshurica* makes a large tree with large leaves as compared to *P. sibirica*. The fruit is small. *P. sibirica* is considered to be a large shrub or small tree with small fruits of low quality, which are often inedible, even almond-like.

Both Kostina (1969) and Holub (1959) recognize *Armeniaca davidiana* (Carr.), yet both comment that it is similar to, and practically just a form of, *P. sibirica*. Holub gives the present range of *A. davidiana* as the mountains of northeast China, and eastern and southeastern Manchuria, extending southward to North Korea. We would suggest that if this form described as *A. davidiana* is to be recognized as a distinct form it should be described as *P. sibirica* var. *davidiana*. Otherwise when we include apricots in the genus *Prunus* there is a possibility of confusion in thinking about the peach-like *P. davidiana* (Carr.) Franch. This species, *P. davidiana*, was described from material collected near Peking. Holub (1959) states that this species extends southward into the warmer and wetter areas in the mountains, especially in the southern part of north China and in the northern part of central China.

Prunus dasycarpa occurs only as isolated trees, principally in orchards or gardens in peripheral areas of apricot production. Kostina and Riabov (1959) have shown that the species *P. dasycarpa* is made up of natural hybrids between *P. armeniaca* and *P. cerasifera*. The trees bloom later and the flower buds are more resistant to cold than most of the cultivated apricots. They are also more resistant to disease. This is especially conspicuous in comparison with the varieties of the Irano-Caucasian and Central Asian groups. The fruit is somewhat plum-like but with very short pubescence. Most forms are dark purple or red but yellow-fruited forms are known.

The occasional occurrence in cultivated apricots of branches bearing plum-like mutant leaves poses new and intriguing questions with respect to the origin of apricots. Tamassy and Pejovics (1965) and Dr. C. O. Hesse of the University of California (personal communication) have observed naturally occurring leaf mutations. Dr. K. O. Lapins at the research station, Summerland, B.C., Canada, (personal communication) has obtained similar leaf types following thermal neutron irradiation. Such leaves are more or less lanceolate and quite plum-like in their over-all appearance. The branches bearing such mutant leaves have been unfruitful. Limited experience with this mutant type suggests that it is not fully competitive with the normal type. Careful pruning while the trees are in leaf is required to maintain the mutant condition.

History of Improvement

Apricots grow best in mountainous regions having a typically continental climate where there are hot, dry summers and steady, cold winters. Such conditions are not conducive to a build-up of large centers of population. So it has been for centuries that good fresh apricots have not been available to the majority of the world. Apricot breeding programs have been initiated in an effort to develop cultivars, both for fresh use and processing, with wider or different ecological adaptation in order that producing areas can be developed closer to the great centers of populations.

The earliest controlled pollinations between apricots were made in the Soviet Union by Michurin early in this century. During the 1920s there was continued concern in the Soviet Union for the evaluation of existing selections and the development of new forms through growing of seedlings. This was a part of the tremendous program in the 1920s and 1930s, directed by N. I. Vavilov, to collect and to evaluate all plant life in the Soviet Union and in the rest of the world that could be of economic value to the Soviet Union (Vavilov, 1951).

The rich collection of apricots at the Nikitsky Botanic Garden was assembled from Dr. Kostina's part in this early work of exploration. The collection that she assembled was so challenging to a trained fruit breeder that it formed the basis of an exciting lifetime for her. Certainly she has been the most productive apricot breeder in the world, and she has either encouraged or challenged others in the Soviet Union to work, too.

The evaluation of existing selections and a concern for developing new cultivars started in the United States, Canada, and Romania about a half century ago. Since then cultivar evaluation programs, and in some cases a limited amount of breeding work, have been undertaken in other countries where apricots were grown or where there was a possibility they might be grown.

Recently some cultivars have been introduced from breeding programs but the new cultivars have not yet had any significant impact on commercial production.

Modern Breeding Objectives

The most important objective in most of the apricot breeding projects is climatic adaptation. The development of apricots with a long period of winter development of the flower buds (i.e., a long chilling period) that will withstand fluctuating temperatures in late winter is probably the most important objective for extending the region where apricots can be grown. In addition to the slow development of flower buds in midwinter, it is very important that the flower buds respond very slowly (require relatively high amounts of day degrees of heat) to warming temperatures in the spring after their rest has been satisfied. The combination of these two characteristics will give the ultimate in late blooming so that the buds or flowers may escape spring frosts.

Greater midwinter cold hardiness of flower buds and wood in combination with acceptable fruit characters is desired for northern areas. Apricot trees that will grow well and ripen fruit under humid growing conditions are an objective of many breeding programs. Trees with a lower chilling requirement, or trees with a high chilling requirement that is met at a higher threshold, are needed to extend the production of apricots into warmer regions.

Resistance to diseases and good pomological fruit characteristics are needed along with these various types of climatic adaptation. Some of the characteristics important for fresh market are large size, attractive appearance (a bright blush over bright orange or cream), freestone, firm flesh, and good quality. For canning apricots, good orange skin and flesh are wanted. Also important are uniform medium size, regular shape, resistance to pit burn during high temperatures just before harvest, good texture (freedom from fibers and vascular bundles), small pit, and a good balance of acid and sugar. For a drying variety, high soluble solids are needed.

Resistance to some or all of the following diseases, depending on the producing area, should be an important objective:

Shot hole disease	*Coryneum beijerinckii* Oud.
Bacterial leaf spot	*Xanthomonas pruni* (E. F. S.) Dow.
Gummosis	*Cytospora* sp. (the imperfect stage of *Valsa* species)
	Cytosporina (the imperfect stage of *Eutypa armeniaca* Hansf. and Carter)
	Pseudomonas syringae Van Hall
Brown rot	*Sclerotinia laxa* Aderh. and Ruhl.
	Sclerotinia fructigena (Pers.) Schr.
	Sclerotinia fructicola (Wint.) Rehm.
Anthracnose	*Gnomonia erythrostoma* (Pers.) Auersw.

Resistance to "apoplexy" must be mentioned. The term *apoplexy*, widely used in Europe, is applied to a sudden wilting and death of a tree or part of a tree. Several causes, environmental and/or pathological, produce this syndrome. Apoplexy is a useful term to refer to this syndrome but the use of the term should not be considered to infer a diagnosis (Paunovic, 1964). When cultivars fully adapted to a given region are developed, resistance to apoplexy will have been achieved.

Extension of the season of ripening, earlier

and/or later, is wanted in all apricot producing areas.

A consideration of the apricot breeding programs in the world dramatizes the need for local adaptation. Breeding programs in Australia, South Africa, and Argentina have been concerned about better local adaptation and about cultivars for processing. There has also been some concern in Argentina and South Africa for cultivars that will withstand shipment to northern markets as fresh fruit.

California and Washington have also been concerned about the development of better cultivars for processing and long distance shipment.

In all of the above programs, the principal effort has been hybridization among recognized cultivars of Kostina's West European subgroup.

In British Columbia, Canada, there has been an effort to collect some cultivars from the Irano-Caucasian Group to extend the genetic base for breeding. In eastern United States and Canada the effort to get adaptation has emphasized the use of locally-adapted selections. Recently we at Rutgers have found additional sources of adaptation in selections derived from *P. mandshurica*.

Similarly in Europe, the principal concern is better adaptation, and resistance to apoplexy. Poland is especially concerned with cold hardiness. Romania has the aggravating situation that prevails in so much of Europe—there are very good cultivars both for fresh market and for processing but they are not dependably productive and long-lived in enough sites. This same problem continues eastward in the European republics of the Soviet Union.

The objectives of any one breeding program will depend on the needs of that region and on the resources available. Some stations are very limited in the objectives they dare or care to undertake. Others have accepted greater challenges for adaptation and development of new combinations of fruit characters. These latter programs have been projected through several to many generations.

Breeding Techniques

Floral Biology

The apricot has a perfect, perigynous flower with a single pistil. The petals are usually white, though some are tinted and, rarely, some are even deep pink in color. Pollen sterility occurs. It is inherited as a single recessive (Hesse, personal communication).

Cultivars may be either self-compatible or self-incompatible. Careful breeding work in the future may well demonstrate an allelic series for self-incompatibility like that known for sweet cherries, since Dr. T. Toyama, USDA, Prosser, Washington, (personal communication) has observed that some crosses between self-incompatible selections have been unsuccessful.

The apricot is a diploid ($2n = 16$, $x = 8$) but occasional tetraploid mutants have been reported.

Pollination

The most efficient way to obtain a volume of pollen is to collect flowers in the field within two to three hours after sunrise. The flowers may be in the balloon stage or beginning to open so long as the anthers have not begun to dehisce. The flowers must be kept at 1–3 C from the time of picking until the anthers can be stripped from the filaments. Consequently if flowers are to be picked from several trees in the morning, it is desirable to carry the bags of flowers in an insulated and refrigerated box until they can be placed in the laboratory refrigerator. Such flowers should not be held more than 24 hours before the anthers are stripped, since even under refrigeration the anthers soon begin to dehisce. The flowers are rubbed over a sieve with mesh openings large enough for the anthers to fall through without crushing but small enough to catch petals and other debris. The anthers are then placed in a tray made of smooth paper and left at room temperature (about 25 C) to dry for 24–36 hours. They should not be placed in direct sunlight. When dehisced and dry, pollen and anther sacs are placed in a straight-sided glass vial, corked and held at −20 C until used. If storage for a long period of time is desired the vial should be cotton-stoppered and kept in a desiccator at 25–50% relative humidity at −20 C. If pollen must be kept at or just above freezing, 16% relative humidity is optimum. When pollen is to be used repeatedly in the field over a period of several days, better viability is maintained when the pollen is taken to the field in a refrigerated container. In this case, immediately after the pollen vial is removed from the refrigerated container, the vial should be warmed in the hand momentarily before the cork is removed in order to avoid condensation inside the vial. Pollen can be shipped anywhere in the world by putting dried pollen in a polyethylene bag, sealing with tape, and sending as letter air mail.

It is helpful, however, to have pollen collected and ready before bloom time. This also makes it possible to use pollen of late blooming sorts on early blooming selections. Branches about 1.5–2 cm in diameter may be brought into a greenhouse for forcing up to a month before bloom. When the buds start to swell, about 1–2 cm of the branch must be cut off under water every other day to insure good water uptake. In comparison with flowers in the field, the anthers on flowers from forced branches dehisce more rapidly. Therefore flowers should be picked at the balloon stage, as often as three times a day, and kept at 1–3 C. Pollen is prepared from such flowers in the manner described above.

When flowers develop normally, emasculation should be delayed as late as practical before the flower opens to expose the stigma. When there is a prolonged period of very cool weather during the blooming season the anthers of some cultivars may dehisce before the petals open. But buds that are just starting to show white are too immature for emasculation. It is mechanically difficult to emasculate them. Furthermore, the pistil may not develop normally if there is a period of unfavorable weather after emasculation.

Emasculation can be accomplished by breaking the calyx cup just below the region where the bases of the sepals, petals, and stamens join. Several techniques have been used. The authors prefer fingernails. Fine-pointed forceps or specially-made or adapted V-notched emasculating tools that will cut the calyx cup but not shear the style can be used. Such tools are best for inexperienced technicians. Apricot flowers are more difficult to emasculate than flowers of other *Prunus* species. Most apricot flowers are bunched together on short shoots or spurs and do not open simultaneously so that they are difficult to manipulate. The most exasperating aspect of emasculation is the fact that the pedicel is short and brittle so that, whether they occur singly or in bunches, many flowers are broken off during emasculation.

Another problem is that, compared with other fruit, it is often much colder during the season of emasculation and pollination making it more difficult for human fingers to work.

It is not necessary to protect the emasculated flowers when the only objective of the cross pollination is to obtain a large segregating population for selection. In general, insects will not visit flowers after the petals have been removed. The occasional bee that does find nectar in the base of the emasculated cup will attempt to collect the nectar by lighting on the twig. Such insects will almost never touch the stigma of the flower from which they are collecting the nectar. On the other hand, protection of the emasculated flowers would be essential in the case of controlled pollinations for a genetic study where a single accidental cross pollination might distort the interpretation of the experiment. Protecting the flowers with opaque materials may give some frost protection by reducing direct radiation from the pistil.

We feel that it is more efficient to apply pollen to the stigma with the tip of the index finger or the end of the cork from the pollen vial than to use a small, fine-haired brush. When the forecast is for fair weather we prefer to apply the pollen the day after emasculation. For good germination the air temperature should be at least 10 C for several hours after pollination. If the forecast is for cold or rainy weather we try to pollinate immediately after emasculation. When the weather is unfavorable or when the pollen has low germination, if the stigmas are still receptive, a second pollination will increase fruit set. Since it is usually cool during apricot bloom, we regularly plan to double pollinate.

Seed Handling

Fruit may be left on the trees as long as possible without rotting. Where animals (squirrels) will eat the sweet-seeded selections, the fruit must be protected somehow several weeks before it ripens.

Pits (seeds within the stony endocarp) from midseason or late-ripening female parents may be stored dry after harvest. Such pits would normally be placed in moist stratification at temperatures just above freezing for two to four months until it is convenient to plant them in the spring.

Very early ripening female parents (those that ripen 70–90 days after bloom) may not have fully mature seeds. Such seeds must have special treatment to give maximum germination. Crossa-Raynaud (1966) has shown that seeds from early-ripening female parents have better germination when placed on an agar-sucrose medium for stratification and germination. It may well be that seeds from many early-ripening female parents would have good germination without special media if they were stratified immediately after harvest so that there was no opportunity for desiccation.

We prefer to stratify all our seeds immediately after harvest (Smith et al., 1969). The pits are removed from the fruit by hand, treated with phenol or carbolic acid (at 150 ml: 3000 ml H_2O for five minutes) and then cracked. The seeds are surface sterilized with aqueous merthiolate (Na ethylmercurithio-salicylate) at 1:2000 dilution for ten minutes, rinsed three times with sterile H_2O, and put in polyethylene bags. The bags must be sealed and labeled. They are then placed in cold storage (1 C) for stratification for two to three months. Additional sterile H_2O should be added to the bags with a hypodermic needle as needed during the stratification period.

Our work schedule is such that we must wait until January to germinate the seeds in a greenhouse. The seedlings are grown in sterilized soil in small plant bands (6 x 6 x 13 cm) without bottoms. We have found that the fungicide ferbam, ferric dimethyldithiocarbamate, has a very low phytotoxicity for seeds and seedlings. Consequently we use it freely and frequently to deter damping-off during the midwinter when light intensity is low, days are short, and the young seedlings are succulent. We plant the seedlings directly from the greenhouse into the orchard as soon as practical in the spring.

Since apricots will germinate after a short stratification period, where there is a long growing season it may often be practical to harvest the seeds about 90–100 days after bloom, stratify them immediately at 4–5 C for about 30 days and then germinate them. Such a procedure makes it possible for the seedlings to be grown in the nursery during the late summer and fall of the same year the cross was made.

Seedling Evaluation

We know of no literature concerning seedling screening or correlation between seedling and adult plant characters with apricots. Breeders working with other fruit crops are finding that such seedling/mature plant correlations are helpful. We would urge breeders working with apricots to develop this sort of information.

Dr. V. Cociu of the Research Institute for Pomology, Pitesti, Romania, (personal communication) has been screening seedlings in the nursery

FIG. 2. Type of seedling apricot tree to keep (left) and type to discard (right). This picture was taken in the nursery at the Institute for Research in Pomology, Pitesti, Romania, in 1972.

on the basis of tree type (Fig. 2). His experience suggests that the trees in the nursery with large leaves and vigorous, relatively-unbranched shoots are more likely to produce medium or large-sized fruits and be better adapted in an ecological condition similar to that of the nursery site. On the other hand, the plants with much branching, very thin shoots, and small leaves are likely to be late in coming into fruiting, small-fruited and unadapted. Consequently, where land for first-test seedlings is at a premium, Dr. Cociu believes he can afford to discard the second type.

It should be recognized, though, that with the many challenges for recombination that are before us in apricot breeding there will frequently be occasion to make "wide" crosses between very different parental types where the desired recombination may well not be expected to be recovered in the first generation. For such objectives it might not be wise to screen the progeny rigorously for local adaptation.

Screening seedlings for disease resistance is certainly important when efficient inoculation and screening methods are available. Drs. R. E. C. Layne and B. N. Dhanvantari, of the Research Station at Harrow, Ontario, Canada, (personal communication) are screening seedlings for resistance to *Xanthomonas pruni*.

Crossa-Raynaud (1969) reported screening seedling selections for resistance to blossom rot (*Sclerotinia laxa*) after artificial inoculation.

Dr. C. N. Clayton, Department of Plant Pathology, North Carolina State University, Raleigh, (personal communication) is attempting to evaluate selections for their resistance to *Cytospora* following artificial inoculation.

Each breeder will evaluate his progenies for the characters important for his program on the basis of the objectives defined for his region. Different progenies may well be evaluated using different criteria. For example where two or more generations are needed to obtain selections with healthy stems and resistance to cold and subsequent canker infection, a lesser level of good horticultural characteristics would be tolerated in the early generations.

Records

In most cases, in a breeding program committed principally to the development of new cultivars, detailed records need only be taken for the seedlings selected, not for every seedling in the progeny.

When records for genetic analysis are a part of the program, special crosses will usually have to be made to effectively elucidate the mode of inheritance. In such cases the data to be obtained and the method of analysis should be planned before the cross is made so that the accumulated data can be readily analyzed, by a computer if possible. Often detailed data are obtained for a sample (the first 50 or 100 seedlings) of a progeny made for cultivar improvement. Whenever it can be obtained, knowledge about the inheritance of characters in apricots would be valuable for planning future hybridization. Since health, longevity, and productivity are very important factors limiting the extension of apricot growing in all areas, all breeders should be concerned with accumulating and disseminating information about these characters.

Testing and Introduction

Naming and release of a new cultivar of apricots by a responsible research institution presumes that, in the mind of the introducer, the new cultivar is superior with respect to certain pomological characters that will make it profitable to grow. There is, however, always an element of chance. On the one hand the desired new characteristic which the introducer has recognized may not add a significant amount of profit. And until the new cultivar has been widely tested in commercial production in different producing regions, there is always the chance that it may show certain weaknesses which the introducer had not previously recognized.

Where the breeding program is working closely with a production group, it is often in the best interests of the producers to have new selections tested in limited volume by several growers before the final decision is made to name and release for extensive commercial planting.

When there is considerable likelihood that the new selections may be much more profitable than the established cultivars, grower groups may be willing to take chances in testing new selections in order to have the privilege of getting information about the new selections more quickly. The ultimate in such an arrangement would be for the growers to assume practically all of the expense for second test evaluation of new selections. In such a situation the fruit breeder, or a competent and interested extension specialist participating in the second test evaluation of new selections, would coordinate the testing.

Such an arrangement makes it possible for growers to be informed at the earliest possible moment after the selections have been recognized in the breeding plots. It allows the breeder to focus all his resources—land, money, and time—on the actual breeding work, which is the production of segregating populations and the selection of superior individuals. It assumes that the industry can and will invest heavily in this second testing of promising selections. The extent of this second testing by growers will vary from selection to selection and from grower to grower depending on the anticipated potential of the selection and on the nature of the grower's operation. Obviously there will be concern about productivity and ecological adaptation, especially with respect to disease resistance, winter and spring cold hardiness, special fruit characters, adaptability for fresh market and/or processing, time of ripening, etc. Finally, dependence on second testing solely by growers assumes that there will be rather conspicuous differences in some or many of these characteristics, so that detailed records over many years are not necessary for the industry to be willing to establish commercial plots of a new selection.

There are situations where such a program is inappropriate. In some places the industry is not able or not willing to undertake any of the risks associated with the introduction of a new cultivar. This is usually true in the Socialist countries in Eastern Europe where the pomologist is responsible for making official cultivar recommendations to government and large government-controlled orchard units. Such a person, therefore, must have a carefully-planned cultivar testing program which has fully replicated plots in each of the important producing regions of his country. He must have production records over a sufficient period of time, five to ten years as a minimum, so that he can be confident his perspective covers most of the climatic variation that will occur. His recommendations must be valid for the plantings and production that the country will have for the second and third decade after he makes a recommendation.

Dr. J. Kalasek, National Fruit Trials, Zelesice u Brna, Czechoslovakia, and Dr. P. Tomcsanyi of the National Institute for Variety Testing, Budapest, Hungary, have very sophisticated and comprehensive programs of cultivar and selection testing which are providing the sort of information called for in the above paragraph.

Another situation arises when a breeder must depend on revenue from patents for much of his support. In such a situation the breeder must do adequate second testing before he dares to decide to patent. He must maintain the reputation for developing patented cultivars that will be truly profitable, so that his new introductions will be accepted readily and in volume by the industry.

Breeding Systems

Varietal and Selection Crossing

For the most part, on the basis of our experience with apricots, the conclusions of Hansche et al. (1972) with sweet cherries, peaches, and walnuts are also appropriate for apricots:

> . . . the straightforward procedure of selecting parents on the basis of their own performance, and their subsequent mating *inter se*, should affect relatively rapid genetic gains for several generations. More complex selection procedures and breeding methods (such as those that employ progeny testing and inbreeding that are designed to exploit non-additive genetic effects) appear to be unnecessary in these stocks. In fact, such procedures would most likely result in a lower rate of genetic gain since with such methods, selection can usually be practiced only every other generation.

This statement should provide the basic perspective for apricot breeders. In any breeding program at any one time it is necessary to budget the total resources of the project among the several objectives. We have felt that it increases our chances of achieving a given objective, when there has been no progeny testing, to use two or three phenotypes which seem equally promising for that objective. Consequently after we have decided on the budget (the total number of seedlings that *we can afford* for the given breeding objective), we prefer to raise several intermediate-sized progenies, rather than trying to choose a single pair of parents for one large population to fulfill the budget.

We would agree that, from the point of view of time, progeny testing often is not necessary.

But where progeny test data are available, as is the case for many of the cultivars in the collection of the Nikitsky Botanic Garden (Kostina, 1972), they should be considered in the choice of parents.

Modified Backcrossing

There are certain characters, such as disease resistance, cold hardiness, and late blooming, that can be effectively incorporated with other desirable pomological characters through a modified backcrossing program. Important considerations for the success of such a program are: 1) an adequate screening procedure so that the desired phenotypes may be identified efficiently in each generation, and 2) the use of different high quality backcross parents in successive generations. This practice allows the breeder to incorporate a greater variety of pomological characters into the breeding lines and at the same time to avoid the deleterious effects of inbreeding that are associated with repeated backcrossing to a single cultivar.

Interspecific Hybridization

There is much that can and should be done in apricot improvement by means of the two procedures described above. However there are characteristics in other *Prunus* species, such as later blooming, greater disease resistance, cold hardiness, and modified tree types, that would be of value for regional adaptation if they could be incorporated into the apricot.

Kostina (1969) and her colleagues have been making interspecific crosses between apricots and plums including *P. cerasifera*, *P. salicina* Lindl., *P. dasycarpa*, and *P. besseyi* Bailey. Uljanishev (1956) specifically mentions work with *P. besseyi* and *P. besseyi* hybrids. He also indicated that three or four generations are necessary to obtain acceptable cultivars for severe climates. Many sterilities are observed. Some fruitful hybrids have been obtained although none are satisfactory commercial forms. Clones of *P. dasycarpa*, however, are grown locally for their fruits in regions too extreme for apricots. The great and difficult task of increasing the yield and improving the quality of the fruit of these hybrids still stands (Kostina, 1969).

Hybridization with other *Prunus* species should also be considered. In this connection we suggest that *Prunus davidiana* should be explored as a genetic bridge to facilitate the transfer of characters from other species to apricots and vice versa. For example, we have made limited pollinations of apricots with pollen of a (peach x *P. davidiana*) hybrid and obtained seedlings with genetic markers proving the seedlings were hybrids. While using *P. davidiana* as a bridge it would be good if we could explore the genetic variability in this species. The specimens of *P. davidiana* in North America are the result of limited importations of seeds from the vicinity of Peking. Holub (1959) indicates that Peking would be in the northern range of the species and that it extends south at least into the mountains of central China including the eastern slopes too wet for most apricots. It seems likely that some desirable aspects of adaptability might even be obtained from *P. davidiana*.

Mutation Breeding

'Early Blenheim' was selected following thermal neutron irradiation and was introduced by Lapins (1972) as an early-ripening cultivar for local markets. Lapin's work is the only report of achievement from mutation breeding with apricots. On the other hand, considerable experience has been accumulated in the irradiation of other tree fruits which can be applied to the irradiation of apricots.

Three concepts are basic for mutation breeding when working with vegetatively-propagated fruit crops: 1) efficient screening; 2) treatment at appropriate stage of plant development; and 3) awareness of chimeras.

An objective should be chosen that will allow for efficient screening of a volume of material at an early stage. This often requires manipulation of the irradiated material to allow for early recognition of the desired mutants. It is costly to grow large volumes of fruit plants to maturity before screening for mutations.

Stages of plant development at which irradiation treatments are most effective were determined by Lapins et al. (1969) working with apple and peach. They concluded that the irradiation of physiologically active meristems gave higher mutation frequencies than the treatment of less active tissue. It was necessary to have conditions favoring extensive cell division immediately after irradiation for the best growth of mutated cells in competition with non-affected cells. This experience should be appropriate, too, for work with apricots.

A somatic mutation originates a chimera. Irradiated plant material must be manipulated so that it is possible to select out stable forms, that is, periclinal chimeras or totally mutant shoots. Ultimately it must be possible to increase the mutant character throughout successive generations of asexual propagation (Hough, 1965).

Breeding for Specific Characters

Adaptation

One of the most critical characters for the increase of apricot culture in an existing area or the spread of apricots to new areas is increased adaptation. In most areas increased cold hardiness is the primary concern.

Cold hardiness is usually interpreted in terms of flower bud survival at the end of the dormant season. There are different stages during the dormant season when the plant responds differently to warm temperature and to cold temperature, and breeders must realize that ultimate bud survival is often not a definitive criterion to describe the nature of the cold injury that occurred to a given cultivar during a given dormant season.

There are three stages of flower bud development when more hardiness is needed: 1) in early winter and midwinter (the period of "rest" of the apricot flower buds); 2) in late winter and early spring in areas where there are fluctuating winter temperatures; and 3) just before or during blossoming. Draczynski (1959), Sholokhov (1961), and Elmanov et al. (1964), working in climates where there was adequate chilling for all the genotypes studied, showed that there is a significant physiological difference in the ability of the flower buds to withstand cold during the course of the winter. This difference is associated with the stage of cell differentiation in the anthers. During the rest period, differentiation of the pollen mother cells continues when temperatures are above 0 C. If temperatures are too hot, 18–20 C, for an extended period of time during this rest period, differentiation stops and bud drop may occur. (This latter is often a serious problem in warmer climates for cultivars with too high a chilling requirement.) The length of time from the beginning of differentiation of pollen mother cells to reduction division is a function of the cultivar and of temperature. Flower buds may become most resistant to cold during this period. Cultivars with a slow rate of winter development (prolonged rest period) are more resistant to fluctuating winter temperatures. Some of the cultivars identified at the Nikitsky Botanic Garden that have this characteristic are 'Oranzhevokrasnyi', 'Zard', 'Supkhany', and 'Kok-pshar' from central Asia, and the *zerdel* type 'Tokshian' (Kostina, 1969).

The onset of meiosis marks the end of the rest period. From this time on, warm temperatures, especially above 5–6 C, favor the development of pollen grains. When the temperature remains cold there is no development. It is during this stage of development that the buds lose hardiness following a period of warm weather. Subsequent cold weather that does not damage the buds still will not recondition the buds to withstand minimum temperatures such as they could have survived at the beginning of the post-meiotic development period. During this period, too, individual cultivars differ in their response to the accumulation of heat units or degree hours above freezing. The cultivars requiring greater total amounts of heat (slow rate of release of the growth processes) are later blooming and consequently may escape spring frosts. Some of such cultivars identified at the Nikitsky Botanic Garden are 'Zard', 'Oranzhevokrasnyi', 'Central Asian No. 35', 'Khurmai tsitrusonyi', and 'August' (Kostina, 1969).

The most hardy cultivars for areas with fluctuating winter temperatures and with spring frosts are those with both a long period of winter development ("rest") and with a long period of development from meiosis to fully formed pollen grains. 'Zard' and 'Oranzhevokrasnyi', occurring in both of the above groups, are two such cultivars. The work at the Nikitsky Botanic Garden (Kostina, 1969) has shown that these two stages of bud development can be inherited separately, with some evidence that cultivars hardy in a given set of conditions will transmit hardiness for those conditions. Breeders wanting to incorporate greater hardiness into their material will need to determine just what kind of hardiness they need. They will also need to determine what kind of hardiness a given "hardy" cultivar has before using it as a parent. Artificial freezing tests throughout the winter and spring, along with the determination of the stage of development of the pollen mother cells at each time of the freezing tests, could help in determining the kind of hardiness of cultivars and seedlings.

The picture is not as clear for adaptation to areas of mild winters. There still may be a hardiness problem in areas where frosts do occur. But in the selection or development of cultivars for a warm climate the principal problem is that there be sufficient chilling for annual flowering and foliation. One specific requirement is tolerance to high winter temperatures that cause abortion of the flower buds. Dr. A. Carraut of the Institut National de la Recherche Agronomique de Tunisie, Ariana, (personal communication) has observed that there was practically no flowering following an unusually warm November and December even though the total chilling for the winter was normal. Thus, the calculation of hours below 7 C is not adequate to predict the response of a given cultivar in a given season. Erez and Lavee's (1971) concept of weighted chilling hours is helpful. This concept considers that the response of leaf buds to the duration of chilling temperatures varies with the actual temperature above freezing. In warm climates it seems to be the amount, duration, and interrelation of chilling and heat during the premeiotic stage of bud differentiation that determine whether or not this stage may proceed to meiosis.

There are three species, *P. brigantiaca*, *P. ansu*, and *P. mume*, that should provide valuable additional sources of adaptation even though in most instances it will require at least two backcross generations before adapted cultivars with acceptable fruit can be recovered.

We assume that *P. brigantiaca* has been developed from the earliest importation of apricots into the valley of the Rhone. For several centuries it has been undergoing natural selection for adaptation on the slopes of the French Alps. Even though mountain slopes are more favorable for apricots than most of the commercial orchard sites in Europe, it seems likely that some of the pomologically most desirable specimens of this species should be useful in breeding for disease resistance and climatic adaptation (resistance to apoplexy) in western Europe.

Prunus mume and the cultivars of *P. ansu* are widely grown in the humid areas of China and Japan where the winters are mild. There are also cultivars of *P. ansu* that are grown north of the range for *P. mume*. On the basis of our limited knowledge of the species, *P. mume* should be the best source of disease resistance and adaptation to humid growing areas. There should also be forms of *P. mume* that would be useful in breeding for adaptation to regions with very mild winters. Since *P. ansu* is made up of cultivars this group offers the most promise for early progress in breeding for adaptation to humid growing areas. A serious effort should be made to survey the variability in both *P. mume* and *P. ansu* and get this material established by apricot breeders in several countries.

Disease Resistance

Resistance to diseases is important for profitable production but, until now, satisfactory screening procedures have not been developed for most diseases. Consequently where disease resistance has been considered, selection for resistance has depended upon field infection in most cases.

Blossom infection from *Sclerotinia laxa* is a serious problem in several areas. Resistance exists in the Tunisian cultivar 'Hamidi' and the Californian cultivar 'Nugget'. Crossa-Raynaud (1969) has developed an artificial inoculation technique for young shoots. The amount of infection in these shoots is correlated with the amount of blossom infection.

Resistance to bacterial leaf spot, *Xanthomonas pruni*, is necessary for production in humid growing areas. Following a natural epiphytotic, Layne (1966) observed resistance in several local selections from Ontario. Unfortunately these have small fruit size and poor quality. 'Curtis' from Michigan and 'Alfred' from New York were also reported to be resistant. An artificial inoculation technique for screening young seedlings for resistance of the leaves to bacterial leaf spot has been worked out by Layne and Dhanvantari (personal communication). Even though a field epiphytotic of bacterial leaf spot results in dramatic disfiguration of apricot fruits, and although the extent of fruit infection is not always correlated with the extent of leaf infection, a high level of resistance of the foliage to infection would seem to be the most important aspect of resistance. When there is little infection on the leaves during the growing season there will be very little inoculum available for fall nodal infection. Consequently there will be very few overwintering cankers to reestablish the disease the following season in either leaves or young fruits.

Fruit Characters

Very little has been reported about the inheritance of most characters in the apricot.

Lapins et al. (1957), in crosses between American cultivars (part of Kostina's European Group), found that the average performance of the progeny could be predicted on the basis of the phenotype of the parents. For freedom of pit and firmness of fruit the mean value of the progenies was close to the average of the parents. For size and edible quality of the fruit there was a shift toward smaller size and poorer quality. However, breeders generally are not interested just in the average performance of progenies. They are more interested in obtaining seedlings with exceptional combinations of desired characteristics.

For a majority of traits, crosses between cultivars of the Central Asian Group and cultivars of the European Group give seedlings that are intermediate in character in the first generation (Kostina, 1969). However, certain characteristics of the central Asian cultivars predominate, such as small fruit size, relatively large pit size, increased sugar content, decreased bitterness of the flesh, longer period of winter dormancy, later flowering, increased winterhardiness, and less resistance to fungal diseases.

Limited data on inheritance of season of ripening suggest that inheritance may be comparable to some of the patterns we know for peaches. Certainly inheritance of season of ripening is not simply controlled. Kostina (1969) reports that the widest range of season of ripening (from May to September) exists among cultivars of the Central Asian Group. 'Kok-pshar', 'Ahrori', and 'Nahichevanski Early' have transmitted early ripening. She also reports having obtained late selections from the cultivars 'August' and 'September'. Very early ripening cultivars are present in Tunisia. These include 'Hamidi', 'Amor Leuch', and some local selections that are less well-defined as distinct clones. It must be remembered, though, that in the warmer growing regions early ripening is a function of both time of bloom and number of days from bloom to maturity. We have recently identified very early ripening seedlings with good size. 'NJA13' and 'NJA19' ripen 80 days or less after bloom and produce fruit 6 cm or more in diameter.

Large size is desirable for fresh market fruits. It is also desirable for high quality parents to be used in crossing with small-fruited apricots, or even plums, which possess desirable characters for adaptation. In breeding for large size we have found that maximum fruit size is more important than average fruit size in the choice of a parent. Cultivars that can produce very large fruit are: 'Montedoro' from Italy, 'Mari de Cenad' from Romania, 'Jitrenka' from Czechoslovakia, 'Perfection', and 'Goldrich', which was recently named by the USDA.

The definition of high edible quality differs with the use to which a given cultivar will be put. The high edible quality of tree-ripened 'Royal' as grown in California consists of a balance of sugar, acid, and aroma. This combination has only been recovered in seedlings of 'Royal'. Unfortunately 'Royal' also transmits its characteristics for lack of widespread adaptability.

Fruits with high sugar and lower acidity would be pleasant to eat over a wider range of maturity so they would be desirable fresh market fruits for distant shipments. Such cultivars would have to be firm-fleshed. The cultivars we now know with high soluble solids and aroma, 'Shalah' and 'Supkhany', have the texture of juicy *P. salicina* plums and are not particularly suitable for long distance shipping.

Canned apricots should retain their shape as processed halves. They should have a good sugar/acid balance and typical apricot aroma. None of these attributes of the processed fruit, texture, flavor, or aroma, can be reliably predicted from the fresh fruit.

Pit burn is the term used for softening and discoloration around the pit as the fruit ripens. In the warm apricot growing regions this may be a serious problem when there are several days of hot weather (over 40 C) just prior to harvest. 'Canino' and the new cultivar 'Castleton' have been reported to be resistant to pit burn.

Apricot flesh varies from white and very light cream through yellow to deep orange and even red. In crosses of Irano-Caucasian cultivars with European cultivars the light color, characteristic of the majority of Irano-Caucasian cultivars, predominated over the yellow-orange color of European cultivars in the first generation (Kostina, 1969, 1972).

Skin color varies also from light cream to deep orange. The amount of overcolor may vary from nothing to practically a full red blush. Limited experience would indicate that flesh color, skin color, and overcolor or blush are inherited independently.

Apricot cultivars with glabrous fruit were observed in China by Holub (1958). Glabrous-fruited apricots also occur in central Asia especially in the Samarkand region (Kostina, 1936).

The tender, juicy consistency of the flesh that is characteristic of many cultivars used for fresh fruit of the Central Asian and Irano-Caucasian groups is transmitted to the progeny in crosses with cultivars of the European Group having firmer, drier flesh.

Sweet seeds are dominant over bitter seeds (Kostina, 1969).

Achievements and Prospects

The first apricot cultivars from controlled pollinations were introduced in Russia. Recently cultivars have been introduced in North America, South Africa, Australia, Argentina, Romania, and Czechoslovakia. A large proportion of these are from open pollination. Until now such new cultivars have made no appreciable impact on production in established areas. Some of the cultivars from eastern North America, as well as some from the breeding programs in European Russia, seem to be more winter-hardy and resistant to disease. Thus they give promise that the range of commercial apricot production can be extended into the humid temperate fruit growing regions which are close to concentrations of population.

The variability existing in desirable fruit characters assures the breeder that new cultivars can be produced that will be readily accepted in competition with other fruits, and the range in ecological adaptation indicates that apricots can be grown much more widely, so they certainly can become a greater part of the world's fruit production. But the limited ecological adaptation of any one genotype is the challenge to apricot breeders. Cultivars must be built for each producing area and for each marketing opportunity. It is exasperating to realize that whenever it is desired to introduce a new fruit character from another region into a breeding program, that character will likely be associated with very specific local adaptation to its original region.

With this perspective in mind, it becomes obvious that ambitious and persistent breeding programs must be carried on if we are to develop cultivars that will make it possible to grow high-quality apricots in many places in the temperate fruit growing regions.

The fruit qualities acceptable to consumers in the great centers of population will be quite similar, so the breeding programs will have similar objectives. Also, the ecological inflexibility of any genotype will require parallel breeding programs. Certainly, it will be most efficient to have coordinated interregional and international breeding programs using a common gene pool and, in some cases, even common seedling populations, but with selection being carried out under the ecological conditions in each region—it is unlikely that there will soon be a single apricot clone like the 'Redhaven' peach or the 'Delicious' apple that can be planted practically everywhere.

As significant progress is made, undoubtedly additional breeding programs will be initiated in potential producing areas. The opportunity to participate in international cooperation should encourage the initiation of such programs.

Eremin (1970) suggested that there are great opportunities for still wider adaptation that can be achieved through interspecific hybridization in *Prunus*. It seems likely that much could be done to develop new types of trees and fruits, as well as to develop good apricots adapted to more extreme growing conditions, through hybridization of apricots with other *Prunus* types. This will require a sustained effort.

Surely this coming generation of breeders will be able to extend the range of commercial production of apricots with a much greater variety of fruit types. If some of this coming generation of fruit breeders will undertake seriously to develop the potentials that are offered through interspecific hybridization, then the next generation of breeders will be able to add even greater increments to the opportunities for commercial production of this most delectable of tree fruits—truly bring apricots out of the mountains!

Literature Cited

The authors wish to acknowledge assistance from the Gerber Products Company in the preparation of this chapter, especially Figure 1.

Blaha, J., L. Luza, and J. Kalasek. 1966. Apricots (in Czech). *In* J. Blaha et al. *Peaches, apricots, almonds.* Prague: Czechoslovak Acad. Sci. p.227–414.

Crossa-Raynaud, P. H. 1966. Techniques of hybridizations and of embryo culture with apricot (in French). *Doc. Tech. Inst. Natl. Rech. Agron. Tunisie* 22:1–18.

———. 1969. Evaluating resistance to *Monilinia laxa* (Adherh. and Ruhl.) Honey of varieties and hybrids of apricots and almonds using mean growth rate of cankers on young branches as a criteria of susceptibility. *J. Amer. Soc. Hort. Sci.* 94:282–284.

Draczynski, M. 1958. The course of pollen differentiation in time in almond, peach, and apricot and the influence of bud temperatures on these processes (in German). *Gartenbauwissenschaft* 23:327–341.

Elmanov, S. I., E. A. Jablonskij, A. M. Sholokhov, and Ju. E. Sudakevic. 1964. The winterhardiness of the floral organs of peach, apricot, and almond in relation to their developmental characteristics (in Russian). *Trud. Nikit. Bot. Sad.* 37:237–255.

Eremin, G. V. 1970. Certain features of the inheritance of characters of the parental species in distant hybrids of plum (in Russian). p. 86–92. In *Wide hybridization of plants and animals.* Vol. 2. Moscow: Kolos. (*Plant Brdg. Abstr.* 42:6068. 1972).

Erez, A., and S. Lavee. 1971. The effect of climatic conditions on dormancy development of peach buds. 1. Temperatures. *J. Amer. Soc. Hort. Sci.* 96:711–714.

Goor, A., and M. Nurock. 1968. The peach, the apricot, the plum and the pear. In *Fruits of the Holy Land.* Jerusalem: Israel Universities Press. p.202–227.

Hansche, P. E., C. O. Hesse, and V. Beres. 1972. Estimates of genetic and environmental effects on several traits in peach. *J. Amer. Soc. Hort. Sci.* 97:76–79.

Holub, J. 1958. The ecological basis for speciation of several fruit genera in China (in Chinese). Ph.D. dissertation, National Agricultural University, Peking.

———. 1959. On growing of apricots in China (in Czech). *Ovocinar a Vinohradnik* 7:82, 104, 125–126, 151, 175.

Hough, L. F. 1965. Summary. 4. Induced mutations and breeding methods in vegetatively-propagated species, p. 793–794. In *The use of induced mutations in plant breeding.* Elmsford, N. Y.: Pergamon Press.

Kostina, K. F. 1936. The apricot (in Russian). *Bul. Appl. Bot. Gen. Plant Brdg.* Suppl. 83. (Inst. of Plant Industry, Leningrad).

———. 1969. The use of varietal resources of apricots for breeding (in Russian). *Trud. Nikit. Bot. Sad* 40:45–63.

———. 1972. Introduction and breeding of apricots (in Russian). *Sel'skokhozyaostvennaya Biol.* 8:86–91.

Kostina, K. F. and I. N. Riabov. 1959. An experiment on distant hybridization of fruit trees (in Russian). *Trud. Nikit. Bot. Sad.* 29:113–137.

Lapins, K. O. 1972. New fruits from Summerland, B.C., 1956–1970. *Can. Dept. Agr. Publ.* 1471, Summerland, B.C.

Lapins, K. O., Catherine H. Bailey, and L. F. Hough. 1969. Effects of gamma rays on apple and peach leaf buds at different stages of development 1. Survival, growth, and mutation frequencies. *Rad. Bot.* 9:379–389.

Lapins, K. O., A. J. Mann, and F. W. L. Keane. 1957. Progeny analysis of some apricot crosses. *Proc. Amer. Soc. Hort. Sci.* 70:125–130.

Layne, R. E. C. 1966. Susceptibility of apricots to bacterial spot infection of foliage and fruit. *Plant Dis. Rptr.* 50:112–115.

Löschnig, H. J. and F. Passecker. 1954. *The apricot and its culture* (in German). Vienna: Austrian Agronomic Publishers.

Paunovic, S. A. 1964. Apricot growing, its problems. Measures for their solution. *16th Int. Hort. Cong.* 1962. Vol. 5:492–501.

Pieniazek, S. A. 1967. Fruit production in the socialist countries of Central and Eastern Europe. *Proc. 17th Int. Hort. Cong.* 4:241–272.

Sholokhov, A. M. 1961. Winterhardiness of apricot in relation to the morphogenesis of flower buds (in Russian). *In* Morfogenezu Rast. Trud. 1(2):283–286.

Smith, C. A., Catherine H. Bailey, and L. F. Hough. 1969. Methods for germinating seeds of some fruit species with special reference to growing seedlings from immature embryos, *N. J. Agr. Expt. Sta. Bul.* 823.

Tamassy, I. and B. Pejovics. 1965. Atavism, or the development of new forms in apricot (*A. vulgaris* L.) (in Hungarian). *Kertesz. szolesz. Foiskol. Kozl.* 29:91–104. (*Plant Brdg. Abstr.* 37:3016, 1967).

Uljanishev, M. M. 1956. Apricot breeding in the southern Voronesh region (in Russian). p. 163–192. *In* V. K. Zeats (ed.) *Selektsiya kostochkovich kultur (Improvement of cultivated stone fruits.)* Moscow: State Publishing House for Agr. Lit.

U.S. Department of Agriculture, Foreign Agric. Service. 1972. *World agricultural production and trade statistical report.*

Vavilov, N. I. 1926. The mountainous districts as the home of agriculture. p. 218–220. *In* N. I. Vavilov. Studies on the origin of cultivated plants (Russian and English). *Bul. Appl. Bot. Plant Brdg.* 16(2).

———. 1951. Phytogeographic basis of plant breeding. p. 13–54. *In* K. S. Chester (trans.) The origin, variation, immunity and breeding of cultivated plants. *Chronica Botanica* 13(1/6).

Temperate Nuts

Almonds

by Dale E. Kester and Richard Asay

The almond is one of the oldest tree nut crops used by man but its exacting environmental requirements have restricted its commercial production to specific areas of the world. Horticulturally, it is classed as a "nut" in which there is an edible seed (the kernel or "meat") which is the commercial product. The kernel is surrounded by a shell which is enclosed in a hull that dehisces at maturity.

Almonds are an important food crop, varying in use from local consumption as an edible nut in its natural state to inclusion as a major ingredient in manufactured food products. Woodroof (1967) has described many aspects of its industrial use. The edible kernel is eaten raw or cooked, either blanched or unblanched. (Blanching involves removal of the pellicle by hot water or steam). Kernels may be roasted dry but are usually cooked in oil followed by salting with various kinds of seasoning. Kernels are combined with chocolate in confectionery. Almond kernels can be sliced and diced into various shapes to be used in pastry, ice cream mixes, and combinations with vegetables. The kernels may be ground and made into a paste which is used in making various bakery products.

Almond kernels are concentrated energy sources due to their high oil content (Table 1). The oil is primarily unsaturated, mostly oleic (67%) and linoleic (24%) fatty acids. The nuts also contain considerable protein, minerals, and some vitamins. The hull contains about 25% sugar and can be used for livestock feed (Cruess, 1951; Weir, 1951).

The almond tree is usually the earliest deciduous fruit crop to bloom in spring because of its relatively low chilling requirement and ready response to warm growing temperatures. Likewise, with favorable growing conditions in the fall (warm temperature, abundant soil moisture), almond

TABLE 1. Chemical analysis and food value of almond kernels (from Watt and Merrill, 1963; Leverton, 1959)

Constituent	Amount in 100 g fresh weight
Water (%)	5
Calories	598
Protein (g)	19
Fat (g)	54
Total Carbohydrate (g)	20
Fiber (g)	3
Ash (g)	3
Calcium (mg)	234
Phosphorus (mg)	500
Iron (mg)	5
Sodium (mg)	4
Potassium (mg)	770
Magnesium (mg)	625
Vitamin A (units)	0
Thiamin (mg)	0.24
Riboflavin (mg)	0.92
Niacin (mg)	3.50
Ascorbic acid (mg)	trace

trees continue to grow and tend to be slow to go dormant and develop hardiness. Therefore, distribution is limited in many areas of the world and production decreased in many seasons by spring frosts that injure blossoms and developing fruits. In the milder winter areas almond trees are limited by having a distinct chilling requirement.

Almond blossoms are genetically self-incompatible and require cross-pollination by insects, primarily bees, from another tree that is cross-compatible. Cool, rainy weather during blossoming may limit production by restricting pollination and inducing fungus diseases.

Because of these climatic limitations the principal production areas for almonds have been the central valleys of California, the areas bordering the Mediterranean Sea, southeast and central Asia, and limited areas in Chile, South Africa, and Australia. Most ideal conditions exist where there is a Mediterranean-type climate which has a rainy, relatively mild winter combined with a warm, rainless spring and summer.

The ability to establish almond culture in various countries has depended upon the genetic material available. Consequently, selection of local cultivars adapted to the environment of specific production areas has been a necessary first step in the development of almond growing. For example, the Spanish, Italian, Australian, and California industries have been built around unique combinations of cultivars or seedling plants originating within those countries. Modern breeding procedures have had a minor impact so far on almond growing but the opportunity for improvement is rather high.

Origin and Early Development

The cultivated almond (*Prunus amygdalus* Batsch) is in the Rosaceae family, subfamily Prunoideae, typified by a drupe fruit structure. Some authors have listed almonds as a separate genus *Amygdalus*. Rehder (1947) classified almond and peach along with several other *Prunus* species as a subgenus Amygdalus and this system will be used here.

Morphologically the almond fruit has the three distinct parts of a drupe fruit (see Fig. 1). The exocarp is more or less pubescent. The mesocarp (hull) is fleshy but becomes dry and leathery at maturity as it loses moisture. The endocarp (shell) varies from being very hard to very thin and soft. The seed contains a full sized embryo surrounded by seed coats. Fruits develop in two stages, differing from other stone fruits in that there is no second period of enlargement preceeding ripening (Brooks, 1943); rather, the mesocarp dehisces and opens to expose the nut.

Wild Species

Almonds representing a wide range of morphological and geographical forms grow wild throughout southwest and central Asia, from Turkey and Syria into the Caucasus Mountains of southern USSR through Iran, and into the deserts of Tian-Shan and Hindu Kush Mountains of southeast USSR and Afghanistan. More than 30 species have been described but because of apparent polymorphism and existence of geographical races, the distinctness of such species is not certain. Fig. 2 shows four geographical areas where major wild species are predominant. The list of named species has been divided into four groups (Cociu, 1967; Evreinoff, 1952; Grasselly, 1972).

1) Section Euamygdalus: Medium-sized trees of 3–4 m, not normally with spines.

Prunus fenzliana Fritch. grows at about 1500 m elevation in Turkey east into northwest Iran. A small tree with elliptical leaves, it has a small flat nut with a hard shell that tends to be sculptured. It is very hardy and is found in Region B.

Prunus argentea (Lam.) Rehd. (*orientalis*) is a shrub that grows in mountains of Syria and northwest Iran (Region B). The leaves are silvery; the shell netted and not pitted.

Prunus bucharica Fedtchenko is the dominant species in slopes of Russian Turkistan (Region D) where it grows at 800 to 2500 m in large forests. A small tree with a small, pointed, flat nut that is smooth and polished, it has kernels that are normally bitter but some sweet forms occur.

Prunus ulmifolia Franchet grows in same region as *P. bucharica* (Region D). This small tree is very bushy. Its shoots and young leaves are pubescent; the adult leaves are pubescent on the

FIG. 1. Almond fruiting habits, fruit, and leaves. Upper left: kernel (seed) and in-shell nut (endocarp). Lower left: typical almond leaves, showing long petiole in relation to leaf blade. Center: spur system with dehiscing almond fruit attached. Short spur at apex will develop lateral flower buds. Right: fruiting spur with leaves removed, showing persistent peduncle of fruit.

upper surface and strongly toothed. The fruits are small and indehiscent at maturity.

Prunus communis Fritsch (*Amygdalus communis* L. var. *spontanea* Korschinsky) is the wild form of cultivated almond, occurring in southeastern Russia and Afghanistan (Region D) and in central Iran (Region C). Believed to have originated from hybridization among *P. fenzliana*, *P. bucharica*, and *P. ulmifolia*, it has larger tree and nut size than other species and includes bitter and sweet, soft-and hard-shelled types.

2) Section Chamaeamygdalus

Prunus nana Stokes (*P. tenella*) is widespread and abundant in steppes of southern and central Russia extending from the Balkans to the Turkistan area (Region A). A small shrub to 1–1.5 m, it is very bushy and dense. The leaves are lanceolate, glabrous, and dentate; the fruits small; the nuts flat and very bitter. It is very polymorphic; some types have very hard shells.

Prunus ledebouriana (Schlechtendal) Grasselly is related to *P. nana* but tree, leaves, and nut are larger. It occurs in the east part of the range of *P. nana* (Region A).

Prunus georgica (Desfontaine) Grasselly is also related to *P. nana* but occurs in mountains of Georgia and west Transcaucasia (Central Region A). Ranging in size from a shrub to 1 m it has lanceolate leaves, large flowers and large flat nuts.

Prunus petunnikowii (Litwin.) Rehd. occurs in the southeastern part of the range (Region A) at slopes of Tian-Shan at 1400–1800 m on the steppes of Turkistan. Varying from a shrub to 1 m, it has asymmetrical nuts.

3) Section Spartiodes

Prunus spartiodes (Spach) Schneid. is a shrub to 1.5 m with long, green, slender, pliant branchlets with small ephemeral leaves. Resembling scotch broom, it is found in the arid mountains of central Iran at 1500 m elevation.

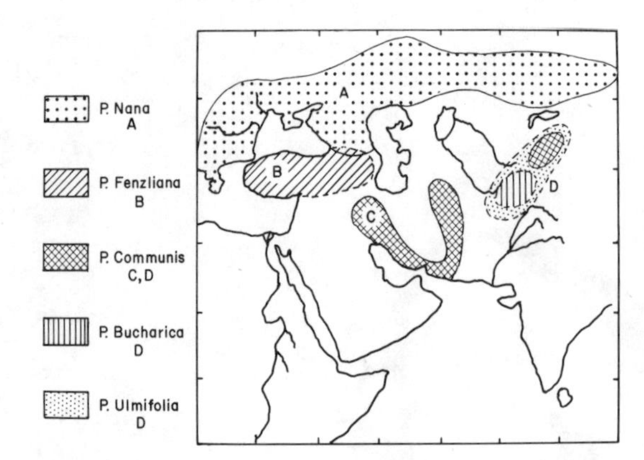

FIG. 2. Geographical distribution of major almond species.

Prunus scoparia (Spach) Schenid. is found in Iran as a small tree to 3–4 m. It is similar to *P. spartiodes* but has narrower leaves.

Prunus arabica (Olivier) Grasselly is found in the deserts of Arabia.

Prunus agrestis (Boissier) Grasselly is found in Syria. Similar to *P. spartiodes* Schneid, it has narrower leaves.

4) Section Lycioides

Prunus spinosissima (Bunge) Franchet occurs in Iran at 300–1500 m elevation. A small tree to 2–2.5 meters with very spiny, very abundant, finely netted, very large fruits, its kernels have a high oil content (60%). The species is xerophytic and hardy.

Prunus brachuica (Boissier) (*Prunus turcomania* (Linc.) Grasselly, *P. lycioides* Spach., *P. eburnea* Ait. and Hem. occur in Iran from Turkistan to Afghanistan. The shrub is 1–2 m tall, and very spiny, with small leaves and small pointed nuts. It is very xerophytic.

Prunus nairica (Fedtchenko) Grasselly is endemic to a small area in Armenia at 1500 m elevation. It is a small bush to 1 m, very spiny, and having very small leaves; its fruits are very large.

Prunus webbi (Spach) (*A. salicifolia* Boiss., *P. webbi salicifolia*) is a small shrub with pubescent leaves and smooth nuts.

Additional species have been described but may be variations of species already named. Sabeti (1966) lists the following under the genus *Amygdalus* as occurring wild in Iran in addition to those described above. In parentheses are listed the probable section to which they belong.

A. erioclada Bornm.	unknown
A. haussknecktii Schneider	(Section Spartiodes)
A. leiocarpa Boiss.	unknown
A. reuteri Boiss. and Bh. (*A. horrida* Spach)	(Section Lycioides)
A. stocksiana Boiss.	(Section Euamygdalus)

In addition to this list from Central Asia, *P. tangutica* Batal. is an almond-like species from China (Rehder, 1947). A group of almond-like *Prunus* species are also found in the deserts and low rainfall areas of the southwestern United States and northern Mexico (Mason, 1913). They include *P. fasciculata* Gray, *P. minutiflora* Engelm, *P. macrophylla* Hens., and *P. havardii* (Wright). These species probably represent adaptations of the North American plum group to a desert environment.

Cultivated Species

Selection of cultivated types appears to have taken place from *P. communis* since before recorded history. *P. amygdalus* Batsch can be considered as the collection of almond cultivars that have been grown, and probably should be considered synonymous with *P. communis* (*Amygdalus communis*). Almond growing is mentioned in the Bible and was known by the Assyrians. The almond accompanied the spread of civilization along the shores of the Mediterranean Sea into Greece, Italy, France, Spain, Portugal, and North Africa in areas where it was adapted to the Mediterranean-type climate. In some cases, trees escaped to the wild and were rediscovered as new species. Columella first described almond growing in the Roman period and Oliver de Serres described it in Spain in 1529 (Candolle, 1886; Evreinoff, 1952).

Seeds and scions were imported to the United States in the early colonial period but their culture was unsuccessful until they were brought to California about 1836 (Wood, 1925). Likewise, seeds or scions were carried to Australia, South Africa, and South America where limited planting has taken place in areas where Mediterranean climate exists.

Cytology

The 2n chromosome number of *Prunus amygdalus*, *P. fenzliana*, and *P. nana (tenella)* is 16 (Darlington and Ammal, 1945; Almeida, 1945; Warfield, 1968) which is the same as many other *Prunus* species. Almeida (1945) examined cytologically cultivars of almond that had histories of reduced fertility, and found various abnormalities that he associated with reduced fertility. Cytological studies of meiosis have been made with individual seedlings of F_2 and backcross generations (Armstrong, 1957) involving hybrids of peach and almond. Pairing during regular 8—8 segregation of chromosomes occurred with an occasional laggard as the only irregularity. Several triploid almond plants were found among this hybrid population. These plants are characterized by very low fertility, much vigor, and floriferousness.

Tetraploid branches were produced on diploid F_1 hybrids of peach and almond by colchicine treatments (Salesses and Bonnet, 1971). Cytological analysis indicated that there was no fundamental structural differentiation between the genomes of the two species.

History of Improvement

Cultivars being grown throughout the world can be considered in three main groups representing stages of evolution in cultivar improvement (Grasselly, 1972).

The first group contains seedling plants propagated by seed, presumably from selected parents; culture by seed propagation has been widespread since the beginning of almond cultivation and still exists in Iran, Turkey, Afghanistan, Greece, Crete, the Balearic Islands, and other areas. Such lines may be maintained by separate land owners or villages and may represent distinct seed-propagated cultivars.

The second represents groups of cultivars vegetatively propagated by budding to rootstocks but selected and maintained by individual land owners, in local areas. Such cultivars may have local names at the orchard but the nuts, when marketed, are grouped together and given collective names based either on type or where they are produced. This situation prevails in Italy, Sicily, Spain, and Portugal, and occurred at one time in France.

Third are the vegetatively-propagated cultivars that are grown extensively in specific production areas and have been adopted as standard for that industry; these are grafted to selected rootstocks. They may be cultivars originating from the second group, as chance seedlings, or from breeding programs. Representing the most advanced stage of evolution in cultivation, this type of development is characteristic of intensive industrial production areas. It has occurred in California, to some extent in Australia, and is being adopted in France and Spain. Selection and evaluation may be made by orchardists, nurserymen, or governmental agencies. Much cultivar evaluation is in progress throughout the world. Controlled breeding programs have been underway or are beginning in California, France, USSR, Italy, and Israel but the impact on almond industries so far has been small.

California Breeding Programs

The evolution of breeding programs and cultivar selection procedures is well documented in California (Wood, 1925; Kester and Jones, 1970). The migration into California and the initiation of agriculture from about 1840 included the importation of almond seeds and scions from the Mediterranean area followed by extensive planting. The origin of these importations has been tabulated as follows (Grasselly, 1972):

Number of cultivars	Area
27	Southern Spain (Alicante, Grenada)
5	Balearic Islands
6	France
2	Italy
2	Northern Spain
1	Southern Portugal

Essentially none of the imported cultivars were adapted, primarily because of susceptibility to disease and spring frosts, and most of the orchards were unsuccessful. From 1840 until about 1900 or slightly later, intensive selection and evaluation of individual seedlings in orchards by nurserymen and almond growers occurred. The most successful of these originators in terms of importance on California horticulture was A. T. Hatch of Suisun. He selected from a large group of unbudded rootstocks four plants which he named 'Nonpareil', 'Ne Plus Ultra', 'I.X.L.' and 'La Prima'. The first three of these became major cultivars in California and 'Nonpareil' has become the most important. 'Peerless' was selected from Colusa County and 'Drake's

Seedling' by Mr. H. C. Drake, also of Suisun. 'Texas Prolific', now known as 'Mission' and sometimes as 'Texas', originated during this period but the origin is unknown.

Dr. W. P. Tufts (1922), University of California, clarified the pollen compatibility relationships among almond cultivars and Milo N. Wood (1925), United States Department of Agriculture (USDA) examined their characteristics. As a result of these efforts and through grower experiences, a system of planting evolved in which 90% of acreage up to fairly recent years revolved around 'Nonpareil' as a main cultivar combined with 'Ne Plus Ultra' and 'Peerless' as earlier blooming pollinizers and 'Mission' (Texas') and 'Drake' as later-blooming pollinizers.

The USDA and University of California (UC) research programs were combined into a formal breeding project in 1923 and intensive efforts were made to develop specific marketing types adapted to California. A major breeding line involved 'Nonpareil' and 'Harriott', from which 'Jordanolo' and 'Harpareil' were introduced in 1937 (Wood, 1939). 'Jordanolo' quickly became a major cultivar in California but by about 1945 the widespread incidence of noninfectious bud-failure (Wilson and Schein, 1956) in the cultivar curtailed its continued planting. Subsequently, its acreage has decreased and it is gradually going out of existence as an important cultivar. Bud-failure susceptibility, apparently inherited from 'Nonpareil', has continued to characterize the developmental programs in California (Kester and Jones, 1970).

'Davey' was introduced from the program as a pollinizer for 'Nonpareil' (Serr et al., 1953) but it has not been successful because of undesirable tree characteristics. The USDA-UC program was separated in 1948. 'Kapareil', a product of a second major breeding line involving 'Nonpareil' and 'Eureka', was introduced from the UC program in 1963 (Kester et al., 1963) and 'Milow' from the same line in 1974 (Kester, unpublished). Both were introduced because of their tendency to produce small kernels for use by industry. 'Vesta' was introduced from the USDA program in 1968 (Jones, 1968) and 'Titan' in 1971 (Jones, 1971).

The major contributions toward cultivar improvement, however, continued to come from chance seedlings or bud mutations observed by growers and distributed by nurserymen (Kester, 1966). 'Merced' and 'Thompson' became the dominant pollinizing cultivars of 'Nonpareil' and 'Mission' ('Texas') in commercial orchards. Between 1955 and 1972, 15 of 27 new cultivars introduced from private sources were chance seedlings, mostly of 'Nonpareil x Texas' hybrids (Kester and Jones, 1970). Three were bud mutations of 'Nonpareil' and it appears that the 'Nonpareil' parentage dominates the group of 15 chance seedlings selected and introduced during this period.

F. W. Anderson, Le Grand, California, began an almond breeding program in the 1950s emphasizing late bloom, well-sealed shells, and high production. 'Texas-Nonpareil' hybrids and advanced generations from these two cultivars or offspring of a late blooming mutant of 'Nonpareil' dominate the breeding lines. Nine cultivars have been introduced since 1960. Development of self-fertile cultivars also has been an objective and a cultivar named 'LeGrand' with this characteristic has been produced.

Programs in Other Countries

Cultivar development in Australia resembles that of California. Imported cultivars had not been satisfactory when introduced and selection of local cultivars from seedling population has occurred. The major cultivar 'Chellaston', as well as most locally developed cultivars, have been seedlings of 'Jordan' (Moss, 1962).

In France, attempts to revive the almond growing industry began in 1950 with a program of cultivar evaluation and controlled crossing. In the first phase of the program, a large collection of cultivars from all over the world was planted at Bordeaux and Nîmes. These were extensively studied, and recommendations from this group were made and crosses between selected parents from the collection were produced (Grasselly, 1967, 1972).

From this collection the following groups of cultivars were identified as representing a common origin and possibly representing specific geographic populations or ecotypes with different genetic characteristics and backgrounds:

1) Mutants of known cultivars. Examples are various mutants of French cultivar 'Sultana' and of American cultivar 'Nonpareil'.

2) Seedling offspring of a single cultivar. In Spain, 'Desmayo' has produced a number of types, all characterized by early flowering, enlarged mesocarp, late maturity, and elongated leaves with long petiole.

3) Cultivars originating from 'Nonpareil'. Many such cultivars have arisen in California with

characteristic tree shape, soft or paper endocarp, and high shelling percentage.

4) Spanish group. This includes a group of cultivars originating in the province of Barcelona. These cultivars have short, spherical fruits with very hard shells, eg., 'Marcona', 'Planeta', 'Pascuala', 'Pastanzela'.

5) French group. 'Dorée' and 'Flots' are similar as are 'Flour-en-bas', 'Fourcoronne', 'Tardine de la Verdiere', 'Tournefort'. These have no double kernels. They tend to leaf out and bloom late.

6) Cultivars from southern Italy. This group contains many local cultivars all of which are characterized by a high percentage of double kernels. A high percentage produce two flowers per bud, are moderately late flowering and have a characteristic bushy growth habit.

7) Cultivars from northern Portugal. 'Verdeal', 'Marcelina grada', 'Prada', are large and flat with hard shells and very bushy growth habit.

8) Tunisian cultivars from Sfax 'Constantini'. These produce relatively thick nuts and leaf out and bloom very early.

Controlled crosses were made among selected parents beginning in 1960 emphasizing late bloom, blossom hardiness, resistance to disease and good nut quality. Two late blooming cultivars, 'Ferrudual' and 'Ferragnes', have resulted.

Russian scientists have contributed to much of the knowledge of the distribution of wild almond species (Kosloff and Kostina, 1935). Almond breeding has been carried on at the Nikitski Botanical Gardens, Yalta, since the 1930s by Dr. A. A. Rickter (1964) (see accounts by Mukjergee, 1967; and Scaramuzzi, 1964). Emphasis has been on developing late-blooming, frost-tolerant almonds by crossing and backcrossing with *P. mira*, *P. bucharica*, and *P. nana*. A number of selections from this group are being grown in Russia and are being tested in Europe. Self-fertility from *P. mira* has also been transferred to almond (Riabov, 1969). *P. davidiana* (Carr) Franch. has been used to incorporate brown rot (*Monilia*) resistance in crosses with almond (Scaramuzzi, 1964).

Modern Breeding Objectives

Objectives of a fruit or nut breeding program can fall into three categories depending on the state of the industry and the use planned for the improved cultivars. First, if the wish is to extend production into areas nonadapted environmentally, specific characteristics not present in available cultivars need to be incorporated by specific breeding procedures. The main limiting factors in almond production are embodied in the following categories and program objectives would include incorporating such factors into acceptable commercial cultivars:

1) Resistance to spring frosts. This may be achieved by later blooming or by greater blossom hardiness.

2) Greater winterhardiness. This character may be associated with earlier fall dormancy.

3) Adaptation to mild winter areas. Winter chilling is required for normal bloom and shoot growth (Chandler and Brown, 1951), although requirements tend to be lower than those of most other deciduous fruit crops.

4) Resistance to fungus disease. This is especially important where considerable rain and high humidity occur in the spring and summer.

Second, the wish may be to select or produce a cultivar for an area which is more or less adapted environmentally. The immediate objective would be to test adaptability of available cultivars and to select those most immediately suitable, followed by development of breeding program. In addition to the previous objectives which may be important, the following are minimum requirements:

5) High consistent yield as determined by number of nuts per tree and average nut size.

6) Ease of nut harvest, separation of hulls from shells, and shell cracking without kernel damage.

7) Moderate shelling percentage approximately equivalent to a semi-soft shell.

8) Shells that are well sealed.

9) Kernels that are uniform in shape and thickness, few or no double kernels, none or medium pubescence, relatively smooth surface, well filled-out for thickness, no sharp or irregular tips, no defects, i.e., gum, blanks, callus, etc., no creases, and good flavor.

This type of program has been the first step in introducing almond growing into California, Australia, and other areas.

Third, one may wish to replace present cultivars in areas where an industry now exists and where standardization has been achieved of a specific group of cultivars for use by that industry. Objectives of a breeding program under these conditions could include all of the above categories, but should include objectives for an improved product or increased efficiency. In this case, new cultivars must compete directly with cultivars already in established use. Among such objectives could be the following:

10) Development of cultivars that interact in relation to bloom and harvest for more efficient production.

a) Cultivars that bloom together but harvest separately in sequence so they can be kept separate and will use the harvesting equipment over a longer period more efficiently.

b) Cultivars that bloom together and are sufficiently alike in harvesting, nut, and kernel characteristics so that they can be harvested and processed together as a single cultivar.

c) Development of self-fertile cultivars that can be planted in solid blocks without the need for cross-pollination.

11) Development of cultivars with kernels directed towards specific market outlets, depending upon proportions of specific sizes and shapes, blanching and roasting qualities, processing uses, etc.

12) Improvement in kernel quality, as determined by organoleptic tests and appearance.

13) Change in nutritional characteristics of the kernel as food by increasing protein levels, increasing vitamins, and change in type and amount of oil produced.

14) Improved resistance to various diseases and insects and other maladies that may reduce dependence on chemical control.

15) Tree structure that requires little or no pruning, and which can be most efficiently harvested, such as by shortened ripening period, ease of knocking, etc.

Almond breeding in California represents the latter category where a highly developed industry has evolved which has been more or less standardized to specific cultivars and where environmental control and management techniques have raised productivity to high levels. Figure 3 is an analysis of the almond production system under these conditions where the necessary economic and production factors are arrayed against the biological characteristics of the almond species.

A first objective of a modern breeding program should be to identify the biological characteristics that control most directly the established horticultural and industrial requirements. A second objective should be to determine the relative genetic-environmental components of these characters and to obtain information about their inheritance patterns. With this information it should be possible to devise strategies by which the required characteristics can be combined to increase the likelihood that a cultivar superior in horticultural qualities can be produced.

Breeding Techniques

The almond has a typical peryginous *Prunus* flower with a single ovary enclosed in the floral cup that bears 30 to 34 stamens. Nectaries are present in the base of this cup. Flower buds are produced laterally on short spurs and sometimes laterally on longer shoots.

Pollination

Pollen is collected from flowers just before they open (popcorn stage). Flower buds of later-blooming cultivars can be forced in a warm room to obtain pollen if necessary. To avoid contamination with unwanted pollen, sterilize hands with alcohol to kill stray pollen before handling flowers. This precaution should be taken with the various implements used and at each step of the procedure. Anthers can be removed from the stamens by scraping with tweezers, squeezing flowers between fingers, or rubbing them across a wire screen. Anthers should be dried by exposing them to air for 24 hours. Examine anthers under a magnifying glass and, if dehisced, transfer to a vial. Stopper with nonabsorbent cotton.

Dry pollen will remain viable for several weeks

FIG. 3. Relationships among biological traits of almond and specific horticultural characteristics and industrial features of the almond production system in California.

Biological	Horticultural	Industrial		
BLOSSOM				
time of bloom and leafing	frost control			
blossom hardiness	disease, insect control	Production costs		
pollen compatibility	pruning			
	orchard management			
	harvesting		Management costs	
TREE				
growth habit	hulling			
vigor	cracking	Processing costs		
susceptibility to disease, insect, nutritional problems	sorting			
size	removal of defects			
shape				
blossom bud density				
precocity of production				
FRUIT				
time of maturity	tree size			
hull dehiscence pattern	blossom density			
gum production	% set	Nuts/tree		Commercial value (net income/acre)
	earliness of bearing			
	regularity			
NUT				
shell hardness	shelling percent			
seal of shell	defects	% good kernels	Productivity lbs/acre	
	gum and callus			
	worm damage			
	undeveloped			
	mold			
KERNEL				
size	size	Kernel weight		
shape	shape			
tendency to double			Market value price/lb.	
uniformity	appearance			
smoothness	uniformity			
color	processing quality			
production of creases	flavor			
pubescence				
bitterness				
chemical constituents				

at room temperature but may also be stored in a desiccator of calcium chloride for 48 hours to dry. For longer storage, pollen should be stored at low temperature, about 0 C or lower. Placing stored pollen in a freezer is an effective method for long-term storage.

Flowers are emasculated by removing the petals, stamens, and parts of the floral cup. This can be done with the fingers. A tweezer modified to make a circular knife edge that will cut through the blossom without injuring the pistil is helpful.

Pollen can be applied with a camel's hair brush, glass rod, or the finger. Be sure to sterilize the hands and implements with alcohol prior to use.

In almond, covering the blossoms is not necessary to prevent cross-pollination by bees if the floral cup and petals are removed. One should obtain 10–50% set with cross-pollinated flowers, whereas with emasculated, unpollinated flowers less than 1% will occur and usually it will be zero. Late-opening flowers should be removed. An alternate procedure is to cover an entire branch with a cloth or cheesecloth bag and pollinate unemasculated flowers. Since almond cultivars are self-incompatible, no self-pollination will occur.

One can usually establish that an individual

fruit is fertilized three to five weeks after bloom when the ovary begins to swell beyond the size of the calyx cup. However, a later drop, six to seven weeks after bloom, may also occur and reduce the final set (Kester and Griggs, 1953).

Seed Handling

Fruits are collected when the hull dehisces, the nuts (consisting of seed and surrounding shell) are extracted, and the seeds dried. Almond seeds require moist-chilling (stratification) for three to four weeks or more which may be done either with the nut (shell intact), with the seed alone, or even with the hull intact. Stratification is carried out in a standard manner, such as in moist, sand-peat media in a plastic bag or other container. Fungicide treatment of seeds is desirable since almonds are susceptible to "damping-off." Seeds may be stratified by planting directly out-of-doors in the fall or by storage in a refrigerator. Stratifying seeds in a small amount of water or fungicide slurry in a glass Erlenmeyer flask is very satisfactory. Seeds should be soaked 12 to 24 hours in water before placing in stratification; 10 C (50 F) is optimum for most rapid after-ripening during stratification. However, after removal from a refrigerator, germination temperature should not be more than 18 C (68 F); if higher, individual embryos that are not well sprouted will not germinate.

As an alternate procedure, remove seeds (with seed coat intact) from the developing fruit several weeks after the embryo has attained full size; collection time is mid-July in California. Seeds are germinated at 10 C (50 F) and transplanted to germinating media after the radicles of the embryos have started to elongate. Avoid high germination temperatures, 20 C (68 F), or severe rosetting of the seedling plant may result. If it does occur, pinching out the tip of the rosetted shoot may force lower laterals to grow which are usually normal.

Seedling Growth

The following sequence of operations for growing seedling plants to an evaluation stage has been used successfully in California. Its basic purpose is to allow the seedling plant to develop naturally in order to indicate its potential bearing habit as early as possible:

1) First year. Make the cross in spring and collect seeds in the fall to be stratified and germinated in a greenhouse.

2) Second year. Transplant seedling plants to a nursery to grow for one season.

3) Third year. Transplant to a seedling block where plants are grown in hedgerow at spacing of one to two m (three to six feet). Spacing could be closer but handling is more difficult and individual plants become too crowded for good tree growth. Early evaluation of tree growth and productivity depends on proper handling at this stage. Plants should grow as well as possible to produce a basic framework. Spring pruning by judicious pinching of growing shoot tips can assist in getting a proper framework. Otherwise, pruning should be held to a minimum.

4) Fourth year. Most seedling plants, if well grown, should begin to initiate flower buds during this season. Again keep pruning to a minimum.

5) Fifth year. Plants should begin to produce some flowers and some fruit. A preliminary evaluation of tree and nuts can be made and in some instances obviously poor-growing trees can be eliminated. However, an evaluation of tree and nut characteristics may not be completely reliable for this first crop and a second year's observation is desirable, depending on the purpose of selection.

6) Sixth year. Good bloom and crop should be produced if weather conditions are favorable. Selection as to tree potential and nut characteristics for further testing should be possible at the conclusion of this year. As soon as desired items are propagated, the remainder can be discarded.

Several modifications designed to shorten the breeding cycle are possible. Nuts may be stratified and then planted directly into an orchard row, omitting the nursery step. This procedure could shorten the developing period one year, but may result in greater losses of seedling plants. A further modification would be to germinate immature seeds (as described above) and to grow the young seedling plants during the same year in a container in a greenhouse or lath house and then transplant to the orchard at the end of the year.

A further shortening by an additional year is possible by germinating the immature seed, and removing buds from it in the fall or next spring to be top-budded into established rootstock plants. If well handled, grafts should produce some flowers during the second season and more during the third (total four-year cycle). In this procedure, rootstock plants from virus-tested stock should be used to avoid contamination of a future introduction.

The procedures described to this point have been designed to obtain information on inheritance, to select parental genotypes of the next breeding cycle, or to identify a superior genotype as a candidate cultivar. Once such objectives are attained, surplus progeny can and should be quickly eliminated. Trees in close plantings become crowded and the value of continued performance evaluations on the seedling plant is questionable even if the block is thinned down to selected plants. Candidate cultivars should be propagated as soon as possible into a test planting where their performance as an orchard tree can be evaluated. The program to be followed does not differ in principle from that of other tree fruit species. Evaluation procedures should be more simple than some of the others, however, in that horticultural manipulations of thinning or special pruning to control crop and quality is not required. Thus, three phases of development can be distinguished: 1) identification of cultivar candidates (selections) from the seedling population; 2) the growing of a large number of selections in special blocks, each selection consisting of a large number of relatively few individual trees compared against standard cultivars and other selections; and 3) planting of best selections in commercial orchards.

Release Procedures

A new almond cultivar should not be released to the public until it has been ascertained that the propagating material distributed is free of known viruses. Indexing procedures to identify such viruses are known (USDA *Handbook*, 1951), and it would be desirable to identify the potential for specific "genetic disorders" such as in noninfectious bud-failure (BF). At the present time, no direct test exists for this purpose. Identification can only occur in some indefinite future when symptoms actually appear in some trees of the clone. The inability to identify such "genetic disorders" should be recognized as a limitation of virus control programs. At the California Agricultural Experiment Station, a selection with commercial promise is proposed as a candidate tree in the California Registration and Certification Program (California State Department of Agriculture). Scions from a source tree are first indexed on 'Shirofugen' cherry to identify the presence of *Prunus* ring spot virus (PRSV). Such a test is completed within about one month. If the test is negative, budwood is supplied to plant pathologists who will conduct a "long index" to seven additional hosts: 'Nonpareil' almond, 'Shiro' plum, 'Italian' prune, 'Elberta' peach, 'Royal' apricot, 'Bing' cherry, and 'Montmorency' cherry. Propagating material from the same source is supplied at the same time to the Foundation Plant Materials Service (Department of Viticulture and Enology, UC Davis, Calif.) to be budded to a registered rootstock seedling plant. One year later the nursery tree will be transplanted to the Foundation Orchard as a candidate tree. Observations of the indexing host plant continues for two growing seasons. If the indexing tests are negative, the tree in the Foundation Orchard is eligible for registration in the registration and certification program. Distribution of propagating material for commercial propagation, however, will not be made until the genetic identity of the candidate plant has been verified.

However, if the indexing is positive, indicating the presence of one or more viruses, the candidate tree is eliminated from the orchard. Most viruses can be successfully eliminated from infected almond stock by heat treatment (Nyland and Goheen, 1969).

Most almond cultivars in California which have originated from private developers are patented, giving the originator the exclusive rights to its propagation for 17 years. He may license other individuals to propagate the cultivar under some licensing arrangement invariably involving the payment of royalties to the originator. To obtain information about propagation, one must deal with the patent holder.

Breeding Systems

The cultivated almond species is a variable, heterozygous group of individual cultivars and seedling lines. Only a few important horticultural traits, e.g., bitterness, have so far been identified as controlled through single major genes. Shell hardness, self-fertility, and late bloom may be under control of a major gene but quantitative genes also appear to be involved. Incompatibility alleles are single genes and their manipulation may be highly significant in identifying and showing relationships among cultivars.

Most of the important horticultural traits such as yield, time of bloom, time of ripening, nut size, shelling percentage, tree size, kernel size, shape and quality factors, etc., are largely controlled through quantitative genes. Studies that have so far been published (Kester, 1965; Grasselly and Gall, 1967; Grasselly, 1972), experiences and observations made over a period of years, and the inspection of the data now at hand indicate that these genes are largely additive. Phenotypic and genotypic heritabilities of most of these traits should be relatively high. The considerable environmental effects that do occur on phenotypic measurements of these traits both in different seasons and in different locations should be capable of being minimized by the statistical techniques utilized by Hansche, et al. (1966, 1968, 1972*a*, *b*), in analyzing heritabilities of economic traits of cherry, strawberry, peach, and walnut. It thus appears that the method of parental selection according to phenotype should be the most efficient and effective method in achieving improvement in almond. General parental selection of phenotype has been the dominant procedure used by almond breeders to date. The great influence to fruit tree improvement of chance seedling selection by orchardists and nurserymen probably reflects the fact that most chance seedlings arise from commercial orchards containing the most horticulturally desirable cultivars available at that time. These had in turn been the best screened from the previous generation of cultivars.

Such a breeding "line" may represent a fairly narrow genetic base, although little is actually known about the degree of inbreeding that is represented among cultivars arising in such manner or whether it is really significant. In California, the rather wide range of genotypes that were apparently initially introduced in the late 1800s has been narrowed significantly to certain breeding lines including 'Nonpareil' x 'Texas' ('Mission'), 'Nonpareil' x 'Eureka', and 'Nonpareil' x 'Harriott'. The high degree of dependence upon 'Nonpareil' as a parent stems from its predominance in commercial orchards, its high nut quality, high shelling percentage, and dependable production. On the other hand, current and future difficulties being experienced by the California almond industry in the existence of noninfectious bud-failure (BF) problem appears to be due to the somatic variability of the BF genotype inherited from 'Nonpareil' and being carried in the 'Nonpareil' breeding lines (Kester and Jones, 1970).

Another possible procedure of breeding would be to develop various lines by inbreeding that would become more or less homozygous for specific characteristics. Such a procedure has been utilized and studied to some extent by R. W. Jones, USDA, Fresno, California (unpublished data). Jones and Kester (unpublished data) produced several lines of sib crossing and have studied them through three consecutive generations. Such lines became quite uniform in nut and tree characteristics without obvious reduction in size or vigor. Most striking in many of the inbreeding lines has been the appearance of seedlings with unstable genotypes analogous to the BF disorder (Kester, 1971). This breeding program might make it possible to concentrate certain desirable characteristics and identify specific undesirable characteristics. Recombination of desirable lines could perhaps be used to increase probability of obtaining desirable characteristics, eliminate undesirable genes, and restore any lost vigor.

Interspecific Hybridization

Almond species: The cultivated almond crosses with many other *Prunus* species depending on the closeness of the relationship. Within the area of southwest Asia where many wild almond species exist, interspecific hybridization apparently occurs naturally (Evreinoff, 1952). In hand pollinations made in California, cultivated almond crossed equally well with individual plants of eight related almond species, including *P. fenzliana*, *P. webbi*, *P. argentea*, *P. tangutica*, *P. bucharica*, and a number of imported seedling species of unknown identification (Warfield and Kester, 1968). Individuals in F_1 populations were fertile and tended to be intermediate between the parental species in most morphological characteristics. Further generations with *P. fenzliana* show a segregation of growth characteristics of parental types.

Peach: Almond crosses readily with peach whenever the two species have come into contact. These hybrids may be spontaneous (Bernhard, 1949) or result from deliberate crossing (Jones, 1969; Kester and Hansen, 1966; Kozahonane, 1970). The F_1 hybrid invariably shows great hybrid vigor and the seedling population is more or less uniform. However, differences have been observed among almond cultivars used as parents that influence the vigor, uniformity, and pistil sterility of the hybrid offspring. A character producing rough-bark, dwarfish hybrid offspring appears to be transmitted by some almond parents, including 'Nonpareil' and 'Tardy Nonpareil'. Flowering habit of peach recurs in the hybrid. The dominant single gene for small-flower in peach is inherited in almond x peach similar to peach x peach. The leaves of the hybrid are intermediate in size, length and petiole length/blade length.

The fruit of the hybrid tends to be intermediate in size and fleshiness but resembles the almond in its growth pattern and dehiscence. The endocarp (stone) is like the peach. Bitterness of the seed is similar to peach or intermediate.

First generation peach x almond hybrids have now been studied extensively in the United States and France as rootstocks for almond, peach, and for other *Prunus*. Characteristics of value have been vigor, resistance to calcareous soil, and nematode resistance, in some cases. The difficulty has been their propagation.

Second generation and backcross populations of peach x almond crosses segregate to produce a wide range of types recovering both peach and almond parental types and wide differences in vigor (Armstrong, 1957; Kester, unpublished). Very small, weak individuals to exceedingly vigorous plants appear. Variation in sterility exists with considerable amounts of pistil sterility. Anomalous flower types have appeared, particularly one with a cluster of pistils with no petals or anthers.

Other Prunus: Almond will cross with plum species but with difficulty. Resulting hybrids are sterile (Bernhard, 1958, 1962). Likewise, F_1 hybrids of apricot x almond have been produced in low percentages (Jones, 1968). Hybrids tend to be intermediate, are pistil sterile, and tend to be weak in growth.

E. F. Serr and H. Forde (unpublished data) crossed almond with *P. andersoni* Gray, the desert peach of Nevada. Both individuals produced were sterile, one very weak, the other vigorous. Almond crosses readily with *P. davidiana* and *P. mira*.

Bud-sports and Bud Selection

A number of naturally occurring bud-sports have been found in almond.

A late blooming mutant of 'Nonpareil', evidently a chimera, was found in three separate orchards. It has been named as 'Tardy Nonpareil' and patented (Brooks and Olmo, 1972).

A late blooming, double flowering mutant, evidently a chimera, was found in 'Nonpareil' (F. L. Anderson, unpublished). This character appears in about half of the offspring. Double flowering seedlings have segregated from peach x almond populations where 'Nonpareil' was a parent.

A sport limb with red flowers and red pellicle was discovered in a 'Nonpareil' tree growing in Wasco, California (Brooks and Olmo, 1972). It is evidently a chimera and has somewhat unstable branches sometimes reverting back to branches with normal flowers.

Leaf variegation has been found in 'Nonpareil'. Variegated individuals also occasionally appear in seedling populations.

A peculiar unproductive, very rough-bark limb has been observed to develop on normal 'Nonpareil' trees on several occasions. These have sometimes been propagated to produce nonproductive orchard trees.

Mutant limbs bearing very narrow, strap-like, crinkly leaves have been found in 'Nonpareil' and 'Peerless'. Such trees are unproductive. The mutant is very unstable since normal and crinkle sectors can be found on the same branch. Individual seedling plants with this unstable crinkle leaf pattern have also segregated from peach x almond populations when 'Nonpareil' was one of the cultivars used in developing the lines.

Mutant limbs with bitter kernels on sweet kerneled trees have occasionally been reported by orchardists but not often authenticated. However, what appears to be true development of a bitter sport on a 'Merced' has been found. The sport appeared to have poorer set than the rest of the tree. A bitter mutant of 'Nonpareil' has also been found which sets as well as the parental tree (R. W. Jones, unpublished data).

Breeding for Specific Characters

A number of characteristics have been used at the California Agricultural Experiment Station as having value in identifying superior almond cultivars either to use as parents in further crosses or to place into selection blocks for tests of potential use by industry.

Blossom Density

Like other *Prunus,* almond plants initiate and develop flower buds laterally on spurs and shoots during one growing season which continue to develop during the following dormant period and, with adequate chilling, flower early the next spring.

The majority of flower buds of most cultivars are produced laterally and singly on spurs. These spurs may be as short as 2 cm or less but could be as long as 13 cm. The number of buds per spur, the number of spurs per branch, and the distribution of these spurs within the tree vary with cultivar, age, and environment. Inasmuch as total number of nuts per tree is one of the components of yield, all of these variables are important factors in selection.

In some cultivars and species, considerable numbers of flower buds are produced laterally on shoots. These may be singly at a node or in combinations of one or two flower buds with a leaf bud. Some cultivars, such as 'Eureka', tend to produce double pistils that grow into double fruits on a single peduncle.

Bud numbers (density) on individual plants in seedling populations can be designated subjectively on a scale of 1 (lowest) to 9 for both shoots and spurs. This scale has been arbitrarily chosen to facilitate computer programming. If trees are properly handled in the orchard with a minimum of pruning, flowering buds may appear by the beginning of the third year, although the fourth year is more reliable. On a mature tree, subjective ratings as described are difficult to estimate accurately and consistently. Actual bud and spur counts would give a more accurate measurement of density, particularly as the plant being tested grows older.

The absolute density of buds gives an indication of production potential and the relative density of buds produced on shoots as compared to spurs (shoot/spur density) characterizes the fruiting habit of many *Prunus* species.

Considerable variation also exists among individual cultivars in shoot/spur density ratios, flower bud density, and precocity of bearing. These characters appear to be heritable but inheritance is polygenic.

In interspecific hybrid populations of cultivated almond a high shoot/spur density tends to dominate. F_1 hybrids of peach and almond invariably show the vigorous and characteristic growth habit of the peach. Segregation in the F_2 and backcross generations shows a complete range in bearing habit representative of the parents.

Grasselly (1972) examined three seedling populations where 'Ardechoise' was a common parent, rating flower density on a scale of 0–5, in consecutive years. A high correlation was found in flower density among the third, fourth, and fifth years; that is, high density in the first year was followed by a high density in the fourth and fifth years. Distribution was polygenic. The precocious high-density cultivar 'Marcona' transmitted higher density and precocity to its seedling offspring than did the less-dense, less-precocious cultivar 'Bartre'. Results indicated that reasonably reliable selection for these characters can be made by at least the fourth and fifth years.

Time of Bloom

Time of bloom is one of the most important characteristics in almond cultivar selection because it is an indicator of chilling requirement during dormancy, modifies vulnerability to frost, and determines how a cultivar is used in an orchard situation in combination with others for cross-pollination.

Blossom opening of an almond tree follows a typical sigmoid growth curve (Fig. 4). The actual time of bloom and the shape of this curve depends upon the interaction of specific genotypes and the temperature patterns of both the previous dormant season and that immediately preceding and during bloom. The duration may be as little as two to three days in some seasons, or as long as ten days or more. In California, the range in time of bloom among different cultivars is from late January to mid-March. The sequence of bloom remains relatively constant from year to year but the actual dates may vary by several weeks in different seasons. Consequently, bloom date should be given in relation to the bloom date of standard cultivars in comparisons among different seasons.

In general, the greater the number of chilling hours during winter, the earlier and faster the bloom; likewise, blossom opening is more rapid the warmer the temperatures are immediately prior to and during bloom. The rate of blossom opening tends to be rapid in the early part of the curve and slows down in the latter part. Designating "time of bloom" of an individual plant as "first bloom," i.e., the date when some fraction of a tree is open, such as 5–10%, is a more accurate way to define this character in genetic studies than to attempt to pinpoint a given date when the tree is in "full bloom"; that is, when all of the blossoms are open. Tabuenca (1972) correlated the maximum and minimum temperatures during specific intervals of the dormant season of a series of almond cultivars in Spain. On this basis she was able to separate the cultivars studied into classes depending on the amount of chilling during winter and the amount of warm temperatures preceding bloom. By this analysis, the number of chilling hours below 5 C varied from 90 to 427, and 422 to 940 below 12 C (Tabuenca et al. 1972).

Time of bloom is quantitatively inherited (Kester, 1965; Grasselly and Gall, 1967). A genetic heritability value of 0.83 has been obtained in a study (Kester et al., 1973) where the number of parental combinations was 20. These values are consistent with observations made with several hundred unanalyzed families grown at the California Experiment Station. Inheritance of bloom in almond follows essentially the same pattern as that for cherry, peach, and walnut (Hansche et al., 1968, 1972 *a*,*b*).

Kester (1965) found a separate case where inheritance from 'Tardy Nonpareil' (a late-blooming mutant of 'Nonpareil') differs from that of other almonds studied in that the offspring segregated into an early blooming group of seedlings and a late blooming group. The mutation was attributed to a dominant major gene controlling time of bloom. There was, however, a correlation between late bloom and low productivity of this group. It seems likely that genetic factors in addition to late bloom were also changed. Grasselly (1972) has also reported that a correlation was found between low vigor and late blooming in two populations but not involving the late-bloom mutant.

Relationships have been found to occur among time of bloom, time of leafing, and chilling requirements (stratification) for germination (Kester et al., 1973). In studying seed production on a parental tree pollinated by different cultivars, the chilling requirement of a population of almond seeds, as measured by the length of time between initiation of chilling at 10 C and radical emergence, i.e., germination (Kester, 1970), was correlated to the bloom time of both parents. Pollen from parents with different blooming dates used on the same seed tree significantly changed the chilling requirements of the separate seed population produced. In studying the resulting offspring, the chilling requirement of an individual seed was not necessarily related to blooming time of the plant that it produced. The correlation coefficient between seed vs. plant was highly significant but low in magnitude. Thus it is possible to separate populations at the germination stage but identifying individual seedling plants for time of bloom at the germination stage is not sufficiently reliable as a routine selection procedure. Correlations did exist between time of bloom and time of leafing-out, and indicate that selection for leafing date at the seedling stage is somewhat more effective.

Fruit Set

A high percentage set of flowers is an obviously desirable horticultural trait. Many environmental factors affect set but a number of genetically controlled traits are undoubtedly involved. Some almond cultivars appear to set more heavily than others even under essentially the same environmental conditions.

Pistil sterility is another cause of variation in set and is particularly characteristic of some interspecific hybrid plants. A fruit set rating made subjectively on a scale of 1 to 9 similar to that for fruit bud density can give an estimate of productivity in young trees and can identify obvious cases of pistil sterility. If one rates on a basis of spurs or shoot density, factors affecting set of individual flowers are paramount; if one rates on whole tree, total tree productivity (yield) as affected by spur and blossom density is involved. Estimating the yield of a mature tree by this method should probably only be a supplement to actual yield records or nut counts.

An accounting of various factors of set requires ratings to be made at three critical times (Kester and Griggs, 1959). Sterile flowers with abortive pistils will usually drop within a few days after bloom without any swelling at the base of the pis-

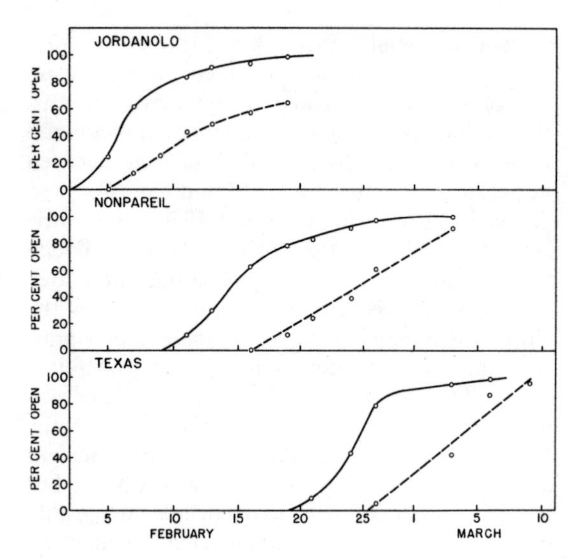

FIG. 4. Blossom opening (solid line) and petal fall (dashed line) pattern of three almond cultivars in California.

til. A rating made seven to ten days after bloom will identify this group. Squeezing the base of a flower at this time will identify sterile flowers since normal blossoms that are unpollinated or unfertilized will produce enlarged pistils up to about 5 mm in diameter but normally do not split the "jacket" or calyx cup. Unfertilized flowers drop or become discernible due to reduced size about three weeks after bloom. A rating made at this time will identify the relative size of this group. Another group of nuts may drop at about six to seven weeks (the so-called "June drop"), after enlarging to about 10 mm in diameter. In this group, as in mature almonds, the abscission layer forms between the end of the peduncle and fruit so that the peduncle persists. The record of fruitfulness in a mature almond tree over a period of years could be observed at any later date by counting persisting peduncles in the different production years.

Fruitfulness is a complex character which, along with flower bud density, determines yield. Observations indicate both characteristics are polygenic. Heritability is not known.

Pistil sterility exists within populations of interspecific hybrids of peach and almond. Different hybrid populations show a wide range of fruitfulness depending on the parental combination. Many are highly fruitful, others have much pistil sterility. Likewise, much variability in pistil sterility within second generation and backcross populations exists.

Tree Characteristics

Trees of different almond cultivars vary considerably in size, shape, vigor, and growth habit. These characteristics affect productivity, ability to train and prune a tree, and the adaptation of the tree to harvesting procedures and orchard conditions. All of these characteristics affect the desirability of a cultivar as an orchard tree.

Tree shape is determined by branching angle, vigor, and relative apical dominance. Trees that tend to be upright with moderately apical, dominant branches are somewhat more preferred by orchardists than those that are rounded or spreading with much lateral branching.

Trees have been rated on shape at the California Experiment Station as very spreading, spreading, medium, upright, and very upright, which provides a scale of 1 to 5, respectively, for analysis. These shapes are compared to type varieties of 'Ne Plus Ultra' = 2, 'Nonpareil' = 3, and 'Texas' ('Mission') =4.

The subjective scale for tree "desirability" can be evaluated as: A = outstanding, B = satisfactory, and D = undesirable. Plus (+) and minus (−) used with B grade provide an overall scale of 1 to 5. Ratings are best made during the dormant season and again during the growing season. Such a tree rating can be used to identify desirable selections at a three to four year age in seedling populations. Tree size, shape, and growing habits are highly specific with a particular cultivar and are evidently closely controlled genetically.

Grasselly (1972) studied the growth habits of six cultivars in France, both as one-year-old nursery plants and as mature trees. These parallel the general range of tree phenotypes observed in California. The specificity of the growth habit for a given cultivar can usually enable an experienced observer to readily identify a given cultivar in an orchard. Comparing the two scales, 'AI' would probably be identified with class 2; 'Marcona' and 'Cristomorto', class 3; 'Texas' and 'Ardechoise', class 4; and 'Bartre', class 5, although even more upright genotypes exist, such as that characteristic of the 'Davey'.

Establishing a correlation between growth habits in a young tree with the structure of the mature tree is particularly important since it would enable a breeder to make early judgments on potential value particularly in selecting parents in consecutive generations. Grasselly (1972) has identified four traits that can be used to measure various aspects of growth characteristics: 1) the number of one-year-old branches on a two-year-old seedling, 2) length of three most vigorous shoots, 3) the angle of insertion of principal lateral branches, and 4) the number of secondary laterals on one-year-old shoots. In the first three of these characteristics the phenotypes of the offspring were intermediate between the parents in both almond x almond and peach x almond progeny. The tendency to produce secondary laterals on current season's shoots was transmitted to their progeny from 'Ardechoise' and 'AI'.

Tree size varies among different cultivars and species but not many specific inheritance data have been analyzed and reported. Among almond x almond progeny, variation in tree size is relatively small although differences occur. Within certain progeny, for example, in a cross of 'Texas' x 'Merced', a number of dwarfed, approximately half-sized phenotypes have appeared. 'Merced' is a 'Nonpareil' x 'Texas' hybrid and the cross represents a backcross to one parent.

Interspecific hybridization involves much size variability. *Prunus fenzliana*, *P. nana*, *P. argentea*, and most others are smaller than *P. amygdalus*. First generation hybrid populations of several of these species with almond are intermediate in size between parents. Hybrids of *P. amygdalus* and peach (*P. persica*), however, are very large in size, apparently an expression of hybrid vigor (Kester and Hansen, 1965). Second generation and backcross populations of almond and peach show an extremely diverse range not only in size but in tree habits. Very dwarfed plants with short internodes have been recovered from such populations.

Tree size, shape, and growing habit are integrated to produce a characteristic tree phenotype that is recognizable as to type. Observations show that such "types" are inherited and parental and intermediate types are recovered in offspring progeny. These traits are evidently polygenic in nature but are highly heritable.

Foliage

Almond cultivars have characteristic differences in density of foliage, size, and, to some extent, color of leaves. Leaf characteristics are also affected by specific environmental conditions and the general health and vigor of the plants. Consequently, foliage appearance can be an indication of weak vigor or the inability of the plant to withstand some unfavorable factor in an environment, such as an attack by various kinds of mites or diseases.

Density of foliage is related to leaf size and number, distribution, and closeness of nodes. Variation of density, however, can be recognized among cultivars and has been rated on a scale of 1 (very sparse) to 5 (very dense). Certain cultivars, such as 'Harriott' and 'Jordanolo', have very dense foliage partly due to their large dark green leaves. This foliage characteristic is heritable to their offspring. At the other extreme, low density may result not only from genotype but also defoliation due to adverse growth conditions or to weak growth capacity. As such, it may be an indicator of unfavorable conditions of the plant or environment.

Leaf size depends upon location of the leaf on the plant. Spur leaves tend to be small; shoot leaves tend to be large. Both can vary with environment so that comparisons should be made only between plants of comparable age and location. Nevertheless, characteristic differences in leaf size of small, medium, or large do occur and can be identified by comparison to preassigned standards. Grasselly (1972) has shown that in four populations of almond x almond and three of almond x peach, the average leaf size of the offspring was directly correlated to the average leaf size of the parents.

Besides the usual "normal" green (G), a dark green (DG) leaf characteristic is observed in some cultivars, such as 'Harriott' and 'Jordanolo'. A light olive green (LG) has sometimes been identified, as in 'Davey'. However, reduced or pale leaf color may also be an indicator of an unhealthy condition although the exact cause may not always be known. An unstable characteristic referred to as yellow-blotch has been found in some cultivars and there is evidence that the condition is inherited (Kester, 1971). "Tip-scorch" or "shot-holing" can occur from unhealthy conditions in the root zone or from disease, but the exact cause is not always known.

Flower Type

Almond cultivars may differ in flower type. A small-petaled phenotype occurs in some cultivars, such as 'Eureka', and is inherited by many of the seedlings of the 'Nonpareil' x 'Eureka' breeding line. Grasselly (1972) found that the small flowered characteristic of 'Marcona' was recessive to large flower types found in 'Ardechoise' and 'Bartre'.

The number of stamens may vary among different cultivars from 20 up to 40 with the usual number being 30 to 33 (Grasselly, 1972). The distribution of stamen number within seedling populations from parents of different stamen number indicate quantitative inheritance.

Maturation of Fruit and Nut

Ripening of the almond fruit involves dehiscence of the hull and loss of moisture to dryness. The pellicle of the kernel changes color from white to brown, and the kernel loses moisture. Ripening begins in the nuts on the periphery of a tree and proceeds to the center. It may continue for three or four weeks from first evidence of dehiscence.

Time of ripening is characteristic of specific cultivars. The sequence among cultivars is repeatable from year to year, but considerable year to year seasonal variation occurs as does variation from location to location. Inheritance appears to be polygenic and observations indicate high heritability. Grasselly (1972) reports similar results with seven populations studied.

Nut Characteristics

The almond nut includes the kernel surrounded by the shell after the hull has been removed. The important shell characteristics from a horticultural standpoint include the hardness of shell, shelling percentage, and adherence of the outer shell layer, and the degree to which the suture edge is sealed. Variation also exists in the kind of outer shell markings, the width of the wing, and other morphological features which are useful in identification and may have value in genetic analyses. Shell types vary from being very hard, thick, and bony to being very thin and papery.

TABLE 2: Characteristics of almond shells

Category	Approximate shelling, %	Description
Very hard shell	<35	Very difficult to break, needs hammer
Hard shell	35–45	unable to break by hand
Semi-soft	45–55	broken by hand with effort
Soft	55–65	broken by hand
Paper	>65	very thin, easily removed

Hardness of shell can be defined descriptively as in Table 2. Shelling percentage can be defined and measured in two slightly different ways. First, the weight of sound, edible kernels in a sample of given weight gives the commercial crack-out and is useful in determining yield of sound nuts. Second, the weight of a given number of well-developed kernels in the same number of nuts gives a biological measurement of shell vs. kernel. In the absence of defective kernels, the two values are the same.

Shell hardness and shelling percentage are the important biological factors. Poorly sealed shells and poorly retained outer corky layers are correlated to high shelling percentages and soft or paper shells.

Grasselly (1972) has identified two major genes affecting shell hardness: D = very hard shell (shelling percentage of 15 to about 35), and d = soft (shelling 35 to 65%). D was dominant over d. He found three genetic groups:

DD = homozygous hard, as in 'Cristomorto'
Dd = heterozygous hard, as in 'Marcona,' 'Bartre'
dd = homozygous soft, as in 'Ardechoise,' 'I.X.L.,' and 'Texas' ('Mission')

Considerable phenotype variation occurs among individuals of the *dd* genotype where the shelling percentage may range from about 40% to as high as 70–80%. It seems probable that there are additional polygenic modifiers that determine hardness of the shell.

FIG. 5. Variation in width of suture opening on nut of a soft- or paper-shelled cultivar. Left to right: well-sealed; small crack, large enough to insert a knife blade; very open surface.

The size of the suture opening (see Fig. 5) of the shell in poorly-sealed nuts may vary according to the following scale:

1) excellent seal; no opening
2) cracked along suture line but no opening; however, a knife blade can be inserted
3) most nuts slightly open; less than 2 mm width
4) mixture of slightly open and wide open
5) most open; 2 mm or more
6) very wide; all very open.

The percentage of nuts in grade 1 (well sealed) is the most important selection characteristic of this group. The remainder are descriptive.

The retention of outer corky layer of shell may be designated as follows: 1) all retained; 2) less than ⅓ of cork missing; 3) about ½ of cork missing; 4) more than ⅔ missing; and 5) all missing.

Double and Twin Kernels

The almond flower contains two ovules. One or both may develop into a seed. If both develop, two kernels are produced that are usually misshapen and undesirable because of difficulty in screening and sizing. The tendency to produce only a single well-formed kernel is a highly desirable characteristic.

Examination of various distribution curves of seedling populations for double kernels shows it to be essentially normal, the mean being intermediate between parents, but the range usually transgresses considerably beyond that of the parents. Selection for low tendency toward doubling has been highly effective.

Nuts with double kernels are produced in varying percentages depending upon cultivar and upon sample. The level of doubling can vary from essentially none in certain cultivars to as high as 40 to 50% or more. Occasionally individual seedlings are observed that consistently produce 100% double kernels. However, the percentage of double kernels of a given cultivar does vary by season which may reflect the general favorability of fruit set for that year over all cultivars.

If the percentage of doubles in a representative sample is used as a criterion of "tendency to double," the distribution curve within a seedling population represents a part of a normal curve. Zero doubling evidently is below a threshold value of internal physiological conditions favoring doubling. Individuals falling below that level cannot be distinguished phenotypically except by breeding behavior. Shifts in the distribution curve in relation to this threshold can account for seasonal and locational variations in percentage of doubling.

A "twin" is a seed in which more than one embryo occurs. These can be detected by the outline of the small embryo showing through the seed coat. If pronounced, it may adversely affect the appearance of the kernel. Twins are uncommon but have been found in some cultivars more than others; sometimes up to 10 to 15% occur. These represent a potential source of haploids but none have been identified.

FIG. 6. Characteristics of almond kernels showing various kinds of defects. Top row, left to right: typical kernel of 'Nonpareil' cultivar showing smooth, uniform, well-shaped kernel with no defects; incompletely developed kernel, shriveled at basel end; damage from Navel orange worm (*Paramyelosis transitella* walk.); callus growth on surface of kernel which grew from a crack on inside of shell. Bottom row, left to right: "crease" in surface of kernel, producing a major depression; "twin" embryos resulting from two embryos within the same seed coat; "split pellicle" in which embryo grows through the skin or pellicle; "black spot" resulting from insect sting given by various plant bugs.

FIG. 7. Variation in grades of smoothness (veininess) of the pellicle of a kernel. Very wrinkled on left to very smooth on right.

Defective Kernels

Defective kernels (see Fig. 6) can be observed in various samples:

1) Blanks and shrivels. These occur if embryos abort and do not develop although the seed coats remain. If abortion occurs after the kernel is partially developed, a shriveled kernel is produced. There are undoubtedly a number of causes of abortive kernels in almond cultivars, some genetic, some environmental. Occasional seedlings are found that tend to produce high percentages of blank kernels.

2) Crease. Some cultivars develop a deep crease or fold on the flat side of the kernel. The number of kernels within a sample with this characteristic may vary from very few to many. Little information is available except that it is inherited.

3) Callus growth. This condition is manifested as a scabby white to brown layer on the surface of the pellicle that dries and may produce an inedible kernel at maturity. This material is callus and grows from cracks that may develop on the inside of the endocarp in early May when the endocarp is still growing but the inside surface begins to harden. The condition is analogous to the split-pit condition that occurs in peaches (Kester, 1957). It can result from a combination of environmental conditions and an inherent tendency in specific cultivars.

4) Gummy nuts. Nuts are sometimes found with a clear gum over the kernel, which nearly fills the cavity. Gumming is a response to physiological injuries that can result from a variety of conditions. These can include the same conditions as those producing callus growth, such that the vascular system of the fruit is injured. Some cultivars appear to have more tendency to produce gum than others.

5) Split pellicle. Occasionally kernels are produced in which the white part of the embryo shows through a rupture of the seed coat. No information is available as to causes.

Each of these defects can be designated as a percentage of a given sample as a measure of the tendency to produce that problem. No specific information is available on inheritance except that they result from a combination of genetic and specific environmental factors.

Kernel Color and Surface

Almond kernels of different cultivars vary in color from very light yellow through yellow, to brown, tan and very dark brown. Pigment develops in the pellicle during maturity so that relative lightness or darkness may depend to some extent on stage at harvest. The kernel becomes darker on drying, with age, and on exposure to light.

The following grades have been observed: 1) very light; 2) light, as in 'Nonpareil'; 3) medium, as in 'Ne Plus Ultra'; 4) dark, as in 'Mission' ("Texas'); and 5) very dark.

Color of pellicle is highly heritable and selection for color can be effective.

Almond kernels may have very short hairs or pubescence on the surface of the pellicles. Where abundant, the surface is rough and unpleasant when eaten. A moderate amount is acceptable but an excessive amount is undesirable.

Kernels can be matched against a set of standard kernels as follows: 1) no pubescence; 2) slight pubescence, as in 'Nonpareil'; 3) moderate pubescence; 4) marked pubescence; and 5) very pubescent.

This characteristic is highly specific to cultivar and selection can be highly effective.

The almond pellicle is derived from integuments which have a network of veins or vascular bundles terminating at the chalazal end.

The surface of the kernel may be quite smooth or it may be variously wrinkled due to variation in prominance of veins and surface depressions between them (see Fig. 7). A tendency toward roughness or veininess decreases attractiveness and may contribute to difficulties in some processing techniques such as blanching.

For measurement, compare sample to specific standards in the following grades: 1) completely smooth; 2) slightly wrinkled, as in 'Nonpareil'; 3) moderately wrinkled; 4) marked wrinkling; and 5) very wrinkled.

Although smoothness can vary somewhat with season and location as a reflection of the relative ability of nut to fill out, it tends to be specific for a cultivar. Smoothness is characteristic of specific breeding lines and selection can be effective.

Kernel Size and Shape

These two characteristics of an almond kernel are considered together because they are intimately interrelated. Genetic analysis is complex because each involves components that appear to be determined by separate genetic and environmental factors. Size can be designated by linear dimensions or by weight but proper analysis requires interaction of both methods (Kester, 1965). Individual nut size as measured by weight is a component of yield. Smaller almonds produce lower yields per acre than larger almonds. Equally important, different kernel sizes, as determined by linear dimension, have different industrial uses, but this use also depends on shape and thickness. Almond kernels in commerce are graded in sizes that range from about 15–16/oz (1.18 g) to 40–50/oz (0.60 g). Almond kernels of different sizes may vary somewhat in their roasting and processing requirements. Small, flat kernels of about 30 or more/oz (0.95 g) are the so-called "bar-types" in demand for confectionary manufacture.

FIG. 8. Width/length ratios of kernels. Left to right: very narrow, W/L about 40; medium narrow, W/L = 47; medium, W/L = 53; broad, W/L = 65.

Considering first the variability within a *single sample* of a single cultivar, there is a wide range in size, as measured by length and width or weight (see Fig. 8). A high correlation exists between length and width among various sizes of kernels, indicating that the shape is relatively constant with change in size. There is also a high correlation between kernel weight and both width and length, and separation into commercial size grades can be made by passing kernels through round-hole screens of various diameters. Further separation of each group for thickness can be made by passing kernels through slot-hole screens of different widths. However, kernel thickness of different-sized kernels is not very well correlated to width and thickness. As size is reduced by width and length dimensions, kernel thickness does not necessarily reduce in a proportional rate. Thus, a small kernel may be just as thick or thicker in actual measurement than a large kernel.

If one compares samples from separate locations (environments) or different years, the average size (width, thickness, weight) for the samples may vary but the ratio between width and length remains relatively more constant than the relationship to thickness. In general, the greater the number of nuts per branch, the smaller the size. Thickness is characteristic of a cultivar but may vary considerably in different samples taken from different years and locations and can affect the width: weight relationship used to screen out different nut sizes.

Biologically, the linear relationship between length, width, and thickness are established in the first developmental period of fruit growth and is complete by May (in California). Weight accumulates in the last phase of fruit development and determines how completely filled the kernel will be at harvest. If improperly filled due to adverse growing conditions or genetic factors, the kernel may be considerably shriveled and reduced in thickness.

As a consequence of this analysis, the following characters are proposed as describing any single almond kernel sample: average weight, width/length (W/L) x 100, and average thickness. These are used to compare kernel samples of different cultivars and seedling plants. In addition, some descriptive characters may be needed. Thus, shapes can be compared readily by their W/L relationship (side view) as being long and narrow W/L = 40–48; medium W/L = 49–55; or broad W/L =

> 55. However, viewed from this way the sides of a kernel may not necessarily be uniformly wide but may be rounded, oval, ovate, oblong, or straight on one side and rounded to various degrees on the other. Kernel thickness may vary from being flat, medium, or plump in relation to size. Likewise thickness may not be constant from base to tip. If the basal part of the kernel is thicker than the apical, a "tear-drop" appearance results. Unequal thickness can result in unequal roasting during processing.

Inheritance of the various components of size and shape is quantitative, complex, and varies with parents used. Comprehensive genetic analysis has not been made. Grasselly (1972) reports that certain cultivars transmit their specific kernel characteristics to offspring. 'Bartre' transmits very large kernel size; 'Texas' and 'Ardechoise', smaller sizes. Very broad 'Marcona' transmits this characteristic. Thicker kernels are inherited by 'Marcona' and 'Texas' and thinner ones by 'Bartre' and 'A la Dame'. He reports a correlation between kernel and comparable endocarp (shell) dimensions but not necessarily to that of the mesocarp (hull).

Incompatibility and Self-fertility

The cultivated almond is self-sterile and must be cross-pollinated by pollen of a different genotype. Sterility results from a gametic incompatibility (Gagnard, 1952), in which a pollen grain with a particular S-allele will not grow down or function in a style of a cultivar whose genotype includes that same allele, as in tobacco (East and Mangelsdorf, 1925) or cherry (Crane and Brown, 1937).

A number of different incompatibility groups of California cultivars have been identified (Olmo and Kester, unpublished data):

1) 'Nonpareil', 'Long I.X.L.', 'I.X.L.', and 'Profuse';

2) 'Languedoc', 'Texas' ('Mission'), and 'Ballico';

3) 'Jordanolo' and 'Harpareil';

4) 'Reams' and 'Jubilee';

5) 'Rivers Nonpareil', 'Kutsch', 'Sultana', and 'Bigelow'; and

6) 'Smith XL' and 'Drake'.

'Harriott' has a common allele with 'Nonpareil' but 'Mission' ('Texas') and 'Eureka' do not. The four possible S-allele genotypes from 'Nonpareil' x 'Mission' and 'Nonpareil' x 'Eureka' progeny have been identified.

Identification of specific S-alleles in different cultivars would make it possible to establish parentages of specific genotypes, to trace origins and to determine the degree of inbreeding. No source of self-fertility has been identified within the cultivated almond species. However, self-fertility can be transmitted to almond from self-fertile species, such as peach and *P. mira*. It is likely that other related species, such as *P. nana*, might also be sources (Warfield and Kester, 1968). Such programs of interspecific hybridization to obtain self-fertility have been carried out at the California Agricultural Experiment Station (first carried out by Dr. H. P. Olmo, later by D. Kester), and in the breeding program of F. W. Anderson, Le Grand, California. A self-fertile almond variety, 'Le Grand', has been placed on the market from the latter program (Brooks and Olmo, 1972).

The first generation hybrids of almond with self-fertile species are also self-fertile. Self-fertile individuals can be recovered in consecutive backcross populations to almond. An analysis of limited data from such crosses (Kester, 1968) suggests that the self-fertility system (sf) is dominant over self-incompatibility (si) system. However, there is apparently considerable variation in degree of self-fertility, possibly due to modifiers of expression, so that selection of individual clones with consistently high self-fertility is required.

One of the difficulties of studying this problem has been the inconsistency and unreliability of obtaining self-pollination results following artificial hand pollinating or covering limbs with bags. The degree of self-fertility seems to vary from year to year and suggests that the ability to set with self-pollination of many cultivars is quite sensitive to environmental influences. To be commercially successful a self-fertile cultivar should produce the same degree of set and quality of kernel with its own pollen as if it were cross-pollinated and such consistent results have not yet been produced.

Organoleptic Qualities

Almond kernels of different cultivars vary somewhat in organoleptic characteristics and commercial almond handlers recognize such differences. Distinct flavor characteristics have been claimed for Mediterranean-produced almonds (Horoschah, 1971) but differences may represent environmental or processing procedures. Not enough information is available on specifically characterizing differences and how to qualitatively and quantitatively measure them.

Differences involve both flavor and texture or physical qualities and can be detected by taste tests and physical measurements. Baker et al. (1961, 1962) found that randomly selected untrained testers could identify specific preferences among a group of cultivars previously selected for "desirability" although the ability to identify preferences varied greatly by individual taster and conditions of the test. In taste panel tests (Johnson and Kester, unpublished data) differences could be readily identified, preference stated, and some characterization in relation to relative sweetness or bitterness described. However, identifiable differences in the raw kernel may change after roasting (Simone and Kester, unpublished data). Roasting requirements with regard to time and temperature may vary with cultivar, size, and shape (Pittman, 1930).

A desirable texture is somewhat difficult to define. Sterling and Simone (1953) showed that significant differences existed among cultivars in "fragmentability," or ability to break up into small pieces and in "hardness" or "crispness," a tendency to break sharply into larger pieces.

A major difference in flavor revolves around the presence of glucoside amygdalin that gives the "bitter" principle. Lower levels or the absence of amygdalin produces the "sweet" flavor. Amygdalin breaks down to benzaldehyde and cyanide (Woodroof, 1967).

A "mild bitter" flavor can be detected to some cultivars that sometimes is considered pleasing. The relative significance of this factor has a great deal to do with detection of flavor differences.

Bitterness in almond has been identified as a recessive single gene s, (Heppner, 1923, 1926). Thus, a bitter almond has a genotype of ss. Most cultivars of almond are heterozygous for bitter Ss, but a number of homozygous sweet SS have been identified through breeding tests.

The fact that the almond kernel is the offspring generation and results from genes of both maternal parent and the pollen parents suggests that expression of bitterness might vary within a tree depending on genotype of individual embryo. Crane and Lawrence (1947) report that when the almond cultivar 'Marie Dupuy' was pollinated by pollen from a bitter almond, the resulting seeds were decidedly bitter. Pollinations carried out for many years at the California Agricultural Experiment Station have not produced any case where a distinct pollen influence (xenia) for bitterness has been identified. Furthermore, the major commercial cultivars ('Nonpareil' and 'Mission', for example) used in California are heterozygous for bitter (Ss). If segregation occurred immediately in the embryo, one-fourth of all kernels produced should be bitter. Segregation does occur in the next generation, however, and approximately one-fourth of the seedling offspring produced are bitter.

Inheritance of bitterness in peach x almond hybrid populations is more complex and shows quantitative variation (McCarty et al., 1952).

Resistance to Disease, Insects, and Other Environmental Disorders

Different cultivars of almond are attacked by various pathogenic organisms in the environment which may become limiting to production under particular situations and be a limiting characteristic of certain cultivars.

Fungus diseases: Certain fungus and bacterial diseases that attack the flowers, young shoots, leaves, and developing fruit in the spring can be destructive and may require costly chemical controls. They are particularly destructive in years and areas where spring rains and high humidity occur. Much variation among cultivars occurs in relative susceptibility to these various pathogens, and selection for resistance can be effective if sufficient observations are made in appropriate environments.

Grasselly (1972) analyzed the relative susceptibility of a large number of cultivars from various parts of the world in their response to *Fusicladium carpophilum* (Thöm.) Oud., *Monilinia laxa* (Aderh. & Ruhl.) Honey, shot-hole fungus (*Coryneum beijerinckii* Oud.), anthracnose (*Gloeosporium amygdalinum* Brizi), and rust (*Tranzschelia pruni-spinosae* [Pers.] Diet.) in high humidity areas of Bordeaux, France. He particularly identified the cultivar 'Ardechoise' as showing high resistance to all of the fungus diseases and suggested that resistance to various fungus diseases might be controlled by similar genetic factors. He also analyzed a number of seedling progeny of 'Ardechoise' compared to progeny of susceptible cultivars as 'Cristomorto' or 'AI'. He concluded that resistance was controlled by quantitative genes. The distribution curve suggests high heritability in that the offspring population showed a wide range from resistant to susceptible with the average being intermediate between those of the parents.

Resistance to the condition known as "crown rot" caused by *Phytophthora* is particularly important in rootstocks. Experience has shown that almond trees on almond seedling rootstocks are more susceptible to this problem than are most other *Prunus* species. High losses can possibly be associated with the short dormancy requirement in almond and the tendency to delay going dormant in the fall if the weather stays warm. Considerably higher losses are often noted in nurseries in those falls when warm temperatures prevail and there is high humidity produced by early rains.

Field observations suggest that resistance to crown rot is associated with the cultivar used as the scion. After one particularly serious winter associated with high crown rot losses, approximately one-half of the 'Nonpareil' trees of a particular test plot involving two different rootstocks—almond seedling and peach x almond hybrid seedlings—died whereas all of the 'Texas' ('Mission') trees on the same two rootstocks survived. The percentages of tree loss of these cultivars in the remainder of the orchard were comparable. Likewise, 'Thompson' trees showed similar susceptibility to crown rot. Analogous observations among various almond cultivars were made that same year. Day (1953) has earlier reported similar observations.

Grasselly (1972) reports also that seedling plants with 'Ardechoise' as a parent showed resistance to *Phytophthora*. On the other hand, when the cultivar 'AI' was in one of the parental pedigrees, either pollen or seed, considerable seedling mortality (about 25%) occurred.

Insects: Several major insect pests of almond occur and differences in susceptibility have been observed among different cultivars. The brown almond mite (*Brobia arbaea* M. & A.) and several kinds of red spider mites (*Panonychus ulmi* K. and *Tetranychus pacificus* McG.) are particularly destructive although their effect is not in immediate loss of a crop. Rather it has a weakening effect over a period of time caused by defoliation and decreased photosynthetic efficiency. Some variation in resistance to mites have been noted but no systematic analysis has been made.

The Navel orangeworm (*Paramyelois transitella* Walk.) and peach twig borer (*Anarsia lineatella* Zell.) are particularly destructive and destroy almond nuts. Reasonably effective control measures are available for the second but not for the first. Differences in susceptibility are primarily related to hardness of the shell and how well the shell is sealed. The widespread use of soft- and paper-shelled almond cultivars has contributed to the problem. Early, efficient, and rapid harvesting before the insect population builds up is also conducive to insect control. Improvements in genetic control for these insects should follow along these lines of development.

Frost resistance: Spring frosts are a major limiting factor in many parts of the world. Selection for later-blooming cultivars is one major source of resistance since the susceptibility to freezing increases rapidly as the flower buds open and is particularly high after full bloom when the small nuts are developing. Thus, the later the bloom period, the greater the chance of escaping frost.

In addition, there appear to be differences in hardiness among cultivars at the same stage of development (Grasselly, 1972; Griggs, 1953; Meith and Rizzi, 1972). The differences appear to be greatest during the period before full bloom. 'Ne Plus Ultra', 'Peerless', and 'Drake' are more sensitive than 'Nonpareil'; 'Texas' ('Mission') is probably intermediate. Kester (unpublished data) has noticed an association between sensitivity to frost and early leafing-out in relation to bloom. Time between leafing-out and blooming varies greatly from season to season but late leafing *in relation to bloom* is closely related to genotype. Most of the native almond species observed so far have this characteristic and it is transmitted to the offspring (Kester, unpublished data). Grasselly (1972) has noted differences in frost resistance among cultivars. For example, 'Desmayo' and 'Ardechoise' are more resistant than 'Marcona', 'Ne Plus Ultra', and 'Peerless'. The early-blooming, early-leafing cultivars 'Cavaliera' and 'Constantini' are quite susceptible. He notes that leafing and blooming dates are not necessarily similar and that late leafing is transmitted to offspring.

Genetic Disorders

Some almond cultivars are subject to specific disorders which have been classed as "genetic" or "hereditary" since no transmissible pathogen has been identified. The condition is perpetuated by vegetative propagation from particular parts of the clone. It has been found only in certain cultivars, and the condition is inherited from them.

The most important of these is noninfectious bud-failure (BF) which is significant because it occurs in 'Nonpareil' and certain other important commercial cultivars of California (Kester and Jones, 1970). The disorder is recognized by specific symptoms, principally the failure of leaf buds to grow out in the spring (see Fig. 9). Loss in viability had actually occurred in these growing points the summer before (Hellali, Kester, and Ryugo, 1971). Necrotic areas and bands of rough bark may develop on shoots. Flower buds are not affected but over a period of years the reduction in bearing surface results in significant losses in yield.

A progressive shift from normal to BF occurs within the clones so that symptoms may develop on a given tree after a period of time (sometimes four or five years or more after planting) even though the propagating material came from a seemingly normal tree. Wilson and Schein (1956) reported that shift in BF level was associated with increasing age of the clone. Experimental evidence has since been obtained that the disorder is temperature sensitive and is gradually induced in somatic cells with continuous exposure of the growing shoot to high growing temperatures (Kester and Hellali, 1972).

FIG. 9. Branch of almond tree showing severe noninfectious bud-failure (BF) symptoms. Note the bare one-year-old shoots and vigorous new shoot growing from a basal bud, and the severe rough bark on an older shoot.

The factor causing BF disorder is transmitted from parent to offspring by both male and female gametes and different levels of BF in somatic tissue are heritable (Kester, 1968). Almond x almond crosses have been compared in all combinations involving normal and BF plants of known BF potential genotypes. When both parents had BF symptoms, BF appeared in high percentages at an early age in the seedling offspring. BF appeared later and with lower percentages at a given age when one parent had BF symptoms. When neither parent showed obvious symptoms, BF appeared much later and at low percentages at a given age. The use of normal-appearing BF cultivars as parents appears to account for the development of BF cultivars, such as 'Jordanolo', 'Harpareil', 'Jubilee', and 'Merced'. These cultivars when selected and released for commercial use did not produce evidence of BF until after they had been grown and propagated for a number of years. Such serious outbreaks have occurred in some of these cultivars as to eliminate them from recommended planting.

Kester (1968) attempted to identify a BF factor in progeny tests using a severe BF plant as a recurrent parent. Since some BF offspring were found in essentially all of 22 populations studied, it was concluded that the BF factor acted as a major dominant factor which was repressed to varying degrees by some unknown cellular mechanism. This behavior suggested a type of somatic variation similar to the genetic phenomenon of paramutation proposed by Brink (1968), or similarly one of the many other types of unstable somatic phenomena that have been described (Pearson, 1970). An alternate hypothesis is that the disorder is caused by a nontransmissible pathogenic agent within the cell that reproduces and grows in response to high temperature. However, no evidence for either a transmissible or nontransmissible agent has been found.

A number of other analogous disorders are believed to occur in the breeding material at the California Experiment Station. These include a leaf-blotch condition, a narrow-leaf somatic variation, a rough-bark phenotype and possibly others (Kester, 1971).

Rootstock Breeding

Rootstock selection and breeding is an important component of almond breeding.

Objectives

There are three main objectives of a rootstock improvement and selection program. The first is graft compatibility interaction with the major scion cultivars, which establishes the limits within which a breeder can work. The second is productivity. Medium to high vigor in the rootstock and the ability to withstand high productivity are desirable in modern almond growing. Because nuts are harvested mechanically and maximum nut yield is the principal point of production, there seems to be little advantage to direct efforts toward dwarfing. The third is adaptation. Individual characteristics include adaptation to the entire range of soil and insect and disease problems that exist in various areas.

Types of Rootstock

There are four principal types of almond rootstocks including: 1) almond seedlings, 2) peach seedlings, 3) plum seedlings or clones, and 4) peach x almond hybrids.

Almonds: Almond seedling rootstocks are invariably deep rooted and drought tolerant and have traditionally been used when orchards are grown under well-drained, deep soil conditions without irrigation. They are tolerant of calcareous soil and more resistant to high boron than many other *Prunus* species (Hansen, 1948). On the other hand, almond seedlings are susceptible to root knot nematodes (*Meloidgyne* sp.), crown gall (*Agrobacterium tumefaciens*), crown rot (*Phytophthora* sp.), and some other soil organisms such as oak root fungus (*Armillaria mellea* [Vahl.] Guel.). Nematode-resistant almond plants have been selected from almond populations in Israel (Kockba and Spiegel-Roy, 1972).

Compatibility relationship to almond scion cultivars is good and gives no problem. Almond trees on almond rootstocks tend to be long-lived under favorable conditions but are so subject to the disease conditions cited above that their survival in irrigated orchards can be erratic. As a rootstock for other *Prunus* species, almond is only partially satisfactory. For peach cultivars, almond shows various degrees of incompatibility, depending on the particular combinations. Unions that appear normal for many years sometimes break off cleanly, although the reciprocal combination of almond scion on peach seems satisfactory in most instances.

Bitter almond seeds have been traditionally used by nurserymen as a preferred stock but there seems little reason to believe that this criterion alone imparts superiority over sweet almond types (Day, 1953; Bernhard and Grasselly, 1969). Seeds collected from standard cultivars are used as a source. For example, seedlings of 'Mission' ('Texas') provide uniform, vigorous nursery plants and have been used. 'Mission' is more susceptible to sodium in high salt areas and the use of this cultivar has sometimes been questioned. However, there has been little interest in selecting among almond sources for rootstocks because of the trend toward peach in most intensively-cultivated areas where irrigation is practiced.

Peach: Peach seedling rootstocks have come into use for almond growing whenever the industry comes under extensive irrigation. Almond trees with peach rootstocks grow more vigorously when young, come into bearing somewhat sooner, and tend to survive better than comparable trees on almond rootstocks. The reason for greater tree survival seems to be due to greater tolerance to crown rot and higher water levels and lower susceptibility to crown gall. This shift from almond to peach rootstock in commercial planting has been noticeable in California, France, and Spain, as the management system changed. On the other hand, peach is not tolerant of soils that are calcareous, droughty, or where boron levels are high. Also, trees with a peach root system may not be as long-lived or as productive when they are old. How significant this difference is has not been demonstrated under adequately controlled conditions.

The principal method of selecting or developing peach rootstock material is to identify clones that produce uniform seedlings with the desired characteristics. Consequently, any of the products of peach rootstock breeding programs should be immediately useful for almond. One of the most widely-used seed sources has been 'Lovell' peach, which produces uniform vigor in almond plants. Various nematode-resistant or -immune peach rootstocks are being used. 'Nemaguard', widely used for peaches in soils infested with root knot nematodes (*Meloidgyne incognito acrita* and *M. javanica*), gives well-growing almond trees.

Plum: Interest in plum rootstocks stems primarily from their tolerance of heavy, poorly-drained soils. In addition, the clonal stock 'Marianna 2624' (*P. cerasifera* Ehrh. x *P. munsoniana* Wight and Hedr.) has been used for its relative tolerance to oak root fungus (*Armillaria mellea*). Plums have a potential advantage in showing considerable capacity for vegetative reproduction as cuttings so that clonal stocks are possible.

Incompatibility problems arise with almond/plum combinations. In general, trees tend to be smaller in size than those of almond/almond (which may be an advantage), but many combinations are short-lived, produce inadequate tree growth, and develop large overgrowth at the union. Kester et al. (1965), found that the almond cultivars tested on 'Marianna 2624' plum separated into a more or less distinctly compatible and an incompatible group. With the incompatible group, differences in degree were identified by the timing in the fall of the onset of defoliation symptoms both within the season and at the age when growth inhibition occurred. Incompatibility was associated with 'Nonpareil' and cultivars related to it. The compatible group was related to 'Mission' ('Texas') and related cultivars. Seedling offspring of incompatible x incompatible cultivars were largely incompatible (Kester, 1970). Crosses of compatible x incompatible produced all compatible offspring when 'Mission' was a parent but not when 'Merced' (compatible) was a parent. ('Merced' is a 'Nonpareil' x 'Mission' hybrid.) Results are interpreted that differences in incompatibility in this combination may be controlled by a relatively few dominant genes. Variability does exist within each of the populations indicating quantitative as well as qualitative effects.

The defoliation symptom which was particularly useful in early identification of incompatibility was associated with a breakdown in phloem function and could be observed morphologically at the almond/plum union. In 'Nonpareil'/'Mission'/'Marianna 2624' combinations, the breakdown factor was translocatable through the compatible almond interstock. However, a special selection of *Prunus insititia* L. or a hybrid of it, known as 'Havens 2B', which is compatible with almond, appeared to react differently as an interstock and to overcome breakdown. This difference in reaction has considerable practical and theoretical interest as it indicates that there are prospects for individual interstock selections of value (Kester et al., 1956). Various *P. insititia* selections, tested for almond at the California Agricultural Experiment Station, have not made satisfactory orchard trees although complete failure does not occur.

Bernhard and Grasselly (1969) have tested a number of different combinations of various plum hybrids with almond in France. Results ranged from combinations showing complete failure to other combinations with promising results. Among the plum materials developed and tested were clones of myrobalan, *P. cerasifera* x *P. salicina* Lindl., 'Marianna' seedling selections, clones of various hybrids involving *P. cerasifera, P. salicina,* 'Marianna', *P. amygdalus,* and *P. persica.*

The breeding and selection of special plum stocks for almond would appear to have possibilities. Primarily, the breeding objective would involve selection for compatibility in addition to special rootstock characteristics. Careful selection would be required not only of the specific rootstock genotype, whether seedling or clonal, but also of the genotype to be used as the scion. Selection of interstock genotype is also a possibility in this case.

Peach x almond hybrids: The first generation (F_1) hybrid between peach and almond can be readily made and has considerable potential as a rootstock for almond. With a suitable parental combination the population of F_1 seedlings is uniform, shows extensive vigor, and, in general, is capable of a wide adaptation to soil conditions (Kester and Hansen, 1966).

As a rootstock, the peach x almond hybrid has shown considerable potential for almond and possibly other species, such as peach, plum, and prune. The full range of characteristics are not known and differences among selection can be expected. In general, the rootstock produces very vigorous grafted trees that grow well over a wide range of soil types (Kester and Hansen, 1966).

Propagation has been a major stumbling block. Two methods are possible: seed and vegetative propagation.

Seed propagation can be accomplished either by making hand crosses or by growing the peach and almond parents in sufficient isolation so that natural hybridization occurs. Selection of hybrid seedling plants is made in the nursery row so that it is necessary to identify them from the peach or almond seedlings in the same population. Selection of proper peach and almond parents is necessary both to get high yield of hybrid seed and to obtain a desirable population of rootstock plants.

Kozahonane (1970) found that a higher percentage of sets could be produced on pollen-sterile peach cultivars with almond pollen compared to various other peaches used as seed parent. He attributed this to the longer period of receptivity. Peach cultivars seem to vary in their ability to set fruit on almond, and selection for high setting ability is possible.

Three general lines of hybrid selection have been underway. The first took place in France and involved the identification of naturally occurring hybrids followed by selection of clones for vegetative propagation. Two clones, 'GF 557' and 'GF 677', have been selected (Bernhard, 1955) which can be propagated by leaf-bud cuttings (Grasselly, 1955).

A second line of selection involves hybrids of late-blooming almond cultivars with 'Nemaguard' peach. Jones (1969) reported on populations produced either by hand pollination or by placing bouquets of peach flowers in almond trees. Populations produced by the open-pollinated method contained 50 to 60% hybrids which could be identified in the nursery. Root systems produced showed the entire range from spreading (like peach), through intermediate, to deep growing (like almond). Nematode galls were not completely absent but were much less pronounced than on known susceptible plants. Jones (1972) later introduced 'Titan', a late-blooming almond to cross-pollinate with 'Nemaguard' to produce uniform, vigorous hybrid seedlings in high percentages.

Production of hybrid seeds from seed orchards where 'Nemaguard' peach and 'Nonpareil' almond were growing together was discovered by Arthur Bright, Le Grand, California. He has produced hybrid rootstocks commercially from this source by growing seeds of these selected almond trees and roguing out almond seedlings in the nursery row.

A third line of breeding for peach-almond hybrid material has been carried out at the California Agricultural Experiment Station involving several lines of peach selections. Some of these are immune to the two species of root knot nematode (Sharpe et al., 1969) and immunity is transmitted when crossed with almond clones. Some of the almond clones had been previously selected for crown gall and bacterial canker resistance from large populations of irradiated seedlings (DeVay, 1965). Nematode immunity from the peaches tested appears to be transmitted into the hybrid offspring as in peach x peach crosses (Kester et al., 1970). Selection of individual clones capable of vegetative propagation is a feature of this program, although seed production is not ruled out. Commercial hardwood cutting propagation can be achieved by proper selection of clone, proper timing, and treatment with growth regulator and fungicide (Hansen and Hartmann, 1968; Mercado-Flores and Kester, 1966). In earlier studies, almond has been found difficult to root, the peach relatively easy (with proper technique), and the hybrid intermediate with a wide range from easy to difficult (Kester and Sartori, 1966). Consequently, selection for ease of rooting is particularly important.

Selection of almond parent must be part of this hybrid seed production program. Some almond cultivars have been found to transmit factors that produce some proportion of stunted, rough-barked offspring that are unlikely to be suitable for rootstock. The effect of these variable types on rootstock performance is unknown.

Other species: There are possibilities for rootstock breeding among other almond species, most of which have not been investigated. Gentry (1956) reports that *P. spinosissima* and *P. spartinoides* growing wild in Iran are used by local growers to topwork almond cultivars to them under very arid conditions. *P. bucharica* has been reported to be a useful rootstock for apricot, almond, and peach under nonirrigated conditions. *P. petunnikowii* is reported to be exceptionally resistant to drought, crown gall, and *Capnodis* (Evreinoff, 1952). Serr and Forde (unpublished date) crossed *P. andersoni* with almond to produce seven seedlings. One clone survived, 51–263, which was found to root moderately well by cuttings and, when used as a rootstock, performed as well as almond seedlings.

Achievements and Prospects

Limitations imposed by characteristics of the almond species and its cultivars have traditionally restricted almond growing to rather specific climatic regions in the world. Furthermore, the adaptation of the almond tree to drought and to concomitant susceptibility of the almond root system to excess water has traditionally tended to restrict almond growing in many of the almond regions of the world to marginal production areas where yields are low and inconsistent. A major shift in the concept of almond growing has occurred with the removal of many of the limitations to production by the control of environmental hazards, such as disease, lack of water, inadequate nutrition, frost, etc., and the shift to more resistant rootstocks, notably the peach. Where these changes have occurred, such as in California, almond growing has tended to expand to the most favorable climatic areas and shift to most productive soil sites. Under these conditions orchardists find that the almond tree is highly adapted to the intensive practices, including mechanization, that characterize modern agriculture, with resulting high yields per acre.

Increased use of the almond on the market through effective merchandizing and the development of new uses has paralleled these production changes. Marketing has been aided by the increased buying power of the consuming public all over the world. The result of these factors has been spectacular increases in per-acre yields, expansion to new areas, and an increase in total production. The prospects are that almond growing will continue to intensify in various parts of the world. However, successful commercial production will require that orchards be situated on high-producing locations of soil and that efficient management techniques be used to make the enterprise economically sound.

Development of almond industries in various parts of the world has been accompanied by the selection of local cultivars adapted to that region. As the production methods and marketing requirements have developed, new demands have been placed on the requirements of the cultivars used. In California production has been standardized around specific cultivar combinations but there has been a continuous search on the part of growers for cultivar combinations better than the existing ones. Consequently, a continuous shifting from one new cultivar to another has occurred as the limitations of a cultivar becomes apparent with extensive cultivation.

Most cultivars used commercially to date have appeared as chance seedlings. An individual new cultivar may possess one or more outstanding traits but none has so far been completely satisfactory. Most discouraging has been the occurrence of specific defects, such as noninfectious bud-failure of individual cultivars that could not have been forseen in the original test trees studied.

Specific breeding programs have not had a major impact on the cultivar picture but considerable progress has been made. There is a wide range of variability within the almond species and its relatives, much of which has not yet been utilized or systematically exploited. Breeding procedures do not appear to be complicated as long as they are based on careful criteria of selection and evaluation.

Future prospects for improvement can be listed in the following manner:

1) Productivity. Consistent, high production will be essential as a first requirement of cultivar improvement. This means that the tree not only must be adapted to the particular environmental conditions imposed but that it possess those particular growth characteristics that will respond to intensive culture and management in modern agricultural techniques. One of the major trends will be to select for cultivars that come into production rapidly (by the third or fourth year), maintain high production for a period of time but perhaps be removed at an age earlier (20 to 30 years) than the traditional age of almond orchards. Specific cultivar types that fit into these categories appear to be available but more information is needed so as to characterize them for their early identification.

2) Desirable market product. The high production being enjoyed by the almond industry in some areas and the prospects for continued increases leading to possible excess production requires that an expanding market be maintained. This means that a high quality product must be produced that is adapted to specific uses. Up to now, the most desirable cultivars from the orchard standpoint are not necessarily the best from the market standpoint. The California almond industry was built

around the high quality of the 'Nonpareil' kernel with its uniformity, smoothness, bright yellowish-brown color, and blanchability. However, its main weakness is a very thin, poorly-sealed shell which is not particularly adapted to mechanization and is subject to worm and bird hazards. Most of these characteristics appear to be highly heritable and should be able to be combined with suitable tree characteristics. However, a wide range of kernel types involving kernel size, flower, and shape occur and specific types should be combined with desirable tree characteristics. A further possibility is to combine the desirable kernel from almond and the peach to produce a dual-purpose fruit crop with edible flesh and seed.

3) Efficiency of orchard management. The requirement to plant separate cultivars side-by-side for cross-pollination produces difficulties in orchard management, involving timing of sprays, frost protection, harvesting, and irrigation schedules. Combinations are chosen primarily for the most efficient blossoming pattern but also should be chosen to provide the most efficient harvesting operation. The large number of new cultivars that have recently been introduced into California makes possible to the grower the opportunity to be selective in choosing various cultivar combinations on both of these counts. The following patterns are now becoming available:

a) Combinations of cultivars that bloom in sequence and harvest in sequence. This system, which may involve three to five cultivars in the same orchard, has been in use for many years. It had been thought to be the most useful for cross-pollination but it has serious drawbacks for management, particularly in the modern mechanized orchard.

b) Combinations of cultivars that bloom together and harvest in sequence. This system allows reduction to a two-cultivar orchard and appears to be as efficient from the cross-pollination standpoint as the multiple cultivar orchard, providing that suitable pollinating cultivars are used. The 'Nonpareil' x 'Merced' and 'Thompson' x 'Mission' combinations are of this type. This system provides the most immediate prospect for advances in almond orchard planning.

c) Combinations of cultivars that bloom together and harvest together. This system requires that the two cultivars be nearly identical in nut and kernel characteristics so that they can be mixed together as a single product on the market. Cultivars can be planted as separate trees or be grafted into the same tree either in the nursery or in the orchard. The 'Nonpareil' x 'Kapareil' and 'Nonpareil' x 'Milow' combinations are examples of this system. Prospects for continuing this kind of selection are promising but careful selection of cultivars is necessary. Careful tree management is necessary if one cultivar is to be grafted into the other.

d) Self-fertile cultivars. The development of commercially-acceptable cultivars with this characteristic would be a major advance in almond culture since single cultivar orchards would be possible. Experimental cultivars now exist primarily through the transfer of the self-fertility system from other self-fertile species, such as peach. Considerable testing will be required on an orchard basis to be as sure that such a cultivar will perform as satisfactorily when self-pollinated as when cross-pollinated and that kernal quality will not be affected. However, it is likely that future developments in almond breeding will be directed towards the incorporation of self-fertility into other cultivars.

4) Freedom from problems. The most discouraging aspects of almond breeding has been the existence of unrecognized characteristics or defects of almond cultivars that may not appear until the vegetatively-propagated cultivar is grown for a period of time in different environments. Some of these defects involve specific adaptations to environments, susceptibility to disease or other pathogens, or inability to perform adequately under intensive culture. These characteristics may be recognized through improvements in the evaluation schemes and the enhanced recognition of the significance of traits in the young tree correlated to traits when older.

Other characteristics such as the noninfectious bud-failure phenomenon appears to represent specific unstable systems, presumably genetic, that result in persistent shifts within the propagating material of the clone. Although improvements in selection procedures during propagation offer promise for "living with" this problem, the long-term success in almond breeding requires that this problem be eliminated through breeding cultivars that do not carry this system. Although the noninfectious bud-failure phenomenon is the most significant, other apparently unstable characters have been observed in breeding material suggesting that there may be other phenomenon currently unrecognized that could be significant under particular situations.

All of the above characteristics are based primarily on the almond as an industrialized agricultural commodity and are based on optimizing the favorable factors for production. In other situations, the ability to grow almonds under less favorable conditions may be desirable to enhance the economy or the food supply of a local area. For example, there are considerable prospects for producing later-blooming and hardier types of almonds than are now grown. Breeding material is available among commercial cultivars to delay the blossom season several weeks later than now occurs. At the other extreme, short chilling, early-blooming cultivars exist and could be used to extend the almond into warmer winter areas providing spring frosts aren't also a hazard.

Finally, non-commerical cultivars as backyard and home orchard trees may be highly useful in areas not particularly adapted commercially. Some of the related almond species that possess dwarf tree characteristics, late bloom and early harvest, possibly combined with self-fertility, could be the basis of such improvement by incorporating desirable nut characteristics into them.

Literature Cited

Almeida, C. R., Marques de. About the unproductivity of almond (in Portuguese). *Anais Inst. sup. Agron. Lisboa* 15:1–186.

Armstrong, D. L. 1957. A cytogenetic study of some derivatives of the F_1 hybrid *Prunus amygdalus* x *Prunus persica*. Ph.D. dissertation, University of California, Davis.

Baker, G. A., M. A. Amerine, and D. E. Kester. 1961. Dependency of almond preference on consumer category and type of experiment. *J. Food Sci.* 26: 377–385.

———. 1962. Consumer preference on a rating basis for almond selections with allowance for environmental and subject induced correlations. *Food Tech.* 16(7):121–123.

Bernhard, R. 1949. Peach x almond and its utilizations (in French). *Revue Hort.* 2164:97–101.

———. 1958. Hybrids of prune x peach and prune x almond: principal characteristics useful as eventual rootstocks of peach. *Adv. in Hort. Sci.* 2:74–86.

Bernhard, R., and Ch. Grasselly. 1969. Rootstocks of the almond (in French). *Bul. Tech. Inform.* 241: 543–549.

Brink, R. A., E. D. Styles, and J. D. Axtell. 1968. Paramutation: directed genetic change. *Science* 159: 161–170.

Brooks, R. M. 1943. A growth study of the almond fruit. *Proc. Amer. Soc. Hort. Sci.* 37:193–197.

Brooks, R. M., and H. P. Olmo. 1972. *Register of new fruit and nut varieties.* 2nd ed. Berkeley: Univ. of Calif. Press.

Candolle, A. P. de. 1886. *Origin of cultivated plants.* New York: Hafner.

Chandler, Wm. H., and D. S. Brown, 1951. Deciduous orchards in California winters. *Calif. Agr. Ext. Serv. Cir.* 179.

Cociu, V. 1967. Pomology of the Socialist Republic of Romania (in Romanian). *Bucarest Acad. de la Repub. Socialiste de Roumanie.*

Crane, M. B., and W. J. C. Lawrence. 1947. *The genetics of garden plants.* London: Macmillan.

Crane, M. B., and A. J. Brown. 1938. Incompatibility and sterility in the sweet cherry (*Prunus avium* L.). *J. Pom. Hort. Sci.* 15:86–116.

Cruess, W. V. 1949. Almond meats and hulls. *Calif. Agr.* 3(12):6, 12.

Darlington, C. D., and E. K. Ammal. 1945. *Chromosome atlas of cultivated plants.* London: Allen and Unwin.

Day, L. H. 1953. Rootstocks for stone fruits. *Calif. Agr. Expt. Sta. Bul.* 736.

DeVay, J. E., G. Nyland, W. H. English, F. J. Schick, and G. D. Barbe. 1965. Effects of thermal irradiation on the frequency of crown gall and bacterial canker resistance in seedlings of *Prunus* rootstocks. *Rad. Bot.* 5:197–204.

East, E. M., and A. J. Mangelsdorf. 1925. A new interpretation of the hereditary behavior of self-sterile plants. *Proc. Natl. Acad. Sci.* 11:166–171.

Evreinoff, V. A. 1952. Some biological observations on the almond (in French). *Rev. Intern. de Bot. App. et d'agric. Trop.* 32:446–459.

———. 1952. Some biological and pomological observations on the almond (in French). *Bul. Soc. Hist. Nat. de Toul.* 87:23–43.

———. 1957. Contributions to the study of the almond (in French). *Ecole. Nat. Sup. Agri. Toulouse.* Vol. 5.

Gagnard, J.M. 1954. Research on the systematic characters and on the sterility phenomenon in almond varieties cultivated in Algeria (in French). *Ann. Inst. Agr. Algerie* T. 8(2).

Gentry, H. S. 1956. Almond culture in southern Iran. *Almond Facts* 21:6–7.

Grasselly, Ch. 1967. Choice of almond varieties and their rootstocks. *In*: L'Amandier-Comité National Interprofessionel de l'amande. Aix en Provence.

———. 1972. The almond: morphological and physiological characters of varieties, measurements, and their transmission to the first generation hybrids (in French). Thesis, University of Bordeaux.

Grasselly, Ch., and H. Gall. 1967. Study of the possibility of combining some agronomic characters of the almond Cristomorto hybridized with three others (in French). *Ann. Amel. Plantes* 17:83–91.

Griggs, W. H. 1949. Effects of low temperature on blossom survival and fruit set of 19 varieties of almonds. *Proc. Amer. Soc. Hort. Sci.* 53:125–128.

Hansen, C. J. 1948. Influence of the rootstock on injury from excess boron in French (Agen) prune and president plum. *Proc. Amer. Soc. Hort. Sci.* 51: 239–244.

Hansen, C. J., and H. T. Hartmann. 1968. The use of indolebutyric acid and captan in the propagation of clonal peach and peach-almond hybrid rootstocks by hardwood cuttings. *Proc. Amer. Soc. Hort. Sci.* 92:135–140.

Hansche, P. E., V. Beres, and R. M. Brooks. 1966. Heritability and genetic correlation in the sweet cherry. *Proc. Amer. Soc. Hort. Sci.* 88:173–183.

Hansche, P. E., R. S. Bringhurst, and V. Voth. 1968. Estimates of genetic and environmental parameters in the strawberry. *Proc. Amer. Soc. Hort. Sci.* 92: 338–345.

Hansche, P. E., C. O. Hesse, and V. Beres. 1972. Estimates of genetic and environmental effects on several traits in peach. *J. Amer. Soc. Hort. Sci.* 97: 76–79.

Hansche, P. E., V. Beres, and H. I. Forde. 1972. Estimates of quantitative genetic properties of walnut and their implication for cultivar improvement. *J. Amer. Soc. Hort. Sci.* 97:279–284.

Hellali, Rachid, D. E. Kester, and K. Ryugo. 1971. Noninfectious bud-failure in almonds: observations on the vegetative growth patterns and concurrent changes in growth regulating substances. *HortScience* 6:283. (Abstr.)

Heppner, M. J. 1923. The factor for the bitterness in the sweet almond. *Genetics* 8:390–391.

———. 1926. Further evidence on the factor for bitterness in the sweet almond. *Genetics* 11:605–606.

Horoschak, S. 1971. The almond industries of Italy and Spain. *USDA* FAS-M228:1–18.

Jones, R. W. 1968. Hybridization of apricot x almond, *Proc. Amer. Soc. Hort. Sci.* 92:29–33.

———. 1969. Selection of intercompatible almond and root knot nematode resistant peach rootstock as parents for production of hybrid rootstock seed. *J. Amer. Soc. Hort. Sci.* 94:89–91.

———. 1972. Titan, a seed source for F_1 almond x Nemaguard peach hybrids. *Fruit Var. Hort. Dig.* 26:18–20.

Kester, D. E. 1957. Corky growth of kernels is recurring Jordanolo problem. *Almond Facts* 22(4):8–9.

———. 1963. California almond varieties. *Calif. Agr. Expt. Sta. and Ext. Serv. Leaflet* 152:1–8.

———. 1965. Size, shape and weight relationships in almond kernels. *Proc. Amer. Soc. Hort. Sci.* 87: 204–213.

———. 1965. Inheritance of time of bloom in certain progenies of almond. *Proc. Amer. Soc. Hort. Sci.* 87:214–221.

———. 1968. Noninfectious bud-failure, a nontransmissible disorder in almond. I. Pattern of phenotype inheritance. *Proc. Amer. Soc. Hort. Sci.* 92:7–15.

———. 1968. Noninfectious bud-failure, a nontransmissible disorder in almond. II. Progeny tests for bud-failure transmission. *Proc. Amer. Soc. Hort. Sci.* 92:16–28.

———. 1969. Pollen effects on chilling requirements of almond and almond-peach hybrid seeds. *J. Amer. Soc. Hort. Sci.* 94:318–321.

———. 1970. Transfer of self-fertility from peach (*Prunus persica* L.) to almond (*Prunus amygdalus* Batsch.). Berkeley: *Abstr. West. Sec. Amer. Soc. Hort. Sci.*

———. 1970. Graft incompatibility of almond seedling populations to Marianna 2624 plum. *HortScience* 5:349. (Abstr.).

———. 1971. Abnormal phenotypes with viruslike appearance within seedling populations of almond. *HortScience* 6:292. (Abstr.).

Kester, D. E., R. Asay, and E. F. Serr. 1963. The Kapareil almond. *Calif. Agr. Expt. Sta. Bul.* 798:1–8.

Kester, D. E., and W. H. Griggs. 1959. Fruit setting in the almond: The pattern of flower and fruit drop. *Proc. Amer. Soc. Hort. Sci.* 74:214–219.

Kester, D. E., and C. J. Hansen. 1966. Rootstock potentialities of F_1 hybrids between peach (*Prunus persica* L.) and almond (*Prunus amygdalus* Batsch.). *Proc. Amer. Soc. Hort. Sci.* 89:100–109.

Kester, D. E., C. J. Hansen, and B. Lownsbery. 1970. Selection of F_1 hybrids of peach and almond resistant and immune to root knot nematodes. *HortScience* 5:349. (Abstr.).

Kester, D. E., C. J. Hansen, and C. Panetsos. 1962. Plum rootstocks for almonds. *Calif. Agr.* 16(6): 10, 11.

———. 1965. Effect of scion and interstock variety on incompatibility of almond on Marianna 2624 rootstocks. *Proc. Amer. Soc. Hort. Sci.* 86:169–177.

Kester, D. E., and Hellali Rachid. 1972. Variation in the distribution of noninfectious bud-failure (BF) in almonds as a function of temperature and growth. *HortScience* 7:322. (Abstr.).

Kester, D. E., and R. W. Jones. 1970. Noninfectious bud-failure from breeding programs of almonds (*Prunus amygdalus* Batsch.). *J. Amer. Soc. Hort. Sci.* 95:492–496.

Kester, D. E., P. Raddi, and R. A. Asay. 1973. Correlation among chilling requirements for germination, blooming, and leafing in almond (*Prunus amygdalus* Batsch). *Genetics* (suppl.) 74: 5135. (Abstr.).

Kester, D. E., and Elvino Sartori. 1966. Rooting of cuttings in populations of peach (*Prunus persica* L.), almond (*Prunus amygdalus* Batsch.), and their F_1 hybrid. *Proc. Amer. Soc. Hort. Sci.* 88:219–223.

Kockba, J., and P. Spiegel-Roy. 1972. Resistance to root-knot nematode in bitter almond progenies and almond x Okinawa peach hybrids. *HortScience* 7: 503–504.

Kostoff, N. V., and K. F. Kostina. 1935. A contribution of the study of the genus *Prunus*. Focke, Leningrad. (in Russian, summary in English) quoted by Evreinoff, 1952.

Kozakanane, H. D. 1970. Improvement in methods of production of plant hybrids between peach and almond species (in French). Thesis, University of Bordeaux.

Leverton, R. M. 1959. Recommended allowances. *In* Food, *USDA Yearbook of Agriculture*. Washington: Supt. of Documents. 1959:227–230.

Mason, W. A. 1913. The pubescent fruited *Prunus* of North America. *J. Agr. Res.* 1:147–178.

McCarty, C. D., J. W. Leslie, and H. B. Frost. 1952. Bitterness of kernels of almond x peach F_1 hybrids and their parents. *Proc. Amer. Soc. Hort. Sci.* 59: 254–258.

Meith, C., and A. D. Rizzi. 1970. Almond production. 1. Establishing the orchard. *Univ. of Calif. Agr. Ext.* AXT-29. 2. Care of the orchard. AXT-83.

Mercado-Flores, I., and D. E. Kester. 1966. Factors affecting the propagation of some interspecific hybrids of almonds by cuttings. *Proc. Amer. Soc. Hort. Sci.* 88:224–231.

Moss, O. E. 1965. The almond industry in South Australia. *J. Agr., S.A.* 69:38–45.

Mukherjee, S. K. 1967. Horticultural research in the USSR. *Indian J. Hort.* 24:1–10.

Nyland, G., and A. C. Goheen. 1969. Heat therapy of virus diseases of perennial plants. *Ann. Rev. Plant Path.* 7:331–354.

Pearson, O. H. 1968. Unstable gene systems in vegetable crops and implications for selection. *HortScience* 3:271–274.

Pitman, G. 1930. Further comparison of California and imported almonds. *Ind. Eng. Chem.* 22:1129–1131.

Riabov, I. N. 1969. Trials on self-pollination in some interspecific hybrids of the common peach and common almond with smooth-pit peach (in Russian). *Bjull. Gos. Mikitsk. Bot. Sada*, No. 3(10):24–28 (*Hort. Abst.* 41:539.)

Rehder, A. 1947. *Manual of cultivated trees and shrubs*. 2nd ed. New York: Macmillan.

Richter, A. A. 1964. Results of practical and theoretical work of hybridization and varietal studies in almond (in Russian). (quoted by Grasselly, 1972). *Trud. Gos. Nikit. Bot. Sada* 37.

———. 1969. Objectives and methods of selection of almond. (in Russian) (quoted by Grasselly, 1972). *Trud. Gos. Nikit. Bot. Sada* 40.

Sabeti, H. 1966. *Trees and shrubs of Iran*. University Press of Teheren.

Salesses, G., and A. Bonnett. 1971. Morphological and cytological study of almond x peach hybrid and the 4x form induced by colchicine treatment (in French). *Ann. Amelior. Plantes* 21:75–82.

Scaramuzzi, F. 1964. Aspects of fruit culture in the Soviet Union (in Russian). *Accad. Econ. Agrar. dei. Georg.* XI:1–70.

Sharpe, R. H., C. O. Hesse, B. F. Lownsbery, V. G. Perry, and C. J. Hansen. 1969. Breeding peaches for rootknot nematode resistance. *J. Amer. Soc. Hort. Sci.* 94(3):209–212.

Serr, E. F., D. E. Kester, M. N. Wood, and R. W. Jones. 1954. The Davey almond. *Calif. Agr. Expt. Sta. Bul.* 741:1–8.

Sterling, C., and M. Simone. 1954. Crispness in almonds. *Food Research* 19:276–281.

Tabuenca, M. C. 1972. Winter chilling requirements in almond. *An. Aula Dei.* 11:325–329.

Tabuenca, M. C., M. Mut, and J. Herrero. 1972. Influence of temperature on the flowering time of almonds. *An. Aula Dei.* 11:378–395.

Tufts, W. P., and C. L. Philp. 1922. Almond pollination. *Calif. Agr. Sta. Bul.* 346:1–36.

U. S. Department of Agriculture. 1951. Virus diseases and other viruslike symptoms of stone fruits in North America. *USDA Handbook* No. 10.

Warfield, D. L., and D. E. Kester, 1968. Some characteristics of *Prunus* introductions of interest in almond breeding. *HortScience* 3:92. (Abstr.).

Warfield, D. L. 1968. An investigation into characteristics of a Yugoslavian *Prunus* introduction of potential value in almond breeding. M. S. thesis, University of California, Davis.

Watt, B. K., A. Merrill, and M. L. Orr. 1959. A table of food values. In *USDA Yearbook of Agriculture*, Washington: Supt. of Documents. 1959:231–265.

Weir, Wm. C. 1951. Almond hulls as feed. *Calif. Agr.* 5(9):13.

Wilson, E. E., and R. D. Schein. 1956. The nature and development of noninfectious bud-failure of almonds. *Hilgardia* 24:519–542.

Wood, M. N. 1925. Almond varieties in the United States. *USDA Bul.* 1282:1–142.

Wood, Milo N. 1939. Two new varieties of almond: The Jordanolo and the Harpareil. *USDA Cir.* 542: 1–12.

Woodroof, J. G. 1967. *Tree nuts: Production, processing, products*. Vol. I. Westport, Conn.: AVI Co.

Pecans and Hickories

by George D. Madden and Howard L. Malstrom

The pecan, most cultivated member of the hickory group, ranks as an important nut crop in the world although its commercial production has been confined primarily to the southern United States. Improvement of native trees began as early as 1850, first by seedling selection and later by hybridization. Most of the commercial trees grown today were developed between 1915 and 1930 when considerable expansion of the pecan industry created a demand for improved cultivars. Following a period of inactivity in cultivar development, the United States Department of Agriculture established an extensive breeding program and began releasing new cultivars in the mid-1950s. Breeding objectives have changed considerably over the past 50 years and a review of the progress made in pecan breeding will serve to underline the major objectives and methods involved with improving quality and production of this crop today.

Pecan and hickory belong to the *Juglandaceae* family. The generic names have been listed variously as *Juglans* and *Hicoria* in the past (Madden et al., 1969). The appellation *Carya*, an ancient Greek name of the walnut, has now been made a conserved generic name by the International Botanical Congress. The pecan, *Carya illinoensis* (Wang.) K. Koch, is the most important of the *Carya* group and will be discussed in detail in this chapter.

Except for *Carya carthayensis* Sarg., which is native to eastern China, *Carya* spp. are native to North America (Crane et al., 1937). The natural range extends from the north central and eastern United States into northern Mexico. The fact that the genus is found over such a vast geographical range with different temperatures and humidity patterns suggests a number of genetic and physiological adaptations. Some species apparently hybridize and perhaps this has led to their unusual vigor and adaptation.

The hickories are large trees occurring naturally in mixed hardwood stands along with ashes, maples, and oaks. Fifteen species of hickory have been described but there has been some taxonomic confusion over the years due to the close relation and possible hybridization of some species that have resulted in integrated forms. Despite an extensive search, botanists, naturalists, and horticulturists have been unable to salvage a single species except the pecan from this genus for the commercial value of its nuts. The recent average annual production (1967–1972) of pecans in the United States exceeded 90 million kg (in shell) with a value over $80 million. The economic value of most other hickories lies primarily in the timber that is noted for its strength and hardness and that has been used extensively in the furniture industry. The flavor imparted by the smoke of burning hickory is among the best for the smoking of meats.

Some hickory species are used as ornamental trees because of their symmetrical form and luxuriant foliage. Yet others produce highly flavored nuts and, despite the thick shell and poor kernel quality, are considered a delicacy in many rural areas. Only limited work has been done with native hickory trees to develop improved cultivars through selection of improved seedlings and controlled crosses. Hickories grow slowly and are nonprecocious, and the poor cracking quality reduces the flavored nuts to little economic importance. Despite poor nut production, some species show a high degree of disease resistance (McKay, 1961; McKay and Romberg, 1962) and others are apparently not attacked by some harmful insects (McKay and Romberg, 1962). The greatest contribution of hickory species may be to furnish genes to the pecan cultivars, which will enable them to resist damaging disease and insects (McKay, 1961).

The total number of identified hickory species varies considerably, depending on the taxonomist. The following species are relatively popular or have breeding potential.

Carya ovata (Mill.) K. Koch, the shagbark hickory, occasionally grows 40 m in height and is distinguished by the shaggy trunk bark found on older trees. The large, glossy leaves usually contain five to seven leaflets, 10 to 15 cm long (Rehder, 1960). The dormant buds, large and overlapping, break dormancy later in spring than other species and shoot growth usually ceases after seven to eight weeks (Reed, 1944). Reed (1944) states that most Americans consider the flavor and quality of the shagbark to be the choicest of native hickory nuts. The nuts are small, 3.5 to 6 cm long, whitish, distinctly angular, and thin but hard-shelled (Rehder, 1960). They form in thick hulls that open entirely to the base along suture seams, and the shell has several longitudinal ridges protruding along its length. The species is difficult to graft or transplant and grows so slowly that it often takes nurserymen up to seven years to produce salable trees (Reed, 1944).

The geographical range overlaps that of the pecan. The species is found from southern Canada to the southern Gulf Coast and west to eastern Oklahoma, Kansas, and Nebraska (Reed, 1944). The shagbark is generally found on upland soil in contrast to many other hickories (MacDaniels, 1969).

Carya laciniosa (Michx. f.) Loud., the shellbark hickory, so closely resembles the shagbark hickory in size and appearance that the two are often difficult to distinguish (Reed, 1944). Leaves of the shellbark are frequently up to 30 cm in length, usually with seven leaflets per leaf (Reed, 1944; Rehder, 1960; MacDaniels, 1969). In contrast to the shagbark, it grows mostly on river bottoms commonly subject to flooding (MacDaniels, 1969).

The hulls or shucks are large, with dimensions of 7.5 by 3 cm not uncommon. The nut, yellowish white and obscurely four-angled, does not usually crack open along suture lines (Bailey, 1960). The kernel, though often not completely filled, tastes sweet. According to Reed (1944), some shellbark x pecan hybrids develop plump, well-filled kernels and probably produce the sweetest meat of all hickories. This species has not been commercially planted, and marketable nuts have come from scattered trees in forests or on farms.

Carya cordiformis (Wang.) K. Koch, the bitternut, is the most hardy of the hickories and is most widespread geographically (Reed, 1944). It is found along bottomlands rich in humus and is commonly referred to as "swamp hickory" (Bailey, 1960). Many trees of this species have been found along northern shores of Lakes Erie and Ontario (Reed, 1944).

The tree grows to about 30 m in height and the foliage is finer than that of either shagbark or shellbark and more like that of pecan (Reed, 1944). Bailey (1960) describes the leaf buds as bright yellow in summer and darker in winter; the leaves contain five to nine leaflets. The fruit, in clusters of two to three, is roundish and 2 to 4 cm in length. The sutures flange toward the apical end and hull sections split only about half the length of the nut (Reed, 1944). Both shell and hull are relatively thin, but the kernels are of little value because of their extreme bitterness and astringency (Reed, 1944; MacDaniels, 1969).

Reed (1944) states that the bitternut develops a dense root system and can be transplanted more successfully than most other hickories. Root sprouts are common, especially when a tree has been removed. The bitternut can easily become a "weed tree" once it is established. Because of its vigorous root system, this species might have promise as a rootstock.

Carya glabra (Mill.) Sweet., the pignut, and *Carya ovalis* (Wang.) Sarg., the sweet pignut or red hickory, have trees and nuts of similar appearance. Reed (1944) claims that the principal difference lies in the taste of the kernels; those of *C. ovalis* are always sweet, whereas those of *C. glabra* are rarely so. Wiegand and Eames (1924) and

MacDaniels (1969) believe that *C. ovalis* is a hybrid of *C. ovata* and *C. glabra*. The trunk of the red hickory becomes shaggy with age but rarely as shaggy as that of shagbark or shellbark (Reed, 1944). The red hickory usually has five to seven leaflets per leaf, whereas sweet pignut usually has three to seven (Bailey, 1960). Both trees produce small, thin-shelled nuts encased by thin shucks; those of *C. glabra* split only half way to the base, whereas those of *C. ovalis* split nearly to the base. The nuts of *C. ovalis* are often mistaken for those of shagbark (Reed, 1944).

Reed (1944) states that both species grow within the native range of shagbark and prefer upland soils that are gravelly or sandy. They often grow in areas where apple and peach orchards are planted. He reports that red hickory has been propagated and several cultivars produced, but the nuts are not of the quality of those of shagbark cultivars in the same area.

Carya tomentosa (Lam.) Nutt., the mockernut, has a number of common names, among them white hickory, black hickory, bullnut, bigbird hickory, hard back hickory, hognut, pignut, and square nut (Reed, 1944). The mockernut is a large (up to 30 m in height), well-shaped tree that produces dense shade; the bark of the trunk is finely grooved and without scales (Rehder, 1960). The dark-green, pubescent leaves range from 8 to 18 cm in length, with usually seven to nine leaflets per leaf.

The thick hulls split entirely to the base, and Rehder (1960) depicts the nuts as pear shaped. 3.5 to 5 cm in length, brown, sharply ridged, and thick-shelled. However, these nuts vary in form and appearance and are often difficult to distinguish from those of shellbark (Reed, 1944). Kernels of mockernut are almost impossible to shell out as complete halves, but are sweet and edible (Reed, 1944; Rehder, 1960).

Carya myristicaeformis (Michx. f.) Nutt., the nutmeg hickory, grows to 30 m in height and the branches are coated with yellow-brown scales that become reddish-brown the second year (Rehder, 1960). The leaflets, five to 11 per leaf, are short stalked, almost sessile, and from 7.5 to 13 cm in length. The staminate catkins are peduncled but the fruit is usually solitary (Bailey, 1960). The nut is reddish-brown with irregular spots and stripes and a thick shell; kernels are sweet.

Carya cathayensis Sarg., the Cathay hickory, is indigenous to eastern China. This cultivar is related to *C. myristicaeformis*, according to Bailey (1960), who describes this species as somewhat smaller, reaching a maximum height of 20 m, with five to seven leaflets, 10 to 14 cm in length, per leaf. The fruit is ovoid, four-angled and small, 2 to 2.5 cm long; kernels are sweet.

Carya aquatica (Michx. f.) Nutt., the bitter pecan or water hickory, is smaller than most hickories, rarely reaching a height of 30 m (Bailey, 1960; Rehder, 1960). The bark is light brown and separates into long, thin plates, and the buds are dark reddish-brown in winter. The leaflets, seven to 13 per leaf, are sessile and 8 to 12 cm long (Bailey, 1960). Both Bailey (1960) and Rehder (1960) describe the fruit, usually three to four per cluster, as ovoid, 2.5 to 3.5 cm long, with a thin husk that splits entirely. The nut is four-angled, irregularly and longitudinally wrinkled, and dull reddish-brown; kernels are bitter.

Carya pallida (Mill.) Engl. and Graebn. has rough pale bark with purple-brown branches (Rehder, 1960). Usually seven, rarely nine, leaflets, 8 to 12 cm in length, are found per leaf. The fruit is obovoid, 2 to 3 cm in length, with a thin husk that usually splits to the base. The nut is flattened and relatively thin shelled.

Rehder (1960) lists two related species. *C. buckleyi* Durand is a smaller tree native to Texas and Oklahoma, and *C. buckleyi* var. *arkansana* (Sarg.) Sarg., a larger tree often reaching a height of 25 m, is native to Missouri.

Carya illinoensis (Wang.) K. Koch., the pecan, is one of the largest of the hickories, and often attains a height of 50 m. The bark is deeply furrowed, grayish-brown and the dormant buds are pubescent and yellow. The leaflets, 11 to 17 per leaf, are short-stalked, oblong lanceolate and 10 to 18 cm in length. The fruit, oblong and ranging from 3.5 to 8 cm in length, is borne, three to ten in number, on spikes. The nut is ovoid or oblong, 2.5 to 7 cm in length, smooth, and light brown with darker striations extending longitudinally. The kernel is sweet.

Origin and Early Development

L. D. Romberg, a long-time pecan breeder, made an extensive study (1968) of the probable evolution of the pecan. Based on fragmentary reports and communications with other pecan historians, he assumes that the glaciers that moved south from Canada in recent geological time probably moved the pecan from northern areas. In the first three interglacial periods, the common North American crow, *Corvus brachyrhynchos* Brehm., probably did more to spread seed nuts than any other animal. When the last ice cap was receding, Indians carried nuts northward and eastward into areas now considered the native pecan habitat. They most likely selected the largest and thinnest-shelled nuts available for planting. The early European settlers found the pecan growing on fertile, alluvial soils along the bottomland of the Mississippi River and its tributaries. However, in the past 60 years large commercial plantings have been established both east and west of the natural origin. Today these areas lead the world in pecan production.

Romberg (1968) postulates that the juvenile state of seedling trees was a strong factor in the evolution of pecan trees. Juvenile trees produce extensive root systems and are apparently more resistant to cold and to insect and disease injury (Romberg and Madden, unpublished data). In the forest environment, the trees grew large, spreading branches over competing trees before nut production commenced. Bearing of fruit required considerable stored food (Worley, 1971; Sparks and Brack, 1972), and probably trees that were capable of eliminating competition by establishing extensive tree structures were the survivors.

Pecan trees have probably always been plagued by fungi that attack tender tissues of leaves, shoots, or nuts. Natural selection must have played a major role in the development of naturally resistant strains, and in the early selection of pecan seedlings by man, disease resistance was a prime objective. 'Stuart' and 'Schley' comprise most of the acreage in the southeastern states, and were originally highly disease resistant. However, present-day cultivars, bred or selected for improved nut characters, are more severely affected by disease organisms than native trees. Chemical fungicides which permit the profitable culture of susceptible cultivars have reduced the dependency of disease resistance in breeding. Nevertheless, without either genetic resistance or chemical control, pecans certainly could not be economically grown (Moznette, 1934; Cole 1968; KenKnight, 1968).

History of Improvement

The word "pecan" was reportedly used by the American Indian at the time America was discovered (Woodard et al., 1929). According to Crane et al. (1937), the earliest written report of pecans is that of the Spanish explorer Cabeza DeVaca in 1520, who described thin-shelled "walnuts" found along the streams in the southern United States. Selection and propagation of superior trees began as early as 1846 when scions of a cultivar later named 'Centennial' were successfully grafted on seedling rootstock by a Louisiana slave named Antoine; subsequent propagations include 'Van Deman' in 1877, 'Rome' in 1882, 'Hollis' in 1884, and 'Pabst' in 1890 (Taylor 1905).

Because of lack of budded trees, most early orchards were started from seed and selections propagated from the resulting seedlings. H. A. Halbert selected 'Halbert' from a native orchard near Coleman, Texas, and S. H. James (1895) named 'Carman', 'James', and 'Moneymaker' from seedling selections in Louisiana. At about this time, J. H. Burkett of Clyde, Texas, named the 'Burkett', selected primarily for its high quality nut (Crane et al, 1937). However, low yield, susceptibility to disease, and sprouting in the shuck have gradually decreased its popularity and today the 'Burkett' is not being planted. J. B. Curtis, Orange Heights, Florida, introduced 'Curtis', 'Hume', 'Kennedy', and 'Randall' in 1896 from a seedling orchard he planted in 1886, and the 'Moore' also originated in Florida at about that time. The 'Mahan', discovered in Mississippi, is believed to be a seedling of 'Schley' (Crane et al., 1937). 'Cape Fear', from North Carolina, was introduced 40 years ago but has only recently gained prominence because of its apparent suitability for high-density plantings (Madden, 1972*b*).

Between 1880 and 1910, Theodore Bechtel, C. E. Pabst, W. R. Stuart, A. C. Delmas, I. P. Delmas, and F. H. Lewis introduced 'Delmas', 'Pabst', 'Schley', and 'Stuart' (Taylor, 1905). This group also released 'Alley', 'Candy', 'Russell', and 'Success' cultivars (KenKnight, 1971). 'Schley', 'Stuart', and 'Success' are the predominant cultivars in the Southeast today, and most of the trees are over 50 years old (R. L. Livingston, personal communication).

E. E. Risien of San Saba, Texas, named several cultivars resulting from selections of wild seedlings. His earliest release was 'San Saba' in 1882 (Wolfe, 1937). Risien was also a pioneer in the field of propagation and began topworking large seedling trees to his selections in the 1890s. He introduced 'Attwater', 'Kincaid', and 'Sloan' from random seedling selections and 'Colorado', 'Jersey', 'Liberty-Bond', 'Onliwon', 'San Saba Improved', 'Texas Prolific', 'Squirrels Delight', 'Supreme', and 'Western Schley' from his planting of 1,000 'San Saba' nuts (Crane et al., 1937).

The first recorded pecan nursery was established at New Orleans, Louisiana, by William Nelson. He began by selling seedlings in 1874 and by 1879 was selling grafted trees (Crane et al., 1937). However, between 1905 and 1925, a great number of new cultivars were introduced and trees propagated in nurseries. During that period, each area of the "Pecan Belt" introduced its favorite cultivars of local origin. The 'Aggie', 'Cowley', 'Hayes', 'Mount', 'Oakla', 'Patrick', and 'Texhan' ('Mahan' seedling) were named as a result of local nut contests and pecan shows.

Forkert (1914) was apparently the first hybridizer of pecans. He introduced the 'Admirable' ('Russell' x 'Success') and 'Dependable' ('Jewett' x 'Success'), neither of which has retained its former popularity. His one meritorious introduction, 'Desirable', is of unknown parentage (Crane et al., 1937) and has found renewed interest in recent years, especially in the Southeast, because it bears large, high-quality crops, is medium in precocity, and has better than average disease resistance. E. E. Risien, who also began making controlled crosses, shortly thereafter released 'Banquet' ('Texas Prolific' x 'Attwater'), 'Commonwealth' ('Long Fellow' x 'Texas Prolific'), 'Kincaid Improved' ('Onliwon' x 'Kincaid'), 'Sloan Improved' ('San Saba' x 'Sloan'), and 'Venus' ('San Saba' x 'Attwater').

Of all the introductions up to 1930, the 'Western Schley' probably has more desirable characters than any other. It was planted extensively in Texas and Oklahoma shortly after introduction and later in New Mexico and Arizona. This cultivar is still one of the most precocious and prolific available and is being planted intensively today in most areas where high-density groves are being established (Madden, 1972*b*).

Selections from native stands of trees in the northern United States began in the 1890s (Crane et al., 1937). Native pecans in the North are adapted to a growing season of 160 to 180 days, as compared to 190 to 220 days in the South. According to Taylor (1905), the first recorded cultivar selected in the North was found in Illinois and named 'Hodge's Favorite' or 'Illinois Mammoth'. Nuts of the original tree were sold for seed (Crane et al., 1937). W. N. Roper of Virginia introduced the 'Major' in 1907 from a selection of seed nuts shipped from southeastern Indiana. Roper later traced the nuts to the parent tree in northern Kentucky. Persistent searches for new sources in northern areas yielded 'Busseron', 'Butterick', 'Greenriver', 'Indiana', 'Kentucky', and 'Posey' between 1910 and 1915. For a long interval following, few cultivars were selected. However, in 1933 interest stimulated in new cultivars through pecan contests resulted in 'Peruque' and 'Starking Hardy Giant', presently important northern cultivars. Very few commercial pecan plantings exist in the northern areas, therefore the northern cultivars are of minor economic importance.

The first carload shipment of pecans from named cultivars, graded according to standards of size and quality, was made in Georgia in 1917. By 1920 large quantities of nuts from new cultivars were being marketed, but growers at that time did not receive a premium price for higher quality nuts. Today nuts of improved cultivars graded on the bases of size, kernel, and other quality factors bring a price premium.

From 400 to 500 cultivars were probably named in the first 50 years of development. However, light bearing, uncertain filling, susceptibility to insects and diseases, and other factors have eliminated most of them. A few withstood the test of time, and now probably only 20 to 30 are commercially propagated.

The USDA Bureau of Plant Industry established a pecan hybridization program in 1915 but very little came from several thousand seedlings because the work was terminated at various times. However, 'Caddo', from a 'Brooks' x 'Alley' cross made about 1920 by C. A. Reed, was selected and propagated at Philema, Georgia, but not until 1968 was this cultivar named and released to the public.

The most notable pecan breeder of present time is L. D. Romberg of Brownwood, Texas (Fig. 1), who was active in pecan improvement from 1923 to 1968 with the Texas Department of Agriculture and the USDA. In all, 12 cultivars resulting from Romberg's crosses have been released: 'Barton' ('Moore' x 'Success'), 1953; 'Comanche' ('Burkett' x 'Success'), 1955; 'Choctaw' ('Success' x 'Mahan') and 'Wichita' ('Halbert' x 'Mahan'), 1959; 'Apache' ('Burkett' x 'Schley') and 'Sioux' ('Schley' x 'Carmichael'), 1962; 'Mohawk' ('Success' x 'Mahan'), 1965; and 'Shawnee' ('Schley' x 'Barton'), 1968. Others released by Madden from crosses made by Romberg include 'Cheyenne' ('Clark' x 'Odom'), 1970; 'Cherokee' ('Schley' x 'Evers'), 1971; 'Chickasaw' ('Brooks' x 'Evers') and 'Shoshoni' ('Odom' x 'Evers'), 1972; and 'Tejas' ('Mahan' x 'Risien #1'), 1973. These cultivars possess widely divergent characters and some of them do not lend themselves to commercial production. However, 'Wichita', 'Sioux', 'Cheyenne', 'Cherokee', 'Chickasaw', and 'Shoshoni' are in demand beyond the supply because their precocity, prolificacy, and compact tree growth make them well suited to the new concept in pecan culture, high-density planting. 'Wichita' is probably the most widely planted cultivar of the last ten years.

The USDA presently is testing about 150 hybrid selections in cooperation with over 200 cooperators in 18 states. Many of these were made by Romberg. In addition to the USDA program, research and extension personnel in many states have maintained active programs to seek out promising seedling pecans for release as cultivars.

FIG. 1. L. D. Romberg of Brownwood, Texas, a pioneer USDA pecan breeder from 1931 through 1968.

Modern Breeding Objectives

New trends in production, marketing, and processing since 1940 have had an important role in changing the traditional objectives of pecan improvement. Early breeders were concerned mainly with disease resistance, high production, and attractive in-shell nuts. They did not have to consider characters desirable for mechanical harvesting, mechanical shelling, or adaptability to high-density plantings.

Most old pecan cultivars did not produce a first crop until about the eighth year but the industry is now demanding more precocious selections that bear commercial crops by the fourth year and that can be planted at a greater density. Trees that bear more uniformly over the years are desired to alleviate fluctuation in market price brought about by the irregular bearing habit and to provide a steady supply of nuts for marketing. Cultivars that mature nuts earlier in the season are of importance from the standpoint of better price and more even distribution of seasonal processing. Factors such as thinness of shell and nut size have decreased in importance because most pecans are marketed as shelled kernels. Smaller nuts, which generally fill out better than large nuts, and thick-shelled nuts, resistant to mechanical harvest damage, are more in demand. Because many chemical pesticides control insects and diseases satisfactorily today, breeding for inherent resistance has not been a critical objective. However, the possibility of restrictions on chemical usage dictates that genetic resistance be considered in future breeding objectives.

Breeding Techniques

Floral Biology

Morphology: The pecan is monoecious. The apical bud is mixed with catkin and vegetative buds enclosed by separate bud scales. The central bud normally gives rise to a shoot that may differentiate terminal pistillate flowers. While the pistillate flowers are completing differentiation, leaves, nodes, and internodes develop rapidly, and rudimentary buds form in the axils of the leaves. The remaining buds produce catkins if they develop (Isbell, 1928). With conditions favorable for growth in the spring, both the vegetative center of the bud and the rudimentary catkins continue their development but the vegetative center develops more rapidly. Frequently, the vegetative shoot will abort completely, leaving a short green stub; or if it grows, it will abort its apical bud about four to six weeks after bud break (Sparks, 1967). Some of the latter shoots can develop a new or false apical bud in late summer. Buds from long one-year-old shoots produce more catkins and clusters of pistillate flowers and set more nuts than buds of short shoots of the same type (Isbell, 1928) and in fact, there is a direct relation between flower formation and fruit set up to 25-cm shoot length (Sparks and Brack, 1972).

Catkins or aments arise from buds developed the preceding year, generally from the apex to the base of one-year-old shoots. They are generally borne in clusters of three, but up to nine and as few as one have been noted on rare occasions (Woodroof and Woodroof, 1930). Catkins are initiated in leaf axils in spring and early summer, a full year before they produce pollen (Woodroof, 1924; Isbell, 1928). According to Isbell (1928), many of the well-developed mixed buds contain three or four catkin buds by the end of the growing season. The stamens are in tiny florets, each having from three to nine stamens tipped with rosy or creamy anthers.

Shuhart (1927) found four distinctly different types of terminal buds in the pecan. Type 1 is a strictly vegetative, or true terminal bud, and the only type found in a young nonbearing tree. Type 2 (a false apical bud) is similar in size, shape, and structure to the lateral buds and is subtended by a full-size leaf. In early spring it is accompanied by a cluster of pistillate flowers that failed to develop, and the bud occupies a terminal position. Type 3 (a false terminal bud) is located at the base of the nut cluster and subtended by a leaf, but differs from Type 2 in that its flower cluster sets and matures nuts. Type 4 (also a false terminal bud) appears to be borne on the peduncle of the nut cluster, but is not subtended by fully developed leaves. Mature trees may have all four classes of terminal buds (Sparks, personal communication), but trees more than 35 years old have only types 2, 3, and 4 (Shuhart, 1927).

Morphological evidence of pistillate flower bud differentiation is seen in early spring after growth has started (Shuhart, 1927) and proceeds rapidly after the bud scales abscise and growth starts (Isbell, 1928). According to Mullenax and Young (1972), cultivars vary in the time of initiation as well as rate of development of terminal and subterminal inflorescences.

The pistillate flowers, normally one to nine in number, are borne sessile on spikes that terminate the current year's growth. An individual flower measures from 5.5 to 8.0 mm in length, about half of which is enclosed in a light-green pubescent calyx; the rest is stigma (Isbell, 1928). The calyx is marked by four ridges that divide it into four sections. Each section is terminated by a tapering bract from 3 to 5 mm in length, varying with cultivar. These may be erect, extended laterally, or reflexed downward along the sides of the flower. Receptive stigmas are sessile, variously shaped and colored, and a viscid fluid covers the uneven surfaces. There are distinct differences among cultivars in the size, color, and shape of the stigmas (Woodroof and Woodroof, 1926).

The vegetative regions of the flower resemble those of the young growing nut with vascular systems fairly complete (Woodroof and Woodroof, 1926). The ovule is about two-thirds enclosed by the integument when flowers become receptive and the vascular system of the septum has become quite distinct. The vascular bundles of the septum develop independently from vascular traces and later become connected with the series of bundles next to the shell region as described by Benson and Welsford (1909) for *Juglans regia* L. Each of the two parallel bundles of the septum becomes divided near the apex into a semicircle distinguishable in cross sections of the integument.

Dichogamy: The more or less complete dichogamy (the differential maturation of staminate and pistillate flowers) found in the pecan is often a major difficulty in pollination, especially in cases of isolated plantings of one or a few cultivars (Madden and Brown, 1973). Smith and Romberg (1940), have observed that the major pollen-shedding period and the major receptive period of many cultivars are not closely concurrent and in several cases they do not overlap at all. Cultivars fall into two groups with reference to the relative time at which the pollen sheds as compared with the time the stigmas are receptive. In one group (protandrous) the pollen is shed before stigma receptivity; in the other group (protogynous) pollen is shed after stigma receptivity. Staminate flowers of protandrous cultivars develop more rapidly than those of protogynous cultivars, but the pistillate flowers develop in the reverse order. As a general rule, pollen is shed earlier and stigma receptivity occurs later in protandrous than in protogynous cultivars.

Variation in degree of dichogamy occurs among cultivars, locations, and from year to year as a result of genetic and environmental effects on relative rates of pistillate and staminate flower development (Stuckey, 1916; Adriance, 1931; Romberg, 1931; Smith and Romberg, 1940; Wolstenholm, 1970*b*; Mullenax and Young, 1972; Madden and Brown, 1973). Nevertheless, Wolstenholm (1970*b*) found that protandrous and protogynous types always remained in discrete classes although the degree of dichogamy was variable. Romberg (1931), however, reported that a few genotypes occasionally self pollinate.

In general, a moist, warm spring favors earlier maturity of staminate flowers, whereas a cool, dry spring advances maturity of pistillate flowers (Adriance, 1931). Warm seasons of heavy rainfall advanced maturity of both. Staminate flowers within a cultivar are less variable in heat unit requirements.

Because in many cases complete dichogamy occurs in pecan, cultivars of the opposite blooming type are required in a planting to insure good pollination (Madden and Brown, 1973). The period of pollen shedding is five to 20 days and that of stigma receptivity is six to 14 days (Smith and Romberg, 1940; Madden and Brown, 1973), depending on cultivar, temperature, and humidity. For effective pollination, Woodroof and Woodroof (1927) suggest that pollinizer rows should not exceed 150 m from trees to be pollinated, but have reported pollen dispersal up to 300 m.

Pollination and Fertilization

Pollen is produced in great abundance, with about 2½ million pollen grains produced by a single catkin (Madden et al., 1969). Because pollen is distributed only by gravity and wind or rain, this apparent extravagance is necessary to insure effective pollination. Pollen sheds only under dry conditions and none is shed when relative humidity is above 85% (Woodroof, 1930). Rain followed by slow drying conditions can prevent shedding for up to 24 hours (Woodroof and Woodroof, 1927). Because of heavy dew in most pecan-growing areas, very little shedding occurs at night; most shedding occurs during daylight with a peak in the afternoon. Catkins normally abscise a few days after pollen shedding. The degree of defective pollen produced varies among cultivars and between years and is apparently related to factors that cause a failure of development of staminate flowers (Woodroof, 1930). Because staminate flowers begin development a full year before their appearance, there is not normally a problem of incomplete stamen differentiation.

McKay (1947) reported that pistillate flowers of 'Schley' contain mature embryo sacs in the nucellus of the ovule at the time of stigma receptivity. Four megaspores appear in nearly all of the ovules at pollination time, three of which have begun to disintegrate (Woodroof, 1928; McKay, 1947). At this time the nucleus of the mother cell has divided once and in some cases twice (Woodroof and Woodroof, 1926). The pollen tube may enter the ovarian cavity as early as the 4-megaspore stage. Adriance (1931) noted there was a considerable time lapse between pollination and fertilization and Shuhart (1932) placed the interval at two weeks. According to Woodroof et al. (1928), the egg and second male nucleus fuse during the fifth to sixth week after pollination; the zygote first divides about two months after pollination.

Chalazogamy (the growth of the pollen tube through the chalaza instead of the micropyle) occurs in pecan (Billings, 1903; Shuhart, 1927; Adriance, 1931). According to Billings (1903), the pollen tube passes down the axial tissue, past the ovary wall, close to the margin of the cavity. After leaving the style, the pollen tube does not have conducting tissue through which to pass and grows through isodiametric cells. At a point slightly below the funiculus, it curves and passes through a region of cells that stain deeply. At a point under the ovule, the pollen tube turns upward into the embryo sac.

Although the quantity of pollen is seldom limiting and wind distribution to all nutlets is quite thorough (Woodroof, 1924, 1930), fertilization in pecans may fail to occur, resulting in "nut dropping." There are three periods of "nut drop." One occurs shortly after bloom and is attributed to a lack of pollination. A second, or "May drop" occurs one to four weeks after pollination and seems associated with conditions that prevent fertilization of the ovule (Woodroof et al., 1928; Adriance, 1931; Sparks and Heath, 1972). The third drop, termed the "summer drop," continues until shell hardening and includes nuts that were damaged or contained aborted embryos. Woodroof et al. (1928) found this third drop more severe on heavy-bearing trees. All authors indicated that the magnitude decreased with successive drops. Sparks and Heath (1972) observed an inverse relationship between shoot vigor and degree of abscission, and the correlation was highest during the first and third drops.

Several studies have been made to determine the factors responsible for failure of fertilization of the egg cell or development of the embryo. Stuckey (1916) suggested that parthenogenesis might occur in the pecan, but Woodroof and Woodroof (1928, 1930) found that the "May drop" included nuts that were apparently not fertilized. Adriance (1931), who observed flower drop in bagged and unpollinated flowers, also concluded that the "May drop" appeared to be the result of a lack of fertilization due primarily to lack of pollination.

Most of the pollination techniques used in breeding today are modifications of the early work of Traub and Romberg (1933) and Smith and Romberg (1940).

For pistillate flowers, the preferred system of bagging flowers uses bags of sausage casing of 3 cm diameter (Smith and Romberg, 1940). The casing, cut into 10 cm lengths, is placed over the flowers well before receptivity. The loose end is tied around a band of cotton on the shoot below the inflorescence. With protogynous cultivars, the bag must be applied quite early; in some cases about as soon as the stigmas appear from beneath the leaves. A protandrous cultivar can be bagged a week to ten days after the stigmas are clearly visible.

Pollen is collected by removing and drying catkins after pollen shedding has commenced. They can be placed between two layers of paper or in a large, sealed paper bag. Gentle shaking will dehise the dried pollen, which is then sieved through organdy cloth to remove the coarse particles (Romberg, 1957).

Often pollen is shed from protogynous cultivars after other protogynous pistillate parents have passed receptivity. Such pollen can be stored for one year in a desiccator at -10 C and 25% relative humidity (Romberg and Smith, 1950) and pollination effected the next year.

The apparatus for pollen transfer consists of a syringe bulb attached to a coiled glass tube terminated by a hypodermic needle (Smith and Romberg, 1940). Pollen is placed in the bulb, the needle inserted through the cotton plug on the shoot, and the pollen gently blown into the sausage casing enclosing the inflorescence (Fig. 2).

Pollen must be applied when stigmas are receptive. Receptivity normally extends over a six to 14-day period (Smith and Romberg, 1940), depending on the cultivar and climatic conditions. However, pollen should be applied, as described by Romberg (1957), during the more limited range of peak receptivity to insure complete pollination and fertilization. Generally a protogynous pistillate parent should be pollinated about three days before its catkins begin shedding pollen and before its uncovered stigmas begin to turn brown. Receptivity cannot be determined by the basic red or green color of the stigma (Romberg, personal observation), but an experienced individual can

FIG. 2. Method of controlled pollination and bagging of pistillate inflorescence.

detect incipient receptivity by noting a glossy appearance of the stigmatic surface. Alternately, one can apply a light coat of pollen to the stigma, blow gently across it, and check for pollen adherence. A limited amount of pollen should be applied to protogynous cultivars during early receptivity because an over-abundance of pollen grains can produce a heavy nutlet drop (Romberg and Smith, 1950). Termination of receptivity cannot as readily be determined by sight. Unpollinated stigmas do not completely dry until long after they have passed receptivity (Woodroof, 1926; Shuhart, 1928).

Seed Germination

It is generally assumed that pecan nuts of southern origin have no well-defined rest period and will germinate at any time after harvest (Adriance, 1960). Lipe et al. (1969), in studies of growth substances in the fruit from pollination to harvest, found that growth inhibitor level reached a peak about 20 days before harvest, after which it dropped to its lowest point at about harvest. These data may help explain pre-harvest sprouting in the shuck (or germination on the tree), which is not uncommon for several cultivars. Finch (1933) reported premature sprouting to be associated with moisture in the shuck in contact with the seed and found a close association between vegetativeness of the tree and the amount of premature germination.

Barton (1936) and Shuhart (1926), testing nuts of southern origin, indicated that as long as they are handled carefully and not permitted to dry or spoil, they apparently can be planted at any time from harvest until late winter or early spring. Bailey and Woodroof (1932) concluded that optimum germination and uniformity of growth were obtained when nuts were stratified in a medium at pH 6.8 and stored at high humidity and 0 C. They did not find that nuts of any particular seedling or cultivar produced more consistently uniform or vigorous seedlings than others. Recent work by Sparks (unpublished data) has also indicated that seed of southern adapted cultivars stratified at 4 C will germinate earlier and more uniformly than unstratified seed. In fact, seed nuts may be stored for one year at 0 to 5 C for nursery planting without lowering the percentage of germination (Smith et al., 1933). The time of initial germination can be reduced and the uniformity of germination enhanced by breaking or removing the shell (Sparks and Pokorny, 1967). Apparently the size of nut and the degree of filling have no influence on germination and growth (Bailey and Woodroof, 1932; Szynoinak, 1932). Nuts from controlled pollinations are collected in early fall and stored dry at 22 C until planting. In January, the nuts are planted in large containers or barrels as described by Madden (1972c).

Nuts of northern origin, when planted in the south, must be stratified and chilled to germinate normally. In fact, many of the northern seed will not germinate at all without chilling (Madden, unpublished data). Nuts of 'Major' and 'Peruque' require stratification at 1 C for ten weeks for optimum germination and seedling growth.

Propagation of Seedlings

After 5½ months of growth, buds of the seedlings are grafted to branches of large, bearing trees. Bearing trees will aid in overcoming juvenility and hasten fruiting and evaluation (Romberg, 1944; Romberg and Smith, 1950; Madden, 1971). Each branch must be carefully labeled; one large tree may receive buds of 30 to 60 seedlings. Branches are developed from the seedling buds by using after-care techniques similar to those commerically used in topworking (Hutt, 1913).

Branches resulting from the buds exhibit typical juvenile characters: erect growth; dry bark at an early age; red-tinted young leaves; and pubescence of the bark epidermis of current season shoots, the leaf rachises, and the veins, with no flowers produced (Romberg, 1944). Nut evaluation and selection normally begin four to eight years after budding. In 1973 the USDA had more than 15,000 seedlings of known parentage propagated in this manner.

Numerous methods of budding and grafting have proved successful in the propagation of pecan, including shield or "T" bud, flute bud, ring bud, patch bud, "H" bud, chip bud, cleft graft, veneer graft, side graft, whip graft, and inlay bark graft (Corsa, 1895; Oliver, 1902; Madden, 1968a, 1971; Romberg and Madden, 1968.) Larger seedling trees may be whip grafted in their second growing season. Smaller seedling trees are usually patch budded in late summer of their second season or the beginning of their third season.

Harvest, Evaluation, Selection

Because of the size of old bearing pecan trees, the breeder must use ladders or hydraulic lifts to evaluate the seedling branches and collect the fruit. A net attached to a pole, with a hook for stripping the nuts located immediately above the net, is used to harvest the seedling branches. The following measurements are recorded for all nut samples: nut volume, nut weight, specific gravity, shell weight per nut volume, kernel weight per nut volume, percent kernel by weight, percent filling, nut shape, relative depth of dorsal parallel kernel grooves, amount of central partition septum, adherence of internal shell, presence of lateral grooves, kernel color and appearance, and commercial shelling characters. Photographs scaled to actual nut and kernel size are maintained for all seedlings; these are useful for inheritance studies.

In the USDA program branches of inferior seedlings are removed each year. Wood is collected from promising selections (75 to 100 per year) and budded in mature trees. Promising selections are evaluated for two or three years and sent to cooperators in other locations for trial.

An alternate method of evaluation and selection was initiated in 1971. Nuts from controlled crosses are planted immediately after harvest and the seedlings grown under greenhouse conditions at 27 C, then planted to the field in the spring at a spacing of 0.8 x 5.3 m. If properly cared for, seedlings may fruit by the seventh to tenth season from transplanting.

In this procedure evaluation of growth and nut characters is simplified and more representative compared with topworked limbs because the seedling trees do not suffer from severe shading as do some branches on topworked trees. This method also eliminates the possibility of virus transfer that could occur in topworking.

To evaluate adaptability to different climatic, soil, or environmental conditions, promising selections are sent to grower organizations and state experiment stations in all areas in the United States where pecans are commercially grown. When a seedling selection shows enough merit to justify its general propagation, the seedling is named and introduced. The USDA has adopted names of Indian tribes for cultivars. The number of testing sites has been increased greatly in the past five years and now numbers 200 in 18 states. A replicated planting of 130 cultivars and selections is being established at the USDA research station at Byron, Georgia. In addition, new testing sites in 1973 are located in Alabama, Arizona, Arkansas, California, Florida, Georgia, Louisiana, Mississippi, New Mexico, and Texas.

Breeding Systems

The pecan is normally cross pollinated in nature as a result of dichogamy. Dichogamy is maintained by a single gene pair with protogynous types *PP* or *Pp* and protandrous genotypes *pp*. As a result of this enforced cross-pollination, pecans have been maintained in a highly heterozygous condition.

In numerous instances where pecan seedlings resulting from both cross- and self-pollinations have been evaluated, plants from the self-pollinated seed were less vigorous (Romberg and Smith, 1950). Romberg and Smith (1946) studied the effects of cross-pollination, self-pollination, and sib-pollination and found no evidence of incompatability of pollen and pistil in self- or cross-pollination and no apparent selectivity of pollen in the fertilization of egg cells in two cultivars pollinated with a mixture of two pollens. There was good nut set on both cultivars as long as pollen was applied when flowers were receptive. This agrees with results of Adriance (1931) who obtained good fruit set by using pollen from many sources when the flowers of eight cultivars were receptive. However, Wolstenholm (1969) and Romberg and Smith (1946) have shown that self-pollination was followed by a much greater early nut drop than was cross-pollination. Further, the percentage of cross-pollinated nuts harvested was greater and the percentage of large clusters was higher than that of self-pollinated nuts.

The quality of matured self-pollinated nuts was much lower than that of cross-pollinated nuts. Although Romberg and Smith (1946) report pollen influences (metaxenia) on kernel weight and nut volume, this was not confirmed by Wolstenholm (1969).

Selection and Hybridization

Improvement of native pecans began about 1850 by selection from seedling stands of trees. Most older pecan cultivars originated in this manner and many present commercially important cultivars were originated by simple selection. Selection from native stands is still practiced to some extent, particularly by amateur breeders and pecan enthusiasts.

Controlled hybridization for pecan improvement was begun early in the twentieth century. Most cultivars which have been introduced into cultivation since the mid 1950s have resulted from hybridization programs. Careful selection of desirable parental material followed by hybridization has proved to be the best method of obtaining superior genotypes. As the demands of the pecan industry become more specific, it becomes increasingly necessary to exert more control in producing seedling genotypes. Thus, future cultivars must be more genetically engineered for adaptability to mechanical harvest, and must be more precocious, prolific, and more consistently productive than in the past. These achievements are best attained through scientifically controlled hybridization programs.

While selection in native stands for potential cultivars is not as efficient and effective as controlled hybridization, selection should continue to be practiced in feral material for specific characters of value in breeding. The greater apparent disease resistance of many selections from native trees when compared with modern cultivars indicates that genes for greater disease resistance can be found in native material.

The use of induced mutations and polyploidy have been of no value in pecan improvement (Romberg, personal communication). However, only limited studies have been conducted and definite conclusions must await further investigations.

Interspecific Hybridization

The taxonomic confusion concerning hickory species may be partially due to natural hybridization and resultant intergrading forms (McKay 1961). According to Woodworth (1930) *C. cordiformis*, *C. laciniosa*, and *C. ovata* are diploid with chromosome numbers of $2n = 32$, $x = 16$ while *C. tomentosa*, *C. glabra*, and *C. ovalis* are tetraploid, ($2n = 64$). McKay (1961) believes that *C. aquatica* and *C. illinoensis* (pecan) are both diploid ($2n = 32$) and indicates that the chromosome numbers of the other species are in doubt.

The natural hybrids between pecan and hickory are called "hicans" (McKay, 1961). Hican seedlings are extremely variable in all characters, but are generally intermediate in tree and nut characters between pecan and the hickory parental species. Although thick, heavy hulls are usually associated with thick-shelled nuts (McKay 1964), some hican seedlings have large, thin-shelled nuts with thick hulls.

The most notable hicans arise from pecan hybridization with *C. laciniosa* (shellbark hickory). However, most of these hybrids are infertile and seldom produce mature fruits. McKay (1961) reports that because of low nut production, the trees tend to be vigorous and make desirable shade trees. In an evaluation of 400 hican seedlings of pecan x *C. laciniosa*, McKay (1964) found that more than half had nine to eleven leaflets per leaf, intermediate between those of the parents. A few had seven to nine broad leaflets, resembling the hickory parent. Hican cultivars of this hybridization include 'McCallister', 'Gerordi', 'Bixbyi', 'Burlington', and 'Koon', but because of poor production most of these have been abandoned (Reed and Davidson, 1958).

Pollen of most pecan and hickory cultivars shows little irregularity in size or content (McKay, 1961). In a study of pollen characters of 222 pecan x shellbark hybrids, about half of the seedlings appeared to be normal with few empty pollen grains and resembled the pollen of most pecan cultivars (McKay and Romberg, 1962). The other seedlings showed a wide range of empty and off-size pollen. Some seedlings produced pollen grains in which occasional cells seemed to be double whereas others produced misshapen grains.

Hybrids between pecan and *C. cordiformis* (bitternut) such as 'Lacey' and 'Pleas' (McKay, 1961) have the characteristic astringent kernel of bitternut but in other characters are intermediate between the parents. These hybrids appear to be more productive than pecan x shellbark hybrids.

The 'Burton', 'Henke', and 'Pixley' are hybrids of pecan x *C. ovata* (shagbark) (MacDaniels, 1969) and generally have a higher degree of cold hardiness. They are grown in many northern states (McKay, 1961).

The water hickory (*C. aquatica*) shows promise of furnishing disease-resistance genes for improving pecan. The water hickory has a high degree of resistance to scab caused by *Fusicaldium effusum* Wint., currently the most damaging disease of the pecan. *C. glabra* and *C. tomentosa* exhibit considerable resistance to a leaf spot disease caused by *Gloeosporium caryae* Ell. and Dearn (McKay, 1961). Because both species are tetraploid, they likely would not hybridize with the pecan. The incorporation of resistance to these major diseases in pecan will probably require the development of tetraploid pecan material (McKay, 1961).

Breeding for Specific Characters

Few characters of horticultural importance have been studied genetically. Observations on breeding behavior by Romberg and Smith (1950) indicate that characters such as nut size, shape and marking, relative shell and kernel weights, and kernel appearance are generally affected by environment whereas characters such as type of catkin, form and color of stigma, dichogamy type, shuck thickness and type of suture, lenticel type and distribution, and branching type are relatively free of environment interaction. Characters shown to be under single gene control include dichogamy and stigma color.

In dichogamy, protogyny is controlled by a single dominant gene (*PP* or *Pp*) with protandry recessive (*pp*). The protogynous cultivar 'Mahan' appears to be homozygous (*PP*) while 'Schley', 'Stuart', 'Curtis', 'Odom', 'Brooks', 'Burkett', 'Nugget', 'Evers', and 'Candy' are heterozygous. The fact that many of the naturally-produced protogynous cultivars are heteroxygous suggests that cross pollination of protandrous and protogynous types takes place normally. Since the degree of dichogamy among trees varies considerably, modifying genes may also affect this character.

Unpollinated stigmas of some cultivars are green whereas others are red. Selfed seedlings of 'Success' (red stigma) segregate three red stigmas: one green stigma, suggesting that red color is controlled by a single dominant gene. Other characters which suggest a dominant pattern of inheritance include a protruding ridge of the four shuck sutures as exhibited by 'Carmichael' and relatively thick shoots and large leaves.

Tree Characters

Presently there is great demand for early-bearing cultivars because growers want an early return on their investment (Madden, 1972*b*; Meadows, 1972). Many of the old standard cultivars do not bear until the eighth to tenth year and this represents a significant loss in return per unit area.

Meadows (1972) has estimated that a precocious, prolific cultivar, at a planting density of 50 trees per acre (0.4 ha), can yield as high as 2000 lb. (900 kg) per year at a value of almost $700 by the eighth to tenth year. However, most precocious cultivars tend to produce excessive crop in relation to leaf area after about the twelfth year, resulting in small, poor quality nuts. Trees of this type must be pruned to control crop as well as total size.

'Stuart', 'Schley', 'Elliott', 'Desirable', 'Success', and 'Farley' lack precocity in bearing (Madden, 1968*b*). Parents used in the USDA breeding program that have induced precocity include 'Evers', 'Candy', 'Brooks', 'Shoshoni', 'Cherokee', 'Mahan', 'Moore', and 'Western' (Madden, 1968*b*).

Most precocious cultivars by nature have good branching characters (Madden, 1972*b*). They should produce wide-angle crotches at the junction to the central leader that resist breakage due to wind or weight of crop and leaves. The more precocious cultivars generally produce short, spur-type growth (fine lateral branching) that bears nuts

(Madden, 1972*b*). This character might make the cultivar more adaptable to mechanical "hedge-row" type pruning because a significant part of the fruit buds are inside the zone removed by pruning.

Some crosses of 'Mahan' x 'Risien #1', 'Schley' x 'Evers', and 'Brooks' x 'Evers' have produced the desired strong crotch angles with a good branching character. The 'Wichita' ('Halbert' x 'Mahan') shows a susceptibility to limb splitting.

In native groves and in many improved cultivar plantings, large pecan trees tend to bear irregularly, presumably because of the variability acquired over the years. Apparently the irregular bearing habit is a physiological effect brought about by the production of excess fruit in relation to the functional leaf surface (Worley, 1971; Sparks and Brack, 1972). Proper training and pruning of precocious, prolific cultivars might produce more consistent yields from year to year.

Precocious and prolific selections begin bearing irregularly by the sixth to eighth year even though they produce more leaf surface than less precocious types (Madden, 1972*d*; Malstrom, unpublished data). Less precocious, prolific trees do not begin such cycles until they are about 15 years old. This problem may not be correctable by breeding alone and presumably will require physiological manipulation to bring productive leaf surface in balance with fruit level.

Regular harvest of improved cultivars usually does not begin until mid-October in most southern regions. Earlier-maturing nuts that can be processed in time for the winter holiday market will bring a higher price. Earlier harvesting also permits the grower to make more efficient use of labor and equipment by spreading his harvest over a longer period. There is less chance of bad weather during early fall, and nuts that have dropped on the ground are less likely to spoil. Early-maturing cultivars also tend to bear more regularly. Sparks (unpublished data) contends that the predominant direction of photosynthates is to the ripening fruit until nut drop after which it is basipetal, presumably to storage sites. Worley (1971) and Worley et al. (1972) found that the earlier the defoliation in Georgia up to August 1, the less the shoot growth and the lower the nutlet count per shoot the next year. Thus, a greater interval between nut harvest and leaf fall should result in more complete development of fruit buds for the next year. In the northern pecan-growing regions, early-ripening cultivars are necessary to escape early frost damage.

There are some disadvantages of early-maturing nuts. Birds and rodents often carry off significant numbers of them before ample food is available from later-maturing nuts. Also, some insects apparently prefer the earlier nuts (Madden, 1972*a*).

Of the popular cultivars grown, 'Mahan' and 'Schley' mature nuts relatively late. Selections with 'Curtis', 'Stuart', and 'Mahan' parentage tend to be late in nut maturity. 'Wichita' and 'Cheyenne' are medium-maturing cultivars and 'Chickasaw' and 'Cherokee' mature nuts in late September. Selections with 'Candy', 'Barton', 'Moore', 'Evers', and northern cultivar parentage have produced early-maturing nuts (Madden, 1968*b*).

Cultivars that are exceptionally vigorous and heavy producers constitute a major objective in pecan breeding. Rarely does a tree lacking vegetative vigor produce heavy crops although occasionally vigorous trees are also poor nut producers (Romberg and Smith, 1950). Unfortunately, vigorous foliage is more susceptible to attack by some disease organisms than less vigorous foliage (KenKnight, 1968). Often more disease-resistant cultivars lack the vigor to make them suitable nut producers (Romberg and Smith, 1950).

Maintenance and retention of foliage is important for the production of high yield and quality crops the next year (Worley, 1971). Some cultivars are apparently physiologically equipped to hold foliage later in the fall than others (Madden, unpublished data). The foliage of some pecan cultivars is also inherently more resistant to attack by insects and disease organisms than that of others (KenKnight and Crow, 1967). Leaf retention is favored by freedom from insect and disease attack. Parents that produce progeny with vigorous growth are similar to those that induce good branching.

Cultivars differ genetically in susceptibility to various diseases. The most damaging and most common disease to attack pecans, especially in the humid Southeast, is scab, caused by the fungus *Fusicladium effusum* Wint. Scab is difficult to control because of numerous races of the fungus, many of which attack all cultivars (KenKnight and Crow, 1967). KenKnight (1968) reported that 'Candy', 'Desirable', 'Mahan', 'Odom', 'Pensacola Cluster', 'Peruque', 'Barton', 'Choctaw', 'Mohawk', 'Hastings', 'Schley', and 'Success' were severely infected by scab; 'Davis', 'Lewis', 'Moneymaker', 'Moore', and 'Stuart' mildly infected; and 'Curtis', 'Farley', 'Elliott', and 'Gloria' only slightly infected by scab. Several expensive fungicide applications are necessary in southeastern states to control the common diseases (Cole, 1968).

Madden (1972a) has reported that pecan cultivars differ in their susceptibility to attack by the numerous insect pests, especially aphids (*Monellia* spp. and *Tinocallis caryaefoliae* Davis), pecan nut casebearer (*Acrobasis caryae* Grote), southern green stink bug (*Nezara viridula* L.), and hickory shuckworm (*Laspeyresia caryana* Fitch).

Delayed spring growth is desirable in areas where spring frosts are prevalent. During the springs of 1971, 1972, and 1973, late freezes severely damaged trees at Brownwood, Texas. Protandrous cultivars lost most of the catkin crop, whereas protogynous cultivars lost a considerable number of the pistillate flowers. Madden (unpublished data) has determined that the pistillate flower cluster is susceptible to frost damage as soon as it protrudes from the surrounding leaf sheath. Lateness in spring growth is necessary in northern pecan zones because of the prevalence of frosts.

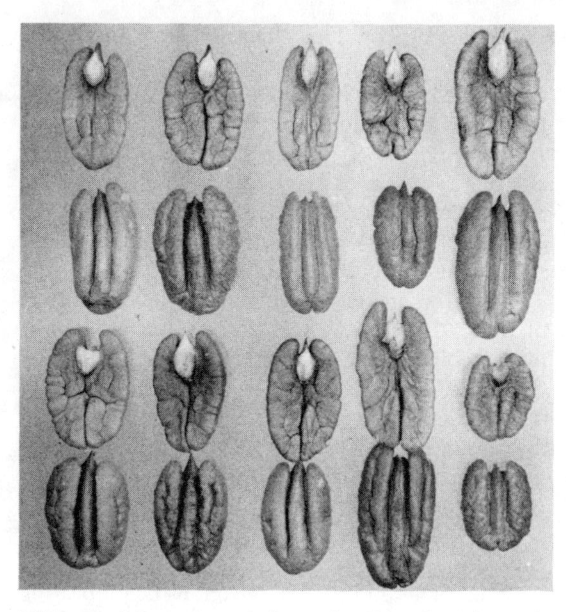

FIG. 3. Variations in kernal color, dorsal paralleled grooves, ventral groove, and lobe connection.

Nut Characters

Nuts spaced evenly on a long peduncle are desirable. Loosely spaced nuts are also less susceptible to attack by some disease and insect pests because good air circulation reduces moisture (Romberg and Smith, 1950).

Delayed shuck opening is predominantly a genetic trait in that a thin shuck will normally open earlier than a thick shuck (Romberg and Smith, 1950). Late shuck opening delays harvest and may result in sprouting of the nut in the shuck. Uniform shuck opening would facilitate harvesting and eliminate the expensive practice of several harvests.

Premature dropping is apparently controlled genetically to some extent (Romberg and Smith, 1950). Failure to fill or mature nuts may be due to an inherent physiological weakness or susceptibility. This problem has been studied at length but as yet the cause is not clear. Premature shuck opening is a character of the 'Burkett' (Smith and Romberg, 1948; Romberg and Smith, 1950).

Pecan kernels should have high oil content, good flavor, little fiber, light color, a smooth surface, and good keeping quality. They should be plump, firm, not crumbly, free from air cavities inside, and free from adherence of corky shell parts. Romberg and Smith (1950) state that the preferred kernel has shallow, open paralleled dorsal grooves, little or no central partition septum, a thin, smooth partition wall, and a small kernel lobe connection (Fig. 3). These features are associated with good cracking and shelling qualities.

The kernel should fall within the shelling color standards. Undesirable dark kernels are usually associated with rancidity development and have poor eye appeal (Romberg and Smith, 1950).

Retention of quality in storage and freedom from rancidity is controlled predominantly by the type and quality of oil in the kernel (Romberg and Smith, 1950). Cultivars vary greatly in this character.

For home cracking, eating out of hand, or in-shell sales, a large pecan comprising 18 to 22 nuts/kg, is preferred. A medium to small pecan, more than 22 nuts/kg, is generally preferred by commercial shellers because the smaller nut generally has a higher oil content which imparts a better flavor, and solid, firm kernels (Romberg and Smith, 1950). The nut should be symmetrical with an elliptical shape. Extremely long or round nuts are difficult to shell. Sharp protrusions or depressions at either the apex or basal end of the nut are undesirable.

Most growers prefer pecans that shell 50% to 60% kernel. The shell should be thick with strong sutures to withstand mechanical harvesting and cleaning damage. Cracked shells are portals of entry for disease organisms that may destroy the kernel in storage (Heaton, 1972).

Rootstock Breeding

The USDA initiated a program in 1965 to breed for more desirable rootstocks (Madden, 1970). Parent material that produces vigorous and uniform seedlings will be established in areas isolated from other pollen sources for rootstock seed production. Pistillate parent trees will also be caged and desired pollen introduced for rootstock seed production. Pecan, hickory, and hican in all combinations are being tested as rootstocks and interstocks for compatability, dwarfing, and effect on scion growth, nut production, and quality.

Often soil-connected problems are a result of poor choice of site by the grower. Pecan trees will not survive in water-logged soils. Research by the Texas Agricultural Experiment Station has shown that seedling rootstocks differ in tolerance to zinc deficiency and chloride toxicity (Hanna and Storey, 1972). These problems are predominantly root controlled and emphasize the importance of the rootstock. Because all pecan cultivars are presently propagated on open-pollinated seedling rootstocks, a breeding program for rootstock adaptability is an important, and up to the present a largely neglected approach to pecan culture.

Nurserymen have always desired vigorous seedlings for the root system of their propagated trees. Vigorous seedlings can be used in propagation sooner and the resultant scion grows faster than on weak seedling stocks allowing them to produce larger trees. Nurserymen asexually propagate pecan by budding or grafting cultivars onto open-pollinated seedling rootstocks of named cultivars and native seedlings (Adriance, 1960). The seed sources for rootstocks are usually selected from trees that produce vigorous, relatively uniform seedlings (Camp, 1927; Traub and Gray, 1931; Burkett, 1933). Nurserymen are currently urged to use open-pollinated seed nuts of 'Riverside', 'Mahan', 'Apache', 'Moore', 'Hollis', 'Western', and other cultivars for rootstock purposes. However, the pecan being predominantly cross-pollinated is extremely heterozygous and open-pollinated seed nuts are variable genetically even though the phenotypes may appear uniform. Seedlings are not rogued in the nursery; consequently, rootstocks in pecan groves lack uniformity. Yarnell (1933, 1934) emphasized the importance of the pollen parent for seedling rootstocks of good vigor. Sitton and Dodge (1938) reported extreme variations in yield and growth of 'Mobile', 'Schley', and 'Moneymaker' budded onto 'Moore' and 'Waukeenah' seedling rootstocks. Seedlings selected for uniformity and those originating from known parentage will produce more uniform trees than randomly selected seedlings (Wolstenholme, 1970a). Uniform rootstocks are particularly important for nutritional and physiological research where genetic consistency is essential. Genetic variability of the rootstock may be a primary reason for lack of uniformity in pecan orchards and explains why researchers are often unable to clearly assess differences in yield and growth in response to applied treatments.

In recent years there have been a number of studies initiated in clonal pecan rootstock production (Sparks and Pokorny, 1966; Romberg, 1967; Allen et al., 1968; Wolstenholme, 1969, 1970a). Pecan wood is difficult to root (Wolstenholme, 1970a) and despite numerous attempts, little success has been achieved. Sparks and Pokorny (1966) have attained probably the best results by air-layering to establish roots and Sparks (unpublished data) has obtained a high degree of survival under field conditons with air-layers. However, most of the work is in its infancy and a true determination of the feasibility in establishing clonal rootstocks by asexual means is largely undetermined.

Hickory and Hican Understock and Interstock

The growing interest in high-density pecan plantings and maintenance of smaller trees has resulted in experimental use of hickory and hican rootstocks or interstocks for dwarfing. Pecan can be successfully grafted on certain hickory species but the resulting trees do not produce satisfactory crops (Reed, 1944; McKay, 1964). Reed (1944) indicates that species most often tested for rootstock are *C. tomentosa* (mockernut) and *C. aquatica* (water hickory). *C. tomentosa* grows slowly and retards growth of pecan tops but seldom produces a tree that bears well or yields large nuts. *C. aquatica* apparently does not even retard pecan top growth and it too decreases nut production. Camp (1927) reported that seedlings of *C. aquatica* were more vigorous than seedlings of a number of native pecan cultivars. He could not ascertain the identity of the pollen parents but suspected they may have been pecan. According to Reed (1944), *C. glabra* and *Pseudocarya stenoptera* (National Arboretum), Chinese wingnut are not compatible with pecan. The former is tetraploid, whereas the latter has a laminated pith such as that of walnut, rather than the solid type found in pecan and hickory (Madden, 1970). *C. ovata* (shagbark) is used primarily as an understock for shagbark cultivars, but the pecan appears to be more suitable because of its faster growth. In general, pecan seedlings are more suitable rootstocks for hickories than the reverse (Reed, 1944).

Smith (1932) used combinations of species and hybrids in attempts to produce improved cultivars on wild hickory species. Shagbark and shellbark cultivars were incompatible with native trees of *C. tomentosa, C. glabra, C. ovalis,* and *C. cordiformis*. He later used hican hybrids and interpieces between the pecan scion and shellbark rootstock and obtained good compatibility. McKay (1964) suggests that hican seedlings with short internodes, slow growth, and enough pecan characters to render them graft compatible should offer the most promise.

Achievements and Prospects

The objectives and methods of pecan breeding have changed over the past century. Early breeding emphasized disease resistance, good foliage, high production, and attractively large, thin-shelled nuts. The industry now desires precocious, prolific trees that bear well-filled nuts early. Agricultural mechanization has dictated the development of smaller trees that can be planted at higher densities, and nuts adaptable to mechanical harvest, cleaning, and shelling. Increasing shortages of ideal agricultural land will necessitate development of cultivars adaptable to a wide range of potential pecan-producing areas.

Breeding programs are long term and some of the efforts begun 35 years ago by L. D. Romberg are only now beginning to produce results. Although programs of the USDA and state experiment stations have produced a few cultivars in the past ten years that are adaptable to the changed culture, we anticipate that a number of new cultivars will be released in the immediate future that will be primarily suited to this dynamic industry. Improvement by hybridization is relatively new in pecans, as compared with other horticultural crops. As more is known about mode of inheritance in the pecan, breeders will be better able to select parental material that will provide the desired type of tree.

Literature Cited

Adriance, G. 1931. Factors influencing fruit setting in the pecan. *Bot. Gaz.* 91:144–166.

———. 1960. Pecan propagation. *Proc. Texas Pecan Growers Assn.* 39:84–98.

Allan, P., M-D. Brutsch, J. C. LeRoux, B. N. Wolstenholme, and D. Cormack. 1968. Preliminary studies on rooting pecan cuttings. *Proc. Texas Pecan Growers Assn.* 47:73–76.

Bailey, J. E., and J. G. Woodroof. 1932. Propagation of pecans. *Ga. Agr. Expt. Sta. Bul.* 172:4–19.

Bailey, L. H. 1960. *Standard cyclopedia of horticulture.* New York: Macmillan. p. 675–678.

Barton, L. V. 1936. Seedling production in *Carya ovata* (Mill.) K. Koch, *Juglans cinerea* L. and *Juglans nigra* L. *Contr. Boyce Thompson Inst.* 8:1–5.

Benson, J., and E. J. Welsford. 1909. The morphology of the ovule and female flower of *Juglans regia* and of a few allied genera. *Ann. Bot.* (London) 23:623–633.

Billings, F. H. 1903. Chalazogamy in *Carya olivaeformis*. *Bot. Gaz.* 35:134–135.

Burkett, J. H. 1933. Seedage project. *Proc. Texas Pecan Growers Assn.* 13:50–52.

Camp, A. F. 1927. Report of Assn. Horticulturist. *Fla. Agr. Expt. Sta. Ann. Rpt.*

Cole, J. 1968. Four fungicides controlled pecan scab

(*Fusicladium effusum* [Wint.]). *Proc. S.E. Pecan Growers Assn.* 61:106–140.

Corsa, W. P. 1895. Nut culture in the United States. *USDA Div. Pomology Special Rpt.—Pecans.* 53–58.

Crane, H. L., C. A. Reed, and M. N. Wood. 1937. Nut breeding. *In* Better plants and animals. In *USDA Yearbook of Agriculture.* 1937:827–856.

Finch, A. H. 1933. Notes on pecan filling and maturity. *Proc. Amer. Soc. Hort. Sci.* 30:387–391.

Forkert, C. 1914. Twelve years' experience in hybridizing pecans. *Proc. Natl. Nut Growers Assn.* 13:28–30.

Hanna, J. D., and J. B. Storey. 1972. New stock promising. *Pecan Quart.* 6(1):17.

Heaton, E. K. 1972. Moisture and molding of pecans. *Proc. S.E. Pecan Growers Assn.* 65:135–138.

Hutt, W. N. 1913. Topworking seedling pecan trees. *Proc. No. Nut Growers Assn.* 4:32–43.

Isbell, C. L. 1928. Growth studies of the pecan. *Ala. Agr. Expt. Sta. Bull.* 226.

James, S. H. 1895. A pecan grove in Louisiana. *Rural New Yorker.* 54:258–259.

KenKnight, G. 1968. Resistance of pecan to disease. *Proc. S.E. Pecan Growers Assn.* 61:168–171.

———. 1971. Pecan varieties "happen" in Jackson County, Mississippi. *Pecan Quart.* 4(3):6–7.

KenKnight, G., and J. H. Crow. 1967. Observations on susceptibility of pecan varieties to certain diseases at the Crow farm in DeSoto Parish, La. *Proc. S.E. Pecan Growers Assn.* 60:48–51.

Lipe, J. A., P. W. Morgan, and J. B. Storey. 1969. Growth substances and fruit shedding in the pecan, *Carya illinoensis. J. Amer. Soc. Hort. Sci.* 94:668–671.

MacDaniels, L. H. 1969. Hickories. *In* R. A. Jaynes (ed). *Handbook of North American nut trees.* Geneva, N.Y.: Humphrey Press, Inc. p. 190–202.

McKay, J. W. 1947. Embryology of pecan. *J. Agr. Res.* 74:263–283.

———. 1961. Interspecific hybridization in pecan breeding. *Proc. Texas Pecan Growers Assn.* 40:66–76.

———. 1964. Hican seedlings, their origin and value. *Proc. Texas Pecan Growers Assn.* 43:67–70.

McKay, J. W., and L. D. Romberg. 1962. The importance of hican seedling characteristics in pecan breeding. *Proc. Texas Pecan Growers Assn.* 41:102–111.

Madden, G. D. 1968*a*. Recent advancements in pecan propagation. *Proc. Texas Pecan Growers Assn.* 47: 44–48.

———. 1968*b*. Potential varieties for the southeast through the USDA pecan breeding program. *Proc. S.E. Pecan Growers Assn.* 61:27–52.

———. 1970. Progress report on the research at the U.S. Pecan Field Station, Brownwood, Texas. *Proc. Texas Pecan Growers Assn.* 49:37–40.

———. 1971. Budding technique—for propagating young seedlings. *Pecan Quart.* 5(1):4–5.

———. 1972*a*. Research notes. Pecan varieties differ as to susceptibility to insect attack. *Pecan Quart.* 6(1): 23.

———. 1972*b*. Pecan varieties for high density plantings. *Pecan Quart.* 6(3):9–10.

———. 1972*c*. Barrel-grown seedlings. *Pecan Quart.* 6(3):16–17.

———. 1972*d*. Comparative production of 'Choctaw' and 'Sioux' topworked to large trees. *Pecan Quart.* 6(2):15.

Madden, G. D., F. R. Brison, and J. C. McDaniel. 1969. Pecans. *In* R. A. Jaynes (ed). *Handbook of North American nut trees.* Geneva, N.Y.: Humphrey Press, Inc. p. 163–189.

Madden, G. D., and E. J. Brown. 1973. Blossom dates of selected pecans. *Pecan Quart.* 7(1):17–19.

Meadows, W. A. 1972. High density hikes early yield. *Pecan South, Ga. Farm.* March:10–11.

Moznette, G. F. 1934. Experiments in control of the pecan black aphid under orchard conditions. *Proc. S.E. Pecan Growers Assn.* 28:55–61.

Mullenax, R. H., and W. A. Young. 1972. Female flower development of the pecan. *Proc. S.E. Pecan Growers Assn.* 65:83–92.

Oliver, G. W. 1902. Budding the pecan. *USDA Bur. Plant Ind. Bul.* 30.

Reed, C. A. 1944. Hickory species and stock studies at the Plant Industry Station, Beltsville, Maryland. *Proc. No. Nut Growers Assn.* 35:88–115.

Reed, C. A., and J. Davidson. 1958. The hickory group. *In* C. A. Reed and J. Davidson (eds.). *The improved nut trees of North America and how to grow them.* New York: Devin-Adair Co. p. 107–157.

Rehder, A. 1960. *Manual of cultivated trees and shrubs.* New York: Macmillan. p. 120–124.

Romberg, L. D. 1931. The present status of the pecan pollination problem. *Proc. Texas Pecan Growers Assn.* 11:56–61.

———. 1944. Some characteristics of the juvenile and the bearing pecan tree. *Proc. Amer. Soc. Hort. Sci.* 44:255–259.

———. 1957. Research work at U.S. Pecan Field Station, Brownwood, Texas. *Proc. Texas Pecan Growers Assn.* 35:33–40.

———. 1967. Clonal pecan rootstocks. *Proc. Texas Pecan Growers Assn.* 46:72–75.

———. 1968. Pecan varieties in 1968. *Proc. S.E. Pecan Growers Assn.* 61:56–65.

Romberg, L. D., and G. D. Madden. 1968. A method for bark grafting pecan stocks of high curvature. *USDA Correspondence Aid* CA-34-154.

Romberg, L. D., and C. L. Smith. 1946. Effects of cross-pollination, self-pollination, and sib-pollination on the dropping, the volume, and the kernel development of pecan nuts and on the vigor of the seedlings. *Proc. Amer. Soc. Hort. Sci.* 47:130–138.

———. 1950. Progress report on the breeding of new pecan varieties. *Proc. Texas Pecan Growers Assn.* 29:12–21.

Shuhart, D. V. 1926. The pecan in Oklahoma. *Okla. Agr. Expt. Sta. Cir.* 59:3–15.

———. 1927. The morphological differentiation of the pistillate flowers of the pecan. *J. Agr. Res.* 34:687–696.

———. 1928. Fruiting habits of the pecan. *Proc. Texas Pecan Growers Assn.* 8:37–38.

———. 1932. Morphology and anatomy of the fruit of *Hicoria pecan. Bot. Gaz.* 93:1–20.

Sitton, B. G., and F. N. Dodge. 1938. A comparison of the behavior of the 'Mobile' variety of pecan on

Hicoria pecan and *Hicoria aquatica* rootstocks. *Proc. Amer. Soc. Hort. Sci.* 36:121–125.

Smith, J. R. 1932. Double-topworking hickory trees. *Proc. No. Nut Growers Assn.* 23:115–118.

Smith, C. L., and L. D. Romberg. 1940. Stigma receptivity and pollen shedding in some pecan varieties. *J. Agr. Res.* 60:551–564.

———. 1948. Pecan varieties and pecan breeding. *Proc. Texas Pecan Growers Assn.* 27:18–26.

Smith, C. L., C. J. B. Thor, and L. D. Romberg. 1933. Effect of storage conditions on the germination of seed pecans. *Proc. Texas Pecan Growers Assn.* 13: 68–71.

Sparks, D. 1967. Shoot and apical bud abortion on nonbearing mature 'Stuart' pecan trees as a function of time. *Proc. S.E. Pecan Growers Assn.* 60:135–143.

Sparks, D., and C. E. Brack. 1972. Return bloom and fruit set of pecan from leaf and fruit removal. *HortScience* 7:131–132.

Sparks, D., and J. L. Heath. 1972. Pistillate flower and fruit drop of pecan as a function of time and shoot length. *HortScience* 7:402–404.

Sparks, D., and F. A. Pokorny. 1966. Investigations into the development of a clonal rootstock of pecans by terminal cuttings. *Proc. S.E. Pecan Growers Assn.* 59:51–56.

———. 1967. Effect of the shell on germination of pecan nuts *Carya illinoensis*, Koch cv. Stuart. *HortScience* 2:145–146.

Stuckey, H. P. 1916. The two groups of varieties of the *Hicoria pecan* and their relation to self sterility. *Ga. Agr. Expt. Sta. Bul.* 124:127–148.

Szymoniak, G. 1932. Stocks for pecans. *Proc. Natl. Pecan Growers Assn.* 31:97.

Taylor, W. A. 1905. Promising new fruits. In *USDA Yearbook of Agriculture* 1904:406–416.

Traub, H. P., and O. S. Gray. 1931. The variation of seedlings from selected pecan trees. *Proc. Texas Pecan Growers Assn.* 11:23–25.

Traub, H. P., and L. D. Romberg. 1933. Methods of controlling pollination in the pecan. *J. Agr. Res.* 47:287–296.

Wiegand, K., and A. J. Eames. 1924. The flora of the Cayuga Lake Basin, New York. *Cornell Univ. Expt. Sta. Memoir* 92:172–173.

Wolfe, R. R. 1937. Life of E. E. Risien. *Proc. Texas Pecan Growers Assn.* 16:57–58.

Wolstenholme, B. N. 1969. Effects of self and cross-pollination on fruit set and nut drop of the pecan at Pietermaritzburg. *Agroplantae* 1:189–194.

———. 1970*a*. The need for and propagation of clonal rootstocks for pecans. *Proc. Texas Pecan Growers Assn.* 49:51–53.

———. 1970*b*. Dichogamy observations on pecans at Pietermaritzburg with particular reference to a cultivar-cultural experiment. *Agroplantae* 2:33–38.

Woodard, J. S., L. D. Romberg, and F. J. Willmann. 1929. Pecan growing in Texas. *Texas Dept. Agr. Bull.* 95.

Woodroof, J. G. 1924. The development of pecan buds and the quantitative production of pollen. *Ga. Agr. Expt. Sta. Bul.* 144.

———. 1926. The fruit-bud, the flower, and then the pecan nut. *Proc. Natl. Pecan Growers Assn.* 25: 81–92.

———. 1930. Studies of the staminate inflorescence and pollen of *Hicoria pecan*. *J. Agr. Res.* 40:1059–1104.

Woodroof, J. G., and N. C. Woodroof. 1926. Fruit-bud differentiation and subsequent development of the flowers in the *Hicoria pecan*. *J. Agr. Res.* 33:677–685.

———. 1927. Distance pecan pollen is carried by wind for practical purposes. *Proc. Ga.-Fla. Pecan Growers Assn.* 21:43–44.

———. 1928. The dropping of pecans. *Proc. Natl. Pecan Growers Assn.* 27:28–34.

———. 1930. Abnormalities in pecans. 1. Abnormalities in pecan flowers. *J. Hered.* 21:39–44.

Woodroof, J. G., N. C. Woodroof, and J. E. Bailey. 1928. Unfruitfulness of the pecan. *Ga. Agr. Expt. Sta. Bul.* 148.

Woodroof, N. C. 1928. Development of the embryo sac and young embryo of *Hicoria pecan*. *Amer. J. Bot.* 15:416–421.

Woodworth, R. H. 1930. Meiosis of microsporogenesis in *Juglandaceae*. *Amer. J. Bot.* 17:863–869.

Worley, R. E. 1971. Effects of defoliation date on yield, quality, nutlet set, and foliage regrowth for pecan. *HortScience* 6:446–447.

Worley, R. E., S. A. Harmon, and R. L. Carter. 1972. Correlation among growth, yield, and nutritional factors for pecan (*Carya illinoensis* W. cv. Stuart): correlations with yield, quality, and terminal shoot growth. *J. Amer. Soc. Hort. Sci.* 97:511–514.

Yarnell, S. H. 1933. The pecan seedling rootstock test to date. *Proc. Texas Pecan Growers Assn.* 13:47–50.

———. 1934. Pecan seedling rootstock studies. *Proc. Texas Pecan Growers Assn.* 14:15–17.

Walnuts

by Harold I. Forde

Species of walnuts, genus *Juglans*, are found in the wild in much of the eastern and southern parts of the United States, in Mexico and Central America, the Andean part of South America from Columbia to Argentina, the West Indies, Japan, China, southern Asia from India to Turkey, and southeastern Europe to the Carpathian Mountains of Poland.

Wild walnuts have been utilized as food for humans and animals since prehistoric times. Some walnut species have been used for timber probably for as long as man has used wood as a building material.

Only eight *Juglans* species are considered here, although several other species are of at least local importance for timber and nuts.

Origin and Early Development

Juglans regia L., the Persian walnut, is probably native to a wide region extending from the Carpathian Mountains across Turkey, Iraq, Iran, Afghanistan, and southern Russia to northern India. *J. regia* trees are large and round-headed, about 27 m tall. The rather smooth light-colored large nuts are free of the hull at maturity. Shell thickness varies from paper-thin to very thick and hard. The Greeks probably obtained *J. regia* from Persia. From Greece, the seeds were transported to Rome where they were known as *Jovis Glans*, or Jupiter's acorn, from which comes *Juglans*. From Italy, *J. regia* spread to what is now France, Spain, Portugal, and southern Germany. Trees of the species were in England by 1562, and were brought to America by the earliest settlers. The American colonists are said to have called the species English walnut to distinguish it from the American black walnut. Most horticulturists prefer the name Persian walnut as associated with the place of origin of the species.

J. sieboldiana Maxim., the Japanese walnut, is native to Japan. Now grown to a limited extent in many countries, the trees reach about 25 m and have a broad head. The leaves are very pubescent on the lower surface. The nuts are usually smooth to rugose and elongated to nearly round, with a hard shell. The heartnut, *J. sieboldiana cordiformis*, Maxim-Mak., is a botanical variety of J. *sieboldiana*.

J. nigra L., the eastern American black walnut, is native to an area extending from the Atlantic Ocean to Texas, Oklahoma, Kansas, and Nebraska, and from the southern Great Lakes region almost to the Gulf of Mexico. This large tree grows as high as 45 m and has a trunk diameter of 2 m. The nuts have a thick, hard, rough shell with a thick, adhering hull.

J. cinerea L., the butternut, is native from Georgia and Arkansas to New Brunswick. This is the most cold-resistant species of American *Juglans*. The slow-growing trees can attain a height of 30 m. The nuts are elongated and have a very hard, rough shell with an adhering hull.

J. microcarpa Berlander grows wild in Texas and New Mexico. This small tree is rarely over 10 m and the nuts are small and round with a very hard, smooth shell and an adhering hull.

J. major Heller grows from Colorado through Arizona and New Mexico and into Mexico. Although usually under 15 m tall, with ideal conditions the trees can attain 25 m or more. The nuts are nearly round, almost smooth, and hard shelled with an adhering hull.

J. californica S. Wats, the southern California black walnut, is native to coastal southern California. They usually grow near small streams, most of which have water only when it rains. The tree is shrubby, usually branching at or near the ground. Most wild trees are 5 to 6 m tall, but under good growing conditions the trees can attain 15 m. The nuts are small, smooth, and hard shelled with an adhering hull.

J. hindsii Jeps., the northern California black walnut, is native to a small area in the central part of northern California. The trees have a round head on a high trunk, growing to about 30 m in height. The nuts are round, smooth, and hard shelled with an adhering hull.

History of Improvement

Historically, *J. regia* trees have been grown from seed and are so grown in many parts of the world. Undoubtedly man has improved *J. regia* by selecting seed from what was considered to be superior trees. Also, when trees were cut for timber the poorer producers of nuts would be selected. A gradual selection for improved nut quality and production must thus have occurred over a long period.

In recent years more and more of the Persian walnut trees have been propagated by budding or grafting, a method first used in western Europe, particularly France. Chosen as scion wood sources were superior individual seedlings, which became the cultivars used by growers. Some of the French cultivars such as 'Franquette' are over 200 years old. Grafted or budded trees have been used in California for over 80 years (Smith, 1912). More recently, grafting or budding of walnuts has become more common in other countries, too. As soon as growers learned to bud or graft, they chose as scion sources the trees that were superior for their area.

In recent years there have been surveys of thousands of seedling trees in several countries including Yugoslavia (MacDaniels, 1961), Poland and Czechoslovakia (Millikan and Cochran, 1966), and Bulgaria (Valev, 1969). Workers in Yugoslavia are studying 80 of the most promising trees. In Poland, five have been selected from 150 superior trees studied. In Czechoslovakia, ten have been selected for further study from 20,000 trees. In Bulgaria, four cultivars were selected from about 20,000 trees. The United States Department of Agriculture Bureau of Plant Introduction is testing selections from Russia, Hungary, and Korea (Smith and Alanger, 1969).

In California, Persian walnuts were first grown in the days of the early Spanish missions. These "mission walnuts" were small and rather hard-shelled.

In 1867 Joseph Sexton planted a sack of walnuts supposedly from Chile. Seed from Sexton's trees were planted in the southern California coastal areas and from these were developed the Santa Barbara soft-shell walnuts. In time, several cultivars were named, including 'Placentia', 'Erhardt', 'Wasson', and 'El Monte'.

In 1871 Felix Gillet, a nurseryman, began to import scions and nursery trees from France, introducing most of the French cultivars. These were planted extensively in northern California along with their seedlings, and all cultivars presently planted extensively in California are descended at least partially from them, originating either as chance seedlings or in a breeding program of the University of California.

In the northeastern United States, immigrants from Germany brought walnuts with them as early as the first part of the eighteenth century.

Descendants of these German walnuts are still growing in the eastern United States. In 1932 seed of walnuts from the Carpathian Mountains of Poland was introduced to Ontario and the northeastern United States by the Reverend Paul C. Crath. A great many seedling trees have descended from these Carpathian walnuts, and a number of named cultivars have resulted. (*Annual Reports of the Northern Nut Growers Association* from 1940 on have articles about Carpathian walnuts.)

J. nigra trees were found in the native forest in much of the eastern United States. These trees are of great value because the wood is prized for gunstocks, furniture, and cabinet making. The nuts are valued for their food value and flavor.

Since about 1945, a commercial shelling industry has been developed. In good years, 23 million kg of hulled black walnuts are shelled, producing about 2.7 million kg of kernels (Webber, 1963). In a poor crop year, only about 9 million kg are cracked (Cavender, 1959).

Zarger (1946) found that common black walnuts of the Tennessee Valley were quite variable in cracking quality, size, kernel percentage, and yield. Data from 132 trees showed an average annual yield of 15 kg per tree (the best tree averaged 72 kg) and kernel yields ranging from 13 to 25% (average of 20%). Recovery of marketable kernels with a hand cracking machine averaged only 17%.

Many cultivars of *J. nigra* have been named (Berhow, 1962). Cultivars have been selected for superior yield and nut qualities such as size, shell thinness, and cracking quality.

The other native eastern American walnut, the butternut (*J. cinerea*), is much less important than *J. nigra*. The shell is very rough and hard, and kernel percentage is low. During this century, a number of cultivars have been selected from the seedling population, largely for the same characters as for *J. nigra* cultivars (yield, cracking quality, and shell thinness).

J. sieboldiana, the siebold or Japanese walnut, was introduced in the United States in about 1870 by a nursery in the Santa Clara Valley of California. A varietal type of *J. sieboldiana* with heart-shaped nuts is called the heart nut, *J. sieboldiana cordiformis*. Seed of either ordinary *J. sieboldiana* or heart nut will produce trees with either heart-shaped nuts or the more round type.

After 1870 the Japanese walnuts were distributed throughout the United States and southern Canada. These walnuts are of interest because of their hardiness and ornamental value. A number of cultivars of *J. sieboldiana* have been named, including several of the heart nut type. Nowhere has the Japanese walnut attained much importance.

The northern California black walnut, *J. hindsii*, was found wild in only a few small groves. The trees are now very common as shade trees in yards and especially along roadways in California, but their most important use is as rootstock; most of the Persian walnuts of the Pacific Coast are grafted onto *J. hindsii* rootstock because of its disease resistance and vigor.

Most *J. hindsii* nuts are harvested and sold to shellers. The nuts are smaller than *J. nigra* nuts but compare favorably in some other respects. Nut size and other characters vary greatly, of course. Average kernel yields are probably similar to those of *J. nigra* nuts. Some better ones have an in-shell weight of 12 or 13 g and a kernel weight of about 4 g. The kernels have a pleasant flavor but not as strong as that of *J. nigra*. No cultivars have been selected for nut production; the only selection of *J. hindsii* has been selection of seed trees for nurseries.

J. hindsii also has value as a timber tree. Since nearly all of the existing trees have grown without being crowded or pruned, they tend to have too many relatively low branches to be ideal timber trees; nevertheless many trees have been sold at good prices, usually for shipment to Europe. The stumps of some Persian walnut trees that had been grafted high on *J. hindsii* have also brought good prices.

The other three North American species (*J. microcarpa*, *J. major*, and *J. californica*) are of little importance except as shade trees and as food for wildlife. The nuts tend to be small and hard and the trees are generally small for a walnut, although *J. major* trees can be quite large under favorable conditions. There are no cultivars of any of these species.

Interspecific Hybrids

All *Juglans* have a diploid chromosome number of 32 (Woodworth, 1930), and it is generally agreed that *Juglans* species cross readily with each other. A case of apparent intersterility, however, was found by E. F. Serr (unpublished data). He took pollen from 12 *J. hindsii* trees that had failed to produce hybrids even though surrounded by *J. regia* trees shedding pollen at the right time. Still, no nuts were set when this pollen was mixed and applied to 200 bagged *J. regia* flowers. When pollen from a *J. hindsii* tree that had produced hybrids was used to pollinate 100 bagged *J. regia* flowers, 26 seeds resulted from which seven trees were grown. The seven trees were all hybrids.

Natural hybrids are known to occur between several of the species. Luther Burbank produced hybrids between *J. regia* and *J. hindsii* (which he called "Paradox") and between *J. hindsii* and *J. nigra* (which he called "Royal") (Howard, 1945). Hybrids are also known between *J. regia* and *J. nigra*, *J. sieboldiana* and *J. cinerea*, *J. sieboldiana* and *J. regia*, *J. californica* and *J. regia*, and probably many other combinations. *J. nigra* and *J. cinerea* probably are intersterile. They grow together over much of the eastern United States, yet no verified hybrids exist (Funk, 1969). The name "paradox" today may apply to the hybrid produced by crossing *J. regia* with any black species. In California, paradox is usually *J. regia* and *J. hindsii*. This hybrid is now produced in large numbers and used as a rootstock for *J. regia* cultivars. Paradox trees are very vigorous but relatively unfruitful, most of them producing only a few nuts. This low nut production makes them even more vegetatively vigorous.

Paradox trees look more like *J. regia* than *J. hindsii;* the bark is light-colored and the leaflets are large like *J. regia*, although more pointed. At maturity the hull dehisces from the shell. The nut shells are light in color like *J. regia* but are thick and hard like a black walnut. Unfortunately, there is no black walnut flavor in any known paradox nuts.

Paradox trees have great potential value as timber because of their rapid growth and wood quality. A number of large paradox trees have been cut for timber. One such tree (Anonymous, 1937), planted by General John Bidwell in Chico, California, in 1871, was cut by the Wood Mosaic Company of Louisville. Kentucky, when it was about 65 years old. It was 30 m high and contained more than 11.8 cu m of lumber (5000 board-feet). The trunk and limbs weighed 29,500 kg. From the stump, trunk and limbs (35 cm in diameter and larger) were cut more than 14,800 sq m of veneer.

The USDA has had a *J. nigra* x *J. regia* breeding project since 1937 (McKay, 1965). The objective has been to develop trees with the good points of both species, such as good cracking quality and annual bearing from the Persian, and late start of annual growth from the *J. nigra*. Because of the near sterility of the F_1 trees, only 2,641 nuts were planted from these trees; 497 F_2 trees had fruited. These trees were quite variable: most produced thick-shelled inferior nuts, some were intermediate, and two had thin shells.

In a similar trial, E. F. Serr and H. I. Forde (unpublished data) grew seed of a paradox tree on the Davis campus of the University of California. The seed germinated rather poorly, and only 30 of the F_2 seedlings were grown. Variation in character was great: one was difficult to distinguish from *J. regia;* two had leaves almost like those of *J. hindsii* (one tree with light bark and the other with dark bark); and the other trees were intermediate in characters. All the nuts were poor in quality, variable in shell thickness, and lacking black walnut flavor.

One of these F_2 trees was backcrossed to the 'Payne' cultivar of *J. regia*. Only 11 such backcross trees were grown, ten of which produced nuts. The nuts were all small (6.5 to 10 g per nut) and of poor quality. Kernel percentage varied from 41 to 55%.

These two tests show that it is possible to get thin shells in descendants of black walnuts. The process would be long and difficult, however, since very large numbers would be needed to have a chance of getting anything good, while large numbers would be difficult to attain because of the semisterility of these hybrids.

Royal hybrids (*J. hindsii* x *J. nigra*) do not seem to have the sterility that paradox does. The many royal hybrids growing in California are quite vigorous, though not as vigorous as paradox. Their heavy crops may be the reason they appear less vigorous than paradox. If they are grown to produce a long straight trunk, royal hybrids should be good timber trees.

Modern Breeding Objectives

A very important objective of any breeding program is greater yield. Yield of nuts is affected by three factors: the number of pistillate flowers produced, the percent set, and the size of the nuts. Although tree size might be thought of as a fourth factor, many small trees might produce as much as a few large trees per unit area—or perhaps even more.

The number of pistillate flowers produced is determined by the number of buds that grow each year and by the number of flowers per shoot. Some cultivars such as 'Franquette' produce flowers only on shoots from terminal buds and subterminals. A subterminal is the bud just below where the nut or nuts were borne on the shoot the previous season. The 'Franquette' usually produces no more than two flowers per terminal or subterminal, and many of these shoots have only one or no pistillate flowers, with the result that even 100% set would not produce a heavy crop.

On other cultivars, such as 'Payne', in addition to terminals and subterminals, 80 to 90% of the lateral buds produce pistillate flowers. These more fruitful cultivars also tend to have two or three flowers on most shoots. Such cultivars with a high percent of lateral bud fruitfulness come into commercial production about twice as fast as cultivars that are not fruitful on laterals.

Not only does large nut size contribute to yield but large nuts usually have higher kernel percentage. The most important measure of nut size is kernel weight. Most commercial nut kernels now weigh between 5 and 7 g, and some new cultivars have kernels of 7 to 9 g. Since genetic material is available with kernels larger than 10 g, an objective of 10 or 11 g is reasonable.

Nut Quality

The several important factors affecting kernel quality are all influenced by environment as well as by genetic factors. Commercial walnut meats are graded into two main categories: sound (edible) kernels and off-grade (inedible) kernels. Off-grade are kernels that are shriveled, insect-damaged, moldy, rancid, black, or dirty. Black kernels are those that are darker than dark amber on a standard color chart. The sound kernels are separated into grades based largely on color, with light color the most valuable. Kernel shrivel is affected by such factors as insect-damage (i.e., by aphids or spider mites), disease-damage (i.e., defoliation by anthracnose), lack of water, and heat damage. Susceptibility to heat damage is influenced by genetically controlled factors. Some walnut cultivars produce more of their nuts in the shade than do other cultivars, but shade is not the only factor since some cultivars have fewer of their exposed nuts damaged.

Some cultivars are more susceptible to mold damage than are others. This is partly associated with heat damage, for even slight heat damage makes nuts more susceptible to mold.

Cultivars also differ in resistance to insects (see Disease and Insect Resistance, below). Dirty kernels may be caused by shell defects, such as perforated or broken shells.

Cultivars differ greatly in ability to produce light-colored kernels. This is due partly, though not entirely, to ripening time: late-maturing cultivars tend to have lighter kernels, probably because they mature when the weather is cooler.

A high kernel percentage, probably always considered by selectors of cultivars, is important but may be overemphasized. Probably a more important factor is total kernel yield/ha or, even more important, kg of sound kernels/ha. Factors affecting percent kernel are nut size, kernel plumpness, shell fill, and shell thickness. A well-filled shell tends to yield a higher kernel percentage, but the kernel may then be so tight in the shell that it is hard to extract. Such kernels also tend to break in the cracking machine. Therefore, a good walnut should be well filled but still have some space between shell and kernel.

Persian walnuts vary from shells as thick and hard as a black walnut to shells that are paper thin. The shells of commercial walnuts must be strong enough to endure harvest and hulling by machine and shipment in bulk or in sacks to nut buyers. Kernels in broken nuts become damaged mechanically and, more important, may get dirty. In view of the above, 60% kernel is about the maximum one could reasonably expect. That is about the percentage of the 'Serr', the California cultivar with the highest kernel percentage, and of the 'Hansen', the Carpathian type with the highest kernel percentage known to the author.

Some Persian walnuts have a poor shell seal

at the suture, even though the shell may be thick and strong. Such nuts break apart and are as bad as perforated or weak shells. It is desirable to have an attractive shell, reasonably smooth and light-colored, when walnuts are sold in the shell. Today, however, more and more nuts are shelled and sold as kernels.

Tree Characters

One of the most pressing needs is to get good cultivars that leaf out later in the spring. In California, the cultivars in common use start growth over a period of about one month. The earliest important cultivar, 'Payne', starts growth at Davis, California, from about March 13 to March 21, depending on the season; 'Franquette', the latest cultivar, starts leafing out about April 15 to 20. All the other cultivars currently being planted start growth during the first half of this period, most in the first third. Frost danger is still considerable in the last half of March in most California walnut-growing areas, and in a few districts the danger extends through the middle of April.

FIG. 1. Pistillate flowers at terminal of current season's first flush of growth. The catkin from the lateral bud is on the previous season's growth. Note that the pistillate flowers are receptive, but the catkin is not yet shedding pollen.

Earlier cultivars are also more likely to be affected by spring rains during blossom time, which can interfere with pollination. Furthermore, rain during and after bloom increases the incidence of walnut blight, caused by *Xanthomonas juglandis* Pearce, which can result in heavy nut losses. Good cultivars are needed that start growth from about 15 days before to a little later than 'Franquette'.

With good later cultivars a grower could spread his risk by having both early and late cultivars, so inclement weather at any one time would be less likely to wipe out his crop. Harvest would also be spread, relieving pressure on labor and equipment.

The winter-hardy Carpathian walnuts grown in the northeastern United States and southern Canada also present a problem by starting growth too early (McKay, 1956). They leaf out early in California also; at Davis some of them have been as early or earlier than 'Payne'. Carpathian walnuts are often severely damaged by spring frost in locations where *J. nigra*, starting growth two to three weeks later, is not damaged. Thus, Carpathian walnut cultivars that start growth two to three weeks later than the present ones would be very desirable. Such cultivars would probably also extend the useful range farther south, where Carpathians have been unsuccessful because of spring frosts. Later flushing Carpathian walnuts can be obtained by crossing or selfing the later ones now available (Dabb, 1968).

Walnuts are monoecious with unisexual wind-pollinated flowers. The staminate flowers are borne on catkins that develop from lateral buds on the previous-season's growth. The pistillate flowers are borne terminally on current-season's growth (Fig. 1).

Although *J. regia* cultivars are self-fertile, most have some degree of self-unfruitfulness because of dichogamy. Most cultivars are protandrous, a few are protogynous. Forde and Griggs (1972) showed that all the common California cultivars are dichogamous, some more than others. They also showed that the degree of dichogamy will vary somewhat from year to year, depending on the weather. In most years, all of the 22 cultivars they studied would have enough overlap of pistillate and staminate bloom to set a crop. Set could be improved on all these cultivars in most years by cross-pollination from a cultivar that sheds at the right time. McDaniel (1957) reported that most Carpathian walnuts grown in Illinois are protandrous, some are protogynous.

An objective of a breeding program could be to get a cultivar with a better overlap of pistillate and staminate bloom, or to get pairs of cultivars, one protandrous and one protogynous, that would overlap each other. The second alternative would probably be easier to accomplish.

Disease and Insect Resistance

A serious disease of Persian walnuts is bacterial blight, caused by *Xanthomonas juglandis* Pearce. While this disease causes some lesions on twigs, its more serious effect is the destruction of pistillate flowers and immature nuts. In general in California, this disease is worse on cultivars that start growth early, incurring the risk of rains which favor the disease. Cultivars also seem to have some real difference in susceptibility. Although 'Payne', an early cultivar, is generally considered susceptible to blight, 'Jensen', a cultivar starting growth ten days later, has been more susceptible at Davis. In contrast, PI 159568, an introduction from Afghanistan, has never had blight at Davis even though it starts growth as early as 'Payne'. Some of the seedlings of PI 159568 also seem to have less blight than other walnuts.

Phloem (or deep bark) canker caused by *Erwinia rubifaciens* Wilson, is confined almost entirely to 'Hartley' and 'Placentia'; nothing is known about the heritability of resistance to this disease.

Several other diseases of walnut include anthracnose, caused by *Gnomonia leptostyla* Ces. & de Not., found throughout the natural range of *J. nigra*. This disease also attacks Persian walnuts. *J. nigra* cultivars are reported to differ in susceptibility (Berry, 1960). Heritability of resistance does not appear to be high (Funk, 1972).

Resistance would be desirable to such serious pests as codling moth (*Carpocapsa pomonella* L.), walnut aphid (*Chromaphilis juglandicolis* Kalt.), husk fly (*Rhagoletis complela* Cress and *R. suavis* Loew), twig girdler (*Oncideres singulata* Say.), and several other insects.

Tree Habit and Vigor

J. regia trees (at least in California) tend to spread more than is desirable and initiate bud growth on the lower side of the lowest main branches. If these low branches are not pruned every year or so they grow down until they are near or on the ground; therefore, more upright trees would be desirable. Walnut trees also grow too large, making it inconvenient to prune, spray, and harvest. The ideal tree would grow vigorously to moderate size and then slow down and produce nuts.

The French types of walnuts grown in California and western Oregon withstand temperatures only down to about −9 to −11 C, whereas Carpathian types grown in the northeast withstand −37 to −40 C. Although greater resistance to winter cold is not much needed in California, the trees, particularly young trees, are often damaged by fall freezing before they are thoroughly dormant. Winterhardiness is not the limiting factor even in the California mountains, for spring frosts will be the factor eliminating walnuts before altitude is great enough for winter killing to result.

Where Carpathians are grown, greater hardiness would probably extend their range farther north until another factor becomes limiting, such as short growing season and insufficient heat to mature the nuts, or too short a season for the wood to mature and harden.

Ornamental and Timber Qualities

Persian walnuts are large handsome trees and used quite often as yard trees, especially by those who want both an ornamental shade tree and nuts for home use. For ornamental use, a Persian walnut should be fairly upright in growth habit and not produce a great many branches, requiring much pruning. Moderate crops would be preferable to heavy crops in most cases because heavy-bearing trees tend to require more pruning and have more limb breakage. There are cut leaf forms of both *J. regia* and *J. nigra* and a purple leaf *J. nigra* available for ornamental use (Jacobs, 1968).

Persian walnut is valued as a timber tree in Europe but not in the United States. The reason is partly that we have very few old Persian walnuts in this country and partly that they tend to be in orchard form, as single trees in yards or as roadside trees. In all cases they are grown for nut production, not timber, and therefore are widely spaced and profusely branched. Persian walnuts grown for timber should be either crowded or pruned so as to produce a long straight trunk. Crowded Persian walnut trees are poor nut producers.

Probably the only way we would ever produce any Persian trees for timber would be by pruning orchard trees to a moderately high trunk, 1.3 to 2.6 m to the first branch. A valuable log could then be available when the orchard is pulled out. This would probably be done by some growers if log buyers should ever offer high enough prices consistently. Many of the younger California orchards have their first branch 1½ to 2 m from the ground now. Such logs would be more valuable if the trees were on black walnut rootstock and grafted high enough to produce a black walnut log 1.3 or 2.6 m long.

In Europe some cultivars are known to be more valuable than others because the wood is of higher quality. In California it seems highly unlikely that any cultivar would be selected for anything but nut production.

Breeding Rootstocks

There are many problems with the rootstocks now in use. In the United States, *J. regia* cultivars have been grown mainly on four rootstock seedlings of *J. nigra*, *J. hindsii*, *J. regia*, and on first-generation paradox. Some have been grown on *J. californica* and royal hybrid, and a few have probably been grown on other species.

The first walnut orchards in California were seedling orchards. When budding and grafting were first practiced, the rootstocks used were *J. regia* seedlings. By about 1900 the California nurseries had shifted to *J. hindsii*, which had better disease resistance and vigor. It is more resistant to crown gall, caused by *Agrobacterium tumefaciens* Smith and Town., and to oak root fungus, caused by *Armillaria mellea* Vahl., and tends to be more vigorous and uniform the first year in the nursery. Early horticulturists (Smith, 1912) held that trees on *J. regia* were more severely affected by any adverse condition such as poorer soil or lack of water. In recent years many trees have been propagated on paradox rootstocks, because of their greater vigor and greater tolerance of the root lesion nematode, *Pratylenchus vulnus* Allen and Jensen.

Blackline disease has renewed interest in *J. regia* as a rootstock. Blackline, a disorder of unknown cause, results in failure of the graft union in trees, usually after they are 12 to 15 years old or older. Persian trees on all rootstocks except *J. regia* may have their tops killed by blackline.

No matter which rootstocks are used, they could be improved in vigor and uniformity by selection or breeding. Since diseases and nematodes which attack the roots are difficult to control, resistant rootstocks would be very desirable. If *J. regia* stocks are used to avoid blackline, it would be desirable for them to have resistance to such things as crown gall, oak root fungus, and root lesion nematode. Clones resistant to these problems, if such exist, would need to be located first and a breeding program designed to incorporate such resistance into a single stock. Unless we learn how to propagate *J. regia* by cuttings, such a stock would have to come true from seed. Another problem would be the possibility of encountering strains of the disease or nematode that would attack the resistant rootstock. Such difficulties would probably make progress slow in a rootstock-breeding program.

Breeding Techniques

Floral Development

The staminate flowers of walnut are borne on catkins that develop from lateral buds on the previous-season's growth. Pistillate flowers are borne terminally on current-season's growth (Fig. 1). Persian walnut commonly has one or two buds in the axil of each leaf; some of the other species commonly have three or more. Catkin buds differentiate early in the growing season as the first flush of growth occurs. If the new growth is examined about when the first flush stops, two kinds of buds are apparent: the leaf buds with scales and the naked catkin buds. The catkin bud is commonly the lower bud in leaf axils with two buds.

The following spring each of these catkins produces a great deal of pollen. Impiumi and Ramina (1967) reported that each flower on the catkin had 13 to 18 anthers, each anther with about 900 pollen grains, about 1.8 million pollen grains per catkin.

During late winter, part of the leaf buds develop pistillate flowers thereby becoming mixed buds. Nast (1935), working with the 'Concord' cultivar of *J. regia*, found that the first indication of pistillate flower development in material collected at Davis, California, was on February 25. The pistillate flower of *J. regia* consists of an involucre composed of a bract and two bracteoles, a four-lobed perianth, and a one-celled bicarpellate ovary terminated by a bilobed stigma. The involucre and perianth develop into the husk. The shell develops from the ovary wall. The kernel is the embryo.

In cultivars that are not very fruitful, relatively few of the buds, mostly terminal buds, become mixed buds, whereas in the most fruitful cultivars nearly all leaf buds, including lateral ones, become mixed and bear pistillate flowers.

Pollen Collection and Storage

Since catkins develop very little after they are removed from the tree, they must be ready or almost ready to shed pollen when gathered. Probably the best catkin for pollen production is one that has already had a few of its anthers dehisce. Most of the anthers will dehisce in about 24 hours if the catkins are spread thin on paper or on fine-screen trays over paper, at room temperature. The pollen can then be separated from the rest of the catkin by sieving through cheesecloth or fine screen.

Lack of a reliable laboratory method for testing pollen has limited work on storage of *J. regia* pollen. Early nut breeders (Crane et al., 1937) concluded that subfreezing temperatures would kill walnut pollen. This idea has persisted. Walnut pollen seems to lose its viability in a week or less at room temperature. At the University of California at Davis we have stored *J. regia* pollen at 0 C and 40% relative humidity. Over a period of 23 years pollen stored this way has always produced nut set when used within one month; however, it did not remain viable until the next season. Recent work (Griggs et al., 1971; Forde and Griggs, 1972) shows that *J. regia* pollen remained viable for at least one year in a freezer at −19 C.

Hall and Farmer (1971) found that black walnut pollen germinated on liquid and agar-based media containing 20% sucrose. Boron (100 ppm boric acid) added to the medium slightly enhanced germination. Germination occurred only in dense (150+ grains/mm^2) populations of the pollen on the agar medium. Wide tree-to-tree variation in percentage of germination was observed. Pollen was stored at −30 C and −19.6 C for three months without major loss of viability.

For controlled crosses, the pistillate flowers must be covered during the time they are receptive to protect them from wind-blown pollen. Wood (1934) covered the tips of branches, including the pistillate flowers, with cotton batting. Serr and Forde (1956) used 9 x 15 cm bags of tightly woven white cotton gabardine. Preliminary work indicated that shoots of Persian walnuts are likely to be severely burned if covered with clear plastic, at least under California weather conditions. Shoots covered with dark-colored cloth also were badly damaged by heat. White cloth bags do not accumulate enough heat to damage walnut shoots or pistillate flowers. High-quality cotton gabardine was found to be sufficiently pollen-proof. Many other apparently tightly woven cloth samples allowed walnut pollen to pass through. A small amount of plugging cotton is placed around the shoot, and the bag is then attached over the cotton with rubber bands. Using two bands minimized loss of bags. Pollen is blown into the bag through a hypodermic needle attached to a plastic bottle. The pollen is metered by a cheesecloth disc in the neck of the bottle.

Hand pollination experiments (Merrill, 1969; Serr and Forde, 1956; Wood, 1934) have determined that pistillate flowers begin to become receptive as soon as there is any division between the two lobes of the stigma (Fig. 2) or any red at the tip of the pistillate flower. Percentage of nut set is highest from pollen applied when the stigmatic lobes are well developed and separated but standing at an angle of approximately 45° to the longitudinal axis of the ovary. The more mature stigmas give less set, becoming practically zero by the time brown specks appear on the stigmatic lobes and the glandular surface dries. Therefore, bags to prevent pollination must be applied before the stigmatic lobes begin to separate. Pollen should be applied when the stigmas are most receptive, and the bags may be removed after they are no longer receptive. Leaves inside the bag must be reduced to about four leaflets, or the bag will be so full of leaves that it would be difficult to get pollen on the stigmas.

Seed Germination and Tree Growth

J. regia seed is mature internally several days before the nuts are hullable. To prevent windfall losses, seed should be harvested before the nuts are completely hullable.

Persian walnut seed needs about six to eight

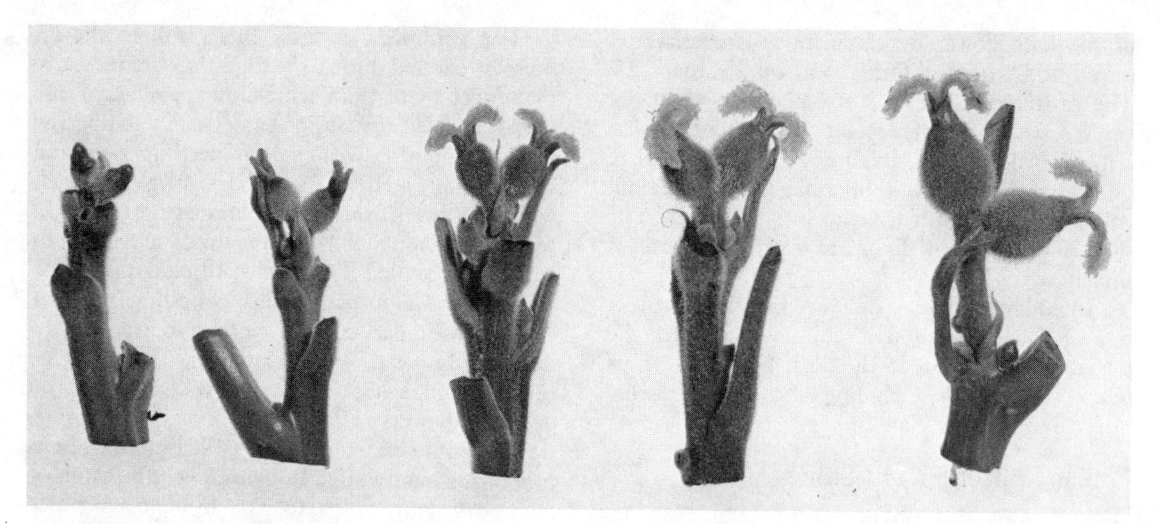

FIG. 2. Walnut pistillate flowers showing progressive stages of development, from left: immature flowers not receptive to pollen, flowers in early receptive stage, flowers in most receptive stage, flowers nearing end of receptive stage, and flowers no longer receptive.

weeks of stratification. In the University of California breeding program seeds are stratified for about eight weeks at 2 C. The seeds are then planted in the greenhouse in flats 20 cm deep. About one month after the seeds have germinated the flats are transferred to a lathhouse, where they remain until the trees are dormant. During the dormant season the trees are lined out in a nursery row. After one year in the nursery the trees can be transferred to a test orchard. In the California program scions are taken from trees that have been in the nursery for one year and grafted onto one-or-two-year-old rootstocks in the test orchard. This speeds the program, saving one year per generation.

Seedling Evaluation

In the University of California walnut breeding program (Serr and Forde, 1956) notes are taken the third, fourth, and fifth years seedlings are field grown as follows:

Tree characteristics
- Leafing date (average terminal bud 2.5 cm long)
- Blight (1–6 scale)
- Sunburn damage (1–6 scale)
- Estimated yield (1–6 scale)

Flowering characteristics
- Pollen shedding
- First catkin
- Last catkin
- Full pistillate bloom

Nut characteristics (on 10 nut sample)
- Shell roughness (rough, medium, smooth)
- Shell color (light, medium, dark)
- Shell seal (1–6 scale)
- Length
- Suture diameter
- Cheek diameter
- Nut weight
- Kernel weight
- Shell fill (1–6 scale)
- Shrivel (number blank, badly shriveled, moderately shriveled)
- Kernel color (number light, amber, black)
- Kernel veins (number with light and dark veins)
- Kernel spots (number with light and dark spots)
- Kernel flavor (good, fair, poor).

Data from the first 14 years of this project were analyzed statistically for 18 traits (Hansche et al., 1972). The population studies consist of 38 parents and approximately 2400 progeny trees resulting from crosses among the parents. Parents were either on *J. hindsii* seedlings or on paradox (*J. hindsii* x *J. regia*) seedling rootstocks at Davis, California. Progenies were on paradox seedling rootstocks spaced 4.6 x 4.6 m. Measurements were obtained as each offspring came into production and were repeated for two to five years. The analysis, however, used only measurements taken during the first two years of production.

Yearly differences in weather have a considerable effect on the year-to-year performance of many commercially important traits of tree fruit crops. When these effects are large they sharply reduce the confidence that a breeder can place on comparisons among genotypes observed in different years. That is generally why he postpones decisions or selections until measurements have been taken for several years. Yearly differences in climate have large effects on many traits of walnut. For example, during the 14 years data were analyzed, the average leafing date of the population had a range of more than 16 days, the average nut weight varied 5 g/nut, and kernel weight varied 3.7 g/kernel. Such effects need not obscure the distinction of genetic differences in segregating populations, however, for year effects can be eliminated from comparisons among genotypes by applying appropriate statistical techniques (Searls and Henderson, 1961). In walnut as in other tree crops, such techniques should greatly increase the precision with which genetic differences among seedlings can be discriminated, reducing the number of years required to identify superior seedlings with confidence. Accompanying such reduction will be major reductions in per-unit costs of genetic improvement and a significant increase in rate of genetic improvement.

In recent years in the University of California project, after two nut samples are evaluated seedlings are either discarded or transferred to a selection block. Each seedling transferred to the selection block is grafted on four rootstock trees at a spacing of 9.2 x 9.2 m. These few progenies in the selection block can be studied further to determine whether they are good enough to be released as new cultivars. Those considered as possible new cultivars are tested in other walnut-growing areas also.

With four trees of each selection there are usually enough pistillate flowers by the fifth year for these trees to be used as parents. At Davis, trees at 9.2 x 9.2 m begin to become crowded in about 12 years. If further data are needed, half of the trees can be removed.

Breeding Systems

Heritability

Table 1 (from Hansche et al., 1972) lists heritability estimates with standard deviations for 18 traits of Persian walnut. Note that heritabilities are extremely high for leafing date, receptive date of pistillate flowers, first shedding of pollen, and harvest date. These results are consistent with heritability estimates of analogous traits in populations of peach (Hansche et al., 1972) and cherry (Hansche et al., 1966; Hansche and Brooks, 1965) and probably reflect a relationship common among a wide variety of tree fruit and nut species.

Shell thickness, nut length, suture diameter, cheek diameter, nut weight, and kernel weight also have very high heritabilities. Since the heritabilities of all these traits are above 0.8, selection of superior seedlings as parents on the basis of performance during their first two years of production, and subsequent random mating *inter* se, should be effective in improving subsequent generations. Repetitive applications of this simple selection technique and mating system should give very rapid genetic improvement of these traits in this population.

Heritability is moderate for fruitfulness of laterals (a trait highly correlated with both early commercial yields of juvenile trees and large yield of mature trees [Serr, 1959]), last shedding of pollen, shell seal, light-colored kernels, kernel veining, and size-adjusted kernel spots. Hence, selection of superior seedlings as parents on the basis of these traits during the first and second years of production should give a good genetic gain from generation to generation, though less spectacular than the gain in traits with heritabilities above 0.8.

Crop rating (estimated yield) and percentage of good kernels (not affected by shrivel, mold, etc.) have heritabilities that are not significantly greater than zero. This indicates: 1) that the breeding stock analyzed is genetically homogeneous with respect to these traits; 2) the effect of segregating genes on expression of these traits is entirely epistatic; 3) change differences in environment are almost entirely responsible for observed phenotypic differences; 4) variability in estimating crop on juvenile trees obscures heritable differences in this trait; or 5) some combination of those phenomena prevails. Whatever the cause, the

TABLE 1. Estimates of Persian walnut progeny and mid-parent means, heritabilities (h^2), and their standard deviations (SD).

Trait	Mid-parent		Progeny			
	Mean	SD	Mean	SD	h^2	SD
Leafing date	27.9	5.5	32.5	7.5	0.96	0.02
First shedding of pollen	39.6	4.5	42.4	9.9	0.91	0.06
Last shedding of pollen	49.9	5.2	48.4	7.2	0.68	0.05
Receptive date of pistil	51.1	5.5	55.8	7.8	0.93	0.02
Fruitfulness of laterals	41.4	20.6	15.4	24.4	0.39	0.02
Crop (estimated yield)	3.9	0.6	2.1	0.8	0.07	0.03
Harvest date	38.2	4.3	43.3	6.1	0.85	0.02
Shell seal	2.4	0.4	2.7	0.8	0.38	0.04
Shell thickness	4.0	0.3	4.0	0.7	0.91	0.04
Nut length	39.6	1.9	40.5	3.5	0.82	0.03
Nut suture	33.6	1.3	34.6	2.5	0.89	0.04
Nut cheek	34.5	1.5	36.0	3.0	0.97	0.04
Nut weight	12.6	1.1	13.8	2.4	0.86	0.04
Kernel weight	6.3	0.6	6.9	1.3	0.87	0.04
No. of good kernels	10.0	0.5	7.1	4.0	−0.08	0.06
No. of light colored kernels	5.6	1.1	4.6	2.5	0.52	0.05
Kernel veins	3.0	1.4	3.0	2.8	0.49	0.04
Size-adjusted kernel spot	5.2	2.1	5.1	4.7	0.41	0.05

practical effect on a breeding program would be the same: crop rating and percentage of good kernels cannot be genetically improved by selecting and mating seedlings on the basis of these traits.

Solving this problem will depend primarily on which of the five above-mentioned factors is the major cause of the low heritability. The proportion of total variability in this population that arises from both genetic and environmental causes should be compared with the proportion of variability arising from differences between trees of the same genotype, that is, variability due to environment. The importance of genes can thus be compared with the importance of environment in determining crop rating and percentage of good kernels in this population. The variance estimates obtained total 1.13 for crop rating, with 1.03 for environment; and 2.09 for percentage of good kernels, with 2.00 for environment. Over 90% of the total variance of both crop rating and percentage of good kernels can be attributed to environmental effects, apparently leaving less than 10% to genetic effects. These results rule out the possibility that epistatic effects of genes are the major cause of low heritability, since the maximum possible contribution to observed differences could be only 10% of the total. It can be concluded that breeding methods designed to take advantage of epistatic effects (inbreeding tests of specific combining ability and the development of hybrids) would be no more successful in improving crop rating and percentage of good kernels than *inter se* random mating of parents selected on the basis of their own performance. Apparently the low heritability of crop rating and percentage of good kernels is due either to genetic homogeneity of this population with respect to these traits, to effect of random environmental variability, or to some combination of these two phenomena. Therefore, no selection procedure or breeding method would be expected to lead to improvements of these traits until the genetic variability of the population is increased and the random effects of environment on crop rating and percentage of good kernels are greatly decreased.

There is considerable evidence that the variability associated with crudeness of the subjective measurements involved in crop rating are very large and may be a major cause of low heritability. Furthermore, it is likely that juvenile walnut trees (in the first two years of production) are more subject to environmental effects on yields than are mature trees. Taken together, these two sources of nongenetic variation could easily account for the low heritability of crop rating. Unfortunately, reducing such variability would seem to be both prohibitively expensive and time consuming. Determining seedling yield by weight, for instance, might be much more accurate than visual estimates; however, this procedure would be expensive. Delaying selection decisions until seedlings are more mature, and presumably more phenotypically stable, not only would add additional maintenance expense but also would significantly lengthen the selection cycle, and thus seriously limit the potential rate of genetic improvement aimed for by increasing the precision of yield measurements.

Probably considerably less expensive and quite possibly more effective would be an alternative solution: selection of traits with moderate to high heritabilities that are presumably highly correlated with yield, such as fruitfulness of laterals, nut suture diameter, length and weight, and kernel weight. Clearly, selection of seedlings as parents on the basis of superior performance of these traits during the first and second years of production (and their subsequent random mating *inter se*) should yield rapid genetic improvement. If, as seems apparent, they are highly correlated with yield, this procedure should also affect genetic improvement of yield potential without adding either to the expense of the breeding program or to the length of selection cycles. The trait of key significance with respect to improving yield is probably the fruitfulness of laterals. Therefore, efforts to improve estimates of its heritability by improving measurement techniques seem to provide the most effective and efficient means of improving the rate of correlated improvement in yield. The few seedlings that are transferred to selection blocks can be studied over a longer period of time and more accurate yield determinations made before any of them are released as cultivars.

Apparently less than 5% of the estimated total variance in percentage of good kernels can be attributed to genetic effects. A little more than half of the rest can be attributed to differences in the environment of the orchard, and the remainder to the size of the sample from which the percentage of good kernels is estimated. Variability associated with sample size could be reduced by about a quarter by doubling the sample size, though that would increase heritability only by 0.01 to 0.05. There thus appears little hope of finding a practical means of genetically improving this population with respect to percentage of good kernels unless additional genetic variability is introduced into the breeding stock.

Phenotypic Correlations

Table 2 shows a complete list of the phenotypic correlations among the 18 traits studied with those underscored that the authors (Hansche et al., 1972) considered strong enough to be of practical significance to a breeding program.

Strong phenotypic correlation among traits can probably be attributed, at least in part, to pleiotropic effects or to close linkage among genes affecting different traits; such phenomena could enhance or inhibit progress in selection of traits, depending on the direction of the correlation. Therefore, they should be taken into account by the breeder.

The correlation −0.44 between shell seal and shell thickness is actually a positive correlation since the two scales of measurement are in opposite directions. That is, a high number for shell seal is a poor seal, whereas a high number for shell thickness is a thick shell. This is a case in which one trait affects the other directly. Other things being equal, a thick shell will appear to be better sealed than a thin one, for two reasons. A thick shell tends to have a greater area of contact at the suture, and thin shells are also more likely to come apart at the suture because they can flex without breaking the shell halves, thereby putting more strain on the suture.

As is usual with tree fruits (Hansche et al., 1966), a high positive correlation exists between leafing, flowering, and harvest dates. Fruitfulness of laterals shows a fairly strong positive correlation with crop rating despite the large measurement error associated with crop rating.

All measurements of nut and kernel size are very highly correlated (0.62, 0.72, 0.78, and 0.82, respectively, for nut weight with nut length, suture diameter, cheek diameter, and kernel weight). In fact, they are probably high enough to negate the practical value of measuring more than one of these traits. They suggest that this breeding program could probably be improved in overall efficiency by diverting the time required to measure nut size to increasing the size of the sample from which kernel characteristics are estimated, that is, obtaining more precise kernel measurements without expending more resources.

Selecting a Breeding System

Selection of a breeding system usually depends on the ratio of additive to nonadditive genetic effects and the degree of dominance in breeding stock. If there are nonadditive genetic traits, breeding methods designed to take advantage of them might be employed, such as a search for unique

TABLE 2. Estimates of phenotypic correlations among characteristics of Persian walnuts.[z, y]
(Correlation coefficients considered to be of practical importance by the authors are underlined.)

No.	Trait	18	17	16	15	14	13	12	11	10	9	8	7	6	5	4	3	2
1.	Leafing date	—.07	—.14	.23	—.03	—.05	.00	.08	.07	.09	.00	.10	.56	—.25	.06	.80	.62	.65
2.	First shedding of pollen	—.10^{x}	—.08^{x}	.26^{x}	.05^{x}	.14^{x}	.09^{x}	.21^{x}	.20^{x}	.15	.15^{x}	.19^{x}	.38^{x}	—.14^{x}	.23^{x}	.46^{x}	.82	
3.	Last shedding of pollen	—.08^{x}	—.04^{x}	.25^{x}	.04^{x}	.12^{x}	.04^{x}	.20^{x}	.19^{x}	.13^{x}	—.17^{x}	.17^{x}	.31^{x}	.03^{x}	.26^{x}	.38^{x}		
4.	Receptive date of pistil	—.02	—.16	.20	—.04	—.04	.05	.04	.06	.12	.08	.09	.58	—.33	.06			
5.	Fruitfulness of laterals	.10	—.10	—.06	.02	.03	—.09	—.01	—.02	—.01	—.18	.21	.07	.41				
6.	Crop	.06	.00	—.03	—.07	—.14	—.20	—.09	—.05	—.13	—.11	.00	—.28					
7.	Harvest date	—.02	—.12	—.04	—.03	.08	.17	.06	.04	.15	.14	.01						
8.	Shell seal	.12	—.04	—.03	—.03	.05	—.08	.21	.14	.11	—.44							
9.	Shell thickness	—.05	—.01	.03	.04	—.17	.27	.01	—.04	.09								
10.	Nut length	.09	.07	.05	.03	.48	.62	.48	.49									
11.	Nut suture	.03	.06	.10	.06	.66	.72	.82										
12.	Nut cheek	.00	.08	.07	.08	.66	.78											
13.	Nut weight	—.03	.09	.08	.20	.82												
14.	Kernel weight	—.02	.09	.05	.23													
15.	No. of good kernels	—.07	.11	.29														
16.	No. of light kernels	—.29	—.11															
17.	Kernel veins	—.19																
18.	Kernel spot with size																	

[z] All measurements were previously adjusted for year effect.
[y] 95% confidence intervals about correlation estimates were obtained via Fisher's "Z" statistic. Unless otherwise indicated, all confidence intervals ranged between ±0.01 and ±0.05.
[x] 95% confidence interval was between ±0.06 and ±0.09.

parents with a high specific combining ability of methods that employ inbreeding and subsequent production of hybrids. If most of the variability in the population results from segregating genes with additive effects, the most effective and efficient means of improving the population genetically is to select individuals as parents in each generation on the basis of their own performance, and mate these selected parents at random *inter se*.

The only Persian walnut population known by the author to have received a genetic study is the population at the University of California, described earlier in this chapter. Their traits are known to be mostly additive and to have moderate to high heritability, so the indicated method would be the second one: selecting superior individuals in each generation on the basis of their own performance and crossing them at random. Inbreeding should be avoided among the selected parents because inbreeding in outbreeding species, such as walnut, usually results in progenies with depressed reproductive potential. The University of California project has studied most of the important walnut tree and nut characteristics except disease and insect resistance. Some that were not analyzed directly are functions of others that were analyzed; for example, kernel percentage is a function of nut weight and kernel weight, kernel plumpness and nut fill are also related to nut and kernel weight, and since sunburn causes nuts to shrivel or become dark, sunburn is reflected in the number of sound kernels and the number of light-colored kernels. The notes taken on walnut blight were not analyzed because there was so little blight on the young trees available. Blight (Smith, 1912) is usually not very serious on young trees in the interior valley of California and it would be quite expensive to keep all seedlings long enough to get good blight readings.

In plants in general, resistant cultivars are the ideal way of combating diseases and insect pests. In Persian walnuts, breeding for such resistance would present some serious difficulties. First of all, no cultivars are known that are immune or highly resistant to most of the problems, though there seem to be definite differences in susceptibility to some diseases. In annual plants, a parent with heritable resistance to a given disease may prove useful despite many defects because many generations can be grown in a few years, allowing the resistance to be combined with the good characters of other parents. In walnut, however, if a potential parent with resistance to disease is a poor specimen in other respects, it may be practically worthless because of the several generations required to get other good characters combined with the resistance. Thus, a breeder would be unlikely to try to use as a parent an apparently immune or resistant individual that is poor in other respects. An example would be PI 159568, an introduction from Afghanistan with good sized nuts with about 53% kernel and with light to light amber kernel color, used in California because it appears blight-resistant. One serious fault, however, keeps it from being a commercial cultivar: it is not fruitful enough. When a parent of this type is crossed with a highly fruitful parent, good fruitful progenies are possible in the first generation. When PI 159568 was crossed with 'Payne' in the California project, several fruitful progenies resulted. Two of them kept in the selection block appear to have some blight resistance, though not as much as PI 159568. One of these, 'Serr', has been released as a cultivar.

Achievements and Prospects

Ten new cultivars have been released from the walnut breeding program in California (Serr and Forde, 1968). Other selections under test may be released in the future. Since breeding stock has been improved, good progress is expected in the future.

In all probability, heritabilities in Carpathian walnuts would be very similar to those in the California population. Therefore, if superior clones were selected as parents, crosses were made among them, and superior seedlings were selected as parents, good progress should be made in each generation in improving this strain of *J. regia*. Since there are many thousands of seedling *J. regia* in several countries spread over a wide area of the world, and their representatives in the United States are relatively few, it seems probable that our breeding stock could be improved by searching for superior trees in some of those other countries. Valuable work of this kind has been done by the USDA

Bureau of Plant Introduction and others, though perhaps more could be done.

The high cost of breeding programs and the relatively low value of the in-shell nuts make it seem unlikely that any breeding program will be started in the foreseeable future to improve any of the other species of *Juglans* as nut producers, although significant genetic improvement could undoubtedly be made.

With the increasing scarcity and high price of walnut lumber, a potential exists for improving black walnut as a timber tree. It would be desirable to increase such factors as rate of growth and apical dominance, producing good logs more quickly. C. F. Bey (1968) reported moderately high heritability for some traits associated with these two factors.

Trees grown from seed from southern sources tend to start growth earlier, continue growth until later in the season, and make more growth than trees from more northern sources (Funk, 1969; Bey, 1970; Bey et al., 1971). If southern strains are moved too far north spring frost or winter kill could be a problem. Bey et al., (1971) suggests moving seed north 150 miles or less.

Much of the selection in *J. nigra* has been for nut quality and yield. This may tend to hinder the improvement of trees for timber. A nut-producing tree should have a large crown and much branching for abundant fruiting wood, whereas a timber tree should have a narrow crown and a long straight trunk. Growth is slowed by a heavy crop on any kind of tree. However, if timber trees are grown so they produce more nuts, this would mean growing them relatively far apart and pruning to get a long straight trunk. Part of the cost of growing the trees can be recovered by selling the nuts. Improvement in seedling timber trees can be achieved by maintaining seedling seed orchards. Trees in a seedling seed orchard are grown from seed from selected trees (Funk, 1969).

Literature Cited

Anonymous. 1937. A Claro walnut treasure. *Flexwood News and Views* 2(1).

Berhow, Seward. 1962. Black walnut varieties, a reference list. *53rd Ann. Rpt. No. Nut Growers Assn.* p. 63–69.

Berry, Fredrich H. 1960. Etiology and control of walnut anthracnose. *Md. Agr. Exp. Sta. Bull.* A-113.

Bey, Calvin F. 1968. Genotypic variation and selection in *Juglans nigra*. Ph.D. thesis, Iowa State Univ.

———. 1970. Geographic variation for seed and seedling characters in black walnut. *U.S. Forest Service Res. Note* NC-101.

Bey, Calvin F., John R. Toliver, and Paul L. Roth. 1971. Early growth of black walnut trees from twenty seed sources. *U.S. Forest Service Res. Note* NC-105.

Cavender, C. L. 1969. Black walnuts from the shellers' point of view. *50th Ann. Rpt. No. Nut Growers Assn.* p. 44–50.

Crane, H. L., C. A. Reed, and M. N. Wood. 1937. Nut breeding. *In* Better plants and animals 2. *USDA Yearbook of Agriculture.* 1937:827–889.

Dabb, Clifford H. 1968. Late vegetating Persian walnuts. *59th Ann. Rpt. No. Nut Growers Assn.* p. 93–94.

Forde, Harold I., and William H. Griggs. 1972. Pollination and blooming habits of walnuts. *Agr. Expt. Univ. Calif.* AXT-N24.

Funk, David T. 1969. Genetics of black walnut. *U.S. Forest Service Res. Note* WO-10.

———. 1972. *Prog. Rept. Forest Science Lab.*, Carbondale, Ill. (mimeo.).

Griggs, William H., Harold I. Forde, Ben T. Iwakiri and Richard N. Asay. 1971. Effect of subfreezing temperatures on the viability of Persian walnut pollen. *HortScience* 6:235–237.

Hall, Geraldine C., and Robert E. Farmer, Jr. 1970. *In vitro* germination of black walnut pollen. *Can. J. Bot.* 49(6):799–802.

Hansche, P. E., and R. M. Brooks. 1965. Temporal and spatial repeatabilities of a series of quantitative characters in sweet cherry *Prunus avium* L. *Proc. Amer. Soc. Hort. Sci.* 86:120–128.

Hansche, P. E., V. Beres, and R. M. Brooks. 1966. Heritability and genetic correlation in the sweet cherry. *Proc. Amer. Soc. Hort. Sci.* 88:173–183.

Hansche, P. E., C. O. Hesse, and V. Beres. 1972. Estimates of genetic and environmental effects on several traits in peach. *J. Amer. Soc. Hort. Sci.* 97:76–79.

Hansche, P. E., V. Beres, and H. I. Forde. 1972. Estimates of quantitative genetic properties of walnut and their implications for cultivar improvement. *J. Amer. Soc. Hort. Sci.* 97:279–285.

Howard, W. L. 1945. Luther Burbank's plant contributions. *Calif. Agr. Expt. Sta. Bull.* 691.

Impiumi, G., and A. Romina. 1967. Floral biology and fruiting of the walnut (*J. regia*) floral morphology and pollen transportation. *Riv. Orthoflorofruttic Ital.* 51:538–543.

Jacobs, Homer L. 1968. Nut culture at Holden Arboretum. *59th Ann. Rpt. No. Nut Growers Assn.* p. 28–36.

MacDaniels, L. H. 1961. Nut growing in Jugoslavia. *52nd Ann. Rpt. No. Nut Growers Assn.* p. 26–31.

McDaniel, J. C. 1957. The pollination of Juglandaceae

varieties—Illinois observations and review of earlier studies. *47th Ann. Rpt. No. Nut Growers Assn.* p. 118–132.

McKay, J. W. 1956. Walnut blossoming studies in 1956. *47th Ann. Rpt. No. Nut Growers Assn.* p. 79–82.

———. 1965. Progress in black x Persian walnut breeding. *56th Ann. Rpt. No. Nut Growers Assn.* p. 76–80.

Merrill, Robert. 1969. 'Franquette' pollination techniques. *Diamond Walnut News* 51(2):8, 15.

Millikan, D. E., and L. H. Cochran. 1966. Selection and propagation studies on *J. regia* in east Europe. *57th Ann. Rpt. No. Nut Growers Assn.* p. 81–82.

Nast, Charlotte G. 1935. Morphological development of the fruits of *Juglans regia*. *Hilgardia* 9:345–381.

Searls, S. R., and C. R. Henderson. 1961. Computing procedures for estimating components of variance in the two way classification mixed model. *Biometrics* 17:607–616.

Serr, E. F. 1969. Persian walnuts in the western states. *In* R. A. Jaynes (ed.). *Handbook of North American nut trees.* Geneva, N.Y.: W. F. Humphry Press, Inc. p. 240–263.

Serr, E. F., and H. I. Forde. 1956. Walnut breeding. *Proc. Amer. Soc. Hort. Sci.* 68:184–194.

———. 1968. Ten new walnut varieties released. *Calif. Agr.* 22(4):8–10.

Smith, Ralph E. 1912. Walnut culture in California—walnut blight. *Univ. Calif. Agr. Expt. Sta. Bull.* 231.

Smith, R. L., and Henry Alanger. 1969. Evaluations of Persian walnuts introduced from Russia, Hungary, and Korea. *60th Ann. Rpt. No. Nut Growers Assn.* p. 61–65.

Valev, K. 1969. Promising new walnut varieties. *Grad. Lozar Nauk* 6(8):3–8.

Webber, George C. 1963. The eastern black walnut. *54th Ann. Rpt. No. Nut Growers Assn.* p. 67–70.

Wood, Milo N. 1934. Pollination and blooming habits of the Persian walnut in California. *USDA Tech Bul.* 387.

Woodworth, R. E. 1930. Meiosis of microsporogenesis in the Juglandaceae. *Amer. J. Bot.* 17:863–869.

Zarger, Thomas G. 1946. Yield and quality of common black walnuts in the Tennessee Valley. *37th Ann. Rpt. No. Nut Growers Assn.* p. 118–124.

Filberts

by Harry B. Lagerstedt

Corylus avellana L., the European filbert, or hazel, as it is also called, is a member of the birch family, Betulaceae. After the Pleistocene glaciations, from about 8,000 to 5,500 B.C., the genus *Corylus* was the dominant native vegetation throughout most of northern Europe. The genus now comprises only about ten species, all indigenous to the Northern Hemisphere. Their habitat ranges from the high Himalayas (*C. ferox* Wall.) to the northern reaches of Canada (*C. cornuta* Marsh.) where short growing seasons and temperatures down to −50 C prevail. There are timber trees (*C. colurna* L. and *C. chinensis* Franch.) native to the Middle East and Asia, respectively, that reach heights of 30–40 m. The habit of most filberts is that of a multistemmed shrub (Fig. 1), but in the United States they are grown as a single-trunk tree. In all the world, the commercial production of edible filberts is located in limited areas of Turkey, Italy, Spain, and the United States. In the United States, 95% of the filbert production occurs in the maritime climate of the Willamette Valley of Oregon, an area about 100 x 250 km.

The history of *Corylus* has frequently been associated with the occult and supernatural. The original divining rod, dowsing rod, or witching wand, used to locate hidden treasures, veins of metals, and water, was made of hazel because of its imputed mystic powers. Ancient Chinese manuscripts assert that the filbert is one of the five sacred nourishments God bestowed upon mankind. The burning of hazel nuts upon the altars of the gods was supposed to give temple priests a power of insight or clairvoyance (Reed and Davidson, 1958). The Greek physician Dioscorides (first century A.D.) used filberts in several of his remedies to cure everything from the common cold to baldness. According to Fuller (1896), the filbert was celebrated in prose by Virgil, who stated it was more honored "than the vine, the myrtle, or even the bay itself." The filbert nut was used as an emblem of fruitfulness in marriage ceremonies. In northern Europe the tree was thought to be a protection against lightning and a charm against witches. It was considered to be sacred to the mythological god Thor (Bunyard, 1920).

The breeding potential of this extremely unusual and interesting plant is tremendous and has barely been touched. The dependency on cross-pollination within most of the *Corylus* species provides for a high degree of heterozygosity. This in turn makes available a wide range of different plant characters and provides the basis for a productive breeding program.

FIG. 1. One of the oldest filbert trees in the northwestern United States. Planted about 1898, it measures about 8m in height and has a spread of 10m. It shows the multistemmed character.

Origin and Early Development

Most of the world's filbert production is grown on seedling trees. Most of the existing named cultivars, some of which have been maintained as clones for nearly 200 years, were selected from seedling populations. Knowledgeable amateurs have contributed to the crossing and selection of desirable filbert cultivars, but their efforts were most often the result of open-pollinated or unprotected crosses. Most cultivars existing today are the result of natural selection and selection by man.

The Filbert Name

Members of the genus *Corylus* have most commonly been called both filberts and hazel nuts. For botanists and horticulturists there has been no controversy regarding these two popular terms since 1942, when the American Joint Committee on Horticultural Nomenclature agreed that the name filbert should represent the genus. However, common usage and colloquial terminology are slow to change, and such terms as hazel, cob, cobnut, lambert nut, Lombardy nut, and Spanish nut remain with us.

The filbert's shape, appearance, husk, and origin have all contributed to its many names. According to Bunyard (1920), Theophrastus recognized only two sorts, trees producing round nuts and trees producing oblong nuts. They were then known as Heracleatic nuts, having been introduced from Heraclea on the Black Sea. To Hippocrates they were known as *Carya thusia*. Pliny was first to use the species name *avellana*, from Abellana in Asia from which the nuts were introduced. This may also be the source for the name of the Italian town Avellino, where the filbert has been grown for centuries. Fuller (1896) finds that the Italian name of Avellana was adopted in Britain and eventually taken by Linnaeus as the specific name of the indigenous species. Evelyn's *Sylva* of 1664 distinguishes between Pontic nuts with long husks and "bald hazelnuts, which, doubtless, we had from abroad, bearing the name Avelan or Avelin." Pontic nuts mentioned above refers to the name given filberts by the Romans, "Nux Pontica," because they were introduced from Pontus.

In the eastern United States, the term "hazel" is most commonly used, especially for the native types *C. americana* Marsh. and *C. cornuta* Marsh. In western United States where *C. avellana* L., the European filbert, is commercially cultivated, "filbert" is the most commonly used name. Yet, when it comes to discussing *C. colurna* L. and *C. chinensis* Franch., the Turkish and Chinese filbert respectively, they are both referred to as "tree hazels" because they do not produce suckers.

The name "filbert" was supposed to have originated from the term "full beard," being used for nuts having a long husk that completely covered the nut. Bunyard (1920) believes that a more plausible origin of the name filbert is taken from St. Philbert, a Frankish abbot whose feast day falls on August 22, coinciding with the ripening of this nut.

The name hazel is said to have come from the Anglo-Saxon *haesel*, a hood or bonnet, describing a husk that was shorter than the nut. Reed and Davidson (1958) trace the root of haesel to the older Anglo-Saxon *haes* or German *heissen*, to give orders, for the hazel rod was a sign of authority for shepherd chieftans. *Corylus*, the genus name, comes from *korys* which also refers to hood, helmet, or bonnet.

For centuries nuts with long or short husks were generally called filberts and hazels respectively. This is no longer true. Since both types, *C. maxima* Mill. and *C. avellana* L., cross readily, present-day cultivars are a mixture. These two species have been crossed so extensively, because of man's culturing them together, that the husk length characteristic is, by itself, no longer valid for distinguishing between the two species (Trotter, 1947).

On the cultivar level, Bunyard (1920) illustrates some of the nomenclatural confusion in this plant group with an extant German nut type called 'Lambertsnuss'. It was once thought that the name originated as *Langbart* or long beard, referring to the husk length. Another point of view contended that 'Lambertsnuss' was derived from Longobards or Lombards of Italy, and it was they who introduced it to Britain in their westward migration. By coincidence, the best-known British filbert cultivar, 'Kentish Cob', was introduced in 1812 by a Mr. A. R. Lambert and became known as "Lambert's nut." To add even more to this confusion, the popular 'Kentish Cob' (Mr. Lambert's introduction) is grown in the United States under the name 'Du Chilly'.

These examples merely indicate that names of species and cultivars in the genus *Corylus* must be accepted cautiously by serious workers in this field. Interspecific crossing, self-sterility, the common use of seedling orchards, and, until recent years, a limitation of adequate nursery practices, all increase genetic heterogeneity and inaccurate terminology in the genus.

The Filbert Plant

Two filberts, *C. colurna* and *C. chinensis*, are tree types that have a central leader. All other species are shrubs or large bushes; that is, they tend to be round headed and to grow with multiple stems, which are renewed continually by the production of numerous sprouts or suckers. In the United States, where the European filbert *C. avellana* is trained to a single trunk, the removal of suckers is a major annual problem. Accordingly, there is a great need to develop a rootstock that does not sucker. All filberts are deciduous, though members of some species retain their leaves into the winter in areas with a mild climate.

Filberts are not large plants, as orchard trees go, but under fertile conditions they may reach heights of 8 to 10 m. They have been planted more than 9 m apart, but it may take 20 years before they use their allotted space. Due to economics and land values, there is a great deal of interest in high-density orchards in which trees number more than 400/ha. There is also a great need for growth-controlling or dwarfing rootstock, one which, in addition to reducing plant size, could control biennial bearing, bear precociously, and yield heavily.

The Filbert Nut

The involucre or husk that covers the filbert nut varies greatly in appearance. It is often used to describe or distinguish between species and cultivars. Husks of *C. avellana* and *C. americana* are generally short and sometimes flared, permitting the nut to drop out readily. This characteristic is called "free husking" and is a highly desirable aid to early harvesting. Husks of *C. colurna* are extremely thick and fleshy at their base. Distally the bracts are deeply forked into lance-like lobes. Upon drying, these lobes curl back so that a cluster of drying husks takes on the appearance of a spiny cactus. Husks of *C. chinensis* are constricted above the nut, and the deeply lobed bracts recurve to resemble antlers (Kasapligil, 1963). Husks of *C. ferox* and *C. tibetica* Batal. are spiny and resemble those of a chestnut bur. *C. maxima* has a long husk, and *C. cornuta* has one that is so constricted above the nut that the nut becomes extremely difficult to remove.

Nuts and kernels are of a variety of sizes and shapes. Shell color varies from tan to a rich deep brown, with stripes more or less distinct. Shell pubescence also varies and has an important influence on nut appearance. When mechanically harvested from the ground, highly pubescent nuts often take on a dull appearance from the dirt trapped under these numerous short, stiff, curved hairs. Shell pubescence is considered an undesirable trait because it tends to hide the rich brown coloration of the nut. Shell thickness is another highly variable characteristic. The Turkish tree hazel has a rather small nut with an extremely thick shell, compared to that of the European filbert. Even so, 'Barcelona', a cultivar of the latter, averages only 40% kernel. Thin-shelled nuts are being sought by means of breeding and selection because the use trend of filberts is towards the kernel trade. In this trade they must compete with walnuts, almonds, and pecans, which yield 45 to 60% kernel. Filbert selections with 55% kernel have been achieved, but none of these has as yet been satisfactory in a sufficient number of characteristics to warrant introduction.

The desirable kernel should be plump and have a small central cavity so that it resists shriveling upon drying. In the United States a round kernel is preferred to a long one. The kernel surface should be smooth, and the brown pellicle should be easy to remove. This would enhance the appearance of the shelled nut and greatly facilitate its processing. Kernel quality should be high, with desirable storage characteristics. Rapid breakdown in storage, causing discoloration, shrivel, or flavor changes would be undesirable. The kernel must be pleasantly flavored as a raw nut, and the changes that occur during roasting should enhance flavor even more.

Geographic Distribution

It is estimated that the latest glacial period waned during a warming trend that occurred from about 17,000 to 8,000 B.C. During this period, known as the Subarctic, glaciers gradually retreated from the British Isles, Northern Europe, and Scandinavia. As the temperatures moderated, *Corylus avellana* was one of the first trees to move northward, along with birch, pine, oak, and alder. At the end of the pre-Boreal Period (8,000 to 7,500 B.C.) there was a sharp rise in amounts of filbert pollen deposited in strata associated with that time. Filbert became the dominant vegetation during the Boreal period which lasted for 2,000 years, that is, until about 5,500 B.C. In this period, mean summer temperatures were considerably higher than those that now occur in the British Isles. The number of filbert pollen grains preserved in certain peat strata during this time exceeded that of all other trees combined by 75%! This enormous "hazel maximum" has been found repeated in the British Isles and several other parts of Europe (Tansley, 1939). Since its period of glory, the filbert has gradually given way to other types of vegetation, and its numbers have been greatly reduced.

However, the distribution and native range of *Corylus avellana* has not diminished greatly since that time. One of the most recent and detailed accounts of the range of the European filbert was published by Kasapligil (1964), from whom we quote in part:

> *Corylus avellana* is distributed throughout Europe from Cintra on the west coast of Portugal, Ireland and Orkney Islands to the southern part of the Ural mountains through Bessarabia, Crimea and Kazakistan. It is abundant throughout the Balkan countries including the coastal regions of eastern Thrace. The northern distribution range extends to 68° northern latitude along the west coast of Norway to 64° northern latitude in Sweden and 60° northern latitude in Russia along the southern shores of Ladoga Lake. In southern Europe it occurs in Spain, Sicily and Greece. In Asia it extends from Turkey through Caucasia to Iran in the east and from the Anti-Taurus Mountains of Anatolia to Syria and Lebanon in the south.

Within this enormous range it seems that the filbert thrives best in areas with very moderate climates. The four major production areas are all influenced to some degree by large bodies of water: Turkey by the Black Sea; Italy and Spain by the Mediterranean; and Oregon by the Pacific Ocean. The distribution of *Corylus maxima*, the giant filbert, is described by Kasapligil (1964) as follows:

> *Corylus maxima* is native to southeastern Europe, i.e., Thrace, Macedonia, Croatia and to the northeastern Anatolia. This species is recorded from the following

> localities of Asia Minor: Giresun, Trabzon, and Gumushane. It represents the characteristic plant of the bush forest of Turkey up to an elevation of 1,300 m where it is also widely cultivated for its nuts.

From this account it can be seen that the native range of *C. maxima* is encompassed by that of *C. avellana*. The overlap of ranges and the lack of genetic blocks between these two species makes plausible the belief that many of the currently grown cultivars are the result of this interspecific cross.

C. tibetica Batal., the Tibetan filbert, is common in much of China and native to western China, especially the provinces of Yunnan, southeastern Szechwan, western Hupei, eastern Kansu, and Shensi (Shun-Ching, 1935). Its outstanding characteristic is the spiny husks, which cause the nut clusters to resemble the burs of the chestnut. Rehder (1940) describes it as a shrubby tree, growing to 8 m. Wilson (1916) considered it a bush, 3 to 7 m in height.

C. ferox Wall., the Himalayan filbert, is thought to be the nearest relative to the above on the basis of husk characteristics. Its nut clusters also resemble chestnut burs. However, the range is different. *C. ferox* is indigenous to northern India, Nepal, and Sikkim, and grows at altitudes of more than 3,000 m. It is the only *Corylus* species we have seen with an ovate-lanceolate leaf. The nuts are small and fairly thick-shelled. They resemble the European filbert in taste.

C. Jacquemontii Dcne. is a tree form, usually listed as a species (Wilson, 1916; Rehder, 1940), but considered by Kasapligil (1963) to be a variety of *C. colurna* L. Its habitat is similar to that of *C. ferox* above, though mostly west of it. It has also been called the Himalayan filbert. Other names are the India tree hazel and *C. colurna* var. *lacera* A. Dc.

C. colurna, the Turkish tree hazel, is native to an area extending from southeastern Europe through northern Turkey, Caucasia, northern Iran, and the Himalayas to China (Kasapligil, 1964). The forms that grow in the latter two areas are considered by most other authors to be *C. Jacquemontii* and *C. chinensis* Franch. The bark on young stems of this species is light gray and deeply furrowed, yet velvet soft. Upon aging, it becomes rough and fissured, exfoliating in vertical plates. Nuts are small and thick-shelled. The tree grows from 20 to 40 m in height.

C. chinensis, the Chinese tree hazel, grows to 40 m. It is characterized by a husk strongly contracted above the nut into a tube, which divides into linear lobes that fork deeply at their apex. It is distributed in the provinces of Yunnan, Hupei, and Szechwan at altitudes of more than 2,000 m (Shun-Ching, 1935).

C. heterophylla Fisch. is known also as the Siberian filbert, miscellaneous filbert, and Japanese hazel. Its range extends from the provinces of Lianoning, Hupei, Shensi, Kansu, and Shansi to Manchuria, Korea, and Japan. The varieties *sutchuenensis* and *yunnanensis* are well recognized. Most of them are shrubs 2 to 3 m in height; some have distinct truncate leaves (Shun-Ching, 1935).

C. sieboldiana Bl. is known alternately as the Manchurian and the Japanese filbert, names that indicate its range. The species is well distributed in Japan, but one variety, *manchurica*, is more common in Korea and China. It is a shrub growing 3 to 5 m in height. The husks constrict into a tube above the nut. This characteristic makes it resemble the beaked filbert of North America so much that it has come to be regarded as the Asiatic representative of *C. cornuta* (Wilson, 1916).

C. cornuta Marsh. and *C. rostrata* Ait. are the species names used for the beaked filbert. It ranges from Quebec to Georgia on the East Coast through Saskatchewan to Missouri. The confusion that reigns around these two species names has existed for nearly 200 years and should be put to rest. As pointed out by Woodworth (1929), Marshall described the species *C. cornuta* and published his work in 1785. Aiton's publication, which called the same species *C. rostrata*, was dated 1788. According to the rules of priority, the accepted name for this species should be *C. cornuta*. On the West Coast, the variety *californica* is distributed from British Columbia to northern California (Rehder, 1940). Members of the species have been hardy at temperatures of −50 C in the Hudson Bay area of Canada. It is primarily a shrub up to 3 m in height, and the *californica* form is somewhat taller.

C. americana Marsh., the American filbert, has a medium-sized, edible nut, which is enclosed in a husk that is twice the length of the nut, but not constricted. It ranges from Saskatchewan to Maine, southward to Florida, westward to Oklahoma, Missouri, and Minnesota (Rehder, 1940). Its range is not as far north or west as that of the beaked filbert. Both species are shrubby, of similar size and habit, but they are easily distinguished by husk shape.

History of Improvement

United States

The two Corylus species indigenous to the United States and Canada, *C. americana* and *C. cornuta,* have contributed far more to the sustenance of wildlife than that of man. Indians and early settlers made some use of these nuts, especially of *C. americana,* but found it easier to seek other larger nuts from the wild. Though the native filberts are abundant and cover an extremely wide range, relatively few selections have been made from the native types.

The early settlers brought filbert cultivars with them from Europe, especially from England. On the eastern seaboard, these cultivars would grow for several years and then succumb to the eastern filbert blight, caused by *Cryptosporella anomala* (Pk.) Sacc. The native filberts not only tolerate this fungus disease, they serve as host and source of inoculum. The European filbert has been introduced to nearly every state east of the Rocky Mountains, but it has generally failed either from disease susceptibility or from a lack of adequate hardiness. The primary objective of filbert breeding in the eastern United States and Canada has been to combine the high quality and large size of the European filbert with the hardiness and disease resistance of the American filbert.

The pioneer filbert breeder in the East was John F. Jones of Lancaster, Pennsylvania. He used the *C. americana* cultivar 'Rush' as the pistillate parent in his crosses. As a staminate parent he used the European cultivar 'Italian Red' to produce the hybrid 'Bixby', and 'Barcelona' to produce 'Buchanan' (MacDaniels, 1964; Reed, 1936; Reed and Davidson, 1958).

Two other selections from the native *C. americana* population were named 'Littlepage' and 'Winkler'. The latter cultivar grows only to 1½–2 m but has large nuts. Both of these native selections, along with 'Rush', were used as pistillate parents in crosses with *C. avellana* made by C. A. Reed of the United States Department of Agriculture from 1928 to 1932 (Crane, Reed, and Wood, 1937). Two selections made from this progeny in 1941 and 1943 have since been named 'Reed' and 'Potomac' (Reed and Davidson, 1958; Slate, 1961).

In Minnesota, Carl Weschcke used a *C. americana* selection of his own and the 'Winkler' cultivar as pistillate parents, crossing them with a mixture of *C. avellana* pollens, thus the staminate parentage is uncertain. The progeny, for which he coined the name "hazilberts," yielded several selections, of which at least three were named: 'Carlola', 'Delores', and 'Magdalene' (Reed and Davidson, 1958; Weschcke, 1954).

Since 1925, the New York State Experiment Station at Geneva has maintained a variety trial to evaluate hardiness in the European cultivars, to select desirable specimens from a *C. americana* collection, and to evaluate *C. americana* x *C. avellana* crosses. Several severe winters with temperatures of −37 C have helped to cull all but the hardiest progeny. To date only a few cultivars have been recommended on a trial basis for home gardeners (Slate, 1947, 1961). Plant material for evaluation at Geneva was obtained from several sources. The original planting was established with 40 cultivars of *Corylus* obtained from New York and Oregon nurseries. In 1928 an additional 40 cultivars were imported from Germany. In 1932, 535 seedlings were received from W. G. Bixby of Long Island, New York, the results of crosses made by C. A. Reed of the USDA. In all, nearly 2,000 seedlings were fruited and evaluated at Geneva by 1947.

A series of crosses made in New York during the years 1930–1933 used the cultivars 'Rush' and 'Barcelona' as pistillate parents, with both *C. americana* and *C. avellana* cultivars as staminate parents. Crosses involving only *C. americana,* such as 'Rush' x 'Winkler' and 'Rush' x 'Littlepage', yielded progeny of no value. Crosses involving only *C. avellana,* such as 'Barcelona' crossed with either 'Medium Long', 'Red Lambert', 'Daviana', or 'Purple Aveline', yielded very few seedlings worthy of further evaluation under New York conditions. The interspecific hybrids produced from 'Rush' and *C. avellana* cultivars proved to be the most promising (Slate, 1947).

In these interspecific crosses, 'Rush' was used to provide hardiness and resistance to the eastern filbert blight. The European filberts were selected to contribute nut size, kernel quality, and the free-husking characteristic. In evaluating seedlings, emphasis was placed on nut size, color, and lack of shell pubescence. Most of the progeny had relatively thin shells. Kernels were evaluated for lack of shrivel, smoothness, and freedom of fiber. The primary plant characteristics considered were productivity and catkin hardiness.

Several of the best interspecific hybrids with 'Rush' were considered to be superior to 'Barcelona'. Some traits of cultivars used as pollen parents in these crosses were recorded: 'Cosford' generally transmitted high productivity, while 'Kentish Cob' ('Du Chilly') did the opposite. Crosses with 'Red Lambert' had the hardiest catkins, followed by those of 'Cosford', 'Bolwiller', 'Italian Red', and 'Barcelona'. Of 1,232 'Rush' hybrids raised to maturity, 39 or 3.2% had qualities that merited further propagation.

No hybridization has been done at the Geneva Experiment Station since the early 1930s. Of the seedlings evaluated, none has produced enough nuts to warrant recommending for commercial planting. The best of the hybrids have produced nuts as large as the European cultivars and with cleaner kernels. Unfortunately, most of these nuts stick in the husk and have a reduced production of male catkins (Slate, 1969).

However, the search for a filbert satisfactory for New England conditions has not ceased. For many years more than a hundred kilograms of nuts from the Geneva seedlings and cultivars were planted by the Soil Conservation Service in various parts of the state. These second-generation seedlings could supply a huge source of segregating *Corylus* plant material, but due to their scattered nature, evaluation of them would be difficult (Slate, 1947).

Also in New York, Samuel H. Graham of Ithaca introduced the cultivars 'Morningside' (1945) and 'Graham' (1950). These were the result of *C. americana* x *C. avellana* crosses. 'Morningside' originated from 'Rush' x 'Du Chilly', and 'Graham' came from 'Winkler' x 'Longfellow'. The cultivar 'Morningside' was lost to the eastern filbert blight, which killed many of Graham's selections (Slate, 1969).

The European filbert was introduced to the West Coast by Felix Gillet, a French barber who started a nursery in the gold-mining town of Nevada City, California, in 1871. Gillet correctly predicted that the filbert would do well in the maritime climate of the coastal valleys of Oregon and Washington. The Willamette Valley of Oregon proved to be ideal for filbert culture and now produces 95% of the United States tonnage, the other 5% being grown in the neighboring state of Washington. In the Northwest, cultivar improvement has occurred primarily through chance seedlings that interested and observant growers have selected from their orchards or nursery rows.

Some of these early selections and their originators follow:

Cultivar	Year Introduced	Originator	Location
Alpha	1925	A. A. Quarnberg	Vancouver, WA
Big Schaad		A. B. Scherf	Newberg, OR
Brixnut	1919	C. T. Brixey	McMinnville, OR
Chaparone	1915	C. T. Brixey	McMinnville, OR
Clackamas	1917	H. A. Kruse	Wilsonville, OR
Fitzgerald	1936	R. Turk & D. Fitzgerald	Washougal, WA
Freehusker		A. B. Scherf	Newberg, OR
Gasaway	1926	C. Bush	Barton, OR
Gem		R. Turk & D. Fitzgerald	Washougal, WA
Henneman	1926	H. A. Henneman	Portland, OR
Kruse	1917	H. A. Kruse	Wilsonville, OR
Longfellow	1925	N. Mosier	Dayton, OR
Mosier	1932	N. Mosier	Dayton, OR
Nibler	1926	A. M. Gray	Milwaukie, OR
Nonpareil		R. Turk & D. Fitzgerald	Washougal, WA
Nooksack	1918	H. E. Altman	Nooksack, WA
Ogden	1926	A. M. Gray	Milwaukie, OR
Roberta	1915	H. A. Kruse	Wilsonville, OR
Royal	1934	K. Pearcy	Salem, OR
Scherf	1936	A. B. Scherf	Newberg, OR
Sicily	1915	H. A. Kruse	Wilsonville, OR
Woodford		E. W. Woodford	Forest Grove, OR

As can be seen from the dates listed, the major selections were made during the 20-year span after 1915. This was a period of rapid expansion of the filbert industry in the Northwest. Interest in cultivar selection was heightened by the fact that some of the original growers had planted seedling orchards, so a heterozygous population was available to select from. As nursery plantings increased and more clonal stocks became available, seedling orchards were replaced, predominantly by 'Barcelona' trees.

Dissatisfaction with the 'Barcelona' has gradually been increasing, and by 1960 growers were supporting a breeding program at Oregon State University with grants from their state filbert commission. Dr. Maxine Thompson, who is currently in charge of this breeding program, has made controlled crosses of selected parents for several years. There are now more than 4,000 progeny in the field growing to maturity (Thompson, 1972). The first phase of the breeding program was to collect, classify, and evaluate most of the important clonally propagated cultivars available in the major growing districts of the world (Thompson, 1969). Most of the first-phase work, involving more than 200 cultivars, has been completed. The second phase, selecting parents for combination of desirable traits and making the necessary controlled crosses, has begun.

Since the first results from this breeding program are still several years away, promising grower selections are being considered in the interim. The selections 'Butler', 'Ennis', 'Jemtegaard No. 5, 'Lansing', and 'Ryan' are being increased and placed in performance trials with growers and nurserymen. 'Butler' and 'Lansing' appear most promising as pollinizers for 'Barcelona'. They have a larger nut of similar type and are expected to be more productive.

Olgar Jemtegaard of Boring, Oregon, has been carrying on his private filbert-breeding program for more than 30 years. He has grown progenies resulting from open-pollinated crosses of superior pistillate parents such as 'Nonpareil'. The seedlings were allowed to fruit in the nursery row. The most promising selections were then grafted into his orchard, and the poor selections were culled out. In this way he has gradually developed an orchard of more than 10 ha, composed largely of superior selections and containing a wide array of desirable characteristics from which to choose cultivars or breeding stock. This gene bank Jemtegaard has developed may be more important to Oregon than the cultivar collection made by the state, since all of the selections are from cultivars developed in the Northwest. As Thompson (1969) points out, the chances are greatest that filbert improvement will come from cultivars developed in and suited for the particular growing conditions of that locality.

Canada

J. U. Gellatly of Westbank, Canada, was a nut-tree nurseryman with a strong desire to produce hardy, late-blooming, early-maturing filbert trees that could survive and produce in the harsh climate of his country. Gellatly was also interested in producing a desirable rootstock for the filbert. He made crosses between the European filbert and species of the so-called tree hazels, designating the resulting progenies as "trazels." He also made selections directly from selfed *C. colurna*, choosing those that had large nuts. The cross-bred trazels were selected for nut characteristics and rootstock possibilities (Gellatly, 1955).

Another series of Gellatly crosses produced a group of hybrids that he termed "Filazels." The seed parent in these crosses was selected from native *C. cornuta* plants growing in the Peace River district of northern Canada. This Peace River hazel could survive − 50 C. Gellatly believed that combining the European filbert with the above would produce a hardy plant with a large nut, requiring a short growing season, and one that could even be used for a windbreak on the Canadian prairies (Gellatly, 1956).

All of Gellatly's crosses were open-pollinations, a questionable practice with a wind-pollinated crop. Catkins were removed from the seed parents to prevent self-pollination, a desirable practice even though most members of the genus *Corylus* are self-sterile. No record was kept of the pollen parents. Gellatly (1966) believed that "the tree hazels cross readily with the filberts," an opinion that has not been substantiated under controlled conditions. He did find that most trazels inherited the tree type of growth and had nuts similar to the tree-hazel parent, but that there was a great deal of variation. Nut samples of several trazel selections obtained from Mr. Gellatly bear strong resemblance to the *C. colurna* parent.

Seeds and seedlings of first and second generation filazel crosses were distributed widely throughout Canada and eastern United States. In addition, numerous selections were named by Gellatly, the names ending in the suffix "oka" to commemorate the Okanagan Valley in which he lived. We have compiled a list of 30 names of filazels and trazels that have been given names ending in "oka." To this can be added numerous numbered selections and his first named cultivars, 'Craig', 'Brag', 'Comet', 'Holder', and 'Gellatly', which are listed in the *Register of New Fruit and Nut Varieties*. Gellatly also was responsible for locating and introducing the cultivar 'Bawden' on Vancouver Island in 1936. Its parentage is unknown, and it has not received much attention.

Europe and Adjacent Countries

Each of the major filbert-producing countries maintains some breeding, selection, or cultivar-improvement program. In addition, interest also exists in the Slavic countries, Scandinavia, central Europe, and countries surrounding the Mediterranean. Test plantings and trials exist in Norway,

Denmark, Sweden, Germany, Austria, France, Corsica, England, The Netherlands, Tunisia, Greece, Yugoslavia, Bulgaria, Czechoslovakia, Rumania, Lithuania, Australia, New Zealand, Japan, and Korea. While an international interest in filbert culture exists, the environment required by the filbert usually restricts its commercial development and so limits extensive research and breeding programs.

The literature covering stations where active breeding or selection programs are in existence is mostly in Italian, Spanish, or Turkish. Some of these publications have been translated, a few contain English summaries, and others are represented only in condensed form as abstracts. Therefore, it must be acknowledged at the outset that language differences are an information barrier. Coverage under this heading will be fragmentary.

Turkey, the foremost nation in filbert production (61%), has an active research group of agriculturists located at Giresun. A research program was initiated in 1936. In 1963, with the acquisition of 384 ha of land and buildings, it became known as the Giresun Filbert Research Institute. The institute was combined with an agricultural technical school in 1967 to strengthen their program further.

Research is directed primarily toward cultural problems, but the institute progress report for 1965–1970 also indicated that self- and cross-pollinations have been made that show that the wild-type filberts are effective pollinizers for the major Turkish cultivars. There is a selection program to locate cultivars from the native *Corylus* stands. Apparently, most of the Turkish crop is produced on seedling trees, which would constitute an unplanned heterozygous population from which to select.

Turkish filberts are harvested by hand from the tree. For handpicking, a freehusking selection, with a short open husk, is not important or even desirable. Nuts which drop to the ground are usually lost due to lack of clean cultivation. Accordingly, Turkish cultivars usually have a long husk that is slightly constricted over the nut. This long husk is characteristic of the native trees.

While cultivars have been selected, the bulk of the crop is classified on the basis of nut shape and the district from which the nuts originate. The shapes, round and long, are called "tombul" and "sivri," respectively. The round is the more popular. The main districts are Giresun and Trabzon. The former is considered to produce the best nuts (Schreiber, 1953).

From the number of publications and workers involved, Italy apparently has the most active filbert-research program of any nation. Yet even in this country, where filberts have been known for thousands of years, and 24% of the world's production is grown, a serious breeding program is just getting started.

Romisondo (personal communication) has described 13 characteristics of the nut and of the tree as being of value for breeding purposes. The inheritance of these characteristics is being studied in intraspecific hybridizations of *C. avellana* and *C. maxima*. The objective of these crosses is to improve the cultivar 'Tonda Gentile della Langhe' (TGDL), which is most popular in the Piedmont growing area. The desirable characteristics of 11 cultivars and two species are listed as being of potential usefulness in a breeding program with 'TDGL'.

Romisondo (personal communication) states that 'TGDL' is almost male-sterile, bears its nuts next to catkins, suckers excessively, and is susceptible to the big bud mite (*Phytocoptella avellanae* Newkirk and Keifer) and what is suspected to be a virus disease. Crosses have been made with other *C. avellana* cultivars, especially 'Cosford', which is a heavy pollen producer, is self-fertile, and is cross-fertile with 'TGDL'. 'Cosford' bears thin-shelled nuts, which are well distributed along the stem.

The first progeny from the above cross have now fruited. These seedlings are highly variable in vegetative characteristics, pollen viability, and self-fertility. It would appear at this time that the oblong shape and thin shell of 'Cosford' is dominant over the round shape and thicker shell of the pistillate parent.

Each of the growing areas of Italy favors somewhat different cultivars because of differences in climate and growing conditions. 'Gentile Romana', 'TGDL', and 'Tonda di Giffoni' are considered to be the most promising cultivars on which to base a breeding program, since they are already well adapted to Italian growing conditions. The cultivar 'Mortarella' has been described as combining most of the qualities desired by both growers and consumers in much of Italy. Filberts are the third most important crop in Sicily, yet growers there distinguish only between early and late types, not cultivars. There is a need for a selection program in that part of the country.

Both Romisondo (1965) and Paglietta (1970) have done a great deal of basic work on floral biology, pollination, and establishing cross-compatibility between cultivars. The latter has also occupied the attention of Tombesi (1966), who determined that cultivars long established in Italy had a large percentage of nonviable pollen grains, many of which were small and shrivelled in comparison to those that were viable. Pollen grains from cultivars introduced from the United States had smaller sized grains, but a much higher proportion of them were viable.

Spain grows about 10% of the world's filbert production on about 30,000 ha located primarily in the northeastern corner of the country bordering the Mediterranean Sea. About 70% of this production comes from the cultivar 'Negreta', a medium-long nut. The balance of the production is called "Comuna" or unclassified common.

Like the United States, Spain bases its production primarily on a single cultivar. Thus the faults of that cultivar become the major problems of the industry. Irregular cropping has been one problem, partly due to environmental factors, but perhaps also due to poor pollination. Large solid plantings of 'Negreta' have been made without pollinizers, even though these are recommended. Since yields are said not to have suffered, it could be that 'Negreta' is partly self-fertile. Cultivars such as 'Negretta primerenca de la Selva', 'Grossal de Constanti', 'Ribeta', and 'Artelleta de Alforja' are recommended as pollinizers. No filbert-breeding program is being carried out in Spain.

The Garrone Valley of southeastern France has the largest filbert plantings, yielding a few hundred tons of nuts annually. Filberts are a popular nut, and more are imported than are produced locally. The Ministry of Agriculture subsidizes new plantings in an effort to popularize filbert culture, increase production, and satisfy the national demand. A centrally located tree-fruits research station initiated a filbert program in 1960, which includes breeding, variety trials, and cultural research. The major cultivars of *C. avellana* grown in France, Spain, Italy, Turkey, Belgium, England, and Germany have been collected for evaluation in the Garrone Valley. According to Germain (personal communication), the cultivars of greatest interest in France are: 'Fertile de Coutard' ('Barcelona'), 'Negret', and 'Tonda Gentile della Langhe'.

Crosses involving the aforementioned and 'Mereville de Bolwiller', 'DuChilly', 'Cosford', and other cultivars have been made. The resulting hybrids have not yet fruited. They will be evaluated for precocious bearing, season of bloom, hardiness, productivity, nut shape and size, shell thickness, and suitability of the kernel for the confectionery trade. 'Daviana' has not proved to be a satisfactory pollinizer for 'Barcelona' in France. A pollinizer having a round nut is high on their list of breeding objectives.

There appears to be a great deal of interest in the possibilities of growing filberts in several widely separated regions of the USSR. Reports have been collected from ten different experiment stations mentioning that variety trials are established and that selections have been made from native stands. Crosses that have been made are primarily involved with incorporating more cold-hardiness into selections with desirable nut characteristics. Seedlings obtained from crosses of the native filbert with *C. avellana* cultivar 'Pontica' were found to be more cold-hardy than those crosses with *C. maxima*, *C. colurna*, and *C. cornuta*.

Numerous selections made from native stands have been named. They are considered to be more frost-resistant, cold-hardy, and disease-resistant, and to have larger nuts than some of the named cultivars. In breeding, it has been considered preferable to use the cultivar as the female parent. For example, the cross 'Barcelona' x native *C. avellana* was made in southern Russia, using northern pollen, and the progeny grown in northern Russia to screen for hardiness.

Breeding Objectives

In the United States, the 'Barcelona' cultivar accounts for about 85% of the annual filbert production. In effect, this industry is based on a single cultivar, a cultivar not without its faults. The 'Barcelona' must be acknowledged as having stood the test of time, as it has been the leading cultivar in Oregon for more than 70 years. This is partly because of certain good characteristics and partly because of the difficulty, cost, and time involved in increasing a new cultivar. The method of propagation, layerage, is slow. It would require from five to eight years for a nurseryman to build up sufficient stocks to offer plants of a new cultivar for sale. Use of a rootstock and grafting may improve this situation in the future (Lagerstedt, 1970, 1971).

The faults of 'Barcelona' dictate that it must gradually yield its leading position. These faults are: moderate productivity, poor kernel percentage (average 40%), high susceptibility to brown stain (a physiological disorder), unattractive kernel (pellicle adheres), lack of early maturity, high blank production (seedlessness), moderate tree vigor, and moderate nut flavor. Each of these characteristics can be improved, since the improvement already exists in some other selection.

The increasing costs of land, of production, and of changes in marketing dictate the major filbert-breeding objectives as outlined by Thompson (1969, 1972): Nuts will be bred for three uses: the kernel trade, the inshell trade, and a multipurpose nut suitable for either use.

With these uses in mind, there follows a list of desirable characteristics sought in new filbert cultivars:

1) Increased yield: increased flower set, more nuts per tree and per cluster, larger nut, lack of biennial bearing.

2) Increased kernel percentage: thin-shelled, well-filled nut.

3) Kernel quality: clean, smooth kernel, no shrivel, good flavor, small cavity, good shelf life, and good processing characteristics.

4) Nut characteristics: round shape, early maturity, minimal pubescence, rich brown color, well-sealed shell.

5) Husk characteristics: medium to short length, early maturing, not constricted.

6) Reproductive characteristics: good pollenizer, synchronized flowering, low blank percentage, good self- or cross-fertility.

7) Resistance: bacterial filbert blight, big bud mite, brown stain.

8) Tree habit: upright, wide crotch angles, nonsuckering, deep fibrous roots.

Cultivars designed for the kernel trade should be thin-shelled. They should have a moderate-sized kernel, which is plump and highly flavored, with a small cavity and no adhering pellicle. The seed coat should separate easily from the kernel on blanching or roasting. It should resist shrivel and have a superior storage quality.

Cultivars for the inshell trade must be large nuts with an attractive external appearance. Large nuts generally have kernels with large cavities. Such kernels shrivel excessively. This fault should be bred out. Thin shells would also be desirable from the packers' and consumers' point of view. To the producer, thin shells mean losses to birds, rodents, and mechanical breakage.

Cultivars for multipurpose use should be medium size, possess as many desirable characteristics as possible, and be able to serve as a pollenizer.

To improve cultural management, tree growth, and reproduction, there is a need for precocious selections that set large crops annually and are resistant to insect, disease, and physiological problems. There is a need for a nonsuckering rootstock, one that is well anchored by a deep, fibrous root system and possesses growth-controlling characteristics. Pollenizer selections with the same type nut as the maincrop cultivar would lead to increased pollenizer-tree numbers and improved pollenizer distribution in the orchard.

Breeding Techniques

In Oregon, the filbert tree is said to never go dormant because it exhibits some growth and development every month of the year. It blooms during the winter from November to March, reaching a peak during January and February. Pollination may occur as early as December, yet fertilization does not occur until June, a time span of half a year between these natural events! In fact, at the time of pollination, the pistillate flower has no ovary, only two elongating stigmatic styles. These persistent styles and the apex of the nut are initiated in one year, but develop the next, so that the apex of the nut is a year older than its base.

The Filbert Flowers

All *Corylus* species are monoecious and bear unisexual anemophilous flowers. The staminate blooms (male catkins) become evident during the growing season as early as June. Pistillate (female) flowers are borne in tight clusters and do not become visible until early winter in climates that are mild. At pollination time, these inconspicuous, apetalous pistillate flowers merely consist of bright red stigmatic strands, extruding from the tip of a bud.

Blooming in filberts is dichogamous, so appearance of staminate and pistillate flowers may not be synchronized. Protandary (staminate bloom first), protogyny (pistillate bloom first), and homogamy (synchronized flowering) occur, depending on cultivar, growing location, and season. Homogamy is of least importance, since most filberts are self-sterile. They benefit from and are usually obligated to cross pollination for adequate nut set and proper kernel development. Pollinizer trees must be provided for the main crop cultivar, and their catkin bloom must be matched to the peak appearance of main-crop pistillate flowers. Attention to dichogamy is of greatest importance in growing areas with cold climates, since blooming is delayed until spring and occurs over a relatively short period of time.

The conspicuous "bloom" of the filbert is most commonly associated with the extension growth of the staminate catkins on leafless twigs, with subsequent release of clouds of pollen. This release signifies the end of a development period that was initiated early in the previous year.

In Oregon the red tip of the first tiny 'Barcelona' catkin can be seen in the latter part of June. It arises in the axils of basal leaves on the current season's stems. By July catkins are easy to find, and in early August they are 11.3 mm long and 100 of them weigh an average of 7.53 g. We have followed their rate of growth by means of biweekly length and weight measurements, starting August 10 and ending after anthesis. Catkins grow rapidly through August and most of September (Fig. 2). At the end of September the growth curve reaches a plateau, indicating a stage of rest or a period of growth inactivity. Then in mid-December, catkin growth is resumed until anthesis in January or February. At this time their length continues to increase, but their weight decreases because of loss of pollen.

Considering the tremendous energy expenditure involved in catkin and pollen production and the nutritive loss it represents annually to the plant, partial or complete male sterility might be an advantage for the main crop cultivar. There is one tree in Oregon that annually sheds its crop of catkins before they mature. This tree is extremely vigorous and bears heavy nut crops annually, perhaps because less of its food reserve is expended by not producing catkins.

Trotter (1947) observes that catkins are present in great numbers in July, but are then only rudimentary flowers. At this time the androecium is actively meristematic. Each individual flower is subtended by a small protective bract which, when combined with 150–200 or more similar bracts, forms the imbricated surface of the catkin. Trotter claims that the androecium is not derived from the primitive meristem that gives rise to the bract, but from a secondary meristem situated at the base of the bract. Towards autumn, two minute

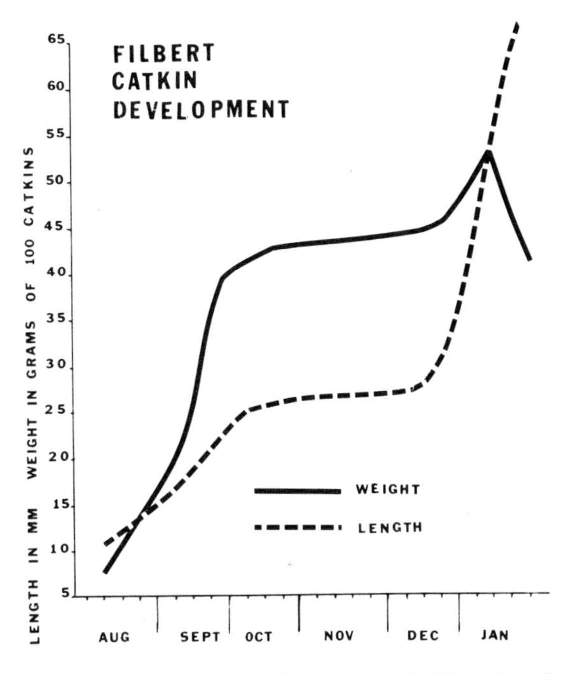

FIG. 2 Seasonal development of 'Barcelona' catkins in Oregon. Dehiscence is signified by the mid-January point, at which the curves for catkin length and weight diverge.

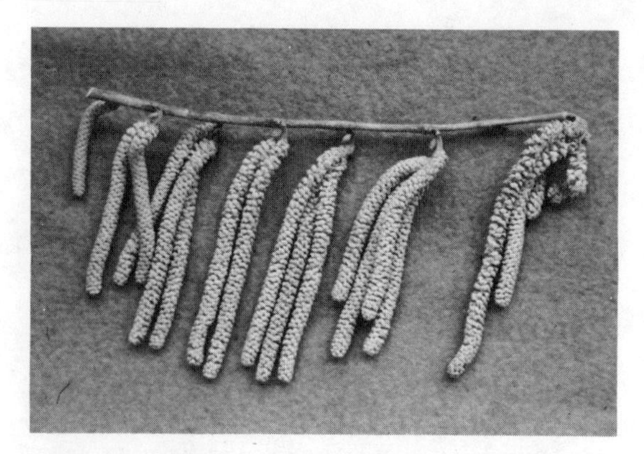

FIG. 3. Varying stages of catkin elongation on a single stem before pollen dehiscence.

bracteoles differentiate and fuse for two-thirds of their length with the central bract, the whole of which functions as a nutritional reserve for the androecium and the pollen.

Four meristematic mammillae within each bract gradually differentiate into four bilocular stamens. The filament of each stamen forks to form a Y, the tip of each fork bearing a unilocular anther. Thus there are eight anthers borne on four filaments within each catkin bract (Trotter, 1947).

When the catkins first appear they are short and stiff, often growing upward. As they elongate and increase in diameter, they become pendulous. When anthesis approaches, the catkins elongate again. As they begin to stretch, the imbricated surface is broken up (Fig. 3). The catkins lose their rigidity and flex readily in the wind, facilitating the release of pollen.

Pollen dehiscence usually occurs with a drop in relative humidity. It is accomplished by successive contractions of the anther walls and a longitudinal separation along a suture between the two lobes of the anther. Pollen that is dehised in still air will drop on the top of the bracts below and be held there until blown off by the wind (Trotter, 1947).

Filbert pollen, which is about 25–30 μ in size, is triangular or ellipsoidal in shape, swelling to spheroidal or slightly ellipsoidal when it takes up water. It contains three germinative pores, each situated over a slight prominence of the pollen wall, or in some cases, over a small vacuole. The nutritive reserves of these pollen grains have been analyzed as: 5% starch, 15% sucrose, 30% protein, 0.2% nucleic bases, 5% water, and other substances such as fats, pigments, various elements, and cellulose (Trotter, 1947).

The inconspicuous "bloom" of the filbert is associated with the appearance of numerous threadlike red styles protruding from a compound bud. Each pair of styles constitutes a pistillate filbert flower. From four to 18 of these flowers can be present in a single flower bud, depending on species and cultivar (Fig. 4). In 'Barcelona' the average number of flowers per cluster is eight (Thompson, 1967), yet the average number of nuts set per cluster is about 2.1 (Lagerstedt, 1972). Cultivars that have a higher average number of flowers per cluster and a greater capacity to set also have a higher yield potential than 'Barcelona'.

The time of flower initiation has been difficult to establish. According to Trotter (1948), who studied the cultivar 'San Giovanni' in Italy, the pistillate flower begins to develop within a bud in a leaf axil during the end of June or beginning of July. The flower arises as a simple meristematic mass or mammilla, which becomes trilobal. During development, the middle lobe disappears. The side

FIG. 4. Pistillate flower cluster on a filbert stem.

lobes grow toward the exterior, forming two rudimentary carpels differentiated at the top into stigmatic style-like appendages. At the bottom they are surrounded by a minute perigynous ring. Each flower is joined to the axil of a small bract connected to the receptacle. Several such flowers make up the compound flower bud or flower cluster (Trotter, 1948).

In Hagerup's (1942) review of floral biology, he mentions that Baillon in 1875 found the first delicate primordium of the pistillate flower on June 15 in France. A month later the styles were forming. By August, the pistillate flowers had terminated their development for the season. At this time, the flower buds are located in the leaf axils and are indistinguishable from vegetative buds. Like the staminate catkin, the pistillate flowers apparently enter a period of rest, resuming growth during the winter before anthesis in January and February. The earliest visible stage of anthesis is called the "red dot stage." This occurs as the styles emerge from the end of the bud, but before they have had a chance to elongate or reflex. Arikan (1963) states that in Turkey, emerging flowers are visible in October. The earliest this stage has been observed on 'Barcelona' in Oregon is October 30. Functional pistillate flowers have also been observed as late as mid-March. While there is a peak of bloom lasting two to three weeks, in areas with mild climates some pistillate flowers may be observed during four to five months of the year. These flowers have been shown to be receptive at all stages of development, and nuts have been set from December to March (Thompson, 1965).

FIG. 5. Pistillate flower cluster with bud scales removed. Each pair of styles constitutes a flower.

The approximate time of flower initiation has not yet been studied in the United States. The above reports, from Italy and France, are the only ones we know of that have described flower initiation. Most other work on pistillate flower development begins at anthesis and carries on through megasporogenesis.

At the time of anthesis, the gynoecium consists only of a pair of long styles (Fig. 5), closely adhering and joined at their base by a small mass of tissue, which will ultimately develop into the ovary (Hagerup, 1942; Trotter, 1948). During anthesis and the time of pollination, there is apparently very little further development of the rudimentary ovarian tissue. However, the stigmatic styles do elongate from about 2 mm, gradually reflexing as they grow. Their terminal epidermal cells differentiate into club-like papillae, which help to retain the pollen grains (Trotter, 1948). This epidermal differentiation progresses down the styles as they grow and become exposed above the bud scales by which they are surrounded. This differentiation also appears to correspond to the stigmas' receptivity (Thompson, personal communication).

Thus the filbert flower is poorly developed. It lacks the showy petals, colors, nectar, and fragrance common to insect-pollinated flowers. Yet for all its deficiencies and small size, it is well designed to intercept wind-blown pollen.

Ultimately the stigmas wither. This may be after pollination, which stops their growth, or be caused by frost, abrasion, or desiccating winds. When the styles wither due to injury, they continue their growth to expose more receptive tissue, adequate to achieve satisfactory pollination. Pollen-tube growth may reach the base of the style in as little as two days at room temperature (Paglietta, 1970), but more slowly under natural conditions. There the generative nuclei will remain quiescent, encased in callose, until the time preceding fertilization.

Growth of the Ovary and Ovule

After pollination (January and February for the 'Barcelona' in the United States), the gynoecium begins its growth very slowly. According to Thompson (personal communication), size increases are first detected during February or March with little visible differentiation. During March (May in Denmark, according to Hagerup, 1942), the developing ovules appear on their placenta at the base of a fissure formed by the two styles. This ovule and placenta differentiates from an intercalary meristem, which extends across the apex of the floral axis at about the point where the two styles are joined.

As intercalary development continues, a rudimentary perianth grows outside the ovary wall and the placenta grows inside the ovary wall. The ovary contains two parietal placentae, each of which usually bears two ovules. Generally, only a single ovule of the four develops into a kernel (Hagerup, 1942).

As growth of the 'Barcelona' gynoecium continues slowly but steadily through April and May, the functional ovaries reach an average size of 2.5 mm, about a ten-fold increase in size since January. Many nonfunctional ovaries (75–80%) cease their development and never grow beyond the 1–2 mm stage (Thompson, 1967). Ovary growth accelerates extremely rapidly during June. A measured 90% increase in diameter results during this month. The ovary increases 5,000% in volume during the five-week period beginning in late May. This is one of the steepest growth curves known for any tree fruit or nut.

During May and June, the ovules change their position within the ovary from basal to apical, being lifted upward on the central vascular stalk and on cells proliferating from the inner ovary wall. The ovules also change their position relative to the horizontal. Originally they were orthotropous (having the chalaza, hilum, and micropyle in the same axial line) and directed obliquely downward, but ultimately they become anatropous, with the micropyle directed upward. Toward the end of this period, the embryo sacs mature, forming functional organs ready to be fertilized some four to five months after pollination.

Thompson (personal communication) found that within the 'Barcelona' ovule there was no synchronization of either meiotic or mitotic divisions leading up to the development of a functional embryo sac. The usual degeneration of three daughter cells following meiosis, which occurs in most seed plants, did not occur in filberts. Instead, they were seen either to degenerate, to enlarge slightly, or to enlarge greatly to become typical macrospores, which in turn could form embryo sacs having two, four, or eight nuclei. In functional, eight-nucleate embryo sacs, three inconspicuous antipodal nuclei migrated to the chalazal end, two larger polar nuclei joined to form the fusion nucleus, and three medium-sized nuclei became embedded in cytoplasm at the micropylar end. Before fertilization, the antipodals migrated down one side of the embryo sac and ultimately disintegrated.

These internal changes take place by the time the 'Barcelona' nut is 8–10 mm in diameter. Less than half of the rapid growth of the ovary has been completed, and it has the shape of the mature nut. This enlarging nut is soft, green in color, and enclosed in its involucre or husk. After reaching full size in early July, the ovary wall begins to lignify to form the hard shell, first at the apex, and gradually progressing to the base of the nut. Color changes are not apparent until late in August, and these also begin at the nut's apex. At this same time, the nut begins to separate from the involucre due to abscission of soft cells located at the nut's base. Once this separation is complete, the nut could drop, but with the 'Barcelona', it is usually held by the enveloping involucre, which is still green. As the involucre matures and dries, it gradually opens to release the nut. Nutfall in Oregon usually proceeds over a six-week period from September 1 to October 15.

It is unusual that the initiation of flowers in June of one year, resulting in styles and rudimentary perianth that persist at the nut's apex, is so far separated in time from ovary development which occurs at its base in the following year. Thus within a single filbert nut, its tip is more than a year older than its base.

Fertilization

At fertilization in about mid-June, two ovules about 1 mm in length are embedded in a pithy parenchymous endocarp, large white spongy cells that fill the interior of the enlarging ovary. But what of the pollen grain that had germinated on the stigma five months previously?

The tip of the pollen tube had enlarged to form

several protuberances or swollen areas; in some cases, it even branched. Romisondo (1965) considered the pollen grains to be encased in callose, which he hypothesized had both a protective and nutritive value to the generative nuclei during their rest. Thompson (personal communication) observed these irregularly shaped tube tips in the base of the style as late as May 17. The tube tip was located at the base of the style or in the adjacent ovary wall, where it remained while the ovary differentiated. These style bases were now located at the apex of the developing ovary. Once the embryo sacs were mature, the pollen tubes resumed their growth. A secondary tube left the apex of the ovary, passed through the ovary wall to the funiculus, then through the funiculus along the integument, curving toward the chalazal end of the nucellus (Trotter, 1948). Pollen tubes were never seen to enter the nucellus in sectioned material, but in squashes a single tube was observed to penetrate the micropylar end of the embryo sac (Thompson, personal communication).

Actual fertilization has not been observed, though Thompson (personal communication) has seen the typical vermiform-shaped sperm before union with the fusion, or endosperm, nucleus. She found that the first indication of fertilization was an increase of acellular endosperm nuclei which, within a few days, formed a protoplasmic strand of several hundred nuclei within the embryo sac. Concurrent to endosperm multiplication is embryo-sac enlargement, which crushes the nucellus and any partly developed embryo sacs in the ovule. Cell division and growth of the zygote appears to be delayed until 400 to 500 endosperm nuclei have developed. Both Romisondo (1965) and Trotter (1948) have noted this delay.

The time lapse between initiation of endosperm and zygotic divisions is estimated to vary from four to seven days. Following double fertilization, the egg apparatus can be seen to differentiate: The egg cell enlarges and forms a more conspicuous cell wall, and the synergids gradually disintegrate. These events have not occurred before endosperm fertilization. When fertilization of the egg does not occur, there is no development of the egg apparatus. This would indicate that fertilization is essential to such development, and that in all likelihood it probably occurs near the time of fertilization of the endosperm (Thompson, personal communication).

Embryo Development

Similar to ovary growth, embryo growth is rapid, once it gets started. Of the two obvious ovules located at the apex of a central vascular strand, usually one aborts while its mate enlarges. This rapidly growing seed gradually fills the ovarian cavity by displacing the spongy parenchyma cells in which it is embedded. These cells, which are the interior of the ovary wall, are gradually crushed against the shell, or outer ovary wall. Upon seed maturity, these cells appear as a brown mat of fibers. Ideally these fibers stay with the shell when the nut is cracked, but in some cultivars they cling to the kernel and become an objectionable characteristic. The enlarging embryo also displaces the central vascular strand so that it, too, is pushed against the ovary wall. This strand resists crushing, and its imprint can be seen on the mature kernel as a groove or furrow.

On occasion, both ovules will be fertilized, and twin seeds will result. We have never seen more than two kernels in a filbert. Hagerup (1942) indicates that three or four kernels, although very rare, do occur. Unfortunately, it is much more common to have nuts without kernels, or blanks. McKay (1966) found partial female sterility (up to 88% blanks) in selections from a *C. americana* x *C. avellana* cross. Male sterility in this progeny was shown to be caused by lack of chromosome pairing (asynapses) during meiosis, and the same was thought to occur during megasporogenesis. The high degree of sterility was believed to be associated with a lack of chromosome homology between the two parent species.

Thompson (personal communication) observed that, in normally developing embryos, the acellular nuclei of the liquid endosperm gradually become cellular, first near the embryo and last at the chalazal end. The endosperm cells become large and vacuolate. They are then absorbed by the cotyledons within a few weeks of fertilization. At that time the embryo has attained about half its total length, it has a well-defined shoot apex, it has the first two leaf primordia differentiated, and it has an obvious radicle. Embryo growth is complete five to six weeks after fertilization, but differentiation continues about two weeks more. Then the shoot has four to eight leaf primordia, and the root cap is distinct from the root meristem. Also during the latter part of this time, food reserves in the cotyledons change from carbohydrate to oil. Nuts harvested from the tree at this stage of development are capable of germination, as the embryo has not entered rest (Lagerstedt, 1968).

Filbert Pollination

For commercial nut production the filbert must be cross-pollinated. This condition imposes certain restrictions on the grower and breeder. For example, the breeder cannot grow F_2 populations and increase homozygosity of certain genetic characteristics.

Flowering dichogamy can further complicate filbert pollination, particularly in areas with cold winters, where flowering occurs over a relatively short period of time in the spring. In the major production areas where there are mild winters, dichogamy is probably less important because the blossom period is extended and pollen is available over a long period of time.

The first filbert plantings in the Northwest were made by A. A. Quarnberg of Vancouver, Washington, in 1894. He observed a single high-yielding 'Barcelona' tree among several others that performed poorly and concluded that its productivity was due to pollen from a nearby 'Du Chilly' tree (Bush, 1941). Quarnberg was among the first to raise questions about the need for cross-pollination, because until then it had been tacitly assumed that the filbert was self-fruitful. In England, Bunyard (1920) wrote of filberts, "No work has yet been done to test the self-sterility or otherwise of nuts, but judging from large orchards, one would assume that the commonly grown varieties are quite self-fertile."

It was Schuster's classic work (1924) that verified the 'Daviana' cultivar as a superior pollen parent for the 'Barcelona'. 'Daviana' has retained that position for 50 years. Schuster selfed 20 different cultivars and found them all essentially self-sterile. In crosses, he concentrated his efforts on 'Barcelona' during three consecutive years, using a total of 30 different pollen parents. He concluded that the numbers of nuts set in a cluster was a characteristic of the seed parent, but that the number of clusters set depended on the pollen parent.

Considering the efficiency with which wind-pollination may occur, the prevention of outcrossing becomes a serious problem for the filbert breeder. Isolation by distance is of doubtful value. The use of muslin bags to cover the pistillate blooms is also questionable because of the extremely small size of filbert pollen. Johansson (1934) used cloth bags impregnated with paraffin.

FIG. 6. Whitewashed plastic cages over filbert trees to be used in making crosses. Catkins are visible on tree in foreground.

Clear plastic bags have been used, but they tend to tear and to heat up too rapidly inside. Finely woven nylon bags have been used satisfactorily, but even these present problems with rubbing blossoms and whipping in the wind. The most satisfactory method has been to construct plastic tents on a wooden framework that completely covers the entire tree (Fig. 6). Clear plastic must be whitewashed, or white plastic used, to prevent heating during the long time the tents must remain closed, December to April in Oregon.

Emasculation is usually no problem except that when a whole tree is involved all the catkins may be difficult to find if they are picked only one time. Tents or bags should be checked a second time just before catkin growth begins, because even part of a catkin will produce pollen. While selfing may not be a problem in filberts, complete emasculation will eliminate the possibility.

An access door in the framework of the tent facilitates emasculation and crossing. Where a single cross is to be made, pollen can be blown into the tent on a jet of compressed air by opening the plastic slightly where it overlaps. If several crosses are to be made within a tent, branches must be bagged individually. Pollination can be achieved by injecting pollen or by opening a sealed bag of catkins after they are placed inside the individual limb bag.

When individual limbs are bagged without the use of tents, the bags are removed and pollen applied with a brush. Most breeders agree that bag removal should be done on a calm, windless, or rainy day. Even under these conditions, there is a chance for cross-contamination. Supply houses serving foresters list a white nonwoven bag made of terylene and equipped with a sealed-in polyvinyl chloride window. These bags reflect heat, are permeable to air and moisture, and are rot resistant. Using needle and syringe, pollen can be "injected" or blown onto receptive stigmas as observed through the window. The needle hole made in the bag must be sealed after pollination. These bags are stiff and subject to blowing off and to rubbing flowers.

Pollen Production and Freezing

Pollen production varies greatly among cultivars with some being consistently poor producers; their grains are few in number and frequently small or lens shaped, signifying that they are empty. Tombesi (1966) found a significant positive correlation between pollen grain size and viability, the smaller ones being nonviable. Trotter (1947) noted that in comparison to the wild *C. avellana*, cultivars had a greater percentage of defective pollen grains that were incapable of germination. In 1966 Tombesi reported a germination range of from 0.5 to 62% for the 20 cultivars he studied. The high degree of pollen sterility is apparently caused by an unequal distribution of chromosomes during microsporogenesis (Woodworth, 1929; Kasapligil, 1968). This type of aberration results in pollen grains with deficient or supernumerary chromosomes which influence both their size and viability.

Pollen production can also vary from year to year in a single cultivar, depending on the previous year's nut production, stem growth, and cultural practices. Schuster (1924) classified several cultivars as to pollen production and observed that those that were classed as heavy pollen producers were light bearers of nuts and vice versa. We have observed a seedling tree that annually drops its catkin crop late in summer, a characteristic that would be considered a great saving of the tree's reserves. This tree also happens to be the largest one in the orchard and a consistent yielder of heavy nut crops.

Slate (1947, 1961) reports that filbert production in New York is seriously limited, partly because of winter catkin injury. He evaluated catkin hardiness of 31 cultivars in 1928 and made crosses between *C. americana* and *C. avellana* to combine the catkin hardiness of the former with the desirable nut characteristics of the latter. Some success was achieved. Interspecific crosses between 'Cosford', 'Bolwiller', 'Italian Red', 'Barcelona', and the *C. americana* cultivar 'Rush' produced selections with moderately hardy catkins. Crosses made only between *C. avellana* cultivars nearly always produced selections with catkins highly susceptible to freeze injury.

Rimoldi (M.S. thesis, Oregon College of Agriculture, 1921) subjected pollen, both in and out of the anther, to heat and cold to determine their effect on viability. Pollen from ten cultivars was frozen at −18 C for 96 hours, causing an average reduction in germination of 12% as compared to unfrozen controls. Subjecting pollen to high temperatures decreased viability rapidly. Temperatures of 27, 32, and 38 C for 48 hours caused the average germination to drop from 53% for the control to 24, 6, and 0.4% respectively for the treatments.

Schuster (1924) indicates that it takes temperatures of −24 to −18 C to injure pollen. Exposure of pollen to −18 C for an extended period resulted in killing only a part of the pollen under laboratory conditions. The same temperature in the orchard during January 1924 when catkins were elongating killed nearly all of them. Freezing rains that encased catkins in ice caused no injury to catkins or pollen.

Oldén (1952) found the critical temperature to be between −18 and −20 C during the rest period, with severe catkin injury occurring at −23 C. For a given low temperature, injury increased with time of treatment and with time of year (January treatment as compared to March). Native *C. avellana* types (Sweden) were hardier than the cultivars tested. Catkins of *C. avellana*, frozen at −31 C, yielded pollen grains that retained their viability but at −36 C, pollen grains also succumbed.

On the American continent *C. cornuta* has been hardy at −50 C in the province of Alberta (Gellatly, 1941).

Pollen Handling, Storage, and Germination

A knowledge of the methods of collecting, extracting, testing, and storing filbert pollen is essential to the serious breeder. To test for viability, pollen is most easily and quickly germinated in a sucrose solution by the hanging-drop technique. There is some disagreement as to the best conditions for germination, but by approximating the environmental conditions normal in the orchard, satisfactory results can usually be obtained.

Filbert pollen is easily collected by bringing branches with elongated catkins into a cool (10–15 C), draft-free, dry room. The branches are placed in water and the container placed on a large sheet of paper. Anthers will usually dehisce overnight, covering the paper with pollen. Elongated catkins can also be cut from the branches and laid on sheets of paper to await dehiscing. If the pollen is to be used in crossing, only one cultivar should be in the room at one time. Johansson (1934) placed small branches with catkins in paper bags, which were tightly closed to prevent cross-contamination. Pollen is stored in vials.

Zielinski (1968) was able to maintain pollen in a viable condition for 360 days by storing it at −18 C in 74 and 92% relative humidity (rh). Pollen kept at 20 or 11 C or at less than 74% rh lost most of its viability after 30 days. Storage at 0 C maintained viability longer, but not up to 360 days. Pollen of *C. americana* has been kept viable for eight months when stored at 0 C in 30 and 40% rh (Cox 1943). When stored at either 4 or 10 C, all viability was lost after eight months. In general, the higher the temperature or the lower the relative humidity after pollen collection, the more rapid the decline in viability.

A great deal of work has been done to determine the best culture solution or medium for filbert pollen. Schuster (1924) germinated pollen in solutions ranging from tap water to 40% cane sugar. He concluded that 12 to 15% solutions gave the best results using Van Tiegham hanging drops, though results varied at different times of the season. Pollen grains tend to burst when sugar concentrations are either too low or too high. Johansson (1934) standardized his germination tests using a 20% sucrose solution, the hanging-drop technique, and a 12-hour time lapse before evaluating germination.

The temperature for best pollen germination was reported by Zielinski (1968) to be from 15 to 20 C at 40 to 46% relative humidity. Pollen failed to germinate at 23 C. Both agar plates and the Van Tiegham hanging-drop method with 20% sucrose were used for pollen germination. Branches with elongating catkins, kept at an air temperature of 23 C for varying lengths of time, showed reduced pollen germination after four hours.

Johansson (1934) cultured pollen of several cultivars at temperatures of 1, 7, 10, and 17 C. There was little or no germination at the lower temperatures after 12 hours. These low temperatures did not affect pollen viability, because when cultures were brought to room temperature, germination proceeded rapidly, even in cultures that had been maintained at 1 C for 36 hours.

Cox (1943), as opposed to Trotter (1947), observed that pollen never germinated in the catkin. By adding a water extract of ground mature catkin tissue to the culture solution, he caused a decrease in pollen germination from 91 to 50%. This germination was abnormal, the pollen tubes being short, stubby, swollen in places, and with some tubes growing upward, away from the agar surface rather than on or into it.

We have germinated pollen with a portion of stylar tissue in the culture solution hoping to develop a quick test for cross-compatibility. We assumed that since selfed pollen rarely germinates on its own stigma, it may contain a water soluble inhibitor. If styles of a different cultivar permitted pollen germination, it would be possible to develop a quick test for inhibition. However, we observed that pollen germinated in both selfed and crossed cultures.

In vitro tests of pollen viability offer fairly reliable data on its performance when applied to compatible stigmas. However, *in vivo* tests are a more accurate estimate of germinability, and they are essential to determine cross-compatibility (Thompson, 1965). Romisondo (1965) stained fixed styles in aniline blue and observed them with fluorescence microscopy. Pollen grains and pollen tubes fluoresce yellow, thus showing their progress through the stylar tissue, which does not fluoresce (Fig. 7). This valuable technique has also been used by Thompson (1965, 1969) and Paglietta (1970) to show that pollen from selfed plants does not germinate on its own stigma. In controlled cross pollinations, numerous pollen tubes deep in the stylar tissue appear to indicate a high degree of compatibility between cultivars.

Post-pollination Events

After pollination, functional pollen grains caught in the papillae of the receptive stigmatic surface will germinate, forming a tube that grows down the style. Pollen grains do not generally germinate on nonreceptive styles. The rate of growth of the pollen tube depends largely on temperature. Paglietta (1970) found that two days were enough for the pollen tube to grow to the base of the style under laboratory conditions. Romisondo (1965) found two to three days enough sometimes, and ten days always enough. Thompson considers four to seven days enough under orchard conditions.

Romisondo (1965) determined that 80% of the length of the style consisted of stigmatic surface, and that pollen could germinate on any part of it. At first the tube went straight in, then curved downward, following an intercellular path through the walls of the stylar tissue. As the pollination season progressed, he observed more numerous pollinations that originated near the base of the styles. With advanced maturity, the apex of the styles became necrotic, and this withering progressed toward the base. Pollen quality was far more important than quantity. Even where mass pollinations were made, very few pollen tubes actually reached the base of the style. No tubes were observed in 48 to 72% of the styles. By contrast, Thompson (1967) found that 90% of the open-pollinated 'Barcelona' flowers she examined had pollen tubes. Of 50 flower clusters examined, all had at least one flower with pollen tubes.

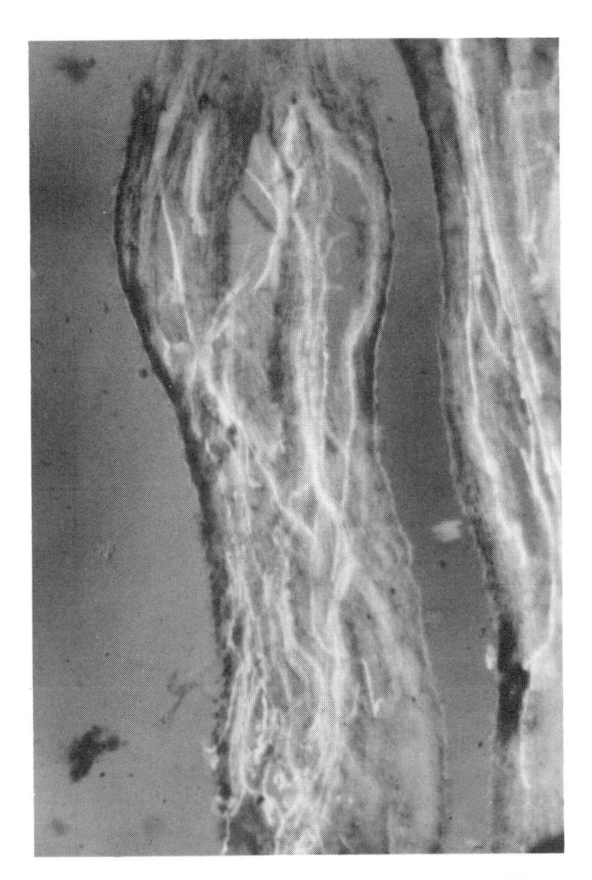

FIG. 7. Pollen tubes fluorescing in a filbert style.

Ovaries of flowers that are not pollinated attain a diameter of 0.5 mm and ultimately drop. Even within the pollinated flowers, Thompson (1967) has found a 75 to 80% drop of individual flowers after they had attained a diameter of 1 to 2 mm. This amounted to a 35 to 45% cluster drop. If at least one of the flowers in a cluster develops normally, the cluster will not drop (Arikan, 1963). The remaining flowers in the cluster will wither or be absorbed by the other enlarging tissues. By mid-May, Thompson (1967) could detect a size difference between normally developing flowers and those she called "developmental dropouts." The developmental dropouts represent a large reduction in the potential yield of the 'Barcelona' cultivar. The extent of similar nut drops in other cultivars needs to be determined, and the possible use of hormonal sprays to reduce this drop should be investigated.

Seedage

A knowledge of how to germinate filbert seed satisfactorily and grow the resulting seedlings is essential to breeders and nurserymen, yet relatively little is published in detail on this subject. Like seeds of most temperate-zone plants, filbert seeds have an internal dormancy or rest period that must be overcome before satisfactory germination can be achieved. This rest period of the filbert seed is easily overcome by cold-moist storage, for example, stratification at 4 C for three months or more. Percentage of germination improves with time as storage is lengthened (Lagerstedt, 1968). Dry cold storage or dry warm storage of seeds is unsatisfactory. Filbert seed germination is usually highly variable. We have rarely obtained more than 70% germination with treatments designed to induce germination, and we have rarely obtained 0% germination in controls.

It is common for filbert seeds to germinate over a period of several years. Seeds that do not germinate in the first year may do so in the following. This is especially true with the tree hazels, such as *C. colurna*, which has a very thick shell. While the shell is highly permeable to water, it presents a mechanical barrier that the swollen seed cannot always overcome. With an additional year in the soil, the action of microorganisms on the shell may weaken it enough to allow the seed to break its outer covering. When seeds are treated with growth regulators to accelerate germination, the shell should be removed first. For stratifying, the shell is left on; otherwise, the seeds will mold during the long storage period.

The rest period of filbert seeds can be circumvented by germinating the seeds before they enter rest (Lagerstedt, 1968). By hand-harvesting nuts from the tree in August, after embryo maturity, we have obtained 23% germination after only five weeks. This increased to 48% after 29 weeks. Seeds harvested late in September had a total germination of only 17%, indicating that more had entered rest at this time. The shell was removed from all of these hand-harvested seeds, but they received no other special treatment.

While it is possible to bypass the rest period with this technique, it is not a recommended procedure. About 60% of the resulting seedlings produced growth distortions resembling galls on their stems. Where these galls were in a terminal position, growth of the seedling was stopped, or death resulted. In some cases one or more new growing points developed from these galls. Once normal seedlings reached a height of 15 to 25 cm, they required a period of chilling before additional growth would occur.

The most practical way of accelerating filbert seed germination is with the use of gibberellic acid (GA_3). Soaking seed in GA_3 solutions of 50 to 100 ppm for six to 18 hours will significantly increase germination of nonstratified seeds, regardless of the length of time they have been stored. The resulting seedlings will also be two to three times as tall as control seedlings during the first growing season (Lagerstedt, 1968).

This technique has been used to hasten development of progeny from both breeding and rootstock studies. Nuts harvested in October were immediately cracked out, and the seeds were treated with GA_3 and placed in the greenhouse for germination. Most of the seeds germinated in three to nine weeks, with a few stragglers and some that completely failed to germinate. Seedlings could be 30 to 45 cm tall by January with proper care. They will continue growth in the greenhouse, especially if provided with supplemental light, fertilizer, and transplanting to larger containers. Seedlings that are permitted to grow until time for field planting (April or May) will usually cease growth upon transplanting. Such late transplanting has proved costly, in that expensive greenhouse facilities are used to provide a season's growth, but then no additional growth is obtained in the nursery. The nursery-planted trees also tend to defoliate during the summer, and some die. As a result, the greenhouse-grown seedlings should be hardened-off in January, chilled, and then transplanted to the nursery as early in the season as possible. Only in this way can the extra time and labor of greenhouse care be justified.

Several different soil media, pot sizes, and pot shapes have been tested. Our standard medium is now equal parts of peat moss, perlite, and vermiculite. We have found that the vermiculite is especially important, probably due to minor-element balance. It has been our experience that filbert trees grow best in light media mixtures, that is, those that are well-aerated and do not compact readily. Soluble, complete, high-analysis fertilizers are used periodically in the irrigation water to maintain correct nutrient balance and sustain active growth.

Shallow flats are unsatisfactory for filbert seed germination and subsequent growth. Deep flats should be used, and a light medium is best to prick out the tender seedling. We have germinated individual seeds in pots 25 cm tall with a 6-cm diameter at the top. These tall, narrow pots were also suitable for growing the seedlings for several months, which eliminated some transplanting.

The least expensive method of filbert seed germination, and the one most practical for nurserymen, is seedbed culture. We have prepared raised beds by adding 50% sand to the existing soil and tilling it in along with the fertilizers. Other types of media can be added and the whole fumigated to eliminate pathogens and reduce weed seeds. Stratified seed can then be planted about 8 cm apart and grown with intensive irrigation, fertilizers, and care. The resulting seedlings are tall and straight as compared to those grown under less competitive conditions.

Breeding Systems

Interspecific Crosses

A wealth of characteristics are available in the various species, awaiting the patience and foresight of the nut-tree breeder willing to combine them. Filbert research centers have been so involved in solving cultural problems that much of the exciting work has been left to enthusiastic amateurs whose facilities are limited, and who, in some cases, possess superficial knowledge of genetics and breeding techniques.

Corylus species are scattered throughout the Northern Hemisphere. The collection of their many forms, getting them established, growing them to maturity, and determining their authenticity is a large undertaking. The confusion in *Corylus* terminology adds to this task. To date, most workers interested in filbert improvement have concentrated their efforts among the cultivars of *C. avellana* and *C. maxima*. Some notable exceptions to this are listed below.

The cultivars, 'Rush' and 'Winkler', selected from the native *C. americana* population, have been crossed with pollen of at least 15 different *C. avellana* cultivars (Slate, 1947; Weschcke, 1954; Reed and Davidson, 1954). These crosses have resulted in the hybrids 'Carlola', 'Dolores', 'Magdalene', 'Bixby', 'Buchanan', 'Reed', 'Potomac', 'Graham', and 'Morningside'. The reciprocal of this cross, *C. avellana* x *C. americana*, has been tried numerous times, but has failed in every instance (Reed, 1936; Weschcke, 1954).

The beaked filbert, *C. cornuta*, has been given less attention in breeding programs, as its nuts have been considered inferior to those of *C. americana*. Schuster (1924, 1927) used pollen of the western variety, *californica*, on the cultivars 'Barcelona' and 'Du Chilly', with complete lack of success. His objective was to use the native western form as a source of pollen in commercial orchards of the European filbert.

Reed (1936) used pollen of *C. colurna*, *C. heterophylla*, and *C. maxima* on the *C. americana* cultivar 'Littlepage', apparently with good success. He described the hybrids as being more hardy than the pollen parents, less inclined to sucker, more vigorous, and more upright in habit. A small number of nuts were obtained from crosses with *C. colurna*. A few of these seedlings expressed *colurna* characteristics, though some had the color and bark character of the pollen parent. The limited number of nuts obtained from this progeny were mostly without kernels.

In 1936, Gellatly (1966) started crossing *C. avellana* x *C. colurna*. He has named several selections (coining the word "trazelnuts") resulting from this cross. He mentions that the hybrids segregate for both the tree and bush type, with many intermediate characteristics expressed for such things as hardiness, nut size, shell thickness, husk type, and bark type. He also claims that many hybrids resulted from the reciprocal cross, namely *C. colurna* x *C. avellana*. The European cultivars used as pistillate parents were 'Craig', 'Holder', and 'Brag'.

In addition to the above, Gellatly mentions a selection named 'Chinoka' as resulting from a cross with *C. chinensis*, but he does not specify the pollen parent. He also developed a progeny line that he called "filazels." This line resulted from crosses between *C. cornuta* x *C. avellana*, achieved by removing all the catkins from *C. cornuta*. His objective in this cross was to incorporate the extreme winter hardiness (−50 C) of the native hazel from the Peace River district of Alberta into a group of trees that could be used as windbreaks on the Canadian prairies. Many of Gellatly's promising selections and named cultivars now serve as a nucleus for a rootstock breeding program at the USDA Research Center, Corvallis, Oregon. The results of this program will be, at least in part, a living legacy to Gellatly's aspirations and industry. Some of his named cultivars will, in time, be backcrossed to both Turkish and European selections and perhaps yield a clue to their hybrid origin which is now in doubt since all the crosses were not carefully controlled.

To the best of our knowledge, the *C. avellana* x *C. colurna* cross has never been made under controlled conditions. Since the odds are obviously heavily against its success, large numbers of crosses and cultivar combinations must be tried, a goal difficult to achieve under controlled conditions. Kasapligil (1963, 1964) does not believe that the cross is possible.

Schuster (1927) used *C. colurna* pollen on the *C. avellana* cultivars, 'Barcelona', 'Du Chilly', 'Daviana', and 'White Aveline', and made the reciprocal crosses as well. All such crosses were failures. He and Kasapligil (1963, 1964) both mention locating individual Turkish trees, sometimes surrounded by *C. avellana* trees, that consistently bear annual crops of empty shells. We have located such trees also and cracked several thousands of these empty shells. However, we have found that not all are empty. In fact, most of our seed sources contain an average of 0.7% kernels. Only one year's lot of seed has been grown out, and they have the appearance of pure Turkish seedlings.

Johansson (1927) reported using pollen of *C. colurna* on the *C. avellana* cultivars 'Apolda', 'Cosford', and 'Multiflorum' with negative results. He thought the poor results could be partly caused by defective pollen, because *in vitro* germination was consistently low.

The hybrids *C. colurnoides* and *C. Jacquemontii* are two old cultivars, listed by Rehder (1940) and other taxonomists as having originated from *C. avellana* x *C. colurna*. Rehder also lists *C. vilmorinii* as a hybrid of *C. avellana* x *C. chinensis*. There is no record on how these cultivars originated, but presumably they were selected from open-pollinated material and named on the basis of their intermediate characteristics. Kasapligil (1963) and Zielinski and Thompson (1967) have observed specimens of *C. colurnoides* at the École Dubneuil in Paris. They claim that it appears to be identical to *C. avellana*, which is in itself a highly variable species.

Fuller (1896) mentions that several hundred seedlings have been raised from crosses between "the two common European species, *C. avellana* and *C. colurna*, and their hybrids." Unfortunately, Fuller was confusing *C. colurna* with *C. maxima*, the Giant filbert. Other authors (Trotter, 1947; Schuster, 1927) mention that *avellana* and *maxima* have been so extensively crossed that it is no longer a simple matter to determine the primary botanical characteristics of either. It is simply assumed that most of the named extant cultivars have resulted from a mixture of these two species.

Farris (1970) is a serious amateur filbert breeder who has collected numerous species and varieties and is investigating their cross-compatibility. He can be credited with making the following crosses:

- *C. americana* x *C. heterophylla sutchuenensis*
- *C. americana* x 'Indoka' (*C. Jacquemontii* x *C. avellana*)
- *C. americana* x 'Royal', 'Holder', 'Luisen' (*C. avellana*)
- *C. heterophylla sutchuenensis* x 'Holder', 'Bellhusk' (*C. avellana*)
- NY 1401 ('Rush' x 'Cosford') x 'Royal' (*C. avellana*)
- 'Manoka' (*C. cornuta* x *C. avellana*) x 'Royal' (*C. avellana*)
- 'Laroka' (*C. colurna* x *C. avellana*) x *C. colurna*
- 'Morrisoka' (*C. colurna* x *C. avellana*) x *C. colurna*

In addition to the above successful crosses, Farris has also supplied information on interspecific crosses that he has been unable to effect:

- *C. avellana* (27 selections) x *C. colurna*
- *C. americana* x *C. colurna*
- *C. heterophylla sutchuenensis* x *C. colurna*
- *C. heterophylla sutchuenensis* x 'Erioka' (*C. colurna* x *C. avellana*)
- *C. americana* x 'Erioka' (*C. colurna* x *C. avellana*)
- *C. avellana* 'Bellhusk' x 'Buchanan' ('Rush' x 'Italian Red')
- *C. avellana* 'Bellhusk' x 'Morrisoka', 'Erioka' (*C. colurna* x *C. avellana*)
- *C. avellana* 'Bellhusk' x *C. chinensis* (hybrid?)

To the above list of failures can be added previously mentioned *C. avellana* x *C. americana* (Reed, 1936; Weschcke, 1954) and *C. avellana* x *C. cornuta* (Schuster, 1924).

Polyploidy

Woodworth (1929) examined ten species, four cultivars, and three hybrids of *Corylus* and concluded: "There is at hand no evidence which would indicate polyploidy in this genus." This is interesting, since he found polyploidy common in the family *Betulaceae*. He listed somatic chromosome numbers of 14, 28, 35, and 42 for *Betulus*, 14 and 28 for *Alnus*, and 8 and 32 for *Carpinus*. In addition to *Corylus*, no polyploidy was found in its closely related genera *Ostrya* and *Ostryopsis*.

Danielsson (1946) and Danielsson-Santesson (1951) described the first and only *Corylus* polyploids arising from diploid *C. avellana* seedlings that were treated with 0.5 to 2.0% solutions of colchicine. Growing points of seedlings were repeatedly painted or submerged in colchicine solutions of varying concentration. Best results were obtained by germinating stratified seeds in a 2% colchicine solution. While many seedlings died or were retarded in growth, most survivors exhibited symptoms of polyploidy. Many of the resulting plants were mixoploids, but several complete tetraploids also occurred. In addition to a somatic chromosome number of 44, the tetraploids had thicker, rounder leaves, with pronounced serrations and enlarged stomata. Like their diploid counterparts, the tetraploid plants showed a great deal of variability in plant characters and vigor.

The only occurrence of a triploid *Corylus* is mentioned by Danielsson (1946). This unique individual tree was found in a population of colchicine-treated seedlings. It was selected for cytologic examination because certain phenotypic characteristics were similar to those of the tetraploids. However, it had a somatic chromosome number of 33 and was therefore believed to have resulted from the combination of a reduced with a nonreduced gamete. Presence of trivalents during first metaphase suggests this selection as an auto- rather than an allo-triploid. There were generally five trivalents per cell, with bivalents and univalents also present. Meiosis was found to be highly irregular. During first anaphase, the univalents often lagged and might not be included in the resulting tetrads. These univalents remained in the cytoplasm and occasionally formed micronuclei. Tetrads of the triploid plant varied, and instead of forming four nearly equal-sized cells, from four to nine smaller pollen grains developed. There was a large variation in pollen-grain size, yet an estimated 50% of the triploid pollen appeared morphologically sound. This polyploid work, done in Sweden, has not been continued.

Breeding for Specific Characters

Plant Characters

Cold hardiness is a characteristic sought in most fruit and nut trees, yet it is rarely required in regions where the filbert is grown commercially. Filberts are grown primarily in regions with climates that are modified by large bodies of water. These regions normally have mild winters, those which are not conducive to acclimating plants to withstand severe freezes. In the Willamette Valley of Oregon damaging freezes have occurred in 1919, 1937, 1950, 1955, and 1972. Each of these freezes differed in timing, intensity, and duration. Each of the freezes injured different plant parts depending on time of the freeze and the stage of development of the plant part. It is not practical to breed for hardiness in areas that lack a pattern of cold acclimation and generally lack a winter sufficiently cold to test for hardiness. However, the extent of freeze injury of seedlings was noted in the Oregon filbert breeding program following the 1972 freeze.

In New York, hardiness was one of the most important plant characteristics considered (Slate, 1947). Evaluations of catkin hardiness of seedlings from crosses between cultivars of *C. avellana* determined that winter injury of catkins was nearly always high. 'Red Lambert' had the hardiest catkins of any *C. avellana* cultivar at Geneva and, in crosses, produced a higher proportion of catkin-hardy seedlings than any other cultivar. 'Cosford' was also fairly good in this respect. When 'Bollwiller', 'Italian Red', or 'Barcelona' were crossed with 'Rush' (*C. americana*), seedlings with moderately hardy catkins were produced.

Weschcke (1954) obtained supposedly hardy seedling filberts from J. U. Gellatly in British Columbia. After seven years, all but two plants had succumbed to the harsh winters of Minnesota. The pollen from these plants was utilized in crosses with the native *C. americana* and ultimately resulted in Weschcke's first "hazilberts." Weschcke's objective was to produce a selection suitable for commercial filbert production in the northeastern United States, and he recognized hardiness as an essential character in achieving this goal.

Farris (1970) has selected seedlings for hardiness in Michigan. Gellatly (1956) utilized the Peace River hazel (*C. cornuta*) in his crosses with *C. avellana* to obtain a hardy filbert with a large nut and short growing season. His objective was to develop a plant which could withstand the Canadian winters while serving as a windbreak and a source of food.

Fecundity is one of the greatest advantages a new cultivar could contribute to the filbert industry. Abundant blossoms, high fertility, and consistently large crops are the prime objectives of every grower. The 'Barcelona' cultivar has endured because it has been considered relatively productive in comparison to other cultivars. Any new selection considered for introduction must be more productive than 'Barcelona'. Mature orchards of 'Barcelona' develop a biennial bearing habit, alternating high with low production years. A new selection should have a more consistant bearing habit. The 'Barcelona' normally sets an average of 2.1 nuts per cluster (Lagerstedt, 1972), but its potential is much greater. Other cultivars are known which consistently set an average of three or four nuts per cluster. There is also a need for a pollinizer cultivar with abundant large catkins containing a high percentage of viable pollen.

Tree vigor is another plant character which varies tremendously in all seedling filbert populations. Controlled crosses resulting in seedlings with uniform hybrid vigor have not yet been established. If such a population could be developed along with the nonsuckering characteristic, a rapidly growing rootstock would be of great importance to the nurseryman. From his point of view, rapid germination, seedling vigor, and rapid callusing for grafting success represent the characteristics of a superior rootstock.

Hybrid vigor in a self-rooted cultivar might be desirable from the standpoint of a tree reaching bearing maturity quickly and utilizing its allotted space in the orchard. However, if vigor persists for several years, it might retard maturity and decrease the bearing potential of the cultivar during its early life. Superior tree vigor may not be desirable in high-density orchards. Such trees must be pruned sooner, more frequently, and more severely to be contained. This, in turn, has a deleterious effect on nut production.

Cultivars such as 'Du Chilly' and 'Imperial de Trebizonde' appear to lack vigor and have been considered for use as dwarfing rootstocks. 'Du Chilly' was found to be too slow growing and too poorly rooted to serve in this capacity, while 'Imperial de Trebizonde' was never adequately tested.

Filbert trees vary considerably in growth habit. 'Daviana' is vase shaped and one of the most erect-growing cultivars. It has thinner and more supple branches as compared to 'Barcelona', which is a round-headed tree and has heavier branches. 'Brixnut' is termed a "viny" tree due to its drooping growth habit. In the mechanized, close-spaced orchards of Oregon, the upright character is a decided advantage. It is also much easier to prune and maintain.

Other growth habits such as branch angles, amount of lateral branching or twigginess, and leaf size influence production and orchard management. Selections with wide branch angles can support larger crop loads and are less subject to breakage during wind, ice, or snowstorms. Twiggy trees take longer to prune and their short stem growth produces fewer nuts. Leaf size has not yet been objectively evaluated, but the characteristic is evident in seedling filbert populations and may prove to be of importance to nut yield.

Nut Characters

Filberts are graded on size with a premium paid for the largest (jumbo) grade. Some of the largest filberts in the world are grown in Oregon and a portion of these are even exported to Europe. Large filberts are utilized as in-shell nuts while the smaller grades are cracked for the kernel trade. Several cultivars having large size have been developed in the Northwest. These are 'Fitzgerald', 'Royal', 'Nonpareil', and 'Pearcy'. Most of these cultivars have thick-shelled nuts and all have kernels which shrivel upon drying. While small nuts retain kernel plumpness, large nuts do not.

In addition to size, the in-shell trade demands that filberts have an attractive appearance. They should have a rich brown color and little pubescence. The stiff, short hairs, which occur predominantly toward the tip of the nut, collect dirt during mechanical harvest and cause a dull appearance, even after bleaching. In a bowl of mixed nuts, there is no place for other than richly toned filberts.

The 'Barcelona' nut is round and sets the shape standard for the trade. The 'Daviana' and 'Du Chilly' are long nuts and are screened out for use in the kernel trade. Since the latter two cultivars are pollinzers for 'Barcelona', an important breeding objective is to develop a pollinizer with a round shape; one which can be freely interplanted with 'Barcelona'. Three such selections have been located and are being increased for this purpose. Selections with round nuts are predominantly used for breeding in Oregon.

The 'Barcelona' averages about 40% kernel by weight. It is in direct competition with walnuts, almonds, and pecans, which have as high as 50, 55, and 60% kernel respectively. Thin-shelled selections of filberts are common, some with as high as 55% kernel. While none are suitable for introduction, they are being utilized in breeding. While desirable to the consumer and processor, thin-shelled nuts could be a problem to the grower. For example, the thin-shelled 'Daviana' is selectively removed from the orchard by birds and rodents, a cause of major yield losses. Thin-shelled nuts may also be more subject to breakage during mechanical harvest and to splitting during drying.

A quality factor being given great consideration in the Oregon breeding program is kernel cleanliness. As the filbert kernel enlarges, it crushes the cells of the inner ovary wall. In species such as *C. americana* these tissues, called the pellicle when they are dry, adhere to the shell and leave a clean kernel. In *C. avellana*, the pellicle usually adheres in part to the kernel. The kernel of the principal cultivar in Italy, 'Tonda Gentile della Langhe', retains too much pellicle and is unattractive and less palatable as a result. Both Slate (1947) and Weschcke (1954) have pointed out the *C. americana* x *C. avellana* progeny generally have nuts with clean kernels. Regardless of whether the filbert is to be used for the in-shell or kernel trade, a pellicle-free kernel would be a desirable character. This character is available in the *C. avellana* cultivars 'Daviana' and 'Ryan' as well as in several Turkish cultivars, and it is being used in breeding.

Some filberts are found to occur without kernels and are called "blanks." The 'Barcelona' cultivar has produced over 25% blanks in certain years, resulting in serious yield reduction. Extra labor is also required to eliminate these from the crop. Blank nuts, occurring as singles or doubles, tend to drop early and must be tilled into the soil or flailed to break them up. Late-dropping blanks, or those occurring in larger clusters, are partly removed during harvesting and field-cleaning processes. However, some always remain with the crop nuts and their presence is charged against the grower. Thus blanks are both a nuisance and expense, and reducing or eliminating the problem would save the grower money. Cultivars are known which produce very few blanks. It is also known that the use of some cultivars as pollinizers for 'Barcelona' results in a greater number of blanks as compared to use of 'Daviana' (Zielinski and Thompson, 1967). The reduction of blanks by developing pollinizers which are highly compatible with the main crop cultivar is an essential breeding objective.

Early nut maturity is especially important in Oregon because rains, which usually start in October, increase the difficulty of mechanical harvest. Harvest operations are two to five times more rapid prior to rains as compared to following them. The rains promote germination and growth of weeds which increase the harvest problem as the season progresses. Wet foliage becomes difficult for blowers to move during mechanical sweeping and windrowing of nuts. Wet foliage is also more difficult for the mechanical harvester to remove during cleaning. Nuts harvested prior to the fall rains are cleaner, require less drying, and receive less mechanical abuse than those harvested after the rains have started.

Early nut drop is considered to be a characteristic of both nut and husk in Oregon. For mechanical harvest it is imperative that commercial cultivars be free husking, that is, that the nuts fall free from the husk, not in it. Thus early husk maturation is important to early nut drop.

The normal pattern of maturation in Oregon begins with abscission of the nut from the base of the husk about the third week of August. The nut is mature and ready for harvest once it has abscissed, but at this time the husk is green and its bracts enclose the nut even though it is loose within them. Harvest is delayed until the husk matures and its bracts reflex and allow the nut to drop out. Husk maturation occurs over a six-week period following the nut's abscission from the base of the husk. Since mechanical harvest is a once-over process, growers await a minimum of 90% nut drop before starting harvest operations. This goal is usually achieved about mid-October and coincides with the fall rains.

Early cultivars and selections are known and are being used in the Oregon breeding program. The Italian cultivar 'Tonda Gentile della Langhe' matures three weeks before 'Barcelona' and is being used in crosses with 'Ryan', an early cultivar which has a thinner shell and cleaner kernel (Thompson, 1972). The long-husked Turkish 'Tombul' is four weeks earlier than 'Barcelona' and has been used in crosses with the selection 'Jemtegaard #20', which is mid-season but has a husk with very short bracts that flare back at maturity. Crosses of this type are planned to combine several desirable characteristics from each of the parents.

Filbert flavor has not received as much attention as other more obvious characteristics, but it does vary greatly among cultivars. While ranking lower in importance during early screening of seedlings, it gains importance as selections approach final evaluation. Roasting filberts enhances their flavor markedly. This characteristic has not been adequately researched among existing cultivars because the majority of the United States production has gone to the in-shell trade where the kernel is eaten raw. Use of the processed kernel is increasing and as it does, response of the filbert to roasting will be more seriously considered. Storage life of the raw or processed filbert is another quality factor which has received relatively little attention from the breeder.

Thompson (1969) summed up breeding objectives for nut characters as follows:

1) Improved kernel quality
   a. Clean kernel
   b. Lack of kernel shrinkage
   c. Superior flavor
   d. Long shelf life
2) Higher production
   a. Reduced blank percentage
   b. Increased kernel percentage
   c. Increased nut size
3) Earlier maturity than 'Barcelona'.

Insect and Disease Resistance

One of the reasons why the 'Barcelona' is the principal cultivar of commerce in Oregon is its resistance to bud mite, *Phytocoptella avellanae* Newkirk and Keifer (previously called *Phytoptus avellanae* Nal. and *Eriophyes avellanae* Nal.). The 'Daviana', used as a pollinizer for 'Barcelona', is susceptible to bud mite, which reduces its yields annually. 'Royal', a cultivar introduced in the Northwest, is also susceptible. This cultivar enjoyed a brief popularity because of its large, round, attractive nut. However, when grown in solid blocks, 'Royal' yields drop off drastically because of the rapid increase in mite infestation. In the Oregon filbert-breeding program, all seedlings are screened for susceptibility to the bud mite to prevent an error such as occurred with 'Royal'.

Ourecky and Slate (1969) have reported on the susceptibility of some of the *Corylus* species, cultivars, and selections at Geneva to the filbert bud mite. This mite was probably introduced from Europe along with *C. avellana* cultivars. It forms galls that completely distort the infested buds of susceptible cultivars (Fig. 8). Where these buds contain the pistillate flower, nut production is eliminated, resulting in serious yield reductions. To date, the only effective control is by resistant cultivars.

FIG. 8. The top bud is galled by *Phytocoptella avellanae* Newkirk and Keifer, the filbert big bud mite. The susceptible cultivar is 'Royal'.

The nature of the resistance is not known, but in the Geneva crosses, susceptibility appeared dominant to resistance. *C. americana* selections appear to lack resistance to the filbert bud mite, as do nearly all of the interspecific crosses with *C. avellana*. This factor alone may be a significant deterrent to getting a satisfactory hybrid established in the New England states. European cultivars such as 'Barcelona', 'Cosford', 'Italian Red', and 'Purple Aveline' appear resistant to the bud mite at Geneva (Ourecky and Slate, 1969).

Another serious problem of *C. avellana* cultivars grown in the eastern United States is a fungal blight caused by *Cryptosporella anomala* (Pk.) Sacc., commonly known as the eastern filbert blight. Filberts, introduced to the Atlantic Coast as early as 1858 (Fuller, 1896), grew for a while, but ultimately succumbed to this blight. It remains as a serious problem for successful growth of European filberts in the East. Morris (1931) claims that purple-leaved filberts are immune to its attack and Fuller (1896) never saw the disease on *C. cornuta*. Host for this fungus is the native filbert, *C. americana*, which is considerably more tolerant to its attacks. One of the original objectives of crosses between *C. americana* and *C. avellana* was to get the disease tolerance of the former combined with the nut size and quality of the latter. Most of the hybrids produced by Graham in New York were lost to this disease (Slate, 1969). We are unaware of any hybrids with tolerance to the eastern filbert blight. The northwest states have a quarantine against importation of filbert plants from any area east of the states of Idaho, Utah, and Arizona because of this disease.

Filberts in the West also have a problem with a blight, but it is caused by a bacteria, *Xanthomonas corylina* Dowson. After several years' observation of *C. colurna* trees inoculated with the bacteria, it would appear that this species is fairly tolerant to the disease. The pollinizer cultivar 'Daviana' is also tolerant as it survives in young plantings while being surrounded by 'Barcelona' trees which have succumbed to the blight. Partly for this reason, the 'Daviana' is temporarily being considered as a rootstock for 'Barcelona'. The western filbert blight is injurious to young trees, especially those which are already under stress due to sunburn, lack of moisture, or cultural neglect. Such trees usually die in the second year after planting and frequently represent considerable losses. Once trees are four to five years old, they are rarely killed by the bacterial blight, but smaller twigs are killed with resulting crop loss. The inheritance of resistance to this blight is unknown, but seedling populations are screened for its presence.

Cameron (1970) reported the presence of a new filbert disease in Oregon, but was unable to determine its nature. Symptoms consist of severe reduction in leaf size, tree growth, and nut production. Most severe symptoms were observed on 'Barcelona'. Leaf symptoms have been slight on 'Du Chilly' and none have been observed on interplanted 'Brixnut' or 'Daviana'. While some tolerance may be readily available toward this new malady, it is as yet of little economic importance and has not attracted the breeder's attention.

Rootstock Breeding

Filbert cultivars occur on their own root systems because they are propagated by the age-old technique of layerage. Layerage is used because other propagating techniques such as cuttage and graftage have never been consistently nor commercially possible. Improved techniques have increased grafting success (Lagerstedt, 1970, 1971), and with it, renewed interest in using rootstocks for filberts.

Dutch nurserymen are using *C. avellana* seedlings as rootstocks for the 'Corkscrew' filbert *C. a. contorta*. The Turkish tree hazel *C. colurna* makes a better rootstock for ornamental types and specimen plants in arboreta and botanical gardens because it does not sucker. The Canadian nurseryman J. U. Gellatly sold trees grafted on *C. colurna*, but in Oregon the last orchards established on this rootstock date back to 1940.

The normal habit of the European filbert is that of a large multistemmed shrub, which annually produces suckers from buds that arise at the base of the trunk. These suckers are the natural replacement for old stems as they deteriorate. To facilitate mechanization of orchard practices, growers in the United States train filbert trees to a single trunk. This training system makes it necessary to eliminate the continually appearing suckers several times each growing season. In large orchards and high-density plantings, sucker removal becomes a major cultural problem.

The Turkish Tree Hazel

Corylus colurna is the best known of the so-called tree hazels. Its synonyms are: Turkish tree hazel, Byzantine filbert, and Constantinople filbert. As the common names imply, it is native to Turkey and western Asia. Kasapligil (1963) considers varieties of it extending its range to India and China. It was first introduced to Europe by Clusius in 1582 and brought to British gardens in 1665. It is used as a timber tree in its native lands, growing to heights of 40 m. Under cultivation it has been used as an ornamental and as a rootstock for the European filbert. It has been distributed throughout the temperate zone of the Northern Hemisphere and proved to be hardy in Canada and the states of New York, Vermont, and Maine. The nut is small, thick-shelled, and difficult to remove from its fleshy husk. Though it is edible, it is not of major importance to man as a crop or food. The Turkish tree hazel apparently has some resistance to both the eastern and western forms of filbert blight (Morris, 1931). As a specimen plant it is usually a tall, pyramidal tree displaying winter catkins and an attractive scaly bark.

Its nonsuckering characteristic and potential use as a rootstock were pointed out in 1841 (Anon.). Since then many authors have commented on it, but the most experience with this species as a rootstock comes from Oregon (Bush, 1941; Gellatly, 1941, 1955, 1956, 1966; Kasapligil, 1963; Lagerstedt, 1970, 1971; Lagerstedt and Byers, 1969). *C. colurna* calluses readily to form a graft union rapidly; an important factor in establishing the scion variety. Due to differences in bark color and texture, the union between the Turkish and European filbert is readily evident (Fig. 9). Once established, it grows vigorously and becomes a productive tree. Grafted to 'Brixnut' or 'Halls Giant', which are vigorous cultivars, it forms a smooth union. Its most outstanding characteristic is the nonsuckering habit. The continual suckering of the layered European filbert necessitates a great deal of annual care and is one of the factors limiting the size of orchard a grower can manage. The nonsuckering characteristic of the Turkish tree hazel is being used in a breeding program designed to produce a filbert rootstock.

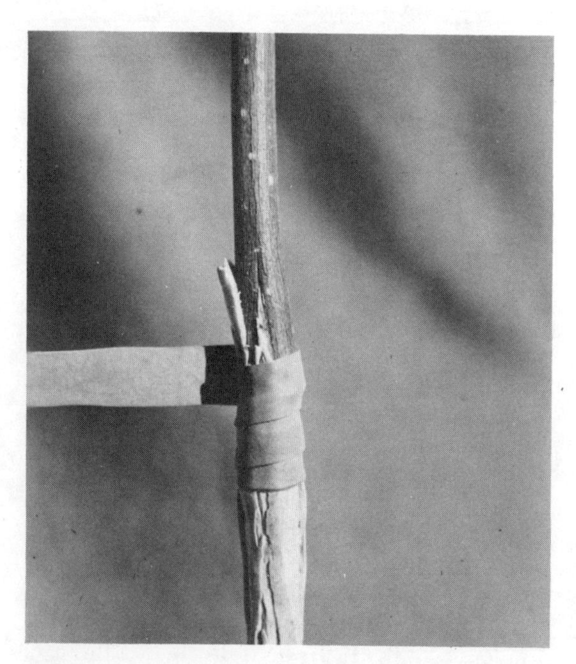

FIG. 9. *C. avellana*, whip grafted on *C. colurna* rootstock.

The first filbert rootstocks used in Oregon were seedlings of the European filbert, as it was believed they suckered less than trees that originated as layers. The natural suckering habit of the European type soon asserted itself and other stocks were sought. Seeds were collected from especially vigorous trees and faster growing cultivars, such as 'Halls Giant'. These attempts also met with failure. In the late 1930s a few nurserymen tried seedlings of *C. colurna* as a rootstock, and several present orchards were established on this stock. At that time, the limiting factor in the use of this rootstock was successful grafting techniques. While grafting was possible, consistent results and uniform stands of trees were difficult to obtain, so layerage has persisted as the method of propagation and with it the suckering problem.

In orchards where the Turkish filbert is used as a rootstock, it tends to overgrow cultivars like 'Barcelona', and productivity is said to decline after 20 to 25 years. These observations have not been consistent, so performance may vary with cultural practices and scion cultivar. As a nursery tree, *C. colurna* has several disadvantages. Its seeds may take two or more years to germinate, though treatment with giberellic acid can hasten this process (Lagerstedt, 1968). The seedling tree will often require two additional years of growth before attaining graftable size. The tree produces a strong tap root with a few stiff, lateral roots. This makes digging, handling, shipping, and transplanting difficult. This poor root system could adversely influence transplanting survival, as well as lengthen the time before the tree is established in the orchard and begins growth.

Turkish Tree Hazel Hybrids

The Turkish filbert is difficult to propagate asexually. Cuttings have been unsuccessful, and layered trees root poorly. These factors reduce its chances for being propagated as a clonal rootstock and for use as a rootstock in commercial nursery practice.

Natural hybridization between the European and Turkish filberts is rare, but Rehder (1940) has described examples: *C. colurnoides* and *C. Jacquemontii*, which arise from *C. avellana* x *C. colurna*. Kasapligil (1963, 1964) doubts the validity of these crosses, stating that the hybrid has never been obtainable through controlled crosses, and the "hybrids" have been selected on the basis of observed intermediate characters. To the best of our knowledge, this is true; yet we have personally observed such intermediate plants in a population of *C. colurna* seedlings.

Gellatly (1956, 1966) found that most of his numerous crosses between the European and Turkish filberts resulted in seedlings of intermediate characters. Many of these trees were grown to fruiting size. A range of variation was obtained for characteristics such as yield, suckering, upright habit, hardiness, nut size, shell thickness, and husk type. When the European filbert was used as the pollen parent, the resulting trees produced nuts that were thick-shelled, but larger than those of the Turkish filbert. They had tree-type growth, were of intermediate size, and exhibited the bark characteristics of the European parent.

More than 20 years ago Olgar Jemtegaard of Boring, Oregon, selected a tree with striking intermediate characters from a *C. avellana* seedling population. This tree has been given the name 'Filcorn' because the nuts it produces resemble an acorn (Fig. 10). These fertile nuts, we assume, were pollinated by the surrounding *C. avellana* trees. Thus, we believe the nuts of the 'Filcorn' represent a backcross as follows:

C. avellana x *C. colurna*
↓
'Filcorn' x *C. avellana*
↓
'Filcorn' seedlings

'Filcorn' seeds have been collected annually for several years, and seedlings are being grown and evaluated for their suckering habit. The majority of the seedlings vary greatly in this characteristic, but several have been selected for little or no suckering and are being increased as clones for further evaluation as rootstocks (Lagerstedt, 1972).

In addition to 'Filcorn' seedlings, other chance seedlings, *Corylus* species, and tree hazels are being collected and placed in layering beds to establish them on their own root systems. Once this is done, they are topworked with a common scion variety and evaluated for suckering and other growth characteristics (Lagerstedt, 1972).

The above approach will result in a nonsuckering rootstock, which must be asexually produced by layerage before it can be grafted. This rootstock would be expensive to produce as compared to a seed source that could breed true for the nonsuckering characteristic. Ultimately, a nonsuckering seedling rootstock would be the ideal solution to this problem. Such a rootstock might be obtained by backcrossing 'Filcorn' to *C. colurna*.

FIG. 10. Top center, nuts and husks of the selection 'Filcorn', believed to be the result of a *C. avellana* x *C. colurna* cross.

The Chinese Tree Hazel and its Hybrids

Corylus chinensis is another tree hazel. It was collected in China in 1888 and introduced to British gardens at the turn of the century, according to Osborn (1930). Kasapligil (1963) considers this species to be a variety of *C. colurna* and states that it has a greater tendency to produce suckers than does *C. colurna*. Gellatly (1956, 1966) mentions it as a tree type with a smoother bark than its Turkish relative. He used it successfully as a rootstock and noted that it formed smoother unions when topworked with the European filbert. However, *C. chinensis* has not been widely used as a rootstock.

Rehder (1940) lists *C. vilmorinii* as an interspecific cross between *C. avellana* x *C. chinensis*. The Vilmorin filbert is a smooth-barked tree of limited distribution and use. There are no reports on its use as a rootstock. Three trees of a similar cross have been selected by the author in Oregon for use as potential rootstocks. Seedlings of these three trees are also being grown so that they may be evaluated for the suckering characteristic.

Additional Rootstock Possibilities

Both Gellatly (1955, 1956) and Morris (1931) used *C. colurna*, *C. chinensis*, and *C. Jacquemontii* as rootstocks for several different species and found them completely compatible with *C. avellana*. *C. avellana*, *C. cornuta*, and *C. americana* have also been used as rootstocks for various species without problems. Species and variety collections have been established in Oregon by grafting on either *C. colurna* or *C. avellana* seedlings. Hundreds of distinct genetic combinations are represented here, all with apparent graft compatibility.

Until a satisfactory nonsuckering rootstock is developed or discovered, we are investigating the possibilities of using the pollinizer cultivar 'Daviana' for this purpose. Its availability from nurserymen as clonal material makes it particularly useful. Though it does produce suckers, they are not as numerous as those from 'Barcelona'. Trees of both cultivars have been dug and 'Daviana' was ascertained to have the best root system of the two in respect to anchorage. 'Daviana' is tolerant to the western bacterial filbert blight and might therefore decrease mortality of young trees during their first few years in the orchard. It should also be fully graft compatible with other European filbert cultivars. For these reasons, we have established 'Daviana' rootstock trials at two locations comparing the scion selections 'Butler', 'Ennis', 'Jemtegaard #5', 'Lansing', and 'Ryan' against 'Barcelona'. These trials will be the first to objectively evaluate different filbert cultivars on a common rootstock and should yield valuable performance data for both cultivars and rootstock.

FIG. 11. An intergeneric graft of *Corylus avellana* 'Barcelona' on *Carpinus betulus*. The scions leafed out, but no functional unions were formed.

Lagerstedt (1970) attempted intergeneric grafting between several cultivars of *Carpinus* and *Ostrya* and *C. avellana* (Fig. 11). The latter was used as both rootstock and scion with the two former. *Carpinus* and *Ostrya* are genera closely related to *Corylus* in the family *Betulaceae*. The objective of intergeneric grafts was to make use of the nonsuckering habit of the proposed rootstocks. All intergeneric grafts failed. Though the genera are closely related, their chromosome numbers differ.

Once the major objective of developing a satisfactory nonsuckering rootstock for the European filbert has been solved, the primary candidate selections will also be evaluated for their growth-controlling characteristics. For an industry that only recently has made the first use of a source of clonally-propagated rootstocks (Lagerstedt, 1971), growth-controlling stocks have hardly been considered, although there is a need for them just as with other tree crops. Their use in high-density orchards and their potential for inducing precocious bearing, minimizing the biennial bearing habit, increasing nut set and yields, and many other characteristics, must be explored.

Achievements and Prospects

The pioneer filbert breeders were C. A. Reed of the USDA; G. L. Slate of the New York Experiment Station; and amateurs such as J. F. Jones of Pennsylvania, R. T. Morris of Connecticut, S. H. Graham of New York, Carl Weschcke of Minnesota, and J. U. Gellatly of Canada. These men, located outside of the principal commercial growing regions, concentrated their efforts on interspecific hybridization to combine the quality of the *C. avellana* nut with desirable characteristics found in other species. Several hybrid cultivars developed by these men were named and increased, but their acceptance and distribution has been somewhat limited. At Geneva, N.Y., the last filbert crosses were made in 1930. The progeny from these and earlier crosses have been evaluated through the years and the most promising selections are still in existence. Filbert breeding at Geneva has received relatively little support, but it has produced a nucleus of selections which could serve as a basis for an expanded breeding program. The continued use of *C. avellana* cultivars for breeding in the East is of doubtful benefit. A more promising approach would be to utilize the interspecific hybrids already selected for hardiness, disease resistance, nut size, and kernel characteristics. Growing large seedling populations arising from controlled crosses from these hybrids should produce more suitable selections for eastern conditions.

The type of breeding done in Oregon and Washington following the turn of the century could at best be considered mass selection. Fortunately, there were a few observant growers with an interest in filbert improvement who investigated seedling orchards and planted seedling nurseries in search of superior selections. Men such as A. A. Quarnberg, C. T. Brixey, A. B. Sherf, H. A. Kruse, A. M. Gray, H. A. Henneman, D. Fitzgerald, and K. Pearcy should be recognized for their contributions of cultivars for use in the Northwest. As growers or nurserymen, they were sufficiently concerned with cultivar improvement to search through the available seedling populations that had originated from open-pollinated seed. The most popular selections named were 'Brixnut', 'Fitzgerald', 'Gem', 'Nonpareil', 'Nooksack', 'Royal', and 'Woodford'. However, these and other western cultivars never seriously challenged the position of 'Barcelona' or 'Daviana' and currently there is little demand for them.

For more than 30 years, Olgar Jemtegaard of Boring, Oregon, has planted seedlings of *C avellana*, mostly from open-pollinated 'Nonpareil'. From these populations he has made several selections which possess desirable characters such as thin shell, clean kernel, earliness, and high productivity. His selection No. 5 is currently being increased and evaluated for suitability to grow with 'Barcelona'. It is of similar size, shape, and appearance; it could serve as a pollinizer for 'Barcelona' and be used for either the in-shell or kernel trade. The 'Jemtegaard #5' selection has also proved to be easy to graft and seems to be a fairly vigorous growing tree. Other selections developed by Jemtegaard are being utilized in the Oregon breeding program for their best qualities.

The filbert breeding program in Oregon is directed by Dr. Maxine M. Thompson and is supported by a grant from the Oregon Filbert Commission. The first seeds obtained from controlled cross-pollinations were planted in 1965. They are now in bearing and are being evaluated. Since that time, superior parent selections have been located and their desirable characteristics catalogued so that they may be utilized in planned crosses (Thompson, 1972). Several thousand seedlings are being grown, but to date no selections have been named.

In Michigan, Cecil Farris has collected most of the *Corylus* species and is actively involved in determining their cross compatibility. While other species collections exist in the United States and Europe, we do not know of anyone else utilizing them for filbert improvement. Farris' objectives include improved nut size and quality, hardiness, and developing filbert rootstocks and ornamental trees.

In Italy, Dr. P. Romisondo has started a breeding program to improve 'TGDL'. The first seedlings have come into bearing and appear highly variable. No selections have been made as yet. This is an established, government-supported program and should result in cultivar improvement in Italy.

In other countries where filberts are grown, cultivar trials are in existence and selections are being made from the wild, but little if any planned breeding is done.

It must be concluded that to date no significant filbert-breeding achievements have been made anywhere in the world. The most popular cultivars currently in use were available at the turn of the century or long before. Filberts, though ancient and widely cultivated, have not enjoyed the general popularity or attention given to certain other fruits and nuts. With commercial production limited to relatively few countries, research interest and support is localized. Other crops compete for research attention, which leaves proportionately less staff and funds for filberts. In addition, there have been more immediate cultural problems to solve.

We have now reached a stage of knowledge where many of the cultural problems have been solved, and long-range research such as tree-nut breeding can be considered. Both tree planting and nut production are increasing worldwide, as is consumption and general popularity of the filbert. Thus it is safe to predict that this crop has a bright research future in which breeding should play a major part. The diversity of characteristics available in the genus *Corylus* presents a tremendous opportunity and challenge to plant breeders.

Literature Cited

Anon. 1841. Filberts. *Gardeners' Chron.* 1:69.

Arikan, F. 1963. The possibilities of improving filbert growing (in Turkish). Ankara: Gũzel Sanatlar Matbaasi. p. 1–64.

Bunyard, E. A. 1920. Cob-nuts and filberts. *J. Royal Hort. Soc.* 45:224–232.

Bush, C. D. 1941. *Nut growers handbook.* New York: Orange Judd.

Cameron, H. R. 1970. An undetermined disease of filbert. *Plant Dis. Rptr.* 54:69–72.

Cox, L. G. 1943. Preliminary studies on catkin forcing and pollen storage of *Corylus* and *Juglans*. *Ann. Rpt. No. Nut Growers Assoc.* 34:58–60.

Crane, H. L., C. R. Reed, and M. N. Woods. 1937. Nut breeding. In *USDA Yearbook of Agriculture*. 1937:827–889.

Danielsson, B. 1946. Polyploid types of hazel (in Swedish). *Sveriges Pom. Fören. Årsskrift* 1945:116–122.

Danielsson-Santesson, B. 1951. Continued investigations of polyploid hazels (in Swedish). *Sveriges Pom. Fören. Årsskrift* 52:84–94.

Farris, C. W. 1970. Inheritance of parental characteristics in filbert hybrids. *Ann. Rpt. No. Nut Growers Assoc.* 61:54–58.

Fuller, A. S. 1896. *The nut culturist.* New York: Orange Judd.

Gellatly, J. U. 1941. Report from Canada. *Ann. Rpt. No. Nut Growers Assoc.* 32:69–72.

———. 1955. Tree hazels as hardy rootstocks. *Ann. Rpt. No. Nut Growers Assoc.* 46:90–92.

———. 1956. Tree hazels—*Corylus colurna*. *Ann Rpt. No. Nut Growers Assoc.* 47:110–112.

———. 1966. Tree hazels and their improved hybrids. *Ann. Rpt. No. Nut Growers Assoc.* 57:98–101.

Hagerup, O. 1942. The morphology and biology of the *Corylus* fruit. *Det. Kgl. Dansk. Vidensk. Selskab. Biol. Meddeleser* 17:3–32.

Johansson, E. 1927. Floral biology trials with hazel at Alnarp. 1924–1926 (in Swedish). *Sveriges Pom. Fören. Årsskrift* 1927:3–20.

———. 1934. Pollination trials with hazel at Alnarp. 1927–1933 (in Swedish). *Sveriges Pom. Fören. Årsskrift* 1934:262–274.

Kasapłigil, B. 1963. *Corylus colurna* and its varieties. *Cal. Hort. Soc. J.* 24:95–104.

———. 1964. A contribution to the histotaxonomy of *Corylus* (*Betulaceae*). *Adansonia* 4:43–90.

———. 1968. Chromosome studies in the genus *Corylus*. *Sci. Rept. Faculty Sci., Ege Univ.* No. 59.

Lagerstedt, H. B. 1968. Germination of filbert seed. *Proc. Nut Growers Soc. Ore. Wash.* 54:46–51.

———. 1970. Filbert propagation techniques. *Ann. Rpt. No. Nut Growers Assoc.* 61:61–67.

———. 1971. Filbert tree grafting. *Ann. Rept. Ore. St. Hort. Soc.* 62:60–63.

———. 1972. Nut cluster size in the 'Barcelona' filbert. *Proc. Nut Growers Soc. Ore. Wash.* 57:71–75.

Lagerstedt, H. B., and D. R. Byers. 1969. Filbert research—progress and results during 1969. *Proc. Nut Growers Soc. Ore. Wash.* 55:43–51.

MacDaniels, L. H. 1964. Hazelnuts and filberts. *Horticulture* 42(10):44–45, 53.

McKay, J. W. 1966. Sterility in filbert (*Corylus*). *Proc. Amer. Soc. Hort. Sci.* 88: 319–324.

Morris, R. T. 1931. *Nut growing*. New York: Macmillan.

Oldén, E. J. 1952. Freezing trials with hazel twigs at Balsgård during the winters of 1950–51 and 1951–52 (in Swedish). *Sveriges Pom. Fören. Arsskrift* 1952: 13–29.

Osborn, A. 1930. The tree *Coryluses*. *Gardeners' Chron.* 2250:106–107.

Ourecky, D. K., and G. L. Slate. 1969. Susceptibility of filbert varieties and hybrids to the filbert bud mite, *Phytoptus avellanae* Nal. *Ann. Rpt. No. Nut Growers Assoc.* 60:89–91.

Paglietta, R. 1970. Filbert nut cultivars tested as pollinizers with the fluorescent light method (in Italian). *Ann. Fac. Sci. Agrar. Univ. Torino* 5:253–272.

Reed, C. A. 1936. New filbert hybrids. *J. Heredity* 27: 427–431.

Reed, C. A., and J. Davidson. 1958. *The improved nut trees of North America*. New York: Devin-Adair. p. 227–242.

Rehder, A. 1940. *Manual of cultivated trees and shrubs*. New York: Macmillan.

Rimoldi, F. J. 1921. A study of pollination and fertilization in the filbert. M. S. Thesis. Oregon Agricultural College, Corvallis.

Romisondo, P. 1965. Some aspects of the floral biology of the filbert cultivar 'Tonda Gentile della Langhe' (in Italian). *Ann. Accad. Agricol. Torino* 170:1–60.

Schreiber, W. R. 1953. Filberts in Turkey, *USDA Foreign Agr. Rpt. No.* 73.

Schuster, C. E. 1924. Filberts: 2. Experimental data on filbert pollination. *Ore. Agr. Expt. Sta. Bul.* 208.

———. 1927. Sterility in filberts. *Mem. Hort. Soc. New York* 3:209–211.

Shun-Ching, L. 1935. Shanghai, China: *Forest botany of China*. Commercial Press.

Slate, G. L. 1947. Some results with filbert breeding at Geneva, N.Y. *Ann. Rpt. No. Nut Growers Assoc.* 38:94–100.

———. 1961. The present status of filbert breeding. *Ann. Rpt. No. Nut Growers Assoc.* 52:24–26.

———. 1969. Filberts—including varieties grown in the East. Chap. 20 in *Handbook of North American nut trees*, ed. R. A. Jaynes, Geneva, N.Y.: Publ. Northern Nut Growers Assoc., Humphrey Press.

Tansley, A. G. 1939. *British Islands and their vegetation*. London: Cambridge Univ. Press.

Thompson, M. M. 1965. A progress report on pollination studies in filberts. *Proc. Nut Growers Soc. Ore. Wash.* 51:49–54.

———. 1967. Role of pollination in nut development. *Proc. Nut Growers Soc. Ore. Wash.* 53:31–36.

———. 1969. Improvement of filberts through breeding. *Proc. Nut Growers Soc. Ore. Wash.* 55:80–84.

———. 1972. Progress of the filbert breeding project. *Proc. Nut Growers Soc. Ore. Wash.* 57:43–46.

Tombesi, A. 1966. Biosystematic studies on the pollen of 20 hazel varieties (Abstract, original in Italian). *Ann. Fac. Agr. Perugia* 21:219–230.

Trotter, A. 1947. Characteristics and phenomena of reproduction in filbert: pollination (in Italian). *Rend. Accad. Lincei.* ser. 8, 2:745–749.

———. 1948. Evolution of the female flower and the maturation of the fruit of filbert (*Corylus avellana* L.) (in Italian). *Rend. Accad. Naz. Lincei.* ser. 8, 4:659–666.

Weschcke, C. 1954. *Growing nuts in the north*. St. Paul: Webb, p. 24–38.

Wilson, E. H. 1916. *Plantae Wilsonionae*. vol. 2, pt. 3, *Corylus* L. Cambridge, Mass.: The University Press. p. 443–455.

Woodworth, R. H. 1929. Cytological studies in the *Betulaceae*. 2. *Corylus* and *Alnus*. *Bot. Gaz.* 88:383–399.

Zielinski, Q. B. 1968. Techniques for collecting, handling, germinating, and storing pollen of the filbert (*Corylus* spp.) *Euphytica* 17:121–125.

Zielinski, Q. B., and M. M. Thompson. 1967. Self- and cross-pollination experiments in filberts evaluating potential pollinizers. *Proc. Amer. Soc. Hort. Sci.* 91: 187–191.

Chestnuts

by Richard A. Jaynes

The chestnut has been an important food for man in Asia, Asia Minor, Europe, and North America for centuries. The nuts have been harvested in Japan and China since before written history. The Greeks reportedly introduced the tree to southern Europe from Sardis in Asia Minor before 50 B.C. (Loudon, 1838), and in North America the nuts were used by the Indians before the continent was discovered by Columbus. Chestnut was a staple in the diet and a necessity for survival during famine. It was also an important food for domestic and wild animals. In fact, the demise of the wild turkey in the eastern United States has been attributed in part to the loss of the native chestnuts. In America, as well as in Europe, farmers fattened their hogs by turning them loose in the woods in early fall to feed on the mast of chestnut and oak.

Three species of chestnut are economically important as nut producers. They are all highly variable, and selection for high-yielding, large-fruiting types has been practiced for hundreds of years. However, the impetus for advances in breeding has largely been a reaction to insect and fungal damage to chestnut trees.

Origin and Early Development

Chestnut (*Castanea*) is in the family Fagaceae which also includes oak (*Quercus*) and beech (*Fagus*). The 13 species (Camus, 1929) are native to the temperate zones of Asia, Asia Minor, southern Europe, and the eastern United States. The United States species, other than the American chestnut, *C. dentata* Borkh., are called chinkapins. They are an imprecisely defined group of shrubs and small trees found in the Southeast; the Allegheny, *C. pumila* Mill., and the Ozark chinkapin, *C. ozarkensis* Ashe, are the two most common. The others are Ashe chinkapin, *C. ashei* Sudw.; Trailing chinkapin, *C. alnifolia* Nutt.; Florida chinkapin, *C. floridana* Ashe; and *C. pauscispina* Ashe. (See Graves, 1961, for taxonomic keys.)

The geological record of *Castanea* sheds little light on the evolution of the genus. We do know that it once had a wide range in North America. According to Sargent (1896), "Before the middle tertiary period *Castanea* existed in northern Greenland, and in Alaska, where traces of the leaves and fruit of *Castanea Ungeri* Heer have been distinguished; and impressions of the leaves of one and perhaps of two species found in the miocene rocks of Oregon, and in those of the upper miocene of the Colorado parks, show that *Castanea* which already existed in Europe in the cretaceous period, once inhabited western North America whence it has now disappeared."

Chestnut probably originated in the Orient.

There we find the Chinese chestnut, *C. mollissima* Bl., as well as three other, less important species: the seguin, *C. seguinii* Dode; the Henry, *C. henryi* Rehd. & Wils.; and the David, *C. davidii* Dode. In Japan and Korea we find *C. crenata* Sieb. & Zucc.

The westward extension of the genus (perhaps carried by man) gave rise in Europe to the European chestnut (*C. sativa* Mill.) and the eastward migration gave rise to the several species in North America (Fig. 1). The Asian species have evolved in the presence of the chestnut blight fungus, *Endothia parasitica* (Murr.) P. J. and H. W. Anderson, and hence possess field resistance to the disease. If the progenitor(s) of the American chestnut were ever resistant, they lost this characteristic as they evolved on the American continent in the absence of the chestnut blight fungus. The highly susceptible European chestnut may have evolved in a similar manner.

FIG. 1. Distribution of chestnut. In North America, the striped area represents *Castanea dentata* (American chestnut) and the stippled areas, chinkapins. To the west is *C. ozarkensis* and to the east, *C. pumila*. In Europe and Asia Minor, *C. sativa* (European chestnut) is shown. In Asia are *C. mollissima* (Chinese chestnut) and *C. crenata* (Japanese chestnut).

China

There is little or no information on chestnut research being conducted in China. From the explorations of Meyer (Galloway, 1926) and others it is known that the Chinese chestnut had a wide range (Fig. 1) and that nut size and other tree characteristics were highly variable. Improved nut types were developed over the centuries by seed selection. Thus trees grown from different importations may produce nuts weighing as little as 4 or as much as 20 g each.

The chestnut blight fungus is a threat even to the "resistant" Chinese chestnut. One of the means used to lessen the effect of the disease was to shave the outer layers of the bark on older trees to give a smooth surface (Galloway, 1926). This treatment was presumably designed to eliminate or reduce natural damp crevices that serve as inoculum centers for germinating spores.

Korea

The Japanese chestnut is native to southern Korea but the Chinese chestnut has been introduced and the two species freely hybridize. Research on selection and breeding of chestnut began in 1961 at Seoul (Kim et al., 1965) and more recently at the Institute of Forest Genetics, Suwon. Efforts center on developing good nut producers that are resistant to attacks of the gall wasp, *Dryocosmus kuriphilus* Yasumatsu. This insect, introduced in 1961, attacks the vegetative buds, disfigures the tree, and greatly reduces fruiting.

There are approximately 7,500 ha of chestnut plantings that are susceptible to the gall wasp and which produce about 2,000 tons of nuts per year (1967–1970) (S. K. Hyun, personal communication). Since 1968, the area planted to gall wasp resistant clones has increased by 26,000 ha and by 1977 an additional 23,000 ha will be planted.

Japan

Chestnuts were cultivated in Japan at least 1,000 and possibly as far back as 4,000 years B.C. and varietal selections were known by 750 A.D. (Eynard et al., 1966). More than 145 distinct cultivars were known in 1935 (Motte, 1935). Morphologically, the Japanese chestnut appears to be most closely related to the European chestnut. They both have similar glands on the undersides of the leaves and the pellicle over the nut kernel is often convoluted and becomes caught by cotyledon tissue, a disadvantage in shelling and cleaning. In general, nuts of the European and Japanese species have fewer soluble sugars than the Chinese chestnut and are not as sweet to the taste when eaten raw. The Chinese chestnut was introduced into Japan in 1922 and is cultivated to a limited extent.

The first chestnut breeding program started in 1947 at the Horticultural Research Station, Hiratsuka (Shimura et al., 1971a). Some 3,600 hybrid seedlings were obtained and four clones were selected for their superior growth characters, nut production, and resistance to the chestnut gall wasp. Propagation of these clones is by grafting.

Nut production for the six years between 1959 and 1964 stayed at about 28,000 tons per year, despite an increase of 35% in the area devoted to bearing trees; by including the young non-bearing orchards the actual area in chestnut doubled from 9,700 to 22,600 ha (Eynard et al., 1966). The lack of increase in nut production, despite the increase in producing trees, was due to the progressively more serious damage caused by the gall wasp, a relatively new pest first reported in Japan in 1941.

Europe

Chestnut has been an important crop in the hilly, rural areas of southern Europe for centuries. Camus in 1929 described more than 200 clones grown in France and another 200 cultivars grown in Italy. Chestnut has been a major crop in Portugal, Spain, Switzerland, and Turkey as well. In the 1950s the United States imported 7,500 tons of chestnuts a year from these countries, with most coming from Italy.

Two fungus diseases seriously threaten nut production in these Mediterranean countries. One, the ink or phytophthora root disease, caused by *Phytophthora cambivora* (Petri) Buissi and *P. cinnamomi* Rands, is believed to have been present in Spain since 1726 (Crandall, 1950) and in Italy since at least 1840 (Borelli and Pettina, 1956). It is particularly serious at lower elevations where the soil does not have good drainage. The second disease, the chestnut blight fungus, was first identified in Italy in 1938 and in the other countries more recently, such as France in 1956. The seriousness of these two diseases could hardly be exaggerated. They are in large measure responsible for drastically reduced nut production within the last century and especially within the last 20 years. Production of nuts in Italy diminished 85% in the years from 1909 to 1965, from 700,000 to 100,000 tons per year (Moser, 1966). In France there was also an 85% reduction from 1890 to 1957, from 500,000 to less than 100,000 tons per year (Gaïssa, 1966). Greece, Portugal, Spain, Switzerland, and Turkey have also experienced declining production.

Considerable research on chestnut has been undertaken in southern Europe. Much of the effort has been on selection with *C. sativa* but intra- and inter-specific hybridization has also received attention. One of the more productive chestnut research centers was established in Italy in 1951 under the direction of A. Pavari at Florence and continued until 1959. Publications from the center report a wide range of research on the genetics, biology, culture, and pathology of chestnut (e.g., Breviglieri, 1951, 1955). A decrease in chestnut research occurred in Italy in the early 1960s but subsequently there has been a renewed interest at the universities and horticultural stations.

In France, chestnut research has been largely supported by the National Institute of Agronomy Research (INRA) since 1948 and a station for this effort was established at Malemort, Correze. In 1971 this work was moved to the Station de Recherches d'Arboriculture Fruitière, Pont de la Maye, near Bordeaux. Research has centered on breeding, selecting, and propagating improved chestnut cultivars. Several clones of the European chestnut as well as hybrids with the Japanese chestnut have been clonally propagated by stooling and are undergoing commercial testing (Schad et al., 1952; Solignat, 1958, 1962, 1964, 1966).

In Spain, breeding work was initiated by Gallastegui (1926) at the Misión Biológica of Galicia and later (1944) by P. Urquijo-Landaluze at the National Agronomy Institute, La Coruna. Urquijo-Landaluze and E. Vieitez, at the University of Santiago, have reported success in stooling and rooting cuttings of selected clones (Molina and Vieitez, 1966). Between 1954 and 1961 over 220,000 plants of *C. sativa* and various *C. sativa*-*C. crenata* hybrids were inoculated in a screening program for ink disease resistance. The most resistant clones are still under test. Although the chestnut blight fungus occurs in Spain it has not yet proved to be a serious pest.

In Switzerland there has been considerable study of host-parasite relations of the chestnut blight fungus at the Federal Institute of Forest Research, Zurich (G. Bazzigher, 1957*a*, 1957*b*; Bazzigher & Schmid, 1962). Chestnut breeding, as well as host-parasite studies and research on related problems, has also been undertaken in Greece, Portugal, and Turkey.

Increased industrialization has accounted in part for the decrease in chestnut production in southern Europe. Less manual labor is available for the traditional, but laborious, means of harvesting. Thus, if chestnut production is to be increased, not only will cultivars highly resistant to the ink and canker diseases be required, but also mechanization of the growing, harvesting, and handling of the nuts will be necessary.

United States

In contrast to the countries just discussed, chestnut in the United States was valued primarily as a timber and tannin source and only secondarily as a nut tree. Ironically, it was at about the time the chestnut blight fungus was introduced into the United States that clones of the native chestnut were selected for their fruit size and quality (Buckout, 1896; Heiges, 1896). Indeed, a few commercial orchards had been established for nut production, but before they reached production age they were destroyed by the chestnut blight fungus.

The fungus was first noted in New York in 1904. It had apparently been introduced on nursery stock from Japan about 1890. From its discovery until about 1940 it spread approximately 20 miles a year in roughly concentric circles through the entire range of *C. dentata* destroying all the large trees and eliminating the American chestnut as a commercial species. More than a fourth of all hardwood sawtimber in the Southern Appalachian Mountains was chestnut and it was unsurpassed in the number and diversity of uses (Hepting, 1971).

The ink disease has not been a major problem in the United States, though damage from it in the South is known and recession of the American chestnut from the Piedmont in the Southeast in the 1800s was likely a result of root attack by *P. cinnamomi* (Hepting, 1971).

The first controlled pollination of chestnut trees was begun in 1888 with a cross of the American and Japanese chestnut. Shortly thereafter, Luther Burbank embarked on a hybridizing program using American, Japanese, and European chestnut trees. Van Fleet began breeding chestnuts for better nut cultivars in 1894 and continued until 1921. Unfortunately about the time his hybrids were coming into bearing, the chestnut blight fungus spread into his plantings. Most of the hybrids became seriously infected because they were of the highly susceptible American and European chestnut species. (See reviews on breeding chestnut by Crane et al., 1937; Jaynes, 1964; and McKay and Jaynes, 1969.)

The work of Van Fleet was taken over and expanded in 1925 by G. F. Gravatt and R. B. Clapper of the Bureau of Plant Industry. Numerous first and second generation crosses between species were successfully completed and some 20,000 nuts were obtained from the controlled crosses. The objective of this work, and that later initiated by A. H. Graves, was to develop a timber-type tree with characteristics much like the American chestnut but with the added quality of resistance to the chestnut blight fungus. Towards this goal Clapper emphasized first and second generation crosses of the American and Chinese chestnut. Many of these hybrids are still under test in plots scattered throughout the eastern United States (Diller and Clapper, 1969; Nichols et al.

1971). One promising forest-type tree has been named the 'Clapper' chestnut.

A. H. Graves began his chestnut breeding at the Brooklyn Botanic Garden in 1929 and later became associated with the Connecticut Agricultural Experiment Station which is continuing the work, making it one of the oldest continuous tree breeding programs in existence. Graves's first crosses were between the American chestnut and the Japanese chestnut and later these F_1 hybrids were crossed to the Chinese chestnut. Selections from these three-way crosses of Chinese x Japanese-American, known as CJA's, have shown promise for nut, game, forest, and ornamental plantings.

Chestnut stands apart from the other nut trees in that a number of people have successfully completed first and second generation crosses among the different species. The reasons for this are not entirely clear but may be connected with the relatively short generation time—as little as three years from seed to fruiting, the accessibility of the flowers, and not least of all the early success of workers like Van Fleet and Burbank in producing hybrids.

Since *Endothia parasitica* was discovered in the United States, the primary breeding objectives have been blight resistance and timber form, but the multiple use value of chestnut has always been recognized. Although there are few large chestnut orchards in the United States, small orchards, consisting largely of seedling Chinese chestnuts, exist in the Midwest, East, and Southeast. At least one grower in Georgia has over 3,000 bearing trees. Stimulus for greater production requires a strong market and a strong market depends on a reliable, constant source of nuts. For instance, the market for chestnut products, such as a chestnut soup, is virtually untapped, and will remain so unless potential growers can be assured of a demand by processors in the future.

Chinese chestnut seedlings are listed by most mail-order nurseries and, in addition to these trees, Christisen (1969) found in a survey conducted in 1966 that close to 100,000 chestnut seedlings were being produced annually in state nurseries for distribution to landowners for wildlife and other planting purposes. Thus there is an awareness of the present real value of the Chinese chestnut and its hybrids for use as orchard and ornamental trees and as food producers for wildlife.

Breeding Techniques

Floral Biology

Chestnuts are monoecious and the flowers are borne on the current year's growth. Two types of inflorescence are found, the unisexual staminate catkins, which are located on the lower parts of the shoot, and the bisexual catkins towards the terminal end of the shoots. Staminate flowers are spirally arranged along the axis of the catkin in clusters of from three to seven. Pistillate inflorescences appear alone or in clusters of two or three at the base of the bisexual catkins. Three pistillate flowers surrounded by the many-bracted involucre are normally found in the true chestnut section of the genus, whereas the chinkapins, e.g., *C. pumila*, have only a single pistillate flower in each involucre. All flowers normally have seven to nine styles (Nienstaedt, 1956).

Flowering occurs after the first leaves have fully expanded and varies somewhat according to species, clone, and season. Most trees flower in June in Connecticut and the nuts mature in the fall of the same year. The lower, unisexual staminate catkins are the first to start opening. The pistillate flowers are next to open and not until eight to ten days after anthesis of the unisexual catkins do the staminate flowers of the bisexual catkins begin to open. Stout (1928) described this sequence of ♂, ♀, then ♂ maturation of flowers as duodichogamy.

Pollen Collection, Storage, and Viability

Dry pollen can be stored at low temperatures. Catkins, previously bagged and dehiscing pollen, are dropped onto waxed paper. The pollen is scraped off with a small stick inserted in a cork, which is then fitted into a glass vial. Most of the pollen remains attached to the stick. Room is left in the bottom of the vial below the stick for a small amount of desiccant (silica gel). After four to eight hours of drying, the vial containing pollen is placed in storage at −15 C. Chestnut pollen has remained viable when stored for a year in this manner (Jaynes, 1964). It is a valuable technique when the flowering times of parent trees are widely separated or the parents are distant from each other.

Pollen viability can be checked *in vitro*; however, many researchers experienced difficulty in obtaining germination percentages above 60% (Breviglieri, 1955; Nienstaedt, 1956; Jaynes, 1961). Incubation at high temperatures is important. Good pollen routinely yields 50 to 80% germination after incubation for 1 hour at 30 C in 0.5% aqueous sucrose solution.

Emasculation and Pollination

Pistillate flowers are not receptive until five days after anthesis of the first staminate flowers and remain receptive up to the 17th day (Clapper, 1954; Nienstaedt, 1956). Best results with controlled crosses can be expected if the pollinations are made between the tenth and 13th day after anthesis. Solignat (1958) found that the pistillate flowers are receptive eight days after full development of all stigmas and remain so for three weeks. Shimura et al. (1971*b*) found the receptive period occurred 16–20 days after the first appearance of stigmas. Prior to bagging, the staminate catkins are removed and the staminate portion of the bisexual catkins pinched off. This eliminates the possibility of self-pollination. Emasculation does not injure the pistillate flowers and may even have a beneficial effect, while bagging does reduce nut set.

Bagging with kraft paper bags is used to ensure isolation of the pistillate flowers. The bag encloses the entire branch tip, and the neck of the bag is usually fastened securely to the past year's growth to prevent breakage of the new growth. Closure is with a piece of wire drawn tightly and given a single twist so it can be readily removed. Staminate flowers are bagged in a similar manner. When the pistillate flowers are receptive, the bag on the pistillate flowers is removed just long enough to perform the cross pollination. A staminate catkin shedding pollen is removed from a bag on the tree that is to serve as the staminate parent and drawn across the stigmas of the pistillate flowers. When the danger of outcrossing has passed (two to three weeks), the bags are removed from the developing burs and replaced with coarse-mesh cloth bags. These bags serve to mark the location of pollinations, prevent the loss of identifying tags, and retain the hybrid nuts when they mature (Jaynes, 1974).

Seed Harvesting and Treatment

Fresh chestnuts contain 40 to 45% carbohydrates, mostly in the form of starch, about 5% oil and 50% moisture. They are highly perishable because they lose water rapidly when exposed to the normal humidity of ambient air; drying causes the kernel to harden and become ungerminable. Stratification for one to three months at 0 to 4 C is required to break embryo dormancy and ensure uniform germination.

As soon as a few burs on a tree begin to crack open and some of the enclosed nuts begin to turn from a pale green to brown color, the burs can be knocked off or picked. For the tree breeder this ensures harvest of nuts at one time and minimizes losses to animals. The alternative is daily harvesting of nuts from the ground for a period of about two weeks. In the future, economical orchard management, at least in the United States, will require mechanical harvesting of the fallen nuts and possibly mechanical shaking to get the nuts down at one time. It may be feasible to initiate dehiscing of the burs at the proper time by spraying the trees with a growth regulator such as ethephon ([2-chloroethyl] phosphonic acid).

The burs are stored in wire crates in a cool (18 C), humid room, such as an earth cellar. The nuts continue to mature and within five to ten days the burs open. Nuts are separated from the debris and placed in plastic bags, along with some nearly dry peat moss, for cold storage (1 to 2 C). A slight loss of moisture (curing) is desirable before storing. As long as free water from condensation does not develop in the bags, growth of mold organisms is limited. Normal storage is from fall harvest until spring planting time. However, if optimal storage conditions are maintained, nuts can be kept viable for up to 3½ years. Seed storage for 18 months, or until the second planting season after harvest, appears to be practical (Jaynes, 1969a).

Germination and Growth of Seedlings

Nuts should be sown in a light, well-drained soil with a pH of about 5.5. Fall planting is satisfactory if rodents, especially squirrels, are not a threat. Nuts are covered with 2.5 to 5.0 cm of soil and, before the ground freezes, mulched with 5 cm of straw or similar material to prevent damage from low winter temperatures. For spring planting, nuts are stored as described and planted when the soil can be worked. Seedlings should be allowed 225 sq cm of space for the first year's growth. Under good conditions, one-year-old seedlings should attain a height of at least 30 cm and a basal diameter of 0.6 cm. As with walnuts, the shoot and root of a germinating seedling will expend less energy and be straighter if the nut is planted on its side instead of having the embryo pointing directly down or up. The use of residual (simazine) and contact (paraquat) herbicides has proved valuable in controlling weeds in nurseries and young plantations (Ahrens, 1969).

Seedling Evaluation

Much research remains to be done on the evaluation and selection of various traits among seedlings. Some progress has been made in this direction in screening for blight resistance (Bazzigher and Schmid, 1962; Clapper, 1952; Solignat, 1962), but in general the correlation between growth of individual inoculations and seedling resistance has not been high. The following data are typical (Jaynes, unpublished). Seventeen three-year-old Chinese and the same number of American chestnut seedlings were inoculated in July and fungal growth was measured one month later. Average canker diameter was 2.0 cm on the 17 *C. dentata* seedlings and only 1.2 cm on the *C. mollissima* trees, but the broad ranges overlap, 0.6 to 3.9 and 0.2 to 2.6, respectively. Variation in response to inoculation is so great that single tree selections are not reliable. The problem becomes especially acute when dealing with hybrid populations where there is virtually a continuous gradation from highly resistant to highly susceptible individuals.

Selection criteria for vigor and form could probably be refined so that meaningful selections could be made by the third year. Seedling vigor is correlated with nut size; hence, care must be taken not to confuse vigor from seed reserves with genetically controlled vigor. Jaynes and Graves (1963) showed that the form of seedlings from different crosses varied dramatically. They used a ratio of height to crown diameter for three-year-old seedlings. American chestnut seedlings had a height/crown ratio of 2.0 compared to 1.3 for the more spreading types. This ratio tends to obviate the differences in mass due to the effects of seed size that often affect direct comparisons of height and stem diameter. However, the vigor and habit of a tree are greatly affected by how soon it begins to flower and fruit. Early fruiting varieties are less vigorous (Solignat, 1964).

Irradiation

Uchiyama (1967) treated nuts of *C. crenata* in the fall with gamma rays to prevent sprouting in storage. By February 15 dosages of 10,000 and 20,000 r had increased sprouting compared to the control, but 35,000 and 50,000 r completely inhibited sprouting and most of these nuts were still salable in April. These treatments were for nuts to be marketed and the irradiation effect on germination ability was not reported.

MacDonald et al. (1962) and Singleton (1969) have irradiated American chestnut seed to induce mutations for increased disease resistance. The LD_{50} has not been precisely determined, but for nuts treated in the spring just prior to germination it is about 6,000 r (Thor and Barnett, 1969). Singleton proposed to use a method he found successful in obtaining mutations in corn: irradiate the seed, allow the resulting seedlings to interpollinate, plant this seed and self-pollinate individuals, plant these nuts and screen the seedlings for mutations. He states that 15 to 20 years will be needed to determine if a blight resistant chestnut can be produced by this means. There are restrictions in following such a course: three generations are required and one of these depends on the ability to self-pollinate these highly self-incompatible trees.

Chromosome Number and Colchiploidy

The somatic chromosome number for chestnut is $2n = 24$. Ten species and several interspecific hybrids were examined (Jaynes, 1962) and all had $2n = 24$ except for two hybrids: one was triploid and the other aneuploid ($2n = 25$).

Techniques for successfully doubling the chromosomes of chestnut with colchicine have been described (Dermen and Diller, 1962; Genys, 1963; Schad et al., 1952). These authors suggested that tetraploids formed by such treatments should be valuable to cross with normal diploids to form vigorous triploids. Selections for disease resistance, fertility, nut size, and/or rate of growth could then be made at the triploid level. Since fertility of such trees might be greatly reduced, their value as nut producers could be limited.

Breeding Systems

Self and Cross Fertilization

Self fertilization is rare in chestnuts. This is not due to the differential maturation of the flowers, because there is ample overlap in the flowering periods of the staminate and pistillate flowers. The cause is self-incompatibility (Clapper, 1954; Nienstaedt, 1956; Solignat, 1958).

There is disagreement as to whether chestnuts are predominantly wind- or insect-pollinated. Groom (1909) stated that staminate catkins are predominantly designed for insect pollination: they are long, conspicuous, and grouped; they are scented and nectar is present; pollen grains cling together, and flying insects visit the flowers. However, the pistillate flowers are best suited for wind pollination: the flower is inconspicuous, devoid of odor, and the styles are long like those of wind-pollinated species. Groom concluded that the chestnut is in a transitional stage, changing from an insect- to a wind-pollinating mechanism. Clapper (1954) and Ohata and Sato (1961) found that wind-borne pollen effected pollination through an insect-proof screen and they, as well as Breviglieri (1955), demonstrated that wind-borne pollen could be trapped on greased slides at a distance of at least 60 m from the trees. Thus wind is apparently the more important factor in pollination, insects not being necessary.

There have been many reports and observations of isolated trees that have not borne fruit (e.g. Graves, 1937; Sargent, 1896), and self pollinations have indicated that chestnut trees are often self-sterile (McKay, 1942; Clapper, 1954). Yet self fertilization does occur in at least some chestnut trees (Breviglieri, 1951; Solignat, 1958). Schad et al. (1955) suggested that the incompatibility might be a polyallelic series such as occurs in *Nicotiana*. have indicated that chestnut trees are often self-fertilization occurs regularly after selfing, but generally at a low level (Solignat, 1958). *Castanea mollissima* and *C. crenata* are somewhat less self-fertile and clones of *C. dentata* and the chinkapin species are usually self-incompatible (Jaynes, 1961).

Sterility, Apomixis, and Inbreeding

A tendency towards either staminate or pistillate sterility in certain trees has been observed by many and has been interpreted as a tendency towards dioecy (Breviglieri, 1951). Complete female sterility is rare but male-sterile trees are well known. In the most common type of male sterility, found within species as well as among species hybrids, no anthers are formed. The sterility therefore appears to be the result of a premeiotic event. Data from Jaynes (1961) indicate that the male sterility in at least a few trees is cytoplasmically controlled. Male-sterile trees are often heavy fruit bearers and many of the European cultivars selected for heavy yield are pollen sterile. Female or completely sterile trees would be desirable as street trees where fruit would be a nuisance.

Apomixis and possible parthenocarpy were reported by McKay (1942) in Chinese chestnuts and by Breviglieri (1951) in European chestnuts. Apomixis is less common than self fertilization but if it is triggered by self pollination then the experimental distinction between the two events becomes difficult. If seedlings from selfing showed inbreeding depression and were not uniform, it could be concluded that they resulted from self fertilization and not apomixis. Jaynes and Graves (1963) found no evidence for apomictic fruit set among hybrids. Apomixis and parthenocarpy appear to occur at such low frequencies that they are not a significant means of seed set and reproduction with most chestnut trees.

Inbreeding depression has been observed by the author from the first generation of selfing and it is doubtful if a line could be kept vigorous enough to allow for several generations of inbreeding. Published data are not available.

Hybrids and Interspecific Crosses

Since the first chestnut breeding work began, heterosis or hybrid vigor has often been attributed to certain interspecific crosses (e.g., Detlefsen and Ruth, 1922; Clapper, 1954). Because of the difficulty in inbreeding, heterosis cannot be tested by crossing inbred chestnut lines as is commonly done by the corn breeder. However, interspecific crosses can be compared with intra-specific crosses. Jaynes (unpublished) found hybrid vigor in several crosses of *C. crenata* with *C. dentata* as compared to intraspecific crosses of the parents. From several comparisons it was apparent that vigor of the progeny (seedling height) was highly dependent on the individual parents used in the cross. Not all *crenata* x *dentata* parental combinations were equally vigorous.

Overdominance or heterosis for catkin length in interspecific crosses has been reported (Jaynes, 1961). Average catkin length of *C. crenata* was 22 cm, for *C. dentata* 18 cm, and for the F_1 hybrid 26 cm, or 30% greater than expected based on an average of the two parents. In crosses of one of these F_1 hybrids (catkins 32 cm) exhibiting heterosis with *C. mollissima* (catkins 19 cm) the 19 flowering seedlings had catkins averaging 17 cm or about 30% less than the expected 25 cm average of the parents.

Solignat (1966) determined from controlled crosses that nut size was highly dependent on the pollen used, but was not directly correlated with the nut size of the pollen parent. He concluded the largest nuts were a result of heterosis.

Crosses have been completed between most species of the three subgenera (Fig. 2). Partial genetic incompatibilities, revealed by poor fruit set, male sterility, and other abnormalities, do exist between certain species (Jaynes, 1963). However, the barriers are incomplete and do not follow a consistent pattern. Speciation has apparently not yet resulted in the development of completely incompatible systems. This allows the geneticist and plant breeder to make use of the total variation among all the species for breeding improved cultivars.

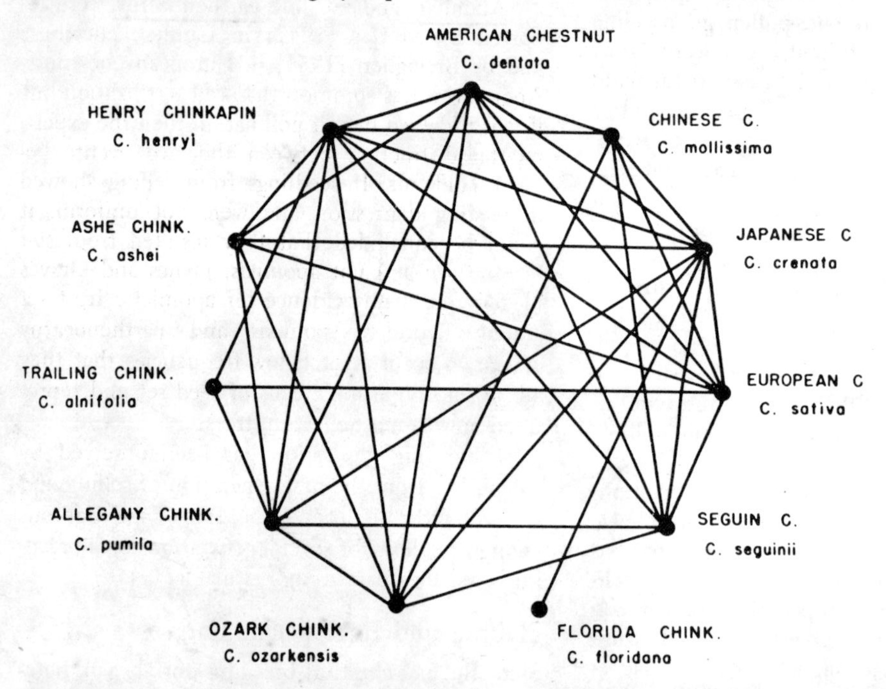

FIG. 2. Summary of controlled crosses that have been successfully completed between species of chestnut (Jaynes, 1969c).

Mendelian Inheritance

Genetic analysis has not been extensive enough to determine the inheritance of more than one or two specific traits. However, preliminary observations of numerous characters suggest inheritance patterns.

Everbearing or continuous flowering, exhibited by some *C. seguinii* trees, is apparently under the control of one or two major genes. Clapper (1954) suggested it was a single dominant gene but substantiating data were not presented. Jaynes (1961), on the other hand, reported that the inheritance of the everbearing trait fits the hypothesis of control by two unlinked recessive genes based on analysis of 79 flowering F_1, F_2, and backcross hybrids with *C. mollissima*.

Flowering at an early age, another *C. seguinii* characteristic, is a dominant trait in crosses with *C. mollissima* and possibly other species (Jaynes, 1961). One hybrid seedling of *C. seguinii* x *C.*

ozarkensis, in fact, produced a staminate catkin within eight months of germination.

Shimura and Yasuno (1969) studied the inheritance of the orange kernel character in *C. crenata*. Orange-kernel is controlled by a single recessive gene. Of 14 Japanese cultivars studied seven were heterozygous Yy but none was yy. Seedlings from orange kernels are albino and die unless grown at low light intensities where they develop light green leaves.

Polygenic Inheritance

McKay (1960), using *C. mollissima*, studied various seed and seedling traits: orange kernel, snow kernel, albino, red shoot pigmentation, and multiseed nuts. All the traits were assumed to be under polygenic control, but detailed analysis was not made. Several nut characters, commonly thought to be under maternal control, are under biparental control, including time of maturation, size, and the dormancy requirement of the mature nuts (Jaynes, 1963; Solignat, 1966). Such male parent effects on seed traits are sometimes referred to as xenia.

Analysis of the genetic control of nut size has not been reported. Personal observations indicate a great deal of variation from crosses, and it is suspected that nut size may be under the influence of codominant polygenes.

Solignat (1962) and Clapper (1952, 1954) concluded from analysis of various species hybrids that at least two recessive genes were involved in blight resistance. The apparent continuous gradation of classes from resistant to susceptible indicates the inheritance is likely polygenic rather than oligogenic.

Many other traits such as form and vigor are undoubtedly under polygenic control. The conclusions of Hansche et al. (1972) from estimates of genetic and environmental effects of several traits in peach, sweet cherry, and walnut are probably applicable to chestnut: "... the straightforward procedure of selecting parents on the basis of their own performance, and their subsequent mating *inter se*, should affect relatively rapid genetic gains for several generations." Jaynes (1969*b*) had earlier suggested a seed orchard technique utilizing similar straightforward selection principles to develop improved chestnut trees.

Breeding for Disease and Insect Resistance

The Chestnut Blight Fungus and Host Resistance

The chestnut blight fungus (*Endothia parasitica*) is a wound parasite which attacks the phloem of susceptible trees, generally forming a girdling canker on the main trunk. The pathogen is an ascomycete that can be readily cultured on artificial media. Conidia are produced under such conditions but perithecia and ascospores, products of sexual reproduction, have not yet been obtained on artificial media. However, in recent studies (Puhalla and Anagnostakis, 1971) on the genetics and nutritional requirements of the blight fungus, perithecia and ascospores were formed in the laboratory on chestnut stem segments inoculated with genetically marked strains. Ascospore analysis suggested that the fungus is homothallic but will preferentially outcross. Heterokaryons and putative diploids were formed.

Two toxins produced by the fungus, diaporthin and skyrin, have been implicated as contributing to the disease but it has not been shown that they are essential to disease development. Neither toxin has apparently been isolated from diseased tissue (Owens, 1969).

Stands of the European chestnut in France and Italy have suffered severely from the chestnut blight, but its spread has been slower than that witnessed among stands of the American chestnut (Biraghi, 1966) even though European chestnut trees grown in the United States are as susceptible to the blight fungus as the American chestnut. Perhaps the slower spread of the disease in southern Europe is related to climatic factors, such as a lower mean relative humidity, which might inhibit spore germination. The role in chestnut stands of a hypovirulent form of the fungus discovered in Italy by Grente (1965) is not completely understood. Recent studies (Grente, 1971) indicate that the hypovirulent form is spreading and causing localized regression of the disease. The Chinese and Japanese chestnuts are the only two species grown in the United States with notable field resistance to the canker fungus. However, they are not uniformly resistant, a fact often overlooked by those writing about breeding blight-resistant chestnuts. Indeed, blight cankers are common on some trees of these "resistant" species (Diller et al., 1964; Jaynes, 1967; Uchida, 1964). Field resistance of hybrids is understandably difficult to achieve, therefore, when resistance factors of the Chinese and Japanese species are diluted with genes from susceptible species such as the European or American chestnut.

Graves (1950) and Clapper (1952) ranked the different chestnut species from most susceptible to most resistant as follows: *C. dentata, C. sativa, C. ozarkensis, C. pumila, C. seguinii, C. crenata,* and *C. mollissima*. Variability of host reaction was common. Since the discovery of *E. parasitica* in New York in 1904, diligent efforts have been made to locate and propagate American chestnut trees showing strong evidence of resistance. So far no tree of *C. dentata* has been demonstrated to be blight resistant enough to warrant strong optimism (Diller and Clapper, 1965). Bingham et al. (1971) suggest that enough resistant genes exist within the species *C. dentata* so that, by appropriate breeding procedures, these genes could be recombined to produce a blight-resistant American chestnut. Solignat (1962) suggested a similar approach to obtain blight-resistant European chestnut trees.

The basis for host resistance is not known although Nienstaedt (1953) suggested that relative resistance among species was, at least in part, a result of the differential solubility and qualitative differences among the tannins. This work was not confirmed by Bazzigher (1957a).

Resistance may prove to be a combination of mechanical and chemical factors. Wound periderm does form in advance of the fungal hyphae (Bramble, 1936), but the cork and callus tissues are seldom able to permanently stop the host pathogen (Bazzigher, 1957*b*). Biochemical resistance alone is not enough for survival (Grente, 1961). Resistance is probably not passive, but active and apparent only after invasion of host tissue.

The chestnut blight fungus is a threat to certain other trees, especially live oak, *Quercus virginiana* (Batson and Witcher, 1968; Stipes and Phipps, 1971). However, barring a change in virulence of the fungus, there is little likelihood that damage to live oak will ever approach that caused to the American chestnut.

Screening for Resistance to the Chestnut Blight

A reliable test is needed to determine field resistance of trees to the chestnut blight fungus without having to wait five to 25 years for natural infection to occur. Young trees are susceptible to the disease and can be inoculated. Generally, however, such procedures use trees which are at least two years old and they must be observed for another one to two years to determine the effect of the inoculation (Bazzigher and Schmid, 1962; Clapper, 1952; Graves, 1950). There is no strict correlation between initial growth of the disease and field resistance. In fact there is a danger of "overkill" from artificial inoculations; trees that are almost girdled the first year may in subsequent years wall off the fungus and heal. A rapid screening method is needed for one-year-old trees that will allow all susceptible seedlings to be culled. An efficient method would enable thousands of seedlings to be screened in a nursery and only the highly resistant plants would be kept for field planting and further evaluation of form, vigor, nut quality, etc.

Ink Disease, or Phytophthora Root Rot

Some of the problems of breeding and screening for ink disease resistance are similar to those encountered with the chestnut blight fungus. Fungicidal control of the disease has not been practical. Screening one-year-old seedlings is inefficient and requires a second inoculation the following year to avoid having susceptible plants "escape." Some resistance apparently exists in *C. sativa*, but *C. crenata* and *C. mollissima* and hybrids of them have greater resistance (Schad et al., 1955; Molina and Vieitez, 1966). The inheritance of resistance still has to be worked out. Grafting of European chestnut cultivars onto resistant Asiatic stock has not been successful because of stock-scion incompatibility problems. Crandall (1950) has discussed the distribution and significance of the chestnut root rot phytophthoras, *Phytophthora cinnamomi* and *P. cambivora*.

Other diseases are, of course, of local importance, and it can be expected that genetic resistance will be available within either the endemic or exotic species of chestnut. Blossom-end rot, *Glomerella cingulata* (Ston.) Spauld. and Schrenk, which causes decay of the mature nut in Chinese chestnut, is such a disease. Resistant and suspectible trees have been noted. In a seedling orchard the removal of susceptible trees greatly reduced damage from the disease (Fowler and Berry, 1958).

Insect Resistance

The most serious insect pests in the United States are the chestnut weevils, *Curculio proboscideous* F. and *C. auriger* Casey, which attack the nuts. There is some variability among trees in their susceptibility but no breeding has been undertaken to enhance this resistance. Another weevil, *Conotrachelus carinifer* Casey, has recently been reported as a problem on chestnut in southeastern United States (Payne et al., 1972). In Europe a shuckworm, *Laspeyresia splendana* Hubner, and a weevil, *Balaninus elephas* Gull., are problems.

The gall wasp, *Dryocosmus kuriphilus*, threatens the chestnut industry of Japan and Korea. Resistant trees have been obtained by breeding and selection, but Shimura (1972) reported that another strain of the wasp has developed which attacks these trees. Late-ripening varieties of *C crenata* tend to be more resistant; there is little resistance in *C. mollissima*. Since the pest is of recent introduction it is possible that in addition to selecting resistant cultivars, natural predators of the gall wasp may ultimately be more effective in exerting control.

Achievements and Prospects

Research for improvement of chestnut has been both discouraging and encouraging. The breeder has had to cope with a catastrophic bark disease in the endemic chestnuts of the United States and Europe, a debilitating root rot in Europe, and most recently, a gall wasp that threatens tree growth and nut production in Asia. Tree improvement programs are difficult enough without the obstacles of such major pests. Chestnut is a valuable crop and one where vast improvement is still possible. Variation within and between species for pest resistance, form, vigor, and productivity is so great that a well-thought-out, long-range breeding program will succeed. On a world-wide basis, endemic species have received the greatest attention, but breeders will be forced to use species hybrids to introduce desirable genes that are unavailable in the native trees. For breeding programs to succeed, much more information on inheritance of many characters will be needed, especially on disease and insect resistances. Efficient methods of vegetative propagation and improved screening techniques are also required.

A. H. Graves, an active chestnut worker from 1911 until 1962, summarized the importance of breeding chestnut trees in 1914, 15 years before he began his own hybridizing work: "Work of this kind is extremely valuable and, although slow in yielding results, may eventually prove to be the only means of continuing the existence in our land of a greatly esteemed tree."

Literature Cited

Ahrens, J. F. 1969. Herbicides. In *Handbook of North American nut trees*. R. A. Jaynes, ed., p. 51–59. Knoxville, Tenn.: No. *Nut Growers* Assoc.

Batson, W. E., Jr., and W. Witcher. 1968. Live oak cankers caused by *Endothia parasitica*. *Phytopathology* 58:1473–1475.

Bazzigher, G. 1957a. Fermentation of tannins and phenols by three parasitic fungi (in German). *Phytopath*. Z. 29:299–304.

———. 1957b. Susceptibility and resistance of different hosts to *Endothia parasitica* (in German). *Phytopath*. Z. 30:17–30.

Bazzigher, G., and P. Schmid. 1962. A method to test for *Endothia* resistance in chestnut (in German). *Phytopath*. Z. 45:169–189.

Bingham, R. T., R. J. Hoff, and G. I. McDonald. 1971. Disease resistance in forest trees. *Ann. Rev. Phytopath*. 9:433–452.

Biraghi, A. 1966. Phytopathological aspects of saving the chestnut (in Italian). In *Atti Conegno Internazionale sul castagno*, p. 120–128. Cuneo, Italy: Camera di Commercio Industria Artigianato e Agricoltura.

Borelli, O., and D. Pettina. 1956. The disease of chestnut (in Italian). *In* Centro di Studio sul Castagno. *Suppl. Ricerca Scientifica*. p. 169–172.

Bramble, W. C. 1936. Reactions of chestnut bark to invasion by *Endothia parasitica*. *Amer. J. Bot.* 23: 89–99.

Breviglieri, N. 1951. Research on the flower and fruiting

biology of *Castanea sativa* and *Castanea crenata* in the district of Vallombrose (in Italian). *In* Centro di Studio sul Castagno. *Suppl. Ricerca Scientifica* 2:5–25.

———. 1955. Research on the dissemination and germination of chestnut pollen (in Italian). *In* Centro di Studio sul Castagno. *Suppl. Ricerca Scientifica* 1:15–49.

Buckhout, W. A. 1896. Chestnut for fruit. *Pa. St. College Agr. Expt. Sta. Bul.* 36.

Camus, A. 1929. *The chestnuts: monograph of* Castanea *and* Castanopsis (in French). Paris: Paul Lechevalier.

Christisen, D. M. 1969. Nut tree plantings for wildlife. In *Handbook of North American nut trees*. R. A. Jaynes, ed., p. 365–375. Knoxville, Tenn.: No. Nut Growers Assoc.

Clapper, R. B. 1952. Relative blight resistance of some chestnut species and hybrids. *J. Forestry* 50:453–455.

———. 1954. Chestnut breeding, techniques and results. *J. Heredity*. 45:106–114; 201–208.

Crandall, B. S. 1950. The distribution and significance of the chestnut root rot phytophthoras, *P. cinnamomi* and *P. cambivora*. *Plant Disease Rptr*. 34: 194–196.

Crane, H. L., C. A. Reed, and M. N. Wood. 1937. Nut breeding. In *USDA Yearbook of Agriculture*. 1937: 827–890.

Dermen, H., and J. D. Diller. 1962. Colchiploidy of chestnuts. *Forest. Sci.* 8:43–50.

Detlefesen, J. A., and W. A. Ruth. 1922. An orchard of chestnut hybrids. *J. Hered.* 13:305–314.

Diller, J. D., and R. B. Clapper. 1965. A progress report on attempts to bring back the chestnut tree. *J. Forestry* 63:186–188.

———. 1969. Asiatic and hybrid chestnut trees in the Eastern United States. *J. Forestry* 67:328–331.

Diller, J. D., R. B. Clapper, and R. A. Jaynes. 1964. Cooperative test plots produce some promising Chinese and hybrid chestnut trees. *U.S. Forest Service Res. Note* NE-25.

Eynard, I., N. Honda, and S. Koma. 1966. The culture and fruiting of chestnut in Japan (in Italian). In *Atti Convegno Internazionale sul Castagno*, p. 90–106. Cuneo, Italy: Camera di Commercio Industria Artigianato e Agricoltura.

Fowler, M. E., and F. H. Berry. 1958. Blossom-end rot of Chinese chestnuts. *Plant Disease Rptr*. 42:91–96.

Gaïssa, M. J. 1966. A brief account of chestnut in France (in French). In *Atti Convegno Internazionale sul Castagno*, p. 115–119. Cuneo, Italy: Camera di Commercio Industria Artigianato e Agricoltura.

Gallastegui, C. 1926. Breeding hybrid chestnuts (in Italian). *Bol. Re. Soc. Esp.* Enero 26:88–94.

Galloway, B. T. 1926. The search in foreign countries for blight-resistant chestnuts and related tree crops. *USDA Cir.* 383.

Genys, J. B. 1963. One-year data on colchicine-treated chestnut seedlings. *Chesapeake Sci.* 4:57–59.

Graves, A. H. 1914. The future of the chestnut tree in North America. *Popular Sci. Monthly*. 84:551–566.

———. 1937. Breeding new chestnut trees. *Ann. Rpt. No. Nut Growers Assoc.* 29:93–100.

———. 1950. Relative blight resistance in species and hybrids of *Castanea*. *Phytopathology* 40:1125–1131.

———. 1961. Keys to chestnut species. *Ann. Rpt. No. Nut Growers Assoc.* 61:78–90.

Grente, J. M. 1961. Observations of the behavior of chestnut trees after inoculation with *Endothia parasitica*. *Ann. Epiphyt.* 12:65–70.

———. 1965. A hypovirulent form of *Endothia parasitica* and the hope of control against the chestnut blight fungus (in French). *Compte. Rend. Hebd. Séances Acad. Agr. France.* 51:1033–1037.

———. 1971. Hypovirulence and biological control of *Endothia parasitica* (in French). Abst. *Ann. Phytopath.* 3:409–410.

Groom, P. 1909. *Trees and their life histories*. London: Cassell. p. 180–187.

Hansche, P. E., C. O. Hesse, and V. Beres. 1972. Estimates of genetic and environmental effects on several traits in peach. *J. Amer. Soc. Hort. Sci.* 97:76–79.

Heiges, S. B. 1896. *Nut culture in the United States*. USDA, Div. Pomology (unnumbered pub.).

Hepting, G. H. 1971. Diseases of forest and shade trees of the United States. *USDA Forest Service Handbook* 386.

Jaynes, R. A. 1961. Genetic and cytological studies in the genus *Castanea*. Ph.D. dissertation, Yale Univ.

———. 1962. Chestnut chromosomes. *Forest Sci.* 8: 372–377.

———. 1963. Biparental determination of nut characters in *Castanea*. *J. Hered.* 54:84–88.

———. 1964. Interspecific crosses in the genus *Castanea*. *Silvae Genetica* 13:146–154.

———. 1967. Natural regeneration from a 40-year-old Chinese chestnut planting. *J. Forestry* 65:29–31.

———. 1969a. Long-term storage of chestnut seed and scion wood. *Ann. Rpt. No. Nut Growers Assoc.* 60:38–42.

———. 1969b. Seed orchards for better chestnut trees. *J. Forestry* 67:453.

———. 1969c. Breeding improved nut trees. In *Handbook of North American nut trees*. R. A. Jaynes, ed., p. 376–399. Knoxville, Tenn.: No. Nut Growers Assoc.

———. 1974. Genetics of chestnut. *USDA, Forest Ser. Res. Paper* (in press).

Jaynes, R. A., and A. H. Graves. 1963. Connecticut hybrid chestnuts and their culture. *Conn. Agr. Expt. Sta. Bul.* 657.

Kim, K. S., S. K. Park and M. D. Lee. 1965. Selection of superior varieties of chestnut. 2. On the superior chestnut Chyung-pu 6 and the three species. *Res. Rpt. Office Rural Dev.*, Suwan 8:49–62.

Loudon, J. C. 1838. *Arboretum et Fruiticetum Britannicum*. Vol. 3, p. 1983–2004. London.

MacDonald, R. D., E. Thor, and J. O. Andes. 1962. American chestnut breeding program at the University of Tennessee. *Ann. Rpt. No. Nut Growers Assoc.* 53:19–21.

McKay, J. W. 1942. Self-sterility in the Chinese chestnut (*C. mollissima*). *Proc. Amer. Soc. Hort. Sci.* 41:156–160.

———. 1960. Seed and seedling characters as tools in speeding up chestnut breeding. *Proc. Amer. Soc. Hort. Sci.* 75:322–325.

McKay, J. W., and R. A. Jaynes. 1969. Chestnuts. In *Handbook of North American nut trees.* R. A. Jaynes, ed., p. 264–286. Knoxville, Tenn.: No. Nut Growers Assoc.

Molina, F., and E. Vieitez. 1966. Protection of chestnut against diseases in Spain (in French). In *Atti Covegno Internationale sur Castagno*, p. 107–114. Cuneo, Italy: Camera di Commercio Industria Artigianato e Agricoltura.

Moser, L. 1966. The preservation and improvement of fruiting chestnut trees (in Italian). p. 138–160. Cuneo, Italy: Camera di Commercio Industria Artigianato e Agricoltura.

Motte, Jean. 1935. *Castanea crenata*, Sieb. et Zucc. Outline of a list of Japanese varieties (in French). *Bul. Maison Franco-Japonaise.* Tokyo.

Nichols, C. R., R. E. Schoenike, and W. Witcher. 1971. Evaluation of the 24-year-old Table Rock (S. Carolina) chestnut plot for growth and disease resistance and a proposed breeding program for its use. *Ann. Rpt. No. Nut Growers Assoc.* 62:62–68.

Nienstaedt, H. 1953. Tannin as a factor in the resistance of chestnut, *Castanea* spp., to the chestnut blight fungus, *Endothia parasitica* (Murr.) A. and A. *Phytopathology* 43:32–38.

———. 1956. Receptivity of the pistillate flowers and pollen germination tests in genus *Castanea*. *Z. Forstgenetik Forstpflanzenzüchtung* 5:40–45.

Ohata, T., and T. Sato. 1961. Studies on the pollinations of chestnut trees. *Bul. Nat. Inst. Agr. Sci.*, ser. E., 9:185–193.

Owens, L. D. 1969. Toxins in plant disease: Structure and mode of action. *Science* 165:18–25.

Payne, J. A., L. S. Jones, and H. Lowman. 1972. Biology and control of a nut curculio, *Conotrachelus carinifer* Casey, a new pest of chestnuts. *Ann. Rpt. No. Nut Growers Assoc.* 63:76–78.

Puhalla, J. E., and S. L. Anagnostakis. 1971. Genetics and nutritional requirements of *Endothia parasitica*. *Phytopathology* 61:169–173.

Sargent, C. S. 1896. *Silva of North America* 9:10.

Schad, C., J. Grente, and G. Solignat. 1955. Improvement of chestnut (in French). *Ann. Amélior. Plantes* 5:303–325.

Schad, C., et al. 1952. Research on chestnut at the Brive Station. *Ann. Inst. Nat. Res. Agron.* 3:369–458.

Shimura, I. 1972. Studies on the breeding of chestnut 2 (in Japanese with Eng. Sum.). *Bul. Hort. Res. Station* ser. A, no. 11.

Shimura, I., and M. Yasuno. 1969. Studies on the genetics and breeding of kernel quality in the chestnut 1 (in Japanese with Eng. Sum.). *Jap. J. Breed.* 19:413–418.

Shimura, I., et al. 1971a. Studies on the breeding of chestnut. *Castanea* spp. 1 (in Japanese with Eng. Sum.). *Bul. Hort. Res. Station* ser. A, no. 10.

Shimura, I., M. Yasuno, and C. Otomo. 1971*b*. Studies on the breeding behaviors of several characters in chestnuts, *Castanea* spp. 2. Effects of the pollination time on the number of nuts in the burr. (in Japanese with Eng. Sum.) *Japan J. Breed.* 21:17–20.

Singleton, W. R. 1969. Mutations induced by treating maize seeds with thermal neutrons. In *Induced mutations in plants*, p. 479–483. Vienna: Int. Atomic Energy Ag.

Solignat, G. 1958. Observations on the biology of chestnut (in French). *Ann. Amélior. Plantes* 8:31–58.

———. 1962. Observations on the resistance of chestnuts to *Endothia parasitica* (in French). *Ann. Amélior. Plantes* 12:59–65.

———. 1964. Suitability of chestnut clones for fruit or timber production (in French). *Ann. Amélior. Plantes* 14:67–85.

———. 1966. Xenie, early manifestation of heterosis in chestnut (in French). *Ann. Amélior. Plantes* 16:71–80.

Stipes, R. J., and P. M. Phipps. 1971. Current status of *Endothia* canker of live oak in Virginia. *Plant Disease Rptr.* 55:201.

Stout, A. B. 1928. Dichogamy in flowering plants. *Bul. Torrey Bot. Club.* 55:141–153.

Thor, E., and P. E. Barnett. 1969. Breeding for resistance to chestnut blight. Second World Consultation for Tree Breeding. (unpublished symposium).

Uchida, K. 1964. Studies on the control of *Endothia* canker of Japanese chestnut (*Castanea crenata* Sieb. et Zucc.). *Hort. Expt. Sta.*, Ibaragi Prefecture 1:13–18.

Uchiyama, Y. 1967. Effect of gamma irradiation on sprout inhibition and its physiological mechanism in chestnuts. *J. Jap. Soc. Hort. Sci.* 36:348–56.

Urquijo-Landaluze, P. 1944. Methods to obtain chestnut hybrids resistant to disease (in Spanish). *Bol. Path. Veg. Ent. Agr.* 13:447–462.

Subtropical Fruits

Citrus

by Robert K. Soost and James W. Cameron

The importance of citrus to agriculture and the world's economy is demonstrated by its wide distribution and large-scale production. Citrus is grown throughout the world in tropical and subtropical climates where there are suitable soils and sufficient moisture to sustain the trees and not enough cold to kill them. The producing regions roughly occupy a belt spreading approximately 35° N and S of the equator.

The main commercial areas are in the subtropical regions at latitudes more than 20° N or S of the equator. The total planted area of the world's 49 citrus-producing countries is estimated at slightly over 1.6 million ha. Much of the world's citrus is grown for local use, and considerable plantings are scattered or away from commercial transport. The Mediterranean area and North and Central America contain about 80% of the commercial plantings. The remaining 20% is distributed in the Far East (10%), South America (6%), and in other Southern Hemisphere countries (4%), including South Africa and Australia. The Mediterranean area supplies about 80% of the total fresh-fruit export and 8% of the processed products. North and Central America supply only about 9% of the fresh-fruit export, but over 80% of the processed products; most of them are from the United States. Oranges constitute over 75% of the total production with lemons and grapefruit each accounting for approximately 10% (Burke, 1967).

Citrus fruits range in size from very small to very large. Among the smallest are the kumquats (*Fortunella* spp. Swingle) and the limes (*Citrus aurantifolia* [Christum.] Swing.), the greatest dimension of which may scarcely exceed 3 cm. At the other extreme are the pummelo (*C. grandis* [L.] Osbeck) and citron (*C. medica* L.), which may attain 30 cm in diameter or length.

Other characters also show a great range of variability. Fruit rind color ranges from the yellow-green of the limes to the red-orange of some of the mandarins (*C. reticulata* Blanco). Fruit shape also shows a full range from oblate to pyriform. At maturity fruits of some cultivars are still very high in acid while others have almost none. Tree size also exhibits tremendous range. Although all of the species of the genus *Citrus* are evergreen, the related genus *Poncirus* is deciduous.

Altogether there is a tremendous amount of variability within the genus with which the breeder can work, and closely related genera provide even a wider selection of characters. However, there are several barriers to the full utilization of this variability. The attempts of breeders to tap this reservoir of characters for the production of improved cultivars are presented in the following material. Considerable progress has been made; some barriers have been removed, but others remain.

Origin and Early Development

The genus *Citrus* and its close relatives are members of the family Rutaceae, subfamily Aurantioideae. *Citrus* is considered to be native to Southeast Asia, and especially to eastern India, but it shows phylogenetic relationships which extend through the East Indies, Australia, central China, Japan, and even Africa. Many present-day citrus cultivars have been cultivated since ancient times and their wild progenitors are usually not definitely known.

The most comprehensive taxonomic treatment of the subfamily Aurantioideae in recent times is that of W. T. Swingle (Swingle and Reece, 1967). In this classification, *Citrus* and five other genera, which constitute a group within the subtribe *Citrinae*, comprise the true citrus fruit trees. Two of these genera, which are notably cold-tolerant, are *Poncirus* (the deciduous, trifoliate orange) and *Fortunella* (the kumquat). These can be crossed with *Citrus* and with each other, and various proven hybrids exist. Two other genera, *Eremocitrus* and *Microcitrus*, are found in the wild, almost exclusively in Australia. *Eremocitrus* is a pronounced xerophytic plant. These two genera also have been successfully crossed with *Citrus* and *Poncirus* respectively. Various suspected natural hybrids among members of these five genera have also been described, especially by Swingle (Swingle and Reece, 1967). The sixth genus of the group, *Clymenia*, is known only from the Pacific island of New Ireland. It has not been hybridized with the others, to our knowledge.

The Japanese botanist, Tyozaburo Tanaka (1954), after exhaustive studies of *Citrus* in the Orient, proposed a theoretical line of demarcation in Southeast Asia (the Tanaka Line) which separates areas of probable development and spread of certain *Citrus* species and close relatives. This line runs southeastwardly from the northeast border of India, passing just above Burma and through a point south of the island of Hainan. Species which include the lemon, lime, citron, and pummelo appear to have arisen on the continent to the south of this line. This region was probably also the home of early forms of the sweet and sour orange. However, among the broad group of mandarin types a much wider area of early development is indicated. A chain-like progression can be traced northeast of the Tanaka Line, along the east China coast, through Formosa, and to Japan. The satsuma and certain other mandarins are believed to have developed there. *Poncirus* and *Fortunella*, the two genera most closely related to *Citrus*, are found in a third chain which crosses south-central China in an east-west direction.

In Europe, the citron was evidently the first citrus fruit to be known. Its culture in Persia was described by Theophrastus of Eresus in about 300 B.C. and probably was established there not later than 500 B.C. It has been an important part of Jewish religious rituals since at least the first century A.D. The Jews were largely responsible for its distribution around the eastern Mediterranean. Citron culture was later extended to Italy and other warmer areas of Europe (Webber, 1967).

The sour orange (*C. aurantium* L.) was known to the Arabs and they were instrumental in expanding its culture in the tenth century in the eastern Mediterranean and later into Africa and southern Europe. Apparently the lemon (*C. limon* [L.] Burm. f.), the lime, and the pummelo were distributed in a similar fashion by the first half of the twelfth century.

The sweet orange (*C. sinensis* [L.] Osbeck) was introduced from China and distributed in Europe by the Portuguese, probably in the sixteenth century. However, sweet oranges were cultivated in Europe prior to the Portuguese introduction but their use was not widespread. They were largely used as condiments and probably were inferior to the cultivar introduced by the Portuguese. The Portuguese cultivar became an important economic factor for Portugal and rapidly spread to other Mediterranean countries and became known as the 'Portugal' orange.

Mandarins were not introduced into Europe until modern times but were cultivated in China and Japan at a very early date. The first mandarin tree was brought to England in 1805 and the cultivar spread from there to the Mediterranean.

Citrus was unknown in the Western Hemisphere until the coming of Columbus, but there is written record that he brought seeds of oranges, lemons, and citrons to Haiti on his second voyage in 1493. Additional introductions were made by the Portuguese and Spanish into the Americas in the sixteenth century.

Most present-day cultivars which are widely

grown in the main commercial producing areas of the world were derived by selection, largely of sports, from the cultivars introduced into Europe and the Americas. The 'Washington' navel orange appears to have arisen as a sport in Brazil. The grapefruit originated in the West Indies, probably as a hybrid of *C. grandis*. Most of the present mandarin cultivars originated in China, Japan, or Southeast Asia. The satsuma mandarins have been developed entirely in Japan.

The status of species within the genus *Citrus* is presently in a state of contradiction. The system of Swingle establishes 16 species. In contrast, Tanaka (1954) has proposed 145 species and later 159 (Swingle and Reece, 1967). This lack of agreement reflects two basic problems: 1) what degrees of difference justify species status; and 2) whether supposed hybrids among naturally occurring forms should be assigned species rank, although hybrids of known origin are often classed as subspecific cultigens. Presently accepted *Citrus* species of economic importance have many characters in common and are generally interfertile; their hybrids are also often fertile. Genetic sterility and self- and cross-incompatibility occur but they are not primarily related to species limits. Asexual seed reproduction (nucellar embryony) is prominent in many species but it also is not exclusively a species characteristic.

A compromise classification of *Citrus* species has been proposed by R. W. Hodgson (1961). It includes 36 species, consisting of the 16 admitted by Swingle and an additional 20 recognized by Tanaka. Among the latter, for example, are *C. limettioides* Tanaka (the Palestine sweet lime); *C. limonia* Osbeck (the sour mandarin-limes), and several species which set apart well-known mandarin forms.

Certain group names which indicate the parentage of hybrids have become established in the literature. The most common of these include tangelo (tangerine x grapefruit); tangor (tangerine x orange); orangelo (orange x grapefruit); citrange (*Poncirus* x sweet orange); citradia (*Poncirus* x sour orange); citrangequat (citrange x kumquat); and a few others. These terms are definitely useful in referring to certain types of hybrids; however, it is impractical to extend such a series indefinitely.

History of Improvement

United States Department of Agriculture

Systematic breeding has been carried out at only a few institutions in the world. The first organized program was begun by the USDA in Florida in 1893 by W. T. Swingle and H. J. Webber (Cooper et al., 1962). Its principal goal was in relation to disease problems, but a severe freeze in the winter of 1894–95 destroyed most of the seedlings. Impressed by the freeze injury, Swingle proposed using the cold-resistant trifoliate orange (*Poncirus trifoliata* [L.] Raf.) as a parent in crosses to produce hardy hybrids with good fruit. This project was initiated, together with a program to obtain loose-rinded fruits resembling orange or grapefruit and intermediate cultivars.

Several important features of citrus genetics, especially the general interfertility of wide crosses and the high variability among F_1 hybrids, were discovered in the course of this work, and some distinct new cultivars of horticultural value were produced.

Two tangelos, the 'Sampson' and the 'Thornton', hybrids between the grapefruit and the 'Dancy' tangerine, indicated promise for this type of cross. These cultivars were grown commercially on a small scale, but have been replaced by newer cultivars. The 'Sampson' has also been used as a rootstock because of its vigor, tolerance to gummosis, and high degree of nucellar embryony.

The crossing of the trifoliate orange with the sweet orange produced many nucellar seedlings, and some hybrids. The hybrids were highly variable, and several produced fruit with certain good characters, but none was satisfactory. The best of them proved rather sour and all had the disagreeable rind oil of the trifoliate parent, as well as an unpleasant flavor. They were hardy enough, however, to succeed hundreds of miles north of the main Florida citrus region. Several of these citranges, including the 'Rusk', 'Morton', 'Savage', and 'Cunningham' have been used as rootstocks, but have been replaced by more recent introductions.

Beginning in 1908, the USDA group produced large numbers of hybrids from many combinations of cultivars and species. Additional citranges were produced, and citranges were backcrossed to sweet orange to produce citrangors. None was obtained which combined a high degree of hardiness with good fruit quality. However, the 'Troyer' citrange has become an important rootstock cultivar.

Kumquats were also used in crossing because of their exceptional winterhardiness. Hybrids between the kumquat and the trifoliate orange were not promising, but crosses with both citranges and limes were named and released. These were hardier than existing cultivars but have not been planted widely because of lack of acceptance of their fruit characteristics.

Additional hybrids were produced between the grapefruit and mandarin groups. Several were named and described. Two of them, the 'Orlando' and 'Minneola', are grown to a considerable extent.

In 1942 another extensive series of crosses was made at the USDA Horticultural Field Station at Orlando, Florida. Progeny from one cross in particular, 'Clementine' x 'Orlando', yielded a number of promising hybrids. Three of these, 'Robinson', 'Osceola', and 'Lee', were introduced in 1959. 'Robinson', a very early maturing cultivar, has been widely planted in Florida. A fourth hybrid from the same cross, 'Nova', was introduced in 1964. The 'Page', a hybrid of 'Minneola' x 'Clementine', was introduced in 1963.

Beginning in 1948, the USDA undertook an expanded program at Orlando and also at Indio, California. Thousands of seedlings from more than 155 different crosses were evaluated at Indio (Furr, 1969). 'Clementine' and 'Temple', both having good characters and producing only zygotic seedlings, were most often used as seed parents and yielded the best of the selections. The bulk of these had mandarin or tangelo as the pollen parent. Three hybrids, 'Fairchild', 'Fremont', and 'Fortune', hybrids of the 'Clementine' mandarin by 'Orlando' tangelo, 'Ponkan' mandarin, and 'Dancy' mandarin respectively, were released in 1964.

Extensive work was done on resistance to chloride injury and Phytophthora root rot. Work on cold hardiness was initiated and is still being actively pursued.

University of Florida

The University of Florida Citrus Experiment Station undertook limited citrus breeding in 1924, with emphasis on the production of acid cultivars. This program was later discontinued. A new program, initiated in 1956, initially emphasized the selection and establishment of nucellar clones. Emphasis has been shifted to the improvement of grapefruit, oranges, and tangerines through hybridization. Basic studies on means of identification of species and hybrids are also being conducted. Beginning in 1951, extensive screening and selection for resistance to the burrowing nematode (*Radopholus similis* [Cobb] Thorne) was initiated (Ford, 1969). In 1956 the program was expanded in cooperation with the USDA.

Two nematode-resistant cultivars, 'Ridge Pineapple' orange and 'Milam' lemon, and one tolerant cultivar, 'Estes' rough lemon, were released in 1964. A previously-released cultivar, 'Carrizo' citrange, was also determined to be resistant and was recommended for use. Additional tolerant or resistant selections were also determined but have not been released because they have not been superior to the cultivars previously released. Hybridization of resistant selections has been initiated.

University of California

Breeding was begun at the University of California Citrus Research Center, Riverside, in 1914 by H. B. Frost. In the first group of crossings, the parent cultivars most often represented were the 'King', 'Willowleaf', and satsuma mandarins; 'Dancy' tangerine; 'Imperial' grapefruit; 'Eureka' and 'Lisbon' lemons; and 'Ruby', 'Valencia', and 'Maltese Oval' oranges. The 'King' proved to be an outstanding parent. Three 'King' hybrids, the 'Kara', 'Kinnow', and 'Wilking', were introduced in 1935. These and other hybrids from the first series of crosses have been used in further breeding.

In the 1930s and 1940s, additional progenies were obtained from three principal groups of crosses. The parentages were: 1) mandarins and mandarin hybrids such as tangors and tangelos; 2) tetraploids crossed with diploids; and 3) pummelos crossed with mandarins, grapefruit, and other pummelos.

Within the first group, 'King', 'Dancy', 'Willowleaf', and 'Clementine' produced hybrids of good flavor. The new hybrids 'Wilking', 'Dweet', 'Mency', and 'Honey' were also useful parents. Two hybrids, 'Encore' and 'Pixie', were introduced in 1965. Both are unusually late maturing and 'Pixie' is almost seedless.

Triploids were obtained by using tetraploid 'Lisbon' lemon and a tetraploid seedy grapefruit as seed parents, crossed with pollen parents 'Kinnow', 'Dancy', and certain lemon hybrids. Many of the hybrids with pummelo cultivars were very seedy, and many were bitter or of poor flavor. Mild early-ripening hybrids were obtained from crosses with a very low-acid pummelo. The hybrid pummelo 'Chandler' was introduced in 1961.

The program was further expanded after 1947, and crossing to produce triploids has been emphasized. The study of nucellar clones initiated by H. B. Frost was continued and several new nucellar seedling selections were developed. Breeding for rootstock purposes was initiated; the progenies are still under test. Additional crosses with the low-acid pummelo cultivar and some of its selected hybrids were made. Many additional crosses to produce orange-like hybrids and mandarin hybrids were made. Efforts to produce vigorous lemon cultivars from intercrossing 'Lisbon' and 'Eureka' lemon strains have not succeeded.

Other Countries

Citrus breeding programs were underway in several areas of the world by about 1930, but World War II caused the interruption or abandonment of many of them. In Java about 1920, H. J. Toxopeus began hybridization directed toward rootstocks resistant to *Phytophthora* spp. (Toxopeus, 1936). At the same time J. P. Torres (1932) in the Philippine Islands initiated a breeding project to produce scion cultivars. Large-scale crossing was carried out in the Transcaucasian region of Russia during the 1930s with the main objective of producing hardy and early-ripening cultivars.

The projects in Java and the Philippines apparently have not been resumed since World War II. Frequent reports on breeding and selection have continued to appear in Russian journals. Much of the work is directed toward development of cold hardiness. Various hybrids have been reported, but interpretation of methods and results is difficult. Unshu (satsuma) and lemon cultivars have been used frequently as parents. Nucellar seedlings that have some characteristics of the pollen parent are reported. Until recently, several Russian researchers have reported vegetative hybrids that were not chimeras but the most recent reports mainly argue against their occurrence.

In Japan, programs of selection among satsuma variants have long been maintained. Many cultivars differing in maturity and other characteristics have been described and named.

Modern Breeding Objectives

Important breeding goals exist in citrus with respect to both scions and rootstocks. Many needs are of long standing and are similar to those in other tree fruits. The needs may be general, or may be closely related to the particular geographical area involved.

Scions

In scion cultivars, vigor and longevity of tree and sufficient amount and regularity of crop are important. Reduced tree size without reduction in yield per unit volume is becoming of increased interest as picking costs increase. With the majority of the commercial acreage in areas subjected to occasional damaging freezes, cold hardiness of the scion cultivars is desirable. Fruit size is a critical factor and many hybrids of otherwise good quality are discarded because the fruit is too small. Fruit shape, exterior appearance, and flavor are of major importance. Acceptable shape and appearance are in part determined by characteristics of existing cultivars. A globose lemon would not be easily accepted. Deep orange or red-orange rind color is desired for oranges and mandarins but not for lemons or grapefruit. Ease of peeling is desirable for eating out-of-hand, but loose rinds may be damaged easily. Although thick rinds are objectionable, fruits with excessively thin rinds do not ship well. Generally, fruits with smooth rinds and without stem-end necks or blossom-end nipples are desired. Desirable flavors are difficult to define and also depend on previously-acquired taste preference. Thus far, only repeated organoleptic testing is reliable for determining acceptability. The ratio of soluble solids to acid is an important factor. The ratio has to be much higher for oranges and mandarins than for fruit types such as lemons, which are used as acid fruits and processed drinks. Seed content is also important. Many of the cultivars now grown for fresh fruit have few seeds, with the notable exception of many mandarin types and some juice oranges. Seedlessness is difficult to obtain by regular breeding methods, however, since mechanisms related to seed production often are related to fruit development.

Season of ripening, storage life, and adaptability to specific environments often determine the success or failure of new cultivars. In Japan, the satsuma is a major cultivar partly because the fruits can be harvested before damaging cold weather begins. In South Africa, storage life is important, since much of the fruit is exported long distances. In desert areas, high foliage density and fruits resistant to sunburn are essential.

Unusually early-maturing or late-maturing cultivars are always of interest because they may fill a need in a pattern of production or marketing. The increasing use of mechanical harvesting methods in many areas will also influence the selection of certain characteristics related to the handling of the fruit.

Disease and pest resistance of the scion cultivars is desired but is difficult to accomplish. The long life cycle of the host and the many and varied pests and diseases combine to reduce the probabilities of combining resistance with other desirable characteristics.

Rootstocks

The need for dependable new rootstocks is of primary concern. Climate, soil composition, diseases, and physiological incompatibility all affect rootstock behavior, and cultivars which are successful in one region may be quite unsatisfactory in another. To be successful the rootstock must lead to longevity, high yield, and good fruit quality in the scion. Reduction of tree size without affecting yield or scion health is desirable. Rootstocks adapted to difficult soil situations such as poor availability of micronutrients or high salt levels are needed. Several of the available cultivars used for rootstocks are limited in their use because of stock-scion incompatibilities, including adverse reactions to certain diseases. A rootstock with the tolerance to *Phythophthora* possessed by sour orange without its susceptibility to the virus tristeza is needed. Lemon cultivars develop bud-union difficulties with most of the available rootstocks that have other desirable characters. Better resistance to *Phytophthora* and nematode species and the combination of resistance to these two pests are needed.

In contrast to scion cultivars, successful rootstocks must produce many seeds and be highly nucellar in order to provide uniformity. Cold hardiness is desirable, particularly if it enhances cold hardiness of the scion. Rapid growth and lack of branching are desirable characters for convenient and economical nursery production of rootstock seedlings.

Underlying the breeding objectives is the need to gain genetic information on the inheritance of specific characters and the combining ability of available parents.

Breeding Techniques

Floral Biology

In subtropical zones with cool winter temperatures, most citrus species flower once a year during the early spring. In such climates flower induction appears to be initiated by early January. In the tropics and in coastal areas, flowering may occur several times a year or the blooming season may be greatly prolonged. Various factors, such as disease and rainfall or heavy irrigation preceded by a dry period, may promote flowering. *Poncirus* may bloom earlier or later than *Citrus*, depending upon the amount of winter chilling received. Other related genera such as *Fortunella* and *Severinia* bloom during summer or fall rather than in the spring. Even in areas with cool winters, Eureka-type lemons tend to bloom throughout the year, but more heavily in the spring than at other times.

In the genus *Citrus*, flowers are borne singly in the axils of the leaves or in short axillary, corymbose racemes. The genera most closely related to *Citrus* (*Microcitrus*, *Fortunella*, *Poncirus*, and *Eremocitrus*) bear one to few flowers in the leaf axils. Some distantly related genera such as *Murraya* bear flowers in panicles. In *Citrus*, inflorescences develop from dormant axillary vegetative buds. The vegetative apical meristem is transformed into a terminal flower bud. Leaves may develop on the shoots from primordia present in the dormant buds. Axillary flower buds develop later than the terminal bud. Therefore, the several flowers on one inflorescence differ greatly in time of development, requiring the removal of the less well developed buds if controlled cross-pollination is to be carried out.

The flowers typically have four to eight thick, linear petals and a four- to five-lobed calyx. The petals are imbricate in the bud but are strongly reflexed at maturity. There are usually four times as many stamens as petals but in some species there may be six to ten times as many. Stamens and petals are borne on the receptacle, immediately below a short, annular disk. The anthers surround the pistil at or near the level of the stigma; they consist of two locules which dehisce longitudinally. In some cultivars, anthers may dehisce prior to opening of the flower.

The ovary of *Citrus* is composed of some six to 14 carpels joined to each other and to a central axis. At a very early stage the pistil is not closed at the top, but consists of a circular wall or ring and a protuberance within the ring (Ford, 1942). The wall is composed of the young carpels which arise as a whorl of crescent-shaped primordia. The carpels grow upward and their margins project inward to meet the central protuberance, which also grows upward, producing the axis or core of the fruit and uniting with the carpel margins.

At the inner angle of the locule of each carpel is developed the placenta, a region of thickened tissue which bears the ovules. These at first grow straight outward, but later develop the anatropous form with the micropyle facing the axis of the ovary. The mature ovule consists of the funiculus, the nucellus, an eight-nucleate embryo sac, and two surrounding integuments. The micropylar opening extends through the nucellus and the integuments at the free end of the ovule.

At flowering the ovary may be subglobose and sharply distinct from the much narrower style, as in the oranges, or subcylindrical and merging gradually into a broad style. The style is usually cylindrical, and expands into the subglobose stigma. The stigma is receptive from one to a few days before anthesis, and for several days afterward. Modified epidermal cells on the stigma secrete a viscous fluid that aids in retention and germination of the pollen. Canals extend from each locule through the style, opening on the surface of the stigma. In cultivars with a navel structure, a large canal also may connect the navel with the surface of the stigma.

The flowers of fertile taxa may be either both perfect and staminate, or regularly perfect. In those taxa with staminate flowers the development of the pistil is curtailed at various stages. In the first group, which includes the genus *Poncirus*, *C. limon*, *C. aurantifolia*, *C. medica*, and related types, the pistil is underdeveloped or absent in a large percentage of the flowers. In the second group, including *C. sinensis*, *C. grandis*, *C. paradisi*, and *C. reticulata*, abortion of the pistil is much less frequent. The proportion of staminate flowers is highly variable, according to the cultivar and growth conditions. In all cultivars tested, the proportion of staminate flowers is higher in later-opening flowers. Low temperature during flower formation is reported to increase the percentage of aborted pistils. In 'Shamouti' orange, zinc-deficient trees had 11% staminate flowers while normal trees had approximately 6%. Leafless inflorescences have higher percentages of staminate flowers than leafy inflorescences.

A large percentage of perfect flowers also may fail to set fruit. Studies in lemon and orange show that 45 to 50% of the flowers set fruit. Heavy loss of set fruit also occurs, much of it shortly after initial set. In lemons only 7% of the buds developed mature fruit. In sweet oranges this varied from a low of 0.2% to over 5%, depending on the cultivar (Erickson, 1967).

The stamens, unlike the pistil, show little tendency to general failure of development. Stamens may infrequently be petaloid and certain hybrids have generally aborted anthers. Imperfect development of part or all of the pollen mother cells or the pollen is common, however. Normal anthers are bright yellow at maturity. In cultivars with much defective pollen, the anther color is considerably lighter. Anthers containing no pollen are pale cream or white and usually do not dehisce.

Pollen is of the sticky, adherent type characteristic of entomophilous plants, and wind is therefore a minor factor in its transfer from flower to flower. Self-pollination can easily occur because of the nearness of anthers to stigma. Four characteristics of the flowers make them attractive to insects: conspicuous corolla, strong perfume, pollen, and abundant nectar. Thrips are present in great abundance, some apparently feeding on the pollen. Honey bees and others work the flowers for both nectar and pollen. Mites are sometimes found in the flowers and may carry pollen. Most pollination is undoubtedly accomplished by bees. In some cases at least, self-pollination occurs without the agency of insects.

Pollen development generally follows the usual course for dicotyledonous plants. The microspore enlarges and develops a complex wall. Before the anther dehisces the microspore nucleus divides, forming the vegetative and the generative nuclei. The pollen remains bi-nucleate until after it germinates when the generative nucleus divides to form the two microgametes. Pollen tube growth through the style appears to be entirely in the stylar canals. The pollen tube enters the embryo sac through the micropyle. One microgamete enters the egg and the second unites with the two polar nuclei to initiate the triploid ($3n = 27$) endosperm. Under favorable environmental conditions, fertilization occurs within eight days after pollination, but periods as long as four weeks have been reported.

Percentage of functional pollen varies greatly among species and cultivars. Some of the most widely-used cultivars are deficient in this respect. The 'Washington' navel orange produces no viable pollen, and the satsuma mandarin very little. Seedless grapefruit cultivars like the 'Marsh' have from 5 to 15% well-developed pollen. Most commercial lemon and orange cultivars have intermediate percentages of viable pollen with some as low as 25%. Most mandarin and pummelo cultivars have high percentages of functional pollen. In the 'Washington' navel orange, related cultivars, and some hybrids, sporogenous tissue degenerates before meiosis. In the satsuma the number of pollen mother cells produced is small and most pollen produced degenerates. In lemons, limes, and low-seeded orange cultivars, many meiotic abnormalties have been reported (Naithani & Raghuvanshi, 1958; Iwamasa, 1966). Temperatures below 19 C have increased meiotic abnormalities in lemon cultivars and resulted in increased sterility in satsuma cultivars. Cultivars with some non-functional pollen generally show comparable ovule abortion. However, the pollen-sterile 'Washington' navel has some functional ovules.

Pollen Collection, Application, and Storage

Controlled pollination in citrus is relatively easy to achieve. When certainty of parentage is required for genetic or taxonomic studies, seed and pollen parent flowers should be protected against contaminating pollen. Trees and branches especially likely to set fruit should be used; those with an excessively heavy bloom may set poorly. The earlier and larger flowers of a cluster are preferable, but only those with full-sized pistils should be used.

If twigs are covered by insect-proof bags before buds open, open flowers with dehisced anthers can be collected from the bags for direct application to emasculated flowers. Without bagging, flowers to provide pollen are collected just before opening, petals and pistil are removed, and anthers allowed to dehisce. Dehiscence usually occurs within 12 to 24 hours at room temperatures. Pollen may be applied by use of this modified flower, or dehisced anthers may be placed in a vial and pollen applied with a brush.

Special storage of pollen is seldom necessary for crosses within the genus *Citrus*, except for long-distance transportation, since most cultivars have similar and rather long blooming periods. But *Fortunella* blooms much later than *Citrus* in many areas, and *Poncirus* may bloom earlier or later. For storage, pollen should be kept over a drying agent in a sealed container, preferably at 4 C or lower. Hybrids have been obtained from pollen stored as long as five weeks (Soost and Cameron, 1954). Longer storage may be possible by using controlled humidities and lower temperature, or by freeze-drying.

Emasculation

Emasculation is easier and less injurious if the flowers are nearly ready to open (Fig. 1); flowers with dehiscing anthers must be avoided. Dehiscence usually occurs soon after the petals separate, but in cultivars such as the pummelos it often occurs in the closed bud. Several flowers may be emasculated on each selected twig. Setting of fruits appears to be improved by removing other blooms and fruit from the surrounding area. Care should be taken to avoid removing leaves because fruit-set is better on leafy inflorescences and twigs. Emasculation is usually accomplished by gently separating the petals with forceps and pulling off the anthers, avoiding contact with the stigma. Forceps should be sterilized in alcohol periodically and when changing trees. Various mechanical aids for emasculation have not been successful. After removal of unemasculated buds, open flowers, and any fruit, twigs with emasculated buds are usually enclosed by flat-bottomed paper bags, tied with twine or pliable wire. Plastic or air-tight bags should be avoided. To exclude small insects and minimize motion of the bag, cotton pads may be wrapped around the twig at the point where the

FIG. 1. Emasculation of a citrus flower, from left: bud nearly ready to open, with petals pulled apart, with anthers removed, and with entire stamens removed.

bag is tied. For selfing, bagging alone is often sufficient, without manipulation of the flowers.

Pollination

Pollination may be carried out immediately after emasculation, or up to several days later, depending upon the condition of the seed parent flowers. Receptivity of the stigma is indicated by a sticky secretion, and unpollinated flowers remain receptive for several days. Self-pollination of self-incompatible cultivars may be successful if it is done three to four days before anthesis. Paper bags should be replaced after pollination but should be removed approximately two weeks later to avoid any build-up of pests and diseases. All pollinated twigs should clearly be marked with appropriate labels. After several weeks, twigs with developing fruit can be covered with open-mesh bags, which often prevents loss of fruit before harvest.

Cross-pollination can be accomplished more rapidly if the twigs are not covered with bags. Some workers, finding that bees seldom visit flowers without petals, remove the petals when emasculating, and pollinate immediately without subsequent bagging.

The proportion of definitely aborted pollen can be determined by its failure to stain in acetocarmine or iodine solutions; however, grains which stain are not always viable. Viability is probably better indicated by germination in suitable culture media than by stainability. Most procedures have used 20% sucrose solutions at 20 C to 25 C. Cultivar differences in response to sucrose concentration have been reported. Some germinate best in sucrose solutions as low as 15% and others as high as 50%. The hanging drop technique has most often been used but satisfactory results have also been obtained using 1% agar in covered dishes.

Seed Harvesting, Treatment, and Storage

Fruit are harvested at maturity and kept identified. Sound fruit can be stored at room temperature for a week or more without damage to the seed. Refrigeration extends the period to several weeks. However, invasion by fungi can be serious even under refrigeration. Seeds are usually extracted by cutting through the rind around the equator of the fruit and twisting the fruit apart. Seeds are then squeezed into a sieve and washed free of pulp. To eliminate brown rot fungi (*Phytophthora citrophthora* [SM. & SM.] Leonian and *P. parasitica* Dastur) seeds should be immersed for ten minutes in well-agitated water held at 51.5 C. After treatment seeds should be dried sufficiently to remove surface moisture before being treated with Arasan (tetramethylthiuramdisulfate) or a comparable fungicide to prevent seedling albinism. If the seeds are not to be stored more than a few weeks, a dip in 1% 8-hydroxyquinoline sulfate may be substituted. Surface-dried, treated seeds can be stored in polyethylene bags at 1.5 C to 7.5 C. The length of successful storage varies but most cultivars can be stored for as long as eight months. *Poncirus trifoliata* seed has been stored for over two years with little reduction in germination. Most citrus seeds give the highest germination if planted shortly after extraction. *P. trifoliata* seed may benefit from chilling for one or two months at 40 C, particularly if harvested early in the season.

Citrus seeds are very sensitive to drying. When preparing seeds for treatment, storage, or planting, they should be dried only enough to remove surface moisture. Germination in some cultivars may be reduced by drying even for a few hours. However, seeds should not be kept in their own juice or immersed in water.

Viability of seed lots can be determined within a few days by removing seed coats from a sample of seeds and germinating them in petri dishes or other containers suitable for germination tests. Seed can also be cut in half longitudinally and placed with the surface down in a 1% solution of 2, 3, 5-triphenyl tetrazolium chloride (TTC). Viable seeds should turn bright pink within 24 hours.

Seed Morphology

The mature seed consists of one or more embryos, surrounded by two seed coats. The endosperm and nucellus are no longer present, except as they may have contributed to the formation of the inner seed coat, the tegmen. This thin membrane closely invests the embryo or embryos; it is largely formed from the inner integument of the ovule. The chalazal end of the tegmen is characteristically colored, which aids in the identification of cultivar groups. The outer seed coat, or testa, is usually grayish white or cream and is tough and woody. It is often ridged or wrinkled and usually extends beyond the embryo at one or both ends to form a beak or flat plate. In some polyembryonic seeds, the surface may be corrugated because of unequal development of the embryos. Each embryo consists largely of fleshy cotyledons, which are attached to a very short hypocotyl. At one end of the hypocotyl, lying between the cotyledons, is the very small plumule; at the other end is a rudimentary radicle. The one to several embryos form a fairly solid, rounded mass, in which the radicles normally point toward the micropylar end of the seed. In many cultivars a sexual embryo may or may not be present, and all other embryos are ordinarily of asexual (nucellar) origin. In monoembryonic seeds the two cotyledons are usually about equal in size and shape, but cotyledon size varies greatly when more than one embryo is present. Multiple embryos are usually crowded together at the micropylar end; often some are small with poorly-developed cotyledons. Small embryos may also be located on the outer side of large cotyledons or between two large cotyledons (Fig. 2).

Seed shape is characteristic for certain species and cultivars, but seeds of a cultivar may vary greatly in size. Both size and shape are altered by the number of seeds per fruit and the number of embryos per seed.

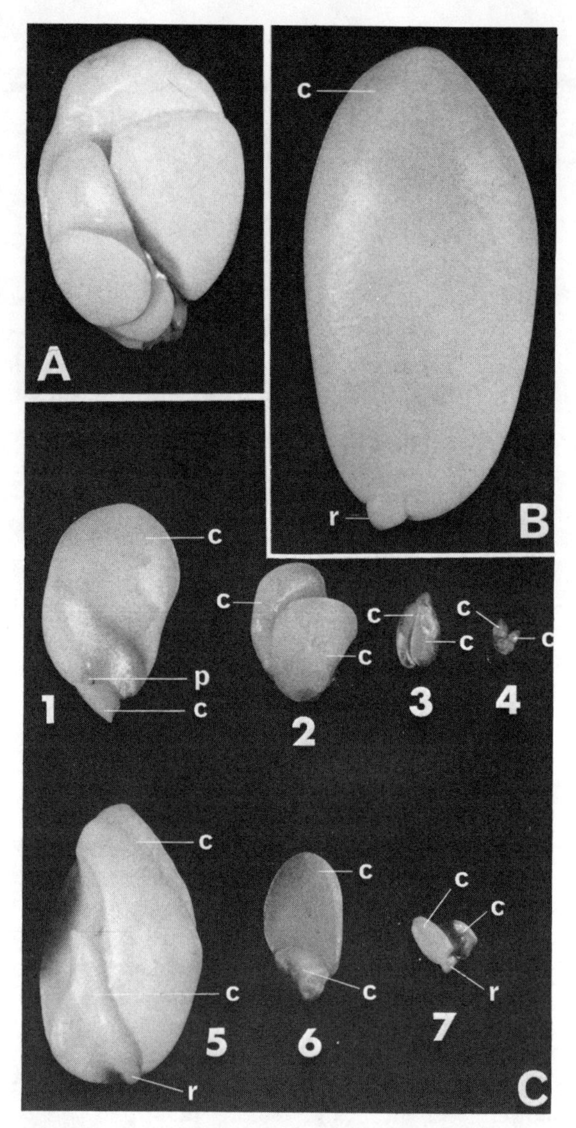

FIG. 2. (A) Multiple embryos in one seed of 'Ponkan'. (B) Monoembryonic seed of *C. grandis*. (C) Seven embryos dissected from one seed of 'Ponkan' with cotyledons (c), plumule (p), and radicle (r).

Germination

Germination of the seeds is hypogeous in the subgenus *Eucitrus*. In species of the subgenus *Papeda* for which data are available, it is epigeous. Both the radicle and plumule-epicotyl normally emerge from the micropyler end of the seed. The radicle emerges first, growing rapidly downward without immediate formation of secondary roots. In polyembryonic seeds particularly, the testa may obstruct the growth of the radicle, causing it to become bent before it can emerge. The epicotyl is bent as it emerges from the seed coat but straightens during subsequent growth. The first two true leaves are situated opposite one another on the stem and are usually unlike the later leaves in shape, sometimes being mistaken for cotyledons.

Germination is slow, requiring two to four weeks at temperatures between 20 C and 25 C; it is slower and more erratic at lower temperatures, and most rapid at 25 C or slightly higher. Germination can be hastened by removing the seed coats or cutting off the chalazal end of the outer coat. Seeds should be planted at a depth of 1½ to 2 cm. Various media that provide adequate moisture and aeration are suitable for germination, but continued growth will depend on the availability of mineral nutrients.

Pre- and post-germination injury by fungi can be a serious problem. Preplant seed treatments are given earlier in this chapter. At planting, sterile containers and planting media should be used, and precautions should be taken against later infection. To aid in the prevention of "damping-off" by *Rhizoctonia* fungi, the upper 3 to 5 cm of soil may be acidified by incorporation of 35 g of aluminum sulfate per 900 cm^2 of soil surface. If damping-off occurs, drenching the soil with fungicides will help arrest the disease.

Sterile nutrient culture can be utilized successfully for the germination of immature or small embryos. Seeds should be surface-sterilized, and the embryos then removed from the seed coats under sterile conditions and transferred to the culture medium. Zygotic embryos have also been recovered from highly polyembryonic cultivars by extracting the presumed zygotic embryo approximately three to four months after pollination. For such very immature embryos it is necessary to add casein hydrolysate to modified nutrient media (Rangan et al., 1969). Subsequent transfer to media without casein hydrolysate and sucrose is essential for continued development. Nucellar seedlings can be developed by similar *in-vitro* culture of nucellar tissue from cultivars which are monoembryonic and do not produce nucellar embryos *in vivo*.

Cytological Methods

Citrus chromosomes are small (about 2 μ long at first metaphase of meiosis) and not very favorable for extensive studies, but usable preparations for routine examination can be made (Fig. 3). Squashes can be made from pollen mother cells, root tips, shoot tips, and embryos. The material is pretreated with 25 ppm *O*-isopropyl-N-phenylcarbamate (IPC) for two to three hours, fixed in 2½:1(V:V) ethyl alcohol:propionic acid for four to 24 hours, hydrolyzed in 1*N* HCl, stained in 1% lacto-propiono-orcein and squashed. Pre-treatment with IPC may not be desirable with all tissues. If material is to be stored, it should be transferred to 70% ethyl alcohol after fixation. Hydrolysis is best done at approximately 60 C with the length of time adjusted for the type of tissue;

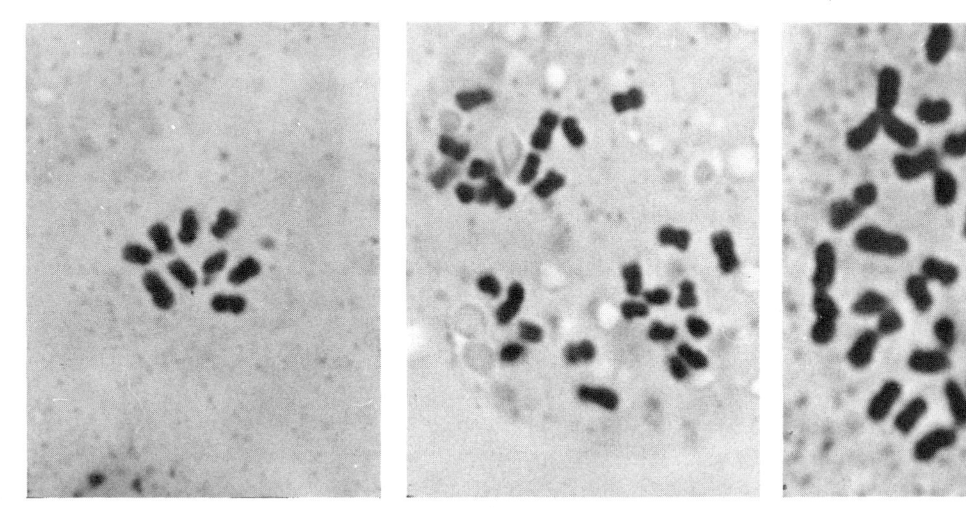

FIG. 3. From left, haploid, triploid, and tetraploid cells prepared from root tip smears at 3654, 5046, and 3480 x, respectively.

usually three to ten minutes is sufficient. Root tips and other tissues can also be prepared by conventional paraffin-embedding methods. Heidenhain's hemotoxylin at 0.25% is the most suitable stain for chromosomes in sectioned material.

Seedling Evaluation

Dependable techniques for the identification of zygotic seedlings among populations containing high percentages of nucellar seedlings would be valuable. When the pollen parent is known and has morphological characteristics that clearly distinguish it from the seed parent, zygotic seedlings can be identified with some degree of accuracy. However, when the pollen parent is either unknown or when its morphological characteristics are identical or similar to the seed parent, as in self-pollination or in "close" crosses, it becomes extremely difficult to distinguish between zygotic and nucellar seedlings. Various chemical tests of bark or leaves have not been sufficiently selective when parents are genetically similar. Infrared spectroscopy and gas chromatography have also been tried but have not been completely satisfactory for routine use. Recent Japanese work (Watanabe et al., 1970) indicates that it may be possible to use radioactive labeled pollen to mark the zygotic embryo.

Culture of partially-developed embryos may also provide a means of recovering zygotic embryos from highly polyembryonic cultivars. Until definitive tests that can be used easily on a large scale are developed, the breeder must either rely on his ability to select on the basis of seedling characters or grow the populations to fruiting.

There are no known correlations between identifiable seedling characters and mature characters that would provide a means for the selection or elimination of seedlings. Seedlings are discarded because of the lack of sufficient vigor. Inheritance of a few specific foliage characters can be determined during the first year of growth.

Disease and Pest Screening

Seedling populations can be tested for resistance to one or more diseases or pests. Some of the tests result in the death of susceptible seedlings and can only be used if resistance to the disease is the primary objective. Most testing has been for one or more species of *Phytophthora* root rot or for various nematode species. For the *Phytophthora* tests, seedlings have been most commonly exposed to water suspensions of the fungi, incubated in beds of peat-vermiculite for two to three months, and the survivors planted in soil beds which are also inoculated (Carpenter and Furr, 1962). A series of control seedlings of cultivars of known performance in relation to *Phytophthora* resistance is included in each test to check the effectiveness of the test and for comparison with the unknown seedlings. Although Carpenter and Furr (1962) tested the hybrid populations directly, the present USDA program grows the hybrids to fruiting and tests the nucellar seedlings of the hybrids. This procedure allows the retention of the hybrids for other purposes and permits replicated tests. For those hybrids that produce low percentages of nucellar seedlings, the zygotic seedlings can be tested to determine the value of the hybrids as parents.

Testing has also been done by growing individual seedlings in separate containers constructed so that they can be flooded from below with a suspension of zoospores. After a period of approximately two weeks, the seedlings are removed from the quartz sand used as the growing medium and the roots are immersed in a 1% solution of 2,3,5-triphenyltetrazolium chloride for 24 hours. Dead roots do not stain. Damage can be expressed as percentage of a given root mass that remains unstained. For testing for resistance to one or more of the nematodes that injure citrus, the young seedlings are grown in containers or soil beds inoculated before or after transplanting the seedlings. With the burrowing nematode, *Radopholus similis* (Cobb) Thorne, it is necessary to inoculate before transplanting and to increase the nematode population by growing a suitable host plant (Ford and Feder, 1958). Ford and Feder (1958) also devised a supplementary laboratory test in which young seedlings are inoculated in petri dishes. The effect of the nematodes in segregating populations can be determined by examining roots under a dissecting microscope or by counting the nematodes extracted from a root sample (Cameron et al., 1954). Supplemental staining techniques may be needed (Ford and Feder, 1958). Ratings on resistance can be based on the number of nematodes per g of roots. Testing is non-destructive so the plants can be field planted for additional evaluation. The field planted seedlings should be checked for nemotode populations by root sampling over a period of years. With the citrus nematode, *Tylenchulus semipenetrans* Cobb, some seedlings with low initial populations develop high populations after several years.

Tristeza virus has its most serious effects in certain scion-rootstock combinations. To test for potential rootstocks in segregating populations is not practical although it could be accomplished. The present procedure is to test for tristeza tolerance only those selections that have tolerance or resistance to other diseases and produce a high percentage of nucellar seedlings. Nucellar seedlings of these selections are then budded with a common scion and inoculated with the virus.

Testing for resistance or tolerance to other diseases and pests has not been done routinely. Some populations being evaluated for use as scion cultivars have also been evaluated for apparent susceptibility or resistance to attack by insects. Ratings for red-scale infestations in seedling populations in California were based on the number of scales apparent in a two-minute examination of each tree. Trees were rated from 0 (apparently scale-free) to 5 (very heavy infestation) (Cameron et al., 1969). Ratings were made for a ten-year period. Selections were made only from trees that remained in grades 0 or 1.

Resistance to Environmental Factors

Tests for tolerance to chloride injury were conducted by the USDA (Furr and Ream, 1969) by irrigating seedlings with water of known high chloride level for a period of two years. Symptoms were rated visually and the chloride content of leaf tissue was determined. Selections were based on the level of leaf chlorides. In the USDA program, various tests have been tried for evaluating cold hardiness. The most reliable procedure is to determine injury following exposure of whole plants to freezing conditions. For evaluation, Young and Peynado (1967) used an index based on the diameter of wood at which 50% of the branches are killed. To determine the index, ratings of the percentage of branches killed in each of several diameter categories from 5 mm up are made. Young and Hearn (1972) used eight categories based on percentage of leaf and stem kill. Seedlings are first preconditioned by holding at temperatures slightly above freezing and then subjected to below-freezing temperature in controlled chambers.

Evaluation for Scion Use

Various planting spacings have been utilized in an effort to increase the number of trees per land area without adversely affecting cultural operations and seedling evaluation. The closest spacing utilized in the California program has been 1.8 x 3.6 m. This has been too close except for some small-sized mandarin populations. Other breeders have utilized row spacing as close as 1 m. Many plantings have been approximately 3 x 3 to 3 x 3.6 m. Even this spacing has been too close for easy operation with many progenies. The best alternative appears to be to retain close tree spacing in the row, 3 m or less, and increase row spacing to approximately 4.8 m.

The question of whether to plant seedlings on their own roots or to propagate them on a common rootstock has been troublesome. Propagation on a common rootstock may provide a more uniform basis for tree growth and therefore a more uniform basis for evaluation. However, the cost of propagation and the additional time required are significant factors. Difficulties in relation to susceptibility to soil fungi, virus diseases, and incompatibilities also eliminate many rootstocks from use with unknown scion progenies. In situations where *Phytophthora* root rot is not a problem, planting the seedlings on their own roots probably is best. It is cheaper, faster, and avoids any problems associated with rootstocks. However, in areas where *Phytophthora* is a problem, the seedlings must be propagated on tolerant rootstocks to avoid excessive loss of seedlings.

Evaluation of Fruit and Tree Characters

Ratings on tree vigor, tree shape, yield, and crop retention are usually made on a visual basis. Ratings on fruit size, shape, exterior rind characters, and interior characters, including peel thickness, pulp characteristics, and seediness are also made on a visual basis. A rating of palatability is made organoleptically, judging the level of sugar and acid. If possible, these ratings are made several times a season in order to determine the period of maturity. For each character a rating scale of several classes has been established. Originally these classes were defined by word description; for example, for peel color yellow, yellow-orange, orange, orange-red, and red were used. For rapidity in evaluation and for statistical handling, these can be converted to numerical scales.

The method of record-keeping varies with the expected utilization of the populations and their size. Because of repeated samplings it is most convenient to have a system of field records that provides columns of each character being rated and sufficient vertical space for samplings at different dates. Cards which can be edge-punched for retrieval of desired characters by means of hand spindles are convenient for relatively small populations. For larger populations or a more critical review and analysis of characters it is best to transfer the original data to a computer acceptable form.

Quantitative measurements and laboratory analyses for fruit quality are made only on those seedlings that show some promise in the initial evaluations or on populations of particular interest for specific quality characters.

Selection of Parents

The selection of parents in citrus breeding is governed by many considerations. Cultivars which are entirely zygotic are to be preferred as the seed parents. The pummelos are such a group. Many known or putative hybrids that are entirely zygotic are now recognized among other citrus groups. For rootstock breeding, at least one parent must produce nucellar embryos because of the need for a high degree of nucellar embryony in rootstock cultivars. Only monoembryonic progeny are produced when both parents are monoembryonic. If the available monoembryonic cultivars are not suitable as parents, cultivars which produce some zygotic embryos must be used. The production of zygotic embryos by some cultivars is so low that they cannot be used effectively as seed parents.

Most citrus cultivars are very heterozygous. Therefore, selfing would appear to be a useful technique for breeding. However, selfing has produced mostly weak or otherwise inferior progeny. Selfing is also restricted by the high degree of nucellar embryony of many cultivars and by self-incompatibility in others.

There is no information on general or specific combining ability on which to base the selection of parents. Few characters have shown simple inheritance. Most characters studied are inherited quantitatively. Several undesirable characters such as small fruit size, seediness, and paleness of color appear to be dominant, making the prediction of results difficult. Information is available for selecting parents for breeding for some specific characters such as resistance to citrus nematodes.

Induction of Genetic Changes

In recent years several Russian workers have reported on chemical treatments of citrus for the production of mutations. Kerkadze (1970) indicated that treatments with several alkylating agents and 8-ethoxy caffeine produced albinos, dwarfs, and other variants. Treatments were made at various times on the developing fruits. Another Russian group (Goliadze and Talakvadze, 1968) reported an increase in the variability of seedlings as a result of treating seeds with N-nitroso compounds.

X-ray treatments of seeds and budwood by researchers in the United States and Russia indicate that the effective dosage level is between 5 and 10 kr. Treatment with thermal neutrons requires between six and 15 hours exposure.

Tetraploids of a few cultivars have been produced by colchicine treatment. H. Barrett (personal communication) has produced tetraploids of several monoembryonic cultivars by treating axillary buds with 1% colchicine solution for five days. Treatment of 'Valencia' seeds for four to eight hours at 0.025% produced tetraploids.

Breeding Systems

Polyembryony

Asexual reproduction by seed is an important characteristic in citrus, and it strongly influences the methods and results of breeding. Most citrus species can produce asexual progeny, which arise from somatic cells of the nucellus; in many cultivars very few sexual seedlings are ever recovered. Virtually all evidence indicates that the initiation of nucellar embryos in nature requires not only pollination but probably also the fertilization of the egg or the polar nuclei. The general lack of seed production in self-incompatible cultivars, even when pollen tubes grow some distance into the styles, suggests that pollen-tube growth alone is not a sufficient stimulus.

Fertilization of the egg cell in citrus has been reported to occur from two days to three or more weeks after pollination, and cell division of the zygote may commence soon after. At this time the endosperm is already multicellular, and partly fills the embryo sac; it is a temporary tissue and later disappears. Except in strictly sexual cultivars, the zygotic embryo usually competes for space and nutrients with one or more nucellar embryos. Embryo sacs may contain a few embryos developing normally, together with others that are partially suppressed. The results no doubt depend on the number, location, and time of initiation of the nucellar embryos, and on their genetic vigor relative to that of the sexual embryo. Table 1 shows data indicating that many embryos fail to germinate and reach the seedling stage. Some cultivars have several embryos in most of their seeds, but few seeds produce more than two or three seedlings. Lower numbers of embryos per seed usually result in larger embryo size and a greater probability that a sexual embryo will survive.

Data on the frequencies of zygotic and nucellar seedlings are shown in Table 1. All tested cultivars of *C. grandis* (the pummelo) are monoembryonic, producing in nature only zygotic seedlings, as do the cultivars of the citron (not listed). Some cultivars of the mandarin group also produce only zygotic seedlings, while others produce mainly nucellar ones. Most cultivars of lemon and lime yield substantial percentages of zygotic seedlings, but among the grapefruits and oranges the percentages are low. The available number of monoembryonic, zygotic cultivars has been much increased in recent hybrid progenies.

The percentage of zygotic seedlings can vary considerably from fruit to fruit within a cultivar, but attempts to control this by treatment of the ovaries have not been successful. However, the pollen parent can sometimes influence the proportions of zygotic seedlings. Thus rough lemon, when self-pollinated, can produce almost exclusively nucellar seedlings, but in one experiment where pollen of *Poncirus trifoliata* was used, 46% of the seedlings were zygotic (Table 2).

TABLE 1. Mean numbers of embryos per seed, and seedlings per seed, in *Citrus* cultivars and species (after Frost and Soost, 1968)

Cultivar or species	Embryos per seed	Seedlings	
		Per seed	Per embryo
King (tangor?)	1.10	1.03	0.94
Ellen grapefruit	1.32	1.20	0.91
Citrus aurantium	1.50	1.03	0.69
McCarty grapefruit	1.70	1.23	0.72
Rangpur lime	1.70	1.08	0.63
Batangas mandarin	2.01	1.23	0.61
China mandarin	2.72	1.42	0.52
Citrus sinensis	2.70	1.31	0.47
Rough lemon	2.90	2.00	0.69
Calamondin	3.59	1.70	0.47

The total number of embryos per seed varies greatly within a tree, as well as among cultivars, and there is little consistency within species having nucellar embryony. Large differences have been observed in the same cultivar between localities and between years, and some differences related to rootstock have been reported. Undoubtedly, many internal and external factors affect this character. However, there are also certain rather clear-cut inherited effects.

Additional embryos in a seed are not always of nucellar origin. Cultivars which are typically monoembryonic occasionally produce two or more embryos per seed. In the study summarized in Table 3, such embryos were shown to be zygotic twins and triplets, identifiable by the dominant trifoliate leaf character from their pollen parent. A later field study of many of these individuals showed that they were in fact genetically identical twins and triplets (Fig. 4), and had evidently originated by fission of the zygotic embryo. Cultivars that produce nucellar seedlings also occasionally produce zygotic twins, but typically monoembryonic cultivars have not been shown to produce nucellar seedlings under natural conditions.

Nucellar embryos and viable nucellar seedlings have recently been induced in monoembryonic citrus forms by tissue culture (Rangan et al., 1969). In the first successful experiments, seedlings of pummelo, 'Ponderosa' lemon, and 'Temple' (tangor?) were obtained. In later trials, nucellar seedlings were produced from several other cultivars. In first tests, the seedlings appear to be free of certain viruses carried by the mother trees. This tissue culture technique provides a method for reproduction by seed of normally strictly zygotic citrus clones, and probably for freeing them from viruses. It will probably be found that various juvenile characters, such as thorniness and slowness to fruit, are present in such seedlings. Using tissue culture, the authors cited above were also able to isolate and grow the zygotic embryo from cultivars in which normally it seldom survives.

Juvenility

A long period of juvenility is characteristic of citrus seedlings and their immediate budded progeny.

TABLE 2. Numbers of zygotic and nucellar seedlings from various seed-parent species and varieties (after Frost and Soost, 1968)

Seed parent	No. of pollen-parent cultivars[z]	No. of seeds	Seedlings per seed		Nucellar seedlings as percentages of all seedlings
			total	zygotic	
LEMON					
Eureka	5	210	1.06	0.71	33
Lisbon	6	703	1.05	0.71	32
Feminello	2	23	1.43	—	69
Monachello	2	281	1.09	—	24
Rough	1	819	1.96	0.05	98
Rough	1[y]	127	1.24	0.57	54
LIME					
Mexican	1[y]	14	1.29	0.29	78
Kusaie	1[y]	87	1.09	0.80	26
Red	1[y]	130	1.17	0.66	43
MANDARIN GROUP					
Dancy	6	193	1.37	0.00	100
Willow Leaf	7	771	1.28	0.18	86
Kishiu	6	333	1.00	1.00	0
Ponkan	1	79	1.42	0.02	98
Kara	1	83	1.71	0.00	100
Szinkom	7	1683	1.50	0.25	83
King	7	387	1.01	0.80	21
Satsuma	11	323	1.44	0.14	90
GRAPEFRUIT GROUP					
Marsh	6	207	1.08	0.05	96
Imperial	8	626	1.26	0.42	66
Imperial	Selfing	99	1.22	0.10	92
Panuban	1	70	1.11	0.83	26
Sukega	2	116	1.00	1.00	0
PUMMELO					
Four cultivars	14	1954	1.001	1.001	0
Seven cultivars	8	610	1.00	1.00	0
SWEET ORANGE					
Maltese Oval	8	167	1.09	0.66	39
Paperrind	5	82	1.56	0.26	84
Valencia	8	106	1.35	0.21	85
Washington navel	6	24	1.33	0.04	97
Washington navel	1	32	2.00	0.62	69
SOUR ORANGE	1	319	1.21	0.18	85
TANGELO					
Orlando	1	525	1.31	0.20	83
Minneola	1	279	1.49	0.04	97
Sunshine	1	401	1.74	0.00	100
SPECIES					
C. taiwanica	1[y]	594	1.27	0.68	47
C. macrophylla	1[y]	139	1.42	0.01	99
C. amblycarpa	1[y]	54	1.28	0.00	100
C. ichangensis	1[y]	11	1.00	1.00	0
C. pennivisiculata	1	110	1.36	0.05	96
P. trifoliata	1	228	1.03	0.89	13
P. trifoliata	Selfing	80	1.26	0.10	73
OTHER					
Yuzu	1[y]	90	1.24	0.41	67
Yuzu	1[y]	198	1.22	0.05	95
Ichang Hybrid	1[y]	231	1.58	0.00	100

[z] Except where selfing is indicated.
[y] *Poncirus trifoliata* pollen.

Juvenility is evidenced by thorniness, vigorous and upright growth habit, slowness to fruit, alternate bearing, and certain physical characters of the fruit. This behavior occurs in both sexual and nucellar seedlings but it can be most clearly demonstrated in nucellar progeny, where genetic recombination is not a factor.

Thorniness is especially prominent. On trees of most old budded cultivars, thorns are few and small, although in some lemons they are notably persistent. Among young nucellar seedlings and budlings, however, thorniness is usually much greater than in the seed parent trees. Thorniness of sexual seedlings is variable but generally high. Thorniness is much reduced after repropagation from the upper parts of seedlings several years old; the lower parts of seedling trees appear to retain indefinitely the ability to produce thorny shoots. H. B. Frost, in an early study of 15 cultivars in California, found that thorniness was much greater on young trees of the second budded generation from nucellar seedlings than on trees of the same age, budded from the old seed parent budlines (Frost, 1943). Table 4 shows some of these comparisons, and also shows that first fruiting was very much less on the trees of the nucellar budlines. Fig. 5 contrasts the thorniness of a 'Minneola' tangelo tree budded from an old budline with that from a nucellar budline only a few years of age from seed.

FIG. 4. Two sets of sexual twins (1, 2 and 3, 4) from the cross 'Clementine' x trifoliate orange. Each set is from a single seed. (After Cameron and Garber, 1967)

TABLE 3. The production of multiple sexual embryos in seeds of typically monoembryonic seed parents (after Ozsan and Cameron, 1963)

Variable	Seed parent[z] Clementine	Wilking	Pummelo 2421
Total seeds examined	1545	213	39
Number of polyembryonic seeds	42	5	3
Total embryos in polyembryonic seeds	112	12	7
Proven sexual seedlings from polyembryonic seeds[y]	58	6	5

[z] The pollen parent was the trifoliate orange, which imparts a dominant trifoliate leaf to its hybrid progeny.
[y] No seedling was shown to be nucellar; embryos not proved sexual did not produce seedlings large enough to be judged for trifoliate leaf.

Greater thorniness has also been shown on trees propagated from budwood taken from low positions, as contrasted to high positions, on young nucellar or other seedling source trees. In such cases, selection of thornless budwood from high positions often results in progeny trees of low average thorniness. However, certain thorny genetic types cannot necessarily be modified by this procedure. It is noteworthy that rooted cuttings and trees from rootsprouts also often show more initial thorniness than do budded repropagations from the same clones.

Growth rate of seedlings and seedling budlines, as evidenced by trunk cross section and top vol-

FIG. 5. Minneola tangelo trees of equal age from budding. From left, old budline, nucellar budline.

ume, is usually greater than that of old budlines of the same clones for many years; however, virus and virus-like diseases may often influence this behavior. Virus diseases in citrus seldom pass through the seed, so that young seedlings, both sexual and nucellar, are commonly initially virus-free. Murashige et al. (1972) also established virus-free sources of old budlines by micro-grafting shoot apices, using sterile nutrient culture techniques. In a study where nucellar-budline 'Eureka' lemon trees were inoculated with buds from old budlines and compared with uninoculated sister trees, growth was much depressed in the inoculated group (Calavan and Weathers, 1959). Exocortis virus introduced from the old budline was considered to be responsible. In a later study with the very vigorous 'Lisbon' lemon, a much lesser effect of inoculation from an old budline was observed (Calavan, personal communication).

First flowering in citrus seedlings may require five or more years, depending upon the cultivars involved. Efforts to reduce this period, whether by budding, topworking, girdling, or stunting, have been largely unsuccessful. Clear differences exist among cultivars; nucellar seedlings from some lemons and limes flower early from seed, while grapefruit and especially oranges are slow. Flowering usually occurs first in the distal portions of the higher shoots. Furr (1961) compared pairs of budlings propagated from basal and top shoots taken from seedlings which had already fruited. The tested budlings of lemon and lime all fruited at about the same time, but those of grapefruit showed much more early fruiting from top buds than from basal ones.

In the first fruiting years, the fruit characters of budlines from seedlings may be noticeably inferior. The fruits often show more elongation, rind puffing, and hollowness of core, and may have thicker rinds than fruits of older budlines. Other characteristics are lower seed number, and in navel orange cultivars, smaller average size of navel structure. These differences tend to disappear as the trees and budlines grow older, and fruit quality, considered over a series of cultivars, is usually the same as in old budlines. Nevertheless, in certain cultivars the differences can be surprisingly persistent. In California, 5- to 15-year-old budded trees of nucellar budlines of 'Washington' navel orange and 'Marsh' grapefruit last reproduced from seed

TABLE 4. Thorniness and fruiting of 3-year-old budlings from old budlines and younger nucellar seedling selections (after Cameron and Frost, 1968)

Cultivar	Budwood source within clone[z]	Degree of excess thorniness in nucellar budline	Average number of fruits per budling
Eureka lemon	Old budline		8.7
	Nucellar	moderate	0.2
Marsh grapefruit	Old budline		8.2
	Nucellar	moderate	0
King mandarin	Old budline		18.0
	Nucellar	high	3.0
Valencia orange	Old budline		16.8
	Nucellar	very high	0

[z] Six trees from each budwood source. Nucellar selections originated from seed in 1915–1917; this experiment budded in 1930.

in 1915 still showed some of those trends in 1965. Table 5 shows representative data from three field plantings. Climate and amount of crop, within years, can influence these characters, but an intrinsically more vegetative behavior in young nucellar budlines must play a part. There were no acute disease symptoms in the field plots of Table 5, but mild, undefined virus forms which may be present in old budlines could indirectly modify fruit characters by depressing growth vigor of the trees.

The effects described above must be distinguished from differences shown by sexual progeny and differences resulting from mutation and chimerism which frequently occur in citrus. Nucellar selections have sometimes been described as showing other differences such as greater frost hardiness, larger fruit size, and added resistance to disease, but unless genetic changes have also been involved, these effects are probably indirect or temporary results of seedling vigor.

Some authors (e.g. Tutberidze, 1966; Kapanadze, 1969) have reported that certain seedlings occurring in assumed nucellar progenies from controlled crosses are indeed nucellar in origin, yet still show genetic characters from the pollen parent. But various crosses can yield a mixture of nucellar and sexual seedlings with intergrading characters, and evidence for such unusual occurrences should be fully documented.

TABLE 5. Fruit characters of nucellar and old budlines of Washington naval orange and Marsh grapefruit in California[z] (from Soost et al., 1965)

Field experiment no.	Year	Budline	% of fruit with stem-end taper	thick rind	hollow core
		Washington navel orange			
1	1963	Old	17	6	8
		Nucellar	37	12	12
	1964	Old	9	12	14
		Nucellar	10	19	15
2	1963	Old	12	5	1
		Nucellar	11	6	22
	1964	Old	13	18	16
		Nucellar	17	25	51
Mean yearly excess % in nucellar			6.0	5.3	15.3
		Marsh grapefruit			
3	1961	Old	44	36	11
		Nucellar	63	51	15
	1962	Old	22	22	0
		Nucellar	47	64	16
Mean yearly excess % in nucellar			22	28.5	10

[z] Based on 20 fruits from each of 10 trees of each budline, at each sampling. Compared budlines were usually of the same age and rootstock.

The most valuable horticultural characters of nucellar selections have been high tree vigor, initial freedom from virus diseases, and, usually, high yields (Cameron et al., 1959). Recent studies by workers in several countries, including Egypt, Israel, and Russia, substantiate this high yield potential.

Incompatibility

In addition to absolute gametic sterility, self- and cross-incompatibility are present in *Citrus*. All tested cultivars of *C. grandis* have been self-incompatible. Several known or suspected hybrid cultivars are incompatible. These include 'Clementine', 'Orlando', 'Minneola', 'Sukega', 'Robinson', 'Nova', 'Page', and 'Fairchild'. Self-incompatibility has been reported for several cultivars in Japan. Other researchers have reported self-incompatibility for some cultivars of the species *C. limon*, *C. limettioides*, and *C. sinensis* (Frost and Soost, 1968). Several of the cultivars in these latter species also have a high degree of pollen sterility making the determination of incompatibility doubtful.

Early work in Japan indicated that some self-incompatible varieties were also cross-incompatible (Nagai and Tanikawa, 1928). 'Orlando' and 'Minneola', both hybrids of 'Duncan' grapefruit with 'Dancy' tangerine, are cross-incompatible. 'Robinson', 'Nova', and 'Page' also are suspected of being cross-incompatible (Hearn et al., 1969).

Hybrids between self-incompatible cultivars have also been self-incompatible, and sometimes cross-incompatible. These compatibility relationships are evidently determined by a series of gametophytic, oppositional alleles, but the extent of their distribution is not known (Soost, 1969). At least four alleles are present in the cultivars tested, but as many as eight may be present. On this basis, it is predicted that all progeny resulting from intercrossing self-incompatible cultivars will be self-incompatible and either one-fourth or one-half of them will be cross-incompatible, depending upon the constitution of the parents. *C. grandis* is the only species in which the system is clearly widespread. Most of the other known self-incompatible cultivars are known or suspected hybrids. Several have *C. paradisi* (grapefruit) as one parent. This species is thought to have been derived from *C. grandis* and may have received self-incompatibility genes from it. The source of the second allele in some grapefruit hybrids must be from a parent which is not closely related to *C. grandis* as indicated by the parentage of 'Orlando' and 'Minneola' (above).

Hybrid Vigor and Inbreeding

There is evidence for hybrid vigor and inbreeding depression in citrus, although critical data are difficult to obtain. Within the genus *Citrus*, narrow crosses are likely to produce many weak offspring. H. B. Frost (1943), in limited early tests, reported weak hybrids from the crosses of 'Ruby' x 'Valencia' orange and 'Eureka' x 'Lisbon' lemon. J. P. Torres (1936), in the Philippine Islands, found that even the wider cross, sour orange x rough lemon, gave hybrids which were predominantly weaker than either parent. Among 100 hybrid field trees of 'Eureka' x 'Lisbon' recently studied at Riverside none were as large or vigorous as comparable nucellar seedling trees of the two parents; 47 of the 100 produced no flowers even at the age of 11 years.

Selfing, especially in forms long propagated by nucellar embryony, is often deleterious, but most such cultivars give too few sexual seedlings to provide extensive data. Selfing of a self-fertile form of the 'Clementine' mandarin, which is probably a hybrid, has given many weak seedlings. Selfing of the 'Temple' (tangor?) has given some moderately vigorous ones.

Wider crosses within *Citrus* often produce favorable proportions of vigorous offspring. Mandarin x grapefruit hybrids frequently average better in vigor than mandarin x mandarin or mandarin x orange. Intercrossing of selected F_1 clones usually yields some vigorous advanced generation individuals. Crosses involving the pummelo are especially fast-growing, reflecting marked hybrid vigor. Table 6 summarizes recent studies at Riverside where trunk diameters of hybrid populations with varying proportions of pummelo ancestry were compared. In each planting, F_1 families with pummelo as one parent had significantly larger trunk size than comparable families with less or no pummelo germplasm. Top volumes of the trees were also obviously larger in most cases. The pummelo has apparently existed as a strictly sexual species for a long period, and it may not be carrying the load of unfavorable genes which seems to be present in many highly apomictic forms.

One of the comparisons in Table 6 involves *Poncirus*. Crosses made with this genus as long as 70 years ago showed that vigorous hybrids with *Citrus* could be obtained. Some vigorous citrus rootstocks, including 'Troyer' citrange, have *Poncirus* as one parent. However, many weak or sterile hybrids have also been produced.

Polyploidy

Diploidy is the general rule in *Citrus* and its related genera, the gametic (n) chromosome number being 9. *Fortunella, Poncirus, Microcitrus, Eremocitrus, Citropsis, Murraya,* and other genera are regularly diploid. However, forms with increased numbers have been identified or produced. The Hong Kong wild kumquat, *Fortunella hindsii* (Champ.) Swing. was the first form to be identified as a tetraploid, but a diploid form has subsequently been reported. Many spontaneous tetraploids in *Citrus* and *Poncirus* subsequently have been identified. Triploids have been identified in zygotic progenies of diploid parents. Some aneuploids have been reported and a very few individuals with higher ploidy levels (Esen and Soost, 1972c).

Most spontaneous tetraploids have been obtained as nucellar seedlings in the course of breeding and varietal studies. They have occurred in frequencies ranging from under 1% to 6% in all major groups of cultivars, including orange, lemon, and grapefruit (Cameron and Frost, 1968). They have also been obtained from other cultivars and individuals that produce nucellar seedlings in sufficient quantities. They often occur in seeds which produce diploid seedlings and in seeds from fruits which mainly produce diploids, indicating that doubling occurs in the ovule or ovary.

Tetraploid hybrids from crossing diploid parents have rarely been reported. Tetraploid hybrids were obtained in unexpectedly high frequencies when monoembryonic diploid seed parents were crossed with tetraploid pollen parents. Esen and Soost (1972a) have determined that this high percentage results from the failure of many triploid embryos to survive while most of the tetraploids survive. Failure is related to the ratio of the chromosome number of the endosperm to the chromosome number of the embryo. In all cases verified cytologically, the tetraploid embryo is accompanied by hexaploid endosperm, giving an embryo

TABLE 6. Effect of pummelo germplasm on tree size in field plantings of citrus hybrids near Riverside, California[z]

Field plot no.	Cross	Generation relative to pummelo parent	% pummelo germplasm in hybrids	Number of trees	Mean trunk diameter (cm)[y]
1	Pummelo x (mandarin x orange)	F_1	50	77	14.4***
	(Mandarin x orange) x (mandarin x orange)	—	0	125	9.4
2	Pummelo x grapefruit	F_1	50	227	13.2***
	(Pummelo x [mandarin x orange]) x (mandarin x orange)	BC_1	25 variable	194	12.4
	(Mandarin x orange) x (mandarin x orange)	—	0	56	10.5
3	Pummelo x (mandarin x orange)	F_1	50	142	15.4***
	(Pummelo x [mandarin x orange]) x Self	F_2	variable	78	10.6
4	Pummelo x *Poncirus*	F_1	50	21	11.9**
	(Mandarin x orange) x *Poncirus*	—	0	87	9.9

[z] Trees 6 to 14 years of age, on their own roots. Trees within each field plot planted in the same year, in adjacent rows.
[y] Trunk diameters measured 30 cm above ground. Asterisks indicate significance of the difference, calculated from logs of the diameters, of the marked mean from other means in that plot (** significant at 1%; *** at 0.1%).

to endosperm chromosome number ratio of 2:3 (0.67). The triploid embryos result from the union of the usual haploid egg with diploid male gametes from the tetraploid pollen parent. With triploid embryos, tetraploid endosperm is invariably present, giving an embryo to endosperm ratio of 3:4 (0.75). The embryo to endosperm ratio with diploid embryo and triploid endosperm is identical to the ratio with tetraploid embryo and hexaploid endosperm (0.67), in contrast to the ratio of 0.75 obtained with triploid embryos in the cross 2n x 4n. Embryo and endosperm are initiated and appear to develop normally, but in most cases the endosperm fails before seed maturity and the embryo does not survive. In these crosses (2n x 4n), the tetraploids result from the production of diploid gametes by the seed parents in low frequencies. The percentage of diploid gametes produced is a function of the particular seed parent. The five cultivars tested (Esen and Soost, 1972a) varied from less than 1% to approximately 20%. Because of the failure of triploid embryos, the percentage of tetraploids in the surviving progenies is much larger than the percentage of diploid gametes. The cultivars tested varied from approximately 6% to 95% in their production of surviving tetraploids; the percentage of survivors was correlated with the percentage of diploid gametes produced. Although the failure of the triploid embryos eliminates 2n x 4n crosses as an effective means of producing triploids, the survival of the tetraploids provides a means of generating new tetraploids for breeding.

Effects of tetraploidy in citrus can be rather accurately evaluated in nucellar selections. Except for possible bud variation, such tetraploids differ genetically from sister nucellar diploids only in chromosome number. Tetraploids grow more slowly, are more compact in habit, and are generally less fruitful than diploids of the same cultivar. Among selections at Riverside only the tetraploids of 'Lisbon' lemon and some grapefruits have been productive, often giving heavy yields. Leaves of the tetraploids are broader, thicker, and darker in color (Fig. 6). Excessive dying of twigs and branches has been a problem in most tetraploids.

The grapefruit tetraploids have been strikingly more vigorous and healthy than any of the others.

Tetraploids on their own roots are usually smaller than when budded on a rootstock. Furusato (1953) found that tetraploid seedlings of *Poncirus* had a thicker main root and fewer lateral roots than diploids. As a rootstock, tetraploid *Poncirus* has been very variable in performance. In most cultivars, tetraploid fruits have been less elongate than diploid ones and are commonly smaller. In 'Lisbon' lemon, tetraploid fruit size has been larger than in the diploid. Irregular fruit shape is characteristic, and the rinds are usually thicker (Fig. 7). Fruits tend to color later, but the content of acid and soluble solids is similar to that in diploids. Seed number in tetraploids seems to depend upon the cultivar; it has been low in some but in 'Lisbon' lemon it has been higher than in the diploid. The number of embryos per seed tends to be fewer in tetraploids than in diploids but there is little evidence on the proportion of sexual embryos in seeds from selfing.

Cameron and Scora (1968) compared rind oil components of diploids and tetraploids of seven cultivars by gas chromatography. Percentages of oil constituents were markedly similar between diploids and their respective tetraploids. The few significant differences tended to occur with the same constituents among different varieties.

The characters of citrus tetraploids show that these forms have little economic value in themselves. However, they are of interest as breeding material, since they can be used in crosses with diploids to produce triploids.

FIG. 6. Diploid (upper row) and tetraploid (lower) leaves from six citrus cultivars. From left: 'Lisbon', 'Paperrind' orange, 'Hall's Silver' grapefruit, 'Dancy' tangerine, 'King' mandarin, 'Owari' satsuma. Note the broader leaf shape of the tetraploids. (After Cameron and Frost, 1968.)

FIG. 7. From left: whole and cut fruits of tetraploid sweet orange parent, triploid hybrid, and diploid 'Kinnow' mandarin parent, showing seedlessness of the triploid and thick rind typical of tetraploids.

The characteristics of citrus tetraploids described above have been determined entirely from comparisons of nucellar tetraploids from existing cultivars. Their usual lack of vigor could be related to the particular genotypes involved. However, the large number of hybrid tetraploids obtained in recent years from crossing diploid seed parents by tetraploid pollen parents have also been lacking in vigor. In populations consisting of hybrid triploids and tetraploids, the tetraploids can be consistently identified by their lack of vigor, indicating that poor growth is associated with the tetraploid condition rather than with tetraploidy of certain genotypes.

Triploids occasionally occur spontaneously in progenies from diploid by diploid crosses. Unlike the spontaneous tetraploids, they are always of zygotic origin. As high as 5% triploids have been identified in some progenies.

Triploids can also be systematically produced by controlled crossing of tetraploids with diploids, although the incidence of viable seeds is low. The first triploids to be identified cytologically were produced by crossing diploid limequat by pollen of tetraploid *Fortunella hindsii* (Champ.) Swing. (Longley, 1926). Frost (1943) obtained a considerable number of triploids from crosses of tetraploid grapefruit and lemon pollinated by several diploid cultivars. Russo and Torrisi (1953) produced triploid lemons by use of a tetraploid pollen parent. Tachikawa et al. (1961) obtained triploids both from the cross 2n x 4n and from its reciprocal. Gecadz (1967) reported triploids from crosses of tetraploid lemon x diploid *C. ichangensis* Swing.

Additional triploids have been obtained at Riverside, using eight tetraploid cultivars and 24 diploids. When the pollen parent was tetraploid, the proportion of developed but empty seed coats was usually high, even if the seed parent produced some nucellar embryos. Subsequent research (Esen and Soost, 1972*a*) shows that this is caused by the failure of the endosperm and triploid embryos to develop normally. The low recovery rate of triploids when the tetraploid is used as the pollen parent makes it desirable to use the tetraploid as the seed parent. However, most tetraploids available produce moderate to high percentages of nucellar embryos. To expedite the production of triploids, monoembryonic tetraploid cultivars are needed. Barrett (personal communication) has recently produced tetraploids of monoembryonic cultivars by colchicine treatments. Esen and Soost (1972*b*) have shown that the recovery of triploids from diploid by diploid crosses can be improved by selecting the small seeds and germinating them under very favorable conditions. Although the embryo develops normally, its size appears to be restricted in later stages of development by early termination of endosperm development. The restriction in development appears to be related to deviation of the embryo to endosperm ratio in chromosome number from the normal ratio of 2 to 3 (0.67). The seeds with triploid embryos invariably have pentaploid endosperm giving a 3:5 (0.60) embryo to endosperm ratio. The percentage of triploids produced is a function of the seed parent used and varies from less than 1% to approximately 25% in three cultivars tested.

The triploids show considerably more average vigor than tetraploids. They also show much genetic variability characteristic of sexual citrus progenies. Triploids can usually be identified as polyploid by their thick, rounded leaves. Measurements show that the leaves are usually intermediate in shape and thickness between those of their 2n and 4n parents. Much sterility is to be expected in triploids and all of those which have fruited at Riverside have had a few or very few seeds (Fig. 7). Fruitfulness has been highly variable: some triploids have set fruit well while others are very unproductive (Soost and Cameron, 1969). This tendency to low yields is a serious problem, but since parthenocarpy is characteristic of some diploid cultivars it may also be present in a reasonable proportion of triploids. Low seed content is horticulturally highly desirable so that a continued breeding effort and study of triploids is worthwhile.

Mutations

Spontaneous mutations: Spontaneous mutation, as evidenced by sudden change in heritable character, occurs frequently in citrus. Mutations are often observed as limb sports or sectors on fruits; they are also detected occasionally in nucellar seedlings or their budded progeny. Among sexual seedlings, mutations can seldom be distinguished from gene recombinations. A. D. Shamel and associates, beginning in about 1909, documented many somatic variations occurring in commercial citrus cultivars, and verified their heritability by bud propagation (Shamel, 1943). These changes were for the most part unfavorable, showing poor yield, atypical leaf characters, or abnormal fruits. A few had characters which could be useful under certain conditions.

Several highly valuable spontaneous mutations have occurred in citrus. The 'Washington' navel orange and the 'Marsh' grapefruit, cultivars of primary importance and both nearly seedless, are each considered to have arisen from a closely related seedy form. The 'Shamouti', which originated in Palestine by bud mutation, is a superior orange when grown in certain environments. Other more recent mutants of interest include the 'Salustiana' orange of Spain, and the 'Marrs' of Texas, somatic variants which ripen early in the areas in which they were discovered. A variant of the navel orange called 'Skaggs Bonanza', described as early ripening and heavy bearing, has recently been introduced in California. In 1913 in Florida the 'Thompson' pink grapefruit was discovered as a limb sport from the white 'Marsh.' The 'Thompson', in turn, has repeatedly produced valuable red-fleshed sports such as the 'Redblush'. These changes evidently involve both mutation and a chimeral system.

Cultivars of the satsuma mandarin are another citrus group in which mutation has been very important. A number of distinct types have been described in Japan during the last 50 years, and investigation of additional variants is continuing there. Changes in rind color and time of ripening have been of particular interest. Many of the satsuma variants also seem to involve chimerism, as described later.

Wood pocket disease of semi-dense 'Lisbon' lemons appears to be due to an unstable mutation, and may reflect chimeral interrelations. Its symptoms include leaf variegation, fruit rind sectoring, and wood cankers. It recurs in lemon trees grown from diseased buds, and is transmitted through a considerable proportion of both sexual and nucellar seedlings, but not transmitted to healthy trees by tissue grafts. Degree of symptom expression is variable; the disease often increases in severity in successive budded or seedling generations.

Induced mutations: An experiment to induce mutation in citrus by x-rays was reported as early as 1935, and several short-term studies using various agents have been published recently (e.g. Kerkadze, 1970). Morphological and cytological changes have been observed, but stable, favorable results have not been clearly demonstrated. Nucellar embryony, natural sterility, and spontaneous mutation complicate such studies in citrus. Hensz (1960) obtained several thousand plants from irradiated seeds and budwood of grapefruit and Valencia orange, and some persistent changes were later observed. A low-seeded, red-fleshed grapefruit, the 'Star Ruby', was found among the seedlings of a seedy parent as part of the study, but there is little evidence that it was produced by the irradiation. Gregory and Gregory (1965) have searched for changes toward winterhardiness in citrus, following irradiation of budwood. So far, results appear to be uncertain.

Chimeras: The existence of chimeral plants has long been recognized in citrus. Chimeral conditions are particularly evident in leaves and fruit. Detailed descriptions of citrus chimeras appeared in the literature as early as the seventeenth century, when a tree which was a mixture of citron and sweet orange was recorded by Ferrari. Synthetic chimeras such as this usually originate at the point of graftage of a stock and scion, and include tissues of both. Some Russian authors have claimed that such chimeras can produce nucellar offspring which inherit characters from both tissue components. This is doubtful, and Majsuradse (1966), after studying pollen and obtaining seedlings from controlled cross and self-pollinations of some of these chimeras, showed that he could account for nearly all of the seedlings on the basis of normal sexual and asexual mechanisms.

FIG. 8. Variegated pink lemon chimera, showing irregular white leaf margins and green and white stripes on fruit. (After Cameron and Frost, 1968).

More commonly, chimeras in citrus result from somatic mutation followed by persistence of both the old and new cell types. A periclinal arrangement of the two genetic types is the most common; this can be relatively stable or highly unstable especially in the fruit. Fig. 8 shows variegated leaves and fruit of a clone of lemon called 'Variegated Pink', grown at Riverside. The leaf patterns of this cultivar indicate it to be a white-over-green chimera, with a subepidermal histogen (Layer II) which produces white tissue and an inner core (Layer III) which produces normal green. The leaves have an irregular, white marginal zone, the white regions often being dwarfed. Several shades of green occur from the midrib outward, indicating that the cell layers involved are arising in variable fashion from the two histogens. The immature fruits have a thin, whitish rind with irregular, narrow longitudinal stripes of green. Even the bark of young shoots can show a white and green pattern.

Citrus appears to be one of many dicotyledonous plants in which the meristems are organized into three histogenic layers; studies of nucellar seedlings from well-characterized old budlines can sometimes show whether a histogenic layer in the old budline carried a given genetic factor. L-II evidently produces the nucellus and the male and female gametes, and in some white-over-green leaf chimeras nucellar seedlings are regularly white leafed, indicating that L-II lacked the chlorophyll factor. Nucellar seedlings have recently provided evidence that the 'Thompson', 'Foster', and 'Burgundy' pigmented grapefruits are periclinal chimeras (Table 7). These cultivars all arose by somatic mutation from white grapefruit clones. 'Thompson' and 'Burgundy' show color in the juice vesicles but not in the rind, while 'Foster' has color in both tissues. 'Thompson', by further somatic variation, has repeatedly given rise to redder-fleshed variants such as 'Redblush', which do express color in the rind. Nucellar seedlings of 'Thompson' produce only white fruit, as do seedlings of 'Burgundy', while nucellar seedlings of 'Redblush' produce red fruit typical of the parent clone. Evidently, the color factors in 'Thompson' and in 'Burgundy' are present in L-I, which contributes to the juice vesicles, but not in L-II. In 'Redblush', the factor must be present in L-II as well as L-I, no doubt as a layer substitution, and the nucellar seedlings must carry it in all layers.

In the case of the 'Foster', fruit of nucellar seedlings has deeper flesh color than that of the parent clone, and continues to show color in the rind. This implies that 'Foster' carries the color factor in L-II, and that when all layers carry it (as in the seedlings) the flesh color is deeper. The parent 'Foster' may also have the factor in L-I; if it does not, some interaction among tissues may produce the color in its flesh.

A similar explanation has recently been proposed to account for the behavior of the 'Suzuki Wase' satsuma, an early-ripening variant which occurred as a limb sport on a common satsuma in Japan, and which has frequently shown somatic reversion. Iwamasa and Nishiura (1970) reported that 25 nucellar seedling trees of 'Suzuki Wase' all bore the later-ripening fruit of the common type, instead of parent-type 'Suzuki Wase' fruit. Three other Wase clones, which had not shown somatic reversion, produced nucellar seedlings which continued to express the early ripening character. 'Suzuki Wase' apparently carried an early-ripening factor in L-I, where it increased sugar content and decreased acidity of the juice; it should not carry it in L-II, since nucellar seedlings lack it. The three other clones presumably carried the factor in both layers. The early-ripening character also involves early coloring of the rind (from green to orange). Since this rind tissue is believed to arise from L-II, the authors postulated an interaction depending upon enzymes or hormones originating in the L-I juice vesicles to account for the early coloring.

TABLE 7. Color expression in parental clones and derived nucellar seedlings of certain grapefruit cultivars (taken principally from Cameron et al., 1964)

Cultivar	Parental clones[z] Juice vesicles (largely L-I)	Parental clones[z] Rind and septa[y] (L-II; partly L-III?)	Nucellar seedlings Juice vesticles, rind, and septa
Marsh	white	white	white
Thompson pink	pink	white	white
Burgundy	red	white	white
Redblush types	red	red	red
Foster pink	pink	red	red

[z] L-I, L-II, and L-III indicate histogenic layers.
[y] Excluding epidermis.

Breeding for Specific Characters

In various citrus breeding programs, groups of specific characters have been sought. However, sexual progenies from hybridization are commonly highly variable; single-gene inheritance is rarely indicated, and F_1 families often display a wide quantitative range of character expression (Furr, 1969). Thus, crosses of oranges with mandarins can yield progenies whose rind and flesh colors range from pale yellow to deep red-orange. In crosses of grapefruit with mandarins, the extremes of various fruit characters may approach the norm of either parent. In many crosses, a comparable range of variability occurs with respect to tree vigor and habit, and leaf characters. Occasionally a hybrid exceeds the limits of its parents in some character. Thus, a new mandarin called 'Fremont' 'Clementine' mandarin x 'Ponkan' mandarin) shows less granulation of the flesh than either parent under hot desert conditions in California. Similarly the hybrid mandarin 'Encore' has a longer summer season of use in some climates than do its parents, 'King' and 'Willowleaf'.

Occasionally there is segregation of a character in citrus progenies which indicates the action of one or a few genes. From certain crosses made both in Java and the United States, extreme dwarf plants occurred. A cross of pummelo x citron yielded only dwarfs, while the reciprocal produced 50% dwarfs. A particular pummelo crossed by 'Clementine' mandarin also produced exclusively dwarfs; crosses of this pummelo by 'Temple', 'Kara' mandarin, and 'Valencia' orange gave partly dwarf and partly normal plants. But in a progeny of 176 plants of the same pummelo x 'Pearl' tan-

FIG. 9. Dwarf seedlings (in circles) seven years old and about 30 cm high, and a normal sibling. Examples are from a cross involving pummelo.

gelo, all plants were normal. Fig. 9 shows two dwarf plants and a larger sister plant at seven years of age.

The trifoliate leaf character of *Poncirus* shows essentially complete dominance over the monofoliate condition of *Citrus;* patterns of segregation in advanced generations suggest that two principal genes (perhaps duplicate factors) could be involved. At Riverside, among 1716 F_2 seedlings from selfing or intercrossing of 33 strictly sexual F_1 hybrids of 'Clementine' x *Poncirus,* 7.2% were monofoliate; among 326 F_2 seedlings from 7 hybrids of an orangelo x *Poncirus,* 5.8% were so. A few progenies from backcrossing such F_1 hybrids to monofoliate *Citrus* have had from 20 to 40% monofoliate plants. Plants with mixed leaf types occur, however, among both F_2 and backcross progenies.

Hybrids among monofoliate *Citrus* species also occasionally show trifoliate leaves, often only transiently. However, H. Toxopeus (1962) found that about one-third of 50 two-year-old seedlings from the cross, *C. grandis* x *C. hystrix,* (both monofoliate) were trifoliate leaved.

Toxopeus also reported that about 50% of 598 seedlings from *C. grandis* x 'Meyer' lemon showed in young leaves the purple anthocyanin coloration typical of lemons. A recent study by M. Malik (Ph.D. dissertation, University of California, Riverside) of 100 inter-clonal hybrids between 'Eureka' and 'Lisbon' lemons showed that 19 had green young-leaf color, and 81 had various shades of brown-red. The data fit the assumption that these lemon cultivars are heterozygous for a dominant gene controlling this reddish color. Color segregation in the young stems of these same 100 hybrids occurred in a different proportion, however.

Iwamasa (1966) found segregations for undeveloped anthers and for sterile pollen in crosses between the pollen-sterile satsuma and other pollen-fertile cultivars. Eight out of 20 hybrids between satsuma and several pollen-fertile *Citrus* clones produced aborted anthers; 26 out of 46 hybrids between satsuma and *Poncirus* had essentially complete pollen abortion.

Nucellar Embryony

The attribute of nucellar embryony seems to be inherited in a fairly simple fashion. In the study shown in Table 8, crosses between entirely sexual, monoembryonic parents within *Citrus* yielded hybrids which were all sexual and monoembryonic.

TABLE 8. Ratios of polyembryonic to monoembryonic hybrids, and range of embryo numbers per seed, in crosses among citrus cultivars (after Parlevliet and Cameron, 1959)

Seed parent cultivar	Pollen parent: Cultivar	Pollen parent: Embryo number per seed	Hybrid progeny trees: Ratio P:M[y]	Hybrid progeny trees: Range of embryo numbers in polyembryonic trees[x]
Crosses among monoembryonic varieties[z]				
Pummelo 2347	Clementine		0:12	
Pummelo 2341	Temple		0:12	
Wilking	Sukega		0:12	
Clementine	Sukega		0:12	
Crosses of monoembryonic cultivars by polyembryonic cultivars				
Pummelo 2347	Honey	2.7	13:14	1.06- 1.4
Pummelo 2347	Ponkan	48.0	19: 3	1.14- 2.9
Pummelo 2240	Kara	8.1	3: 6	1.16- 2.4
Pummelo 2240	Frua	18.0	8: 0	2.40-10.7
Clementine	Honey	2.7	16:22	2.80-16.0
Clementine	Ponkan	48.0	8: 3	6.70-16.0
Wilking	Mency	15.0	10: 2	6.10-38.0

[z] These cultivars occasionally have multiple embryos, which are apparently sexual.
[y] P:M = polyembryonic:monoembryonic.
[x] quantitative determinations were not made on all the polyembryonic trees.

Crosses of monoembryonic cultivars x polyembryonic nucellar ones gave hybrids of both types in ratios sometimes approaching 1:1. In some cases, two polyembryonic parents have produced strictly sexual, monoembryonic hybrids; thus, 'King' x 'Willowleaf' produced 'Wilking' and 'Encore', both monoembryonic. A recent study in Japan (Iwamasa et al., 1967) substantiated this general pattern of inheritance. Five crosses involving four monoembryonic parents, all different from those of Table 8, produced 81 hybrids which were all monoembryonic. Monoembryonic x polyembryonic parents again gave many hybrids of each kind. One or two dominant genes may be conditioning polyembryony in these cases, seemingly with modifiers which affect the degree of polyembryony. However, in crosses of *Poncirus*, which is polyembryonic, with certain monoembryonic *Citrus* cultivars, very high proportions of monoembryonic hybrids have occurred, as if additional genes were involved.

Acidity

Levels of acidity in citrus fruits can be inherited in a semiquantitative manner in certain crosses. In many cultivars both acid and very low-acid forms are known. At Riverside hybrids of relatively low acidity were obtained from crosses of a very low-acid pummelo by medium-acid parents of other species and cultivars; when pummelos of medium acidity were crossed by these same medium-acid pollen parents, the progenies had acid levels which were mostly above those of either parent (Table 9). A selfed F_2 progeny was later obtained from

TABLE 9. The influence of acid and very low-acid pummelos on the acidity of hybrid citrus progenies (Soost and Cameron, 1961, and later data)

Selections tested[z]	Number of plants tested	Range of acidity (%) in 1957	1958	1959
PARENTS				
Very low-acid pummelo seed parent	1	0.1	0.05	0.1
Medium-acid pummelo seed parents	4	1.6-1.9	1.7-1.9	1.0-1.7
Medium-acid citrus pollen parents[y]	4	1.0-2.2	1.0-2.0	0.8-1.9
F_1 HYBRIDS				
Very low-acid pummelo x medium-acid citrus	30	0.7-2.6	0.7-2.3	0.8-1.6
Medium-acid pummelos x medium-acid citrus	40	1.9-5.0	1.7-5.2	1.4-3.5
		Range of acidity (%) in 1970		
F_2 GENERATION FROM SELFING				
Very low-acid pummelo x medium-acid mandarin, selfed	32	0.1-2.7[x]		

[z] Fruits tested between January and March.
[y] Including an orange, a mandarin, a tangor, and a pummelo.
[x] One other individual had a high acid content of 4.0%.

one of the very low-acid pummelo hybrids. Among 32 plants tested, nine had acidity levels almost as low as the very low-acid grandparent. But in other crosses, in which a very low-acid orange was hybridized with medium-acid cultivars, no such predominance of low acidity was found. Among some 200 F_1 individuals, most were as high or higher in acidity than the medium-acid parent.

Pests and Disease

Resistance to pests and diseases in citrus depends sometimes upon the scion, sometimes upon the rootstock, and occasionally upon interactions between the two. Root rots caused by the fungi, *Phytophthora citrophthora* (SM. & SM.) Leonian and *P. parasitica* Dastur, are serious and widespread. Two rootstock species which are notably resistant to these fungi are the sour orange and the trifoliate orange. H. J. Toxopeus in Java (Toxopeus, 1955) by 1940 obtained some hybrids resistant to *Phytophthora*, as tested by bark inoculations. The parents included sour orange, mandarin, and Japanese citron. One of the first hybrids produced by the USDA, the 'Sampson' tangelo, was used as a rootstock because of tolerance to *Phytophthora*. Recently, the USDA has screened many cultivars and hybrids for tolerance to *P. parasitica*. There was some evidence for tolerance from widely-varying sources, including especially *Poncirus* and some of its hybrids. Some pummelos and a few hybrids of 'Clementine' mandarin x pummelo were tolerant, as was the citrus relative, *Severinia*. Inheritance is no doubt quantitative.

At Riverside, L. J. Klotz and others recently tested hybrids and other cultivars as rootstocks in the field, by inoculation of the trunk cambium with *P. citrophthora*, and *P. parasitica* separately. Many of the clones which had shown no evidence of trunk infection while growing in heavily-infected field soils developed serious lesions after the bark inoculation. Certain *Poncirus* selections were resistant, but others were susceptible. Two important *Poncirus* hybrids, the 'Troyer' and 'Carrizo' citranges, were relatively resistant to both pathogens. The order of susceptibility among the cultivars as a group was not the same for the two pathogen species.

At Riverside, recent new hybrids of *Poncirus* x sweet orange, mandarin, or an orange-grapefruit selection have shown high resistance to natural trunk infection by *Phytophthora* when growing as field trees in infested soil. These same clones, tested as young nucellar seedlings under the severe conditions of sand culture root inoculation, were much more variable in their reactions. It is clear, nevertheless, that *Poncirus* is a valuable parent for breeding *Phytophthora*-resistant rootstocks.

Many selections of *Poncirus* and some *Poncirus* hybrids show high resistance to the citrus nematode, *Tylenchulus semipenetrans* Cobb, which is a serious root pest in California. Most of the common *Citrus* rootstock cultivars are susceptible, and many cases of reduced tree vigor and yields of scion orchards have been reported. In a study by J. W. Cameron and others in 1954, nearly all of 484 young hybrid seedlings from crosses of *Poncirus* with five susceptible *Citrus* species showed high initial resistance to root infestation in greenhouse pot tests. Many of these seedlings were later placed in field plantings in naturally infested soil. After several years, about half of this group continued to have low infestation while comparable trees of susceptible *Citrus* clones were generally highly infested (Cameron et al., 1969). A few of these hybrid selections, especially those of *Poncirus* x sweet orange, show promise as rootstocks since they also have resistance to *Phytophthora*, and in some cases to the tristeza virus (see below). Biotypes of the citrus nematode have been identified, however, and hybrid rootstock clones show variable reactions to them, as do selections of *Poncirus* itself.

In Florida the burrowing nematode, *Radopholus similis* Thorne, has been very damaging to citrus. It spreads through the soil in widening circles from tree to tree. Among more than 1200 *Citrus* selections and relatives tested by H. W. Ford and others by 1962, only a few specific source trees were found to be tolerant or resistant (Ford, 1969). Several of these are sweet orange types, but the others include the 'Carrizo' citrange and three selections with some lemon parentage. Such scattered instances of resistance suggest rather specific genetic factors occurring in low frequencies within the species involved. Crosses between nematode resistant selections have been made by Dr. Ford at the University of Florida. Selections from these crosses are being tested for resistance to the citrus nematode as well as to the burrowing nematode. Dr. J. C. Hearn of the USDA at Orlando, Florida, has also had a breeding program for nematode resistance since 1966 (personal communication). F_1 hybrids from crosses between selections resistant to *R. similis* and *T. semipentrans* have been produced.

During the last 25 years, virus and virus-like disease have become a critical factor in citrus culture throughout the world. The most damaging citrus viruses include tristeza, psorosis, exocortis, and sometimes cachexia. Tristeza and exocortis act primarily on specific rootstock-scion combinations, while psorosis affects many scion varieties, irrespective of rootstock. One of the most widely-injurious virus reactions known has been the destruction of the sweet orange on sour orange rootstock by tristeza; in many regions, use of this otherwise valuable stock has had to be abandoned. Stubborn disease, now considered to be caused by a mycoplasma, is also causing grave injury to citrus in many areas, and appears to be spreading rapidly. Strains varying in virulence are known or suspected in most of these agents.

The trifoliate orange as a rootstock is tolerant to tristeza, in addition to its other favorable characters. However, its hybrids show much variability in their reaction to the disease. Studies by W. P. Bitters at Riverside have shown that some long-established citranges from early breeding work are susceptible, while others, including 'Troyer' and 'Carrizo', are tolerant, at least to the more common forms of tristeza in California. New trifoliate hybrids now being evaluated show similar variability. In addition to citranges, some hybrids of *Citrus sunki* Hort. ex Tan. x *Poncirus*, and some hybrids of rough lemon x other *Citrus*, for example, show tolerance.

Little has been reported on inheritance in citrus in relation to susceptibility to pests such as aphids, mites, and scale insects. In some cases where differential resistance appears to exist, the effect may be only an indirect result of other cultivar differences. Thus, in regions with rather low winter temperatures, aphid injury can be slight on the young leaves of cultivars which produce early spring growth flushes; among cultivars which begin growth later, when aphid populations are high, damage is much more severe. There is evidence, however, that infestation of citrus hybrids by the California red scale, *Aonidiella aurantii* Mask., is significantly related to parentage. Among long-established parent cultivars, lemons are highly susceptible, oranges and grapefruit appear somewhat less so, and mandarin types are often only lightly infested. At Riverside, a nine-year study of scale infestation on about 1800 hybrids supported this order of susceptibility in relation to parentage. Eighty-one of the hybrids, representing selected proportions of mandarin, orange, grapefruit, and pummelo ancestry, together with a group of 'Eureka' lemon trees, were then repropagated in a replicated trial and again classified for natural scale infestation over a six-year period (Cameron et al., 1969). Infestation was significantly lowest in the groups having the highest proportion of mandarin parentage. It was higher in groups with greater proportions of mixed orange, grapefruit, and pummelo parentage, and highest in the lemon control trees. The nature of the resistance is not known.

Cold Resistance

Cold resistance in scions and rootstocks has always been a goal of citrus breeders and growers. In any area which is not frost-free, a difference in cold tolerance of a very few degrees can be highly important to the success of a citrus cultivar. New breeding efforts toward cold hardiness have been carried out by the USDA since the 1950s. For scion hardiness the satsuma and 'Changsha' mandarins, the 'Meyer' lemon, and the kumquat were among the parents used. Mandarins as a group, and especially the 'Changsha', have been recognized as cold-tolerant, and many hybrids involving them have been tested. Furr et al. (1966) found that under natural freeze conditions in the field, mandarin progenies with 'Changsha', 'Owari' satsuma, or 'Wilking' mandarin as a parent averaged much more cold-tolerant than those with pummelo or grapefruit parentage. There was much variation, however, and progenies involving the 'Clementine' mandarin ranged from tolerant to susceptible. Controlled freezing experiments with young citrus hybrids in the field, by the use of large portable freezer boxes, have also been carried out (Young and Peynado, 1967). Again, mandarin hybrids showed more average hardiness than hybrids involving pummelo, grapefruit, or lemon. On the basis of cumulative tests, Young and Hearn (1972) rated over 30 cultivars and hybrid selections for cold hardiness. The 'Owari' satsuma and 'Changsha' mandarin are rated excellent. 'Clementine' and some of the newer hybrids, including 'Robinson', 'Page', and 'Nova' are rated good. Using controlled testing conditions, they are also able to effectively screen segregating seedlings for the selection of cold-hardy individuals. The degree of dormancy present at the time of testing is an important factor in such studies. Controlled freezing of citrus leaves has also been reported.

In Japan, where citrus undergoes low winter temperatures in many areas, cold hardiness is of much concern. *Poncirus* is the most widely-used rootstock, and cultivars of the early-ripening and cold-hardy satsuma are the most important scion types. Crosses for new rootstocks have been made between *Poncirus* and the 'Yuzu', 'Natsudaidai', and other cultivars. 'Natsudaidai', which is apparently a pummelo hybrid, has been a standard late-season cultivar, and it has been crossed with pummelos, sweet oranges, and grapefruit, with the aim of obtaining improved fruit quality and retaining cold hardiness and late ripening (Nishiura, 1964). Satsumas have also been crossed with various parents; a seedless hybrid of satsuma x pummelo showed promise as late-ripening.

Cold hardiness has long been stressed in breeding and selection studies in southern Russia. Large-scale crossing was carried out during the 1930s with the main objective of producing hardy and early-ripening cultivars. During the last few years there have been several new reports of frost-resistant hybrids obtained from breeding (e.g. Mikautadze, 1968). Satsuma has often been one of the parents involved, but inter- and intra-specific crosses with lemon were also cited. Some frost resistant crosses of *Citrus* x *Poncirus*, suitable for juice, were described; F_1 hybrids with *Poncirus* usually have unpleasant flavors which make them undesirable for eating or drinking.

Chloride Resistance

Evidence on the tolerance of rootstock cultivars and hybrids to high levels of chlorides in the soil has been obtained in recent studies by the USDA (Furr and Ream, 1969). Tolerant parents, such as the 'Rangpur' lime, 'Cleopatra' mandarin, and 'Shekwasha' mandarin, were crossed with trifoliate orange, sour orange, and other cultivars, and the progeny seedlings (hybrids) were grown in soil irrigated with water high in chlorides. A wide quantitative range of injury occurred within crosses, but there was a definite trend for more progeny plants to show tolerance when both parents were tolerant, as in the cross, 'Rangpur' x 'Cleopatra'. Selected tolerant hybrids were carried to fruiting and their nucellar seedlings were used in replicated tests as rootstocks with orange scions. Chloride analyses of the scion leaves indicated that some of the hybrids were as effective as 'Cleopatra' in limiting chloride accumulation. Their value with respect to disease resistance is still under test.

Leaf and Rind Oils

A study (Scora et al., 1970) of leaf and rind oils of six triploid hybrids from each of five crosses and their diploid and tetraploid parents showed much variability of individual components among hybrids from the same cross. When sets of hybrids had a common parent there was no constant effect of that parent. There also was no overall dominating effect of the tetraploid parents. Transgression in relation to individual components occurred commonly and was particularly prominent with decanal, linalool, and linalyl acetate. It is clear that these hybrids show many inherited differences in several oil components. The differences are quantitative and probably reflect genetic control of intermediate biochemical pathways.

Achievements and Prospects

Future advances in citrus breeding and genetics will no doubt continue to be slow and will require long-range planning and effort, due to the quantitative nature of citrus inheritance, the long generation time, and the presence of asexual embryony. Information is now available, however, on several of the more important problems. Thus, the inheritance of sexual embryony is partly understood, and many potential new seed parents which are entirely sexual have become available through breeding. The rather widespread occurence of self- and cross-incompatibility is becoming recognized, together with its implications for fruit setting and for further breeding. Data are accumulating on the hybrid vigor to be expected from the crossing of various parent groups and species. A groundwork of knowledge has been laid concerning the range and nature of inheritance of such desirable characters as cold resistance, *Phytophthora* resistance, and nematode resistance; while all these characters appear to be quantitative, it may now be possible to make further progress with selected parents.

Further experiments can be aimed at more specific goals than in the past, in the areas just mentioned and in studies where particular mechanisms such as triploidy, self- and cross-incompatibility, or the incorporation of cytologically visible chromosome differences may produce clear-cut effects in F_1 or advanced generations. Improvements in fruit characters such as size, rind color, or flavor have been achieved in particular hybrids, but systematic advancement from generation to generation will not necessarily follow.

Two major problems to which solutions have not been found arc the slowness to fruit of most citrus propagated from seed, and the difficulty of separating nucellar seedlings from sexual ones in the young seedling stage. However, the analysis of leaf or bark tissue by chromatography, spectrophotometry, electrophoresis, or other techniques may eventually provide better identification of hybrid seedlings. First fruiting seems to be only slightly hastened by the budding of seedlings onto rootstocks. Tissue culture techniques may provide a means of accomplishing crosses in which the embyro would otherwise fail to reach maturity. *In vitro* somatic hybridization and fertilization of ovules are also being investigated.

Literature Cited

Burke, J. H. 1967. The commercial citrus regions of the world. p. 40–189. *In* Walter Reuther, H. J. Webber, and L. D. Batchelor (eds.) *The citrus industry*. Vol. 1, rev. ed. Berkeley: Div. Agr. Sci., Univ. Calif.

Calavan, E. C., and L. G. Weathers. 1959. Transmission of a growth-retarding factor in Eureka lemon trees, p. 167–177. *In* J. M. Wallace (ed.) *Citrus virus diseases*. Berkeley: Div. Agr. Sci., Univ. Calif.

Cameron, J. W., R. C. Baines, and O. F. Clarke. 1954. Resistance of hybrid seedlings of the trifoliate orange to infestation by the citrus nematode. *Phytopathology* 44:456–458.

Cameron, J. W., R. C. Baines, and R. K. Soost. 1969. Development of rootstocks resistant to the citrus nematode by breeding and selection. p. 949–954. *In* H. D. Chapman (ed.) *Proc. 1st Int. Citrus Symp.* Vol. 2. Riverside: Univ. Calif.

Cameron, J. W., G. E. Carman, and R. K. Soost. 1969. Differential resistance of *Citrus* species hybrids to infestation by the California red scale, *Aonidiella aurantii* (Mask). *J. Am. Soc. Hort. Sci.* 94:694–696.

Cameron, J. W., and H. B. Frost. 1968. Genetics, breeding, and nucellar embryony. p. 325–370. *In* Walter Reuther, L. D. Batchelor, and H. J. Webber (eds.) *The citrus industry*. Vol. 2, rev. ed. Berkeley: Div. Agr. Sci., Univ. Calif.

Cameron, J. W., and M. J. Garber. 1968. Identical-twin hybrids of *Citrus* x *Poncirus* from strictly sexual seed parents. *Am. J. Bot.* 55:199–205.

Cameron, J. W., and R. W. Scora. 1968. A comparison of rind oil components of diploid and tetraploid citrus by gas-liquid chromatography. *Taxon* 17:128–135.

Cameron, J. W., R. K. Soost, and H. B. Frost. 1959. The horticultural significance of nucellar embryony in citrus. p. 191–196. *In* J. M. Wallace (ed.) *Citrus virus diseases*. Berkeley: Div. Agr. Sci., Univ. Calif.

Cameron, J. W., R. K. Soost, and E. O. Olson. 1964. Chimeral basis for color in pink and red grapefruit. *J. Heredity* 55:23–28.

Carpenter, J. B., and J. R. Furr. 1962. Evaluation of tolerance to rootrot caused by *Phytophthora parasitica* in seedlings of *Citrus* and related genera. *Phytopathology* 52:1277–1285.

Cooper, W. C., P. C. Reece, and J. R. Furr. 1962. Citrus breeding in Florida—past, present, and future. *Proc. Fla. St. Hort. Soc.* 75:5–13.

Erickson, L. C. 1967. The general physiology of citrus. p. 86–126. *In* Walter Reuther, L. D. Batchelor, and H. J. Webber (eds.) *The citrus industry*, Vol. 2, rev. ed. Berkeley: Div. Agr. Sci., Univ. Calif.

Esen, A., and R. K. Soost. 1972*a*. Tetraploid progenies from 2x X 4x crosses of *Citrus* and their origin. *J. Am. Soc. Hort. Sci.* 97:410–414.

———. 1972*b*. Unexpected triploids in *Citrus*: their origin, identification, and possible use. *J. Heredity* 62:329–333.

———. 1972*c*. Aneuploidy in *Citrus*. *Am. J. Bot.* 59: 473–477.

Ford, E. S. 1942. Anatomy and histology of Eureka lemon. *Bot. Gazette* 104:288–305.

Ford, H. W. 1969. Development and use of citrus rootstocks resistant to the burrowing nematode, *Radopholus similis*. p. 941–948. *In* H. D. Chapman (ed.) *Proc. 1st Int. Citrus Symp.* Vol. 2. Riverside: Univ. Calif.

Ford, H. W., and W. A. Feder. 1958. Procedures used for rapid evaluation of citrus for resistance to certain endoparasitic nematodes. *Proc. Am. Soc. Hort. Sci.* 71:278–284.

Frost, H. B. 1943. Genetics and breeding. p. 817–913. *In* H. J. Webber and L. D. Batchelor (eds.) *The citrus industry*, Vol. 1. Berkeley: Univ. of Calif. Press.

Frost, H. B., and R. K. Soost. 1968. Seed reproduction: development of gametes and embryos. p. 290–324. *In* Walter Reuther, L. D. Batchelor, and H. J. Webber (eds.). *The citrus industry*, Vol. 2, rev. ed. Berkeley: Div. Agr. Sci., Univ. Calif.

Furr, J. R. 1961. Earliness of flowering and fruiting of citrus trees propagated from top and basal shoots of young fruiting seedlings. *J. Rio Grande Val. Hort. Soc.* 15:44–49.

———. 1969. Citrus breeding for the arid southwestern United States. p. 191–197. *In* H. D. Chapman (ed.). *Proc. 1st Int. Citrus Symp.* Vol. 1. Riverside: Univ. Calif.

Furr, J. R., R. T. Brown, and E. O. Olson. 1966. Relative cold tolerance of progenies of some citrus crosses. *J. Rio Grande Valley Hort. Soc.* 20:109–112.

Furr, J. R., and C. L. Ream. 1969. Breeding citrus rootstocks for salt tolerance. p. 373–380. *In* H. D. Chapman. (ed.). *Proc. 1st Int. Citrus Symp.* Vol. 1. Riverside: Univ. Calif.

Furasato, K. 1953. Tetraploidy in Citrus. *Ann. Rpt. Natl. Inst. Japan* (Japan) 3:51–52.

Gecadze, G. N. 1967. Caryology of the first generation of hybrids obtained from a cross of tetraploid lemon with the diploid *C. ichangensis* (in Russian). *Subtrop. Crops* 3:97–105.

Goliadze, D. K., and S. M. Talakvadze. 1968. Comparative characteristics of different methods of using chemical mutagens in the breeding of citrus crops (in Russian). *Subtrop. Crops* 5:43–49.

Gregory, W. C., and M. P. Gregory. 1965. Induced mutations in quantitative characters: experimental basis for mutations to hardiness in citrus. *Proc. Soil Crop Sci. Soc. Fla.* 25:372–396.

Hearn, C. J., P. C. Reece, and R. Fenton. 1969. Self-incompatibility and the effects of different pollen sources upon fruit characteristics of four *Citrus* hybrids. p. 183–187. *In* H. D. Chapman (ed.) *Proc. 1st Int. Citrus Symp.* Vol. 1. Riverside: Univ. Calif.

Hensz, R. A. 1960. Effects of X-ray and thermal neutrons on citrus propagating material. *J. Rio Grande Val. Hort. Soc.* 14:21–25.

Hodgson, R. W. 1961. Taxonomy and nomenclature in citrus. p. 1–7. *In* W. C. Price (ed.) *Proc. 2nd Conf. Int. Org. of Citrus Virologists.* Gainesville: Univ. of Florida Press.

Iwamasa, M. 1966. Studies on the sterility in genus *Citrus* with special reference to the seedlessness. *Bul. Hort. Res. Sta. Japan,* (Ser. B) 6:1–77.

Iwamasa, M., and M. Nishiura. 1970. Evidence for the chimeral nature of vegetative reversion in 'Suzuki Wase,' an early-ripening satsuma mandarin. *Sabrao Newsletter* 2:109–114.

Iwamasa, M., I. Ueno, and M. Nishiura. 1967. Inheritance of nucellar embryony in citrus. *Bul. Hort. Res. Sta. Japan* (Ser. B) 7:1–8.

Kapanadze, I. S. 1969. Development of citrus endosperm and its mutagenic effect on the nucellar embryos (in Russian, English summary). *Soobshch. Akad. Nauk. Gruz. SSR.* 56:193–196.

Kerkadze, I. G. 1970. Type of obtained mutants in citrus by induced mutagenesis (in Russian, English summary). *Genetika* 6:26–32.

Longley, A. E. 1926. Triploid citrus. *J. Wash. Acad. Sci.* 16:543–545.

Majsuradse, N. I. 1966. Intergeneric chimeras and their significance for breeding (in Russian). *Genetika* 11:69–82.

Mikautadze, E. S. 1968. Frost-resistant forms of mandarin-pummelo hybrids (in Russian). *Bul. Vavilov All-Un. Inst. Plant Industr.* 12:23–27.

Murashige, T., W. P. Bitters, T. S. Rangan, E. M. Nauer, C. N. Roistacher, and P. B. Holliday. 1972. A technique of shoot apex grafting and its utilization towards recovering virus-free Citrus clones. *HortScience* 7:118–119.

Nagai, K., and T. Tanikawa. 1928. On citrus pollination. *Proc. 3rd Pan Pacific Sci. Cong.* 2:2023–2029.

Naithani, S. P., and S. S. Raghuvanshi. 1958. Cytogenetical studies in the genus *Citrus*. *Nature* 181:1406–1407.

Nishiura, M. 1964. Citrus breeding and bud selection in Japan. *Proc. Fla. St. Hort. Soc.* 77:79–83.

Ozsan, M., and J. W. Cameron. 1963. Artificial culture of small citrus embryos, and evidence against nucellar embryony in highly zygotic varieties. *Proc. Am. Soc. Hort. Sci.* 82:210–216.

Parlevliet, J. E., and J. W. Cameron. 1959. Evidence on the inheritance of nucellar embryony in citrus. *Proc. Amer. Hort. Sci.* 74:252–260.

Rangan, T. S., T. Murashige, and W. P. Bitters. 1969. *In vitro* studies of zygotic and nucellar embryogenesis in citrus. p. 225–229. *In* H. D. Chapman (ed.) *Proc. 1st Int. Citrus Symp.* Vol. 1. Riverside: Univ. Calif.

Russo, F., and M. Torrisi. 1953. Problems and objectives of citrus genetics. 1. Selection of hybrids, nucellar embryos, and triploids and the artificial production of mutations (in Italian). *Ann. Sper. Agrar.* (Rome) 7:883–906.

Shamel, A. D. 1943. Bud variation and bud selection. p. 915–952. *In* H. J. Webber and L. D. Batchelor. (eds.). *The citrus industry,* Vol. 1. Berkeley: Univ. Calif. Press.

Scora, R. W., J. W. Cameron, and J. A. Berg. 1970. Rind and leaf oils of triploid interspecific *Citrus* hybrids and their diploid and tetraploid parents. *Taxon* 19:752–761.

Soost, R. K. 1969. The incompatibility gene system in citrus. p. 189–190. *In* H. D. Chapman (ed.) *Proc. 1st Int. Citrus Symp.* Vol. 1. Riverside: Univ. Calif.

Soost, R. K., and J. W. Cameron. 1954. Production of hybrids by the use of stored trifoliate orange pollen. *Proc. Am. Soc. Hort. Sci.* 63:234–238.

———. 1961. Contrasting effects of acid and non-acid pummelos on the acidity of hybrid citrus progenies. *Hilgardia* 30:351–357.

———. 1969. Tree and fruit characters of *Citrus* triploids from tetraploid by diploid crosses. *Hilgardia* 39:569–579.

Soost, R. K., J. W. Cameron, R. H. Burnett, and B. England. 1965. Fruit characters of nucellar Washington navel orange and Marsh grapefruit budlines. *Calif. Citrograph* 51:3, 18, 20, 22, 24.

Swingle, W. T., and P. C. Reece. 1967. The botany of *Citrus* and its wild relatives. p. 190–430. *In* Walter Reuther, H. J. Webber, and L. D. Batchelor (eds.) *The citrus industry,* Vol. 1, rev. ed. Berkeley: Div. Agr. Sci., Univ. Calif.

Tachikawa, T., Y. Tanaka, and S. Hara. 1961. Investigations on the breeding of triploid citrus varieties. *Bul. Shizuoka Citrus Exp. Sta.* 4:33–44.

Tanaka, T. 1954. *Species problem in citrus.* Ueno, Tokyo: Japanese Society for the Promotion of Science.

Torres, J. P. 1932. Progress report on citrus hybridization. *Philipp. J. Agr.* 3:217–229.

———. 1936. Polyembryony in citrus and study of hybrid seedlings. *Philipp. J. Agr.* 7:37–58.

Toxopeus, H. 1962. Notes on the genetics of a few leaf characters in the genus *Citrus. Euphytica* 11:19–25.

Toxopeus, H. J. 1936. The breeding of rootstocks for *Citrus sinensis* Osb. immune against *Phytophthora parasitica,* the cause of 'gum-disease' in Java (in German). *Züchter* 8:1–10.

———. 1955. Breeding of stocks for *Citrus* resistant to gum-disease and 'Mal Tristeza.' p. 1423–1427. *In* J. P. Nieuwstraten (ed.) *Report 14th Int. Hort. Cong.* Wageningen, Netherlands: H. Veenman and Zonen.

Tutberidze, B. D. 1966. The genetic characteristics of the sexual generation of nucellar hybrids of lemon and *Citrus junos* (in Russian). *Subtrop. Crops* 2: 101–106.

Watanabe, H., H. Yamagata, and K. Syakudo. 1970. Studies on the citrus generic polyembryony in relation to breeding: 2, Discrimination of the embryo fertilized by H-labeled pollen grains (in Japanese). *Jap. J. Breeding* 20:141–145.

Webber, H. J. 1967. History and development of the citrus industry. p. 1–39. *In* Walter Reuther, H. J. Webber, and L. D. Batchelor (eds.). *The citrus industry,* Vol. 1, rev. ed. Berkeley: Div. Agr. Sci., Univ. Calif.

Young, R. H., and A. Peynado. 1967. Freeze injury to 3-year-old citrus hybrids and varieties following exposure to controlled freezing conditions. *J. Rio Grande Val. Hort. Soc.* 13:80–88.

Young, R. H., and C. J. Hearn. 1972. Screening citrus hybrids for cold hardiness. *HortScience* 7:14–18.

Avocados

by B. O. Bergh

The avocado (*Persea americana* Mill.) is unique among tree fruits. It is neither sweet nor acid, but of a bland nature; superior cultivars have a mild nutty or anise-like flavor. It can be served in many ways. Over most of the world it is generally eaten alone or with salt, lemon or lime juice, vinegar, or as part of a salad. In Brazil its flesh is commonly sweetened and eaten as a dessert, or even used to make primitive soap.

Fruits of the better cultivars will store on the tree long past palatable maturity, but the fruit ripens (with concomitant softening) only after it is picked. Its nature and flavor are such that most adults encountering it for the first time need to acquire a liking for it. The usual progression of impressions is from varying degrees of aversion at first taste, through steadily increasing acceptance, to the point at which it is prized as a distinctive delicacy.

The English word *avocado* is a corruption, dating back to the seventeenth century, of the Spanish *ahuacate* (usually spelled *aguacate*), derived from the Aztec *ahuacaquahuitl,* usually shortened to *ahuacatl. Aguacate* has been the common name for the fruit since soon after the Spanish conquest, in Mexico, Central America, and the Caribbean. Other New World regions developed various minor local names, notably *palta* in western South America.

Persea is one of about 50 genera in the aromatic laurel family (Lauraceae). Nearly all of the 1000 or more laurel species are tropical, but a few are subtropical and there are even a very few of mild temperate adaptation—such as the classic laurel (*Laurus*) of the Mediterranean area, and *Sassafras* of the eastern United States. In addition to *Persea,* the only genus appreciably cultivated is *Cinnamomum;* species of the latter yield commercial cinnamon and camphor. Linnaeus described the avocado as a species of *Laurus;* Miller's valid designation of *P. americana* dates from 1754.

Mexico is rapidly becoming the leading producer of avocados (Gustafson, 1972). There are also major commercial industries in the United States (California, Florida, and Hawaii, with a few in the lower Rio Grande Valley of Texas); in Israel, chiefly the coastal areas (Gustafson, 1970); in South Africa, chiefly the Lowveld of northeastern Transvaal (Smith et al., 1970); and in Chile, chiefly in the lower Aconcagua Valley (Schmidt, 1965). In terms of monetary value, United States production easily ranks first.

At least incipient commercial enterprises have been established in Australia (Shepard, 1971) and New Zealand (Fletcher, 1972); in Argentina (Palacious, 1965), Brazil, Venezuela (Serpa, 1968), Colombia, Peru (Morin, 1965), Guatemala, El Salvador (Popenoe, 1958), and most other countries of South and Central America; in Cuba, Jamaica (Davidson, 1967), Puerto Rico, St. Croix (Bond and Frederiksen, 1968), Martinique, and various other islands of the West Indies; in Ghana

(Godfrey-Sam-Aggrey, 1969), Cameroun (Anonymous, 1968), Mozambique, Madagascar (Moreuil, 1971), Kenya, and other tropical African nations; in the Canary (Cabezon, 1965) and Madeira Islands, Morocco (Chavanier, 1967), Sicily, Greece (Zamenes, 1967), Turkey (Chapot, 1967), and other Mediterranean countries; and in India (Raman and Balaram, 1967), Malaysia, Indonesia, the Philippines (Rodrigo, 1968), and other parts of tropical Asia and Oceana.

Prospects for the prompt commercial utilization of successful avocado breeding projects are brightest in the countries that already have major industries—their growers have the knowledge and ability to profit financially from improved cultivars. But all adapted regions can benefit from the development of superior cultivars. For example, Brazil is perhaps second only to Mexico in total number of avocado trees, but most of them are highly variable, nondescript seedlings. Brazil and various other tropical regions around the world have equally good soil, and a climate that is probably superior for avocados to that of the countries with large commercial orchards. Thus far they lack the needed cultivars to extend the marketing season with standardized, high quality fruit that has adequate storing and shipping ability, and their crops are largely dumped on local or nearby markets over a relatively short period at low prices and with considerable spoilage.

As yet only Israel and South Africa have large export markets; these are in Western Europe, especially France and Great Britain. The considerably larger production of United States and Mexico is practically all consumed internally. Most of the world, especially the more affluent areas, is largely an undeveloped market for the avocado. For example, while California alone now has an annual gross return for avocados of over $35 million, Americans average only about one avocado per person per year. Average consumption is already over twice as high in Israel, where the fruit was unknown until the late 1950s (Arkin, 1967), and in Chile, per capita consumption is considerably higher yet (Schmidt, 1965). Worldwide, the chief factors limiting avocado consumption are lack of familiarity and high cost; these are the result of low total production; production is in turn limited by the avocado's sensitivity to climatic extremes (Oppenheimer, 1947), its vulnerability to root rot, and delay in developing the superior cultivars possible on the basis of available germplasm. While it is a luxury food in much of the world, the avocado has been a staple in the diet of the natives of Central America and nearby regions since pre-Columbian times. In Mexico it is still referred to as "the butter of the poor" (Wilkins, 1965).

Its mineral, vitamin, and protein content are remarkably high for a fruit (Hodgson, 1950). Its yields are potentially high, especially in tropical areas. Other advantages include palatability, beneficial effects on the digestive system (Hume, 1951), year-round availability, and high caloric values for regions with chronic food shortage (Raman and Balaram, 1967).

The avocado fruit is botanically a berry. The edible portion is the mesocarp, which is an oily pulp.

Origin and Early Development

A few *Persea* species are indigenous to the Old World, notably *P. indica* L. Sprengel from the Canary and nearby islands, but most are from the New World. The only modern monograph on the genus (Kopp, 1966) was limited to the Western Hemisphere and described 81 indigenous species, plus a number of botanical varieties. She divided the species into two subgenera, *Eriodaphne* and *Persea*, and noted a very sharp demarcation between them. This morphological judgment is supported by graft and hybridization studies at the University of California; the species so far tested fit into the respective subgenera based on morphology, with intra-compatibility complete and inter-compatibility nil.

The commercial avocado is in the *Persea* subgenus. Kopp (1966) described three additional species in that subgenus. However, my observations (unpublished) suggest that all members of *Persea* subgenus *Persea* may best be classified as belonging to a single polytypic species, *P. americana* Mill. All *Persea* species examined have a chromosome number of $2n = 24$.

Horticultural Races

Three distinguishable horticultural races have long been recognized in the avocado, known respectively from their presumed centers of origin as the Mexican, Guatemalan, and West Indian races. Kopp (1966) followed the usual practice of uniting the Guatemalan and West Indian races into one botanical variety distinct from the Mexican race variety, which is sometimes considered a separate species. But my observations indicate that the three races are each about equally distinct from the other two, and that the status of botanical variety best reflects the systematic position of each taxon. Consequently, the Mexican, Guatemalan, and West Indian races have been respectively designated *P. americana* var. *drymifolia*, *P. americana* var. *guatemalensis* (var. nov.), and *P. americana* var. *americana* (Bergh et al., in press). The accumulating evidence that the Mexican race is no more distinct from the two more tropical races than they are from each other, vindicates the astute field judgment of Popenoe (1941) over 30 years ago. Other species in this subgenus are also reduced to varietal status, but since none has as yet demonstrated value for commercial avocado breeding, they will not be discussed further.

Kopp (1966) pointed out that *P. americana* "probably originated in the Chiapas (Mexico)-Guatemala-Honduras area where the wild avocados are found. Its origin is obscure because of its close connection with representatives of early civilizations of the area who valued its fruits. It is, therefore, a problem in ethnobotany." Forms judged chiefly by their smaller fruits and larger seeds to be possible wild progenitors of the three horticultural races have been found at higher elevations in southern Mexico (Mexican); in the highlands of central Guatemala (Guatemalan); and in Colombia (West Indian). A still more primitive form in certain parts of Central America could conceivably be ancestral to all three species (Bergh, 1969).

There are no known sterility barriers among the three races—or among any members of the *P. americana* complex. Hence, hybridization occurs readily wherever trees of different races are growing in proximity, whether indigenously (Popenoe and Williams, 1947) or planted by man (Bergh, 1969). The cultivar that has long led in production in California and most other less tropical avocado regions, 'Fuerte', is apparently a natural Mexican x Guatemalan hybrid. It was found growing in southern Mexico. The second cultivar in most of these regions, and gaining in relative importance, is 'Hass'; it is generally regarded as straight Guatemalan, but progenies produced by self-pollination (Bergh, unpublished) indicate that it is of perhaps one-quarter, or somewhat less, Mexican race germplasm. Guatemalan x West Indian hybrids are now the leading cultivars in Florida, and are very promising for future avocado development in other more tropical areas as will be discussed later.

TABLE 1. Comparison of the three horticultural races

Character	Mexican	Guatemalan	West Indian
TREE			
Climatic adaptation	semitropical	subtropical	tropical
Cold tolerance	most	intermediate	least
Salinity tolerance	least	intermediate	most
Pubescence	more	less	less
Leaf anise	present	absent	absent
Leaf underside	more waxy	less	less
Leaf size	average smaller	larger	larger
Leaf color	medium green	average redder	paler, yellower
FRUIT			
Bloom to maturity	7 months	14 months	7 months
Stem	shorter	average longer	shorter
Size	small	variable	variable
Shape	longer	average rounder	longer
Perianth persistence	greater	less	less
Color	more often dark	usually green	green or reddish
Skin thickness	very thin	thick	thin, leathery
Skin surface	waxy bloom	rough	shiny
Seed size	large	small	large
Seed cavity	often loose	tight	often loose
Seed surface	smooth	smooth	rough
Seed integuments	thinner	thinner	thicker
Oil content	highest	high	low
Pulp flavor	anise-like, rich	often rich	sweeter, milder
Pulp fibre	common	less common	less common
Cold storage	more tolerance	more tolerance	less

In spite of such increasing race mixing, certain morphological and physiological differences among the races remain of great horticultural importance. These, with other distinguishing traits, are listed in Table 1. Rhodes et al. (1971) effectively quantified some of these racial differences by different methods of numerical taxonomy.

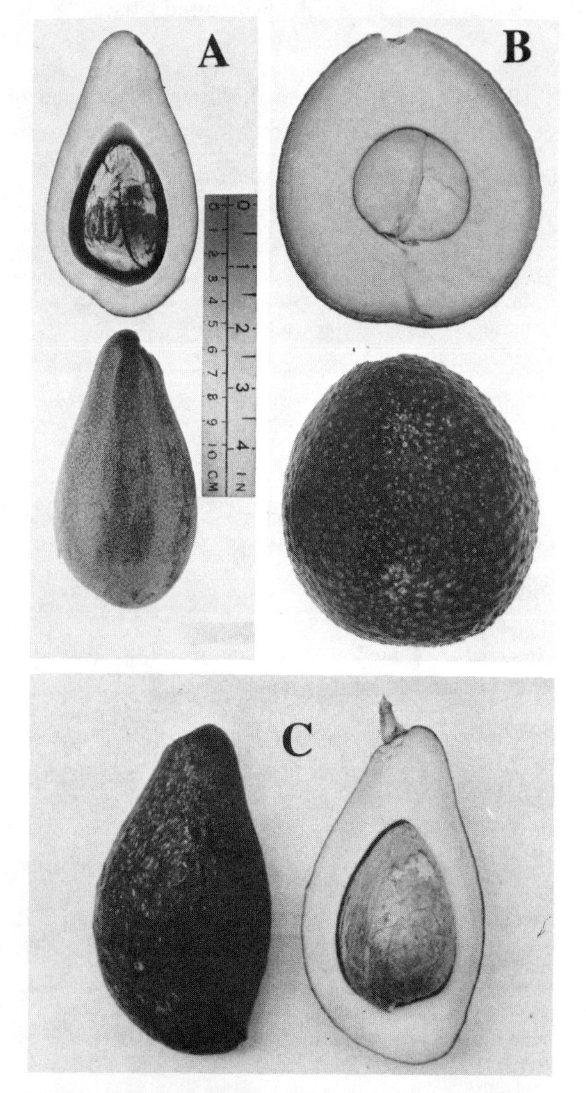

FIG. 1. Fruits of the three horticultural races of avocados. (A) Mexican: small, very thin skin, large and loose seed. (B) Guatemalan: round, thick and rough skin, remarkably small seed. (C) West Indian: leathery skin, large and slightly loose seed. (Courtesy R. H. Whitsell for A and B, and W. B. Storey for C.)

Even without racial hybridization, there is some overlapping for most of these traits. Since most of the major commercial cultivars now are racial hybrids, identification in terms of racial origin becomes much more difficult. But the two races most similar in terms of the fruit, Mexican and West Indian, are most dissimilar in terms of general climatic adaptation. Hence, the problem is largely one of distinguishing between (usually relative amounts of) West Indian and Guatemalan germplasm in tropical regions, and Mexican and Guatemalan germplasm in less tropical areas. In both cases, probably the most useful criterion is season of maturity; other criteria often helpful include skin thickness and surface, and seed size and tightness (Fig. 1). All of these characters appear to be polygenetically determined.

The length of time shown in Table 1 from bloom to fruit maturity is an approximation and varies among cultivars of a given race, and, for a given cultivar, with climatic factors. For example, in Trinidad the spring dry season delays the blooming period of the local nonirrigated West Indian types until early summer; the fruits thus develop during the period most conducive to rapid development, and begin maturing only about three months from set (W. B. Storey, personal communication). In the more rigorous climate of California, Guatemalan cultivars may require 18 months to mature fruit, as compared with nine to 12 months in Florida (Ruehle, 1963).

The presence of anise scent in the foliage and bark is a definite Mexican race character, and is especially useful because it does not require fruit to be present. For distinguishing nonfruiting trees of the other two races, the paleness of West Indian foliage is fairly determinative; and some Guatemalans have quite reddish new leaves.

Climate-adaptational differences may themselves be sufficient to identify trees in certain locations. Only the West Indian race is adapted to a truly tropical climate—trees of the other races may fail to set fruit or even flower in such a climate (Serpa, 1968). Conversely, West Indian trees in a climate like California's set very little or no fruit even when not frost-injured. And in a really cold location, with several degrees of winter frost occurring consistently, only Mexican-race trees can be expected to survive.

The three climate classes of the first line of Table 1 follow the usage of Popenoe (1919) and Ochse et al. (1961). While *semitropical* may be considered a synonym of *subtropical,* in this usage it is considered to be a less tropical climate—with more distinct seasons, including periods of both greater cold and lower humidity. According to Popenoe (1919, 1952, 1958), in the Torrid Zone the tropical region extends from sea level to about 1000 m, and in addition to the West Indian avocado, such fruits as the mango, breadfruit, and tamarind thrive. The subtropical region extends from roughly 1000 to 2000 m, and favors the Guatemalan avocado, plus the Mexican at higher elevations, plus the orange, loquat, cherimoya, and others. The semitropical was considered by Popenoe to extend from perhaps 2000 to 3000 m; only the Mexican race would be expected to stand the winter cold of this region, as suggested by the success of such temperate zone fruits as apple, pear, and peach. But according to Chavanier (1967), all three races thrive and bear well in the climate of Skirat, Morocco.

A given race or cultivar thus can succeed at a considerable range of elevations. Popenoe (1919) estimated that fruit maturity was delayed about one month by each 300 m increase in elevation. Time of maturity is also delayed by increased latitude, and these two forces together with climatic differences at the same elevation and latitude make possible a wide range of harvesting times for the same cultivar. When a fourth factor is added, differences among cultivars or individual seedlings, it is understandable that Mexican-race trees in Mexico have been reported to mature fruit "almost continuously throughout the year" (Turu, 1970), and this has even been reported to be true of West Indian trees, in Venezuela (D. Serpa, personal communication).

Distribution in 1492 and Later

When Europeans reached the Americas, they found the avocado cultivated from northern Mexico to Peru (Hodgson, 1950). Apparent wild ancestors were growing over part of this range, from southern Mexico to present-day Panama, as noted for the different races above. But it had already been cultivated so long that its early history was, and remains, very difficult to trace.

The Spanish conquistadores soon carried the avocado to Venezuela (Serpa, 1968), the West Indies (Hume, 1951), Chile (Schmidt, 1965), the Madeira and Canary Islands (Cabezon, 1965), and similar regions touched by their wide-ranging sovereignty. Eventually the fruit was established in nearly all regions of the world with suitable climates. It reached Ghana in 1750 or soon afterwards (Godfrey-Sam-Aggrey, 1969); Madagascar in 1802 (Moreuil, 1971); Hawaii prior to 1825 (Hodgson, 1950); Florida in 1833 if not considerably earlier (Ruehle, 1963); California by about 1860; and Malaysia by 1900 (Lambourne, 1934).

The race of avocado established would vary with the local climate; for example, in the tropical West Indies, only the tropical-adapted West Indian race became prevalent (Bond and Frederiksen, 1968; Storey, 1968)—hence its rather misleading name. The same race was similarly established in other tropical areas around the world. The Guatemalan and Mexican races were gradually being grown in progressively less tropical areas. But in Chile (Schmidt, 1965), only the Mexican race was brought in by the Spaniards and grown for some 300 years before the adapted and (on the whole) horticulturally superior Guatemalan race was introduced less than 40 years ago.

The first recorded European description of the avocado was in 1519 by a Spanish cartographer, M. F. de Enciso, who had seen the fruit in Colombia (Popenoe, 1941). Several other reports followed soon after, most emphasizing its exceptional food values, long before an understanding of nutrition had been achieved.

History of Improvement

Two steps are usually involved in tree fruit improvement: selection of improved genotypes and their fixation by asexual propagation. The avocado produces only sexual seeds, and flowering dichogamy largely ensures cross pollination. Consequently, the seedlings produced by a single tree (or cultivar) are extremely variable. This is less true of the West Indian (Smith et al., 1970; Storey, 1968) than of the other two horticultural races. But even with the West Indian, seedlings are far too variable to form the basis of a sound commercial industry. Most seedling trees set fruit belatedly, if ever. Those that do set generally have light crops and fruit quality tends to be inferior. The few with good crops of superior fruit vary so much in fruit characteristics from tree to tree that a severe marketing handicap results. The first known asexual propagation of the avocado was in Florida before 1900 (Ruehle, 1963).

Selection of horticulturally improved avocados occurred long before they were asexually propagated. Indeed, in Mexico, Smith (1966) found avocado seeds of varying antiquity (beginning about 7000 B.C.) which seemed to him to suggest selection for larger size beginning about 4000 B.C. But this conclusion was based largely on a comparison of the largest seeds at each level; number of seeds present was greater in the later deposits and sample size was strongly correlated with largest individual size. Still, it seems reasonable that early man would select for smaller proportional seed size as well as larger fruit size (Popenoe, 1919). Smith's data are thus compatible with selection for larger avocado fruits over the past several thousand years, but give no clear evidence for it.

Extensive pre-Columbian selection is evident from the high horticultural quality already present when European man first encountered the avocado. As Popenoe (1952) noted, "In prehistoric times, by the laborious process of selection and propagation by seed, the avocado was developed from small-fruited wild forms of which many are still found in the forests of Mexico and Central America, to the splendid varieties now being propagated vegetatively, varieties which so admirably meet the needs of man that modern science has not yet been able to better them materially." Earlier, Popenoe (1919) suggested some of the probable selection methods: cutting down the poorer seedlings, planting seeds from superior seedlings, selling choice fruit and thereby spreading the better types. In the same publication he suggested that selection has "tended to conceal the racial characters, by bringing the various races to a common level." As a possible example, he adduced the fact of much smoother skin in some Guatemalan lines. The smaller Mexican fruit size apparently indicates less selection within that race (Chandler, 1958).

The sort of procedure involved is strikingly illustrated by the Rodiles groves near Atlixco, Mexico (Anderson, 1950). Here the selection has been perhaps uniquely intensive and consistent; for generations the Rodiles family has planted seeds from the finest available local avocados, resulting in thousands of bearing trees, with unusually high average quality. 'Fuerte', long the world's leading cultivar, was introduced from this grove in 1911, with 'Puebla' which also was a leading California cultivar for many years. The selections were made by Carl Schmidt for the first large California avocado nursery, headed by F. O. Popenoe (Kellogg, 1971).

F. O. Popenoe's son, Wilson, conducted the most zealous program ever attempted to improve the industry by introducing grafting wood from superior seedlings in the avocado's native habitat. It was centered in Guatemala, and culminated in 1917 with the introduction of 'Nabal', 'Benik', 'Panchoy' and several other Guatemalan race cultivars. "Considering the time spent and the top calibre of the man heading the exploration, the results in terms of cultivars directly adapted to California must be regarded as disappointing. Unfortunately, it was not realized at that time that the California consumer of avocados wants a much smaller fruit than is preferred in [many] other avocado-consuming regions of the world. Most of Popenoe's introductions were too large-fruited. Others bore poorly under California conditions, or lacked other desirable characters. It is probably as breeding materials that they have made—and are making—their major contribution. However, for about 20 years 'Nabal' was usually second only to 'Fuerte' in total California production" (Bergh, 1957). 'Nabal' and 'Itzamna' were both recommended for colder parts of Florida for many years. 'Nabal' and 'Benik' are still being grown commercially in Israel, the latter on a small scale to cross-pollinate the former.

Popenoe (1951) later made a few additional selections from the Atlixco grove, chiefly of predominantly Mexican types and intended primarily as additional germplasm for Central American avocado culture. Early introduction of seeds from superior seedlings in Central America and adjoining regions has also given rise to commercial cultivars (Bergh, 1957) in California, Florida, and elsewhere.

Every present Florida cultivar of any importance was selected from locally grown seedlings that were produced by open pollination. These were predominantly West Indian and (more recently) West Indian x Guatemalan hybrids. Other tropical countries have improved their avocado industries by introducing Florida selections, including Madagascar (Moreuil, 1971), Venezuela (Serpa, 1968), Virgin Islands (Bond and Frederiksen, 1968), and the more tropical parts of Mexico (Gustafson, 1972; Turu, 1970).

Several tropical areas have advanced the regional avocado industry also by selecting superior local seedlings. Storey (1968) concluded from the great seedling uniformity in Trinidad and Tobago that perhaps only one or two seeds had been originally introduced by the Spaniards, with subsequent selection for productivity and quality.

In California also, with the exception of 'Fuerte', every major commercial cultivar is a local selection. Other parts of the avocado world with a similar Mediterranean-type climate have built their industries chiefly on 'Fuerte', the rapidly advancing 'Hass', and other California cultivars. This is true of Israel, other Mediterranean countries, South Africa, Australia and New Zealand, Chile, other southern latitudes and also higher elevations in South America, and the extensive avocado groves being planted in Mexico inland from the tropical coastal areas. In part this is a consequence of the established preference. The European preference, in turn, is partly a result of 'Fuerte' being the first cultivar to arrive in quantity (so that the different-appearing 'Hass' is encountering difficulty getting a fair trial), and partly a result of the very high quality of 'Fuerte' (like 'Hass'). The popularity of the 'Fuerte' and 'Hass' thousands of miles from their places of origin is also a reflection of the remarkably wide adaptability and superior performance of these two cultivars, including their ability to remain on the tree for months after reaching maturity, in countries all around the world. Indeed, 'Fuerte' is reported to bear better in some other regions than in either its Mexican "home" or its California "foster home"; for example, of many cultivars tested in western Kenya, "'Fuerte' is outstanding with us because of its colossal bearing capacity" (R. Andersen, personal communication).

Modern Breeding Objectives

Objectives change as consumer preferences in fruit size and other qualities change, and as disease and other problems increase or decrease.

The rootstock needs above all to be conducive to consistent heavy bearing of the scion top. Dwarfing or semi-dwarfing stocks would be desirable (Ticho, 1971) since a major weakness of many avocado cultivars is excessive vigor at the expense of fruit-set. The stock should be easy to propagate, sexually or asexually. It should grow vigorously, and be easy to successfully graft or bud. A stock that enhances scion cold hardiness would be highly desirable. There are marked differences both among and within the three horticultural races in resistance to the salinity injury to which the avocado is unusually susceptible. Genetically superior lines are also available in terms of stock absorption of iron to prevent chlorosis (Kadman and Ben-Ya'acov, 1970). The most sought-after rootstock objective is resistance to *Phytophthora cinnamomi* Rands, the root-rot fungus that is decimating the avocado industry in most regions, although it is unknown in Israel (Gustafson, 1972), Morocco, and a few other countries. Other rootstock diseases to which genetically-induced resistance is known are wilt caused by *Verticillium albo-atrum* Berth & Reinke, and canker caused by *Dothiorella gregaria* Sacc. There are no known rootstock differences in effect on avocado fruit quality.

The avocado tree should have a spreading rather than an upright growth habit (Fig. 2) to reduce wind injury, facilitate spraying, and reduce picking costs. The tree needs to be vigorous, but not to such a degree that fruitfulness is impaired. Most of the major world avocado regions are subject to occasional frost damage, and cold tolerance of both the fruit stem and the tree generally is highly advantageous. There are also marked cultivar differences in tolerance of excessive heat, in terms of both fruit dropping and tree injury. Apart from differences in climatic tolerances, some cultivars thrive and bear well over a much wider range of microclimates than do others, thereby minimizing marketing and other problems that result from numerous commercial cultivars. The most important tree characteristic is good fruit setting ability. Without that, excellence in the preceding qualities becomes meaningless; indeed, the objectives listed previously are important chiefly for their indirect effect on amount and regularity of bearing. Consistency of production may be as important as heavy overall production (Bergh, 1961). Except for fungi that are of primary concern in terms of fruit injury, the only serious disease is "sunblotch" caused by an unidentified virus; no genetic source of resistance is known.

The fruit should mature over a relatively short period, to obviate the need for spot picking. Even more desirable is long storage life on the tree, which gives two major marketing advantages: picking can be given optimum timing, and fewer different cultivars are needed to supply the entire marketing season (the year around in California). While some local tropical markets accustomed to large-fruited West Indian types favor larger fruits (Bond and Frederiksen, 1968; Godfrey-Sam-Aggrey, 1969; Krome, 1967; Le Roux, 1970; Serpa, 1968; Trabucco, 1968), for export and many internal markets the preferred size is about 300 g. The thick-pyriform fruit shape of 'Hass' and 'Teague' is ideal, but the more elongate or "pear" shape of 'Fuerte' is acceptable (Fig. 3). Highly elongate fruits are objectionably awkward to pack. Spherical or slightly oblate fruits simplify packing, but are not as readily identifiable as avocados. Uniformity in shape and size is desirable. The skin should be thick enough to protect the flesh in transit, but not so thick as to add unnecessary waste material or to prevent the consumer from determining ripeness by gentle palm pressure. The skin should peel easily. A dark (red to purple to nearly black) skin color is preferred in a few markets, color seems immaterial in others, and green is preferred in many markets where 'Fuerte' has become established as the standard of quality. Regardless of color, fruit should be bright, attractive, and unblemished; russeting is undesirable. Seed size should be small. If the seed is loose, there is added waste bulk and the outer seed coat is likely to stick objectionably to the flesh. Unless the flesh softens adequately and then has an acceptable flavor, superiority in all other regards is meaningless. Flesh color or the visibility or toughness of fibrovascular bundles can influence taste reaction. Pulp quality should not decline appreciably with chilling storage.

FIG. 2. Desirable (A) and undesirable tree shapes. (A) 'Topa Topa' cultivar about 9m wide by 8m high. (B) 'Bacon' cultivar about 6m at its widest by 22m high. (Courtesy R. H. Whitsell.)

After ripening, the fruit should remain in good edible condition for a reasonable length of time—at least a day at room temperature, at least a week under 5 C refrigeration—without becoming overly "mushy," i.e., susceptible to a condition of semifluid structural breakdown, or developing unpleasant odors or tastes. California law, at present under litigation, prohibits the marketing of avocado fruit with less than 8% oil, which minimizes the sale of local immature fruit but also prevents the importation of West Indian avocados. No other avocado oil level restrictions have been involved anywhere. Consumer preference depends largely on the oil content of the avocados to which one originally became accustomed. Both physiological and fungal diseases can cause fruit breakdown, and genetic resistance to both is known.

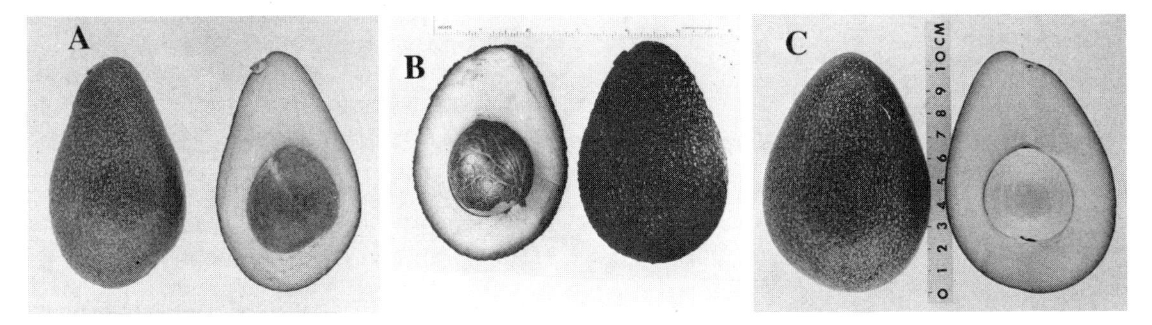

FIG. 3. (A) 'Fuerte', the world's leading cultivar, is Mexican-Guatemalan. (B) 'Hass', of growing importance in less-tropical regions, is largely Guatemalan. (C) 'Teague', the only commercially successful cultivar produced by the University of California breeding program, is largely Mexican. It evinces the early maturity, short season, thin skin, and cold hardiness, but has the Guatemalan type of tight, moderate-size seed, and larger fruit size. (See also Fig. 1A and B.) 'Hass' is black at maturity; the other two are green. (Courtesy R. G. Platt for A, C. A. Schroeder for B, and R. H. Whitsell for C.)

Breeding Techniques

Floral Biology

The avocado flower is perfect, with three members in each whorl (Fig. 4). The perianth parts are pale green to slightly yellow. The sepals average about 5 mm long, with the petals usually a bit longer. Also, the sepals usually reflex a little more widely (Figs. 5, 6). But the only way that they can be differentiated with certainty is the placement of each of the three sepals opposite an inner stamen, while each petal is opposite a staminode (Figs. 4, 5, 6). All stamens are quadrilocular (Figs. 4, 6). The two outer whorls appear identical, and their pollen sacs open inward. Pollen sacs of the inner whorl open outward or laterally, and attached to each filament base is a pair of glands that secrete nectar at the second opening of the flower (Fig. 6). The innermost whorl is sterile, and these staminodes secrete nectar at the first opening (Fig. 5). The simple pistil has an ovary with just one ovule.

Avocado flower behavior is both singular and important. It can be described as protogynous dichogamy with synchronous daily complementarity. A given tree (or cultivar) is classified as "A" type if each flower is functionally female (pistil-receptive, Fig. 5) in the morning and functionally male (pollen-shedding, Fig. 6) the following afternoon; or "B" type if each flower is "female" in the afternoon and then "male" the following morning. An A-type flower opens for the first time in early or mid morning, remains open and pistil-receptive until about noon, then closes and remains closed until about noon of the second day, when it reopens and begins shedding pollen with the pistil no longer receptive, finally closing permanently that night. On a single tree there may be thousands of flowers that open for the first time the same morning and then follow the same behavior pattern synchronously hour after hour for their two-day existence. Flowers on B trees function analogously but with transposed timing.

The flower behavior outlined above is regular only under conditions of warm weather, with diurnal temperatures going above about 26 C and nocturnal temperatures remaining above about 15 C. As day maximum and night minimum temperatures drop, the time of opening becomes later for both the "female" and "male" stages. Eventually the daily timing of the two types will be about reversed from expectation. With further chilling, the second ("male") opening may be delayed one

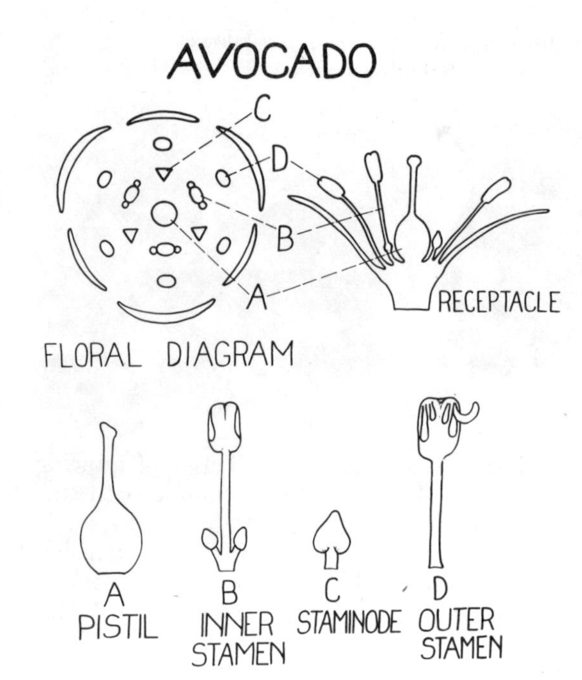

FIG. 4. Diagrammatic representation of the flower parts. The inner stamens also contain four pollen sacs each. (Courtesy C. A. Schroeder.)

or more days, and various abnormalities in flower behavior may be manifest. Either opening may continue through the night and into the next day. The first ("female") opening may fail to occur; this will not necessarily inhibit fruit set, as the pistil appears receptive at the single flower opening. With a daily mean temperature during the blooming period of less than about 15 C, 'Fuerte' set sharply declined (Hodgson and Cameron, 1936); other cultivars seem to respond similarly to this critical temperature, although fruit set is seldom as greatly affected.

Less severe weather chills upset the flower dichogamy so as to bring about partial overlapping of the "male" and "female" stages (Bergh, 1969). Hence, self pollination becomes possible. This is enhanced by variability in flower behavior brought about by variability in sun and wind exposure of flowers on the same tree, and possibly by seedling rootstock effects among trees of the same cultivar. So an isolated tree or a large solid block of one cultivar may set very well, but the opportunity for cross pollination can markedly increase fruit set.

Cultivars of the Guatemalan race usually bloom later than those of the other two races. Under California conditions, each tree or cultivar

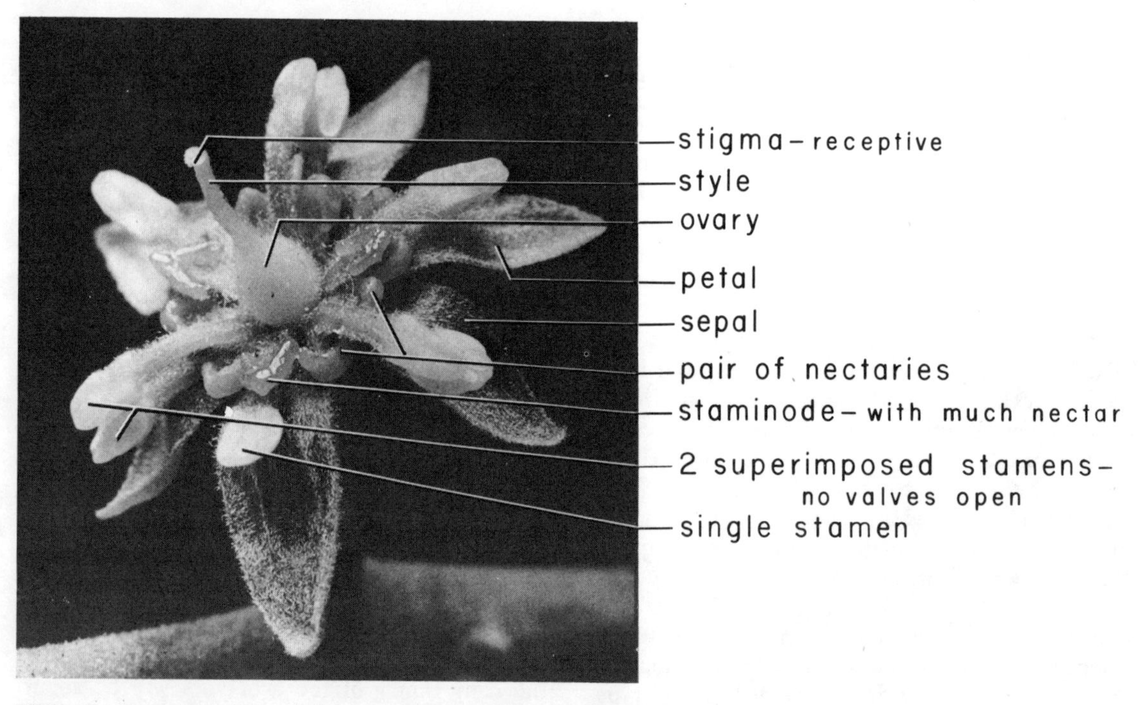

FIG. 5. First opening of flower, functionally pistillate. The stigma is receptive and no pollen is shed. (Courtesy K. L. Middleham.)

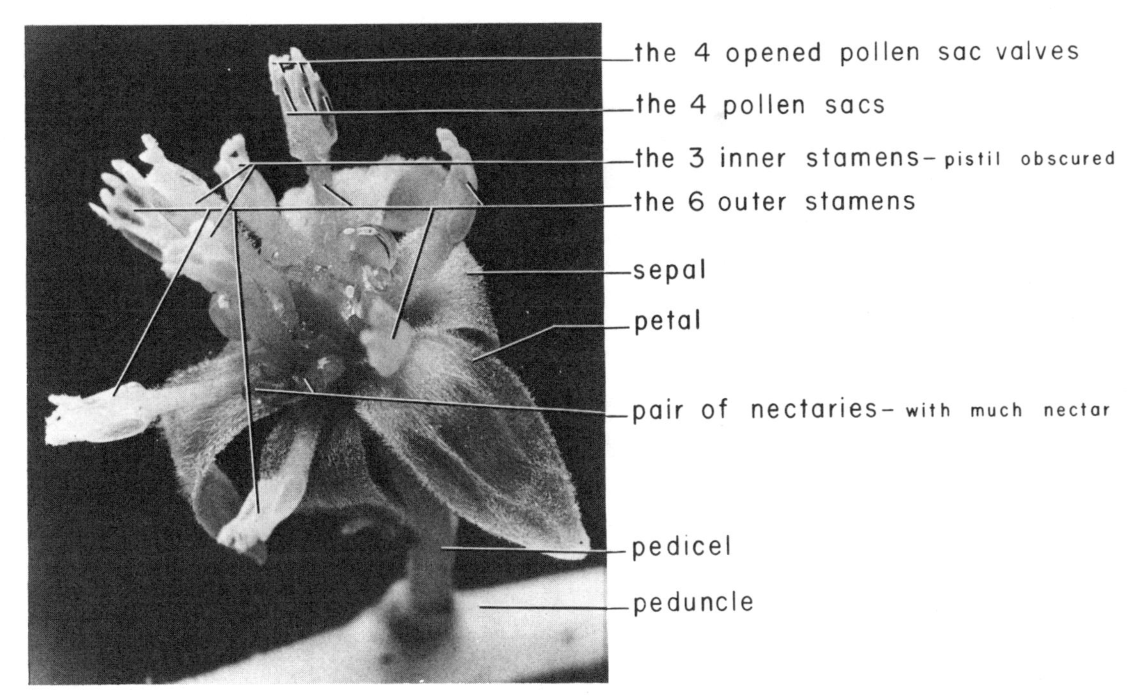

FIG. 6. Second opening of flower, functionally staminate. The stigma is deteriorated and pollen is shedding. (Courtesy K. L. Middleham.)

will bloom continuously for perhaps two months, and it is rare for the earliest to be finished blooming when the latest begins. But in Peru (Morin, 1967), total avocado bloom is spread out over a much longer period, while individual cultivars tend to have a shorter period of bloom; hence, few cultivars have overlapping blooming periods.

The flowers are grouped in usually terminal, highly compound cymes (Fig. 7A) of dozens to hundreds of flowers each. Hence, a single tree may have a million flowers during one blooming period; only a fraction of 1% of this total could ever be held to fruit maturity. Nevertheless, bloom that is less than normal for that cultivar indicates a physiological condition not conducive to good set, while genetically sparse bloom may indicate unusually good fruit-setting ability, perhaps because superabundance of flowers is physiologically exhausting. The remarkably heavy setting new Israeli cultivar 'Tova' forms comparatively few flowers (S. Gazit, personal communication). While the buds may be initiated as long as two months before the flowers open (Schroeder, 1951), failure of the first group of flowers to set may promptly stimulate a new flush to develop (Fig. 7B), especially in lines with Mexican race germplasm.

Pollination

Commerical avocado production requires pollen transfer by large flying insects, primarily the honey bee in California. The pollen requires only about two to four hours to grow from stigma to ovary, depending on the temperature; its progress can easily be followed by the new aniline blue fluorescence method (S. Gazit, personal communication).

The pollen from each sac usually sticks in a clump to the opened valve until it is removed by insects or drops with the flower. Methods of pollen collection successful with other fruits, such as vacuum devices, have not worked well with the avocado. A satisfactory breeding approach is to pick male-stage flowers with clumped pollen and daub stigmas of female-stage flowers. Or, the pollen can be removed by the fingernail (preferably painted black to make the pollen readily discernible) and then applied.

Functional pollen viability after more than a few days of storage (in low temperature and humidity) has not been unequivocally demonstrated. A high proportion of pollen grain germination *in vitro* has recently been achieved (R.

Makino et al., unpublished), although the tubes appear somewhat abnormal.

FIG. 7. (A) Typical inflorescence (the more basal leaves have been removed, showing two vegetative buds in petiole axils). (B) 'Fuerte' inflorescence almost finished blooming without fruit set, and new flower buds forming at the base of the peduncles. (Courtesy C. A. Schroeder.)

FIG. 8. Bee-proof cage for cross-pollination or selfing (bees enclosed). It is 2.44m square by 3.66m high. Larger cages have also proven useful. The ladder inside the cage gives easy access to the flowers, which are predominantly in the tree top. The right half of the lower front section is a hinged door. (Courtesy R. Makino.)

Since, apart from man, large flying insects are the only known pollinating vectors, exclusion of insects by screening material insures that the only fruit set will be from manually applied pollen. Screened and unpollinated trees have set fruits only rarely in California and Israel; flower emasculation is thus unnecessary in avocado breeding.

Increased fruit set due to cross pollination decreases rapidly with distance from the pollen source (Bergh and Gustafson, 1958; Bergh et al., 1966). Hence, self seed can be obtained by: 1) isolated trees; 2) trees in commercial blocks at least 50–100 m from pollinating cultivars; or 3) caged trees containing bees, and a source of fresh water. A satisfactory cage type is a wooden frame bolted together and either a rigid top (Fig. 8) or a loose screening fastened to the framework with a sturdy cord.

Controlled hybridization can be achieved by hand pollination of flowers protected from insects by cages or bags, or by growing the trees in a greenhouse. Seed obtained from two cultivars enclosed in a single cage with bees, or from adjoining trees of two cultivars in an isolated block, will usually be a mixture of hybrids and selfs.

Increasing Seed Set

During early California breeding (Schroeder, 1958), over 10,000 'Fuerte' flowers were cross-pollinated and only four mature fruits were obtained. In contrast, recent Israeli hybridization with their extremely productive new selection 'Tova' as seed parent has given a success rate of 2 to 5% (S. Gazit, personal communication).

A number of techniques can increase seed set:

1) Select trees growing in an optimum location (Bergh, 1967), and provide optimum care, such as irrigation, fertilization, disease and insect control, and windbreak protection if desirable.

2) Avoid excessive shading (Lahav, 1970); use screen material that transmits as much light as possible, and remove limbs shading the breeding tree.

3) Girdling maximizes fruit yield (Lahav et al., 1971; Ticho, 1971), and especially fruit number (Lahav et al., 1972); girdling even after fruit is set may increase its chances of survival (Ticho, 1971).

4) Hybridize or self in the productive ("on") year for that tree, since nearly all avocado trees are alternate bearing.

5) Select a heavy-setting cultivar as seed parent.

6) Handle flowers, especially the pistil, minimally and gently.

7) Pollinate only a few flowers in each cluster.

8) Remove the excess flowers a few days earlier (Torres, 1936).

9) Never pollinate a darkened or abnormal-looking stigma.

10) Never pollinate at the second (pollen-shedding) opening, regardless of stigma appearance.

11) Pollinate before mid-afternoon (Bringhurst, 1951).

12) Pollinate during optimum weather conditions (usually toward the end of the blooming season in a region with cool spring weather like California's).

Germinating the Seedlings

Seeds may germinate in unharvested fruits. Rarely will the embryo show much leaf development, but the roots may grow out through the seed coats and into the flesh, and in some cases, roots grow right out through the skin (Fig. 9). In Fig. 1B, a root has reached the unusually thick and hard skin, and then turned to grow along it. Seeds with manifest root development should be planted immediately.

An avocado seed left at ordinary room temperature and humidity remains viable for only a few days after its removal from the fruit; the embryo has no real protection against desiccation. However, avocado seeds remain viable for several months at a temperature a little above freezing (Bergh, 1961), in the high-humidity conditions of the ordinary cold storage. Humidity can be ensured by storing in such media as slightly damp peat moss, or in polyethylene bags. Before they are placed in storage, seeds should be treated with a fungicide to prevent rot. Seeds that have not been in cold storage are probably best planted without further treatment. But after just 24 hours of chilling, germination rate was increased markedly by peeling off the seed coat (Kadman, 1963). Further increase in germination of chilled seeds can be obtained by cutting the seed surface, but fungicide must then be applied to prevent decay.

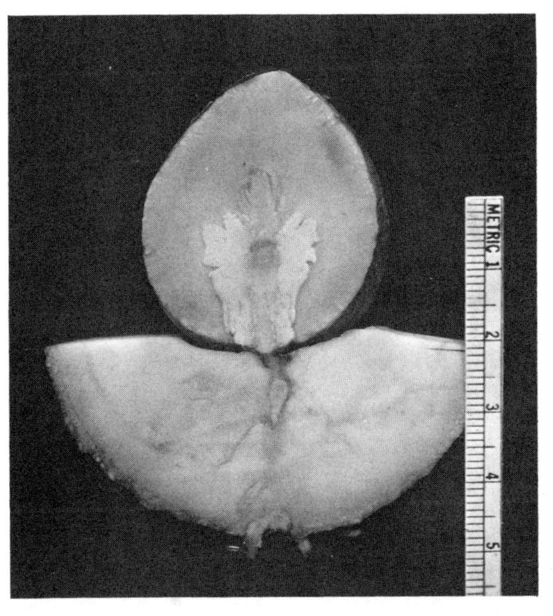

FIG. 9. Seed germination inside fruit on tree. The seed has been split in half to show the embryo (note the incipient leaves). A few roots had traversed both pulp and skin. (Courtesy C. A. Schroeder.)

The embryo in the seed is oriented correctly for planting as it hangs on the tree—the plumule is next to the stem end of the fruit. The seed is broader and flatter at the radicle end. Plant so that the top of the seed is at ground level; when hot weather can be expected, a mulch of peat moss or other available material, about 1 cm thick, will protect the seeds.

In regions with adequate rainfall, the seeds can be planted directly in their field location (Davidson, 1967), but since young seedlings are extremely sensitive to water stress, planting in an irrigated nursery about 40 cm apart is recommended in most areas (Platt and Frolich, 1965; Ruehle, 1963). Seedlings a meter or less in height can be transplanted to their permanent location. In California it is necessary to retain soil around the roots of transplanted seedlings but "bare-root" handling is possible in humid climates.

There are important advantages to growing the seedlings in containers: easier control of *Phytophthora* root rot, reduced space requirement, and reduced transplanting shock. Container-grown seedlings are generally used for commercial avocado propagation (Ruehle, 1963; Ticho and Gefen, 1965).

Growing the Seedlings

Under optimum growing conditions, including a night-day temperature range of about 24–30 C, seeds will germinate in about a month. More vigorous seedlings will attain a height of about a meter in another three months, while slower growers may require six months to reach this plantable size. Cooler temperatures greatly retard growth.

In California soils, container-grown seedlings may require added nitrogen by two months after seeding and about once per month thereafter. Planted trees should be fertilized very cautiously until roots are well established, then a small amount several months apart, and eventually a larger and increasing amount once per year (in spring) as the trees grow larger. Nitrogen needs vary with the avocado cultivar (Embleton and Jones, 1972), and with cultivar progeny (Bergh, unpublished). In California, zinc or iron applications are needed in some locations. Nitrogen is the only element generally applied in Israel (Ticho and Gefen, 1965) but in other regions complete fertilizers have been recommended (Ruehle, 1963; Yee, 1957). Water relations are critical. The avocado is very sensitive to drought, but excess water is conducive to *Phytophthora* root rot if the spores are present (Zentmyer et al., 1967).

At the time of field planting, protect any exposed parts of the stem from sunburn with a coat of whitewash or white latex (water-base) paint; non-latex bases will kill tender plant tissue. Stake the trees for a year, longer in windy areas.

Because of their shallow rooting, the soil around avocado trees should not be disturbed by cultivation. Control weeds by frequent mowing, by herbicides, or by mulches.

Most avocado trees require no pruning. The more erect types may be tipped repeatedly in order to develop a bushier tree. Other pruning is chiefly to keep fruit off the ground, or for easier irrigation and other grove operations. Pruning seedlings tends to delay fruiting.

Evaluating and Propagating the Seedlings

I have found useful a 5 m telescoping aluminum pole with a cutting edged V-blade to hook and sever the fruit pedicel; another essential is a sharp knife.

In the California program, individual records are taken of the following fruit characters: size, shape, color, russet, appearance (on a scale of 1 to 4), depth of skin, flesh appearance (1 to 4), seed cavity, fiber, maturity, yield, ripening qualities, peeling (1 to 4), flavor (1 to 4). The last three characters are rated by laboratory taste panel. All ratings are subjective. Using this system, about 80 seedlings can be evaluated by one person per day, and twice as many if an assistant works ahead of the evaluator, searching for fruit, harvesting several, and cutting them open. Popenoe (1969) gives valuable suggestions for detailed descriptions of selections.

Detailed instructions and illustrations for propagation will be found in Platt and Frolich, 1965; Ruehle, 1963; Soule and Lawrence, 1969; Yee, 1957, and in standard plant propagation texts.

Breeding Systems

Selection of Naturally Occuring Variants

Every avocado cultivar now being grown on a large scale throughout the world originated as a chance seedling; none of importance has been derived from breeding programs. Propagation material from areas with established avocado populations usually formed the initial basis for emerging industries in new regions. 'Fuerte', which for decades has been the leading cultivar world-wide, is a good example. Introduced into California in 1911 as budwood from a seedling tree in Atlixco, Mexico, it gradually gained preeminence. Later it was sent from California to Israel, South Africa, Chile, Australia, Mexico, and other subtropical countries, in each of which it has become the leading cultivar.

Many early California cultivars originated in Mexico or Central America, either from asexual propagation of selections made there, or else from imported seeds (Bergh, 1957). The Guatemalan cultivars 'Benik', 'Itzamna', and 'Nabal' were introduced as budwood; 'Dickinson' came from imported seed. Many of these early cultivars accompanied or followed 'Fuerte' to other regions, and have been grown commercially to a limited degree. In recent years, introductions to California have been almost entirely limited to those collected in a search for resistance to root rot caused by *Phytophthora cinnamomi* (Zentmyer, 1972). These introductions have also been chiefly from Central American and neighboring countries where *Persea* species abound.

Seeds imported from California, Mexico, and Central America formed the basis for the Florida industry (Ruehle, 1963); chance seedlings selected within the state account for well over three-quarters of the total acreage. Florida has become a source of introduction for tropical areas in the same way as California for the subtropics. Floridian cultivars, some entirely of the West Indian race, others Guatemalan-West Indian hybrids, have been successful in various countries of Central America, South America, and the West Indies, in coastal Mexico, and in tropical regions on other continents.

Several hundred chance seedlings selected in California have been named, and three ('Hass', 'Bacon', and 'Zutano') are now recommended for commercial planting. The seed parent is unknown for these cultivars as for most privately produced chance seedlings.

In nearly every region where the avocado is grown, local seedlings have been selected and named. While numerous mediocre-quality cultivars aggravate marketing problems, such selection is to be encouraged to provide new materials for thorough testing. The most promising selections in one region soon become plant introductions to others that have similar climates.

Inbreeding

While self-incompatibility is unknown, the avocado's unique mating system based on synchronous dichogamy insures considerable cross pollination and, consequently, a high degree of heterozygosity. Thus, self pollination usually produces a wide array of segregants, with the great variability of traits among individuals providing a basis for selection.

A program of selfing has been pursued at the University of California, Riverside, based on: 1) ease of producing selfed seedlings for selection, 2) selfing as a rapid method to evaluate breeding potential, and 3) development of superior inbred lines for further breeding. Our experience with about 6000 seedlings derived from a number of self-pollinations as compared with 1200 seedlings from controlled crosses suggest that selfing has potential in avocado as a breeding method. The frequency of promising selections has been about the same (about 4 per 100 seedlings) for our selfed and crossed populations. Although there is some inbreeding depression, the obviously weak, abnormal seedlings which appear with somewhat higher frequency in the S_1 generation than in outcrosses can be discarded in the nursery prior to field planting, and the reduced vigor observed in the second and third generation of selfing may be beneficial because high vegetative vigor tends to be associated with reduced fruit setting.

Selfing and selection have resulted in a number of inbred lines with a high proportion of desirable phenotypes, presumably through the discarding of inferior, deleterious combinations and concentrations of advantageous ones. These lines will be used for subsequent hybridization, while promising selections within these inbred lines are now under test.

FIG. 10. 'Fuerte' is especially prone to bud sporting. (A) Periclinal ("hand-in-glove") chimera produces an odd, reticulate russet pattern. (B) Two examples of sectorial chimeras, which are more frequent in all cultivars. To the left, the mutant tissue is a sunken, yellowish strip, somewhat like an unusually regular "sunblotch" virus symptom. On the right, only a broader central strip is not mutant. This figure also illustrates the marked shape variation in a more elongate fruit like 'Fuerte': (A) is its typical shape in inland regions, while (B–left), at least, is typical along the coast. (Courtesy C. A. Schroeder.)

Outbreeding

Hybridization is the only way to combine the complementary desirable features of different cultivars or to obtain a desirable intermediate trait when the available breeding materials have extreme phenotypes (e.g., breeding lines with clearly superior qualities, but one of which has too large a fruit while the other is too small; or one has too elongate a fruit while the other is round). The commercially important Booth numbered selections in Florida are evidently natural hybrids of the Guatemalan and West Indian races, and are intermediate in harvesting season as well as other useful traits. Parallel interracial hybridizations between Guatemalan and West Indian types have given rise to important intermediate cultivars in Hawaii. Hybridization would also serve to increase heterozygosity should overdominance be involved in commercially important avocado traits.

The University of California breeding program is now shifting its emphasis to hybridization based on the use of lines identified as superior by the self progeny and the use of promising selections obtained directly from selfing.

Interspecific Hybridization

Attempts have been made to solve the serious root rot problem through development of resistant rootstocks by interspecific hybrids of avocado with those *Persea* species immune to the disease. Up to the present time, all such crosses have been unsuccessful; both graft and cross incompatibility appear to be complete between subgenus *Persea* which includes the avocado and subgenus *Eriodaphne* which includes all the known immune species.

The hirsute *Persea floccosa* Mez (probably better designated *P. americana* var. *floccosa*) has the valuable trait of setting much larger numbers of fruits than do other taxa in the *Persea* subgenus. However, its fruits are very small and the seed relatively large. It has been hybridized with several large-fruited, small-seeded cultivars, and a few of the better F_1s selfed or crossed back to commercial cultivars. The results have not been promising; heavy setting ability has been lost at least as rapidly as commercial quality has been approached and flavor at best has been mediocre.

Mutation Breeding and Polyploidy

Occasional spontaneous mutations have long been recognized in the avocado. Tree shape, leaf size, shape, and color, fruit size and shape, or skin surface and thickness have been clearly different on certain "sported" limbs. While trees of a number of cultivars have been affected, 'Fuerte' (Fig. 10) appears unstable, with pronounced proclivity to somatic mutation; in contrast 'Hass' seems to be very stable. No mutation has yet proven horticulturally beneficial, although a tree of the new 'Jim' cultivar has been reported to have a sport limb producing fruit with a longer season and tougher skin.

Mutations affecting fruit yield only are much more difficult to detect, especially because of the highly erratic nature of avocado bearing. Good evidence for genetic yield differences has been obtained only in 'Fuerte' (Hodgson, 1945). The sound nursery practice of taking buds from limbs of demonstrated high-yielding ability guards against detrimental mutations and may pick up beneficial ones.

There is no published information on induced avocado mutations, but a program involving the irradiation of meristems of 'Duke' and certain of the *Persea* species resistant to root rot has been started at the University of California using fast neutrons.

Tetraploidy has been induced with colchicine in the small-fruited 'Mexicola' cultivar. Typical gigas characteristics were observed in the vegetative organs. Fruit set was reduced almost to nil.

Breeding for Specific Characters

Since not one single-gene character is known in the avocado, sources of specific desired traits must be sought from phenotypic evaluation augmented by quantitative inheritance data where available.

Rootstock

The predominant desired trait is adequate resistance to *Phytophthora* root rot. Taxa that are either sexually or asexually compatible with the avocado have little tolerance. 'Duke', a Mexican race cultivar, has some resistance (Zentmyer et al., 1967), but not enough to offer a commercial solution. Also, 'Duke' seedlings germinate erratically and grow slowly (Burns et al., 1968). A major search is now underway by Zentmyer (1972) and his co-workers for additional sources of resistance in taxa compatible with the avocado. A very recent Mexican race introduction designated 'G6' looks highly promising (Zentmyer, personal communication). Even if 'G6' resistance were not greater than that of 'Duke', it might have a different genetic base and so hybridization of the two could combine genes for resistance to provide a real answer to root rot. Extensive attempts to hybridize the avocado (or other members of sub-genus *Persea*) with different *Persea* species of sub-genus *Eriodaphne* that are immune to the root rot organism have failed; special techniques to facilitate crossing are now being tried.

Resistance to salinity is most common in the West Indian race and most rare in the Mexican. There is, however, much variability within each of the three races (Ben-Ya'acov, 1972; Kadman and Ben-Ya'acov, 1970), and even among seedlings from the same tree (Kadman, 1968, and earlier publications). One or more of the Israeli selections (Kadman, 1968, Kadman and Ben-Ya'acov, 1970) may well prove superior to any present rootstock in California or in other regions with salinity problems, or at least be useful as a breeding source of salinity resistance. Chlorosis resistance is also greatest in the West Indian race, and there is considerable intra-racial variability in this trait also (Ben-Ya'acov, 1972). The Mexican race is generally most susceptible to high-lime chlorosis, and Guatemalan lines to poor-drainage chlorosis (Kadman and Ben-Ya'acov, 1970). Thus, West Indians are preferred on both counts. However, their greater cold sensitivity, and their apparently poorer performance in water-logged soils during and after unusually wet winters (Ben-Ya'acov and Oppenheimer, 1970) make it likely that their usefulness for less tropical regions will be limited to breeding. The Mexican race has more genetic resistance to *Dothiorella* canker and to *Verticillium* wilt (Zentmyer et al., 1965).

Breeding for other rootstock traits has been minimal and it is difficult to justify much effort until the overshadowing need for root rot resistance has been met.

'Topa Topa', 'Mexicola', 'Zutano', and most Guatemalan cultivars produce stocks that are readily grafted, since they grow vigorously and take grafts or buds easily. Hence, these might be good parents for crosses with lines having resistance to root rot or salinity.

Possible sources of dwarfing rootstocks include 'Mt4' (Bergh and Whitsell, 1962), 'Jalna', 'Wurtz', and 'Nowels', and *Persea americana* botanical varieties *schiedeana, floccosa,* or *nubigena.* 'Fuerte' and 'Hass' scion tops grow up to twice as rapidly on Guatemalan as compared with Mexican stocks (Bergh, unpublished), with fruit set apparently about proportional to tree size.

Tree

Tree form (Fig. 2), whether of the desired spreading habit as in 'Topa Topa', 'Hass', and 'Fuerte', or erect like 'Bacon', 'Zutano', and 'Reed', has high heritability. However, there is great variability, and many of the progeny from selfing differ markedly from their parents. Nevertheless, if a very upright tree is used as one parent, the progeny would be more likely to have a favorable intermediate tree form if the other parent were sprawling, such as 'Wurtz'.

Excessive vigor and fruitfulness are not fully compatible—exceptionally robust seedlings usually have little or no fruit. Tree vigor declines with inbreeding, and is restored when inbred lines are crossed. There are also polygenic differences in vigor apart from inbreeding.

Outstanding cold hardiness is limited to the Mexican race; members of the West Indian race may be injured by above-freezing chilling. 'Hass' has been considered an extraordinarily cold hardy Guatemalan cultivar, but its progeny suggest that perhaps a quarter of its genes came from the Mexican race; this presumably explains both its cold tolerance and the fact that it is palatable much earlier in the season than fully Guatemalan cultivars. There are differences in hardiness within as well as between the races. For example, among cultivars believed to be pure Guatemalan, 'Nabal' is unusually cold hardy and 'Anaheim' unusually cold tender. The 'Yama' is usually considered one of the hardiest Mexicans, taking −8 C without severe injury (Bergh, unpublished), but its self progeny indicate that it may be a quarter Guatemalan—making it a most propitious parent for cold hardiness breeding. Techniques have been developed for determining relative hardiness by freezing leaves under controlled conditions (Manis and Knight, 1967). Mexican lines appear promising in Florida for breeding commercially acceptable selections with enhanced cold hardiness (Knight, 1971).

Cultivar heat tolerance performance varies even more erratically than cold tolerance, and there appears to be less genetic variability. However, the Mexican race has greater average hardiness, as has been shown especially by 'Mexicola', 'Mayo', and 'Indio'. The Guatemalans 'Frey' and 'Hass' appear to be among the more heat-sensitive cultivars. 'Irving', a Mexican-Guatemalan hybrid, has shown exceptional tolerance of desert heat and low humidity in California. In Morocco, hot and dry wind injured 'Fuerte', 'Hass', and 'Jalna' most severely, 'Choquette' and 'Chavanier #2' least (Chavanier, 1967).

For marginal areas with problematic soil, sudden climatic variations, and perhaps other stress factors, such sturdy cultivars as 'Bacon' and 'Zutano' make good commercial selections and therefore preferred parents. 'Hass' seems especially prone to injury from various environmental stresses, but paradoxically it is remarkable for its wide adaptability. In California, Guatemalan cultivars generally do best near the coast and are less fruitful inland, while Mexican race lines are just the reverse. But 'Hass' does exceptionally well in nearly every microclimate where it has been tried, excepting too cold or really tropical areas. It is, therefore, an indicated parent where performance in a broad spectrum of situations is important. In contrast, the 'Fuerte' bears well only in very limited regions of California.

Fruit set that is precocious, consistent, and heavy is the most important characteristic. Two new cultivars with outstanding yielding ability are 'Reed' and especially 'Tova'. 'Ruehle' (Popenoe, 1963a) is a very productive West Indian selection. 'Zutano' also has unusual productivity, but its progeny resemble their parents more in inferior fruit quality than in superior fruit set.

Fruit

For earlier-season cultivars with short tree life, the great majority of fruits should reach harvestable maturity at about the same time. Early-maturing 'Irving' and mid-season 'Rincon' both reach maturity over an objectionably long period of time.

Nothing is known about the heritability of this trait, and it is more difficult to rate. Although this character may be ignored in the original seedling evaluation it should be evaluated in second testing.

'Hass' is also exceptional for length of successful fruit storage on the tree after maturity has been reached. 'Hass' fruits are harvested every month of the year in California. Partly this is a result of climatic differences, but in a given locality they may be palatable for six months or longer before they begin dropping. Just as remarkable in this respect is 'Fuerte', since it stores on the tree about as long (during a cooler part of the year) and is much earlier maturing—the earlier a cultivar matures, the more condensed its entire fruiting schedule, and thus the shorter its expected tree storage period. Mexican-race fruit will usually store for only two or three weeks, before it starts to drop or the skin discolors, cracks, and breaks down. Most purely Guatemalan cultivars have a much shorter season than 'Fuerte' or 'Hass' in spite of their reaching maturity weeks or months later than even the latter. Since 'Fuerte' and 'Hass' both have genes of both of the less tropical races, racial hybridity may favor long tree storage, but 'Bacon' is an example of hybrids with short fruit life on the tree. 'Hass' and 'Fuerte' also both have exceptionally high oil content at maturity, which could be the major contributing factor; this would point to quite different guidelines in choosing breeding parents.

The present optimum fruit size for most markets of about 300 g may be obtained by different breeding approaches. One can self or hybridize parents with genotypes that result in this optimum phenotype. Or one can hybridize two parents of which one has below-optimum and the other above-optimum genetic size potentialities. Phenotype provides biased estimate of genotype: in every selfed progeny that I have studied, average seedling fruit size was lower than that of parent fruit size. Presumably this is a consequence of human selection for larger fruit size, so that progenies tend to regress toward the natural population mean as dominance and epistatic size effects are reduced by gene segregation. The larger size favored in many tropical regions can be achieved by using some of the very large-fruited cultivars of the West Indian race; large-fruited Guatemalan lines are also available.

Fruit shape segregates extensively in most self progenies. Nevertheless, there is enough heritability to make parental phenotype a useful guide. As with size, the parent(s) can be of about optimum for the trait, or complementary: one too elongate and the other too oblate. The squat-pyriform fruits of 'Hass' or of the new 'Teague' (Fig. 3) are about optimum, as is the ovate fruit of 'Bacon', and comparable shapes. Progeny tend to be more elongate than the parents, presumably due to human selection for a more round shape.

The preferred medium-thick skin (Fig. 3B) can also be achieved by crossing either optimum with optimum, or one extreme with the other; an example of the latter approach might be A and B of Fig. 1. The 'Fuerte' skin has proven too thin for best shipping and handling in California and elsewhere (Pretorius, 1971). Good peeling ability is usually associated with a medium-thick skin as in 'Hass' (Fig. 3B), but the thinner skin of 'Fuerte' and even the very thin skin of 'Stewart' usually peel well. Attractive skin colors are exemplified by the deep green of 'Rincon', the yellower speckling of 'Fuerte' or 'Elsie', and the solid purple of 'Mexicola'. Skins that are part green and part purple are usually less attractive, as is true of a considerable proportion of selfed-'Nabal' progeny. The severe russeting of 'Regina', and the end-spotting of 'Zutano' are unfortunately present in a majority of their respective progeny. Where skin fungus diseases are a problem, genetic resistance is desirable for good appearance, since fungicide spray treatments are expensive and not fully successful. Ruehle (1963) gives some relative cultivar susceptibilities: Cercospora spot or blotch (causal organism, *Cercospora purpurea* Cke.) is much less severe on 'Collinson', 'Fuchsia', and 'Pollock'; the 'Fuerte' is especially susceptible to anthracnose or black spot (caused by *Colletotrichum gloeosporioides* Penz.); 'Fuchsia', 'Pollock', 'Booth 1', and 'Waldin' are quite resistant to avocado scab (caused by *Sphaceloma perseae* Jenkins), while the 'Lula' is highly susceptible. The newer 'Ruehle' cultivar has been reported fairly resistant to all three diseases (Popenoe, 1963a). Under conditions of high humidity, as in the tropics generally, tolerance of anthracnose becomes a principal consideration. The Mexican race is extremely susceptible (N. P. Maxwell, personal communication), although 'Brogden' has shown some resistance in Florida (Popenoe, 1962). 'Hass' has demonstrated high anthracnose resistance in Mexico (Wilkins, 1965).

A seed that is small relative to fruit size and

tight in the pulp cavity is a superior attribute of many Guatemalan lines (Fig. 1B). 'Stewart' and 'Teague' (Fig. 3C) demonstrate that these traits can be transferred into an otherwise largely Mexican genetic background. 'Irving', whose later-maturing fruit and slightly thicker skin indicate more Guatemalan germplasm, has an exceptionally small seed ratio. A high proportion of 'Stewart' and 'Irving' self progeny have seeds comparable to or even smaller than their parents. The West Indian 'Ruehle' (Popenoe, 1963a) has an outstandingly small seed for that race. The Australian selection 'Sharwil' is of mixed Mexican and Guatemalan ancestry and has a tiny seed (Hope, 1963).

'Hass' fruit has a rich, attractive pulp color, as does the fruit of most of its progeny. However, the even yellower flesh color of 'Yama' is passed on to comparatively few of its selfs. In most cultivars, the fibrovascular bundles tend to become more prominent with increasing maturity, but this undesirable development is less marked in most Guatemalan lines. Uniform and adequate softening of the fruit as it ripens is independent of race; the Mexican 'Jalna' and the Guatemalan 'Ryan' are two cultivars with weaknesses in this respect. Little is known about the inheritance of flesh softening. Flavor is probably the most subjective of all traits, but the rich, slightly nutty taste of 'Hass', 'Fuerte', or 'Edranol' is generally preferred to more bland Guatemalan and hybrid flavors. The spicy or anise-like flavor of Mexican types such as 'Mexicola' or 'Duke' is considered still more desirable by many people, while the mild, sweeter taste typical of West Indians may be preferred by people accustomed to such. Flavor differences between races tend to be inherited, but within a race there has seemed to be little correlation between parent and average progeny rating. For example, the 'Mayo' has an unusually rich and pleasing flavor but its selfed seedlings have a poor average rating, with a high proportion bland to unpleasant. Long shelf life is not an attribute of the avocado fruit, although the rapidity with which 'Anaheim' softens into a mushy consistency contrasts with the better-keeping qualities of 'Hass', 'Fuerte', 'Stewart', and others. West Indian lines have a much lower oil content than those of the other two races but there is considerable intra-race variability. While Mexicans and Guatemalans commonly have more than 20% oil (Rounds, 1950), the 'Mayo' and 'Anaheim' cultivars of these two respective races may both have an oil content of under 8% when they begin dropping or deteriorating.

Achievements and Prospects

Subtropical Regions

These are the areas where the less tropical Mexican and Guatemalan races are adapted.

California: The cultivars produced in California form the backbone of the avocado industries of Israel, South Africa, Chile, Mexico, Australia, New Zealand, and other countries with a similar climate.

The first commercially important cultivars were selected from native seedlings in Mexico ('Fuerte' and 'Puebla'), and in Guatemala ('Nabal' and others). Subsequent selections of importance have been from open-pollinated seedlings grown within the state. These include such declining cultivars as the very largely Mexican 'Jalna', the hybrid (intermediate phenotype) 'Rincon', the largely Guatemalan 'MacArthur', and the presumably fully Guatemalan 'Anaheim' and 'Dickinson'.

Local selections of growing importance are the largely Mexican 'Zutano', the hybrid 'Bacon', and the largely Guatemalan 'Hass'. These three plus the hybrid 'Fuerte' are the four cultivars presently recommended for California. 'Fuerte' still constitutes about half of all bearing acreage, but its proportion is declining because total acreage is increasing while 'Fuerte' is being little planted and is suffering losses to root rot, urbanization, and topworking to more productive cultivars. 'Bacon' and 'Zutano' acreage will probably continue to increase somewhat, but both are now grown on a relatively small scale, are recommended only for areas unsuitable for 'Hass' or 'Fuerte' ('Bacon' is very cold hardy), and have serious commercial limitations—tree form, short season, mediocre fruit quality (especially 'Zutano'), and only moderate production ('Bacon').

Commercially successful avocado cultivar development in California has so far involved only the most primitive method, selection from chance seedlings. Even the seed parent is not known for any of the above cultivars. With no racial sterility barriers, later originations within the state are a highly variable mixture of the two races. Relative racial proportions can be estimated imperfectly from the individual phenotype, and better from a progeny test. Analysis of self progeny has indicated that about one-sixth or one-quarter of the genes in 'Hass'—previously thought to be straight Guatemalan—are actually from the Mexican race.

More is known about the ancestry of the recent selections. Of these, the most promising are the following five.

'Reed', from a 'Nabal' seed, is conjectured to have 'Anaheim' as pollen parent because of the proximity of a tree of that cultivar. A Guatemalan phenotype, including late maturity, its fruits are larger than present consumer preference, and have a thick, ovate fruit shape, green skin, and high quality. The trees are erect with precocious, consistent, very heavy production. It is a private selection.

'Jim' is from a 'Bacon' seed, perhaps with the same pollen parent. Its fruits begin maturing with 'Bacon' and store on the tree at least a month longer, but they mature over a longer period so that two pickings are probably necessary. The fruit is excessively long-necked, but suffers less from surface deterioration, and is green. The tree seems equally cold hardy, is less erect, and is apparently more productive. 'Jim' is a private selection.

'Creelman' is a hybrid of 'Fuerte' and 'Hass', so is theoretically about two-thirds Guatemalan and one-third Mexican. Mature about the beginning of March in California, it holds into July. Fruit size is about 275 g when carrying a good crop. The fruit is pyriform with a green skin and very good flavor. A University of California selection, it has a spreading tree and good production.

'Teague' (Fig. 3C) is a hybrid of 'Fuerte' and 'Duke', so is theoretically about one-quarter Guatemalan, but matures in the Mexican season (about mid-September through October in California). Although it is typically Mexican in short tree storage and thin skin, its seeds are tight and exceptionally small for that season. The fruit is about 300 g with a thick, ovate shape, green skin, and a mild, good flavor. It has a vigorous, spreading tree; production has been variable. It is a University of California selection.

'Alboyce', a 'Hass' open-pollinated seedling, is thinner-skinned than 'Hass' and earlier (about January to July—a remarkably long tree storage). The fruit is 'Hass' size (about 240 g, so smaller than optimum), and shape; but has green skin, very high quality, small seed, and excellent flavor. A University of California selection, it is a spreading tree with only moderate production.

Over the past 60 years or so, some 500 named cultivars have been introduced in California, nearly all by private individuals. The great majority of these introductions were a reflection of personal partialities and hopes; in monetary terms they probably did the industry more harm than good. While many soon disappeared, about 200 different cultivars are delivered annually to California packinghouses, creating marketing difficulties and added expense. Repeated campaigns to encourage the topworking of mediocre trees to the superior cultivars have been only partly successful. However, it is only by introducing and testing new selections that superior cultivars can be developed. Indeed, the commercial industry of California (and of the less tropical avocado world generally) is based very largely on cultivars developed by private individuals as described above. The problem is how to encourage the growing and testing of seedlings, and yet discourage dissemination (beyond the few trees needed for testing in different environments) until superiority is proven; unpromising seedlings should be quickly eliminated or topworked.

The California Avocado Society, through its variety committee, has brought order out of an initially chaotic cultivar situation by publicizing the weaknesses of the great majority of introductions and recommending only the few that really excel in overall commercial desirability. This has been a critical factor in the success of the avocado in California. An invaluable complement to an avocado breeding program anywhere is some such evaluating committee or group, whether private or governmental.

Recommendations are expected to change with time. Of the four cultivars at present endorsed for California growers, none is recommended for all five growing districts, and two are suggested for only one district each. In more tropical regions, fruits have shorter tree life and so more cultivars are needed over the marketing season.

Even 'Hass' has enough commercial weaknesses to make a replacement desirable. Its rough skin and dark color are drawbacks in some markets.

More serious is its rather small size, which becomes aggravated with increasing tree age. 'Hass' is sensitive to environmental stress, and although production is outstanding, it is surpassed by 'Reed' and the Israeli 'Tova'.

Perhaps the greatest need of the avocado industry in California and some other regions is a *Phytophthora*-resistant rootstock. Hybridization, irradiation, species graft-union studies, and search among natural populations in Central America are among the approaches being used.

Israel: The Israeli avocado industry has involved a number of stages. The first involved chance seedlings that had come to their attention. 'Scotland' has been grown on a small scale but 'Ettinger' (Ohad, 1965; Ticho and Gefen, 1965) is the only such cultivar of present commercial importance selected in Israel, a situation that seems about to change.

'Ettinger' is probably a 'Fuerte' seedling, but earlier, maturing in October and harvested through December—a long season for so early a cultivar. The fruit is pyriform and green. Trees are erect, resistant to cold and wind, and noted for good production. 'Ettinger' grown in California does not mature noticeably earlier than the 'Fuerte' (about December), and yet by February there is much russeting; it has poor peeling ability and a large seed, but very little fiber and a superior, rich, spicy flavor.

From hundreds of private chance seedlings and thousands of windbreak trees, some 16 selections have been made over the past years (Oppenheimer et al., 1970; Slor and Spodheim, 1972). The following appear to be the most promising:

'Tova' is harvested about December through February (compared with 'Fuerte', about November through March). The fruit is about 250 g, ovate, green, easily peeled, with fairly good flavor. The tree is rather small, perhaps as a result of its uniquely heavy annual yields, apparently in excess of 20,000 kg per ha (S. Gazit, personal communication).

'Horshim', harvested about December through April, has fruit which averages 275 g, and is elongate pyriform, and green. Superb flavor, small seed, slower flesh darkening when cut, and alternate bearing but very good set are characteristic.

'Netaim' is harvested about March through June ('Nabal' season). The fruit averages 300 g with very good flavor. It is alternate bearing but has a very good set.

A breeding program has now been started at the Volcani Institute (S. Gazit, personal communication), involving both self-pollination and hybridization, with 'Tova' understandably the preferred seed parent. Although the program was delayed until the industry was well established, Israeli horticulturists have advanced to the forefront of avocado research, and more effort and funds are directed towards obtaining new cultivars than in any other country (Malo, 1971).

South Africa: Because breeding is so costly and time consuming, cultivar improvement has been based on the intensive study of existing seedlings (Le Roux, 1970). Extensive travels, primarily in Natal, resulted in many selections, including 'Mcfie' (probably a 'Fuerte' seedling); 'Teteluku' and 'Aitken' (resemble 'Fuerte' in appearance); 'Frankish'; 'Clarke'; 'Doornkop' (probably Guatemalan); 'Hillcrest' and 'Ixopo' (perhaps Guatemalan x West Indian hybrids.)

An earlier South African selection, the 'H.L.H. Carton' (van der Meulen, 1970), is the only Mexican race cultivar to be grown there on a commercial scale. It tolerates more cold and also more heat, maturing in late summer in the very hot Lowveld areas. It has a typical Mexican phenotype: somewhat elongate fruit with a large seed and thin purple skin.

Seedling trees in the Nelspruit area have not been promising. A 'Fuerte' progeny of 393 seedlings had no tree with fruit as good as the parent, although one selection was made for further evaluation; and of 1,614 seedlings examined elsewhere, "not a single tree with promising fruit was found" (Smith et al., 1970).

Australia and New Zealand: A number of selections have been made, including the rather 'Fuerte'-like 'Ormond' and the late-maturing 'Hopkins' from New Zealand (Fletcher, 1972), and a superior 'MacArthur' type and a 'Hass' x 'Nabal' hybrid in Australia (Shepard, 1971).

'Sharwil', the really superior Australian selection, ranks with 'Fuerte' and 'Hass' as the recommended cultivars for that country (Chalker and Robinson, 1969). It is considered one of the three top cultivars also for Hawaii (R. A. Hamilton, personal communication). It is intermediate between the Guatemalan and Mexican races in phenotype, maturing a little later than 'Fuerte'. The fruit is pyriform and green, with a small seed and a rich nutty flavor. Its set is moderate but consistent (Hope, 1963).

Other countries: A number of Mexican race selections have been made in Chile, including 'Campeon', 'Cholula', and the 'Peuminas' numbered cultivars (Hodgson, 1959). The more recent introduction 'Black La Cruz' is apparently a Mexican x Guatemalan hybrid. It resembles 'Fuerte' in size, shape and quality, but is two months earlier in Chile; it offers good production (Schmidt, 1965). Also 'Fuerte'-like and with superior yielding ability, but later in maturity, is the increasingly planted 'Corona' (Malo, 1971).

Local selections are also being made in Mexico (Gustafson, 1972; Turu, 1970).

Several selections have been made in Morocco (Chavanier, 1967). A 'Fuerte' x 'Caliente' hybrid (Vogel, 1958) is typically Mexican, pyriform, with high oil content and good quality. Very recently a selection, apparently of Guatemalan ancestry, has been named 'Skirat' (M. Benzit, personal communication).

In Greece, a 'Fuerte'-like seedling with a more pronounced neck has been discovered (Zamenes, 1967). Although thin-skinned (Mexican type), it matures late and ships well, and has very good flavor.

A 'Linda' x 'Lyon' hybrid with superior flavor and heavy production, but with very erect tree form, is under observation in Kenya. Seedlings of all sizes and colors abound there and are eagerly consumed, but none with commercial promise has been found (R. Andersen, personal communication).

Many countries contain tropical, subtropical, and intermediate areas, and so their choice of horticultural race may be mixed. In the Canary Islands (Cabezon, 1965), promising selections include those of presumed all-Guatemalan ancestry (the rather large-fruited 'Orotava'); Guatemalan x West Indian (the very large 'Gema'); and Mexican x West Indian (the small 'Robusta').

The commercial avocado industry of the Hawaiian Islands traces its origins to the introduction of Guatemalan seeds in 1891 (Yee, 1957). Subsequent development was rapid. One seedling became the famous 'MacDonald' cultivar and it gave rise to a number of further selections, a few of which, such as 'Beardsley', are still being grown, and one of which ('Wilder') in turn was the seed parent for several selections that resulted from pollination by nearby West Indian trees. Recent additional selections have been made, and R. A. Hamilton (personal communication) has rated the five most desirable cultivars as 'Sharwil', 'Kahaluu', 'Morishige', Fujikawa', and 'H.A.E.S. 7315'. At least the 'Kahaluu' is part West Indian, as well as part Guatemalan.

Even in the more tropical latitudes, sufficient elevation can create enough frost hazard to make the Mexican the only successful race. W. Popenoe (1950, 1951) at the Escuela Agrícola Panamericana, in Tegucigalpa, Honduras, selected two predominantly Mexican seedlings in the Rodiles grove at Atlixco, Mexico, where the 'Fuerte' originated. Both selections had glossy, attractive green fruits, with thin skins, tight seeds, and rich, nutty, excellent flavor. The 'Aztec' averaged about 310 g, matured October to November at Tegucigalpa, had much leaf anise, and produced very vigorous trees. The 'Toltec' weighed about 250 g, matured November to January, had less leaf anise, and the tree was erect and rather small.

Tropical Regions

The West Indian race is best adapted to the tropics.

Florida: The commercially important Florida cultivars also originated as open-pollinated chance seedlings, but unlike the situation in California, the seed parent of many selectors is known. All were developed by private individuals. They may be classified by approximate season of maturation (Krome, 1967; Malo, 1971; Popenoe, 1962; Ruehle, 1963). All of the following cultivars are green-skinned at maturity.

Those maturing in summer (July and August) are of the West Indian race. 'Pollock', asexually propagated as early as 1901 and long the standard for its season, has fruit which is very attractive and high in quality but very large. Production is low. 'Simmonds' is a seedling of 'Pollock'. It is replacing its parent because of heavier yields and slightly smaller fruits, but its fruits are less attractive, and the tree is weaker. It has performed well in Honduras (Popenoe, 1958). 'Ruehle', a seedling of 'Waldin', is smaller than 'Pollock' and 'Simmonds' (about 425 g) and has a relatively small seed. It has a heavy set and will usually hold through September (Popenoe, 1963a).

Fall crops (September through November) are mostly West Indian x Guatemalan hybrids. 'Waldin' is usually considered straight West Indian, but its later maturity, higher oil percentage, and greater cold hardiness suggest that it may be part Guatemalan, possibly from Central American ancestors. Large by subtropical standards (about 550 g), it has a large seed and is precocious and productive. 'Tonnage' is a seedling of a Guatemalan ('Taylor', which in turn was a seedling of the old California cultivar 'Royal'). Its earliness and slightly loose seed indicate West Indian ancestry. It is about 500 g and cold hardy for its season. 'Booth 8' had a Guatemalan seed parent fertilized by West Indian pollen, and is a little later than 'Tonnage'. Its size varies with fruit number—only about 300 g in clusters, with rather thick skin and fair flavor. It is prolific but alternate bearing and tends to overbear, requiring thinning. Now the leading Florida cultivar, it accounts for about 30% of the total production. 'Booth 7' is of the same origin as 'Booth 8', but has a slightly later maturity. About 425 g in size, it has thick skin, and a moderate seed size. It is prolific, but fruit may drop at 3 C.

Avocados maturing in the winter (about December-January) are of unknown ancestry, but they have both Guatemalan and West Indian characteristics. 'Lula' is second to 'Booth 8' in volume, although now fewer are planted. It may be part Mexican. Fruit is about 500 g, generally necked, large seeded, and susceptible to scab. It has very good flavor and refrigerates unusually well for a tropical cultivar. The tree is too erect, but prolific; it is the most cold hardy of these Florida cultivars. 'Choquette' produces a fruit which is very large (over 800 g) and flattened on one side at the apex. It has a very good flavor and a medium but somewhat loose seed. Production is moderate. 'Monroe', similar to 'Choquette' in size and shape, is handsome and prolific. 'Hall', about 700 g, is handsome but scab-susceptible. It offers heavy set in alternate years.

For the spring season (February–May), 'Kampong' is the best of the later cultivars (Popenoe, 1963*b*). The commercial season is from January into April, with suitability for home use until about the end of May. Fruit size is about 700 g.

Of all the above cultivars, only the 'Ruehle' is of fairly recent origin. However, renewed interest in new cultivars for Florida was reported by Krome (1967), including selection from populations of both self and cross hybridizations. Manis and Knight (1967) announced plans to breed cultivars for areas now considered too cold to support a commercial avocado industry.

West Indies: Most of these islands produce considerable seedling West Indian fruit, of highly variable appearance and quality, and with a harvesting season largely limited to summer and early fall.

In Trinidad and Tobago, Storey (1968) found the West Indian seedlings to be remarkably uniform, suggesting to him that the original Spanish introduction may have been a single seed. Guatemalan lines are being introduced to provide parents for fruits with later maturity and longer tree storage. An extension of the harvesting season by breeding would be a major contribution to the avocado industry of tropical regions.

Cultivars from many areas and including all three races were introduced into the United States Virgin Islands (Frederiksen and Krochmal, 1963; Bond and Frederiksen, 1968). Cultivars with much Mexican ancestry have generally done very poorly. Only a few Guatemalans matured fruit satisfactorily and their quality has been inferior. Nevertheless, an apparently pure Guatemalan of local origin, the 'Gregory', was one of the best, although the skin is too thick to peel and fruit size is smaller than the local preference. Also, the numbered Booth cultivars from Florida (Guatemalan x West Indian hybrids) did well. But the local consumers preferred predominantly West Indian types, especially the local selection 'D.W.I. (Danish West Indian) Bank', which has fruit to about 900 g, medium and apparently tight seed, good peeling, and excellent flavor. Two Puerto Rican selections were also highly regarded: 'Semil 34' (about 450 g, small tight seed, fine flavor), and 'Miguelito' (to 1300 g, with a long slender neck and loose seed, thus non-commercial, yet very popular locally).

A number of local selections have been made in Jamaica, including 'Elgin', 'Gimball', 'Huntley', and 'Stuart' (Davidson, 1967); a program to incorporate Guatemalan germplasm into the better available West Indian lines is planned.

The 'Catalina' is a superior West Indian selection made in Cuba (Popenoe, 1958). A number of other cultivars have been selected there (Cañizares y Zayas, 1937), as well as on other islands of the West Indies.

Other countries: In Venezuela, 'Pollock' is the primary cultivar, with 'Simmonds' and 'Choquette' probably next in importance (Serpa, 1968). There are many local selections, most with pronounced necks and large, often loose seeds. Serpa (personal communication) has suggested that Guatemalans should be tried further, for higher elevations and for hybridization with the West Indians. But West Indian lines perform so well over a wide range of the Venezuelan elevations that fruit is available almost the entire year. The practice of using avocado seedlings to shade cacao plantings provides a ready population from which to select.

The avocado industry in Ghana could be greatly improved by selection of local cultivars (Godfrey-Sam-Aggrey, 1969). More than a dozen seedling selections in Bangalore, South India, have been described (Raman and Balaram, 1967). Most appear to be of the less tropical Guatemalan or Guatemalan x Mexican types. In Thailand (S. Kosiyachinda, personal communication), local seedlings are being examined for possible commercial usefulness, as are seedlings grown from Hawaiian seeds.

Avocado production is increasing in many countries around the world, and new regions are beginning to import cultivars to evaluate the potentialities of a new industry. The development of locally-adapted cultivars with high production could be of great nutritive benefit in many tropical areas. Expanding consumption in more developed nations is providing a healthful source of dietary variation and pleasure. Other uses of the fruit are promising, especially as a tropical oil crop for cosmetic and other purposes; because the West Indians best adapted to the tropics are much the lowest in oil of the three races, further racial hybridization should be explored. For all avocado regions, continued future expansion of the industry would be greatly assisted by superior cultivars, better adapted to each region.

Literature Cited

Anderson, E. 1950. Variation in avocados at the Rodiles plantation. *Ceiba* 1:50–55.

Anonymous. 1968. Fruitful days in Cameroun (in French). *Fruits d' outre mer* 23:329–32.

Arkin, I. H. 1967. Opportunities for marketing Israeli avocados in western Europe. *Calif. Avocado Soc. Yrbk.* 51:73–84.

Ben-Ya'acov, A. 1972. Avocado rootstock-scion relationships: A long-term large-scale field research project. *Calif. Avocado Soc. Yrbk.* 55:158–161.

Ben-Ya'acov, A., and C. Oppenheimer, 1970. Trials with rootstocks and sources of budwood, p. 40–43. In *The Volcani Inst. Agr. Res., Div. Subtrop. Hort. 1960–1969.* (Israel).

Bergh, B. O. 1957. Avocado breeding in California. *Proc. Florida St. Hort. Soc.* 70:284–290.

———. 1961. Breeding avocados at C. R. C. *Calif. Avocado Soc. Yrbk.* 45:67–74.

———. 1966. A Hass open-pollinated progeny set. *Calif. Avocado Soc. Yrbk.* 50:64–77.

———. 1967. Reasons for low yields of avocados. *Calif. Avocado Soc. Yrbk.* 51:161–167, 169–172.

———. 1969. Avocado (*Persea americana* Miller). *In* F. P. Ferwerda and F. Wit, eds. *Outlines of perennial crop breeding in the tropics.* p. 23–51. Misc. Pap. 4. Wageningen, The Netherlands: Landbouwhogeschool.

Bergh, B. O., M. J. Garber, and C. D. Gustafson. 1966. The effect of adjacent trees of other avocado varieties on 'Fuerte' fruit-set. *Proc. Amer. Soc. Hort. Sci.* 89:167–174.

Bergh, B. O., and C. D. Gustafson. 1958. 'Fuerte' fruit set as influenced by crosspollination. *Calif. Avocado Soc. Yrbk.* 42:64–66.

Bergh, B. O., R. W. Scora, and W. B. Storey. A comparison of leaf terpenes in *Persea* subg. *Persea. Bot. Gaz.* in press.

Bergh, B. O., and R. H. Whitsell. 1962. A possible dwarfing rootstock for avocados. *Calif. Avocado Soc. Yrbk.* 46:55–62.

Bond, R. M., and A. L. Frederiksen. 1968. Avocados in St. Croix, U. S. Virgin Islands. *Calif. Avocado Soc. Yrbk.* 52:139–144.

Bringhurst, R. S. 1951. Influence of glasshouse conditions on flower behavior of 'Hass' and 'Anaheim' avocados. *Calif. Avocado Soc. Yrbk.* 1951:164–168.

Burns, R. M., R. J. Drake, J. M. Wallace, and G. A. Zentmyer. 1968. Testing 'Duke' avocado seed source trees for sunblotch. *Calif. Avocado Soc. Yrbk.* 52: 109–112.

———. 1954. Interspecific hybridization in and chromosome numbers in *Persea. Proc. Amer. Soc. Hort. Sci.* 63:239–242.

Cabezon, A. G. 1965. Avocado culture in the Canary Islands. *Calif. Avocado Soc. Yrbk.* 49:47–48.

Chalker, F. C., and P. W. Robinson. 1969. The avocado. *Agr. Gaz. New South Wales* 80:347–354.

Chandler, W. H. 1958. The avocado. In *Evergreen or-*

chards. 2nd ed. p. 205–228. Philadelphia: Lea & Febiger.

Chapot, H. 1967. Avocado culture in Turkey. *Calif. Avocado Soc. Yrbk.* 51:93–96.

Chavanier, G. 1967. New observations on avocado growing in Morocco. *Calif. Avocado Soc. Yrbk.* 51:111–113.

Davidson, M. R. L. 1967. Preliminary notes on avocados. *Inf. Bul. Sci. Res. Council Jamaica* 8(1):22–31.

Embleton, T. W., and W. W. Jones. 1972. Development of nitrogen fertilizer programs for California avocados. *Calif. Avocado Soc. Yrbk.* 55:90–96.

Fletcher, W. A. 1972. Avocado growing in New Zealand. *Calif. Avocado Soc. Yrbk.* 55:152–155.

Frederiksen, A. L., and A. Krochmal. 1963. New avocado varieties for the U. S. Virgin Islands. *Caribbean Agr.* 1:293–301.

Godfrey-Sam-Aggrey, W. 1969. Avocado production in Ghana. *World Crops* 21:271–272.

Gustafson, C. D. 1970. Israel's avocado industry—1969. *Calif. Avocado Soc. Yrbk.* 53:38–41.

———. 1972. Avocado growers' study mission #2 to Mexico—1970. *Calif. Avocado Soc. Yrbk.* 55:61–67.

Hodgson, R. W. 1945. Suggestive evidence of the existence of strains in the 'Fuerte' avocado variety. *Calif. Avocado Soc. Yrbk.* 1945:24–26.

———. 1950. The avocado—a gift from the middle Americas. *Econ. Bot.* 4:253–293.

———. 1959. The avocado industry of Chile. *Calif. Avocado Soc. Yrbk.* 43:45–49.

Hodgson, R. W., and S. H. Cameron. 1936. Temperature in relation to the alternate bearing behavior of the 'Fuerte' avocado variety. *Proc. Amer. Soc. Hort. Sci.* 33:55–60.

Hope, T. 1963. Quality tests identify best avocados. *Queensland Agr. J.* 89:657–660.

Hume, E. P. 1951. Growing avocados in Puerto Rico. *USDA Fed. Expt. Sta. Puerto Rico Cir.* 33.

Kadman, A. 1963. Germination experiment with avocado seeds. *Calif. Avocado Soc. Yrbk.* 47:58–60.

———. 1968. Selection of avocado rootstock suitable for use with saline irrigation water. *Calif. Avocado Soc. Yrbk.* 52:145–147.

Kadman, A., and A. Ben-Ya'acov. 1970. Selection of rootstocks and other work related to salinity and lime, p. 23–40. In *The Volcani Inst. of Agr. Res., Div. Subtrop. Hort. 1960–1969.* (Israel.)

Kellogg, G. 1971. The Fuerte avocado in Yorba Linda. *Calif. Avocado Soc. Yrbk.* 54:47–48.

Knight, R. J. 1971. Breeding for coldhardiness in subtropical fruits. *HortScience* 6:157–160.

Kopp, L. E. 1966. A taxonomic revision of the genus *Persea* in the Western Hemisphere. (*Persea-Lauraceae*). *Memoirs N. Y. Bot. Garden* 14, no. 1.

Krome, W. H. 1967. Avocado varieties I am planting now—and why. *Proc. Florida St. Hort. Soc.* 80:359–361.

Lahav, E. 1970. Localization of fruit on the tree, branch girdling and fruit thinning. a. Localization of fruit on the tree, p. 60–61. In *The Volcani Inst. Agr. Res., Div. Subtrop. Hort. 1960–1969.* (Israel).

Lahav, E., B. Gefen, and D. Zamet. 1971. The effect of girdling on the productivity of the avocado. *J. Amer. Soc. Hort. Sci.* 96:396–398.

———. 1972. The effect of girdling on fruit quality, phenology, and mineral analysis of the avocado tree. *Calif. Avocado Soc. Yrbk.* 55:162–168.

Lambourne, J. 1934. The avocado pear. *Malayan Agr. J.* 22:131–140.

Le Roux, J. C. 1970. The ideal summer-maturing avocado cultivar. *Farming in South Africa* 46:25–27, 29.

Malo, S. E. 1971. Mango and avocado cultivars: present status and future developments. *Proc. Florida St. Hort. Soc.* 83:357–362.

Manis, W. E., and R. J. Knight, Jr. 1967. Avocado germplasm evaluation: Technique used in screening for cold tolerance. *Proc. Florida St. Hort. Soc.* 80:387–391.

Moreuil, C. 1971. Brief notes on some fruit species on the east coast of Madagascar (in French). *Fruits d'outre mer* 26:53–63.

Morin, C. 1967. The avocado (in Spanish). In *Production of tropical fruits* (in Spanish). 2nd ed. p. 84–172. Lima, Peru: Librerias A.B.C.

Ochse, J. J., M. J. Soule, Jr., M. J. Dijkman, and C. Wehlburg. 1961. Avocado in *Tropical and Subtropical Agr.*, 2 vols. p. 617–642. New York: Macmillan Co.

Ohad, R. 1965. More on avocados in Israel. *Calif. Avocado Soc. Yrbk.* 49:61–66.

Oppenheimer, C. 1947. The avocado industry in Palestine. *Calif. Avocado Soc. Yrbk.* 1947:112–122.

Oppenheimer, C., S. Gazit, E. Slor, and H. Lippman. 1970. Varieties, introductions and local selections, p. 54–57. In *The Volcani Inst. Agr. Res., Div. Subtrop. Hort. 1960–1969.* (Israel).

Palacious, J. 1965. Avocado culture in the Argentine Northwest. *Calif. Avocado Soc. Yrbk.* 49:43.

Platt, R. G., and E. F. Frolich. 1965. Propagation of avocados. *Univ. Calif. Agr. Expt. Sta. Ext. Ser. Cir.* 531.

Popenoe, J. 1962. Summer avocado varieties. *Proc. Florida St. Hort. Soc.* 75:358–360.

———. 1963a. The Ruehle avocado. *Florida Agr. Expt. Sta. Cir.* S-144.

———. 1963b. Spring avocado varieties for South Florida. *Amer. Soc. Hort. Sci.: Caribbean section. Proc. Annual Meeting* 7:80–83.

Popenoe, W. 1919. The avocado in Guatemala. *USDA Bul.* 743.

———. 1941. The avocado—a horticultural problem. *Trop. Agr.* 18:3–7.

———. 1950. 'Aztec', a new horticultural variety of avocado. *Ceiba* 1:116–118.

———. 1951. 'Toltec', another new horticultural variety of avocado. *Ceiba* 1:225–227.

———. 1952. The avocado. *In* Central American fruit culture. p. 305–310. *Ceiba* 1:269–367.

———. 1958. A program to increase the production of good fruits in the republic of El Salvador (in Spanish). *Ceiba* 7:44–61.

———. 1969. The value of systematic pomology in tropical fruit culture. *Florida St. Hort. Soc. Proc.* 82:309–313.

Popenoe, W., and L. O. Williams. 1947. The expedition to Mexico of October 1947. *Calif. Avocado Soc. Yrbk.* 1947:22–28.

Pretorius, W. J. 1971. South Africa's avocado industry—the present position. *South Africa Citrus J.* 449:3, 5, 8.

Raman, T. S. V., and P. D. Balaram. 1967. Avocado culture in India. *Calif. Avocado Soc. Yrbk.* 51:97–106.

Rhodes, A. M., S. E. Malo, C. W. Campbell, and S. G. Carmer. 1971. A numerical taxonomic study of the avocado (*Persea americana* Mill.) *J. Amer. Soc. Hort. Sci.* 96:391–395.

Rodrigo, P. A. 1968. Avocado: nutritious neglected fruit. *Philippine Farms and Gardens.* 5:8–9, 17, 30–34.

Rounds, Marvin B. 1950. Check list of avocado varieties. *Calif. Avocado Soc. Yrbk.* 1950:178–205.

Ruehle, G. D. 1963. The Florida avocado industry. *Univ. Florida Agr. Exp. Sta. Bul.* 602.

Schmidt, M. 1965. Avocado growing in Chile. *Calif. Avocado Soc. Yrbk.* 49:45–46.

Schroeder, C. A. 1951. Flower bud development in the avocado. *Calif. Avocado Soc. Yrbk.* 1951:159–163.

———. 1958. The origin, spread, and improvement of the avocado, sapodilla, and papaya. *Indian J. Hort.* 15:116–128.

Serpa, D. 1968. Avocado culture in Venezuela. *Calif. Avocado Soc. Yrbk.* 52:153–168.

Shepard, S. 1971. Avocado culture in Australia and New Zealand. *Calif. Avocado Soc. Yrbk.* 54:110–114.

Slor, E., and R. Spodheim. 1972. Selection of avocado varieties in Israel. *Calif. Avocado Soc. Yrbk.* 55:156–157.

Smith, C. E., Jr. 1966. Archeological evidence for selection in avocado. *Econ. Bot.* 20:169–175.

Smith, J. H. E., J. J. Kruger, and D. H. Swarts. 1970. Selecting avocado seedlings. *Farming in South Africa* 46(4):22–23.

Soule, J., and F. Lawrence. 1969. How to grow your own avocado tree. *Florida Agr. Ext. Ser. Cir.* 340.

Storey, W. B. 1968. The avocado in Trinidad and Tobago. *Calif. Avocado Soc. Yrbk.* 52:148–152.

Ticho, R. J. 1971. Girdling, a means to increase avocado fruit production. *Calif. Avocado Soc. Yrbk.* 54:90–94.

Ticho, R. J., and B. Gefen. 1965. The avocado in Israel. *Calif. Avocado Soc. Yrbk.* 49:55–60.

Torres, J. P. 1936. Some notes on avocado flower. *Philippine J. Agr.* 7:207–227.

Trabucco, E. 1968. Marketing of Florida avocados and limes. *Proc. Florida St. Hort. Soc.* 81:330–332.

Turu, T. 1970. Avocados south of the border. *Calif. Avocado Soc. Yrbk.* 53:31–37.

Van der Meulen, A. 1970. Commercial avocado cultivars. *Farming in South Africa* 45(10):20–22.

Vogel, R. 1958. Characteristics of some avocado cultivars grown in Morocco (in French). *Fruits d'outre mer* 13:507–509.

Wilkins, W. C. 1965. Asexual propagation in Mexico. *Calif. Avocado Soc. Yrbk.* 49:39–42.

Yee, W. 1957. Producing avocado in Hawaii. *Hawaii Agr. Ext. Ser. Cir.* 382.

Zamenes, N. P. 1967. Avocados in ancient Greece. *Calif. Avocado Soc. Yrbk.* 51:91–92.

Zentmyer, G. A. 1972. Expansion of avocado root rot resistance program. *Calif. Avocado Soc. Yrbk.* 55:87–89.

Zentmyer, G. A., A. O. Paulus, and R. M. Burns. 1967. Avocado root rot. *Calif. Agr. Expt. Sta. Ext. Ser. Cir.* 511, rev.

Zentmyer, G. A., A. O. Paulus, C. D. Gustafson, J. M. Wallace, and R. M. Burns. 1965. Avocado diseases. *Calif. Agr. Expt. Sta. Ext. Ser. Cir.* 534.

Figs

by W. B. Storey

The common fig (*Ficus carica* L.; family Moraceae; 2n = 26) has been cultivated as a fruit tree in the eastern Mediterranean regions of Europe and Africa and the southwestern regions of Asia from time immemorial. Even the approximate date of its beginnings as a cultivated crop is lost in antiquity.

Today, the fig is an important fruit crop in many parts of the world. It is especially so in countries bordering on the Mediterranean Sea, the Red Sea, and the Arabian Sea. The people of these countries eat the fruit fresh, dried, as paste, and baked in pastries. Several countries export it dried or as paste. Principal exporting countries, their long-term average annual production of dried figs and paste in metric tons, and exports to the United States in 1970 and 1971 are given in Table 1.

TABLE 1. Long-term average annual production of fig products by principal producing countries, and 1970 and 1971 exports to the United States, in metric tons (California Fig Institute, 1972).

Country	Average annual production (metric tons)	Exports to the U.S. (metric tons)	
		1970	1971
Spain	82,555	1,837	2,839
Italy	76,205	23	27
Turkey	30,845	1,616	437
Greece	21,866	1,272	1,177
Portugal	8,427	3,997	2,909
Others	ca. 20,000	7	4
Total	239,898	8,752	7,393

Figs are cultivated around the world in subtropical and tropical regions and to some extent in moderate climatic regions of the Temperate Zones. Production is limited in most countries, and usually the fruit is consumed fresh locally or in dried, canned, and preserved form.

The only significant plantings outside the Mediterranean-Asian region are in the United States, principally the state of California. In 1972 approximately 8,800 ha of land in the United States were planted with fig trees, 7,000 ha of which were in California, mostly in the San Joaquin Valley. The remainder were in Texas, Louisiana, Florida, other southeastern states, and in southern Oregon. Of California's production, 85% is marketed as dried figs, 12% as canned figs and fig juice, and 3% as fresh fruit. Fruit produced by other states is eaten fresh, canned, or made into preserve.

Total United States utilization of dried figs and fig paste in 1972 was 20,025 mt. California's contribution of 12,720 mt accounted for 63% of the total. Foreign imports of 7,933 mt accounted for the remaining 37%. All but a small fraction of the total of dried figs and paste utilized goes into the making of fig bars.

From 1923 to 1952, the United States ranked

fourth among fig producing nations; now it ranks sixth. California's land area in figs, the amount of fruit produced, and monetary return to growers in 1969 represent the culmination of a steady decline of the industry since its heyday in 1943 when there were 14,269 ha in orchards, production was 40,269 mt, average farm return was $322/mt, and total farm return was $10,716,000. In 1969, California produced 11,930 mt of dried fruit, and 6,713 mt of fresh fruit (2,467 mt equivalent dry weight), 5,444 mt of which were canned. The average farm price was about $240/mt, and the year's farm value was $3,085,350.

Causes for the decline included competition from imported figs from countries with low cost labor, increasing local labor costs, encroachment of residential and industrial developments into the principal producing areas, which happen to be near large cities, and a tax structure which taxes land on market value as determined by adjacent properties rather than on actual agricultural use, making economical production impossible in some localities.

Since 1969 there has been a resurgence of activity in the California fig industry. This was generated by the 1967 Arab-Israeli War which resulted in closing the Suez Canal, disrupting shipping in the Mediterranean Sea, and shutting off supplies from countries east of Suez. Since then, the entire California crop has been marketed annually at satisfactory prices. Completion of the California Aqueduct down the west side of San Joaquin Valley in 1971 brought water from northern California, opening vast areas to irrigated agriculture. Since fig trees grow rapidly and come into bearing quickly, they are being planted to produce income for defraying taxes and stand-by water charges. About 194 ha were planted in 1968, about 190 ha in 1969, about 223 ha in 1970, and as many as 405 ha in 1971.

Revitalization of the fig industry in California has brought a need for stepped-up research in all phases of fig culture, including development of new superior cultivars to enhance the growers' economic position.

Origin and Early Development

Early History

According to Greek mythology, Zeus was pursuing Ge and her son Sykeus in the war of the Titans, when Ge, in order to save Sykeus, metamorphosed him into a fig tree. The city of Sykea in the ancient country of Cilicia derived its name from this myth.

An Athenian myth credits the goddess Demeter with having been the first to reveal the "fruit of autumn" to humans, who named it *fig*. The word *sycophant* is derived from the Greek words *sykon* (fig) and *phanein* (to show). The original connotation is obscure, but is thought to have alluded to a person who accused another falsely. In modern usage it connotes a parasitic or servile hanger-on or flatterer.

According to Aigremont (1908), the fig is a symbol of fertility and propagation in oriental countries. Among the Hellenes it was sacred to the sensuous, flabby, procreative god Dionysius who, in order to keep a promise to Polyhymnos, placed a phallus of fig wood on the latter's grave and kept the promised favor for himself. The phallus carried at Dionysian festivals was carved out of fig wood. The fig became the tree of phallic worshippers of India and Italy where people still use the gesture "fico," with the thumb inserted between the first two fingers. In modern times it has become an obscene gesture of contempt. However, it also represents the letter *T* in a one-hand dactylic alphabet for communicating with deaf persons.

Botanists believe that the fig was first brought into cultivation by inhabitants of the fertile region of southern Arabia, where wild trees still are to be seen. It was mentioned by King Urukagina in the Sumerian Era ca. 2900 B.C., and became known to the Assyrians as early as 2000 B.C.

In time, it spread in cultivation through Asia Minor and into all countries of the Mediterranean Region. The specific epithet *carica* refers to Caria, an ancient region in Asia Minor noted for its figs. It was known on Crete early as 1600 B.C. Homer, ca. 850 B.C., mentions it several times in his *Iliad* and *Odyssey*.

A Roman legend recounts that Romulus and Remus, who founded Rome, (ca. 746 B.C.,) were

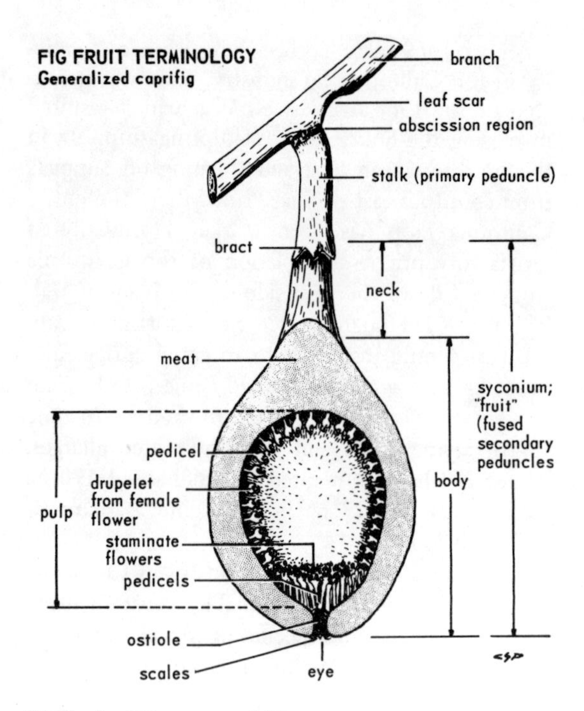

FIG. 1. Diagram of *Ficus carica* syconium.

sheltered in infancy by a sacred fig tree, *ficus Ruminalis,* while being nursed and cared for by a she-wolf and a woodpecker. The tree was named for Ruminia, a goddess who watched over suckling animals.

The spread of fig culture eastward in Asia went slowly, for the species, which thrives in arid subtropical climates, is not well adapted to the humid tropics of India and southeastern Asia. It is said to have reached China no earlier than the Tang period, 618–907 A.D. Its spread southward in Africa was still slower, and it seems not to have reached South Africa until sometime in the nineteenth century.

During the great age of exploration that followed "discovery" of the New World by Christopher Columbus in 1492, the fig spread in cultivation to all parts of the tropical, subtropical, and moderately temperate regions of the Americas. European cultivars were established in the West Indies in 1520, in Peru by 1526, and in Florida about 1575. By 1683 trees had been established in Baja California, Mexico. Franciscan missionaries led by Junipero Serra planted the first figs in California when they established the mission at San Diego in 1769. The commercial cultivar 'Franciscana', or 'Mission' as it is commonly called, owes its name to this event.

Systematic Position

The family *Moraceae* includes 60 genera and more than 2000 species of trees, shrubs, vines, and a few herbs. Familiar members of the family are: *Morus alba,* L., and *M. nigra,* L., the white and black mulberries; *Broussonetia papyrifera* (L.) Vent., the paper mulberry or wauke, a widely used ornamental tree in California; *Artocarpus altilis* (Parkins) Fosb., the breadfruit, and *A. heterophyllus* Lam., the jak fruit; and numerous *Ficus* species, including the banyans. The genus *Ficus* comprises about 1000 species, several of which have edible fruit (Condit, 1969). Corner (1965) divided the genus into 48 subgenera on the basis of characteristics which delineate one group of closely allied species from another.

F. carica belongs to the subgenus *Eusyce,* which is characterized by species having unisexual flowers only and by gynodioecism. Some of the allied species resemble *F. carica* in growth form, and in leaf and fruit characters. Several are hosts to *Blastophaga psenes* Cavolini, the fig wasp, which is the instrument of pollination. Forms intermediate between valid species of *Ficus* suggest natural hybridization, which explains the diversity of opinions among systematic botanists with respect to species delineation and typification. However, the following are regarded as probably being valid

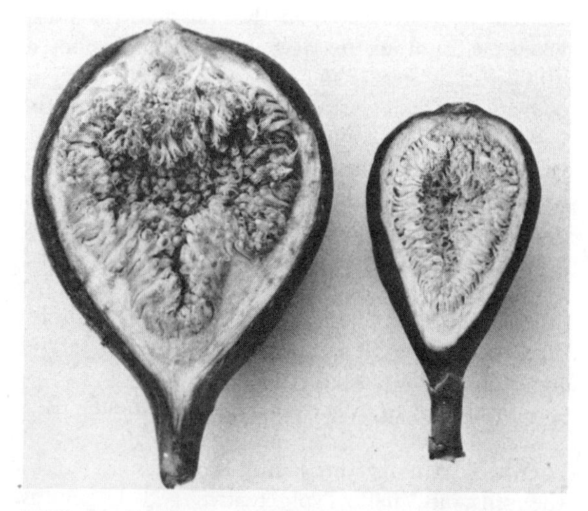

FIG. 2. *F. carica* syconia at time of pollination. Left, monoecious caprifig syconium with staminate flowers in anthesis; right, pistillate fig syconium with flowers in anthesis.

species: *F. geranifolia* Miq.; *F. palmata* Forsk.; *F. pseudo-carica* Miq.; and *F. serrata* Forsk.

The characteristic inflorescence of all *Ficus spp.* is the *syconium* (Figs. 1, 2). It is a complex, enlarged, fleshy, hollow peduncle bearing closely massed flowers on the entire inner wall. It has an apical pore, the ostiole, which forms a passage between the hollow interior and the outer air. The canal of the ostiole is lined with numerous scales, and the external orifice or eye is sheathed by overlapping scale-like bracts.

Pomologically, the fig fruit is the mature ("ripe"), succulent syconium. Botanically, however, it is a spurious fruit consisting entirely of vegetative peduncular tissue. The true fruits are tiny pedicellate drupelets within. In fruit classifications, the fig usually is included with the multiple or collective fruits.

In fig literature, the syconium frequently is called a "hollow receptacle." Although the term *receptacle* is used in the dictionary sense that it is a structure that can receive and contain something, its use to denote the syconium is unfortunate, for persons are known to have interpreted it as being homologous to the receptacle of an aggregate fruit such as the strawberry (*Fragaria* x *ananassa* Bailey, family Rosaceae) and papaw (*Asimina triloba* Dunal., family Annonaceae), the fruits of which consist of the more or less fleshy receptable of an individual flower bearing few to many carpels. Each fig flower within a syconium has its own receptacle.

Horticultural Classification

F. carica is a gynodioecious species, having two distinct forms of trees, the caprifig which is monoecious, and the fig which is pistillate.

Fig flowers are minute. Typically, they are pedicellate, hypogynous, and unisexual, with a single five-parted perianth. There are three kinds of flowers: short-styled pistillate; long-styled pistillate; and staminate (Fig. 3). Both kinds of pistillate flowers are single-carpelled with a usually bifid stigma. The short-styled pistillate flower has a more nearly globose ovary, and a style about 0.70 mm long; it is adapted for oviposition by the fig wasp. Commonly, it is called a gall flower, but Condit and Flanders (1945) considered this term to be a misnomer. The long-styled flower has a more or less ovoid or ellipsoid ovary, and a style about 1.75 mm long; it is not adapted for oviposition by the fig wasp. Both kinds of pistillate flowers

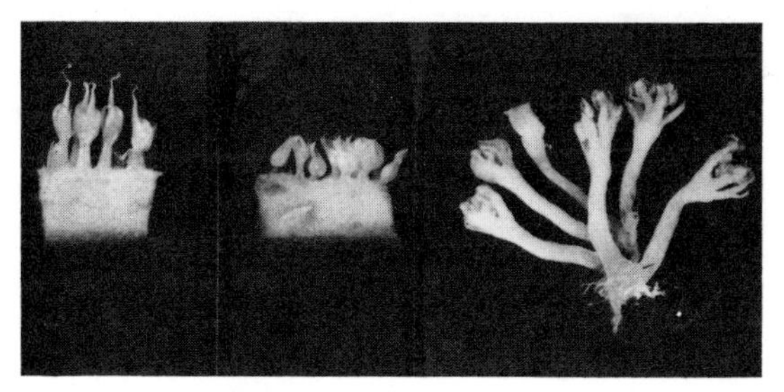

FIG. 3. Fig flowers. From left: long-style pistillate, short-style pistillate, staminate.

are fertile, and, following pollination and syngamy, each develops into a drupelet which is the true fruit of the fig. In vernacular language, these are "fig seeds." The staminate flower is longer pedicelled than the pistillate flowers, has a five-parted perianth, five stamens, and a vestigial pistil.

The caprifig tree is typified by syconia which contain short-styled pistillate flowers and staminate flowers. The pistillate flowers are distributed over most of the inner wall. The staminate flowers are massed around the interior end of the ostiolar canal (Fig. 2). The pistillate fig tree is typified by syconia which contain only long-styled pistillate flowers.

The caprifig produces three crops of syconia annually: 1) the *mamme* crop, which is initiated on current vegetative growth in fall, remains dormant through winter, and matures in spring; 2) the *profichi*, which is initiated in spring from latent buds on the previous season's wood, matures in summer, and supplies the grower with pollen-carrying wasps which pollinate figs of the Smyrna type, described later; 3) the *mammoni*, which is initiated on current growth in summer and matures in fall.

The fig may produce two crops a year or it may produce only one, depending on the cultivar and prevailing climatic conditions. The first crop appears in spring from latent buds on the previous season's growth. It is known as a *breba* crop. It matures in June and July in California. The second crop appears on current growth from May to July, and matures from August to December, depending on the cultivar. This crop has acquired no special name and is known simply as "the main or second crop." The relationship of the caprifig and fig crops to one another, to the life cycle of the fig wasp, and to fig breeding is covered more fully in the section on floral biology.

Horticulturists and breeders regard the caprifig

as a single type with about 20 cultivars which have been grown commercially for producing profichi for use in caprification. Presently, only the following five are grown for this purpose: 'Brawley', 'Milco', 'Roeding 3', 'Samson', and 'Stanford'.

Fig cultivars are classified into three horticultural types, Smyrna, San Pedro, and Common. The types are described below.

Cultivars of the Smyrna type usually do not produce breba crop. A few, such as 'Sari Lop' (syn. 'Calimyrna'), and 'Taranimt' may produce scanty to fair breba crops in some years. The syconia that reach maturity are seedless, flaccid, and insipid in flavor. Adequate main crops are initiated, but the syconia shrivel and fall from the tree at the end of the period of anthesis unless the flowers are pollinated. A fertilized ovule develops into a viable seed within the thin, shell-like, brittle endocarp, with endosperm surrounding the embryo. Condit (1955) described 116 named cultivars of this type. Only 'Sari Lop', which was renamed 'Calimyrna' for promotional purposes, is commercially important in California. In 1971 the area in producing 'Calimyrna' trees in California was 3,366 ha or 48% of the total of 7,000 ha. Important cultivars of the Mediterranean Region that are being evaluated as commercial types or for use in breeding in California are 'Azendjar', 'Cheker Injir', 'Marabout', 'Tameriout', 'Taranimt', and 'Zidi'.

Fig trees of the San Pedro type initiate fair to large breba crops. These persist to maturity without any known stimulus and ripen into palatable fruits. An adequate second crop of syconia is initiated, but usually only those in which the flowers have been fertilized develop to maturity. Condit (1955) described 21 named cultivars in this class. The only ones produced commercially in California for the early fresh fruit market are 'Gentile', 'King', and 'San Pedro'. Three important French cultivars are: 'Dauphine', 'Drap d'Or', and 'Pied de Boeuf'. 'Lampeira' is an important Portuguese cultivar.

Common type fig trees may or may not initiate breba crops which develop to maturity without any known stimulus. Some cultivars, e.g., 'Dottato' (syn. 'Kadota'), 'Beall', 'Archipel', and 'Franciscana' (syn. 'Mission'), mature fair to good annual breda crops regularly; others, such as 'Verdone' (syn. 'Adriatic'), 'Brunswick' (syn. 'Magnolia'), and 'San Piero' (syn. 'Brown Turkey' in California trade), may produce small to fair crops in some years, no crops in other years; and a few, such as 'Marseilles', 'Panachée', and 'Vernino', produce no breba crops at all. Most common fig cultivars produce fair to large main, or second, crops which develop to maturity without pollination or any other known stimulus. Condit (1955) described 470 named cultivars in this class. The important ones commercially or for breeding are 'Archipel', 'Beall', 'Conadria', 'Dottato', 'Genoa', 'Marseilles', and 'Verdone'.

Condit (1955) described 627 named cultivars of the fig. Doubtless, there are many obscure ones in various parts of the world that escaped notice. The synonyms for cultivars in widespread cultivation are legion.

History of Improvement

Excepting the modern hybrid 'Conadria' (Condit and Warner, 1956), all fig cultivars grown today owe their origins to selection by unknown persons in Asia and Europe in past centuries among wild seedling trees and volunteer seedlings of cultivated trees. These have been maintained clonally by rooted cuttings and, in the course of time, have acquired names.

The earliest attempts to create new improved cultivars by breeding were those reported by Swingle (1908, 1912), Hunt (1911, 1912), Burbank (1914), Rixford (1918, 1926), Noble (1922), and Condit (1928).

In 1922 R. E. Smith initiated a fig breeding project at the University of California, Davis. It was transferred to the Citrus Experiment Station in Riverside (now the Citrus Research Center and Agricultural Experiment Station) in 1928, and Ira J. Condit (Fig. 4) was appointed project leader. The project encompasses various studies as indicated by its title: "Genetics, Cytology, Morphology, and Breeding of Fig (*Ficus carica*), with

FIG. 4. Dr. Ira J. Condit (1883–), premier fig breeder, known affectionately to his colleagues as the "High Priest of the Fig."

Emphasis on Improving Varieties." Condit served as project leader from 1928 to 1968. W. B. Storey has been project leader since Condit's retirement in 1968. Storey and Condit (1969) contributed a chapter on fig breeding to *Outlines of Perennial Crop Breeding in the Tropics*, published by the Agricultural University, Wageningen, The Netherlands.

Material presented in this chapter is based largely on Condit's findings from more than 300 hybrid progenies totaling more than 30,000 seedlings. Insofar as is known, 'Conadria', which he and Warner introduced into the trade in 1956, is the only cultivar in commercial production that was developed in a planned breeding program. More than 267 ha are planted with 'Conadria' in Fresno and Merced Counties, California (Fig. 5).

Besides the writer, E. N. O'Rourke, Jr., is carrying on fig breeding at Louisiana State University, Baton Rouge, Louisiana. The only fig breeder listed for any other country besides the United States by the Food and Agricultural Organization's *World List of Plant Breeders* (1961, 1965, 1967) is G. N. Slykov of the All-Union Institute of Plant Industry, Leningrad, U.S.S.R.

FIG. 5. Fig orchard at Chowchilla, California. Trees to the right of the road are a portion of 267 ha planted with 'Conadria'.

Modern Breeding Objectives

Maximum Productivity

Yields per hectare of marketable dry figs differ among cultivars, as well as being affected by location, the vagaries of climate, and orchard management. In San Joaquin Valley, the yield of 'Verdone' ranges all the way from 3.7 to 7.4 mt/ha and of 'Sari Lop' all the way from 1.1 to 5.1 mt/a. A well managed 5.3 ha 'Verdone' orchard consistently produced around 7.2 mt/ha annually over a span of several years.

California's 1972 production of 12,720 mt on 7,000 bearing hectares averaged 1.82 mt/ha. Customary orchard spacing is 10.7 x 10.7 m between trees which provides for 89 trees/ha. Average yield per tree was 21.2 kg. This corresponds closely to the 18.2–22.7 kg per tree average in Turkey's Meander Valley, which is considered to be the world's prime producing region.

Higher yield per unit area is desirable, but this seems more likely to be achieved by closer planting and better orchard management than by breeding. In fact, production in well-managed 'Sari Lop' orchards has to be controlled to prevent overloading, which causes branches to bend or break and reduces the size of the figs. Growers do this by following a system which prorates the number of profichi caprifigs placed in the trees for caprification according to the size of tree (Condit, 1920).

Fruit Quality

Today's objectives in fig breeding are dictated primarily by the requirements of the industry. Growers and processors alike would like a fig for California with all of the good attributes of 'Calimyrna' (i.e., 'Sari Lop') which would not require caprification. The ideal fruit would have golden skin, white meat, amber pulp, the distinctive flavor of 'Sari Lop' which is due in part to the endosperm of the fertile seeds, attractive size and shape, and skin not easily bruised when fresh fruit is handled.

In Louisiana, breeding is directed toward development of an improved type resembling 'Celeste', which is favored by growers (O'Rourke, 1966). Success or failure of a new fig cultivar depends upon its fruit, which must process into a high-quality, readily marketable product, and horticultural characteristics which conform to growers' requirements.

The subject is covered more fully in the section on breeding for specific characters.

Elimination of Caprification

'Sari Lop', a cultivar of the Smyrna type, is the finest fig produced in California, where it is known as 'Calimyrna'. It must be caprified, of course, in order to set a crop. Caprification is the transfer of caprifig pollen to fig flowers by fig wasps. To accomplish it commercially, profichi syconia of the caprifig are placed in branches of fig trees. It is a costly, time-consuming, and inconvenient practice, which growers would like to have eliminated. The procedures followed in caprification have been described in detail by Condit (1920).

Perhaps the worst feature of caprification is that it may result in fruit spoilage by endosepsis, or internal rot caused by *Fusarium moniliforme* Sheld. var. *fici* Caldis, a fungus which is transmitted to figs from infected caprifigs by female blastophagas.

Resistance to Insects and Nematodes

The only serious insect problem in California is entrance into the syconia by species of vinegar flies (*Drosophila* spp.), dried-fruit beetles (*Carpophilus* spp.), and thrips (*Thrips* spp., and *Franklinilla* spp.). They lay eggs which hatch into larvae inside the figs, and also carry in yeasts which cause souring. A solution is to breed figs with small ostiolar eyes which are tightly closed by scales. Tightly closed eyes prevent the insects mentioned above from entering the syconium, but are no barrier to the fig wasp.

A root-knot nematode, *Meloidogyne incognita* (Kofoid and White) Chitwood var. *acrita* Chitwood, causes severe damage to the roots of fig trees in California, Louisiana, and Florida, especially trees on light, sandy soils. Many *Ficus* spp. are virtually immune or highly tolerant to the nematode. One of the objectives of breeding is to obtain interspecific hybrids which will result ultimately in a cultivar combining desirable tree and fruit characters with nematode resistance. Another objective is to determine which of the numerous *Ficus* spp. available may be both nematode resistant and graft-congenial with *F. carica* for possible use as rootstocks for fig cultivars.

Breeding Techniques

Floral Biology

There is a pronounced dichogamy in the development of the flowers of the caprifig tree. The flowers are protogynous, the pistillate flowers coming into anthesis three or more months ahead of the staminate flowers. Ordinarily, only the stamens of the profichi or summer crop of caprifigs develop to full maturity and shed pollen.

As noted previously, the caprifig tree bears three crops of syconia a year: The mamme, or spring crop; the profichi or summer crop; and the mammoni, or fall crop. The times of ripening vary from year to year, and from one locality to another, usually as the result of the vagaries of climate. The annual cycle in Riverside, California, where most of the breeding is done, is essentially as described below.

Syconia of the mamme crop are initiated on current growth in August, September, and October. The pistillate flowers come into anthesis and are receptive to oviposition by the fig wasp in October and November. Staminate flowers are initiated but seldom develop to maturity. Syconia which wasps fail to enter soon cease development and fall from the tree. Syconia which have been entered, and in the ovules of which eggs have been laid and larval development begun, remain on the tree and begin developing. Development is stopped by cold weather, but the syconia remain on the tree through the winter season. They resume development with the coming of warm weather, and ripen in March and April. The male wasps hatch from their enclosing drupelets, and cut the drupelets containing the females, releasing them and mating with them as they emerge. The females leave the syconium through the ostiole, and enter the young syconia of the profichi crop.

The syconia of the profichi crop are initiated on the previous season's wood in February and March. The pistillate flowers are in anthesis when the wasps are emerging from the mamme crop in March and April. Again the syconia drop when not inhabited by wasps, but remain on the tree if inhabited. This crop matures in June and July. At this time, the stamens are fully developed, shedding pollen which adheres to the emerging female wasps. This is the crop which furnishes syconia for caprifying, i.e., pollinating the commercial crop of the otherwise nonpersistent Smyrna type of fig, as well as pollen for use in breeding.

The mammoni syconia are initiated on current growth in June and July and are receptive to the wasps emerging toward the end of the profichi crop. Inhabited mammoni syconia ripen in October and November, when the wasps emerge and enter the developing mamme syconia, thus completing the cycle. Caprifig trees in which the fig wasp can complete the cycle are said to be *colonized*. Inhabited mammoni syconia sometimes contain seeds from flowers which escaped oviposition, but received pollen. The staminate flowers of the mammoni crop seldom develop to maturity.

A breba crop matures on fig trees in June and July. This crop generally escapes caprification because it is initiated at the time the mamme crop, which produces no pollen, is ripening. All fig cultivars produce a main crop or summer crop on current growth, which matures from August to December, depending on the cultivar.

Pollen Collection

The profichi crop which ripens in June and July is the only one of the three caprifig crops in which the stamens develop to anthesis. Ripe caprifigs are harvested and taken to a warm, dry place where there is little air circulation, split open longitudinally, and spread out to dry on kraft or similar paper on a table or laboratory bench. In a few days, the fruit is dry, and all of the anthers have dehisced. The pollen is collected on a piece of bond or similar paper by holding the opened fig over it and thumping it with a finger or giving it some hard taps with a pencil. By shaking the paper, one can winnow parts of stamens, wasps, and other detritus from the pollen, which can then be put into a cork-stoppered or screw-capped vial (Fig. 6–B) and stored until ready for use. Pollen stored in a refrigerator at 8–10 C remains viable for as long as 120 days.

Pollen from different parental caprifig trees preferably is collected in separate rooms or on separate days, for fig pollen is so fine and light and so easily airborne that one runs the risk of contaminating one lot by another if several are collected in the same room, especially if the air is stirring.

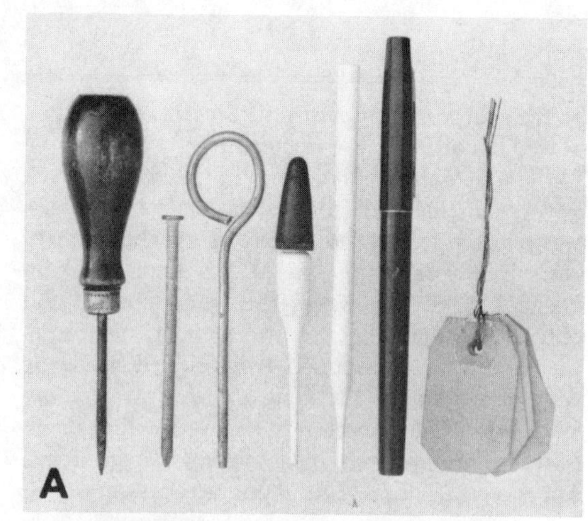

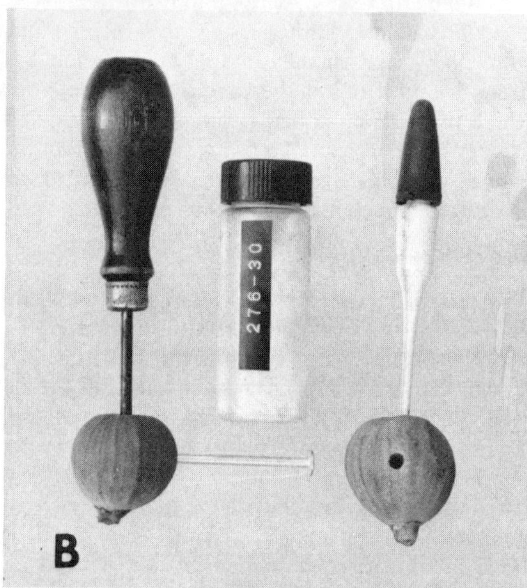

FIG. 6. Fig pollination. (A) Instruments for puncturing, inserting pollen into syconium, and labeling. (B) Fig pollen in vial and procedure for puncturing and pollinating.

Pollination

Only a few simple instruments and supplies are needed for breeding figs (Fig. 6-A). An awl, gimlet, 10-penny nail, or similar sharp-pointed object is needed for puncturing the syconium. Small pipettes or droppers from dropping bottles or medicine bottles are needed for introducing pollen into syconia. Small paper marking tags and a pen with indelible ink are useful for identifying pollinated syconia and recording the parentage and date of the cross.

Fig syconia of the main or summer crop are initiated on current growth from May until growth ceases in August or September on most cultivars. The flowers are in anthesis and receptive to pollination of each syconium successively from the middle of June into August in the order of initiation. Receptive syconia are light green in color and yield slightly to light pressure when squeezed between the thumb and forefinger. Pollination for breeding is done in late June and early July.

Two holes are punched into the interior of the syconium with the pointed instrument. One may be through the ostiole, the other through the side wall (Fig. 6–B), or both through the side wall (Fig. 9). The tapered end of a pipette is dipped in pollen, and inserted into one of the holes. The pollen is puffed in by mouth or by squeezing a rubber bulb on the large end. The other hole serves simply as an air vent which allows for puffing the pollen in easily. Puncturing the syconia does not damage them enough to cause them to drop. The marking tag containing desired data is tied by its string to the branch at the base of the pollinated syconium.

O'Rourke (1966) splits the syconium to be pollinated longitudinally about one third of the distance from apex to base with a knife. He then picks up a bit of pollen on the knife point and blows it over the flowers.

If there are caprifig trees in the vicinity of cultivars to be bred, one should remove all of the caprifigs of the profichi crop to preclude contamination by pollen-carrying wasps. Generally, however, caprifigs are not grown within flight distance of common figs or cultivars grown for use in breeding. No bagging of a pollinated syconium need be done, for blastophagas do not enter syconia that have been pollinated and begun development.

Each progeny of a season's successful crosses is assigned a progeny number. Every seedling of a hybrid progeny that is planted or budded into an old tree for rearing to fruiting is identified by a plant label bearing the progeny number and the seedling number, from 1 to as many as desired for evaluation. These constitute pedigree numbers.

Seed Collection

Ovule development begins a few days after introduction of pollen into the syconium, and effectively plugs the puncture holes. The syconium ripens in the normal season; i.e., August to September. When the figs are in the last stages of maturity prior to ripening, it may be necessary to enclose them in cloth or fine mesh nylon bags to protect them from being eaten by birds. Ripe fruits are broken up and dropped into beakers or jars of water for fermenting. In a few days, drupelets with viable seeds sink to the bottom. Empty drupelets and the fermented flesh float, and are decanted off. After being washed in two or three changes of water and spread out on a piece of paper to dry, the good drupelets may be used for planting. A single syconium yields 300 to 500 drupelets, which are "seeds" when the fleshy portions of the pericarp, i.e., exocarp and mesocarp, have been washed off.

Germination and Growth of Seedlings

Seeds are set for germination soon after collection in August or September. They are sown on a relatively fine grade of vermiculite in plastic trays and covered lightly with the same material. The trays are put into a warm glasshouse. By December when they are 5.1 to 7.6 cm tall, with several leaves, they are transplanted to 345 ml styrofoam tumblers filled with good potting soil, and provided with a punched-out hole for water drainage. By the end of January, most are 10 to 25 cm tall, and 5 to 7.5 mm thick toward the base. At this time, they are set outside in the cold to induce dormancy.

Shortening the Period of Juvenility

Fig seedlings growing on their own roots take five to seven years to come into bearing from the time of germination. The period can be reduced to one to three years by grafting buds or scions from seedlings on the branches of three- to 20-year-old or even older trees. Grafting is done during March and April, when stock trees are beginning their spring flush of vegetative growth and sap is flowing freely. The branches of the stock tree are cut back severely (Fig. 7) and buds or scionwood inserted. The simple bark graft is preferred, because it is the easiest and fastest method. Two or three scions of the same seedling are topworked on each branch. The cut surface of the branch is coated with a pruning compound. The scions are enclosed in a plastic bag to keep them from drying out, and the plastic bag is covered with a larger kraft paper bag to prevent sunburning. Each grafted branch is tagged with an embossed metal label bearing the progeny and seedling number. This is attached to a staple on the branch by a short length of copper wire.

FIG. 7. Hybrid fig seedlings topworked to stock tree to shorten period of juvenility.

Buds begin to break on the scions in two to three weeks, and, as soon as the stems begin to elongate and leaves develop, the bags may be removed. Despite the small caliper of many seedlings, some being no thicker than an ordinary match stick, at least 90% of the scions unite with the stock and grow. Growth is rapid, and some seedlings produce fruit the first season. Most fruit in the second season, and just about all fruit by the third season.

In addition to circumventing juvenility, the method has three other advantages: 1) as many as 20 seedlings can be worked into large trees with many branches, saving considerable land area over what would be required for seedlings growing on their own roots even when as closely spaced as 1 m apart in a row; 2) seedlings have the undesirable characteristic of sending up numerous sucker stems from the crown every year for several years which interfere with maintenance and must be pruned off; 3) as soon as the seedlings have been evaluated, the branches bearing the discards can be cut back and reused for topworking a new set of seedlings the following year.

Seedling Evaluation and Selection

Seedlings are evaluated initially on the basis of obvious fresh fruit characters: i.e., size, shape, skin color, pulp color, flavor, size of eye, and freedom from splitting. Criteria for selection are covered later in the section on breeding for fruit characters. Additional evaluations are made later for dried fig and paste color and flavor.

Three other characters judged are vigor of the seedling, potential productivity, and ripening period in the crop season.

After two or three seasons of evaluation, only a few seedlings usually remain which are considered to be worthy of further evaluation as potential commercial cultivars. In the fourth bearing season all growers and processors who wish to do so may serve on an appraisal committee sponsored by the California Fig Institute, which assists the breeder in making final evaluations and selections, if any, in the field planting. Final selections are tested in the laboratory to see if they meet processors' criteria. Propagation material of selections which seem to meet growers and processors' requirements is distributed to interested growers for growing in a large scale trial and to nursery firms which wish to multiply them as clones. The material is distributed through the agricultural extension service by the farm adviser who includes the fig among the crops he services. It is distributed under its pedigree number, and is given no horticultural cultivar name until its commercial success seems reasonably assured. This tends to preclude proliferating an already enormous directory of cultivars with names of new cultivars which may never go into production or may have a short-time existence.

The philosophy behind having growers and processors participate in selection is based on the fact that as a result of many years of experience, they have rather definite ideas as to just what they want in a commercial fig. It is thought best, therefore, to let them make the final selection of types that they want rather than for the breeder to try to coerce them into accepting something he thinks they ought to have.

Figs for home planting or for possible production as fresh fruit usually are evaluated, selected, and tested by the breeder usually according to his own criteria. Trees are placed with cooperators for growing under various climatic conditions. If performance and quality of a selection are satisfactory in most localities, the tree is given a cultivar name, and propagation material is furnished to nurserymen for multiplication and sale of trees.

Inducing Mutations

Insofar as is known, the only attempts to induce mutations have been with the use of colchicine, mostly on germinating seeds. Mutations have resulted, but these generally have been slow-growing, with malformed foliage, and sparsely fruiting. The fruit on those still growing is late in ripening and very poor in quality.

No natural polyploid is known among named cultivars, and despite the tens of thousands of seedlings that have been reared, no polyploid has ever been found among them. A matter of interest is: what would tetraploid plants and fruits of some of the commonly grown cultivars be like? In 1970, an experiment was initiated on treating flushing axillary buds on a few cultivars with colchicine using customary concentrations and procedures. The appearance of some of the shoots suggests that they may be polyploid, but at this writing they have yet to be examined cytologically.

Breeding Systems

Sex Determination

Sex appears to be determined in fig by two closely linked pairs of alleles on a presently unidentified pair of chromosomes. They are symbolized as follows:

- G, dominant allele for gynoecious flowers with short-styled pistils.
- g, recessive allele for gynoecious flowers with long-styled pistils.
- A, dominant allele for presence of the androecium.
- a, recessive allele for suppression of the androecium.

The caprifig homologue is GA and genotypes of caprifig trees are GA/GA and GA/ga. The pistillate homologue is *ga*, and the trees always have the genotype *ga/ga*.

Table 2 shows the genetics of sex determination in *F. carica*, hypothesized from the results of genetical analysis.

TABLE 2. Sex determination in *F. carica*, with ratios of genotypes shown.

Seed parent	Pollen parent: Homozygous caprifig G A/G A	Pollen parent: Heterozygous caprifig G A/g a
Homozygous caprifig G A/G A	All homozygous caprifig G A/G A	1 Homozygous caprifig G A/G A 1 Heterozygous caprifig G A/g a
Heterozygous caprifig G A/g a	1 Homozygous caprifig G A/G A 1 Heterozygous caprifig G A/g a	1 Homozygous caprifig G A/G A 2 Heterozygous caprifigs G A/ g a 1 fig g a/g a
Fig g a/g a	All heterozygous caprifig G A/g a	1 Heterozygous caprifig G A/g a 1 fig g a/g a

Two crosses of fig x caprifig made with seedling caprifigs from the progeny of a caprifig x caprifig cross yielded progenies 100% caprifig, demonstrating the existence of the homozygous GA/GA genotype. One cross involved 'Mission', which produced a seedling progeny of 79 individuals, all caprifigs. The other cross involved 'Partridge Eye', which produced a seedling progeny of 45 individuals, also all caprifigs.

All commercial caprifigs probably originated from fig seeds and, therefore, are heterozygous. When used for breeding, they produce progenies which segregate into the ratio of 1 caprifig: 1 fig. It follows that, in fig breeding, every cross involves a caprifig. Productivity and fruit characters of caprifigs are important, therefore, in the selection of parents for further breeding.

No way is known for distinguishing caprifigs from figs in the young seedling stage; consequently, one must wait until they produce their first syconia and "declare." Since virtually all caprifigs are worthless from the standpoint of fig fruit selection, one must reconcile himself to the fact that one-half of every seedling population is wasted on this account. Discovering a juvenile sex-linked character or inducing one by radiation or other means would serve the useful purpose of enabling one to discard the caprifigs from the progenies to be evaluated.

Fertility and Sterility

Both the short-styled pistillate flowers of the caprifig and the long-styled pistillate flowers of the fig are sexually functional and capable of setting seed when pollinated. Furthermore, no caprifig tree has ever been found in which the staminate flowers of the profichi are incapable of functioning.

The following situations do occur, however, with respect to pistillate flower development and maturation.

1) Without pollination or a wasp larva developing in the ovule, the flower aborts soon after anthesis. Abortive flowers are the cause of "blank" caprifigs and "seedless" figs.

2) Following pollination and syngamy, the flower develops normally, the ovary maturing into a drupelet containing a viable seed consisting of an embryo surrounded by endosperm. In the vernacular, the drupelet with exocarp and mesocarp removed is the "seed."

3) Without pollination, the flowers on some trees develop parthenocarpically. They mature as empty drupelets or *cenocarps*, commonly referred to in the trade as "seedlike bodies."

4) Following oviposition, the flower develops with a blastophaga larva in the ovule which devours the embryo and endosperm as it grows. Eventually, the entire interior of the drupelet is inhabited by a single mature wasp. Drupelets which contain wasps are called *psenocarps*.

Persistent vs. Caducous Syconia

The three horticultural types of figs are classified primarily by whether their syconia persist on the tree without any known form of stimulus or drop soon after anthesis (Fig. 8). In vernacular usage, the common type of fig is said to be completely parthenocarpic. The San Pedro type is said to have a parthenocarpic breba crop. The Smyrna type is said to be completely nonparthenocarpic.

Use of the term *parthenocarpic* is unfortunate, for mature "ripe" caprifigs and seedless figs consist of a fruit-like vegetative structure entirely

FIG. 8. Caprifig branches. (A) Branch from a caducous type of tree showing lower, undeveloped, light-colored, uninhabited syconia about to drop, and upper, large, dark, inhabited syconia developing normally. (B) Branch from persistent type of tree showing uninhabited syconia developing normally.

devoid of carpellary tissue. The terms *persistent* for syconia that develop to maturity without some form of stimulation and *caducous* for those that drop prematurely would seem to be more accurate botanically. Sometimes, syconia of Smyrna and main crop San Pedro fig trees which otherwise would fall can be made to persist simply by venting them and blowing air into them through one of the vents by means of a pipette. Gibberellin and (2-chloroethyl)phosphonic acid (ethephon) solutions sprayed on them have the same effect. (Crane, Marei, and Nelson, 1970*a*, 1970*b*; Gerdts and Obenauf, 1972).

Parthenocarpy

Parthenocarpy *per se* refers to development of the carpels of the flowers without syngamy, resulting in cenocarps. There are two genetically determined forms of parthenocarpy, stimulative and vegetative.

Stimulative parthenocarpy is induced by a stimulus introduced into the syconium from outside. One such stimulus is entry of the female fig wasp and her ovipositing into the embryo sac. Other effective factors are blowing air into the syconium, introducing dead pollen, and invasion by thrips and other insects. Parthenocarpy by stimulation prevents dropping of the syconium, allowing it to develop to maturity. Ovaries containing wasps develop into psenocarps; other stimulated ovaries develop into cenocarps.

Vegetative parthenocarpy needs no known stimulus. It is characteristic of all common type figs in cultivation and of persistent caprifigs, which almost always have some cenocarps. Trees lacking this trait produce blank syconia in caprifigs and "seedless fruits" in figs.

The tree characters of caducous syconia and persistent syconia are genetically controlled (Saleeb, 1965). Saleeb analyzed the progenies of the four possible combinations of crosses, i.e., caducous x caducous, caducous x persistent, persistent x caducous, and persistent x persistent, comprising 2,473 individuals. He reported the segregation ratios that are given in Table 3.

TABLE 3. Compositions of progenies from crosses of caducous and persistent figs with caducous and persistent caprifigs.

Fig		Caprifig		Progeny
Caducous	x	caducous	→	all caducous
Caducous	x	persistent	→	1 caducous:1 persistent
Persistent	x	caducous	→	all caducous
Persistent	x	persistent	→	1 caducous:1 persistent

The results are determined by a single pair of alleles, one of which is an egg lethal. These are symbolized as follows:

+, wild type allele for caducous syconia and normal ovule development.

P, mutant allele for persistence of syconia, and ovule abortion; dominant to +.

The genotype of trees with caducous syconia is ++; the genotype for trees with persistant syconia is + *P*.

The genetics of the four combinations shown above is as shown in Table 4.

Selection of Parents

Parent trees for fig breeding are selected primarily on the basis of fruit characters. The greatest need

TABLE 4. Genetics of caducous vs. persistent syconia in *F. carica*, corresponding to phenotypes in Table 3.

Fig		Caprifig		
Genotype	Eggs	Genotype	Pollen	Progeny
++	All +	++	All +	All ++
++	All +	+P	1+:1*P*	1++:1+*P*
+*P*	1+:1*P* (lethal)	++	All +	All ++
+*P*	1+:1*P* (lethal)	+P	1+:1*P*	1++:1+*P*

is for a new cultivar that will have all the desirable characters of 'Sari Lop' and will have persistent syconia, obviating the need for caprification. 'Sari Lop', therefore, is one of the cultivars commonly selected as a seed parent. Other light-skinned cultivars, including 'Conadria', 'Verdone', 'Dottato', 'Genoa', 'Marseilles', and 'Verdal Longue', were used in the early stages of the program.

The several caprifigs used for breeding today are types with persistent syconia which trace their pedigrees back to 'Croisic', a type having persistent syconia with greenish yellow skin and white edible pulp, which was brought to California from France by an unknown person and planted at Cordelia, California, sometime prior to 1893.

Size of Progeny

As with many tree fruit crops, the principal factor limiting progeny size is availability of land. In the early days of breeding, seedling trees were planted in orchard rows, with the trees spaced 2 m in the row and the rows 7 m apart. This was as close as seedlings could be lined out for easy maintenance until they fruited after three to seven or more years in the ground. This spacing limited the number of trees to 900 trees/ha.

Stock trees, remnants of a hybrid seedling planting at the University of California Horticultural Field Station in San Joaquin Valley, are spaced 3 m in rows, 7 m apart. These spacings provide for about 538/ha. The numbers of work able branches vary from as few as three on some trees to 20 on others. In 1971, the stock trees in a 0.4 ha planting were supporting more than 1,200 seedlings. Excepting the few that are selected for further evaluation by the end of the third crop year after grafting, there is a complete replacement of seedlings for observation.

A single pollinated syconium yields 300 to 500 seeds, so obtaining an adequate supply for growing seedlings for testing is no problem. The practice at the University of California is to make about ten crosses using five selected figs as seed parents and two selected caprifigs as pollen parents. About 100 seedlings from each progeny, i.e., a total of about 1,000, are tested each year, which is all the space available will allow for.

With emphasis on the common type, i.e., persistent fig, one must accept the fact that only 25% of a seedling progeny will be material for selection. Figs comprise only 50% of the progeny, and 50% of these will be the caducous type.

Breeding for Specific Characters

Fig Tree Characters

With present objectives in breeding, persistence of syconia to ripeness is a prime requisite.

Vigorous trees that grow large are desired. Since fruit for the dried fig industry is harvested from the ground, large size is no object. Except for shaping and repairing wind or other damage, fig trees grown for dried figs are not pruned.

Trees with numerous branches and round heads are preferred to low broadly spreading trees.

The tree must be productive within limits. Some trees overbear, and branches droop or break.

Maturing and dropping the crop early in the season is a desirable character. In August and early September, the weather is hot and dry, and the probability of dew formation, fog, and rain is virtually nil. This is the period of highest quality. By late September, dew formation and "tule" fogs are likely as the weather cools, and the probability of rain showers increases. Temperatures below a critical low of about 13 to 15 C and saturation of the soil which has purposely been allowed to go dry after mid-July cause excessive splitting and souring, as well as a drop in fruit quality.

Fig Fruit Characters

So-called "white figs", i.e., figs with greenish-yellow or golden-yellow skins, are preferred by the industry both for canning and drying. They dry to a light straw color and present a more attractive appearance than dark-skinned figs in fancy packs of dehydrated figs, and in paste which is baked in fig bars and made into preserves.

Figs with white meat and amber pulp are preferable to those with strawberry or reddish colored meat and pulp, because they yield a light amber or straw-colored paste.

Flavor of figs, as of many other fruits, defies accurate description. Growers and processors, however, recognize what might be called a distinctive fig flavor.

All of the commercial cultivars serve the dual purpose of furnishing some fruit to the fresh fruit market and the rest to processors as dried fruit. For this reason fresh fruit flavor is important.

Dried fruit and paste flavor are determined by the flavor of the fresh fruit. Like the fresh fruit, they should have the distinctive fig flavor and no "wild taste." They are sweeter than fresh fruits because of the concentration of sugars by drying.

Medium-sized figs, i.e., figs measuring about 5 cm in diameter and 7.5 cm in length, are preferred. Such figs number 20/kg. Shape may be fully rounded globose, oblate spheroidal, ovoid, and pyriform since these pack easily, uniformly, and attractively in confectionery packages. Overly flattened oblate and elongated pyriform are undesirable from the standpoint of sorting and packing.

Caprifigs

The first requirement in selecting a caprifig tree for use in improving fruit quality by breeding is that it carry the allele for persistence. As shown in Table 3, persistence can be transmitted to the offspring by the caprifig only. As mentioned earlier, all caprifig trees used for this purpose are descendants of 'Croisic'. Otherwise, the looked-for characters are the same as for the fig.

To be selected for use in caprification, a seedling caprifig tree must be of the caducous syconia type. This is dictated by the fact that few commercial fig growers produce their own caprifigs for pollinating 'Sari Lop'. In the main, caprifigs are produced in separate orchards removed many miles from the commercial fig growing region, as a precaution against spreading endosepsis, i.e., the internal fruit rot caused by the fungus *Fusarium moniliforme* var. *fici*. In the spring all mamme caprifigs are harvested at maturity, split open, treated with a fungicide, and stored. They are distributed among the caprifig trees when the profichi crop is receptive to colonization by the wasp. Growers then buy caprifigs from the producer according to their needs for caprifying 'Sari Lop'. However, not all caprifig cultivars are well colonized. Some, such as 'Roeding 1', 'Excelsior', and 'Forbes', produce large numbers of "blanks" which persist to maturity and are not easily distinguished from the syconia inhabited by wasps. The buyer wants to be assured that every caprifig he buys contains wasps; hence, the requirement for caducous types. For the grower, the cultivar must yield a heavy profichi crop of well-inhabited syconia.

Another consideration is time of maturity of the profichi syconia during the season of fig syconia receptivity. Early, medium, and late cultivars are required to spread fruit set and to provide overlapping during the season to compensate for fig and caprifig crops being thrown out of phase by the vagaries of climate. The syconium must be dry for satisfactory hatching of blastophagas and emergence of the females. 'Croisic' type caprifigs are unsuitable, because the pulp is juicy when the syconium is ripe and tends to preclude the wasps' getting out. It is also important that the syconium be hollow for satisfactory hatching and exiting of blastophagas.

Interspecific Hybridization

Trabut (1922) crossed *F. carica* with *F. palmata* and *F. pseudo-carica*, closely related species in the subgenus *Eusyce*, and grew numerous seedlings at the Montpellier Garden in Algeria, but does not appear to have reported later on their behavior or value.

Condit (1947) made the same crosses later, and reported the following: When caprifigs of the two species are used as the pollen parents, the Smyrna type (i.e., the caducous character) is dominant; when persistent *F. carica* caprifigs are used as the pollen parents, the progenies segregate into persistent and caducous phenotypes among both the figs and the caprifigs; the figs are edible, but of poor quality and worthless commercially; the seedlings tend to be precocious, many producing syconia in the same season in which the buds are grafted on mother trees.

FIG. 9. Interspecific hybridization in *Ficus*. (A) *F. pumila*, juvenile leaves on vegetative branch, syconium, and leaf from fruiting branch. (B) *F. pumila* x *carica* F_1 syconium and leaf. The punctures near the apex of the syconium were made for pollinating with *F. carica*. (C) *Ficus carica*, caprifig syconium. (Not to scale.)

'Brawley', a caprifig cultivar which was selected from a cross of *F. carica* x *pseudo-carica* by the hybridizer Francis Heiny of Brawley, California, is still used by growers. It is a good caprifig, but is no longer propagated because many fig growers did not like its small-sized syconia.

Condit (1950) reported hybridizing *F. pumila* with *F. carica*, using pollen from a caprifig of the latter species. I remade the cross in 1965, using pollen from a caprifig tree having light yellow persistent syconia (Fig. 9). Certain characters of the parents make the cross especially intriguing.

F. pumila, which is called "climbing fig," is an evergreen clinging vine with dimorphic branches. Juvenile branches cling to walls, fences, buildings, trees, and other upright structures. They are thin and wiry, and have small simple leaves arranged in the ½ phyllotaxy. Fruiting branches project horizontally for 30 to 60 cm, are thick and woody, and have larger leaves arranged in the ⅖ phyllotaxy. Generally, a fruiting branch has only

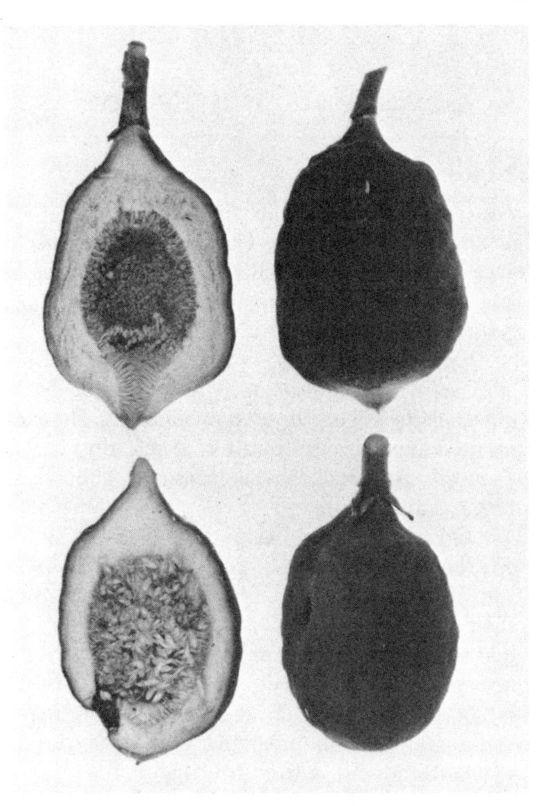

FIG. 10. *F. pumila*. At top, unpollinated syconium; bottom, seed-filled syconium after pollinating with *F. carica*.

one to three syconia. The syconia tend to persist to maturity without pollination. The mature syconium has purple skin color, and tough leathery texture. Pollinated ones are full of seeds (Fig. 10).

F. carica is a moderate-sized deciduous tree with stout, deliquescent, monomorphic branches which produce numerous syconia that are succulent at maturity. The leaves are large, palmately lobed, and arranged on the branches in the ⅖ phyllotaxy.

Some seedlings of the F_1 progeny are deciduous, some evergreen; most are vine-like, but some are weeping; none is tree-like; all bear fleshy, succulent, purple syconia. They segregate sexually into a 1:1 ratio of pistillate plants to monoecious plants.

An F_1 pistillate plant was backcrossed successfully with an *F. carica* caprifig in 1970. Scions of 30 seedlings of the backcross progeny were grafted on mother stock trees in the spring of 1971. Attempts will be made to obtain an F_2 progeny for the promise it would offer in terms of recombinations of characters.

An accomplishment one would hope to achieve ultimately from such a cross is development of vines bearing palatable figs which could be espaliered on walls or trained on trellises—evergreen vines for tropics and subtropics, hardy deciduous vines for temperate regions.

Although *F. pumila* and *F. carica* grow in close proximity to each other in some localities, and inhabited caprifigs of the latter have been placed near syconia on vines of the former, no natural hyrbidization by the agency of the fig wasp has occurred, nor have wasps been found in cut-open syconia.

N. K. Arendt, who was active in fig breeding in the Crimea, reported on pollinating figs with various species in Moraceae, as well as several other families. In her most recent report (1969), she summed up the results by stating viable seeds are plentiful but no hybrids were realized. Instead, foreign pollens stimulate apomixis.

Achievements and Prospects

New Fig Cultivars

To date, the most significant achievement of the fig breeding program in California is the acceptance of four hybrid cultivars by commercial fig growers and by nurserymen catering to the home garden trade. The fruits are described in Table 5.

'Conadria' (Ped. 143–5; Fig 11-A) was the first cultivar in history developed in a planned breeding program to become a commercial fig. Its name resulted from growers continually referring to it as 'Condit's Adriatic', which eventually abbreviated to 'Conadria'.

'DiRedo' (Ped. 143–38; Fig. 11-B), a sibling of 'Conadria', 'Flanders' (Ped. 151–37; Fig. 11-C), and 'Excel' (Ped. 195–36; Fig. 11-D) acquired their names from popular usage. 'DiRedo' was introduced into commercial production by Nicholas and Joseph DiRedo of Fresno, California, father and son cooperating growers in the hybrid seedling evaluation program. 'Flanders' attracted much attention as a fresh fruit fig on the property of Stanley E. Flanders of Riverside, California. 'Excel', a 'Dottato' hybrid, was so named by I. J. Condit because it excels its parent as a fresh fruit cultivar in southern California. It is now being tested as a possible replacement for 'Dottato' in San Joaquin Valley.

In the 1974 crop season, the project leader assisted by growers, processors, agricultural extension service personnel, and a representative of the California Fig Institute and Dried Fruit Advisory Board selected seven seedlings from hybrid progenies totaling more than 1,500 individuals fruiting at that time. Fruit characters are described and tentative cultivar names are given in Table 6.

Nematode Resistance

O'Rourke (1966), Puls, Birchfield and O'Rourke (1967), and Puls and O'Rourke (1967) have reported significant progress in breeding figs for nematode root-knot (*Meloidogyne incognita* var. *acrita*) resistance in Louisiana. A fairly high order of resistance was found in individual seedlings in progenies of the commercially grown cultivars 'Hunt' and 'Celeste'.

No special attention has been given to nematode resistance among hybrid fig seedlings in California because of the practice of grafting the seedlings on established trees. However, tree vigor, and the small size and paucity of root-knots on three

TABLE 5. Fruit characters of four hybrid fig cultivars.

Cultivar	Pedigree number	Skin color	Meat color	Pulp color	Shape	Eye	Avg. no. fresh figs/kg	Wt. (g)	Remarks
Conadria (Fig. 11-A)	143-5	light yellowish-green	white	light strawberry	pyriform with thick neck	very small; tight	20	45–55 avg. 50	Good Verdone-like type for drying in hot interior valleys, & for table use fresh. Resistant to splitting.
DiRedo (Fig. 11-B)	143-38	light yellowish-green	white	amber	globose —short, thick neck	small, tight	18	50–60 avg. 55	Good Verdone-like type for drying in hot interior valleys; dries light in color; some splitting in adverse weather.
Flanders (Fig. 11-C)	151-37	light tawny w/violet stripes & scattered white flecks	white	light strawberry	pyriform —long slender neck	medium tight	20	45–54 avg. 50	Excellent fresh fruit fig for the home garden. Virtually no splitting. Dried figs dark, commercially unattractive.
Excel (Fig. 11-D)	195-36	light yellow	white	amber	ovoid to globose	medium, tight	24	35–50 avg. 42	Excellent Dottato-like type for fresh fruit, canning, and drying. Virtually no splitting.

introduced cultivars namely 'Azendjar', 'Taranimt', and 'Zidi' suggest a high order of tolerance. These are being tested more intensively in a locale with soil highly infested with the root-knot nematode.

Achievements with Caprifigs

Three superior caprifig trees having persistant syconia have been selected as pollen parents for breeding for fruit quality. All bear heavy crops of syconia with green skin, white meat, and amber pulp. The syconia are of medium size and oblate spheroidal shape. The staminate flowers produce copious amounts of fertile pollen. They have the following pedigree numbers: 228–20; 271–1; and 276–31. These three have been responsible for all of the hybrid selections listed in Table 6.

Pedigree 228–20 is a good, productive, medium-sized, fleshy, persistent caprifig with green skin and amber pulp. It was selected from progeny of 'Conadria' x ('Monstreuse' x 'Croisic').

Pedigree 271–1 is a highly productive, medium-sized, fleshy, persistent caprifig. It has green skin and amber pulp and was selected from progeny of 'Beall' x 228–20.

Pedigree 276–31 is an attractive, fleshy, persistent, productive caprifig. Its skin is yellowish green, its pulp amber. It was selected from progeny of 'Sari Lop' x 228–20.

One caprifig tree of the caducous type has been selected for use in caprifying 'Sari Lop'. The syconia are large, obturbinate, hollow, and dry, and generally the profichi crop is abundant. Its chief advantage over commercially grown caprifigs now in use is the later season of the profichi crop. It should prove especially useful in years when synchrony of anthesis in caprifigs and figs is out of phase with the caprifigs maturing before the figs are receptive for pollination. This caprifig bears Pedigree 113–66. It was selected from a progeny of *F. palmata* x ('Brown Turkey' x 'Roedings 3').

The Future

Much has yet to be done in fig breeding, but the prospects for attaining desired goals in the not-too-distant future are bright. Some of the objectives are discussed below.

1) A prime objective is development of a persistent-type fig having the attractive appearance and high quality of 'Sari Lop', a small tight eye, resistance to splitting, and less susceptibility to internal spoilage. Such a type would obviate the need for the time-consuming, expensive practice of

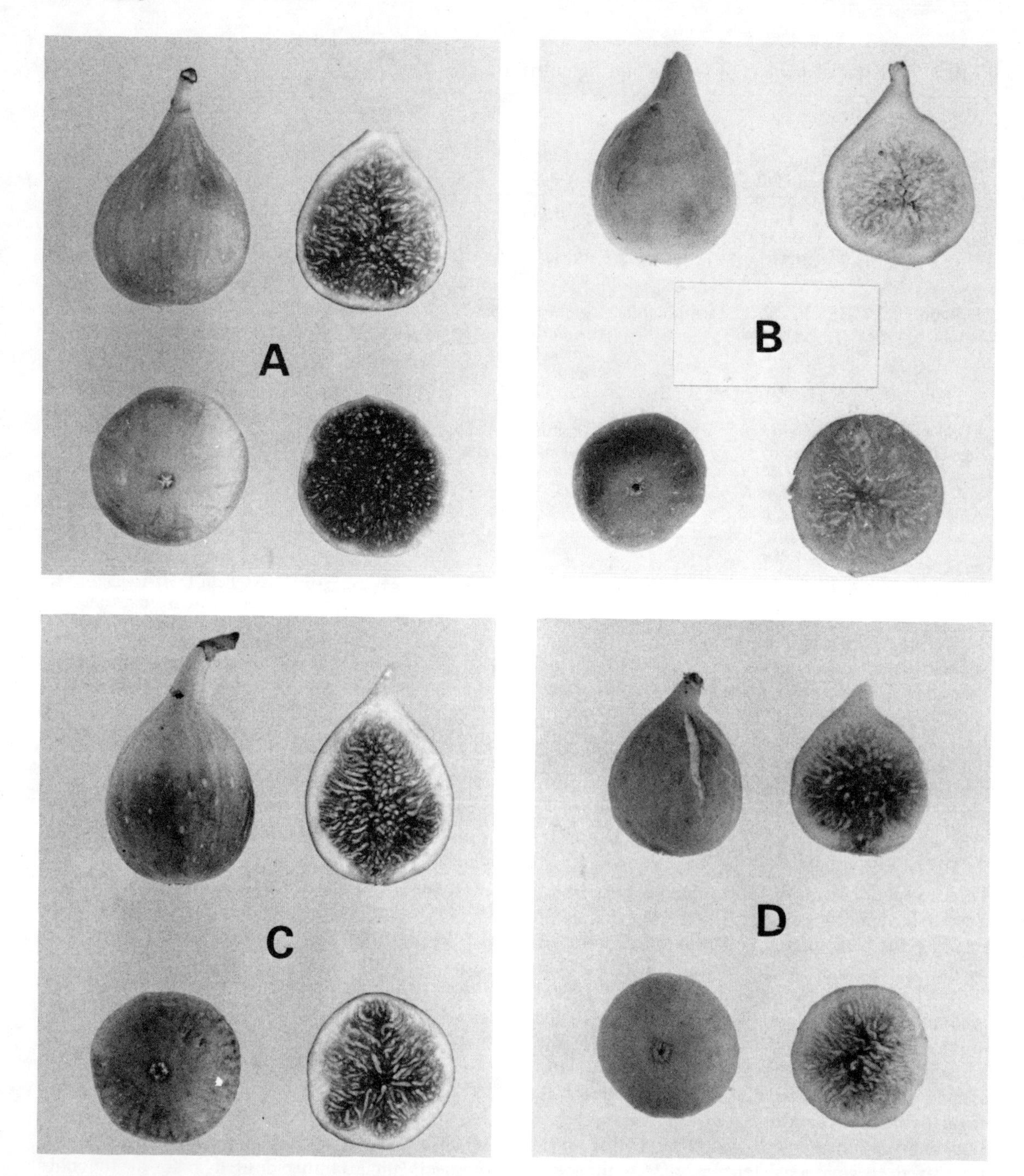

FIG. 11. Hybrid fig cultivars. (A) 'Conadria'. (B) 'DiRedo'. (C) 'Flanders'. (D) 'Excel'.

caprification, as well as splitting and internal spoilage which follow the entrance of contaminated fig wasps. The unique flavor of 'Sari Lop' is due in large measure to the oily nutlike endosperms of the fertile seeds. Ordinarily, the common type figs produce cenocarps only, but some tend to develop parthenogenetic endosperms (Condit, 1932). Possibly, this tendency can be bred into a persistent-syconia 'Sari Lop' derivative. Or it might be induced by the use of a plant growth regulating chemical.

Progenies resulting from parents both of which

TABLE 6. Proposed names and fruit characters of seven hybrid fig seedlings selected in 1971.

Pedigree number	Parentage	Proposed name	Skin color	Meat color	Pulp color	Shape	Eye	Wt. (g)	Avg. no. fresh figs/kg	Remarks
233-10	'Conadria' x ('Dottato' x 'Croisic')	Yvonne	canary yellow	white	light strawberry	obovoid	medium, tight	45–55 avg. 50	20	matures early, all figs dropping in a short time; virtually no splitting
276-7	'Sari-Lop' x 228-20	Saleeb	light green	white	pale pink	oblate	medium, tight	48–60 avg. 53	19	matures early, yields well, no splitting
279-57	158-46 x 271-1	Tena	greenish	white	light strawberry	oblate	medium, tight	45–60 avg. 52	19	mid-season, attractive, no splitting
284-11	276-14 x 276-1	Nardine	light yellow	white	amber	oblate	medium, tight	43–58 avg. 50	20	mid-season, attractive, yields well, few split
291-8	233-10 x 271-1	Deanna	light yellow	white	amber	oblate	medium, tight	45–56 avg. 50	20	mid-season, yields well, attractive, no splitting
291-30	233-10 x 271-1	Gulbun	light yellow	white	light pink	oblate	medium, tight	65–78 avg. 72	14	large, mid-season, yields well, few split
291-50	233-10 x 271-1	Evrem	light yellow	white	greenish	oblate	medium, tight	43–55 avg. 50	20	mid-season, attractive, yields well, no splitting

have 'Sari Lop' in their ancestry are now available, e.g., pedigrees 276 and 278. Such progenies contain persistent-syconia segregants of both fig and caprifig. By judicious selection, sibling pollination, and back-crossing to 'Sari Lop', presumably all of the desirable characters of 'Sari Lop' can be restored through recombination, excepting, probably, the development of endosperm-containing seeds. Sibmating Pedigree 276–14 fig with 276–31 caprifig yielded progeny 284, the first ever to have 'Sari Lop' genes from both parents. A superior 'Sari Lop'-like selection, Pedigree 284–11, has been made from this progeny (see Table 6).

2) Continued selection of well-flavored figs with light green or yellow skin color, white meat, and amber pulp for drying and canning.

3) Figs with small eyes tightly closed by bracts precluding entrance by species of vinegar flies, souring beetles, thrips, and other insects.

4) "Seedless fruit," i.e., fruit lacking fertile drupelets or cenocarps but having good flavor. This kind of fruit would be a boon to persons wearing dentures or dental bridges who like figs but avoid them because they find the hard drupelets distressing. Trees with syconia completely lacking any sort of carpel development are found occasionally in hybrid seedling progenies.

5) Fruit with skin that does not crack readily and that is tough enough to resist bruising or tearing in handling.

6) Induction of polyploidy with the use of colchicine or by some other means. Although polyploid seedlings have been produced by germinating seeds in colchicine, attempts to induce polyploidy in established cultivars have been singularly unsuccessful. The few polyploid seedlings that have fruited have been worthless. It would be interesting to see what polyploids of various cultivars would be like.

7) Development of clones resistant to root-knot nematode. Three cultivars from North African sources, 'Azendjar', 'Taranimt', and 'Zidi',

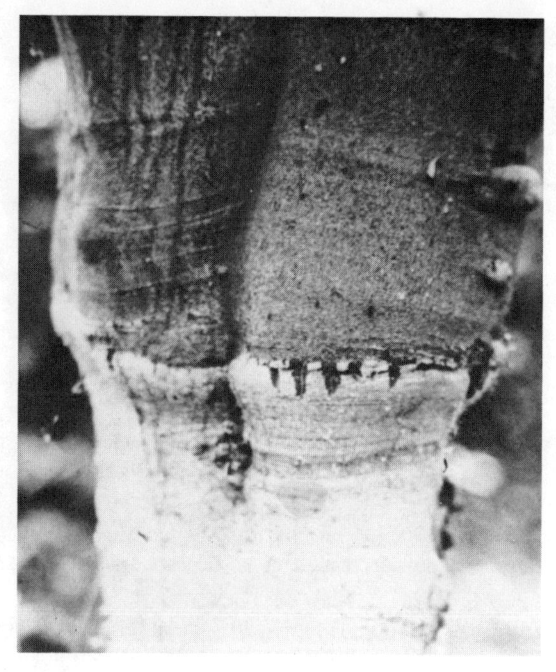

FIG. 12. Ten-year-old bud union of *F. carica* topworked on *F. cocculifolia*.

appear to be highly resistant and are being tested in a heavily infested nematode plot in San Joaquin Valley. O'Rourke (1966), Puls and O'Rourke (1967), and Puls, Birchfield, and O'Rourke (1967) have developed some resistant seedlings in Louisiana.

Some *Ficus* spp., such as *F. cocculifolia* Baker, *F. pumila* L., and *F. racemosa* L. are nematode resistant. Graft-congenial species might well serve as rootstocks for commercial fruit cultivars (Fig. 12). The area of graft-congeniality has scarcely been touched.

8) Production of interspecific hybrids such as the one already achieved between *F. pumila* and *F. carica*. Attempts at interspecific hybridization is an area of fig breeding that is largely untouched. As many as 100 species of *Ficus* are available for the purpose in southern California.

As mentioned previously, the ultimate aim of the *F. pumila* x *F. carica* cross is to develop evergreen vines bearing palatable fruit which could be grown on trellises or espaliered on the walls of buildings and other structures in the tropics and subtropics, and hardy deciduous vines for similar use in temperate regions.

Resurgence of interest in commercial fig production in recent years presages increasing growth of the industry and the concommitant demand by growers and processors for indicated research on problems which arise. In many of these problems, breeding may play a major role.

Literature Cited

Aigremont. 1908. *Volkserotik und Pflanzenwelt*. Halle: S. Trensigner. (Feige. 1:74–78). Cited from Condit, 1947.

Arendt, N. K. 1969. The use of apomixis in fig breeding (in Russian). *Trud. Nikit. Bot. Sad.* [*Trans. Nikita Bot Gdn.* 1969] 40:95–120. Abstr. 8761 in *Plant Br. Abstr.* 41 (4). 1971.

Burbank, L. 1914. *Luther Burbank* 4:297. New York: Burbank Press.

California Fig Institute. 1972. *Statistical review of the California dried fig industry*. Fresno.

Condit, I. J. 1920. Caprifigs and caprification. *Calif. Agr. Expt. Sta. Bul.* 319:341–377.

———. 1928. Fig breeding. *J. Hered.* 19:417–424.

———. 1932. The structure and development of flowers in *Ficus carica L. Hilgardia* 6:443–481.

———. 1947. The fig. Waltham: Chronica Botanica.

———. 1950. An interspecific hybrid in *Ficus*. *J. Hered.* 41:165–168.

———. 1955. Fig varieties: a monograph. *Hilgardia* 11:323–538.

———. 1969. *Ficus: the exotic species*. Berkeley: Univ. of Calif. Div. Agr. Sci.

Condit, I. J., and Flanders, S. E. 1945. "Gall-flower" of the fig, a misnomer. *Science* 102:128–130.

Condit, I. J., and Warner, R. M. 1956. Promising new seedling fig. *Calif. Agr.* 10:4, 14.

Corner, E. J. H. 1965. Check list of *Ficus* in Asia and Australasia. *Gardener's Bul. Singapore* 21:1–186.

Crane, J. C., Marei, Nash, and Nelson, M. M. 1970*a*. Ethrel speeds growth and maturity of figs. *Calif. Agr.* 24(3):8–10.

———. 1970*b*. Growth and maturation of fig fruits stimulated by 2 chloroethyl phosphonic acid. *J. Amer. Soc. Hort. Sci.* 95:367–370.

Food and Agricultural Organization of the United Nations. 1961. *World list of plant breeders. Sup. 1.* 1965. *Sup.* 2. 1967. Rome.

Gerdts, M., and Obenauf, G. 1972. Effects of preharvest applications of ethephon on maturation and quality of Calimyrna figs. *Calif. Agr.* 26(5):8–9.

Hunt, W. B. 1911. Fig breeding. *Univ. of Ga. Bul.* 11: 146–148.

———. 1912. Fig breeding. *Univ. of Ga. Bul.* 12:107–110.

Noble, E. G. The work of the Yuma Reclamation Project Experiment Farm in 1919 and 1920. *USDA Cir.* 221:3–37.

O'Rourke, E. N., Jr. 1966. Breeding new figs for Louisiana. *La. Soc. Hort. Res.* 3:35–41.

Puls, E. E., Jr., Birchfield, W., and O'Rourke, E. N., Jr. 1967. Evidence of root-knot nematode resistance in fig seedlings. *Proc. So. Agr. Workers* 14:227.

Puls, E. E., and O'Rourke, E. N., Jr. 1967. Root-knot nematode resistance in fig seedlings. *La. Agr.* 10(4): 14–15.

Rixford, G. P. 1918. Smyrna fig culture. *USDA Bul.* 732:1–43.

———. 1926. Work with seedling figs. *Calif. Ann. Fig Inst. Proc.* 10:6–9.

Saleeb, W. F. 1965. *Genetics and cytology of sycomium persistence in Ficus carica*. Ph.D. dissertation, Univ. of Calif.

Storey, W. B., and Condit, I. J. 1969. Fig (*Ficus carica* L.). In F. P. Ferwerda and F. Wit, eds. *Outlines of perennial crop breeding in the tropics*. p. 259–267. *Misc. Pap. 4* Wageningen, The Netherlands: Landbouwhogeschool.

Swingle, W. T. 1908. Some points in the history of caprification and its bearing on the Smyrna fig industry in this country. *Calif. Fruit Growers Conv. Proc.* 34:178–187.

———. 1912. Cooperative distribution of new varieties of Smyrna figs and caprifigs. *USDA Bur. Plant Ind. Bul.* 537. (Second rev. of *Bul.* 438. 1909.)

Trabut, L. 1922. On the origins of the fig (in French). *Rev. Bot. Appl. Agr. Colon.* 2:393–396.

Author Index

Subject Index

This index includes species and common names, cultivars, diseases, disorders, pathogens, and pests.